OPERATIONS AND PRODUCTION SYSTEMS WITH MULTIPLE OBJECTIVES

**WILEY SERIES IN SYSTEMS ENGINEERING
AND MANAGEMENT**

Andrew P. Sage, Editor

OPERATIONS AND PRODUCTION SYSTEMS WITH MULTIPLE OBJECTIVES

BEHNAM MALAKOOTI

WILEY

Published by John Wiley & Sons, Inc., Hoboken, New Jersey.
Published simultaneously in Canada.

For general information on our other products and services or for technical support, please contact our Customer Care Department within the United States at (800) 762-2974, outside the United States at (317) 572-3993 or fax (317) 572-4002.

Wiley also publishes its books in a variety of electronic formats. Some content that appears in print may not be available in electronic books. For more information about Wiley products, visit our web site at www.wiley.com.

Library of Congress Cataloging-in-Publication Data:

Malakooti, Behnam, 1953–
  Operations and Production systems with multiple objectives / Behnam Malakooti.
    pages cm
  Includes bibliographical references and index.
  ISBN 978-0-470-03732-4 (cloth)
 1. Production management.   2. Operations research.   I. Title.
  TS155.M33136 2013
  658.5–dc23

                                                    2012035463

Printed in the United States of America

10 9 8 7 6 5 4 3 2 1

To

Mina, Nima, Shahdi, Ramin

&

Shaya Sheikh, Camelia Al-Najjar, G. M. Komaki

BRIEF TABLE OF CONTENTS

CONTENTS

HOW TO USE THIS BOOK

Online Supplements and Solutions to Problems in Microsoft Excel, MATLAB, and LINGO formats

Throughout the book, the reader will note the icon [WWW], which refers to:

- Supplemental material in Microsoft Excel or Word formats;
- Modeling resources using Excel Solver, MATLAB®, and LINGO optimization software.

Each formulation of solved examples can be modified to solve homework or larger practical problems. Currently, student versions of LINGO optimization software for solving small problems can be downloaded free of charge.

www.wiley.com/go/ops

All resources are available at www.wiley.com/go/ops. Click the "More information" tab and select "Student Companion Site" from the drop-down menu for access to the resources.

Instructor's Homework Solutions Manual

A solutions manual for all the homework problems in this book is available for professors and instructors. See the Instructor's site at www.wiley.com/go/ops.

† Graduate and Advanced Sections

Graduate and advanced topics designated by † may be omitted without loss of continuity or generality.

Suggested Outlines for Different Courses and PE Exam

   I. Production Systems
  II. Supply Chain Operations
 III. Quality, Productivity, and Measurement Methods
  IV. Manufacturing and Production Systems
   V. Introduction to Systems, Industrial, Operations, & Management Engineering
  VI. Layout Design, Location, and Cellular Systems
 VII. Professional Engineering (PE) (Principle and Practice) of Industrial Engineering. (Exclude all MCDM and all optimization formulations, exclude all †, include all introductory sections)

Suggested Outlines for Different Courses and PE Exam

Chapters	I	II	III	IV	V	VI	VII- PE Exam
1. Introduction	✓	✓	✓	✓	✓	✓	1.6, 1.7, 1.8, 1.9.1
2. Multicriteria Decision Making	✓	✓	✓		✓	✓	2.2.1, 2.2.3, 2.3.7
3. Forecasting	✓	✓	✓	✓	✓		3.2, 3.3.1, 3.4.1, 3.5.1, 3.6, 3.7.1
4. Aggregate Planning	✓			✓			4.2, 4.3, 4.7
5. Push and Pull (MRP/JIT) Systems	✓			✓			5.2.3, 5.3.2, 5.6, 5.7
6. Inventory Planning and Control	✓			✓	✓		6.2, 6.3, 6.6.1, 6.9.1, 6.10.1
7. Scheduling and Sequencing	✓			✓			7.2.1&2, 7.3.1&2, 7.5.1
8. Project Management	✓	✓		✓	✓	✓	8.2, 8.3.1, 8.5.1
9. Supply Chain and Transportation	✓	✓		✓	✓	✓	9.1, 9.2.1&2, 9.4.2, 9.5.3
10. Productivity and Efficiency	✓	✓	✓				10.2, 10.4.1, 10.7.1
11. Energy Systems: Design and Operation	✓	✓			✓		11.4.1
12. Clustering and Group Technology		✓	✓	✓	✓	✓	12.2, 12.3
13. Cellular Layouts and Networks				✓		✓	13.1, 13.2.1, 13.2.2
14. Assembly Systems	✓			✓		✓	14.2, 14.3, 14.6
15. Facility Layout						✓	15.1, 15.2, 15.3.1, 15.6.1
16. Location Decisions		✓				✓	16.2, 16.3.1, 16.4.1, 16.5.1
17. Quality Control and Assurance		✓	✓				All (except 17.2&3, 17.4.6, 17.6.3&4, 17.7.4, 17.8.2)
18. Work Measurement		✓	✓				All (except 18.5&18.6.4)
19. Reliability And Maintenance		✓	✓	✓	✓		19.2.1, 19.3, 19.4, 19.6.1, 19.7.1
Total Number of Chapters	12	12	8	12	9	9	

PREFACE

My perception of what should be covered in this book was formed by working with different manufacturing industries, corporations, businesses, computer companies, government organizations, and NASA. This book is the result of the following different aspects of my academic and professional life:

1. The use of multiple criteria decision-making (MCDM) approaches in my undergraduate and graduate courses in operations, production, manufacturing, optimization, engineering economics, intelligent networks, decision making, and systems engineering at Case Western Reserve University (CWRU).
2. The successful completion of funded projects from industry and government, dependent on actual, down-to-earth applications using MCDM as a philosophical systems approach and tool for achieving results.
3. The publication of articles in technical journals, which has required me to hone the scientific aspects of MCDM.
4. The relentless search for developing nonlinear systems for modeling descriptive and prescriptive decision making in real life.

SYSTEMS APPROACH

A *system* is a combination of interdependent components (or subsystems) that form a complex whole. Systems encompass a multitude of fields such as operations, production, manufacturing, control engineering/systems, industrial systems, energy systems, biology, ecology, information technology, artificial intelligence, telecommunications, management, health, medicine, and social science. Systems engineering is the art and science of designing, architecting, modeling, engineering, and building complex systems. The systems architecting process creates an architectural systems design, a conceptual model, and an operational system. This design receives and analyzes feedback from multiple parties, including clients and developers of subsystems and components. Each system must be optimized based on the objectives of different systems stakeholders. To understand and build operations and production systems, one needs to know the basic concepts and tools of systems engineering, systems architecting, and multi-objective optimization. In this book, we demonstrate how operations and production systems are architected, engineered, and optimally operated by decomposing and simplifying complex problems. This book provides a new perspective on problem solving for several disciplines such as systems engineering, industrial engineering, decision making, operations research, operations management, engineering management, and production planning.

NEW METHODS FOR SOLVING OPERATIONS AND PRODUCTION SYSTEMS (SINGLE OBJECTIVE)

In this book, we also provides new approaches for solving several classic operations/production systems (single objective) problems. They include the following:

- Assembly systems: Approaches for the design of assembly systems (Ch. 14, Qualitative Assembly Line Balancing)
- Cellular systems: Design and analysis of cellular systems and single-layer networks (All methods of Ch. 13 Cellular Layouts and Networks)
- Energy systems: Design and operations of energy systems (All methods of Ch. 11 Energy System Design and Operation)
- Facility layout: New approaches for single-layer, multiple-layer, and complex layout problems (Ch. 15, Rule Based Layout Method)
- Push and pull systems: Development of hybrid push-pull systems (Ch. 5 Push-and-Pull (MRP/JIT) Systems)
- Scheduling: New head-tail approaches for solving flow-shop and job-shop sequencing problems (Ch. 7)
- Productivity and efficiency: New approaches for measurement and analysis of systems (Ch. 10)

NEW METHODS FOR SOLVING OPERATIONS AND PRODUCTION SYSTEMS (MULTIPLE OBJECTIVES, CRITERIA)

Traditional optimization models focus primarily on one objective, usually cost. However, in real-world problems, more than one objective should be considered. In all real-world problems, the decision maker must consider conflicting objectives when architecting, modeling, and optimizing a system. The objectives of most operations/production/manufacturing systems are as follows:

- Cost (to be minimized)
- Quality (to be maximized)
- Productivity (to be maximized)
- Sustainability (to be maximized)
- Flexibility (to be maximized)

These objectives are often in conflict with each other, especially when resources are limited. For example, quality and cost may be inversely related; that is, better quality is associated with higher costs. A conventional approach for solving multiple objectives problems is to consider only one objective and to treat the remaining objectives as constraints or given requirements. This approach inevitably leads to the selection of inferior solutions. In this book, we provide several methods for solving MCDM and multiple objective optimization (MOO) problems. MCDM/MOO models support the optimization of several objectives simultaneously and the understanding of the nature of tradeoffs among different objectives. This book provides the MCDM/MOO formulations and solutions for all classic operations and production problems.

COMMON MISCONCEPTION ABOUT MULTIPLE CRITERIA DECISION-MAKING (MCDM)

One common misconception about formulating MCDM problems is that using too many objectives (criteria) can help in solving the problem, however, the real-world problems have few meaningful conflicting objectives. Most of the operations and production problems covered in this book have only two or three objectives. Another misconception in MCDM applications is that many meaningful alternatives exist for solving a problem; however, usually only a few good alternatives should be considered carefully. Sophisticated MCDM approaches designed for many objectives and many alternatives may not be helpful for solving the realistic MCDM problems. See Section 2.1 and Table 2.1 for more details.

STUDENTS' ROLES ON THE FORMATION OF THIS BOOK

The MCDM approach presented in this book is the result of my struggle to define a unified, objective, and systems-engineering-based approach for solving decision-making (in general) and operations/production systems (in particular) problems. This book was developed with feedback from my undergraduate and graduate engineering students at Case Western Reserve University. The majority of this book is based on my lectures of the past 10 years. The lecture notes were developed in the process of attempting to simplify advanced and complicated topics to be taught at the junior and senior undergraduate and the master's level graduate courses. To make this transformation, my graduate student advisees were involved in testing and verifying the topics of the book and in providing me with valuable feedback that improved the book substantially. As I developed homework assignments on MCDM approaches to the classic operations systems problems, it became more apparent that the objectives and the problem-solving approaches needed to be simplified. As my students graduated and got jobs in industries, they updated me on their activities. To my surprise and complete satisfaction, they reported that they had not forgotten about my unique MCDM coverage of classic operations/production systems topics; they use these methods in practice and consider them powerful tools for solving everyday business and industrial problems.

THE LONG JOURNEY OF THIS BOOK

As a graduate student at Purdue University, I was introduced to the fields of optimization, MCDM, and decision making by A. Ravindran, Jyrki Wallenius, and Herbert Moskowitz.

At that time, I could not have imagined a book similar to the current book may exist. In 1984, the 6th Multiple Criteria Decision Making (MCDM) Conference was organized by two of my colleagues: Yacov Haimes and Vira Chankong of the Systems Engineering Department at Case Western Reserve University. In that meeting, many MCDM pioneers and visionaries gathered to discuss the future of the field of MCDM. At that time, MCDM theories and methods were not used widely in academia, government, industry, and business. The pioneering book of Keeney and Raiffa (1976) solidified decision-making (under risk) approaches based on expected utility theory, and the MCDM field suffered from not having a competing book. It was a common understanding that the field of MCDM had not matured to the point that one could write a textbook on the subject. During one of the conference breaks (of 1984), Ralph Steuer (1986) and Milan Zeleny (1982) encouraged my idea of writing a book on operations and production systems to illustrate solving real-world problems using MCDM. The idea of writing such a book sounded promising and achievable. But, it took more than 29 years to formulate and apply MCDM to the fields of operations, productions, industrial engineering, and systems engineering as presented in this book. In retrospect, writing this book and (Z Utility Theory papers Malakooti, 2014 and 2014a) has been far more challenging than the combined effort of writing all of my numerous technical papers. This is the first time a textbook of this nature for undergraduate and graduate students has been written, and its success will bring MCDM to the mainstream of academia and industry.

Z UTILITY THEORY: A PLEASANT BY-PRODUCT OF THIS BOOK

Current decision theories and approaches are based on the work of Von Neumann and Morgenstern (1947), who presented axioms and theorems of expected utility theory (EUT) [see Dyer (2005)]. With the exception of goal programming [Charnes and Cooper (1977)], there are two popular schools of thought for assessing utility functions for solving decision problems: an axiomatic approach based on EUT (advocated by decision theorists) and a procedural approach based on analytic hierarchy process (AHP) [Saaty (1980), advocated by a multitude of various disciplines in academia]. We cannot underestimate the substantial contributions of EUT, Prospect Theory, AHP, and goal programming to realizing decision making. These four systems of thinking solve real-world problems in a systematic way and may be used as testing-beds for human behavior. Although EUT and AHP are astronomically distant from each other and do not share a common ground; but unfortunately and ironically, both EUT and AHP suffer from the same terminal disease, namely, linearity. [The linearity of AHP and AHP's critique is covered in Chapter 2, MCDM (Section 2.4.6). The linearity of EUT is well known and has been documented by a variety of decision paradoxes; for example, see Kahneman and Tversky (1979).] Later I realized that, despite its appearance, Cumulative Prospect Theory is also linear. Goal programming remained to be the only functional nonlinear system but it may generate inefficient alternatives as the best alternative (see Section 2.6.2). In retrospect, developing nonlinear utility functions seems to be the focus of my research, for example see Malakooti (1985). During the writing of a chapter on decision making under risk [Malakooti (2014b)], I realized that existing risk theories based on EUT (and also Cumulative Prospect Theory) do not address the bicriteria nature of risk problems. In my attempt to formulate the bicriteria problem for risk, I was inspired by the mean-variance approach of Markowitz (1959); however, Camelia Al-najjar brought to my attention that the mean-variance approach violates the stochastic dominance (efficiency) principle. Z utility theory (ZUT) [Malakooti (2014a)] was developed for nonlinear utility functions to solve bicriteria risk problems using only a one-parameter model, i.e., a "Z," while satisfying the stochastic dominance (efficiency) principle. However, ZUT also solves MCDM problems (under certainty); Chapter 2 (MCDM) presents some of the applications of ZUT for solving MCDM. As ZUT was evolving, I noticed that ZUT also solves all decision (risk) paradoxes including all problems posed by Prospect Theory [Kahneman and Tversky (1979) and their adversaries]. I also realized that all prominent risk theories such as EUT and Cumulative Prospect Theory [Tversky and Kahneman (1992)] are special cases of ZUT. Furthermore, the experimental results of G. M. Komaki and Camelia Al-najjar showed that ZUT substantially outperforms all existing risk methodologies including EUT and Cumulative Prospect Theory. To my complete astonishment, our experiments illustrated that ZUT also very closely (but not exactly) solves descriptive risk problems, resulting in an integrated/unified approach for solving both prescriptive (normative) and descriptive risk models. However, ZUT (and all MCDM and risk approaches) addresses only one dimension of solving decision-making problems: The decision-making problem cannot be addressed completely unless all four dimensions of decision process are considered [Malakooti (2012)].

REFERENCES

Charnes, A., W. W. Cooper, "Goal Programming and Multiple Objective Optimization—Part I." *European Journal of Operations Research*, Vol. 1, 1977, pp. 39–54.

Dyer, J., "Maut—Multiattribute Utility Theory," Multiple Criteria Decision Analysis: State of the Art Surveys. *International Series in Operations Research & Management Science*, Vol. 78, 2005, pp. 265–292.

Kahneman, D., A. Tversky, "Prospect Theory: An Analysis of Decision Making Under Risk." *Econometrica*, Vol. 47, No. 2, 1979, pp. 263–292.

Keeney, R. L., H. Raiffa, *Decisions with Multiple Objectives: Preferences and Value Tradeoffs*. Wiley, New York (1976). (Reprinted, Cambridge University Press, New York, 1993).

Malakooti, B., "A Nonlinear Multiple Attribute Utility Theory," Decision Making with Multiple Objectives, ed. Y. Y. Haimes and V. Chankong, Springer-Verlag, 1985, pp. 190–200.

Malakooti, B., "Decision Making Process: Typology, Intelligence, and Optimization." *Journal of Intelligent Manufacturing*, Vol. 23, No. 3, 2012, pp. 733–746.

Malakooti, B., "Z Utility Theory: Decisions Under Risk and Multiple Criteria." to appear, (2014a).

Malakooti, B. *Decisions and Systems with Multiple Objectives*, J. Wiley, forthcoming, 2014b.

Markowitz, H. M., *Portfolio Selection: Efficient Diversification of Investments*. Wiley, New York (1959). (Reprinted, Yale University Press, New Haven, CT, 1970; 2nd ed., Basil Blackwell, Oxford, U.K., 1991).

Saaty, T., *The Analytic Hierarchy Process: Planning, Priority Setting, Resource Allocation*. McGraw-Hill, London (1980).

Steuer, R. E., *Multiple Criteria Optimization: Theory, Computation, and Application*. Wiley, New York (1986).

Von Neumann J., O. Morgenstern, *Theory of Gaines and Economic Behaviour*. Princeton University Press, Princeton, NJ (1st ed., 1944, 2nd ed., 1947).

Tversky, A., D. Kahneman, "Advances in Prospect Theory: Cumulative Representation of Uncertainty." *Journal of Risk and Uncertainty*, Vol. 5, No. 4, 1992, pp. 297–323.

Zeleny, M., *Multiple Criteria Decision Making*. McGraw-Hill, New York (1982).

ACKNOWLEDGMENT

TO STUDENTS

I would like to express my sincere appreciation to many of my graduate and undergraduate students for their invaluable contributions in various aspects of developing this book over several years. The list that follows is only a partial list of the students whose reviews or comments were used in improving this book. Most of the graduate students in my courses were working full time in various industries while they took my graduate courses. These graduate students substantially enhanced my lecture notes as I was teaching my courses and applied them to industrial-related problems. In particular:

Shaya Sheikh, Ph.D., has made substantial improvements to this book.

Camelia Al-Najjar, Ph.D., provided crucial critiques for new theories presented in this book.

Michael Fuentes, M.S., provided substantial improvements and critiques.

G. Mohammad Komaki, Ph.D., provided improvements to this book.

Hyun Kim, Ph.D., provided excellent substantial improvements for several chapters.

Mathew B. Lehman, Ph.D., read and improved all chapters in this book.

Youchul Jung, M.S., verified and improved the early versions of most chapters.

Jason J. Chung, M.S., verified and improved the more recent version of the book.

Past Students: I would also like to acknowledge the contributions of hundreds of my students who also supported the development of this book; they include: Dr. Hanieh Agharazi, Professor Jumah Alalwani, Satoshi Araki, Faizah Abu Bakar, William Balhorn, Brian Barritt, John Brainerd, Chinglun Cheng, Gerard D'Souza, J. Deviprasad, Shuan Endres, Scott Fassett, Dr. Sunjaya Gajurel, Dr. Zhihao Gou, Michael Griffith, James Heintel, Werle James, Professor Lowell Lorenzo, Camil Martinez, Joel Mathewson, Dennis Milam, S. Rajamani, Aaron Sell, Jerome Steiner, Evan C. Tandler, Siva K. Tanguturi, Dr. Atikan Teber, Ivan Thomas, Akira Tsurushima,

S. Vishnuraman, C. J. Wang, Professor Jun Wang, Timothy Watson, Dr. Ziyong Yang, and Dr. Yingqing Zhou.

TO MENTORS

I have been lucky to have had very brilliant, wise, caring, and inspiring mentors throughout my life. They include Professors A. Ravi Ravindran (Pennsylvania State University), Irving Lefkowitz (Emeritus Professor of Systems Engineering, CWRU), Late Yoh-Han Pao and Late Robert E. Collin (both from the Electrical Engineering Department at CWRU), Arthur H. Heuer (the Department of Materials Science and Engineering), Thomas H. Kicher, and Eli Reshotko (Emeritus Professors of Mechanical and Aerospace Engineering), and the most delightful Dr. Javad Kashani, MD.

TO COLLEAGUES AT CASE WESTERN RESERVE UNIVERSITY

This book reflects in part the inspirations of my colleagues: Professors Marcus R. Buchner, Howard J. Chizeck, the unforgettable Lucien Duckstein, Minguo Hong, Yacov Y. Haimes, Benjamin F. Hobbs, Irving Lefkowitz, Kenneth Leoparo, Wei Lin, Fraink Merat, Mehran Mehregany, Mihajlo D. Mesarovic, Soroosh Sorooshian, Mario Garcia Sanz, and Sree Sreenath. In particular, I must acknowledge the contributions of my dear friend and colleague Professor Vira Chankong.

TO NASA GLENN RESEARCH CENTER AND MY ASSOCIATES

I am indebted to Dr. Kul Bhasin (Chief of the Satellite Networks and Architectures Branch at NASA Lewis Research Center) who introduced me to architecting and engineering of complex systems for NASA missions, which led me to architect the Intelligent Internet Protocol (IIP) for the deep space communication complex system for NASA missions. I am also indebted to the insights of Dr. Bijan Bastani, Dr. Donna Kashani, Mr. Manooher Missaghi, Dr. Farid Sabet, Mr. Sohrab Sadegi, Mr. Nader Salehi, Dr. Naser Pourahmadiand, and the beloved Professor Joseph Rutman.

TO THE STAFF AT WILEY

I would like to thank the excellent support that I have received from the editorial staff at Wiley: Kari Capone (Editor), George Telecki (past Publisher), Kris Parrish (Production Editor), and Professor Andrew Sage (past Series Editor); along with Prakash Naorem (Project Manager, Aptara).

TO MY FAMILY

My wife (Mina), son (Nima), and daughter (Shahdi) have been extremely supportive in various stages of developing this book. I also enjoyed the enthusiasm of my brothers (Behrooz, Behzad, and Ramin) and my sister (Freshteh) who all bring me the sweet memories of my mother and father.

CHAPTER 1

INTRODUCTION

1.1 INTRODUCTION

Over thousands of years, the methods of production, operations, and manufacturing systems have been continually developed and improved. Such systems have been the concern of people from ancient Egyptians along the Nile River to Henry Ford in Detroit; from businessmen in the British Isles during the industrial revolution to current integrated circuits (IC) manufacturers. Manufacturing systems are concerned with the conversion of physical inputs into physical outputs by using workers, machines, and equipment. Production systems are concerned with the modeling and engineering management of converting inputs into outputs using forecasting, inventory, production planning, and scheduling. Operation systems are concerned with a broad view of such conversions; they include strategic decisions such as product design and development, quality policies, logistical systems, facility location decisions, facility layout, human resources, supply chain management, quality control, reliability, and maintenance. Operation systems are also concerned with efficiently delivering quality products and services to customers in a timely and cost-effective manner. Supply chain management involves purchasing, storing, and distributing raw materials and semifinished and finished products through a network of suppliers, production facilities, and distributors. Production, operation, and manufacturing systems are also concerned with developing systems' capabilities and utilizing resources effectively to fulfill the needs of the customers.

In business, the primary objective is to maximize the net profit. However, measuring and predicting net profit are often difficult because of market fluctuations, uncertainties, risks, and the dynamic nature of most businesses. Predicting net profit becomes even more difficult as the planning periods are set further into the future. It is, however, easier to predict the

Note: Advanced material that can be omitted without loss of generality will be indicated by a dagger.

Operations and Production Systems with Multiple Objectives, First Edition. Behnam Malakooti.
© 2014 John Wiley & Sons, Inc. Published 2014 by John Wiley & Sons, Inc.

TABLE 1.1   Possible Objectives (Criteria) in Design, Planning, and Control of Production and Operation Systems

• Minimize total cost	• Maximize employees' job satisfaction
• Minimize risk	• Maximize ease of change of systems
• Maximize quality (products and services)	• Maximize ease of use of systems
• Maximize productivity	• Maximize achievement of just-in-time
• Maximize flexibility	• Maximize agility
• Minimize adverse environmental impact	• Maximize simplicity of systems
• Maximize customer satisfaction	• Minimize variations/fluctuations
	• Minimize use of energy

factors that contribute to net profit. Experience has shown that long-term net profit depends upon some of the objectives shown in Table 1.1, where many organizations (including businesses, government, and nonprofit organizations) must consider these objectives in design and operations of systems. Most businesses cannot simply measure and maximize net profit directly. However, they can measure and optimize some of these objectives, which will lead to maximizing the net profit.

In the design of products, processes, operations, and facilities, one must consider some or all of the above objectives. It is more feasible to measure and optimize some of the above objectives, especially over a long period of time, instead of trying to quantify and optimize the net profit directly. The result of optimizing several objectives often leads in the direction of maximizing net profits. Such objectives, however, often conflict with each other, especially when resources are limited. For example, one cannot expect to buy a high-quality product at a low price. Multiple-criteria decision-making (MCDM) models are based on optimizing several objectives simultaneously, while traditional decision-making models only consider one objective (usually minimizing cost). An understanding of trade-offs among different objectives will allow a decision maker to arrive at the best compromise solution that balances competing objectives.

This book includes a comprehensive study of multiple-objective approaches for production, operation, and manufacturing systems. The most common objectives that are used in this book are the ones listed in the left column of Table 1.1.

This book also covers concepts and methods that can be used for effective decision making in businesses, manufacturing industries, process industries, government organizations, the military, and nonprofit organizations as well as for personal life decision problems.

Business organizations manage three important functions: marketing, operations, and finance/accounting. Decisions related to these three areas can be categorized as strategic, tactical, and operational. Table 1.2 illustrates these three decisions.

1.1.1   Overview of Input and Output Systems

One way to describe production and operation systems is from an input–output point of view. From this point of view production and operation systems transform inputs into useful outputs. For example, a university can be viewed as a production system where incoming freshmen are the inputs and educated students with degrees are the outputs. The educational activities (classes, projects, extracurricular activities, etc.) are the transformation processes. This input–output relation, as shown in Figure 1.1, can be used to describe every form of production, operation, and manufacturing systems.

TABLE 1.2   Strategic, Tactical, and Operation Decisions

Strategic Decisions	Tactical Decisions	Operations Decisions
• Long-term plans (years) • Long-term forecasting • Less structured decisions • Long-term consequences • Focus on entire organization • Cut across departments • More difficult to define and solve	• Medium-term plans (months) • Routine and repetitive • More structured decisions • Medium-term consequences • Focus on departments, teams, and tasks • Interrelated decisions	• Short term (weeks) • Immediate plans • Short-range consequences • Focus on details, workforce, inventory, etc. • Allocation of resources • Quality issues • Reliability issues

The integration of several interrelated production, operation, and manufacturing systems leads to a supply chain system. Supply chain management involves the integration of several phases of operations, including materials procurement, manufacturing processes, product assembly, and distribution systems. Supply chain management (Figure 1.2) integrates all aspects of a business in order to reduce costs and maximize speed of delivery. Generally the supply chain relies on a competitive and customized chain of suppliers. It is a generalization of the input–output model (Figure 1.1), incorporating all phases and cycles of production from the materials supplier to the customer.

1.1.2   Overview of the Chapters of This Book

A view of production and operation systems is presented in Figure 1.3. At the top level, the functional areas of production and operation systems are categorized as production and operations planning, process design and logistics, and quality, reliability, and measurement. the systems approaches, multiple-criteria decision analysis, and problem-solving tools are presented at the lower level of Figure 1.3. Sections 1.3 and 1.4 discuss each of these topics. The top level of Figure 1.3 is the subject of this book and the bottom level of Figure 1.3 is the subject of the forthcoming book (Malakooti, 2014).

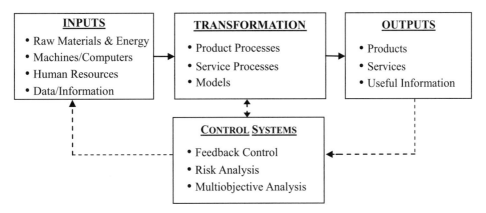

Figure 1.1   Input–output relation in production/operation systems with control systems feedback.

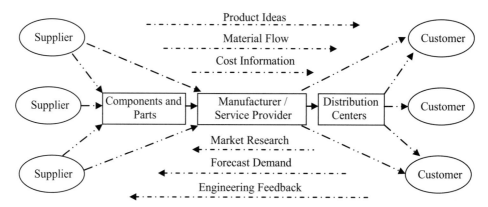

Figure 1.2 View of supply chain system and management.

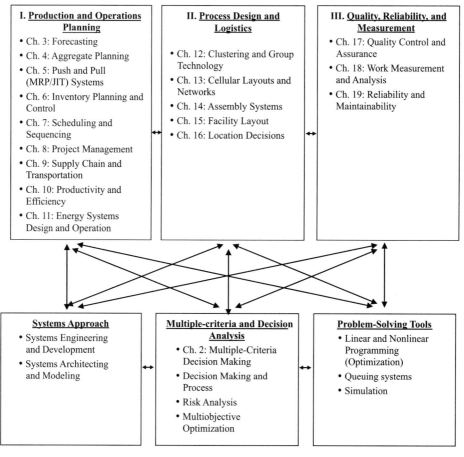

Figure 1.3 Functional areas of production and operation systems, along with systems approaches, multiple criteria and decision analysis, and problem-solving tools.

1.2   PRODUCTION AND OPERATIONS HISTORY AND PERSPECTIVE

The history of production and operation systems began around 5000 B.C. when Sumerian priests developed the *ancient system* of recording inventories, loans, taxes, and business transactions. The next major historical application of operation systems occurred in 4000 B.C. It was during this time that the Egyptians started using planning, organization, and control in large projects such as the construction of the pyramids. By 1100 B.C., labor was being specialized in China; by about 350 B.C., an archaic assembly line was formed in Greece.

In the Middle Ages, kings and queens ruled over large areas of land. Loyal noblemen maintained large sections of the monarch's territory. This hierarchical organization in which people were divided into classes based on social position and wealth became known as the *feudal system*. In the feudal systems, servants produced for themselves and people of higher classes by using the ruler's land and resources. The Renaissance brought about specialization in labor within the *European system*. This system had some of the first significant contributions that led to the industrial revolution. The industrial revolution occurred as the result of two conceptual phenomena: interchangeability of parts and division of labor. Division of labor was developed several centuries before interchangeability of parts. It was in the late eighteenth century when Eli Whitney popularized the concept of interchangeability of parts when he manufactured 10,000 muskets. Up to this point in history of manufacturing, each product (e.g., each gun) was considered a special order, meaning that parts of a given gun were fitted only for that particular gun and could not be used in other guns. Interchangeability of parts allowed the mass production of parts independent of the final products in which they will be used. This changed the factory layout from a job shop to a flow shop, which resulted in a substantial increase in product quality and speed of production.

In 1883, Frederick W. Taylor introduced the stopwatch method for accurately measuring the time to perform each single task of a complicated job. He developed the scientific study of productivity and identifying how to coordinate different tasks to eliminate wasting of time and increase the quality of work. The next generation of scientific study occurred with the development of (a) work sampling and (b) predetermined motion time systems (PMTSs). Work sampling is used to measure the random variable associated with the time of each task. PMTSs allow the use of standard predetermined tables of the smallest body movements (e.g., turning the left wrist by 90°) and integrating them to predict the time needed to perform a simple task. PMTSs have gained substantial importance due to the fact that they can predict work measurements without actually observing the actual work. The foundation of PMTSs was laid out by the research and development of Frank B. and Lillian M. Gilberth around 1912. The Gilberths took advantage of taking motion pictures at known time intervals while operators were performing the given task.

At the beginning of the twentieth century, *American Industry* invented the revolutionary idea of an assembly synchronized system. Henry Ford developed the first auto assembly system where a car chassis was moved through the assembly line while workers added components to it until the car was completed. During World War II, the growth of complex systems led to the development of efficient manufacturing methods and the use of advanced mathematical and statistical tools. This was supported by the development of academic programs in industrial and systems engineering disciplines as well as fields of operations research and management science (as multidisciplinary fields of problem solving). While systems engineering concentrated on the broad characteristics of the relationships between inputs and outputs of generic systems, operations researchers concentrated on solving specific and focused problems. The synergy of operations research and systems engineering allowed for the realization of solving large-scale and complex problems in the modern era. Recently, the development of faster and smaller computers, intelligent systems, and the World Wide Web (WWW) has substantially changed the nature of operations, manufacturing, production, and service systems.

Based upon this evolutionary history of production, operations, and manufacturing systems, it is clear that each era built upon the ideas of its predecessors. Table 1.3 shows a chronological display of this history and the major achievements for each period. The history and future of production and operation systems are divided into five phases. We believe that currently we are at the beginning of phase V.

I. Empiricism (learning from experience)
II. Analysis (scientific management)
III. Synthesis (development of mathematical problem-solving tools)
IV. Isolated systems with single objective (use of integrated and intelligent systems and WWW)
V. Integrated complex systems with multiple objectives (development of ecologically sound systems, environmentally sustainable systems, considering individual preferences)

As stated before, productions, operations, and manufacturing systems have been constantly changing over the past several thousand years. The initial production systems contained a significant amount of human insight and artistic creativity. However, the historical path of production systems has shifted more toward scientific models. Industrial engineering and management science were developed as a result of the industrial revolution, when the size and complexity of manufacturing increased significantly. These new complex systems, as well as the changes in the scale of businesses and organizations, have necessitated more precise and innovative models. The use of these modern models has resulted in highly efficient production and operations systems. Recent developments in computers, the Internet, and enhanced transportation systems led to the world becoming a "smaller place" with a global economy. Countries are no longer self-sufficient, and companies now produce for the world market. Therefore, it is necessary to view production, operation, and manufacturing systems from a global point of view, as opposed to the traditional regional or national perspective. Today, production, operation, and manufacturing systems face many new challenges. As a result, factors such as language barriers must be properly addressed. Large inventories are no longer feasible due to quick changes in demand and increases in holding costs. Designers must find new ways to design, manufacture, assemble, and transport products. As customized products and services replace standardized products and services, flexibility in manufacturing processes becomes more important.

Recent changes in the global market have changed the fields of production, operation, and manufacturing systems to handle the dynamic global needs. Information systems that were used to control, monitor, and maintain manufacturing and production systems no longer stop at the physical boundaries of the enterprise. They often interact with both clients and suppliers in order to more efficiently manage the supply chain system. Modern systems designers must overcome a multitude of challenges and problems. Considering

TABLE 1.3   Production, Operation, and Manufacturing System History and Perspective

Phase I: Empiricism	**Renaissance** • Double-entry bookkeeping and cost accounting **Industrial Revolution** • Began in the eighteenth century in British Isles • New technology, new forms of organization and management of production process • Replacement of people by machines • Improvements in machinery led to early mass production systems • Interchangeability of parts (Eli Whitney) • Specialization of labor (Adam Smith's "Wealth of Nations") • Dramatic reduction in unit cost of consumer products
Phase II: Analysis	**Scientific management** • "The Principles of Scientific Management," developed by Frederick Taylor (1895) • Scientific analysis and measurement of work • Emphasis on analysis (dissolving into basic elements) • Decomposition of a large problem into several smaller problems • Solving each smaller problem independently • Combining optimum results from small problems • Was criticized as an insult to human dignity—view man as a machine • Acknowledgment of human abilities and limitations • Development of more humane definition of role of people as elements in production process
Phase III: Synthesis	**Operations problem solving** • Developed after creation of "special teams" during World War II to solve problems related to war operations • Shifted focus from war tactical problems to industrial problems • Development of deterministic models • Development of stochastic models • Development of simulation models • Use of analytical models • Use of statistical models • Use of computers to solve problems • Synthesis aimed at producing balanced system solutions
Phase IV: Isolated systems with single objective	**Systems approach** • Complexity analysis • Structures, decomposition, and coordination of systems • Systems architecting • Identifying complex problems, variables, and their interactions • Looking at system as a whole with emphasis on integration **Globalization** • Global markets and global corporations • New international laws allowing global competition • Increase in choices and sensitivity to global markets

(continued)

TABLE 1.3 (*Continued*)

	Intelligent systems
	• Shift in emphasis from well-structured to ill-structured problems
	• Ability to integrate experience and intelligence with quantitative models
	• Learning from biological systems
	World Wide Web and Internet
	• Paperless communication and operations
	• Instant ordering and service
	• Worldwide collaboration and trade
	• Integrated supply chain management
Phase V: Integrated complex systems with multiple objectives	**Multiobjectives-based systems**
	• Developing methods for simultaneously optimizing quality, cost, productivity, flexibility, and ecological soundness of systems and products
	• Considering individual decision makers' multiple-criteria preferences
	• Development of customized products for massive-scale customers
	Risk measurement and analysis
	• Considering individual risk preferences in decision making
	• Measurement of risk for individual and organizational decision makers
	• Measurement and understanding of catastrophic risks
	Ecological, biological and energy-aware systems
	• Developing biologically inspired systems and products
	• Developing sustainable production, operations, and manufacturing systems
	• Developing environmentally friendly systems
	• Developing energy-efficient and renewable systems and products
	• Developing reusable/recyclable systems and products
	• Developing pollution and waste-free systems and products

ecological and human issues, a framework designed to optimize multiple objectives is essential.

While technology has been progressing at an amazing pace, centuries of abuse of natural resources, eradication of forests, and pollution of the whole Earth environment (air, soil, and oceans) are the dark sides of human advancements. A solution to these catastrophic human behaviors is to change unidimensional–single-objective consuming behavior to multidimensional–multiobjective behavior to develop environmentally friendly systems. Ecologically sound production and manufacturing systems can be developed to ensure that the natural environment is not negatively affected. "The ecologically sound systems" must be developed to allow life in all its forms to continue and thrive on Earth.

1.3 PRODUCTION AND OPERATIONS MODELS

In this section, we provide an overview of the production and operation problems and models and their objectives that are covered in this book.

1.3.1 Production and Operations Planning

Forecasting   Forecasting can be used to predict future patterns of behavior—for example, predicting demands for a certain product for each of the next 12 months. While it is unreasonable to expect an exact prediction using forecasting models, a good estimate with some degree of confidence of its occurrence can be found. Usually, the forecasting of periods farther into the future is less accurate.

Models	Objectives

• Qualitative models
• Time series models
  (a) Moving averages
  (b) Exponential smoothing
  (c) Cyclical/seasonal analysis
• Regression (cause-and-effects) models
  (a) Linear regression
  (b) Nonlinear regression models
  (c) Multivariable regression models
  (d) Qualitative regression models
	• Minimize risk associated with forecasting
	• Maximize confidence in forecasted values
	• Maximize accuracy in forecasted values
	• Maximize stability of forecasting

Aggregate Planning   Aggregate planning is concerned with medium-range planning for several periods (e.g., months). The basis of this planning model is to simplify the problem formulation by forming cumulative units of several products. An aggregate plan is developed based on forecasted demands to determine the optimal resource utilization to meet a given demand. Resources include regular workforce, facility capacity, inventory level, outsourcing, and subcontracting.

Models	Objectives
• Tabular and graphical approaches • Linear programming approach for single product • Linear programming approach for multiproducts • Disaggregating and master scheduling • Multiple-objective linear programming approaches	• Minimize total cost • Minimize total backlog • Minimize total inventory • Minimize fluctuations in hiring and firing

Push-and-Pull Systems: MRP and JIT   In push systems, the materials are produced and kept in inventory regardless of whether the downstream stations are ready to process them. The material requirements planning (MRP) model is a push scheduling that determines the time and the amount of the components of a product required in order to satisfy its demand for each given period of time. This model is dependent upon an accurate forecast, at which point the exact required quantities for all components or subassemblies can be determined. In material requirements planning, lot sizes for production of each component are determined based on forecasts, setup costs, and inventory costs. Just-in-time (JIT) is a pull system. In JIT, production planning of components is based on whether the downstream stations are ready for processing them. That is, the production scheduling is based on the internal demand of facilities that produce the items. In JIT, ideally no inventory is kept. In pull systems (JIT), materials are not shipped to downstream stations unless the downstream stations are ready to produce them.

Models	Objectives
• Tabular model for material requirements plan	• Minimize total inventory
• Lot-sizing methods	• Minimize total backorder
(a) Economic order quantity	• Minimize carrying costs
(b) Part period balancing	• Maximize achieving JIT
(c) Least-unit cost	
• Material resource planning	
• Just-in-time production system	
• Multicriteria hybrid push-and-pull systems	

Inventory Planning and Control   Inventory systems are used to determine how much inventory should be stored for each period of time in order to meet future needs. Inventories are hedged against uncertainties for possible shortages. There are four types of inventories: raw materials, product components, work-in-process, and finished goods. Inventories tie up precious capital resources and do not produce any direct income. Therefore, inventories must be minimized. However, inventory holding costs must be balanced with the cost of possible backlogs and unsatisfied customers. To be conservative, one may keep more inventory than is needed to avoid possible shortages due to future demand fluctuations and variations in production planning.

Models	Objectives
• Economic order quantity model	• Minimize total inventory cost
• Economic order quantity model with inventory discount	• Minimize possible shortages
• Economic production quantity model	• Maximize achieving JIT inventory plan
• Economic order quantity model allowing shortage	• Minimize risks
• Inventory with probabilistic demand	• Maximize flexibility
• Inventory of perishable items	

Scheduling and Sequencing   In order to produce a product, its components must be processed by a number of processors. Scheduling and sequencing are concerned with the optimal planning of a set of given tasks where each task must be processed by different processors (e.g., machines). Scheduling and sequencing can be used for repetitive products. Scheduling is an important tool in the control of production operations. Factors such as job arrival patterns, the variety of processors, processing time, the number of available processors, the labor force, and flow patterns are considered for determining a feasible schedule that minimizes total processing time.

Models	Objectives
• Sequencing n jobs on one processor	• Minimize tardiness
• Sequencing n jobs on two processor	• Maximize JIT
• Sequencing two jobs on m processors	• Minimize total processing time
• Sequencing n jobs on m processor	• Maximize machine utilization
• Sequencing flow shop problems	• Minimize work-in-progress inventory
• Sequencing job shop problems	

Project Management Project management is concerned with the planning of single large projects that consist of many activities. Projects are unique, that is, not repetitive in nature (e.g., building a stadium). To complete a project, many activities must be planned in a certain sequence. The traditional objective of project management is to sequence all project activities in such a way that the total project duration is minimized while the proper sequencing of activities is maintained.

Models	Objectives
• Critical-path method	• Minimize total cost of project
• Time and cost trade-offs	• Minimize total duration of project
• Project/program evaluation and review technique for probabilistic activities	• Maximize probability of completion
	• Minimize the required number of resources
• Planning with resource constraints	• Minimize risk of not meeting deadlines

Supply Chain and Transportation

Supply Chain As discussed before, the supply chain management system is an integrated system to manage the flow of materials from suppliers to the manufacturer and finally to the customer; see Figure 1.2. In supply chain management, the overall inventory is reduced at all levels, including raw materials inventory, work-in-progress inventory, and finished-goods inventory. Supply chain management is a model for efficiently managing a network of customers, retailers, distributors, suppliers, and manufacturers and effectively moving materials and information through the system.

Assignment The assignment problem is encountered in many applications: for example, to assign contractors to jobs, facilities to locations, and salespersons to territories.

Transportation   The objective of transportation problems is to minimize the total cost of delivering a quantity of goods from a set of origin points (e.g., suppliers) to a set of destination points (e.g., customers) where there are different transportation costs associated with each pair of point of origin and point of destination.

Transshipment   Transshipment is a broadening of the transportation problem where intermediate points (e.g., warehouses) are used between the origin points and destination points, creating a network of transportation problems.

Models	Objectives
• Supply chain management—network	• Minimize transportation costs
• Assignment methods	• Minimize transportation time
• Transportation methods	• Minimize assignment costs
• Transshipment method	• Maximize flexibility and agility
	• Maximize quality

Productivity and Efficiency   Productivity is defined as the amount of output per input for production and service processes. It is commonly measured in terms of output per unit worker per unit hour, which is known as labor productivity. Productivity growth is a rate comparing the productivity of a given period to a base period. Generally, higher productivity (and productivity growth) is associated with a higher standard of living in a country. Efficiency is used when comparing the productivity of comparable processes.

Models	Objectives
• Single productivity indexes	• Minimize cost
• Multiple productivity indexes	• Maximize accuracy
• Single efficiency indexes	• Maximize quantity
• Multiple efficiency indexes–linear programming approach	• Maximize quality
• Productivity and efficiency of system of multiple processors	

Energy Systems: Design and Operation   An energy system is defined based on the relationships of three entities: users, savers, and energy generators. A user is an entity that consumes energy for certain purposes. Examples of a user include cars, home appliances, and factory machinery. A user can receive energy from either a generator or a saver. A saver is a device that is able to store energy for future use. A generator is a device that generates energy. If a generator can produce energy faster than it is consumed, it can be saved by the saver for later use. This allows the saver to provide energy at a later time and prevents the generator from having to constantly produce energy. For different applications different energy systems can be designed and operated optimally.

Models	Objectives
• Energy conversion and types	• Minimize cost per unit of energy
• Multicriteria energy selection	• Minimize environmental impact per
• Energy routing optimization	unit of energy
• Optimization for design and operation of energy systems	• Maximize rate of extraction of renewable energy (per unit)
• Optimization for multiperiod operation of energy systems	• Maximize JIT use of energy
• Multicriteria energy systems	• Maximize efficiency of energy systems

1.3.2 Process Design and Logistics

Clustering and Group Technology   Clustering methods can be used to cluster groups of objects based on their attributes for serving certain objectives. An example of clustering is group technology. Group technology is a strategy that combines the benefits of mass production with the high product variation characteristics of a job shop environment. Group technology strategy reduces production waste, material handling, production time, waiting time, and setup time. This leads to a streamlining of the production operations to achieve a flexible manufacturing system based upon grouping families of parts and clustering similar machines. Parts are generally grouped by similarity in use of production machinery. Machines that process similar parts are placed in the same cluster to maximize efficiency.

Models	Objectives
• Rank order clustering	• Minimize material flow (and cost)
• Similarity coefficient clustering	• Minimize setup time
• P-median mathematical programming	• Minimize cost of duplicated machines
• K-means algorithm—iterative	• Maximize flexibility
• Hierarchical clustering	• Maximize productivity

Cellular Layouts and Networks   Many design (e.g., layout) problems can be formulated as a single-row network (e.g., layout). The materials (or information) flow can be either unidirectional or bidirectional. For practical purposes, manufacturing facilities usually divide their facilities into separate self-sufficient and self-contained cells. These areas are then designed in the form of a single row or U-shaped layout. These layouts are advantageous because they are easy to monitor, maintain, and utilize. The autonomous behavior of cells and their independence improves the workflow and reliability of the system. At times, however, the duplication of certain workstations (machines) may become necessary for the purpose of having independent cells where each cell operates independent of other cells.

Models	Objectives
• Single-row unidirectional layout	• Minimize material flow (and cost)
• Single-row bidirectional layout	• Maximize productivity
• Head–tail heuristic	• Maximize flexibility
• Integer programming—optimization	• Maximize agility

Assembly Systems   Assembly systems are generally used for the assembly of high-volume and low-variation products. Assembly line balancing is used to design an effective arrangement of the workstations needed to assemble a product. This form of production became popular at the beginning of the twentieth century with automobile assembly line production systems. Several objectives must be considered when designing effective and flexible assembly lines.

Models	Objectives
• Single-product line balancing	• Minimize number of stations
• Mixed-product line balancing	• Minimize cycle time (to maximize productivity)
• Stochastic assembly line balancing	• Minimize operational cost

Facility Layout   Layout design is concerned with identifying the optimal locations of machines and departments within a given facility such as factories, offices, schools, banks, airports, and nearly every type of business. Factors influencing the layout of a production facility include available space, material flow costs, mobility, and employee satisfaction with the work environment. Different types of facilities are product, process, fixed position, cellular, and group technology. In designing an effective layout, one should consider several objectives. In construction models a solution is constructed; in improvement models a given solution is improved.

Models	Objectives
• Systematic layout planning—construction	• Minimize total materials flow and cost
• Pairwise exchange model—improvement	• Minimize total transportation time
• Qualitative-based model—construction	• Maximize space utilization
• Head–tail model—construction	• Maximize flexibility
• Relayout model—improvement	• Maximize ease of use
	• Maximize ease of expansion

Location Decisions   The process of determining a geographic site for a facility is an important aspect of strategic decision making. In determining a facility location, factors such as availability, cost of location, technology, labor, utilities, accessibility, environmental conditions, and interactions with suppliers and customers must be considered. Different mathematical and analytical models can be used to make this decision.

Models	Objectives
• Break-even analysis • Assignment model • Rectilinear distance measurement • Euclidean distance measurement • Single-facility location model • Multifacility location model	• Minimize fixed and variable costs such as transportation and distribution • Maximize future flexibility and expansion • Maximize proximity to suppliers, resources, facilities, and customers • Maximize quality of products

1.3.3   Quality, Reliability, and Measurement

Quality Control and Assurance   Quality control methods can be used to identify quality problems during and after the production process. Efficient sampling plans are used to predict the quality of products or processes while using minimal inspection costs. For the determined level of quality, a proper sampling approach should be used to minimize the number of defective products shipped to customers. Quality function deployment is an approach for incorporating the customer voice into the manufacturing process.

Models	Objectives
• Control charts (continuous and discrete measurements) • Operating characteristic curve • Sequential sampling plans • Average outgoing quality • Quality function deployment (QFD) • Multicriteria QFD	• Minimize cost of quality control • Maximize average outgoing quality • Maximize accuracy of measurements • Maximize confidence in results • Minimize risk to consumer • Minimize risk to producer

Work Measurement and Analysis   With many corporations and firms focusing on efficiency gains and cost reductions, statistical based work measurement methods are becoming popular. Work measurement involves the use of time studies to measure, predict, and set standards for the acceptable amount of time that specific tasks and jobs should take.

Work measurement and labor standards have applications in manufacturing operations, service industries, government, and educational institutes.

Models	Objectives
• Stopwatch studies—continuous time	• Minimize cost of measurement
• Work sampling—discrete time	• Maximize accuracy of measurement
• Predetermined time standards	• Maximize confidence in results
• Multicriteria sampling plans	• Minimize risk

Reliability and Maintainability   Reliability refers to the likelihood that a product will perform up to its specifications for an expected period of time. It can also be used to predict future costs and to determine optimal maintenance and replacement policies. Proper maintenance policies can prolong the life of a system or product and minimize the failure rates. Replacement polices are used to identify when a component or a system should be replaced to minimize the total cost and the risk. A balanced decision between maintenance costs and replacement costs must be made while considering the required level of reliability.

Models	Objectives
• Reliability of single components	• Maximize total reliability of system
• Mean time between failure rates	• Minimize total cost
• Reliability measurement of sequential and parallel systems	• Minimize total number of breakdowns
• Maintenance policies	• Minimize total preventative and breakdown costs
• Replacement policies	• Minimize risk of failure

1.4   SYSTEMS APPROACH AND TOOLS

1.4.1   Systems Approach

Systems Engineering and Development   Systems engineers use modeling to plan for the development and construction of a system. The systems engineering process encompasses the design, development, and construction of a system and tracking the progress through the system life cycle. Systems engineering also encompasses integration approaches and management throughout the entire life cycle of a system. Every system is different, but the fundamental steps involved in the systems engineering process remain the same for all systems. The systems engineering process is an iterative process. Based on such objectives, the system is constructed. Systems engineering is an evolving field which continues to grow as new complex problems are encountered.

Models	Objectives
• Physical systems (micro vs. macro) • Biological systems (micro vs. macro) • Man-made systems (conceptual vs. physical) • Environment-based systems (static vs. dynamic) • Conceptual systems (logical vs. intuitive) • Dynamic systems (nonadaptive vs. adaptive) • Mathematical systems (linear vs. nonlinear) • Boundary-based systems (closed vs. open) • Goal-seeking systems (single objective vs. multiobjectives)	• Maximize efficiency of systems • Maximize productivity of systems • Maximize quality of systems • Minimize cost of systems • Maximize ease of use • Maximize ease of modification • Maximize simplicity of models

Systems Architecting and Modeling Systems architecting is the art and science of designing complex systems, which are also called systems of systems. Systems architecting is based on multiple disciplines of art, science, management, and engineering. In the design of a system, the systems architect must consider the objectives of different stake holders. The systems architecting process creates a systems architectural design by using the customer's requirements to produce a conceptual model. This process is performed by the systems architect, but it has feedback from many parties, including clients, systems engineers, and the developers of subsystems and components.

Models	Objectives
• Analog models • Physical models • Analytical models • Formal models • Managerial models	• Maximize customer's satisfaction • Maximize simplicity of design • Maximize quality of systems • Minimize project duration • Minimize project cost • Maximize aesthetic of system

1.4.2 Multicriteria Decision Analysis

Multicriteria Decision Making The first step in solving a MCDM problem is to clearly define the problem and its constraints. Once the problem is understood and formulated, the next step is to determine the objectives associated with the problem. Then one should apply a proper model for solving and analyzing the problem. Then, a set of promising feasible alternatives satisfying constraints should be generated and analyzed. It is important to realize that there may be many different alternatives to a given problem.

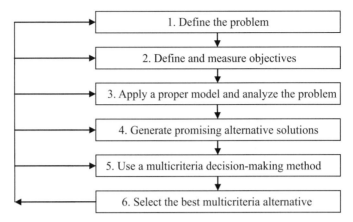

Figure 1.4 Flowchart of multicriteria problem-solving approach.

But, the set of promising alternatives is a small subset of alternatives. A MCDM approach (see Chapter 2) can be used to evaluate alternatives and choose the best multicriteria (also called the most compromised) alternative. This multicriteria process is shown in Figure 1.4. Each phase of decision making can be modified based on feedback from other phases.

Often, the decision-making process is complex and requires analyses of many different aspects, including several different objectives (criteria) and perspectives. The decision maker should always consider different alternatives and evaluate them based on given criteria. The MCDM methods can be used for evaluating, rating, and ranking alternatives while considering several conflicting objectives.

Models	Objectives
• Identifying efficient alternatives	• Minimize efforts of decision-making process
• Ranking by additive utility functions	
• Ranking by Z utility theory functions	• Maximize reliability of approach
• Ranking by closeness to goals	• Maximize simplicity of approach
• Ranking by interactive methods	• Maximize robustness of approach
• Ranking by paired comparisons of alternatives	• Maximize accuracy of approach

Decision Making and Process Understanding the decision process can provide insight into why people make decisions in certain ways. Decision makers who are conscious of their decision process and types can make more effective and balanced decisions. The decision process model is based on four dimensions, where each dimension has two opposing types: information processing (concrete and abstract), alternative generation (adaptive and constructive), alternative evaluation (risk averse and risk prone), and decision closure (decisive and tentative); see Chapter 2. We provide approaches for assessing and optimizing the decision process. In decision making, expected values are used to determine the best alternative while considering the payoffs associated with different probabilistic states of nature (of future events). Decision trees are one of the primary visual tools used in decision analysis. Decision trees can also be used to present multiphase or sequential decision-making problems. To estimate the value of additional information (to improve expected value result), Bayes's approach is used.

Models	Objectives
• Decision-making process and four types • Decision trees and tables • Sequential decision making • Expected value with partial information on probabilities • Worth of additional information (Bayesian approach)	• Maximize optimal value of having more information • Maximize optimal value of having more alternatives • Maximize expected value of payoffs • Maximize optimal timing of decision making • Maximize worth of additional information

Risk Analysis In risk analysis for decision making, consider two cases: decisions under risk (with known probabilities) and decisions under uncertainty (with unknown probabilities). Utility functions are used for presenting and reflecting the decision maker's risk attitude. Expected-utility theory is the classical approach for assessing utility function but it has its limitations. Z utility theory (Malakooti, 2014b) generalizes expected-utility theory and provides mechanisms to quantify risk as an objective measurement independent from the decision maker. Z utility theory functions can be easily assessed, verified, and graphically presented; they can be used for both risk and uncertainty. Z utility theory functions can also be used to identify the type of decision-making behavior: risk averse, risk prone, risk neutral, and mixed risk averse/prone. The picture depicted on the front cover of this book shows Z utility functions as birds that can fly nonlinearly in any direction and having different degrees of freedom.

Models	Objectives
• Decisions under risk • Expected utility theory • Z utility theory • Catastrophic risk measurement • Regret/reward theory • Multiplicative and feasible goal approaches • Decision making under uncertainty	• Maxmize expected value of (e.g., monetary) payoffs • Maximize qualitative payoffs • Minimize potential loss for risk averse • Maximize potential gain for risk prone

1.4.3 Problem-Solving Tools

Linear and Nonlinear Programming (Optimization) Optimization is a mathematical tool used to obtain the optimal solution (that optimizes an objective function) to a set of constraints on decision variables. The constraints are a set of equalities and/or inequalities that restrict the range of feasible alternatives. For example, in production problems, the objective is cost (to be minimized) subject to limited resources such as labor, materials, machines, and capital constraints. Optimization helps the decision maker plan and allocate resources effectively and efficiently. There are four classes of optimization problems:

- Linear programming (LP): linear objective and constraints
- Nonlinear programming (NLP): nonlinear objective and/or constraints
- Integer programming (IP): LP or NLP where some decision variables must assume integer values
- Combinatorial programming (CP): optimization problems where the set of alternatives is combinatorial

Models	Objectives
• Graphical solutions for two variables • Simplex method for solving LP • Feasible direction for solving NLP • Steepest descent methods • Constrained and unconstrained optimization • Computer-based algorithms for LP, NLP, IP, and CP • Sensitivity analysis of coefficients • Economic interpretations • Heuristics for combinatorial problems	• Maximize accuracy of presenting model • Minimize number of constraints, decision variables, and objectives • Maximize simplicity of model • Maximize ease of communication to decision maker • Maximize chance of finding optimal solution • Minimize computational effort to find optimal solution

Queuing Systems   In most queuing problems, objects (customers or products) arrive to the queuing line in a random fashion. Similarly, the service time by the server (processor) to process each object is random. Based on the characteristics of a queuing line, many performance measurements, such as the average service time per object, can be found. There can be one or several channels of arrival to each processor. The service can occur in one phase or several phases. Queuing systems are used extensively in industry as well as service industries and organizations. Minimizing total cost, average service time, and waiting time is an important objective when designing a queuing system.

Models	Objectives
• Single server, single phase • Single server, multiphase • Multiple server, single phase • Multiple server, multiphase	• Minimize number of servers • Minimize waiting and service time • Minimize costs • Maximize quality of service • Maximize long-term productivity

Simulation   Simulation is an important analytical tool for analyzing complex systems. Simulation can be used to assist decision makers in experimenting with different possible alternatives prior to choosing one. Simulation incorporates and performs many "runs" of a given scenario to achieve a realistic performance of the systems under study. Some complex systems can be modeled by simulation.

Models	Objectives
• Generating random numbers	• Maximize accuracy of model
• Continuous-event simulation	• Minimize cost of modeling
• Discrete-event simulation	• Maximize validity of model
• Random-number generation	• Minimize computational time
• Inverse transformation functions	
• Learning in simulation	
• Multiobjective optimization in simulation	

1.5   MULTICRITERIA PRODUCTION/OPERATION SYSTEMS

1.5.1   Pyramid of Multicriteria Production/Operations

There are five long-term objectives that should be considered in operational systems:

Cost   Cost can be measured in terms of monetary units (e.g., dollars). It usually consists of fixed and variable costs. Generally, the present worth of costs can be calculated by considering relevant interest and inflation rates. Cost is the easiest objective to measure for the current time or the committed near future.

Productivity   Productivity can be measured in terms of the number of products produced during a period of time (e.g., 100,000 products per 40 h).

Quality   Quality can be measured in a variety of ways. In the broadest sense, it can be measured by the customer's satisfaction for the intended use of the product. From a production standpoint, quality is a measure in terms of the conformation of a manufactured product to its design specifications. These measurements are discussed in Chapter 17.

Flexibility (Reconfigurable and Agility)   Flexibility can be measured in a variety of ways. It may refer to what extent a system can be changed (reconfigured) to produce a different line of products. It may also refer to the ease of changing the layout for future expansion (or reduction) and technological upgrades. Agility may refer to the speed at which the changes may take place.

Ecological Soundness   Ecological soundness refers to the biological and environmental impacts of the system under study. Ecologically sound systems support energy efficiency, sustainability, minimal hazardous by-products, and long-term buildup of the natural environment.

Multicriteria Pyramid of Production and Operation Systems   Figure 1.5 shows the relationship between the five objectives of production and operation systems. The tip

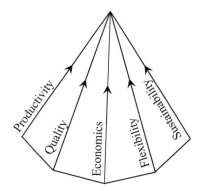

Figure 1.5 Multicriteria pyramid of production and operation systems shows feasible and efficient multicriteria alternatives; tip of pyramid can be achieved at the highest cost.

of the pyramid is associated with the highest productivity, the highest quality, the most economical, the most flexibility, and the most sustainability. The points inside the pyramid are associated with different combinations of the five criteria. The tip of this Pyramid is the ideal point, but it is infeasible. The base of this Pyramid consists of the worst points. Therefore, in practice, one should make a tradeoff among these five objectives to find a feasible solution closer to the tip of the Pyramid. In the next section, we discuss how to identify efficient alternatives which are not dominated by other points. This book provides an overall approach for maximizing these five objectives.

Trade-Off Analysis and Efficiency   Once the target values for costs, productivity, quality, flexibility, and ecological soundness are set, strategic decisions and needed resources to achieve such objectives can be determined. In the following, the trade-offs between cost and the other four criteria are discussed; these are presented in Figures 1.6a, b, c, and d. In order to have consistent notation for maximizing all objectives, consider maximizing the inverse of cost (1/cost), which is the same as minimizing cost. Therefore, all these five objectives are maximized. To illustrate the concept of efficiency, for example, consider flexibility and the inverse of cost (1/cost). An alternative is inefficient (or dominated) if there exists another alternative that is better or equivalent in terms of both objectives. For example, in Figure 1.6a, alternative $(1/C_1, F_1)$ is dominated by alternative $(1/C_1, F_2)$; it is inefficient. The set of alternatives can be reduced to a set of efficient alternatives.

Cost versus Flexibility   Strategies that heavily emphasize minimizing cost may lead to more rigid and inflexible systems. Likewise, the strategy that opts for more flexibility is associated with higher cost. Consider Figure 1.6a. All alternative strategies on the curve labeled as "current trade-off" are efficient with respect to each other. This is an indifference curve meaning that the feasible alternatives in terms of C and F are located on this curve. In order to have an increase in flexibility, the cost can be identified. Suppose that point $(1/C_1, F_1)$ is selected as the operational system by a given industry. The operation system is then designed for this given flexibility and cost. However, with the progress of technology as time passes, improvements will shift the curve upward, labeled as "future trade-off." The

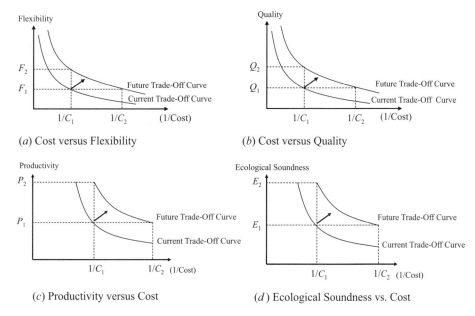

Figure 1.6   Current vs. future trade-off curves. Future trade-off curve will be better than current trade-off curve because of technological and process improvements. Also, one current alternative will lead to many improved alternatives in the future.

current point $(1/C_1, F_1)$ could be replaced by one of the choices between points $(1/C_2, F_1)$ and $(1/C_1, F_2)$ on the future trade-off curve. In making a strategic decision, one should design a system that considers the future improvement in cost and flexibility. That is, the future choices depend on the choice made at the current time. Therefore, at the current time, one should consider an alternative that provides the most choices for cost and flexibility in the future.

Cost versus Quality   Now consider quality versus cost. Figure 1.6*b* demonstrates a similar relationship to Figure 1.6*a*. Once a strategy [e.g., a given cost and quality (C_1, Q_1)] is selected, the future improvements in cost and quality will render more choices on the future trade-off curve, that is, all points between $(1/C_2, Q_1)$ and $(1/C_1, Q_2)$ on the future curve.

Cost versus Productivity   Similarly, Figure 1.6*c* shows the trade-off between cost and productivity. A higher cost (i.e., lower 1/cost) is associated with a higher productivity assuming everything else is constant. For example, for the same quality and flexibility, it will cost more to have a system with higher productivity.

Cost versus Ecological Soundness   Similarly, Figure 1.6*d* shows the trade-off between cost and ecological soundness. There is a trade-off between cost and ecological impact. Currently, it costs more to develop systems that are more ecologically friendly. In the future, it may be possible to develop systems that are less costly and more ecologically friendly.

1.5.2 Hierarchical Multicriteria Planning

In this section, we propose a hierarchical framework for production and operation systems. See Figure 1.7. In this framework, six levels are considered based on the duration of the planning. Level 1 has the longest planning horizon, and level 6 has the shortest one. This hierarchical multilevel system is characterized by vertical (or top-down) decomposition

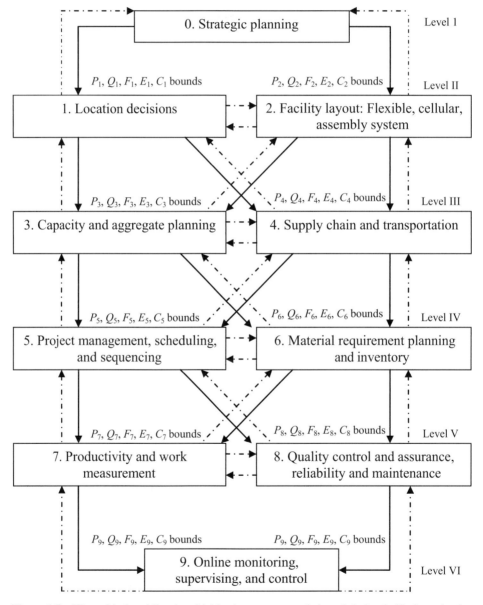

Figure 1.7 Hierarchical multilevel multiobjective system consisting of six levels. Each production planning function communicates to other functions based on their common objectives: productivity level (P), quality level (Q), flexibility (F), ecological soundness (E), and cost (C).

and the interdependence of adjacent levels. Higherlevel planning and decision making are usually associated with longer durations and more uncertainty compared to the lower levels. In this hierarchical formulation of production planning, the initial action starts with higher levels and bounds on objectives are given to lower levels. The objectives are:

- Maximize P (productivity)
- Maximize Q (quality)
- Maximize F (flexibility)
- Maximize E (ecological soundness)
- Minimize C (cost)

The information on objectives is provided in terms of lower and upper bounds where the bounds are iteratively reduced according to the interaction among connected subsystems. Figure 1.7 diagrams the interrelation of the six-level problems. A solid arrow indicates the dictation of bounds on objectives from a parent to children. Children provide feedback or make adjustments using a dashed arrow. For the same child, if the given bounds from two parents are not feasible, then negotiations should occur to find a feasible solution and the best compromise solution. Each level is discussed in the following.

Level I   At the highest level strategic planning, bounds on all five objectives, P, Q, F, E, and C, are determined for a long period of time (e.g., for several years).

Level II   In location decisions, the best locations of facilities are determined based on objectives such as minimizing transportation and distribution cost, maximizing proximity to suppliers, resources, and facilities, and maximizing proximity to customers. In the facility layout problem, the best arrangement and location of departments (or machines) within a given facility are determined. The objectives of the facility layout problem include minimizing the materials movement, maximizing the flexibility, and minimizing the materials movement cost.

Level III   In capacity and aggregate planning, the optimal master schedule is determined. Based on this master schedule, broad planning in terms of aggregated units and the needed capacities is determined. Objectives include minimizing the production cycle or project duration, satisfying priorities on due dates, and minimizing the project cost such as resources, labor, and machinery. In supply chain and transportation, optimal plans to deliver materials to manufacturers and from manufacturers and suppliers to customers are determined. Objectives include minimizing transportation cost, minimizing transportation time, and maximizing the quality of transportation.

Level IV   In project management, scheduling, and sequencing, the best ways to assign jobs to different departments, machines, or contractors are determined. Objectives include minimizing durations satisfying priorities (e.g., due dates) and minimizing cost. In material requirement planning and inventory, optimal plans to acquire necessary components and materials to manufacture the products based on the demands are determined. Objectives include minimizing total inventory, minimizing total backorder, and maximizing achievement of JIT.

Level V   In productivity and work measurement, the best approaches for measuring productivity and work measurements are used. These measurements can help higher level decision-making units to be more realistic about their target objectives. Objectives include maximizing the accuracy of measurements, maximizing the confidence level in determined values, and minimizing the total cost (and time) of making the measurements. In quality control and assurance, the quality of products is measured in accordance with the required specifications. Objectives include maximizing the number of outgoing items, maximizing the quality of outgoing items, and minimizing the cost of quality control. In reliability, reliability of the system or product is measured. In maintenance, the best maintenance policy for the optimal performance of the system or product is determined. Objectives include minimizing cost, maximizing systems' life, and maximizing the quality of products.

Level VI   In online monitoring, supervision, and control, the main concern is to adjust the system parameters automatically with respect to changes in raw materials and manufacturing dynamics. An automated system would simulate the decision maker's behavior (preferences) instantly. Objectives for online control are minimizing the time needed to stabilize the systems' behavior to achieve certain goals, minimizing the deviation from the goals, and minimizing the cost of production, operations, and monitoring.

1.6   PRODUCT AND PROCESS LIFE CYCLE

1.6.1   Product Life Cycle

The product life cycle is commonly used to describe the life of a product over time. This cycle consists of four stages: product introduction, growth, maturity, and decline. A product life-cycle curve is shown in Figure 1.8 where the annual sales volume of a product is plotted against time. The four phases are described below.

Phase I: Introduction   During the introduction of a product, production and manufacturing facilities are designed to produce a small number of products. In this stage, manufacturing costs are high. Flexibility in production methods is needed in order to adapt to changes that will occur in the future. Marketing for the product is just beginning, and the sales volume slowly increases.

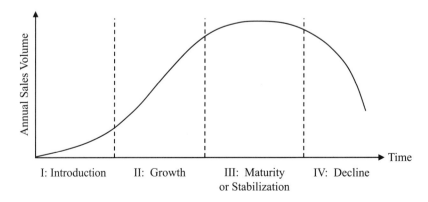

Figure 1.8   Product life-cycle curve.

Phase II: Growth   In the growth phase, the sales volume increases rapidly. Improvement methods are introduced in this stage. Permanent equipment is acquired to plan for a stable system for the next stage. A concerted effort to gain market share is made in this stage as competing products emerge.

Phase III: Maturity   The maturity phase is reached when demand for the product does not increase anymore and there is little room for improving the process. This phase is also referred to as stabilization. The focus for marketing the product relies on competitive pricing. In this phase, the most efficient and least expensive production method is used. The economy of scale substantially helps to keep the market price low and competitive.

Phase IV: Decline   Finally, the product enters the fourth stage of its life cycle, the decline. At this point, the market becomes saturated and sales volume declines. Creative marketing techniques may be used to maximize the sales of products. The product may be modified in order to prolong its life cycle. In some cases, decline may occur after a long period of time.

Every product may have a different life cycle in terms of the duration of each of its four phases. The challenge for decision makers is to make production planning congruent with the product's life cycle. Each phase should be planned in terms of material procurements, labor hiring, production capacity, marketing, and sales. Most industries (such as auto industries) try to predict the life cycle of their products and begin new models even before the maturity of their current models occurs. In Figure 1.9 three curves for three generations of the same type of products are shown. When the decline of product *a* occurs, product *b* is at its maturity, and product *c* is being introduced. Following this strategy, a company will stay competitive and dynamic.

1.6.2   Process Life Cycle

The process life cycle refers to the development of process capability for a given line (type) of products. For example, in the auto industry, once the process capability of producing certain types of cars is stabilized, this process can be used to produce a variety of car models, including the new lines of cars that will be developed in the future. The life cycle of a process will usually outlast many generations of product lines. The process life cycle also consists of four phases; see Figure 1.10.

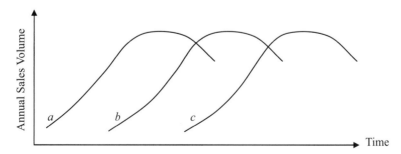

Figure 1.9   Product life cycle of different generation of same type of products, where product *b* replaces product *a*, product *c* replaces product *b*, and so on.

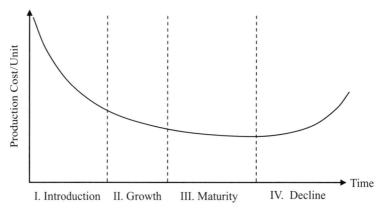

Figure 1.10   Process life cycle.

The basic idea behind the process life cycle is that as time goes on, manufacturing for mass production of a product becomes more efficient, skilled, and also automated, and products are produced more quickly and at a lower cost per unit. This is due in large part to the stabilization of the product's production capabilities and learning curves (discussed in the next section).

1.6.3   Product–Process Matrix

The relationship between the types of products and the types of processes can be shown through the use of the product–process matrix. The product–process matrix is a useful visual tool for providing information regarding the interaction of the product and process volume, flow, and flexibility. This matrix is shown in Figure 1.11. In Figure 1.11, by moving diagonally down from the upper left-hand corner to the lower right-hand corner, different

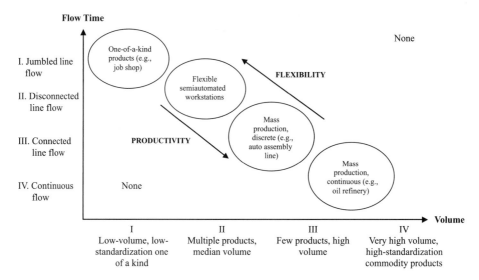

Figure 1.11   Product–process matrix classifying different products and process.

pairings of product and process types occur. In the upper left-hand corner, jobs are unique and manufacturing costs per unit are high. There is much flexibility in the production processes. Moving down the diagonal line, flexibility is reduced, but manufacturing costs per unit decrease. Product design and manufacturing processes become more standardized by moving toward the lower right-hand corner of the matrix. In this direction, production volume also increases. The top right and the bottom left are labeled as "none" because it is technologically and economically impossible to operate in these regions. For example, it is not economically feasible to develop an automated mass production system to only produce a few one-of-a-kind products (the bottom left).

1.7 LEARNING CURVES

Generally, with the passage of time, production systems become more efficient as workers and managers learn to become more effective. The learning process is especially rapid for repetitive systems (e.g., mass production and assembly systems). A famous learning curve is Moore's law, which regards the assembly time of computing hardware. It states that the number of transistors that can be placed on an IC doubles approximately every 18 months. For the past 50 years, this trend has been observed, but it is expected that this rate will start to flatten out in the next 10–15 years. Learning from experience is not the only reason for improved efficiency. Other factors are better machines and tools, streamlined production lines, automation, and improved products and processes. A learning curve is a graph that depicts labor hours required per unit of product versus the total number of units produced. For example, for a 60% learning rate, if the first unit takes 20 h, then the second unit will take $0.6 \times 20 = 12$ h, and the fourth unit will take $0.6 \times 12 = 7.2$ h. Continuing in the same manner, the 1024th unit will take only 0.12 h. As the number of units increase, the curve becomes flatter. This given example (learning rate of 0.6, or 60%) is very fast and not realistic for many industries. In practice, for most industries, the learning rate varies from 0.85 (or 85%) to 1 (or 100%). Three learning curves are presented in Figure 1.12 and Table 1.4.

Neglecting the learning curve for any labor-based organization will lead to overstaffing labor and additional cost. Learning curves are generated based on past experiences, so one

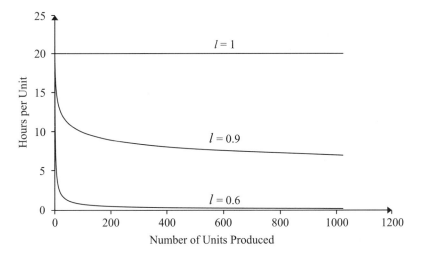

Figure 1.12 Learning curves for 60, 90, and 100% learning rates.

TABLE 1.4 Production Time for *n*th Unit for Learning Rates of 60, 90, and 100%, the Higher Learning Rate Associated with Less Productivity

	60% Rate		90% Rate		100% Rate	
n Units Produced	$0.6 \times T(n/2)$	Production time for *n*th unit, $T(n)$ (h)	$0.9 \times T(n/2)$	Production time for *n*th unit, $T(n)$ (h)	$1 \times T(n/2)$	Production time for *n*th unit, $T(n)$ (h)
1	—	20	—	20	—	20
2	$(0.60) \times 20$	12	$(0.90) \times 20$	18	1×20	20
4	$(0.60) \times 12$	7.2	$(0.90) \times 18$	16.2	1×20	20
8	$(0.60) \times 7.2$	4.32	$(0.90) \times 16.2$	14.58	1×20	20
16	$(0.60) \times 4.32$	2.59	$(0.90) \times 14.58$	13.12	1×20	20
32	$(0.60) \times 2.59$	1.56	$(0.90) \times 13.12$	11.81	1×20	20
1024	$(0.60) \times 0.20$	0.12	$(0.90) \times 7.75$	6.97	1×20	20

should make sure that the recorded information is accurate and relevant. The learning curve should be reevaluated when more recent data become available. Two different approaches, the arithmetic and the logarithmic, can be used to solve learning curve problems, that is, find the learning rate.

1.7.1 Arithmetic Approach

The arithmetic approach uses the following basic equation that defines learning curves:

$$T(2n) = lT(n) \tag{1.1}$$

where

$$n = \text{number of units}$$
$$T(n) = \text{production time for } n\text{th unit}$$
$$l = \text{learning rate, where } 0 < l \leq 1$$

When $l = 1$, no learning occurs, that is, the time to produce an item remains the same. As l decreases, the learning rate increases, and it takes less time to complete the same job.

Example 1.1 Aircraft Industry Using Arithmetic Approach   In an aircraft industry it takes 1000 h to build the first aircraft and 700 h to build the second aircraft.
Find the learning rate l and the time it takes to produce the 4th, 8th, ..., 1024th units.

First, find the learning rate l, where $T(1) = 1000$ and $T(2) = 700$. Consider Equation (1.1) where $n = 1$:

$$T(2) = lT(1) \qquad 700 = 1000\, l$$
$$l = \frac{700}{1000}$$
$$= 0.70$$

Therefore, use $T(2n) = 0.7T(n)$ to find the time it takes to build the products. These times are presented in the table below:

nth (aircraft)	1	2	4	8	16	32	64	128	256	512	1024
$T(n)$	1000	700	490	343	240	168	118	82.4	57.7	40.4	28.3

1.7.2 Logarithmic Approach

The arithmetic approach can be used to calculate the production time per unit for units $n = 2^k$, where k is a positive integer. However, the arithmetic approach cannot be used to predict the production time for any arbitrary number (e.g., 25th or 341th). The logarithmic approach can be used to find the value of $T(n)$ for any given n (the time it takes to produce the nth item) as follows:

$$T(n) = T(1)n^{-b} \qquad (1.2)$$

where

$T(1) =$ production time for 1st unit
$b =$ coefficient exponent of learning curve

Relationship of b to Learning Rate l   Consider two numbers, n and $2n$. From Equations (1.1) and (1.2), the equation can be written as

$$l = \frac{T(2n)}{T(n)} = \frac{T(1)(2n)^{-b}}{T(1)n^{-b}} = 2^{-b} \qquad \text{or} \qquad l = 2^{-b} \qquad (1.3)$$

Solving for b,

$$l = 2^{-b} \qquad \log(l) = -b\log(2) \qquad b = -\frac{\log(l)}{\log(2)} \qquad (1.4)$$

Example 1.2   Logarithmic Approach: Detail Example   Consider Example 1.1, where the learning rate is 70% (i.e., $l = 0.70$) and the time it takes to produce the first unit is 1000 h, that is, $T(1) = 1000$.

(a) Find the time required to produce the 600th unit.
(b) Find the time it takes to produce unit number 50, 100, 500, 1000, 2000, and 5000.
(c) Do parts (a) and (b) if the learning rate is 90%.

Solution

(a) First, find the value of b by using Equation (1.4):

$$b = -\frac{\log(0.70)}{\log(2)}$$

$$= 0.515$$

Using Equation (1.2),

$$T(n) = 1000(n)^{-0.515}$$

Therefore, the production time for the 600th unit is

$$T(600) = 1000(600)^{-0.515} = 37.1 \text{ h}$$

(b) For $l = 0.70$ and $b = 0.515$, use Equation (1.2) to find $T(n)$ for different values of n:

n (aircrafts)	1	50	100	500	1000	2000	5000
$T(n)$	1000	133.36	93.325	40.741	28.51	19.951	12.446

(c) For $l = 0.90$, using Equation (1.4), $b = 0.152$. Using Equation (1.2), we find:

n (aircrafts)	1	50	100	500	1000	2000	5000
$T(n)$	1000	551.76	496.59	388.83	349.95	314.95	274

1.7.3 Deteriorating Curves

In certain applications, the productivity (or quality or efficiency) of some systems decreases as the system gets older. For example, due to older age, a machine may require more shutdowns for needed repairs. Another example is how the output of a battery may decrease due to deterioration with age. In this case, Equations (1.1), (1.2), and (1.4) can be used where the learning rate $l > 1$, that is, over 100%. An example of a deteriorating curve is shown in Figure 1.13.

Example 1.3   Deteriorating Curve   As time passes, a given machine will deteriorate and will require more shutdowns for needed repairs. Suppose that for this machine it takes 1000 h to produce the first unit, and the learning rate is 110% (i.e., $l = 1.1$). Respond to parts (a) and (b) of Example 1.2.

Solution

(a) First, find the value of b using Equation (1.4):

$$b = \frac{\log(1.1)}{\log(2)}$$

$$= -0.138$$

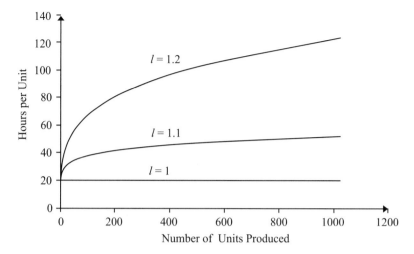

Figure 1.13 Deteriorating curve for 110, 120, and 100% deteriorating rates.

Using Equation (1.2),

$$T(n) = 1000(n)^{0.138}$$

Therefore, to produce the 600th unit, it will take

$$T(600) = 1000(600)^{0.138} = 2418 \text{ h}$$

(b) For $l = 1.1$ and $b = -0.138$, use Equation (1.2) to find $T(n)$ for different values of n:

n (aircrafts)	1	50	100	500	1000	2000	3000	4000	5000
$T(n)$	1000	1716	1888	2358	2594	2855	3019	3141	3239

1.8 CAPACITY PLANNING

1.8.1 Capacity Strategies

Capacity changes (growth or decline) must be continuously reevaluated by different organizations. This decision is based on the long-term forecasting of demand for products and services. Capacity planning is based on considering future (predicted) demands, technological changes, changes in construction costs, interest rate fluctuations, and the prediction of strategies employed by competitors. The relationship between capacity growth and future (predicted) demands can be categorized into three general strategies which are illustrated in Figure 1.14.

(a) *Capacity Leads Demand* In Figure 1.14a, planned capacity growth exceeds the predicted demand. This strategy allows a company to meet the demands without having shortages when there are unexpected increases in demand. The more than needed capacity in terms of facility, manpower, and inventory costs makes this plan

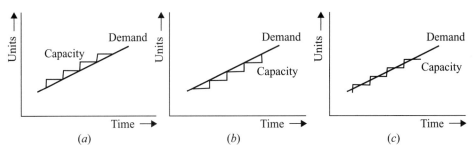

Figure 1.14 Capacity-planning strategies: (*a*) capacity leads demand; (*b*) capacity lags demand; (*c*) capacity matches demand.

less advantageous. This strategy is conservative in terms of consumer satisfaction but risky because of a possible loss of income from the excess capacity.

(b) *Demand Leads Capacity* In Figure 1.14*b*, capacity growth lags the demand. Demand is greater than the capacity for this case. The capacity is being fully utilized, but there will be shortages. In this case, the company will lose some of its future income. The risk of this strategy is that it may result in permanent loss of the market share.

(c) *Capacity Matches Demand* In Figure 1.14*c*, the capacity growth closely matches the demand. Minor shortages or excess accumulations may occur, but generally demands are met. This is a risk-neutral strategy compared to the other two options.

1.8.2 Break-Even Analysis

Economies of scale are the most significant considerations regarding capacity growth. Economies of scale can be used to justify greater capacity growth, because a larger facility will be able to produce units at a lower unit cost when demand is high. As the number of units increases, the average cost per unit decreases. A fundamental problem when considering capacity expansion is to what extent the organization can rely on outsourcing to meet demands. In order to analyze this problem, break-even curves can be used. However, break-even curves are only based on rough estimates of the demand, the time value of money, and learning curves. Breakeven curves can also be used for comparing alternatives when the total cost can be presented as a function of the volume of demand.

Suppose the unit cost is denoted by C_I if it is produced internally and by C_O if it is outsourced. The initial cost of producing the item internally is I_I, and the initial cost of outsourcing is I_O. Let x be the number of units produced (to be determined). The cost of the two alternative methods (producing internally or outsourcing) can be set equal to each other to find x^*, the break-even point.

$$\text{Cost of producing internally} = \text{cost of outsourcing}$$
$$I_I + C_I x = I_O + C_O x \tag{1.5}$$

Equation (1.5) can be solved to find the break-even quantity (x^*):

$$x^* = \frac{I_I - I_O}{C_O - C_I} \tag{1.6}$$

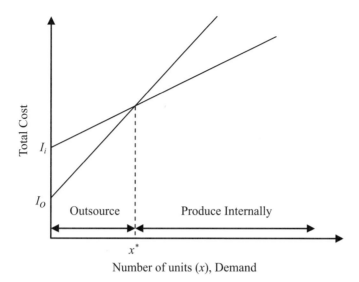

Figure 1.15   Break-even analysis for finding break-even quantity.

Note that generally $I_I > I_O$ and $C_I < C_O$. Figure 1.15 illustrates break-even curves and the break-even point quantity x^*. When the total quantity is less than x^*, one should choose outsourcing; when the total quantity is more than x^*, one should choose producing internally.

Example 1.4   Break-Even Analysis with Three Alternatives   Consider the following three alternatives:

  I. Full internal production at $14 per unit with $20,000 invested
 II. Full outsourcing at $18 per unit with $0 invested
III. Partial internal production and partial outsourcing at $16 per unit with $12,000 invested

Identify which alternatives should be selected for different demand quantities.
   To solve this problem, we need to find the break-even points for each pair of alternatives. The three different break-even points are shown in Figure 1.16. For example,

$$x_1^* = \frac{20,000 - 12,000}{16 - 14} = 4000 \text{ units}$$

$$x_3^* = \frac{12,000 - 0}{18 - 16} = 6000 \text{ units}$$

In this example:

   If the quantity x is less than $x_1^* = 4000$, choose alternative II.
   If the quantity x is more than $x_3^* = 6000$, choose alternative I.
   If the quantity x is between 4000 and 6000, choose alternative III.

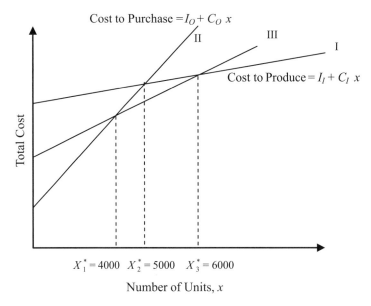

Figure 1.16   Break-even analysis for three alternatives.

Note that the break-even point of alternatives I and II, x_2^*, is not used as it is inferior to alternative III.

1.9   MACHINING/OPERATION OPTIMIZATION

In the last section, the importance of capacity planning as a key strategic decision making was discussed. The capacity of a given process is directly proportional to the speed at which the process is operated. Increasing the production rate, however, usually increases the cost of operations and machinery maintenance and may also impact the quality of the product. Production systems consist of production departments, workstations, and machines. The smallest component of a production system is a machine. In this section, we discuss how to optimize the setup of a machine by considering both production rate and the cost of operation. For the purpose of illustration, the optimization for metal cutting/machining and tool replacements is presented here. However, the same basic idea can be applied to production departments and workstations.

In metal cutting/machining (such as turning, drilling, milling, and grinding), a higher cutting speed costs more because the cutting tools wear faster. Replacing cutting tools more frequently translates into a higher maintenance cost for the company. Alternatively, the company can buy more expensive cutting tools that last longer, but this also translates into a higher cost. Therefore, the production manager must choose an operating speed of the machines that balances the conflicting goals of increasing production rate and minimizing costs given that the quality of products are unchanged for the range of changes in production rate. In this section, we discuss the basics of machine setup and tool life optimization using the well-known Frederick W. Taylor tool life equation developed in the late nineteenth century. We also demonstrate the trade-off between the productivity and cost for machining optimization.

1.9.1 Taylor's Tool Life Equation

Taylor's equation, developed by Frederick W. Taylor (the father of scientific management), shows the basic relationship between machine speed and tool life for machinery processes:

$$vt^n = C \tag{1.7}$$

where

$v =$ speed of cutting (inches per minute)
$t =$ tool life (minutes)

In Taylor's equation, the parameters n and C are found for the given machine, tool, and type of materials being cut. The parameters of this equation (n and C) can be derived by cutting materials at different speeds and recording the tool life for each given speed. Then, a linear regression method (covered in Chapter 3) can be used to estimate n and c. In the following we show how to find these parameters by solving two equations with two unknowns. To verify Taylor's equation, one can experimentally generate several points and check if speeds and tool life match the generated Taylor equation.

Example 1.5   Taylor's Equation Parameters Identification   The table below shows the results of two experiments for a lathe cutting machine using a specific type of cutting tool for a given type of steel material.

(a) Derive the tool life equation for this tool.
(b) Find the tool life if the speed is 18 in./min; also find the speed which has a tool life of 120 min.
(c) Verify if the equation is valid.

	Experiment 1	Experiment 2
Speed (in./min), v	10	13
Tool life (min), t	150	50

Solution

(a) The following two equations can be written based on Equation (1.7) using the data from experiments 1 and 2, respectively:

$$10 \times 150^n = C \qquad 13 \times 50^n = C$$

Equating the left-hand sides of the above two equations yields

$$10 \times 150^n = 13 \times 50^n \qquad \text{or} \qquad \frac{10}{13} = \left(\frac{50}{150}\right)^n$$

Taking the log of both sides gives

$$\ln 0.769 = n \ln 0.333 \qquad \text{or} \qquad n = 0.24$$

Now, the value of C can be obtained by substituting the value of n in one of the above equations:

$$C = 10 \times 150^n = 10 \times 150^{0.24} = 33.29$$

Therefore, the tool life equation is

$$vt^{0.24} = 33.29$$

Figure 1.17 shows Taylor's equation for this example.

(b) Examples of how to find t (or v) for given v (or t) are shown in the table below.

Speed (in./min)	Tool Life (min)	n	C	Solution
18	$t = ?$	0.24	33.29	$18 \times t^{0.24} = 33.29$ or $t = 12.96$ min
$v = ?$	120	0.24	33.29	$v \times 120^{0.24} = 33.29$ or $v = 10.51$ in./min

(c) To verify Taylor's equation, more experiments must be conducted to verify if the derived equation closely matches the data from the experiments. For example, at speed $v = 18$ in./min, use a new tool and measure its life. Also, at speed $v = 10.51$ in./min, run the experiment with a new tool and measure the tool life. According to the obtained Taylor equation, the tool lives should be about 12.96 and 120 min, respectively. If the results of experiments are approximately close to these numbers, the model is valid.

1.9.2 Multicriteria of Machining Operation

Consider the following two objectives for the tool life/machining problem introduced in the last section:

Minimize f_1 = total cost (measured by the total cost of replacing all tools during a production period)

Maximize f_2 = productivity (measured by the total number of parts produced per period)

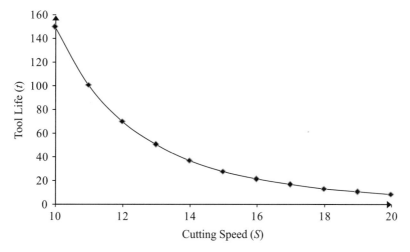

Figure 1.17 Taylor's Tool Life for Example 1.5.

Based on Taylor's equation for tool life, these two objectives can be presented as follows:

$$f_1 = c_t N \tag{1.8}$$

where c_t is the cost of replacing one tool in dollars and N is the number of times the tool is replaced per production period. The value of N can be found by dividing the production period (T) by the tool life:

$$N = \frac{T}{t}$$

To find the productivity, consider the total cutting length required for one part, denoted as L. Then, obtain the total number of parts produced per period for the given speed:

$$f_2 = \frac{vT}{L} \tag{1.9}$$

To generate multiple-criteria alternatives, consider the given minimum and maximum cutting speeds. Then, consider several equally distributed points between the given minimum and maximum speeds. For each given speed, derive the tool life, number of tool changes, total costs, and total production per production period.

Example 1.6   Multicriteria of Maching Operation   Consider Taylor's equation in Example 1.5; that is,

$$vt^{0.24} = 33.29$$

Suppose that the production period is four weeks and there are 40 h per week. Also suppose that the total cost of replacing each tool is $30 (the cost of a new tool and the labor cost of replacing it). Also suppose that each part requires a total of 200 in. of cutting, and the cutting speed v can vary from 10 to 15 in./min. Generate six multicriteria alternatives and identify whether they are efficient. Rank all alternatives.

We know $c_t = \$30$ and $L = 200$. We also know from Equation (1.9) that the total production time is

$$T = 4 \times 40 \times 60 = 9600 \text{ min}$$

We can consider six equally distributed cutting speeds: 10, 11, 12, 13, 14, and 15. The six multicriteria alternatives are stated in the table below. For each given speed v, find the tool life t, the number of tool life changes (N), the total cost f_1, and the total productivity f_2. All these alternatives are efficient in terms of optimizing both objectives simultaneously. More alternatives can be generated by considering additional cutting speeds. The ranking of these alternatives by a hypothetical decision maker is shown in the following table. Therefore, the decision maker will use the third alternative, cutting speed $v = 12$, which will result in a tool life $t = 70$, productivity of 576 parts per period, and total cost of \$4102 per period.

	a_1	a_2	a_3	a_4	a_5	a_6
Cutting speed, v	10	11	12	13	14	15
Tool life, t	150	101	70	50	37	28
Number of tool changes, N	64	95	137	191	260	346
Minimum cost (\$), $f_1 = c_t N$	1919	2855	4102	5726	7797	10,394
Maximum number of parts, $f_2 = vT/L$	480	528	576	624	672	720
Efficient?	Yes	Yes	Yes	Yes	Yes	Yes
Ranking by the decision maker	3	2	1	4	5	6

   See Supplement S1.1.xls.

REFERENCES

General

Abernathy, W. J., and P. L. Townsend. "Technology, Productivity, and Process Change." *Technological Forecasting and Social Change*, Vol. 7, No. 4, 1975, pp. 379–396.

Abernathy, W. J., and K. Wayne. "Limits of the Learning Curve." *Harvard Business Review*, Vol. 52, 1974, pp. 109–119.

Argote, L., and D. Epple. "Learning Curves in Manufacturing." *Science*, Vol. 247, No. 4945, 1990, pp. 920–924.

Bowman, E. H. "Consistency and Optimality in Managerial Decision Making." *Management Science*, Vol. 9, No. 2, 1963, pp. 310–321.

Devinney, T. M. "Entry and Learning." *Management Science*, Vol. 33, No. 6, 1987, pp. 706–724.

Dhalla, N. K., and S. Yuspeh. "Forget the Product Life Cycle Concept." *Harvard Business Review*, Vol. 54, 1976, pp. 102–112.

Hayes, R. H., and S. Wheelwright. "Link Manufacturing Process and Product Life Cycles." *Harvard Business Review*, Vol. 57, 1979, pp. 133–140.

Hiezer, J., and B. Render. *Operations Management*, 9th ed. Prentice-Hall, 2007.

Ravindran, A. R., and D. P. Warsing, Jr. *Supply Chain Engineering: Models and Applications*. Boca Raton, FL: CRC Press, 2012.

Schaller, R. R. "Moore's Law: Past, Present and Future." *Spectrum IEEE*, Vol. 34, No. 6, 1997, pp. 52–59.

Silver, E. A., and R. Peterson. *Decision Systems for Inventory Management and Production Planning*, 2nd ed. New York: Wiley, 1985.

Sivazlian, B. D., and L. E. Stanfel. *Analysis of Systems in Operations Research*. Englewood Cliffs, NJ: Prentice Hall, 1975.

Spence, A. M. "The Learning Curve and Competition." *The Bell Journal of Economics*, Vol. 12, No. 1, 1981, pp. 49–70.

Stevenson, W. *Operations Management*, 11th ed. McGraw-Hill, 2011.

Yelle, Y. E. "The Learning Curve: Historical Review and Comprehensive Survey." *Decision Science*, Vol. 10, 1979, pp. 302–328.

Multiobjective Optimization of Operations & Machining

Ghiassi, M., R. E. Devor, M. I. Dessouky, and B. A. Kijowski. "An Application of Multiple Criteria Decision Making Principles for Planning Machining Operation." *IIE Transactions*, Vol. 16, No. 2, 1984, pp. 106–114.

Ho, W. "Integrated Aanalytic Hierarchy Process and Its Applications—A Literature Review." *European Journal of Operational Research*, Vol. 186, No. 1, 2008, pp. 211–228.

Iwata, Y., K. Taji, and H. Tamura. "Multi-objective Capacity Planning for Agile Semiconductor Manufacturing." *Production Planning & Control*, Vol. 14, No. 3, 2003, pp. 244–254.

Malakooti, B. "A Gradient-Based Approach for Solving Hierarchical Multiple Criteria Production Planning Problems." *Computers and Industrial Engineering, An International Journal*, Vol. 16, No. 3, 1989a, pp. 407–417.

Malakooti, B. "An Interactive Hierarchical Multi-Objective Approach for Computer Integrated Manufacturing." *Robotics and Computer-Integrated Manufacturing*, Vol. 6, No. 1, 1989b, pp. 83–97.

Malakooti, B. "An Interactive On-Line Multiobjective Optimization Approach with Application to Metal Cutting Turning-Operation." *International Journal of Production Research*, Vol. 29, No. 3, 1991, pp. 575–598.

Malakooti, B. *Decisions and Systems with Multiple Objectives*, J. Wiley, forthcoming, 2014.

Malakooti, B. "Double Helix Value Functions, Ordinal/Cardinal Approach, Additive Utility Functions, Multiple Criteria, Decision Paradigm, Process, and Types." *International Journal of Information Technology & Decision Making (IJITDM)*, Forthcoming, 2014a.

Malakooti, B. "Z Utility Theory: Decisions Under Risk and Multiple Criteria" to appear, 2014b.

Malakooti, B., and J. Deviprasad. "An Interactive Multiple Criteria Approach for Parameter Selection in Metal Cutting." *Operations Research*, Vol. 37, No. 5, 1989, pp. 805–818.

Malakooti, B., J. Wang, and E. C. Tandler. "A Sensor-Based Accelerated Approach for Multi-Attribute Machinability and Tool Life Evaluation." *International Journal of Production Research*, Vol. 28, No. 12, 1990, pp. 2373–2392.

Malakooti, B., Y. Q. Zhou, and E. C. Tandler. "In-Process Regressions and Adaptive Multicriteria Neural Networks for Monitoring and Supervising Machining Operations." *Journal of Intelligent Manufacturing*, Vol. 6, No. 1, 1995, pp. 53–66.

Sadiq, R., and F. I. Khan. "An Integrated Approach for Risk-Based Life Cycle Assessment and Multi-Criteria Decision-Making." *Business Process Management Journal*, Vol. 12, No. 6, 2006, pp. 770–792.

Wang, J. "Optimisation of Cutting Parameters Using a Multi-Objective Genetic Algorithm." *International Journal of Production Research*, Vol. 47, No. 21, 2009.

Wang, N. "Multi-Criteria Decision-Making Model for Whole Life Costing Design." *Structure and Infrastructure Engineering*, Vol. 7, No. 6, 2011.

EXERCISES

Introduction (Section 1.1)

1.1 Describe characteristics of each of the following types of decisions:

(a) Strategic decisions

(b) Tactical decisions

(c) Operational decisions

1.2 For a given industry (e.g., automobile):

(a) Provide strategic, tactical, and operation decisions.

(b) What are the inputs, outputs, and transformation processes?

(c) What are the supply chain characteristics?

1.3 Provide an example for each of the following systems. For each example, identify inputs, transformation, and outputs.

(a) Systems with tangible inputs and outputs (e.g., a physical product)

(b) Systems with intangible inputs and outputs (e.g., a service)

(c) Systems with intangible inputs but tangible outputs (e.g., an information system)

Production and operations history and perspective (Section 1.2)

2.1 Describe the major historical events that have changed the way operations, manufacturing, and production systems operate.

2.2 Describe the characteristics and differences of the five phases of operations, manufacturing, and production systems history.

Production and operations Models (Section 1.3)

3.1 Provide an example and its characteristics for each of the following problems. (You may use the Internet to find examples.)

(a) Location decisions

(b) Layout design

(c) Supply chain

   (d) Transportation

   (e) Clustering and group technology

   (f) Assembly systems

3.2 Provide an example and its characteristics for each of the following problems. (You may use the Internet to find examples.)

   (a) Forecasting

   (b) Aggregate planning

   (c) Push-and-pull systems: MRP and JIT

   (d) Inventory planning and control

   (e) Project management

   (f) Scheduling and planning

3.3 Provide an example and its characteristics for each of the following problems. (You may use the Internet to find examples.)

   (a) Quality control and assurance

   (b) Reliability, maintenance, and replacement

   (c) Work measurement and analysis

   (d) Productivity and efficiency

   (e) Energy systems

Systems Approach and Tools (Section 1.4)

4.1 Provide an example and its characteristics for each of the following problems. (You may use the Internet to find examples.)

   (a) Systems engineering and development

   (b) Systems architecting and modeling

   (c) Multi objective perspective

4.2 Provide an example and its characteristics for each of the following problems. (You may use the Internet to find examples.)

   (a) Optimization

   (b) Decision analysis

   (c) Queuing systems

   (d) Simulation

Multiple-Criteria Production/Operation Systems (Section 1.5)

5.1 Consider a decision-making problem (either a personal or a business decision). For this problem, define and apply each of the six steps of solving multiple-criteria problems. Justify why the selected alternative is the best.

5.2 Describe the difference between efficient and inefficient alternatives. Give examples of efficient and inefficient alternatives for exercise 5.1.

5.3 Provide an example (e.g., choose an industry or a business) where each of the five strategic criteria are used. Also discuss the trade-offs for pairs of criteria for the selected example. (You may use the Internet to find examples.)

5.4 Choose a business. *Hint:* You can search the Internet.
  (a) Provide an example where the six levels of production planning are used.
  (b) Describe each level and show how each level interacts with other levels.
  (c) Describe the objectives of each level and the objectives of the whole system.

5.5 Choose a manufacturing industry. (You may use the Internet to find examples.) Respond to questions (a) (b) and (c) of exercise 5.4.

Product and Process Life Cycle (Section 1.6)

6.1 Choose a product. For this product, briefly describe the four stages of the product life cycle. (You may use the Internet to find examples.)

6.2 **(a)** Describe two products that would experience a sharp decline on the life-cycle curve in the fourth stage.
  (b) Describe two types of products for which sales would remain steady for a long period after maturation.

6.3 Choose a process (e.g., an industry or a business). For this process industry, briefly describe the four stages of the product life cycle. (You may use the Internet to find examples.)

6.4 Give an example (the name of the products and the name of the manufacturing) for each of the following product/process types. (You may use the Internet to find examples.)
  (a) Jumbled flow/low volume
  (b) Disconnected flow/multiple products
  (c) Connected line flow/high volume
  (d) Continuous flow/higher volume

Learning Curves (Section 1.7)

7.1 Provide an example for each of the following organizations with respect to learning curves. (You may use the Internet to find examples.)
  (a) Manufacturing company
  (b) Governmental organization
  (c) Nonprofit organization

7.2 Draw learning curves for each of the following learning rates assuming the first unit takes 20 h to produce:
  (a) 25%   **(b)** 50%   **(c)** 75%   **(d)** 100%

7.3 If it takes 10 h to make the 1st car and 8 h to produce the 2nd car:
  (a) Find the learning rate l and the learning curve exponent b.
  (b) Find the time required to produce 5th unit, 16th unit, and 1000th unit.

7.4 The following table shows the time to produce the nth units for three cases:

Unit n	Hours for nth unit		
	Case I	Case II	Case III
1	50	50	50
5	12	10	14

(a) Consider case I. Find the learning rate, the exponent of the learning curve, and the time needed to produce the 8th, 10th, and 90th units.

(b) Consider case II. Find the learning rate, the exponent of the learning curve, and the time needed to produce the 8th, 10th, and 90th units.

(c) Consider case III. Find the learning rate, the exponent of the learning curve, and the time needed to produce the 8th, 10th, and 90th units.

7.5 Suppose that the exponent of the learning curve is $b = 0.24$, and it takes 30 h to produce the 1st unit.

(a) How long will it take to produce the 50th and 1500th units?

(b) Find the learning rate l.

(c) Find the total time it takes to produce 50 and 1500 units.

7.6 The following table shows the times to produce the nth units.

Unit n Produced	Hours for nth unit
1	40
2	x
4	16.9

(a) How many hours will it take to produce the 2nd unit?

(b) Find the learning rate and the exponent of the learning curve coefficient.

(c) How long will it take to produce the 64th, the 70th, and the 3000th units?

7.7 If it takes 33.75 h to produce the 1st unit and 40 h to produce the 5th unit:

(a) Find the learning (deteriorating) rate and the exponent of the learning curve coefficient.

(b) How long will it take to produce the 2nd, 15th, and 200th units?

7.8 If it takes 100 h to produce the 1st unit and 120 h to produce the 30th unit:

(a) Find the learning (deteriorating) rate and the exponent of the learning curve coefficient.

(b) How long will it take to produce the 100th, 1000th, and 10,000th units?

Capacity Planning (Section 1.8)

8.1 Provide an example (e.g., choose an industry or a business) where each of the following cases has occurred. Discuss the risks (advantages/disadvantages) associated with each case. (You may use the Internet to find examples.)

(a) Capacity leads demand.

(b) Capacity lags demand.

(c) Capacity matches demand.

8.2 A fast food restaurant expects to cook 600 hamburgers per hour.

(a) The current grill's capacity is 2 hamburgers per minute. How many grills does the restaurant need?

(b) The owners can replace the current grills with a new model that can cook three hamburgers per minute where each new grill costs $1500. Alternatively, the restaurant can buy supergrills that cook seven hamburgers per minute where each costs $3300. Which one of the new grills is the better alternative?

(c) If demand is increased to 800 hamburgers per hour, does your answer to part (b) change? Explain why.

8.3 Suppose a computer company can internally produce processors for $10 per processor, but this requires some initial investment in setting up the system. An outside vendor can make the same processor for $13 per processor. If the company needs to produce 6000 processors, what is the maximum amount of investment the company should be willing to spend to produce the product internally (instead of outsourcing).

8.4 An electric company can make its own oscillating fans for $4.50 per unit after making an $1800 investment. An outside vendor will make the oscillating fans for $5.25 per unit.

(a) What is the break-even quantity of oscillating fans for the company?

(b) Show this result graphically

Machining Optimization (Section 1.9)

9.1 A cutting tool life is 40 min if the cutting speed is 120 in. /min; its life is 60 min if the cutting speed is 100 in./min

(a) What is the tool life equation for this tool?

(b) What is the tool life if the cutting speed is 110 in./min?

(c) What cutting speed will result in a tool life of 30 min?

9.2 Consider Taylor's equation where $n = 0.40$ and $C = 50$.

(a) What cutting speed should be used to have a tool life of 70 min?

(b) What is the tool life if the cutting speed is 15 in./min?

9.3 Consider the following equations for cost and production time.

$$f_1(v) = 5 + 220v^{-1} + 0.134v^{0.678} \quad \text{production cost}$$
$$f_2(v) = 1.7 + 40v^{-1} + 0.0103v^{0.678} \quad \text{production time}$$

(a) Find the multicriteria alternatives for speeds of 100, 125, 150, 175, and 200.

(b) Identify efficient and inefficient alternatives.

9.4 Consider the following tool life equation:

$$vt^{0.4} = 50$$

Suppose the minimum cutting speed is 80 in./min and the maximum cutting speed is 120 in./min. The total production time is 100 h, the total cutting length per part is 40 in., and the cost to replace each tool is $16.

(a) Generate five equally distributed alternatives based on cutting speed.

(b) Measure total cost and total productivity for each alternative.

(c) Identify inefficient and efficient alternatives. Show the multicriteria alternatives graphically.

9.5 Consider the following tool life equation:

$$vt^{0.65} = 24$$

Suppose the minimum cutting speed is 20 in./min and the maximum cutting speed is 28 in./min. The total production time is 80 h, the total cutting length per part is 20 in., and the cost to replace each tool is $6. Respond to parts (a), (b), and (c) of exercise 9.4.

CHAPTER 2

MULTICRITERIA DECISION MAKING

2.1 INTRODUCTION

Decision making is a central issue in numerous fields of studies, from product design to product manufacturing, from agriculture to aerospace industry, from design of a microchip to design of a worldwide distributed Internet system, from psychology to bioinformatics, and from politics to engineering. At the personal level, decision making can be as simple as deciding what to wear for the day or as complex as choosing a spouse. At the organizational level, decision making can be as simple as deciding where to buy office supplies or as complex as determining which lines of products to develop (or to terminate). In almost all real-world decision-making problems, decision makers must consider several conflicting criteria (objectives) when evaluating alternatives. The MCDM aspects of operations and production systems are covered in different chapters of this book. MCDM must support different decision makers (DMs) to be able to form an informed opinion for evaluating alternatives according to their own individual (or organizational) preferential behavior, decision-making skills, and problem-solving capabilities. *MCDM approaches* should have the following nine characteristics:

Principle oriented (axiom based), convincing, coherent, defendable (justifiable), enlightening (illuminating, informative, supportive), versatile (allows for the use of different preferential behaviors), transparent, systematic, and verifiable (testable and repeatable).

The general steps of the MCDM methodology are

1. Understand and utilize decision process paradigm (Section 2.1.1) that the success of decision making depends on four different dimensions; one of them is MCDM.

Note: Advanced material that can be omitted without loss of generality will be indicated by a dagger. This chapter is adapted from Malakooti 2012, 2014a and 2014b.

Operations and Production Systems with Multiple Objectives, First Edition. Behnam Malakooti.
© 2014 John Wiley & Sons, Inc. Published 2014 by John Wiley & Sons, Inc.

2. Choose an appropriate and valid model to formulate the given problem.
3. Identify alternatives and the criteria that should be used to evaluate the alternatives.
4. Assess criteria values for the given alternatives.
5. Identify the set of efficient and feasible alternatives.
6. Choose an appropriate and valid method for evaluating alternatives for the given DM.
7. Apply the chosen method to identify the best alternative for the given DM.
8. Analyze the consequences of the chosen alternative.
9. If necessary, repeat the steps above.

Step 1 is about the individual DM's information-processing and problem-solving approach (it is subjective and belongs to the internal world of the DM). Steps 2–5 are related to the definition of a problem (they are subjective but belong to the external world of the DM). Steps 6–8 are related to the analyst who helps the DM to solve the problem (they are objective and belong to the external world of the DM). All these steps depend on the DM but they should be defendable and rational. In this chapter, we develop several MCDM methods for ranking all alternatives and identifying the best alternative by using different (multiple criteria) utility functions that represents the DM's rational preferential behavior.

Three Fields of Studies in Decision Making Decision making challenges people in every aspect of their lives. For important decisions, having poor decision-making skills can be very costly. Analyzing and learning from previous experiences enables DMs to recognize and utilize useful decision-making criteria and methods. Before making a decision, one must gather relevant information and perform data analysis. After understanding and formulating the problem, a set of possible alternatives can be generated. From this set, the best, or most preferred, alternative can be selected. Decision making can be classified based on the three fields of study seen in Figure 2.1. These three fields are related and decision making is the product of these three fields:

- Decision process: how decisions are made from a processing point of view.
- MCDM under certainty: how alternatives are evaluated and selected.
- Risk measurement and analysis for decisions under uncertainty: how risk attitudes are measured.

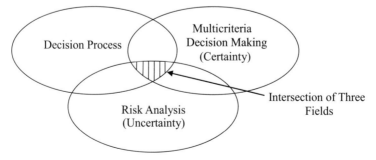

Figure 2.1 Decision making based on relationship between three fields of decision making.

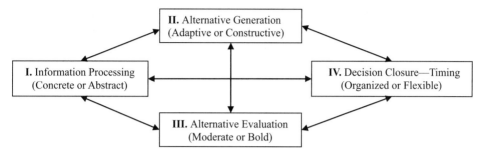

Figure 2.2 The four dimensions of the decision-making process where each dimension has two opposing types.

2.1.1 Decision Process Paradigm

Decision Process (Internal) Humans have the ability to solve complex decision-making problems that are closely related to man's perception, recognition, memory, intelligence, and creativity in solving problems. Decision-making ability allows one to consider different alternatives with the ability to estimate each alternative's consequences. In the scientific realm, the relationships between causes and effects are identified and measured with substantial certainty and such relationships are independent of time and environment. However, in decision making the relationships between causes and effects are very much time and environment dependent, and there is usually a level of uncertainty. Decision-making approaches are often cultural and subjective. In this context, understanding the "decision-making process" and defining a DM's types is a step toward the development of a scientific decision-making theory. Understanding the decision process can provide insight into how and why people make decisions in certain ways. DMs that are conscious of their decision process and types can make more effective and balanced decisions. Malakooti (2012) introduced a decision process model for addressing the process of decision making, seen in Figure 2.2.

The decision process model is based on four dimensions where each dimension has two opposing types; a brief description of these types is presented below.

Information Processing (Concrete or Abstract) This dimension pertains to how people view and process data and information. Concrete processors perceive and view information from its content point of view; they use details and specific information. Abstract processors perceive and view information from a larger perspective and scope; they use aggregate and approximate values.

Concrete (Precise, Specific)	Abstract (Concise, General)
• Explores the depth of information	• Explores the breadth of information
• Uses exact values	• Uses approximate values
• Observes item by item specifically	• Observes the relationships of items

Alternative Generation (Adaptive or Constructive)   This dimension pertains to the DM's patterns of generating alternatives. Adaptive DMs are more creative in modifying and improving existing alternatives to generate new alternatives. Constructive DMs are more creative in generating unique, out of the box, alternatives.

Adaptive (Improver, Reformer)	Constructive (Originator, Inventor)
• Forms ideas by comparing alternatives • Uses existing approaches to find solutions • Uses similar patterns for finding solutions	• Forms ideas by contrasting alternatives • Uses intuition to find solutions • Uses uncommon patterns for finding solutions

Alternative Evaluation (Moderate or Bold)   This dimension pertains to the DM's evaluation approach of alternatives. By taking a more cautious approach, moderate DMs try to avoid potential loss and uncertainty by paying a higher cost. On the other hand, bold DMs may choose riskier alternatives associated with higher possible outcomes and may proceed even if there is uncertainty.

Moderate (Conciliator)	Bold (Challenger)
• Places high emphasis on possible loss when there is uncertainty • Excessively concerned with probable loss • Believes that not taking risks pays off	• Places moderate emphasis on possible loss there is uncertainty • Not excessively concerned with probable loss • Believes that taking some risks pays off

Decision Closure (Organized or Flexible)   This dimension is related to a DM's approach in finalizing a decision. Organized DMs are more realistically aware of timing and duration and are reluctant to change their decisions. They finalize their decisions sooner rather than later. On the other hand, flexible DMs are more inclined to delay making decisions and may change their decisions more often than others. Flexible DMs are less aware of timing and duration.

Decisive (Expediter, Determined)	Flexible (Procrastinator, Indecisive)
• Decides sooner than most people • Definite in selecting alternatives • More aware of timing and duration of tasks	• Decides later than most people • Tentative in selecting alternatives • Less aware of timing and duration of tasks

2.1.2 Decision-Making and Process Pyramid

Decision-Making Approach (External)   Similar to the decision process, the decision-making model consists of four interrelated steps:

1. Problem formulation (PF) (for presenting a problem to be solved)
2. Problem solving techniques (PSTs) (for generating useful solutions)
3. MMCDM (for the selection of the best alternative)
4. Decision realization (DR) (for identifying when a decision closure occurs)

The last step of decision making, DR, occurs when the decision is made and is ready for implementation. Figure 2.3 shows the relationship between the decision process and decision making. Contributing factors (inputs) to internal and external decision process and decision-making approaches are also listed in Figure 2.3.

Decision-Making and Process Pyramid   Figure 2.4 illustrates a pyramid of relating decision-making typology to the four phases of problem solving. The top of the pyramid presents the decision process where each edge of the pyramid represents one of the four dimensions of decision process types. The base of the pyramid presents the four phases of problem solving: problem formulation, alternative generation, alternative evaluation, and DR.

2.1.3 Decision-Making Paradigm

Table 2.1 provides a perspective for formulating, analyzing, and solving decision-making and MCDM problems.

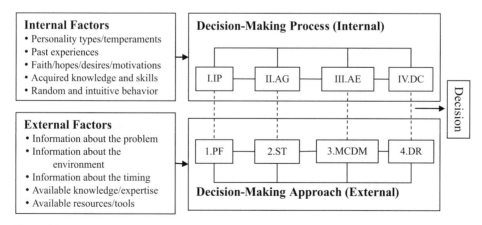

Figure 2.3   Contributing factors to internal and external decision-making process and approach. IP, information processing; AG, alternative generation; AE, alternative evaluation; DC, decision closure; PF, problem formulation; ST, solution techniques; MCDM, multiple-criteria decision making; DR, decision realization.

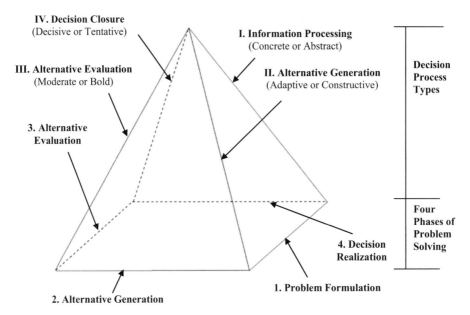

Figure 2.4   Pyramid of the decision making and process: four types of decision process versus four phases of problem solving.

2.2   EFFICIENCY AND ITS EXTENSIONS

In MCDM, the DM is concerned with finding the best (optimal or best-compromised) alternative from a set of multicriteria alternatives. The definitions of efficiency and convex efficiency can be used to classify the set of alternatives so that the set of most promising alternatives can be identified.

The set of alternatives S can be represented as

$$S = [F_1, F_2, \ldots, F_n], \quad \text{or} \quad F_j \text{ for } j = 1, \ldots, n$$

where n is the number of alternatives. An alternative F_j is a k-tuple vector of criteria (objectives) represented by

$$F_j = [f_{1j}, f_{2j}, f_{3j}, \ldots, f_{kj}], \quad \text{or} \quad f_{ij} \text{ for } i = 1, \ldots, k$$

where k is the   number of criteria. Criteria are also referred to as objectives or attributes. Alternatives can be presented by the following matrix:

	Alternatives					
Maximize	F_1	F_2	$\ldots$	F_j	$\ldots$	F_n
Objective f_1	f_{11}	f_{12}	$\ldots$	f_{1j}	$\ldots$	f_{1n}
Objective f_2	f_{21}	f_{22}	$\ldots$	f_{2j}	$\ldots$	f_{2n}
$\ldots$	$\ldots$	$\ldots$	$\ldots$	$\ldots$	$\ldots$	$\ldots$
Objective f_k	f_{k1}	f_{k2}	$\ldots$	f_{kj}	$\ldots$	f_{kn}

TABLE 2.1   A Perspective for Solving Decision-Making and MCDM Problems[a]

I. Problem formulation, modeling, and information processing	• Define criteria to be independent, comprehensive, and conflicting. • Combine similar and related criteria into one criterion to minimize the number of criteria. • First try to formulate bicriteria problems; they can be easily solved, for example, graphically. • Use brainstorming (and feedback from others) to formulate the problem. • Use the "divide and conquer" strategy by dividing the decision problem into smaller problems so that each small problem can be solved independently. • Formulate the problem from different perspectives to gain insight, for example, act as your opposite. • Determine and adapt your decision type in information processing: concrete (precise, specific) or abstract (concise, general).[a]
II. Generation and analysis of alternatives	• Be creative and open-minded in considering more alternatives. • Identify a small set of "promising" alternatives. • Make a simple list of pros and cons for each "promising" alternative. • Do not eliminate "promising" alternatives that are inefficient (dominated); instead search for hidden criteria/reasons that may make "promising" inefficient alternatives efficient. • Use intuition and gut feelings to gain insight regarding the alternatives. • Play devil's advocate and criticize your own decision to gain insight, for example, act as your opposite. • Determine and adapt your decision type in considering or generating alternatives: adaptive (improver, reformer) or constructive (originator, inventor).[a]
III. Evaluation: risk analysis and modeling	• Find out the cost of gathering more information and reducing uncertainty. • Use proper editing rules to maximize the accuracy of paired comparison of risky alternatives.[a] • Use nonlinear models for solving the risk problem.[a] • Identify the uniqueness (the signature) of each individual DM decision behavior.[a] • Also, use expected value and risk value to evaluate different alternatives.[a] • Reduce the possibility of large losses by paying needed premium (e.g., buy insurance). • Consider taking calculated risks. • Realize that biological fight or flight reactions may not render a good decision. • Realize risk and MCDM methods that directly and independently assess single value and single risk utility functions (e.g., methods based on expected utility theory) are limited/handicapped in solving decision problems.[b] • Determine your decision type in risk attitude: risk averse (moderate, conciliator) or risk prone (bold, challenger); and adapt it as needed.[a]

(Continued)

TABLE 2.1   *(Continued)*

IV. Evaluation: multicriteria analysis and modeling	• Use paired comparison of "the most promising" alternatives to choose the best one.

IV. Evaluation: multicriteria analysis and modeling

- Use paired comparison of "the most promising" alternatives to choose the best one.
- Before considering all criteria, first gain insight by analyzing pairs of criteria.
- If possible, first use simpler models/methods (e.g., additive utility functions) but realize that they are restrictive, and then consider nonlinear utility functions.
- Be parsimonious in the number of parameters needed to accurately assess the DM's utility function; that is, use the minimum number of parameters (coefficients) to assess the utility function.
- Use transparent models and methods with simpler assumptions.
- Use different methods to solve the same problem and compare the results.
- Identify the ideal goal/aspiration/zenith alternatives; find the closest alternatives to them.
- Identify the most disastrous/despised/nadir alternatives; find the farthest alternatives from them.
- Determine and adapt your decision type by multicriteria preference attitude: independent (flat/linear), convergent (centrist), divergent (extremist); or hybrid (adaptive).[a]

V. Decision energy, closure, and timing

- Realize that waiting (doing nothing alternative) could be the best or the worst alternative.[c]
- Measure the possible gains or possible losses of waiting or expediting the decision.
- Optimize your decision-making effort in accordance to the importance of the decision problem.[a]
- Determine and adapt your decision type in decision closure and timing: decisive (organized, expediter, determined) or tentative (flexible, procrastinator, indecisive).[a]

---

[a]The details and justifications of some of these guidelines are provided in Malakooti (2012, 2014 and b).

[b]An excellent and exquisite example of methods that directly and independently assess single value and single risk utility functions is Keeney and Raiffa (1993), which epitomizes expected utility theory, an additive single utility function based on the theoretical work of von Neumann and Morgenstern (1947).

[c]In most decision-making problems, the "do nothing" alternative is sometimes ignored. For example, a patient has the choice of not undergoing treatment, or a manager has the choice of not hiring any of the candidates.

2.2.1   Efficiency (Nondominancy/Pareto Optimality)

The definition of efficiency is based on the assumption that one wants to maximize each and every objective. This assumption means that preferences with respect to objectives are increasing; that is, the more is always better.

An alternative F_p is efficient if and only if there does not exist another alternative F_q in the set of alternatives S such that

(a) $f_{iq} \geq f_{ip}$ for all i and
(b) $f_{iq} > f_{ip}$ for at least one i

The above two conditions ensure that the values of all criteria for alternative F_q are as good as and better in at least one objective than the values of the criteria for alternative F_p. Constraint (b) ensures that the two alternatives, F_p and F_q, are not equal. In other words, a point is efficient if and only if it is not dominated by any other point. Inefficient alternatives are also referred to as inferior, dominated, or non-Pareto-optimal alternatives. Efficient alternatives are referred to as noninferior, nondominated, or Pareto optimal.

In some MCDM problems, some of the criteria may need to be minimized. If an objective is to be minimized, maximize the negative value of the objective instead when determining efficient alternatives. That is, if an objective f_i is to be minimized, replace it by maximizing $-f_i$.

When the problem is to find the best alternative, inefficient alternatives can be removed from further consideration. The remaining alternatives are efficient. Note that an inefficient alternative can be the second best alternative, and therefore, when ranking a set of alternatives (to choose a subset of alternatives), inefficient alternatives should not be eliminated.

For example, consider four alternatives ($n = 4$) and two criteria ($k = 2$). The set of alternatives and their associated objective values are shown in the following table and Figure 2.5a.

	F_1	F_2	F_3	F_4
f_1	3	2	4	4
f_2	3	2	1	0

Since both criteria are maximized, the more desirable alternatives will appear in the upper right-hand corner of Figure 2.5, as this area will maximize both f_1 and f_2. In Figure 2.5a, each alternative is presented by a point. In Figure 2.5a, F_1 is closer to the upper right corner than F_2. It can be observed that F_1 dominates F_2, since $f_{11} > f_{12}$ and $f_{21} > f_{22}$. Therefore, F_2 can be eliminated from further consideration. Similarly, F_3 dominates F_4 since $f_{13} = f_{14}$ and $f_{23} > f_{24}$. Hence F_4 can also be eliminated from further consideration. Only bicriteria alternatives can be graphically presented in Cartesian objective space (e.g., see Figure 2.5a).

2.2.2   Polyhedral Graphical Efficiency

The graphical method for determining efficiency can only present two criteria problems; for example, see Figure 2.5a. For more than two criteria, we develop multicriteria polyhedron graphical efficiency where each alternative is presented in a two-dimensional space using k lines all starting from the same point of origin, where at the origin the values of each objective are at their minimum. The value of each objective is plotted on its respective line. Lines are equally distanced from each other starting at the origin; for example, for four axes use a " $+$ " shape, for six use a "*" shape. To show an alternative, its criteria values on different lines are connected. In this approach, a bicriteria alternative is represented by a line, a tricriteria by a triangle, four criteria by a square, and higher dimensions by a k polyhedral graph. If the area created by an alternative is contained within the area of another alternative, then the inner alternative is dominated.

For example, consider the four alternatives presented in Figure 2.5a where each alternative is presented by a point. Figure 2.5b presents the same information but each alternative

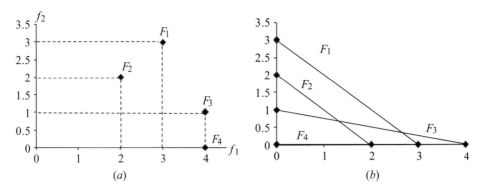

Figure 2.5 A plot of the four alternatives (*a*) as points in Cartesian objective axes and (*b*) as lines in polyhedral graphical efficiency.

is presented by a line. F_2 is dominated by F_1 because its line is completely below the line of F_1 using (0,0) as the point of reference. Similarly, F_4 is dominated by F_3 because the line for F_4 is below that of F_3. The efficient alternatives are therefore F_1 and F_3. Now, consider Figure 2.6 where two alternatives with four criteria are presented. Note that $F_1 = (-2,4,5,-1)$ dominates $F_2 = (-6,3,4,-2)$, which is shown graphically. Multicriteria graphical efficiency is most useful for presenting a few top alternatives for comparison purposes.

2.2.3 Convex Efficiency

We extend the definition of efficiency to convex efficiency as follows:

Any efficient point that is dominated by a convex combination of other efficient points is a nonconvex efficient point; otherwise it is convex efficient.

Therefore, the set of efficient points can be divided into convex and nonconvex efficient points, and so the set of alternatives can be divided into three subsets: convex efficient, nonconvex efficient, and inefficient. For example, consider a bicriteria problem having the

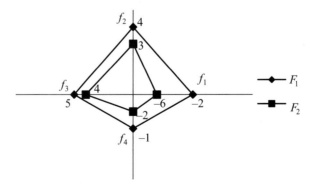

Figure 2.6 Four-dimensional graphic illustrating $F_1 = (-2,4,5,-1)$ dominates $F_2 = (-6,3,4,-2)$.

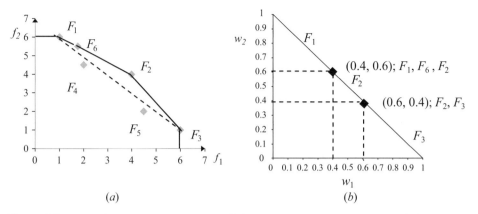

Figure 2.7 Example 2.1. (*a*) F_1 to F_6 are efficient. F_4 and F_5 are nonconvex efficient points. (*b*) Range of weights for which convex efficient points are optimal.

following three alternatives: $F_1 = (0,1)$, $F_2 = (1,0)$, and $F_3 = (0.4, 0.4)$. F_1 and F_2 are convex efficient and F_3 is nonconvex efficient. This definition has several useful applications, especially in identifying nonlinear utility functions. For additive utility functions (covered in the next section), it can be shown that nonconvex efficient alternatives can be eliminated as their utility values are always inferior with respect to convex efficient alternatives. It should be noted that convex efficient points are not necessarily extreme points; for example, in Figure 2.7 F_6 is on the line that connects F_1 and F_2, and it is not an extreme point but it is convex efficient. In the following, we explain these concepts for bicriteria problems. Computationally efficient methods for identifying convex efficient points for multicriteria problems are provided in Malakooti (1989a, c, 2000).

A solution to the following problem is a convex efficient point.

PROBLEM 2.1   GENERATING CONVEX EFFICIENT POINTS

$$\text{Maximize } U = w_1 f_1 + w_2 f_2 + \cdots + w_k f_k$$
$$\text{Subject to the set of alternatives } S \tag{2.1}$$

where $w_1 > 0$, $w_2 > 0, \ldots,$ and $w_k > 0$ and $w_1 + w_2 + \cdots + w_k = 1$. These weights are given parameters. It can be shown that all convex efficient points can be generated by varying the values of weights.

Graphical Approach for Identifying Convex Efficient Alternatives   For bicriteria problems, identifying nonconvex efficient alternatives is simple. First, plot all efficient alternatives (points) in objective space. Consider all efficient points that are farthest away from the origin. Next, connect such efficient points by straight lines to form a convex efficient frontier. Then, identify efficient alternatives that are dominated by the convex efficient frontier. These are nonconvex efficient points. Note that inefficient points are also dominated by the convex efficient frontier.

Example 2.1   Convex Efficiency Example   Consider the following six alternatives having two criteria. Identify convex and nonconvex efficient alternatives.

	F_1	F_2	F_3	F_4	F_5	F_6
Max. f_1	1	4	6	2	4.5	1.75
Max. f_2	6	4	1	4.5	2	5.5

In the above example, both criteria f_1 and f_2 are maximized, and all alternatives, F_1 to F_6, are efficient. However, from a convex efficiency point of view, F_4 is dominated by some of the points on the line connecting F_1 to F_2. Similarly, some points on the line connecting F_2 to F_3 dominate F_5. Therefore, F_4 and F_5 are nonconvex efficient as they are dominated by the line connecting F_1 and F_3. The rest are convex efficient.

2.2.4   Weight Space for Convex Efficient Points

Weight Space for Multicriteria   Consider Problem 2.1. It can be shown that a set of vectors of weights can be used to generate the same convex efficient point. But sometimes, one vector of weights $(w_1, w_2, \ldots, w_k)$ can be associated with several efficient points (as shown in Example 2.1 and Figure 2.7b). In the following, we discuss how to find a range of weights for bicriteria problems.

Weight Space for Bicriteria   For bicriteria problems, $U = w_1 f_1 + w_2 f_2$ is maximized where $w_1 > 0$, $w_2 > 0$, and $w_1 + w_2 = 1$. Note that nonconvex efficient (and also inefficient) points cannot be generated by maximizing $U = w_1 f_1 + w_2 f_2$; that is, nonconvex efficient (and also inefficient) points do not have weight space, but convex efficient points have weight space (see Problem 2.1). For determining weight ranges (space) associated with each convex efficient alternative, consider the pairs of adjacent (graphically) convex efficient points F_q and F_p. To find the weights where both F_q and F_p are optimal, set $U(F_q) = U(F_p)$ and use $w_1 + w_2 = 1$. Do this for all pairs of adjacent alternatives. Then, find the ranges at which different alternatives are optimal where $U = w_1 f_1 + w_2 f_2$ is maximized. A simple approach to show the ranges is by drawing $U(1 - w_2))f_1 + w_2 f_2$ for each alternative where f_1 and f_2 are shown as two vertical lines and w_2 varies from 0 to 1. $w_1 = 1 - w_2$. See Figure 2.8.

Example 2.2   Weight Space for Convex Efficient Alternatives   Consider Example 2.1 where there are six alternatives. Four alternatives, f_1, f_2, F_6, and F_3, are convex efficient. The range of weights can be generated using these four alternatives. For example, consider F_1 and F_2, which are adjacent to each other. Set

$$U(F_1) = U(F_2) \quad \text{or} \quad w_1(1) + w_2(6) = w_1(4) + w_2(4)$$

By substituting $w_2 = 1 - w_1$:

$$w_1 + (1 - w_1)(6) = w_1(4) + (1 - w_1)(4) \quad \text{or} \quad w_1 = 0.4 \text{ and } w_2 = 0.6$$

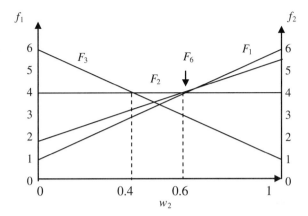

Figure 2.8 Weight space for Example 2.1 corresponding to Figure 2.7a, b.

This means that at $w_1 = 0.4$ and $w_2 = 0.6$, both F_1 and F_2 have the same utility value, and that is the point of demarcation between the two points. Now consider, F_2 and F_3. To find the weights associated with these two points, set

$$U(F_2) = U(F_3) \quad \text{or} \quad w_1(4) + w_2(4) = w_1(6) + w_2(1), \quad \text{or}$$
$$w_1(4) + (1 - w_1)(4) = w_1(6) + (1 - w_1)(1) \quad \text{or} \quad w_1 = 0.6 \text{ and } w_2 = 0.4$$

In this example, point F_6 can only be generated by $w_1 = 0.4$ and $w_2 = 0.6$, because it is the convex combination of F_1 and F_2. The range of weights for this problem is presented in the following table. Note that there is no gap in the weight space; that is, any given weight (w_1, w_2) is associated with a convex efficient point.

w_1 range	0–0.4	0.4	0.4–0.6	0.6	0.6–1
w_2 range	0.6–1	0.6	0.4–0.6	0.4	0–0.4
Best alternative	F_1	F_6, f_1, f_2	F_2	F_2, F_3	F_3

The above range of weights can also be obtained graphically as shown in Figure 2.8. To do this, draw each alternative by a line as shown in Figure 2.8. For example, for $F_1 = (1,6)$, the line is drawn by connecting points $f_1 = 1$ and $f_2 = 6$ in Figure 2.8. Note that $U(F_1) = w_1(1) + w_2(6) = (1 - w_2) + 6w_2 = 1 + 5w_2$; therefore the utility changes based on the value of w_2. Draw the lines of all alternatives. The intersection of each pair of convex efficient alternatives is associated with the weights shown in the above table. For example, $U(F_2) = U(F_3)$ and intersection is at $w_1 = 0.6$, $w_2 = 0.4$. Nonconvex efficient alternatives such as F_4 and F_5 do not have a weight space.

Ranking with Partially Known Weights In many cases, the DM may not be able to assess the exact weights of importance for linear additive utility functions, but the DM can provide or agree with an acceptable range of weights. The lower and upper bounds on each weight can be presented as follows:

$$w_{i,\min} \leq w_i \leq w_{i,\max} \quad \text{for } i = 1, 2, \ldots, k \qquad (2.2)$$

Alternatives can be ranked by using the given lower and upper bounds on weights. For example, consider Example 2.1. Suppose that $0.2 \leq w_1 \leq 0.5$, and $0.5 \leq w_2 \leq 0.8$, then according to Figure 2.7b (or Figure 2.8), alternatives f_1, f_2, F_6 are acceptable, but F_3 is not. See Malakooti (1989a, 2000) and Kirkwood and Sarin (1985) for efficient methods of ranking alternatives for partial information given on weights.

2.3   UTILITY FUNCTIONS

2.3.1   Z Utility Theory

Four Types of Utility Functions   A "utility function" is a function of all criteria (also called objectives or attributes), $U(f_1, f_2, \ldots, f_k)$. The focus of this chapter is the construction and validation of $U(f_1, f_2, \ldots, f_k)$ under certainty that presents the DM's preferences (preferential behavior) for MCDM problem. We define four types of utility functions:

(a) Independent (deviation neutral, flat/linear presented by an additive utility function)

(b) Convergent (deviation averse, centrist presented by a concave utility function)

(c) Divergent (deviation prone, extremist presented by a convex utility function)

(d) Adaptive/hybrid (deviation adaptive, a hybrid of the above three cases)

Depending on the DM and the decision problem, one of these utility functions can be used to solve the problem. In Figure 2.9, the first three types of utility functions are presented by three-dimensional surfaces, where the utility function $U(f_1, f_2)$ is a function of two criteria, f_1 and f_2. In MCDM, all criteria are maximized; therefore all utility functions are increasing functions of all criteria (f_1 and f_2) in the range [0,1]. In Figure 2.9a, the utility function is additive (linear) where the best alternative is (1,1). In Figure 2.9b, the utility function is concave where the best alternative is also (1,1). In Figure 2.9c, the utility function is convex where there are four equally best alternatives (–1,–1), (–1,1), (1,–1), and (1,1).

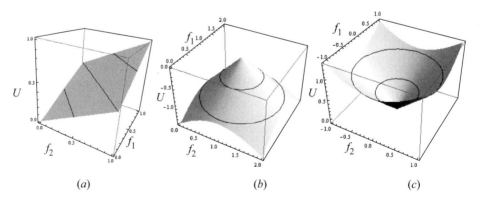

(a) (b) (c)

Figure 2.9   Examples of three types of utility functions $U(f_1, f_2)$, in three dimensions: (a) independent: additive utility, Max $U = 0.5f_1 + 0.5f_2$; (b) convergent: concave utility, Max $U = -((f_1 - 1)^2 + (f_2 - 1)^2)^{0.5} + 0.01\, f_1^{0.4} f_2^{0.4}$; (c) divergent: convex utility, Max $U = (f_1^2 + f_2^2)^{0.5} - 0.01(f_1 + 1)^{0.4}(f_2 + 1)^{0.4}$.

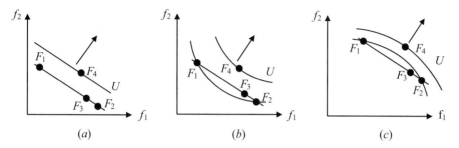

Figure 2.10   Contours of three types of bicriteria utility functions $U(f_1, f_2)$: (a) contours of independent, additive utility; (b) contours of convergent, concave utility; (c) contours of divergent, convex utility; (d) adaptive/hybrid (deviation adaptive, a hybrid of the above three cases).

In reality, ideal alternatives such as (1,1) in Figure 2.9 are not part of the feasible set of alternatives (if they exist, the MCDM problem would have only one solution, (1,1)). Generally, there is a set of efficient alternatives. Therefore, only a small portion of the utility functions in Figure 2.9 are feasible. In Figure 2.10, alternatives F_1, F_2, F_3, and F_4 are all efficient. Because the utility function is defined over the set of efficient alternatives, only utility functions over the efficient area domain need to be considered. Figure 2.10 shows two contours for each of the utility functions over the efficient area where arrows point toward higher (or better) utility values. Note that all points on the same contour have the same utility value; that is, they are equally preferred.

In all three graphs, F_1 and F_2 are equally preferred to each other. F_3 is a convex combination of F_1 and F_2; that is, it is on the line that connects F_1 and F_2. In Figure 2.10a, for independent (additive utility function) decision behavior, a convex combination of F_1 and F_2 (e.g., F_3) is equally preferred to F_1 and F_2. In Figure 2.10b, for convergent (concave feasible goal) decision behavior, any convex combination of F_1 and F_2 (e.g., F_3) is preferred to both F_1 and F_2. In Figure 2.10c, for divergent (convex feasible goal) decision behavior, both F_1 and F_2 are preferred to any convex combination of F_1 and F_2 (e.g., F_3).

Additive Utility Functions (Ordinal/Cardinal Approach, Sections 2.4.1–2.4.4)

If the DM's utility function is additive, it is best identified initially, without imposing the DM to a more challenging task of assessing nonlinear utility functions. In Sections 2.4.1 through 2.4.4, the ordinal/cardinal approach (OCA) is developed that uses preference difference for assessing the weights of importance of additive utility functions. The same questions asked for verification of OCA can be used for the assessment and verification of nonlinear Z utility theory functions (ZUTs).

Additive Utility Functions (Analytic Hierarchy Process, Sections 2.4.5 and 2.4.6)

A well-known method for assessing additive utility functions is analytic hierarchy process (AHP) which is reviewed in Section 2.4.5 and a critique of its effectiveness is provided in Section 2.4.6.

Quasi-linear Additive Utility Functions (Section 2.4.7)

In Section 2.4.7, we develop a new approach that generalizes additive utility functions by allowing the use of concave/convex or quasi-linear value functions; and by evaluating actual alternatives. An optimization approach is used to assess value and additive utility functions simultaneously.

Using Nonlinear (Concave or Convex) Utility Functions (Sections 2.5 and 2.6) Additive utility functions are a special case of nonlinear functions, which can provide additional flexibility when selecting alternatives. In MCDM and multiple objective optimization (MOO) literature, convex utility functions are not considered. In MOO, mathematically, it is more convenient to use concave functions because one may be able to prove convergence to the optimal solution for maximizing concave utility functions. In MCDM, concave functions are used because they present a conservative decision behavior; for example, central points (nonextreme) are preferred to extreme points. We, however, believe that convex functions must also be considered as in some cases, they are the most suitable functions to present the decision behavior. For example, consider choosing a food based on two attributes, taste and nutrition. A person may prefer either very tasty food or very nutritious food to food that is both moderately tasty and moderately nutritious. Another example is when an investor prefers either less risky (with low rate of return) or more risky (with high rate of return) to moderately risky (with medium rate of return). This type of decision behavior can occur in everyday life and in critical decision problems in government, industry, and corporations. Nevertheless, generally, concave utility functions are more justifiable than convex utility functions for many applications, and in this chapter, we will focus and elaborate on them.

ZUT Goal Seeking and Resolving of Goal Programming (Section 2.6) One of the most known and effective methods for solving MCDM/MOO problem is goal programming. In Section 2.6, we review this method and suggest replacing it with Z-goal programming, which overcomes the possible shortcomings of goal programming. Sections 2.7 and 2.8 cover MOO for methods covered in Sections 2.4–2.6.

ZUT: Adaptive (Hybrid) Concave/Convex Nonlinear Utility Functions Classical utility theory (expected utility theory, e.g., see Keeney and Raiffa, 1993; Dyer, 2005) has been influenced by economists' understanding of utility (and value) functions of commodities which have diminishing marginal utility values presented by concave functions (Bernoulli, 1738); for example, see Figure 2.14a. Although this assumption seems to be justifiable and rational, it simplifies, restricts, and contradicts rational human decision behavior. That is, depending on the problem and the alternatives, DMs may have concave or convex or some other forms of utility functions; for example, see Figures 2.14a–d. Alternatively, the utility function may be a hybrid of different (e.g., concave and convex) functions. In fact, one of the advantages of additive utility functions could be that they approximate the combined concave and convex utility functions by an additive function and therefore, they appear to be more successful than nonlinear utility functions. For complex decision problems, combined concave and convex functions are considered in Z utility theory (ZUT) (e.g., see Section 2.6.5). In ZUT, the utility function is a hybrid of an additive function, a concave deviational value, and a convex deviational value (Malakooti, 2014b). There are several types and extensions of ZUTs; a review of two of them is presented in Sections 2.5 (multiplicative) and 2.6 (goal seeking).

ZUT: For Decision Making Under Risk and Uncertainty ZUT employs a new class of nonlinear utility functions that solve decision-making problems under risk, uncertainty, and ambiguity. ZUT provides editing rules (on how to present risky alternatives) that substantially improve the accuracy of paired comparisons of alternatives and the assessment of the single utility function for risk. Experimental results show that ZUT effectively solves all complex risky decision paradoxes (such as Allais, 1953; Kahneman and Tversky, 1979) and has a high accuracy in predicting decision behavior. ZUT provides nonlinear utility functions generalizing expected utility theory of Von Neumann and Morgenstern (1947), mean variance of Markowitz (1959), and cumulative prospect theory of Tversky and Kahneman (1992).

2.3.2 Nonlinear Utility Functions

A utility function $U(F) = U(f_1, f_2, \ldots, f_k)$ represents the DM's preferences for an alternative, $F = (f_1, f_2, \ldots, f_k)$ where each objective f_i is maximized for $i = 1, 2, \ldots, k$. Note that we use the term utility U for all cases of decision making under certainty, uncertainty, and risk. When $U(F)$ is differentiable with respect to objectives F, then monotonicity (efficiency) means that

$$\partial U / \partial f_i > 0 \qquad \text{for all } i = 1, 2, \ldots, k$$

Gradient of a Nonlinear Utility Function   The gradient is defined as the best direction of improvement at a given point (alternative). For MCDM problems the gradient is always positive.

$$\nabla = [\partial U / \partial f_1, \ldots, \partial U / \partial f_k] > 0$$

$\partial U / \partial f_i$ for $i = 1, 2, \ldots, k$ is also defined as the marginal utility (MU_i) with respect to f_i, which is the local rate of change in the utility value as a function of objective f_i, while all other objectives are kept constant. The rate of change is defined as the amount of increase in the utility as a result of one unit increase of the corresponding objective (criterion). Therefore, the elements of the gradient vector correspond to the marginal utilities of all criteria (objectives). A utility function is nonlinear if its gradient evaluated at different points is not constant; that is, it varies.

Examples of Nonlinear Utility Functions   Consider a bicriteria utility function $U(f_1, f_2)$. Contours of two different utility functions, one concave and one convex, are presented in Figure 2.11.

In Figure 2.11a, two gradients are shown at two different points on the same contours. For nonlinear utility functions, gradients are different at different points. For example,

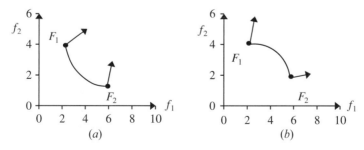

Figure 2.11   Gradients of nonlinear functions are different at different points.

for a convex utility function, suppose that $U = f_1^2 + f_2^2 + 0.1 f_1 f_2$ is maximized where $0 \le f_1 \le 10$ and $0 \le f_2 \le 10$. The gradient of this function is:

$$\nabla U = [2f_1 + 0.1f_2, 2f_2 + 0.1f_1]$$

The gradient at point $F_1 = (2,4)$ is $(4.4, 8.2)$. The gradient at point $F_2 = (6,2)$ is $(12.2, 4.6)$. That is, at different points, the best directions of improving the utility functions are different.

2.3.3 Additive Utility Functions

Additive utility functions are very popular and have been used in many applications. Suppose that all criteria are maximized.

Linear Additive Utility Functions The linear additive utility function is

$$U = w_1 f_1 + w_2 f_2 + \cdots + w_k f_k \tag{2.3}$$

where $w_1 > 0$, $w_2 > 0$, ..., $w_k > 0$. $w_1, w_2, \ldots, w_k$ are the weights of importance of each criterion (objective). The weights of importance are solicited from the DM and reflect the importance of objectives. All weights are positive numbers and therefore the utility function is maximized. Alternatives are ranked in descending order of their utility values U, and the best alternative is the one with the highest utility value. The gradient of a linear utility function with respect to objectives is

$$\nabla U = [\partial U / \partial f_1, \ldots, \partial U / \partial f_k] = (w_1, w_2, \ldots, w_k)$$

This means that for linear utility functions, the gradient is always constant and it is equal to weights of importance of criteria regardless of the values of criteria.

Additive Utility Functions Using Value Functions of Criteria Additive utility functions are a generalization of the linear additive utility function (2.3) where value functions of criteria are used.

$$U = w_1 v_1(f_1) + w_2 v_2(f_2) + \cdots + w_k v_k(fk) \tag{2.4}$$

where $w_1 > 0$, $w_2 > 0$, ..., $w_k > 0$, and $w_1 + w_2 + \cdots + w_k = 1$.

$v_i(f_i)$ is an assessed value function of f_i. The value function $v_i(f_i)$ is an increasing function of f_i. All assessed value functions should use comparable scales, for example, a 0–1 scale where 0 is the worst and 1 is the best. Value functions are discussed in Section 2.3.5. A utility function is additive if its gradient with respect to each value function is constant. Note that (2.4) is a generalization of (2.3) where $v_i(f_i) = f_i$ in Equation (2.3). This additive utility function (2.4) still has a constant gradient with respect to each $v_i(f_i)$.

$$\nabla U = [\partial U / \partial v_1(f_1), \ldots, \partial U / \partial v_k(f_k)] = (w_1, w_2, \ldots, w_k)$$

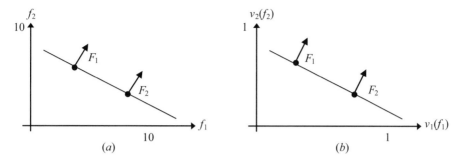

Figure 2.12   Gradients of additive utility functions are constant and equal at all points: (*a*) a contour of a linear additive utility function $U = 2f_1 + 4f_2$; (*b*) a contour of an additive utility function $U = 3v_1(f_1) + 6v_2(f_2)$.

Equation (2.4) can be written in a more general form as (2.5), where both (2.4) and (2.5) can rank alternatives similarly.

$$U = v_1^o(f_1) + v_2^o(f_2) + \cdots + v_k^o(f_k) \tag{2.5}$$

Equation (2.5) is also an additive utility function. For example, $U = 3f_1^3 + 5f_2^5 + 17f_3^2$ is an additive utility function in the form of (2.5) and it can be converted to the form of (2.4).

Graphical Examples of Gradient of Linear and Additive Utility Functions   For linear utility function $U = 2f_1 + 4f_2$ the gradient is (2,4); that is, the best direction of improvement is to increase f_2 by twice as much as the increase in f_1 at any given point. The gradients for points F_1 and F_2 are shown in Figure 2.12*a*. For additive utility function (2.4), see Figure 2.12*b*.

Mutual Preferential Independence (MPI) of Criteria in Additive Utility Functions   We showed that for additive utility functions the gradient of the utility function with respect to $v_i(f_i)$ is constant at all different alternatives (points). Traditionally, the explanation and validation of the additive utility function is accomplished through the definition of MPI. Two criteria p and q are mutual preferential independent from each other if their preferential relationship (the trade-off) does not change based on the value of all other criteria (i.e., the $k - 2$ criteria other than p and q). This means that the preferential relationship (the trade-off) between the two criteria p and q will remain the same regardless of how high or low the values of the other criteria are. For example, consider buying a house based on three criteria of price (cost in dollars), space (in square footage), and locality (e.g., better neighborhoods in scale of 0–10). If one's rate of substitution (trade-off) between price and space is independent of the locality, then these two criteria are mutual preferential independent of the locality criterion. The utility function is additive when each and every pair of criteria is MPI. In the house buying example, if each pair of criteria, (price, space), (price, locality), and (space, locality), is MPI, then the utility function is additive. Note that the last MPI can be concluded from the first two MPIs; that is, when (price, space) and (price, locality) are MPI then (space, locality) is also MPI. In practice, direct verification

of MPI is cumbersome and not meaningful to most DMs. (It is often a futile exercise in self-verification and self-ascertaining.)

Restrictiveness and Deficiency of Linear and Additive Utility Functions

Additive utility functions may be useful for some decision problems, but their linearity (which was explained through MPI assumption) may lead to the elimination of the best efficient alternative. That is, some efficient alternatives (that are nonconvex efficient) cannot be selected as the best alternative (see Section 2.2.3). For example, in Figure 2.7a, F_4 is efficient but convex inefficient, but this alternative cannot be selected as the best regardless of which weights are used in (2.3). However, some DMs may select F_4 as the best alternative. The above deficiency is applicable when value functions are used (2.4). For example, consider a decision problem having the following three alternatives: $F_1 = ((v_1(5), v_2(15))$ $= (0,1)$, $F_2 = ((v_1(25), v_2(7)) = (1,0)$, and $F_3 = ((v_1(12),v_2(9)) = (0.4,0.4)$. F_3 is efficient but convex inefficient. This alternative cannot be selected as the best alternative regardless of which weights are used in (2.4), that is, maximizing $U = w_1v_1(f_1) + w_2v_2(f_2)$. However, it is easy to see that many DMs would select F_3 as the best alternative. To be able to select such efficient (but convex inefficient) alternatives, nonlinear utility functions should be used as presented in Sections 2.5 through 2.9. The consistency tests presented in Section 2.4.2 can be used to identify whether the utility function is additive or nonlinear by evaluating external alternatives (that were not used for assessing the additive utility function).

2.3.4 Additive versus Multiplicative Utility Functions[†]

Consider the following multiplicative utility function.

$$U = v_1^o(f_1) \times v_2^o(f_2) \times \cdots \times v^o k(f_k) \qquad (2.6)$$

By taking the natural logarithm of both sides of this function we have

$$\ln U = \ln v_1^o(f_1) + \ln v_2^o(f_2) + \cdots + \ln v_k^o(f_k) \qquad (2.7)$$

Both multiplicative forms (2.6) and (2.7) can rank alternatives similarly. Therefore, one can utilize additive utility functions to achieve the same ranking of multiplicative utility functions; that is, multiplicative form (2.6) is not genuinely a nonlinear utility function. However, consider (2.8) which is a simple form related to nonlinear ZUTs (discussed in Section 2.5).

$$U = z_{11}(v_1^o(f_1) + v_2^o(f_2) + \cdots + v_k^o(f_k)) + z_{12}(v_1^o(f_1) \times v_2^o(f_2) \times \cdots \times v_k^o(f_k)) \quad (2.8)$$

where z_{11} and z_{12} are given coefficients. The above function can be interpreted as the addition of additive (2.5) and multiplicative (2.6) utility functions. We observe that this function (2.8) cannot be transformed into an additive function, which means that it is genuinely a nonlinear utility function. For example, $U = 2f_1 + 5f_2 + f_3 + 6f_1 f_2 f_3$ cannot be expressed in the forms of (2.4), (2.5), (2.6), or (2.7). However, note that the celebrated multiplicative (multiattribute) utility function of Keeney (1974) and Keeney and Raiffa (1993) has only a slight nonlinearity and it can also be transformed to an additive function by a logarithmic transformation. Our experiments show that it performs at the level of additive functions; and therefore, for practical purposes, one may use additive utility functions in its place.

2.3.5   Value Functions for Multicriteria[†]

Value Functions   Value function refers to the worth of a criterion usually presented on a scale of 0–1. Each DM may have a different value function for a given criterion. Suppose that the value function of each criterion can be assessed independent of other criteria. Value functions can be linear (e.g., see Figure 2.13a) or nonlinear (e.g., see Figure 2.13b).

If a value function is linear, then its marginal values have the same worth at different points. However, in nonlinear value functions, marginal values may have different worth at different points. For example, the worth of $10,000 (as a marginal value) may be much higher for you if your total assets are $25,000 than if your total assets are $10,000,000. The same concept of value function of money may be applied to other criteria such as reliability.

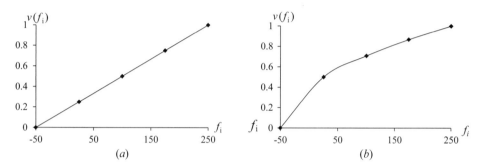

Figure 2.13   (a) A linear value function; (b) a nonlinear value function.

Four Types and Examples of Value Functions   The value functions of different criteria can assume different forms. The most common ones are linear (Figure 2.13a), concave (Figure 2.14a), convex (Figure 2.14b), and quasi-linear (a combination of concave and convex functions). We consider two forms of quasi-linear functions: S shape (first convex then concave, Figure 2.14d); reverse S shape (first concave then convex, Figure 2.14c). Here are some examples of the types of value functions: Figure 2.14a (value of higher energy efficiency, e.g., higher miles per gallon (MPG), gas consumption for cars); Figure 2.14b (value or the joy of winning more money for a gambler); Figure 2.14c (value of appreciating healthy life as one gets older); Figure 2.14d (value or effectiveness of having healthy food/vitamins). We would like to emphasize that concave value functions (Figure 2.14a) are more practical than convex value functions (Figure 2.14b). Similarly, S-shaped value functions (Figure 2.14d) are more practical than reverse S-shaped value functions (Figure 2.14c). For S-shaped value functions, for example, consider buying a car where the value function of MPG criterion may range from 0 to 70 where the point of inflection in Figure 2.14d is 15 MPG. The value function curve on both sides of this inflection point rapidly increases; however, when MPG reaches 40 MPG the increase in the value function becomes gradual. The S shape includes the advantage of using both concave and convex functions.

Increasing Property of Value Functions   A value function (e.g., Figures 2.14a–d) must be an increasing function of criterion (objective) f_i. To quantify value functions, use a one parameter model. For concave or convex value functions use:

$$v_i(f_i) = f_i^\beta \qquad \text{where } \beta_i > 0 \qquad (2.9)$$

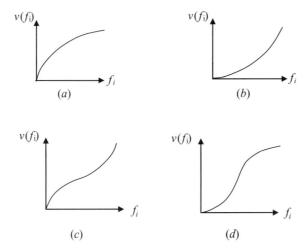

Figure 2.14   Increasing value functions: (*a*) concave, (*b*) convex, (*c*) quasi-linear (concave then convex), (*d*) quasi-linear (convex then concave).

When $0 < \beta_i < 1$, this value function is concave (Figure 2.14*a*) and when $\beta_i > 1$, this value function is convex (Figure 2.14*b*). To quantify quasi-linear value functions, use

$$v_i(f_i) = (f_i^{\gamma_i}/(f_i^{\gamma_i} + (1 - f_i)^{\gamma_i})^{1/\gamma_i}) \quad \text{where } \gamma_i > 0.279 \tag{2.10}$$

When $0.279 < \gamma_i < 1$ this function is reverse S shape (first concave then convex, Figure 2.14*c*); when $1 < \gamma_i < 2.5$ this function is S shape (first convex and then concave, Figure 2.14*d*); and when $\gamma_i > 2.5$ this function is convex. For practical purposes, one may use either concave functions $0 < \beta_i \le 1$ in (2.9); or S-shaped quasi-linear $1 \le \gamma_i$ in (2.10). (Note that there are several functional forms (e.g., sigmoid function) to present quasi-linear functions using one, two, or more parameters. We adopted (2.10) for value functions based on cumulative probability function of Tversky and Kahneman, 1992, for which Ingersoll, 2008, showed that when $\gamma_i < 0.279$, $v_i(f_i)$ is not increasing; so one should use $\gamma_i > 0.279$.) New functions are developed in Malakooti (2014a, b).

Double Helix and Single Helix Value Functions for Complex Problems   Each of the above two value functions have only one parameter (either β or γ). It is possible to use value functions that have two (e.g., both β and γ) or more parameters for solving more complex decision problems. For example, a two parameter value function can present all the above four forms: concave, convex, S, and reverse S shapes. A five parameter value function (called double helix) can present a combination of the above four forms (e.g., the value function may pass the diagonal line twice) to have a double S, double reverse S, or combined S and reverse S; these functions are developed in Malakooti (2014a, b).

Challenge of Assessing Individual Value Functions   The concept of value functions was motivated based on economic value of commodities; that is, as the quantity of a commodity increases its marginal value decreases, for example see Bernoulli (1738). For MCDM problems, although value functions appear to be interesting and meaningful, in practice, their independent assessment may not be possible, may confuse the DM, and may result in an invalid model. Also, the range of criteria values of practical alternatives is relatively small that they can be presented by linear value (LV) functions. In general, it is far more effective to resolve the issue of nonlinearity of (multiple criteria) utility functions (Sections 2.5 through 2.9) than the nonlinearity of single value functions. That is, the effect of nonlinearity of single value functions (of individual criteria) is overshadowed by the nonlinearity of the utility function (of all criteria). However, if it becomes necessary to use nonlinear value functions (as the last resort), then assess all value functions simultaneously by using only one parameter value function for all criteria; this is discussed in the method of Section 2.4.7.

2.3.6   Double Helix (Dual) Value Functions: Prosperity and Mortality

Solving Advantages and Disadvantages Bicriteria Problems   Consider evaluating alternatives under certainty having two criteria: advantages (labeled as expected value (EV)) and disadvantages (labeled as expected risk value (RV)). Suppose that we can measure these two criteria for all alternatives. The value functions presenting each of these criteria are inherently different. In the following we use two different double helix functions for presenting these two criteria. Assuming EV and RV are independent from each other, these two value functions can be assessed independently. Then we can use an additive utility function to solve the bicriteria problems where the utility function is $U = v_1(\text{EV}) + v_2(\text{RV})$. Two examples of value functions of EV and RV are presented below. The first one is the Prosperity (food or fun) curve for the expected value of alternatives. The second one is the mortality (fight or flight) curve of risk value of alternatives. Both of these functions may consist of up to four distinct segments where each segment may have different slope.

Expected Value (EV) and Risk Value (RV) for Risk Problems   Consider evaluating alternatives under risk for a single criterion (attribute). An alternative under risk can be presented by two different quantity, expected value (EV) of payoffs (potential return, e.g., in dollars) and risk value (RV) of payoffs (potential loss, e.g., in dollars). Methods for measuring RV for problems under risk are provided in Malakooti (2014). A risk averse DM maximizes EV and minimizes RV, therefore risk can be presented as a bicriteria problem.

Prosperity (Food or Fun) Curve for Expected Value, $v_1(EV)$

I. *Necessity*   The amount of money needed for living necessities such as food, clothing, and shelter.

II. *Desirous*   The amount of money needed for desired expenditures beyond necessity.

III. *Security*   The amount of money needed for security, protection, and future/retirement supports.

IV. *Aspiration*   The amount of money needed for luxury, the fulfillment of dreams, ambitions, and life-long goals.

I and III can be categorized as food and II and IV as fun. Different curves can be assessed for different people. For example, the priority (order) could be I, II, III, and IV (food first then fun), or the priority (order) can be II, I, IV, and III (fun first then food). Figures 2.15*a, b* present two possible types of prosperity curves.

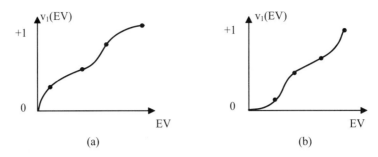

Figure 2.15   Two examples of prosperity (food or fun) curves for expected value (0–1): (*a*) concave–convex–concave; (*b*) convex–concave–convex.

Mortality (Fight or Flight) Curve for Risk Value, v₂(RV)   For risk averse, four levels of RV can be considered.

I. *Immortal Risk*   The value of risk is inconsequential.
II. *Consequential Risk*   The value of risk is consequential.
III. *Mortal Risk*   The value of risk is critical, the loss is mortal.
IV. *Postmortal Risk*   The value of risk is beyond critical (mortal).

$v_2(RV)$ can be explained based on fight or flight position, describing at what point a person changes his/her mind from fight to flight and vice versa. Flight attitude is when the value increases are rather flat, and fight attitude is when the value increases are sharp. One can categorize I and IV as flight and II and III as fight. But different DMs may have different attitudes and priorities with respect to these four levels. $v_2(RV)$ can be concave (first fight then flight), convex (first flight then fight), or different combinations of convex and concave functions. Figure 2.16*a* shows flight, fight, and flight. Figure 2.16*b* shows fight, flight, and fight.

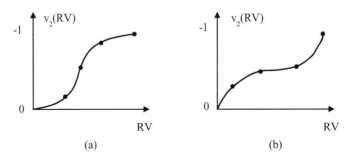

Figure 2.16   Two examples of immortality (fight or flight) curve for risk averse (0 to −1) where 0 is the best and −1 is the worst: (*a*) flight–fight–flight type; (*b*) fight–flight–fight type.

2.3.7   Normalized Criteria and Weights

Normalized Criteria   A simple approach to assess LV functions (e.g., see Figure 2.14*a*) is normalizing all objective values. Using normalized values simplifies solving MCDM problems. Each criterion can be normalized as follows:

Find the range of each criterion (objective), f_i, $i = 1, 2, \ldots, k$.

$$f_{i,\min} = \text{Minimum}\{f_{ij}, \ j = 1, 2, \ldots, n\}$$
$$f_{i,\max} = \text{Maximum}\{f_{ij}, \ j = 1, 2, \ldots, n\}$$

For each given alternative F_j, normalize each f_{ij}:

$$f'_{ij} = (f_{ij} - f_{i,\min})/(f_{i,\max} - f_{i,\min}) \qquad \text{for } i = 1, 2, \ldots, k$$

The additive utility function (2.4) can be presented as

$$U = w_1 f'_1 + w_2 f'_2 + \cdots + w_k f'_k \tag{2.11}$$

$f'_1, f'_2, \ldots, f'_k$ are normalized criteria values of $f_1, f_2, \ldots, f_k$, respectively. $w_1, w_2, \ldots, w_k$ are weights of importance of the normalized criteria, respectively, where the summation of weights is one, $w_1 + w_2 + \cdots + w_k = 1$, and each weight has a positive value, $w_1 > 0, w_2 > 0, \ldots, w_k > 0$.

Alternatives are ranked in descending order of their U values. The alternative with the highest U is the best alternative.

Example 2.3   Ranking Alternatives Using Normalized Criteria   Consider the problem shown in the table below. First normalize all objective values, then assess weights associated with normalized criteria, and finally rank the alternatives.

	F_1	F_2	F_3	F_4	F_5	Min	Max
Maximize f_1	10	8	4	2	3	2	10
Maximize f_2	−200	450	850	900	500	−200	900

The normalized objective values, f'_1 and f'_2, are listed in the following table. For example, $f'_{12} = (8 - 2)/(10 - 2) = 6/8 = 0.75$. For the given normalized values, suppose the DM's weights of importance are $w_1 = 0.75$ and $w_2 = 0.25$, where $0.75 + 0.25 = 1$. Therefore, utility function is: $U(F) = 0.75 f'_1 + 0.25 f'_2$. The ranking of the alternatives from best to worst is f_1, f_2, F_3, F_5, and F_4.

	F'_1	F'_2	F'_3	F'_4	F'_5	w_i
Maximize f'_1	1.000	0.750	0.250	0.000	0.125	0.75
Maximize f'_2	0.000	0.591	0.955	1.000	0.636	0.25
Maximize $U(F)$	0.750	0.710	0.426	0.250	0.253	–
Rank	1	2	3	5	4	–

Normalized Weights In some applications, it is necessary to rank alternatives using the actual criterion values. In this case, it is possible to use (2.12) to directly rank alternatives by using normalized weights of importance, $w_1', w_2' \ldots w_k'$.

$$U = w_1' f_1 + w_2' f_2 + \cdots + w_k' f_k \tag{2.12}$$

Consider Example 2.3, rank the alternatives by using normalized weights $w_1' = 0.75/10 = 0.075$ and $w_2' = 0.25/900 = 0.00028$. Therefore, we use $U = 0.075 f_1 + 0.00028 f_2$ to rank alternatives.

	F_1	F_2	F_3	F_4	F_5	Max	w_i	w_i'
Maximize f_1	10	8	4	2	3	10	0.75	0.075
Maximize f_2	−200	450	850	900	500	900	0.25	0.00028
Maximize U	0.69	0.73	0.54	0.40	0.36			
Rank	2	1	3	4	5			

Relationship of Normalized Weights to Normalized Criteria Note that (2.9) and (2.10) may not be converted to each other; that is, they may rank alternatives differently. However, if $f_{1,\min}$ is set equal to zero for $i = 1, 2, \ldots, k$, that is, $f_{i,\min}$ is not the minimum value but it is a reference point, then it is possible to convert (2.9) and (2.10) to each other. In this case, both methods would rate and rank alternatives similarly. For example, the normalized values of $(2,4,10)$ are $(0.2,0.4,1)$, respectively, and those of $(-5,2,4,10)$ are $(-0.5,0.2,0.4,1)$, respectively.

$$U(F_j) = \sum_{i=1}^{k} w_i f_{ij}' = \sum_{i=1}^{k} w_i \frac{f_{ij} - f_{i,\min}}{f_{i,\max} - f_{i,\min}} \text{ for alternative } F_j$$

By setting $f_{i,\min} = 0$ for all i:

$$U(F_j) = \sum_{i=1}^{k} w_i \frac{f_{ij}}{f_{i,\max}} = \sum_{i=1}^{k} \frac{w_i}{f_{i,\max}} f_{ij} = \sum_{i=1}^{k} w_i' f_{ij}$$

where $w_i' = w_i / f_{i,\max}$ for all i.

For example, suppose that in Example 2.3, the $f_{1,\min}$ and $f_{2,\min}$ are set as zero. Then, by using $w_1' = 0.75/10$ and $w_2' = 0.25/900$, both methods will provide the same rating and ranking of alternatives. Note that the normalized values for Example 2.3 will change by using 0 for both $f_{1,\min}$ and $f_{2,\min}$.

	F_1'	F_2'	F_3'	F_4'	F_5'	w_i
Maximize f_1'	1	0.8	0.4	0.2	0.3	0.75
Maximize f_2'	−0.22	0.5	0.94	1	0.56	0.25
Maximize $U(F)$	0.69	0.73	0.54	0.40	0.36	–
Rank	2	1	3	4	5	–

Minimization and/or Maximization of Criteria and Utility Function   So far
we assumed that all criteria and utility functions are maximized. In contrary cases,
simply change minimizing f_i to maximizing $-f_i$ and then apply the usual procedures for
maximization of the utility function. Mathematically, there is no problem in applying this
concept. However, from a communication point of view with the DM, when all criteria
(or most of them) are minimized it is counterintuitive to maximize the utility function.
For example, suppose that when manufacturing parts, the cost and the number of defective
parts are minimized. In that case it makes more sense to minimize cost, the number of
defectives, and the utility function, such that the best alternative will have the lowest
utility value. Therefore, if the utility function is minimized, change each maximizing f_i
to minimizing $-f_i$ and then apply the procedure for minimization of the utility function for
the given weights of importance.

An Example for Minimize f_1, Maximize f_2, and Minimize U: Consider the four
alternatives presented in the following table where the first criterion is minimized and the
second one is maximized. Consider the normalized values of both of these criteria. Note
that to find f_2' for minimization, one must convert maximizing f_2 to minimizing $-f_2$ and
then normalize this value. Do this by multiplying the second row by -1, and then normalize
the values. For example, consider the second criterion for alternative F_2, $f_{22}' = (-355 - (-440))/(-300 - (-440)) = 0.61$.

Now suppose that for the normalized values, the weight of the second criterion is twice
as important as the weight of the first criterion, that is,

$$\text{Minimize } U = w_1 f_1' + w_2 f_2' = 0.33 f_1' + 0.67 f_2'$$

The ranking is shown in the following table. The first alternative with the lowest utility
is ranked first.

	F_1	F_2	F_3	F_4	Min	Max	Weight
Minimize f_1	9	8	7	6	6	9	–
Maximize f_2	440	355	325	300	300	440	–
Minimize f_1'	1.00	0.67	0.33	0.00	0	1	0.33
Minimize f_2'	0.00	0.61	0.82	1.00	0	1	0.67
Minimize $U(F)$	0.33	0.63	0.66	0.67	–	–	–
Rank	1	2	3	4	–	–	–

See Example 2.3.xlsx.

2.4   ADDITIVE UTILITY FUNCTION: ORDINAL/CARDINAL APPROACH

2.4.1   Ordinal/Cardinal Approach: I. Assessment of Weights

For additive utility function, OCA approach can be used to assess weights of importance
of criteria, assess criteria values, and rank all alternatives. OCA method consists of the
following steps.[1]

---

[1] Ordinal/cardinal approach is adopted from systematic decision process of Malakooti (2011) which is updated
in Malakooti (2014a); also see Malakooti (2000).

(i) Assess the weights of importance of criteria, verify ordinal and cardinal consistency.

(ii) Verify the consistency of assessed weights for additive utility functions by using a set of testing (external) alternatives. If inconsistent, use z-theory (Section 2.5 or 2.6) approach for nonlinear utility functions.

(iii) Assess the criteria values for all alternatives using value functions.

(iv) Use the assessed utility function to rank all alternatives. Verify consistency of ranked alternatives. If inconsistent, adjust and verify above steps.

Qualitative Paired Comparison Ratings: The DM is asked to respond to paired comparisons of weights of importance using either qualitative or numerical ratings. Examples of qualitative ratings for six levels are shown in the following table. More levels can be considered if needed.

Qualitative Ratings	A	E	I	O	U	X
Description	Absolutely important	Especially important	Important	Ordinarily important	Weakly important	Equally important or N/A

Quantitative Paired Comparison Ratings: Three levels of precision can be considered in quantitative ratings where each rating is presented by an integer number.

Quantitative Ratings	Low Precision	Medium Precision	High Precision
An Integer No. in Range of	0–5	0–10	0–100

Depending on the problem and the DM's precision, one of the three levels (or other scales) can be selected. The final ranking of alternatives will be as precise as the selected level of precision. The qualitative ratings can be converted to quantitative ratings by using one of the three levels of precision. We recommend using 0–100 high precision for assessing quantitative values and utility ratings.

Assessment of Weights of Importance   In this section, the method for assessing weights of importance, $w_1, w_2, \ldots, w_k$, are presented. The weight question corresponds to normalized criteria values ranging from 0 to 1. The following question is asked when comparing two different objectives with reference to the selected precision scale, for example, use a scale of 0–10. For two objectives i and q,

By how much is the weight of objective i more important than the weight of objective q?

For example, if w_i is more important than w_q, provide a rating R_{iq} such that

$$w_i - w_q = \alpha R_{iq} \tag{2.13}$$

where R_{iq} is the rating with respect to the given scale, and α is a positive coefficient that relates the given rating scale to the weights scale of 0–1. Specifically, α is 1/(range of rating scale). For example, if range is 0–10 and $R_{12} = 6$, then $w_1 - w_2 = 6\alpha = 6(1/10) = 0.6$.

TABLE 2.2   Comparing Pairs of k Weights

R_{ij}	w_1	w_2	w_3	$\cdots$	w_k	w_0
w_1	0					
w_2		0				
w_3			0			
$\cdots$				0		
w_k					0	
w_0						0

Thus, since we now know that $w_1 - w_2 = 0.6$ and $w_1 + w_2 = 1$ we can solve to find that $w_1 = 0.8$ and $w_2 = 0.2$.

We define w_0 as the weight of importance of a hypothetical worst criterion which has the value of zero, that is, $w_0 = 0$. With respect to this zero value as a reference, all assessed weights will be positive. In comparison to this worst objective, the above question can be stated as:

What is the weight of importance of objective i?

The responses can be inserted into Table 2.2. Since each weight is equally preferred to itself, the diagonal of this matrix always has zero values.

The assessment of weights by OCA consists of the following steps:

1. Construction of Response Matrix

- Ask the DM to rank $k + 1$ weights, $w_1, w_2, \ldots, w_k, w_0$, in descending order of their importance, where $w_0 = 0$.

- Arrange rows and columns of a $(k + 1)$ by $(k + 1)$ matrix (in the form of Table 2.2) in order of their importance; that is, the most important weight will be in the first row and also in the first column. The least important weight, w_0 (which has the value of 0), will be the last row and the last column.

- Ask the DM to assign numerical ratings, R_{ij}, for each pair of weights of the $(k + 1)$ by $(k + 1)$ response matrix. Use 0 when the two weights are equivalent.

2. Weights Assessment

Find the average weight, w_i, by performing the following operations.

- For each column j, find

$$R_{\min, j} = \text{Minimum } \{R_{ij} \text{ for } i = 1, 2, \ldots, k + 1\}$$

- Find adjusted preference ratings

$$R_{ij}^* = R_{ij} - R_{\min, j} \quad \text{for all } i \text{ and } j$$

- Find the sum of R_{ij}^* for each column j; that is, for each j, find

$$R_{\text{sum}, j}^* = R_{1j}^* + R_{2j}^* + \cdots + R_{k+1, j}^*$$

- Find each weight i with respect to each column j:

$$w_{ij} = R_{ij}^* / R_{\text{sum},j}^* \quad \text{for each } i \text{ and } j$$

- Find average weight $\bar{w}_i$ for each row i:

$$\bar{w}_i = \sum_{j=1}^{k+1} (w_{i,j})/(k+1) \qquad \text{for } i = 1, \ldots, k \tag{2.14}$$

3. Ordinal Inconsistency

Consider response matrix **R** (Table 2.2) whose columns and rows are ordered in descending order of weights. Responses are ordinal consistent when

- In each row, the assessed numbers are in ascending order.
- In each column, the assessed numbers are in descending order.
- The upper right triangle of the preference response table is nonnegative, and the lower left triangle of the preference response table is nonpositive.

4. Cardinal Inconsistency

Consider response matrix **R** (Table 2.2). Matrix **R** is cardinal consistent if all columns of matrix **R*** are identical. (Recall that $R_{ij}^* = R_{ij} - R_{\text{min},j}$ for all i and j). That is,

$$R_j^* = R_{j+1}^* \qquad \text{for all } j \text{ (columns)} \tag{2.15}$$

Verifying cardinal consistency, that is, Equation (2.15), is computationally very easy. Note that cardinal consistency means that

$$R_{iq} + R_{qj} = R_{ij} \qquad \text{for all } i, j, \text{ and } q \tag{2.16}$$

However, it is easier to verify (2.15) instead of verifying (2.16).

When responses are cardinal consistent, there is no need to calculate average weights as the weights w_{ij} generated by each column are identical. But if there are some inconsistencies, then cardinal inconsistency ratio (CIR) (2.18) must be calculated to determine the magnitude of inconsistency.

5. Cardinal Inconsistency Ratio

CIR can be used to determine the magnitude of cardinal inconsistency for each row of responses and also for **R** matrix. CIR_i for each weight i is the average of the absolute deviation of weight i from its average value. Therefore, CIR is the same as the well-known mean absolute deviation (MAD), which is used in the context of weight assessment.

$$\text{CIR}_i = \sum_{j=1}^{k+1} \left(\left|\bar{w}_i - w_{i,j}\right|\right)/(k+1) \qquad \text{for } i = 1, 2, \ldots, k \tag{2.17}$$

$$\text{CIR} = \sum_{i=1}^{k} \text{CIR}_i \tag{2.18}$$

The theoretical maximum value of CIR is 1 and its minimum is 0. If CIR is more than a threshold ε, for example, $\varepsilon = 0.05$ (or 5%), the cardinal inconsistency is significant and the additive utility function should not be used, instead use z-theory nonlinear utility function (Sections 2.5 and 2.6).

Example 2.4 A Location Selection Problem: OCA Weight Assessment   Consider the selection of the best location for a new manufacturing plant. After the initial analysis, five cities (A, B, C, D, and E) are identified as the best possible alternatives. The following four criteria are considered to evaluate these alternatives.

1. Total cost including land cost, building cost, operational cost, local taxes, and so on.
2. Proximity index, measured based on the closeness of the location to major highways, airports, resources, raw materials, vendors, and the market.
3. Labor availability index (skilled labor and staff).
4. Community support index based on school systems, higher education institutes, cultural institutes, support of local authorities, and so on.

In the first step of OCA, the weights of importance of the four criteria are assessed as follows.

Construction of Response Matrix: Ask the DM to rank the four weights, that is, w_1, w_2, w_3, w_4, in descending order where $w_0 = 0$ is the reference point.

Suppose that the DM's ranking of weights is: w_2, w_4, w_3, w_1, and w_0. Note that w_0 is always ranked last.

Suppose the DM assesses preference differences for each pair of weights, R_{ij}, on a scale of 0–10. The ratings are shown in Table 2.3a. For example, w_2 is better than w_3 by 4 units on a scale of 0–10. In this example, the maximum difference is 8, which is the difference between w_2 (the best) and w_0 (the worst equals 0).

Weight Assessment: Table 2.3c shows the application of OCA to find weights; note that because all responses are cardinal consistent, then all weights generated in each column are also identical.

TABLE 2.3   Application of OCA to Find Weights for Example 2.4

R_{ij}	w_2	w_4	w_3	w_1	w_0
w_2	0	2	4	6	8
w_4	−2	0	2	4	6
w_3	−4	−2	0	2	4
w_1	−6	−4	−2	0	2
w_0	−8	−6	−4	−2	0
$R_{\min,j}$	−8	−6	−4	−2	0

a. Assessed Difference Ratings for Each Pair

R_{ij}^*	w_2	w_4	w_3	w_1	w_0
w_2	8	8	8	8	8
w_4	6	6	6	6	6
w_3	4	4	4	4	4
w_1	2	2	2	2	2
w_0	0	0	0	0	0
Sum	20	20	20	20	20

b. Transformation to Nonnegative Values of Weights and Ordinal Consistency Values and Cardinal Consistency

Weights	w_2	w_4	w_3	w_1	w_0	$\bar{w}_i$	CIR_i	Consistent?
w_2	0.4	0.4	0.4	0.4	0.4	0.4	0	Yes
w_4	0.3	0.3	0.3	0.3	0.3	0.3	0	Yes
w_3	0.2	0.2	0.2	0.2	0.2	0.2	0	Yes
w_1	0.1	0.1	0.1	0.1	0.1	0.1	0	Yes
Sum	1	1	1	1	1	1	0	Yes

c. Weight Assessment
   See Example 2.4.xlsx

Ordinal Consistency   Table 2.3*a* shows that all responses R_{ij} are ordinal consistent.
Cardinal Consistency   Table 2.3*b* shows all column of R_{ij}^* matrix are identical, therefore all responses are cardinal consistent.
Cardinal Inconsistency Ratio   Since all responses are consistent, CIR = 0.

Example 2.5   OCA Weight Assessment with Inconsistent Responses   For purposes of illustration, suppose the responses of Example 2.4 are inconsistent. We randomly increased or decreased responses of Table 2.3a by about 20%. Table 2.4*a* presents these inconsistent responses. Tables 2.4*b*, *c* show the OCA solution.

TABLE 2.4   Example 2.5: Inconsistent Responses, Weights Assessment by OCA, and CIR Measurement

R_{ij}	w_2	w_4	w_3	w_1	w_0	R_{ij}^*	w_2	w_4	w_3	w_1	w_0
w_2	0	0	6	8	10	w_2	10	8	12	12	10
w_4	0	0	4	6	8	w_4	10	8	10	10	8
w_3	−6	−4	0	0	6	w_3	4	4	6	4	6
w_1	−8	−6	0	0	4	w_1	2	2	6	4	4
w_0	−10	−8	−6	−4	0	w_0	0	0	0	0	0
$R_{\min,j}$	−10	−8	−6	−4	0	Sum	26	22	34	30	28
	a. Assessed Difference Ratings						*b*. Transformation to Nonnegative Values				

Weights	w_2	w_4	w_3	w_1	w_0	$\bar{w}_i$	CIR_i	Consistent?
w_2	0.38	0.36	0.35	0.41	0.36	0.37	0.018	No
w_4	0.38	0.36	0.29	0.33	0.29	0.33	0.032	No
w_3	0.16	0.19	0.18	0.13	0.21	0.17	0.024	No
w_1	0.08	0.09	0.18	0.13	0.14	0.13	0.030	No
Sum	1	1	1	1	1	1	0.104	> 0.05, No

c. Weight Assessment

See Example 2.5.xlsx.

Weights Assessment: The assessed average weights are $\overline{W} = (w_2, \ w_4, \ w_3, \ w_1) = (0.37, 0.33, 0.17, 0.13)$.

Ordinal Inconsistency Test: By examining Table 2.4*a*, it can be observed that the responses are ordinal consistent.

Cardinal Inconsistency Test: Columns of Table 2.4*b* are not identical, therefore responses are cardinal inconsistent. Also, because of cardinal inconsistencies, the columns of the weight matrix are not equal.

Calculating CIR: We need to measure CIR. First, find CIR for each weight, for example CIR_1 for w_1 is

$$CIR_1 = (|0.08 - 0.13| + |0.09 - 0.13| + |0.18 - 0.13| + |0.13 - 0.13| + |0.14 - 0.13|)$$
$$/5 = 0.030$$

Similarly, the CIR for w_2, w_3, and w_4 are 0.018, 0.024, and 0.032, respectively; see Table 2.4. The total CIR is

$$CIR = 0.030 + 0.018 + 0.024 + 0.032 = 0.104$$

This CIR is not acceptable because 0.104 is higher than the acceptable threshold error, 0.05. If this inconsistency cannot be resolved (i.e., the DM does not change his/her responses to reduce CIR), then a nonlinear utility function, as presented in Sections 2.5 and 2.6, should be used to rank alternatives.

2.4.2 Ordinal/Cardinal Approach: II. External Verification

Consider the following additive utility function:

$$U = w_1 f_1' + w_2 f_2' + \cdots + w_k f_k' \tag{2.19}$$

where $f_1', f_2', \ldots, f_k'$ are normalized criteria values of $f_1, f_2, \ldots, f_k$, respectively. $w_1, w_2, \ldots, w_k$ are weights of importance with respect to the normalized criteria values where $w_i > 0$ for all i and $w_1 + w_2 + \cdots + w_k = 1$. For example, the additive utility function for Example 2.4 is: $U = 0.1 f_1' + 0.4 f_2' + 0.2 f_3' + 0.3 f_4'$.

Identity Alternatives: Consider k alternatives based on a k by k identity matrix as presented in Table 2.5. Alternative $k + 1$(labeled as null or zero) has 0 values; it represents the worst criteria values.

Convex Alternatives for External Consistency Test: Consider k convex combination of identity alternatives as presented in Table 2.5 for external consistency verification. Each

TABLE 2.5   General Identity and Convex Alternatives for Testing

	Identity Alternatives				Null	Testing: Convex Alternatives				
	I_1	I_2	...	I_k	I_0	$I_{1,2}$	$I_{2,3}$	...	$I_{k-1,k}$	$I_{k,k}$
f_1	1	0	...	0	0	0.5	0	...	0	$1/k$
f_2	0	1	...	0	0	0.5	0.5	...	0	$1/k$
f_3	0	0	...	0	0	0	0.5	...	0	$1/k$
...	...	...	...	...	...	...	...	...	0.5	$1/k$
f_k	0	0	...	1	0	0	0	...	0.5	$1/k$

alternative has two components of 0.5 in the consecutive order of criteria. Also, in order to have more precise fitting, alternative $I_{k,k}$ with equal criteria values is added.

Now apply the assessed additive utility function to each column (alternative) of Table 2.5.

Identify Alternatives: For identity alternatives, the result is shown in (2.20). Note that the utility value of each alternative is the same as the value of the weight whose corresponding row has the value of 1 in Table 2.5. For example, $U(I_1) = 1w_1 + 0w_2 + 0w_3 + \cdots + 0w_k = w_1$; and in general,

$$U(I_1) = w_1, U(I_2) = w_2, \ldots, U(I_k) = w_k, U(I_0) = w_0 = 0 \qquad (2.20)$$

External Weight Assessment Consistency Verification: Generate the utility value of each testing alternative by using the assessed additive utility function. For example,

$$U(I_{1,2}) = 0.5w_1 + 0.5w_2$$

Then rank all alternatives. Ask the DM if he/she agrees with the assessed ratings and rankings of alternatives.

Mean Absolute Deviation for Utility Assessment   MAD can be used to measure the accuracy of the assessed utility function. It is the average of absolute deviation of the computed utility values of Q alternatives (U) using assessed utility function versus the utility values given by the DM (U^*).

$$\text{MAD} = \left(\sum_{j=1}^{Q} |\text{Computed } U_j - \text{DM's } U_j^*| \right) \bigg/ Q \qquad (2.21)$$

Example 2.6   External Weight Assessment Verification: Consistent Case   Consider Example 2.4. The set of alternatives for testing is presented in Table 2.6.

Now ask the DM to verify the ranking and rating by the assessed utility function $U = 0.1f_1 + 0.4f_2 + 0.2f_3 + 0.3f_4$. The following table shows the comparison of the DM's assessment versus the assessments by OCA for the given alternatives. Because

TABLE 2.6   Identity and Convex Alternatives for $k = 4$

	Identity Alternatives				Testing: Convex Alternatives			
	I_1	I_2	I_3	I_4	$I_{1,2}$	$I_{2,3}$	$I_{3,4}$	$I_{4,4}$
f_1	1	0	0	0	0.5	0	0	0.25
f_2	0	1	0	0	0.5	0.5	0	0.25
f_3	0	0	1	0	0	0.5	0.5	0.25
f_4	0	0	0	1	0	0	0.5	0.25

MAD $= 0.01$ is less than the threshold of 0.05, the assessed utility function is valid to be used.

		Used for Assessment				Used for Test				Ave
		I_1	I_2	I_3	I_4	$I_{1,2}$	$I_{2,3}$	$I_{3,4}$	$I_{4,4}$	
DM's utility	U^*	0.1	0.4	0.2	0.3	0.24	0.32	0.24	0.25	–
	Ranking	8	1	7	3	5	2	5	4	–
Assessed	U	0.1	0.4	0.2	0.3	0.25	0.30	0.25	0.25	–
	Ranking	8	1	7	2	4	2	4	4	–
Error =		0	0	0	0	0.01	0.02	0.01	0	MAD =
$\lvert U^* -$ Assessed $U\rvert$										0.01

See Example 2.6.xls.

$$MAD = (\lvert 0.1 - 0.1\rvert + \lvert 0.4 - 0.4\rvert + \lvert 0.2 - 0.2\rvert + \lvert 0.3 - 0.3\rvert + \lvert 0.24 - 0.25\rvert$$
$$+ \lvert 0.32 - 0.30\rvert + \lvert 0.24 - 0.25\rvert + \lvert 0.25 - 0.25\rvert)/8 = 0.01$$

Example 2.7   External Weight Assessment Verification: Inconsistent Case   Consider Example 2.4. The set of alternatives for testing is presented in Table 2.6. Now ask the DM to verify the ranking and rating by the assessed utility function $U = 0.1f_1 + 0.4f_2 + 0.2f_3 + 0.3f_4$.

		Used for Assessment				Used for Test				Ave
		I_1	I_2	I_3	I_4	$I_{1,2}$	$I_{2,3}$	$I_{3,4}$	$I_{4,4}$	
DM's utility	U^*	0.1	0.4	0.2	0.3	0.4	0.1	0.45	0.45	–
	Ranking	7	3	6	5	3	7	1	1	–
Assessed	U	0.1	0.4	0.2	0.3	0.25	0.3	0.25	0.25	–
	Ranking	8	1	7	3	4	2	4	4	–
Error =		0	0	0	0	0.15	0.2	0.2	0.2	MAD =
$\lvert U^* -$ Assessed $U\rvert$										0.0938

See Example 2.7.xlsx.

$$\text{MAD} = (|0.1 - 0.1| + |0.4 - 0.4| + |0.2 - 0.2| + |0.3 - 0.3| + |0.4 - 0.25| \\ + |0.1 - 0.3| + |0.45 - 0.25| + |0.45 - 0.25|)/8 = 0.0938$$

Because MAD $= 0.0938$ is higher than the threshold of 0.05, the assessed utility function is not valid to be used. Therefore, a nonlinear utility function should be used (see Sections 2.4.7, 2.5, and 2.6).

2.4.3 Ordinal/Cardinal Approach: III. Assessment of Qualitative Criteria

There are two types of criteria: quantitative (objective) and qualitative (subjective). Quantitative (objective) criteria can be directly measured by the known conventional units (e.g., for cost, use dollars). However, qualitative (subjective) criteria (such as taste, convenience, comfort, aesthetics) have no commonly acceptable units of measurement. Measurements of qualitative criteria are subjective and complex. OCA assesses qualitative criteria values using the strengths of preferences scales (such as A, E, I, O, U, and X) and then converts these ratings to their associated integer ordinal values. Then all criteria values can be normalized, for example, on a scale of 0–1, or use the value functions as discussed in Section 2.3.5.

Example 2.8   Location Selection: Criteria Assessment   Consider Example 2.4 with five possible alternative cities: A, B, C, D, and E. Each of the five cities can be presented by a vector of four criteria: F_A, F_B, F_C, F_D, and F_E.

The first criterion is quantitative and the other criteria are qualitative. For the first criterion, the cost values are given in dollars, as presented in the first row of Table 2.7. Now assess the second criterion, the closeness to resources and markets. Convert qualitative criteria ratings to integer numerical values; in this example, we use 0, 1, 2, 3, 4, and 5 to present X, U, O, I, E, and A, respectively. The DM is asked to assess this criterion for each alternative. The responses are shown in the second row of Table 2.7. Similarly, each of the remaining criteria is assessed. The assessed values are shown in Table 2.7 where indices of actual alternatives are $j = $ A, B, C, D, and E. Table 2.8 shows the normalized values of criteria.

2.4.4 Ordinal/Cardinal Approach: IV. Ranking of Alternatives

In this step, the assessed utility function is used to rank all alternatives. A small set of alternatives, for example, the $k + 1$ best ranked alternatives, can be presented to the DM for final external verification. That is, ask the DM to confirm the method's rankings and utility values of the best $k + 1$ alternatives. This finalizes the external verification for the assessed criteria values and the assessed weights of the additive utility function.

TABLE 2.7   Assessment of Each of Four Criteria for Each of the Five Alternatives (A, B, C, D, and E) Where the Nadir Alternative (N) Has the Worst Values for All Criteria

Criterion Description: Actual Assessment	f_{ij}	F_A	F_B	F_C	F_D	F_E	$f_{i,\min}$	$f_{i,\max}$
1. Total cost in million dollars per year (minimize)	f_1	4	20	12	16	4	4	20
2. Closeness to resources and markets (maximize)	f_2	0	3	4	5	2	0	5
3. Labor availability (maximize)	f_3	5	0	3	2	3	0	5
4. Community support (maximize)	f_4	0	5	2	3	4	0	5

TABLE 2.8   Ranking of the Five Alternatives of Example 2.4 by Additive Utility Function

Criterion Description—Normalized Assessment	f_i'	F_D	F_C	F_B	F_E	F_A	Weight, w_i
2. Normalized closeness (maximize)	f_2'	1	0.8	0.6	0.4	0	0.4
4. Normalized community support (maximize)	f_4'	0.6	0.4	1	0.8	0	0.3
3. Normalized labor availability (maximize)	f_3'	0.4	0.6	0	0.6	1	0.2
1. Normalized total cost (maximize)	f_1'	0.25	0.5	0	1	1	0.1
$U = 0.4f_2' + 0.3f_4' + 0.2f_3' + 0.1f_1'$	–	0.69	0.61	0.54	0.62	0.30	–
Rank by Maximize U	–	1	3	4	2	5	–
Consistency verification by the DM for ranking and calculated utility values?	–	Yes	Yes	Yes	Yes	Yes	Consistent

See Example 2.8.xlsx.

Example 2.9   A Location Selection: Ranking Alternatives   Consider Examples 2.4 and 2.8 where there are four criteria, $i = 1, 2, 3$, and 4; and five alternatives, $j = $ A, B, C, D, and E. In step I, the weights were assessed; they are shown in the last column of Table 2.4. In step II, criteria values for each of the alternatives were assessed as shown in Table 2.7. All criteria are then normalized to a scale of 0–1 as shown in Table 2.8. Note that because the first criterion is minimized, its normalized values are calculated using $-f_1$ rather than f_1. For example, for alternative C, the normalized cost is $f_1' = (-12 - (-20))/(-4 - (-20)) = 8/16 = 0.5$. The following additive utility function is used to rank all alternatives.

Maximize $U = 0.1f_1' + 0.4f_2' + 0.2f_3' + 0.3f_4'$

Table 2.8 shows the resulting utility values and the rankings of all alternatives. The last row shows the responses from the DM that agree with the given rankings and ratings (the utility values) of all alternatives. Because all responses are yes, the assessed utility function is externally verified.

2.4.5   Analytic Hierarchy Process[†]

AHP Weight Assessment   AHP assesses weights and criteria values for additive utility functions and provides a ranking of alternatives. AHP, developed by Saaty (1980, 1990, 2008), has gained substantial popularity. AHP uses a simple and structured approach in communicating with the DM, and it has an appealing systematic approach for solving MCDM problems. Assessing weights using the AHP method is presented in this section. To assess weights in AHP, the DM provides pairwise comparisons of weights in the form of preference ratios. AHP asks the DM to choose the degree of preference ratios for pairs of all criteria, from a scale of 1–9, where 1 and 9 signify equal preference and highest preference, respectively. Integer numbers 1–9 or their reciprocals are used. Let P_{iq} represent the preference ratio between criteria i and q. For instance, $P_{23} = 5$ indicates that the DM prefers criterion 2 to criterion 3 by a factor of 5. P_{32} would simply be a reciprocal of P_{23}, that is, $P_{32} = 1/P_{23} = 1/5$.

The AHP procedure for determining weights can be stated as follows:

1. Ask the DM to state the preference ratio for each pair of weights.
2. Normalize each column by dividing each entry by the sum of each column.
3. Find the average of each row to determine the average weight of each criterion.
4. Measure inconsistency of assessed weights.

TABLE 2.9 Presentation of k Identity Alternatives for AHP

	I_1	I_2	$\ldots$	I_k
f_1	1	0	$\ldots$	0
f_2	0	1	$\ldots$	0
$\ldots$	$\ldots$	$\ldots$	$\ldots$	$\ldots$
f_k	0	0	$\ldots$	1

The AHP method of comparing pair of weights is equivalent to the paired comparisons of a $k \times k$ identity matrix where k is the number of criteria. The general form of this matrix is presented in Table 2.9. AHP assumes an additive utility function and uses it to rank alternatives. Applying the additive utility function for the identity alternatives of Table 2.9:

$$U(I_i) = w_i \qquad \text{for } i = 1, 2 \ldots, k \tag{2.22}$$

The preference ratio of alternatives I_i to I_q is

$$P_{iq} = U(I_i)/U(I_q) = w_i/w_q \qquad \text{(InAHP)} \tag{2.23}$$

That is, comparing pairs of weights is the same as comparing pairs of identity alternatives when additive utility functions are used. AHP uses the preference ratio for all pairs of weights and then generates an average value for each weight.

Example 2.10 Weight Assessment by AHP Consider Example 2.4 with four criteria. Ask the DM to provide the preference ratio information for pairwise comparisons of all weights.

1. The responses are listed in Table 2.10. Find the sum of the entries of each column. For example, the sum of the first column is $1 + 4 + 2 + 3 = 10$.
2. Each entry, P_{ij}, is normalized by dividing it by the sum of the column it belongs to. Assess weights based on each column. See Table 2.11. For the first column, weights are calculated as follows: $w_1 = 1/10 = 0.10$, $w_2 = 4/10 = 0.40$, $w_3 = 2/10 = 0.20$, $w_4 = 3/10 = 0.30$.
3. Now determine the average of each row; the result is the assessment of each weight.

Therefore, weights obtained using AHP are $W = (0.10, 0.38, 0.23, 0.30)$. To respond to the AHP questions for this example, we used $(0.1, 0.4, 0.2, 0.3)$ as the weights of importance and rounded the numbers to the closest integers for a 1–9 scale. AHP solution is relatively close to the exact values of assumed weights.

TABLE 2.10 DM's Preference Ratio Responses; Finding the Sum of Each Column

P_{ij}	w_1	w_2	w_3	w_4
w_1	1.00	0.25	0.50	0.33
w_2	4.00	1.00	2.00	1.00
w_3	2.00	0.50	1.00	1.00
w_4	3.00	1.00	1.00	1.00
Sum	10.00	2.75	4.50	3.33

TABLE 2.11   Weight Assessment by Each Column (Step 2) and the Average of Each Weight (Step 3)

P'_{ij}	w_1	w_2	w_3	w_4	Average
w_1	0.10	0.09	0.11	0.10	0.10
w_2	0.40	0.36	0.44	0.30	0.38
w_3	0.20	0.18	0.22	0.30	0.23
w_4	0.30	0.36	0.22	0.30	0.30
Sum	1	1	1	1	1

See Example 2.10.xlsx.

2.4.6   Effectiveness of AHP[†]

Rank Reversal: Criteria Dependence   To assess the value functions of each criterion, AHP uses the same AHP procedure for assessing the weights of importance. To do this, normalized criteria values of all alternatives (f_{ij}, the value of criterion i for alternative j) are assessed with respect to each other (all alternatives). That is, a matrix with n columns and n rows (where n is the number of alternatives) is assessed for each criterion. The approach for assessing normalized criteria values in AHP may cause inconsistency in ranking alternatives. The reason is that in AHP, the assessment of normalized criteria for each alternative depends on other alternatives. A symptom of this approach for criteria assessment is rank reversal of alternatives. Rank reversal was observed by Belton and Gear (1983), who showed that by adding a copy of an existing alternative, the ranking of alternatives is reversed. Dyer (1990a, b) extended the work of Belton and Gear (1983) and also showed that by adding new alternatives, the ranking of alternatives can be reversed. This phenomenon is explained through the following example.

Example 2.11   AHP Rank Reversal in Criteria Assessment   Consider the example presented in Table 2.12 where there are three alternatives and four criteria. Suppose a scale of 4–18 is used to present criteria values. The DM assesses the values of the first objective for alternatives 1, 2, and 3 as 4, 18, and 18, respectively. To find the normalized values for the three alternatives, first find the sum of assessed criteria values, $4 + 18 + 18 = 40$. Then calculate the normalized values: $4/40 = 0.100$, $18/40 = 0.450$, and $18/40 = 0.450$. See the first row of Table 2.12. The same procedure is repeated to assess the normalized values of the remaining three criteria. Suppose that the weights of importance of criteria

TABLE 2.12   Ranking of Three Alternatives by AHP for Example 2.11

	F_1	F_2	F_3	Sum	F'_1	F'_2	F'_3	Sum
f_1	4	18	18	40	0.100	0.450	0.450	1
f_2	18	4	4	26	0.692	0.154	0.154	1
f_3	4	18	10	32	0.125	0.563	0.312	1
f_4	8	4	10	22	0.364	0.181	0.455	1
U	–	–	–	–	0.320	0.337	0.343	1
Rank	–	–	–	–	3	2	1	–

TABLE 2.13   Rank Reversal of Table 2.12 Alternatives by AHP for Example 2.11

	F_1	F_2	F_3	F_4	Sum	F_1'	F_2'	F_3'	F_4'	Sum
f_1	4	18	18	10	50	0.08	0.360	0.360	0.200	1
f_2	18	4	4	4	30	0.6	0.133	0.133	0.133	1
f_3	4	18	10	18	50	0.08	0.360	0.200	0.360	1
f_4	8	4	10	10	32	0.25	0.125	0.313	0.313	1
U	–	–	–	–	–	0.253	0.245	0.251	0.251	1
Rank	–	–	–	–	–	1	4	2	2	–

for the additive utility function are $w_1 = 0.25$, $w_2 = 0.25$, $w_3 = 0.25$, and $w_4 = 0.25$. The ranking of alternatives using AHP method is shown in Table 2.12 by Maximize $U = 0.25 f_1'$ $+ 0.25 f_2' + 0.25 f_3' + 0.25 f_4'$.

Now, suppose that a new alternative $F_4 = (10,4,18,10)$ is considered. The set of updated alternatives is shown in Table 2.13. The same normalization approach is used to find normalized values. Note that the normalized values of the alternatives in Table 2.13 compared to Table 2.12 are changed. The ranking of alternatives by AHP using the same utility function is shown in Table 2.13. Note that F_3 was preferred to F_1 (in Table 2.12) but after considering the new alternative, F_4 (in Table 2.13), F_1 is preferred to F_3. That is, the ranking of alternatives is reversed.

Rank Reversal: Weights of Nonlinear Utility Functions   If the utility function is nonlinear and the DM's responses to the AHP questions are consistent, then AHP incorrectly concludes that responses are consistent with the assessed additive utility function and uses the assessed additive utility function. This confusion occurs because consistency in response to weight questions does not mean the utility function is additive. Implicitly, AHP makes its consistency conclusion based on evaluating k vectors of identity alternatives (see Table 2.9) for assessing k unknown weights, that is, assessing k unknown variables by using k nonredundant (including $w_1 + w_2 + \cdots + w_k = 1$) linear equations. To verify that the utility function is additive, more than k alternatives (e.g., at least $k + 1$ or preferably $2k + 1$ alternatives) should be used.

Example 2.12   AHP Rank Reversal in Weight Assessment   Consider a problem with two criteria. Consider alternatives: $I_1 = (1,0)$, $I_2 = (0,1)$, and suppose that the DM ranks these alternatives equally. That is, $U(I_1) = U(I_2)$, and therefore the response to the AHP ratio question is one, that is, $w_1/w_2 = 1$. Using $w_1/w_2 = 1$ and $w_1 + w_2 = 1$, the solution is $(w_1, w_2) = (0.5, 0.5)$. The assessed additive utility function is $U = 0.5 f_1 + 0.5 f_2$. In this case, AHP declares there is no inconsistency and uses the assessed additive utility function. However, suppose that the DM's actual utility function is $U = 0.5 f_1 + 0.5 f_2 - 0.5 f_1 f_2$. With respect to this utility function, the DM's response is still the same, that is, $P_{12} = U(I_1)/U(I_2) = w_1/w_2 = 1$. Now, consider a third alternative, $F_3 = (0.6, 0.6)$, for external consistency verification. See Table 2.14. According to AHP's additive utility function, F_3 is preferred to I_1 and I_2. But, according to the actual nonlinear utility function, I_1 and I_2 are preferred to F_3. This occurs because AHP incorrectly assumes that consistent responses to identity alternatives means the assessed utility function is additive.

TABLE 2.14   Rank Reversal of AHP Due to Nonlinearity of Utility Function

	I_1	I_2	F_3
AHP assessed additive utility, $U = 0.5f_1 + 0.5f_2$	0.5	0.5	0.6
Ranking by AHP	2	2	1
DM's actual utility, $U = 0.5f_1 + 0.5f_2 - 0.5f_1f_2$	0.5	0.5	0.42
Ranking by the DM	1	1	2

Excessive Number of Questions to Assess Criteria Values in AHP   In AHP, the value of each criterion for each alternative is assessed with respect to all other alternatives. For assessing the values of one criterion for all n alternatives, AHP asks at least $(n^2 - n)/2$ questions, and for all k criteria, at least $k(n^2 - n)/2$ questions. The first number in each cell of Table 2.15 shows the number of questions that is asked by AHP for assessing criteria. By a simple observation of this table, one can conclude that AHP is impractical for solving problems that have more than 10 alternatives or 5 criteria. For example, for a problem with 5 criteria and 1000 alternatives, about 2.5 million questions should be responded by the DM.

AHP Hierarchical Approach versus Multiple objective Optimization   AHP advocates using a hierarchical structure for presenting criteria, that is, identifying several criteria, then breaking down each criterion into a number of subcriteria, and continuing this process. A hierarchical multiobjective structure is an attractive concept. However, the assessment process can be extremely confusing when different criteria are functions of the same decision variables. The proper way to formulate such problems is to define each objective as a function of decision variables as presented in MOO; see Sections 2.7 and 2.8. Different objective functions can be functions of the same decision variables but have different coefficients. Most of the MOO problems presented in different chapters of this book have this type of structure, (e.g., see MOO formulations in Chapter 2). Furthermore, each objective function can be a nonlinear function of decision variables, which cannot be presented using the AHP assessment process; also see Ravindran and Warsing (2013).

The Inaccuracy of Assessing Additive Utility Functions   AHP assumes that DMs cannot assess utility values precisely and therefore it uses 1–9 ratio ratings (also see Muther, 1973; Malakooti, 1985). For additive utility functions, our experimental results show that using ratio information of preferences to assess the utility function results in inaccurately assessed utility function, that is, the function is not precisely assessed. Experimental evidence supports that DMs can provide relatively accurate (precise) numerical ratings of

TABLE 2.15   Number of Questions Asked to Assess the Criteria Values in AHP versus OCA

k	n 10	20	50	100	1000
2	90/20	380/40	2450/100	9900/200	999,000/2000
5	225/50	950/100	6125/250	24,750/500	2,497,500/5000
10	450/100	1900/200	12,250/500	49,500/1000	4,995,000/10,000

utility values of alternatives and using such information will result in a far more accurately assessed additive utility function. In contrast, OCA (Section 2.4.1) recommends using more precise direct numerical (e.g., 0–100) ratings of weights and alternatives (by comparing the utility of alternatives to the null alternative which has a zero utility value). The method of the next section allows for direct ratings of actual alternatives for assessing the additive utility function.

AHP Appeal and Remedy   As discussed AHP has an appealing systematic approach for solving MCDM problem. The critique of AHP (provided in this section) can be remedied by some simple modifications to AHP. In fact, the OCA (Sections 2.4.1–2.4.4) provides such remedies and the interested user of AHP may use preference ratios of AHP instead of preference difference of OCA provided in Sections 2.4.1 through 2.4.4.

2.4.7   Quasi-linear Double Helix Value and Utility Functions[†]

Simultaneous Assessment of Value and Utility Functions   The additive utility function can be assessed more accurately using the following approach. First choose a sample of actual alternatives that are the best representative of all alternatives. Suppose Q number of alternatives are selected. Ask the DM to rank the Q selected alternatives. Given that the utility rating of the absolute minimum alternative $(f_{1,\min}, f_{2,\min}, \ldots, f_{k,\min})$ is 0, and the utility rating of the absolute maximum alternative $(f_{1,\max}, f_{2,\max}, \ldots, f_{k,\max})$ is 1, ask the DM to rate the given Q alternatives on a scale of 0–1. Suppose the ratings are U_j^* for $j = 1, 2, \ldots, Q$. Use these rating in the following optimization problem to find the unknown weights of the additive utility function. Use the normalized values of all criteria when solving the optimization problem.

Using OCA to Assess Utility Ratings of Alternatives, U_j^*: Instead of directly assessing utility values (ratings) U_j^* for $j = 1, 2, \ldots, Q$, it is possible to employ the OCA approach of Section 2.4.1 to assess these ratings. To do this, compare the utilities of each pair of Q alternatives and use the method of Section 2.4.1. The OCA solution will be the assessed utility values, U_j^* for $j = 1, 2, \ldots, Q$.

One Value Function for All Criteria: Consider each criterion value function as $v_i(f_i) = f_i^\beta$ where β is a positive coefficient; also see (2.9) in Section 2.3.5. When $0 < \beta < 1$, $v_i(f_i)$ is concave; when $\beta > 1$, $v_i(f_i)$ is convex; and when $\beta = 1$, $v_i(f_i)$ is linear. In the following optimization, one may require $0 < \beta \leq 1$ to have a concave utility function (2.24).

$$U = \sum_{i=1}^{k} w_i f_i^\beta \tag{2.24}$$

In (2.24), we use only one β for all criteria value functions to be consistent with the parsimonious principle of using the minimum number of parameters for assessment of utility functions. However, (2.24) can be generalized by using $v_i(f_i) = f_i^{\beta_i}$ where β_i is a positive coefficient; in this case, the following optimization problems will have $2k + 1$ parameters. Note that w_k can be written as $1 - (w_1 + w_2 + \cdots + w_{k-1})$; therefore the utility function of the following optimization problem has only $k + 1$ parameters.

Concave and Convex Utility Functions: By using value functions (2.9), the utility function (2.24) is concave when $0 < \beta < 1$, and it is convex when $\beta > 1$.

PROBLEM 2.2 OPTIMIZATION FOR ASSESSING ADDITIVE UTILITY FUNCTION AND VALUE FUNCTIONS

$$\text{Minimize SSE} = \sum_{j=1}^{Q} (e_j^+)^2 + \sum_{j=1}^{Q} (e_j^-)^2 \tag{2.25}$$

Subject to:

$$\sum_{i=1}^{k} w_i f_{ij}^{\beta} + e_j^+ - e_j^- = \alpha U_j^* \qquad \text{for } j = 1, 2, \ldots, Q \tag{2.26}$$

$$\sum_{i=1}^{k} w_i = 1 \tag{2.27}$$

$$w_i \geq 0 \qquad\qquad \text{for } i = 1, \ldots, k \tag{2.28}$$

$$e_j^+ \geq 0, e_j^- \geq 0 \qquad\qquad \text{for } j = 1, 2, \ldots, Q \tag{2.29}$$

$$\alpha \qquad\qquad \text{Unrestricted in sign} \tag{2.30}$$

$$\beta > 0 \qquad\qquad \text{for } i = 1, \ldots, k \tag{2.31}$$

The above problem can be solved using a nonlinear optimization; for example, use LINGO or Isqnonlin in MATLAB. $k + 1$ parameters are used to assess this utility function; therefore, at least $k + 2$ (and preferably more than $2k$) alternatives should be used to assess this utility function.

Quasi-linear Value Functions: Instead of using (2.24), we can use quasi-linear (2.10) in the additive utility function which results in (2.24)'. Then the utility function can take a more flexible form of combination of concave or convex functions.

$$U = \sum_{i=1}^{k} w_i (f_i^{\gamma} / (f_i^{\gamma} + (1 - f_i)^{\gamma})^{1/\gamma}) \tag{2.24a}$$

Note that (2.24a) can be generalized by using (2.10) for each of the objective function, to have a total of $2k + 1$ parameters (this generalization is not recommended for practical purposes).

Quasi-linear Value Functions (Double Helix: S&S; 1/S&1/S; or S &1/S): Instead of using (2.24) one can use more general Double Helix functions, see Malakooti (2014).

Example 2.13 Assessment of Concave Additive Utility Function by Optimization
Consider Table 2.16, a four-criteria problem with 11 alternatives. Ranking and ratings of these alternatives are also provided in Table 2.16. Use the first eight alternatives for assessing the unknown weights; and use the last three alternatives for verification purposes.

The optimization problem 2.2 for this problem can be written as

$$\text{Minimize SSE} = \sum_{j=1}^{8} (e_j^+)^2 + \sum_{j=1}^{8} (e_j^-)^2$$

Subject to:

$w_1 \times 1^\beta + e_1^+ - e_1^- = \alpha \times 0.38$	$w_1 + w_2 + w_3 + w_4 = 1$
$w_2 \times 1_\beta + e_2^+ - e_2^- = \alpha \times 0.10$	$w_i \geq 0 \quad$ for $i = 1, \ldots, 3$
$w_3 \times 1^\beta + e_3^+ - e_3^- = \alpha \times 0.38$	$e_j^+ \geq 0, e_j^- \geq 0$
$w_4 \times 1^\beta + e_4^+ - e_4^- = \alpha \times 0.10$	$\quad$ for $j = 1, 2, \ldots, 8$
$w_1 \times 0.5^\beta + w_2 \times 0.5^\beta + e_5^+ - e_5^- = \alpha \times 0.30$	$\alpha \quad$ Unrestricted in sign
$w_2 \times 0.5^\beta + w_3 \times 0.5_\beta + e_6^+ - e_6^- = \alpha \times 0.27$	$\beta > 0$
$w_3 \times 0.5^\beta + w_4 \times 0.5^\beta + e_7^+ - e_7^- = \alpha \times 0.30$	
$0.25^\beta(w_1 + w_2 + w_3 + w_4 + e_8^+ - e_8^- = \alpha \times 0.31)$	

The solution to this problem using MATLAB is $w_1 = 0.39$, $w_2 = 0.11$, $w_3 = 0.39$, $w_4 = 0.11$, $\alpha = 1.02$, and $\beta = 0.80$. The obtained solution is optimal and the MAD (using all 11 alternatives, F_1 to F_{11}) is 0.012 (which is acceptable); see Table 2.17. Therefore, $U = 0.39 f_1^{0.8} + 0.11 f_2^{0.8} + 0.39 f_3^{0.8} + 0.11 f_4^{0.8}$. The sum of square of errors (SSE) of this solution is 0.0011.

TABLE 2.16   Example 2.13: Assessment of Additive Utility Function

f_{ij}	Alternatives for Reference		Alternatives for Assessment								Alternatives for Test		
	F_{min}	F_{max}	F_1	F_2	F_3	F_4	F_5	F_6	F_7	F_8	F_9	F_{10}	F_{11}
f_1	0	1	1	0	0	0	0.5	0	0	0.25	0.5	0	0.5
f_2	0	1	0	1	0	0	0.5	0.5	0	0.25	0	0.5	0
f_3	0	1	0	0	1	0	0	0.5	0.5	0.25	0.5	0	0
f_4	0	1	0	0	0	1	0	0	0.5	0.25	0	0.5	0.5
Ranking	NA	NA	2	10	2	11	5	8	5	4	1	9	5
DM's U^*	0	1	0.38	0.10	0.38	0.1	0.3	0.27	0.3	0.31	0.44	0.12	0.30

Comparisons to (2.24a): For comparison purposes, using (2.24)' instead of (2.24) in solving the above optimization problem, the solution is: $W = (0.38, 0.12, 0.38, 0.12)$, $\gamma = 0.88$, $\alpha = 0.93$, and SSE is 0.0051 which is worse than using (2.24) whose SSE = 0.0011. That is, for the above utility values, using (2.24) provides a better solution than using (2.24)', and therefore, in practice we need to try both (2.24) and (2.24)' approaches.

TABLE 2.17   Solution of Example 2.13

f_{ij}	Alternatives for Assessment								Alternatives for Test			MAD
	F_1	F_2	F_3	F_4	F_5	F_6	F_7	F_8	F_9	F_{10}	F_{11}	
DM's U^*	0.38	0.1	0.38	0.1	0.30	0.27	0.30	0.31	0.44	0.12	0.30	–
U	0.39	0.11	0.39	0.11	0.29	0.29	0.29	0.33	0.45	0.13	0.29	–
$|U - U_j^*|$	0.01	0.01	0.01	0.01	0.01	0.02	0.01	0.02	0.01	0.01	0.01	0.012

See Examples 2131.m, 2132.m, 2.13-3.xlsx.

TABLE 2.18   Solution of Example 2.14 Using $v(f_i) = f_i^\beta$

f_{ij}	Alternatives for Assessment								Alternatives for Test			MAD
	F_1	F_2	F_3	F_4	F_5	F_6	F_7	F_8	F_9	F_{10}	F_{11}	–
DM's U^*	0.38	0.11	0.41	0.1	0.21	0.22	0.24	0.14	0.34	0.09	0.21	–
U	0.37	0.11	0.41	0.11	0.19	0.21	0.21	0.16	0.31	0.09	0.19	–
$\lvert U - U_j^* \rvert$	0.01	0	0	0.01	0.02	0.01	0.03	0.02	0.03	0.00	0.02	0.014

See Examples 2141.m, 2142.m, 2.14-3.xlsx.

Example 2.14   Comparison of Concave and Quasi-linear Value Functions   Consider the alternatives used in Example 2.13. The DM's utility, U^*, of these alternatives is presented in both the following tables. Assess the additive utility using (2.24) and (2.24)′ and compare their accuracies.

For (2.24), that is, using $v_i(f_i) = f_i^\beta$, the optimization problem is similar to the optimization problem presented in Example 2.13, but using the new DM's utility, U^*. Solving this optimization problem, we have $\beta = 1.33$ (a convex value function), $W = (0.37, 0.11, 0.41, 0.11)$, $\alpha = 0.97$, and SSE is 0.0015. The details are shown in Table 2.18.

For (2.24)′, we use $v_i(f_i) = f_i^\gamma/(f_{i+}^\gamma(1 - f_i)^\gamma)^{1/\gamma}$ in the optimization problem 2.2 and Example 2.13. The solution to this optimization problem is: $\gamma = 1.54$ (an S-shaped quasi-linear value function), $W = (0.37, 0.11, 0.41, 11)$, $\alpha = 0.99$, and SSE is 2.8×10^{-4}. The details are shown in Table 2.19. The SSE of this example (2.24a) is much lower than the SSE of (2.24); therefore (2.24a) is a better fit.

2.5   MULTIPLICATIVE ZUT

Basics of All ZUTs

We introduce a class of nonlinear utility functions having $k + 1$ parameters where k is the number of criteria. All ZUTs have the form $U = LV + z_M DV$ where LV is linear and distance value (DV) is a nonlinear function. z_M is a coefficient that ranges from -1 to $+1$. When z_M is negative, the utility function is concave; when z_M is positive the utility function is convex; and when z_M is zero the utility function is additive (linear). ZUTs can be interpreted as a hybrid of additive (LV) and nonadditive (nonlinear) (DV) which are balanced by the value of z_M. Therefore, LV and DV can be considered as two aggregate objective functions. For a concave utility functions, LV can be interpreted as the merit (goodness) of a given

TABLE 2.19   Solution of Example 2.14 Using $v_i(f_i) = f_i^\gamma/(f_i^\gamma + (1 - f_i)^\gamma)^{1/\gamma}$

f_{ij}	Alternatives for Assessment								Alternatives for Test			MAD
	F_1	F_2	F_3	F_4	F_5	F_6	F_7	F_8	F_9	F_{10}	F_{11}	–
DM's U^*	0.38	0.11	0.41	0.1	0.21	0.22	0.24	0.14	0.34	0.09	0.21	–
U	0.37	0.11	0.41	0.11	0.21	0.23	0.23	0.15	0.35	0.10	0.21	–
$\lvert U - U_j^* \rvert$	0.01	0.00	0.00	0.01	0.00	0.01	0.01	0.01	0.01	0.01	0.00	0.006

See Examples 2144.m, 2145.m, 2146.xlsx.

alternative and DV as its drawback (demerit). In this case, *the problem of maximizing LV and minimizing DV can be presented as*

$$\text{Maximize } U = \text{LV} + z_\text{M}\text{DV} \qquad \text{where} -1 < z_\text{M} \leq 0$$

Details and extensions of ZUTs for solving MCDM problems are presented in Malakooti (2014b); here we provide an overview of multiplicative (this section) and goal-seeking (Section 2.6) ZUTs.

Absolute (Universal) Minimum and Maximum Values for Objectives: For a decision problem, suppose that for each given criterion i, f_i presents the normalized value (0–1) with respect to the absolute universal minimum and the absolute universal maximum for all possible values of the given criterion. Therefore, each f_i of any alternative will always remain within the absolute universal minimum and maximum criteria values.

2.5.1 Multiplicative ZUT

Linear Value Function: Consider $w_1, w_2, \ldots, w_k$ as the weights of importance of criteria $f_1, f_2, \ldots, f_k$, respectively, where $w_1 > 0$, $w_2 > 0, \ldots, w_k > 0$ and $w_1 + w_2 + \cdots + w_k = 1$. LV is

$$\text{LV} = w_1 f_1 + w_2 f_2 + \cdots + w_k f_k \tag{2.32}$$

Multiplicative Value Function:

$$\text{MV}_\text{M} = \left(f_1^{w_1} \times f_2^{w_2} \times f_3^{w_3} \times \cdots \times f_k^{w_k} \right) \tag{2.33}$$

(2.33) is a concave functions. Maximizing MV_M presents an extreme convergent (concave) behavior in ZUT. MV_M (2.33) for k factors is a generalization of the Cobb–Douglas (1928) function, which is used in production economics modeling of two factors.

Additive Transformation of Multiplicative Value Functions: By taking natural logarithm (ln) of MV_M (2.33), the function becomes an additive function in the form $\ln(\text{MV}_\text{M}) = w_1\ln(f_1) + w_2\ln(f_2) + \cdots + w_k\ln(f_k)$, where w_i is the importance of $\ln(f_i)$. Therefore, (2.33) presents a transformed additive utility function which is defined by k parameters. In some applications, it is more justified to use (2.33) instead of using additive utility functions, LV (2.32).

Multiplicative Deviational Value Function: Now, we derive deviation value (DV) with respect to MV_M which results in multiplicative DV_M:

$$\text{DV}_\text{M} = |\text{LV} - \text{MV}_\text{M}| = \text{LV} - \text{MV}_\text{M} \tag{2.34}$$

Since all f_i for $i = 1, 2, \ldots, k$ are nonnegative, $\text{LV} \geq \text{MV}_\text{M}$; therefore $|\text{LV} - \text{MV}_\text{M}| = \text{LV} - \text{MV}_\text{M}$. We can verify that at the ridge, $f_1 = f_2 = \cdots = f_k$, the value of $\text{DV}_\text{M} = 0$.

Multiplicative ZUT (A Generalized Arithmetic Mean and Geometric Mean Function): Multiplicative ZUT (M-ZUT) (2.35) generalizes the well-known arithmetic mean and geometric mean functions.

$$U = \text{LV} + z_\text{M}\text{DV}_\text{M} \qquad -1 < z_\text{M} \leq 0 \tag{2.35}$$

Alternative Presentation of M-ZUT: By substituting (2.34) in (2.35), U can be written as

$$U = (1 + z_M)LV - z_M MV_M \qquad -1 < z_M \leq 0 \qquad (2.35a)$$

This means that U is an additive function of LV (linear) and MV_M (multiplicative) where both LV and MV_M are maximized (because both $(1 + z_M)$ and $-z_M$ are positive). Note that when $z_M = -1$, $U = MV_M$ and therefore the generalized Cobb–Douglas is a special case of M-ZUT (2.35).

Increasing and Concavity Properties: For $-1 < z_M \leq 0$, ZUT (2.35) is an increasing function of objectives $(f_1, f_2, \ldots, f_k)$ and is also a concave function; see proofs in Malakooti (2014b).

Use of Value Functions, $v_i(f_i) = f_i^\beta$: Similar to the approach of Section 2.4.7, we can use value functions $v_i(f_i) = f_i^\beta$ instead of f_i. That is, replace each f_i by f_i^β where β is a positive coefficient; therefore (2.35) can be generalized by having one more parameter β. A more general function can be obtained by replacing each f_i by $f_i^{\beta i}$ where each β_i is a positive coefficient (we do not recommend using this latter general utility function as it is not parsimonious, i.e., it has too many parameters).

The Calibration and Generalization of M-ZUT: M-ZUT can become more flexible and generalized; see Malakooti (2014b).

Illustrative Example: Consider a bicriteria where weights are equal, $w_1 = w_2 = 0.5$. Figure 2.17a shows the contours of LV and U that pass through point $(f_1, f_2) = (0.4, 0.4)$. At $(0.4, 0.4)$, $LV = 0.5(0.4 + 0.4) = 0.4$, and $MV_M = (0.4^{0.5} \times 0.4^{0.5}) = 0.4$, $DV_M = 0$; therefore, $U = LV = MV_M = 0.4$. But at the point $(0.6, 0.2)$, $LV = 0.5(0.6 + 0.2) = 0.4$; for $z_M = -1$, $U = (0.4 - 1(0.4 - (0.6^{0.5} \times 0.2^{0.5})) = 0.35$. For $z_M = -0.5$, $U = (0.4 - 0.5(0.4 - (0.6^{0.5} \times 0.2^{0.5})) = 0.37$. Figure 2.17$b$ shows the LV and MV_M functions in three dimensions.

Figure 2.18a shows contours of a bicriteria problem for where weights are equal, $w_1 = w_2 = 0.5$ and $z_M = -0.5$. Figure 2.18b shows contours of a bicriteria problem for where weights are not equal, $w_1 = 0.7$ and $w_2 = 0.3$ and $z_M = -0.5$.

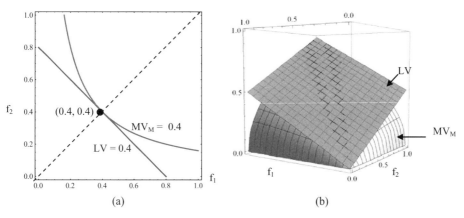

(a)   (b)

Figure 2.17 A concave multiplicative ZUT for $z_M = -1$, $U = MV_M$: (a) contours of LV and MV_M; (b) functions of LV and MV_M.

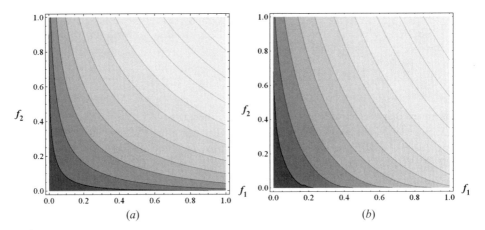

$$(a) \qquad\qquad\qquad (b)$$

Figure 2.18   Contours of concave multiplicative ZUT with $z_M = -0.5$ using (a) $w_1 = w_2 = 0.5$; (b) $w_1 = 0.7$, $w_2 = 0.3$.

Example 2.15   Multiplicative ZUT   Consider the 10 alternatives presented in Table 2.20. Suppose that $w_1 = 0.5$, $w_2 = 0.3$, $w_3 = 0.2$. Find utility values and show rankings for M-ZUT where $z_M = -0.3$.

For alternative F_1, $MV_M = 0.6^{0.5} \times 0.5^{0.3} \times 0.2^{0.2} = 0.46$, and $U = LV - 0.3|LV - MV_M|$ $= 0.49 - 0.3(0.49 - 0.46) = 0.48$. The solutions for all alternatives are shown in Table 2.20.

2.5.2   Direct Assessment of M-ZUT[†]

To assess $k + 1$ parameters, $w_1, w_2, \ldots, w_k, z_M$, in $U = LV + z_M\,DV_M$ (2.35), the DM is asked to assess the utility values for k identity alternatives. The assessed utility value $U(I_i)$ of an identity alternative I_i results in the following equation:

$$U(I_i) = LV\,(I_i) + z_M\,DV_M\,(I_i) = (1 + z_M)LV(I_i) - z_M MV_M(I_i)$$

TABLE 2.20   Example 2.15: Multiplicative Z Utility Function

Objectives	w_i	Alternatives									
		F_1	F_2	F_3	F_4	F_5	F_6	F_7	F_8	F_9	F_{10}
f_1	0.5	0.6	0.4	0.3	0.5	0.4	0.2	0.1	0.5	0.1	0.9
f_2	0.3	0.5	0.1	0.2	0.3	0.4	0.5	0.6	0.1	0.9	0.1
f_3	0.2	0.2	0.8	0.7	0.6	0.4	0.5	0.6	0.7	0.5	0.1
LV		0.49	0.39	0.35	0.46	0.40	0.35	0.35	0.42	0.42	0.50
MV_M		0.46	0.30	0.31	0.44	0.40	0.32	0.24	0.33	0.27	0.30
$DV_M = |LV - MV_M|$		0.03	0.09	0.04	0.02	0.00	0.03	0.11	0.09	0.15	0.20
$U = LV - 0.3DV_M$		0.48	0.36	0.34	0.45	0.40	0.34	0.32	0.39	0.38	0.44
Rank by Max. ZUT		1	7	8	2	4	8	10	5	6	3

Where $LV = 0.5f_1 + 0.3f_2 + 0.2f_3$ and MV_M is based on $W = (0.5, 0.3, 0.2)$

TABLE 2.21   Presentation of k Nonzero Alternatives for Testing Multiplicative Assessment

	N_1	N_2	$\ldots$	N_k
f_1	$2/(k+1)$	$1/(k+1)$	$\ldots$	$1/(k+1)$
f_2	$1/(k+1)$	$2/(k+1)$	$\ldots$	$1/(k+1)$
$\ldots$	$\ldots$	$\ldots$	$2/(k+1)$	$\ldots$
f_k	$1/(k+1)$	$1/(k+1)$	$\ldots$	$2/(k+1)$

$MV_M(I_i) = 0$, since any criteria with zero value will cause MV_M to be 0.

$$U(I_i) = (1 + z_M)w_i \tag{2.36}$$

For any number of objectives k, the parameter z_M can be found by adding all k of the assessed utility values for the identity alternatives $U(I_i)$, where $i = 1,2,\ldots,k$. Summation of Equation (2.36) over all $i = 1,2,\ldots,k$ is

$$\sum_{i=1}^{k} U(I_i) = (1 + z_M)\sum_{i=1}^{k} w_i = (1 + z_M) \quad \text{or} \quad z_M = \left(\sum_{i=1}^{k} U(I_i)\right) - 1 \tag{2.37}$$

Then, each of the weights w_i can be found by solving for w_i in Equation (2.36) for the corresponding i:

$$w_i = U(I_i)/(1 + z_M) \tag{2.38}$$

A Sample of Alternatives for Testing Multiplicative ZUT: The MV_M of identity and convex alternatives (Table 2.21) is zero, because any $f_i = 0$ will cause MV_M to be zero. Therefore, we should consider alternatives that have no zero objective values. Table 2.21 presents k alternatives that are all positive. The ith component of each of the k alternatives is $2/(k-1)$ but all other components are equal to $1/(k + 1)$.

Approximation of z_M and Verification of M-ZUT: After all weights are assessed, use assessed weights W to find LV, DV_M of all Q alternatives. Then find the $z_{M,j}$ of each alternative j by using its known U_j^*, LV_j, $DV_{M,j}$, that is, find $z_{M,j} = (U_j^* - LV_j)/DV_{M,j}$. Then find the average value of z_M of all Q alternatives; if MAD of this assessed z_M value is acceptable, then the assessed M-ZUT is valid. Details are shown in the following example.

Example 2.16   Assessment for M-ZUT   Consider a four-criteria problem for the symmetric case $\alpha_1 = \alpha_2 = \alpha_3 = \alpha_4 = 1$. The assessed utility values U^* (by the DM) of four identity alternatives I_i and four nonzero alternatives are shown in Table 2.22. Note that $k = 4$. Therefore, for nonzero alternatives, use $2/(k + 1) = 2/5 = 0.4$ and $1/(k + 1) = 1/5 = 0.2$.

Assess and verify the multiplicative utility function.

TABLE 2.22   Assessed Utility Values U^* and Four Nonzero Alternatives

	Alternatives for Assessment				Alternatives for Testing			
	I_1	I_2	I_3	I_4	N_1	N_2	N_3	N_4
f_1	1	0	0	0	0.4	0.2	0.2	0.2
f_2	0	1	0	0	0.2	0.4	0.2	0.2
f_3	0	0	1	0	0.2	0.2	0.4	0.2
f_4	0	0	0	1	0.2	0.2	0.2	0.4
DM's U_j^*	0.20	0.05	0.20	0.05	0.25	0.25	0.27	0.25

Solution   Using Equation (2.37) and substituting the assessed utility values, we find z_M as

$$z_M = \left(\sum_{i=1}^{k} U(I_i) \right) - 1 = 0.5 - 1 = -0.5$$

Now, using Equation (2.38) $w_i = U(I_i)/(1 + z_M)$ for $i = 1$ and substituting $U(I_i)$ with $U^*(I_i)$, we get

$$w_1 = 0.2/(1 - 0.5) = 0.4$$

Similarly, using Equation (2.38) for $i = 2,3,4$:

$$w_2 = U(I_2)/(1 + z_M) = 0.05/(1 - 0.5) = 0.1$$
$$w_3 = U(I_3)/(1 + z_M) = 0.2/(1 - 0.5) = 0.4$$
$$w_4 = U(I_4)/(1 + z_M) = 0.05/(1 - 0.5) = 0.1$$
$$w_2 = U(I_2)/(1 + z_M) = 0.05/(1 - 0.5) = 0.1$$
$$w_3 = U(I_3)/(1 + z_M) = 0.2/(1 - 0.5) = 0.4$$
$$w_4 = U(I_4)/(1 + z_M) = 0.05/(1 - 0.5) = 0.1$$

Therefore, the weight solution is $(0.4,0.1,0.4,0.1)$ with $z_M = -0.5$ where $\mathrm{MV_M} = f_1^{0.4} f_2^{0.1} f_3^{0.4} f_4^{0.1}$.

Approximation of z_M and Verification: Using assessed weights $(0.4,0.1,0.4,0.1)$,

$$\mathrm{LV} = 0.4 f_1 + 0.1 f_2 + 0.4 f_3 + 0.1 f_4$$

$$\mathrm{MV_M} = f_1^{0.4} f_2^{0.1} f_3^{0.4} f_4^{0.1}$$

Now find the z_M for each of $2k$ alternatives. The solution is shown in Table 2.23.

$\mathrm{MAD} = 0.01 < 0.05$ is acceptable. Therefore, the assessed ZUT is $U = \mathrm{LV}{-}0.5\mathrm{DV_M}$; or alternatively $U = 0.5\mathrm{LV} + 0.5\mathrm{MV_M}$. For example, $U(N_4) = 0.22{-}0.50 \times 0.01 = 0.215$ or $U(N_4) = 0.5 \times 0.22 + 0.50 \times 0.21 = 0.215$. This MAD is less than 0.05 and therefore the assessed M-ZUT is acceptable.

TABLE 2.23   Example 2.16: Assessment for Z Utility Function with DV_M

	I_1	I_2	I_3	I_4	N_1	N_2	N_3	N_4	Average
LV	0.4	0.1	0.4	0.1	0.28	0.22	0.28	0.22	–
MV_M	0.00	0.00	0.00	0.00	0.26	0.21	0.26	0.21	–
$DV_M = |LV - MV_M|$	0.40	0.10	0.40	0.10	0.02	0.01	0.02	0.01	–
$z_M = (U_j^* - LV)/DV_M$	–0.50	–0.50	–0.50	–0.50	–	–	–	–	$z_M = -0.50$
$U = LV - 0.50 DV_M$	0.20	0.05	0.20	0.05	0.27	0.22	0.27	0.215	
U^*	0.2	0.05	0.20	0.05	0.25	0.25	0.27	0.25	
$|U - U^*|$	0.00	0.00	0.00	0.00	0.02	0.03	0.00	0.03	MAD = 0.01

See Example 2.16.xls.

2.5.3   Assessment of M-ZUT by Nonlinear Equations[†]

In this section, we discuss how to use a Q sample of actual alternatives to assess M-ZUT. For a given Q sample of alternatives, let the DM's utility be denoted by U_j^* for $j = 1$, $2, \ldots, Q$. To find unknown parameters, $w_1, w_2, \ldots, w_k$, and z_M, the following nonlinear optimization problem can be formulated where the SSE is minimized.

PROBLEM 2.3   ZUT ASSESSMENT PROBLEM

$$\text{Minimize SSE} = \sum_{j=1}^{Q} (e_j^+)^2 + \sum_{j=1}^{Q} (e_j^-)^2 \tag{2.39}$$

Subject to:

$$(1 + z_M) \sum_{i=1}^{k} w_i \times f_{ij} + z_M (f_{1,j}^{w_1} \times f_{2,j}^{w_2} \times \cdots \times f_{k,j}^{w_k}) + e_j^+ - e_j^- = U_j^*$$
$$\text{for } j = 1, 2, \ldots, Q \tag{2.40}$$

$$\sum_{i=1}^{k} w_i = 1 \tag{2.41}$$

$$w_i = 0 \qquad\qquad \text{for } i = 1, \ldots, k \tag{2.42}$$

$$e_j^+ = 0, e_j^- = 0 \qquad\qquad \text{for } j = 1, 2, \ldots, Q \tag{2.43}$$

$$-1 < z_M \le 0 \tag{2.44}$$

Levenberg–Marquardt Method for Solving Simultaneous ZUT Nonlinear Equations: Problem 2.3 cannot be solved by conventional nonlinear optimization methods. We recommend using Levenberg–Marquardt method (Levenberg, 1944; Marquardt, 1963) for solving the set of equations of Problem 2.3. This method minimizes SSE (the objective function of Problem 2.3). The Levenberg–Marquardt method significantly outperforms the gradient descent methods by using an improved iteration step; however, there is no guarantee of convergence. This method has also been used in nonlinear regression which has a similar format as in Problem 2.3. This approach is available in MATLAB (http://www.mathworks.com/help/optim/index.html) as the Isqnonlin function.

TABLE 2.24 **Example 2.17: Sample of Alternatives, the DM's Utility, and Levenberg–Marquardt Method Solution**

f_{ij}	Alternatives for Assessment							Alternatives for Test			MAD
	F_1	F_2	F_3	F_4	F_5	F_6	F_7	F_8	F_9	F_{10}	–
f_1	1	0.5	0.5	0.5	0.5	0.25	0.25	0.5	0	0.5	–
f_2	0.5	1	0.5	0.5	0.5	0.5	0.25	0	0.5	0	–
f_3	0.5	0.5	1	0.5	0.25	0.5	0.5	0.5	0	0	–
f_4	0.5	0.5	0.5	1	0.25	0.25	0.5	0	0.5	0.5	–
DM's U^*	0.63	0.55	0.60	0.69	0.35	0.32	0.40	0.16	0.18	0.20	–
U solution	0.643	0.545	0.612	0.683	0.340	0.324	0.397	0.157	0.139	0.201	–
$|U-U_j^*|$	0.013	0.005	0.012	0.007	0.010	0.004	0.003	0.003	0.041	0.001	0.010

See Examples 2.17.xls, 2171.m, 2172.m.

Example 2.17 **Assessment of ZUT Using Simultaneous Equations** Consider Table 2.24 for a four-criteria problem. The first seven alternatives are used for assessing the unknown coefficients, and the last three alternatives are used for verification. Assess the ZUT and verify the accuracy of the assessment. The solution is shown in Table 2.24.

The solution to this problem using MATLAB is: $w_1 = 0.3085$, $w_2 = 0.0927$, $w_3 = 0.2049$, and $w_4 = 0.3939$, and $z_M = -0.4286$. Therefore, $U = 0.5714\text{LV} - 0.4286\text{MV}_M$ where LV $= 0.3085f_1 + 0.0927f_2 + 0.2049f_3 + 0.3939f_4$ and $\text{MV}_M = f_1^{0.3085} f_2^{0.0927} f_3^{0.2049} f_4^{0.3939}$. The obtained solution is optimal and the MAD (using all 10 alternatives $F_1 - F_{10}$) is 0.010 which is acceptable.

2.6 GOAL-SEEKING ZUT

In this section, we develop a new goal-seeking utility function of the form $U = \text{LV} + z_G\text{DV}_G$ where LV is additive and DV_G is a rectilinear distance to a given goal alternative. In Section 2.6.2, we review goal programming (GP) and show that it violates efficiency principle by preferring inefficient alternative to efficient ones. In Section 2.6.3, we develop the goal-seeking method that resolves GP and finds the closest efficient feasible alternative to the goals (even if the given goal alternative is inefficient). In Section 2.6.4, we show how to assess goal-seeking utility function and in Section 2.6.5, we consider binary goals (poles) that generalizes single goal-seeking method.

2.6.1 Goal-Seeking ZUT

Since all objectives are maximized, generally a goal alternative (goals), $F_G = (f_{1,G}, f_{2,G}, \ldots, f_{k,G})$, should be efficient with respect to the set of alternatives (or possibly dominate some efficient points). However, we relax this requirement to allow for more flexibility.

Goal-Seeking Method: In goal-seeking methods, the DM identifies the goal, F_G, and the utility values of alternatives are based on their weighted closeness to the goals and their additive utility function LV.

Goal-Seeking ZUT: Goal-seeking ZUT is presented in (2.45) where $w_1 > 0$, $w_2 > 0, \ldots, w_k > 0$.

$$\text{DV}_G = \sum_{i=1}^{k} w_i|f_i - f_{iG}| \tag{2.45}$$

$$U = \text{LV} + z_G\text{DV}_G \qquad \text{for} \qquad -1 < z_G < +1 \tag{2.46}$$

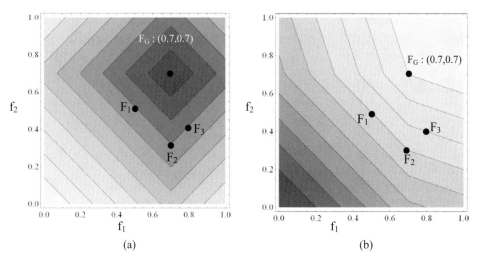

Figure 2.19 Contours in f_1, f_2 space for (a) $DV_G = 0.5|f_1 - 0.7| + 0.5|f_2 - 0.7|$; (b) goal-convergent ZUT contours, $U = 0.5f_1 + 0.5f_1 - 0.5DV_G$.

Increasing Property of Goal-Seeking ZUT and Concavity/Convexity of ZUT: Goal-seeking ZUT (2.46) is an increasing function of objectives. Therefore, even if the goal alternative is inefficient, an inefficient alternative will not be preferred to an efficient alternative when using U.

ZUT (2.46) is concave when $z_G \leq 0$ and convex when $z_G \geq 0$; see proofs in Malakooti (2014b).

If the goal alternative is defined as a desired point, the DM would minimize DV_G, and therefore z_G should be a negative number.

$$U = LV + z_G DV_G \qquad \text{for} \qquad -1 < z_G \leq 0 \quad \text{(goal convergent)} \qquad (2.47)$$

Graphical Bicriteria Example for Goal Seeking   For the goal-convergent (concave) case, consider Figure 2.19 where goals $F_G = (0.7, 0.7)$. Suppose that $w_1 = w_2 = 0.5$, therefore $DV_G = 0.5|f_1 - 0.7| + 0.5|f_2 - 0.7|$. Figure 2.19a shows the contours of DV_G (2.45) in f_1 and f_2 space. Note that a point such as (0.8,0.8) that dominates $F_G = (0.7, 0.7)$ is less preferred to it. Figure 2.19b shows contours of goal-convergent ZUT (2.47) where $z_G = -0.5$, that is, $U = 0.5f_1 + 0.5f_2 - 0.5DV_G$. In this case, point (0.8,0.8) that dominates $F_G = (0.7, 0.7)$ is preferred to it because both LV and DV_G are used in the utility function. To explain this concept more clearly, consider three points $F_1 = (0.5, 0.5)$, $F_2 = (0.7, 0.3)$, and $F_3 = (0.8, 0.4)$. Consider only minimizing DV_G; then all three of these alternatives would be ranked equally: $DV_G(F_1) = DV_G(F_2) = DV_G(F_3)$, contradicting the fact that F_2 is dominated by F_3. But for $LV = 0.5f_1 + 0.5f_2$, we can see that $LV(F_3) > LV(F_1)$ $= LV(F_2)$. When using ZUT (2.47), both LV and DV_G are used and F_3 is preferred to both F_1 and F_2.

Example 2.18 Goal-Seeking Utility Function   Consider the alternatives shown in Table 2.25 where $W = (0.4, 0.4, 0.2)$ are weights of importance for the three objectives. Rank these alternatives for the following cases:

TABLE 2.25 Four Examples for Ranking Alternatives with Respect to Two Goals and a Nadir

Objectives	Alternatives							Goals a, b F_G	Goals c F_G	Weight w_i
	F_1	F_2	F_3	F_4	F_5	F_6	F_7			
f_1	0.3	0.4	0.2	0.2	0.5	0.1	0.9	0.7	0.2	0.40
f_2	0.2	0.4	0.5	0.2	0.1	0.9	0.1	0.65	0.2	0.40
f_3	0.7	0.4	0.5	0.2	0.7	0.5	0.1	0.3	0.2	0.20
LV	0.34	0.40	0.38	0.20	0.38	0.50	0.42	0.60	0.2	–
a. DV_G	0.42	0.24	0.30	0.40	0.38	0.38	0.34			
a. Rank by Min DV_G	7	1	2	6	5	4	3			
b. $U = LV - 0.6DV_G$	0.09	0.26	0.20	–0.04	0.15	0.27	0.22			
b. Rank by Max U	6	2	4	7	5	1	3			
c. DV_G	0.14	0.20	0.18	0.00	0.26	0.38	0.34			
c. $U = LV - 0.6DV_G$	0.26	0.28	0.27	0.20	0.22	0.27	0.22			
c. Rank by Max U	4	1	2	7	5	2	6			

See Example 2.18.xls.

(a) Only use distances to goals (i.e., GP) where goals are (0.7,0.65,0.3).

(b) Goal-seeking (convergent case) where goals are (0.7,0.65,0.3) and z_G is –0.6, that is,

$$\text{Maximize } U = LV - 0.6DV_G \quad \text{using goals } (0.7, 0.65, 0.3)$$

(c) Convergent case where goal alternative is a dominated point, (0.2,0.2,0.2), and z_G is –0.6, that is,

$$\text{Maximize } U = LV - 0.6DV_G \quad \text{using goals } (0.2, 0.2, 0.2)$$

The solutions are shown in Table 2.25. For example, for alternative F_1:

(a) $DV_G = 0.4|0.3–0.7| + 0.4|0.2 - 0.65| + 0.2|0.7–0.3| = 0.42$.

(b) $U = LV - 0.6DV_G = 0.34 - 0.6 \times 0.42 = 0.09$.

(c) $U = LV - 0.6DV_G = 0.34 - 0.6(0.4|0.3 - 0.2| + 0.4|0.2 - 0.2| + 0.2|0.7 - 0.2|) = 0.26$.

In question c, purposely an inefficient goal point is given to illustrate that goal-seeking ZUT finds an efficient point F_3, instead of choosing F_4 which is inefficient but the closest point to the goals.

2.6.2 Review of Goal Programming and Its Effectiveness[†]

GP (Charnes and Cooper, 1961, 1977) is the most known and applied MOO method; also see Zeleny (1982), Steuer (1986), and Chankong and Haimes (2008) for related works. MOO is usually applied in the context of a continuous set of alternatives presented by an optimization problem (such as linear programming (LP) with multiple objectives). GP approach can also be applied to a discrete set of multicriteria alternatives. GP is based

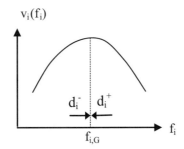

Figure 2.20 A concave unimodal value function used in goal programming.

on minimizing a variation of DV_G denoted by $DV_{G,G}$ as its utility function. The main disadvantage of GP is that it may select an inefficient alternative as the best alternative (see examples below and also in Section 2.8).

Unimodal Value Function and Use of 2k Objectives in GP In GP, implicitly, a unimodal value function for each objective is assumed, for example, Figure 2.20 where value is increasing up to point $F_{i,G}$, and then it is decreasing. For example, suppose that in Figure 2.20, f_i is the intake of sugar (or salt or needed medication) where $v_i(f_i)$ is its value. One needs to use the optimal value of f_i; that is, $f_{i,G}$.

Note that this definition of an objective f_i contradicts the definition used in MCDM that all objectives are maximized. Therefore, GP can be understood as having $2k$ objectives which are all "minimized."

$$\text{Min } d_i^- = f_{i,G} - f_i \qquad \text{when } (f_i - f_{i,G}) < 0 \text{ for } i = 1, \ldots, k$$
$$\text{Min } d_i^+ = f_i - f_{i,G} \qquad \text{when } (f_i - f_{i,G}) > 0 \text{ for } i = 1, \ldots, k$$

In GP, the DM sets the goals for each objective and then specifies the priorities for underachievement (d_i^-) and overachievement (i^+) of each goal. That is, the deviations from each given goal are minimized differently depending on whether the deviation is an underachievement or an overachievement. Two well-known classes of GP methods are on-preemptive (Archimedean) and preemptive (lexicographic ordering). In the following, we give an overview of these two methods.

Non-preemptive (Archimedean) GP In Archimedean GP, priorities are provided in the form of weights of importance for achieving goals. In the Archimedean (unlike the preemptive GP), trade-offs in achieving goals are allowed and vary depending on the values of weights. Then, alternatives are ranked by minimizing $DV_{G,G}$ (2.48).

PROBLEM 2.4   NON-PREEMPTIVE GP

$$\text{Minimize } DV_{G,G} = \sum_{i=1}^{k} (w_i^+ d_i^+ + w_i^- d_i^-) \tag{2.48}$$

$$\text{Subject to } f_i + d_i^- - d_i^+ = f_{i,G} \quad \text{for} \quad i = 1, 2, \ldots, k \tag{2.49}$$

$$d_i^+ \geq 0, d_i^- \geq 0 \quad \text{for} \quad i = 1, 2, \ldots, k \tag{2.50}$$

And the set of constraints on all alternatives; e.g., LP constraints $\quad$ (2.51)

In order to scale the problem properly, use the normalized values of all objectives and goals; for example, use a scale of 0–1 for maximizing all objectives.

House Buying Example for Non-preemptive (Archimedean) GP Consider buying a house based on three criteria of price (cost in dollars), space (in square footage), and locality (e.g., quality of neighborhoods in scale of 010). Suppose that ranges are from $100,000 to $300,000 for price; from 1500 to 3000 for square footage; and from 0 to 10 for locality. Suppose the goals are: $F_G^o = (\$200,000, 2500, 9)$. Consider two houses: $F_1^o = (\$200,000, 2200, 9)$ and $F_2^o = (\$200,000, 2300, 6)$. The normalized values are $F_G = (0.50, 0.67, 0.90)$, $F_1 = (0.50, 0.47, 0.90)$, and $F_2 = (0.50, 0.53, 0.60)$. Suppose that for each objective both over- and underachievements have the same weights of importance, that is, both are minimized equally where $(w_1^+, w_2^+, w_3^+) = (w_1^-, w_2^-, w_3^-) = (0.5, 0.3, 0.2)$. Therefore, utility function (minimized) $U = \text{DV}_{G,G}$ (2.48) can be written in the form of DV_G (2.45), where weights are $w_1 = 0.5$, $w_2 = 0.3$, and $w_3 = 0.2$.

$$U(F_1) = 0.5|0.5 - 0.5| + 0.3|0.67 - 0.47| + 0.2|0.9 - 0.9| = 0.06$$
$$U(F_2) = 0.5|0.5 - 0.5| + 0.3|0.67 - 0.53| + 0.2|0.9 - 0.6| = 0.10$$

Therefore, F_1 is the closer to goals and is the best alternative.

Preemptive (Lexicographic Ordering) GP: Successive LP   Preemptive GP is an extension of non-preemptive GP where achieving each objective goal $f_{i,G}$ for $i = 1, \ldots, k$ are in preemptive (lexicographical) order presented by $P_1 \succ\succ P_2 \succ\succ \cdots \succ\succ P_k$ where $\succ\succ$ signifies infinitely more important. For example, for three objectives, $P_1 \succ\succ P_2 \succ\succ P_3$ means that you should first achieve the first objective's goal regardless of the second and third objectives' goals. If the first objective's goal is achieved, then consider achieving the second objective goal. If the second objective goal is achieved, then consider the third objective's goal. In lexicographical ordering, no trade-offs among objectives are allowed. In preemptive GP problem, $P_1 \succ\succ P_2 \succ\succ \cdots \succ\succ P_k$ are given before solving the problem.

PROBLEM 2.5   PREEMPTIVE GP

$$\text{Minimize } U = \sum_{i=1}^{k} P_i(w_i^+ d_i^+ + w_i^- d_i^-) \tag{2.52}$$

Subject to (2.49), (2.50), and (2.51).

House Buying Example for Preemptive (Lexicographic Ordering) GP Consider the above house buying example where the goals are $F_G = (\$200,000, 2500, 9)$ and priorities are $P_1 \succ\succ P_2 \succ\succ P_3$. Now, rank the two houses: $F_1 = (\$200,000, 2200, 9)$ and $F_2 = (\$200,000, 2300, 6)$. In this example, the first goal is achieved in both alternatives; therefore consider the second goal. In this case, F_2 is the best alternative because it is closer to the goal of higher square footage. Therefore, the third goal is not considered.

Simplicity and Nonlinearity of Utility Function in GP   The main advantage of GP is that it uses LP for solving MOO problems. Furthermore, the utility function of GP is nonlinear and can select (and generate in MOO problems) nonconvex efficient points. As discussed in Sections 2.2.3 and 2.3.3, these nonconvex efficient points cannot be selected (or generated) when additive utility functions are used. The fact that GP uses LP for solving nonlinear utility functions cannot be underestimated. GP is also a powerful tool for trial and error analysis of different possible alternative scenarios.

GP Selection of Inefficient Alternatives In the context of MCDM/MOO, the main disadvantage of GP is that it may select (or generate) an inefficient alternative as the best alternative. For example, suppose that $w_1 = 0.5$, $w_2 = 0.5$, $F_1 = (20,10)$, and $F_2 = (23,12)$, where the goals are $F_G = (19,13)$. In this example, $U(F_1) = 0.5|19-20| + 0.5|13-10| = 2$, and $U(F_2) = 0.5|19-23| + 0.5|13-12| = 2.5$. F_1 is closer than F_2 to F_G; therefore, F_1 is preferred to F_2. But F_2 dominates F_1; that is, F_1 is inefficient. Note that F_G is efficient with respect to both F_1 and F_2. However, by using goal-convergent ZUT (2.47), an inefficient point will never be selected.

Minimizing Overachievements of GP Contradicts Efficiency in MOO In the context of MCDM/MOO, maximizing each objective contradicts minimizing overachievements, d_i^+. By minimizing overachievements, when an objective value is more than the stated goal, that objective is minimized. This contradicts efficiency principle. In some situations, the minimization of overachievement can be justified. For example, taking 500 mg of vitamin D per day can be stated as a goal to stay healthy, whereas both over- and under-intake of it should be minimized. But the proper way to address such objectives is by using unimodal value functions as functions of decision variables. For example, consider maximizing $v_i(x_i)$ where x_i is the decision variable (not the objective). For the vitamin example, x_i is the amount of vitamin, where $v_i(0\text{ mg}) = 0$, $v_i(500\text{ mg}) = 1$, and $v_i(1000\text{ mg or more}) = 0$. Then, maximizing the value function, $v_i(x_i)$, is consistent with the efficiency principle (see Section 2.3.5 for explanation of value functions).

Disadvantage of Preemptive GP The above disadvantages listed for non-preemptive GP are also applicable to preemptive GP. For example, by using this method an inefficient point can also be selected as the best alternative. Furthermore, preemptive GP contradicts the principle of considering trade-offs because no compromises are allowed. It is based on all or nothing decisions. To make this more clear, in the above house buying example, suppose that there are three houses: $F_1^o = (\$200{,}000, 2200, 9)$, $F_2^o = (\$200{,}000, 2300, 6)$, and $F_3^o = (\$200{,}000, 2301, 0)$ where $P_1 \succ\succ P_2 \succ\succ P_3$. In this case, alternative F_3 will be preferred to F_1 and F_2. That is, for only one additional square foot, F_3^o, which has the worst locality rating of 0, is selected.

2.6.3 Goal Seeking: Z-GP[†]

In this section, we discuss the relationship of non-preemptive GP goal seeking of Z-theory (Section 2.6.1). In the non-preemptive formulation of GP, there are $2k$ weights of importance: $W_G = (w_1^+, w_2^+, \ldots, w_k^+; w_1^-, w_2^-, \ldots, w_k^-)$ associated with over- and under-achievement variables $(d_1^+, d_2^+, \ldots, d_k^+; d_1^-, d_2^-, \ldots, d_k^-)$, respectively. These weights, W_G, are nonnegative, that is, ≥ 0; they are given values prior to solving the problem. The summation of overachievement weights, and also underachievement weights, can be set to be equal to one, that is, $(w_1^+ + w_2^+ + \cdots + w_k^+) = 1$, $(w_1^- + w_2^- + \cdots + w_k^-) = 1$. By solving the following Z-GP problem, efficient solutions will be generated for non-preemptive GP where all objectives are maximized.

PROBLEM 2.6   GOAL SEEKING: Z-GP-LP (Z-GP)

Maximize $U = \text{LV} + z_G \text{DV}_{G,G}$ (2.53)

Subject to : $\text{LV} = \sum_{i=1}^{k} (w_i^+ f_i + w_i^- f_i)$ (2.54)

$\text{DV}_{G,G} = \sum_{i=1}^{k} (w_i^+ d_i^+ + w_i^- d_i^-)$ (2.55)

$f_i + d_i^- - d_i^+ = f_{i,G}$ for $i = 1, 2, \ldots, k$ (2.56)

$d_i^+ \geq 0, d_i^- \geq 0$ for $i = 1, 2, \ldots, k$ (2.57)

And the set of constraints on all alternatives; e.g., LP constraints   (2.58)

In a Z-GP problem, z_G is a given parameter such that $-1 < z_G \leq 0$, and w_i^-, $w_i^+ > 0$ for all i are also given. In the solution of Z-GP problem, either d_i^+ or d_i^- will be positive (but both can be zero). After the problem is solved, the values of d_i^+, d_i^-, and f_i for $i = 1, 2, \ldots, k$, will be known.

If $f_{i,G} < f_i$ then $f_{i,G} - f_i = -d_i^+$ (overachievement) where $d_i^+ \geq 0$.

If $f_{i,G} > f_i$ then $f_{i,G} - f_i = d_i^-$ (underachievement) where $d_i^- \geq 0$.

In GP terminology, for each objective f_i, w_i^- is the importance of underachievement d_i^-, and w_i^+ the importance of overachievement d_i^+. That is, weights of importance for under- and overachievements can be different in $\text{DV}_{G,G}$ but they are the same in DV_G. If we replace $(d_i^- - d_i^+)$ by $|d_i|$, and set $w_i^- = w_i^+ = w_i$ for $i = 1, 2, \ldots, k$, then we have the following result.

$$\text{DV}_{G,G} = \sum_{i=1}^{k} w_i(|d_i|) = \sum_{i=1}^{k} w_i(|f_{i,G} - f_i|) = \text{DV}_G$$ (2.59)

Increasing Property of Z-GP: It can be shown that $U = \text{LV} + z_G \text{DV}_{G,G}$ is an increasing function of objectives and the solution to Z-GP problem is always efficient; see proofs in Malakooti (2014b).

Solving Preemptive GP by ZUT: The same above results are applicable to preemptive GP when LV is added to the last objective function of LP where LP problems are solved sequentially (i.e., add LV to the last active term of preemptive ordering).

Bicriteria Optimization Example for Z-GP Consider a bicriteria problem where the set of alternatives is continuous and presented by $0 \leq f_1 \leq 1$, $0 \leq f_2 \leq 1$, and $f_1 + f_2 \leq 1.5$. Suppose that $w_i = w_i^- = w_i^+ = 0.5$ for i = 1,2 (i.e., $w_1 = w_2 = 0.5$) and $z_G = -0.5$. Suppose given goals are $F_G = (0.3, 0.4)$. Solving Z-GP problem by LP, the solution is $f_1 = 0.5$ and $f_2 = 1$, where LV = 0.75, $\text{DV}_{G,G} = 0.4$, and $U = 0.75 - 0.5 \times 0.4 = 0.55$. This solution is efficient and dominates goals $F_G = (0.3, 0.4)$. But if only $\text{DV}_{G,G}$ is minimized as the objective function of Z-GP problem (which is the case in GP), the solution would be (0.3,0.4) which is an inefficient point. As a comparison, suppose that the given goals $F_G = (0.8, 0.9)$, which is infeasible. Solving Z-GP problem by LP, the solution is $f_1 = 0.6$

TABLE 2.26   Presentation of $k + 1$ Alternatives for Assessment of Weights

	$F_{\text{Mid},1}$	$F_{\text{Mid},2}$	$\cdots$	$F_{\text{Mid},k}$	F_0
f_1	f_{Mid}	0	$\cdots 0 \cdots$	0	0
f_2	0	f_{Mid}	$0 \cdots$	0	0
$\cdots$	$\cdots$	$\cdots$	$\cdots f_{\text{Mid}}$	$\cdots$	0
f_k	0	0	0	f_{Mid}	0

and $f_2 = 0.9$. This solution is feasible and efficient. Section 2.8 provides details of solving optimization problems such as LP problems.

2.6.4 Assessment of Goal-Seeking ZUT[†]

Consider goal-seeking ZUT (2.46) or goal-convergent ZUT (2.47). In this section, we show how to find weights and z_G when the goal (F_G) is known. Choose an alternative $F_{\text{Mid}} = (f_{\text{Mid}}, f_{\text{Mid}}, \ldots, f_{\text{Mid}})$ such that $f_{\text{Mid}} \leq \text{Min } \{f_{i,G} \text{ for all } i \text{ objectives}\}$. F_{Mid} is the minimum acceptable alternative which is dominated by the goals F_G. For practical purposes, one can use $F_{\text{Mid}} = (0.5, 0.5, \ldots, 0.5)$. Consider Table 2.26 that presents k alternatives whose ith component is f_{Mid} and all other components are zero. The $k + 1$th alternative, F_0, has (all) zero components.

The DM is asked to provide the utility of each of these alternatives; therefore $k + 1$ equations can be generated. We use these equations to assess all k weights and z_G. The LV of F_0 is zero; therefore, the utility of F_0 is

$$U(F_0) = z_G(w_1 f_{1,G} + w_2 f_{2,G} + \cdots + w_i f_{i,G} + \cdots + w_k f_{k,G}) \qquad (2.60)$$

For each alternative of Table 2.26, $F_{\text{Mid},i}$, the utility is

$U(F_{\text{Mid},i}) = f_{\text{Mid}} w_i + z_G(w_1 f_{1,G} + w_2 f_{2,G} + \cdots + w_i |f_{\text{Mid}} - f_{i,G}| + \cdots + w_k f_{k,G})$
Since $f_{i,G} \geq f_{\text{Mid}}$:
$U(F_{\text{Mid},i}) = f_{\text{Mid}} w_i + z_G(w_1 f_{1,G} + w_2 f_{2,G} + \cdots + w_i(f_{1,G} - f_{\text{Mid}}) + \cdots + w_k f_{k,G})$, or
$U(F_{\text{Mid},i}) = f_{\text{Mid}} w_i(1 - z_G) + z_G(w_1 f_{1,G} + w_2 f_{2,G} + \cdots + w_i f_{i,G} + \cdots + w_k f_{k,G})$

Using (2.60) in the above equation, we have $U(F_{\text{Mid},i}) = f_{\text{Mid}} w_i(1 - z_G) + U(F_0)$, or

$$w_i(1 - z_G) = (U(F_{\text{Mid},i}) - U(F_0))/f_{\text{Mid}} \qquad (2.61)$$

The summation of Equation (2.61) over all i results in

$$\sum_{i=1}^{k} w_i(1 - z_G) = \sum_{i=1}^{k} (U(F_{\text{Mid},i}) - U(F_0))/f_{\text{Mid}}, \text{ or } (1 - z_G)$$

$$= \sum_{i=1}^{k} (U(F_{\text{Mid},i}) - U(F_0))/f_{\text{Mid}}, \text{ or} \qquad (2.62)$$

$$z_G = 1 - \sum_{i=1}^{k} w_i(U(F_{\text{Mid},i}) - U(F_0))/f_{\text{Mid}}$$

After z_G is found in (2.62), use it in (2.61) to find all weights as follows:

$$w_i = (U(F_{\text{Mid},i}) - U(F_0))/(f_{\text{Mid}}(1 - z_G)) \quad \text{for } i = 1, 2, \ldots, k \qquad (2.63)$$

Approximation of z_G and Verification of G-ZUT: After all weights are assessed, use assessed weights W to find LV, DV_G of all Q alternatives. Then find the $z_{G,j}$ of each alternative j by using its known U_j^*, LV_j, $DV_{G,j}$; that is, find $z_{G,j} = (U_j^* - LV_j)/DVG_{M,j}$. Then find the average value of z_G of all Q alternatives; if MAD of this assessed z_G value is acceptable, then the assessed G-ZUT is valid. Details are shown in Example 2.19.

Assessment of Goal-Seeking ZUT by Nonlinear Equations (Based on Section 2.5.3): To assess goal-seeking utility function, a problem similar to Problem 2.3 (the assessment method of Section 2.5.3) can be formulated and solved by Levenberg–Marquardt method using MATLAB.

We can consider three different variations of goal-seeking problem: (a) When goals are given then find weights (W) and z_G; in this case at least $Q = k + 2$ alternatives should be evaluated. (b) When weights (W) are given (or use equal weights), then find goals and z_G; in this case at least $Q = k + 2$ alternatives should be evaluated. (c) When both goals and weights (W) are unknown, then find goals, W, and z_G; in this case at least $Q = 2k + 2$ alternatives should be evaluated.

Supplement MC-S.6.2 shows how to find goals and weights simultaneously.

Example 2.19   Assessment of Weights of Goal-Convergent Utility Function   Consider a four-criteria problem and suppose that all values are normalized and $F_G = (0.6,0.8,0.5,0.5)$. For the set of nine alternatives, the assessed utility values by the DM, U^*, are shown below. Use the four nonzero alternatives, $N_1 - N_4$, for testing purposes. Use the assessment method to find weights and z_G and verify the assessments using testing alternatives.

We first find z_G using (2.62).

$$z_G = 1 - \sum_{i=1}^{k} (U(F_{\text{Mid},i}) - U(F_0))/f_{\text{Mid}} = 1 - (0.62 + 0.14 + 0.61 + 0.18) = -0.55$$

Now, find weights using (2.63); for example, the first weight is

$$w_1 = (U(F_{\text{Mid},1}) - U(F_0))/(f_{\text{Mid}}(1 - z_G)) = 0.62/(0.5(1 + 0.55)) = 0.4$$

The solutions are shown in the following table. Using assessed weights $(0.4,0.09,0.39,0.12)$, LV and DV_G of each alternative can be obtained. For example, for the first alternative:

$$LV(F_{\text{Mid},1}) = 0.4 \times 0.5 + 0.09 \times 0 + 0.39 \times 0 + 0.12 \times 0 = 0.20$$
$$DV_G(F_{\text{Mid},1}) = 0.4|0.5 - 0.6| + 0.09|0 - 0.8| + 0.39|0 - 0.5| + 0.12|0 - 0.5| = 0.37$$

	$F_{\text{Mid},1}$	$F_{\text{Mid},2}$	$F_{\text{Mid},3}$	$F_{\text{Mid},4}$	F_0	N_1	N_2	N_3	N_4	Find
DM's U_j^*	0.02	−0.22	0.016	−0.20	−0.29	0.30	0.55	0.20	0.28	
$(U(F_{\text{Mid},i}) - U(F_0))/f_{\text{Mid}}$	0.62	0.14	0.61	0.18	–	–	–	–	–	$z_G = -0.55$
Assessed w_i	0.40	0.09	0.39	0.12	–	–	–	–	–	
LV	0.20	0.05	0.20	0.06	0.00	0.47	0.60	0.33	0.40	–
DV_G	0.37	0.52	0.37	0.51	0.57	0.34	0.20	0.32	0.19	–
$U = $ LV−0.552DV_G	0.00	−0.24	0.00	−0.22	−0.31	0.28	0.49	0.15	0.30	
Error $= |U-U^*|$	0.02	0.02	0.02	0.02	0.02	0.02	0.06	0.05	0.02	MAD $= 0.03$

See Example 2.19.xlsx.

This MAD $= 0.03$ is acceptable. Therefore, the assessed ZUT is $U = $ LV−0.55DV. For example, $U(N_4) = 0.40 - 0.55 \times 0.19 = 0.30$. MAD is 0.03 which is acceptable.

2.6.5 Binary Pole ZUT: Goal Seeking/Nadir Avoiding

People may make their decisions by simultaneously considering two reference points: goals and nadir.

- Goals F_G (flexible ideal, aspired, zenith, target) is a point that the DM identifies as desired alternative, and therefore the distance to this point is minimized.
- Nadir F_N (flexible nadir, pitfall, despised) is a point that the DM identifies as undesired and therefore the distance to this point is maximized.

The DM would like to be as close as possible to the goals, and to be as far as possible from the nadir. It is a common mistake to assume that the utility function of the DM is symmetric with respect to these two reference points; that is, minimizing distance to goals would maximize the distance from the nadir and vice versa. See Figure 2.21. Section 2.8.3 explains that the value function of each objective of binary pole utility functions has a quasi-linear S-shaped valued function.

Before presenting binary pole, we first discuss ZUT by only considering nadir.

Nadir-Divergent ZUT: In this method, the DM identifies the nadir, F_N, alternative. The nadir-divergent ZUT is

$$U = \text{LV} + z_N \text{DV}_N \qquad \text{for} \quad 0 < z_N < +1 \quad \text{(nadir divergent)} \qquad (2.64)$$

$$\text{DV}_N = \sum_{i=1}^{k} w_i |f_i - f_{iN}| \qquad (2.65)$$

U is increasing for $-1 < z_N < 1$; however, since DV_N is maximized, we use $0 \leq z_N < +1$. Since DV_N is convex, for $0 < z_N < +1$, this function is convex.

Binary Pole ZUT: The binary pole ZUT is

$$U = \text{LV} + z_G \text{DV}_G + z_N \text{DV}_N \quad \text{for} -1 < z_G \leq 0, 0 \leq z_N < 1, \text{ and } |z_G| + z_N < 1 \qquad (2.66)$$

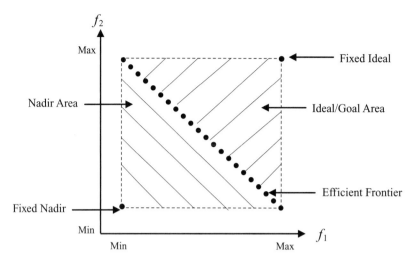

Figure 2.21   Efficient frontier, goal, and nadir areas, fixed ideal, and nadir.

U is an increasing function of all objectives. Therefore, for any given two reference points (regardless of whether they are efficient or inefficient) (2.66) is increasing and generates (chooses) efficient points. When $z_N = 0$, the function is concave and the DM minimizes the distance to the goals. When $z_G = 0$, this function is convex, and the DM maximizes the distance from the nadir point. When z_G is negative and z_N is positive (2.66) is neither concave nor convex; proofs are given in Malakooti (2014b).

Multiple Pole ZUT: People may make their decisions by simultaneously considering several reference points (poles). In this case, two reference points, goals and nadir, can be expanded to several reference points (poles); see Malakooti (2014b).

Example 2.20   Goal-Convergent/Nadir-Divergent (Binary Pole) ZUT   Consider the alternatives shown in Table 2.27 where $W = (0.4, 0.4, 0.2)$ are weights of importance for the three objectives. Rank these alternatives for the following cases:

(d) Divergent case where nadir point is $(0.2, 0.3, 0.2)$ and z_N is $+0.3$, that is,

$$\text{Maximize } U = \text{LV} + 0.3\text{DV}_N.$$

TABLE 2.27   Example 2.19: Assessment of Weights of Goal-Convergent Utility Function

	Alternatives for Assessment					Alternatives for Testing			
	$F_{\text{Mid},1}$	$F_{\text{Mid},2}$	$F_{\text{Mid},3}$	$F_{\text{Mid},4}$	F_0	N_1	N_2	N_3	N_4
f_1	0.5	0	0	0	0	0.3	0.9	0.1	0.5
f_2	0	0.5	0	0	0	0.1	0.4	0.3	0.9
f_3	0	0	0.5	0	0	0.8	0.5	0.6	0.2
f_4	0	0	0	0.5	0	0.2	0.1	0.2	0.3
DM's U_j^*	0.02	−0.22	0.016	−0.20	−0.29	0.30	0.55	0.20	0.28

TABLE 2.28   Four Examples for Ranking Alternatives with Respect to Two Goals and a Nadir

Objectives	Alternatives							Goals	Nadir	Weight
	F_1	F_2	F_3	F_4	F_5	F_6	F_7	F_G	F_N	w_i
f_1	0.3	0.4	0.2	0.2	0.5	0.1	0.9	0.7	0.2	0.40
f_2	0.2	0.4	0.5	0.2	0.1	0.9	0.1	0.65	0.3	0.40
f_3	0.7	0.4	0.5	0.2	0.7	0.5	0.1	0.3	0.2	0.20
LV	0.34	0.40	0.38	0.20	0.38	0.50	0.42	0.60	0.24	–
a. DV_G	0.42	0.24	0.30	0.40	0.38	0.38	0.34			
d. DV_N	0.18	0.16	0.14	0.04	0.30	0.34	0.38			
d. Rank by Max DV_N	4	5	6	7	3	2	1			
d. $U = LV + 0.3DV_N$	0.39	0.45	0.42	0.21	0.47	0.60	0.53			
d. Rank by Max U	6	4	5	7	3	1	2			
e. Bipole U (a, d)	0.14	0.30	0.24	–0.03	0.24	0.37	0.33			
e. Rank for dual pole	6	3	4	7	5	1	2			

See Example 2.20.xls.

(e) Dual pole case where both goals (0.7,0.65,0.3) and nadir (0.2,0.3,0.2) are used, that is,

$$\text{Maximize } U = LV - 0.6DV_G + 0.3DV_N.$$

The solutions are shown in Table 2.28. For example, for alternative F_1:

$$DV_G = 0.4|0.3 - 0.7| + 0.4|0.2 - 0.65| + 0.2|0.7 - 0.3| = 0.42$$
$$DV_N = 0.4|0.3 - 0.2| + 0.4|0.2 - 0.3| + 0.2|0.7 - 0.2| = 0.18$$
$$U = LV - 0.6DV_G + 0.3DV_N = 0.34 - 0.6 \times 0.42 + 0.3 \times 0.18 = 0.14$$

In question d, purposely an inefficient goal point is given to illustrate that ZUT finds an efficient point F_3 instead of choosing F_4 which is inefficient but the closest point to the goals.

2.7   MULTIPLE OBJECTIVE OPTIMIZATION

MOO is concerned with solving MCDM problems where the set of multicriteria alternatives is presented by a set of mathematical constraints. In MOO, objectives and constraints are functions of decision variables. Therefore, MOO problems are more complex to solve than discrete MCDM problems where the set of alternatives is enumerable and directly presented in the objective space. The optimization methods used for solving MOO problems are based on LP, nonlinear programming (NLP), integer programming (IP), and combinatorial methods for solving single objective optimization problems. The reader must have an elementary knowledge of optimization before studying MOO. In this section, we will focus on solving multiple objective linear programming (MOLP) problems because they are the most useful and the simplest of all MOO problems. The approaches for solving MOLP problems can be extended to solve multiple objective nonlinear programming, IP, and combinatorial problems.

2.7.1 Formulation of Multiple Objective Optimization[†]

In LP, the optimal solution is found for only one objective function (e.g., minimizing cost). MOLP is a generalization of LP where several linear objective functions are simultaneously optimized (each objective is either maximized or minimized) subject to a set of linear constraints.

A standard MOLP formulation is as follows.

PROBLEM 2.7   MOLP

Objectives	Subject to: Objective Equations	Constraints
Maximize f_1 Maximize f_2 Maximize f_k	$f_1 = c_{11}x_1 + c_{12}x_2 + \cdots + c_{1n}x_n$ $f_2 = c_{21}x_1 + c_{22}x_2 + \cdots + c_{2n}x_n$ $f_k = c_{k1}x_1 + c_{k2}x_2 + \cdots + c_{kn}x_n$ $f_1, f_2, \ldots, f_k$ unrestricted in sign	$a_{11}x_1 + a_{12}x_2 + \cdots + a_{1n}x_n = b_1$ $a_{21}x_1 + a_{22}x_2 + \cdots + a_{2n}x_n = b_{2:}$ $a_{m1}x_1 + a_{m2}x_2 + \cdots + a_{mn}x_n = b_m$ $x_1, x_2, \ldots, x_n \geq 0$

The notations of the MOLP are

k: number of objectives

n: number of decision variables

m: number of constraints

f_i for $i = 1, \ldots, k$: objective variables (unknown)

x_j for $j = 1, \ldots, n$: decision variables (DVs) (unknown)

c_{ij} for $i = 1, \ldots, k$ and $j = 1, \ldots, n$: coefficients of DVs in objective functions (known)

a_{pj} for $p = 1, \ldots, m$ and $j = 1, \ldots, n$: coefficients of DVs in constraints (known)

b_p for $p = 1, \ldots, m$: constraint (resource) limitations (known)

In the MOLP model, there are k objective functions which are maximized with respect to all constraints. Decision variables are nonnegative (i.e., $x_1, x_2, \ldots, x_n \geq 0$), but objective variables ($f_1, f_2, \ldots, f_k$) are unrestricted in sign and can assume negative or nonnegative values.

The above MOLP problem 2.7 formulation can be presented in a compact form.

$$\text{Maximize } F(\mathbf{x}) \text{ Subject to } \mathbf{x} \in S$$

where $F \in R^k$ and $\mathbf{x} \in R^n$. S is the set of constraints presented in MOLP problem 2.7, $F = (f_1, f_2, \ldots, f_k)$, and $\mathbf{x} = (x_1, x_2, \ldots, x_n)$. All objectives are assumed to be maximized in MOLP. If an objective function has to be minimized, it can be represented by maximizing its negative. That is, minimizing f_i is equal to maximizing $(-f_i)$, that is,

Minimize $f_i = c_{k1}x_1 + c_{k2}x_2 + \cdots + c_{kn}x_n$ or

Maximize $(-f_i) = -c_{k1}x_1 - c_{k2}x_2 - \cdots - c_{kn}x_n$

The purpose of solving MOLP problems is to find the optimal values for unknown objectives, $F = (f_1, f_2, \ldots, f_k)$, and decision variables, $\mathbf{x} = (x_1, x_2, \ldots, x_n)$, for a given DM. The optimal solution to the MOO problem is also called the best compromise or the best solution.

Finding the Maximum Value for Each Objective Function: To find the maximum value for each objective function, f_i, solve the following LP problem.

Maximize f_i Subject to the constraints of Problem 2.7.

Finding the Combination of All Objective Functions Based on Equal Weights: Consider equal weights of $1/k$ given to each objective function.

$$\text{Maximize } (1/k)f_1 + (1/k)f_2 + \cdots + (1/k)f_k \quad \text{Subject to the constraints of Problem 2.7.}$$

Payoff Matrix: For each of the k objectives, find its maximum value and record the values of the other objectives. Present these k alternatives in the form of a $k \times k$ matrix. This matrix is the payoff matrix. The payoff matrix shows the extreme objective alternatives. The payoff matrix can be used to identify both the maximum and the (efficient) minimum of each objective, and to show the range for each objective.

Finding the Minimum Value of Each Objective Function Associated with Efficient Alternatives: The minimum value of a given objective associated with all efficient alternatives is usually higher than the minimum value of the given objective associated with all alternatives. This value is the minimum for the given objective in the payoff matrix.

Example 2.21   Multiobjective Diet Problem   The university hospital dietary staff is required to satisfy the minimum nutritional needs of the patients while minimizing total food costs. Each food can consist of a combination of five ingredients as presented in Table 2.29. Each food must contain the minimal nutritional needs of 38 units of vitamin C and 44 units of vitamin D. There is a maximum limit of 50 units of all ingredients to be used in a food. The three objective functions are

$f_1 = $ Minimize cost
$f_2 = $ Maximize food taste
$f_3 = $ Maximize food nutrition

The problem is to decide the best combination of ingredients to optimize the three objectives.

TABLE 2.29   Food Ingredients and Their Characteristics

	Ingredient 1	Ingredient 2	Ingredient 3	Ingredient 4	Ingredient 5	Minimum Required
Vitamin C	4	1	2	1	2	Minimum of 38
Vitamin D	2	2	3	0.5	1	Minimum of 44
Ingredients	1	1	1	1	1	Maximum of 50
Cost/unit	0.3	0.15	0.2	1	0.1	–
Taste/unit	1	2	1	10	1	–
Nutrition/unit	15	2	1	1	2	–

The decision variables, $x_1, x_2, \ldots, x_5$, represent the number of units of each ingredient. The MOLP problem formulation is as follows:

Objectives	Subject to: Objective Equations	Constraints
Minimize food cost: f_1	$f_1 = 0.3x_1 + 0.15x_2$ $+ 0.2x_3 + x_4 + 0.1x_5$	$4x_1 + x_2 + 2x_3 + x_4 + 2x_5 \geq 38$
Maximize food taste: f_2	$f_2 = 1x_1 + 2x_2 + 1x_3$ $+ 10x_4 + 1x_5$	$2x_1 + 2x_2 + 3x_3 + 0.5x_4 + x_5 \geq 44$
Maximize food nutrition: f_3	$f_3 = 15x_1 + 2x_2 + 1x_3$ $+ 1x_4 + 2x_5$	$x_1 + x_2 + x_3 + x_4 + x_5 \leq 50$
	f_1, f_2, f_3 unrestricted in sign	$x_1, x_2, x_3, x_4, x_5 \geq 0$

The three LP formulations to generate the optimal solution for each objective are as follows. The weights for the combined problem (Problem 4) are $(1/k) = 1/3$.

Problem 1	Minimize f_1	Subject to: MOLP constraints of Example 2.21
Problem 2	Maximize f_2	Subject to: MOLP constraints of Example 2.21
Problem 3	Maximize f_3	Subject to: MOLP constraints of Example 2.21
Problem 4	Maximize $U = -(1/3)f_1 + (1/3)f_2 + (1/3)f_3$	Subject to: MOLP constraints of Example 2.21

For Problems 1, 2, and 3, each of the associated LP problems is solved separately and independently from the other problems. Problem 4 combines all three objectives into one utility function. While the first three problems generate three extreme solutions in terms of the three objectives, Problem 4 generates a solution where equal importance is given to the three objectives. See Table 2.30 for a summary of the solutions. Each row represents the solution to each problem in terms of decision variables and objective functions. Each row is an alternative solution to the MOLP problem. All three alternatives are efficient. In reality, one probably will not choose one of these solutions, but will choose a combination of these solutions; the methods for generating such solutions will be discussed in the remaining sections of this chapter.

TABLE 2.30   Summary of Different Solutions for Example 2.21

Problem	Decision Variables					Objective Functions		
	x_1	x_2	x_3	x_4	x_5	f_1	f_2	f_3
P1: Minimize f_1 S.T. MOLP	0	0	12.5	0	6.5	$3.15	19	25.5
P2: Maximize f_2 S.T. MOLP	0	0	7.6	42.4	0	$43.92	431.6	50
P3: Maximize f_3 S.T. MOLP	50	0	0	0	0	$15.00	50	750
P4: Maximize U S.T. MOLP	0.04	0	12.5	12.82	0	15.33	140.73	26
Minimum value	0	0	0	0	0	$3.15	19	25.5
Maximum value	50	0	12.5	42.4	6.5	$43.92	431.6	750

See Examples 2.21.xlsx, 2.21.lg4.

The payoff matrix for this example is the 3×3 matrix of the objective function values associated with P1, P2, and P3. The maximum and the minimum of each objective are presented in the last two rows.

2.7.2 Generation of Efficient Extreme Alternatives[†]

In this section, we define efficiency (or Pareto optimality) and show how to generate efficient extreme points for MOO problems.

Efficiency (Nondominancy) in MOO: The efficiency definition is the same as the one presented in Section 2.2.1 where the set of multicriteria alternatives, S, is presented by MOO problem. For MOLP problems, S is the set of all linear constraints of Problem 2.7.

Extreme versus Nonextreme Alternatives: In MOLP problems, there are two types of points: extreme and nonextreme points. The optimal solution of an LP problem is always an extreme point. For example, corner points f_1, f_2, F_3, F_4, and F_5 in Figure 2.22b are extreme points. All other feasible points to LP are nonextreme points. For example, all the points on the lines that connect F_1 and F_2, or F_2 and F_3, or F_3 and F_4, or F_4 and F_5 in Figure 2.22b are nonextreme points.

Generally, for the MOLP problem, the set of efficient extreme points is finite and the set of efficient nonextreme points is infinite. The optimal solution for an additive (linear) utility function is always an efficient extreme point; and for a nonlinear utility function it could be an efficient nonextreme point (see Section 2.8.1).

Generating Efficient Extreme Points to MOLP Problem: Here, we discuss how efficient extreme points (alternatives) are generated. Consider the weighted additive utility function:

$$U = w_1 f_1 + w_2 f_2 + \cdots + w_k f_k$$

where $w_1, w_2, \ldots, w_k$ are weights of importance of each objective function. All weights are positive numbers (i.e., $w_1 > 0, w_2 > 0, \ldots, w_k > 0$), and $w_1 + w_2 + \cdots + w_k = 1$. The weights can be represented by a vector $W = (w_1, w_2, \ldots, w_k)$. The following formulation converts the MOLP problem into an LP with a single composite objective function, which we will call weighted MOLP (W-MOLP).

The efficient frontier is the set of all efficient points including efficient extreme and nonextreme points. All efficient extreme points can be generated by the W-MOLP and all efficient nonextreme points can be generated by using the method presented in Section 2.8.1.

PROBLEM 2.8 WEIGHTED-MOLP PROBLEM

Maximize $U = w_1 f_1 + w_2 f_2 + \cdots + w_k f_k$
Subject to: MOLP constraints of Problem 2.7

where $w_1 > 0, w_2 > 0, \ldots, w_k > 0$ are given values.

It is assumed that there exists at least one feasible solution to the W-MOLP problem and that all objective functions are being maximized. Consider the following three important results.

Result 1: For any given vector of weights W, the solution to the W-MOLP (Problem 2.8) is an efficient extreme point.

Result 2: All efficient extreme points of a MOLP problem can be generated by varying the weights in Problem 2.8.

Result 3: For additive utility functions, such as (2.67), the optimal (the best compromise) solution is an efficient extreme point (because the objective function of Problem 2.8 is linear).

Example 2.22   Two Objectives and Two Variables—Graphical Solution

Objectives	Subject to: Objective Equations	Constraints	
Maximize f_1	$f_1 = x_1 + 12x_2$	$x_1 + x_2 \leq 13$	$2x_1 + x_2 \leq 19$
Maximize f_2	$f_2 = 10x_1 + 2x_2$	$x_1 + 2x_2 \leq 22$	$3x_1 + x_2 \leq 27$
	f_1, f_2 unrestricted in sign	$x_1, x_2 \geq 0$	

Generate five efficient extreme points by varying weights using Problem 2.8.
The W-MOLP problem for this example is

Maximize $U = w_1 f_1 + w_2 f_2$
Subject to: MOLP constraints of Example 2.22

The solutions for maximizing each objective separately are presented in Table 2.31 and are found by solving the two independent LP problems P_1 and P_2.
By randomly varying the weight values, efficient alternatives can be generated.
Consider, for example, the following weights: (0.99,0.01), (0.5,0.5), (0.33,0.67), (0.2,0.8), and (0.01,0.99). The following five alternatives ($F_1 - F_5$) are generated by solving five different W-MOLP problems (Problem 2.8) by

Maximizing $U = w_1 f_1 + w_2 f_2$
Subject to: MOLP constraints of Example 2.22

TABLE 2.31   Solution for Maximizing Each Objective Function

W-MOLP Problem	Decision Variables		Payoff Matrix	
	x_1	x_2	f_1	f_2
P_1: Max $U = 0.99f_1 + 0.01f_2$ S.T. Example 2.22 constraints	0	11	132	22
P_2: Max $U = 0.01f_1 + 0.99f_2$ S.T. Example 2.22 constraints	9	0	9	90
Minimum	0	0	9	22
Maximum	9	11	132	90

For example, the following LP problem is solved to generate alternative F_3.

Objectives	Subject to: Objective Equations	Constraints	
Maximize $U = 0.33f_1 + 0.67f_2$	$f_1 = x_1 + 12x_2$ $f_2 = 10x_1 + 2x_2$ f_1, f_2 unrestricted in sign	$x_1 + x_2 \leq 13$ $x_1 + 2x_2 \leq 22$	$2x_1 + x_2 \leq 19$ $3x_1 + x_2 \leq 27$ $x_1, x_2 \geq 0$

The solution to this LP problem is $x_1 = 6$, $x_2 = 7$, $f_1 = 90$, and $f_2 = 74$. Note that

$$f_1 = 6 + 12(7) = 90$$

$$f_2 = 10(6) + 2(7) = 74$$

The LP solutions for Example 2.22 using the five different weights are presented below.

Figure 2.22 presents (a) the decision variable space and (b) the objective space. The range of weights (weight space) for this example is shown in the last two rows of Table 2.32. The weight space presents the range of weight values for which the same optimal solution will be generated. For example, for any $0.42 \leq w_1 \leq 0.64$ and $0.36 \leq w_2 \leq 0.58$, where $w_1 + w_2 = 1$, the W-MOLP problem generates alternative F_2. In the next section, we show how to find such weight ranges.

In this example, there are only five efficient extreme points, F_1–F_5 (note that alternative F_0 shown in Figure 2.22b is inefficient). Note that there are six extreme points in the decision space, including ($x_1 = 0$, $x_2 = 0$) whose objective values are $f_1 = 0 + 12(0) = 0$, and $f_2 = 10(0) + 2(0) = 0$. This point, $F_0 = (0,0)$, is an inefficient extreme point and cannot be generated by W-problem 2.8.

2.7.3 MOO of Additive and Multiplicative Utility Functions[†]

Consider the weighted additive utility function:

$$U = w_1 f_1 + w_2 f_2 + \cdots + w_k f_k \qquad (2.68)$$

Suppose that the DM assesses the weights of importance of all objectives (e.g., using the method of Section 2.4). In this case, a W-MOLP problem (Problem 2.7) can be solved

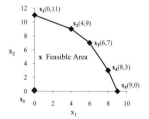

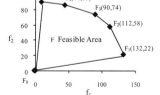

(a) **Decision Variable Space:**
Feasible Region x_1 vs. x_2

(b) **Objective Space:**
Efficient Frontier f_1 vs. f_2

Figure 2.22 Feasible region (decision variables) and efficient frontier (objective space) for Example 2.22.

TABLE 2.32   LP Solutions for Example 2.22 Using Five Different Weights

Alternative	F_1	F_2	F_3	F_4	F_5
w_1, w_2	0.99,0.01	0.50,0.50	0.33,0.67	0.2,0.8	0.01,0.99
f_1	132	112	90	44	9
f_2	22	58	74	86	90
x_1	0	4	6	8	9
x_2	11	9	7	3	0
Weight range: $w_1 = 1 - w_2$	0.64–1	0.42–0.64	0.21–0.42	0.1–0.21	0–0.1
Weight range: w_2	0–0.36	0.36–0.58	0.58–0.79	0.79–0.90	0.90–1

to find the best alternative for the DM. Note that it is not required that the sum of weights be equal to one in (2.68) but all weights are assumed to be positive, that is, $w_1 > 0$, $w_2 > 0, \ldots, w_k > 0$. Similarly, one can use the assessed multiplicative utility function from Section 2.5 as the objective function for MOO problem.

Example 2.23   Additive Utility Function   Consider Example 2.22. Suppose the DM's weights are $w_1 = 0.5$ and $w_2 = 0.5$. The W-MOLP problem 2.7 is formulated as follows.

Objective	Subject to: Objective Equations	Constraints	
Maximize $U = 0.5f_1 + 0.5f_2$	$f_1 = x_1 + 12x_2$ $f_2 = 10x_1 + 2x_2$ f_1, f_2 unrestricted in sign	$x_1 + x_2 \leq 13$ $x_1 + 2x_2 \leq 22$ $2x_1 + x_2 \leq 19$	$3x_1 + x_2 \leq 27$ $x_1, x_2 \geq 0$

See Example 2.23.lg4.

Solving this W-MOLP yields the following solution:

x_1	x_2	f_1	f_2	U
4	9	112	58	85

Example 2.24   Multiplicative ZUT   Consider Example 2.22. Suppose that the DM's utility function is multiplicative, where $LV = 0.5f_1 + 0.5f_2$ and $MV_M = f_1^{0.5} f_2^{0.5}$, and $z_M = -0.9$. Therefore, maximize $U = (1-0.9)LV + 0.9MV_M$. Find the best solution for this problem.

Objective	Subject to: Objective Equations	Constraints	
Maximize $U = (0.1)(0.5f_1 + 0.5f_2)$ $+ 0.9(f_1^{0.5} f_2^{0.5})$	$f_1 = x_1 + 12x_2$ $f_2 = 10x_1 + 2x_2$ f_1, f_2 unrestricted in sign	$x_1 + x_2 \leq 13$ $x_1 + 2x_2 \leq 22$ $2x_1 + x_2 \leq 19$	$3x_1 + x_2 \leq 27$ $x_1, x_2 \geq 0$

Maximizing this U as the objective function subject to the constraints of Example 2.22 yields the following solution. Note that the solution of Example 2.23 that maximizes $U = 0.5f_1 + 0.5f_2$ is different than Example 2.24 that uses $(0.1)(0.5f_1 + 0.5f_2) + 0.9(f_1^{0.5} f_2^{0.5})$ in its utility function. The solution is

x_1	x_2	f_1	f_2	U
5.3	7.7	97.6	68.5	81.9

See Example 2.24.lg4.

2.8 GOAL-SEEKING MULTIPLE OBJECTIVE OPTIMIZATION

2.8.1 Generation of Efficient Nonextreme Alternatives[†]

In the goal-seeking MOLP (GS-MOLP) method, the DM can identify targets or goals (also called aspiration levels or ideals) for the objectives. The DM also gives the weights of importance for achieving such goals. If the weights are not given, equal weights are used. The GS-MOLP seeks to optimize the value of all the objective functions. The goal-seeking method finds the closest feasible efficient alternative (called feasible goal) to the given goal. An objective f_i is considered achieved, overachieved, or underachieved if it is equal to, greater than, or less than its goal, $f_{i,G}$, respectively.

Basic Concept of Goal Seeking: The purpose of this method is to find a feasible efficient solution (called feasible goal) with respect to the given goal regardless of whether the goal itself is efficient or inefficient. If the given goal dominates some of the efficient points, then the obtained efficient feasible goal will be at a minimum rectilinear distance from the goal. If some of the efficient points dominate the given goal, then the obtained efficient feasible goal will be at the maximum rectilinear distance from the goal.

Comparison to GP: The GS-MOLP method is closely related to GP (Section 2.6.2), a well-known multiobjective method with the following differences:

1. GS-MOLP always generates efficient points while GP may not. That is, when a goal is inefficient, GP generates an inefficient solution (see Section 2.6.2).
2. GP requires $2k$ number of weights (for under- and overachievements) but GS-MOLP requires only k weights (see Section 2.6.2).

Z-GP (Section 2.6.3, Problem 2.6) can be simplified by only having k number of weights as shown in Problem 2.9. Suppose the goal is presented by $F_G = (f_{1,G}, f_{2,G}, \ldots, f_{k,G})$ and weights of importance for achieving them by $W = (w_1, w_2, \ldots w_k)$, the goal-seeking problem is formulated as follows.

PROBLEM 2.9   GS-MOLP

$$\text{Maximize} \quad U = \text{LV} + z_G \text{DV}_G \tag{2.69}$$

$$\text{Subject to :} \quad \text{LV} = \sum_{i=1}^{k} w_i f_i \tag{2.70}$$

$$\text{DV}_G = \sum_{i=1}^{k} w_i(d_i^+ + d_i^-) \tag{2.71}$$

$$f_i + d_i^- - d_i^+ = f_{i,G} \qquad \text{for} \quad i = 1, 2, \ldots, k \tag{2.72}$$

$$d_i^+ \geq 0, d_i^- \geq 0 \qquad \text{for} \quad i = 1, 2, \ldots, k \tag{2.73}$$

And the set of constraints on all alternatives; e.g., LP constraints (2.74)

Problem 2.9 can be solved by LP because all the constraints and the objective function are linear.

Result 4: For any given vector of weights W and goals F_G, the solution to the GS-MOLP (Problem 2.9) is efficient.

Result 5: All efficient nonextreme alternatives can be generated by varying weights W and goals F_G, in GS-MOLP (Problem 2.9).

Example 2.25   GS-MOLP Generating Efficient Nonextreme Alternatives   Consider maximizing f_1 and f_2 subject to $f_1 + 2f_2 \leq 4$ and $f_1, f_2 \geq 0$. Using the additive function (W-MOLP problem 2.8), only two efficient extreme solutions can be generated for this problem; they are $(4,0)$ and $(0,2)$. However, by using GS-MOLP problem 2.9, all efficient points to this problem can be generated; two of such nonextreme points are shown in Figures 2.23a, b. For example suppose that weights are equal $w_1 = w_2 = 0.5$, $F_G = (3.5, 1.5)$, and $z_G = -0.5$. The feasible goal solution to this problem is: $F_2 = (3.5, 0.25)$; see Figure 2.23a. This solution is both an efficient and nonextreme point to the original problem.

Objectives (Maximize)	Subject to: Objective Equations	Constraints
$U = \text{LV} - 0.5\text{DV}_G$	$\text{LV} = 0.5 f_1 + 0.5 f_2$ $\text{DV}_{G,G} = 0.5\, d_1^+ + 0.5\, d_1^-$ $\qquad + 0.5\, d_2^+ + 0.5\, d_2^-$ $f_1 + d_1^- - d_1^+ = 3.5$ $f_2 + d_2^- - d_2^+ = 1.5$	$f_1 + 2f_2 \leq 4$ $f_1, f_2 \geq 0$ $d_i{}^+ \geq 0, d_i^- \geq 0$ for $i = 1,2$

Now suppose that given goals is $F_G = (1, 0.25)$ (which is an inefficient solution). The feasible goal solution to this problem is $F_2 = (3.5, 0.25)$ which is efficient; see Figure 2.23b.

See Example 2.25.lg4.

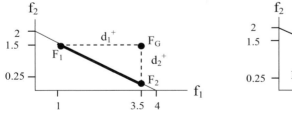 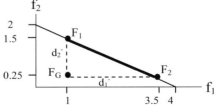

Figure 2.23   Graphical presentation of Problem 2.9, the relationship of the goal to the solutions.

2.8.2   Goal-Seeking MOO[†]

Problem 2.9 can be used to solve the goal-seeking MOO for both LP and NLP problems. To do this, the DM should provide the desired goals and the weights of importance of objectives. Three examples are shown below.

Example 2.26   GS-MOLP Example: Two Objectives   Consider Example 2.22 (which is also presented in the below table).

Case I Example: This is an example where the goal dominates some efficient alternatives and it is infeasible. Suppose $F_G = (100,95)$, and the weights of importance to achieve the goal are $W = (0.7,0.3)$. Suppose that $z_G = -0.5$. Find the best alternative for the DM. The GS-MOLP problem is formulated as follows:

Utility Function	S.T.: Goals	Objectives	Constraints	
Maximize $U = 0.7f_1 + 0.3f_2$ $-0.5((0.7d_1^+$ $+0.7d_1^-)$ $+(0.3d_2^+$ $+0.3d_2^-))$	$f_1 + d_1^- - d_1^+ = 100$ $f_2 + d_2^- - d_2^+ = 95$ $d_1^-, d_1^+, d_2^-, d_2^+ \geq 0$	$f_1 = x_1 + 12x_2$ $f_2 = 10x_1 + 2x_2$ f_1, f_2 unrestricted	$x_1 + x_2 \leq 13$ $x_1 + 2x_2 \leq 22$ $x_1, x_2 \geq 0$	$2x_1 + x_2 \leq 19$ $3x_1 + x_2 \leq 27$

Solving the above problem by LP, the following solution is obtained.

	x_1	x_2	f_1	f_2	d_1^-	d_1^+	d_2^-	d_2^+
GS-solution	4	9	112	58	0	12	37	0

See Example 2.26.lg4.

Therefore, the feasible solution is $f_1 = 112$ and $f_2 = 58$. See Figure 2.24a. The solution shows the goal of f_1 was overachieved and the goal of f_2 was underachieved. That is, the given goal dominates part of the set of efficient alternatives.

f_1 is achieved because $d_1^- = 0$, $d_1^+ = 112 - 100 = 12$.

f_2 is underachieved because $d_2^- = 95 - 58 = 37$, $d_2^+ = 0$.

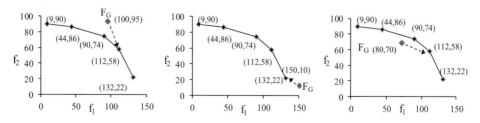

Figure 2.24   The three cases of feasible goal for Example 2.26.

Case II Example: This is an example where the goal is efficient and infeasible. Suppose the goal is $F_G = (150,10)$. The feasible (efficient) solution to this problem is $(132,22)$, see Figure 2.24b. This problem is formulated as follows:

Utility Function	S.T.: Goals	Objectives and Constraints
Maximize $U = 0.7 f_1 + 0.3 f_2$ $-0.5((0.7d_1^+ + 0.7d_1^-)$ $+ (0.3d_2^+ + 0.3d_2^-))$	$f_1 + d_1^- - d_1^+ = 150$ $f_2 + d_2^- - d_2^+ = 10$ $d_1^-, d_1^+, d_2^-, d_2^+ \geq 0$	Objectives and constraints of Case I

Solving this LP, the following solution is obtained.

	x_1	x_2	f_1	f_2	d_1^-	d_1^+	d_2^-	d_2^+
GS-solution	0	11	132	22	18	0	0	12

See Example 2.26-Case II.lg4.

Therefore, f_1 is underachieved by 18, and f_2 is overachieved by 12.

Case III Example: This is an example where the goal is dominated by some efficient alternatives and it is feasible. Suppose the goal is $F_G = (80,70)$. The feasible (efficient) solution to this problem is $(112,58)$; see Figure 2.24c. This problem is formulated as follows. Note that in Figure 2.24c, the rectilinear distance is maximized rather than minimized (horizontal distance is preferred over the shorter vertical distance; this generates the greatest distance from the inefficient goal).

Utility Function	S.T.: Goals	Objectives and Constraints
Maximize $U = 0.7 f_1 + 0.3 f_2$ $-0.5((0.7d_1^+ + 0.7d_1^-)$ $+ (0.3d_2^+ + 0.3d_2^-))$	$f_1 + d_1^- - d_1^+ = 80$ $f_2 + d_2^- - d_2^+ = 70$ $d_1^-, d_1^+, d_2^-, d_2^+ \geq 0$	Objectives and constraints of Case I

Solving this LP, the following solution is obtained:

	x_1	x_2	f_1	f_2	d_1^-	d_1^+	d_2^-	d_2^+
GS-solution	4	9	112	58	0	32	12	0

See Example 2.26-Case III.lg4.

Therefore, f_1 is overachieved by 32 and f_2 is underachieved by 12. This is an example of Case IV, where the goal is feasible, but is inefficient to the alternatives. See Figure 2.24c.

2.8.3 Binary Pole (Goal-Seeking/Nadir-Avoiding) MOO[†]

In this section, we extend the concept of Section 2.6.5 for solving MOO when considering binary poles: goals F_G (to be close to) and nadir F_N (to be far from). Suppose that the

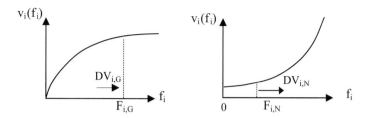

Figure 2.25   The value function for goal seeking is convex and for nadir avoiding is concave.

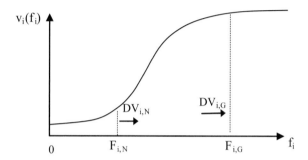

Figure 2.26   The value function of goal seeking/nadir avoiding is quasi-linear; S shape.

following information is given before solving the problem: weights of importance of objectives: $W = (w_1, w_2, \ldots, w_k))$, $F_G = (f_{1,G}, f_{2,G}, \ldots, f_{k,G})$, $F_N = (f_{1,N}, f_{2,N}, \ldots, f_{k,N})$, z_G, and z_N.

Value Functions of Binary Pole in MOO   Consider the following two problems.

Maximizing LV and Minimizing DV_G (Figure 2.25a; solved by Problem 2.9).

Maximizing LV and Maximizing DV_{N-} (Figure 2.25b; solved by Problem 2.10, when $z_G = 0$).

Now consider combining the above two MCDM problems; they result in a three-criteria problem:

Maximizing LV, Minimizing $DV_{G,-}$ and Maximizing DV_N (Figure 2.26; Problem 2.10).

In binary pole goal seeking and nadir avoiding, objective function (2.75) is maximized which optimizes LV, DV_G, and DV_N. Figure 2.26 shows that this value function has S shape; therefore this value function increases rapidly as the distance from point F_N increases; but when F_G is achieved, the increase in value function becomes gradual. This value function is an increasing function of the objective f_i.

PROBLEM 2.10   BINARY POLE (GOAL-SEEKING/NADIR-AVOIDING) MOO (GS/NA-MOO)

$$\text{Maximize } U = \text{LV} + z_G\, DV_G + z_N\, DV_N \qquad \text{or}$$

$$\text{Maximize } U = \sum_{i=1}^{k} w_i f_i + z_G \sum_{i=1}^{k} w_i |f_i - f_{iG}| + z_N \sum_{i=1}^{k} w_i |f_i - f_{iN}| \quad (2.75)$$

Subject to:

The set of constraints on all alternatives; for example, constraints of Problem 2.7. (2.76)

Note that z_G and z_N are given values such that $-1 < z_G \leq 0, 0 \leq z_N < 1$, and $| z_G | + z_N < 1$. Problem 2.10 has a nonlinear objective function and linear constraints.

Result 6: For any given vector of weights W and goals F_G, nadir, F_N, the solution to Problem 2.10 is efficient.

Result 7: All efficient nonextreme alternatives can be generated by varying weights W and goals F_G, nadir, F_N, in Problem 2.10.

Example 2.27   GS/NA-MOO Binary Poles Example: Two Objectives   Consider Example 2.26. Suppose the weights of importance of objectives are $W = (0.7, 0.3)$, goals is $F_G = (100, 95)$ with $z_G = -0.4$, and the nadir point is $F_N = (10, 20)$ with $z_N = 0.5$. Solve this problem by (a) using only the nadir point and (b) using both goals and nadir points.

(a) The nadir-avoiding MOO problem is formulated as follows: Maximize $U = 0.7 f_1 + 0.3 f_2 + 0.5 DV_N = 0.7 f_1 + 0.3 f_2 + 0.5(0.7|f_1 - f_{1,N}| + 0.3|f_2 - f_{2,N}|)$.

Utility Function	Subject to: Objective Equations	Constraints	
Maximize $U = 0.7 f_1 + 0.3 f_2$ $+ 0.5(0.7|f_1 - 10|$ $+ 0.3|f_2 - 20|)$	$f_1 = x_1 + 12x_2$ $f_2 = 10x_1 + 2x_2$ f_1, f_2 unrestricted in sign	$x_1 + x_2 \leq 13$ $x_1 + 2x_2 \leq 22$ $x_1, x_2 \geq 0$	$2x_1 + x_2 \leq 19$ $3x_1 + x_2 \leq 27$

The solution to this optimization problem is as follows and shown in Figure 2.27.

	x_1	x_2	f_1	f_2
NA-solution	0	11	132	22

See Example 2.27.lg4.

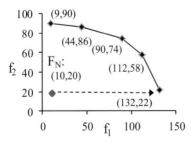
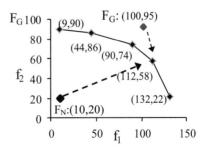

Figure 2.27   The three cases of feasible goal for Example 2.27.

(b) The GS/NA-MOLP problem is formulated as follows: Maximize $U = 0.7f_1 + 0.3f_2 - 0.4DV_G + 0.5DV_N = 0.7f_1 + 0.3f_2 - 0.4(0.7|f_1 - f_{1,G}| + 0.3|f_2 - f_{2,G}|) + 0.5(0.7|f_1 - f_{1,N}| + 0.3|f_2 - f_{1,N}|).$

Utility Function	Subject to: Objective Equations	Constraints	
Maximize $U = 0.7f_1 + 0.3f_2$ $-0.4(0.7\|f_1 - 100\|$ $+ 0.3\|f_2 - 95\|) + 0.5(0.7\|f_1$ $- 10\| + 0.3\|f_2 - 20\|)$	$f_1 = x_1 + 12x_2$ $f_2 = 10x_1 + 2x_2$ f_1, f_2 unrestricted in sign	$x_1 + x_2 \leq 13$ $x_1 + 2x_2 \leq 22$ $x_1, x_2 \geq 0$	$2x_1 + x_2 \leq 19$ $3x_1 + x_2 \leq 27$

The solution to this optimization problem is as follows and shown in Figure 2.27.

	x_1	x_2	f_1	f_2
GS/NA-solution	4	9	112	58

See Example 2.27b.lg4.

2.9 PAIRED COMPARISON AND INTERACTIVE METHODS

2.9.1 Paired Comparison: Exhaustive Search

Paired comparison of actual alternatives is one of the easiest and most reliable ways of communicating with DMs. For paired comparisons of alternatives, the DM's response can be preferred, not preferred, or indifferent. The only assumption about the DM's preferential behavior is transitivity of preferences. That is, if F_j is preferred to F_p, and F_p is preferred to F_q, then F_j is preferred to F_q. No other assumption is made about the utility function; therefore this utility function is the most flexible (least restrictive) of all methods. If the number of alternatives is discrete and small, the exhaustive search can be used to find the best alternative.

In the exhaustive interactive search method, the DM must enumerate all efficient alternatives to find the best one. In this method, for n alternatives, $n-1$ paired comparisons are required. In each iteration, the best selected alternative is compared to the next alternative. The search continues until all alternatives are enumerated.

Example 2.28   Exhaustive Search Interactive Paired Comparison   Consider 10 alternatives $F_1 - F_{10}$ as presented in Table 2.33. The problem is presented in the standard

TABLE 2.33   Set of 10 Alternatives with Three Criteria

	F_1	F_2	F_3	F_4	F_5	F_6	F_7	F_8	F_9	F_{10}
Max f_1	60	55	52	50	48	45	43	40	37	35
Max f_2	51	49	43	39	32	44	50	56	62	67
Max f_3	34	37	44	48	55	65	53	47	40	35

TABLE 2.34 Exhaustive Search Results to Determine the Best Alternative in Table 2.33

Iteration	1	2	3	4	5	6	7	8	9
Compare	F_1–F_2	F_1–F_3	F_1–F_4	F_4–F_5	F_5–F_6	F_6–F_7	F_6–F_8	F_6–F_9	F_9–F_{10}
Response	F_1	F_1	F_4	F_5	F_6	F_6	F_6	F_9	F_9

MCDM form, but here it is not assumed that f_1 is the most important criterion. But alternatives are in descending order of f_1.

In the exhaustive search method, the DM is asked to make comparisons for pairs of alternatives, one at a time. The responses are listed Table 2.34, and the best alternative is F_9.

2.9.2 Paired Comparison: Basic Idea of Interactive Methods

If the set of alternatives is small and enumerable, then the above method can be directly used. However, in many real-world problems, the set of alternatives is either large, continuous, or combinatorial, and therefore interactive paired comparison methods can be used. The DM's preferential behavior can be represented by a utility function. In interactive methods, it is assumed that the complete assessment of utility functions can be difficult, time consuming, and sometimes impossible. In interactive methods, the utility function does not have to be completely assessed; only paired comparisons of alternatives are used to partially assess the utility function and find the best alternative.

In interactive methods, the best or optimal alternative is found with the minimum partial information needed about the DM's preferences. The convergence to the optimal alternative is guaranteed while the number of paired comparisons of alternatives is minimized. We classify the DM's preferential behavior to be independent (additive utility function), convergent (concave utility function), or divergent (convex utility function); see Section 2.3. In interactive methods, it is possible to identify the DM's preferential behavior based on the responses of the DM to a set of paired comparison questions. Then one of the three types of utility functions that best represents the DM's preferential behavior is used. While from the DM's point of view, interactive paired comparison methods are simple and flexible, from the analyst's (or from an algorithmic) point of view, interactive methods are complex and involved.

In interactive methods, the optimal solution is found with respect to partially known preferences, usually presented by a utility function. Section 2.2.4 shows that for different range of weights of an additive utility function, different alternatives can be optimal. For example, consider Figure 2.28, maximizing an additive utility function $U = w_1 f_1 + w_2 f_2$. Suppose alternative F_2 is preferred to F_1 and F_3. Therefore, the following two constraints can be formed: $U(F_2) - U(F_1) > 0$ and $U(F_2) - U(F_3) > 0$. By using paired comparison questions, the weights are implicitly and partially assessed, for example, $\{0.3 \leq w_1 \leq 0.5$ and $0.5 \leq w_2 \leq 0.7\}$; see Section 2.2.4. In this case, for all different utility functions ranging from $U = 0.3f_1 + 0.7f_2$ to $U = 0.5f_1 + 0.5f_2$, alternative F_2 is the optimal point. Therefore, the best alternative can be found without knowing the exact values of weights, w_1 and w_2.

The above concept can be extended for nonlinear utility functions (e.g., to ZUTs). That is, it is possible to find the best alternatives with respect to a partially assessed concave or

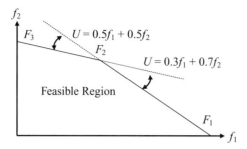

Figure 2.28 F_2 is optimal with respect to a family of additive utility functions.

convex utility function, that is, without completely assessing the nonlinear utility function. Convergence for bicriteria problems is shown in the following section.

2.9.3   Paired Comparison: Interactive Bicriteria Method[†]

Bicriteria problems are the most applicable and used MCDM problems in the real world. Many systems and operations problems in this book are also formulated as a bicriteria problem. In this section, we have developed an interactive method in which the DM responds to questions of paired comparisons of bicriteria alternatives until the best alternative is obtained.

Nonlinear Value and Utility Functions: Paired Comparison Exhaustive Search
In general any bicriteria problem with any nonlinear value and utility functions can be solved if all alternatives can be graphically presented in (f_1, f_2) Cartesian space. Then, the DM can be asked to choose the best alternative. That is, $U(v_1(f_1), v_2(f_2)) = U(f_1, f_2) =$ the DM's preferences over the set of alternatives; this problem can be solved directly by the DM choosing the best alternative. Therefore, in this case, there is no need to assess value or utility functions. For example, in Figure 2.29 all alternatives are presented graphically, and the DM can directly choose the best alternative.

Interactive Paired Comparison Search   A different approach for solving bicriteria problem is by presenting to the DM only a small subset of bicriteria alternatives in each interaction session. In this case, the following bicriteria interactive approach can be used to find the best alternative. It is assumed that the utility function is an increasing function of both objectives and it is quasi-concave (which is more general than concave function). In each interaction session (iteration) of the following interactive method, a total of five equally distributed alternatives are presented to the DM, and the DM is asked to select the best alternative. Using this information, we may eliminate up to half of the total remaining alternatives in each iteration (without presenting them to the DM). The algorithm is based on nonderivative single variable optimization (e.g., see bisectional or golden section search methods which are covered in optimization and NLP textbooks). In bicriteria interactive approach, alternatives are placed in ascending order of one of the two criteria, either the first or the second criterion. The method is as follows.

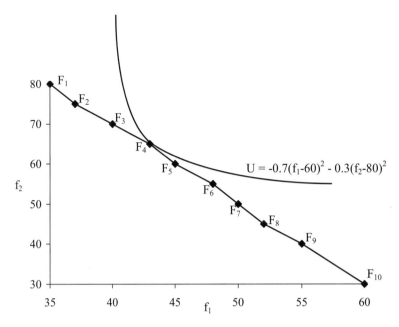

Figure 2.29 Graphical representation of Example 2.29.

Initial Step: Arrange alternatives in ascending order of the first objective (or the second objective).

Iterative Steps:

1. Re-label alternatives as F_p from p_{min} to p_{max}.
2. Select five alternatives that are equally distributed in terms of their index p from p_{min} to p_{max}. The indices of the five alternatives are 1, $(1 + (n + 1)/2)/2$, $(n + 1)/2$, $(n + (n + 1)/2)/2$, and n, where n is the remaining number of alternatives. (If the generated number is not an integer number, round it to a lower integer). Label these indices as p_{min}, $p_{mid,1}$, p_{mid}, $p_{mid,2}$, and p_{max}, respectively.
3. Ask the DM to choose the best of the five alternatives.
4. Eliminate alternatives for which the following "Rule of Elimination" applies.
5. Stop when the set of alternatives is five or less; choose the best alternative from this set of alternatives. Otherwise, go to Step 1.

Rules of Elimination of Bicriteria Alternatives When the DM ranks F_i higher than F_j, it means that $U(F_i) > U(F_j)$, this information is used to eliminate alternatives. In bisectional search, after the first iteration, in each iteration five alternatives are presented from which two are new alternatives.

Consider two alternatives, F_i and F_j. Suppose $i < j$ in terms of their index p.

Rule A: If F_i is preferred to F_j, then eliminate all alternatives on the right-hand side of j; that is, eliminate all F_p such that $p > j$, that is, for $p = j + 1, j + 2, \ldots, n$.

Rule B: If F_j is preferred to F_i, then eliminate all alternatives on the left-hand side of i; that is, eliminate all F_p such that $p < i$, that is, for $p = 1, 2, \ldots, i{-}1$.

Rule C: If F_i is equal to F_j, then consider alternative F_{i-1} versus F_j and apply rules A and B.

The proof of the above rules of elimination is based on single variables optimization (see single variable unconstrained optimization in NLP/optimization textbooks) and the quasi-concave utility function properties (see Malakooti, 1989b, c, 2010).

In this method, all alternatives are arranged in ascending order of one of the two objectives (e.g., the first objective). Since alternatives are ordered in ascending order of the first objective, they are also ordered in descending order of the second objective. Then, this problem can be presented as a one-dimensional search in terms of the indices p.

A Summary of the Elimination Rules: Choose the best alternative of the given five alternatives. Keep the two segments adjacent to it. Eliminate all alternatives on the other remaining segments.

If best alternative is:	p_{min}	$p_{mid,1}$	p_{mid}	$p_{mid,2}$	p_{max}
Alternatives to be eliminated	$p_{mid,1}-p_{max}$	$p_{mid}-p_{max}$	$p_{min}-p_{mid,1}$ and $p_{mid,2}-p_{max}$	$p_{min}-p_{mid}$	$p_{min}-p_{mid,2}$

Example 2.29   Interactive Method for Bicriteria Problem   Suppose there are 10 alternatives. First organize them in ascending order of f_1; see Table 2.35. In this example, for illustrative purposes, maximizing $U = -0.7(f_1-60)^2 - 0.3(f_2-8\ 0)^2$ was used to simulate a DM's responses for ranking alternatives. In reality, different DMs have different utility functions that will result in different rankings.

In this example, $p_{max} = n = 10$. The indices for five equally distributed points are as follows:

$$p_{max} = 10, p_{min} = 1, p_{mid} = (10+1)/2 = 5.5 \text{(round to lower integer, i.e., 5)},$$
$$p_{mid,1} = (1+5.5)/2 = 3.25 \text{ or 3, and } p_{mid,2} = (5+10)/2 = 7.5 \text{ or 7.}$$

Consider the following five alternatives associated with $p_{min}, p_{mid1}, p_{mid}, p_{mid2}$, and p_{max}, that is, F_1, F_3, F_5, F_7, and F_{10}. The DM is asked to rank the following five alternatives (presented in ascending order of f_1).

In the first iteration, using the above rules of elimination, the two segments adjacent to the best solution, F_5, are kept and the other two segments are eliminated. That is, F_1 and F_2 on the left of F_3 are eliminated. Also, F_8, F_9, and F_{10} on the right side of F_7 are eliminated. The remaining alternatives are from F_3 to F_7.

TABLE 2.35   Example 2.29 in Ascending Order of the First Objective

	F_1	F_2	F_3	F_4	F_5	F_6	F_7	F_8	F_9	F_{10}
Maximize f_1	35	37	40	43	45	48	50	52	55	60
Maximize f_2	80	75	70	65	60	55	50	45	40	30

In the second iteration, the five equally distributed alternatives are F_3, F_4, F_5, F_6, and F_7.

	First Iteration					Second Iteration				
	F_1	F_3	F_5	F_7	F_{10}	F_3	F_4	F_5	F_6	F_7
f_1	35	40	45	50	60	40	43	45	48	50
f_2	80	70	60	50	30	70	65	60	55	50
The DM's rank	4	2	1	3	5	4	1	2	3	5

Since there are no other alternatives to be considered, the best alternative is F_4. Figure 2.29 illustrates the set of alternatives and the utility function contour that passes through F_4.

2.9.4 Paired Comparison: Advanced Interactive Methods[†]

Interactive Bicriteria for LP (MOLP) Problems The method for solving bicriteria discrete problems (covered in Section 2.9.3) can be used to solve bicriteria LP (MOLP) problems. A simple approach to solve Bicriteria LP problems is to generate the efficient frontier (in the two-dimensional space) and then apply Section 2.9.3 method to find the best solution. See Sections 2.7.2 and 2.8.1 for generation of efficient alternatives.

Interactive Multicriteria for LP (MOLP) and Discrete Problems Several researchers (such as Geoffrion et al., 1972 (for concave); Zionts and Wallanious, 1976 (for linear); and Korhonen et al., 1984 (for quasi-concave)) have developed interactive methods for MOO problems; also see Steuer (1986). Malakooti (1988, 1989b, c) provided exact paired comparison methods for multiple objective discrete programming, LP, and NLP problems for quasi-concave utility functions. Quasi-convex utility functions were considered by Malakooti (1989c), Malakooti and Al-alwani (2002), and Malakooti (2010). (See Wallenius et al. (2008) and Zopounidis and Pardalos (2010) for a comprehensive list of references.

2.9.5 MCDM Validation and Extension to Clustering[†]

Verification of MCDM Methods Using Risk, Economics, and Regression How can we measure the accuracy, validity, and effectiveness of an MCDM method? For most parts, the literature lacks to respond to these questions, especially because of the subjectivity involved in MCDM. That is, it is difficult to objectively test MCDM methods by field studies of human behavior. For example, it is difficult to find a set of standard questions for a given MCDM problem that can be tested on different people so that some patterns of decision making for MCDM can be identified. Also, the existence of utility functions presenting DM's preferential behavior and its concavity (or convexity) need to be tested by field studies. In this chapter, we developed and advocated prescriptive (rational) models that are sufficiently flexible which can also be validated as descriptive models. This marriage of prescriptive and descriptive models may sound impossible to achieve, yet there appears to be hope for resolving these challenges by learning from objective experimental data already available for risk, consumer behavior, and economics problems. Malakooti (2014b) shows how to convert risk problems to a form that can be used for a preliminary testing of MCDM problems, for which MCDM ZUT models can simultaneously satisfy both descriptive and prescriptive models with high precision of accuracy. A different aspect of identifying decision-making patterns is the identification and verification of decision types; four dimensions (types) of decision making have been identified and verified by field studies (Malakooti, 2012). Also, a closely related approach for solving MCDM problem is multiple variable regression which can also be used for validating MCDM methods; see Chapter 3 (Forecasting) of this book. Also artificial neural networks can be used to solve MCDM problems, for example, see Malakooti and Zhou (1998, 2000).

Multicriteria Clustering (Multiperspective) A closely related problem to MCDM is clustering, where objects (labeled as alternatives in MCDM) are presented by a set of attributes (labeled as criteria in MCDM). Several approaches can be used to cluster alternatives based on their similarities. Then, each cluster can be analyzed based on its unique features. We call this multiperspective multicriteria clustering, where each cluster is viewed from a different perspective and its alternatives are evaluated differently. Clustering can significantly simplify complex decision problems, by allowing different MCDM methods to be applied to different clusters. This problem is also closely related to the selection of groups of alternatives. See Z-Theory Multi-Criteria Clustering (Section 8) in Chapter 12 (Clustering and Group Technology).

REFERENCES

Allais, M. "Le comportement de l'hommerationneldevant le risque: critique des postulatsetaxiomes de l'écoleAméricaine." *Econometrica*, Vol. 21, No. 4, 1953, pp. 503–546.

Arthur, J. L., and A. Ravindran. "A Partitioning Algorithm for Linear Goal Programming Problems." *ACM Transactions on Mathematical Software*, Vol. 6. No. 3, 1980, pp. 378–386.

Belton, V., and T. Gear. "On a Short-Coming of Saaty's Method of Analytic Hierarchies." *Omega*, Vol. 11, No. 3, 1983, pp. 228–230.

Belton, V., and T. J. Stewart. *Multiple Criteria Decision Analysis: An Integrated Approach*. Boston, MA: Kluwer Academic, 2002.

Bernoulli, D. "Exposition of a New Theory on the Measurement of Risk." *Econometrica*, Vol. 22, (1738 original) 1954, p. 23.

Bordley, R., and C. W. Kirkwood. Multiattribute Preference Analysis with Performance Targets. *Operations Research,* Vol. 52, 2004, pp. 823–835.

Buchanan, J. T., and J. L. Corner. "Capturing Decision Maker Preference: Experimental Comparison of Decision Analysis and MCDM Techniques." *European Journal of Operational Research*, Vol. 98, No. 1, 1997, pp. 85–97.

Chankong, V., and Y. Y. Haimes. *Multi-objective Decision Making: Theory and Methodology*. Minneola, NY: Dover, 2008.

Charnes, A., and W. W. Cooper. *Management Models and Industrial Applications of Linear Programming*. New York: Wiley, 1961.

Charnes, A., and W. W. Cooper. "Goal Programming and Multiple Objective Optimization—Part I." *European Journal of Operations Research*, Vol. 1, 1977, pp. 39–54.

Cobb, C. W., and P. H. Douglas. "A Theory of Production." *American Economic Review*, Vol. 18, 1928, pp. 139–165.

Dyer, J. S. "Remarks on the Analytic Hierarchy Process." *Management Science*, Vol. 36, No. 3, 1990a, pp. 249–258.

Dyer, J. S. "A Clarification of Remarks on the Analytic Hierarchy Process." *Management Science*, Vol. 36, No. 3, 1990b, pp. 274–275.

Dyer, J. "MAUT-Multiattribute Utility Theory, Multiple Criteria Decision Analysis: State of the Art Surveys." *International Series in Operations Research & Management Science*, Vol. 78, 2005, pp. 265–292.

Edwards, W., R. Miles, and D. von Winterfeldt, eds. *Advances in Decision Analysis*. New York: Cambridge University Press, 2007.

Gal, T., T. J. Stewart, and T. Hanne. *Multicriteria Decision Making and Advances in MCDM Models, Algorithms, Theory, and Applications*. Boston, MA: Academic, 1999.

Geoffrion, A. M., J. S. Dyer, and A. Feinberg. "An Interactive Approach for Multi-criterion Optimization, with an Application to the Operation of an Academic Department." *Management Science*, Vol. 19, No. 4, 1972, pp. 357–368.

Ingersoll, J. "Non-monotonicity of the Tversky Kahneman Probability-Weighting Function: A Cautionary Note." *European Financial Management*, Vol. 14, No. 3, 2008, pp. 385–390.

Kahneman, D., and A. Tversky. "Prospect Theory: An Analysis of Decision Making Under Risk." *Econometrica*, Vol. 47, No. 2, 1979, pp. 263–292.

Keeney, R. L. "Multiplicative Utility Functions." *Operations Research*, Vol. 22, No. 1, 1974, pp. 22–34.

Keeney, R. L., and H. Raiffa. *Decisions with Multiple Objectives: Preferences and Value Tradeoffs*. New York: John Wiley & Sons, 1976, and New York: Cambridge University Press, 1993.

Kirkwood, C. W., and R. K. Sarin. "Ranking with Partial Information: A Method and an Application." *Operations Research*, Vol. 33, No. 1, 1985, pp. 38–48.

Korhonen, P., J. Wallenius, and S. Zionts. "Solving the Discrete Multiple Criteria Problem Using Convex Cones." *Management Science*, INFORMS, Vol. 30, No. 11, 1984, pp. 1336–1345.

Lagrèze, J., and J. Siskos. "Assessing a Set of Additive Utility Functions for Multicriteria Decision-Making, the UTA Method." *European Journal of Operational Research*, Vol. 10, No. 2, 1982, pp. 151–164.

Levenberg, K. "A Method for the Solution of Certain Non-linear Problems in Least Squares." *The Quarterly of Applied Mathematics*, Vol. 2, 1944, pp. 164–168.

Malakooti, B. "Assessment Through Strength of Preference." *Large Scale Systems: Theory and Applications*, Vol. 8, No. 2, 1985, pp. 169–182.

Malakooti, B. "An Exact Interactive Paired Comparison Method for Exploring the Efficient Facets of MOLP Problems with Underlying Quasi-Concave Utility Functions." *IEEE Transactions on Systems, Man, and Cybernetics*, Vol. 18, No. 5, 1988, pp. 787–801.

Malakooti, B. "Identifying Nondominated Alternatives with Partial Information for Multiple Objective Discrete and Linear Programming Problems." *IEEE Transactions on Systems, Man, and Cybernetics*, Vol. 19, No. 1, 1989a, pp. 95–107.

Malakooti, B. "Theories and an Exact Interactive Paired Comparison Approach for Discrete Multiple Criteria Problems." *IEEE Transactions on Systems, Man, and Cybernetics*, Vol. 19, No. 2, 1989b, pp. 365–378.

Malakooti, B. "Ranking Multiple Criteria Alternatives with Half-Space, Convex, and Non-convex Dominating Cones." *Computers and Operations Research: An International Journal*, Vol. 16, No. 2, 1989c, pp. 117–127.

Malakooti, B. "Ranking and Screening Multiple Criteria Alternatives with Partial Information and Use of Ordinal and Cardinal Strength of Preferences." *IEEE Transactions on Systems, Man, and Cybernetics Part A*, Vol. 30, No. 3, 2000, pp. 355–369.

Malakooti, B. "Independent, Convergent, and Divergent Decision Behaviour for Interactive Multiple Objectives Linear Programming." *Engineering Optimization*, Vol. 42, No. 4, 2010, pp. 325–346.

Malakooti, B. "Systematic Decision Process for Intelligent Decision Making." *Journal of Intelligent Manufacturing*, Vol. 22, No. 4, 2011, pp. 627–642.

Malakooti, B. "Decision Making Process: Typology, Intelligence, and Optimization." *Journal of Intelligent Manufacturing*, Vol. 23, No. 3, 2012, pp. 733–746.

Malakooti, B. "Double Helix Value Functions, Ordinal/Cardinal Approach, Additive Utility Functions, Multiple Criteria, Decision Paradigm, Process, and Types." *International Journal of Information Technology & Decision Making (IJITDM)*, Forthcoming, 2014a.

Malakooti, B. "Z Utility Theory: Decisions Under Risk and Multiple Criteria" to appear, 2014b.

Malakooti, B., and J. E. Al-alwani. "Extremist vs. Centrist Decision Behavior; Quasi-convex Utility Functions for Solving Interactive Multi-objective Linear Programming." *Computers & Operations Research*, Vol. 29, 2002, pp. 2003–2021.

Malakooti, B., and Y. Zhou. "An Adaptive Feedforward Artificial Neural Network with Application to Multiple Criteria Decision Making." *Management Science*, Vol. 40, No. 11, 1994, pp. 1542–1561.

Malakooti, B., and Y. Zhou. "Approximating Polynomial Functions by Feedforward Artificial Neural Networks: Capacity, Analysis, and Design." *Applied Mathematics and Computation*, Vol. 90, 1998, pp. 27–52.

Markowitz, H. M. *Portfolio Selection: Efficient Diversification of Investments*. New York: John Wiley & Sons, 1959. (Reprinted by London: Yale University Press, 1970; 2nd ed. Oxford, UK: Basil Blackwell, 1991.)

Marquardt, D. "An Algorithm for Least-Squares Estimation of Nonlinear Parameters. *SIAM Journal on Applied Mathematics*, Vol. 11, No. 2, 1963, pp. 431–441.

Masud, A. S. M., and A. Ravindran. "Multiple Criteria Decision Making." In *Operation Research Methodologies*, A. R. Ravindran, ed. Boca Raton, FL: CRC Press, Chapter 5, 2009.

Muther, R. *Systematic Layout Planning*, 2nd edn. Boston: Cahners, 1973.

Peters, M., and S. Zelewski. "Pitfalls in the Application of Analytical Hierarchy Process to Performance Measurement." *Management Decision*, Vol. 46, No. 7, 2008, pp.1039–1051.

Ravindran, A. R., and D. P. Warsing, Jr. *Supply Chain Engineering: Models and Applications*. Boca Raton, FL: CRC Press, 2013.

Saaty, T. *The Analytic Hierarchy Process: Planning, Priority Setting, Resource Allocation*. London: McGraw-Hill International, 1980.

Saaty, T. "An Exposition of the AHP in Reply to the Paper: Remarks on the Analytic Hierarchy Process." *Management Science*, Vol. 36, No. 3, 1990, pp. 259–268.

Saaty, T. "Decision Making with the Analytic Hierarchy Process." *International Journal of Services Sciences*, Vol. 1, No. 1, 2008, pp. 83–98.

Soleimani-Damaneh, M. "On Some Multi-objective Optimization Problems Arising in Biology." *International Journal of Computer Mathematics*, Vol. 88, No. 6, 2011, pp. 1103–1119.

Steuer, R. E. *Multiple Criteria Optimization: Theory, Computation, and Application*. New York: Wiley, 1986.

Stewart, T. J. "Evaluation and Refinement of Aspiration-Based Methods in MCDM." *European Journal of Operational Research*, Vol. 113, No. 3, 1999, pp. 643–652.

Tabucanon, M. T. *Multiple Criteria Decision Making in Industry*. New York: Elsevier, 1988.

Triantaphyllou, E. *Multi-criteria Decision Making Methods: A Comparative Study*. Boston, MA: Kluwer Academic, 2000.

Tsetlin, I., and R. Winkle. "Decision Making with Multiattribute Performance Targets: The Impacts of Changes in Performance and Target Distributions." *Operations Research*, Vol. 55, 2007, pp. 226–233.

Tversky, A., and D. Kahneman. "Advances in Prospect Theory: Cumulative Representation of Uncertainty." *Journal of Risk and Uncertainty*, Vol. 5, No. 4, 1992, pp. 297–323.

Velazquez, M. A., D. Claudio, and A. R. Ravindran. "Experiments in Multiple Criteria Selection Problems with Multiple Decision Makers." *International Journal of Operational Research*, Vol. 7, No. 4, 2010, pp. 413–428.

von Neumann, J., and O. Morgenstern. *Theory of Gaines and Economic Behaviour*. Princeton, NJ: Princeton University Press, 1944; 2nd edn. 1947.

Wallenius, J., J. S. Dyer, P. C. Fishburn, R. E. Steuer, S. Zionts, and K. Deb. "Multiple Criteria Decision Making, Multiattribute Utility Theory: Recent Accomplishments and What Lies Ahead." *Management Science*, Vol. 54, No. 7, 2008, pp. 1336–1349.

Yoon, K., and C. L. Hwang. *Multiple Attribute Decision Making: An Introduction*. Thousand Oaks, CA: Sage, 1995.

Zeleny, M. *Multiple Criteria Decision Making*. New York: McGraw-Hill, 1982.

Zionts, S., and J. Wallenius. "An Interactive Programming Method for Solving the Multiple Criteria Problem." *Management Science*, Vol. 22, No. 6, 1976, pp.652–663.

Zopounidis, C., and P. M. Pardalos. *Handbook of Multicriteria Analysis*, 1st edn. New York: Springer, 2010.

EXERCISES

Introduction (Section 2.1)

1.1 List, describe, and characterize three important applications of MCDM. Discuss the advantages and disadvantages of considering multiple criteria versus using a single criterion.

1.2 For each of the three problems stated in exercise 1.1, apply the guidelines presented in Table 2.1 to formulate and analyze the problem. At least show three possible alternatives.

1.3 Consider an important decision problem that you are facing now (or that you faced in the past). Apply the guidelines presented in Table 2.1 to formulate and analyze the problem. At least show three possible alternatives.

Efficiency and Its Extensions (Section 2.2)

2.1 Suppose that you are planning to get a job. Two criteria are used for evaluating alternatives: f_1 is salary in $10,000 and f_2 is overall job satisfaction (including location and all other criteria except salary) which ranges from 0 to 10.

	F_1	F_2	F_3	F_4	F_5
Max f_1	4	6	9	10	12
Max f_2	10	8	9	7	5

(a) Identify inefficient and efficient alternatives.

(b) Identify convex efficient and inefficient alternatives. (Show the solution graphically.)

(c) Identify the weight space associated with these alternatives.

2.2 Suppose that you are in charge of purchasing a piece of production equipment. Two criteria are used for evaluating alternatives: f_1 is production capacity in 100,000 units of production per period (4 weeks) and f_2 is the cost of the equipment in $10,000. Respond to parts (a)–(c) of exercise 2.1.

	F_1	F_2	F_3	F_4	F_5
Max f_1	33	19	18	15	14
Min f_2	52	40	39	31	25

2.3 Suppose that you are planning to continue your education and get a higher degree. Three criteria are used for evaluating alternatives: f_1 is the potential salary that you will have after graduating in $10,000; f_2 is the overall quality of education of such a school which ranges from 0 to 100; and f_3 is your quality of life (including location and all other criteria except salary and quality of education) while you are studying in the given school. It ranges from 0 to 31. Identify inefficient and efficient alternatives.

(a) Identify inefficient and efficient alternatives.

	F_1	F_2	F_3	F_4	F_5	F_6	F_7	F_8	F_9	F_{10}
Max f_1	9	7	13	5	20	16	15	19	15	19
Max f_2	70	84	65	90	95	92	73	100	85	95
Max f_3	30	15	26	15	10	19	20	18	31	18

2.4 Suppose that you are in charge of selecting a location for construction of a new plant for your company. Four criteria are used for evaluating alternatives: f_1 is tax credit benefits in million dollars per year, f_2 is proximity to major highways and better transportation ranging from 0 to 1000, f_3 is availability of skilled labor, and f_4 is community support.

(a) Identify inefficient and efficient alternatives.

	F_1	F_2	F_3	F_4	F_5	F_6	F_7	F_8	F_9	F_{10}
Max f_1	5	7	10	11	13	4	12	18	15	9
Max f_2	900	700	800	700	400	1000	950	750	550	650
Max f_3	5	-20	-46	-25	15	0	-15	-10	-20	-8
Max f_4	-5	-15	-10	-40	2	20	40	-20	-10	-6

Utility Functions (Sections 2.3.2–2.3.3)

3.1 Consider the following utility functions:

(i) $U = f_1 + 2f_2 + 3f_3$

(ii) $U = (f_1 - 5)^2 + (f_2 + 2)^2 - 2(f_3 - 4)^2$

For each of the above utility functions determine:

(a) If this function is additive or nonadditive.

(b) If this function is linear or nonlinear.

(c) If it is an increasing function of the objectives.

3.2 Consider the following utility functions:

(i) $U = (f_1 - 5)^2 + (f_2 + 2)^2 - 2(f_3 - 4)^2 - 7f_1f_2f_3$

(ii) $U = LV - 0.5DV$

where $LV = 0.3f_1 + 0.2f_2 + 0.5f_3$ and

$DV = \sqrt{0.3(f_1 - LV)^2 + 0.2(f_2 - LV)^2 + 0.5(f_3 - LV)^2}$ Respond to parts (a)–(c) of exercise 3.1.

Value Functions (Section 2.3.5)[†]

3.3 Consider exercise 2.1. Assess your value functions of the two objectives. Then assess your weights of importance for an additive utility function and rank alternatives.

3.4 Consider exercise 2.3. Assess your value functions of the three objectives. Then assess your weights of importance for an additive utility function and rank alternatives.

Normalized Criteria and Weights (Section 2.3.7)

3.5 Consider exercise 2.1. For an additive utility function, weights associated with normalized criteria values are (0.7,0.3).

(a) Rank alternatives using normalized criteria.

(b) Find equivalent normalized weights for this problem and rank alternatives based on them.

3.6 Consider exercise 2.2. For an additive utility function, weights associated with normalized criteria values are (0.2,0.8). Respond to parts (a) and (b) of exercise 3.5.

3.7 Consider exercise 2.3. For an additive utility function, weights associated with normalized criteria values are (0.3,0.2,0.5). Respond to parts (a) and (b) of exercise 3.5.

3.8 Consider exercise 2.4. For an additive utility function, weights associated with normalized criteria values are (0.25,0.20,0.35,0.20). Respond to parts (a) and (b) of exercise 3.5.

OCA for Additive Utility (Sections 2.4.1–2.4.4)

4.1 Consider exercise 2.1. The following table shows the responses of a DM for this problem. Hint: You should normalize criteria values of exercise 2.1.

R_{ij}	w_1	w_2	w_0
w_1	0	4	10
w_2	-4	0	6
w_0	-10	-6	0

(a) Assess weights using OCA.

(b) Verify ordinal and cardinal consistency. How much is CIR? What does it mean?

(c) Verify external consistency (you can act as the DM).

(d) Apply OCA to rank the first five alternatives.

4.2 Consider exercise 2.3. The following table shows the responses of a DM for this problem. Hint: You should normalize criteria values of exercise 2.3. Respond to parts (a)–(d) of exercise 4.1.

R_{ij}	w_3	w_1	w_2	w_0
w_3	0	3	3	9
w_1	−3	0	3	6
w_2	−3	−3	0	3
w_0	−9	−6	−3	0

4.3 Consider exercise 2.3. The following table shows the responses of a DM for this problem. Hint: You should normalize criteria values of exercise 2.3. Respond to parts (a)–(d) of exercise 4.1.

R_{ij}	w_1	w_3	w_2	w_0
w_1	0	2	4	8
w_3	−2	0	2	6
w_2	−4	−2	0	2
w_0	−8	−6	−2	0

4.4 Consider exercise 2.4. The following table shows the responses of a DM for this problem. Hint: You should normalize criteria values of exercise 2.4. Respond to parts (a)–(d) of exercise 4.1.

R_{ij}	w_3	w_1	w_2	w_4	w_0
w_3	0	3	5	7	8
w_1	−4	0	1	3	4
w_2	−5	−1	0	3	4
w_4	−7	−3	−2	0	1
w_0	−8	−4	−4	−1	0

4.5 Consider an important decision problem that you are facing now or you faced in the past—this could be personal, professional, business, financial, health related, and so on.

(a) Define your criteria (consider at least three criteria).

(b) List at least three alternatives.

(c) Assess weights by OCA.

(d) Verify ordinal and cardinal consistency of the assessed weights. If your inconsistency is significant, try to modify your responses.

(e) Assess criteria values for each alternative.

(f) Use your assessed additive utility function to rank alternatives. Verify the ranking is consistent with your preferences.

(g) Comment on how accurate OCA is reflecting your actual preferences.

4.6 Consider a social, environmental, or governmental-related problem (e.g., choosing a site for a nuclear power plant). Act as the DM. Respond to parts (a)–(g) of exercise 4.5.

Analytic Hierarchy Process (Sections 2.4.5 and 2.4.6)[†]

4.7 Consider exercise 2.3 (with three criteria). AHP ratio responses are presented in the following table.

(a) Use AHP method to assess weights.

(b) Are these responses consistent?

(c) Rank the first five alternatives of exercise 2.3.

P_{ij}	w_1	w_2	w_3
w_1	1	0.5	1
w_2	2	1	1
w_3	2	1	1

4.8 Consider exercise 2.4 (with four criteria). AHP ratio responses are presented in the following table.

(a) Use AHP method to assess weights.

(b) Are these responses consistent? What is the consistency ratio?

(c) Rank the first five alternatives of exercise 2.4.

P_{ij}	w_1	w_2	w_3	w_4
w_1	1	1	0.33	0.33
w_2	1	1	0.2	0.5
w_3	2	5	1	0.14
w_4	3	2	7	1

Optimization for Assessing Weights and Value Functions (Section 2.4.7)[†]

4.9 Consider the following set of alternatives for the bicriteria problem (of job selection using $f_1 =$ salary and $f_2 =$ job satisfaction). The DM's utility values of alternatives based on 0–1 ratings are provided in the following table.

	F_1	F_2	F_3	F_4	F_5	F_6
f_1	1.00	0.47	0.11	0.00	0.19	0.36
f_2	0.00	0.23	0.86	1.00	0.53	0.42

(a) Assess the utility function and measure its accuracy (without using value function).

(b) Assess the utility function and measure its accuracy by using one value function.

f_{ij}	F_1	F_2	F_3	F_4	F_5	F_6
DM's U_j^*	0.45	0.34	0.30	0.25	0.28	0.38

4.10 Consider the following set of alternatives for the tricriteria problem (of getting a higher educational degree using f_1 = potential salary, f_2 = overall quality of education, and f_3 = quality of life). The DM's utility values of alternatives based on 0–1 ratings are provided in the following table.

	F_1	F_2	F_3	F_4	F_5	F_6	F_7
f_1	0.35	0.53	0.67	1.00	0.27	0.00	0.15
f_2	1.00	0.00	0.43	0.86	0.14	0.71	0.50
f_3	0.40	0.80	0.50	0.00	1.00	0.65	1.00

(a) Assess the utility function and measure its accuracy (without using value function).

(b) Assess the utility function and measure its accuracy by using one value function.

	F_1	F_2	F_3	F_4	F_5	F_6	F_7
DM's U_j^*	0.5	0.32	0.45	0.55	0.35	0.41	0.47

4.11 Consider the following set of alternatives for the four-criteria problem (of choosing a new plant location, where f_1 = tax credit benefits, f_2 = proximity to public transportation, f_3 = availability of skilled labor, and f_4 = community support). The DM's utility values of alternatives based on 0–1 ratings are provided in the following table.

	F_1	F_2	F_3	F_4	F_5	F_6	F_7	F_8
f_1	0.00	0.27	0.06	0.77	0.33	1.00	0.31	0.56
f_2	1.00	0.47	0.93	0.13	0.43	0.00	0.44	0.26
f_3	0.00	0.36	0.03	1.00	0.48	0.93	0.41	0.74
f_4	0.81	0.75	1.00	0.14	0.58	0.00	0.67	0.39

(a) Assess the utility function and measure its accuracy (without using value function).

(b) Assess the utility function and measure its accuracy by using one value function.

	F_1	F_2	F_3	F_4	F_5	F_6	F_7	F_8
DM's U_j^*	0.5	0.48	0.51	0.65	0.46	0.65	0.48	0.56

Multiplicative ZUT (Section 2.5.1)

5.1 Consider exercise 4.9 alternatives. Suppose that the weights of importance are 0.6 and 0.4 for objective numbers 1 and 2, respectively. Show the M-ZUTs of each of the following cases and rank alternatives. Rank alternatives by maximizing MV_M. Identify the type utility function and z_M value.

(a) $z_M = -0.8$.

(b) $z_M = -0.3$.

5.2 Consider exercise 4.10 alternatives. Suppose that the weights of importance are 0.3, 0.4, and 0.3 for the three objectives, respectively. Show the M-ZUTs of each of the following cases and rank alternatives. Rank alternatives by maximizing MV_M. Identify the type utility function and z_M value.

(a) $z_M = -0.8$.

(b) $z_M = -0.3$.

5.3 Consider exercise 4.11 alternatives. Suppose that the weights of importance are 0.4, 0.1, 0.2, and 0.3 for the four objectives, respectively. Show the M-ZUTs of each of the following cases and rank alternatives. Rank alternatives by maximizing MV_M. Identify the type utility function and z_M value.

(a) $z_M = -0.8$.

(b) $z_M = -0.3$.

Assessment of M-ZUT (Section 2.5.2)[†]

5.4 Consider a bicriteria problem. The assessed utility values U^* (by the DM) are shown in the following table. Assess and verify multiplicative utility function.

	Alternatives for Assessment		Alternatives for Testing	
f_1	1	0	0.67	0.33
f_2	0	1	0.33	0.67
DM's U_j^*	0.195	0.255	0.44	0.50

5.5 Consider a four-criteria problem (use alternatives presented in Example 2.16, Table 2.22). The assessed utility values U^* (by the DM) are shown in the following table. Assess and verify multiplicative utility function.

	Alternatives for Assessment				Alternatives for Testing			
	I_1	I_2	I_3	I_4	N_1	N_2	N_3	N_4
DM's U_j^*	0.12	0.2	0.07	0.27	0.25	0.25	0.27	0.25

Assessment of M-ZUT by Nonlinear Equations (Section 2.5.3)[†]

5.6 Consider exercise 4.9. Solve it using nonlinear equations for M-ZUT.

5.7 Consider exercise 4.10. Solve it using nonlinear equations for M-ZUT.

5.8 Consider exercise 4.11. Solve it using nonlinear equations for M-ZUT.

Goal-Seeking ZUT (Sections 2.6.1 and 2.6.5)

6.1 Consider exercise 4.9. Suppose that the weights of importance for objectives 1 and 2 are (0.7,0.3), respectively. The goal is (0.9,0.15) and the nadir point is (0.1,0.15). Rank alternatives and identify the best alternative using the following cases:
(a) Goal-convergent case where z_G is –0.4.
(b) Nadir-divergent case where z_N is + 0.3.
(c) Consider both above goal and nadir points.

6.2 Consider exercise 4.10. Suppose that the weights of importance for objectives 1, 2, and 3 are (0.5,0.2,0.3), respectively. The goal is (0.92,0.75,0.77) and the nadir point is (0.2,0.12,0.18). Rank alternatives and identify the best alternative using the following cases:
(a) Goal-convergent case where z_G is –0.5.
(b) Nadir-divergent case where z_N is + 0.4.
(c) Consider both above goal and nadir points.

6.3 Consider exercise 4.11. Suppose that the weights of importance for the four objectives are (0.4,0.3,0.2,0.1), respectively. The goal is (0.86,0.64,0.65,0.58) and the nadir point is (0.11,0.05,0.08,0.13). Rank alternatives and identify the best alternative using the following cases:
(a) Goal-convergent case where z_G is –0.5.
(b) Nadir-divergent case where z_N is + 0.2.
(c) Consider both above goal and nadir points.

Multiple Objective Optimization (Sections 2.7.1–2.7.3)[†]

7.1 Consider the following MOLP problem.

Objectives	Subject to: Objective Equations	Constraints	
Maximize f_1	$f_1 = 5x_1 - x_2$	$x_1 + 3x_2 \le 6$	$x_1 < 1$
Maximize f_2	$f_2 = -x_1 + 3\,x_2$	$4\,x_1 + x_2 \le 5$	$x_2 \le 1.9$
	f_1, f_2 unrestricted in sign		$x_1, x_2 \ge 0$

(a) Find the maximum and minimum for each objective and generate a payoff matrix.

(b) Generate different efficient extreme points by using five different sets of weights.

(c) Present the problem in objective space graphically.

7.2 Consider the following problem. Respond to parts (a)–(c) of exercise 7.1.

Objectives	Subject to: Objective Equations	Constraints	
Maximize f_1	$f_1 = -2x_1 + 3x_2$	$5x_1 + 2x_2 \le 13$	$x_1 + x_2 \le 5$
Maximize f_2	$f_2 = 4x_1 - x_2$	$6x_1 + x_2 \le 13$	$x_2 \le 4.5$
	f_1, f_2 unrestricted in sign		$x_1, x_2 \ge 0$

7.3 Consider the following problem. Respond to parts (b) and (c) of exercise 7.1.

Objectives	Subject to: Objective Equations	Constraints	
Maximize f_1	$f_1 = 3x_1 + 2x_2 - 4x_3$ $-2\,x_4$	$2x_1 + 3x_2 + 4x_3$ $+ x_4 \le 35$	$x_1 + 3x_3 \le 12$ $3x_2 + 2x_4 \le 20$
Maximize f_2	$f_2 = -2x_1 - x_2 + 3x_3$ $+ 4x_4$	$x_1 + 4x_2 + 3x_3$ $+ 6x_4 \le 29$	$x_1, x_2, x_3, x_4 \ge 0$
	f_1, f_2 unrestricted in sign	$4x_3 - x_4 \le 18$ $x_1 + 4x_2 \le 15$	

7.4 Consider the following problem. Respond to parts (a) and (b) of exercise 7.1.

Objectives	Subject to: Objective Equations	Constraints	
Maximize f_1	$f_1 = 3x_1 + 2x_2$	$2x_1 + x_2 \le 6$	$x_1 \le 2$
Maximize f_2	$f_2 = x_1 + 4x_2$	$4x_1 + x_2 \le 9$	$x_1 + 3x_2 \le 15$
Maximize f_3	$f_3 = x_1 - x_2$		$x_1, x_2 \ge 0$
	f_1, f_2, f_3 unrestricted in sign		

7.5 Consider the following problem. Respond to parts (a) and (b) of exercise 7.1.

Objectives	Subject to: Objective Equations	Constraints	
Maximize f_1	$f_1 = 2x_1 + 3x_2 - x_3 + 7x_4$	$2x_1 - 3x_2$	$2x_2 + 3x_3 + 2x_4$
Maximize f_2	$f_2 = x_1 - x_2 + 4x_3 + 2x_4$	$+ 4x_4 \leq 30$	≤ 10
Maximize f_3	$f_3 = 3x_1 + x_2 - 5x_3 - 3x_4$	$-x_1 + 2x_3$	$x_1 + x_2 + x_3 \geq 6$
Maximize f_4	$f_4 = -6x_1 + 3x_2 - 3x_3 - 4x_4$	$+ 5x_4 \leq 20$	$x_1, x_2, x_3, x_4 \geq 0$
	f_1, f_2, f_3, f_4 unrestricted in sign		

Goal-Seeking MOO (Z-GP) (Sections 2.8.1–2.8.3)[†]

8.1 Consider exercise 7.1, where weights are $W = (0.4, 0.6)$, goals are $F_G = (3,4)$, and nadir point is $F_N = (0.5, 0.4)$. Suppose that z_G is –0.4 and z_N is + 0.3.

(a) Solve MOO problem using goal-seeking method. Is the solution efficient?

(b) Solve MOO problem using nadir-avoiding method. Is the solution efficient?

(c) Consider both above goal and nadir. Solve the MOO problem.

(d) Show all the above solutions graphically in objective space f_1 and f_2.

8.2 Consider exercise 7.2, where weights are $W = (0.25, 0.75)$, goals are $F_G = (3,3)$, and nadir point is $F_N = (0.3, 1)$. Suppose that z_G is –0.3 and z_N is + 0.6. Respond to parts (a)–(d) of exercise 8.1.

8.3 Consider exercise 7.3, where weights are $W = (0.32, 0.68)$, goals are $F_G = (12,8)$, and nadir point is $F_N = (2, 2.5)$. Suppose that z_G is –0.6 and z_N is + 0.3. Respond to parts (a)–(d) of exercise 8.1.

8.4 Consider exercise 7.4, where weights are $W = (0.3, 0.3, 0.4)$, goals are $F_G = (10, 15, 8)$, and nadir point is $F_N = (1.5, 2, 0)$. Suppose that z_G is –0.6 and z_N is + 0.3. Respond to parts (a)–(c) of exercise 8.1.

8.5 Consider exercise 7.5, where weights are $W = (0.25, 0.25, 0.25, 0.25)$, goals are $F_G = (21, 32, 15, 25)$, and nadir point is $F_N = (3, 2, 1.5, 4)$. Suppose that z_G is –0.7 and z_N is + 0.25. Respond to parts (a)–(c) of exercise 8.1.

Paired Comparison: Exhaustive Search (Section 2.9.1)

9.1 Consider 10 alternatives F_1–F_{10} as presented below.

	F_1	F_2	F_3	F_4	F_5	F_6	F_7	F_8	F_9	F_{10}
Maximize f_1	35	37	40	43	45	48	50	52	55	60
Maximize f_2	67	62	56	50	44	32	39	43	49	51

Perform the exhaustive search method. Respond to the paired comparison questions using the following utility functions (to be maximized) for presenting different DMs.

(a) The utility function is an additive utility $U = 0.8f_1 + 0.2f_2$.

(b) The utility function is $U(f_1, f_2) = 0.5f_1^2 + 0.6f_2 - 0.3f_1f_2$.

(c) The utility function is $U(f_1, f_2) = f_1^2 + 0.3f_2^2 - 0.7\ f_1^{0.5}f_2^{0.5}$.

9.2 Consider the following 12 alternatives.

Alternative	F_1	F_2	F_3	F_4	F_5	F_6	F_7	F_8	F_9	F_{10}	F_{11}	F_{12}
Max f_1	19	21	24	27	29	31	33	36	38	41	44	45
Max f_2	59	56	51	46	42	40	37	35	32	27	20	18

Perform the exhaustive search method. Respond to the paired comparison questions using the following utility functions (to be maximized) for presenting different DMs.

(a) The utility function is an additive utility $U = 0.3f_1 + 0.7f_2$.

(b) The utility function is $U(f_1, f_2) = 7f_1^2 + 2f_2^2 - 0.8f_1f_2$.

(c) The utility function is $U(f_1, f_2) = 10f_1^2 + 20f_2^2 - 0.4\ f_1^{0.5}f_2^{0.5}$.

Paired Comparison: Interactive Methods (Section 2.9.2)

9.3 Consider the following five alternatives, where an additive utility function is used.

	F_1	F_2	F_3	F_4	F_5
Maximize f_1	35	37	40	50	55
Maximize f_2	80	60	20	15	10

(a) Show the range of weight values of additive utility function for which each of the above five alternatives are optimal. (Hint: Use Section 2.2.4 method)

Paired Comparison: Interactive Bicriteria Method (Section 2.9.3)[†]

9.4 Consider exercise 9.1. Solve this problem by the interactive method. Do maximum of three iterations.

(a) Respond to the paired comparison questions using $U(f_1, f_2) = 0.5f_1^2 + 0.6f_2 - 0.3f_1f_2$. Is the obtained final solution optimal for the given utility function. Why?

(b) Respond to the paired comparison questions using $f_1^2 + 0.3f_2^2 - 0.7\ f_1^{0.5}f_2^{0.5}$. Is the final solution optimal for the given utility function. Why?

9.5 Consider exercise 9.2. Solve this problem by the interactive method. Do maximum of three iterations.

(a) Respond to the paired comparison questions using $U(f_1, f_2) = 7f_1^2 + 2f_2^2 - 0.8f_1f_2$. Is the obtained final solution optimal for the given utility function. Why?

(b) Respond to the paired comparison questions using $U(f_1, f_2) = 10f_1^2 + 20f_2^2 - 0.4\ f_1^{0.5}f_2^{0.5}$. Is the obtained final solution optimal for the given utility function. Why?

CHAPTER 3

FORECASTING

3.1 INTRODUCTION

Forecasting approaches are used to predict future values of certain variables such as market demands, prices, supplies, weather temperatures, economic factors, social factors, medical-related factors, population, productivity, quality, and reliability of systems. While it is almost impossible to determine the exact future values of a given variable, it is possible to estimate its value with some level of precision using a proper forecasting method.

There are two types of forecasting methods:

- *Time Series Methods*   In time series methods, the value of the variable being predicted depends on the past values of the variable as a function of time in chronological order. For example, the average sale price of houses can be predicted based on the average monthly sale price of houses in the past 36 in chronological order.
- *Causal Methods*   In causal methods, the variable being predicted does not necessarily depend on time; it can be predicted based on the values of some other independent variables. For example, the average sale price of houses can be predicted based on the inflation rate. If the inflation rate is high, house prices will be high; and if it is low, house prices will be low. In this case, house prices depend not on the chronological order of time but on the inflation rate.

Both time series and causal methods will be covered in this chapter. Time series methods are more applicable to operation planning problems for demand forecasting while causal

Note: Advanced material that can be omitted without loss of generality will be indicated by a dagger.

Operations and Production Systems with Multiple Objectives, First Edition. Behnam Malakooti.
© 2014 John Wiley & Sons, Inc. Published 2014 by John Wiley & Sons, Inc.

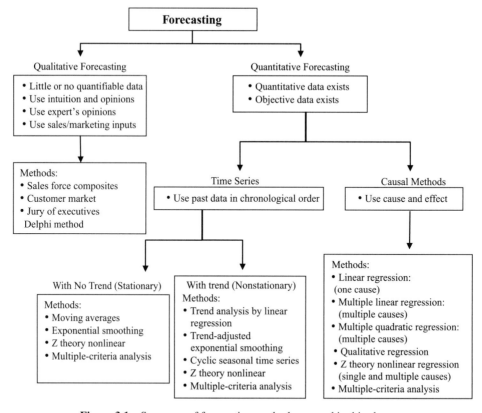

Figure 3.1   Summary of forecasting methods covered in this chapter.

methods can be used for a variety of applications, including scientific problems. For example, causal methods can be used to measure the effects of medications on patients or to predict global warming as a function of CO_2 emissions and other variables. For causal methods, linear and quadratic regression, autoregressive moving-average (ARMA), and autoregressive integrated moving-averages (ARIMA) models are used. In this chapter, we also develop a new nonlinear approach for both times series and regression methods based on Z theory.

In forecasting, one should try to rely on objective numerical data based on relevant information. The objective numerical data could be in terms of dollars, hours, or actual units of products. If relevant numerical data are not available, then qualitative or subjective methods are used. These methods predict future variable values based on subjective and nonquantifiable information such as the opinions of salespeople, marketing staff, customers, and experts. The subjective methods are covered in Section 3.2.

Figure 3.1 shows the key concepts and methods related to the forecasting methods that are covered in this chapter. In this chapter, we introduce MCDM approaches for solving forecasting problems. In multicriteria methods, forecasting alternatives are evaluated by considering different criteria such as risk, the likelihood of occurrence, and the accuracy of the forecast. The multicriteria methods help the planner choose the best alternative among different forecasting alternatives while considering conflicting criteria.

3.1.1 Time Horizons in Time Series Forecasting

Forecasting methods can be used to predict (a) the next (one) period or (b) a number of future periods. As the number of future periods increases, so does the inaccuracy of the prediction. In this chapter, we will mostly focus on the prediction of the next period, but some of the methods developed in this chapter (the trend-adjusted exponential smoothing and the causal methods) can be used to predict a number of future periods.

The time horizon is the length of the forecasting period. Forecasting periods can be long term, intermediate term, short term, or immediate term. The selection of time horizon can have a major effect on business and/or personal decisions depending on its accuracy. Long-term forecasts normally involve more complex issues than short-term forecasts. As the prediction period moves further into the future, uncertainties become more pronounced, leading to less accurate forecasting:

Long Term This type of forecasting usually ranges in years (e.g., from 2 to 10 years). Examples of long-term decisions are demand for new products, capital expenditures, expansion (or reduction), and capacity planning.

Intermediate Term This type of forecasting usually ranges in months (e.g., from 2 to 24 months). Examples of decisions related to intermediate terms are equipment planning, budgeting, staffing, and project scheduling.

Short Term Short-term forecasting usually ranges in days (e.g., from 1 to 60 days). Examples of short-term decisions are job assignments, product scheduling, inventory planning, production planning, and purchasing.

Immediate Term The decision for immediate terms may range in minutes or hours (e.g., 1 min to 24 h. An example of an immediate-term decision is the daily trade of stocks in the stock market.

3.1.2 Principles of Forecasting

The following principles can be considered when applying forecasting methods:

1. Forecasting is about uncertain or probabilistic future events.
2. A forecasting method should be supportable by evidence and be documentable.
3. Forecasts are usually inaccurate because there could be several unknown factors that may contribute to future events.
4. Forecasts for the near future are usually more accurate than for the far future. As the prediction period moves further into the future, the number of unknowns and the level of uncertainty increase.
5. When a forecasting method applied to the past data has very little error, it is usually the case of over-fitting. The results are too good to be true and the forecasting will have a very poor predictability.
6. Evaluate forecasted alternatives by considering multiple criteria such as risk and accuracy.

3.2   FORECASTING APPROACHES

3.2.1   Forecasting Process

The forecasting process consists of the following steps:

I. Choosing Outputs and Inputs (Data) of Forecasting   Establishing the desired outputs to be forecasted is the first step in forecasting. The output will determine the amount of effort for data gathering and the selection of an appropriate forecasting method. However, the quality of forecasted output is only as good as the quality of the input. Related input data may be obtained from different sources such as industry experts, planners, marketing managers, or distributors as well as records such as shipping reports and purchase orders. Data may also be available from past performances, quality reports, machine capacities, and various economic indicators. These data may be internal or external to the company. The DM must filter out any bad data to improve the accuracy of the system. Bad data may be the result of problems with the data collection process or the use of outliers that do not represent general trends. The level of expertise within the organization and the availability of resources for gathering data affect the accuracy of the forecast. One may ask the question: Is the gathered data based on estimates of past performances or is it based on objective data? Using an estimate of past performances is riskier than using objective data.

II. Establishing Time Constraints for Forecasting   The time frame must include information such as (a) how far into the future to forecast, (b) how far into the past the data are relevant, and (c) how much time is available and is needed to apply the chosen forecasting method. The selection of time frames is crucial in achieving accurate forecasting. Often, time frames may be changed based on new data, perspectives, and the objectives of the forecasting.

III. Choosing Forecasting Method   A summary of forecasting methods and their assumptions are presented in Figure 3.1. Qualitative and/or quantitative analysis may be performed on the data. As a general insight, qualitative forecasting is based on human judgment to utilize subjective data, whereas quantitative methods are based on analytical techniques to utilize objective data. These methods will be discussed in more detail later in this chapter. A simple forecasting approach is to plot the past data and predict the future demand by graphing it. This is called the freehand method.

Supplement S3.1 explains forecasting by the freehand method.

IV. Choosing Accuracy Measurement and Error-Tracking Methods   In choosing an accuracy measurement method, one must consider the definition of the error involved in forecasting. The error can be an overestimation or an underestimation of a forecast. Error-tracking methods are used to find the pattern of changes in errors and also to identify if certain data points should be discarded as they may not be representative of the data. The accuracy measurements and tracking methods will be covered in Section 3.2.3 and 3.2.4.

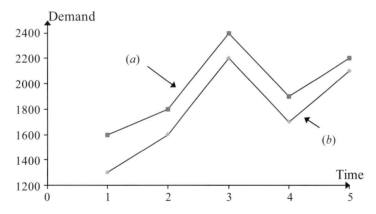

Figure 3.2   Two alternatives: (*a*) minimizing only the error of estimation; (*b*) maximizing the desired underestimation.

V. Selecting Forecast by Use of Multiple-Criteria Analysis   There is usually a range of possible forecasting alternatives that can be selected. A number of criteria can be used to evaluate different possible forecasting alternatives. In multicriteria regression analysis, depending on the problem, the DM may desire either an overestimation or an underestimation in the error. Risk attitude can be either conservative (risk averse) or risk taking (risk prone). In multicriteria forecasting problems, we can consider the following two objectives:

f_1 = minimize error	Error of estimation (to have most accurate forecasting)
Either f_2 = maximize confidence level for risk averse	Being conservative in forecasting to maximize the confidence level of occurrence for desired output
Or f_2 = maximize confidence level for risk prone	Being aggressive in forecasting to maximize the confidence level of occurrence for desired output

Depending on the problem and the point of view, a risk-averse (or risk-3 prone) person may consider either overestimation or underestimation to be the desired output. The DM may determine the level of risks that he or she is willing to take. A multiple-criteria method can be used to choose the proper alternative for DMs. See Figure 3.2.

VI. Identifying Responsible Stakeholders   Once forecasted values are selected, they must be analyzed for accuracy and feasibility by several departments such as marketing, sales, scheduling, and human resources. The forecasted values must also be shared by the responsible departments with external sources who will be involved in the decision-making process.

3.2.2 Qualitative–Subjective Forecasting

Qualitative forecasting is also known as subjective or judgmental forecasting. This method is based on human judgment. It is used when:

- There is not enough data to make the prediction.
- It is too costly to collect the data.
- The data cannot be collected.
- The data do not exist.

The four most common qualitative forecasting methods are summarized below.

I. Sales Force Composites   One way to forecast the product demand is to interview the sales force associated with the product. The sales force has the most direct access to the customer. They are in an up-front position to directly foresee trends for the current period. Sales forces can also provide beneficial estimates for the upcoming period based on their insights to customer needs. Unfortunately, the sales force may not necessarily possess objective estimates. Hoping that the demands for the next period will be higher, the sales force may overestimate the demand. A salesperson's judgment can be also affected by his or her attitude toward work or people and may be further influenced if the company requires quotas. If a quota is required, it is more likely that the salesperson would be influenced by it and their estimate would be invalid.

II. Customer (Market) Surveys   Customer, or market, surveys are obtained by directly interviewing the intended customers. These can be carried out through questionnaires, interviews, or other forms of information gathering in the marketplace. These data can be gathered and analyzed to make a forecast. For example, many car dealerships will send a customer satisfaction survey to a car buyer. This survey helps the dealership and the parent company determine what the buyer likes and dislikes about the car and the dealership. Also, when the buyer brings the car to the dealership for maintenance, the dealership may obtain a survey for the service and performance of the car. The parent company can then compile data from all the surveys to determine the wants and needs of the consumer.

III. Jury of Executive DMs   A Jury of executive DMs is based on gathering the opinions of experts through interviews. In this method, accounting, marketing, engineering, and production representatives must all be represented. This method is usually carried out on new products where no history exists. The opinions may be gathered individually or as a group, with an individual responsible for preparing the interviews. For example, the engineering department may design a new product to fill a niche that has been identified by the marketing department. Therefore, representatives from accounting, marketing, engineering, and production could have an assessment meeting. The purpose of the meeting would be to determine how to market the product, what the product should cost, and how to produce the product. Allowing many departments to join in the assessment meeting will allow more thoughts and ideas to be generated. A better prediction may be developed as a result of the cross-functional team sharing opinions, ideas, and knowledge.

IV. Delphi Method   The purpose of this method is to allow an objective process of inter-action among a team of experts to take place. Very often, open meetings are dominated by certain personalities. Some individuals have the ability to convince others to their way of thinking, which could be totally flawed. Major project disasters such as the NASA *Challenger* confirm that the objective views of a few who predicted disaster were overshadowed by certain judgmental dominant personalities. The Delphi method allows each individual to think for themselves and form their opinion based on written feedback from others. The Delphi method is similar to the jury of executive decision makers. The difference is that in the Delphi method the questionnaire is updated and returned to the participants after the results have been compiled. The participants will continue to modify their decisions until, ideally, a group consensus is formed. The biggest advantage to the Delphi method is that the repeated modifications and the removal of personal interaction will allow the participants to be more objective.

3.2.3 Accuracy Measurement

The accuracy of a forecasting method can be measured by using the error of forecasts for the existing data. The error for each data point e_t is defined as the forecasted value F_t minus the actual demand value D_t where there are N data points, $t = 1, 2, \ldots, N$:

$$e_t = F_t - D_t \quad \text{for} \quad t = 1, 2, \ldots, N \tag{3.1}$$

There are three well-known methods for measuring accuracy: mean average deviation, mean-square error, and mean absolute percentage error. These are discussed below.

> *MAD*   The mean average deviation (MAD) is the most common method used for measuring the forecasting error. The MAD is found by calculating the average of the absolute value of the errors:

$$\text{MAD} = \frac{1}{N} \sum_{t=1}^{N} |e_t| \tag{3.2}$$

> *MSE*   The mean square error (MSE) is determined by finding the average of the squares of the errors. This method places a higher penalty on larger errors:

$$\text{MSE} = \frac{1}{N} \sum_{t=1}^{N} e_t^{\,2} \tag{3.3}$$

The standard deviation (SD) is the square root of the MSE:

$$\text{SD} = \sqrt{\text{MSE}}$$

MAPE   The mean absolute percentage error (MAPE) is a percentage based on the ratio of errors to their corresponding demand values:

$$\text{MAPE} = \frac{1}{N} \left(\sum_{t=1}^{N} \frac{|e_t|}{D_t} \right) \times 100\,\% \tag{3.4}$$

Example 3.1   Three Accuracy Measurement Methods   The following information presents demand and forecasted values for five periods. Calculate the MAD, MSE, and MAPE for the forecasted values:

Period, t	1	2	3	4	5
Demand, D_t	100	190	200	220	300
Forecast, F_t	116	159	202	245	288

The following table shows the details of the calculations for each measurement method:

Period, t	Demand, D_t	Forecast, F_t	$e_t = F_t - D_t$	$|e_t|$	e_t^2	$|e_t| / D_t$
1	100	116	16	16	256	0.16
2	190	159	−31	31	961	0.16
3	200	202	2	2	4	0.01
4	220	245	25	25	625	0.11
5	300	288	−12	12	144	0.04
SUM	1010	1010	0	86	1990	0.48
Avg. = SUM / 5	202	202	0	17.2	398	0.097
				MAD	MSE	MAPE
				= 17.2	= 398	= 9.7%

In this example, the average forecast is equal to the average demand, that is, 202. The results of calculating the MAD, MSE, and MAPE are presented in the second column of the following below. It is possible to compare these three methods by converting them to comparable units. In the third column of the table below, the average units of the three methods are shown. For example, the second root of the MSE is its standard deviation: SD $= \sqrt{398} = 19.95$. This number is comparable to the MAD value of 17.2. The fourth column compares these units in terms of their percentages. For example, in terms of percentage, MAD $= 17.2$ can be compared to the average demand 202, that is, $(17.2/202) \times 100\% = 8.52\%$. This number is comparable to (but not the same as) the MAPE of 9.7%. In terms of the percentage, the MSE can be presented as $(19.95/202) \times 100\% = 9.88\%$, which is also comparable to the MAPE of 9.7%. Therefore, the MAD, MSE, and MAPE provide similar methods of calculating the error, and it is sufficient to use one of the three methods.

	Actual Units	Comparable Units	Percent Units
MAD	$MAD = \dfrac{1}{N}\sum\limits_{t=1}^{N} |e_t|$ $= 86/5 = 17.2$	$MAD = 17.2$	$(17.2/202)$ $\times\ 100 = 8.52\%$
MSE	$MSE = \dfrac{1}{N}\sum\limits_{t=1}^{N} e_t^2$ $= 1990/5 = 398$	$SD = \sqrt{398} = 19.95$	$(19.95/202)$ $\times\ 100 = 9.88\%$
MAPE	$MAPE = \left(\dfrac{1}{N}\sum\limits_{t=1}^{N}\dfrac{|e_t|}{D_t}\right)\times 100$ $= (0.48/5)\times 100$ $= 9.7\%$	$MAPE\ unit = (9.7)$ $\times\ (202)/100$ $= 19.59$	9.7%

3.2.4 Quality Control and Tracking Error Charts

Quality Control Chart Based on Normal Distribution   Over a period of time, the forecasting error may increase, decrease, or be stationary. If the error is increasing or decreasing, it means that there is something wrong with the forecasting method. The upper and lower acceptable bounds for the forecasting errors are labeled as upper control limit (UCL) and lower control limit (LCL), respectively. The UCL and LCL can be determined by the desired level of accuracy and confidence level. Generally, when some of the errors are outside of the UCL and LCL bounds, the forecasting will be inaccurate. In Figure 3.3a, for example, the error is increasing and the last point is above the UCL. In Figure 3.3b, all the errors are within the bounds except one point, which is completely out of the normal range. This point should be removed and the forecasting should be reapplied.

The lower and upper control limits can be set up by using

$$UCL = +z\ SD \tag{3.5}$$

$$LCL = -z\ SD \tag{3.6}$$

where z is based on the normal distribution. See the standard normal distribution table at the back of the book for finding the probabilities associated with different z values. For example, $z = 1, 2, 3$ are associated with 68.26, 95.44, and 99.74% confidence levels that the data will fall within the given UCL and LCL, respectively. The subject of quality control charts is covered in depth in Chapter 17.

Example 3.2   Quality Control Chart Based on Normal Distribution   Consider Example 3.1. Show the error on the control charts for a 99.74% confidence level. Is the process under control?

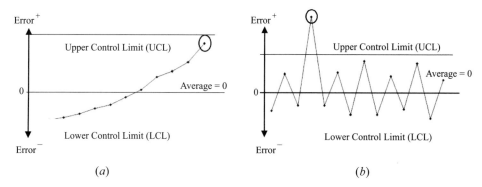

Figure 3.3 (*a*) Example of increasing error over time. (*b*) Data point 4 is out of range and should be removed.

For the 99.74% confidence level, use $z = 3$. Since $\text{SD} = \sqrt{\text{MSE}} \approx \sqrt{398} \approx 20$, the control limits are

$$\text{UCL} = 3 \times 20 = 60 \qquad \text{LCL} = -3 \times 20 = -60$$

The errors and the control chart are presented in Figure 3.4. In this example, the error pattern is acceptable because errors of all periods are within the acceptable range.

Tracking Signal Control Chart A different approach for monitoring errors is the tracking signal method. The tracking signal is a measurement based on dividing the cumulative value of errors by the MAD:

$$\text{Tracking signal} = \sum_{t=1}^{N} \frac{e_t}{\text{MAD}_t} \tag{3.7}$$

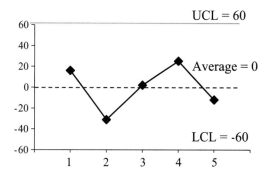

Figure 3.4 Quality control chart of errors from Example 3.2.

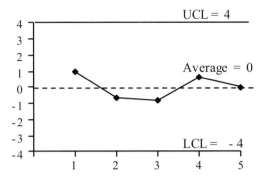

Figure 3.5   Control chart of tracking signal errors from Example 3.3.

Generally, tracking signals vary from ± 0 to ± 8. In practice, a tracking signal between -4 and 4 is equivalent to using $z = 3$ (i.e., 99.74% confidence level). If errors are between -4 and 4, then the forecasting error measurement is under control.

Example 3.3   Tracking Signal Control Chart   Consider Example 3.2. Find the tracking signal for each period and identify if the process is under control.

The tracking signal solution is as follows:

Period, t	Demand, D_t	Forecast, F_t	$e_t = F_t - D_t$	$\|e_t\|$	$\sum_{t=1}^{N} \|e_t\|$	$\sum_{t=1}^{N} e_t$	MAD for period t	Tracking Signal at Period t
1	100	116	16	16	16	16	$16/1 = 16$	$16/16 = 1$
2	190	159	-31	31	47	-15	$47/2 = 23.5$	$-15/23.5 = -0.64$
3	200	202	2	2	49	-13	$49/3 = 16.3$	$-13/16.3 = -0.80$
4	220	245	25	25	74	12	$74/4 = 18.5$	$12/18.5 = 0.65$
5	300	288	-12	12	86	0	$86/5 = 17.2$	$0/17.2 = 0$

In this example, the tracking signals are 1, -0.64, -0.80, 0.65, and 0 for periods 1–5, respectively. All these values are between -4 and $+4$, and therefore the error is under control. See Figure 3.5.

3.2.5   Over- and Underfitting in Forecasting

As discussed before, forecasting is always associated with some level of error due to the randomness of events. By using a highly nonlinear forecasting model such as artificial neural network models, it is possible to find a highly nonlinear function that perfectly fits all existing data points, but such a model may have a poor predictability power for future data points. The above case is called overfitting. Underfitting is the reverse of overfitting, that is, not using a more nonlinear model when it is proper to use it.

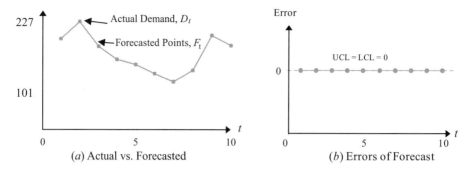

(a) Actual vs. Forecasted   (b) Errors of Forecast

Figure 3.6 Forecasting by a highly nonlinear function which has a perfect (over) fit with very little errors but has poor future predictability.

Example 3.4 Overfitting Forecasting Model Consider the following 10 points where two forecasting methods are used for forecasting: (I) a highly nonlinear function and (II) a linear function. Analyze their predictability based on their error analysis.

Product, t	1	2	3	4	5	6	7	8	9	10
Demand, D_t	191	227	175	148	137	118	101	125	198	177
I. Nonlinear F_t	191	227	175	148	137	118	101	125	198	177
Error	0	0	0	0	0	0	0	0	0	0
II. Linear F_t	180	175	170	165	160	155	150	145	140	135
Error	−11	−52	−5	17	23	37	49	20	−58	−42

The results of models I (highly nonlinear fitting) and II (linear fitting) are shown in Figures 3.6 and 3.7, respectively. The MSE for (a) a highly nonlinear function is zero and for (b) a linear one is 1611.90.

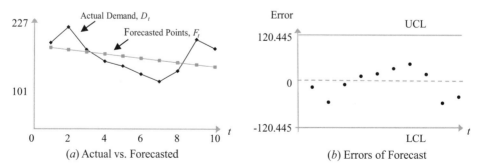

(a) Actual vs. Forecasted   (b) Errors of Forecast

Figure 3.7 Forecasting by a linear function which does not have a perfect fit but has acceptable error. It may be more accurate than a nonlinear function for future predictions.

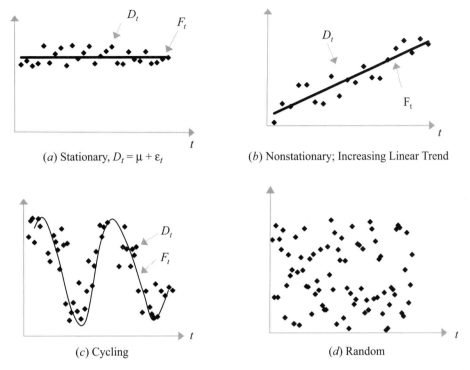

(*a*) Stationary, $D_t = \mu + \varepsilon_t$

(*b*) Nonstationary; Increasing Linear Trend

(*c*) Cycling

(*d*) Random

Figure 3.8   Time series stationary and nonstationary data patterns; actual data D_t versus forecast data F_t shown for each point.

3.2.6   Basics and Types of Time Series

A very important class of problems in forecasting is the time series. A time series problem refers to revealing a pattern in a set of data points measured at equally spaced time intervals. There are four types of time series patterns:

(a) Stationary: The time series has a line pattern with no slope; see Figure 3.8*a*.

(b) Linear trend: The trend line can be either upward or downward; see Figure 3.8*b*.

(c) Cyclic trend: There is a predictable deviation from the expected demand usually in the form of a cycle; see Figure 3.8*c*. For example, seasonal variations are cyclic.

(d) Random variations: These have irregular patterns (usually due to completely random behavior); see Figure 3.8*d*. There is no recognizable pattern in random variations.

Stationary Assumption   The underlying assumption of stationary time series problems is that each past observation can be presented by a constant μ plus a random number ε_t. The random number ε_t represents the noise or fluctuations of each data point t, where

$$D_t = \mu + \varepsilon_t \quad \text{for } t = 1, 2, \ldots, N \tag{3.8}$$

where μ is the mean value which is constant for all periods and ε_t is the random variation associated with the data at period t and ε_t has a mean value of zero and a variance of σ^2.

The Stationary assumption implies that only one parameter, μ, needs to be identified. That is, the demand for the next period $N + 1$ is $F_{t+1} = \mu$. Furthermore, it means that data points do not have a trend (increasing or decreasing pattern). The prediction of the past and the future periods are all the same, that is,

$$F_1 = F_2 = \cdots = F_N = F_{N+1} = F_{N+2} = \cdots = \mu \qquad (3.9)$$

In this chapter, we cover two time series methods that are based on a stationary assumption: moving averages and exponential smoothing. Then, we cover two methods that are not based on the stationary assumption: trend-adjusted exponential smoothing and linear regression.

3.3 TIME SERIES: MOVING AVERAGES

3.3.1 Moving Averages

Suppose at the current period N the past known demands are denoted by $D_1, D_2, \ldots, D_N$. We wish to forecast the next period demand denoted by F_{N+1}. Because distant past information may not be as useful to the forecasting of the next period, in this method, only the latest n periods of data need to be considered, where $n \leq N$:

$$F_{N+1} = \frac{1}{n} \sum_{i=1}^{n} D_{N-i+1} \qquad (3.10)$$

where

$$\begin{aligned}
F_{N+1} &= \text{forecast for period } N+1 \\
D_{N-i+1} &= \text{actual demand values for period } N - i + 1 \\
n &= \text{number of most recent } n \text{ observations} \\
N &= \text{current period number}
\end{aligned}$$

Moving Averages Using One Past Period (Naive Method) This method is called naive because it is based on the naive assumption that the next-period forecasted value is the same as the last-period actual value. This is the simplest form of time series forecasting where it is assumed that $n = 1$. That is,

$$F_{N+1} = D_N$$

This forecasting is based on the assumption that only the most recent data point is relevant to the next period of forecasting.

Example 3.5   Naive Method (Moving Averages with One Period) Forecasting   Consider the number of sales of a given automobile. The data (in 1000 units) for the past 16 weeks are presented in Table 3.1 as D_t for weeks $t = 1, 2, \ldots, 16$. What is the prediction for sale in period 17, that is, F_{17}? Use the naive method. Also, show the forecasting for periods $t = 1, \ldots, 16$.

TABLE 3.1   Moving Averages with $n = 1, 2, 3, 4, 5$ period for Example 3.6

Period, t	Demand, D_t	$n = 1$	$n = 2$	$n = 3$	$n = 4$	$n = 5$
			Forecasting, $F_t(n)$, using n-Period Moving Average			
1	62					
2	66	62				
3	72	66	64			
4	74	72	69	66.67		
5	78	74	73	70.67	68.5	
6	92	78	76	74.67	72.5	70.4
7	106	92	85	81.33	79	76.4
8	118	106	99	92	87.5	84.4
9	126	118	112	105.33	98.5	93.6
10	120	126	122	116.67	110.5	104
11	128	120	123	121.33	117.5	112.4
12	112	128	124	124.67	123	119.6
13	90	112	120	120	121.5	120.8
14	80	90	101	110	112.5	115.2
15	92	80	85	94	102.5	106
16	98	92	86	87.33	93.5	100.4
17	$F_{17} = ?$	98	95	90	90	94.4
MAD		9.60	12.64	15.28	18.67	21.71

In this example, $N = 16$ is the current period, and there are $N = 16$ data points, that is, D_t is known for $t = 1, 2, \ldots, 16$. The prediction of the next period, 17, is equal to the demand of the last period, 16. Since $F_{N+1} = D_N$, then

$$F_{17} = D_{16} = 98$$

The forecasting is shown as F_t for weeks $t = 2, \ldots, 17$ in Table 3.1 under $n = 1$ column. For example, at period 15, the forecast for period 16 would be $F_{16} = D_{15} = 92$.

The actual demand and the prediction for all periods are also presented in Figure 3.9 (denoted by $n = 1$). Note that the prediction follows the exact pattern of the past sale with one period elapsed.

Moving Averages Using n Past Periods   In this case, data of the n past periods are used to make the forecasting. The key question is how many past periods n should be used. Choosing an optimal n may not be a trivial problem.

Example 3.6   Moving-Average Forecasting   Consider Example 3.5. Forecast all periods using two, three, four, and five past periods. Which n value should be selected?

The results for $n = 1, 2, 3, 4, 5$ are presented in Table 3.1. For example, for $n = 5$, we use the past five periods. The calculations for forecasting periods 16 and 17 are shown below:

$$F_{16} = \frac{1}{5}(D_{15} + D_{14} + D_{13} + D_{12} + D_{11}) = \frac{1}{5}(92 + 80 + 90 + 112 + 128) = 100.4$$

$$F_{17} = \frac{1}{5}(D_{16} + D_{15} + D_{14} + D_{13} + D_{12}) = \frac{1}{5}(98 + 92 + 80 + 90 + 112) = 94.4$$

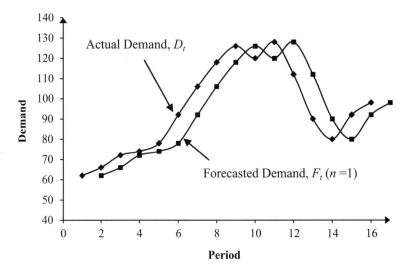

Figure 3.9 Naive method: Moving-average forecasting based on only one past period, that is, $n = 1$. Notice how the forecasted demand closely shadows the actual demand.

By using more periods, the predicted value becomes "smoother" and is less subject to fluctuations in the actual data. Figure 3.10 shows the forecast values for $n = 2$ and $n = 5$.

Comparing $F_t(n = 1)$ (Figure 3.9) to $F_t(n = 2)$ and $F_t(n = 5)$ (Figure 3.10), it can be observed that as the number of periods n increases, the prediction function becomes less responsive to recent changes, that is, it becomes more stable. Which n must be selected? We will answer this question in Section 3.3.3 by the use of a multicriteria approach for moving averages.

3.3.2 Weighted Moving Averages

In the moving-average methods, the weight of importance of the past n data points are assumed to be equal. For n points, each weight is $1/n$ where the total of all weights is unity.

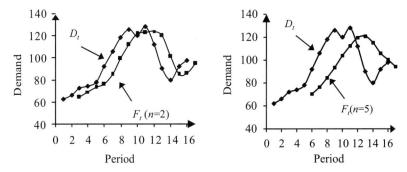

Figure 3.10 Example 3.6 moving-average (MA) forecasting vs. actual demand (D) for $n = 2$ and $n = 5$.

In the weighted moving-average method, each past data point can have a different weight of importance. The forecast is

$$F_{N+1} = \sum_{i=1}^{n} w_{N-i+1} D_{N-i+1} \tag{3.11}$$

where w_{N-i+1} is the weight of importance associated with the demand of period $N\text{-}i+1$ and

$$\sum_{i=1}^{n} w_{N-i+1} = 1 \tag{3.12}$$

In this method, assigning weights is subjective. The weight assignment may become difficult when there are many past periods. The exponential smoothing method of the next section resolves both of these problems.

Example 3.7   Weighted Moving Averages   Consider Example 3.5. Use the past four periods (i.e., $n = 4$). Suppose the weights of importance associated with the past four periods are (0.4, 0.3, 0.2, 0.1) for demands of periods 16, 15, 14, and 13, respectively.
  The weighted moving average is

$$\begin{aligned} F_{17} &= w_{16} D_{16} + w_{15} D_{15} + w_{14} D_{14} + w_{13} D_{13} \\ &= 0.40 \times 98 + 0.3 \times 92 + 0.2 \times 80 + 0.1 \times 90 = 91.8 \end{aligned}$$

In this example, the highest weight of importance (0.4) is given to the most recent period, and the lowest weight of importance (0.1) is given for the earliest period.

Lagging Effect of Moving Averages   By examining Figure 3.10 (for moving averages), it can be observed that the forecasted demand pattern lags behind the actual demand pattern. This occurs because the moving-average method (and exponential Smoothing, which will be covered in the next section) cannot incorporate increase or decrease trends because of the stationary assumptions. Incorporating trends are covered later in this chapter.

Supplement S3.2 explains details of multicriteria for moving averages.

3.4   TIME SERIES: EXPONENTIAL SMOOTHING

3.4.1   Exponential Smoothing

Exponential smoothing is a stationary time series forecasting method. It is a special case of the weighted moving-average method where, instead of assigning n weights $(w_1, w_2, \ldots, w_n)$ only one parameter α, where $0 \leq \alpha \leq 1$, is assessed. A larger value of α signifies that a greater weight is given to more recent periods and less weight is given to earlier periods. The exponential smoothing equation can be written as

$$F_t = \alpha D_{t-1} + (1 - \alpha) F_{t-1} \quad \text{for } t = 2, 3, \ldots, n \tag{3.13}$$

where

$$F_t = \text{forecast for period } t$$
$$D_{t-1} = \text{demand for period } t - 1$$
$$F_{t-1} = \text{forecast for period } t - 1$$

Equation (3.13) can be rewritten as

$$F_t = F_{t-1} - \alpha(F_{t-1} - D_{t-1}) \qquad (3.14)$$

where $F_{t-1} - D_{t-1}$ represents the error between the forecasted demand and the actual demand for the past period $t - 1$. This means the forecast in period t, that is, F_t, is equal to the forecast of the last period minus a fraction (α) of the error of the last period.

Now consider the forecasts for period $t - 1$, that is, F_{t-1}.

$$F_{t-1} = F_{t-2} - \alpha(F_{t-2} - D_{t-2}) \qquad (3.15)$$

By substituting F_{t-1} of Equation (3.15) into Equation (3.14), we have

$$F_t = [F_{t-2} - \alpha(F_{t-2} - D_{t-2})] - \alpha([F_{t-2} - \alpha(F_{t-2} - D_{t-2})] - D_{t-1}),$$

or

$$F_t = \alpha D_{t-1} + \alpha(1 - \alpha)D_{t-2} + (1 - \alpha)^2 F_{t-2} \qquad (3.16)$$

If we continue expanding Equation (3.16), F_t can be expressed as

$$F_t = \sum_{i=0}^{n} \alpha(1 - \alpha)^i D_{t-i-1} \qquad (3.17)$$

where

$$\sum_{i=0}^{n} \alpha(1 - \alpha)^i = 1 \quad \text{as } n \to \infty$$

In Equation (3.13), the forecast value for the first period, F_1, is not known. It is customary to set the first forecasting value equal to the value of the initial demand, that is, set $F_1 = D_1$. This approach, however, may bias the final result of the forecast, especially if there are few periods. An alternative approach is to set its value equal to the average of several initial demands (e.g., average of the first 10 demands).

Relationship of α and Weights of Weighted Moving-Average Method   The weight coefficient of the weighted moving-average method [Equation (3.2)], w_t, based on each $i = 0, 1, \ldots, (n-2), (n-1)$, can be expressed as

$$w_i = \alpha(1 - \alpha)^i \qquad \text{where } t = n - i \quad \text{for} \quad i = 0, 1, \ldots, (n-2), (n-1)$$

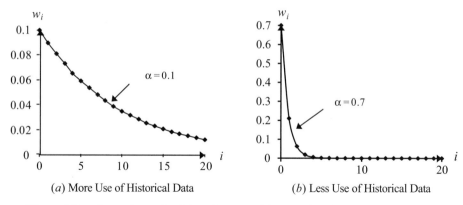

Figure 3.11   Comparison of weights of exponential smoothing for $\alpha = 0.1$ and $\alpha = 0.7$.

where w_i is the weight for the demand data at period $t = n - i$, n being the number of periods. Note that $i = n - t$. As i increases, the value of this weight decreases exponentially, which is the reason for naming this method exponential smoothing. Note that in every case

$$w_i > w_{i+1}$$

that is, the weight of more recent data is higher than that of less recent data. The examples for two different values of $\alpha = 0.1$ and $\alpha = 0.7$ are given in Figure 3.11.

See Supplement S3.4.xlsx.
   As can be seen in Figure 3.11, when α is close to 1 (e.g., $\alpha = 0.7$), weights of distant past periods are zero. In most applications, the chosen α is between 0.1 and 0.3 to put more weights on older historical data. Table 3.2 shows the value of weight w_t for three different values of α. Note that the value of weights in each time period decrease as α decreases.

Supplement S3.5.doc shows an example for exponential smoothing forecasting for 11 periods.

Example 3.8   Exponential Smoothing Forecasting for 21 Periods   Consider demand D_t for 20 periods as shown in Table 3.3. Find the forecasting for period 21. Consider five different values of α, $\alpha = 0.2, 0.3, 0.5, 0.7, 0.9$. Compare the results.
   The solution is provided in Table 3.3.

TABLE 3.2   Equivalent Weight Values of Weighted Moving Averages to Exponential Smoothing for $n = 20$ Periods and $\alpha = 0.1, 0.7, 1$ where $i = 20 - t$

		Weight, w_t									
		$t = 0$	$t = 1$	$t = 2$	$t = 3$	$t = 4$	$t = 5$	$t = 6$	$t = 10$	$t = 20$	Total
$n = 20$		20	19	18	17	16	15	14	10	0	—
$\alpha = 0.1$	w_t	0.100	0.090	0.081	0.073	0.066	0.059	0.053	0.035	0.012	1
$\alpha = 0.7$	w_t	0.700	0.210	0.063	0.019	0.006	0.002	0.001	0	0	1
$\alpha = 1$	w_t	1	0	0	0	0	0	0	0	0	1

TABLE 3.3   Five Exponential Smoothing Forecasts with $\alpha = 0.2, 0.3, 0.5, 0.7, 0.9$

Period, t	Demand, D_t	Five Forecast using Exponential Smoothing, $F_t(\alpha)$				
		$F_t\,(0.2)$	$F_t\,(0.3)$	$F_t\,(0.5)$	$F_t\,(0.7)$	$F_t\,(0.9)$
1	22	22.0	22.0	22.0	22.0	22.0
2	24	22.0	22.0	22.0	22.0	22.0
3	26	22.4	22.6	23.0	23.4	23.8
4	25	23.1	23.6	24.5	25.2	25.8
$\vdots$	$\vdots$	$\vdots$	$\vdots$	$\vdots$	$\vdots$	$\vdots$
19	49	43.9	46.1	46.6	46.1	45.4
20	47	44.9	47.0	47.8	48.1	48.6
21	F_{21} ?	45.4	47.0	47.4	47.3	47.2
MAD		6.27	4.86	3.74	3.28	3.30

Supplements S3.6.doc, S3.6.xlsx shows details of Table 3.3.

For purpose of illustration, consider $\alpha = 0.3$. The following two equations present the exponential forecasts for $\alpha = 0.3$:

$$F_2 = \alpha D_1 + (1 - \alpha)F_1 = [0.3 \times 22] + [(1 - 0.3) \times 22] = 22 \quad \text{(for period 2)}$$

$$F_3 = \alpha D_2 + (1 - \alpha)F_2 = [0.3 \times 24] + [(1 - 0.3) \times 22] = 22.6 \quad \text{(for period 3)}$$

For $\alpha = 0.3$, the above calculations are carried over for periods 4–21. The solutions are presented under the column $F_t(\alpha = 0.3)$ in Table 3.3. The calculation for $F_{21}(0.3)$ is

$$F_{21} = \alpha D_{20} + (1 - \alpha)F_{20} = [0.3 \times 47] + [(1 - 0.3) \times 47] = 47$$

Figure 3.12 shows the comparison of forecasts for $\alpha = 0.2, 0.5, 0.9$. The comparison of the three graphs shows that the forecasting associated with small values of α (e.g., $\alpha = 0.2$) is more stable than with the high value of α (e.g., $\alpha = 0.9$). If we choose the forecasting method based on minimizing MAD, then forecasting associated with $\alpha = 0.9$ should be selected. The proper way to choose the forecast parameter (α) is presented in the next section using the multicriteria analysis.

See Supplements S3.6.doc, S3.6.xlsx.

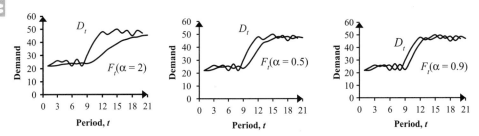

Figure 3.12   Comparison of forecasting F_t for $\alpha = 0.2, 0.5, 0.9$ vs. actual demand D_t for Example 3.9.

3.4.2 Multicriteria Exponential Smoothing[†]

Multicriteria exponential smoothing can be performed by using α as the decision variable. Different values of α will result in different estimation of the demand. Therefore, the search is performed on the value of α. Suppose the given range for α is $\alpha_{min} \leq \alpha \leq \alpha_{max}$. The two objectives for choosing the best value for α are:

Minimize $f_1 = \text{MAD}$ (to minimize mean absolute deviation error of forecasting)
Minimize $f_2 = F_{N+1}$ (to maximize desirable underestimation)

Or

Maximize $f_2 = F_{N+1}$ (to maximize desirable overestimation)
Subject to $\alpha_{min} \leq \alpha \leq \alpha_{max}$

By minimizing $f_2 = F_{N+1}$, more weight is given to past data for presenting underestimation. Depending on the decision problem, this objective may be associated with being risk averse or risk prone. On the other hand, by maximizing $f_2 = F_{N+1}$, more weight is given to past data for presenting overestimation. Depending on the decision problem, this objective may be associated with being risk averse or risk prone. It should be noted that, depending on the problem, the number of efficient points may vary. In the above example, as α increases, $f_1 = \text{MAD}$ may not decrease. This may not be the case in some other examples.

A set of multicriteria forecasting alternatives can be generated by considering different values of α. To generate a set of alternatives, choose a set of discrete values of α. For example, choose five different values of α equally distributed from α_{min} to α_{max}. Find the forecast $F_t(\alpha)$ for each α. Then measure MAD for each generated alternative. The best multicriteria alternative can then be selected. It is also possible to perform a one-dimensional search (optimization) on α to find the optimal α.

Example 3.9 Multicriteria of Exponential Smoothing   Consider Example 3.8. Suppose that the DM is risk averse and that underestimation of demand is the desired output. Use $\alpha = 0.2, 0.3, 0.5, 0.7, 0.9$ to generate bicriteria alternatives and identify efficient alternatives and ask the DM to rank alternatives.

The solution is shown in Table 3.4; which is based on Table 3.3. To solve this problem, for each α value, find the absolute error ($|e_t| = |F_t - D_t|$) for each forecasted value. The sum of these errors is used to find the MAD. The details are shown in Table 3.4. For example, consider period 20; for alternative $\alpha = 0.2$, the forecasted demand is 44.9 (this was shown in Table 3.3), and the actual demand is 47. Therefore, its absolute error is $|44.9 - 47| = 2.1$ as presented in Table 3.4. Ranking by a hypothetical DM is also shown in Table 3.4. Therefore, the exponential smoothing solution with $\alpha = 0.9$ is selected as the best forecasting method. The selected forecasted value for period 21 is 47.2.

See Supplement S3.8.xls.
Supplement S3.9.doc shows details of Table 3.4.

TABLE 3.4   Absolute Error $|e_t|$ and MAD for Each Forecast of Table 3.3

Period, i	Demand, D_t	$\alpha = 0.2$, $\|e_t\|$ for F_t (0.2)	$\alpha = 0.3$, $\|e_t\|$ for F_t (0.3)	$\alpha = 0.5$, $\|e_t\|$ for F_t (0.5)	$\alpha = 0.7$, $\|e_t\|$ for F_t (0.7)	$\alpha = 0.9$, $\|e_t\|$ for F_t (0.9)
1	22	0.0	0.0	0.0	0.0	0.0
2	24	2.0	2.0	2.0	2.0	2.0
3	26	3.6	3.4	3.0	2.6	2.2
4	25	1.9	1.4	0.5	0.2	0.8
⋮	⋮	⋮	⋮	⋮	⋮	⋮
18	45	1.3	1.6	3.3	3.6	3.8
19	49	5.1	2.9	2.4	2.9	3.6
20	47	2.1	0.0	0.8	1.1	1.6
Total $\|e_t\|$	—	125.30	97.22	74.75	65.63	65.97
Min. $f_1 =$ MAD	—	6.27	4.86	3.74	3.28	3.30
Min. F_{21} (α)	—	45.4	47.0	47.4	47.3	47.2
Rank by DM	—	5	4	3	2	1
Efficient	—	Yes	Yes	No	Yes	Yes

3.5   TIME SERIES: TREND-BASED METHODS

A trend is a gradual upward or downward pattern in data that can be used to predict subsequent data points. There are generally three types of trends: linear, cyclic, and nonlinear. In this section, only linear trends are considered. Linear regression and adjusted exponential smoothing can be used to predict linear patterns. Cyclic trends are discussed in Section 3.6 and nonlinear trends are discussed in Section 3.10.

For example, the price of scarce resources such as oil has been generally increasing with some fluctuations (upward trend). On the other hand, the price of semiconductors and computer-related components has been decreasing over time (downward trend). The Dow Jones Index, a U.S. stock market composite index, has generally been increasing with some fluctuations. In Figure 3.13a, the actual data for the Dow Jones Index along with an approximate linear trend is presented for the past 100 years. Figure 3.13b shows the Dow Jones Index for a shorter period of time (10 years). It can be observed that although it is easier to see the increasing linear trend pattern over the 100-year period, as in Figure 3.13a, it is more difficult to observe the linear trend pattern for a short period of time, as in Figure 3.13b. The accuracy of the forecasting method, therefore, can greatly depend on the length of observation and on how periods are defined.

It is possible to consider segments of time and use linear trend analysis for that segment. For example, consider the four stages of the product life cycle: birth, growth, maturity, and decline. Each of these phases can be forecasted using linear trend analysis. See Figure 3.14.

3.5.1   Time Series Trend Analysis by Linear Regression

Linear regression can be used to find the linear trend in time series forecasting. The complete coverage of linear regression and its theoretical justification are given in Section 3.7.1 for causal forecasting problems. Linear regression for solving time series forecasting is a special case of causal forecasting and is explained in Section 3.7.2.

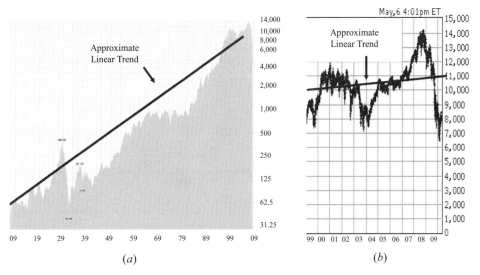

Figure 3.13 Illustration of approximately (*a*) 100 years and (*b*) 10 years of the Dow Jones Composite Index, where the trend is approximated by a linear function.

3.5.2 Trend-Adjusted Double-Exponential Smoothing[†]

Trend-adjusted (also called "double") exponential smoothing is an extension of exponential smoothing which includes a trend adjustment factor. Consider the demand for the past n periods denoted by D_t for $t = 1, 2, \ldots, n$. The problem is to find the adjusted forecasting AF_{t+1} for period $t+1$. The adjusted forecasting (AF) for a given period is the sum of the forecasting (F) and the trend (T) adjustment for that period. Forecasting in this method consists of predicting the stationary parameter (AF) and the trend parameter (T); see Equation (3.18). The stationary prediction F is found by exponential smoothing

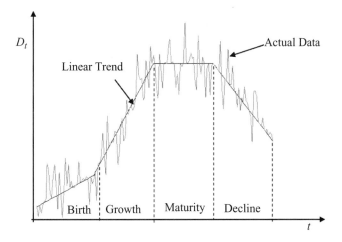

Figure 3.14 Linear trend in the four stages of a life cycle.

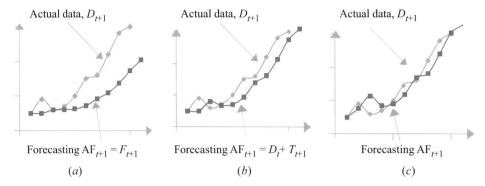

Figure 3.15 Comparison of using α, β, or both (α and β) in adjusted forecasting: (*a*) forecasting using only exponential smoothing, $\alpha = 0.25$, $\beta = 0$, $T = 0$; (*b*) forecasting using only trend analysis, $\alpha = 1$, $\beta = 0.3$, $T > 0$; (*c*) adjusted forecasting using $\alpha = 0$, $\beta = 0.3$, and $T > 0$.

using Equation (3.19) and the trend value T is found by a simple linear function using Equation (3.20). That is,

$$AF_{t+1} = F_{t+1} + T_{t+1} \tag{3.18}$$

$$F_{t+1} = \alpha D_t + (1 - \alpha)F_t \quad \text{(exponential Smoothing for stationary portion)} \tag{3.19}$$

$$T_{t+1} = \beta(F_{t+1} - F_t) + (1 - \beta)T_t \quad \text{(trend factor for trend portion)} \tag{3.20}$$

where

$T_t =$ trend factor for period t
$\alpha =$ exponential Smoothing factor, $0 \leq \alpha \leq 1$
$\beta =$ trend smoothing factor, $0 \leq \beta \leq 1$

A positive T_t means an upward trend, and a negative T_t means a downward trend. To start the forecasting, the initial values for the first period for both F_1 and T_1 must be known. If they are not given, one can simply set $F_1 = D_1$ and $T_1 = 0$.

By setting β less than α, the stability of the prediction increases as more weight is given to find the stationary portion of the prediction than its trend portion. As a rule of thumb, it is recommended to set α and β such that $\alpha + \beta \leq 1$. Figure 3.15 shows the relationship between α and β. Three cases can be considered:

Case (a)   If $\alpha > 0$ and $\beta = 0$, then the initial trend value as a constant is added to each value of F_t. Furthermore, if T_t is also zero, then $AF_{t+1} = F_{t+1}$. In this case, use Equation (3.19), that is, exponential smoothing.

Case (b)   If $\alpha = 1$ and $\beta > 0$, then $AF_{t+1} = D_t + T_{t+1}$.

Case (c)   If $\alpha > 0$, $\beta > 0$, and $T > 0$, then use Equation (3.18).

Note that if $\alpha > 0$, $\beta = 1$, and $T > 0$, then Equation (3.20) will become $T_{t+1} = F_{t+1} - F_t$, and Equation (3.18) will become $AF_{t+1} = F_{t+1} + F_{t+1} - F_t$ or $AF_{t+1} = 2F_{t+1} - F_t$.

TABLE 3.5   Summary of Results for Adjusted Forecast and Trend for Example 3.10 for $\alpha = 0.25$ and $\beta = 0.3$

Period, t	Demand, D_t	Unadjusted Forecasted Demand, F_t ($\alpha = 0.25$)	Trend, T_t ($\beta = 0.3$)	Adjusted Forecast, AF_t ($\alpha = 0.25$, $\beta = 0.3$)
1	5500	5500	300	5500
2	5900	5500	210	5710
3	5600	5600	177	5777
4	5700	5600	124	5724
5	6000	5625	94	5719
6	6500	5719	94	5813
7	6600	5914	124	6039
8	7200	6086	139	6224
9	7800	6364	181	6545
10	8000	6723	234	6957
11	—	7042	260	7302
MAD	—	616.9		519.4

Therefore, Equation (3.19) will result in

$$AF_{t+1} = 2F_{t+1} - F_t = 2\alpha D_t + 2(1 - \alpha)F_t - F_t = 2\alpha(D_t - F_t) + F_t.$$

Example 3.10   Adjusted Exponential Smoothing   Consider the demand of a product, D_t, for the past 10 periods as presented in Table 3.5. Suppose $\alpha = 0.25$, $\beta = 0.3$, and the initial forecast values for the first period are $F_1 = D_1 = 5500$. Also, the trend factor for the first period is 300, that is, $T_1 = 300$ units. Forecast all periods.

The solution for the forecast for each period is presented in the last column of Table 3.5. The details of calculations are as follows. For period 2, we have

$$F_2 = 0.25D_1 + (1 - 0.25)F_1$$

$$= 0.25 \times 5500 + (1 - 0.25) \times 5500 = 5500 \quad \text{[using Equation (3.19)]}$$

$$T_2 = 0.3(F_2 - F_1) + (1 - 0.3)T_1$$

$$= 0.3(5500 - 5500) + (1 - 0.3)300 = 210 \quad \text{[using Equation (3.20)]}$$

Therefore, the adjusted forecasting for period 2 is

$$AF_2 = F_2 + T_2 = 5500 + 210 = 5710 \quad \text{[using Equation (3.18)]}$$

Now consider period 3, and find F_3, T_3, and AF_3 as follows:

$$F_3 = 0.25 \times 5900 + (1 - 0.25) \times 5500 = 5600 \quad \text{[using Equation (3.19)]}$$

$$T_3 = 0.3(5600 - 5500) + (1 - 0.3)210 = 177 \quad \text{[using Equation (3.20)]}$$

$$AF_3 = 5600 + 177 = 5777 \quad \text{[using Equation (3.18)]}$$

Table 3.5 shows the calculations for the rest of the periods.

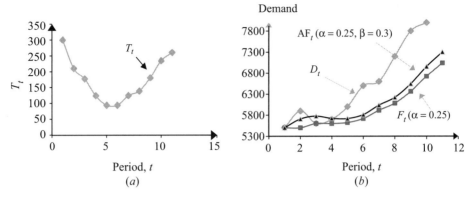

Figure 3.16   Presenting F_t, T_t and AF_t for Example 3.10 when $\alpha = 0.25$, $\beta = 0.3$; (*a*) trend amount, T_t, (*b*) adjusted forecasting factor AF_t.

Figure 3.16 shows a graph of the actual demand (D_t), forecast value by exponential smoothing only (F_t), and AF_t for this example.

See Supplement S3.10.xls.

Supplement S3.11 shows an example for impact of different β values on trend-adjusted double-exponential smoothing. It also presents a heuristic method for finding optimal α and β.

3.6   TIME SERIES: CYCLIC/SEASONAL

Cyclic or seasonal time series is applied to forecasting problems which have predictable regular cycles of highs and lows. For example, a cycle can be one year, where each year may consist of 12 periods. In some situations, the cycle duration may be very short (days) or very long (years). A season may be any length of time and does not necessarily refer to spring, summer, fall, or winter. For example, a restaurant may be busy during lunch time and dinner time. Other examples of seasonal events are Christmas, Independence Day, and Mother's Day.

There are two types of seasonal patterns: (a) with no long term trend (see Figures 3.17*a*) and (*b*) with a long-term trend (see Figure 3.17*b*). The trend can be upward or downward.

The procedure for solving seasonal time series forecasting problems consists of the following steps:

1. Estimate the long-term trend (e.g., use linear regression to estimate the trend).
2. Predict the aggregate value for each cycle.
3. Calculate the seasonal pattern for each period.
4. Combine the trend and the seasonal pattern to predict each period's forecasting.

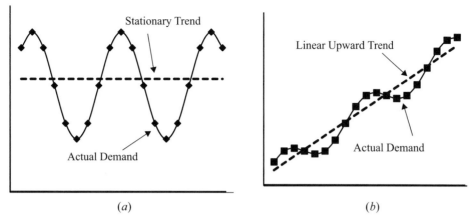

Figure 3.17   Seasonal pattern with (*a*) no long-term trend and (*b*) upward long-term trend.

The notation for this problem is presented below.

Given: Information on past K cycles:

$$N = \text{no. of periods in each cycle}$$
$$i = \text{index of each period, } i = 1, \ldots, N$$
$$K = \text{no. of cycles}$$
$$j = \text{index of each cycle, } j = 1, \ldots, K$$
$$D_{ij} = \text{given demand of period } i \text{ of cycle } j$$

Find: Information for next cycle, $K + 1$

$$D_j = \text{demand for cycle } j$$
$$F = \text{forecast for the next cycle, } K + 1$$
$$F_i = \text{forecast for period } i \text{ of next cycle, } K + 1$$

The cyclic (seasonal) approach consists of two phases:

  I. Forecast the aggregate value of the next cycle (K+1), denoted by F_{K+1}, by using the past cyclic aggregate demands D_j for $j = 1, \ldots, K$.

  II. Forecast each period of the next cycle (K+1), denoted by $F_{i,(K+1)}$, for $i = 1, \ldots, N$ periods.

Below are the steps of phase II:

1. $\bar{D}_i = (1/K) \sum_{j=1}^{K} D_{i,j}$ Average demand of period i over past K cycles for $i = 1, \ldots, N$
2. $\bar{D} = (1/N) \sum_{i=1}^{N} \bar{D}_i$ Average demand per period for all cycles
3. $SI_i = \bar{D}_i / \bar{D}$ Seasonal index for period i for $i = 1, \ldots, N$

TABLE 3.6 Example 3.11 Given Demand for Product for Past 36 Months

Month, i	1	2	3	4	5	6	7	8	9	10	11	12	Total	Average
Sales in year 1	45	51	57	59	58	55	52	48	46	50	54	59	634	52.83
Sales in year 2	59	66	69	72	71	68	66	62	58	62	67	72	792	66.00
Sales in year 3	73	79	81	84	85	83	79	76	73	74	80	86	953	79.40
Sales in year 4	?	?	?	?	?	?	?	?	?	?	?	?	1122	93.50

4. $\bar{F}_{K+1} = F_{K+1}/N$ Average per period for next cycle $(K+1)$ to be forecasted
5. $F_{i,(K+1)} = SI_i \bar{F}_{K+1}$ Forecast for period i for $i = 1, \ldots, N$ for next cycle $(K+1)$

Example 3.11 Seasonal Analysis Consider the demand for a product for the past three cycles (years), where each cycle has 12 months. Table 3.6 shows the demand for the past 36 months.

To find total sales for year 4, one can use linear regression where total sales of years 1–3, that is, 634, 792, and 953, respectively, are used to estimate the fourth year. Suppose the forecasted demand for year 4 is $F_4 = 1122$. Find the forecasted demand for each month using the seasonal index method.

The summary of the solution is presented in Table 3.7. The steps are as follows:

Step 1 Find $\bar{D}_i$ for each month. For example, average demand for the first month, $\bar{D}_1$, is

$$\bar{D}_1 = \frac{1}{3}(D_{1,1} + D_{1,2} + D_{1,3}) = \frac{1}{3}(45 + 59 + 73) = 59$$

See the first row of Table 3.7 for calculated demand averages per month.

Step 2 Find

$$\bar{D} = \frac{1}{N}\sum_{i=1}^{N}\bar{D}_i = \frac{1}{12}(59.00 + 65.33 + 69.00 + 71.67 + 71.14 + 68.76 + 65.65$$

$$+61.92 + 59.00 + 62.00 + 66.94 + 72.33) = \frac{1}{12}793 = 66.08$$

Step 3 Find the seasonal index for each month. For example, the seasonal indices for the first and second months are

$$SI_1 = \frac{\bar{D}_1}{\bar{D}} = \frac{59.00}{66.08} = 0.89$$

$$SI_2 = \frac{\bar{D}_2}{\bar{D}} = \frac{65.33}{66.08} = 0.99$$

Step 4 Find the average forecasted demand per period where the next cycle forecast is $F_4 = 1122$.

$$\bar{F}_4 = \frac{F_4}{N} = \frac{1122}{12} = 93.5$$

TABLE 3.7 Example 3.11 Solution: Forecasted Demand for Each Period of Year 4

Month, i	1	2	3	4	5	6	7	8	9	10	11	12	Total	Averages
Avg. demand for each period, $\bar{D}_i$	59.00	65.33	69.00	71.67	71.33	68.67	65.67	62.00	59.00	62.00	67.00	72.33	793.00	66.08
Seasonal index, SI_i $= \bar{D}_i / \bar{D}$	0.89	0.99	1.04	1.08	1.08	1.04	0.99	0.94	0.89	0.94	1.01	1.09	12	1
Seasonal forecast, $F_{i,4} = SI_i \times 93.5$	83	92	98	101	101	97	93.5	88	83	88	95	102	1122	93.5

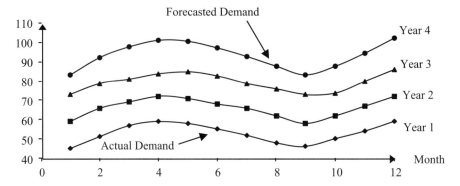

Figure 3.18   Example 3.11 Given past demands for years 1, 2, and 3; the demand for year 4 is forecasted using the seasonal index method.

Step 5   Find the forecasted demand for every period i by using $F_{i,4} = SI_i \bar{F}_4$. For example, the forecasted sales for the first and second periods of year 4 are

$$F_{1,4} = 0.89 \times 93.5 = 83$$
$$F_{2,4} = 0.99 \times 93.5 = 92$$

The seasonal forecast for every month of year 4 is presented in the last row of Table 3.7. The complete results of the calculations are given in the Table 3.7 and also presented in Figure 3.18.

3.7   LINEAR REGRESSION

3.7.1   Linear Regression

Causal forecasting represents the relationship between causes and effects through a mathematical function. In the time series forecasting methods covered in the last sections, forecasting was a function of time (e.g., year, month, day). In causal forecasting, prediction can be based on any variable including time. For example, the total sales of cars (as a dependent variable) can be forecasted using the gross national product (GNP) (as an independent variable). Also, the amount of wear on a tire (as a dependent variable) is a function of the number of miles that a car has traveled (as the independent variable). In this section, we cover linear regression based on one cause (independent variable) and one effect (dependent variable). In Section 3.8, we discuss multiple linear regression where there are multiple decision variables (independent variables) that contribute to a single output (dependent variable). In Section 3.9, we discuss multiple quadratic (additive) regression, and in Section 3.10, we develop multiple nonlinear regression using Z theory. A polynomial (additive) regression of one independent variable x to a dependent variable y is given as

$$\hat{y} = b_0 + b_1 x + b_2 x^2 + \cdots + b_n x^n \tag{3.21}$$

where $b_0, b_1, \ldots, b_n$ are coefficients whose values must be assessed. The simplest form of regression function (3.21) is a single linear regression in the form

$$\hat{y} = a + bx \tag{3.22}$$

where

    $y =$ dependent variable, e.g., demand values
    $x =$ independent variable, e.g., time periods
    $a =$ initial value of y
    $b =$ rate of change in y per unit of x
$a, b =$ coefficients that need to be estimated for given data

Equation (3.22) represents a straight line. Consider a set of given data points (x_1, y_1), $(x_2, y_2), \ldots, (x_n, y_n)$. These points can be represented as (x_i, y_i) for $i = 1, 2, \ldots, n$. Consider Equation (3.22). The problem is to find (a, b) such that the sum of the square errors (deviations) in Equation (3.23) is minimized:

$$\text{Minimize } T(a, b) = \sum_{i=1}^{n} [y_i - (a + bx_i)]^2 \tag{3.23}$$

The optimal values of a and b can be obtained by taking the partial derivative of $T(a, b)$ with respect to a and b and setting them equal to zero:

$$\frac{\partial T}{\partial a} = -2 \sum_{i=1}^{n} [y_i - (a + bx_i)] = 0$$

$$\frac{\partial T}{\partial b} = -2 \sum_{i=1}^{n} x_i [y_i - (a + bx_i)] = 0$$

By solving the above two sets of linear equations, the following solutions are found:

$$b = \frac{\sum_{i=1}^{n} x_i y_i - n\bar{x}\bar{y}}{\sum_{i=1}^{n} x_i^2 - n\bar{x}^2} \tag{3.24}$$

$$a = \bar{y} - b\bar{x} \tag{3.25}$$

where

$$\bar{x} = \frac{1}{n} \sum_{i=1}^{n} x_i \quad \text{and} \quad \bar{y} = \frac{1}{n} \sum_{i=1}^{n} y_i$$

Alternatively, we can represent b as

$$b = \frac{S_{XY}}{S_{XX}} \tag{3.26}$$

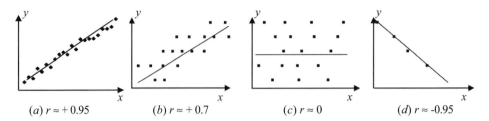

(a) $r \approx +0.95$ (b) $r \approx +0.7$ (c) $r \approx 0$ (d) $r \approx -0.95$

Figure 3.19   Four r values for linear correlation coefficient for x and y: (a) very strong positive; (b) strong positive; (c) no correlation; (d) very strong negative.

where

$$S_{xy} = \sum_{i=1}^{n} x_i y_i - n\bar{x}\bar{y} \tag{3.27}$$

$$S_{xx} = \sum_{i=1}^{n} x_i^2 - n\bar{x}^2 \tag{3.28}$$

$$S_{YY} = \sum_{i=1}^{n} y_i^2 - n\bar{y}^2 \tag{3.29}$$

The variance error of estimation is defined as the MSE s_e^2 and the standard error of estimation is defined by S_e:

$$S_e^2 = \frac{S_{XX} S_{YY} - S_{XY}^2}{S_{XX}(n-2)} \tag{3.30}$$

Correlation Coefficient   The correlation coefficient r measures the degree or strength of the relationship between x and y. The correlation coefficient is expressed as

$$r = \frac{S_{XY}}{\sqrt{S_{XX} S_{YY}}} \tag{3.31}$$

The correlation coefficient ranges from -1 to 1. Values closer to 1 indicate a strong positive (increasing) relationship between x and y, while values closer to -1 indicate a strong negative (decreasing) relationship between x and y. When the correlation coefficient is close to zero, the relationship between x and y is weak. For a linear regression model to be valid, a high $|r|$ is required; usually $|r| > 0.75$ is an acceptable range. Figure 3.19 shows four examples of correlation coefficients for linear regression.

An equivalent alternative approach to calculate correlation is Equation (3.32), but this equation does not provide information about negative or positive correlation; the sign of r should be determined afterward:

$$r = \pm \sqrt{\sum_{i=1}^{n} (\hat{y}_i - \bar{y})^2 / \sum_{i=1}^{n} (y_i - \bar{y})^2} \tag{3.32}$$

TABLE 3.8   Summary of Solution for Example 3.12

Data point, i	1	2	3	4	5	Total	Average
x_i	25	28	15	11	20	99	19.8
y_i	322	395	246	235	268	1,466	293.2
x_i^2	625	784	225	121	400	2,155	–
$x_i y_i$	8,050	11,060	3,690	2,585	5,360	30,745	–
y_i^2	103,684	156,025	60,516	55,225	71,824	447,274	–
Forecasted sales, $\hat{y}_i$	339.07	365.53	250.86	215.58	294.96	1,466	293.2
Error, $e_i = \hat{y}_i - y_i$	17.064	−29.476	4.864	−19.416	26.964	MAD	19.56

where $\bar{y}$ is the average of all y_i and $\hat{y}_1$ is the estimated value for y_i using the linear regression function. The sign of the correlation coefficient can be determined as follows:

If y is an increasing function of x, then r is positive [i.e., the slope of the regression line (b) is positive].

If y is a decreasing function of x, then r is negative [i.e., the slope of the regression line (b) is positive].

Example 3.12   Linear Regression and Correlation   Suppose the sale of a product is a function of the number of salespeople working for a company. The history of sales for the past five months is shown in the table below. It is predicted that in the next month (the sixth month) the number of salespeople will be 18. The company wishes to plan for its production based on the prediction of the sixth month total sales (in \$10,000). Find the estimated mean of sales, the mean standard error of estimation, and the correlation between the total sales and the number of salespeople. Note that this is not a time series problem because the total sales is not a function of time; it is a function of the number of salespeople.

Date point, i	1	2	3	4	5	6
No. of salespeople, x_i	25	28	15	11	20	18
Total sales, y_i	322	395	246	235	268	?

The summary of the solution is presented in Table 3.8.

The details of finding the linear equation coefficients a and b are as follows:

$$S_{XY} = \sum_{i=1}^{5} x_i y_i - 5\bar{x}\bar{y} = 30,745 - 5(19.8)(293.2) = 1718.2$$

$$S_{XX} = \sum_{i=1}^{5} x_i^2 - 5\bar{x}^2 = 2155 - 5(19.8)^2 = 194.8$$

$$S_{YY} = \sum_{i=1}^{5} y_i^2 - 5\bar{y}^2 = 447274 - 5(293.2)^2 = 17442.8$$

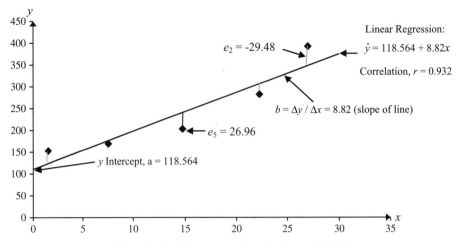

Figure 3.20   Linear regression for Example 3.12.

Hence,

$$b = \frac{S_{XY}}{S_{XX}} = \frac{1718.2}{194.8} = 8.82$$

$$a = \bar{y} - b\bar{x} = 293.2 - 8.82 \times 19.8 = 118.56$$

Therefore, the causal linear regression equation for this example is

$$\hat{y} = 118.56 + 8.82x$$

We can use the above equation to find the forecasted sales $\hat{y}$ for any given number of salespeople x. The forecasted values $\hat{y}$ for the given five points are presented in the Table 3.8. In linear regression, the average of actual value y and the average of forecasted values $\hat{y}$ are always the same, (e.g., 293.2 in this example). Also, the total error is zero in all regression solutions.

This function is presented in Figure 3.20. Note that the order of the data points (i.e., months) is not in the order of increasing value of x. For example, month 2 appears as the last point in Figure 3.20. If the number of salespeople was 18 (i.e., $x = 18$), the total sales would be

$$\hat{y}_6 = 118.564 + 8.82 \times 18 = 277.324$$

The standard error of estimation using Equation (3.30) for this example is

$$S_e^2 = \frac{S_{XX}S_{YY} - S_{XY}^2}{S_{XX}(n-2)} = \frac{194.8 \times 17442.8 - 1718.2^2}{194.8(5-2)} = 762.57 \quad \text{or} \quad S_e = \sqrt{762.57} = 27.61$$

The correlation coefficient using Equation (3.31) is

$$r = \frac{S_{XY}}{\sqrt{S_{XX}S_{YY}}} = \frac{1718.2}{\sqrt{194.8 \times 17442.8}} = 0.932$$

TABLE 3.9   Demand Data for Past Five Periods

Period $t = x_t$	1	2	3	4	5
Demand, y_t	1650	2600	2700	3600	3700

Alternatively, the correlation coefficient obtained using Equation (3.32) is

$$r = \pm \sqrt{\frac{\sum_{i=1}^{n} (\hat{y}_i - \bar{y})^2}{\sum_{i=1}^{n} (y_i - \bar{y})^2}} = \pm \sqrt{\frac{\sum_{i=1}^{5} (\hat{y}_i - 293.2)^2}{\sum_{i=1}^{8} (y_i - 1.34)^2}} = \pm \sqrt{\frac{15,155.09}{17,442.8}} = \pm 0.932$$

The sign of ± 0.932 is positive because $b = 8.82$ is positive. The correlation $r = 0.932$ means that the total sales strongly depends on the total number of salepeople. This is a very high correlation, which justifies the use of linear regression.

See Supplement S3.12.xlsx.

3.7.2   Linear Regression for Time Series

To use linear regression for causal forecasting, we use the following notation. Denote each period t by x_t for $t = 1, 2, \ldots, n$. This is a special case of linear regression where $x_t = t$.

$$X_1 = 1, \quad X_2 = 2, \ldots, \quad X_n = n$$

Also, denote the actual demand D_t by y_t for $t = 1\ 2, \ldots, n$:

$$y_1 = D_1, \quad y_2 = D_2, \ldots, \quad y_n = D_n$$

Also denote the forecasted demand for each period F_t as $\hat{y}_t$.

The linear regression method $\hat{y} = a + bx$ (3.22) finds a straight line that is the best fit for the data.

Supplement S3.13.doc explains multicriteria for trend analysis with linear regression.

Example 3.13   Linear Regression for Time Series   Consider demand data for the past five periods as shown in Table 3.9. Find the forecasted demand for periods 6 and 10.

Table 3.10 summarizes the calculations needed to determine the values of a and b of Equations (3.24) and (3.25).

Now use the calculations from Table 3.10 to find the linear regression parameters b and a:

$$b = \frac{\sum_{t=1}^{n} x_t y_t - n \bar{x} \bar{y}}{\sum_{t=1}^{n} x_t^2 - n \bar{x}^2} = \frac{47,850 - 5(3)(2850)}{55 - 5(3)^2} = 510 \quad \text{[using Equation (3.24)]}$$

$$a = \bar{y} - b\bar{x} = 2850 - 510(3) = 1320 \quad \text{[using Equation (3.25)]}$$

TABLE 3.10 Calculation of Values Used in Linear Regression Example 3.13

						Total	Average
Period $t = x_t$	1	2	3	4	5	15	$\bar{x} = 3$
y_t	1,650	2,600	2,700	3,600	3,700	14,250	$\bar{y} = 2850$
x_t^2	1	4	9	16	25	55	
$x_t y_t$	1,650	5,200	8,100	14,400	18,500	47,850	

Therefore, the linear regression equation for this example is

$$\hat{y} = 1320 + 510x$$

This equation can be used to find the forecast for any given period x_t.

To find the forecast for period 6, use $x = 6$ in the above equation and then calculate the value of y:

$$F_6 = \hat{y}_6 = 1320 + 510 \times 6 = 4380$$

Thus, the forecasted demand for period 6 is 4380. All future periods can be forecasted using the linear regression equation. The forecasted demand for period 10 is

$$F_{10} = \hat{y}_{10} = 1320 + 510 \times 10 = 6420$$

Figure 3.21 shows the linear regression line that predicts the demand along with the actual demand points.

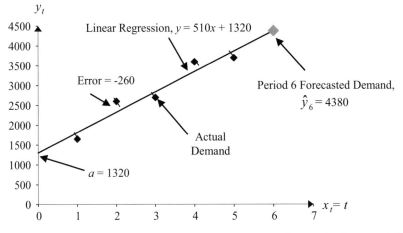

Figure 3.21 Time series linear regression forecasting for Example 3.13 for five actual demands where slope of line is $b = \Delta y/\Delta x = 510$.

Linear regression can be used to predict future demands for any given period. The table below shows forecasted values for periods 1–5 using linear regression:

	1	2	3	4	5	Sum	Average
Actual demand, $D_t = y_t$	1,650	2,600	2,700	3,600	3,700	14,250	2850
Forecasted values, $F_t = \hat{y}_t$	1,830	2,340	2,850	3,360	3,870	14,250	2850
Error, $\lvert F_t - D_t \rvert = \lvert \hat{y}_t - y_t \rvert$	180	260	150	240	170	1,000	MAD = 200

Note that the average actual demand is the same as the average forecasted demand. Also, the summation of errors is zero. The above two conditions are always true for linear regression.

See Supplement S3.14.xls.

3.7.3 Multicriteria Linear Regression[†]

In most real-world problems, forecasting is about making certain commitments based on the forecasted value. For example, based on the prediction of the sales of its products, a company may plan for its production, staffing, needed equipment, and other required resources. In multicriteria linear regression, the DM can choose the best multicriteria alternative. For a risk-averse DM, depending on the problem, the DM may wish either to underestimate (e.g., to minimize possible surpluses) or to overestimate (e.g., to minimize possible shortages).

Consider a normal probability distribution. The predicted quantity (e.g., demand) corresponding to a given z value of the normal distribution can be found by

$$\hat{y}_z = \hat{y} + zS_e \qquad (3.33)$$

where $\hat{y}$ is the estimated mean that minimizes the total error according to Equation (3.23), that is, it is the solution to the linear regression problem. The predicted value $\hat{y}_z$ is associated with a given z. Different alternatives can be generated by considering different values of z where z ranges from -3 to $+3$. When $z < 0$, then zS_e is the amount of shortage. When $z > 0$, then zS_e is the amount of surplus. The probability $P(y \geq \hat{y}_z)$ is the probability that the actual demand y will be greater than or equal to the planned quantity $\hat{y}_z$.

The following two bicriteria linear regression problems can be defined:

Bicriteria Problem When Underestimation Is Preferred (to Maximize Shortages)   Choose a $\hat{y}_z$ which is less than $\hat{y}$, that is, $z < 0$:
Minimize f_1 = deviation from estimated mean $e = \lvert \hat{y} - \hat{y}_z \rvert = \lvert zS_e \rvert$.
Maximize $f_2 = P(y \geq \hat{y}_z)$ (this maximizes shortages because $z < 0$).

Bicriteria Problems When Overestimation Is Preferred (to Maximize Surpluses)   Choose a $\hat{y}_z$ which is greater than $\hat{y}$, that is, $z > 0$.
Minimize f_1 = deviation from estimated mean $e = \lvert \hat{y} - \hat{y}_z \rvert = \lvert zS_e \rvert$.
Minimize $f_2 = P(y \geq \hat{y}_z)$ (this maximizes surpluses because $z > 0$).

TABLE 3.11 Probability That Actual Quantity y Will Be Higher Than Predicted Values $\hat{y}_z$ for Example 3.12 ($\hat{y} = 171.46, \hat{y}_z = \hat{y} + zS_e$)

	Lower Than Estimated Mean			Estimated Mean,	Higher than Estimated Mean		
	$z = -3$	$z = -2$	$z = -1$	$z = 0$	$z = 1$	$z = 2$	$z = 3$
Predicted sales, $\hat{y}_z$ ($)	88.60	116.22	143.84	171.48	199.08	226.70	254.32
Min. $f_1 = $ |Error| $= \|zS_e\|$	|−82.84|	|−55.23|	|−27.62|	0	|27.62|	|55.23|	|82.84|
Max. $f_2 = P(y \geq \hat{y}_z)$, %	99.87	97.72	84.13	50.00	15.87	2.28	0.13
Efficient for underestimation	Yes	Yes	Yes	Yes	No	No	No

Then, a multicriteria method can be used to choose the best alternative. The details of the method are presented in the following example.

Example 3.14 Bicriteria of Linear Regression When Underestimation Is Desired
Consider Example 3.12 where sales y is a function of the number of salespeople x. Identify multicriteria alternatives associated with $z = -3, -2, -1, 0, 1, 2, 3$ (i.e., seven alternatives) for this problem. Also, suppose that the DM is risk averse (conservative) and the DM maximizes shortages, that is, $z < 0$, and maximizes $f_2 = P(y \geq \hat{y}_z)$. Find the production quantity for which the company should plan.

In Example 3.12, the total production (sales) was estimated to be $171.48 (in $1000) for 18 salespeople where the estimated error $S_e = 27.62$. In Table 3.11, the probability of the sales being higher than the predicted value of $171.48 is presented. For example, $z = -2$ corresponds to a 97.72% confidence that the predicted value will be higher than 116.24:

$$\hat{y}_z = \hat{y} + zS_e = 171.48 - 2 \times 27.62 = 116.24$$

Also, there is a 99.87% probability that the sales will be higher than $171.48 - 3 \times 27.62 = 88.62$.

The predicted values for seven different z values are presented in Table 3.11.

Now a multicriteria approach can be used to choose the best multicriteria alternative. For example, if the alternative associated with $z = -1$ is selected, it means that the company prefers to choose $143.84 as the predicted sales. Therefore, the company faces shortages 84.13% of the time. The company prefers the alternative with 84.13% probability confidence that the total sales will be higher than $143.84 (in $10,000). Based on this sales information, the company can make a production plan for the given scenario. Figure 3.22 shows the range of sales values y associated with the normal probability distribution for this example: (a) prediction based on minimizing the least-squares error (use of linear regression), $\hat{y}_{z=0} = 118.56 + 8.82x$ (for $z = 0$); (b) maximum underestimation, $\hat{y}_{z=-3} = 118.56 + 8.82x - 82.86$ (for $z = -3$); and (c) maximum overestimation, $\hat{y}_{z=+3} = 118.56 + 8.82x + 82.86$ (for $z = 3$).

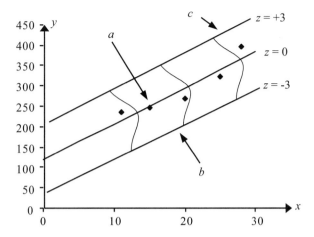

Figure 3.22 Range of forecasted values associated with probability distribution for Example 3.13.

3.8 MULTIPLE-VARIABLE LINEAR REGRESSION

In multiple-variable linear regression, there is more than one cause (independent variable) but there is always one effect (dependent variable). For example, the number of homes sold (as the dependent variable) may depend upon two independent variables: the 15-year fixed mortgage rate and the gross national product (GNP). A lower interest rate and higher GNP is associated with a higher number of homes sold. Multiple linear regression can be written as a generalization of a single linear regression equation (3.22):

$$\hat{y} = a + b_1 x_1 + b_2 x_2 + \cdots + b_k x_k \tag{3.34}$$

where

$\hat{y} =$ forecasted (dependent variable), e.g., demand
$x_k =$ independent variable (cause k) for $k = 1, \ldots, K$
$a =$ estimated y intercept of regression line
$b_k =$ estimated coefficient associated with variable x_k for $k = 1, 2, \ldots, K$

There are many commercial statistical analysis software packages that can be used to solve the above problem. Conventional regression methods use derivative-based methods similar to the method of the last section (see the references at the end of the chapter) to solve this problem. Here (and in Section 3.9), we solve regression problems by optimization as the reader is familiar with optimization methods. Therefore, we propose the following multiple linear regression by quadratic optimization (MLR-QO) to solve multiple linear regression problems:

3.8.1 Multiple Linear Regression by Quadratic Optimization

The multiple linear regression (MLR) problem can be formulated as the following optimization problem.

PROBLEM 3.1 MLR-QO

Minimize:

$$\text{Total square of errors (TSE)} = \sum_{i=1}^{n} \left(e_i^+\right)^2 + \sum_{i=1}^{n} \left(e_i^-\right)^2 \tag{3.35}$$

Subject to:

$$a + b_1 x_{1,i} + b_2 x_{2,i} + \cdots + b_K x_{K,i} + e_i^+ - e_i^- = y_i \quad \text{for} \quad i = 1, 2, \ldots, n \tag{3.36}$$

$$e_i^+ \geq 0 \quad e_i^- \geq 0 \qquad \text{for} \quad i = 1, 2, \ldots, n \tag{3.37}$$

$$a \text{ and } b_k \text{ unrestricted in sign} \qquad \text{for} \quad k = 1, 2, \ldots, K \tag{3.38}$$

where e_i^+ and e_i^- are the positive and negative errors of the predicted $\hat{y}$ from the actual y.

Solve the above problem using a quadratic optimization software package to find the optimal values for decision variables $a, b_1, b_2, \ldots, b_k$. Then use these values in Equation (3.34).

The details of this procedure are shown in the following example.

Example 3.15 Multiple Linear Regression by (Quadratic Optimization)   Two main competitors provide the market share of high-speed Internet in the greater Cleveland area in Ohio: AT&T high-speed service and Time Warner Cable. Both companies advertise to attract customers. The total income in a given year for AT&T high-speed Internet service in Ohio is related to its advertising expenditure in a year and the ratio of its advertising to Time Warner Cable's advertising expenditure. The data for the past eight years are presented in Table 3.12. It is estimated that AT&T's advertising budget for year (data point) 9 will be $3.02 million. Also, it is estimated that the ratio of AT&T's advertising budget to Time

TABLE 3.12 Data and Solution for Example 3.15

Data Point/ Year, i	Problem: Given Data for Past 8 years			Forecasted Solution: MLR-QO	
	AT&T Advertising (in $Million), $x_{1,i}$	Advertising Ratio to Time Warner, $x_{2,i}$	AT&T Income, (in $100 Million), y_i	AT&T Income, (in $100 Million), $\hat{y}_i$	Error, $e_i = y_1 - \hat{y}_i$
1	1.30	1.25	1.15	1.04	0.11
2	1.80	1.34	1.05	1.14	−0.09
3	1.00	1.22	1.22	0.99	0.23
4	1.20	1.78	1.14	1.21	−0.07
5	1.10	2.15	1.15	1.32	−0.17
6	1.60	2.43	1.29	1.49	−0.20
7	1.30	2.87	1.62	1.59	0.03
8	2.06	3.54	2.10	1.93	0.17
9	3.02	3.59	?	2.08	MAD = 0.13

Warner Cable's will be about 3.59 for year (data point) 9. Predict AT&T's total income for data point/year 9.

Table 3.12 gives AT&T's advertising budget and the ratio of its advertising budget to Time Warner Cable's advertising budget for the past eight years.

Based on the data in Table 3.12, determine the multiple linear regression equation that best fits the data using the MLR-QO method.

The MLR-QO problem formulation based on Problem 3.1 is presented below:

Minimize:

$$TSE = \sum_{i=1}^{8} \left(e_i^+\right)^2 + \sum_{i=1}^{8} \left(e_i^-\right)^2$$

Subject to:

$a + b_1 1.3 + b + 21.25 + e_1^+ - e_1^- = 1.15$ $a + b_1 1.1 + b_2 2.15 + e_5^+ - e_5^- = 1.15$

$a + b_1 1.8 + b_2 1.34 + e_2^+ - e_2^- = 1.05$ $a + b_1 1.6 + b_2 2.43 + e_6^+ - e_6^- = 1.29$

$a + b_1 1 + b_2 1.22 + e_3^+ - e_3^- = 1.22$ $a + b_1 1.3 + b_2 2.87 + e_7^+ - e_7^- = 1.62$

$a + b_1 1.2 + b_2 1.78 + e_4^+ - e_4^- = 1.14$ $a + b_1 2.06 + b_2 3.54 + e_8^+ - e_8^- = 2.10$

a, b_1 and b_2 unrestricted in sign

$e_i^+ \geq 0$ $e_i^- \geq 0$ for $i = 1, \ldots, 8$

Solving the above problem using a quadratic optimization software package (LINGO was used here), the solution is

$$a = 0.441 \quad b_1 = 0.137 \quad b_2 = 0.340$$

The multiple linear regression equation is

$$\hat{y} = 0.441 + 0.137x_1 + 0.340x_2$$

Therefore, the forecasted AT&T income for year 9 is

$$\hat{y}_9 = 0.441 + 0.137 \times 3.02 + 0.340 \times 3.59 = 2.08$$

The forecasted AT&T income for the past eight years, based on the above equation, is also presented in Table 3.12.

Figure 3.23 shows a three-dimensional plot of the multiple linear regression equation for this example. Notice that the regression function closely resembles the actual data.

See Supplements S3.15.lg4, S3.15.xlsx.

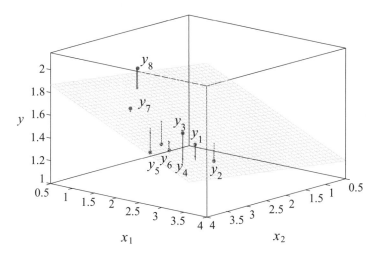

Figure 3.23 Three-dimensional plot of predicted plane of values based on multiple linear regression and actual values for Example 3.15.

3.8.2 MSE and Correlation for Multiple Linear Regression[†]

The sum-of-square error (SSE) measures the amount of variation that is not explained by the proposed regression model. The SSE for single and multiple linear regression is defined by the equation

$$SSE = \sum_{i=1}^{n}(y_i - \hat{y}_i)^2 \tag{3.39}$$

For a single-variable linear regression, Equation (3.30) was used to find the standard error of estimation S_e. There is another way to find the MSE, S_e^2 and standard error of estimation S_e for single linear regressions, shown in Equation (3.40). The MSE S_e^2 is also applicable for multiple linear regression:

$$S_e^2 = \frac{SSE}{n - p} = \frac{\sum_{i=1}^{n}(y_i - y_i)^2}{n - p} \tag{3.40}$$

where n is the number of data points and p is the number of estimated parameters. Note that for single linear regression [Equation (3.22)] two parameters a and b are estimated; therefore $p = 2$. Then, Equation (3.40) is equivalent to Equation (3.30).

For multiple linear regression [Equation (3.34)], there are k estimated coefficients $(b_1, \ldots, b_k)$ and one estimated intercept (a); therefore, $p = k + 1$.

The correlation coefficient r for multiple linear regression is computed using the same equation that was used for single linear regression [see Equations (3.32)].

Example 3.16   Finding MSE and Correlation   Consider Example 3.15 and its solution presented in Table 3.12.

(a) Find the MSE and the correlation for this regression equation.

(b) Compare the results of the following three models for predicting y:

    (i) Single linear regression (SLR) considering only the first independent variable x_1

    (ii) Single linear regression considering only the second independent variable x_2

    (iii) Multiple linear regression (MLR) considering both independent variables (x_1 and x_2)

To compare single linear regression to multiple linear regression, find the MAD, S_e^2 and correlation r for all three models.

(a) In this example, $n = 8$ and $p = 3$. The details for computing the SSE, S_e^2, and correlation are shown below where $\bar{y} = 1.34$:

$$\text{SSE} = \sum_{i=1}^{8} (y_i - y_i)^2 = (1.15 - 1.04)^2 + (1.05 - 1.14)^2 + (1.22 - 0.99)^2$$

$$+ (1.14 - 1.21)^2 + (1.15 - 1.32)^2$$
$$+ (1.29 - 1.49)^2 + (1.62 - 1.59)^2$$
$$+ (2.10 - 1.93)^2 = 0.18$$

$$S_e^2 = \frac{\text{SSE}}{n - p} = \frac{0.1767}{8 - 3} = 0.035$$

$$r = \sqrt{\frac{\sum_{i=1}^{8} (\hat{y}i - \bar{y})^2}{\sum_{i=1}^{8} (yi - \bar{y})^2}} = \sqrt{\frac{0.70}{0.87}} = 0.90$$

$$\sum_{i=1}^{8} (\hat{y}_i - \bar{y})^2 = \sum_{i=1}^{8} (\hat{y}i - 1.34)^2 = (1.04 - 1.34)^2 + (1.14 - 1.34)^2$$

$$+ (0.99 - 1.34)^2 + (1.21 - 1.34)^2$$
$$+ (1.32 - 1.34)^2 + (1.49 - 1.34)^2$$
$$+ (1.59 - 1.34)^2 + (1.93 - 1.34)^2 = 0.7$$

$$\sum_{i=1}^{8} (yi - \bar{y})^2 = \sum_{i=1}^{8} (yi - 1.34)^2 = (1.15 - 1.34)^2 + (1.05 - 1.34)^2$$

$$+ (1.22 - 1.34)^2 + (1.14 - 1.34)^2$$
$$+ (1.15 - 1.34)^2 + (1.29 - 1.34)^2$$
$$+ (1.62 - 1.34)^2 + (2.10 - 1.34)^2 = 0.87$$

Note that $S_e^2 = 0.035$ is a small number, that is, the method is valid from the MSE point of view. The correlation is about 0.9. This correlation shows a strong linear relationship between the two independent variables and the dependent variable. Therefore, this multiple linear regression equation is a valid model for this problem.

(b) Table 3.13 shows the summary of the final results for the three models: (i), (ii), and (iii). The multiple linear regression model (iii) has the highest correlation and a low MSE; it is the best model for this example.

TABLE 3.13   Summary of Three Models for Example 3.15

Model	Method	Regression Equation	MAD	S_e^2	r
(i) Single linear regression using only, x_1	SLR (Section 3.7.1)	$\hat{y} = 0.57 + 0.54x_1$	0.22	0.099	0.57
(ii) Single linear regression using only, x_2	SLR (Section 3.7.1)	$\hat{y} = 0.57 + 0.37x_2$	0.13	0.031	0.89
(iii) Multiple linear regression using x_1 and x_2	MLR (Section 3.8.1)	$\hat{y} = 0.441 + 0.137x_1 + 0.340x_2$	0.13	0.035	0.90

Note: Model (iii) has the best correlation and fitting.

Note that in this example the multiple linear regression model (iii) is a slight improvement to the single linear regression model (ii), but it is much better than model (i). Depending on the data, the multiple linear regression may be substantially better than all different single-variable models.

See Supplement S3.16.xls.

3.9   QUADRATIC REGRESSION

Quadratic Forecasting   In some applications, linear regression may not provide accurate forecasting. For example, consider a problem with one independent variable as presented in Figure 3.24. In this example, the graphed data have a quadratic shape. Two forecasting functions are presented:

(a) A linear function which appears to be a poor fit
(b) A quadratic function which appears to be a better fit

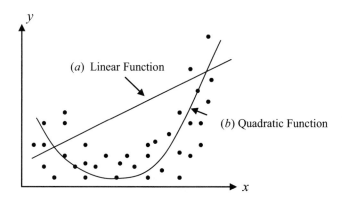

Figure 3.24   Comparison of linear regression and quadratic regression functions: (*a*) linear forecasting regression function (inappropriate to use); (*b*) quadratic forecasting regression function (appropriate to use).

In this section, we develop our approach for solving a quadratic regression equation. The optimal solution for this problem can be obtained by solving a quadratic optimization problem that will be defined in the next section. It should be noted that by using a quadratic regression function a better forecast may not necessarily be achieved, even if the total forecasting error is zero. That is, overfitting of the data may occur when using quadratic regression resulting in a poor model (see Section 3.2.5 for an explanation). Therefore, quadratic regression should be used carefully, usually when the quadratic pattern can be verified. For example, a quadratic function is useful to find the pattern of cycles (see the seasonal time series Section 3.6).

3.9.1 Regression by Quadratic Optimization

A well-known quadratic (in terms of the variables) linear (in terms of the parameters) regression model (3.41) which is a special case of (3.21) is

$$\hat{y} = b_0 + b_1 x + b_2 x^2 \tag{3.41}$$

In the same spirit of quadratic optimization in Problem 3.1, we develop the following quadratic optimization approach for solving quadratic single-variable problems. In this formulation, we use (3.41) because it is much easier to solve an optimization problem with linear constraints. Therefore constraints (3.43) are generated based on (3.41), where e_i^+ and e_i^- are the positive and negative errors of the predicted y from the actual y.

PROBLEM 3.2 QUADRATIC LINEAR REGRESSION BY QUADRATIC OPTIMIZATION (QLR-QO)

Minimize

$$\text{TSE} = \sum_{i=1}^{n} \left(e_i^+\right)^2 + \sum_{i=1}^{n} \left(e_i^-\right)^2 \tag{3.42}$$

Subject to:

$$b_0 + b_1 x_i + b_2 x_i^2 + e_i^+ - e_i^- = y_i \qquad \text{for } i = 1, \ldots, n \tag{3.43}$$

$$e_i^+ \geq 0 \quad e_i^- \geq 0 \qquad \text{for } i = 1, \ldots, n \tag{3.44}$$

b_0, b_1, b_2 unrestricted in sign

Example 3.17 Quadratic Linear Regression   Consider a problem with one independent variable x and one dependent variable y. Information for the past nine periods is given in the first three columns of Table 3.14.

(a) Use QLR to forecast the value for the 10th period.

(b) Use linear regression (Section 3.7.1) to forecast the value for the 10th period.

(c) Compare the above two solutions and identify which one is more accurate; also calculate the MSE.

TABLE 3.14   Example 3.17 Problem and Solution for Linear and Quadratic Regression

Data Point/ Year, i	Problem: Given Data		(a) Solution: Using Single QLR (Problem 9.3.2), Estimated Value, $\hat{y}_i$	(b) Solution: Using Linear Regression (Section 3.7.1), Estimated Value, $\hat{y}_i$
	x_i	y_i		
1	12	34.46	36.19	32.34
2	23	30.16	31.55	34.27
3	16	35.14	33.26	33.04
4	18	33.28	32.32	33.39
5	34	37.91	37.69	36.20
6	21	32.84	31.59	33.92
7	32	36.21	35.77	35.85
8	35	39.58	38.79	36.38
9	29	31.14	33.56	35.32
10	20	?	31.79	33.74
—	—	—	MAD $= 1.17$	MAD $= 2.11$

(a) The single QLR problem is formulated as follows:

Minimize

$$\text{TSE} = \sum_{i=1}^{9} \left(e_i^{+}\right)^2 + \sum_{i=1}^{9} \left(e_i^{-}\right)^2$$

Subject to:

$b_0 + 12b_1 + (12)^2 b_2 + e_1^{+} - e_1^{-} = 34.46$
$b_0 + 23b_1 + (23)^2 b_2 + e_2^{+} - e_2^{-} = 30.16$
$b_0 + 16b_1 + (16)^2 b_2 + e_3^{+} - e_3^{-} = 35.14$
$b_0 + 18b_1 + (18)^2 b_2 + e_4^{+} - e_4^{-} = 33.28$
$b_0 + 34b_1 + (34)^2 b_2 + e_5^{+} - e_5^{-} = 37.91$
$b_0, b_1,$ and b_2 unrestricted
$e_i^{+} \geq 0 \qquad e_i^{-} \geq 0 \quad$ for $i = 1, \ldots, 9$

$b_0 + 21b_1 + (21)^2 b_2 + e_6^{+} - e_6^{-} = 32.84$
$b_0 + 32b_1 + (32)^2 b_2 + e_7^{+} - e_7^{-} = 36.21$
$b_0 + 35b_1 + (35)^2 b_2 + e_8^{+} - e_8^{-} = 39.58$
$b_0 + 29b_1 + (29)^2 b_2 + e_9^{+} - e_9^{-} = 31.14$

Solving this problem using LINGO optimization software, the solution is

$$b_0 = 53.59 \qquad b_1 = -1.99 \qquad b_2 = 0.045$$

Using these values in Equation (3.45), the QLR function is

$$\hat{y} = 53.59 - 1.99x + 0.045x^2$$

TABLE 3.15 Comparison of Quadratic and Linear Regression

	Function	MAD	S_e^2	r
(a) Quadratic linear regression	$\hat{y} = 53.59 - 1.99x + 0.045x^2$	$\dfrac{10.55}{9} = 1.17$	$\dfrac{17.7}{9-3} = 2.95$	$\sqrt{\dfrac{57.42}{75.13}} = 0.874$
(b) Linear regression	$\hat{y} = 30.23 + 0.1756x$	$\dfrac{18.97}{9} = 2.11$	$\dfrac{57.78}{9-2} = 8.25$	$\sqrt{\dfrac{17.34}{75.13}} = 0.481$

Therefore, for period 10, the forecasted value is

$$\hat{y}_{10} = 53.59 - 1.99(20) + 0.045(20)^2 = 31.79$$

Table 3.14 shows the forecasted $\hat{y}$ values using quadratic regression for all nine periods.

(b) The solution for linear regression (Section 3.7.1) is

$$\hat{y} = 30.23 + 0.1756x$$

Table 3.14 shows the forecasted values $\hat{y}$ using linear regression for all nine periods.

(c) To compare the two methods, we find MAD, S_e^2, and the correlation r for the two methods. The results are shown in Table 3.15.

The variance error of estimation, Equation (3.40), of multiple linear regression can also be used for quadratic regression. For linear regression (single variable) there are two estimated parameters ($p = 2$). For quadratic regression, Equation (3.41), three parameters (b_0, b_1, and b_2) are estimated. Therefore, $p = 3$ as in Equation (3.45):

$$S_e^2 = \frac{\text{SSE}}{n-3} = \frac{\sum\limits_{i=1}^{n} (y_i - \hat{y}_i)^2}{n-3} \tag{3.45}$$

Figure 3.25 compares the actual value to the solutions of the two methods for solving this problem. To find out which of the two methods is more accurate, the correlation, S_e^2, and the MAD of both methods are calculated in Table 3.15.

See Supplements S3.17.lg4, S3.17.xls.

The correlation for linear regression is weaker than the correlation for quadratic regression and the MAD and S_e^2 are lower for quadratic regression. This supports the use of the QLR model in this problem.

For the purpose of illustration, we compare the results of linear regression and quadratic regression for this problem in Figure 3.25. The result further supports the use of quadratic regression for this problem.

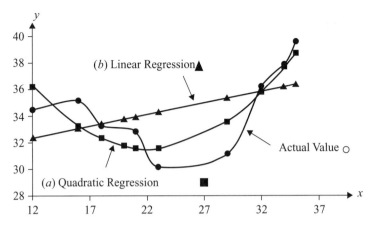

Figure 3.25 Plots of (•) actual demand and forecasted demands by (■) quadratic regression and (▲) linear regression for Example 3.17.

3.9.2 Multiple Variable by Quadratic Optimization[†]

In this section, we extend the optimization method that we developed in the last section for single variable (3.41) to solve the quadratic function for several independent variables as

$$\hat{y} = b_0 + b_{11}x_1 + b_{12}x_1^2 + b_{21}x_2 + b_{22}x_2^2 \cdots + b_{k1}x_k + b_{k2}x_k^2 \qquad (3.46)$$

where the parameters to be determined are $b_0, b_{11}, b_{12}, b_{21}, b_{22}, \ldots, b_{k1}, b_{k2}$. The multiple quadratic regression quadratic optimization (MQLR-QO) problem is formulated as follows:

PROBLEM 3.3 MULTIPLE QUADRATIC REGRESSION QUADRATIC OPTIMIZATION

Minimize

$$\text{TSE} = \sum_{i=1}^{n} \left(e_i^+\right)^2 + \sum_{i=1}^{n} \left(e_i^-\right)^2 \qquad (3.47)$$

Subject to:

$$b_0 + b_{11}x_{1i} + b_{12}x_{1i}^2 + \cdots + b_{k1}x_{ki} + b_{k2}x_{ki}^2 + e_i^+ - e_i^- = y_i \quad \text{for} \quad i = 1, \ldots, n \quad (3.48)$$

$$e_i^+ \geq 0 \quad e_i^- \geq 0 \quad \text{for} \quad i = 1, \ldots, n \qquad (3.49)$$

$$b_0, b_{11}, b_{12}, \ldots, b_{k1}, b_{k2} \text{ unrestricted in sign} \qquad (3.50)$$

where e_i^+ and e_i^- are the positive and negative errors of the predicted y from the actual y.

TABLE 3.16   Problem and Solution of Example 3.18

Data Point/ Year, i	Problem: Given Data			(a) Solution: Multiple Quadratic Regression (Problem 3.3), Estimated Value, $\hat{y}_i$	(b) Solution: Multiple Linear Regression (Problem 3.1), Estimated Value, $\hat{y}_i$
	$x_{1,i}$	$x_{2,i}$	y_i		
1	9	31	15.72	16.03	14.80
2	5	42	20.16	20.16	18.95
3	1	35	15.3	15.30	15.30
4	6	31	14.34	14.01	14.34
5	4	48	24.84	24.84	21.39
6	7	51	31.57	31.03	23.15
7	2	45	23.47	22.82	19.78
8	4	35	?	15.00	15.76

Example 3.18   Multiple Quadratic Regression Quadratic Optimization   Consider a problem with two independent variables x_1 and x_2 and one dependent variable y. Information for the past seven periods is given in Table 3.16.

(a) Use multiple quadratic regression to forecast the value for period 8.

(b) Use multiple linear regression (of Section 3.8) to forecast the value for period 8.

(c) Compare the above two solutions and identify which one is more accurate.

(a) Using the multiple quadratic regression method, Problem 3.3 is formulated as follows:

Minimize

$$\text{TSE} = \sum_{i=1}^{7} \left(e_i^+\right)^2 + \sum_{i=1}^{7} \left(e_i^-\right)^2$$

Subject to:

$$b_0 + b_{11}9 + b_{12}(9)^2 + b_{21}31 + b_{22}(31)^2 + e_1^+ - e_1^- = 15.72$$
$$b_0 + b_{11}5 + b_{12}(5)^2 + b_{21}42 + b_{22}(42)^2 + e_2^+ - e_2^- = 20.16$$
$$b_0 + b_{11}1 + b_{12}(1)^2 + b_{21}35 + b_{22}(35)^2 + e_3^+ - e_3^- = 15.3$$
$$b_0 + b_{11}6 + b_{12}(6)^2 + b_{21}31 + b_{22}(31)^2 + e_4^+ - e_4^- = 14.34$$
$$a + b_{11}4 + b_{12}(4)^2 + b_{21}48 + b_{22}(48)^2 + e_5^+ - e_5^- = 24.84$$
$$a + b_{11}7 + b_{12}(7)^2 + b_{21}51 + b_{22}(51)^2 + e_6^+ - e_6^- = 31.57$$
$$a + b_{11}2 + b_{12}(2)^2 + b_{21}45 + b_{22}(45)^2 + e_7^+ - e_7^- = 23.47$$
$$b_0, b_{1,1}, b_{1,2}, b_{2,1}, b_{2,2} \text{ unrestricted}$$
$$e_i^+ \geq 0 \qquad e_i^- \geq 0 \quad \text{for } i = 1, \dots, 7$$

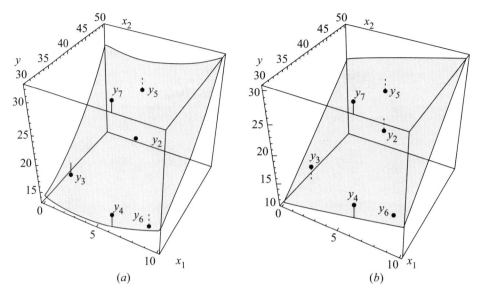

Figure 3.26   Three-Dimensional Plot of actual values with (*a*) predicted multiple quadratic regression values and (*b*) predicted multiple linear regression values for Example 3.18.

The solution is shown in Table 3.16. Solving this problem using optimization software (LINGO was used here), the optimal solution is

$$b_0 = 33.003 \qquad b_{1,1} = -0.554 \qquad b_{1,2} = 0.082 \qquad b_{2,1} = -1.467 \qquad b_{2,2} = 0.028$$

Using these values in Equation (3.42), the multiple quadratic regression is

$$\hat{y} = 33.003 - 0.554x_1 + 0.082x_1^2 - 1.467x_2 + 0.028x_2^2$$

Therefore, for period 8, the forecasted value is

$$\hat{y}_8 = 33.003 - 0.554(4) + 0.082(4)^2 - 1.467(35) + 0.028(35)^2 = 15.05$$

Table 3.16 shows the forecasted $\hat{y}$ values for multiple quadratic regression for all eight periods. Also, Figure 3.26 shows a three-dimensional plot of this solution.

(b) The solution for linear regression is shown in Table 3.17 and Figure 3.26. The solution for multiple linear regression (using Problem 3.1) is

$$\hat{y} = -12.87 + 0.520x_1 + 0.770x_2$$

Therefore, for period 8, the forecasted value is

$$\hat{y}_8 = -12.87 + 0.520 \times 4 + 0.770 \times 35 = 16.6$$

TABLE 3.17   Comparison of Quadratic Regression and Linear Regression

	Function	MAD	S_e^2	r
(a) Multiple-variable quadratic regression	$\hat{y} = 33.003$ $-0.554x_1 + 0.082x_1^2$ $-1.467x_2 + 0.028x_2^2$	0.47	$\dfrac{2.58}{7-5} = 1.29$	$\sqrt{\dfrac{237.15}{237.64}} = 0.999$
(b) Multiple-variable linear regression	$\hat{y} = -12.87$ $+0.520x_1 + 0.770x_2$	0.91	$\dfrac{8.72}{7-3} = 2.18$	$\sqrt{\dfrac{229.18}{237.64}} = 0.982$

Table 3.16 shows the data values of y and the forecasted $\hat{y}$ for both methods for all eight periods.

See Supplement S3.18.lg4.

(c) To compare the two methods, find the MAD, S_e^2, and the correlation for the two methods. The results are shown in Table 3.17.

For the MQLR-QO model, there are five estimated parameters $(b_0, b_{1,1}, b_{1,2}, b_{2,1}, b_{2,2})$. Therefore, the MSE is computed as

$$S_e^2(\text{MQLR-QO}) = \frac{\text{SSE}}{n-p} = \frac{2.54}{7-5} = 1.27$$

For the MLR model, there are three estimated parameters (a, b_1, b_2). Therefore, the MSE for multiple linear regression is

$$S_e^2(\text{MLR}) = \frac{\text{SSE}}{n-p} = \frac{8.72}{7-3} = 2.18$$

The correlation for multiple linear regression is as strong as the correlation for quadratic regression. However, the MAD and S_e^2 for quadratic regression are lower, which makes MQLR-QO a better model for this problem.

By comparing the results of linear regression and multiple quadratic regression for this problem in Figure 3.26, it can be seen that the multiple quadratic regression better fits the actual data.

See Supplement S3.19.xls.

3.9.3   Qualitative Regression[†]

In Section 3.2.2, we discussed qualitative–subjective forecasting methods where a number of experts or juries of executives express their opinions for the rating of the data. In this section, we show how to use linear and quadratic regression when there is such qualitative information. Suppose that the experts or the jury of executives agree on providing Q_i as the qualitative rating of each given data point i where Q_i can be interpreted as the accuracy or reliability or importance of each given point i. We can use any scale of numerical ratings or percentages to represent Q values. Here we use values ranging from 1 to 5 where 1 is the least accurate rating and 5 is the most accurate rating. Furthermore, the approach developed here can also be used to handle fuzzy or imprecise data. The DM can assess the precision ratings (which are similar to weights of importance, or confidence level) for each data point. The classical forecasting methods cannot be used to solve these types of problems.

The following optimization problem can be used to solve qualitative linear regression problems as presented in Section 3.7. The notation in this problem is similar to the notation used for multiple linear regression by quadratic optimization (MLR-QO) (Problems 3.1).

PROBLEM 3.4   LINEAR REGRESSION QUADRATIC OPTIMIZATION WITH QUALITATIVE RATINGS

Minimize

$$\text{Total weighted square of errors} = \sum_{i=1}^{n} Q_i \left(e_i^+\right)^2 + \sum_{i=1}^{n} Q_i \left(e_i^-\right)^2 \tag{3.51}$$

Subject to:

$$a + bx_i + e_i^+ - e_i^- = y_i \quad \text{for } i = 1, \dots, n \tag{3.52}$$

$$e_i^+ \geq 0 \quad e_i^- \geq 0 \quad \text{for} \quad i = 1, \dots, n \tag{3.53}$$

$$a \text{ and } b \text{ unrestricted in sign} \tag{3.54}$$

The regression optimization methods that we developed in Sections 3.8 and 3.9 can be extended for qualitative regression by using objective function (3.51) subject to the same set of constraints of the developed optimization problems. The regression optimization problems can be solved by using optimization software packages.

3.10   Z THEORY NONLINEAR REGRESSION AND TIMES SERIES

3.10.1   Nonlinear Multiple-Variable Regression by Z Theory[†]

The quadratic multivariable regression function used in Section 3.9.2 is an additive function, and it is not truly nonlinear; for example, interactions among variables are not considered. This problem can be resolved by using the Z theory developed in Chapter 2 (MCDM) and by Malakooti (2014b). In this section, we illustrate the use of Z theory to solve this problem. The quadratic regression function (3.46) can be represented as

$$\begin{aligned} LV &= b_0 + (b_{11}X_1 + b_{12}x_1^2) + (b_{21}x_2 + b_{22}x_2^2) + \dots + (b_{k1}X_K + b_{K2}X_K^2) \\ LV &= b_0 + v_1(X_1) + v_2(X_2) + \dots + v_K(X_K) \end{aligned} \tag{3.55}$$

That is, each variable k is represented by a value function

$$v_k(X_k) = b_{k1}X_k + b_{k2}X_k^2 \qquad \text{for } k = 1, 2, \ldots, K$$

In Z theory, the interactions among variables are presented as the deviation value (DV). Different forms of DV and nonlinear functions $\hat{y} = f(\text{LV}, \text{DV})$ are presented in Malakooti (2014b). Here, we use DV based on the Euclidean distance to the mean (the SD) denoted by

$$\text{DV}_{L=2} = \sqrt{[v_1(x_1) - \text{LV}]^2 + [v_2(x_2) - \text{LV}]^2 + \cdots + (v_K(x_K) - \text{LV}]^2} \quad (3.56)$$

The nonlinear regression function using Z theory is

$$\hat{y} = \text{LV} + z\,\text{DV}_{L=2} \tag{3.57}$$

PROBLEM 3.5   NONLINEAR MULTIPLE-VARIABLE REGRESSION OPTIMIZATION

Minimize:

$$\text{TSE} = \sum_{i=1}^{n} \left(e_i^+\right)^2 + \sum_{i=1}^{n} \left(e_i^-\right)^2 \tag{3.58}$$

Subject to:

$$\text{LV}_i + z\,\text{DV}_i + e_i^+ - e_i^- = y_i \quad \text{for} \quad i = 1, \ldots, n \tag{3.59}$$

$$e_i^+ \geq 0 \qquad e_i^- \geq 0 \quad \text{for} \quad i = 1, \ldots, n \tag{3.60}$$

$$b_0, b_{11}, b_{12}, \ldots, b_{K1}, b_{K2}, z \text{ unrestricted in sign} \tag{3.61}$$

where LV_i and DV_i are defined in (3.55) and (3.56), respectively; e_i^+ and e_i^- are the positive and negative errors of the predicted y from the actual y. Problem 3.5 contains nonlinear constraints and is computationally difficult to solve. Two possible methods for solving this problem are as follows:

1. *Levenberg–Marquardt algorithm (see Levenberg, 1944)*   The Levenberg–Marquardt method significantly outperforms gradient descent methods by using an improved iteration step. This approach is available in MATLAB (http://www.mathworks.com/help/toolbox/optim/ug/brnoyhf.html) as the solve function, in SAS as NLPLM Call, and in Origin (scientific graphing and data analysis software).

2. *Z-Nonlinear Heuristic*   The Z-nonlinear heuristic consists of the following steps:
   (a) Find $b_0, b_{11}, b_{12}, b_{21}, b_{22}, \ldots, b_{K1}, b_{K2}$ using the conventional quadratic regression method (covered in Section 2.9.2).

TABLE 3.18 Example 3.19 Data and Z-Nonlinear Regression Solution

i	$x_{1,i}$	$x_{2,i}$	y_i	Quadratic Regression (Problem 3.3), $\hat{y}_i = \text{LV}$	$v_1(x_1)$	$v_2(x_2)$	DV	z	$\hat{y}$	Error
1	9	31	15.72	16.03	1.66	−18.57	37.47	−0.008	16.19	0.47
2	5	42	20.16	20.16	−0.72	−12.22	38.53	0.000	20.32	0.16
3	1	35	15.3	15.3	−0.47	−17.05	35.99	0.000	15.45	0.15
4	6	31	14.34	14.01	−0.37	−18.57	35.61	0.009	14.16	0.18
5	4	48	24.84	24.84	−0.90	−5.90	40.10	0.000	25.01	0.17
6	7	51	31.57	31.03	0.14	−1.99	45.22	0.012	31.22	0.35
7	2	45	23.47	22.82	−0.78	−9.32	39.87	0.016	22.99	0.48
								Avg. z = 0.004		MAD = 0.28

(b) Find LV_i and DV_i of each data point $i = 1, 2, \ldots, n$ using (3.55) and (3.56), respectively.

(c) For each data point $i = 1, 2, \ldots, n$, find z_i using Equation (3.57). Then find the average of z.

(d) Use the average z in $\hat{y} = \text{LV} + z\,\text{DV}_{L=2}$. This is the nonlinear regression function.

Example 3.19 Nonlinear Multiple-Variable Regression Using Z Theory Consider Example 3.18 with two independent variables x_1 and x_2 and one dependent variable y. Information for the past seven periods is also given in Table 3.18. Compare the quality of the solution obtained by Z theory versus the solution obtained by the conventional quadratic regression method.

The quadratic solution obtained in Example 3.18 was $b_0 = 33.003$, $b_{1,1} = -0.554$, $b_{1,2} = 0.082$, $b_{2,1} = -1.467$, and $b_{2,2} = 0.028$. Therefore,

$$\text{LV} = 33.003 + (-0.554x_1 + 0.082x_1^2) + (-1.467x_2 + 0.028x_2^2)$$

where

$$v_1(X_1) = (-0.554x_1 + 0.082x_1^2) \qquad v_2(x_2) = (-1.467x_2 + 0.028x_2^2)$$

$$\text{DV}_{L=2} = \sqrt{(-0.554x_1 + 0.082x_1^2 - \text{LV})^2 + (-1.467x_2 + 0.028x_2^2 - \text{LV})^2}$$

Table 3.18 shows the LV and DV for each data point. For example, for point $i = 1$,

$$\text{LV}_1 = 33.003 + (-0.554 \times 9 + 0.082 \times 9^2) + (-1.467 \times 31 + 0.028 \times 31^2) = 16.09$$

$$\text{DV}_1 = \sqrt{(-0.554 \times 9 + 0.082 \times 9^2 - 16.03)^2 + (-1.467 \times 31 + 0.028 \times 31^2 - 16.03)^2}$$

$$= 37.54$$

$$z_1 = \frac{16.03 - 15.72}{37.54} = -0.008.$$

The average $z = 0.004$. Therefore, the regression function is

$$\hat{y} = \text{LV} + 0.004\,\text{DV}_{L=2}$$

Now, we can compare the nonlinear ZUF to the quadratic function solution by comparing their MAD. To compare the two methods, find the MAD. The following results show that the ZUF provided a better solution.

	Function	MAD
Quadratic regression	$\text{LV} = 33.003 - 0.554x_1 + 0.082x_1^2 - 1.467x_2 + 0.028x_2^2$	0.48
Z utility nonlinear regression	$\hat{y} = \text{LV} + 0.004\,\text{DV}$	0.28

See Supplement S3.20.xlsx.

3.10.2   Nonlinear Time Series by Z Theory[†]

The approach for solving time series forecasting is based on Section 3.10.1. For this problem, only two variables are considered. The first independent variable is time t, where $x_t = t$ for $t = 1, 2, \ldots, n$. The second independent variable is predicted F_t for $t = 1, 2, \ldots, n$, by using any given time series methods [e.g., using exponential smoothing (Section 3.4.1), linear regression (Section 3.7.2), or quadratic regression (Section 3.9.1)]. These two independent variables are evaluated by the actual output y_t, for example, the actual demand $y_t = D_t$.

First, we find F_t for $t = 1, 2, \ldots, n$ using a forecasting method. Then, we find the quadratic regression (3.62) for the given two independent variables x_t and F_t, that is,

$$\text{LV} = b_0 + (b_{11}t + b_{12}t^2) + (b_{21}F_t + b_{22}F_t^2) = b_0 + v_1(t) + v_2(F_t) \qquad (3.62)$$

Example 3.20   Nonlinear Regression Time Series Using Linear Regression   Consider Example 3.13 as shown in Table 3.9. Solving this problem by the linear regression method of Section 3.7.2 (alternatively, one can use the quadratic regression method of Section 3.9.1), the solution is $F_t = 1320 + 510\,t$. Table 3.19 shows the details.

The quadratic solution for (10.8) is $b_0 = 0$, $b_{1,1} = 0$, $b_{1,2} = -280.486$, $b_{2,1} = -0.438$, and $b_{2,2} = 0.0008$. Therefore,

$$\text{LV} = -280.486t^2 - 0.438F_t + 0.0008F_t^2$$

$$v_1(t) = -280.486t^2 \qquad v_2(F_t) = -0.438F_t + 0.0008F_t^2$$

$$\text{DV}_{L=2} = \sqrt{(-280.486t_1^2 - \text{LV})^2 + (-0.438F_t + 0.0008F_t^2 - \text{LV})^2}$$

TABLE 3.19 Example 3.20 Data and Z-Nonlinear Regression Solution

$x_{1,i} = t$	$x_{2,i} = F_t$	y_i	LV_i	$v_1 (x_{1,i})$	$v_2 (x_{2,i})$	DV_i	z_i	$\hat{y}_i$	Error $=$ $\hat{y}_t - y_t$
1	1830	1650	1597.09	−280.49	1978.05	1915.84	0.028	1680.49	30.49
2	2340	2600	2233.6	−1121.96	3519.83	3593.63	0.102	2390.04	209.96
3	2850	2700	2725.29	−2524.41	5493.38	5934.78	−0.004	2983.65	283.65
4	3360	3600	3072.16	−4487.84	7898.69	8969.34	0.059	3462.63	137.37
5	3870	3700	3274.21	−7012.25	10735.77	12707.72	0.034	3827.43	127.43
	MAD $=$						Avg $z =$		MAD $=$
	200						0.044		157.78

Table 3.19 shows the LV and DV for each data point i. For example, for point $i = 1$,

$$LV_1 = -280.486 \times 1^2 - 0.438 \times 1830 + 0.0008 \times 1830^2 = 1597$$

$$DV_1 = \sqrt{(-280.486 \times 1^2 - 1597)^2 + (-0.438 \times 1830 + 0.0008 \times 1830^2 - 1597)^2}$$

$$= 1915.84$$

$$z_1 = \frac{1650 - 1830}{1915.84} = 0.028.$$

The average $z = 0.044$. Therefore, the regression function is

$$\hat{y} = LV + 0.044 \, DV_{L=2}$$

For linear regression, using $F_t = 1320 + 510 \times t$, Example 3.13, MAD $= 200$. Using the ZUF, $\hat{y} = LV + 0.044 \times DV$ and MAD $= 158$, which is better than the MAD of linear regression.

See Supplements S3.21.lg4, S3.21.xlsx.

Example 3.21 Nonlinear Regression Time Series Using Exponential Smoothing
For the past 10 periods, demand D_t is presented as y_t in Table 3.20 (this is the same as Example 3.10). For the exponential smoothing method, suppose $\alpha = 0.25$ and the initial forecast value for the first period is $F_1 = y_1 = 5500$. The forecast value based on exponential smoothing, F_t, is shown in Table 3.20.

Using t and F_t as independent variables, the solution to (3.62) quadratic regression is $b_0 = 9904.32$, $b_{1,1} = -251.50$, $b_{1,2} = 65.77$, $b_{2,1} = -0.084$, and $b_{2,2} = -0.0001$. Therefore,

$$LV = 9904.32 - 251.50t + 65.77t^2 - 0.084F_t - 0.0001F_t^2$$

$$v_1(t) = -251.50t + 65.77t^2 \qquad v_2(F_t) = -0.084F_t - 0.0001F_t^2$$

$$DV_{L=2} = \sqrt{(-251.50t + 65.77t^2 - LV)^2 + (-0.084F_t - 0.0001F_t^2 - LV)^2}$$

TABLE 3.20 Z-Nonlinear Regression Solution Using Exponential Smoothing Method

$x_{1,i} = t$	$x_{2,i} = F_t$	y_i	LV_i	$v_1(x_{1,i})$	$v_2(x_{2,i})$	DV_i	z_i	$\hat{y}_i$	Error $=$ $\hat{y}_t - y_t$
1	5500	5500	6231.59	−185.73	−4406.72	12424.00	−0.06	5682.82	182.82
2	5500	5900	6177.40	−239.92	−4334.93	12316.29	−0.02	5633.39	266.61
3	5600	5600	6135.35	−162.57	−4279.62	12171.09	−0.04	5597.75	2.25
4	5600	5700	6344.24	46.32	−4557.85	12590.45	−0.05	5788.12	88.12
5	5625	6000	6654.51	386.75	−4987.23	13221.75	−0.05	6070.50	70.50
6	5719	6500	7011.95	858.72	−5505.74	13948.29	−0.04	6395.85	104.15
7	5914	6600	7372.23	1462.23	−6054.25	14669.65	−0.05	6724.27	124.27
8	6086	7200	7886.44	2197.28	−6882.05	15826.39	−0.04	7187.38	12.62
9	6364	7800	8383.56	3063.87	−7732.63	16971.48	−0.03	7633.93	166.07
10	6723	8000	8881.72	4062.00	−8634.55	18167.26	−0.05	8079.27	79.27
	MAD $=$						Avg. $z =$		MAD $=$
	616.9						−0.044		109.67

Table 3.20 shows the LV and DV for each data point. Therefore, the nonlinear regression function is $\hat{y} = LV - 0.044\,DV_{L=2}$ with $z = -0.044$ and MAD $=109.67$, which is much better than the exponential smoothing MAD $= 616.9$ and double-exponential smoothing MAD $= 519$ (see Table 3.5)

See Supplements S3.22.lg4, S3.22.xlsx.
Supplement S3.23.doc presents methods and examples for qualitative regression for linear and quadratic multiple variable regression problems.

REFERENCES

General

Armstrong, J. S. "Forecasting by Extrapolation: Conclusions from Twenty-Five Years of Research." *Interfaces*, Vol. 14, No. 4, 1984, pp. 52–66.

Box, G. E. P., and G. M. Jenkins. *Time Series Analysis Forecasting and Control.* San Francisco: Holden-Day, 1976.

Brown, R. G. *Statistical Forecasting for Inventory Control.* New York: McGraw-Hill, 1959.

Brown, R. G. *Smoothing, Forecasting, and Prediction of Discrete Time Series.* Englewood Cliffs, NJ: Prentice Hall, 1962.

Chambers, C., S. K. Mullick, and D. D. Smith. "How to Choose the Right Forecasting Technique." *Harvard Business Review*, Vol. 65, 1971, pp. 45–74.

Draper, N. R., and H. Smith. *Applied Regression Analysis.* New York: Wiley, 1968.

Fisher, M. L., J. H. Hammond, W. R. Obermeyer, and A. Raman. "Making Supply Meet Demand in an Uncertain World." *Harvard Business Review,* 1994, pp. 221–240.

Fogler, H. R. "A Pattern Recognition Model for Forecasting." *Management Science*, Vol. 20, No. 8, 1974, pp. 1178–1189.

Gass, S. I., C. M. Harris, eds. *Encyclopedia of Operations Research and Management Science*, Centennial Edition, Dordrecht, The Netherlands: Kluwer, 2000.

Gross, C. W., and R. T. Peterson. *Business Forecasting*, 2nd ed., New York: Wiley, 1983.

Holt, C. C. "Forecasting Seasonals and Trends by Exponentially Weighted Moving Averages." *International Journal of Forecasting*, Vol. 20, No. 1, 2004, pp. 5–10.

Levenberg, K. "A Method for the Solution of Certain Non-Linear Problems in Least Squares." *Quarterly of Applied Mathematics*, 1944, pp. 164–168.

"Levenberg-Marquardt Method." Available: http://en.wikipedia.org/wiki/Levenberg%E2%80%93 Marquardt_algorithm.

Makridakis, S., S. C. Wheelwright, and R. J. Hyndman, *Forecasting: Methods and Applications*, 3rd ed., John Hoboken, NJ: Wiley, 2008.

Makridakis, S., and R. L. Winkler. "Averages of Forecasts: Some Empirical Results." *Management Science*, Vol. 29, No. 9, 1983, pp. 987–996.

Montgomery, D. C., L. A. Johnson, and J. S. Gardiner. *Forecasting and Time Series Analysis*, 2nd ed., New York: McGraw-Hill, 1990.

Wiener, N. *Extrapolation, Interpolation, and Smoothing of Stationary Time Series.* Cambridge, MA: MIT Press, 1st M.I.T. paperback ed. 1964.

Witners, P. R. "Forecasting Sales by Exponentially Weighted Moving Averages." *Management Science*, Vol. 6, No. 3, 1960, pp. 324–342.

Multiple-Objectives Forecasting

Charnes, A., W. W. Cooper, and T. Sueyoshi. "Least Squares/Ridge Regression and Goal Programming/Constrained Regression Alternatives." *European Journal of Operational Research*, Vol. 27, No. 2, 1986, pp. 146–157.

Greco, S. "Ordinal Regression Revisited: Multiple Criteria Ranking Using a Set of Additive Value Functions." *European Journal of Operational Research*, Vol. 191, No. 2, 2008, pp. 416–436.

Leung, M. T., H. Daouk, and A.-S. Chen. "Using Investment Portfolio Return to Combine Forecasts: A Multiobjective Approach." *European Journal of Operational Research*, Vol. 134, No. 1, 2001, pp. 84–102.

Malakooti, B. "Regression and Forecasting by Single and Multiple Objective Optimization and Z Theory." *Case Western Reserve University, Cleveland, Ohio 2013*.

Malakooti, B. "Z Utility Theory : Decisions Under Risk and Multiple Criteria", to appear, 2014b.

Narula, S. C., and J. F. Wellington. "Multiple Criteria Linear Regression." *European Journal of Operational Research*, Vol. 181, No. 2, 2007, pp. 767–772.

Özelkan, E. C., and L. Duckstein. "Multi-Objective Fuzzy Regression: A General Framework." *Computers & Operations Research*, Vol. 27, No. 7–8, 2000, pp. 635–652.

Reeves, G. R., and K. D. Lawrence. "Combining Multiple Forecasts Given Multiple Objectives." *Journal of Forecasting*, Vol. 1, No. 3, 1982, pp. 271–279.

Sakawa, M., and H. Yano. "Multiobjective Fuzzy Linear Regression Analysis and Its Application." *Electronics Communication Japan*, Vol. 73, 1990, pp. 1–9.

Tran, L. "Multiobjective Fuzzy Regression with Central Tendency and Possibilistic Properties." *Fuzzy Sets and Systems*, Vol. 130, No. 1, 2002, pp. 21–31.

EXERCISES

Introduction (Section 3.1)

1.1 Identify and explain the key entities that must be forecasted along with their time horizons for the following applications:

(a) A given manufacturing industry

(b) A given service industry

(c) A government organization

(d) A nonprofit organization

1.2 Give an example or justification for each principle of forecasting.

Qualitative–Subjective Forecasting (Section 3.2.2)

2.1 For a given industry, discuss and explain each of the six steps of the forecasting process.

2.2 Compare and contrast the four different qualitative forecasting methods. List advantages and disadvantages of each method.

2.3 Give an example of a situation in which a company would use each of the following qualitative forecasting methods:

(a) Sales force composition

(b) Customer (market) survey

(c) Jury of executive decisions

(d) Delphi method

Accuracy Measurement (Section 3.2.3)

2.4 Describe the assumptions, advantages, and disadvantages of the mean average deviation, mean square error, and mean absolute percentage error methods in measuring accuracy. Compare and contrast the three methods.

2.5 Consider the demand for a product for the past seven periods. Two different methods, 1 and 2, are used to forecast the demand for each period. Show demand versus the two forecasting methods graphically.

Period, t	1	2	3	4	5	6	7
D_t	410	420	390	450	510	510	520
Method 1, F_t	420	440	450	450	500	490	460
Method 2, F_t	380	410	420	390	450	510	510

(a) Using method 1 forecasts, measure the mean average deviation, mean square error, and mean absolute percentage error.

(b) Using method 2 forecasts, measure the mean average deviation, mean square error, and mean absolute percentage error.

(c) Compare the results of (a) and (b) and indicate which method (1 or 2) was more accurate. Also, plot actual demand versus the forecasting values of methods 1 and 2.

2.6 Consider the demand for a product for the past eight periods. Two different methods, 1 and 2, are used to forecast the demand for each period. Show demand versus the two forecasting methods graphically.

Period, t	1	2	3	4	5	6	7	8
D_t	84	115	121	126	152	104	119	112
Method 1, F_t	80	112	125	130	168	99	113	105
Method 2, F_t	91	127	128	142	160	84	119	102

(a) Using method 1 forecasts, measure the mean average deviation, mean square error, and mean absolute percentage error.

(b) Using method 2 forecasts, measure the mean average deviation, mean square error, and mean absolute percentage error.

(c) Compare the results of (a) and (b) and indicate which method (1 or 2) is more accurate. Also, plot actual demand versus the forecasting values of methods 1 and 2.

Quality Control and Tracking Error Charts*† (Section 3.2.4)

2.7 Consider exercise 2.5.

(a) For method 1 forecasted values, show the control charts for a 99.74% confidence level using mean square error values (standard deviation).

(b) For method 2 forecasted values, show the control charts for a 99.74% confidence level using mean square error values (standard deviation).

(c) Compare the results of (a) and (b). Which method is more under control in terms of its data?

2.8 Consider exercise 2.5.

(a) For method 1 forecasted values, show the control charts for a 99.74% confidence level using tracking signals.

(b) For method 2 forecasted values, show the control charts for a 99.74% confidence level using tracking signals.

(c) Compare the results of (a) and (b). Which method is more under control in terms of its data?

2.9 Consider exercise 2.6.

(a) For method 1 forecasted values, show the control charts for a 99.74% confidence level using tracking signals.

---

* This section is based on the original contribution by the author, see Malakooti (2013).
† Advanced material that can be omitted without loss of generality.

(b) For method 2 forecasted values, show the control charts for a 99.74% confidence level using tracking signals.

(c) Compare the results of (a) and (b). Which method is more under control in terms of its data?

2.10 Give a business example where stationary, trend, seasonal, or random data might be observed. Explain the meaning of stationary and nonstationary assumptions for each example.

2.11 Consider exercise 2.5. By plotting the data on a graph identify whether the data are stationary, trend, seasonal, or random.

2.12 Consider exercise 2.6. By plotting the data on a graph identify whether the data are stationary, trend, seasonal, or random.

Time Series: Moving Averages (Section 3.3)

3.1 The following table gives the number of customers in thousands who called customer service over the past seven months:

Month, t	1	2	3	4	5	6	7
No. of customers	65	54	72	43	68	56	61

Use moving averages to predict the number of customers who will call in month 8. Show four predictions based on one, two, three, and seven past periods. Which forecast do you suggest to use?

3.2 The table below gives the demand values for a product for periods 1–8. Forecast the value of demand for period 9 using the moving-average method for two, three, four, and eight past periods. Which forecast do you suggest using?

Period, t	1	2	3	4	5	6	7	8	9
Demand, D_t	203	198	210	234	186	175	214	194	?

3.3 Consider exercise 3.1. Suppose the weights of importance for the data for period 1–7 are (0.05, 0.05, 0.1, 0.1, 0.2, 0.2, 0.3), respectively. Find the demand for day 8.

3.4 Consider exercise 3.2. Suppose the weights of importance for the data too periods 1–8 are (0.00, 0.05, 0.05, 0.1, 0.1, 0.2, 0.25, 0.25). Find the demand for day 9.

Exponential Smoothing (Section 3.4.1)

4.1 Consider exercise 3.1. Use exponential smoothing with three different values of α (0.2, 0.4, and 0.6) to forecast the number of customers for period 8. Assume the forecast for the first period is 65. Which forecast should be used?

4.2 Consider exercise 3.2. Use exponential smoothing with three different values of α (0.1, 0.25, and 0.5) to forecast the value of demand for period 9. Assume the forecast for the first period is 203. Which forecast should be used?

Multicriteria for Exponential Smoothing (Section 3.4.2)†

4.3 Consider exercise 4.1. Solve the MCDM problem for three possible alternatives of α (0.2, 0.4, and 0.6).

(a) Identify efficient alternatives.

(b) Rank the three alternatives where the weights of importance for the two normalized objective functions are 0.4 and 0.6, respectively. What forecasting should be used?

4.4 Consider exercise 4.2. Solve the MCDM problem for three possible alternatives of α (0.1, 0.25, and 0.5).

(a) Identify efficient alternatives.

(b) Rank the three alternatives where the weights of importance for the two normalized objective functions are 0.7 and 0.3, respectively. What forecasting should be used?

Time Series Trend Analysis by Linear Regression (Section 3.5.1)

5.1 Consider the following data for seven periods (same as exercise 3.1):

Period, i	1	2	3	4	5	6	7	8
No. of customers	65	54	72	43	68	56	61	?

(a) Use linear regression analysis to predict the number of customers in period 8.

(b) What would be the number of customers in period 30? Is it justified to use this linear regression to predict the demand in period 30? Explain. Graphically show the linear regression versus the actual demand.

(c) How much error does the forecast have for period 7? Is it reasonable? Why?

5.2 Consider the following data:

Period, i	1	2	3	4	5	6	7	8	9
Demand, D_i	203	198	210	234	186	175	214	194	?

(a) Forecast the value of demand for periods 5, 9, and 22 using linear regression.

(b) Graphically show the linear regression versus the demand.

(c) How much error does the forecast have for period 5? Is it reasonable to have this error? Why?

5.3 Consider the data presented in the table below for the past 12 periods.

(a) Use linear regression to predict demand for periods 5, 13, and 20.

(b) How much error does the forecast have for the forecasting period 5? Is this error reasonable? Why?

Period, i	Demand, D_i	Period, i	Demand, D_i
1	2479	7	3483
2	1999	8	3705
3	2544	9	3562
4	2704	10	3813
5	3229	11	3798
6	2924	12	4239
		13	?

Trend-Adjusted Double-Exponential Smoothing[†] (Section 3.5.2)

5.4 Consider exercise 5.1. Use exponential smoothing to forecast the value of demand for period 8. Suppose the trend factor for period 1 is -2. Use the following parameters. Suppose that the initial forecast value for the first period is $F_1 = D_1 = 65$.

   (a) Use $\alpha = 0.4, \beta = 0.1$.
   (b) Use $\alpha = 0.3, \beta = 0.1$.
   (c) Use $\alpha = 0.1, \beta = 0.5$.
   (d) Compare and contrast the above results. Plot the above three alternatives. Which one is the most accurate?

5.5 Consider exercise 5.2. Use exponential smoothing to forecast the value of demand for period 9. Suppose the trend factor for the first period is $+20$. Use the following parameters. Suppose that the initial forecast value for the first period is $F_1 = D_1 = 203$.

   (a) Use $\alpha = 0.2, \beta = 0.3$.
   (b) Use $\alpha = 0.4, \beta = 0.1$.
   (c) Use $\alpha = 0.5, \beta = 0.5$.
   (d) Compare and contrast the above results. Plot the above three alternatives. Which one is the most accurate?

Time Series: Cyclic/Seasonal (Section 3.6)

6.1 A manager of a special clothing store wants to forecast the store's total sales for each month of the next year based on the three previous years sales data. See the table below. Total sales are in $1000's. The demand for this special clothing peaks around the winter. It is predicted that the total demand for year 4 is 5340.

   (a) Find the forecasted demand for each month of year 4.
   (b) Find the forecast for each quarter (every three months) for year 4 using seasonal analysis. Graphically show the demand versus forecasted values.

Month, i	Year 1	Year 2	Year 3	Month, i	Year 1	Year 2	Year 3
1	635	701	685	7	102	126	96
2	602	586	642	8	245	289	199
3	575	458	500	9	368	370	399
4	425	412	389	10	546	530	576
5	280	189	196	11	610	601	600
6	185	145	98	12	680	650	729

6.2 For a given product, the past three years' demand per month is presented in the following table:

Month	1	2	3	4	5	6	7	8	9	10	11	12
Year 1	203	198	210	234	186	175	214	194	200	180	195	300
Year 2	224	232	241	262	189	180	225	220	240	192	237	314
Year 3	243	239	251	260	194	192	240	236	253	200	249	327

(a) Use linear regression to predict total demand in year.

(b) Find forecasted demand for every month of year 4 using seasonal analysis.

(c) Find forecasted demand for every three months (quarter) of year 4 using seasonal analysis. Graphically show the past demand-versus-forecasted values.

Linear Regression (Section 3.7.1)

7.1 An energy company wishes to predict total electricity usage for a town based on the following population (in 1000s) versus total electricity in megawatts consumed in the past five years.

Year	1	2	3	4	5	6
Population	230	350	430	420	460	500
Total electricity usage	340	353	451	442	482	?

(a) If the population in the town becomes 500, in year 6, find the total electricity using causal linear regression forecasting where electricity usage depends on the population size.

(b) Find the standard error of estimation and the mean square error. What do they mean?

(c) Find the correlation coefficient. Describe the meaning of this coefficient value.

(d) Use linear regression to predict total demand in year. Also, graphically show demand versus forecasting.

7.2 For a company the total defective products that were returned by customers is a function of the total units sold. The information for seven periods are presented below.

Month	1	2	3	4	5	6	7	8
Unit sales (in 100,000)	24	32	21	19	28	27	30	35
No. of units returned	479	500	452	439	488	490	501	?

(a) If the company sells 35 units (in 100,000s) in the next period 8, find the total number of products that will be returned. Use causal linear regression forecasting.

(b) Find the standard error of estimation and the mean square error. What do they mean?

(c) Find the correlation coefficient. Describe the meaning of this coefficient value.

(d) Use linear regression to predict total demand in year 4. Also, graphically show demand versus forecasting.

7.3 A company believes that its total sales depends on the number of its salespersons. The information for the past 10 months is provided below.

Month	Salespersons	Total Sales in $100,000	Month	Salespersons	Total Sales in $100,000
1	143	363	6	175	430
2	158	380	7	185	480
3	170	411	8	190	510
4	164	400	9	195	530
5	188	433	10	200	550
			11	121	?

(a) The company is planning to have 121 salespersons in the next period. What will be the total sales of the next period using causal linear regression forecasting.

(b) Find the standard error of estimation and the mean square error for this problem. What do they mean?

(c) Calculate the correlation coefficient. Describe the meaning of this coefficient value.

(d) Use linear regression to predict total demand in year 4. Also, graphically show demand versus forecasting.

Multicriteria Linear Regression (Section 3.7.3)[†]

7.4 Consider exercise 7.1. Use bicriteria linear regression to generate seven bicriteria alternatives.

(a) To minimize the underestimation problem, identify the efficient alternatives.

(b) To minimize the overestimation problem, identify the efficient alternatives.

7.5 Consider exercise 7.2. Use bicriteria linear regression to generate seven bicriteria alternatives.

(a) To minimize the underestimation problem, identify the efficient alternatives.

(b) To minimize the overestimation problem, identify the efficient alternatives.

Multiple-Variable Linear Regression (Section 3.8)

8.1 Consider the table below for seven points of data where x_1 and x_2 are two independent variables and y is a dependent variable.

i	y	x_1	x_2
1	15.62	2	25
2	17.92	5	40
3	14.2	4	15
4	17.66	8	26
5	18.94	9	31
6	16	1	20
7	15.88	3	24
8	?	6	27

(a) Find the multiple linear regression equation for this problem.

(b) If the values of x_1 and x_2 are 6 and 27, respectively, what is the predicted value of y?

(c) How accurate is this method and why?

(d) Compare the results above using single variable linear regression (of Section 3.7.1) for x_1 only. Do the same analysis for x_2 only. Which model is the most valid to use?

8.2 Consider the below table where x_1, x_2 and x_3 are three independent variables and y is a dependent variable.

i	y	x_1	x_2	x_3
1	49.4	13.1	24.0	18.1
2	94.9	7.7	11.3	15.5
3	125.0	9.2	9.6	19.9
4	108.9	13.9	15.1	15.0
5	60.3	8.5	15.9	18.6
6	99.4	10.2	11.7	18.7
7	138.3	14.7	12.8	17.3
8	83.5	10.8	13.4	17.5
9	76.0	5.2	15.1	17.5
10	45.0	9.2	19.8	19.7
11	?	12.0	20.0	14.0

(a) Find the multiple linear regression equation for this problem.

(b) If the values of x_1, x_2, and x_3 are 12, 20, and 14, respectively, what is the predicted y value?

(c) How accurate is this method and why?

(d) Compare the results above using single-variable linear regression (of Section 3.7.1) for x_1 only. Do the same analysis for x_2 only. Which model is the most valid to use?

Single-Variable Quadratic Regression by Quadratic Optimization (Section 3.9.1)

9.1 Consider exercise 7.1.

(a) Respond to question (a) of exercise 7.1 using QLR-QO.

(b) Is it justified to use quadratic regression for this problem? Why? Compare your results with the linear regression exercise 7.1.

(c) Graphically show the demand versus forecasting for both methods.

9.2 Consider exercise 7.2.

(a) Respond to question (a) of exercise 7.2 using QLR-QO.

(b) Is it justified to use quadratic regression for this problem? Why? Compare your results with the linear regression exercise 7.2.

(c) Graphically show the demand versus forecasting for both methods.

Multiple Variable by Quadratic Optimization[†] (Section 3.9.2)

9.3 Consider exercise 8.1.

(a) Respond to parts (a) and (b) of exercise 8.1 using multiple quadratic regression.

(b) Is it justified to use multiple quadratic regression for this problem? Why?

(c) Compare your results to the results obtained in exercise 8.1 using multiple linear regression.

9.4 Consider exercise 8.2.

(a) Respond to parts (a) and (b) of exercise 8.2 using multiple quadratic regression.

(b) Is it justified to use multiple quadratic regression for this problem? Why?

(c) Compare your results to the results obtained in exercise 8.2 using multiple linear regression.

Nonlinear Multiple Variable Regression by Z Theory (Section 3.10.1)[†]

10.1 Consider exercise 8.1.

(a) Respond to parts (a) and (b) of exercise 8.1 using Z theory nonlinear regression.

(b) Is it justified to use Z theory nonlinear regression for this problem? Why?

(c) Compare your results to the results obtained in exercise 9.3 using multiple quadratic regression.

10.2 Consider exercise 8.2.

   (a) Respond to parts (a) and (b) of exercise 8.2 using Z theory nonlinear regression.

   (b) Is it justified to use Z theory nonlinear regression for this problem? Why?

   (c) Compare your results to the results obtained in exercise 9.4 using multiple quadratic regression.

Nonlinear Time Series by Z Theory (Section 3.10.2)[†]

10.3 Consider exercise 5.1.

   (a) Use Z theory nonlinear regression and compare its solution to the solution of exercise 5.1 using linear regression.

   (b) Compare Z theory nonlinear regression to exponential smoothing using $\alpha = 0.5$.

   (c) Is it justified to use Z theory v Regression for this problem? Why?

10.4 Consider exercise 5.2.

   (a) Use Z theory nonlinear regression and compare its solution to the solution of exercise 5.2 using linear regression.

   (b) Compare Z theory nonlinear regression to exponential smoothing using $\alpha = 0.4$.

   (c) Is it justified to use Z theory nonlinear regression for this problem? Why?

CHAPTER 4

AGGREGATE PLANNING

4.1 INTRODUCTION

The primary goal of most production industries is to meet their production demands by efficiently utilizing their resources, such as facilities and manpower. Aggregate planning supports the development of optimal production planning to meet the forecasted future demands for an intermediate time horizon, which is usually between 2 and 24 months. Aggregate planning constraints include meeting the forecasted demand, workforce availability, facility capacity, and desired inventory levels. In aggregate planning, it is assumed that the forecasted demands are known with certainty. Although this assumption appears to be restrictive, errors in forecasted demands may not have severe consequences due to the use of averages and aggregate units for several periods of time. That is, overestimation and underestimation of the demands of several units and periods may result in relatively accurate aggregate demand forecasts.

Aggregate planning uses information such as forecasted demands, raw material availability, workforce levels, and capacity constraints. This information is used to provide optimal aggregate planning in terms of production plans, inventory levels, and workforce plans. Figure 4.1 illustrates three levels of the time horizon of production planning. In level 1, strategic planning decisions are made for the long term (e.g., years) regarding the product types and strategic issues such as facility location, layouts, and capacities. In level 2, for a given intermediate time horizon (e.g., months), demands are predicted and overall aggregate production plans are determined. In level 3, detailed production planning for short periods (e.g., days) is decided. In this level, the production plan for all components of products is planned. This level consists of scheduling, materials requirement planning, and inventory planning. These topics are covered in subsequent chapters of this book.

Note: Advanced material that can be omitted without loss of generality will be indicated by a dagger.

Operations and Production Systems with Multiple Objectives, First Edition. Behnam Malakooti.
© 2014 John Wiley & Sons, Inc. Published 2014 by John Wiley & Sons, Inc.

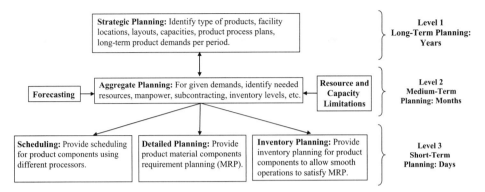

Figure 4.1 Relationship of aggregate planning to long-term strategic planning and short-term planning such as scheduling, material requirement planning, and inventory planning.

This chapter deals with aggregate planning and the disaggregation process. Disaggregation is the process of breaking the aggregate units and longer periods into smaller units. An important objective in aggregate planning is not to have fluctuating production, that is, to have constant and continuous production. This is a difficult objective to achieve because demand may vary from period to period. There are a number of production planning options that can be used to cope with fluctuating demands. They include allowing excess inventory to meet possible future high demands, allowing backlogs (shortages), changing the workforce by hiring or firing workers, allowing overtime work, and subcontracting the work to outside sources (outsourcing). All of these options can be measured in terms of their cost, and then the most economical option or combination of options can be selected. The "traditional" aggregate planning approach emphasizes the minimization of total cost for determining the optimal aggregate plan. In this chapter, however, we introduce a multiple-objective optimization aggregate planning approach that considers several objectives, such as minimizing costs, backlogs, workforce fluctuations (hiring and firing), and inventory fluctuations.

In the classical aggregate planning approach (Sections 4.2–4.4 and 4.7), different products are first aggregated and represented by aggregate units. Then, the aggregate planning problem is solved. In the next step, aggregate units are decomposed into their actual units through the disaggregation process, and finally the master schedule is developed. In modern aggregate planning (Section 4.5), an integrated approach is used to solve all three problems (the aggregate planning, disaggregation, and master scheduling) simultaneously by solving one large-scale linear programming (LP) problem for all products and their components.

This chapter consists of seven sections. Section 4.2 introduces the input–output model for aggregate planning. It also covers a simple graphical solution to the aggregate planning problem. Section 4.3 covers tabular production planning, which uses spreadsheets. Section 4.4 explains the LP optimization approach to solve the aggregate planning problem where only one type of product is considered. Section 4.5 explains the LP formulation for considering several types of products. Section 4.6 covers the multiple-objective linear programming approach for aggregate planning. Section 4.7 covers the traditional aggregation process, disaggregation, and master scheduling.

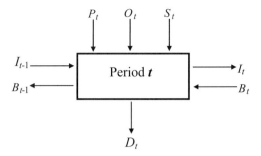

Figure 4.2 Input–output model for period t.

4.2 GRAPHICAL APPROACH

4.2.1 Input–Output Model

In the input–output model, all production activities including hiring and firing are assumed to occur at the start of each period. Resources are also assumed to be available at the beginning of each period. Suppose that there are n production planning periods denoted by t, where $t = 1, 2, \ldots, n$. The relationship of different variables and the given forecasted demand for period t is presented in Figure 4.2.

The notation used in Figure 4.2 is defined as follows:

P_t = regular production for period t

H_t = hiring for period t

F_t = firing for period t

O_t = overtime production for period t

S_t = subcontracted production for period t

I_{t-1} = inventory level at the end of period $t-1$

B_{t-1} = backorder level at the end of period $t-1$

I_t = inventory level at the end of period t

B_t = backorder level at the end of period t

D_t = given forecasted demand for period t

In the input–output model, the value of the demand for each period t is known. Also, the inventory of the last period, I_{t-1}, and the backorder of the last period, B_{t-1}, are known. All other values are unknown and should be determined. For each period t, the following two equations present the relationship among decision variables and the given demand. Equation (4.1) is known as the "hire–fire equation," and Equation (4.2) is known as the "input–output equation." The relationship between the decision variables for each period t, where the values for the previous period $t-1$ are known, is given below:

$$P_t - P_{t-1} = H_t - F_t \quad \text{or} \quad P_t - P_{t-1} - H_t + F_t = 0 \tag{4.1}$$

$$I_{t-1} + P_t + O_t + S_t + B_t = I_t + B_{t-1} + D_t \quad \text{or} \quad I_{t-1} - I_t + P_t + O_t + S_t + B_t - B_{t-1} = D_t \tag{4.2}$$

$$\text{All decision variables are nonnegative (i.e., } \geq 0) \tag{4.3}$$

In Equation (4.1):

(a) If the net change in regular production is positive for two consecutive periods, that is, $P_t - P_{t-1} > 0$, then there will be a hiring, that is, $H_t > 0$. In this case, there would be no firing, that is, $F_t = 0$.

(b) If the net change in regular production is negative, that is, $P_t - P_{t-1} < 0$, then there will be a firing, that is, $F_t > 0$, and consequently there will be no hiring, that is, $H_t = 0$.

That is, in Equation (4.1) both F_t and H_t cannot be positive in the same period.

Equation (4.2) assures that the total input is equal to the total output in each period. The third equation indicates that all decision variables are nonnegative, for example, the inventory value cannot be negative. But, if there are shortages, then backordering $B_t > 0$ and consequently $I_t = 0$. On the other hand, if the net inventory is positive, that is, $I_t > 0$, then consequently $B_t = 0$. Both I_t and B_t cannot be positive in the same period. The inventory of a given period is available for the next period. Similarly, backorders can be used to postpone fulfilling demand in a given period, but usually it is not allowed in the last period.

Depending on the requirements of the aggregate plan, some of the decision variables may be set equal to zero. For instance, if subcontracting and backorders are not allowed, then set $S_t = 0$ and $B_t = 0$ for all periods.

Production Units versus Actual Units In this chapter, all activities (such as regular production, overtime production, subcontracting production, hiring, and firing) are presented by their equivalent production units. For example, suppose that in period t production by hiring is 10 units, that is, $H_t = 10$. This means that the number of products produced by the hired workers is 10 units, but it does not mean that 10 workers are hired in period t. By using production units, the relationships between variables are simplified. It is possible to convert the production units of each variable to the actual unit of that variable. Consider regular production in period t, P_t. Let q be the number of products produced by one worker per period. Therefore, the actual number of workers in period t is

$$W_t = \frac{P_t}{q} \tag{4.4}$$

For example, suppose that the production by regular workers is 10 units and each worker produces 5 units per period; therefore $10/5 = 2$. That is, there are two regular workers in period t. Suppose that the regular worker's payment per period is $C_{R,W}$; then the production cost for one unit, C_R, is

$$C_R = \frac{C_{R,W}}{q} \tag{4.5}$$

Similarly, the cost per unit for overtime and subcontracting can be calculated.

Rounding of Actual Units In Equation (4.4), the actual number of workers W_t may not be an integer number. A conservative approach is to round the number of workers to the next highest integer number. For example, consider 900 production units in period t. If each worker can produce 160 production units in one period, the number of needed workers would be

$$W_t = \frac{P_t}{q} = \frac{900}{160} = 5.625$$

Since it is not possible to actually hire 5.625 workers, then the number can be rounded up to 6 workers to produce at least 900 production units. Similarly, hiring H_t and firing F_t solutions may need to be converted and rounded to find the actual number of workers. To avoid shortages, the hiring number should be rounded up, and the firing number should be rounded down.

As will be shown in Section 4.4, the rounding approach may not result in an optimal solution. One may try to improve the generated solution by trial and error, that is, trying higher and lower integer values. Alternatively, one may use the integer linear programming (ILP) approach, covered in Section 4.4.2 of this chapter, to obtain the optimal integer solution. In ILP, W_t will be required to assume only integer values.

4.2.2   Graphical Approach: One Variable

In a simplified aggregate planning, only one decision variable is considered. All other variables are assumed to be known, that is, they have given values. To find the optimal value for one selected variable, one can use a single-variable optimization approach such as the derivative and one-dimensional search methods, which are covered in conventional linear and nonlinear programming (NLP optimization) textbooks. For example, if the unknown variable is the inventory level, then all other variables, such as workforce, overtime, and subcontracting, are assumed to be known. Then, the optimal inventory for each period is derived such that the demand for each period is met. Similarly, if the unknown variable is the workforce level, then all other variables (such as inventory, subcontracting, and overtime) are assumed to be known. Once the optimal value of one variable is found, its value can be set constant, and another variable can be considered. A simplified plan can provide easy and fast aggregate planning, but it is unlikely that it generates an optimal solution in terms of all variables. It can, however, be used to analyze and improve upon a good final solution by trial and error. It can also be used to respond to if–then questions. In this case spreadsheets such as Excel can be used to evaluate different possible scenarios. The trial-and-error approach consists of the following steps:

1. Choose the most important variable.
2. Vary the value of the variable while keeping the values of all other variables constant. Measure the values of important objectives of the aggregate plan.
3. Find the best value for the given variable. Make it a constant.
4. Choose the next most important variable and repeat steps 2–4.

In the above heuristic approach, better alternatives may be obtained by varying one variable at a time. However, as stated before, the solution may not be optimal. The LP method of Section 4.4 provides an optimal solution to this problem.

The two most important single-variable aggregate planning methods are discussed below:

(a) *Constant Inventory Plan But Variable Workforce*   In the constant inventory plan, an aggregate plan is found where the inventory level is kept constant in all periods. Usually the constant inventory level is zero. This is accomplished by adjusting the workforce levels (by hiring or firing). The zero inventory plans provide aggregate plans that keep the inventory levels at zero and do not allow backordering. Therefore, in zero inventory plans, backorders are set equal to zero.

(b) *Constant Workforce Plan But Variable Inventory*   In constant workforce plans, hiring and firing are not allowed, except at the beginning of the first period. Therefore, a constant workforce must be kept from the beginning of the first period through the end of the last period. The initial hiring or firing in the first period should be made in such a way that the total demand at the end of the final period is satisfied. The inventory level for each period is decided such that the workforce for all periods would remain constant.

Graphical Approach   When dealing with only one variable, the graphical approach can be used effectively. Suppose the workforce is constant but inventory can vary. In this case, set the regular production level of each period equal to the value of the average demand over the planning period. This method produces a feasible solution if backorders are allowed. The total demand is met at the end of the planning period, that is, there will be no backorders at the end of the last period. In constant workforce plans, three cases can occur in terms of meeting the total demand by the end of the last planning period:

I. The cumulative demand is less than the cumulative production over all periods; see Figure 4.3*a*. In this case the demand is satisfied in each period, but there is inventory in some periods.

II. The cumulative demand is more than the cumulative production over all periods; see Figure 4.3*b*. In this case the demand is not satisfied in each period, but it is satisfied in the last period and there are backorders in some periods.

III. The cumulative demand is sometimes less and sometimes more than the cumulative production; see Figures 4.4*a,b*. The demand is not satisfied in some of the periods, but it is satisfied in the last period. In this case there is inventory in some periods and backorders in some other periods.

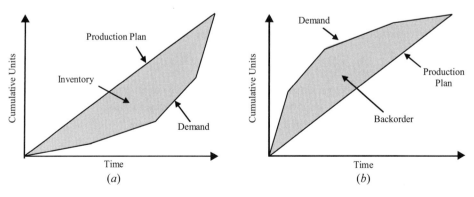

Figure 4.3   (a) Inventory accumulation with constant production. Total demand is satisfied in all periods. (b) Backorder accumulation with constant production. Total demand is satisfied in the last period.

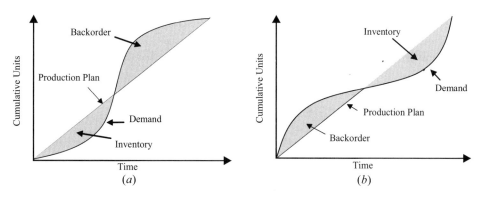

Figure 4.4 (a) Initial demand is less than average production. (b) Initial demand is more than the average production.

Example 4.1 Graphical Approach with Four Periods Consider an aggregate plan with demand requirements of 500, 400, 700, and 400 units for periods 1, 2, 3, and 4, respectively. The initial workforce is 300 production units. The production cost of the workforce is \$5/unit. The inventory cost is \$3/unit per period, and the backlog cost is \$7/unit per period. The hiring cost is \$6/unit, and the firing cost is \$8/unit. Show aggregate plans for (a) constant inventory of zero (where no backlog is allowed) and (b) constant workforce.

Solution

(a) To have a constant inventory of zero, hire and fire to meet the exact demand in each period. The solution is presented below.

t	0	1	2	3	4	Total
Demand	—	500	400	700	400	2000
Workforce	300	500	400	700	400	2000
Hire (fire)	—	200	(100)	300	(300)	500, (400)
Cumulative inventory	0	0	0	0	0	0

The total cost for this plan is $2000 \times 5 + 500 \times 6 + 400 \times 8 = \$16{,}200$.

(b) To have a constant workforce per period, divide the cumulative demand by four periods, that is, $2000/4 = 500$. Then hire 200 units in the first period to produce 500 units in each period by workforce.

t	0	1	2	3	4	Total
Demand	—	500	400	700	400	2000
Workforce	300	500	500	500	500	2000
Hire (fire)	—	200	0	0	0	200
Cumulative inventory (Backorder)	0	0	100	(100)	0	100, (100)

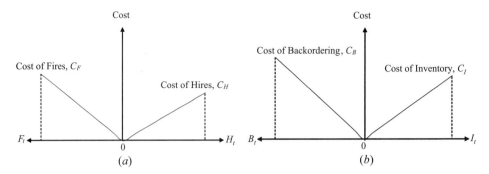

Figure 4.5 (a) Cost of hiring and firing per unit. (b) Cost of inventory and backordering per unit.

Therefore, the total cost for this plan is $2000 \times 5 + 200 \times 6 + 100 \times 3 + 100 \times 7 = $12,300$.

4.3 TABULAR METHOD

Aggregate planning problems can be solved by a tabular approach using spreadsheets. In the tabular method the values of different variables (such as production by regular workforce and inventory levels) can be calculated by using the costs associated with production, overtime, subcontracting, hiring, firing, inventory, and backorders. Generally the cost of firing is higher than the cost of hiring, and the cost of backordering is higher than the cost of inventory per unit; see Figures 4.5a,b.

The tabular method is widely used because it is easy to understand and utilize. The downside of this method is that the generated solution may not be optimal and many trials and errors may be needed to find the optimal solution. For example, it could be necessary to reapply the method many times by varying different variables while minimizing the total cost. The following notation is used in the tabular aggregate planning method.

Known Demands and Initial Values	Known Cost Coefficients
D_t = production demand for period $t = 1, 2, \ldots, n$	C_R = regular workforce cost per unit of production
P_0 = initial production by current workforce at $t = 0$	C_S = subcontracting cost per unit of production
I_0 = initial inventory in period $t = 0$	C_O = overtime cost per unit of production
B_0 = initial backorder in period $t = 0$	C_H = hiring cost per unit of production
B_n = final backorder in period $t = n$	C_F = firing cost per unit of production
	C_I = holding cost per unit of inventory per period
	C_B = backorder cost per unit per period

Using the above known values and cost coefficients, find the optimal values of the following decision variables:

Decision Variables

P_t = production for period $t = 1, 2, \ldots, n$

S_t = subcontracting for period $t = 1, 2, \ldots, n$

O_t = overtime for period $t = 1, 2, \ldots, n$

I_t = inventory for period $t = 1, 2, \ldots, n$

B_t = backorder for period $t = 1, 2, \ldots, n$

H_t = hiring production for period $t = 1, 2, \ldots, n$

F_t = firing production for period $t = 1, 2, \ldots, n$

The total cost of aggregate planning is measured as follows:

$$\text{Total cost} = \sum_{t=1}^{n} [(C_R P_t) + (C_O O_t) + (C_S S_t) + (C_I I_t) + (C_B B_t) + (C_H H_t) + (C_F F_t)] \tag{4.6}$$

The problem is to find values of the decision variables such that Equation (4.6) is minimized while all constraints of the problem are satisfied. Generally, it is assumed that all backorders must be fulfilled in the last period.

In the tabular approach, a table is formed as presented in Table 4.1. The initial values of regular production, inventory, and backorder, that is, P_0, I_0, and B_0, are known for the initial period $t = 0$. These are presented in the first row of Table 4.1 for period 0. Also, all forecasted demands $D_1, D_2, \ldots, D_n$ are known.

The method for finding the values of each cell of the table is a forward approach by finding the values of each row from period 1 to period n where the values for period 0 are given. First find the unknown values for the row for period 1. Then, decide on the given decision variables for the second period. Repeat this process until all periods are considered. The total values at the last row are the totals for periods $1, \ldots, n$. Note that period 0 is not used for the calculation of total values.

TABLE 4.1   Generic Aggregate Plan Table Used in (Tabular) Spreadsheet Approach

t	D_t	P_t	O_t	S_t	I_t	B_t	H_t	F_t	Total Cost ($)
0	0	P_0	0	0	I_0	B_0	0	0	
1	D_1								
2	D_2								
$\vdots$	$\vdots$								
n	D_n								
Total									Equation (4.6)

Example 4.2   Tabular Aggregate Planning Method: Three Periods   Consider the following problem:

Known Demands and Initial Values	Known Cost Coefficients
$D_1 = 110$	Production cost with regular workforce $C_R = \$10/\text{unit}$
$D_2 = 120$	Production cost using subcontractor $C_S = \$12/\text{unit}$
$D_3 = 80$	Production cost using overtime $C_O = \$12/\text{unit}$
Initial work force $P_0 = 90$ units	Hiring cost $C_H = \$12/\text{unit}$
Initial inventory $I_0 = 10$ units	Firing cost $C_F = \$15/\text{unit}$
Initial backorder $B_0 = 0$	Inventory-carrying cost $C_I = \$8/\text{unit}$
Final backorder $B_n = 0$	Backorder cost $C_B = \$14/\text{unit}$

Solutions of different strategies to solve this problem are presented below.

Constant Inventory Plan I (Allow Changes in Hiring and Firing)   In this plan, keep the inventory at zero and change the workforce to satisfy the demand in each period. Overtime and subcontracting are not allowed. The solution is shown in the table below.

t	D_t	P_t	O_t	S_t	I_t	B_t	H_t	F_t	Total Cost
0	0	90	0	0	10	0	0	0	
1	110	100	0	0	0	0	10	0	$1120
2	120	120	0	0	0	0	20	0	$1440
3	80	80	0	0	0	0	0	40	$1400
Total	**310**	**300**	**0**	**0**	**0**	**0**	**30**	**40**	**$3960**

In period 1, the demand of 110 is satisfied by using 90 units of regular force, 10 units from the inventory of last period, and 10 units from hiring, that is, $90 + 10 + 10 = 110$. The cost for period 1 is

$$100(\$10) + 0(\$12) + 0(\$12) + 0(\$8) + 0(\$14) + 10(\$12) + 0(\$15) = \$1120$$

In period 2, the demand is satisfied by hiring 20 units. In period 3, the demand is satisfied while 40 units are reduced by firing.

Now consider the last row for total values. Note that the initial production ($P_0 = 90$) should not be used to calculate the total production. The total demand of 310 is satisfied by 300 production units plus 10 units from the initial inventory, $310 = 300 + 10$. Also, note that H_t and F_t are used to calculate P_t. The above plan has a total cost of $3960. In the following example, it will be shown that this cost can be reduced by considering more variables.

Constant Inventory Plan II (Allow Changes in Hiring, Firing, and Subcontracting)   In this plan, the inventory is kept constant at zero, but subcontracting, hiring, and firing are allowed. Use the variable that has a lower cost to produce the needed additional units. This plan is shown in the table below.

t	D_t	P_t	O_t	S_t	I_t	B_t	H_t	F_t	Total Cost
0	0	90	0	0	10	0	0	0	
1	110	90	0	10	0	0	0	0	$1020
2	120	90	0	30	0	0	0	0	$1260
3	80	80	0	0	0	0	0	10	$950
Total	**310**	**260**	**0**	**40**	**0**	**0**	**0**	**10**	**$3230**

The total cost for period 1 is $90(\$10) + 10(\$12) = \$1020$. The total demand of 310 is satisfied by 260 units from regular production, 40 units from subcontracting, and 10 units from the initial inventory, that is, $260 + 40 + 10 = 310$. The table shows the effects of using subcontractors instead of hiring and firing. If extra workers were hired for the high demands in periods 1 and 2, they would be fired in period 3. The unit cost of hiring in one period and firing in a later period is $\$12 + \$15 = \$27$ per unit, which is greater than the cost of subcontracting at $12 per unit.

Constant Workforce Plan III (Vary Inventory and Subcontracting)   This plan is presented as follows:

t	D_t	P_t	O_t	S_t	I_t	B_t	H_t	F_t	Cost
0		90			10				
1	110	90	0	10	0	0	0	0	$1020
2	120	90	0	30	0	0	0	0	$1260
3	80	90	0	0	10	0	0	0	$980
Total	**310**	**270**	**0**	**40**	**10**	**0**	**0**	**0**	**$3260**

The cost for the first period is $90(\$10) + 10(\$12) = \$1020$. The total production is $270 + 40 + 10 = 320$, which is higher than the total demand 310. There is an inventory of 10 at the end of the third period because it is less expensive to have inventory ($8/unit) than to have firing ($15/unit).

Mixed Plan IV (Allow Changes in All Variables)   In this plan, all variables (i.e., inventory, workforce, overtime, and subcontracting) can be varied to find the lowest total cost. This plan is shown in the table below with a total cost of $3200. This plan has a lower cost than the other plans. By allowing all variables to change, the optimal solution can be obtained. Due to the complexity of this problem, it is necessary to solve this problem by linear programming, which will be covered in Section 4.4.

t	D_t	P_t	O_t	S_t	I_t	B_t	H_t	F_t	Cost
0		90			10	0			
1	110	90	0	10	0	0	0	0	$1020
2	120	90	0	20	0	10	0	0	$1280
3	80	90	0	0	0	0	0	0	$900
Total	**310**	**270**	**0**	**30**	**0**	**10**	**0**	**0**	**$3200**

The total demand is satisfied at the end of the last period, that is, $310 = 270 + 30 + 10$. Note that in period 3 there are 90 units of production for the 80 units of demand and the 10 units of backorder from period 2. In this example, allowing backorders is cheaper than firing workers. Note that the cost of subtracting and overtime are the same ($12/unit), so there is an alternative optimal solution. The alternative optimal solution is to produce 10, 20, and 0 units from overtime and 0, 0, and 0 units from subtracting in periods 1, 2, and 3, respectively. This alternative plan also has a total cost of $3200.

Example 4.3   Tabular Aggregate Planning Method: Four Periods   Consider the following aggregate planning problem for four periods:

Known Demands and Initial Values	Known Cost Coefficients
$D_t = 500, 400, 700, 400$ for $t = 1, 2, 3, 4$, respectively	Production cost with regular workforce $C_R = \$5/\text{unit}$
Initial workforce $P_0 = 300$ units	Production cost with overtime $C_O = \$8/\text{unit}$
Initial inventory $I_0 = 0$ units	Production cost using subcontractor $C_S = \$10/\text{unit}$
Initial backorder $B_0 = 0$	Inventory-carrying cost $C_I = \$3/\text{unit}$
Final backorder $B_n = 0$	Backorder cost $C_B = \$7/\text{unit}$
	Hiring cost $C_H = \$6/\text{unit}$
	Firing cost $C_F = \$8/\text{unit}$

Note: There is a limit of 500 units of regular production per period, $P_t \leq 500$ for $t = 1, 2, 3, 4$. Overtime is limited to 20% of regular production in each period, $O_t \leq 0.20\, P_t$ for $t = 1, 2, 3, 4$.

Constant Inventory and Workforce Plan I (Allow Changes in Subcontracting and Overtime)   Find the lowest cost solution that maintains a constant workforce and does not allow backorders. Since the cost of overtime is less than subcontracting, first use overtime and then subcontracting to meet the demand. This plan is shown in the table below.

t	D_t	P_t	O_t	S_t	I_t	B_t	H_t	F_t	Cost
0		300			0	0			
1	500	300	60	140	0	0	0	0	$3,380
2	400	300	60	40	0	0	0	0	$2,380
3	700	300	60	340	0	0	0	0	$5,380
4	400	300	60	40	0	0	0	0	$2,380
Total	**2000**	**1200**	**240**	**560**	**0**	**0**	**0**	**0**	**$13,520**

The cost for the first period is

$$300(\$5) + 60(\$8) + 140(\$10) + 0(\$3) + 0(\$7) + 0(\$6) + 0(\$8) = \$3380$$

Note that total demand is satisfied, that is, $2000 = 1200 + 240 + 560$.

Constant Inventory and Workforce Plan II (Allow Changes in Subcontracting and Overtime)

A solution is provided below by using a constant workforce of 400 in each period. This solution has a constant inventory of zero. The cost of both hiring and firing is greater than the cost of subcontracting. However, the cost of hiring without having to fire the workers later is cheaper than the cost of subcontracting. In addition, due to the lower cost of overtime production, use the maximum amount of overtime up to the limit of 20% of the regular production ($0.2P_t$). When the maximum overtime is achieved, use subcontracting to meet demand. This plan is shown in the table below.

t	D_t	P_t	O_t	S_t	I_t	B_t	H_t	F_t	Cost
0		300	≤ 80		0	0			
1	500	400	80	20	0	0	100	0	$3,440
2	400	400	0	0	0	0	0	0	$2,000
3	700	400	80	220	0	0	0	0	$4,840
4	400	400	0	0	0	0	0	0	$2,000
Total	**2000**	**1600**	**160**	**240**	**0**	**0**	**100**	**0**	**$12,280**

The total cost for the first period is

$$400(\$5) + 80(\$8) + 20(\$10) + 0(\$3) + 0(\$7) + 100(\$6) + 0(\$8) = \$3440$$

Note that total demand is satisfied, that is, $2000 = 1600 + 160 + 240$.

Constant Workforce Plan III (Allow Changes in Hiring, Firing, Inventory, and Backordering)   In this plan, hire 200 units at the beginning of period 1. Therefore, there are 500 units of regular production per period. The following table shows the solution to this problem. Period 3 requires 700 units, so either inventory has to be carried from period 2 or backorder needs to be allowed in period 3. Set 100 units of inventory in period 2 and set 100 units of backlog in period 3.

t	D_t	P_t	O_t	S_t	I_t	B_t	H_t	F_t	Cost
0		300			0	0			
1	500	500	0	0	0	0	200	0	$3,700
2	400	500	0	0	100	0	0	0	$2,800
3	700	500	0	0	0	100	0	0	$3,200
4	400	500	0	0	0	0	0	0	$2,500
Total	2000	2000	0	0	100	100	200	0	$12,200

The total cost for the first period is

$$500(\$5) + 0(\$8) + 0(\$10) + 0(\$3) + 0(\$7) + 200(\$6) + 0(\$8) = \$3700$$

The total demand of $2000 = 2000 + 100 - 100$ is satisfied.

Mixed Plan IV (Allow Changes in All Variables)   In this plan, all variables can be changed to reach the lowest cost solution. To find the optimal solution to this problem, ILP (to be discussed in Section 4.4.2) is used.

t	D_t	P_t	O_t	S_t	I_t	B_t	H_t	F_t	Cost
0		300			0	0			
1	500	469	31	0	0	0	169	0	$3,607
2	400	469	0	0	69	0	0	0	$2,552
3	700	469	93	0	0	69	0	0	$3,572
4	400	469	0	0	0	0	0	0	$2,345
Total	2000	1876	124	0	69	69	169	0	$12,076

The total demand of $2000 = 1876 + 124$ is satisfied.

Comparison of Four Plans   The following table shows a summary of the totals (last rows) of the four aggregate planning solutions. The mixed plan has the lowest cost solution of $12,076 compared to the costs of other production plans.

Total	D	P	O	S	I	B	H	F	Cost ($)
Constant inventory and workforce plan I	2000	1200	240	560	0	0	0	0	13,520
Constant inventory and workforce plan II	2000	1600	160	240	0	0	100	0	12,280
Constant workforce plan III	2000	2000	0	0	100	100	200	0	12,200
Mixed plan IV	2000	1876	124	0	69	69	169	0	12,076

4.4   LINEAR PROGRAMMING METHOD

The graphical and tabular approaches for solving the aggregate planning problem use trial and error in an attempt to find the optimum plan. The aggregate planning problem can be easily formulated and solved by LP, where the solution is always optimal. The reader is encouraged to review the LP formulation (see LP and NLP textbooks) and using an appropriate software package to solve the LP problem.

4.4.1   Linear Programming Formulation

Consider the notation and equations presented in Sections 4.2 and 4.3 for input and output models and the cost information for all variables. In formulating aggregate planning by LP, the decision variables are regular production, overtime, subcontracting, inventory levels, backorder levels, production gained from hires, and production lost from fires per period represented by P_t, O_t, S_t, I_t, B_t, H_t, and F_t, respectively. The LP objective function is to minimize the total cost, which is the sum of the total cost for all periods ($t = 1, 2, \ldots, n$) as presented in Equation (4.7). The LP problem is subjected to the following constraints:

1. The hiring and firing equations as presented in Equation (4.8).
2. The input–output constraint as presented in Equation (4.9).
3. No backordering allowed at the end of the production planning as presented by Equation (4.10).
4. The allowable limits are presented in Equations (4.11), (4.12), and (4.13), where L_t, M_t, and N_t are given limits for regular production, overtime production, and production by subcontracting, respectively. That is, for each period:
   (a) Regular production is within capacity limit: $P_t \leq L_t$.
   (b) Overtime production is within capacity limit: $O_t \leq M_t$.
   (c) Subcontracted production is within allowable limit: $S_t \leq N_t$.

5. All variables are nonnegative (≥ 0), as given in Equation (4.14).
6. Any other relevant linear equality or inequality constraints are given in Equation (4.15).

Formally, aggregate planning LP can be formulated as follows where the right-hand side of each equation is a given constant.

PROBLEM 4.1 (LP AGGREGATE PLANNING PROBLEM)

Minimize:

$$\text{Total cost} = \sum_{t=1}^{n} [(C_R P_t) + (C_O O_t) + (C_S S_t) + (C_I I_t) + (C_B B_t) + (C_H H_t) + (C_F F_t)] \tag{4.7}$$

Subject to:

$P_t - P_{t-1} - H_t + F_t = 0$	for $t = 1, \ldots, n$	(4.8)
$I_{t-1} - I_t + P_t + O_t + S_t + B_t - B_{t-1} = D_t$	for $t = 1, \ldots, n$	(4.9)
$B_n = 0$		(4.10)
$P_t \leq L_t$	for $t = 1, \ldots, n$	(4.11)
$O_t \leq M_t$	for $t = 1, \ldots, n$	(4.12)
$S_t \leq N_t$	for $t = 1, \ldots, n$	(4.13)
$P_t, O_t, S_t, I_t, B_t, H_t, F_t \geq 0$	for $t = 1, \ldots, n$	(4.14)
Other linear constraints as required		(4.15)

Note that in the LP formulation both F_t and H_t are minimized due to their costs. In the optimal solution, either F_t or H_t will be positive for each given period t (but not both). Similarly, either I_t or B_t will be positive (but not both) in each period. These conditions are automatically achieved when solving the LP problem. Note that constraint (4.10), $B_n = 0$, means that all orders must be fulfilled by the end of the planning period.

Example 4.4   Linear Programming Formulation with Three Periods   Consider Example 4.2. Solve Example 4.2 by LP to find the optimal solution.
The LP is formulated as follows:

Minimize:

$$\begin{aligned}
\text{Total cost } f_1 = {}& 10P_1 + 12O_1 + 12S_1 + 8I_1 + 14B_1 + 12H_1 + 15F_1 && \text{(cost for period 1)} \\
& +10P_2 + 12O_2 + 12S_2 + 8I_2 + 14B_2 + 12H_2 + 15F_2 && \text{(cost for period 2)} \\
& +10P_3 + 12O_3 + 12S_3 + 8I_3 + 14B_3 + 12H_3 + 15F_3 && \text{(cost for period 3)}
\end{aligned}$$

Subject to:

$$P_1 - P_0 - H_1 + F_1 = 0 \qquad \text{Hiring and firing equation for } t = 1$$
$$P_2 - P_1 - H_2 + F_2 = 0 \qquad \text{Hiring and firing equation for } t = 2$$
$$P_3 - P_2 - H_3 + F_3 = 0 \qquad \text{Hiring and firing equation for } t = 3$$
$$I_0 - I_1 + P_1 + O_1 + S_1 + B_1 - B_0 = 110 \qquad \text{Input-output equation for } t = 1$$
$$I_1 - I_2 + P_2 + O_2 + S_2 + B_2 - B_1 = 120 \qquad \text{Input-output equation for } t = 2$$
$$I_2 - I_3 + P_3 + O_3 + S_3 + B_3 - B_2 = 80 \qquad \text{Input-output equation for } t = 3$$
$$B_3 = 0 \qquad \text{Final backorder}$$
$$P_0 = 90 \qquad \text{Initial regular production}$$
$$I_0 = 10 \qquad \text{Initial inventory}$$
$$B_0 = 0 \qquad \text{Initial backorder}$$

$$P_1, P_2, P_3, O_1, O_2, O_3, S_1, S_2, S_3, I_1, I_2, I_3,$$
$$B_1, B_2, B_3, H_1, H_2, H_3, F_1, F_2, F_3 \geq 0$$

Note that $B_3 = 0$, $P_0 = 90$, $I_0 = 10$, and $B_0 = 0$ can be directly inserted in the above LP formulation, but they are shown as equations for the purpose of illustration. For example, the first constraint in the above LP can be stated as $P_1 - H_1 + F_1 = 90$ because the production in period zero, $P_0 = 90$, is known. Also, note that $B_3 = 0$ means that all orders must be fulfilled by the end of the planning period.

Solving this LP problem using an LP software package, the optimal solution is found as presented in the table below:

t	D_t	P_t	O_t	S_t	I_t	B_t	H_t	F_t	Cost
0		90	0		10	0			
1	110	90	0	10	0	0	0	0	$1020
2	120	90	0	20	0	10	0	0	$1280
3	80	90	0	0	0	0	0	0	$900
Total	310	270	0	30	10	10			$3200

This solution has a total cost of $3200.

See Supplements S4.1.xlsx, S4.1.lg4.

Example 4.5   Linear Programming Formulation with Four Periods   Consider Example 4.3. Solve Example 4.3 by LP.

The LP formulation of Example 4.3 is presented below.

LP Problem

Minimize:

$$f_1 = 5P_1 + 8O_1 + 10S_1 + 3I_1 + 7B_1 + 6H_1 + 8F1$$
$$+5P_2 + 8O_2 + 10S_2 + 3I_2 + 7B_2 + 6H_2 + 8F_2$$
$$+5P_3 + 8O_3 + 10S_3 + 3I_3 + 7B_3 + 6H_3 + 8F_3$$
$$+5P_4 + 8O_4 + 10S_4 + 3I_4 + 7B_4 + 6H_4 + 8F_4$$

Subject to:

$$P_1 - P_0 - H_1 + F_1 = 0 \qquad \text{Hiring and firing equation for } t = 1$$
$$P_2 - P_1 - H_2 + F_2 = 0 \qquad \text{Hiring and firing equation for } t = 2$$
$$P_3 - P_2 - H_3 + F_3 = 0 \qquad \text{Hiring and firing equation for } t = 3$$
$$P_4 - P_3 - H_4 + F_4 = 0 \qquad \text{Hiring and firing equation for } t = 4$$
$$I_0 - I_1 + P_1 + O_1 + S_1 + B_1 - B_0 = 500 \qquad \text{Input−output equation for } t = 1$$
$$I_1 - I_2 + P_2 + O_2 + S_2 + B_2 - B_1 = 400 \qquad \text{Input−output equation for } t = 2$$
$$I_2 - I_3 + P_3 + O_3 + S_3 + B_3 - B_2 = 700 \qquad \text{Input−output equation for } t = 3$$
$$I_3 - I_4 + P_4 + O_4 + S_4 + B_4 - B_3 = 400 \qquad \text{Input−output equation for } t = 4$$
$$P_1 \leq 500 \qquad \text{Production limit for period 1}$$
$$P_2 \leq 500 \qquad \text{Production limit for period 2}$$
$$P_3 \leq 500 \qquad \text{Production limit for period 3}$$
$$P_4 \leq 500 \qquad \text{Production limit for period 4}$$
$$O_1 - 0.2P_1 \leq 0 \qquad \text{Overtime limit for period 1}$$
$$O_2 - 0.2P_2 \leq 0 \qquad \text{Overtime limit for period 2}$$
$$O_3 - 0.2P_3 \leq 0 \qquad \text{Overtime limit for period 3}$$
$$O_4 - 0.2P_4 \leq 0 \qquad \text{Overtime limit for period 4}$$
$$B_4 = 0 \qquad \text{No backordering for } t = 4$$
$$P_0 = 300 \qquad \text{Initial regular production}$$
$$I_0 = 0 \qquad \text{Initial inventory}$$
$$B_0 = 0 \qquad \text{Initial backorder}$$

$$P_1, P_2, P_3, P_4, O_1, O_2, O_3, O_4, S_1, S_2, S_3, S_4, I_1, I_2, I_3, I_4,$$
$$B_1, B_2, B_3, H_1, H_2, H_3, H_4, F_1, F_2, F_3, F_4 \geq 0$$

Note that $B_4 = 0$, $P_0 = 300$, $I_0 = 10$, and $B_0 = 0$ can be directly inserted in the above LP formulation, but they are shown as equations for the purpose of illustration. For example, the first constraint in the above LP can be stated as $P_1 - H_1 + F_1 = 300$ because the production in period zero, $P_0 = 300$, is known.

By solving this LP problem using an LP software package, the optimal solution can be found. The optimal solution is presented in the table below.

t	D_t	P_t	O_t	S_t	I_t	B_t	H_t	F_t	Cost ($)
0		300			0	0			
1	500	468.75	31.25	0	0	0	168.75	0	3,606.25
2	400	468.75	0	0	68.75	0	0	0	2,550
3	700	468.75	93.75	0	0	68.75	0	0	3,575
4	400	468.75	0	0	0	0	0	0	2,343.75
Total	**2000**	**1875**	**125**	**0**	**68.75**	**68.75**	**168.75**	**0**	**12,075**

This optimal solution requires a constant regular production (P_t) of 468.75 units per period and overtime production (O_t) of 31.25 and 93.75 in periods 1 and 3, respectively. Subcontracting is not necessary. To cover the high demand of 700 units in period 3, 68.75 units are inventoried in period 2 and an additional 68.75 units are backordered in period 3.

The 68.75 units backordered in period 3 are completely satisfied in period 4. There is no inventory remaining in period 4. The total cost of this optimal plan is $12,075.

See Supplements S4.2.xls, S4.2.lg4.

4.4.2   Integer Linear Programming

The LP formulation for aggregate planning covered in Section 4.4.1 does not require decision variables to be integer. In many applications, some decision variables must assume only integer values and rounding them to integer values will not render an optimal solution. In this case the LP problem should be formulated as an ILP problem. However, substantial computational time will be required to solve larger ILP problems, while very large LP problems can be solved very fast.

Example 4.5 with Integer Variables   Consider Example 4.5, where integer values were not required. For example, there are 68.75 inventory units in period 2. To generate integer values, ILP can be used. The solution to Example 4.5, found using ILP software where all decision variables are required to be integers, is presented below. The total cost of this plan is slightly higher than the total cost of Example 4.5, but this is just a coincidence; generally, requiring integer values considerably increases the total cost as presented in the next example.

t	D_t	P_t	O_t	S_t	I_t	B_t	H_t	F_t	Total Cost ($)
0		300			0	0			
1	500	469	31	0	0	0	169	0	3,607
2	400	469	0	0	69	0	0	0	2,552
3	700	469	93	0	0	69	0	0	3,572
4	400	469	0	0	0	0	0	0	2,345
Total	**2000**	**1876**	**124**	**0**	**69**	**69**	**169**	**0**	**12,076**

See Supplement S4.3.lg4.

The ILP can be used to find the actual values of some decision variables, such as the actual number of regular workers, number of hires, and number of fires.

Finding Actual Number of Regular Workers by ILP   Consider the number of regular workers, W_t, as presented in Equation (4.4). The LP problem of Section 4.4.1 can be reformulated to find the optimal integer values of the number of workers per period t, W_t.

PROBLEM 4.2   (ILP AGGREGATE PLANNING)

Minimize:

$$\text{Total cost} = \left(\sum_{t=1}^{n} (C_R P_t) + (C_O O_t) + (C_S S_t) + (C_I I_t) + (C_B B_t) + (C_H H_t) + (C_F F_t) \right)$$
$$+ \sum_{t=1}^{n} M W_t \qquad (4.16)$$

Subject to: Constraints (4.8)–(4.15) of Problem 4.1 and

$$W_t \geq \frac{P_t}{q} \qquad \text{for } t = 1, \ldots, n \tag{4.17}$$

$$W_t \qquad \text{for } t = 1, \ldots, n \text{ are integers} \tag{4.18}$$

In Problem 4.2, M is the fixed cost per regular worker regardless of the quantity that the regular worker produces and q is the total number of products that a regular worker produces per period. Therefore, the total cost of all regular workers in a given period t consists of fixed and variable costs as follows:

$$M W_t + C_R P_t \tag{4.19}$$

For example, suppose $M = \$6000$, $W_t = 8$, $C_R = 20$, $P_t = 9200$, and $q = 1000$. According to Equation (4.17), $W_t \geq (9200/1000) = 9.2$, that is, at least 9.2 workers are needed. The number of workers W_t must be an integer number. If 9.2 is rounded up to 10, then the cost for the given period for regular work based on Equation (4.19) would be $6000 \times 10 + 20 \times 9200 = \$244,000$.

The formulation for finding the actual integer values for regular workers was shown in Problem 4.2, but a similar approach can be used to find actual integer values for hire, fire, overtime, and subcontracting workers.

Example 4.6   Integer Linear Programming Formulation with Four Periods   Consider Example 4.5. Suppose the fixed cost per regular worker is $150 per period, and each worker can produce 73 units per period, that is, M = $150 and q = 73. In this problem, all decision variables are required to be integers. Solve Example 4.5 by ILP and identify the optimal number of regular workers.

The ILP problem is formulated as follows:

Minimize:

$$
\begin{aligned}
\text{Total cost } f_1 = \; & 5P_1 + 8O_1 + 10S_1 + 3I_1 + 7B_1 + 6H_1 + 8F_1 && \text{(cost for period 1)} \\
& + 5P_2 + 8O_2 + 10S_2 + 3I_2 + 7B_2 + 6H_2 + 8F_2 && \text{(cost for period 2)} \\
& + 5P_3 + 8O_3 + 10S_3 + 3I_3 + 7B_3 + 6H_3 + 8F_3 && \text{(cost for period 3)} \\
& + 5P_4 + 8O_4 + 10S_4 + 3I_4 + 7B_4 + 6H_4 + 8F_4 && \text{(cost for period 4)} \\
& + 150W_1 + 150W_2 + 150W_3 + 150W_4 && \text{(fixed cost of} \\
& && \text{regular workers)}
\end{aligned}
$$

Subject to: All LP constraints of Example 4.5, and

$$W_1 \geq P_1/73 \qquad \text{or} \qquad 73W_1 - P_1 \geq 0$$

$$W_2 \geq P_2/73 \qquad \text{or} \qquad 73W_2 - P_2 \geq 0$$

$$W_3 \geq P_3/73 \qquad \text{or} \qquad 73W_3 - P_3 \geq 0$$

$$W_4 \geq P_4/73 \qquad \text{or} \qquad 73W_4 - P_4 \geq 0$$

W_1, W_2, W_3, W_4 are integer values

Solving the above ILP problem using ILP software (we used LINGO), the solution is presented below.

t	D_t	P_t	O_t	S_t	I_t	B_t	H_t	F_t	W_t	Total Cost ($)
0		300			0	0				
1	500	365	73	62	0	0	65	0	5	4,169
2	400	365	35	0	0	0	0	0	5	2,855
3	700	365	73	262	0	0	0	0	5	5,779
4	400	365	35	0	0	0	0	0	5	2,855
Total	**2000**	**1600**	**160**	**240**	**0**	**0**	**100**	**0**	**20**	**15,658**

See Supplement S4.4.lg4.

To illustrate the difference between LP and ILP solutions, Example 4.6 is solved below by LP, that is, not requiring that W_1, W_2, W_3, and W_4 be integers.

t	D_t	P_t	O_t	S_t	I_t	B_t	H_t	F_t	W_t	Total Cost ($)
0		300			0	0				
1	500	400	80	20	0	0	100	0	5.48	4,169
2	400	400	0	0	0	0	0	0	5.48	2,855
3	700	400	80	220	0	0	0	0	5.48	5,779
4	400	400	0	0	0	0	0	0	5.48	2,855
Total	**2000**	**1600**	**160**	**240**	**0**	**0**	**100**	**0**	**20**	**15,567.67**

Note that the cost of the LP solution (not requiring integer values) is less than the cost of the ILP solution ($15,658) for the same problem (Example 4.6).

See Supplement S4.5.lg4.

4.4.3 LP Excel Solver

In the following, we show how to use Microsoft Excel to set up the LP for Example 4.5 and then solve it using Excel Solver to obtain the optimal solution. We also illustrate how to solve the problem by ILP.

In Excel, we can set up the problem as shown below in Table 4.2.

In this example the decision variables are P_t, O_t, S_t, I_t, B_t, H_t, and F_t for $t = 1, 2, 3, 4$. The problem is to find the optimal values of these variables. The above spreadsheet presents the problem before it is solved by Excel Solver. The cells that have zero values

TABLE 4.2 Excel Linear Programming Setup for Example 4.4

	A	B	C	D	E	F	G	H	I	J
1	Period (t)	Demand (D_t)	P_t	O_t	S_t	I_t	B_t	H_t	F_t	Cost (\$)
2	0		300			0	0			
3	1	500	0	0	0	0	0	0	0	0
4	2	400	0	0	0	0	0	0	0	0
5	3	700	0	0	0	0	0	0	0	0
6	4	400	0	0	0	0	0	0	0	0
7	Total	2000	300	0	0	0	0	0	0	0
8										
9								Constraints		
10	Alternative	Cost/Unit	Limit in t				Hiring & firing	0	=	0
11	Regular time prod.	5	500				Hiring & firing	0	=	0
12	Overtime prod.	8	See const.				Hiring & firing	0	=	0
13	Subcontracting	10	1000				Hiring & firing	0	=	0
14	Inventory	3	No limit				Inventory	0	=	500
15	Backordering	7	No limit				Inventory	0	=	400
16	Hiring	6	No limit				Inventory	0	=	700
17	Firing	8	No limit				Inventory	0	=	400
18							Overtime	0	<=	0
19							Overtime	0	<=	0
20							Overtime	0	<=	0
21	Target cost =	\$0					Overtime	0	<=	0

are calculated by Excel assuming all the initial values for decision variables are zero and constraints are not yet implemented. For example, cells H10 through H21 are calculated by Excel assuming the values of all decision variables are zero.

The target cell, B21, presents the objective function. It should be defined in Excel as follows:

$$B21 = \$B\$11*SUM(C3:C6) + \$B\$12*SUM(D3:D6) + \$B\$13*SUM(E3:E6)$$
$$+ \$B\$14*SUM(F3:F6) + \$B\$15*SUM(G3:G6) + \$B\$16*SUM(H3:H6)$$
$$+ \$B\$17*SUM(I3:I6)$$

The cells in the column under "Constraints" refer to the constraints involving more than one variable as in the LP formulation of Example 4.5. The first four constraints are for hiring and firing [Equation (4.8)] and will be required to have a value of zero. Note that before using the Solver function, the initial values may not be equal to zero:

$$\begin{aligned}
H10 &= C3 - C2 - H3 + I3 &\quad \text{for} \quad& P_1 - P_0 - H_1 + F_1 \\
H11 &= C4 - C3 - H4 + I4 &\quad \text{for} \quad& P_2 - P_1 - H_2 + F_2 \\
H12 &= C5 - C4 - H5 + I5 &\quad \text{for} \quad& P_3 - P_2 - H_3 + F_3 \\
H13 &= C6 - C5 - H6 + I6 &\quad \text{for} \quad& P_4 - P_3 - H_4 + F_4
\end{aligned}$$

For example, cell C3 is the value of P_1. This sets the left-hand side of Equation (4.8). Because the right-hand side of the equation is zero, in the Solver window, the constraint is set so these cells are zero.

The next four constraints in the Excel spreadsheet (Table 4.2), are for the input–output constraints [Equation (4.9)]. As in Equation (4.9), the right side is equal to the demand for that period. In this case, cells J14–J17 are set as follows:

$$J14 = B3 \quad \text{for } D_1 \quad J15 = B4 \quad \text{for } D_2 \quad J16 = B5 \quad \text{for } D_3 \quad J17 = B6 \quad \text{for } D_4$$

The left-hand side of Equation (4.9) is filled into cells H14–H17 for periods 1–4, respectively, as follows:

$$
\begin{aligned}
H14 &= F2 - F3 + C3 + D3 + E3 + G3 - G2 \quad \text{for} \quad I_0 - I_1 + P_1 + O_1 + S_1 + B_1 - B_0 \\
H15 &= F4 - F3 + C4 + D4 + E4 + G4 - G3 \quad \text{for} \quad I_1 - I_2 + P_2 + O_2 + S_2 + B_2 - B_1 \\
H16 &= F5 - F4 + C5 + D5 + E5 + G5 - G4 \quad \text{for} \quad I_2 - I_3 + P_3 + O_3 + S_3 + B_3 - B_2 \\
H17 &= F6 - F5 + C6 + D6 + E6 + G6 - G5 \quad \text{for} \quad I_3 - I_4 + P_4 + O_4 + S_4 + B_4 - B_3
\end{aligned}
$$

Now the overtime capacity constraints must be entered. In this example, overtime capacity is not a fixed number but a function of the production. The last four constraint cells are used to represent these constraints and will be required to have a value less than or equal to zero. These constraints are defined as follows:

$$
\begin{aligned}
H18 &= D3 - 0.2 * C3 \quad \text{for} \quad O_1 - 0.20P_1 \\
H19 &= D4 - 0.2 * C4 \quad \text{for} \quad O_2 - 0.20P_2 \\
H20 &= D5 - 0.2 * C5 \quad \text{for} \quad O_3 - 0.20P_3 \\
H21 &= D6 - 0.2 * C6 \quad \text{for} \quad O_4 - 0.20P_4
\end{aligned}
$$

Explanation of Constraint Values   In Table 4.2, there are three sets of constraints under the column labeled "Constraints." The left-hand column contains the equations for the constraint. For example, the contents of cell H10 is defined as equal to C3 – C2 – H3 + I3. The right-side is the constant that is used for the given constraint. For example, the content of cell J10 is 0. Now that the spreadsheet has been set up, it can be solved by Solver. Go under the Tools menu and select Solver. If you cannot find it, click on Add-Ins and check on the Solver Add-In box. In the Solver dialog box, set up the parameters as follows. A screen capture of the Solver follows the description.

"Set Target Cell" is B21	Objective function
"Equal To" is Min	Since the objective function is to be minimized
"By Changing Cells" is C3:I6	Decision variables

"Subject to the Following Constraints"

H10:H17 = \$J\$10:\$J\$17	Hiring/firing and input–output constraint cells
C3:C6 <= C11	Regular production capacity limits
H18:H21 <= 0	Overtime constraint cells
E3:E6 <= C13	Subcontracting capacity limits
C3:I6 >= 0	Nonnegativity of all variables
C3:I6 integer	For integer variables, not always necessary

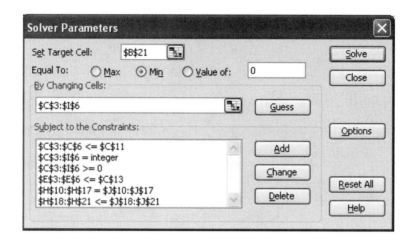

Now click on Solve. The answer should look like the following table. This is the same table format presented initially before solving the problem, except that the cell values have been calculated by the LP software. The solution is presented below.

	A	B	C	D	E	F	G	H	I	J
1	Period (t)	Demand (D_t)	P_t	O_t	S_t	I_t	B_t	H_t	F_t	Cost ($)
2	0		300			0	0			
3	1	500	469	31	0	0	0	169	0	3607
4	2	400	469	0	0	69	0	0	0	2552
5	3	700	469	93	0	0	69	0	0	3572
6	4	400	469	0	0	0	0	0	0	2345
7	**Total**	**2000**	**1876**	**124**	**0**	**69**	**69**	**169**	**0**	**12076**
8										
9								Constraints		
10	Alternative	Cost/Unit	Limit in t				Hiring & firing	0	=	0
11	Regular time prod.	5	500				Hiring & firing	0	=	0
12	Overtime prod.	8	See const.				Hiring & firing	0	=	0
13	Subcontracting	10	1000				Hiring & firing	0	=	0
14	Inventory	3	NL				Inventory	500	=	500
15	Backordering	7	NL				Inventory	400	=	400
16	Hiring	6	NL				Inventory	700	=	700
17	Firing	8	NL				Inventory	400	=	400
18							Overtime	31	<=	93.8
19							Overtime	0	<=	93.8
20							Overtime	93	<=	93.8
21	Target cost =	$12,076					Overtime	0	<=	93.8

From the solution above, the minimum cost is $12,076. This solution was presented in Example 4.5.

See Supplement S4.6.xlsx.

4.5 INTEGRATED LP AGGREGATE PLANNING

Consider m products that must be considered simultaneously where these products may use common resources. See Figure 4.6 for illustration of m products at a given period t.

"Classical" aggregate planning is based on the definition of the hypothetical "aggregate unit" as the combination of m number of units. A major difficulty occurs in the disaggregation process of translating the aggregate units into actual different product units while maintaining feasibility with respect to the constraints. That is, the disaggregated solution must also be feasible and optimal with respect to the original constraints of the problem. Classical aggregate planning for multiproducts, as it is currently practiced, often may generate infeasible or suboptimal solutions. More importantly, there is substantial difficulty in the disaggregation of an aggregate plan. In this section, we develop a large-scale integrated LP approach for solving the aggregate planning problem. This LP problem is much larger than the LP problem if aggregate units were used. However, due to the development of more powerful computers and linear programming optimization software packages, large-scale LP problems can be easily solved. Therefore, the integrated LP approach presented in this section is feasible to use and much easier to apply than the "traditional" aggregate planning approach. This approach eliminates the most confusing aspects of aggregate planning, that is, aggregation, disaggregation, and the development of a master schedule. By solving the large-scale LP problem, there will be no need for aggregation, disaggregation, and a master schedule. The solution of the large-scale LP problem is the master schedule. The solution is both optimal and feasible.

4.5.1 Multiproduct LP Aggregate Planning[†]

The notation and equations presented in this section are based on the Problem 4.1 aggregate planning formulation of Section 4.4, but expanded for m products denoted by j for $j = 1, 2, \ldots, m$. In the LP formulation of this section (Problem 4.3), the regular production for each product type is dedicated for that given product type, but this assumption is relaxed in the next section to allow the use of common resources.

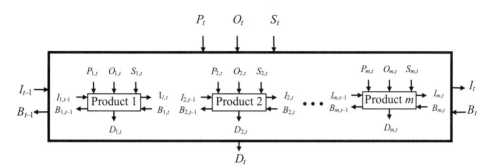

Figure 4.6 Integrated production planning for m products for a given period t.

Given Information	Decision Variables
n = total no. of periods, $t = 1, 2, \ldots, n$	$P_{j,t}$ = no. of units of product j produced in period t
m = total no. of products, $j = 1, 2, \ldots, m$	$O_{j,t}$ = no. of units of overtime of product j in period t
$D_{j,t}$ = forecasted demand for product j in period t	$S_{j,t}$ = no. of units of product j subcontracted in period t
$C_{R,j}$ = cost of regular production of one unit of product j	$I_{j,t}$ = no. of units of product j in inventory at the end of period t
$C_{S,j}$ = cost of subcontracting one unit of product j	$B_{j,t}$ = no. of units of product j backordered in period t
$C_{O,j}$ = overtime cost of producing one unit of product j	$H_{j,t}$ = no. of units of product j gained from hiring in period t
$C_{I,j}$ = cost of inventory of one unit of product j in one period	$F_{j,t}$ = no. of units of product j lost from firing in period t
$C_{B,j}$ = cost of backordering one unit of product j	
$C_{H,j}$ = hiring cost per unit of product j	
$C_{F,j}$ = firing cost per unit of product j	
$P_{j,0}$ = initial production for product j in period $t = 0$	
$B_{j,0}$ = initial backorder for product j in period $t = 0$	
$I_{j,0}$ = initial inventory for product j in period $t = 0$	
$B_{j,n}$ = final backlog for product j in period $t = n$	

The integrated LP aggregate planning is formulated as follows. This formulation is a generalization of the LP problem presented in Section 4.4, [Equations (4.7)–(4.15)].

PROBLEM 4.3 (MULTIPRODUCT LP AGGREGATE PLANNING PROBLEM)

Minimize:

$$\text{Total cost} = \sum_{t=1}^{n} \sum_{j=1}^{m} \left[(C_{R,j} P_{j,t}) + (C_{O,j} O_{j,t}) + (C_{S,j} S_{j,t}) + (C_{I,j} I_{j,t}) + (C_{B,j} B_{j,t}) \right.$$
$$\left. + (C_{H,j} H_{j,t}) + (C_{F,j} F_{j,t}) \right] \tag{4.20}$$

Subject to:

$$P_{j,t} - P_{j,t-1} - H_{j,t} + F_{j,t} = 0 \qquad \text{for } t = 1, \ldots, n, \, j = 1, 2, \ldots, m \tag{4.21}$$

$$I_{j,t-1} - I_{j,t} + P_{j,t} + O_{j,t} + S_{j,t} + B_{j,t} - B_{j,t-1} = D_{j,t}$$
$$\text{for } t = 1, \ldots, n, \, j = 1, 2, \ldots, m \tag{4.22}$$

$$B_{j,n} = 0 \qquad \text{for } j = 1, 2, \ldots, m \tag{4.23}$$

$$P_{j,t} \leq L_{j,t} \qquad\qquad \text{for } t = 1, \ldots, n, \, j = 1, 2, \ldots, m \quad (4.24)$$

$$O_{j,t} \leq M_{j,t} \qquad\qquad \text{for } t = 1, \ldots, n, \, j = 1, 2, \ldots, m \quad (4.25)$$

$$S_{j,t} \leq N_{j,t} \qquad\qquad \text{for } t = 1, \ldots, n, \, j = 1, 2, \ldots, m \quad (4.26)$$

$$P_{j,t}, O_{j,t}, S_{j,t}, I_{j,t}, B_{j,t}, H_{j,t}, F_{j,t} \geq 0 \qquad \text{for } t = 1, \ldots, n, \, j = 1, 2, \ldots, m \quad (4.27)$$

Other linear constraints including common resources $\qquad\qquad\qquad$ (4.28)

In Problem 4.3, all products are independent from each other. That is, by solving m different LP Problem 4.1, the same results can be achieved as by solving Problem 4.3. However, in the next section, we expand Problem 4.3 to consider the use of common resources as well as the use of common items used in different products. This will be shown in Problem 4.4, but first we need to discuss Problem 4.3 for independent products.

Example 4.7   Three-Product LP Aggregate Planning   Consider a company that produces three types of light bulbs, LB_1, LB_2, and LB_3. Currently there are 80, 110, and 90 (in thousands) regular production unit workforces per month for products LB_1, LB_2, and LB_3, respectively. The costs for the three bulbs are as follows:

Product Costs	$LB_1, j = 1$	$LB_2, j = 2$	$LB_3, j = 3$
Cost to produce (regular) one unit, $C_{R,j}$	$0.07	$0.05	$0.06
Cost to produce one unit during overtime, $C_{O,j}$	$0.08	$0.07	$0.08
Cost to subcontract one unit, $C_{S,j}$	$0.08	$0.08	$0.07
Cost to store one unit in inventory, $C_{I,j}$	$0.02	$0.05	$0.04
Cost to backorder one unit, $C_{B,j}$	$0.03	$0.01	$0.05
Cost to hire one unit, $C_{H,j}$	$0.13	$0.12	$0.10
Cost to fire one unit, $C_{F,j}$	$0.15	$0.13	$0.12

The resource limits (in thousands) are listed below.

Resource Limit per Product	LB_1	LB_2	LB_3
$P_{j,t}$ for $t = 1, 2, 3$	90	100	95
$O_{j,t}$ for $t = 1, 2, 3$	8	10	9
$S_{j,t}$ for $t = 1, 2, 3$	10	15	12

Determine the production plan for the next three months where the forecasted demand for each product (in thousands) for each month as follows:

		Demand	
Product	LB_1	LB_2	LB_3
Period 1	80	100	100
Period 2	90	115	95
Period 3	85	120	80
Total	255	335	275

The integrated LP formulation for this example is as follows:

Minimize:

$$\text{Total cost} = \sum_{t=1}^{3} (0.07P_{1,t} + 0.05P_{2,t} + 0.06P_{3,t} + 0.08O_{1,t} + 0.07O_{2,t} + 0.08O_{3,t}$$
$$+ 0.08S_{1,t} + 0.08S_{2,t} + 0.07S_{3,t} + 0.02I_{1,t} + 0.05I_{2,t} + 0.04I_{3,t}$$
$$+ 0.03B_{1,t} + 0.01B_{2,t} + 0.05B_{3,t} + 0.13H_{1,t} + 0.12H_{2,t} + 0.10H_{3,t}$$
$$+ 0.15F_{1,t} + 0.13F_{2,t} + 0.12F_{3,t})$$

Subject to:

$$P_{1,t} - P_{1,t-1} - H_{1,t} + F_{1,t} = 0 \quad \text{for } t = 1, 2, 3$$
$$P_{2,t} - P_{2,t-1} - H_{2,t} + F_{2,t} = 0 \quad \text{for } t = 1, 2, 3$$
$$P_{3,t} - P_{3,t-1} - H_{3,t} + F_{3,t} = 0 \quad \text{for } t = 1, 2, 3$$

$P_{1,0} = 80$	$B_{1,3} = 0$
$P_{2,0} = 110$	$B_{2,3} = 0$
$P_{3,0} = 90$	$B_{3,3} = 0$
$B_{1,0} = 0$	$I_{1,0} = 0$
$B_{2,0} = 0$	$I_{2,0} = 0$
$B_{3,0} = 0$	$I_{3,0} = 0$

$P_{1,t} \leq 90$	for $t = 1,2,3$
$P_{2,t} \leq 100$	for $t = 1,2,3$
$P_{3,t} \leq 95$	for $t = 1,2,3$
$O_{1,t} \leq 8$	for $t = 1,2,3$
$O_{2,t} \leq 10$	for $t = 1,2,3$
$O_{3,t} \leq 9$	for $t = 1,2,3$
$S_{1,t} \leq 10$	for $t = 1,2,3$
$S_{2,t} \leq 15$	for $t = 1,2,3$
$S_{3,t} \leq 12$	for $t = 1,2,3$

$$I_{1,0} - I_{1,1} + P_{1,1} + O_{1,1} + S_{1,1} + B_{1,1} - B_{1,0} = 80$$
$$I_{1,1} - I_{1,2} + P_{1,2} + O_{1,2} + S_{1,2} + B_{1,2} - B_{1,1} = 90$$
$$I_{1,2} - I_{1,3} + P_{1,3} + O_{1,3} + S_{1,3} + B_{1,3} - B_{1,2} = 85$$
$$I_{2,0} - I_{2,1} + P_{2,1} + O_{2,1} + S_{2,1} + B_{2,1} - B_{2,0} = 100$$
$$I_{2,1} - I_{2,2} + P_{2,2} + O_{2,2} + S_{2,2} + B_{2,2} - B_{2,1} = 115$$
$$I_{2,2} - I_{2,3} + P_{2,3} + O_{2,3} + S_{2,3} + B_{2,3} - B_{2,2} = 120$$
$$I_{3,0} - I_{3,1} + P_{3,1} + O_{3,1} + S_{3,1} + B_{3,1} - B_{3,0} = 100$$
$$I_{3,1} - I_{3,2} + P_{3,2} + O_{3,2} + S_{3,2} + B_{3,2} - B_{3,1} = 95$$
$$I_{3,2} - I_{3,3} + P_{3,3} + O_{3,3} + S_{3,3} + B_{3,3} - B_{3,2} = 80$$
$$P_{j,t}, O_{j,t}, S_{j,t}, I_{j,t}, B_{j,t}, H_{j,t}, F_{j,t} \geq 0 \quad \text{for } t = 1,2,3, \text{ and } j = 1,2,3$$

Solving the above LP problem, the following solution is generated (units are in thousands):

Period	Production, $P_{j,t}$		
	LB_1	LB_2	LB_3
1	80	100	90
2	80	100	87.5
3	80	100	87.5
Total	240	300	265

Period	Overtime, $O_{j,t}$			Subcontracted, $S_{j,t}$		
	LB_1	LB_2	LB_3	LB_1	LB_2	LB_3
1	0	0	0	0	0	10
2	8	10	0	2	5	0
3	5	10	0	0	10	0
Total	13	20	0	2	15	10

Period	Inventory, $I_{j,t}$			Backorders, $B_{j,t}$		
	LB_1	LB_2	LB_3	LB_1	LB_2	LB_3
1	0	0	0	0	0	0
2	0	0	0	0	0	7.5
3	0	0	0	0	0	0
Total	0	0	0	0	0	7.5

Period	Hiring, $H_{j,t}$			Firing, $F_{j,t}$		
	LB_1	LB_2	LB_3	LB_1	LB_2	LB_3
1	0	0	0	0	10	0
2	0	0	0	0	0	2.5
3	0	0	0	0	0	0
Total	0	0	0	0	10	2.5

The total cost of this solution is $54,175. The solution clearly specifies the numbers needed for each product and for each period. Note that for each product the total demand is satisfied.

For example, for product 1, the demand of 255 is satisfied: $(80 + 80 + 80) + (0 + 8 + 5) + (0 + 2 + 0) = 255$ (in thousands).

See Supplement S4.7.lg4.

Integer Linear Programming   To find the number of workers and costs for regular production, hiring, firing, overtime, and subcontracting, Equations (4.17) and (4.18) can be used for each product type. The integrated LP problem can be modified so that the number of workers and some of the other decision variables will assume only integer values. This is an extension of the ILP covered for a single product in Section 4.4.2.

4.5.2   Multiproduct LP with Common Resources[†]

In this section, we expand Problem 4.3 when common resources are used to produce different products. For example, suppose that the same regular workforce can be used for the production of different products. In this case, the hiring, firing, subcontracting, or use of overtime for different products can be reduced because the common regular workforce can be used for all products. In the following LP formulation, we present the use of a common regular workforce but the approach can be extended for the use of other common resources. Common regular workforce production to produce different products is represented by Equation (4.29). Problem 4.4 is the same as Problem 4.3, except that the set of constraints (4.29) replaces the set of constraints (4.21) of Problem 4.3.

PROBLEM 4.4   (MULTIPRODUCT LP WITH COMMON RESOURCES)

Minimize:

$$\text{Total Cost} = \text{Equation (4.20)}$$

Subject to:

Constraints (4.22)–(4.28), and

$$\sum_{j=1}^{m} P_{j,t} - \sum_{j=1}^{m} P_{j,t-1} - \sum_{j=1}^{m} H_{j,t} + \sum_{j=1}^{m} F_{j,t} = 0 \quad \text{for} \quad t = 1, \ldots, n \qquad (4.29)$$

Example 4.8   Three-Product LP with Common Regular Workforce   Consider Example 4.7 Suppose that the same regular workforce can be used to produce all three products. Find the total cost and the values of the decision variables. Compare this solution to the solution of Example 4.7.

Consider the LP formulation of Example 4.7 Remove the first three sets of constraints (i.e., a total of nine constraints) which are based on Equation (4.21), that is, remove

$$P_{j,t} - P_{j,t-1} - H_{j,t} + F_{j,t} = 0 \quad \text{for} \quad t = 1, 2, 3, \quad j = 1, 2, 3$$

Instead, use the following set of three constraints based on Equation (4.29):

$$\sum_{j=1}^{3} P_{j,t} - \sum_{j=1}^{3} P_{j,t-1} - \sum_{j=1}^{3} H_{j,t} + \sum_{j=1}^{3} F_{j,t} = 0 \quad \text{for} \quad t = 1, 2, 3$$

Now solve the problem using an LP software package. The solution is presented below (units are in thousands).

| Period | Production, $P_{j,t}$ | | |
	LB_1	LB_2	LB_3
1	80	100	95
2	85	100	90
3	90	100	85
Total	255	300	270

| Period | Overtime, $O_{j,t}$ | | | Subcontracted, $S_{j,t}$ | | |
	LB_1	LB_2	LB_3	LB_1	LB_2	LB_3
1	0	0	0	0	0	5
2	0	10	0	0	5	0
3	0	10	0	0	10	0
Total	0	20	0	0	15	5

| Period | Inventory, $I_{j,t}$ | | | Backorders, $B_{j,t}$ | | |
	LB_1	LB_2	LB_3	LB_1	LB_2	LB_3
1	0	0	0	0	0	0
2	0	0	0	5	0	5
3	0	0	0	0	0	0
Total	0	0	0	0	0	0

| Period | Hired, $H_{j,t}$ | | | Fired, $F_{j,t}$ | | |
	LB_1	LB_2	LB_3	LB_1	LB_2	LB_3
1	0	0	0	0	0	5
2	0	0	0	0	0	0
3	0	0	0	0	0	0
Total	0	0	0	0	0	0

The total cost is $53,000, which is less than the $54,175 of Example 4.7. The cost of this problem is lower because common resources of workers were utilized. In this solution, more products are produced by regular production: $255 + 300 + 270 = 825$ versus the solution of Example 4.7, where there were $240 + 300 + 265 = 805$ produced by regular production.

See Supplement S4.8.lg4.

4.5.3   Multicomponent Multiproduct Planning[†]

Each product can consist of several components (items). Usually, some of the components of different products are the same. For example, the brake pads of several different cars may be of the same type. Problem 4.4 can be expanded to include planning for all common components of all products.

Suppose each product j consists of K components denoted by k for $k = 1, \ldots, K$. The regular production can be expressed as follows:

$$\sum_{k=1}^{K} P_{j,k,t} = P_{j,t} \qquad \text{for} \quad t = 1, \ldots, n, \quad j = 1, \ldots, m \qquad (4.30)$$

$$P_{j,k,t} \geq 0 \qquad \text{for} \quad \text{all } j, k, \text{ and } t \qquad (4.31)$$

The relevant cost information for different components can also be included in Problem 4.4.

Furthermore, each period t can consist of subperiods (e.g., a month may consists of four weeks). For this case, suppose each period t consists of R subperiods denoted by r for $r = 1, \ldots, R$:

$$\sum_{r=1}^{R}\sum_{k=1}^{K} P_{j,k,t,r} = P_{j,t} \qquad \text{for all } j, t \qquad (4.32)$$

$$P_{j,k,t,r} \geq 0 \qquad \text{for all } j, \ k, \ t, \ \text{and } r \qquad (4.33)$$

4.6   MULTIOBJECTIVE LP AGGREGATE PLANNING

4.6.1   LP with Linear Objectives

Consider the LP formulation of Section 4.4 (Problem 4.1). In the multiobjective linear programming (MOLP) problem of aggregate planning, several objectives can be considered. The most important linear objectives are:

Minimize total cost:

$$f_1 = \sum_{t=1}^{n} [(C_R P_t) + (C_O O_t) + (C_S S_t) + (C_I I_t) + (C_B B_t) + (C_H H_t) + (C_F F_t)]$$

Minimize total backorders:

$$f_2 = \sum_{t=1}^{n} B_t$$

Minimize total firing:

$$f_3 = \sum_{t=1}^{n} F_t$$

These three objectives are all linear functions and can be used in linear programming. The first objective, f_1, minimizes the total cost. The second objective, f_2, minimizes the customer's dissatisfaction because of delay in meeting due dates. The third objective, f_3, minimizes the total firing, which maximizes job stability.

Bicriteria LP Problem   The two most important objective functions are total cost and total backorders. As stated before, backorders lead to customer dissatisfaction and will be detrimental to a business in the long term. Here, we discuss the first two objectives (f_1 and f_2), but the method can be applied to more objectives. The MOLP problem is:

> Minimize total cost, f_1
> Minimize backorders, f_2
> Subject to constraints of Problem 4.1, that is, Equations (4.8)–(4.15)

The method for solving this MOLP problem for additive utility functions is presented below. First, find the minimum and maximum values of the two objectives, so that both objective functions can be normalized. Then, formulate the MOLP problem to be solved by LP.

Finding the Extreme Point Associated with Minimum Cost

> Minimize total cost, f_1
> Subject to LP constraints (4.8)–(4.15)

Solve this problem by LP. Label the optimal objective function value as $f_{1,\min}$. For this optimal solution, calculate the f_2 value and label it as $f_{2,\max}$.

Finding Extreme Point Associated with Minimum Backlog   Since the minimum backorder is zero, label $f_{2,\min} = 0$. That is, set all B_t values to zero. Then solve the following LP problem to find the cost associated with this requirement:

Minimize:

> Total cost $= f_1$

Subject to:
> $B_t = 0$ for $t = 1, 2, \ldots, n$
> LP constraints (4.8)–(4.15)

Solve this problem by LP. Label its objective function value as $f_{1,\max}$.

Now, one can find the best alternative for a given additive utility function. Suppose the weights of importance for normalized objective values are w_1 and w_2, where $w_1 + w_2 = 1$. The normalized values of the two objective functions are f_1' and f_2', which are also minimized. Both normalized objective functions are linear functions as presented in the following LP problem. The following linear programming problem can be used to find the optimal solution for minimizing the additive utility function:

Minimize:

$$U = w_1 f_1' + w_2 f_2' \quad \text{(additive utility function)}$$

Subject to:

$$f_1' = \frac{f_1 - f_{1,\min}}{f_{1,\max} - f_{1,\min}} \quad \text{(normalized equation for } f_1)$$

$$f_1' = \frac{f_2 - f_{2,\min}}{f_{2,\max} - f_{2,\min}} \quad \text{(normalized equation for } f_2)$$

$$f_1 = \sum_{t=1}^{n} [(C_R P_t) + (C_O O_t) + (C_S S_t) + (C_I I_t) + (C_B B_t) + (C_H H_t) + (C_F F_t)]$$

$$f_2 = \sum_{t=1}^{n} B_t$$

LP constraints (4.8)–(4.15)

Supplement S4.9.doc contains two additional examples for MOLP problems.

Example 4.9   MOLP for Two Objectives and Four Periods   Consider Example 4.5. Suppose the weights of importance for the normalized values for the first and the second objectives are 0.55 and 0.45, respectively. Find the best alternative to this problem.

The MOLP formulation for this problem is presented below:

Minimize f_1
Minimize f_2
Subject to:

$$f_1 = 5P_1 + 8O_1 + 10S_1 + 3I_1 + 7B_1 + 6H_1 + 8F_1$$
$$+ 5P_2 + 8O_2 + 10S_2 + 3I_2 + 7B_2 + 6H_2 + 8F_2$$
$$+ 5P_3 + 8O_3 + 10S_3 + 3I_3 + 7B_3 + 6H_3 + 8F_3$$
$$+ 5P_4 + 8O4 + 10S_4 + 3I_4 + 7B_4 + 6H_4 + 8F_4$$
$$f_2 = B_1 + B_2 + B_3 + B_4$$

All constraints of Example 4.5.

Finding the Extreme Point Associated with the Minimum Cost

Minimize total cost f_1
Subject to Example 4.9 LP constraints

The solution to this problem is the same as the solution produced in Example 4.5 for f_1 with integer variables. That is,

$$f_{1,min} = \$12{,}076 \qquad\qquad\qquad \text{(total cost)}$$

For this given solution, the value of f_2 is

$$f_{2,max} = B_1 + B_2 + B_3 + B_4 = 0 + 0 + 69 + 0 = 69 \quad \text{(total backorder)}$$

Finding Extreme Point Associated with Minimum Backorders Now, find the total cost by requiring $f_2 = 0$. That is, set $B_1 = B_2 = B_3 = B_4 = 0$ and solve the LP problem:

Minimize cost f_1
Subject to $B_t = 0$ for $t = 1, 2, 3, 4$ and Example 4.9 constraints.

The optimal solution to this problem is

$$f_{1,max} = \$12{,}280 \quad \text{(total cost)}$$
$$f_{2,min} = 0 \qquad\qquad \text{(total backorder)}$$

The ranges for both objective functions are as follows:

	f_1	f_2
Maximum	12,280	69
Minimum	12,076	0

Use the above range information to define the normalized objective function constraints for f_1' and f_2'.

$$f_1' = \frac{f_1 - 12{,}076}{12{,}280 - 12{,}076} \qquad \text{or} \qquad f_1 - 204 f_1' = 12{,}076$$

$$f_2' = \frac{f_2 - 0}{69 - 0} \qquad\qquad \text{or} \qquad f_2 - 69 f_2' = 0$$

Formulate the following linear programming to find the best multicriteria alternative that minimizes the additive utility function:

Minimize:

$$U = 0.55 f_1' + 0.45 f_2'$$

Subject to:

$$f_1 - 204 f_1' = 12{,}076 \qquad f_2 - 69 f_2' = 0$$

and Example 4.9 constraints.

Solving this LP problem using an LP software package yields the optimal multicriteria solution shown in the following table. The total cost f_1 is $12,192 and the total number of backorders f_2 is 20.

Period, t	D_t	P_t	O_t	S_t	I_t	B_t	H_t	F_t	Cost ($)
0		300			0				
1	500	420	80	0	0	0	120	0	3,460
2	400	420	0	0	20	0	0	0	2,160
3	700	420	84	156	0	20	0	0	4,472
4	400	420	0	0	0	0	0	0	2,100
Total	**2000**	**1680**	**164**	**156**	**20**	**20**	**120**	**0**	**12,192**

See Supplements S4.10.xls, S4.10.lg4.

The following table shows the bicriteria planning solutions for the following range of weights: (0.99, 0.01), (0.6, 0.4), (0.55, 0.45), (0.4, 0.6), and (0.01, 0.99).

Weights	D	P	O	S	I	B	H	F	f_1	f_2
(0.99, 0.01)	2000	1876	124	0	69	69	169	0	12076	69
(0.60, 0.40)	2000	1872	125	3	68	68	168	0	12078	68
(0.55, 0.45)	2000	1680	164	156	20	20	120	0	12192	20
(0.40, 0.60)	2000	1660	166	174	15	15	115	0	12208	15
(0.01, 0.99)	2000	1600	160	240	0	0	100	0	12280	0

4.6.2   Multicriteria Problem with Nonlinear Objectives[†]

In addition to the linear objective functions presented in Section 4.6.1, nonlinear objective functions can also be considered. Suppose that the ideal inventory level for each period is

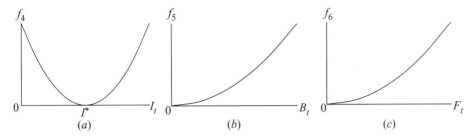

Figure 4.7   Three quadratic objective functions (*a*) inventory; (*b*) backorders; (*c*) firing fluctuation.

I^*. By minimizing the square of the distance from the ideal inventory, $(I_t - I^*)^2$, much higher penalties will be given for higher deviations from the ideal I^*. This is presented in Figure 4.7a. For simplicity, one can set $I^* = 0$. Similarly, quadratic functions for minimizing the backorder and firing can be defined, where the ideal or goal could be zero; see Figures 4.7b,c, respectively. The following three objective functions are nonlinear, more specifically quadratic, functions:

Minimize fluctuations in inventory:

$$f_4 = \sum_{t=1}^{n} (I_t - I^*)^2$$

Minimize fluctuations in backorders:

$$f_5 = \sum_{t=1}^{n} B_t^2$$

Minimize fluctuations in firing:

$$f_6 = \sum_{t=1}^{n} F_t^2$$

These three objective functions are presented in Figure 4.7.

Because the first three objective functions (f_1, f_2, and f_3) are linear, the MOLP problem can be solved by LP. For objective functions f_4, f_5, and f_6, the approach for finding the minimum and maximum values of each objective and the normalized functions is similar to the procedure in Section 4.6.1. For objective functions f_4, f_5, and f_6, the problem can be solved using a quadratic programming method where the utility objective function is quadratic and the constraints are linear. This can be done by incorporating objective functions and their normalization equations directly into the utility objective function which will be used as the objective function for this problem. Currently, many optimization software packages can solve quadratic programming problems. For simplicity of presentation, in the following example, we use the previous section problem formulation, which in this case includes nonlinear constraints.

Example 4.10   MOLP for Three Objectives and Four Periods   Consider the following three objectives for Example 4.5:

Minimize total cost, f_1.
Minimize fluctuations in inventory, f_4.
Minimize fluctuations in backlogs, f_5.
Subject to constraints of Example 4.5.

Suppose the weights of importance are $w_1 = 0.4$, $w_2 = 0.3$, and $w_3 = 0.3$ for the normalized values of the three objectives, respectively. Suppose the ideal inventory is zero, that is, $I^* = 0$. That is, the ideal inventory for each period is zero. Constraints of Example 4.10 are presented below.

Constraints of Example 4.10

$$
\begin{aligned}
f_1 = &\; 5P_1 + 8O_1 + 10S_1 + 3I_1 + 7B_1 + 6H_1 + 8F_1 \\
&+ 5P_2 + 8O_2 + 10S_2 + 3I_2 + 7B_2 + 6H_2 + 8F_2 \\
&+ 5P_3 + 8O_3 + 10S_3 + 3I_3 + 7B_3 + 6H_3 + 8F_3 \\
&+ 5P_4 + 8O_4 + 10S_4 + 3I_4 + 7B_4 + 6H_4 + 8F_4 \quad \text{(total cost)} \\
f_4 = &\; I_1^2 + I_2^2 + I_3^2 + I_4^2 \qquad\qquad \text{(total fluctuation in inventory)} \\
f_5 = &\; B_1^2 + B_2^2 + B_3^2 + B_4^2 \qquad\qquad \text{(total fluctuation in backorders)}
\end{aligned}
$$

And LP constraints of Example 4.5.

Finding Solution Associated with Minimum Cost

Minimize f_1.
Subject to constraints of Example 4.10.

The optimal solution to the above problem is $f_1 = \$12{,}076$, $f_4 = 4761$, and $f_5 = 4761$.

Finding Solution Associated with the Minimum Inventory Fluctuations

Minimize f_1.
Subject to $I_1 = 0$, $I_2 = 0$, $I_3 = 0$, $I_4 = 0$ and constraints of Example 4.10.

The optimal solution to the above problem is $f_1 = \$12{,}276$, $f_4 = 0$, and $f_5 = 162$.

Finding Solution Associated with Minimum Backorder Fluctuations

Minimize f_1.
Subject to $B_1 = 0$, $B_2 = 0$, $B_3 = 0$, $B_4 = 0$ and constraints of Example 4.10.

The optimal solution to the above problem is $f_1 = \$12{,}280$, $f_4 = 0$, and $f_5 = 0$.

Based on the solutions of the above three problems, the resulting ranges for the three objectives are shown in the table below.

	f_1	f_4	f_5
Minimum	12,076	0	0
Maximum	12,280	4761	4761

Using this information, the three equations for normalized objective functions are

$$f_1' = \frac{f_1 - 12{,}076}{12{,}280 - 12{,}076} \quad \text{or} \quad f_1 - 204 f_1' = 12{,}076$$

$$f_4' = \frac{f_4 - 0}{4761 - 0} \quad \text{or} \quad f_4 - 4761 f_4' = 0$$

$$f_5' = \frac{f_5 - 0}{4761 - 0} \quad \text{or} \quad f_5 - 4761 f_5' = 0$$

Now, solve the following optimization problem:

Minimize:

$$U = 0.4 f_1' + 0.3 f_4' + 0.3 f_5'$$

Subject to:

$$f_1 - 204 f_1' = 12{,}076 \quad f_4 - 4761 f_4' = 0 \quad f_5 - 4761 f_5' = 0$$

and constraints of Example 4.10.

Solving this nonlinear programming problem yields the plan shown in the following table. We used LINGO to solve this problem.

Period, t	D_t	P_t	O_t	S_t	I_t	B_t	H_t	F_t	Cost ($)
0		300			0				
1	500	400	80	20	0	0	100	0	3,440
2	400	400	0	0	0	0	0	0	2,000
3	700	400	80	220	0	0	0	0	4,840
4	400	400	0	0	0	0	0	0	2,000
Total	2000	1600	160	240	0	0	100	0	12,280

The solution in terms of the three objective functions is $f_1 = 12{,}280$, $f_4 = 0$, and $f_5 = 0$.

See Supplement S4.11.xlsx.

4.7   DISSAGGREGATION AND MASTER SCHEDULE

In traditional aggregate planning, all different products are aggregated and presented in terms of a single aggregate unit. After solving the aggregate planning problem, the aggregate unit of products must then be disaggregated into the original products in order to create a master schedule. See Figure 4.8. The master schedule shows how many units of each product must be produced in each period while considering the needed resources. In the master schedule, periods are typically smaller than those used in the aggregate plan. For example, the aggregate plan may specify the production planning for each month, while the master schedule would specify production planning on a weekly basis.

Long-Term and Short-Term Aggregate Planning   In long-term aggregate planning, decisions regarding the final products are made. For example, plan 100, 125, and 95 units of product A and 55, 60, and 80 units of product B for months 1, 2, and 3, respectively. Based on this plan, long-term planning in terms of needed resources, such as manpower and inventory levels, is decided for long-term periods (e.g., months). Then, the production can be disaggregated for shorter periods of time, for example, weeks or days. In short-term aggregate planning, planning for the components of each product occurs for periods with smaller duration. The push-and-pull (MRP/JIT) systems discussed in Chapter 5 show how the detailed planning is accomplished.

Family, Items, and Conversions Factor   In aggregate planning the following definitions are used to show the relationship of items to families of products and create aggregate units:

- *Product*   A product consists of items (also called components).
- *Item*   An item is the smallest unit that is considered in the aggregate planning problem.
- *Family*   A family is a group of items whose production similarities allow them to share the same resources, facilities, and labor. Clustering methods can be used to identify families of similar items; See Chapter 12.
- *Aggregate*   Unit An aggregate unit is a common unit for a family of items for planning purposes. Aggregate units can be presented in terms of period, resources, or any appropriate physical units.
- *Conversion Factor*   A conversion factor is a scalar that converts a given item to the aggregate unit. A factor of 1 represents a normal (e.g., an average) product. Each item has its own factor.

Conversion factors are used for aggregation and disaggregation purposes. For example, consider three products A, B, C with factors 1.4, 1.1, and 0.5 respectively. This means product A uses 1.4 units of resources compared to the hypothetical aggregate unit, which

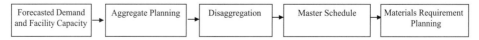

Figure 4.8   Aggregate planning relationship to disaggregation, master schedule, and MRP.

uses one unit of resources. Now suppose the demands for items A, B, and C are 100, 150, and 80 units, respectively. The demand for the aggregate unit will be

$$1.4 \times 100 + 1.1 \times 150 + 0.5 \times 80 = 345$$

Simple Disaggregation Method Different lot sizing and optimization methods are covered in Chapters 5 and 6. Here, for the purpose of demonstration, we show a simple-lot-sizing method for aggregate planning for a multiproduct problem.

Suppose the same resources are used to produce different products, and only one product can be produced at a time, that is, in each period only one type of product can be produced. The first step is to disaggregate the aggregate unit into the actual units of production using the given conversion factors. The aggregate production plan is then broken down into production requirements for each product within the production periods. Then, determine whether or not the generated production plan is feasible with respect to the capacity limits. Notation used in disaggregating are:

$D_{j,t}$ = demand for product j in period t (given demand value)
$P_{j,t}$ = production plan for product j in period t (to be decided)
L_j = maximum lot size for product j (given value)
T_j = the number of periods needed to produce product j (to be decided), $j = 1, \ldots, m$
n = total number of production periods (given value)
m = the number of products

The number of periods required for product j is

$$T_j = \sum_{t=1}^{n} \lceil D_{j,t}/L_j \rceil \quad \text{for } j = 1, \ldots, m$$

where $\lceil \; \rceil$ indicates that noninteger numbers are rounded up to the next highest integer value. For example, 6.2 is rounded up to 7. The numbers are rounded up to avoid shortages. An aggregate plan is feasible if

$$\sum_{j=1}^{m} T_j \leq n$$

If the schedule is feasible, then there is a sufficient number of periods to produce all the needed lot sizes. Otherwise, there will be some shortages at the end of the last period.

The following heuristic can be used to assign lots of different products to different periods:

1. For the first period choose the product that has the highest demand.
2. Consider the next period. Choose the product that has the minimum inventory. Break ties arbitrarily.
3. Calculate inventory $I_{j,t}$ for all products for the given period.

Repeat steps 2 and 3 until all periods are enumerated.

It may not be possible to always satisfy the production demand of all products on time. In this lot-sizing planning, it is assumed that the products being produced in a period can be used to satisfy the demand of the same period.

Example 4.11 Master Scheduling for Two Products and Eight Periods   Consider the following demands for two products for eight biweekly periods:

Biweekly period, t	1	2	3	4	5	6	7	8	Total
Actual $D_{1,t}$	25	25	20	20	30	35	40	35	230
Actual $D_{2,t}$	40	30	25	35	40	45	50	55	320

In this example, the aggregate unit is simply the summation of the two products; that is, a factor of 1 is used for each product. The capacity constraints for the maximum lot size are 70 for product 1 and 90 for product 2 for each period.

Find a feasible solution to this plan. Present the master schedule.

First, check the feasibility. The number of periods required for each product one as follows:

For product 1

$$T_1 = \sum_{t=1}^{8} \lceil D_{1,t}/L_1 \rceil = \lceil 230/70 \rceil = \lceil 3.29 \rceil = 4$$

For product 2

$$T_2 = \sum_{t=1}^{8} \lceil D_{2,t}/L_2 \rceil = \lceil 320/90 \rceil = \lceil 3.56 \rceil = 4$$

Therefore, the minimum number of periods required is $4 + 4$, where $n = 8$. That is,

$$T_1 + T_2 \le n \quad \text{i.e.} \quad 4 + 4 \le 8$$

Since there are eight available production periods, it is feasible to produce the required products.

The following is a master schedule for this example:

	Demand by Period								Total
	1	2	3	4	5	6	7	8	Demand
Product 1 to produce, $P_{1,t}$	0	70	0	70	0	70	0	70	280
Product 2 to produce, $P_{2,t}$	90	0	90	0	90	0	90	0	360
Inventory, $I_{1,t} = P_{1,t} - (D_{1,t} - I_{1,t-1})$	−25	20	0	50	20	55	15	50	
Inventory, $I_{2,t} = P_{2,t} - (D_{2,t} - I_{2,t-1})$	50	20	85	50	100	55	95	40	

The first production period is used to produce product 2 as it has a higher demand than product 1 in the first period. For the rest of the periods, a lot is produced for the product whose inventory is the lowest. In this plan, product 1 has shortages in period 1 (where the inventory is negative). However, its total demand is met at the end of the planning period. In this example, the whole lot size is produced in each period. In this plan, the final inventories (period 8) are 50 for product 1 and 40 for product 2.

REFERENCES

General

Axsater, S. "Aggregation of Product Data for Hierarchical Production Planning." *Operations Research*, Vol. 29, 1981, pp. 744–756.

Ebert, R. J. "Aggregate Planning with Learning Curve Productivity." *Management Science*, Vol. 23, 1976, pp. 171–182.

Elsayed, E. A., and T. O. Boucher. *Analysis and Control of Production Systems*, 2nd E. Englewood Cliffs. NJ: Prentice Hallm 1994.

Heizer, J., and B. Render. *Operations Management.* Upper Saddle River, NJ: Prentice Hall, 2001.

Jain A., and U. S. Palekar. "Aggregate Production Planning for a Continuous Reconfigurable Manufacturing Process." *Computers & Operations Research*, Vol. 32, No. 5, 2005, pp. 1213–1236.

Kanyalkar, A. P., and G. K. Adil. "An Integrated Aggregate and Detailed Planning in a Multi-site Production Environment Using Linear Programming." *International Journal of Production Research*, Vol. 43, No. 20, 2005, pp. 4431–4454.

Khoshnevis, B., P. M. Wolfee, and M. P. Terrell. "Aggregate Planning Models Incorporating Productivity—An Overview." *International Journal of Production Research*, 1982, Vol. 20, No. 5, pp. 555–564.

Masud, A. S. M., and C. L. Hwang. "An Aggregate Production Planning Model and Application of Multiple Objective Decision Methods." *International Journal of Production Research*, Vol. 118, 1980, pp. 115–127.

Nahmias, S. *Production and Operation Analysis*, 6th ed. New York: McGraw-Hill, 2008.

Sipper, D., and R. L. Bulfin, Jr. *Production: Planning, Control, and Integration.* New York: McGraw-Hill, 1997.

Stevenson, M., L. C. Hendry, and B. G. Kingsman. "A Review of Production Planning and Control: The Applicability of Key Concepts to the Make-to-Order Industry." *Journal of Production Research*, Vol. 43, No. 5, 2005, pp. 869–898.

Taubert, W. H. "A Search Decision Rule for the Aggregate Scheduling Problem." *Management Science*, Vol. 14, 1968, pp. 343–359.

Vánczaa, J., T. Kisa, and A. Kovács. "Aggregation—The Key to Integrating Production Planning and Scheduling." *CIRP Annals—Manufacturing Technology*, Vol. 53, No. 1, 2004, pp. 377–380.

Vollmann, T. E., W. L. Berry, and D. C. Whybark. *Manufacturing, Planning, and Control Systems*, 4th ed. New York: McGraw-HillIrwin, 1997.

Zoller, K. "Optimal Disaggregation of Aggregate Production Plans." *Management Science*, Vol. 17, 1971, pp. 533–549.

Multiple-Objective Optimization

Chen, Y. K., and H. C. Liao. "An Investigation on Selection of Simplified Aggregate Production Planning Strategies Using MADM Approaches." *International Journal of Production Research*, 2003, Vol. 41, No. 14, pp. 3359–3374.

Deckro, R. F., and J. E. Hebert. "Goal Programming Approaches to Solving Linear Decision Rule Based Aggregate Production Planning Models." *IIE Transactions*, Vol. 16, No. 4, 1984, pp. 308–315.

Goodman D. A. "A Goal Programming Approach to Aggregate Planning of Production and Work Force." *Management Science*, Vol. 20, No. 12, 1974, pp. 1569–1575.

Leung, S. C. H., and S. S. W. Chan. "A Goal Programming Model for Aggregate Production Planning with Resource Utilization Constraint." *Computers & Industrial Engineering*, Vol. 56, No. 3, 2009, pp. 1053–1064.

Leung, S. C. H., Y. Wu, and K. K. Lai. "Multi-Site Aggregate Production Planning with Multiple Objectives: A Goal Programming Approach." *Production Planning & Control*, Vol. 14, No. 5, 2003, pp. 425–436.

Malakooti, B. "A Gradient-Based Approach for Solving Hierarchical Multi-Criteria Production Planning Problems." *Computers & Industrial Engineering*, Vol. 16, No. 3, 1989, pp. 407–417.

Malakooti, B. and E. C. Tandler. "An Overview of a Decision Support System for Solving Discrete MCDM Problems under Certainty and Uncertainty with Applications to Multi-objective Production Planning." *International Journal of Information and Management Sciences*, Vol. 2, No. 1, 1991, pp. 61–81.

Malakooti, B. "Multiple Objective Aggregate Planning and Optimization." *Case Western Reserve University, Cleveland, Ohio 2013*.

Malakooti, B. "Double Helix Value Functions, Ordinal/Cardinal Approach, Additive Utility Functions, Multiple Criteria, Decision Paradigm, Process, and Types." *International Journal of Information Technology & Decision Making (IJITDM)*, Forthcoming, 2014a.

Malakooti, B. "Z Utility Theory: Decisions Under Risk and Multiple Criteria" to appear, 2014b.

Masud, A. S. M., and C. L. Hwang. "An Aggregate Production Planning Model and Application of Three Multiple Objective Decision Methods." *International Journal of Production Research*, Vol. 18, No. 6, 1980, pp. 741–752.

Mezghani M., A. Rebai, A. Dammak, and T. Loukil. "A Goal Programming Model for Aggregate Production Planning Problem." *International Journal of Operational Research*, Vol. 4, No. 1, 2009, pp. 23–34.

Silva, C. G. D., J. Figueira, J. Lisboa, and S. Barman. "An Interactive Decision Support System for An Aggregate Production Planning Model Based on Multiple Criteria Mixed Integer Linear Programming." *Omega*, Vol. 34, No. 2, 2006, pp. 167–177.

Vercellis C. "Multi-Criteria Models for Capacity Analysis and Aggregate Planning in Manufacturing Systems." *International Journal of Production Economics*, Vol. 23, Nos. 1–3, 1991, pp. 261–272.

EXERCISES

Graphical Approach (Section 4.2)

2.1 Consider five periods with the following demands: 1021, 1963, 1555, 2196, and 1438 for periods 1–5. Suppose the initial workforce is 1500. The cost of the regular workforce is $102/unit—$100/unit for hiring and $120/unit for firing. The cost of inventory is $118/unit per period. The cost of backorder is $132/unit. Overtime and subcontracting are not allowed.

   (a) Show the input–output equations for this problem.

   (b) Consider a constant inventory plan where no backorder is allowed. Identify the cost of this plan. Show your solution graphically (no overtime or subcontracting is allowed).

   (c) Consider a constant workforce plan where only inventory is allowed to vary. Identify the cost of this plan. Show your solution graphically.

   (d) Allow both the inventory and workforce to vary; find the best plan and show it graphically.

2.2 Consider six periods with the following demands: 121, 383, 155, 196, 383, and 168 for the next six months. Suppose the initial workforce is 225. The costs of the regular workforce are \$80/unit, \$159/unit for overtime, and \$190/unit for subcontracted work. The cost for hiring is \$145/unit and \$120/unit for firing. The cost to store inventory is \$118/unit and the cost to backorder is \$132/unit.

Respond to parts (a)–(d) of exercise 2.1.

D_1	D_2	D_3	D_4	D_5	D_6
121	383	155	196	383	168

P_0	I_0	B_0	C_R	C_O	C_S	C_I	C_B	C_H	C_F
225	0	0	80	159	190	118	132	145	120

Tabular Method (Section 4.3)

3.1 A company has expected demands of 1510, 1830, 1580, and 1760 for the next four months, respectively. The initial workforce and inventory of the company are 1650 and 50, respectively. The costs for the workforce are \$12/unit for regular, \$25/unit for overtime, and \$24/unit for subcontracted work. The cost of backorder is \$10/unit. The inventory carrying cost is \$8/unit. The cost for hiring is \$6/unit and \$10/unit for firing.

D_1	D_2	D_3	D_4
1510	1830	1580	1760

P_0	I_0	B_0	C_R	C_O	C_S	C_I	C_B	C_H	C_F
1650	50	0	12	25	24	8	10	6	10

(a) Maintain zero inventory, but allow only hiring and firing. Find the minimum cost.
(b) Maintain a constant workforce but allow only inventory and backorders to vary. Find the minimum cost.
(c) Maintain a constant workforce but allow only inventory and subcontracting to vary. Find the minimum cost.
(d) Try to find the minimum total cost by considering all variables to vary (the best mix plan). Only use the spreadsheet approach; that is, do not use linear programming.
(e) Compare and contrast the solutions found in parts (a)–(d).

3.2 A company has expected demands of 1021, 1963, 1555, 2196, and 1438 for the next five months, respectively. The initial workforce is 1650 and the initial inventory is 50. The production level is limited to 2200. Production cost with a regular workforce is \$1/unit, production cost using subcontractors is \$2.9/unit, and production cost using overtime is \$3.5/unit. The cost for hiring is \$1.5/unit and \$1.2/unit for firing. The backorder cost is \$2.4/unit and the inventory-carrying cost is \$2.5/unit.

Respond to parts (a)–(e) of exercise 3.1.

D_1	D_2	D_3	D_4	D_5
1021	1963	1555	2196	1438

P_0	I_0	B_0	C_R	C_O	C_S	C_I	C_B	C_H	C_F
1650	50	0	1	3.5	2.9	2.5	2.4	1.5	1.2

3.3 A company has expected demands of 121, 383, 155, 196, 383, and 168 for the next six months. The initial workforce of the company is 225 and the company cannot employ more than 400 workers at a time but the labor laws permit the company to allow overtime to 10% of the workforce. The costs for the workforce are \$102/unit for regular, \$159/unit for overtime, and \$190/unit for subcontracted work. The cost for hiring is \$145/unit and \$120/unit for firing. The storage capacity for inventory is limited to 25 units. The cost to store inventory is \$118/unit and the cost to backorder a unit is \$132.

Respond to parts (a)–(e) of exercise 3.1.

D_1	D_2	D_3	D_4	D_5	D_6
121	383	155	196	383	168

P_0	I_0	B_0	C_R	C_O	C_S	C_I	C_B	C_H	C_F
225	0	0	102	159	190	118	132	145	120

Linear Programming Method (Section 4.4)

4.1 Consider exercise 3.1.

(a) Use linear programming to determine the minimum cost. Identify the optimal value for all decision variables.

(b) Use integer linear programming to determine the minimum cost. Identify the optimal value for all decision variables.

(c) In exercise 3.1, also consider the following constraints: Overtime should be less than or equal to 20% of regular production and backorders should be less than or equal to 5% of the demand. Solve by LP and ILP.

4.2 Consider exercise 3.2.

(a) Use linear programming to determine the minimum cost. Identify the optimal value for all decision variables.

(b) Use integer linear programming to determine the minimum cost. Identify the optimal value for all decision variables.

(c) In exercise 3.2, also consider the following constraints: Overtime should be less than or equal to 15% of regular production, backorders should be less than or equal to 5% of the demand, and subcontracting should be less than 25% of regular production. Solve by LP and ILP.

4.3 Consider exercise 3.3.

(a) Use linear programming to determine the minimum cost. Identify the optimal value for all decision variables.

(b) Use integer linear programming to determine the minimum cost. Identify the optimal value for all decision variables.

(c) In exercise 3.3, also consider the following constraints: Overtime should be less than or equal to 10% of the regular workforce, backorders should be less than or equal to 15% of the demand, and subcontracting should be less than 12% of regular production. Solve by LP and ILP.

4.4 Consider exercise 3.1. Suppose the cost per regular worker is $250 per period, and each worker produces 15 units per period.

(a) Solve the problem by integer linear program and identify the optimal number of regular workers.

(b) Solve this problem by LP. How does the solution differ from ILP?

4.5 Consider exercise 3.2. Suppose the cost per regular worker is $500 per period, and each worker produces 12 units per period.

(a) Solve by integer linear programming and identify the optimal number of regular workers.

(b) Solve this problem by LP. How does the solution differ from ILP?

4.6 Consider exercise 3.3. Suppose the cost per regular worker is $50 per period and each worker produces 5 units per period.

(a) Solve by integer linear programming and identify the optimal number of regular workers.

(b) Solve this problem by LP. How does the solution differ from ILP?

Integrated Aggregate Planning (Section 4.5)[†]

5.1 Consider a company that produces three types of car tires: CT_1, CT_2, and CT_3. Currently there are 75, 90, and 80 units (in thousands) regular production workforce per month for products CT_1, CT_2, and CT_3, respectively. The costs for the tires are as follows.

Each Product Cost	CT_1	CT_2	CT_3
Cost to produce one unit by regular worker, $C_{R,j}$	$0.17	$0.16	$0.15
Cost to subcontract one unit, $C_{S,j}$	$0.18	$0.19	$0.17
Cost to produce one unit during overtime, $C_{O,j}$	$0.19	$0.16	$0.18
Cost to store one unit in inventory, $C_{I,j}$	$0.2	$0.5	$0.4
Cost to backorder one unit, $C_{B,j}$	$0.3	$0.5	$0.7
Cost to hire one production unit, $C_{H,j}$	$0.23	$0.22	$0.20
Cost to fire one production unit, $C_{F,j}$	$0.25	$0.23	$0.22

The resource limitations (in thousands) of the company are listed below.

Resource Limitation per Product	CT_1	CT_2	CT_3
$P_{j,t}$ for $t = 1, 2, 3$	85	100	95
$O_{j,t}$ for $t = 1, 2, 3$	10	2	9
$S_{j,t}$ for $t = 1, 2, 3$	15	5	10

Determine the production plan for the next three months, where the forecasted demand (in thousands) for each product for each month as follows:

Each Product	CT_1	CT_2	CT_3
Period 1	80	90	100
Period 2	90	85	95
Period 3	85	90	80

5.2 Consider a company that produces four types of products, A_1, A_2, A_3, and A_4. Currently there are 7.5, 8.5, 8, and 9 units (in thousands) of regular production workforce per period for products A_1, A_2, A_3, and A_4, respectively. The costs for the products are as follows.

Product Cost	A_1	A_2	A_3	A_4
Cost to produce one unit, $C_{R,j}$	$0.37	$0.35	$0.34	$0.32
Cost to subcontract one unit, $C_{S,j}$	$0.39	$0.38	$0.37	$0.36
Cost to produce one unit during overtime, $C_{O,j}$	$0.28	$0.29	$0.28	$0.28
Cost to store one unit in inventory, $C_{I,j}$	$0.2	$0.5	$0.4	$0.24
Cost to backorder one unit, $C_{B,j}$	$0.3	$0.5	$0.7	$0.21
Cost to hire one unit, $C_{H,j}$	$0.41	$0.42	$0.40	$0.40
Cost to fire one unit, $C_{F,j}$	$0.43	$0.43	$0.42	$0.42

The resource limitations (in thousands) of the company are listed below.

Resource Limits per Product	A_1	A_2	A_3	A_4
$P_{j,t}$ for $t = 1, 2, 3$	10	9	8.5	8
$O_{j,t}$ for $t = 1, 2, 3$	1	1	0.9	0.8
$S_{j,t}$ for $t = 1, 2, 3$	1.1	1.2	1.1	1

Determine the production plan for the next three months, where the forecasted demand (in thousands) for each product for each period as follows:

Product	A_1	A_2	A_3	A_4
Period 1	8.5	9	10	9
Period 2	7	8	8	8.5
Period 3	9	10	9	9

5.3 Consider exercise 5.1. Suppose that regular production can be conducted by the same workforce for different products. Find the total cost and the value of decision variables. Compare this solution to the solution of exercise 5.1.

5.4 Consider exercise 5.2. Suppose that the regular production can be conducted by the same workforce for different products. Find the total cost and the value of decision variables. Compare this solution to the solution of exercise 5.2.

LP with Linear Objectives (Section 4.6.1)[†]

6.1 Consider exercise 3.1.

  (a) Consider a bicriteria problem where the objectives are to minimize total cost (f_1) and minimize total backorders (f_2). Suppose the weights of importance for the normalized objective values are (0.4, 0.6), respectively. Use an additive utility function for the MOLP problem to determine the optimal plan.

  (b) Consider three linear objectives: total cost, total backorder, and total firing. Suppose the weights of importance for the normalized objective values are (0.2, 0.3, 0.5), respectively. Use an additive utility function for the MOLP problem to determine the optimal plan.

6.2 Consider exercise 3.2.

  (a) Consider a bicriteria problem where the objectives are to minimize cost (f_1) and minimize backorders (f_2). Suppose the weights of importance for the normalized objective values are (0.5, 0.5). Use an additive utility function for the MOLP problem to determine the optimal plan.

  (b) Consider three linear objectives: total cost, total backorder, and total firing. Suppose the weights of importance for the normalized objective values are (0.4, 0.4, 0.2). Use an additive utility function for the MOLP problem to determine the optimal plan.

6.3 Consider exercise 3.3.

  (a) Consider a bicriteria problem where the objectives are to minimize cost (f_1) and minimize backorders (f_2). Suppose the weights of importance for the normalized objective values are (0.5, 0.5). Use an additive utility function for the MOLP problem to determine the optimal plan.

(b) Consider three linear objectives: total cost, total firing, and total backorders. Suppose the weights of importance for the normalized objective values are (0.3, 0.4, 0.3). Use an additive utility function for the MOLP problem to determine the optimal plan.

Multiobjective LP Aggregate Planning (Section 4.6.2)[†]

6.4 Consider exercise 3.1. Consider three objectives—total cost, total inventory fluctuations, and total backlog fluctuations—where the ideal inventory and ideal total backorders are both zero. Suppose the weights of importance for the normalized objective values are (0.4, 0.3, 0.3). Solve the nonlinear optimization problem to determine the optimal cost.

6.5 Consider exercise 3.2. Consider three objectives—total cost, total inventory fluctuations, and total backorder fluctuations—where ideal inventory and ideal total backorders are both zero. Suppose the weights of importance for the normalized objective values are (0.1, 0.3, 0.6). Solve the nonlinear optimization problem to determine the optimal cost.

6.6 Consider exercise 3.3. Consider three objectives—total cost, total inventory fluctuations, total firing fluctuations—where the ideal inventory is 15 and the ideal firing is zero. Suppose the weights of importance for the normalized objective values are (0.2, 0.5, 0.3). Solve the nonlinear optimization problem to determine the optimal cost.

Disaggregation and Master Schedule (Section 4.7)

7.1 Suppose an aggregate product is based on two different products for eight periods. Consider production demand requirements for two products as presented below.

Product	\multicolumn{8}{c	}{Demand per Period}	Total Demand
	1	2	3	4	5	6	7	8	
1	250	300	250	200	300	400	450	350	2500
2	400	300	350	400	450	500	400	300	3100

That is, the conversion factors are 1 for both products. The capacity constraints are 700 for product 1 and 850 for product 2 for each given period. Only one type of product can be produced per period.

(a) Determine how many lots are required for each product and check the feasibility of the plan. Which products at what periods are backordered?

(b) If the capacity constraints are 500 for product 1 and 350 for product 2, what is the production plan?

7.2 Consider the following table for three products for 10 production periods. The capacity constraints for products 1, 2, and 3 are 750, 500, and 400, respectively.

Product	\multicolumn{10}{c}{Demand by Production Period}	Total Demand
	1	2	3	4	5	6	7	8	9	10	
1	100	150	80	250	320	80	100	130	100	120	1430
2	130	120	70	100	180	120	120	50	50	200	1140
3	90	100	105	95	150	110	100	80	100	120	1050

(a) Determine how many lots are required for each product and check the feasibility of the plan. Which products at what periods are backordered?

(b) If the capacity constraints are 600, 500, and 300 for products 1, 2, and 3, respectively, what is the capacity planning?

CHAPTER 5

PUSH-AND-PULL (MRP/JIT) SYSTEMS

5.1 INTRODUCTION

This chapter covers detailed production planning methods when product demands are known. The two most important production planning approaches are push systems (known as materials requirement planning, MRP) and pull systems (known as just-in-time, JIT). Figure 5.1 presents an overview of the production planning process as related to MRP and/or JIT. Production planning is performed by using information from the master production schedule (MPS), the product processing plan, bills of material (BOMs), available inventory, outstanding purchase orders, and lead times for components (items) of the product.

Push-and-pull systems have two contrasting points of view on how to organize the production plan to satisfy the given demands. The push system (MRP) is based on forcing different production facility processes to produce the needed components to satisfy the needed demands. An analogy is opening the flood gates of a dam and allowing a desired amount of water to flow through. The water, in this case, is the input materials. Ideally, the pressure of this push (which is based on the external demand) forces all of the downstream workstations to work to their maximum capacity to produce the needed items.

The pull system (JIT) is based on internal planning, meaning that each downstream production facility announces when it is ready to provide the needed services so that a specific amount of materials can be sent down to it. An analogy of this process is drip irrigation, where water is sent through a series of pipes to needed areas and is released drop by drop as needed. This system is based on the internal capacity of the downstream workstations. As a result, if the downstream workstations are not demanding items to be processed, the system will be idle. Due to the pulling (on-demand) nature of JIT systems, the system may not be fully utilized if the individual downstream stations are not demanding to produce items to their full capacities. On the other hand, if the individual downstream

Operations and Production Systems with Multiple Objectives, First Edition. Behnam Malakooti.
© 2014 John Wiley & Sons, Inc. Published 2014 by John Wiley & Sons, Inc.

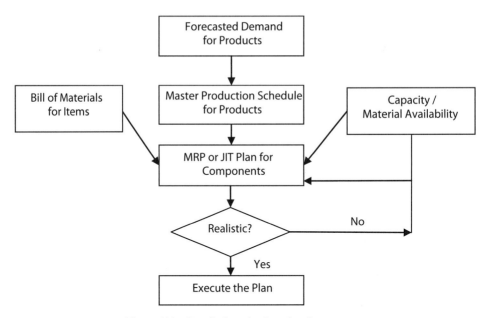

Figure 5.1   Detailed production planning process.

stations are efficient and demanding, then less efficient upstream stations can be identified quickly. The JIT approach is based on a chain effect (free will of workers), while MRP is based on a hierarchical structure (enforcement system). The JIT system can quickly identify the problem areas if the workers are interested in making the system efficient.

The MRP approach as it is known today was developed mostly after World War II by expanding the concept of a BOM, which was used for several decades before MRP. The BOM is a simple tree structure that shows the components (items) of a product and how long it takes to produce each component or part. The MRP takes the BOM information one step further. It identifies the production planning for given lots of products for given periods. The MRP is a multistage planning approach that identifies the quantity and timing for producing different components to satisfy the final product demands for each period. The basic calculation of this system is called MRP explosion calculus, which is based on simple arithmetic for finding production requirements in the form of one table for each component. Based on the required demand and lead time for each component, the MRP table is built to calculate the needed quantities of each component for each period.

There are two approaches in satisfying the demand for components for a given period:

(a) Produce the exact number of components (items) needed per period; this is called the lot-for-lot or variable-lot-size approach. This approach is consistent with the JIT philosophy.

(b) Produce an economic order quantity called the lot size where additional items produced are accumulated to be used in future periods. This is consistent with the MRP system.

In the first approach, JIT, the total inventory is minimized to zero. In the second approach, MRP, additional inventory (or shortage) can be carried over for future periods, but the method minimizes the total cost of inventory and setup.

Just-in-time was originally developed in Japan by the Toyota Motor Company for the purpose of reducing the in-process inventory, which usually ties up substantial amounts of investment and space. By reducing the inventory, several important results were achieved. The most important one is that reducing inventories helps to identify the troubled or bottle-necked production processes (e.g., workstations). In JIT, there is no in-process inventory; the whole production line stops when one of the workstations is not functioning as planned. This perpetuates a domino effect that shuts down the whole system. To avoid system shut-down, all workstations must be smoothly operational with high reliability and low variance in their production cycle times.

JIT systems are usually based on implementation of a kanban system. *Kanban* is the Japanese word for "card." There are usually two types of cards: (a) cards that demand the upstream station for new input parts to be processed and (b) cards that inform the downstream stations that the requested demands are fulfilled. Kanban is a simple manual information processing system which has made production planning very effective and smooth. The other achievement of JIT is its strategy to increase the agility of the system by developing strategies that allow for quick setup of stations for different lots. In particular, the development of Single Minute Exchange of Dies (SMED) at Toyota aimed at reducing the setup time from approximately 24 h to 1 min. SMED was achieved by realizing that 99% of the machine setup and exchanging tools can be accomplished offline while the machines are operating and producing needed parts. By minimizing the total time for tool exchange and machine setup, substantial productivity improvements have been achieved not only at Toyota but also in many different industries. JIT emphasizes smaller lot sizes and quick turnovers, leading to what is called lean manufacturing.

Traditional push-and-pull production systems have fixed views of production planning, which means that a factory or plant commits to one of the two extreme philosophies of push or pull systems. In this chapter, we introduce a third approach for production planning, called multiobjective hybrid push-and-pull production planning. In multiobjective (or multicriteria) production planning, we develop a strategy that is based on the advantages of both push (MRP) and pull (JIT) systems. We allow different levels of push-and-pull systems based on the multiple objectives of the given problem at hand. In multiple-objective push-and-pull production planning, depending on the product, the problem, and the attitude of the workers, a hybrid push–pull system can be built to make the system work very efficiently. The level of push and pull is determined by the multiobjective planning. This will be discussed in Section 5.8.

5.2 MATERIALS REQUIREMENT PLANNING: PUSH SYSTEM

Material requirement planning is a detailed production planning process which uses inputs from all functional areas of a business, including management, purchasing, production, marketing, finance, human resource, and engineering. Consider a product that consists of several components or subproducts. Furthermore, each subproduct may consist of other subproducts called components, items, or parts. MRP is used in dependent demand sit-uations. Dependent demand means that the requirement of one item is dependent on the requirements of other items or subproducts. The final product quantities and their scheduled deliveries are based on the forecasted future demands for final products. In a dependent demand system, once the demand forecast for the final product is decided, the required quantities for all dependent components (subassemblies) can be determined. MRP

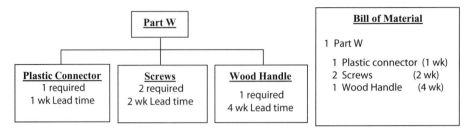

Figure 5.2 Product structure diagram (tree on left) and indented BOM (table on right).

generates a schedule that specifies the type and exact quantity of each component. It also specifies the time required to produce each component. The input to MRP is the MPS, which is a plan for both the quantity and scheduled time of the end product. MRP provides the plan for the components of the end product to satisfy the demands of MPS.

5.2.1 Bill of Materials

Bills of materials are used to illustrate exactly what is required to produce one unit of the end product. A BOM is a list of the components and their quantities needed to assemble or manufacture the end product. This information can be presented as either a product structure diagram (in the form of a tree) or an indented list (in the form of a table). Thus, a BOM provides a detailed disaggregated picture to be used in the production plan.

For illustration purposes, consider a corporation that produces a part called "W." Part W consists of one plastic connector, two screws, and one wood handle. The plant can produce and assemble part W at a rate of 100 per week. The screws need to be ordered at least two weeks prior to the assembly. The plastic connectors need to be completed one week prior to the assembly. The wood shop must receive the order for wood handles four weeks in advance.

The product structure diagram and the indented BOM are presented in Figure 5.2.

5.2.2 Time-Phased Product Structure

A time-phased product structure can be built by using the BOM of the given product. The BOM lists the required quantities and lead times of parts and subassemblies for one final product assembly. The time-phased diagram illustrates the time needed to build or acquire the needed components to assemble the final product. For each product, the time-phased product structure shows the sequence and duration of each operation. For example, consider the assembly of part A, whose time-phased product structure diagram is shown in Figure 5.3. Part A takes eight weeks to be completed. In this example, part A is assembled using subcomponents B and C. The numbers in the parentheses indicate the number of components needed for the assembly. In this example, to produce one part, A, two units of B and four units of C are required. To produce each unit of C, three units of E and one unit of F is required. Note that components D and E are used in different subassemblies. Therefore, to assemble one unit of A, the material requirements for B, C, D, E, F, and G are 2, 4, $2 \times 3 + 3 \times 1 \times 4 = 18$, $3 \times 2 + 3 \times 4 = 18$, $4 \times 1 = 4$, and $2 \times 1 \times 4 = 8$, respectively. In the time-phased product structure diagram, the lead time for each component is presented. For example, it takes four weeks to produce or acquire part F. The lead time for parts A, B, C, D, E, F, and G are 1, 1, 1, 1, 2, 4, and 2, respectively.

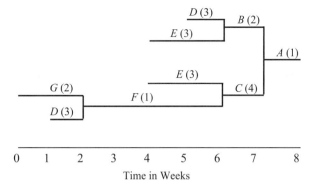

Figure 5.3 Time-phased product structure diagram for part A.

5.2.3 Material Requirement Planning

The MRP is a set of tables which show the number of units and the time each component should be ready to satisfy the given demand for the final product in each period of time. Once the demand, the BOM, and the time-phased product structure diagram for a finished product are known, it becomes a simple arithmetic problem to determine the quantity and the time when each component will be needed.

In the MRP table, as presented in the following example, the following terminology is used.

- The *gross requirement* is the demand for each component for each period.
- The *net requirement* is the number of components needed to be provided in each period to satisfy the gross requirement. The net requirements for a component are found by subtracting any scheduled receipts and any existing inventory from the gross requirements.
- The component's *release date* is determined by subtracting its lead time from the product due date. Below is an example to illustrate the MRP approach.

Example 5.1 MRP Schedule Consider Figure 5.3. For the purpose of illustration, only the calculation of the MRP tables for components A, B, and E are shown here. The given gross requirements and the scheduled receipts for the final product A are presented in the table below. The known values are presented in **bold**. All other numbers are calculated. Find the MRP for parts A, B, and E.

Part Id: A			Lead Time (weeks): 1		Number Needed: 1		Current Period: 4			
	By Period, t (weeks)									
	4	5	6	7	8	9	10	11	12	13	**Total**
1. Gross requirements	**0**	**0**	**30**	**36**	**40**	**36**	**42**	**40**	**48**	**32**	304
2. Scheduled receipts	**0**	**0**	**6**	**0**	**0**	**6**	**0**	**0**	**0**	**0**	12
3. Inventory on-hand	**0**	**28**	4	0	0	0	0	0	0	0	32
4. Net requirement	0	0	0	32	40	30	42	40	48	32	264
5. Time-phased net requirement	0	0	32	40	30	42	40	48	32	0	264
6. Planned order release (lot size)	0	0	32	40	30	42	40	48	32	0	264

Part A

The table above shows the gross requirements for part A beginning in period 4, and it illustrates how scheduled receipts (row 2) and existing inventory (row 3) are treated. Note that part A takes one week to produce; therefore, its lead time is one week.

Row 1: Gross Requirements This is the required quantity of part A that must be fulfilled. For example, in period 6, 30 units must be ready to be shipped out to the customer.

Row 2: Scheduled Receipts These are items that have been ordered and are expected to be received at these periods. For example, in periods 6 and 9, six items are scheduled to be received, but in periods 7, 8, 10, 11, 12, and 13, zero items will be received.

Row 3: Inventory On-Hand This is the existing number of part A available in the inventory. For example, in period 5, the existing inventory is 28.

Now, calculate the inventory on-hand for the remaining periods. In period 6, the inventory on-hand is

> Inventory on-hand of last period + scheduled receipts of this period − gross
> requirements of this period
> = inventory on-hand of this period
> or
> 28 (period 5) + 6 (period 6) − 30 (period 6) = 4 (period 6)

The inventory on-hand is either a positive number or zero. For example, the inventories on-hand for periods 7–13 are all zero.

Row 4: Net Requirements This is the amount needed to be produced. It is only considered for periods in which the inventory is zero. For example, there is no net requirement in periods 5 and 6, as they both have positive inventory (meaning that their gross requirement was satisfied). For period 7, the calculation is

> Gross requirement of this period − inventory on-hand of last period − scheduled
> receipts of this period
> = net requirement of this period
> or
> 36 − 4 − 0 = 32

Similarly, the net requirement for period 8 is 40 − 0 − 0 = 40, and the net requirement for period 9 is 36 − 0 − 6 = 30.

Row 5: Time-Phased Net Requirement This is the same as the net requirement except that it is moved back for the lead time period. For part A, the lead time is one week. Therefore, to satisfy the net requirement of 32 units in period 7, the production must start one week earlier. That is,

> Time-phased net requirement (of this period)
> = net requirement (of future period minus its lead time)

Row 6: Planned Order Release (for Lot-for-Lot Strategy) To satisfy the time-phased net requirement, one can release (meaning order or produce) exactly what is needed. This is called the lot-for-lot strategy. For example, the planned order release of period 6 is 32. Alternatively, one can produce different quantities of lots because it could be more economical to produce larger lot sizes. The process of finding optimal lot size is called lot sizing, and it is covered in the next section. To verify the calculations, one can check the totals, where

$$\text{Total gross requirement} = \text{total receipts} + \text{initial inventory} + \text{total planned orders}$$
$$304 = 12 + 28 + 264$$

Part B

With the material requirement plan established for part A, now the plan for part B can be developed. The planned release (lot size) from part A's MRP from the above table becomes the gross requirement for part B's MRP, as shown below. There are two part B's needed to make one part A. The input to part B is based on the planned order release (row 6) of the table for part A. To calculate the gross requirement for part B (row 1 of the following table), each value of the planned order release of part A (row 6 of the table for part A) must be multiplied by 2. For example, $32 \times 2 = 64$ units of part B are needed in period 6. Also, it is given that there are 16 planned receipts of part B in period 7.

Part Id: B				Lead Time (weeks): 1 Number Needed: 2						Current Period: 4
	By Period, t (weeks)									
	4	5	6	7	8	9	10	11	12	13	**Total**
1. Gross requirements	**0**	**0**	**64**	**80**	**60**	**84**	**80**	**96**	**64**	**0**	528
2. Scheduled receipts	**0**	**0**	**0**	**16**	**0**	**0**	**0**	**0**	**0**	**0**	16
3. Inventory on-hand	**0**	**0**	0	0	0	0	0	0	0	0	0
4. Net requirement	0	0	64	64	60	84	80	96	64	0	512
5. Time-phased net requirement	0	64	64	60	84	80	96	64	0	0	512
6. Planned order release (lot size)	0	64	64	60	84	80	96	64	0	0	512

Now, it can be verified that the totals match. That is, $528 = 16 + 0 + 512$.

Continuing the example, part B is made from parts D and E. For the purpose of illustration, consider part E.

Part E

Three part E's are required for the production of one unit of part B. From the time-phased product structure diagram above, one order of part E takes two weeks. Also, it is given that the on-hand inventory of part E in period 4 is 100. The planned release of part B (times 3) becomes the gross requirements of part E. For example, $64 \times 3 = 192$ units of part E are needed in period 5. See the following table for the remaining calculations.

Part Id: E		Lead Time (weeks): 2		Number Needed: 3					Current Period: 3	
		By Period, t (weeks)								
	3	4	5	6	7	8	9	10	11	12	**Total**
Gross requirements	**0**	**0**	**192**	**192**	**180**	**252**	**240**	**288**	**192**	**0**	1536
Scheduled receipts	**0**	**0**	**0**	**0**	**0**	**0**	**0**	**0**	**0**	**0**	0
Inventory on-hand	**0**	**100**	0	0	0	0	0	0	0	0	100
Net requirement	0	0	92	192	180	252	240	288	192	0	1436
Time-phased net requirement	92	192	180	252	240	288	192	0	0	0	1436
Planned order release (lot size)	92	192	180	252	240	288	192	0	0	0	1436

The MRP for the other parts of Figure 5.3 (i.e., C, D, F, and G) can be similarly calculated.

5.3 LOT-SIZING APPROACHES

In many industries, the same facility, equipment, space, and manpower are used to produce different types of products. For example, in the auto industry, the same forming and pressing equipment can be used to produce a variety of car body parts, such as doors, fenders, and hoods for a variety of cars. However, each piece of equipment must be set up with the appropriate tools and dies for the given part. The process of changing from one set of tools and dies to another is time consuming and expensive. Therefore, it is usually preferred to run a production line for the maximum possible time. Furthermore, usually by having longer runs of the same parts, operational problems can be diagnosed and resolved. The process of identifying how many parts should be produced (before changing the setup for another part) is called the lot-sizing problem. There are two types of lot sizing: static and dynamic. In static lot sizing, the production quantities (lots) are equal for each period. In this case, the lot size is easily determined. In the dynamic lot sizing, the production quantity may vary from period to period depending on the fluctuating product demands and other constraints. This section covers the dynamic lot sizing.

One way to meet the net requirement is to produce exactly what is needed at each given period. This is called lot-for-lot sizing. The lot-for-lot scheme is usually the most expensive method of production. There are different lot-sizing methods, where each scheme has different production costs associated with it. The lot-sizing cost consists of factors such as setup cost, holding cost, and backlog (shortage) cost. In each period, a different lot size of items can be produced that results in a different total cost. The variable-lot-size approach will result in different levels of inventory which require inventory space usage and the capital investment associated with holding the inventory. The objective of lot sizing is to minimize the total lot-sizing cost while fulfilling the customer demands on time. Five lot-sizing methods are discussed in the subsequent sections. In all these methods, the short-term future demands are known, but the long-term future demands are unknown. Before choosing a method, one should identify how well the assumptions of each method apply to a given problem.

5.3.1 Lot-for-Lot

In lot-for-lot, the number of units scheduled for production during each period is the same as the net requirement for that period. In the MRP approach of Section 5.2, a lot-for-lot production schedule was used. As stated before, due to the setup cost for production, lot-for-lot lot sizing is usually the most expensive method of production. Lot-for-lot is closest to the JIT philosophy of production planning.

Example 5.2 Lot-for-Lot Sizing   Consider the following net requirements:

Period, t (weeks)	4	5	6	7	8	9	10	11	12	Total
Net requirement (r_t)	0	0	32	40	30	42	40	48	32	264

Suppose that one technician at \$44/h for 3 h is needed to set up the equipment and there is \$14 worth of maintenance materials. Therefore, the setup cost is $44(3) + 14 = \$146$. The holding costs for the inventory are at approximately 20% per part per year, where each part is worth \$150. Assume there are 52 weeks/year. Therefore, the holding cost is $150(0.20)/52 = \$0.58$ per part per week.

Find the total cost associated with this lot size plan.

The table below presents the lot-for-lot sizing scheme for this problem.

	By Period, t (weeks)									
	4	5	6	7	8	9	10	11	12	**Total**
Net requirement, r_t	**0**	**0**	**32**	**40**	**30**	**42**	**40**	**48**	**32**	264
Lot size, Q_t	0	0	32	40	30	42	40	48	32	264
Net inventory, I_t	0	0	0	0	0	0	0	0	0	0

For this example, there are seven setups for seven lots. The table below shows the total cost of this lot-for-lot sizing policy.

	Number of Setups (Y)	Setup Cost (C_p)	Inventory Total (I^+)	Holding Cost (C_h)	Total Cost $= (Y)(C_p) + (I^+)(C_h)$
Lot for lot	7	\$146	0	\$0.58	$7 \times \$146 + 0 \times \$0.58 = \$1022$

5.3.2 Economic Order Quantity

The economic order quantity (EOQ) lot-sizing technique finds the best lot size that minimizes the total lot-sizing cost assuming a constant rate of demand per period. For example, it uses the average annual demand to find the weekly constant rate of demand. The total cost consists of the setup or ordering cost and the holding cost.

Economic order quantity lot sizing minimizes the total cost with respect to the setup and the holding cost assuming a constant demand over a period of time (e.g., one year). The average demand rate may be calculated from past annual production or from the given net requirements in a given MRP. The formula for finding the optimal lot size [Equation (5.1)] is derived in Chapter 6.

To use the EOQ, the following information is needed: the holding cost rate per period, C_h; the setup cost per period, C_p; and the average demand rate per period, D. The EOQ lot size Q is given by

$$Q = \sqrt{\frac{2DC_p}{C_h}} \qquad (5.1)$$

In this method, schedule to produce the lot in the period where

Net Inventory of last period < net requirement of the given period

Example 5.3   Economic Order Quantity Lot-Sizing Example   Consider Example 5.2 where the setup cost (C_p) is \$146 per setup and the inventory holding cost (C_h) is \$0.58 per part per period. Assume that the average annual demand is 2000 parts.

Find the economic order quantity lot size and its associated lot-sizing costs.

Use Equation (5.1) to find the EOQ:

$$Q = \sqrt{\frac{2(2000/52)(146)}{0.58}} = 139.15$$

The lot size, 139.15, can be rounded to the closest integer number, 139. The resulting MRP for the product is as follows:

| | By Period, t (weeks) | | | | | | | | | |
	4	5	6	7	8	9	10	11	12	**Total**
Net requirement, r_t	0	0	32	40	30	42	40	48	32	264
Lot size, Q_t	0	0	139	0	0	139	0	0	0	278
Net inventory, I_t	0	0	107	67	37	134	94	46	14	499

The lot size of 139 units is produced in period 6, but only 32 units of it are used for the net requirements. Then, no lots are produced in periods 7 and 8. But, in period 9, the lot size of 139 units is produced again. Period 9 was chosen because the inventory of 37 units from the previous period is less than the net requirement of 42 for period 9. The total cost is given below.

	Number of Setups (Y)	Setup Cost (C_p)	Inventory Total (I^+)	Holding Cost (C_h)	Total Cost $= (Y)(C_p) + (I^+)(C_h)$
EOQ	2	\$146	499	\$0.58	$2 \times \$146 + 499 \times \$0.58 = \$581.42$

In this example, two setups were used. Recall that the total cost using lot-for-lot was $1022, but the EOQ cost is $581.42. In both lot-for-lot and the EOQ method, the total net requirements must be fulfilled by the end of the planning period. As stated before, the EOQ method minimizes the total cost assuming a constant rate of demand per period. In reality, the EOQ may not be applicable if there are fluctuating demands.

5.3.3 Part Period Balancing

The EOQ policy finds the optimal lot size by considering both the setup and inventory holding costs while assuming a constant demand rate per period. In EOQ all lot sizes are the same. Part period balancing (PPB) attempts to balance the holding and setup costs when the demand rate is variable; therefore, lot sizes may be different in each period.

The PPB approach can be summarized as follows. Accumulate demand period by period until the total holding cost exceeds the setup cost. The lot size is then the accumulated value of the items to be produced. The first lot size is produced in the first period at which accumulation was started. Shortages will not occur in this method.

Consider the following notation:

i = period number (for iterative calculation purpose), ith period in a given iteration where $i = 1, 2, \ldots, T$

t = period for a given lot size where $t = 1, \ldots, N$

r_i = net requirement associated with the ith period of a given lot

Example 5.4   Part Period Balancing   Consider Example 5.2. Calculations for the PPB lot-sizing method are presented in the table below.

Period, t	Net Requirement, $r_t = r_i$	i	Tentative Lot Size, Q_i	Tentative Periods That a Lot Is Held (y)	Tentative Additional Inventory Costs = $r_i y \times 0.58$	Tentative Total Inventory Costs for period i	Is Total Inventory Cost Higher Than Setup Costs, $146?	Lot Size, Q_t
6	32	1	$0 + 32 = 32$	0	$0.00	$0.00	No	144
7	40	2	$32 + 40 = 72$	1	$23.20	$23.20	No	0
8	30	3	$72 + 30 = 102$	2	$34.80	$58.00	No	0
9	42	4	$102 + 42 = 144$	3	$73.08	$131.08	No	0
10	40	—	$144 + 40 = 184$	4	$92.80	$223.88	Yes	—
10	40	1	$0 + 40 = 40$	0	$0.00	$0.00	No	120
11	48	2	$40 + 48 = 88$	1	$27.84	$27.84	No	0
12	32	3	$88 + 32 = 120$	2	$37.12	$64.96	No	0

In this example, there are two lots. In period 10, since the holding cost of $223.88 is greater than the setup cost, $146, the lot size is chosen to be 144, which should be produced in period 6. After the first lot is determined, the next lot size is determined, beginning from period 10. In the above table, period 10 is shown in two rows for clarity of calculations. In period 12, the total holding cost is still less than the setup cost. However, since this is the last period, the lot size of 120 is selected, which is to be produced in period 10. For these lot sizes, the MRP table solution is as follows:

	By Period, t (weeks)									
	4	5	6	7	8	9	10	11	12	**Total**
Net requirement, r_t	0	0	32	40	30	42	40	48	32	264
Lot size, Q_t	0	0	144	0	0	0	120	0	0	264
Net inventory, I_t	0	0	112	72	42	0	80	32	0	338

The above MRP has two setups where the total cost of this lot size policy is 2($146) + 338($0.58) = $488.04. The total cost of this PPB policy is less than the cost of the lot-for-lot and EOQ schemes. The table below summarizes the total cost of the PPB lot-sizing approach.

	Number of Setups (Y)	Setup Cost (C_p)	Inventory Total (I^+)	Holding Cost (C_h)	Total Cost = $(Y)(C_p) + (I^+)(C_h)$
PPB	2	$146	338	$0.58	$2 \times \$146 + 338 \times \$0.58 = \textbf{\$488.04}$

The next two lot-sizing methods are similar to the PPB lot-sizing method but may generate different solutions.

5.3.4 Silver–Meal Heuristic

The Silver–Meal heuristic, developed by Silver and Meal (1973), utilizes the average cost per period as a function of the number of periods that the orders cover. At each period, the average cost is calculated. As long as the average cost decreases, the lot size increases. Once the average cost per period increases compared to the previous period, the lot size is set. This heuristic attempts to minimize the total holding cost.

Consider the following notations:

T = last period for a given lot size iteration (different lots may have different T values)

C_T = average cost for a given lot size

For each period, the total holding cost is calculated, and then its average is found by dividing it into the number of periods. The total cost is determined by adding the setup cost C_p and the holding cost C_h multiplied by the net requirement r_i for the given period. This value is divided by the total number of periods considered so far to determine the average cost C_T at period i. The calculation is summarized below. Consider the periods associated with a given lot, $i = 1, 2, \ldots, T$:

$$C_T = \frac{C_p + \sum_{i=1}^{T} (i-1) r_i C_h}{i} \tag{5.2}$$

Assume $C_T > C_{T-1}$. Then the sum of all r_i from 1 to $T-1$ is used as the lot size. After the first lot size is determined, the procedure is repeated at the next available net requirement, r_i, until all net requirements are satisfied for the MRP schedule.

Example 5.5   Silver–Meal Heuristic Example   Consider Example 5.2. The Silver–Meal calculations for each iteration are summarized in the following table:

t	Net Requirement, $r_t = r_i$	i	Tentative Lot Size, Q_i	Average Holding and Setup per Period, C_T Costs	Is $C_T > C_{T-1}$?	Lot Size, Q_t
6	32	1	32	$C_{(1)} = [\$146 + (1-1)(32)(\$0.58)]/1$ $= \$146$	No	102
7	40	2	$32 + 40 = 72$	$C_{(2)} = [\$146 + (1-1)(32)(\$0.58)$ $+ (2-1)(40)(\$0.58)] / 2$ $= \$84.6$	$C_{(2)} > C_{(1)}$? No	0
8	30	3	$72 + 30 = 102$	$C_{(3)} = [\$146 + (1-1)(32)(\$0.58)$ $+ (2-1)(40)(\$0.58)$ $+ (3-1)(30)\,(\$0.58)] / 3$ $= \$68.00$	$C_{(3)} > C_{(2)}$? No	0
9	42	4	$102 + 42 = 144$	$C_{(4)} = [\$146 + (1-1)(32)(\$0.58)$ $+ (2-1)(40)(\$0.58)$ $+ (3-1)(30)(\$0.58)$ $+ (4-1)(42)(\$0.58)] /4$ $= \$69.27$	$C_{(4)} > C_{(3)}$? Yes Stop: Use Q_3	—
9	42	1	42	$C_{(1)} = [\$146 + (1-1)(42)(\$0.58)]/1$ $= \$146$	No	162
10	40	2	$42 + 40 = 82$	$C_{(2)} = [\$146 + (1-1)(42)(\$0.58)$ $+ (2-1)(40)(\$0.58)]/2$ $= \$84.60$	$C_{(2)} > C_{(1)}$? No	0
11	48	3	$82 + 48 = 130$	$C_{(3)} = [\$146 + (1-1)(42)(\$0.58)$ $+ (2-1)(40)(\$0.58)$ $+ (3-1)(48)(\$0.58)]/3$ $= \$74.96$	$C_{(3)} > C_{(2)}$? No	0
12	32	4	$130 + 32 = 162$	$C_{(4)} = [\$146 + (1-1)(42)(\$0.58)$ $+ (2-1)(40)(\$0.58)$ $+ (3-1)(48)(\$0.58)$ $+ (4-1)(32)(\$0.58)]/4$ $= \$70.14$	Stop: Last period, use Q_4	0

The heuristic stops at period 12 because all net requirements have been satisfied. The MRP table is presented below.

	By Period, t (weeks)									Total
	4	5	6	7	8	9	10	11	12	
Net requirement, r_t	0	0	32	40	30	42	40	48	32	264
Lot size, Q_t	0	0	102	0	0	162	0	0	0	264
Net inventory, I_t	0	0	70	30	0	120	80	32	0	332

The following table summarizes the total cost of the Silver–Meal heuristic lot-sizing method:

	Number of Setups (Y)	Setup Cost (C_p)	Inventory Total (I^+)	Holding Cost (C_h)	Total Cost $= (Y)(C_p) + (I^+)(C_h)$
Silver–Meal	2	$146	332	$0.58	$2 \times \$146 + 332 \times \$0.58 = $**\$484.56**

5.3.5 Least-Unit-Cost Heuristic

The least-unit-cost (LUC) heuristic is similar to the Silver–Meal heuristic. In the LUC method, the purpose is to minimize the total holding cost by finding the average cost per part, as compared to the average cost per period of the Silver–Meal method:

$$C_T = \frac{C_p + \sum_{i=1}^{T} (i-1)r_t C_h}{\sum_{i=1}^{T} r_i} \tag{5.3}$$

After the first lot size is determined, the procedure is repeated using the next available net requirement.

Example 5.6   Least-Unit-Cost Heuristic Example   Consider Example 5.2. Use the LUC heuristic to provide the lot sizing for the MRP schedule. The solution is provided below.

t	Net Requirement, $r_t = r_i$	i	Tentative Lot Size, Q_i	Average Holding and Setup Costs per Period, C_T	Is $C_T > C_{T-1}$?	Lot Size, Q_t
6	32	1	32	$C_{(1)} = [\$146 + (1-1)(32)(\$0.58)]/$ $[32] = \$4.56$	No	144
7	40	2	$32 + 40 = 72$	$C_{(2)} = [\$146 + (1-1)(32)(\$0.58)$ $+ (2-1)(40)(\$0.58)]/[32 + 40]$ $= \$2.35$	$C_{(2)} > C_{(1)}$? No	0
8	30	3	$72 + 30 = 102$	$C_{(3)} = [\$146 + (1-1)(32)(\$0.58)$ $+ (2-1)(40) (\$0.58)$ $+ (3-1)(30)(\$0.58)]/$ $[32 + 40 + 30]$ $= \$2$	$C_{(3)} > C_{(2)}$? No	0
9	42	4	$102 + 42 = \mathbf{144}$	$C_{(4)} = [\$146 + (1-1)(32)(\$0.58)$ $+ (2-1)(40)(\$0.58)$ $+ (3-1)(30)(\$0.58)$ $+ (4-1)(42)(\$0.58)]/$ $[32 + 40 + 30 + 42]$ $= \$1.92$	$C_{(4)} > C_{(3)}$? No	0
10	40	—	$144 + 40 = 188$	$C_{(5)} = [\$146 + (1-1)(32)(\$0.58)$ $+ (2-1)(40) (\$0.58)$ $+ (3-1)(30)(\$0.58)$ $+ (4-1)(42)(\$0.58)$ $+ (5-1)(40)(\$0.58)]/$ $[32 + 40 + 30 + 42 + 40]$ $= \$2.01$	$C_{(5)} > C_{(4)}$? Yes Stop: Use Q_4	—
10	40	1	40	$C_{(1)} = [\$146 + (1-1)(40)(\$0.58)]/40$ $= \$3.65$	No	120
11	48	2	$40 + 48 = 88$	$C_{(2)} = [\$146 + (1-1)(40)(\$0.58)$ $+ (2-1)(48) (\$0.58)]/[40 + 48]$ $= \$1.98$	$C_{(2)} > C_{(1)}$? No	0
12	32	3	$88 + 32 = \mathbf{120}$	$C_{(3)} = [\$146 + (1-1)(40)(\$0.58)$ $+ (2-1)(48)(\$0.58)$ $+ (3-1)(32)(\$0.58)]/$ $[40 + 48 + 32]$ $= \$1.76$	Stop: Last period use Q_3	0

The heuristic stops at period 12 because all net requirements have been satisfied. The MRP plan and its associated cost for the LUC are as follows:

| | By Period, t (weeks) | | | | | | | | | |
	4	5	6	7	8	9	10	11	12	**Total**
Net requirement	**0**	**0**	**32**	**40**	**30**	**42**	**40**	**48**	**32**	264
Lot size	0	0	144	0	0	0	120	0	0	264
Net inventory	0	0	112	72	42	0	80	32	0	338

The following table summarizes the total cost of the LUC lot-sizing method.

	Number of Setups (Y)	Setup Cost (C_p)	Inventory Total (I^+)	Holding Cost (C_h)	Total Cost $= (Y)(C_p) + (I^+)(C_h)$
LUC	2	$146	338	$0.58	$2 \times \$146 + 338 \times \$0.58 = \mathbf{\$488.04}$

5.3.6 Comparison of Lot-Sizing Approaches

For the given example, the solutions of the five different lot-sizing methods are presented in the following table:

Lot-Sizing Methods	Number of Setups (Y)	Inventory Total (I^+)	Total Cost $= (Y)(C_p) + (I^+)(C_h)$
Lot-for-lot	7	0	$7 \times \$146 + 0 \times \$0.58 = \mathbf{\$1,022}$
Economic order quantity	2	499	$2 \times \$146 + 499 \times \$0.58 = \mathbf{\$581.42}$
Part period balancing	2	338	$2 \times \$146 + 338 \times \$0.58 = \mathbf{\$488.04}$
Silver–Meal	2	332	$2 \times \$146 + 332 \times \$0.58 = \mathbf{\$484.56}$
Least unit cost	2	338	$2 \times \$146 + 338 \times \$0.58 = \mathbf{\$488.04}$

Part period balancing, the LUC heuristic, and the Silver–Meal heuristic are similar and, in this example, have similar solutions. They generally generate better solutions (in terms of total cost) than the lot-for-lot and EOQ methods. It must be noted that each of the above methods is based on certain assumptions of the problem and may be appropriate for certain problems. The main advantage of the EOQ method is that it simplifies the planning process, as the manufacturer can always plan for exactly the same size quantity. This further simplifies machine setup routines, ordering of materials, consistent quality control procedures, and maintenance plans for equipment. In all of the methods, no backlogging is allowed, and as a result, customers will be equally satisfied with all of them. Lot-for-lot requires substantial agility and flexibility to handle different lot sizes frequently. Each of the above five methods have some practical limitations because the setup cost alone may not be the most determinant factor. The multicriteria lot-sizing method, covered later in this chapter, provides a practical solution to this problem by allowing equal-sized lots to be produced.

Finding Optimal Lot Size   The five methods covered in this section are heuristics (may not provide optimal lot sizing). The optimization method of Section 5.5.1 can be used to find the optimal lot-sizing solution.

5.4 LOT SIZING WITH CAPACITY CONSTRAINTS

5.4.1 Backward–Forward Lot Sizing

In the lot-sizing methods, the capacity constraint was not considered. In this section, we discuss a simple approach to find a feasible (but not necessarily optimal) solution while considering the capacity limitation per period. The optimization method of the next section can be used to find the optimal solution. In this approach, first solve the lot-sizing problem by using one of the five lot-sizing methods discussed in the last section. If the obtained solution is feasible with respect to the given constraint, then that solution can be used. If the solution is not feasible, the following method can be used to generate a feasible solution.

Let r_t be the net requirement and c_t be the capacity for a given period t. If

$$r_t < c_t \quad \text{for all } t = 1, 2, \ldots, N \tag{5.4}$$

then the lot-for-lot approach is feasible but not necessarily optimal.

Now, suppose Equation (5.4) does not hold for some of the periods. Consider the following two cases and the solution for each case:

(a) *Infeasible Case*   If shortage is not allowed at the end of the last period, the following condition holds:

$$\sum_{t=1}^{N} r_t > \sum_{t=1}^{N} c_t \tag{5.5}$$

In this case, the problem is infeasible. To find the minimum shortage at the end of the last period, allow a total shortage of $\sum_{t=1}^{N} r_t - \sum_{t=1}^{N} c_t$ at the end of the last period. The lot-sizing solution is to set

$$r_t = c_t \quad \text{for } t = 1, 2, \ldots, N \tag{5.6}$$

Example 5.7   Lot Sizing with Insufficient Capacity   Suppose $N = 6$. The capacity for each period is 100. The net requirement for each period is given in the table below.

Period, i (weeks)	1	2	3	4	5	6	Total
Net requirement, r_t	80	150	70	150	120	80	650
Lot size, Q_t	100	100	100	100	100	100	600
Net inventory, I_t	20	−30	0	−50	−70	−50	—

Note that a negative inventory indicates a shortage. Since $650 > 600$, the problem is infeasible. Therefore, plan to produce 100 units per period and allow $600 - 650 = -50$ units of net inventory (shortage) in the last period.

(b) *Feasible Case*   Consider the following case:

$$\sum_{t=1}^{N} r_t \leq \sum_{t=1}^{N} c_t \quad \text{for last period, i.e., } k = N \qquad (5.7)$$

That is, in the last period, there is no shortage. But in some periods, there could be some shortages.

The method to solve this problem consists of two phases: backward pass and forward pass.

Backward Pass   The method starts in the last period N. Overflows, that is, where $r_t > c_t$, are moved backward in time by adding $r_t - c_t$ to period $t - 1$. Repeat this process until the first period is reached. If the first period is reached and the solution is not feasible, then consider the forward pass.

Forward Pass   Start at period 1 and pass the overflow to the next period. Repeat this process until the last period is reached. It is important to note that shifting requirements to future periods assumes backlogs are allowed (although it is discouraged). While doing this, additional inventory holding costs must be weighed against the shortage costs.

Note that if the solution is feasible in the backward pass, then there are no shortages and there is no need to proceed with the forward pass. Otherwise (i.e., using forward pass), there will be some shortages.

Consider Q_t, the lot size, with respect to the capacity constraints. This method redistributes excess requirements to earlier periods by overflowing them backward such that $Q_t \leq c_t$ for $t = N, \ldots, 2, 1$. Continue shifting requirements to earlier periods until the first period is reached. If $Q_1 \geq c_1$, redistribute the excess requirements to the next period by overflowing them forward such that $Q_t \leq c_t$ for $t = 1, 2, \ldots, N$.

Example 5.8   Lot Sizing with Capacity Constraints   Consider the net requirement and the capacity information given in the first two rows of the table below. The capacity for each period is 100. Determine a solution to this problem.

Total net requirement, 580, is less than total capacity, 600; therefore, this problem is feasible. The feasible solution is presented below:

	By Period, i						Total	Comment
	1	2	3	4	5	6		
Net requirement, r_t	**60**	**80**	**100**	**140**	**180**	**20**	580	600 >580; feasible
Capacity, c_t	**100**	**100**	**100**	**100**	**100**	**100**	600	
	60	80	100	140	180	20		Move 80 from 5 to 4
Backward pass, Q_t	60	80	100	220	100	20		Move 120 from 4 to 3
	60	80	220	100	100	20		Move 120 from 3 to 2
	60	200	100	100	100	20		Move 100 from 2 to 1
	160	100	100	100	100	20		Infeasible, do forward pass
Forward pass, Q_t	160	100	100	100	100	20		Move 60 from 1 to 2
	100	160	100	100	100	20		Move 60 from 2 to 3
	100	100	160	100	100	20		Move 60 from 3 to 4
	100	100	100	160	100	20		Move 60 from 4 to 5
	100	100	100	100	160	20		Move 60 from 5 to 6
Final lot size, Q_t	100	100	100	100	100	80		Feasible
Final net inventory	40	60	60	20	−60	0		

The above solution is feasible but not necessarily optimal in minimizing the total cost. In the next section, we develop an optimization approach to obtain the optimal feasible solution to this problem.

5.4.2 Multiple-Item Lot Sizing

When there are several parts that must use the same resource for their production, a simple approach is to expand the backward and forward approaches for solving such problems while forcing identical lot size for each part. The following example illustrates a three-part MRP problem where each part has an identical lot size.

Example 5.9   Lot Sizing for Three Parts (Identical Lot Sizes)   Consider three parts, A, B, and C, which are supposed to be produced on the same production line where only one of the three parts can be produced in each period t (week). The capacity of the line and the net requirements for each of the three parts are presented below.

	Net Requirement per Week (Demand)									Total Demand	Machine Capacity (units/week)
	4	5	6	7	8	9	10	11	12		
Part A	0	0	32	40	30	42	40	48	32	264	100
Part B	0	32	32	30	42	40	0	32	48	256	90
Part C	29	64	60	84	30	96	64	0	0	427	150

A lot size solution to this problem is provided below.

	Production Plan per Week (Lot Sizes)									Total Production	Total Demand	End Invent.
	4	5	6	7	8	9	10	11	12			
Part A												
Lot	0	0	100	0	0	100	0	0	100	300	264	36
Inventory	0	0	68	28	−2	56	16	−32	36	—	—	36
Part B												
Lot		0	90	0	0	90	0	0	90	0	270	256	14
Inventory		0	58	26	−4	44	4	4	62	14	—	—	14
Part C												
Lot	150	0	0	150	0	0	150	0	0	450	427	23
Inventory	121	57	−3	63	33	−63	23	23	23	—	—	23

In this solution, there are some shortages (backlogs), presented by negative inventory. For part A, there are inventories of −2 and −32 for periods 8 and 11, respectively. For part B, there is an inventory of −4 in period 7. For part C, there are inventories of −3 and −63 in periods 6 and 9, respectively. The leftover production from the maximum capacity can be used to produce more items in the future, listed as the end inventory. The planned production satisfies the total demand at the end period (although there are some shortages in the intermediate periods).

5.5   LOT-SIZING OPTIMIZATION

In Section 5.3, we covered five classical lot-sizing methods. For each classical method, one can show an example that the given classical method would fail to find the optimal lot size that minimizes the total cost. In this section, we present an optimization method that finds an optimal feasible solution for the lot-sizing problem. Furthermore, the classical lot-sizing methods cannot be used when there are constraints such as capacity limitations, machine availability, manpower availability, maintenance plans, space availability, and capital limits on inventory levels for each period. The method that we introduce in this section can find the optimal lot-sizing solution while considering different types of constraints. We first formulate the lot-sizing problem by single-objective integer linear programming (ILP) subject to a set of linear constraints. The objective is to minimize the total cost. Then, in the next section, we will formulate and solve a multiobjective ILP lot-sizing problem.

5.5.1   Lot-Sizing Optimization with Constraints

In the following, we present an integer linear program for solving a lot-sizing problem with constraints.

PROBLEM   5.1   SINGLE-OBJECTIVE   LOT   SIZING: MIXED-INTEGER LINEAR PROGRAM

Minimize:

$$f_1 = \sum_{t=1}^{N} (C_p a_t + C_h I_t + C_b B_t) \quad \text{(total cost)} \tag{5.8}$$

Subject to:

$$I_t - B_t = I_{t-1} - B_{t-1} + x_t - d_t \quad \text{Input–output Equation for } t = 1, 2, \cdots, N \quad (5.9)$$

$$a_t c_{t,\min} \leq x_t \leq a_t c_{t,\max} \quad \text{Limits on capacity } t = 1, 2, \ldots, N \quad (5.10)$$

$$\sum_{t=1}^{N} a_t = Y \quad \text{Number of setups} \quad (5.11)$$

$$Y_{\min} \leq Y \leq Y_{\max} \quad \text{Limits on no. of setups} \quad (5.12)$$

$$B_N = 0 \quad \text{Number of backorders for last period} \quad (5.13)$$

$$I_{t,\min} \leq I_t \leq I_{t,\max} \quad \text{Limits on inventory for } t = 1, 2, \ldots, N \quad (5.14)$$

$$B_t \leq B_{t,\max} \quad \text{Limits on backorders for } t = 1, 2, \ldots, N \quad (5.15)$$

$$a_t = \{0, 1\} \quad \text{Either/or setup per period for } t = 1, 2, \ldots, N$$

$$(5.16)$$

$$x_t, I_t, B_t, Y \geq 0 \quad \text{Nonnegativity of variables for } t = 1, 2, \ldots, N$$

$$(5.17)$$

Other constraints $\qquad\qquad\qquad\qquad\qquad\qquad\qquad (5.18)$

The given coefficients and decision variables of Problem 5.1 are as follows:

Given Coefficients for Problem 5.1	Decision Variables for Problem 5.1
C_p = setup (order) cost per period	$x_t = Q_t$ lot size in period t
C_h = inventory holding cost per period	I_t = inventory at end of period t
C_b = backlog (shortage) cost per period	B_t = backorder at end of period t
$c_{t,\min}$ = minimum capacity constraint in period t	Y = number of setups
$c_{t,\max}$ = maximum capacity constraint in period t	$a_t = \begin{cases} 1 & \text{if an order is placed in period } t \\ 0 & \text{otherwise} \end{cases}$
d_t = demand (or requirement, r_t) in period t	
$Y_{\min}$ = minimum number of setups	
$Y_{\max}$ = maximum number of setups	I_0 = initial inventory in period 0
$I_{t,\min}$ = minimum inventory in period t	B_0 = initial backorder in period 0
$I_{t,\max}$ = maximum capacity constraint in period t	$B_{t,\max}$ = maximum backorder in period t

Problem 5.1 is a mixed ILP problem because some of its decision variables are integers. Specifically, decision variables a_t are required to be either zero or unity.

The objective function (5.8) is the total cost, which includes holding, ordering, and shortage costs. Constraints (5.9) are the input–output equations, which assure that the total input is equal to the total output in each period. Constraints (5.10) show the minimum production (lot size) and the maximum production (capacity) values, where c_{min} and c_{max} are multiplied by the binary variable a_t. Constraint (5.11) shows the given number of setups. It is the summation of all binary variables a_t. Constraints (5.12) enforce the number of setups to be between Y_{min} and Y_{max} where Y_{min} and Y_{max} are given bounds on the number of setups. If Y_{min} and Y_{max} are not specified, set $Y_{min} = 1$ and $Y_{max} = N$. Constraint (5.13) requires that no backordering be allowed in the last period. Constraints (5.14) show the minimum and the maximum limits on inventory for each period, where I_0 is the given initial inventory. Constraints (5.15) state the maximum limit on the amount of backorder for each period, where B_0 is the given initial backlog. Constraints (5.16) force the either/or condition for setup per period, that is, a_t can be either 0 or 1. Constraints (5.17) show that all decision variables are nonnegative. Constraint (5.18) can be used for additional production options and constraints such as subcontracting and overtime. (See examples of additional constraints in the linear programming formulation of aggregate planning covered in Chapter 4.)

Equal Lot Size   If it is desired to have equal lot size for each period, then add the following set of constraints where Q is the decision variable for the lot size:

$$x_t = a_t Q \quad \text{for } t = 1, 2, \ldots, N \tag{5.19}$$

Constraints (5.19) replace constraints (5.10), which means that the lower and upper bounds on the lot size are equal, that is, $c_{t,min} = c_{t,max} = Q$.

Backlogging   By allowing $B_t \geq 0$ for $t = 1, 2, \ldots, N - 1$, backlogging is allowed during the intermediate periods. If the company's policy is not to allow backlogging in any period, simply remove all B_t variables from the problem formulation.

Integer Lot Sizes   If lot sizes are required to be integers, then set decision variables x_t for $t = 1, 2, \ldots, N$ to be the integers in Problem 5.1.

Example 5.10   Mixed ILP Lot Sizing with Capacity Constraints   Consider a lot-sizing problem for six periods with demands as provided in the table below. Setup cost is \$350, and inventory holding cost is \$2 per unit per period. The shortage cost is \$3 per unit per period. The number of setups must be at least two and no more than five. The demand for each period must be met by the end of the planning period; that is, no shortage is allowed at the end of the last period. Furthermore, the lot size should be between 40 and 80.

Period, t	1	2	3	4	5	6	Total
Demand, d_t	28	35	40	17	50	30	200

Use the integer programming approach to solve the problem to minimize the total cost. The formulation of the problem is as follows:

Minimize:

$$f_1 = \sum_{t=1}^{6}(C_p a_t + C_h I_t + C_b B_t) = 350(a_1 + a_2 + \cdots + a_6) + 2(I_1 + I_2 + \cdots + I_6)$$

$$+ 3(B_1 + B_2 + \cdots + B_6)$$

Subject to:

$$
\begin{aligned}
I_0 - I_1 - B_0 + B_1 + x_1 &= 28 & 2 &\leq Y \leq 5 \\
I_1 - I_2 - B_1 + B_2 + x_2 &= 35 & B_6 &= 0 \\
I_2 - I_3 - B_2 + B_3 + x_3 &= 40 & a_t &= \{0, 1\} & \text{for } t = 1, 2, \ldots, 6 \\
I_3 - I_4 - B_3 + B_4 + x_4 &= 17 & I_t, x_t, B_t &\geq 0 & \text{for } t = 1, 2, \ldots, 6 \\
I_4 - I_5 - B_4 + B_5 + x_5 &= 50 & B_0 &= 0 \\
I_5 - I_6 - B_5 + B_6 + x_6 &= 30 & I_0 &= 0 \\
40a_t \leq x_t \leq 80a_t & \quad \text{for } t = 1, 2, \ldots, 6 \\
a_1 + a_2 + a_3 + a_4 + a_5 + a_6 &= Y
\end{aligned}
$$

Now, solve this problem using a mixed ILP software package. (We used LINGO to solve this problem.)

TABLE 5.1   ILP Solution to Example 5.10

		By Period, t						
		1	2	3	4	5	6	Total
Given	Net requirement	**28**	**35**	**40**	**17**	**50**	**30**	200
ILP solution	Net inventory, I_t	35	0	17	0	30	0	82
	Net backlog, B_t	0	0	0	0	0	0	0
	Lot size, x_t	63	0	57	0	80	0	200
	Setup, a_t	1	0	1	0	1	0	3

The optimal solution to this mixed ILP problem indicates that there are three setups in this problem. The solution is to order 63 units (x_1) in period 1, 57 units (x_3) in period 3, and 80 units (x_5) in period 5. The detailed solution is given in Table 5.1.

The three important values for this problem are measured as follows:

Total cost: $3 \times 350 + 2 \times 82 = \$1214$

Total inventory: $35 + 17 + 30 = 82$ units

Total backlog: 0 units

In this example, even though backlogging was allowed, it was not economical to use.

See Supplement S5.1.lg4.

5.5.2   Bicriteria Push-and-Pull Lot Sizing

In MRP (push) lot-sizing approaches, the objective is to minimize the total cost over the entire planning period. The decision variables are the order quantity (lot size) for each period. In practice, however, one must take into account other objectives. Attaining the minimum total cost may result in poor system performance by having high inventory levels. Inventory not only ties up cash but is also a symptom of an ineffective production system. In the JIT (pull) philosophy, inventories are discouraged and ideally their size must be zero. However, the JIT objective of zero inventory may be too extreme. In this section, we develop push-and-pull lot sizing with limited inventory levels.

Let $Y_{\min}$ be the total number of setups when minimizing the total cost (e.g., use one of lot-sizing methods of Section 5.3 or Section 5.5.1). Label this as the minimum number of setups. Now consider the JIT approach, in which a zero inventory strategy requires a lot-for-lot strategy. Label the JIT number of setups as $Y_{\max}$, which is the maximum number of setups. In bicriteria lot sizing, the decision variable is the number of setups Y such that $Y_{\min} \leq Y \leq Y_{\max}$.

In multicriteria push-and-pull lot sizing, the two conflicting objectives are as follows:

PROBLEM 5.2   BICRITERIA LOT-SIZING PROBLEM

Minimize:

$$f_1 = \sum_{t=1}^{N}(C_p a_t + C_h I_t) \quad \text{(total cost, MRP objective)} \tag{5.20}$$

Minimize:

$$f_2 = \sqrt{\sum_{t=1}^{N} I_t^2} \quad \text{(root of total inventory fluctuations, JIT objective)} \tag{5.21}$$

Subject to:

$$\sum_{t=1}^{N}(a_t) = Y$$

$$Y_{\min} \leq Y \leq Y_{\max}$$

$$a_t = \begin{cases} 1 & \text{if order is placed in period } t \\ 0 & \text{otherwise} \end{cases}$$

Other constraints from Problem 5.1

where I_t = inventory at end of period t.

Equation (5.21) is the sum of squares of inventory for each period. It represents the total fluctuation in inventory. This objective closely resembles the JIT objective of minimizing inventory and its fluctuations at different periods. Figure 5.4 shows the relationship of these two objectives.

In Figure 5.4, all solutions associated with $Y < Y_{\min}$ are inefficient and all solutions associated with $Y \geq Y_{\min}$ are efficient because any bicriteria alternative on the left side of $Y_{\min}$ is dominated by the bicriteria alternative at $Y_{\min}$ where both objectives f_1 and f_2 are minimized. For example, the alternative at $Y = 1$ has a higher cost and higher total inventory fluctuations than the alternative at $Y = 2$.

The procedure for generating efficient multicriteria alternatives is described below. First, we show how to generate an efficient point when the number of setups is given. Then, we show how to generate all efficient alternatives and rank them.

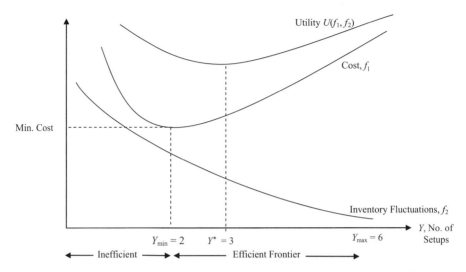

Figure 5.4 Relationship of total cost, total inventory, and multicriteria utility function for Example 5.10.

Finding Best Alternative for Bicriteria Problem The bicriteria approach to solve this problem is based on a one-dimensional search in terms of the number of setups Y. In this method, find the minimum number of setups Y_{min} by using a lot-sizing method; for example, use the optimization method of Section 5.5.1. Then, find the maximum number of setups Y_{max} by using the lot-for-lot method. For small problems, one can generate all efficient alternatives by enumerating all different numbers of setups from Y_{min} to Y_{max}. If the possible number of setups is large, a one-dimensional search approach (e.g., the bisectional search) can be used to find the best number of setups.

Now apply a multicriteria method to choose the best alternative. An additive utility function as a function of f_1 and f_2 can be used to rank alternatives. For example, in Figure 5.1, $U(f_1, f_2)$ is minimized at $Y = Y^*$. An example for using the additive utility function for choosing the best alternative is given in Example 5.11.

Finding Lot Sizing for Given Number of Setups For each given number of setups, find the lot-sizing solution. This can be done heuristically or by use of the optimization method of Section 5.5.1.

Example 5.11   Bicriteria Push and Pull for Lot Sizing Consider the net requirement demand given in the table below. Setup costs are $350, and the inventory holding cost is $2 per unit per period. No shortages are allowed in any periods.

(a) Generate all efficient points by enumerating all number of setups.
(b) Rank all alternatives using the additive utility function where the weights of importance for normalized objectives are 0.4 for total cost and 0.6 for the total fluctuating inventory (i.e., $w_1 = 0.4$, $w_2 = 0.6$), respectively.

Period, t	1	2	3	4	5	6	Total
Net requirements	28	35	40	17	50	30	200

TABLE 5.2   Summary of Six Bicriteria Alternatives in Example 5.11

No. of Setups, Y	Period, t Net Requirement	1 28	2 35	3 40	4 17	5 50	6 30	Total 200	Cost, f_2	Inventory, f_1	Efficient?
1	Lot size	200	0	0	0	0	0	200	1382	255.1	No
	Net inventory	172	137	97	80	30	0	516			
2	Lot size	120	0	0	0	80	0	200	1092	113.6	Yes
	Net inventory	92	57	17	0	30	0	196			
3	Lot size	63	0	57	0	80	0	200	1214	49.1	Yes
	Net inventory	35	0	17	0	30	0	82			
4	Lot size	28	35	57	0	80	0	200	1494	34.5	Yes
	Net inventory	0	0	17	0	30	0	47			
5	Lot size	28	35	57	0	50	30	200	1784	17.0	Yes
	Net inventory	0	0	17	0	0	0	17			
6	Lot size	28	35	40	17	50	30	200	2100	0.0	Yes
	Net inventory	0	0	0	0	0	0	0			

The solution to the problem above is presented below.

(a) In this example, the maximum number of setups is six (one for each period), $Y_{max} = 6$. The minimum number of setups can be obtained by minimizing the total cost using a lot-sizing method. The optimization solution of minimizing total cost requires two setups. Therefore, the minimum number of setups is $Y_{min} = 2$. Therefore, in this example, efficient points are associated with $Y = 2, 3, 4, 5, 6$. For the purpose of illustration, the alternative for $Y = 1$ is also shown.

For each given setup $Y = 1, 2, 3, 4, 5, 6$, solve the lot-sizing problem. The lot-sizing solutions for each given number of setups are shown in Table 5.2.

For $Y = 2$, lot-sizing is presented below, and this (optimal) lot-sizing solution is shown in Table 5.2.

	By Period, t; Y = 2						
	1	2	3	4	5	6	Total
Net requirement	28	35	40	17	50	30	200
Lot size	120	0	0	0	80	0	200
Net inventory	92	57	17	0	30	0	196

Total cost $f_1 = (2)(\$350) + (196)(\$2) = \$1092$

Total inventory fluctuations $f_2 = \sqrt{92^2 + 57^2 + 17^2 + 30^2} = 113.6$

Similarly, when the number of setups is four, $Y = 4$, the lot-sizing solution is shown in Table 5.2, where the two objectives are calculated as follows:

Total cost $f_1 = (4)(\$350) + (47)(\$2) = \$1494$

Total inventory fluctuations $f_2 = \sqrt{17^2 + 30^2} = \sqrt{1189} = 34.5$

The summary of all six lot-sizing solutions and their objective values are presented in Table 5.2.

Multicriteria alternatives associated with the number of setups from 2 to 6, that is, $2 \leq Y \leq 6$, are efficient. The solution for $Y = 1$ is inefficient, as expected.

(b) The multicriteria alternatives and their normalized values are presented below. The additive utility function to rank alternatives using normalized objective values f_1' and f_2' is

$$\text{Minimize } U = w_1 f_1' + w_2 f_2' = 0.4 f_1' + 0.6 f_2'$$

Alternatives	a_1	a_2	a_3	a_4	a_5	a_6	Minimum	Maximum
No. of Setups	1	2	3	4	5	6	1	6
Min. f_1	1382	1092	1214	1494	1784	2100	1092	2100
Min. f_2	255.1	113.6	49.1	34.5	17	0	0	255
Min. f_1'	0.29	0.00	0.12	0.40	0.69	1.00		
Min. f_2'	1.00	0.45	0.19	0.14	0.07	0.00		
Min. U	0.72	0.27	0.16	0.24	0.31	0.40		
Rank	6	3	1	2	4	5		

For this example, a_3 provides the best lot-sizing policy for the given utility function where the total cost is \$1214 and the total inventory fluctuation is 49.1 units. Also see Figure 5.4.

See Supplement S5.2.xls.
Supplement S5.3.docx shows the details of lot sizing by enumeration.

5.5.3   Multiobjective Optimization (MOO)[†]

In addition to minimizing total cost, one may also consider minimizing total inventory, total number of setups, and total backlog. In most cases, one should not plan for backlogging. However, in some situations, such as when the setup cost is very high, the backlog cost is negligible, or due dates are not as important, allowing backlogging may substantially lower the total cost.

The three linear objective functions of the lot-sizing problems are defined as (5.23), (5.24), and (5.25).

PROBLEM 5.3   MOO LOT-SIZING PROBLEM WITH LINEAR OBJECTIVE FUNCTIONS

Minimize:

$$U = w_1 f_1' + w_2 f_2' + w_3 f_3' \qquad (5.22)$$

Subject to:

$$f_1 = \sum_{t=1}^{N} (C_p a_t + C_h I_t + C_b B_t) \quad \text{(total cost, MRP objective)} \tag{5.23}$$

$$f_2 = \sum_{t=1}^{N} I_t \qquad\qquad\qquad \text{(total inventory JIT objective)} \tag{5.24}$$

$$f_3 = \sum_{t=1}^{N} B_t \qquad\qquad\qquad \text{(total backorder customer's objective)} \tag{5.25}$$

Set of constraints of Problem 5.1, that is, (5.9)–(5.18) $\qquad$ (5.26)

Finding Maximum Value of Each Objective To find the practical maximum value of each objective, the procedure is as follows. Consider three different problems where each objective is minimized; note that they are not maximized. From the solution of each problem, record the value of the three objectives. Then, identify the maximum values for each objective. This is shown in the following example.

Example 5.12   Finding Maximum Value of Each Objective Consider Example 5.10. To find the maximum values of the three objective functions, three separate optimization problems should be solved.

For problem 1,

Minimize f_1.
Subject to Example 5.10 constraints.

The solution is $f_1 = 1214, f_2 = 82$, and $f_3 = 0$.
The summary of all three problems and their solutions are shown in the table below.

Problem	Objective	Subject To	Solution f_1	f_2	f_3
1	Min. f_1	Example 5.10 constraints	1214	82	0
2	Min. f_2	Example 5.10 constraints	1634	0	78
3	Min. f_3	Example 5.10 constraints	1862	56	0
	Maximum objective values		1862	82	78
	Minimum objective values		1214	0	0

See Supplement S5.4.lg4.

Example 5.13   MOO for Lot Sizing with Capacity Constraints Consider Example 5.10. Suppose the weights of importance of normalized objectives f_1', f_2', and f_3', are (0.5, 0.3, 0.2), respectively. Find the best multicriteria solution for this problem.

First find the maximum values of f_1, f_2 and f_3. This was shown in Example 5.12. Using this information, the three equations for normalized objective functions are

$$f_1' = \frac{f_1 - 1214}{1862 - 1214} \qquad \text{or} \qquad f_1 - 648 f_1' = 1214$$

$$f_2' = \frac{f_2 - 0}{82 - 0} \qquad \text{or} \qquad f_2 - 82 f_2' = 0$$

$$f_3' = \frac{f_3 - 0}{78 - 0} \qquad \text{or} \qquad f_3 - 78 f_3' = 0$$

Now formulate Problem 5.3:

Minimize:

$$U = 0.5 f_1' + 0.3 f_2' + 0.2 f_3'$$

Subject to:

$$f_1 - 648 f_1' = 1214$$

$$f_2 - 82 f_2' = 0$$

$$f_3 - 78 f_3' = 0$$

$$f_1 = 350(a_1 + a_2 + \cdots + a_6) + 2(I_1 + I_2 + \cdots + I_6)$$
$$+ 3(B_1 + B_2 + \cdots + B_6)$$

$$f_2 = I_1 + I_2 + \cdots + I_6$$

$$f_3 = B_1 + B_2 + \cdots + B_6$$

$$
\begin{array}{ll}
I_0 - I_1 - B_0 + B_1 + x_1 = 28 & 1 \le Y \le 6 \\
I_1 - I_2 - B_1 + B_2 + x_2 = 35 & B_6 = 0 \\
I_2 - I_3 - B_2 + B_3 + x_3 = 40 & a_t = \{0, 1\} \quad \text{for } t = 1, 2, \ldots, 6 \\
I_3 - I_4 - B_3 + B_4 + x_4 = 17 & I_t, x_t, B_t \ge 0 \quad \text{for } t = 1, 2, \ldots, 6 \\
I_4 - I_5 - B_4 + B_5 + x_5 = 50 & B_0 = 0 \\
I_5 - I_6 - B_5 + B_6 + x_6 = 30 & I_0 = 0 \\
40 a_t \le x_t \le 80 a_t \quad \text{for } t = 1, 2, \ldots, 6 \\
a_1 + a_2 + a_3 + a_4 + a_5 + a_6 = Y
\end{array}
$$

Solving this problem using a mixed ILP package (we used LINGO), the solution is found as follows:

		By Period, t						
		1	2	3	4	5	6	Total
Given	Net requirement, x_t	**28**	**35**	**40**	**17**	**50**	**30**	200
	Net inventory, I_t	0	0	17	0	30	0	47
	Net backlog, B_t	28	0	0	0	0	0	28
ILP solution	Lot size, x_t	0	63	57	0	80	0	200
	Setup, a_t	0	1	1	0	1	0	3

The measurements of the three objective function values are:

Total cost $f_1 = 3 \times 350 + 47 \times 2 + 3 \times 28 = \$1228$
Total inventory $f_2 = 0 + 0 + 17 + 0 + 30 + 0 = 47$ units
Total backlog $f_3 = 28$ units

The optimal solution using the MOO additive utility function is to produce 63 units in period 2, 57 units in period 3, and 80 units in period 5.

See Supplement S5.5.lg4.

Extension to Nonlinear Objective Functions   Instead of using linear objective functions for total inventory f_2 and total backlog f_3, one can minimize the total fluctuations in inventory and the total fluctuations in backlog, respectively. In this case, fluctuations in the number of inventory and the number of backlogs are minimized. The nonlinear MOO problem is as follows.

PROBLEM 5.4 MOO LOT SIZING WITH NONLINEAR OBJECTIVE FUNCTIONS

Minimize:

$$U = w_1 f_1' + w_2 f_2' + w_3 f_3' \tag{5.22}$$

Subject to:

$$f_1 = \sum_{t=1}^{N}(C_p a_t + C_h I_t + C_b B_t) \quad \text{(total cost, MRP objective)} \tag{5.23}$$

$$f_2 = \sum_{t=1}^{N} I_t^2 \quad \text{(total inventory fluctuations, JIT objective)} \tag{5.27}$$

$$f_3 = \sum_{t=1}^{N} B_t^2 \quad \text{(total backorder fluctuations, customer's objectives)} \tag{5.28}$$

Constraints (5.9)–(5.18) \tag{5.18}

Example 5.14   MOO Lot Sizing with Nonlinear Objective Functions   Consider Example 5.10. Suppose the weights of importance of normalized objectives f_1', f_2', and f_3' are $(0.2, 0.4, 0.4)$, respectively. Find the best multicriteria solution for this problem.

The approach for solving this problem is exactly the same as the approach we used to solve Example 5.13.

			Solution		
Problem	Objective	Subject To	f_1	f_2	f_3
1	Min. f_1	Example 5.10 constraints	1214	2414	0
2	Min. f_2	Example 5.10 constraints	1567.25	0	1181.69
3	Min. f_3	Example 5.10 constraints	1592	3782	0
	Maximum objective values		1592	3782	1181.69
	Minimum objective values		1214	0	0

See Supplement S5.6.lg4.

Now, Problem 5.4 can be formulated. Solving this problem by a mixed ILP package (we used LINGO), the solution is found as follows:

		By Period, t						
		1	2	3	4	5	6	Total
Given	Net requirement, x_t	**28**	**35**	**40**	**17**	**50**	**30**	200
ILP solution	Net inventory, I_t	27.27	0	17	0	30	0	74.27
	Net backlog, B_t	0	7.73	0	0	0	0	7.73
	Lot size, x_t	55.27	0	64.73	0	80	0	200
	Setup, a_t	1	0	1	0	1	0	3

The measurements of the three objective function values are

Total cost $f_1 = 3 \times 350 + 2 \times 74.27 + 3 \times 7.73 = \$1221.73$

Total inventory $f_2 = 27.27^2 + 17^2 + 30^2 = 1932.49$

Total backlog $f_3 = 7.73^2 = 59.80$

The optimal solution using the MOO additive utility function is to produce 55.27 units in period 1, 64.73 units in period 3, and 80 units in period 5.

See Supplement S5.7.lg4.

5.6 EXTENSIONS OF MRP

Material requirements planning is an effective method of production control and planning that provides a good starting point for operational managers. However, over the years, various drawbacks were observed in the MRP system, and changes were suggested to make MRP a more useful and practical method. Thus, many extensions of MRP exist in the workplace today that increase the knowledge and control over production operations. These extensions of MRP are individually discussed below.

5.6.1 Closed-Loop MRP

A closed-loop MRP system provides additional information about the process through a controlled feedback information loop. The information obtained through this loop is used throughout the master production plan as a means of process optimization.

Two of the main assumptions made in MRP are that the demand is known and the production lead times are consistent. In actuality, forecasted demand could be probabilistic. Such uncertainty in demand may affect the master production schedule and may require updating orders. On the other hand, production lead times may be highly dependent on lot sizes and suppliers' schedules of raw materials. They are also subject to changes depending on factors which are not controlled by the manufacturing company. The effect of these changes can have major implications on the entire production system. Fluctuations in

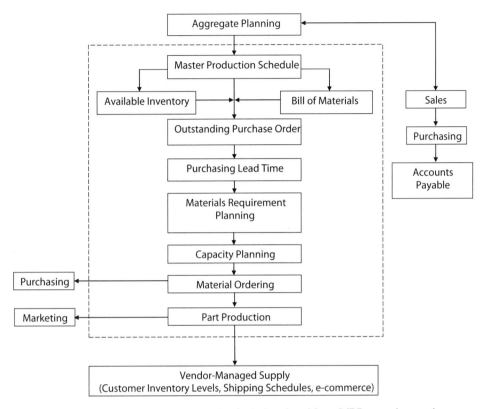

Figure 5.5 Enterprise resource planning that includes closed-loop MRP, capacity requirements, MRP II, and MRP.

production can have a domino effect on all aspects of the production system. By maintaining an information feedback system throughout the process, the master production schedule and the MRP can be altered accordingly. Figure 5.5 shows a graphical example of a closed-loop MRP system (presented in the dashed box).

5.6.2 Capacity Requirements Planning

Another assumption made in basic MRP is the availability of unlimited production capacity. Capacity requirement planning is an addition to the MRP that considers available capacities and resources. For example, when there are not sufficient resources and capacities, this information is communicated to the master production schedule so that corrective actions can be taken. Examples of corrective actions include scheduling overtime, outsourcing by contracting, subcontracting, or revising the master production schedule. If a revision to the master production schedule is necessary, an iterative trial-and-error process can be implemented to identify the best course of action.

5.6.3 Manufacturing Resource Planning

Manufacturing resource planning (labeled as MRP II) incorporates several internal aspects of a business, including marketing, forecasting, and accounting as well as considers different types of required resources. The role of the master production schedule can be used to illustrate the differences between MRP and MRP II. In MRP, the master production schedule is used as the initial input into the process. In MRP II, the master production schedule is generated through a decision-making process and can change depending on the circumstances. For example, the marketing manager would provide the operations scheduler with updated forecast demands while the purchasing manager deals with acceptable suppliers and their required lead times. Using this information, MRP can be used to provide a master production schedule that is more in line with the overall strategy of the company. In addition, MRP utilizes aggregate capacity planning, incorporating all of the benefits discussed in the previous section.

5.6.4 Enterprise Resource Planning

Enterprise resource planning has evolved from MRP to include all internal aspects of a business as well as the external aspects, such as customers and suppliers. This direct interface with the customers and suppliers can rely on electronic data interchange, which is useful in purchasing, ordering, and vendor-managed supply. While the enterprise resource planning system is a tool used to tie together all aspects of a business, the applications of enterprise resource planning may differ drastically for different applications. An example of the enterprise resource planning in a production facility is presented in Figure 5.5.

5.6.5 Distribution Resource Planning

Distribution resource planning is a time-based process used in the development of a material replenishment plan within a distribution environment. It is currently used in supply chain management. A distribution environment often consists of a hierarchy of inventory controllers, each dependent upon each other. The procedure of distribution resource planning is very similar to that of MRP, although many of the inputs into the process are dependent variables. The overall goal of a distribution resource planning system is to manage small but frequent replenishment while maintaining the lowest possible costs. The procedure for using distribution resource planning is to apply the MRP approach while considering the distribution environment structure and plan accordingly.

5.6.6 Supply Chain Management

Supply chain management is the highest level of integration of all aspects of MRP and its extensions, including materials procurement and the various customer delivery methods. As in enterprise resource planning, supply chain management incorporates all aspects of the business into the basic product supply chain. The overall goal of supply chain management is to reduce costs and maximize customer satisfaction by building a competitive and customized chain of suppliers. In supply chain management, the suppliers and customers become key strategic business partners. A more in-depth analysis of supply chain management may be found in Chapter 9.

Supply chain management systems can incorporate the use of e-commerce and Internet ordering and capacities. Depending on the extent of involvement, a business may simply use the Internet as a means of taking orders and issuing invoices or use a server–client system for full-control vendor-managed accounts and inventory control systems.

5.7   JUST-IN-TIME: PULL SYSTEM

Just-in-time was pioneered at Toyota in the 1950s. It was then embraced by Western companies in the early 1980s. Just-in-time has two primary goals: to reduce or eliminate both inventory and equipment idle time. A supplier who could meet frequent delivery schedules is therefore required. Just-in-time also utilizes small batch sizes, which allows an abundance of continuous flow through the system to prevent bottlenecks. Overall, JIT provides a smooth system for planning production schedules, reducing bottlenecks, and excessive material storage costs.

To have a better understanding of JIT systems, it is useful to compare "push" and "pull" production systems. In push systems (MRP), the production is based on a schedule that has been prepared in advanced. In pull systems (JIT), the schedule is based on the production line capabilities to pull partially finished items. In JIT the production releases are based on the authorization and depend on the status of a process (workstation) being available. In MRP systems the materials are moved to the next station regardless of its readiness to process the semifinished parts and materials. In JIT systems, the materials are moved when the next (downstream) workstation is ready. As the result of this policy, the whole system becomes idle if one of the stations stops functioning. Consequently, this system requires that all the stations work smoothly and in a synchronized manner. In JIT, bottlenecks will very quickly become apparent as they will hold up the whole production line. To achieve JIT, one must eliminate the variability in the production process. Also, it requires developing production methods that eliminate safety stocks or buffers, which are usually held in front of a processor.

5.7.1   Principles of JIT

Reducing Inventory Reveals Problems   In push systems, carrying excess inventory typically hides many production problems, as shown in Figure 5.6*a*. The first step for implementing JIT, therefore, is to reduce inventory levels throughout the production process. As inventory is reduced, the production problems that cause delays and variations will become apparent, as illustrated in Figure 5.6*b*.

Once the troubles are identified, attempts can be made to reduce or eliminate them. As long as the variability is reduced, the production flow will become even and smooth.

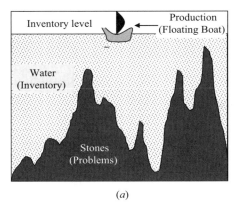 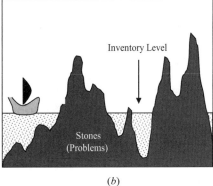

(*a*)   (*b*)

Figure 5.6   (*a*) Excessive inventory hides problems. (*b*) By reducing inventory level, problems are revealed.

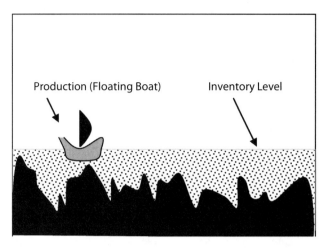

Figure 5.7   By removing major problems (stones), smooth production flow can be achieved.

Inventory is not the only factor that must be reduced to successfully implement JIT. See Figure 5.7. Other factors to be reduced are setup times, work-in-process, batch sizes, scheduled and unscheduled downtime, and materials handling.

Setup Time Reduction Improves Productivity   Because smaller lot sizes are used in JIT, it is necessary that the setup time for lots be reduced substantially. One of the most critical components of JIT production is to reduce the equipment downtime for setups. This is SMED, which advocates performing the majority of the setup tasks offline before the equipment is shut down for die exchange. Ideally the time needed to change the setup of any machinery is 1 min. Working toward quick setup can also reduce setup cost, the EOQ, work-in-process, and lead times.

Waste Elimination Supports Continuous Quality and Productivity Improvement   The JIT philosophy advocates continuous attention to identifying and eliminating waste. Waste is defined as any activity that does not add value to the product. This leads to JIT, continuously improving production quality. For example, materials handling activities, although necessary, are considered a waste as they do not add value to the product. By minimizing materials handling, substantial improvements in production time and cost can be achieved. In JIT, several types of *muda* are identified, where muda is the Japanese term for waste. Seven important muda are listed below.

1. Unnecessary motions in materials handling
2. Waiting in any form and for any reason
3. Overproduction for any reason
4. Unnecessary work-in-process inventory
5. Defects that should be discarded or repaired
6. Underutilization of manpower
7. Underutilization of resources

Allowing Proper Breakdown and Maintenance Times Make the System More Efficient Just-in-time production systems may not perform efficiently due to machine breakdowns, order cancellations, unexpected orders, delayed supply shipments, and so on. Just-in-time is designed to operate very leanly and therefore does not account for these situations. And yet provisions must be made in the real world for such problems. One solution is to schedule the facility to operate less than its total capacity. For example, if a plant can run 24 h per day, it can be scheduled as 8–4–8–4, where for each 8-h production period a 4-h break is provided. These breaks provide sufficient time to take care of irregularities, maintenance, materials shortages, and so on.

Use of Independent Cells Simplifies the Flow Process Many production process flows can be improved by the use of manufacturing cells. The design of each cell is covered in Chapter 13. Each cell then can be run by a JIT approach. The factory can be divided into several independent cells, each being autonomous and responsible for its inputs, outputs, and quality control. Establishing cells facilitates the formation of teams who often take ownership of the product and process and may lead to significant cost reductions and quality improvements. In addition, cross training the employees within a cell can provide substantial flexibility to cope with employment fluctuations and variations in product demand.

For example, consider a system that consists of 11 workstations that produce two families of products. The sequence of workstations to process part 1 is 1, 2, 3, 4, 5, 6, and 7. But the sequence of workstations to process part 2 is 4, 10, 3, 1, 9, 2, 11, 8, and 5. That is, each part goes through a different sequence of machines. The traditional layout approach is to locate all the 11 machines in one area and process the two parts. This layout has a complicated material flow. But, in cellular manufacturing, two cells can be considered. In Figure 5.8, the layouts for the two cells are presented where each cell consists of several different workstations. Each cell may have a different sequence of processing the parts. By organizing the workflow in separate cells, the product flow and process are simplified. Work-in-progress and throughput time can also be decreased. One key benefit of cell layout is the integration of quality control or inspection into each cell. With the ease of reworking or correcting problems, very few defective products will be delivered to the next cell. To reorganize the production processes into cells, a clustering technique can be implemented; this is covered in Chapter 12.

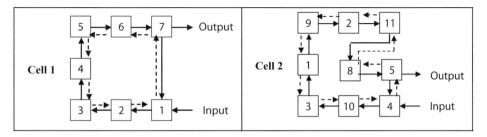

Figure 5.8 Manufacturing cell layout: example of two cells each having a different sequence of using workstations; solid arrows represent move of materials and dashed arrows represent JIT information feedback.

5.7.2 Kanban System

The kanban system was developed independent of the JIT system but soon became the most critical component of the JIT system. Kanban is a manual information system that uses cards for communicating the production status among adjacent workstations. In general, kanban cards are used to signal to an upstream work center to send down more materials or semifinished products to be processed by a downstream work center. Each card represents a request for a small batch of lot.

In a kanban system, adjacent upstream and downstream workstations communicate with each other through their cards, where each container has a kanban associated with it. The two most important types of kanbans are:

Production (P) Kanban   A P-kanban, when received, authorizes the workstation to produce a fixed amount of products. The P-kanban is carried on the containers that are associated with it.

Transportation (T) Kanban   A T-kanban authorizes the transportation of the full container to the downstream workstation. The T-kanban is also carried on the containers that are associated with the transportation to move through the loop again.

There are a number of ways to handle kanban cards. The most well-known kanban system is the dual-card system, where production and transportation systems are independent. If the production and transportation systems are combined, then a single card system can be used. Figure 5.9 shows a dual kanban system:

Production Loop   In Figure 5.9, the input materials arrive at point *I* at station $j - 1$. They are then loaded to station $j - 1$ to be processed. The P-containers, full with materials, are denoted by ▨. After they are processed at station $j - 1$, the containers are emptied at point II. The empty containers, (▢), are then moved through the production loop and wait in the P-kanban post. When an empty P-kanban and its container arrive at point *I*, it is the authorization of station $j - 1$ to produce the needed items and to fill them with the finished products to move through the loop again.

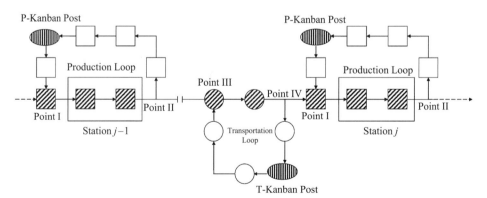

Figure 5.9   Dual-card kanban system: (▢) empty production container; (▨) full production container; (◪) P-kanban post or T-kanban post; (○) empty transportation container; (◪) full transportation container. Each container has a kanban card.

Transportation Loop   In Figure 5.9, at point III, the materials are moved toward station *j*. They are carried in full containers (🌀) with the T-kanban on them. At point IV, the materials are transferred to point *I* of workstation *j*. The empty containers (◯) are moved into the T-kanban post waiting to be moved into the loop. Arrival of an empty T-kanban container at point III authorizes the transportation of full materials to occur.

The methods of load, unload, and exchange of containers depend largely on the type of manufacturing process.

Calculating the Number of Kanban (Containers)   Calculating the number of kanban cards requires several pieces of information. The required percentage of safety stock (*s*) to be held as a buffer and the size of each kanban container (*C*) must be known. In general, the container size is kept small; for example, it can be a small percentage of the cycle (daily) demand. Safety stock is also typically set to a small number, for example, 10% of the product demand for each cycle. Given this information, the following formula is used to determine the number of kanbans (*K*):

$$K = \frac{DL(1+s)}{C} \tag{5.29}$$

where *C* is the container size (in terms of the number of parts), *D* is the total demand per unit of planning time, *L* is the lead time (the production and transportation time for the given station), and *s* is the safety stock (percentage expressed as a decimal). Lead time *L*, safety stock *s*, and container size *C* are given quantities. The result of this calculation can be rounded to the next highest integer to avoid shortages.

Example 5.15   Determining Kanban Size   Suppose there is a daily demand for 90 units of a product, where there is a lead time of three days to produce each unit. Also, suppose that a safety stock of 10% of the produced units is required. Each container can hold 30 units. Given the data below, determine the number of kanbans needed:

$$K = \frac{DL(s+1)}{C} = \frac{(90 \text{ units/day})(3 \text{ days})(1+0.1)}{30 \text{ units}} = \frac{270(1.1)}{30} = 9.9 \text{ kanbans}$$

Hence, 9.9 rounded up to 10 kanbans (or containers) should be used. Over a three-day period, (10 kanbans) × (30 units/kanban) = 300 units will be produced to cover the demand of (3 days) × (90 units/day) = 270. The safety stock is 300 − 270 = 30. This is consistent with the planned 10% safety stock, that is, 30/300 = 0.1.

5.8 MULTICRITERIA HYBRID PUSH-AND-PULL SYSTEMS

Consider the following bicriteria problem when designing production systems:

Maximize push system approach.
Maximize pull system approach.

The above two criteria are conflicting; in the following we discuss hybrid approaches that consider both above objectives.

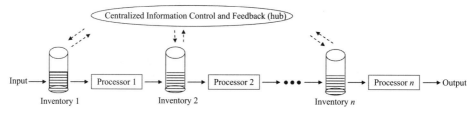

Figure 5.10   Decoupling principle of push MRP system using buffers: (———▶) material flow; (----▶) information flow.

5.8.1 Push-and-Pull Hybrid Systems

Decoupling of Push Systems Versus Interdependence of Pull Systems   The traditional push production systems, such as MRP, are based on using buffers of materials to allow the system to be operational when some stations fail or when there are unwanted fluctuations in the system. Buffers can also allow the system to operate smoothly, even if it is not synchronized (see Figure 5.10). By using buffers, starvation and blocking of machines can be somewhat prevented. It was then realized that such in-process inventories hide the production problems.

The JIT philosophy, contrary to common sense, advocates the reduction and possible elimination of in-process inventory (buffer). This in turn identifies the production problems very quickly. This emphasizes interdependencies of stations instead of decoupling of stations. By using this strategy, JIT made a substantial contribution in developing lean manufacturing. See Figure 5.11.

The JIT philosophy advocates a goal of zero in-process inventory. This philosophy is based on the assumption that once inventories are eliminated, the system will work efficiently and smoothly. There are several problems with this JIT assumption:

- To achieve zero inventory, the machine operators may collaborate to set a lower production rate to match the slowest workstation. This will lead to a lower productivity.
- Due to zero inventory, the system becomes highly interdependent and the system reliability decreases because workstations are serially dependent (see Chapter 19 on serial reliability with and without buffers).
- The JIT system becomes ineffective when multiple products are processed sequentially on the same line because there is no in-process inventory. For example, as one station is being re-setup, the other stations must also be shut down (as there are no in-process inventories for them to run).
- JIT can be a risky operational planning approach as the whole line may shut down due to any problem, such as lack of materials and machine breakdowns.

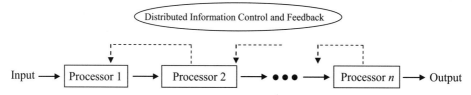

Figure 5.11   Reciprocal Interdependence of JIT without using buffers: (———▶) material flow; (◀--) information flow.

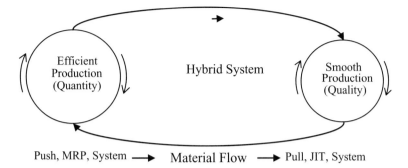

Figure 5.12 Multicriteria hybrid push-and-pull system uses both push-and-pull powers to roll production smoothly and efficiently.

The pure JIT philosophy can be effective for troubleshooting problems for some industries, but it may not be an effective strategy for many industries. The hybrid push–pull production system proposed here uses some principles of JIT to make the system more reliable but also uses some push principles of MRP to make the system more efficient.

The following three hybrid push–pull multicriteria production systems are based on using the advantages of both push and pull systems while avoiding their disadvantages. The principles of these approaches are as follows:

1. Develop a hybrid MRP and JIT plan and decide on the lot sizes and buffer sizes for the push–pull system.
2. Determine the necessary number of kanbans.
3. Push the system based on pushing objectives.
4. Pull the system based on pulling objectives.

In hybrid systems, the lot sizes are decided based on the internal collective assessment of the plant managers, machine operators, and quality control inspectors as well as the materials procurement, marketing, and sales. The lot size can be changed depending on the system's objectives. See Figure 5.12.

5.8.2 Three Types of Hybrid Push-and-Pull Systems

We define three types of hybrid push–pull systems:

I. Push-and-Pull System with Limited Inventory By reducing the inventory level to a specific size, the push system converges toward a pull system. In this system, both push and pull can occur simultaneously. In the event of a breakdown, the downstream stations can use the limited buffer inventory to stay operational; see Figure 5.13. Also, if downstream

Figure 5.13 Type I: push system with limited inventory.

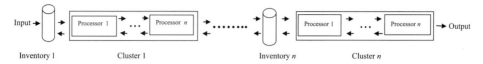

Figure 5.14 Type II; push–pull system with cluster-zoned areas; within each cluster there is a pull system, but the system of clusters is a push system.

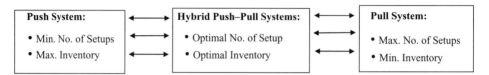

Figure 5.15 Type III: push–pull system with optimal number of setups; see multicriteria approach of Section 5.5.2.

stations are not functional, the upstream stations can produce items and leave them in the buffer spaces.

II. Push-and-Pull System with Cluster-Zoned Areas In this system, several stations are grouped into blocks (or clusters). Inside each block, the subsystem is run by a pure pull system, but the blocks are run as a push system. In this system, the decoupling of MRP and the interdependence of JIT systems are reduced and moderated; see Figure 5.14. This concept is similar to the manufacturing cells covered in last section; see Figure 7.2.

The clustering methods are covered in Chapter 12; Chapter 13 can also be used to design this system.

III. Push-and-Pull System with Optimal Number of Setups In MRP, lot sizing is used to find the number of setups that minimizes the cost. This usually leads to some level of inventory. On the other hand, in JIT the lot-for-lot approach is used, which results in zero inventory. In lot-for-lot, the maximum number of setups is used. In a hybrid push–pull system, the number of setups is in between these two extreme cases. Depending on the multiple objectives and the desired level of MRP and JIT, an optimal number of setups can be selected. The multicriteria lot sizing for this method is similar to Section 5.3. This is presented in Figure 5.15.

REFERENCES

General

Anderson, J. C., R. G. Schroeder, S. E. Tupy, and E. M. White. "Material Requirements Planning Systems: The State of the Art." *Production and Inventory Management*, Vol. 23, 1982, pp. 51–66.

Anily, S., and A. Grosfeld-Nir. "An Optimal Lot-Sizing and Offline Inspection Policy in the Case of Nonrigid Demand." *Operations Research*, Vol. 54, No. 2, 2006, pp. 311–323.

Berry, W. L. "Lot Sizing Procedures for Requirements Planning Systems: A Framework for Analysis." *Institute for Research in the Behavioral, Economic, and Management Sciences*, Vol. 13, No. 2, Second Quarter, 1972.

Bevis, G. E. "A Management Viewpoint on the Implementation of a MRP System." *Journal of Production and Inventory Management*, First Quarter, Vol. 17, No. 1, 1976, pp. 105–116.

DeMatteis, J. J. "An Economic Lot Sizing Technique: The Part-Period Algorithm." *IBM Systems Journal*, Vol. 7, No. 7, 1968, pp. 30–38.

Golhar, D. Y., and C. L. Stamm, "The Just-In-Time Philosophy: A Literature Review." *International Journal of Production Research*, Vol. 29, No. 4, 1991, pp. 657–677.

Gorham, T. "Dynamic Order Quantities." *Production and Inventory Management*, Vol. 9, 1968, pp. 75–81.

Hiezer, J., and B. Render. *Operations Management*, 9th ed. Upper Saddle Rive, NJ: Prentice–Hall, 2008.

Ishiwata, J. *Productivity through Process Analysis*. Portland, OR: Productivity Press, 1997.

Love, S. *Inventory Control*. New York: McGraw-Hill, 1979.

Nahmias, S. *Production and Operations Analysis*, 6th ed. New York: McGraw-Hill, 2008.

Silver, E. A., and H. C. Meal. "A Heuristic for Selecting Lot Size Quantities for the Case of a Deterministic Time-Varying Demand Rate and Discrete Opportunities for Replenishment." *Production and Inventory Management*, Vol. 14, 1973, pp. 64–74.

Sipper, D., and R. L. Bulfin Jr. *Production Planning, Control, and Integrations*. New York: McGraw-Hill, 1997.

Vollman, T. E., W. L. Berry, and D. C. Whybark. *Manufacturing Planning and Control Systems for Supply Chain Management*. New York: McGraw-Hill, 2005.

Wagner, H. M., and T. M. Whitin. "Dynamic Version of the Economic Lot Size Model." *Management Science*, Vol. 5, No. 1, 1958, pp. 89–96.

Zipkin, P. "Does Manufacturing Need a JIT Revolution?" *Harvard Business Review*, Vol. 69, 1991, pp. 40–50.

Muliobjective Push-and-Pull (MRP/JIT) Systems

Andijani, A. A. "A Multi-Criterion Approach for Kanban Allocations." *Omega*, Vol. 26, No. 4, 1998, pp. 483–493.

Gao, Zh., and L. Tang, "A Multi-Objective Model for Purchasing of Bulk Raw Materials of a Large-Scale Integrated Steel Plant." *International Journal of Production Economics*, Vol. 83, No. 3, 2003, pp. 325–334.

Kalpi, D., V. Mornar, and M. Baranovi. "Case Study Based on a Multi-Period Multi-Criteria Production Planning Model." *European Journal of Operational Research*, Vol. 87, No. 3, 1995, pp. 658–669.

Li, X., L. Gao, and W. Li. "Application of Game Theory Based Hybrid Algorithm for Multi-Objective Integrated Process Planning and Scheduling." *Expert Systems with Applications*, Vol. 39, No. 1, 2012, pp. 288–297.

Malakooti, B. "A Gradient-Based Approach for Solving Hierarchical Multiple Criteria Production Planning Problems." *Computers and Industrial Engineering*, Vol. 16, No. 3, 1989, pp. 407–417.

Malakooti, B., and E. C. Tandler. "An Overview of a Decision Support System for Solving Discrete MCDM Problems under Certainty and Uncertainty with Applications to Multiobjective Production Planning." *International Journal of Information and Management Sciences*, Vol. 2, No. 1, June 1991, pp. 61–81.

Mansouri, S. A. "A Multi-Objective Genetic Algorithm for Mixed-Model Sequencing on JIT Assembly Lines." *European Journal of Operational Research*, Vol. 167, No. 3, 2005, pp. 696–716.

Malakooti, B. "Hybrid Material Requirement Planning and Just-in-Time with Multiple Objective and Optimization." *Case Western Reserve University, Cleveland, Ohio 2013*.

Razmi, J., H. Rahnejat, and M. K. Khan. "Use of Analytic Hierarchy Process Approach in Classification of Push, Pull and Hybrid Push-Pull Systems for Production Planning." *International Journal of Operations & Production Management*, Vol. 18, No. 11, 1998, pp.1134–1151.

Shahabudeen, P., R. Gopinath, and K. Krishnaiah. "Design of Bi-Criteria Kanban System Using Simulated Annealing Technique." *Computers & Industrial Engineering*, Vol. 41, No. 4, 2002, pp. 355–370.

Sharma, S., and N. Agrawal. "Selection of a Pull Production Control Policy under Different Demand Situations for a Manufacturing System by AHP-Algorithm." *Computers & Operations Research*, Vol. 36, No. 5, 2009, pp. 1622–1632.

Ustun, O., and E. A. Demirtas. "An Integrated Multi-Objective Decision-Making Process for Multi-Period Lot-Sizing with Supplier Selection." *Omega*, Vol. 36, No. 4, 2008, pp. 509–521.

EXERCISES

Materials Requirement Planning: Push System (Section 5.2)

2.1 A company makes traffic signals. Each traffic signal is composed of a housing and bracket assembly. Each housing assembly is composed of optical and casing sub-assemblies. Each bracket assembly is composed of a hanger part and a wire outlet part. The optical subassembly is composed of four wire lead subassemblies, two lens parts, three bulbs, and four socket parts. Each casing subassembly is composed of four plastic molds and two hardware subassemblies. Each wire outlet subassembly is made up of one conductor part, one insulation part, and four spade connector parts. Each hanger part is made up of five nuts, five bolts, and six washers. Assume the lead time for each component or assemblies is one day.

(a) How many spade connectors are needed to make one traffic signal?

(b) Construct a product structure diagram.

(c) Create a BOM for the traffic signals.

2.2 The time-phased product structure diagram, scheduled receipts, and gross requirement for product A are presented below where the number needed for assembly is shown in parentheses, for example, C (2) means it takes two units of part C to assemble part A. Determine the net requirements for items A, C, and D.

Period, t (weeks)	4	5	6	7	8	9	10	11	12	13
Gross requirements for part A	0	0	10	26	42	15	33	52	40	22
Scheduled receipts for part A	0	14	0	0	0	0	5	0	0	0

Period, t (weeks)	4	5	6	7	8	9	10	11	12	13
Scheduled receipts for item C	0	0	0	100	0	0	0	0	0	0
Scheduled receipts for item D	0	40	180	0	0	0	0	300	0	0

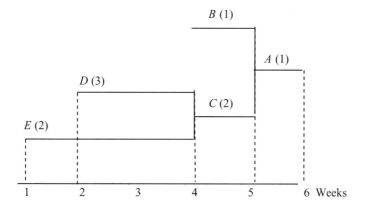

2.3 Consider the following net requirements for product A for weeks 9–15. Lead times are all one week. The product A time-based structure diagram is also presented below.

Week	9	10	11	12	13	14	15	Total
A's net requirement	30	40	20	50	20	30	10	200

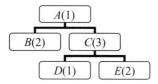

Determine the planned order release for items A, C, and E.

2.4 Consider the following time-phased product structure and net requirements for part A. The lead times are one week for each part. Determine the planned order release for components A, C, F, and G.

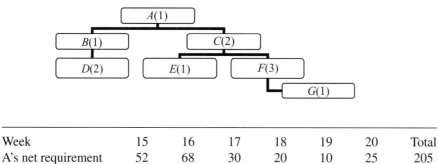

Week	15	16	17	18	19	20	Total
A's net requirement	52	68	30	20	10	25	205

Lot-Sizing Approaches (Section 5.3)

3.1 The net requirements for an electronic component for the next six weeks are presented below.

Week	1	2	3	4	5	6	Total
Net requirement	400	150	350	600	100	400	2000

The setup cost for the production of this component is $500 and the holding cost is $1.2 per component per week. Determine the lot size, inventory for each week, and the total cost using the following methods.

For part (b) assume the average annual demand is 4000:

(a) Lot-for-lot

(b) Economic order quantity

(c) Part period balancing

(d) Silver–Meal heuristic[†]

(e) Least Unit Cost[†]

(f) Compare the total cost for all of the methods above and determine which method is the best.[†]

3.2 The setup cost for the assembly of a product is $1500 per order, and the inventory carrying cost is $4.00 per product per week. The estimated net requirements for the product for the next eight weeks are as follows:

week	1	2	3	4	5	6	7	8
Net requirements	150	200	350	300	250	400	150	600

For part (b) assume the average annual demand is 65,000.
Respond to parts (a)–(f) of exercise 3.1.

3.3 Consider a lot-sizing problem where inventory holding cost is $1.2 per unit per period and the ordering cost per order is $50. Below are the net requirements for 10 periods.

Period (week)	1	2	3	4	5	6	7	8	9	10
Net requirements	50	45	25	30	10	15	29	43	60	23

For part (b) assume the average annual demand is 4000.
Respond to parts (a)–(f) of exercise 3.1.

3.4 A part has an ordering cost of $1000 per order and an inventory holding cost of $2.5 per item per period.

Period	1	2	3	4	5	6	7	8	9	10	11	12
Requirements	110	90	120	160	100	140	110	140	155	135	125	115

For part (b) assume the average annual demand is 20,000.
Respond to parts (a)–(f) of exercise 3.1.

Lot Sizing with Capacity Constraints (Section 5.4)

4.1 Consider exercise 3.1.

(a) If the capacity is 350 units per period, find the lot size for each period.

(b) If the capacity is 340 units per period, find the lot size for each period.

4.2 Consider exercise 3.2.

(a) If the capacity is 300 units per period, find the lot size for each period.

(b) If the capacity is 350 units per period, find the lot size for each period.

4.3 Consider exercise 3.3.

(a) If the capacity is 30 units per period, find the lot size for each period.

(b) If the capacity is 40 units per period, find the lot size for each period.

4.4 Consider exercise 3.4.

(a) If the capacity is 110 units per period, find the lot size for each period.

(b) If the capacity is 130 units per period, find the lot size for each period.

Lot-Sizing Optimization with Constraints (Section 5.5.1)

5.1 Consider exercise 3.1. If backorders are allowed, the cost would be $0.75 per unit per period.

(a) Solve the problem where backorders are not allowed using the optimization method.

(b) Suppose backorders are allowed except for the last period. Use the optimization method and compare the solutions with part (a) and also with the solutions of exercise 3.1.

(c) Suppose backorders are allowed except for the last period and capacity is 340. Solve the problem by optimization and compare to the solution of exercise 4.1.

5.2 Consider exercise 3.2. If backorders are allowed, the cost would be $2.50 per unit per period.

(a) Solve the problem where backorders are not allowed using the optimization method.

(b) Suppose backorders are allowed except for the last period. Use the optimization method and compare the solutions with part (a) and also with the solution of exercise 3.2.

(c) Suppose backorders are allowed except for the last period and capacity is 350. Solve the problem and compare to the solution of exercise 4.2.

5.3 Consider exercise 3.3. If backorders are allowed, the cost would be $1.80 per unit per period.

(a) Solve the problem where backorders are not allowed using the optimization method.

(b) Suppose backorders are allowed except for the last period. Use the optimization method and compare the solutions with part (a) and also with the solution of exercise 3.2.

(c) Suppose backorders are allowed except for the last period and capacity is 40. Solve the problem and compare to the solution of exercise 4.3.

5.4 Consider exercise 3.4. If backorders are allowed, the cost would be $0.50 per unit per period.

(a) Solve the problem where backorders are not allowed using the optimization method.

(b) Suppose backorders are allowed except for the last period. Use the optimization method and compare the solutions with part (a) and also with the solution of exercise 3.4.

(c) Suppose backorders are allowed except for the last period and capacity is 130. Solve the problem and compare to the solution of exercise 4.4.

Bicriteria Push-and-Pull Lot Sizing (Section 5.5.2)

5.5 Consider exercise 3.1.

(a) Suppose the number of setups can vary from 2 to 5. For two objectives of total cost and total inventory fluctuations, identify efficient and inefficient alternatives.

(b) Rank alternatives using an additive utility function with weights of importance of 0.6 and 0.4 for normalized values of objectives.

5.6 Consider exercise 3.2.

(a) Suppose the number of setups can vary from 3 to 7. For two objectives of total cost and total inventory fluctuations, identify efficient and inefficient alternatives.

(b) Rank alternatives using an additive utility function with weights of importance 0.3 and 0.7 for normalized values of objectives.

5.7 Consider exercise 3.3.

(a) Suppose the number of setups can vary from 4 to 8. For two objectives of total cost and total inventory fluctuations, identify efficient and inefficient alternatives.

(b) Rank alternatives using an additive utility function with weights of importance of 0.5 and 0.5 for normalized values of objectives.

Multiobjective Optimization (Section 5.5.3)[†]

5.8 Consider exercise 3.1. Suppose backorder cost is $0.75 per unit per period. Suppose the weights of importance for the normalized objectives of total cost, total inventory, and total number of backlog are 0.3, 0.4, and 0.3, respectively. Use the MOO lot-sizing approach to determine the optimal solution. For this solution, report total cost, lot size for each period, inventory level for each period, and backlog for each period.

5.9 Consider exercise 5.2. Backorder would cost $2.5 per unit per period. Suppose the weights of importance for the normalized objectives of total cost, total inventory, and total number of backlog are 0.6, 0.1, and 0.3, respectively. Use the MOO lot-sizing approach to determine the optimal solution. For this solution, report total cost, lot size for each period, inventory level for each period, and backlog for each period.

5.10 Consider exercise 5.3. Backorder cost is $1.80 per unit per period. Suppose the weights of importance for the normalized objectives of total cost, total inventory, and total number of backlog are 0.2, 0.5, and 0.3, respectively. Use the MOO lot-sizing approach to determine the optimal solution. For this solution, report the total cost, lot size for each period, inventory level for each period, and backlog for each period.

Extensions of MRP (Section 5.6)

6.1 For a given manufacturing industry of a product, explain the meaning or examples of:
 (a) Closed-loop MRP
 (b) Capacity requirements planning
 (c) Manufacturing resource planning
 (d) Enterprise resource planning
 (e) Distribution resource planning
 (f) Supply chain management

Just-In-Time: Pull System (Section 5.7)

7.1 A corporation wants to install a kanban system on one of its assembly lines to reduce inventory levels. The corporation has daily demand of 3600 compressors. The lead time is six days, the safety stock is 16.7% of production (600 compressors), and the container size is 50. Determine the number of kanbans needed.

7.2 A manufacturer of a product is planning to use a kanban system. The manufacturer produces an average of 15,000 items every 30 days. Production lead time is 15 days and the plant would like to have 25% safety stock.
 (a) If each container can hold 200 lamps, how many kanban tickets are required? What is the safety stock for this solution?
 (b) If each container can hold 150, what will be the total number of kanban tickets required? What is the safety stock for this solution?
 (c) Compare the two plans of (a) versus (b).

7.3 A company produces 2400 items monthly. The company would like to decide between two kanban systems to implement. Plan A has a buffer stock of 8%, a container size of 200, and a lead time of 12 days. Plan B has a buffer stock of 12%, a container size of 225, and a lead time of 10 days. Suppose there are 30 days in a month. Compare the two alternatives. Which plan should they choose?

Multicriteria Hybrid Push-and-Pull Systems (Section 5.8)

8.1 List the advantages and disadvantages of the pull system, the push system, and the three different hybrid systems defined in Section 5.8.2.

8.2 Consider the following net requirements. The setup cost for the production is $500 and the holding cost is $1.2 per component per week.

Week	1	2	3	4	5	Total
Net requirement	400	150	350	600	100	1600

Show the production plan for (a) push, (b) JIT, and (c) hybrid push–pull systems.

8.3 Consider the net requirements given in exercise 8.2. Show the production plan for the three multicriteria hybrid systems discussed in Section 5.8.2.

CHAPTER 6

INVENTORY PLANNING AND CONTROL

6.1 INTRODUCTION

Inventory planning and control are crucial in most organizations, including manufacturing, wholesale, grocery stores, retail businesses, the autoindustry, military, and government. The investment tied up in inventory is substantial compared to the total assets of many companies. In the United States, the total annual inventory asset is about 25% of the U.S. gross domestic product. In traditional manufacturing companies, this ratio is about 40% of capital investments. Therefore, it is not surprising that inventory planning is one of the oldest fields of study in production, manufacturing, and operation systems. Around the beginning of the twentieth century, the first inventory models were created, and intensive research on inventory planning was conducted. The economic order quantity model, developed by Ford Harris in 1913, is a fundamental approach for planning and controlling inventory. This basic model served as a basis upon which more encompassing models were developed at the end of twentieth century. The inventory planning and control methods have evolved along with just-in-time (JIT), supply chain management, and Internet-based business systems.

Inventory is the storage of physical resources that can be used for future demands and use. A simple inventory decision problem is to determine the optimal amount to order for an item and the optimal time when each order should be placed. Solving this simple problem becomes challenging in real-world situations because many interrelated factors should be considered. The primary input for inventory planning is the forecasted demand for each specified period of time. It is crucial that demand forecasting be as accurate as possible.

Note: Advanced material that can be omitted without loss of generality will be indicated by a dagger.

Operations and Production Systems with Multiple Objectives, First Edition. Behnam Malakooti.
© 2014 John Wiley & Sons, Inc. Published 2014 by John Wiley & Sons, Inc.

Two possible errors are overestimating the demand and underestimating the demand. Not having enough stock in the inventory results in possible loss of sales and profit and can also create irreversible damage to a company's service reputation. Conversely, having too much stock in inventory can bankrupt a company as it can tie up substantial capital and space. Designing an optimal inventory control system is not the sole responsibility of production and operations managers. Input from marketing, finance officers, demand forecasters, and suppliers contributes to create an inventory system that can handle the stochastic and dynamic nature of demand fluctuations. A large area of research is now focused on creating models that use data from different sources and can handle uncertain and stochastic demands. These models are complex and may require substantial data and statistical information.

In all businesses, effective inventory control is essential. Relying on emergency orders and having stock-outs could be detrimental and extremely costly. Sometimes, consumers demand products in an erratic manner that makes it nearly impossible to meet their demands on a produce-to-order basis. The suppliers of the raw materials also deal with similar inventory challenges. Inventory control can become highly complex when there are thousands, if not hundreds of thousands, of different inventory items. Manually updating the databases, checking, and recording inventory items would be nearly impossible. Computer-based inventory systems can handle an enormous amount of data and perform computations necessary to keep track of vast inventories. Also, for the effective management of inventory systems, it is essential to use state-of-the-art computer-based automated inventory control systems such as barcoding and radio frequency identification tags.

This chapter covers the fundamental concepts and methods for solving inventory problems and determining optimal inventory control strategies. In classical inventory problems, all information, such as ordering, holding, and setup costs, is assumed to be known with certainty. Finding the optimal ordering quantity (which is called economic order quantity, EOQ) is the basic goal of all inventory methods. In more complex inventory models, shortages, production time, quality discounts, and probabilistic demands are considered. Inventory problems often have multiple conflicting objectives. For example, meeting due dates or avoiding possible shortages may be achieved by holding extra inventory, which is associated with a higher total inventory cost. In this chapter, we develop multicriteria planning approaches for all different inventory problems.

There are several reasons not to keep high inventory levels:

- *Obsolescence*   Due to advances in technology and changes in product design, parts procured for future use may become obsolete, leading to a substantial loss. Also, in food-related industries, outdated food products must be thrown away, leading to a waste of resources.
- *Capital Investment*   Inventory ties up a substantial capital of the company and may not allow the company to be agile in response to market fluctuations.
- *Space Usage*   Inventory occupies precious usable space that may be essential for other purposes. Also the space is usually costly to maintain.
- *Complicated Inventory Control Systems*   A higher number of inventory items complicates the control and monitoring, such as identifying where items are and how many items exist. The JIT philosophy (discussed in Chapter 5) advocates the smallest possible inventory level, which is based on relying on reliable suppliers for effective JIT delivery of needed raw materials and inventory items.

Basic Definitions   The basic concepts and terminology associated with inventory planning and control are presented below.

- *Safety Stock*   Safety stock is the extra inventory kept beyond the projected demand to protect against fluctuations due to the uncertainty of demand or failure of production systems.
- *Production Smoothing*   Demand patterns for most products may vary considerably over time; production smoothing allows a company to produce items at a constant rate to deal with the varying demands. The excess units produced during periods of lower demands are held in inventory to satisfy periods of higher demands.
- *Selection Power*   Holding different inventory items enables a customer (or manufacturer) to have a larger selection of possible alternatives.
- *Controlling Inflation and Price Fluctuations*   Inventory can be used as a hedge against inflation and price changes. There could be a substantial profit when raw materials or components are bought at a lower price but sold at a higher sales price as the components of the final product.
- *Utilizing Quantity Discounts*   Purchasing larger quantities usually reduces the cost per item and the cost of delivering the product.
- *Optimizing Economy of Scale*   When producing many items, the production cost per unit usually decreases.

There are several types of inventory:

- *Raw Materials Inventory*   Material required for production that has not been processed.
- *Work-in-Process*   Inventory of materials between phases of production; these are semifinished products.
- *Finished Goods*   Final products that are stored and waiting for shipment.
- *Maintenance Supply*   Supplies needed for maintenance, repair, or operation of machines and processes.

There are several costs associated with inventory:

- *Ordering Cost*   Ordering cost is the average cost per order of given items; this cost includes the order processing cost but excludes the cost of materials. Usually, the ordering cost is independent of the size of the order.
- *Setup Cost*   Items are either produced internally or ordered from outside. When items are produced internally, there exists a setup cost associated with the setup of machines and processes to produce the items.
- *Carrying or Holding Cost*   This is the cost for holding one unit of inventory for one period. This cost is usually determined by two factors: the product size (required floor space for storage) and the product cost (the more expensive the product, the higher the holding cost).
- *Shortage Cost*   This is the cost for not fulfilling a customer's demand for one unit of product for one period, also referred to as stock-out cost due to lost sales because of loss of unhappy customers.

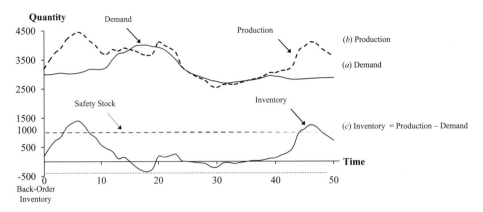

Figure 6.1 Relationship between production, demand, and inventory versus time where inventory = production – demand and the safety stock is 1000.

Variable Demands Due to uncertainty, both production and demand may fluctuate. Therefore, inventory may also fluctuate. An example of fluctuation in inventory is shown in Figure 6.1. Suppose that in every period there is production and demand but production is done based on the previous period's demand prediction. In each period, inventory is defined as production minus demand. For example, in period 0, production is 3200 and demand is 3000. Therefore, at the start of period 0, the inventory is 200. The production size in period 1 is based on the estimated demand. To be conservative, it is recommended that extra items be kept, which is called safety stock. Suppose that the safety stock in this example is 1000 items and the estimated (not actual) demand for period 1 is 2700. Therefore, $2700 + 1000 - 200 = 3500$ items should be produced in period 1. But suppose that the actual demand in period 1 is 3000. Therefore, the actual inventory at the end of period 1 is $3500 - 3000 = 500$.

Figure 6.2 shows an example where the demand rate is uncertain or probabilistic (variable) but the ordering size (quantity order Q) is constant. In this example, it takes four periods for an order to be received. Each order quantity is 5000 units. The reorder point is 1500 units. That is, when the inventory level reaches 1500, an order is placed with the expectation that when the order arrives there will be 250 items available as safety stock. The following table shows the quantities for three cycles. Sections 6.9 and 6.10 cover probabilistic demand.

Period	0	5	6	10	12	14	17	21	23	26
Inventory	5000	2894	1500	250 + 5000	4400	1500	550 + 5000	2052	1500	−300 + 5000
Order	0	0	5000	Received	0	5000	Received	0	5000	Received

Lot-Sizing Methods for Variable Demands The EOQ model, covered in the next section, is based on known constant demand rates and the same order quantity is used for all periods (cycles); see Figure 6.3. In lot-sizing problems, the future demands for all periods are known but they could be different for different periods; see Figure 6.2. In Chapter 5, the following methods are covered to solve the lot-sizing problem for variable demand rates: lot for lot, part period balancing, Silver–Meal heuristic, and the least-unit-cost heuristic.

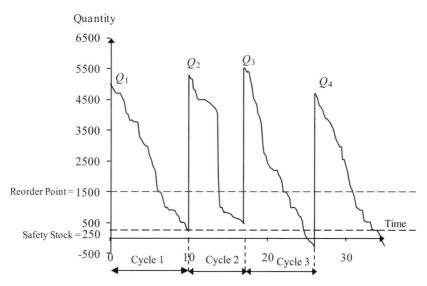

Figure 6.2 Example of uncertain demand rate and constant order quantity (5000) and constant ordering period (four).

6.2 ECONOMIC ORDER QUANTITY

6.2.1 Basic EOQ Model

The EOQ model is a widely used method for determining the order quantity to minimize total inventory cost. Despite the restrictions and simplifications of the EOQ model, it provides a good solution for minimizing total inventory cost. The EOQ method is based on the following assumptions, which significantly simplify the model:

- Demand rate is known and constant.
- Ordered items are delivered in one batch (all will be available at the same time).
- There are no quantity discounts. That is, the cost per unit of items is the same regardless of the quantity of items.
- Lead time (defined as the duration between the order placement time and the arrival time) is known and constant.

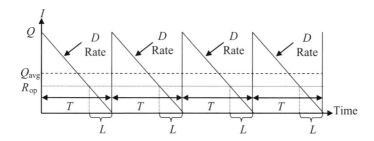

Figure 6.3 EOQ model: inventory versus time where Q, R_{op}, and L are constant.

• Orders, setups, and holding costs are constant and known with certainty.
• Shortage is not allowed.

The following notations are used in the EOQ model:

R_{op} = reorder point, minimum inventory amount at which an order is placed for the future period
L = lead time, length of time between an order placement and its arrival
T = duration of each cycle
Q = economic order quantity
D = demand of units per period (e.g., year)
S_s = safety stock

Figure 6.3 presents a typical inventory planning based on the EOQ model. Since the demand rate and lead time L are constant, the reorder point R_{op} can be calculated directly regardless of the value of the order quantity Q. The average quantity is $Q_{avg} = Q/2$. Cycle period T is represented by a $\longleftrightarrow$ line. When the inventory level reaches the reorder point R_{op}, the order is placed, where it takes L duration when inventory level would have reached zero. Right when the inventory reaches zero, the order arrives. The problem is to find the optimal quantity Q^* that minimizes the total inventory cost. Once Q^* is found, the cycle period T can be calculated. Note that Q^* and T will remain the same for all planning periods.

When inventory is at R_{op}, reordering of the lot occurs, where R_{op} can be obtained as

$$R_{op} = DL \tag{6.1}$$

Note that R_{op} does not depend on the order quantity Q. As discussed before, safety stock is used as a hedge against uncertainty in demand quantity, production time, or ordering time. Safety stock is the minimum amount of inventory kept at all times. When safety stocks are considered, the reorder point R_{op} can be set as in Equation (6.2). Figure 6.4a,b illustrate the inventory planning without and with safety stock:

$$R_{op} = DL + S_S \tag{6.2}$$

In EOQ models, the given known values are as follows:

C_R = cost of each order or setup (\$/order)
C_H = holding cost per unit of inventory per planning period (\$/unit/period)

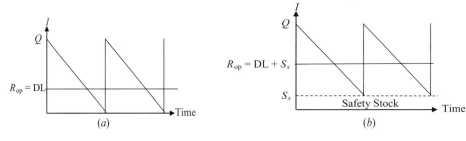

Figure 6.4 (*a*) Inventory without safety stock; (*b*) inventory with safety stock.

In some cases, one can estimate the holding cost by the formula

$$C_H = ic$$

where c is cost per unit and i is the annual interest percentage rate.

In the EOQ model, the decision variable is Q. The problem is to find the optimal Q, labeled as Q^*. Once Q^* is derived, the total cost of the inventory and the cycle period can be calculated.

For a given Q,

$$\frac{D}{Q} = \text{number of orders per period}$$

The cost equations can be derived as follows:

$$\text{Setup cost} = \text{cost per order} \times \text{no. of orders} = C_R \left(\frac{D}{Q} \right)$$

$$\text{Holding cost} = \text{holding cost per unit per period}$$

$$\times \text{ average inventory size per period} = \frac{1}{2} C_H Q$$

The total cost per period, TC, is the summation of the setup cost and the holding cost:

$$\text{Total cost} = \text{setup cost} + \text{holding cost} = \frac{C_R D}{Q} + \frac{C_H Q}{2} \tag{6.3}$$

The graph of holding cost, setup cost, and total inventory cost versus different possible order quantities (Q) is shown in Figure 6.5.

In Figure 6.5, the optimal order quantity Q^* is associated with the minimum total inventory cost.

To find the optimal minimum solution for this problem, find the first derivative of the total cost function with respect to Q:

$$\frac{\partial \text{TC}}{\partial Q} = -C_R D Q^{-2} + \frac{1}{2} C_H \tag{6.4}$$

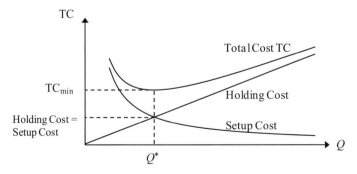

Figure 6.5   Total inventory cost (TC) versus order quantity (Q).

By setting the first derivative [Equation (6.4)] equal to zero, the global optimal solution can be obtained:

$$\frac{\partial \text{TC}}{\partial Q} = 0 \quad \text{or} \quad -C_R D Q^{-2} + \frac{1}{2} C_H = 0 \quad \text{or} \quad Q^2 = \frac{2 C_R D}{C_H}$$

Since $Q > 0$, the optimal order quantity is

$$Q^* = \sqrt{\frac{2 C_R D}{C_H}} \tag{6.5}$$

To prove that the obtained point Q^* is optimal, consider the second derivative of the total cost function as presented below:

$$\text{TC}''(Q) = 2 C_R D Q^{-3} > 0 \quad \text{for } Q > 0$$

Since the second derivative of the total cost function is positive, the total cost is a convex function. Therefore, a global minimum point exists and can be obtained by setting the first derivative equal to zero.

It should be noted that the optimal order quantity Q^* is always at the intersection of the holding cost and the setup cost functions, as shown in Figure 6.5. In other words Q^* can also be found by solving for Q in the equation where setup cost is equal to holding cost, which also results in Equation (6.5):

$$C_R \left(\frac{D}{Q} \right) = \frac{C_H Q}{2}$$

Once the optimal order quantity is known, the total cost can be calculated by substituting the optimal quantity Q^* in Equation (6.3) where the minimum total cost can be calculated as

$$\text{TC} = \sqrt{2 C_H C_R D} \tag{6.6}$$

For a given Q^*, the cycle time per order can be found. It is the reciprocal of the number of orders per year:

$$T = \frac{Q^*}{D} \tag{6.7}$$

Example 6.1   EOQ Example   A manufacturing supply company wishes to determine the optimal order quantity for a given part. Parts are sold at a constant rate of 60 units per week. The cost of each order is $25. Holding costs are based on an annual percentage rate of 12% of the part cost, which includes the administration and insurance cost. Each part costs $4.50. The lead time for order arrival is 14 days. Consider 365 days and 52 weeks in a year.

Find the optimal order quantity, the number of orders per year, the duration of each cycle (in days), and the reorder point.

For a one-year period the problem can be summarized as follows:

Demand per year	$D = 60 \times 52 = 3.120$	Parts/year
Holding cost	$C_H = 0.12 \times 4.50 = \$0.54$	Per part per year
Ordering cost	$C_R = \$25$	Per order
Lead time	$L = 14/365 = 0.038$	Year (for 365 days per year)

Using Equation (6.5), the optimal EOQ can be found as

$$Q^* = \sqrt{\frac{2C_R D}{C_H}} = \sqrt{\frac{2 \times 25 \times 3120}{0.54}} = 537.48 \text{ parts per order}$$

Then

$$\frac{D}{Q^*} = \frac{3120}{537.48} = 5.81 \text{ orders per year}$$

The total annual cost using Equation (6.3) is

$$TC = \frac{DC_R}{Q} + \frac{QC_H}{2} = \frac{3120 \times 25}{537.48} + \frac{537.48 \times 0.54}{2} = \$290.24$$

Alternatively, total cost can be found directly by using Equation (6.6):

$$TC = \sqrt{2 \times 0.54 \times 25 \times 3120} = \$290.24$$

The cycle time per order, using Equation (6.7), is

$$T = \frac{Q^*}{D} = \frac{537.48}{3120} = 0.1723 \text{ years} \qquad \text{or} \qquad 0.1723 \times 365 = 62.88 \text{ days}$$

The optimal solution calls for ordering 537.48 parts every 62.88 days where the minimum total cost of ordering and holding the parts is $290.24 per year.

With the lead time of 14 days, the reordering point, that is, the inventory level at which the company should place the order, can be found by using Equation (6.1):

$$R_{op} = DL = 3120 \times \left(\frac{14}{365}\right) = 119.67 \text{ parts}$$

Rounding Number of Products   In general, after finding the optimal EOQ solution, one can round up Q^* and then generate its associated values. Also, by rounding up the reorder point R_{op} to the next highest integer value, no shortages will occur. For example,

$$Q^* = 537.48 \quad \text{can be rounded to} \quad Q = 538$$

Now, using $Q = 538$, $T = Q^*/D = 538/3120 = 0.1724$ year. In terms of days, $T = 0.1724 \times 365 = 62.93$ days, which can be rounded up to 63 days.

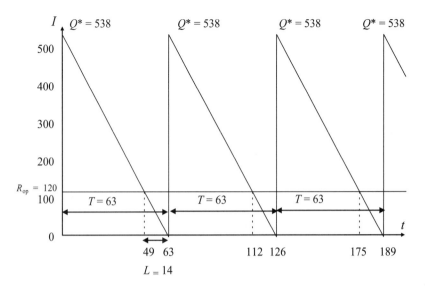

Figure 6.6 Graphical solution to Example 6.1 with rounded numbers.

Also, $R_{op} = LQ^*/T = (14 \times 538)/63 = 119.56$, which can be rounded up to 120 units. Figure 6.6 shows the solution graphically.

The EOQ model does not account for practical constraints that may occur in real-life problems. Also, in addition to cost, other objectives may influence the selection of the order size. Furthermore, by simply rounding up Q^*, the solution may not be optimal. All these problems will be addressed later in this chapter.

6.2.2 Robustness: Sensitivity of EOQ

In the real world, it may not be possible to know the exact values of holding, setup, and ordering costs. Furthermore, the demand could be based on forecasts that are subject to errors. The robustness of the EOQ model can be checked with respect to these inherent uncertainties and the errors in estimation of EOQ coefficients. The question is to what extent the EOQ optimal solution is sensitive to the errors in the estimation of coefficients. In general, the error in order quantity Q is about $\sqrt{\alpha}$, where α is the error in the estimation of a given coefficient. For example, $\alpha = 2$ means that the actual value of the coefficient is two times higher than what it was assumed.

Ordering Cost Sensitivity If the ordering cost increases by a factor of α, the EOQ model can be represented by replacing C_R with αC_R:

$$Q^*_{\alpha C_R} = \sqrt{\frac{2D\alpha C_R}{C_H}} = \sqrt{\alpha}Q^*$$

The total cost, Equation (6.6), is calculated as

$$\mathrm{TC}_{\alpha C_R} = \sqrt{2C_H\alpha C_R D} = \sqrt{\alpha}(TC) \tag{6.8}$$

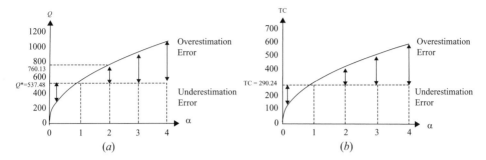

Figure 6.7   Example 6.1: changes in (*a*) order quantity and (*b*) total cost with respect to changes in ordering cost.

For example, if the actual ordering cost will be twice what was estimated,

$$Q^*_{2C_R} = \sqrt{2}Q^* = 1.41Q^* \qquad TC_{2C_R} = \sqrt{2}TC = 1.41\ TC$$

Example 6.2   Robustness and Sensitivity Analysis for 100% Error   Consider Example 6.1. Suppose the ordering cost is \$50 instead of \$25 (i.e., a 100% error or $\alpha = 2$ occurs in the estimation of the ordering cost). Determine the possible value for EOQ and the total cost. Also, graphically show the impact of different values of α.

Since $Q^* = 537.48$ and $\alpha = 2$, the estimated Q^* would be $537.48\sqrt{2} = 760.11$. In other words, the error is

$$\frac{760.11 - 537.48}{537.48} = 0.4142 \quad (\text{or } 41.42\%)$$

Similarly, the total cost would be $290.24\sqrt{2} = \$410.46$. Therefore,

$$\frac{410.46 - 290.24}{290.24} = 41.42\% \text{ error}$$

Figure 6.7 shows the sensitivity analysis for different values of α for Example 6.1.

Holding Cost Sensitivity   Similarly, the following results can be found for an error factor of α in the holding cost:

$$Q^*_{\alpha C_H} = \sqrt{\frac{2DC_R}{\alpha C_H}} = \frac{1}{\sqrt{\alpha}}Q^* \qquad TC_{\alpha C_H} = \sqrt{2\alpha C_H C_R D} = \sqrt{\alpha}(TC)$$

In Example 6.1, if $\alpha = 2$,

$$Q^*_{2C_H} = \frac{1}{\sqrt{2}} \times 537.48 = 380.05 \text{ parts per order} \qquad TC_{2C_H} = \sqrt{2} \times \$290.24 = \$410.46$$

Demand Sensitivity Suppose the demand per period is wrongly estimated and the actual value is αD. Similar to the above, the following results are found:

$$Q^*_{\alpha D} = \sqrt{\frac{2\alpha D C_R}{C_H}} = \sqrt{\alpha} Q^* \qquad TC_{\alpha D} = \sqrt{2 C_H C_R \alpha D} = \sqrt{\alpha}(TC)$$

In Example 6.1, if $\alpha = 2$,

$$Q^*_{2D} = \sqrt{2} \times 537.48 = 760.12 \text{ parts per order} \qquad TC_{2D} = \sqrt{2} \times \$290.24 = \$410.46$$

For $\alpha = 2$, the above results show that when variables change by 100%, the corresponding EOQ and TC change by about 41.42%.

6.3 ECONOMIC PRODUCTION QUANTITY

The EOQ model covered in the last section is based on the assumption that the entire order will be received at one point in time, that is, arrival rate is infinity. But if items are produced in-house, the produced items become available gradually during the production period. Also, it is possible that the ordered items arrive gradually, that is, have a finite arrival rate. In these cases, the arrival rate or the production rate should be considered. The economic production quantity (EPQ) model is an extension of the EOQ model in which the production (or arrival) rate (P) is also used to determine the optimal EOQ. Figure 6.8 illustrates the basic concepts of the EPQ model.

In addition to the EOQ model given values, the following information is also given in the EPQ model:

$$P = \text{production rate per planning period}$$

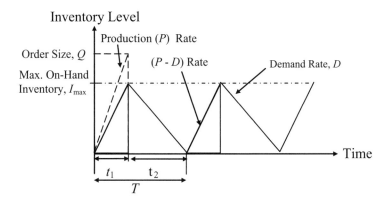

Figure 6.8 EPQ model is a modified EOQ that considers production or arrival rate P where $P - D$ is the Inventory accumulation rate during the production period t_1.

The EPQ decision variables, in addition to EOQ decision variables, are

I_{max} = maximum inventory

t_1 = production (or arrival) duration for Q items in one cycle

t_2 = nonproduction duration in one cycle

T = cycle time, where $T = t_1 + t_2$

The EPQ model assumptions, in addition to the EOQ model assumptions, are

- Production (or arrival) rate P is constant.
- Production (or arrival) rate is greater than the demand rate per period, that is, $P > D$.

In Figure 6.8, the cycle time T consists of the production period t_1 and nonproduction period t_2. During the entire cycle time T, the demand is satisfied. That is, the same constant demand rate is used during both periods t_1 and t_2. In Figure 6.8, at the end of production period t_1, the total production is Q, which is produced at production rate P. During the production period t_1, the inventory accumulation rate is the production rate minus the demand rate $(P - D)$, assuming that $P > D$. Therefore, the order quantity Q is not equal to the maximum inventory because as Q items are produced, a portion of them will be used to satisfy demand during period t_1. The maximum inventory I_{max} is equal to the total products produced during the production period t_1 minus the total products used to satisfy the demand during production period t_1:

$$I_{max} = Pt_1 - Dt_1 = (P - D)t_1 \qquad (6.9)$$

The total quantity produced within a production period is

$$Q = Pt_1 \qquad \text{or} \qquad t_1 = \frac{Q}{P} \qquad (6.10)$$

Equations (6.11) can be derived by substituting t_1 from Equation (6.10) into Equation (6.9):

$$I_{max} = (P - D)t_1 = (P - D)\left(\frac{Q}{P}\right) = Q - D\left(\frac{Q}{P}\right) = Q\left(1 - \frac{D}{P}\right) \qquad (6.11)$$

In the EPQ model, as in the EOQ model, the annual holding cost is the average annual inventory times the holding cost per unit per period. However, the annual holding cost equation for the EPQ model is different from the EOQ model because of production rate, and the inventory holding cost is different.

The holding cost is the average of the inventory (i.e., $I_{max}/2$) times the holding cost per unit, C_H. Substituting I_{max} from Equation (6.11) results in

$$\text{Total holding cost} = \left(\frac{I_{max}}{2}\right)C_H = \frac{Q}{2}\left(1 - \frac{D}{P}\right)C_H$$

The setup cost is the number of orders (D/Q) times the cost per order, C_R:

$$\text{Total setup cost} = \frac{C_R D}{Q}$$

The total inventory cost is equal to the setup cost plus the holding cost:

$$\text{TC} = \frac{C_R D}{Q} + \frac{Q}{2}\left(1 - \frac{D}{P}\right)C_H \tag{6.12}$$

Similar to the EOQ model, the optimal quantity Q^* can be obtained by setting the first derivative of the total cost function (with respect to Q) equal to zero and then solving the equation to find the optimal Q^*:

$$Q^* = \sqrt{\frac{2C_R D}{(1 - D/P)C_H}} \tag{6.13}$$

Equation (6.13) is an extension of the EOQ Equation (6.5), where the holding cost per unit, $(1 - D/P)C_H$, is used.

Example 6.3   Finding the Optimal Economic Production Quantity   Consider Example 6.1. Suppose the company can produce items in house. The production rate is 100 items (parts) per week. The setup cost for each production run is $25 and the holding cost per unit per period is $0.54. Determine the optimal production quantity, the maximum inventory, the total inventory cost, the cycle duration, the production duration, and the number of setups.

The given information for this problem can be summarized as follows:

$D = 3120$ parts/year

$P = 100$ parts/week   or $100 \times 52 = 520$ parts/year (assuming 52 weeks a year)

$C_H = \$0.54$ per unit per year

$C_R = \$25$ per setup

The optimal quantity to be produced in one batch is Q^*. It can be found by using Equation (6.13):

$$Q^* = \sqrt{\frac{2(25)(3120)}{(1 - 3120/5200)(0.54)}} = 849.84$$

The maximum inventory is

$$I_{\text{max}} = Q\left(1 - \frac{D}{P}\right) = 849.84\left(1 - \frac{3120}{5200}\right) = 339.94$$

The annual average total cost is

$$\text{TC} = \frac{C_R D}{Q} + \frac{Q}{2}\left(1 - \frac{D}{P}\right)C_H = \frac{25 \times 3120}{849.84} + \frac{849.84}{2}\left(1 - \frac{3120}{5200}\right) \times 0.54 = \$183.57$$

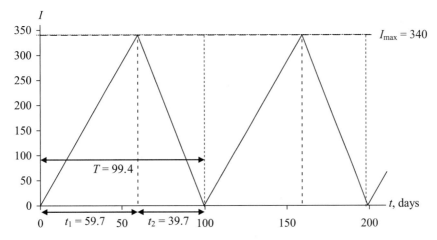

Figure 6.9   Graphical solution of Example 6.3.

The cycle duration, that is, the time between two consecutive productions, is

$$T = \frac{Q^*}{D} = \frac{849.84}{3120} = 0.2724 \text{ years} \qquad \text{or} \qquad 0.2724 \times 365 = 99.43 \text{ days}$$

The production period is calculated using Equation (6.10):

$$t_1 = \frac{Q^*}{P} = \frac{849.84}{5200} = 0.1634 \text{ years} \qquad \text{or} \qquad 0.1634 \times 365 = 59.64 \text{ days}$$

Therefore, the nonproduction duration is

$$t_2 = T - t_1 = 99.43 - 59.64 = 39.79 \text{ days}$$

The total number of setups per year is

$$\frac{\text{Number of setups}}{\text{Year}} = \frac{D}{Q^*} = \frac{3120}{849.84} = 3.67$$

The solution to this example is presented in Figure 6.9.

6.4   EOQ: ALLOWING SHORTAGES

Why Allow Shortages   Most recent books in the fields of operations, production, and inventory systems do not cover inventory planning allowing shortages. Furthermore, most production planning computer packages do not allow the use of shortages. Their reasons are that one should plan to satisfy demands on time and, by planning for shortages, the allowed shortages may get even worse if there are unexpected delays. However, with the advent of more reliable transportation systems, the use of the Internet, JIT, and supply chain management, it is possible to optimize production planning by allowing shortages. In some applications, allowing shortages may result in substantial savings. This especially occurs when the setup or ordering cost is too expensive and the shortage cost is relatively inexpensive.

Other important advantages of planning with shortages include a protection against over-estimated forecasted demand, the canceling of customer's orders, and the obsolescence of items due to technological changes. Furthermore, in the production environment, allowing shortages may allow for planning of smaller lot sizes, which is more consistent with JIT philosophy. By planning, managing, and controlling the amount of shortages, the planning can be accomplished smoothly avoiding large fluctuations. However, there should be a limit on the amount of shortages.

Push systems (material requirements planning, MRP) advocate positive inventories and pull systems (JIT) advocate zero inventory. The third option is to allow negative inventory, which forces smaller lot sizes and makes the system even more agile. Therefore, if systems are well organized and agile, it is possible to plan for shortages. Furthermore, the MRP (push) system plans for additional inventory to cover the future possible shortages. The JIT (pull) system plans for zero inventory to produce the exact number of items needed at given times. The reverse push system (the term we introduce here) allows for shortages to optimize the system. The implementation of this reverse MRP system is more stringent than JIT as it requires JIT to be even more accurate in production planning.

6.4.1   EOQ Model Allowing Shortages[†]

In the EOQ model allowing shortages, the order quantity Q consists of two components: the maximum inventory M and the maximum shortage R where $Q = M + R$. See Figure 6.10. When the order is received, the inventory is at its maximum, M. During period t_1, the inventory is a nonnegative value. In this period, there is no shortage and the demand is satisfied. During period t_2, shortage occurs. In this period, the inventory has a negative value (i.e., the backorder has a positive value) and the demand is not being satisfied. The longest delay to satisfy demand is the length of the second period, t_2. Note that the duration of t_2 is directly related to the value of R. To limit t_2, one can put a limit on the allowable shortage R. The following section shows how to find the optimal order quantity with a given limit on the allowable shortage.

As stated before, the EOQ model assumes no shortages are allowed. In this section, we allow for shortages. All the assumptions of the EOQ model apply to this method except that shortages are allowed. Therefore, the cost of shortages is added to the total cost of inventory. The shortage (or stock-out) cost C_s is the cost of not fulfilling a customer's

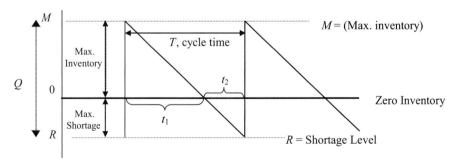

Figure 6.10   Inventory allowing shortage.

demand per unit per period. The cost may be determined by a penalty paid or calculated based on the possible loss of current and future sales.

In this problem, decision variables are

Q = size of each order

M = maximum inventory at beginning of cycle time, where $Q \geq M$

Figure 6.10 illustrates the relationship between the decision variables.

The following values can be calculated once the optimal values for Q and M are determined:

R = maximum shortage, where $R = Q - M$

t_1 = duration when positive inventory exists

t_2 = duration when backlog exists

T = cycle time, where $T = t_1 + t_2$

The total inventory cost consists of holding, shortage, and ordering costs:

$$\text{Holding cost} = \text{holding cost per unit} \times \text{average inventory} = C_H \frac{M^2}{2Q}$$

$$\text{Shortage cost} = \text{shortage cost per unit} \times \text{average shortage} = C_S \frac{(Q - M)^2}{2Q}$$

$$\text{Ordering cost} = \text{ordering cost per unit} \times \text{no. of orders} = C_R \left(\frac{D}{Q}\right)$$

The total cost as a summation of these three costs can be expressed as

$$\text{TC} = C_H \frac{M^2}{2Q} + C_s \frac{(Q - M)^2}{2Q} + C_R \frac{D}{Q} \tag{6.14}$$

This cost function is similar to the EOQ cost function, and it is also a convex function of the two unknown variables Q and M. The optimal values of Q and M, for which the total cost is minimized, can be found as follows. Find the partial derivative of total cost function (6.14) first with respect to Q and then with respect to M. Set the resulting two functions equal to zero, which results in generating two equations. By solving the two equations, the optimal values of Q and M can be found:

$$Q^* = \sqrt{\frac{2C_R D}{C_H}} \sqrt{\frac{C_H + C_S}{C_S}} \tag{6.15}$$

$$M^* = \sqrt{\frac{2C_R D}{C_H}} \sqrt{\frac{C_S}{C_H + C_S}} \tag{6.16}$$

Substituting Q^* and M^* into Equation (6.14), the TC equation can be written as

$$TC = \sqrt{2C_H C_R D} \sqrt{\frac{C_S}{C_H + C_S}} \tag{6.17}$$

It should be noted that if shortages are not allowed, this model becomes the same as the EOQ model. That is, by setting $C_S = \infty$, Equations (6.15) and (6.16) will become the same as the EOQ equation (6.5) that is, $Q^* = M^*$. Also, Equation (6.17) becomes the same as the EOQ total-cost equation (6.6).

Because t_2 is the maximum duration of the shortage, it is important to calculate this value. Using the triangles in Figure 4.1,

$$\frac{t_2}{T} = \frac{R}{Q} \qquad \text{or} \qquad t_2 = \frac{RT}{Q}$$

Example 6.4   Finding Optimal EOQ Allowing Shortage   Consider an inventory problem where demand $D = 5000$ units a year, setup cost $C_R = \$1500$, shortage cost $C_S = \$150$, and holding cost $C_H = \$200$. Find the optimal EOQ, the maximum inventory, the maximum shortage, the maximum shortage time, the cycle time period, and the total cost of this inventory solution. Assume 365 days per year.

Applying Equations (6.15) and (6.16), the solution is as follows:

$$Q^* = \sqrt{\frac{2(1500)(5000)}{200}} \sqrt{\frac{200 + 150}{150}} = 418.33 \text{ units}$$

$$M^* = \sqrt{\frac{2(1500)(5000)}{200}} \sqrt{\frac{150}{200 + 150}} = 179.28 \text{ units}$$

$$R = Q - M = 418.33 - 179.28 = 239.05 \text{ units}$$

$$T = \frac{Q}{D} = \frac{418.33}{5000} = 0.0837 \text{ year} \qquad \text{or} \qquad 0.0837 \times 365 \text{ days} = 30.55 \text{ days}$$

$$\text{Number of orders} = \frac{D}{Q} = \frac{5000}{418.33} = 11.95 \text{ orders per year}$$

The minimum total cost can be calculated as

$$TC = \sqrt{2 \times 200 \times 1500 \times 5000} \times \sqrt{\frac{150}{200 + 150}} = \$35{,}856.86$$

The maximum shortage time t_2 is

$$t_2 = \frac{RT}{Q} = \frac{239.05 \times 30.55}{418.33} = 17.46 \text{ days}$$

Note that $t_1 = T - t_2 = 30.55 - 17.46 = 13.09$.

6.4.2 Inventory Model Allowing Limited Shortage[†]

As stated before, in some cases, allowing shortages may bring about a substantial cost savings; however, the amount of shortage should be closely monitored. In Example 6.4, the maximum shortage is 239.05, where the ratio $R/Q = 239.05/418.33 = 0.57$, or 57% of the total order, may be backordered. In practice, one can limit the maximum shortage amount or its duration. The procedure for finding the optimal EOQ with limited shortage is described below. Suppose $R_{\max}$ is the maximum allowable shortage. Solve the EOQ model allowing shortages:

- If $R \leq R_{\max}$, then the solution is feasible and optimal; use the solution.
- If $R > R_{\max}$,

    Set $R = R_{\max}$; then find optimal Q^* for the given R value.
    Set $M = Q^* - R_{\max}$.

Finding Optimal Order Q^* for Given R   Because R is known and $M = Q - R$, substitute $Q - R$ for M in Equation (6.14). The total cost for a given R is

$$\text{TC} = C_H \frac{(Q-R)^2}{2Q} + C_s \frac{R^2}{2Q} + C_R \frac{D}{Q} \tag{6.18}$$

In this function, there is only one variable, Q. To find the optimal Q value, find its first derivative with respect to Q and set it equal to zero, that is, $\partial \text{TC}/\partial Q = 0$, or

$$C_H \frac{Q^2 - R^2}{2Q^2} - C_s \frac{R^2}{2Q^2} - C_R \frac{D}{Q^2} = 0$$

or

$$C_H Q^2 - C_H R^2 - C_s R^2 - 2C_R D = 0$$

or

$$Q^* = \sqrt{\frac{C_H R^2 + C_s R^2 + 2C_R D}{C_H}} \tag{6.19}$$

Example 6.5   Finding Optimal EOQ with Limited Shortage   Consider Example 6.4. Suppose the maximum allowable shortage $R_{\max} = 250$ units. Determine the optimal EOQ.
   Since $R \leq R_{\max}$, that is, $239.05 < 250$, the solution is feasible and optimal, and EOQ $= 418.33$.

Example 6.6   Finding Optimal EOQ with Limited Shortage   Consider Example 6.4. Suppose the maximum allowable shortage is 100. Determine the optimal EOQ.

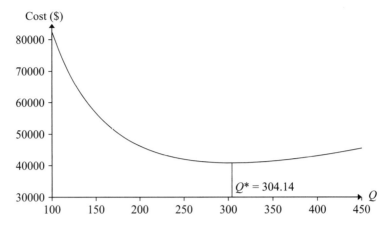

Figure 6.11 Example 6.6: TC function with a given maximum shortage $R_{min} = 100$.

Since $R > R_{max}$, that is, $239.05 > 100$, the previous solution of this example is infeasible. Therefore, set $R = R_{max} = 100$. Now, solve for Q using Equation (6.19):

$$Q = \sqrt{\frac{200(100)^2 + 150 \times 100^2 + 2(1500)5000}{200}} = 304.14$$

$$M = Q - R_{max} = 304.14 - 100 = 204.14 \text{ units}$$

The total cost can be calculated using Equation (6.18):

$$TC = 200\frac{(304.14 - 100)^2}{2 \times 304.14} + 150\frac{(100)^2}{2 \times 304.14} + 1500\frac{5000}{304.14} = \$40,827.63$$

The total shortage time for this solution is

$$t_2 = \frac{RT}{Q} = \frac{100 \times 30.55}{304.14} = 10.05 \text{ days}$$

Alternatively, one can find the solution by graphing the TC function versus different values of Q, where R is a given value. The total cost function for this example is presented in Figure 6.11, where the minimum total cost is \$40,827.63 and $Q^* = 304.14$. Note that this function is convex and it has only one global minimum.

6.5 MULTICRITERIA INVENTORY

6.5.1 Bicriteria Economic Order Quantity

In the EOQ problem, only one objective, total inventory cost, is minimized. The EOQ model is based on the traditional "push" concept of MRP. The JIT strategy is based on the "pull" production planning concept, which advocates the strategy of zero inventory.

In the JIT strategy, the setup and lead times are decreased, allowing for smaller lot sizes which are produced more frequently. Therefore, in JIT, the order quantity Q is reduced. More importantly, JIT advocates minimizing the amount of inventory for the purpose of identifying and resolving the production and management problems. In general, zero inventory implies that the order quantity is reduced while the number of setups is increased. In this section, we introduce multicriteria EOQ to find a compromise solution between MRP and JIT solutions. For different users or applications, a different compromise solution can be identified. Furthermore, other objectives such as space priorities, employee issues, customer preferences, and production priorities can also be identified and considered. The need for considering several criteria is evident because inventory cost does not encompass all of the qualitative and quantitative objectives.

In the bicriteria inventory problem, the following two objectives are considered:

PROBLEM 6.1   BICRITERIA EOQ VERSUS JIT

Minimize: Total inventory cost $f_1 = (C_R D)/Q + (C_H Q)/2$ MRP objective

Minimize: Order quantity $f_2 = Q$ JIT objective

    (to minimize inventory)

Subject to: Problem constraints

Note that in the EOQ model the maximum inventory is equal to the order quantity Q. Minimizing the order quantity will result in a more agile production system which is associated with less inventory. Suppose the feasible and meaningful lower bound on order quantity is Q_{min}.

Two extreme alternatives based on the above two objectives can be generated as follows:

1. For a given Q_{min}, find the corresponding total cost and label it as TC_{max}.
2. Determine Q^* by the EOQ model; find the corresponding cost and label it as TC^*.

If $Q^* \leq Q_{min}$, there exists only one (efficient) solution to this problem. Use Q_{min} as the optimal solution.

When $Q_{min} < Q^*$, all alternatives between Q_{min} and Q^* are efficient and should be considered. For a matter of illustration, five efficient alternatives that are equally distributed in terms of Q are presented in the following example.

Once the bicriteria alternatives and their objective function values are found, a multicriteria method can be used to choose the best alternative. If there are many alternatives, the bisectional one-dimensional search can be used to find the best efficient solution.

Example 6.7   Generating Five Efficient Alternatives for Bicriteria EOQ   Consider Example 6.1. Suppose the minimum order size is 100 items. Generate five equally distributed efficient alternatives in terms of Q.

In this example, $Q^* = 537$. Five different Q values ranging from 100 to 537 can be considered as follows. The first midpoint is $Q = (100 + 537)/2 = 319$. The five values for Q are 100, 209, 319, 427, and 537 are listed in the table below.

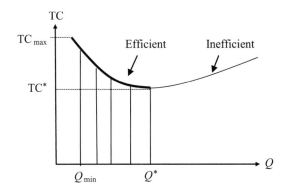

Figure 6.12 Total inventory cost vs. order quantity, efficient and inefficient alternatives.

For each of these given Q values, find the total cost using $f_1 = (C_R D)/Q + (C_H Q)/2$. For example, for $Q = 319$,

$$f_1 = \frac{25 \times 3120}{319} + \frac{0.54 \times 319}{2} = \$331$$

The table below shows the five alternatives and their objective function values.

Alternatives	a_1	a_2	a_3	a_4	a_5
Minimum total inventory cost, f_1	807	430	331	298	290
Minimum order size, $f_2 = Q$	100	209	319	427	537
Efficient?	Yes	Yes	Yes	Yes	Yes

Figure 6.12 shows efficient and inefficient frontiers based on two objectives of total cost and order quantity. Figure 6.13 shows a plot of the five efficient alternatives based on two objectives.

Ranking Alternative by Additive Utility Function For the purpose of illustration, suppose that an additive utility function is used to rank the alternatives. The additive utility function for the two objectives is

$$U = w_1 f_1' + w_2 f_2' \tag{6.20}$$

where f_1' and f_2' are normalized values of objectives f_1 and f_2 and w_1 and w_2 are weights of importance of the two objectives.

To find the optimal solution with more precision, the additive utility function [Equation (6.20)] can be expressed as a function of only Q, that is,

$$U(Q) = w_1 \left(\frac{DC_R/Q + C_H Q/2 - f_{1,\min}}{f_{1,\max} - f_{1,\min}} \right) + w_2 \left(\frac{Q - f_{2,\min}}{f_{2,\max} - f_{2,\min}} \right)$$

f_1 = Total Cost

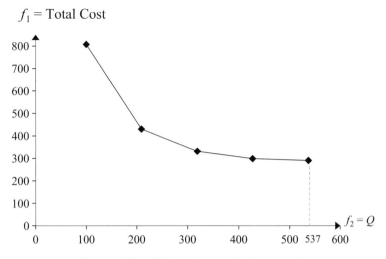

Figure 6.13   Efficient Frontier for Example 6.7.

Find the optimal Q by setting the first derivative of $U(Q)$ equal to zero, that is, $\partial U(Q)/\partial Q = 0$:

$$U'(Q) = w_1 \left(\frac{-DC_R/Q^2 + C_H/2}{f_{1,\max} - f_{1,\min}} \right) + \left(\frac{w_2}{f_{2,\max} - f_{2,\min}} \right) = 0$$

or

$$\frac{DC_R}{Q^2} = \frac{w_2}{w_1} \left(\frac{f_{1,\max} - f_{1,\min}}{f_{2,\max} - f_{2,\min}} \right) + \frac{C_H}{2}$$

Then

$$Q^* = \sqrt{\frac{DC_R}{\dfrac{w_2}{w_1} \left(\dfrac{f_{1,\max} - f_{1,\min}}{f_{2,\max} - f_{2,\min}} \right) + \dfrac{C_H}{2}}} \tag{6.21}$$

Example 6.8   Ranking by Additive Utility Function   Consider Example 6.7. Suppose that 70% importance is given to the cost and 30% to minimizing order quantity (to minimize inventory) for the JIT objective. The weights of importance for normalized objective functions (f_1' and f_2') are $w_1 = 0.7$ and $w_2 = 0.3$, respectively.

   (a) Rank the five alternatives equally distributed in terms of Q.
   (b) Find the exact best alternative.

Alternatives are ranked by the following utility function:

$$\text{Minimize } U = w_1 f_1 + w_2 f_2 = 0.7 f_1' + 0.3 f_1'$$

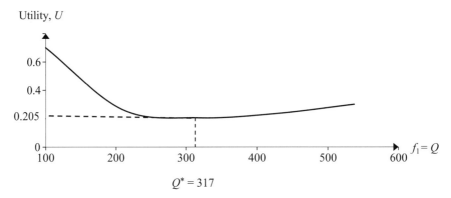

$Q^* = 317$

Figure 6.14 Example 6.8: Ranking of all alternatives by the utility function, the best solution is $Q^* = 317$, associated with the minimum utility $U = 0.205$.

Note that the normalized values can be found by $f_i' = (f_i - f_{min})/(f_{max} - f_{min})$ for each objective i.

(a) *Finding Best of Five Discrete Alternatives*

Alternatives	a_1	a_2	a_3	a_4	a_5	Minimum	Maximum
Min. f_1 = TC	807	430	331	298	290	290	807
Min. f_2 = Q	100	209	319	427	537	100	537
Min. f_1' = TC	1	0.271	0.079	0.015	0	0	1
Min. f_2' = Q	0	0.249	0.501	0.748	1	0	1
Min. U	0.7	0.264	0.206	0.235	0.3		
Rank	5	3	1	2	4		

The best alternative is a_3, which has the lowest utility value.

The optimal point is $f_1 = 331.65$ and $f_2 = Q = 316.83$, where $U = 0.205$. Figure 6.14 shows the utility function for different values of Q.

(b) *Finding Optimal Alternative*   Use Equation (6.21) to find the optimal lot size Q^* for the bicriteria problem:

$$Q^* = \sqrt{\dfrac{3120 \times 25}{\dfrac{0.3}{0.7}\left(\dfrac{807 - 290}{537 - 100}\right) + \dfrac{0.54}{2}}} = 316.83$$

See Supplement 6.1.xlsx.

6.5.2   Bicriteria Economic Production Quantity

The bicriteria EPQ is formulated as follows.

PROBLEM 6.2   BICRITERIA OF EPQ VERSUS JIT

Minimize: Total inventory cost	$f_1 = (DC_R)/Q$ $\qquad + Q(1 - D/P)C_H/2$	MRP objective
Minimize: Order quantity (to minimize inventory)	$f_2 = Q$	JIT objective
Subject to: Constraints of EPQ model		

The solution of the bicriteria EPQ problem is similar to the method of the last section for solving the bicriteria EOQ problem. The additive utility function can be written as

$$U(Q) = w_1 \left(\frac{\frac{DC_R}{Q} + \left(1 - \frac{D}{P}\right)\left(\frac{C_H Q}{2}\right) - f_{1,\min}}{f_{1,\max} - f_{1,\min}} \right) + w_2 \left(\frac{Q - f_{2,\min}}{f_{2,\max} - f_{2,\min}} \right)$$

The optimal solution for the additive utility function is derived by setting $\partial U(Q)/\partial Q = 0$:

$$w_1 \left(\frac{-\frac{DC_R}{Q^2} + \left(1 - \frac{D}{P}\right)\frac{C_H}{2}}{f_{1,\max} - f_{1,\min}} \right) + \left(\frac{w_2}{f_{2,\max} - f_{2,\min}} \right) = 0$$

or

$$\frac{DC_R}{Q^2} = \frac{w_2}{w_1} \left(\frac{f_{1,\max} - f_{1,\min}}{f_{2,\max} - f_{2,\min}} \right) + \frac{C_H}{2} \left(1 - \frac{D}{P}\right)$$

Then

$$Q^* = \sqrt{\frac{DC_R}{\left[\frac{w_2}{w_1} \left(\frac{f_{1,\max} - f_{1,\min}}{f_{2,\max} - f_{2,\min}} \right) + \frac{C_H}{2} \left(1 - \frac{D}{P}\right) \right]}} \qquad (6.22)$$

Example 6.9   Bicriteria EPQ versus JIT   Consider Example 3.3, where

$D = 3120$ parts/year

$P = 100$ parts/week   or $100 \times 52 = 5200$ parts/year

$C_H = \$0.54$ per unit per year

$C_R = \$25$ per setup

Suppose the minimum order size is 100 parts. Find the optimal EPQ for the bicriteria problem where the weights of importance for normalized total cost and order quantity are (0.7, 0.3), respectively.

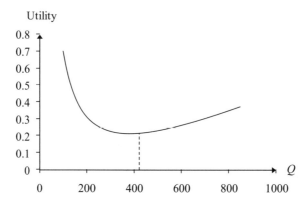

Figure 6.15 Example 6.9: optimal EPQ solution for the additive utility function, $Q^* = 414.01$, $U = 0.21$.

The optimal EPQ, when only minimizing total inventory cost, was found in Example 6.3 by using Equation (6.13):

$$Q^* = \sqrt{\frac{2(25)(3120)}{(1 - 3120/5200)(0.54)}} = 480.38$$

Objectives f_1 and f_2 are minimized using Equation (6.22) where the weights of importance for the two objectives are $w_1 = 0.7$ and $w_2 = 0.3$. The optimal EPQ is

$$Q^* = \sqrt{\frac{3120 \times 25}{\dfrac{0.3}{0.7}\left(\dfrac{790.80 - 183.57}{849.84 - 100}\right) + \dfrac{0.54}{2}\left(1 - \dfrac{3120}{5200}\right)}} = 414.01$$

Note that $f_1 = (3120 \times 25/414.01) + 414.01 \times [1 - (3120/5200)] \times 0.54/2 = 233.11$ and $f_2 = 414.01$. The utility function values of all alternatives are presented in Figure 6.15.

See Supplement 6.2.xls.

6.5.3 Tricriteria EOQ Allowing Shortages[†]

In this section, we discuss the multiobjective optimization (MOO) of inventory planning. The optimization model can be used for single-objective as well as multiple-objective optimization problems for EOQ, EPQ, and EOQ with shortages allowed. Different constraints for all three of these problems can be considered by the following MOO approach. In the following, only the details for EOQ with shortages allowed are shown.

MOO of Inventory Allowing Shortage   Consider an EOQ inventory problem allowing shortages where there are a set of constraints on decision variables. See Section 6.4 and Figure 6.10 for details of the relationship between variables and their definitions for EOQ with shortages allowed.

The MOO of EOQ allowing shortages is as follows:

Minimize: Total cost $\qquad$ $f_1 = C_H \frac{M^2}{2Q} + C_s \frac{(Q-M)^2}{2Q} + \frac{C_R D}{Q}$ $\quad$ MRP objective

Minimize: Order quantity (inventory) $\quad f_2 = Q$ $\qquad\qquad\qquad\qquad$ JIT objective

Minimize: Shortages $\qquad\qquad\qquad f_3 = R$ $\qquad\qquad\qquad\qquad\quad$ Customer
$\qquad\qquad\qquad\qquad\qquad\qquad\qquad\qquad\qquad\qquad\qquad\qquad\qquad\qquad$ objective

Subject to: Constraints of the problem

By minimizing only the total cost f_1, the optimal Q^* and M^* for the single-cost-objective function can be obtained. Use the obtained Q^* as the upper limit for order quantity. The lower limit on the order quantity can be selected by the decision maker, labeled as $Q_{\min}$. Therefore,

$$Q_{\min} \le Q \le Q^*$$

Before solving the problem, the desired lower and upper bounds on Q and R can be identified. The MOO formulation for the additive utility function is presented below.

PROBLEM 6.3   TRICRITERIA OPTIMIZATION INVENTORY PROBLEM

Minimize:

$$U = w_1 f_1' + w_2 f_2' + w_3 f_3' \tag{6.23}$$

Subject to:

$$f_1' = \frac{f_1 - f_{1,\min}}{f_{1,\max} - f_{1,\min}} \tag{6.24}$$

$$f_2' = \frac{f_2 - f_{2,\min}}{f_{2,\max} - f_{2,\min}} \tag{6.25}$$

$$f_3' = \frac{f_3 - f_{3,\min}}{f_{3,\max} - f_{3,\min}} \tag{6.26}$$

$$f_1 = C_H \frac{M^2}{2Q} + C_s \frac{R^2}{2Q} + C_R \frac{D}{Q} \tag{6.27}$$

$$f_2 = Q \tag{6.28}$$

$$f_3 = R \tag{6.29}$$

$$R = Q - M \tag{6.30}$$

$$Q_{\min} \le Q \le Q_{\max} \tag{6.31}$$

$$R_{\min} \le R \le R_{\max} \tag{6.32}$$

$$Q, M, R \ge 0 \tag{6.33}$$

In order to solve the tricriteria inventory problem by multiobjective optimization, each objective function f_1, f_2, and f_3 must be normalized as presented in constraints (6.24),

(6.25), and (6.26) in Problem 6.3. To normalize each objective function, the minimum and maximum values of each objective function $f_{i,\min}$ and $f_{i,\max}$ for $i = 1, 2, 3$ should be determined. These are presented in the following table.

	Objective	Subject To	Solution $f_1 = TC$	$f_2 = Q$	$f_3 = R$
Problem 1	Min. f_1	Constraints (6.27)–(6.33) of Problem 6.3	$f_{1,\min}$	f_2	f_3
Problem 2	Min. f_2	Constraints (6.27)–(6.33) of Problem 6.3	f_1	$f_{2,\min}$	f_3
Problem 3	Min. f_3	Constraints (6.27)–(6.33) of Problem 6.3	f_1	f_2	$f_{3,\min}$
Maximum objective values			$f_{1,\max}$	$f_{2,\max}$	$f_{3,\max}$
Minimum objective values			$f_{1,\min}$	$f_{2,\min}$	$f_{3,\min}$

Example 6.10   MOO of Inventory with Shortages Allowed   Consider Example 6.4 where demand $D = 5000$ units per year, setup cost $C_R = \$1500$ per setup, shortage cost $C_S = \$150$ per unit per year, and holding cost $C_H = \$200$ per unit per year. Suppose the given lower bound on Q is 300. The weights of importance for the normalized values of the three objective functions are $(w_1, w_2, w_3) = (0.4, 0.3, 0.3)$. Determine the optimal total cost, order quantity, inventory, and shortage for this MOO inventory problem.

Before solving the MOO problem, first we must determine the maximum and minimum values of each objective in order to construct the normalized objective function constraints. The summary of all three problems is presented in the table below.

Consider Problem 1. Minimizing f_1 subject to constraints (6.27)–(6.33) of Problem 6.3:

Minimize: f_1.
Subject to:

$$f_1 = \left(200\frac{M^2}{2Q} + 150\frac{R^2}{2Q} + 1500\frac{5000}{Q}\right) \qquad f_2 = Q \qquad f_3 = R$$
$$R = Q - M \qquad 300 \le Q \le 418.33 \qquad Q, M, R \ge 0$$

See Supplement 6.3.lg4.
The solution is $f_{1,\min} = 35{,}856.86$. Also, record the values obtained for the order quantity $f_2 = Q = 418.33$ and the shortage $f_3 = R = 239$.

The summary of solutions for Problems 1, 2, and 3 are presented in the following table:

	Objective	Subject To	Solution $f_1 = TC$	$f_2 = Q$	$f_3 = R$
Problem 1	Min. f_1	Constraints (6.27)–(6.33) of Problem 6.3	35,857.86	418.33	239.05
Problem 2	Min. f_2	Constraints (6.27)–(6.33) of Problem 6.3	49,733.82	300	28.74
Problem 3	Min. f_3	Constraints (6.27)–(6.33) of Problem 6.3	59,761.43	418.33	0
Maximum objective values			59,761.43	418.33	239.05
Minimum objective values			35,857.86	300	0

See Supplements 6.4-1.lg4, 6.4-2.lg4.

Now, the above known values can be incorporated into Equations (6.24), (6.25), and (6.26) of Problem 6.3 to find the normalizing equations:

$$f_1' = \frac{f_1 - 35{,}857.86}{59{,}761.43 - 35{,}857.86} \quad \text{or} \quad f_1 - 23{,}903.57 f_1' = 35{,}857.86$$

$$f_2' = \frac{f_2 - 300}{418.33 - 300} \quad \text{or} \quad f_2 - 118.33 f_2' = 300$$

$$f_3' = \frac{f_3 - 0}{239.05 - 0} \quad \text{or} \quad f_3 - 239.05 f_3' = 0$$

The MOO of inventory is as follows:

Minimize:

$$U = 0.4 f_1' + 0.3 f_2' + 0.3 f_3'$$

Subject to:

$$f_1 - 23{,}903.57 f_1' = 35857.86$$
$$f_2 - 118.33 f_2' = 300$$
$$f_3 - 239.05 f_3' = 0$$
$$f_1 = \left(200\frac{M^2}{2Q} + 150\frac{R^2}{2Q} + 1500\frac{5000}{Q}\right)$$
$$f_2 = Q; \quad f_3 = R; \quad R = Q - M; \quad 300 \le Q \le 418.33; \quad Q, M, R \ge 0$$

Solving the above optimization problem by a nonlinear programming optimization software (we used LINGO), the solution is

$$f_1 = TC = \$40{,}267.57 \qquad f_2 = Q = 300 \qquad f_3 = R = 107.15$$

See Supplement 6.5.lg4.

Solving Example 6.10 with Limited Shortages In Example 6.10, the ratio of shortage to order quantity is $R/Q = 107.15/300 = 0.3572$, or 35.72%. Suppose that the maximum allowable shortage is 10% of the quantity value. That is,

$$R \le 0.1Q \tag{6.34}$$

Now add the above constraint to the set of constraints in Example 6.10 and solve the problem. The solution is

$$f_1 = TC = \$49{,}525 \qquad f_2 = Q = 300 \qquad f_3 = R = 30$$

See Supplement 6.6.lg4.

6.6 QUANTITY DISCOUNT INVENTORY

6.6.1 EOQ with Quantity Discount

In the EOQ model, it is assumed that the price per unit is constant regardless of the quantity purchased. In reality, producers may offer discounts when selling large quantities. This type of sale is particularly appealing to producers because the setup cost is fixed regardless of the order size. There are several different types of quantity discount methods, but the most popular are the incremental and all-units methods. The incremental method involves discounting the price of the volume of materials after certain break points. The all-units method involves discounting the entire order quantity after break point volumes are met. The all-units method is used more extensively by manufacturing companies and wholesale stores.

As an example of the incremental method, consider an order of up to 1000 units costs $1.00 per unit. Every unit after the 1000th unit and before the 2000th unit costs $0.95 per unit. Each unit over 2000 costs $0.85 per unit. For example, if an order is 2500 units, using the incremental method, the cost would be

$$1000 \times \$1.00 + 1000 \times \$0.95 + 500 \times \$0.85 = \$2375$$

For the same example, using the all-units method, orders up to 1000 units cost $1 per unit. Orders between 1000 and 2000 units cost $0.95 per unit, and orders greater than or equal to 2000 units cost $0.85 per unit. In the all-units method, the cost of an order of 2500 units would be

$$2500 \times \$0.85 = \$2125$$

See Figure 6.16a,b for a graphical representation of these two examples where c_j is the price unit cost for level j discount. The all-units method may be preferred because it is easier to understand and for bookkeeping. In this section, only the all-units method is covered.

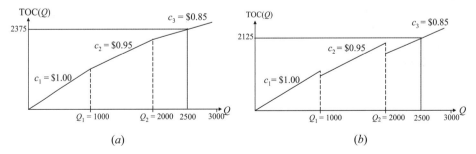

Figure 6.16 (*a*) Incremental method: total order cost as function of order quantity Q. (*b*) All-units method: total order cost as function of order quantity Q.

Because the total inventory cost is a function of the order quantity, each different level of quantity discounts has a different total inventory cost function. The EOQ model can be modified to include quantity discounts. The notation and variables used by the modified EOQ model are shown below. In this model, c_j is the price per unit for the jth level of discount. The decision variable to be found is the optimal order quantity Q^*, where

$$\text{Total cost:} \quad TC_j = \frac{C_H Q}{2} + \frac{C_R D}{Q} + c_j D \quad \text{for given level } j \text{ where } Q_{j-1} \le Q < Q_j$$

(6.35)

$$\text{Optimal order size:} \quad Q_j^* = \sqrt{\frac{2C_R D}{C_H}} \quad \text{for given level } j \text{ where } Q_{j-1} \le Q < Q_j \quad (6.36)$$

The holding cost can be directly given or calculated as

$$C_H = ic_j$$

where i is the approximate interest percentage. For different price levels, the total annual cost can be determined and then the best (lowest) cost solution is selected from the feasible range of the order quantity. For example, Figure 6.16b shows the three levels $0 \le Q_1 < 1000$, $1000 \le Q_2 < 2000$, and $2000 \le Q_3$ where the three price levels are $c_1 = 1$, $c_2 = 0.95$, and $c_3 = 0.85$, respectively. Figure 6.17 shows the total inventory cost for each of three different levels.

To illustrate how to find the optimal order quantity, consider a two-level discount price plan as presented in the following table. The break point, where the price changes from c_1 to c_2, is at quantity Q_1. In the following table, the first order size level ranges from 0 to Q_1 and the second order size level ranges from Q_1 to infinity. The first order size range has a

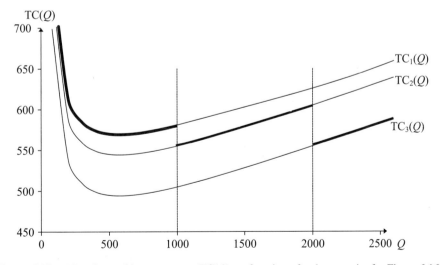

Figure 6.17   All-units total inventory cost TC(Q), as function of order quantity for Figure 6.16b.

unit cost of c_1 and a total cost of TC_1, while the second order size range has a unit cost of c_2 and a total cost of TC_2.

Order Size	Unit cost	Holding Cost, C_H	Total Cost	Optimal Quantity
$0 < Q < Q_1$	c_1	ic_1	TC_1	Q_1^*
$Q_1 \le Q < \infty$	c_2	ic_2	TC_2	Q_2^*

Suppose the optimal EOQ using cost c_1 is Q_1^*, which has a total cost of $TC_1(Q_1^*)$. Also, suppose that the optimal EOQ using cost c_2 is Q_2^*, which has a total cost of $TC_2(Q_2^*)$. The purpose of the algorithm is to find the best feasible quantity Q^* that minimizes the total cost. Suppose $Q_1^* < Q_2^*$. Four possible cases are discussed:

Case (a_1) In this case, both Q_1^* and Q_2^* are greater than the discount break point Q_1, that is, $Q_1 < Q_1^* < Q_2^*$. Therefore, choose $Q^* = Q_2^*$ as the optimal, since it has the lowest feasible total cost (due to a lower unit price c_2). See Figure 6.18a.

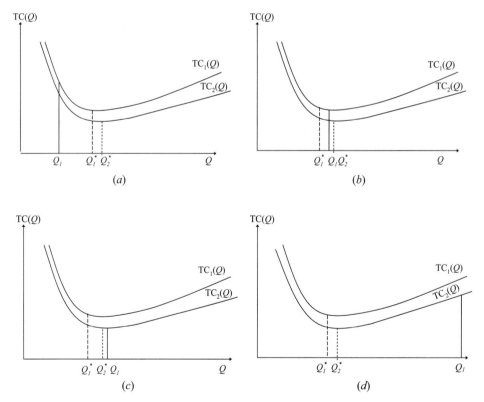

Figure 6.18 (*a*) Graphical representation of case (a_1), $Q_1 < Q_1^* < Q_2^*$; use Q_2^*. (*b*) Graphical representation of case (a_2), $Q_1^* < Q_1 < Q_2^*$; use Q_2^*. (*c*) Graphical representation of case (b_1), $Q_1^* < Q_2^* < Q_1$ and $TC_2(Q_1) \le TC_1(Q_1^*)$; use Q_1. (*d*) Graphical representation of case (b_2), $Q_1^* < Q_2^* < Q_1$ and $TC_1(Q_1^*) \le TC_2(Q_1)$; use Q_1^*.

Case (a₂) In this case, $Q_1^* < Q_1$ and $Q_2^* > Q_1$, that is, $Q_1^* < Q_1 < Q_2^*$. Also due to the lower unit price c_2, since $TC_2(Q_2^*)$ is less than $TC_1(Q_1^*)$, choose $Q^* = Q_2^*$ as the optimal. See Figure 6.18*b*.

Case (b) In this case, both Q_1^* and Q_2^* are less than the break point Q_1, that is, $Q_1^* < Q_2^* < Q_1$.

For this case, there are two possibilities:

Case (b₁) If $TC_2(Q_1) < TC_1(Q_1^*)$, then the optimal point $Q^* = Q_1$ (see Figure 6.18*c*).

Case (b₂) If $TC_2(Q_1) > TC_1(Q_1^*)$, then choose $Q^* = Q_1^*$ as the optimal (Figure 6.18*d*).

Finding EOQ with Discount Quantities for All-Unit Discount

1. Start with the lowest price c_n and compute its EOQ. Continue until a feasible quantity Q_j^* is obtained.
2. Compare the total cost for this feasible quantity with the total cost of each of the break points whose order quantities are greater than or equal to this feasible quantity. The optimal Q is where the lowest total cost occurs.

Example 6.11 EOQ with Two-Level All-Unit Quantity Discount Consider Example 6.1 where the price of parts are discounted as follows:

Discount Number	Discount Quantity, Q	Discount Price, c_j
1	$0 < Q < Q_1$	$c_1 = \$4.50$
2	$Q_1 \le Q < \infty$	$c_2 = \$4.20$

In this example, the average demand is 3120 parts per year. The reorder cost $C_R = \$25$ per order. The holding costs are based on a 22% API (annual percent interest) of the unit price. Find the optimal order quantity for the following three cases:

(a) Suppose the discount quantity is applied when $Q \ge 225$, that is, $Q_1 = 225$.

In this example, $D = 3120$ parts per year, $C_R = \$25$, $i = 0.22$, and $C_H = ic_j$. The optimal quantity and cost for each level j are

$$Q_j^* = \sqrt{\frac{2(25)(3120)}{0.22c_j}}$$

$$TC_j = \frac{25(3120)}{Q_j^*} + \frac{0.22c_j Q_j^*}{2} + 3120c_j$$

Start with the lowest cost which is associated with level 2. The optimal quantity for level 2 is

$$Q_2^* = \sqrt{\frac{2(25)(3120)}{0.22 \times 4.2}} = 411$$

Since $(Q_1 = 225) \leq Q_2^*$, the solution is feasible, and therefore it is optimal: $Q^* = Q_2^* = 411$. The summary is presented below.

Discount Level, j	Discount Quantity, Q_j	Discount Price, c_j	Holding Cost, $0.22c_j$	Optimal Order Quantity, Q_j^*	Feasible?	Total Cost TC_j
2	$225 - \infty$	$4.20	$0.924	411	Yes, stop	$13,484

(b_1) Suppose the price break point is at $Q_1 = 420$. Since $Q_2^* < Q_1$, it is not feasible. Now compute Q_1^*:

$$Q_1^* = \sqrt{\frac{2(25)(3120)}{0.22 \times 4.5}} = 397$$

Since $0 \leq Q_1^* \leq (Q_1 = 420)$, Q_1^* is a feasible quantity. Now, compare the total cost for Q_1 and Q_1^*:

$$TC(Q_1^*) = \frac{25 \times 3120}{397} + \frac{0.22 \times 4.5 \times 397}{2} + 3120 \times 4.5 = \$14,433$$

$$TC(Q_1) = \frac{25 \times 3120}{420} + \frac{0.22 \times 4.2 \times 420}{2} + 3120 \times 4.2 = \$13,484$$

Since $TC(Q_1) < TC(Q_1^*)$, the optimal quantity $Q_1 = 420$.

(b_2) Suppose the price break point is at $Q_1 = 2900$. Since $(Q_2^* = 411) < Q_1$, it is not feasible. Now compute Q_1^*. Since $0 \leq (Q_1^* = 397) \leq (Q_1 = 2900)$, Q_1^* is a feasible quantity. Compare the total cost for Q_1 and Q_1^* using Equation (6.35):

$$TC(Q_1^* = 397) = \frac{25 \times 3120}{397} + \frac{0.22 \times 4.5 \times 397}{2} + 3120 \times 4.5 = \$14,433$$

$$TC(Q_1 = 2900) = \frac{25 \times 3120}{2900} + \frac{0.22 \times 4.2 \times 2900}{2} + 3120 \times 4.2 = \$14,471$$

Since the cost of $Q_1 = 2900$ is more than the cost of $Q_1^* = 397$, the optimal solution is $Q = 397$.

Example 6.12   EOQ with Three-Level Discount   Given the following information:

Demand of product $(D) = 1000$ units per year
Holding cost $(C_H) =$ associated with interest rate of 15% of unit price
Ordering cost $(C_R) = \$25$

The quantity discounts are presented below for three levels.

Order Size	Unit Cost	$C_H = ic_j = 0.15c_j$
$0 < Q < 120$	16	2.4
$120 \leq Q < 200$	12	1.8
$200 \leq Q \leq \infty$	10	1.5

Find the optimal order quantity.

Level 3: Find the EOQ for the lowest price of $10 per unit:

$$Q_3^* = \sqrt{\frac{2(25)(1000)}{1.5}} = 182.57$$

Since $Q_3^* \leq Q_3$, it is not feasible. Therefore, go to the next level of higher cost.

Level 2: For the price of $12 per unit

$$Q_2^* = \sqrt{\frac{2(25)(1000)}{1.8}} = 166.67$$

Since $(Q_2 = 120) \leq Q_2^* < (Q_3 = 200)$, this solution is feasible. Now, compare the total costs for Q_2^* and Q_3:

$$TC(Q_2^* = 166.67) = \frac{1.8(166.67)}{2} + \frac{25(1000)}{166.67} + 12(1000) = \$12,300$$

$$TC(Q_3 = 200) = \frac{1.5(200)}{2} + \frac{25(1000)}{200} + 10(1000) = \$10,275$$

Since $TC(Q_3 = 200) < TC(Q_2^* = 166.7)$, the optimal solution is to order 200 units at a cost of $10 per unit. This solution has a total cost of $10,275, shown in boldface in the following table.

The summary of the above solution is presented in the table below.

Discount Number, j	Discount Quantity, q_j	Discount Price, c_j	Holding Cost, $0.15c_j$	Optimal Order Quantity, Q_j^*	Is $Q_j^* > Q_j$; Feasible?	Total Cost, TC_j
3	$200 - \infty$	$10	$1.50	182.57	No	-
2	$120 - 199$	$12	$1.80	166.67	Yes	$12,300
3	$200 - \infty$	$10	$1.50	200	—	**$10,275**

6.6.2 Bicriteria EOQ with Quantity Discount[†]

The bicriteria quantity discount is similar to the bicriteria EOQ model except that the total cost is calculated differently. Consider the following two objective functions:

Minimize: $f_1 = TC = \dfrac{C_H Q}{2} + \dfrac{C_R D}{Q} + c_j D$ Total cost

Minimize: $f_2 = Q$ Order quantity (to minimize inventory)

This method is based on the method of Section 6.6.1. First, identify the optimal quantity that minimizes the total cost considering the quantity discount problem. Label this optimal quantity Q_{max}. The DM can also identify the minimum acceptable quantity Q_{min}. Then apply the multicriteria method to generate a number of efficient points for Q values ranging from Q_{min} to Q_{max}. For each level lower than Q_{max}, identify the optimal order quantity and the associated costs. Also, consider the break points and find their associated costs. Eliminate all infeasible and inefficient alternatives. Make sure that all break points Q_j are also used as alternatives.

Example 6.13   Efficient Points of EOQ with Quantity Discount   Consider Example 6.11, case (a), where the discount quantity is applied when $Q \geq 225$, that is, $Q_1 = 225$. Suppose that the minimum quantity is 100, that is, $Q_{min} = 100$. Generate a set of multicriteria efficient points.

It was shown in Example 6.11, case (a), that $Q^* = 411$ parts. Therefore, efficient quantities vary from 100 to 411. The three initial quantities are 100, 255, and 411, where the break point $Q_1 = 225$ is also used. Two more order quantities can be generated as the midpoints, that is, $(100 + 255)/2 \approx 177$ and $(255 + 411)/2 \approx 333$. The six quantity values are $Q = 100, 177, 225, 255, 333, 411$. Note that for the 225 break point quantity the lower cost is used because it is feasible. The following table and Figure 6.19 present these efficient alternatives:

Alternatives	a_1	a_2	a_3	a_4	a_5	a_6
Min. order size, $f_2 = Q$	100	177	225	255	333	411
Min. total cost, f_1	14,870	14,568	13,555	13,528	13,492	13,484
Price per unit used	4.50	4.50	4.20	4.20	4.20	4.20
Efficient?	Yes	Yes	Yes	Yes	Yes	Yes

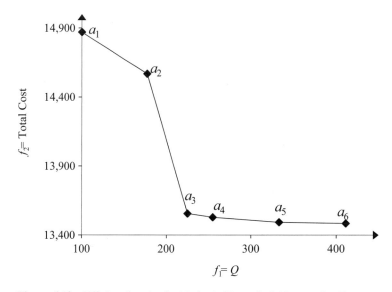

Figure 6.19   Efficient frontier for bicriteria Example 6.13: quantity discount.

All six alternatives in Figure 6.19 are efficient. One may choose a_3 because it is at the elbow of the trade-off function. A more detailed search on the value of Q can be conducted to possibly generate more efficient alternatives.

6.7 MULTI-ITEM INVENTORY

6.7.1 Multi-Item Inventory Optimization

The EOQ model and its extensions that were covered in the previous sections are based on one type of item. In many applications, the inventory planning is based on several items in which several constraints should be considered. In this section, we develop an optimization approach for multi-item inventory while considering different constraints.

PROBLEM 6.4   MULTI-ITEM OPTIMIZATION

Minimize:

$$f_1 = \sum_{i=1}^{n} \left(\frac{C_{R,i} D_i}{Q_i} + \frac{C_{H,i} Q_i}{2} \right) \quad \text{(total inventory cost)} \tag{6.37}$$

Subject to:

$$\sum_{i=1}^{n} a_i Q_i \leq A \quad \text{(space restriction)} \tag{6.38}$$

$$\sum_{i=1}^{n} c_i Q_i \leq B \quad \text{(purchase budget restriction)} \tag{6.39}$$

$$Q_{i,\min} \leq Q_i \leq Q_{i,\max} \quad \text{for } i = 1, \ldots, n \quad \text{(order size restriction)} \tag{6.40}$$

where

$C_{R,i}$ = cost per order for item i

$C_{H,i}$ = cost of holding item i per period

D_i = demand for item i per period

a_i = space required per unit of item i

A = total space available

c_i = purchase price per unit of item i

B = total budget available for purchase

$Q_{i,\min}$ = minimum order size allowed for item i

$Q_{i,\max}$ = maximum order size allowed for item i

The decision variables are Q_i, quantity of item i for $i = 1, 2, \ldots, n$.

In Problem 6.4, the objective function is a summation of n convex functions; therefore, it is a convex function. The set of constraints are all linear. Therefore, optimization (nonlinear programming) can be used to find the optimal solution.

If it is required that the order sizes be integers, then the following constraint should be added to the set of constraints of Problem 6.4:

$$Q_i = \text{integer value for } i = 1, \ldots, n \qquad (6.41)$$

Example 6.14 Minimizing Total Cost for Multi-Item Inventory Planning   Consider an inventory planning problem for three items. Suppose that there are the following limitations for this problem:

Total available space: $A = 2000 \text{ ft}^2$
Total available budget: $B = \$7500$

The cost information and the order quantity restrictions for each item are given below.

Item i	D_i	$C_{R,i}$ ($\$$)	$C_{H,i}$ ($\$$)	a_i (ft^2)	c_i ($\$$)	$Q_{i,\min}$	$Q_{i,\max}$
1	500	20	0.4	5	8	100	200
2	1000	10	0.2	1	10	150	300
3	1200	30	0.3	2	12	200	400

The following optimization problem can be formulated to solve this problem:

Minimize:

$$f_1 = \left(\frac{20(500)}{Q_1} + \frac{0.4(Q_1)}{2}\right) + \left(\frac{10(1000)}{Q_2} + \frac{0.2(Q_2)}{2}\right) + \left(\frac{30(1200)}{Q_3} + \frac{0.3(Q_3)}{2}\right)$$

Subject to:

$$5Q_1 + 1Q_2 + 2Q_3 \leq 2000$$
$$8Q_1 + 10Q_2 + 12Q_3 \leq 7500$$
$$100 \leq Q_1 \leq 200$$
$$150 \leq Q_2 \leq 300$$
$$200 \leq Q_3 \leq 400$$

Solving this problem by an optimization software package (we used LINGO), the solution is

$$Q_1^* = 178.84 \qquad Q_2^* = 203.78 \qquad Q_3^* = 335.95$$

This solution satisfies the 2000-ft^2 space restriction and the total budget of $\$7500$. The total inventory cost for this solution is $f_1 = \$318.685$.

If the order quantities are integers, that is, constraints (6.41) are also used, then the optimal solution to this problem is

$$Q_1^* = 180 \qquad Q_2^* = 204 \qquad Q_3^* = 335$$

See Supplements S6.7.xls, S6.7.lg4.

This solution is feasible and satisfies the 2000-ft^2 space restriction and the total budget of \$7500. The total inventory cost for this integer solution is $f_1 = \$318.688$, which in this example is slightly higher than the noninteger solution.

Sometimes, it is possible to find the optimal EOQ for each item independent from other items (using the EOQ model of Section 6.2). If the optimal solutions for all items satisfy constraints (6.38), (6.39), and (6.40), the EOQ solutions are optimal and feasible, that is, it may not be needed to solve Problem 6.4. However, if the constraints are not feasible with respect to all constraints, Problem 6.4 must be solved. For the purpose of illustration, compare the Example 6.14 optimization solution to the solution of the EOQ model of Section 6.2. The EOQ solution using Equation (6.5) for each item is as follows:

$$Q_1^* = \sqrt{\frac{2(20)(500)}{0.4}} = 224 \quad Q_2^* = \sqrt{\frac{2(10)(1000)}{0.2}} = 316 \quad Q_3^* = \sqrt{\frac{2(30)(1200)}{0.3}} = 490$$

The total inventory cost of this inventory planning is

$$f_1 = \left(\frac{20(500)}{224} + \frac{0.4(224)}{2}\right) + \left(\frac{10(1000)}{316} + \frac{0.2(316)}{2}\right) + \left(\frac{30(1200)}{490} + \frac{0.3(490)}{2}\right)$$

$$= 89.44 + 63.25 + 146.97 = \$299.66$$

As shown in the following table, these optimal solutions are infeasible and violate the restrictions for space, budget, and limits on the order quantity.

Q_i^*	$a_i Q_i^*$ (ft^2)	$c_i Q_i^*$ (\$)
224	1120	1,792
316	316	3,160
490	980	5,880
Total	2416	10,832

6.7.2 Multiobjective Multi-Item Inventory Optimization[†]

When there are inventory constraints, different priorities should be given to different items to reflect their inventory importance. The critical index of item i is denoted by CI_i, which represents its priority. Suppose the critical index for different items can range from 0 to 100 where 0 is the lowest and 100 is the highest priority. That is, it is important to maintain near the maximum level of inventory, $Q_{i,\max}$, for item i, whose priority rating is high.

The total critical index of all items is presented as the objective function (6.42), which is maximized in Problem 6.5. The optimal number of inventory for each item can be found by solving Problem 6.5. In this solution of this problem, the order quantity of items that have a higher critical index will be higher than items that have a lower critical index.

PROBLEM 6.5 SINGLE-OBJECTIVE CRITICAL INDEX OPTIMIZATION PROBLEM

Maximize:

$$f_2 = \sum_{i=1}^{n} \text{CI}_i Q_i \tag{6.42}$$

Subject to: Constraints (6.38)–(6.41)

PROBLEM 6.6 MULTIOBJECTIVE MULTI-ITEM INVENTORY PROBLEM

Minimize: Total inventory cost f_1 (6.37)
Maximize: Total critical index objective f_2 (6.42)
Subject to: Constraints (6.38)–(6.41)

Problem 6.7 can be solved to find the best alternative solution based on an additive utility function, where w_1 and w_2 are the weights of importance for the normalized values of the two objectives. Note that the weights of importance (w_1, w_2) are given values where $w_1 > 0$, $w_2 > 0$, and $w_1 + w_2 = 1$.

PROBLEM 6.7 MULTIOBJECTIVE MULTI-ITEM INVENTORY PROBLEM WITH ADDITIVE UTILITY FUNCTION

Maximize:

$$U = w_1 f_1' + w_2 f_2' \tag{6.43}$$

Subject to:

$$f_1' = -\frac{f_1 - f_{1,\text{max}}}{f_{1,\text{max}} - f_{1,\text{min}}} \tag{6.44}$$

$$f_2' = \frac{f_2 - f_{2,\text{min}}}{f_{2,\text{max}} - f_{2,\text{min}}} \tag{6.45}$$

$$f_1 = \sum_{i=1}^{n} \left(\frac{C_{R,i} D_i}{Q_i} + \frac{C_{H,i} Q_i}{2} \right) \tag{6.37}$$

$$f_2 = \sum_{i=1}^{n} \text{CI}_i Q_i \tag{6.42}$$

Constraints (6.38)–(6.41) of Problem (6.4)

Different efficient multicriteria alternatives can be generated by using different weights (w_1, w_2) in Problem 6.7.

Example 6.15   Multiobjective Multi-Item Inventory Problem   Consider Example 6.14. Suppose the critical index priorities for the three items are as follows:

Item	1	2	3
Critical index	100	5	40

Suppose the weights of importance for the normalized values of the two objectives are (0.6, 0.4).

  (a) Generate the best alternative for the given weights.

  (b) Generate five different efficient alternatives by varying weights.

The solution to this problem follows.

  (a) First find the equation for normalizing each of the two objective functions. To do this, first find the minimum and maximum values of each objective.

  In Example 6.14, the minimum total cost was obtained as $ 318.69, where $Q_1^* = 180$, $Q_2^* = 204$, and $Q_3^* = 335$. For these order quantities, the total critical index f_2 can be calculated using Equation (6.42):

$$f_2 = 100 \times 180 + 5 \times 204 + 40 \times 335 = 32{,}420$$

To find the maximum value of f_2, solve Problem 6.5.

  Maximize: $f_2 = 100Q_1 + 5Q_2 + 40Q_3$
  Subject to: Constraints of Example 6.14 and Problem 6.4

The solution to this problem is $f_2 = 35{,}390$, where $Q_1 = 200$, $Q_2 = 150$, and $Q_3 = 366$. The total cost function using Equation (6.37) for this solution is

$$f_1 \left(\frac{10{,}000}{200} + 0.2 \times 200 + \frac{10{,}000}{150} + 0.1 \times 150 + \frac{36{,}000}{366} + 0.15 \times 366 \right) = 324.93$$

See Supplement S6.8.lg4.

The summary of the solutions for the above two problems is presented in the following table:

	Solution				
	f_1	f_2	Q_1	Q_2	Q_3
Problem 1: Minimize f_1 subject to Example 6.14 constraints	318.69	32,420	180	204	335
Problem 2: Maximize f_2 subject to Example 6.14 constraints	324.93	35,390	200	150	366

The above table shows the range of values for the two objectives. These values are used to find the normalizing objective function equations for the maximization problem. The critical index f_2 is maximized but f_1 is minimized. Therefore, maximize $-f_1$ for normalization purposes. Therefore both normalized values, f_1' and f_2', are maximized:

$$f_1' = -\frac{f_1 - 324.93}{324.93 - 318.69} \qquad f_2' = \frac{f_2 - 32{,}420}{35{,}390 - 32{,}420}$$

Now solve Problem 6.7 to find the optimal solution for the additive utility function:

Maximize: $U = 0.6 f_1' + 0.4 f_2'$

Subject to:

$$f_1' = -\frac{f_1 - 324.93}{324.93 - 318.69} \qquad \text{or} \qquad f_1' = \frac{324.93 - f_1}{6.24}$$

$$f_2' = \frac{f_2 - 32{,}420}{35{,}390 - 32{,}420} \qquad \text{or} \qquad \frac{f_2 - 32{,}420}{2{,}970}$$

where

$$f_1 = -\left(\frac{20(500)}{Q_1} + \frac{0.4(Q_1)}{2}\right) - \left(\frac{10(1000)}{Q_2} + \frac{0.2(Q_2)}{2}\right) - \left(\frac{30(1200)}{Q_3} + \frac{0.3(Q_3)}{2}\right)$$

$f_2 = 100Q_1 + 5Q_2 + 40Q_3$

$5Q_1 + 1Q_2 + 2Q_3 \leq 2000 \qquad 8Q_1 + 10Q_2 + 12Q_3 \leq 7500$

$100 \leq Q_1 \leq 200 \qquad 150 \leq Q_2 \leq 300 \qquad 200 \leq Q_3 \leq 400$

Q_1, Q_2, Q_3 are integers

The solution to the above nonlinear programming (NLP) problem is presented below (we used LINGO to solve this problem).

Q_1	Q_2	Q_3	f_1	f_2	f_1'	f_2'	U
200	188	335	319.70	34,340	0.838	0.647	0.761

See Supplement S6.9.lg4.

(b) To generate five efficient multicriteria alternatives, use a trial-and-error approach by considering different weights and solve Problem 6.7 for each given set of weights. For example, if weights are (0.25, 0.75), then solve the following problem:

Maximize: $U = 0.25 f_1' + 0.75 f_2'$

Subject to: Constraints of Problem 6.7 and Example 6.15

The solution to this problem is presented in the second row of the following table as alternative a_2. Suppose the set of five weights are (0.01, 0.99), (0.25, 0.75), (0.5, 0.5),

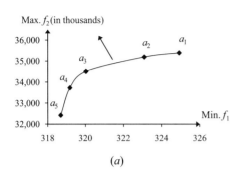

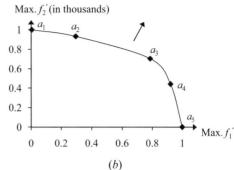

Figure 6.20   Efficient frontier for Example 6.15: (a) actual objective values; (b) normalized objective values.

$(0.75, 0.25)$, and $(0.99, 0.01)$. The optimal solution for each given set of weights is presented below.

Alternatives	Weights (w_1, w_2)	Q_1	Q_2	Q_3	f_1	f_2	f_1'	f_2'
a_1	$(0.01, 0.99)$	200	150	366	324.927	35,390	0.004	1
a_2	$(0.25, 0.75)$	200	158	360	323.091	35,190	0.2947	0.9327
a_3	$(0.5, 0.5)$	200	182	340	320.027	34,510	0.7857	0.7037
a_4	$(0.75, 0.25)$	194	194	334	319.177	33,730	0.9219	0.4411
a_5	$(0.99, 0.01)$	180	204	335	318.688	32,420	1	0

The efficient frontier for this problem is presented in Figure 6.20.

In Figure 6.20a, f_1 is minimized and f_2 is maximized; therefore the efficient frontier is in the northwest direction. In Figure 6.20b, the efficient frontier is in the northeast direction where both normalized objectives are maximized.

6.8   MULTI-ITEM INVENTORY CLASSIFICATION

6.8.1   Cost-Based Multi-Item Classification

Many manufacturing and retail companies have thousands of different types of items in their inventory. Inventory planning and management of a large number of different items are challenging. Classifying different items into groups is an essential aspect of effective inventory planning, control, and management. ABC analysis is a simple and effective inventory classification which is commonly used by management to simplify the inventory policies for a wide variety of inventory items. ABC analysis is based on grouping different types of inventory items into three categories—group A, group B, and group C—based on the percentage of their total dollar value. The classification is based on the Pareto diagram of the nineteenth-century Italian economist Vilfredo Pareto, who observed that for many applications a small portion of the membership has the highest importance. For example, in Italy, 20% of the population owned 80% of the property. Since then, such ratios have

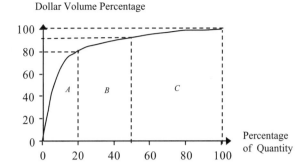

Figure 6.21   An example of ABC classification: group A has the lowest percentage of quantity but has the highest dollar volume percentage.

been widely used to analyze different problems. For example, it is said that 20% of a task force is responsible for 80% of the productivity of a company. In inventory systems, it can also be observed that only about 20% of different inventory items represent about 80% of the total dollar value. By classifying inventory items, more emphasis can be placed on the recordkeeping, planning, and control of the high-value items. The definitions of the three groups are given below:

Group A   This group represents the smallest percentage of the items carried in inventory but has the largest percentage of the annual dollar volume. In practice, it may be about 20% of the total quantity volume but about 80% of the total dollar volume. This group of items should be closely controlled and monitored to minimize the inventory cost.

Group B   This group represents a low-to-mid-range percentage (e.g., 20%–30%) of the quantity of items carried in inventory and accounts for a mid-to-low percentage (e.g., 15%–20%) of the total dollar volume. For this group, an average level of monitoring and controlling inventory is recommended.

Group C   This group represents the highest percentage (e.g., 50%–60%) of the quantity of items carried in inventory but has the lowest percentage in terms of the total dollar volume (e.g., less than 5% of the total inventory cost). For this group, a low level of monitoring and controlling is recommended. It may be cost and time effective to use aggregate ordering and monitoring rules for this group. For this group, one may put the order of many items together, which is known as "joint ordering." Joint ordering reduces paperwork, saves shipping costs, and decreases receiving costs. For example, different sizes of inexpensive nails and screws can be placed into different sizes of containers. Then, the content of each container can be estimated. Also, for these items, large quantities can be ordered to avoid frequent monitoring.

Figure 6.21 illustrates an example of ABC classification.

Procedure for ABC Classification Using Cumulative Cost   The following steps are used to classify items into groups A, B, and C using the cumulative cost of items:

1. For each item, calculate

$$\text{Total cost} = \text{demand } (D_i) \times \text{price per unit } (c_i) \tag{6.46}$$

2. Arrange the list of items in ascending order of total cost of each item.
3. Calculate the cumulative cost starting from the lowest total cost in ascending order.
4. Calculate the percentage of each cumulative cost.
5. Break down items into three (A, B, and C) or more classes of items.

For example, for A, B, and C classification use:

- All inventory items with cumulative cost of less than 5% belong to group C.
- All inventory items with cumulative cost between 5 and 20% belong to group B.
- All inventory items with cumulative cost more than 20% belong to group A.

Example 6.16 ABC Classification Using Cumulative Cost   Consider the following table, which provides information on demand D_i and unit price c_i for item i, where there are 12 items. Group these items into A, B, and C categories.

The ABC analysis steps are performed as follows. Calculate the total cost of each item by multiplying its demand by its unit price as shown in the table below.

Item Name	Demand, D_i	Unit Price, c_i	Total Cost $= D_i c_i$	Rank
S	125	500	62,500	11
XX	30	25	750	3
Y	100	18	1,800	5
U	500	40	20,000	9
V	4,800	2	9,600	8
Z	1,000	1	1,000	4
X	120	25	3,000	6
T	100	310	31,000	10
ZZ	1,000	0.15	150	1
R	2,100	158	331,800	12
YY	5,000	0.12	600	2
W	250	20	5,000	7
Total	15,125	—	467,200	

Now, sort the items in ascending order of their total cost. This is shown in the following table. Then, calculate the cumulative total cost for each item. Now, the cumulative cost percentage for each item can be calculated by dividing its cumulative cost to the total cumulative cost of all items. For example, for item T, the cumulative cost percentage is

$$\frac{72,900}{467,200} = 0.156 \quad (\text{or } 15.6\%)$$

The table below shows the results of the ABC inventory classification.

Item Name	Demand (D_i)	Unit Price (c_i)	Total Cost, TC_i	Cumulative Total Cost	Percent of Cumulative Cost	Group
ZZ	1,000	0.15	150	150	0.03	C
YY	5,000	0.12	600	750	0.16	C
XX	30	25	750	1,500	0.32	C
Z	1,000	1	1,000	2,500	0.54	C
Y	100	18	1,800	4,300	0.92	C
X	120	25	3,000	7,300	1.56	C
W	250	20	5,000	12,300	2.63	C
V	4,800	2	9,600	21,900	4.69	C
U	500	40	20,000	41,900	8.97	B
T	100	310	31,000	72,900	15.60	B
S	125	500	62,500	135,400	28.98	A
R	2,100	158	331,800	467,200	100.00	A
Total	15,125	—	467,200	—	—	—

From the above table, it can be observed that there are only a few items that are critical in terms of the total cost. For example, items R, S, T, and U account for 95% of the total cost. However, in terms of quantity, the majority of the items are in group C.

See Supplement S6.10.xls.

6.8.2 The Critical Index Multi-Item Classification[†]

The critical index (CI) of each item is a subjective rating that represents its qualitative and strategic priority. The CI is based on noncost factors. For simplicity, critical ratings are from 0 to 100, where 0 is the lowest importance and 100 is the highest importance. The ABC classification based on the CI can be conducted using the same procedure of the last section but using the CI instead of the cost of each item. In this procedure, for each item, find

$$\text{Total CI}_i = D_i \text{CI}_i$$

where D_i and CI_i are the demand and the critical index for item i, respectively.

Example 6.17 ABC Classification by Critical Index   Consider Example 6.16. The critical index for each item is given in the column labeled CI_i in the following table. Classify the items using their critical indices.

The method for ABC classification using the critical index is identical to the ABC classification based on cost of Section 6.8.1. The final result of classifying items (after reordering all the rows in ascending order of total critical index) is given in the table below.

Item Name	Demand (D_i)	CI_i	Total CI_i	Cumulative CI_i	Percent	Group
ZZ	1,000	0.2	200	200	0.31	C
XX	30	7	210	410	0.63	C
Y	100	4	400	810	1.25	C
T	100	5	500	1,310	2.02	C
X	120	9	1,080	2,390	3.69	C
W	250	7	1,750	4,140	6.39	B
U	500	4	2,000	6,140	9.48	B
YY	5,000	0.6	3,000	9,140	14.12	B
R	2,100	3	6,300	15,440	23.85	A
Z	1,000	8	8,000	23,440	36.21	A
S	125	100	12,500	35,940	55.51	A
V	4,800	6	28,800	64,740	100.00	A
Total	15,125	—	64,740	—	—	—

Classification of items based on the cumulative critical index is different than the one based on the cumulative total cost presented in Example 6.16.

See Supplement S6.11.xls.

6.8.3 Multicriteria Inventory Classification[†]

In multicriteria inventory classification, we consider the two objectives covered in the previous two sections. In general, the higher the cost or the higher the critical index, the higher the priority given to the item in terms of its classification.

The ABC multicriteria classification problem is:

Classify items based on total cost, TC.

Classify items based on total critical index, TCI.

That is, items are classified based on two criteria: cost and critical index. In this method, for each item, a composite multicriteria utility index (in terms of f_1 and f_2) is calculated. The composite utility indexes are used to classify all items.

This method is as follows:

1. For each item, normalize its two criteria from 0 to 1 using

$$TC_i' = \frac{C_i - C_{i,\min}}{C_{i,\max} - C_{i,\min}} \tag{6.47}$$

$$TCI_i' = \frac{CI_i - CI_{i,\min}}{CI_{i,\max} - CI_{i,i,\min}} \tag{6.48}$$

2. Assess the weights of importance (w_1, w_2) for the normalized values of the two objectives.

3. Generate the composite utility index for each item, $U_i = w_1(TC_i') + w_2(TCI_i')$.

4. Rank the items in ascending order of their U_i values.
5. Find the cumulative value for each item in ascending order of U_i values.
6. Find the percentage of the cumulative values of each item.
7. Apply the ABC classifying method to classify each item based on its percentage of importance.

Example 6.18   Multiple-Criteria ABC Classification   Consider the following data which were presented in Examples 6.16 and 6.17. Suppose the weights of importance for two normalized criteria values are (0.6, 0.4). Classify items by the multicriteria approach.

The following table shows the normalized values for total cost, total CI, and composite utility value for each of the items.

Items are ranked from the lowest to the highest in terms of their cumulative utility value.

Item	Total Cost, TC_i	Total CI, TCI_i	TC'_i	TCI'_i	$U_i = 0.6(TC'_i) + 0.4(TCI'_i)$	Rank
R	2,100	6,300	0.4340	0.2133	0.3457	11
S	1,250	12,500	0.2558	0.4301	0.3255	10
T	100	500	0.0147	0.0105	0.0130	4
U	500	2,000	0.0985	0.0629	0.0843	7
V	4,800	28,800	1.0000	1.0000	1.0000	12
W	250	1,750	0.0461	0.0542	0.0494	6
X	120	1,080	0.0189	0.0308	0.0236	5
Y	100	400	0.0147	0.0070	0.0116	3
Z	1,000	8,000	0.2034	0.2727	0.2311	9
XX	30	210	0.0000	0.0003	0.0001	1
YY	500	3,000	0.0985	0.0979	0.0983	8
ZZ	100	200	0.0147	0.0000	0.0088	2

Now, rearrange the rows in ascending order of their U_i values. Then, find the cumulative U_i values and their percentages for each item i. Classify items into A, B, and C groups.

Item	U_i	Rank	Cumulative U_i	Percent	Multicriteria Group
XX	0.0001	1	0.0001	0.01	C
ZZ	0.0088	2	0.0089	0.41	C
Y	0.0116	3	0.0205	0.94	C
T	0.0130	4	0.0335	1.53	C
X	0.0236	5	0.0572	2.61	C
W	0.0494	6	0.1065	4.86	C
U	0.0843	7	0.1908	8.71	B
YY	0.0983	8	0.2891	13.19	B
Z	0.2311	9	0.5202	23.74	A
S	0.3255	10	0.8457	38.59	A
R	0.3457	11	1.1914	54.37	A
V	1.0000	12	2.1914	100.00	A
Total	2.1914	—	100.00%	—	—

The results of the multicriteria ABC classification are presented in the table above. The resulting classifications are different than the ones obtained by only considering total cost (Example 6.16) or only considering the critical index (Example 6.17).

See Supplement S6.12.xls.

6.9 PROBABILISTIC INVENTORY AND SAFETY STOCK

In the inventory models that were covered in the previous sections, it was assumed that all inventory data are known with certainty. In particular, demand rates were assumed to be constant and known with certainty. In this section, probabilistic inventory planning is considered where both demand rate and lead time are not known with certainty; see Figure 6.2. In the EOQ, the reorder point was defined as

$$R_{\text{op}} = DL$$

In practice, demand rates D and lead times L are rarely constant. For example, suppose that the average demand rate $D = 100$ units per week, and the average lead time $L = 3$ weeks. Also, demand may vary from 80 to 120 units per week, and the lead time may vary from two to four weeks. For this example, three possible options for reorder point are

Average: $R_{\text{op}} = 100 \times 3 = 300$
Lower bound: $R_{\text{op}} = 80 \times 2 = 160$
Upper bound: $R_{\text{op}} = 120 \times 4 = 480$

Using the average value of 300 may be too risky or too conservative depending on the DM's risk attitude and the probability distributions associated with D and L. The approach for solving this problem is based on identifying the probability distributions associated with both D and L and also to consider the risk-taking attitude of the DM.

Suppose $\bar{D}$ is the average demand rate per period and $\bar{L}$ is the average lead time per period. The reorder point for probabilistic inventory planning is

$$R_{\text{op}} = \bar{D}\bar{L} + S_S \tag{6.49}$$

where S_S is the safety stock. The safety stock is used as a precaution against the variability in demand and lead time. The larger the value of the safety stock, the more risk averse is the DM. In the next section, we show how to choose the amount of safety stock using the multicriteria formulation for this problem.

6.9.1 Basic Model of Probabilistic Inventory

To find the amount of safety stock, it is necessary to know the standard deviation for both the lead time and the demand. Using basic statistics, it can be shown that the combined standard deviation of both demand and lead time during the reorder period is defined as

$$\sigma_{DL} = \sqrt{\bar{L}\sigma_D^2 + \bar{D}^2\sigma_L^2} \tag{6.50}$$

TABLE 6.1   Seven Values of z and Their Associated Probabilities

z	0	0.5	1	1.5	2	2.5	3
Confidence level	50.00%	69.15%	84.13%	93.32%	97.72%	99.38%	99.87%

Equation (6.49) can be represented as

$$R_{op} = \bar{D}\bar{L} + z\sigma_{DL} \tag{6.51}$$

where

$\bar{D}$ = average demand per period
$\bar{L}$ = average lead time per period
σ_D = standard deviation of demand per period
σ_L = standard deviation of lead time per period
z = number of standard deviations

In practice, one chooses a z value that provides the needed security for not having shortages. Note that the higher the z value, the lower the stock-out possibility. But, higher values of z are associated with higher inventory costs because a higher amount of safety stock must be kept in the inventory.

The probabilities associated with different values of z for normal distribution are given at the back of this book. A few of these z values are presented in Table 6.1 and some are shown in Figure 6.22.

Example 6.19   Probabilistic Demand and Lead Time   For a product, the average lead time $\bar{L} = 3$ weeks and the average weekly demand $\bar{D} = 100$ units. For this product, the standard deviation for demand is $\sigma_D = 10$ units, and the standard deviation for the lead time is $\sigma_L = 0.5$ weeks. The required confidence that stock-out will not occur is 99.38%. Find the safety stock for this example.

First, find the standard deviation for the combined demand and lead time:

$$\sigma_{DL} = \sqrt{\bar{L}\sigma_D^2 + \bar{D}^2\sigma_L^2} = \sqrt{3(10)^2 + 100^2(0.5)^2} = 52.92$$

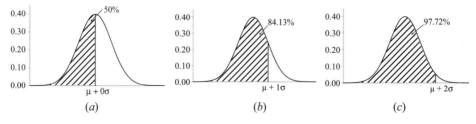

Figure 6.22   Confidence levels for normal distribution for $z = 0, 1, 2$: (a) $X \leq \mu + 0\sigma$; (b) $X \leq \mu + 1\sigma$; (c) $X \leq \mu + 2\sigma$.

The z value associated with 99.38% is 2.5. Therefore, the reorder point associated with a 99.38% confidence level is

$$R_{\text{op}} = \bar{D}\bar{L} + z\sigma_{DL} = 100(3) + 2.5(52.92) = 300 + 132.29 = 432.29$$

That is, 132.29 items (as a safety stock) in addition to the average order size of 300 should be ordered to hedge against variations in the demand rate and lead time. The selection of $z = 2.5$ means that there is only a 0.62 % (i.e., less than 1%) chance of being out of stock.

6.9.2 Multicriteria Probabilistic Inventory[†]

In probabilistic inventory, the amount of safety stock depends on the required confidence level that stock-out will not happen. But different confidence levels are associated with different costs. In multicriteria probabilistic inventory, the DM is provided with a number of choices from which the best alternative can be selected. Consider the following two objectives:

Minimize $f_1 = C_H S_s$ Cost of safety stock per period
Maximize $f_2 =$ probability of having stock (using z)   Confidence in having stock

The probability of having stock is calculated based on the value of z. Using different values of z, a set of efficient bicriteria alternatives can be generated for this multicriteria problem. To simplify the procedure, consider using the seven different values of z given in Table 6.1. Then calculate the safety stock:

$$S_s = z\sigma_{DL}$$

The cost associated with the obtained safety stock can be calculated.

The multicriteria model developed in this section is a risk-averse (conservative) case because only positive safety stocks, that is, $z\sigma_{DL}$, are considered.

Example 6.20 Multicriteria for Selection of Safety Stock   Consider Example 6.19. Suppose the holding cost of inventory per item per period is \$20. Recall that $\sigma_{DL} = 52.92$ in this example. Generate seven different efficient alternatives for this problem.

Consider the seven values of z presented in Table 6.1; the values of the two objectives are shown in the following table. All alternatives are efficient. Now, based on the values of the two objectives f_1 and f_2, the DM can choose the best alternative using a multicriteria method. A ranking by the DM is shown below.

Alternatives	a_1	a_2	a_3	a_4	a_5	a_6	a_7
z	0	0.5	1	1.5	2	2.5	3
$f_2 =$ Confidence level of not having stock-out (%)	50.00	69.15	84.13	93.32	97.72	99.38	99.87
Safety stock $= z\sigma_{DL}$	0	26.46	52.92	79.38	105.84	132.30	158.76
$f_1 =$ Cost (\$) $= 20S_S$	0	529.2	1058.4	1587.6	2116.8	2646	3175.2
Rank by DM	3	1	2	4	5	6	7

6.10 SINGLE-PERIOD MODEL: PERISHABLE

In inventory planning, the assumption is that extra inventory at the end of each period can be used for the next period. That is, by waiting, the accumulated inventory will eventually be used. This assumption is not true for a class of inventory problems labeled "perishable" inventory. The perishable items have a life expectancy and cannot be kept for future use, but in some cases, they may have a salvage value. Examples of perishable items are as follows:

- Newspapers/magazines
- Textbooks
- Perishable food
- Meat/fish
- Flowers
- Calendars

The inventory planning of perishable items is different than nonperishable items, which were covered in previous sections. In perishable inventory planning, there is no reorder point and all items at the end of the period are disposed of or sold at a discounted (salvage) value. Therefore, perishable inventory planning is designed only for one period. Three cases can occur at the end of a period:

(a) There is some excess inventory (loss of income due to excess inventory).
(b) There is some shortage (loss of income due to not having items to sell).
(c) There is no excess inventory or shortage (no loss of income).

The decision problem is to develop a policy that maximizes the potential gains and minimizes the potential losses. To solve this problem, two different costs are defined:

$$C_E = \text{excess inventory cost per unit} = \text{item cost} + \text{disposal cost} - \text{salvage value}$$
$$C_S = \text{shortage cost per unit} = \text{selling price} - \text{item cost}$$

To illustrate this, suppose that a new edition of a textbook sells at $250 per book, the salvage value is $60, the disposal cost is $0, and the cost of publishing the book is $100. Therefore,

$$C_E = (100 + 0 - 60) = \$40 \text{ per book} \quad \text{(excess inventory cost)}$$
$$C_S = (250 - 100) = \$150 \text{ per book} \quad \text{(shortage inventory cost)}$$

If the publisher underestimates the demand and does not print a sufficient number of books, the loss is $150 per book. On the other hand, if the publisher overestimates the demand, the loss is $40 per book. To solve this perishable inventory problem, the following two definitions are used:

- Target service ratio: ratio of shortage cost to total of shortage and excess cost
- Target inventory level: given target inventory level for target service ratio

Total Cost

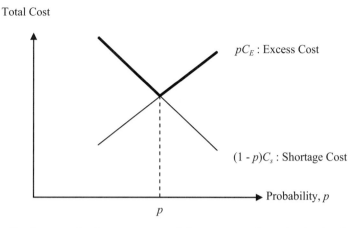

pC_E : Excess Cost

$(1 - p)C_s$: Shortage Cost

Probability, p

p

Figure 6.23 Break-even point for excess cost and shortage cost. The optimal point also minimizes the total cost.

In the following section, it is first shown how to determine the inventory level associated with a target service ratio. For the target inventory level, consider two different cases will be considered: continuous-demand probability and discrete-demand probability.

6.10.1 Target Service Ratio

The target service ratio is the break-even point that can be determined by solving the equation

$$\text{Expected excess cost} = \text{expected shortage cost}$$

Suppose that p is the probability that there is excess inventory at the end of the period and $1 - p$ is the probability that there is a shortage at the end of the period. The expected values for excess and shortage costs are pC_E and $(1 - p)C_S$, respectively. The break-even point, that is, the optimal service ratio, can be found by solving Equation (6.52); also see Figure 6.23. Note that either excess cost or shortage cost can occur; therefore, in Figure 6.23, the total cost is presented by the thick line:

$$pC_E = (1 - p)C_S \qquad \text{or} \qquad p = \frac{C_S}{C_E + C_S} \qquad (6.52)$$

Example 6.21 Finding Optimal Service Ratio for Perishable Inventory   Consider a publisher of a textbook. Find the target service ratio for each of the following three cases:

(a) The excess inventory cost is \$40 and the shortage inventory cost is \$150 per book. The target service ratio is

$$p = \frac{C_S}{C_E + C_S} = \frac{150}{40 + 150} = \frac{150}{190} = 0.79 \approx 0.8 \quad \text{(or 80\%)}$$

This means that the publisher should print enough books to meet the demand 80% of the time. In other words, the publisher should plan to be out of books $100 - 80 = 20\%$ of the time.

(b) Both the excess cost and the shortage cost are $150. The target service ratio is

$$p = \frac{150}{150 + 150} = \frac{150}{300} = 0.5 \quad \text{(or 50\%)}$$

This means that 50% of the time the publisher should meet the demand, that is, 50% of the time, the demand is not met. This makes sense, as the marginal benefit of $150 is equal to the marginal loss of $150.

(c) The excess cost is $350 and the shortage cost is $150:

$$p = \frac{150}{350 + 150} = \frac{150}{500} = 0.3 \quad \text{(or 30\%)}$$

In this case, the publisher should plan to meet the demand only 30% of the time and not to meet the demand 70% of the time. In this case, the excess cost exceeds the shortage cost. Therefore, shortages are preferred to the loss from excess.

6.10.2 Target Inventory with Continuous Probabilities

In the last section, it was shown how to find the desired target service ratio. In the following, it is shown how to find the optimal order quantity to satisfy the probabilistic demand using the given target service ratio. The probabilistic demand can be continuous or discrete:

(a) In the continuous case, demand is represented by a normal probability distribution, where the average μ and standard deviation σ are given. This is covered in this section.

(b) In the discrete case, demand is presented by a discrete probability distribution where probabilities for discrete-demand values are given. This is covered in the next section.

Consider the continuous case, where demand is presented by a normal distribution. The total order quantity, that is, target inventory, is calculated by satisfying the given target service ratio. To do this, find the z value associated with the target service ratio for a normal distribution. Then, for the given mean μ, the given standard deviation σ, and the obtained z, find the target order size using the equation

$$\text{Target inventory} = \mu + z\sigma$$

where $z\sigma$ is the safety stock. Note that it is possible that z will be negative, as presented in case (c) above. In this case there is no safety stock and there will be shortages. An example of the relationship between the target service ratio and the target inventory level to the total cost is presented in Example 6.22.

Example 6.22 Finding Target Inventory with Continuous Probability   Consider the publisher of a book, where the excess inventory cost is $40 and the shortage inventory

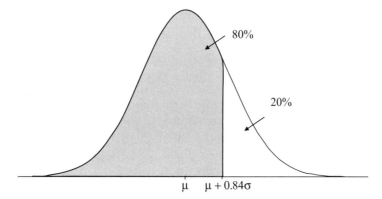

Figure 6.24   Normal distribution where 80% of demand is satisfied.

cost is $150 per book. In Example 6.21, case (a), the service ratio for this problem was calculated. It was 80%. The 80% service target value is associated with a z value of 0.84. (This number can be obtained from the normal distribution table presented at the back of the book.) The normal distribution for this example is presented in Figure 6.24. For different mean values, the target inventory level can be calculated.

Suppose that the publisher can print books in the fall, spring, and summer semesters. The following information is available based on historical data:

Period of Year	Mean Demand, μ	Standard Deviation of Demand, σ
Fall	850	100
Spring	700	80
Summer	100	20

To satisfy the target service ratio of 80%, the publisher should publish the following number of books for each semester. Note that $z\sigma$ represents the safety stock.

Fall $\mu + z\sigma = 850 + 0.84(100) = 934$ Where $0.84 \times 100 = 84$ is targeted safety stock
Spring $\mu + z\sigma = 700 + 0.84(80) \approx 767$ Where $0.84 \times 80 = 67.2$ is targeted safety stock
Summer $\mu + z\sigma = 100 + 0.84(20) \approx 117$ Where $0.84 \times 20 = 16.8$ is targeted safety stock

6.10.3   Target Inventory with Discrete Probabilities[†]

The procedure to find the target inventory when there are probabilistic discrete-demand points is as follows. First, order the data in ascending order of the demand values. Then find the cumulative percentage for each discrete-demand value. Finally, identify the minimum cumulative level that matches the needed target service level.

The details of the method are presented in the following example.

Example 6.23   Finding Target Inventory with Discrete Probability   Consider a textbook publisher which has the following weekly sales (demand) history. How many books should the publisher publish to satisfy the target service of 80%?

The calculations are presented in the following table. The answer is shown in boldface.

Weekly Demand	No. of Weeks	Percentage of Weeks with this Demand	Cumulative Probability
100	0	$0/30 \times 100\% = 0$	0
200	2	$2/30 \times 100\% = 6.7\%$	$0 + 0.067 = 0.067$
300	4	$4/30 \times 100\% = 13.3\%$	$0.067 + 0.133 = 0.200$
400	4	$4/30 \times 100\% = 13.3\%$	$0.200 + 0.133 = 0.333$
500	5	$5/30 \times 100\% = 16.7\%$	$0.333 + 0.167 = 0.500$
600	7	$7/30 \times 100\% = 23.3\%$	$0.500 + 0.233 = 0.733$
700	4	$4/30 \times 100\% = 13.3\%$	$0.733 + 0.133 = \mathbf{0.866}$
800	3	$3/30 \times 100\% = 10.0\%$	$0.866 + 0.100 = 0.966$
900	1	$1/30 \times 100\% = 3.3\%$	$0.966 + 0.033 = 0.999$
1000 or more	0	$0/30 \times 100\% = 0\%$	1
Total	30	$30/30 \times 100\% = 1\%$	

The target service is 80%. In the above table, the weekly demand of 700 is associated with 0.86, or 86%, and is the lowest demand over the 80% threshold. To satisfy the target service of 80%, 700 units should be ordered.

6.10.4   Multicriteria of Perishable Inventory[†]

Continuous Probabilities   Consider the following two objectives:

Minimize $f_1 = $ cost of holding additional safety stock (producer's preference).

Maximize $f_2 = $ probability of not having a shortage, p (consumer's preference, also called the confidence level).

Figure 6.25 illustrates the basic relationship between the two objectives.

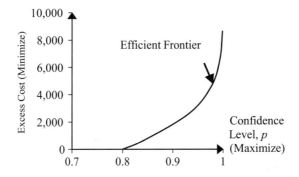

Figure 6.25   Bicriteria for excess cost and probability of not having shortages with continuous probability.

Additional safety stock is the number of items kept in addition to $\mu + z_{min}\sigma$, where z_{min} is the z value associated with the target service rate p^*. By using z values that are larger than z_{min}, additional (excess) safety stock ΔS_S is considered:

$$\Delta S_S = (\mu + z\sigma) - (\mu + z_{min}\sigma) = \sigma(z - z_{min}) \tag{6.53}$$

Now, consider the cost of each excess unit, C_E. The cost of holding additional safety stock is equal to the excess cost per unit multiplied by the additional amount of safety stock:

$$f_1 = C_E \Delta S_s \quad \text{(excess safety stock cost)} \tag{6.54}$$

The second objective is the probability of not having a shortage, which is associated with a given z.

A set of efficient multicriteria alternatives can be generated by varying the value of z, ranging from the minimum service level z_{min} to the highest value, for example, $z = 3$. Then, a multicriteria method can be used to choose the best alternative that satisfies the DM's preferences. In the following example, we will demonstrate the approach for continuous probabilities. Then, the same procedure is applied to discrete probabilities.

Example 6.24 Multicriteria of Perishable Inventory with Continuous Probability   Consider Example 6.22, where $C_E = \$40$, $C_S = \$150$, and the target service level is about 80%. Consider the fall semester (where the mean is 850 and the standard deviation is 100). Generate a set of bicriteria alternatives whose z values are 0.84, 1, 1.5, 2, 2.5, and 3.

In this problem, the target service level is about 80% with a z value of 0.84. Therefore, $z_{min} = 0.84$. The z values that will be used are 0.84, 1, 1.5, 2, 2.5, and 3. Note that only z values higher than $z_{min} = 0.84$ are considered. The set of alternatives is presented in the following table and Figure 6.25.

Alternatives	a_1	a_2	a_3	a_4	a_5	a_6
z	0.84	1	1.5	2	2.5	3
Max. f_2, confidence level, p	0.8000	0.8413	0.9332	0.9772	0.9938	0.9987
Safety stock, $\Delta S_S = \sigma\,(z = 0.84)$	0	16	66	116	166	216
Min. f_1, safety stock cost	0	640	2640	4640	6640	8640
Ranking by DM	2	1	3	4	5	6

See Supplement S6.13.xls.

For example, compare alternatives a_2 and a_3. By moving from $z = 1$ to $z = 1.5$, the confidence level increases from 0.8413 to 0.9332, and cost increases from $640 to $2640. At the very extreme, the confidence level of 0.9987 costs $8,640. A ranking of alternatives by a hypothetical DM is given in the above table.

Figure 6.25 shows the efficient points (frontier) for this bicriteria problem.

Discrete Probabilities   This method is similar to the target inventory with continuous probabilities. The only difference is that each demand level is considered as an alternative. The demand level starts at the target demand level obtained by minimizing the cost for a given target service level. This demand is denoted by D_{min}. Then, all demands higher than D_{min} are considered along with their probabilities. Then, the cost of additional safety stock is calculated. Once a set of alternatives is generated, the best alternative can be selected.

Example 6.25   Multicriteria of Perishable Inventory with Discrete Probability
Consider Example 6.23. Suppose the excess inventory cost $C_E = \$40$ and the target service level is 80%. Generate efficient alternatives for this example.

From Example 6.23, the target demand level is 700; therefore set $D_{min} = 700$. The demand levels to be considered are 700, 800, 900, and 1000.

Alternatives	a_1	a_2	a_3	a_4
Weekly demand	700	800	900	1,000
Excess safety stock, $\Delta S_S = (z\sigma - 0.84\sigma)$	0	100	200	300
Min. f_1, excess safety stock cost: 40 ΔS_s	0	\$4,000	\$8,000	\$12,000
Max. f_2, confidence level, p	0.84	0.97	1	1
Efficient?	Yes	Yes	Yes	No

The alternative associated with the 300 excess safety stock is dominated by the one with 200 excess safety stock. There are three efficient safety stock alternatives of 0, 100, and 200, from which one can be chosen as the best.

REFERENCES

General

Arrow, K. A., T. E. Harris, and T. Marschak. "Optimal Inventory Policy." *Econometrica*, Vol. 19, No. 3, 1951, pp. 250–72.

Barbosa, L. C., and M. Friedman. "Deterministic Inventory Lot Size Models—A General Root-Law." *Management Science*, Vol. 23, 1978, pp. 820–829.

Cohen, M. A., and D. Pekelman. "LIFO Inventory Systems." *Management Science*, Vol. 24, No. 11, 1978, pp. 1150–1162.

Elsayed, E. A., and C. Teresi. "Analysis of Inventory Systems with Deteriorating Items." *International Journal of Production Research*, Vol. 21, No. 4, 1983, pp. 449–460.

Fetter, R. B., and W. C. Dalleck. *Decision Models for Inventory Management*. New York: McGraw-Hill/Irwin, 1961.

Hadley, G. J., and T. M. Whitin. *Analysis of Inventory Systems*. Englewood Cliffs, NJ: Prentice Hall, 1963.

Johnson, L. A., and D. C. Montgomery. *Operations Research in Production Planning, Scheduling and Inventory Control*. New York: Wiley, 1974.

Korhonen, K., and T. Pirttilä. "Cross-Functional Decision-Making in Improving Inventory Management Decision Procedures." *International Journal of Production Economics*, Vols. 81–82, 2003, pp. 195–203.

Larson, C. E., L. J. Olson, and S. Sharma. "Optimal Inventory Policies When the Demand Distribution Is Not Known." *Journal of Economic Theory*, Vol. 101, No. 1, 2001, pp. 281–300.

Lau, H.-S., and A. Zaki. "The Sensitivity of Inventory Decisions to the Shape of the Lead Time Distributions." *IIE Transactions*, Vol. 14, No. 4, 1982, pp. 265–271.

Lovell, M. C. "Optimal Lot Size, Inventories, Prices and JIT under Monopolistic Competition." *International Journal of Production Economics*, Vols. 81–82, 2003, pp. 59–66.

Magee, J. F., and D. M. Boodman. *Production Planning and Inventory Control*, 2nd ed. New York: McGraw-Hill, 1967.

Nahmias, S. *Production and Operation Analysis*, 6th ed. New York: McGraw-Hill, 2008.

Parsons, J. A. "Multi-Product Lot Size Determination When Certain Restrictions Are Active." *Journal of Industrial Engineering*, Vol. 17, 1966, pp. 360–365.

Silver, E. A., and R. Peterson, *Decision Systems for Inventory Management and Production Planning*, 2nd ed. New York: Wiley, 1985.

Starr, M. K., and D. W. Miller. *Inventory Control: Theory and Practice*. Englewood Cliffs, NJ: Prentice Hall, 1962.

Stevenson, W. J. *Operations Management*, 10th ed. New Tork: McGraw-Hill, 2009.

Wagner, H. M. *Statistical Management of Inventory Systems*. New York: Wiley, 1962.

Whitin, T. M. *The Theory of Inventory Management*, revised ed. Princeton, NJ: Princeton University Press, 1957.

Multiobjective Inventory

Agrell, P. J. "A Multicriteria Framework for Inventory Control." *International Journal of Production Economics*, Vol. 41, Nos. 1–3, 1995, pp. 59–70.

Chen, Y., K. W. Li, D. M. Kilgour, and K. W. Hipel. "A Case-Based Distance Model for Multiple Criteria ABC Analysis." *Computers and Operations Research*, Vol. 35, No. 3, 2008, pp. 776–796.

Cheng, L., S. Eswaran, and A. W. Westerberg. "Multi-Objective Decisions on Capacity Planning and Production—Inventory Control under Uncertainty." *Industrial & Engineering Chemistry Research*, Vol. 43, No. 9, 2004, pp. 2192–2208.

Flores, B. E., and D. C. Whybark. "Multiple Criteria ABC Analysis." *International Journal of Operations and Production Management*, Vol. 6, No. 3, 1986, pp. 38–46.

Guvenir, H. A., and E. Erdal. "Multicriteria Inventory Classification Using a Genetic Algorithm." *European Journal of Operational Research*, Vol. 105, No. 1, 1998, pp. 29–37.

Lenard, J. D., and B. Roy. "Multi-Item Inventory Control: A Multicriteria View." *European Journal of Operational Research*, Vol. 87, No. 3, 1995, pp. 685–692.

Mahapatra, N. K., and M. Maiti. "Multi-Objective Inventory Models of Multi-Items with Quality and Stock-Dependent Demand and Stochastic Deterioration." *Advanced Modeling and Optimization (AMO)*, Vol. 7, No. 1, 2005, pp. 69–84.

Malakooti, B. "*Multiple Criteria of Inventory Planning Models Optimization*." Case Western Reserve University, Cleveland, OH, 2013.

Moslemi, H., and M. Zandieh. "Comparisons of Some Improving Strategies on MOPSO for Multi-objective (r, Q) Inventory System." *Expert Systems with Applications*, Vol. 38, No. 10, 2011, pp. 12051–12057.

Padmanabhan, G., and P. Vrat. "Analysis of Multi-Item Inventory Systems under Resource Constraints: A Non-Linear Goal Programming Approach." *Engineering Cost and Production Economics*, Vol. 20, 1990, pp. 121–127.

Puerto, J., and F. R. Fernández. "Pareto-Optimality in Classical Inventory Problems." *Naval Research Logistics*, Vol. 45, No. 1, 1998, pp. 83–98.

Puerto, J., J. Gutiérrez, and J. Sicilia. "Bicriteria Trade-Off Analysis in a Two-Echelon Inventory/Distribution System." *Journal of the Operational Research Society*, Vol. 53, 2002, pp. 468–469.

Ramanathan, R. "ABC Inventory Classification with Multiple-Criteria Using Weighted Linear Optimization." *Computers & Operations Research*, Vol. 33, No. 3, 2006, pp. 695–700.

EXERCISES

Basic EOQ Model (Section 6.2.1) and Robustness: Sensitivity of EOQ (Section 6.2.2)

2.1 A company wishes to determine the optimal order quantity for a given part. Parts are sold at a rate of 1200 per month. Each order costs $1500 and the holding cost for each part is $50 per year. The lead time for order arrival is 0.5 months.

(a) Determine the optimal order quantity.

(b) Determine the number of orders per year.

(c) Determine the duration of each cycle (in days) assuming 365 days per year.

(d) Determine the reorder point.

(e) Graphically show all the above solutions.

2.2 A company wishes to determine the optimal order quantity for its products. The product is sold at a rate of 800 per week. Each order costs $500. On average, the cost of each product is $8 and the holding cost is based on an annual percentage rate of 10% per product. The lead time for order arrival is 1.5 weeks. Respond to parts (a)–(e) of exercise 2.1.

2.3 A manufacturing company wishes to determine the optimal order quantity for a component which is used in the final assembly of a product. Each order costs $7500. The cost of each component is $1000 and the holding cost is based on an annual percentage rate of 15% per component. The demand for components is 10,000 per month. The lead time for order arrival is 10 days. Respond to parts (a)–(e) of exercise 2.1.

2.4 Consider exercise 2.1.

(a) Suppose that holding cost for each part may vary up to $30 per year instead of $50. How much will the optimal order quantity and the total cost vary? Perform the robustness analysis.

(b) Suppose the ordering cost may vary up to $2500 instead of $1500. Compare the $2500 and $1500 ordering costs by performing robustness analysis.

(c) Perform robustness analysis when the demand is 1700 per week instead of 1200 per week.

2.5 Consider exercise 2.2.

(a) Suppose that annual percentage rate may vary up to 14% per product. How much will the optimal order quantity and the total cost vary? Perform the robustness analysis.

(b) Suppose the ordering cost may vary up to $650 instead of $500. Compare the $650 and $500 ordering cost by performing robustness analysis.

(c) Perform robustness analysis when the demand is 1100 per week instead of 800 per week.

Economic Production Quantity (Section 6.3)

3.1 Consider exercise 2.1. Suppose the company can produce parts in their own facility. They can produce the parts at a rate of 1500 parts per month where the setup cost is

$1500. Determine:

(a) Optimal production quantity

(b) Maximum inventory

(c) Total inventory cost

(d) Cycle period in days

(e) Production duration in days

(f) Number of setups per year

3.2 Consider exercise 2.2. Suppose that the company can produce products in its own facility at rate of 1200 products per week where the setup cost is $500. Respond to parts (a)–(f) of exercise 3.1.

3.3 Consider exercise 2.3. Suppose the company can produce 18,000 parts per month in its own facility where the setup cost is $ 7500. Respond to parts (a)–(f) of exercise 3.1.

EOQ Model Allowing Shortages (Section 6.4.1)[†]

4.1 A company wishes to determine the EOQ of a particular part. The company has a demand of 500 units per week, order cost of $250, shortage cost of $10 per unit per year, and holding cost of $15 per unit per year. Determine:

(a) Maximum inventory

(b) Optimal order quantity

(c) Maximum shortage

(d) Cycle period, nonshortage period, and shortage period (all in days)

(e) Total cost

4.2 A product has a demand of 2500 units per month, order cost of $500, shortage cost of $10 per unit per month, and holding cost of $75 per unit per year. Respond to parts (a)–(f) of exercise 4.1.

Inventory Model Allowing a Limited Shortage (Section 6.4.2)[†]

4.3 Consider exercise 4.1. Suppose the maximum allowable shortage is 100. Respond to parts (a)–(e) of exercise 4.1.

4.4 Consider exercise 4.2. Suppose the maximum allowable shortage is 75. Respond to parts (a)–(e) of exercise 4.1.

Bicriteria Economic Order Quantity (Section 6.5.1)

5.1 Consider exercise 2.1. Suppose the minimum order size is 25 parts.

(a) List five equally distributed alternatives in terms of order quantity.

(b) Determine the best alternative using an additive utility function, where the weights of importance for the two normalized objectives are (0.6, 0.4), respectively.

5.2 Consider exercise 2.2. Suppose the minimum order size is 500 parts.

(a) List five equally distributed alternatives in terms of order quantity.

(b) Determine the best alternative using an additive utility function, where the weights of importance for the two normalized objectives are (0.5, 0.5), respectively.

5.3 Consider exercise 2.3. Suppose the minimum order size is 500 parts.

(a) List five equally distributed alternatives in terms of order quantity.

(b) Determine the best alternative using an additive utility function, where the weights of importance for the two normalized objectives are (0.3, 0.7), respectively.

Bicriteria Economic Production Quantity (Section 6.5.2)

5.4 Consider exercise 3.1. Suppose that the minimum order size is 50 parts. Find the optimal order quantity for an additive utility function where the weights of importance for the normalized objectives are (0.7, 0.3), respectively.

5.5 Consider exercise 3.2. Suppose the minimum order size is 1000 parts. Find the optimal order quantity for an additive utility function where the weights of importance for the normalized objectives are (0.6, 0.4), respectively.

Tricriteria EOQ Allowing Shortages (Section 6.5.3)[†]

5.6 Consider exercise 4.1.

(a) Formulate and solve the multiobjective optimization inventory problem where the weights of importance for the three normalized objectives are (0.3, 0.4, 0.3), respectively. Suppose the given lower bound on the order quantity is 1000.

(b) Generate five efficient alternatives to this problem.

5.7 Consider exercise 4.2.

(a) Formulate and solve the multiobjective optimization inventory problem where the weights of importance for the three normalized objectives are (0.6, 0.2, 0.2), respectively. Suppose the given lower bound on order quantity is 500.

(b) Generate five efficient alternatives to this problem.

EOQ with Quantity Discount (Section 6.6.1)

6.1 Consider a product with the following two-level discount prices. The demand is 1560 units per year. The reorder cost is $200 per order. The annual holding cost is 10% of the unit price. Find the optimal order quantity.

Level	Order Size	Unit Cost, $
1	1–449	$12
2	450–∞	$10

6.2 Consider a part with the following three-level discount prices. The demand is 550,000 per year. The reorder cost is $50 per order. The annual holding costs is 13% of the unit price. Find the optimal order quantity.

Level	Order Size	Unit Cost, $
1	$0 < Q < 1500$	150
2	$1500 \leq Q < 2000$	140
3	$2000 \leq Q \leq \infty$	120

6.3 Consider the following four-level discount prices for a product. The demand is 30,000 units per year. The reorder cost is $1500 per order. The annual holding costs is 9% of the unit price. Find the optimal order quantity.

Level	Order Size	Unit Cost, $
1	$0 < Q < 2500$	42
2	$2500 \leq Q < 5000$	41
3	$5000 \leq Q \leq 30000$	40
4	$30,000 \leq Q \leq \infty$	39

Bicriteria EOQ with Quantity Discount (Section 6.6.2)[†]

6.4 Consider exercise 6.1. Suppose that the minimum quantity is 100, that is, $Q_{min} = 100$. Generate five efficient alternatives.

6.5 Consider exercise 6.2. Suppose that the minimum quantity is 1000, that is, $Q_{min} = 1000$. Generate five efficient alternatives.

6.6 Consider exercise 6.3. Suppose that the minimum quantity is 250, that is, $Q_{min} = 250$. Generate five efficient alternatives.

Multi-Item Inventory Optimization (Section 6.7.1)[†]

7.1 Consider an inventory problem for three types of items. All the needed information is presented in the table below. The maximum area is 7500 ft^2 and the maximum total budget is \$15,000. Use optimization to find the optimal order quantity for each item. Also find the total cost.

Item, i	D_i (per year)	$C_{R,i}$ (\$)	$C_{H,i}$ (\$)	a_i (ft^2)	c_i (\$)	$Q_{i,\min}$	$Q_{i,\max}$
1	18,000	200	10	14	18	300	500
2	24,000	300	8	6	21	100	300
3	48,000	95	15	12	13	50	200

7.2 Consider an inventory problem for four types of items. All the needed information is presented in the table below. The maximum area is 1500 ft^2 and the maximum total budget is \$5000. Use optimization to find the optimal order quantity for each item and the total cost for all items.

Item, i	D_i (per year)	$C_{R,i}$ (\$)	$C_{H,i}$ (\$)	a_i (ft^2)	c_i (\$)	$Q_{i,\min}$	$Q_{i,\max}$
1	7,020	10	5	8	23	50	100
2	12,740	15	0.8	4	31	50	75
3	9,880	12	12	25	29	25	50
4	16,120	18	4.5	18	24	15	45

7.3 Consider an inventory problem for five types of items. All the needed information is presented in the table below. The maximum area is 75,000 ft^2 and the maximum total budget is \$150,000. Use optimization to find the optimal order quantity for each item and the total cost for all items.

Item, i	D_i (per year)	$C_{R,i}$ (\$)	$C_{H,i}$ (\$)	a_i (ft^2)	c_i (\$)	$Q_{i,\min}$	$Q_{i,\max}$
1	102,200	950	50	80	230	50	100
2	164,250	840	8	40	310	100	200
3	135,050	1200	120	250	290	50	75
4	76,650	750	45	180	240	100	150
5	120,450	1100	25	120	220	100	150

Multiobjective Multi Item Inventory Optimization (Section 6.7.2)[†]

7.4 Consider exercise 7.1. Suppose the critical indexes for the three items are (40, 70, 80), respectively. Suppose the weights of importance for the two normalized objective values are (0.3, 0.7), respectively. Find the best alternative.

7.5 Consider exercise 7.2. Suppose the critical indexes for the four items are (10, 5, 15, 2), respectively. Suppose the weights of importance for the two normalized objective values are (0.6, 0.4), respectively. Find the best alternative.

7.6 Consider exercise 7.3. Suppose the critical indexes for the four items are (20, 5, 40, 8, 80), respectively. Suppose the weights of importance for the two normalized objective values are (0.5, 0.5), respectively. Find the best alternative.

Cost-Based Multi-Item Classification (Section 6.8.1)

8.1 Consider the following list of items along with their cost and critical index per item:

Item	Annual Demand	Unit Cost, $	Critical Index
1. Cards	3,000	0.40	100
2. Shirts	1,500	1.30	35
3. Gifts	875	4.00	60
4. Clothes	2,000	12.75	20
5. Jewelry	500	6.45	90
6. Cookies	2,500	15.50	85
7. Pants	1,250	2.15	75
8. Hats	975	1.05	70
9. Gloves	1,050	1.25	5

Perform ABC classification using the cost-based approach.

8.2 Consider the following list of items along with their cost and critical index per item:

Item	Annual Demand	Unit Cost, $	Critical Index
1	2100	158	10
2	1250	50	6
3	100	310	7
4	500	40	2
5	4800	2	3
6	250	20	9
7	120	25	5
8	100	18	4
9	1000	1	7
10	30	25	3
11	500	1.2	6
12	100	1.5	8

Perform ABC classification using the cost-based approach.

8.3 Consider the following list of items:

Item	Annual Demand	Unit Cost, $	Critical Index	Item	Annual Demand	Unit Cost, $	Critical Index
1	7426	86	2	9	5988	84	3
2	2543	24	9	10	7715	51	6
3	8569	85	3	11	2955	56	5
4	8520	17	9	12	3493	81	3
5	7088	100	1	13	1378	34	8
6	8352	65	5	14	2006	50	6
7	9933	70	4	15	6838	45	7
8	1911	69	4	16	2967	86	2

Perform ABC classification using the cost-based approach.

Critical Index Multi-Item Classification (Section 6.8.2)[†]

8.4 Consider exercise 8.1. Perform ABC classification using the critical index approach.

8.5 Consider exercise 8.2. Perform ABC classification using the critical index approach.

8.6 Consider exercise 8.3. Perform ABC classification using the critical index approach.

Multiple-Criteria Inventory Classification (Section 6.8.3)[†]

8.7 Consider exercise 8.1. Determine the optimal classification using the multiple-criteria inventory classification approach where the weights of importance for the two criteria are (0.4, 0.6), respectively.

8.8 Consider exercise 8.2. Determine the optimal classification using the multiple-criteria inventory classification approach where the weights of importance for the two criteria are (0.5, 0.5), respectively.

8.9 Consider exercise 8.3. Determine the optimal classification using the multiple-criteria inventory classification approach where the weights of importance for the two criteria are (0.7, 0.3), respectively.

Basic Model of Probabilistic Inventory (Section 6.9.1)

9.1 The average monthly demand of a product is 1000 units with a standard deviation of 75 units. The average lead time to order this product is 2 months with a standard deviation of 0.5 month. The required confidence level is 97.72%. Find the safety stock for this problem.

9.2 To order a product, the average lead time is 12 days with a standard deviation of 3 days. This product has an average daily demand of 85 units with a standard deviation of 15 units. The required confidence level is 99.38%. Find the safety stock for this problem.

9.3 The average monthly demand of a product is 400 units with a standard deviation of 50 units. The average lead time to order this product is four weeks with a standard deviation of 1.5 weeks. The required confidence level is 99.87%. Find the safety stock for this problem.

Multicriteria Probabilistic Inventory (Section 6.9.2)[†]

9.4 Consider exercise 9.1. Suppose the cost of inventory per item per period is $1.5. Generate a set of alternatives associated with $z = 0, 0.5, 1, 1.5, 2, 2.5, 3$ confidence levels. Identify efficient alternatives.

9.5 Consider exercise 9.2. Suppose the cost of inventory per item per period is $25. Generate a set of alternatives associated with $z = 0, 0.5, 1, 1.5, 2, 2.5, 3$ confidence levels. Identify efficient alternatives.

9.6 Consider exercise 9.3. Suppose the cost of inventory per period per item is $10. Generate a set of alternatives associated with $z = 0, 0.5, 1, 1.5, 2, 2.5, 3$ confidence levels. Identify efficient alternatives.

Target Service Ratio (Section 6.10.1) and Target Inventory with Continuous Probabilities (Section 6.10.2)

10.1 A newsstand wishes to determine the number of a given local newspaper to purchase each day. Each newspaper sells for $0.50. Newspapers that are not sold have a salvage value of $0.10. The cost of purchasing each paper is $0.25.

(a) Calculate the excess inventory and shortage costs.

(b) Determine the target service ratio.

(c) If the demand mean is 250 newspapers with a standard deviation of 30 papers per day, determine how many newspapers should be purchased by the newsstand.

10.2 A meat market store must determine how many pounds of meat to purchase every day. A pound of meat sells for $2.50. At the end of the day, the unsold meat has a salvage value of $0.50 per pound. Each pound of meat costs $1.25 to purchase.

(a) Calculate the excess inventory and shortage costs.

(b) Determine the target service ratio.

(c) If the mean demand is 200 pounds per day and the standard deviation is 11.5 pounds per day, determine how much meat the grocery store should buy.

Target Inventory with Discrete Probabilities (Section 6.10.3)[†]

10.3 A book store wishes to determine how many *Newsweek* magazines to purchase each week. Each *Newsweek* magazine sells for $5. *Newsweek* magazines that are not sold have a salvage value of $1.50. The cost of purchasing each magazine is $2.

(a) Calculate the excess inventory and shortage costs.

(b) Determine the target service ratio.

(c) The historical data for demand are shown in the table below. Determine how many magazines should be purchased by the grocery store.

Weekly demand	500	550	600	650	700	750	800 or more
No. of weeks	3	5	6	9	7	4	2

10.4 For a given product, a retail store wishes to determine how many to purchase every week. Each product sells for $115 and has a salvage value of $35. The cost of purchasing each product is $60.

(a) Calculate the excess inventory and shortage costs.

(b) Determine the target service ratio.

(c) Given the historical data below, determine how many products should be purchased by the retail store.

Weekly demand	100	200	300	400	500	600	700	800	900	1000 or more
No. of weeks	1	3	5	7	8	9	7	6	4	2

Multicriteria of Perishable Inventory (Section 6.10.4)[†]

10.5 Consider exercise 10.1. Generate a set of at least four bicriteria alternatives for the confidence levels between z_{min} and $z = 3$.

10.6 Consider exercise 10.2. Generate a set of at least four bicriteria alternatives for the confidence levels between z_{min} and $z = 3$.

10.7 Consider exercise 10.3. Generate a set of at least four bicriteria alternatives for the confidence levels between z_{min} and $z = 3$.

10.8 Consider exercise 10.4. Generate a set of at least four bicriteria alternatives for the confidence levels between z_{min} and $z = 3$.

CHAPTER 7

SCHEDULING AND SEQUENCING

7.1 INTRODUCTION

The purpose of solving scheduling problems is to find the optimal order of processing different tasks on a set of different processors while effectively utilizing the limited available resources. Effective schedules result not only in improved utilization of resources but also customer satisfaction. For example, in manufacturing, different scheduling of products will result in a different utilization of machines and completion times. Scheduling allows production to be accomplished efficiently while optimizing the utilization of resources and satisfying the due dates. Data processing by computers is another example of scheduling in which data are processed, stored, and/or retrieved. Each computer has several devices and processors that need to be properly scheduled to perform the needed tasks. The most important objectives of scheduling are listed below; these objectives are usually in conflict, and therefore MCDM approaches can be used to find a satisfactory solution.

1. Minimize the total time to complete all jobs (to maximize productivity).
2. Maximize meeting due dates (to maximize customer satisfaction).
3. Maximize utilization of resources (to minimize operational cost).
4. Minimize the average number of jobs waiting in the system (to minimize queues).

An important application of scheduling is in service-related systems such as hospitals, restaurants, and banks. In service systems, emphasis is placed on the staff (as processors) and customers (as jobs). Service industries rarely deal with inventory, but they must deal

Note: Advanced material that can be omitted without loss of generality will be indicated by a dagger.

Operations and Production Systems with Multiple Objectives, First Edition. Behnam Malakooti.
© 2014 John Wiley & Sons, Inc. Published 2014 by John Wiley & Sons, Inc.

with customer demand fluctuations which require quick response. Two types of scheduling in service industries are (a) by appointment, such as doctors and lawyers, and (b) by first come, first serve, such as retails and banks. In most scheduling problems the formation of waiting lines is inevitable. The queuing techniques can be used to analyze the formation of queues for different applications.

Scheduling methods can also be used to solve transportation-related problems, such as in the freight, trucking, and airline industries. In airline industries, optimal scheduling must be made by considering the number and types of airplanes, pilots, flight attendants, and passengers while considering unexpected occurrences such as atmospheric and landing conditions. The transportation and assignment methods covered in Chapter 9 can be used to solve some of these problems. Furthermore, single-row layouts (covered in Chapter 15) can be used to solve a set of combinatorial scheduling problems. Because the majority of sequencing problems are combinatorial, finding the optimal solution for a number of sequencing problems is computationally intensive. Therefore, various heuristics are employed to find relatively good solutions. In this chapter, we develop a new method called the head–tail approach for solving combinatorial sequencing problems.

One of the key objectives of scheduling is to improve the productivity of a system. One way to improve the productivity of a system is to identify its bottleneck. A bottleneck is one particular operation that limits the production rate of the entire system. A bottleneck operation has the lowest production rate, and therefore, it sets the limit on the production rate for the whole system.

There are several ways to handle bottleneck problems:

1. Increase the capacity of the bottleneck operation to reduce the bottleneck congestion.
2. Find alternative production routes (if possible) to avoid the bottleneck area.
3. Identify jobs that may cause the slowdown of the bottleneck resource; resolve the causes of such slowdowns before submitting such jobs through the bottleneck area.
4. Rebalance the system so that other machinery may become the bottleneck, but at a better rate of production.
5. Decrease the expected throughput to match the capacity of the bottleneck area.

Since most systems have occurring bottlenecks in different operations, maintaining a smooth balanced operation is hard to achieve. It should be realized that once one bottleneck is fixed, another bottleneck may emerge. Therefore bottleneck problems need to be resolved at the system level. Consistent attention to bottleneck areas is essential for optimizing productivity.

Scheduling problems can be classified based on the job arrivals to the system, which can be either static or dynamic. In static problems, all jobs are known and are ready to be processed at the beginning of the scheduling period. In dynamic problems, jobs may arrive intermittently throughout the system operations, and the sequence of jobs may change as new jobs arrive. Furthermore, the processing times of jobs by processors can be either deterministic or stochastic. Sequencing problems can also be classified as flow shop or job shop. In flow shop, n discrete jobs are processed by m different processors in the same order of processors. But in job shop, each job may be processed by a different sequence of processors. The classes of sequencing problems are $n \times 1$ (where n jobs are processed on one processor), $n \times 2$, $n \times 3$, $2 \times m$, and $n \times m$. Optimal solutions for $n \times 1$, $n \times 2$, and $2 \times m$ problems can be obtained. For $n \times m$ problems, heuristics are used. This chapter covers all of the above problems, including the MCDM analysis for each of these problems.

7.2   SEQUENCING *n* JOBS BY ONE PROCESSOR

The simplest sequencing problem involves *n* jobs and only one processor where the processing time for each job is known and deterministic and the problem is static. An example of sequencing *n* jobs by one processor is a repair shop where *n* repair jobs are performed by one repairman. The repairman can repair jobs in different orders depending on his or her objectives. In this section, we introduce the $n \times 1$ sequencing definitions and measurements, $n \times 1$ sequencing methods, and the MCDM of $n \times 1$ sequencing.

7.2.1   Sequencing Performance Measurements

The sequencing problem can be stated as determining a sequence *S* such that one or several sequencing objectives are optimized. The following sequencing criteria can be used to measure the performance of static sequencing problems. It is assumed that the start times for all jobs are set at zero. Therefore, all due dates are defined with respect to the starting time, zero.

1. *Flow Time*   The flow time $F_{i,S}$ for a given job *i* in sequence *S* is its total waiting time $W_{i,S}$ plus its processing time t_i:

$$F_{i,S} = W_{i,S} + t_i \tag{7.1}$$

where $W_{i,S}$ is the waiting time for job *i* in sequence *S* and t_i is the processing time for job *i*.

2. *Average Flow Time*   The average flow time is the average of the flow times of all *n* jobs in a given sequence *S*:

$$\bar{F}_S = \frac{1}{n} \sum_{i=1}^{n} F_{i,S} \tag{7.2}$$

where *n* is the number of jobs to be processed. Different sequencing solutions may have different average flow times and different total flow times.

3. *Average Lateness*   Lateness for a job is the difference between the completion time of the job and its due date. A job will have positive lateness if it is completed after its due date; it will have negative lateness if it is completed before its due date. The total of absolute values of latenesses is used to calculate the average lateness. The average lateness of a sequence is always a positive number. The average lateness for a given sequence *S* is given as

$$\bar{L}_S = \frac{1}{n} \sum_{i=1}^{n} |L_{i,S}| \quad \text{or} \quad \bar{L}_S = \frac{1}{n} \sum_{i=1}^{n} |(F_{i,S} - d_i)| \tag{7.3}$$

where d_i is the due date for job *i* and $L_{i,S}$ is the lateness of job *i* in sequence *S*.

4. *Average Tardiness*   Tardiness measures if a job is late with respect to its due date. Jobs completed earlier than their due dates are not tardy, so they are excluded from the calculation. Thus, tardiness is the same as positive lateness. Therefore, tardiness is calculated

as the maximum of zero and lateness. Here "0" signifies that the job was either early or on time. The average tardiness of a given sequence S is given as

$$\bar{T}_S = \begin{cases} \dfrac{1}{n} \displaystyle\sum_{i=1}^{n} T_{i,S} \\[2.5ex] \dfrac{1}{n} \displaystyle\sum_{i=1}^{n} \mathrm{Max}\left[0, L_{i,S}\right] \\[2.5ex] \dfrac{1}{n} \displaystyle\sum_{i=1}^{n} \mathrm{Max}\left[0, \left(F_{i,S} - d_i\right)\right] \end{cases} \tag{7.4}$$

where $T_{i,S}$ is the tardiness of job i in sequence S.

5. *Makespan (Total Time)* Makespan for a given sequence is the total time required to complete all n jobs on given processors. For static systems, the makespan of a sequence S is given as

$$M_S = F_{n,S} \tag{7.5}$$

where $F_{n,S}$ is the flow time of the last job n in the given sequence S. Note that for $n \times 1$ problems there is only one processor; therefore the processor is utilized all the time, and the makespan is the same as the summation of the processing times t_i of all jobs. This is not true for other sequencing problems, that is, $n \times 2$, $n \times m$, and $m \times 2$.

6. *Utilization Factor* Utilization refers to how effectively processors are utilized. By maximizing utilization, the amount of idle time in a system is reduced. Utilization for a given sequence S is calculated as the total available processing time divided by the total makespan. For m processors

$$\bar{U}_S = \frac{1}{m \times M_S} \sum_{i=1}^{n} t_i \tag{7.6}$$

For $n \times 1$ problems, since there is only one processor, that is, $m = 1$, the utilization is always 1 regardless of the sequence.

7.2.2 Sequencing Algorithms for $n \times 1$

Different sequencing methods use different priority rules in order to optimize different sequencing objectives. In this section, first we discuss how to use each sequencing rule to find the sequence of jobs. Then we show how to calculate objectives such as average flow time, average lateness, and average tardiness. The same example is used to demonstrate all rules.

1. *First Come, First Serve (FCFS) or First In, First Out (FIFO) Rule* First come, first serve sequences jobs in the order that they arrive at the processor. This rule is often considered the fairest rule to customers.

Example 7.1   Sequencing *n* Jobs by One Processor   Consider sequencing four jobs by one processor. Each job has the following processing times and due dates in days. The starting time is zero for all jobs.

Job, i	1	2	3	4
Process time, t_i	4	9	5	2
Due date, d_i	18	10	7	26

Suppose that before the start time the four jobs arrived in the order of 1, 2, 3, and 4. Apply the FCFS rule to find the order of jobs and measure the objectives.

Job, i	t_i	W_i	F_i	d_i	$|L_i|$	T_i
1	4	0	4	18	14	0
2	9	4	13	10	3	3
3	5	13	18	7	11	11
4	2	18	20	26	6	0
Total	20	35	55	—	34	14

Objectives

Average flow time ($\bar{F}$) = 55/4 = 13.75
Average lateness ($\bar{L}$) = 34/4 = 8.50
Average tardiness ($\bar{T}$) = 14/4 = 3.50

2. Last Come, First Serve (LCFS) or Last In, First Out (LIFO) Rule   Last come, first serve sequences jobs in the opposite order of FCFS. Sometimes, physical circumstances will necessitate using this rule, such as a stack piled up and then taken down from the top.

Job, i	t_i	W_i	F_i	d_i	$|L_i|$	T_i
4	2	0	2	26	24	0
3	5	2	7	7	0	0
2	9	7	16	10	6	6
1	4	16	20	18	2	2
Total	20	25	45	—	32	8

Objectives

Average flow time ($\bar{F}$) = 45/4 = 11.25
Average lateness ($\bar{L}$) = 32/4 = 8.00
Average tardiness ($\bar{T}$) = 8/4 = 2.00

3. *Shortest Processing Time (SPT) Rule*   Using the SPT, jobs are ordered by their processing times where the job with the shortest processing time will be processed first. This rule minimizes the average flow time and the average waiting time.

Job, i	t_i	W_i	F_i	d_i	$|L_i|$	T_i
4	2	0	2	26	24	0
1	4	2	6	18	12	0
3	5	6	11	7	4	4
2	9	11	20	10	10	10
Total	20	19	39	—	50	14

Objectives

Average flow time ($\bar{F}$) = 39/4 = 9.75
Average lateness ($\bar{L}$) = 50/4 = 12.50
Average tardiness ($\bar{T}$) = 14/4 = 3.50

It has been consistently demonstrated that the SPT rule performs well in more complex cases such as when there are multiple processors. One disadvantage of the SPT rule occurs in dynamic cases. If shorter tasks keep arriving, the SPT rule will ignore the longer tasks and perform the shorter ones first. This causes the longer tasks to continue to pile up. This problem can be resolved by using critical ratio rule for dynamic systems as discussed later in this section.

4. *Earliest Due Date (EDD) Rule*   The EDD rule ranks jobs by increasing order of their due dates, so the job with the earliest due date is processed first. This rule provides a schedule that minimizes the average tardiness.

Job, i	t_i	W_i	F_i	d_i	$|L_i|$	T_i
3	5	0	5	7	2	0
2	9	5	14	10	4	4
1	4	14	18	18	0	0
4	2	18	20	26	6	0
Total	20	37	57	—	12	4

Objectives

Average flow time ($\bar{F}$) = 57/4 = 14.25
Average lateness ($\bar{L}$) = 12/4 = 3.00
Average tardiness ($\bar{T}$) = 4/4 = 1.00

5. *Priority-Weighted SPT Rule*   In some applications, different jobs are given different importance or priority, for example, considering the producer and/or the consumer priorities. The importance of a job can be presented by a priority-weighted flow time rule. Consider I_i

as the priority or the weight of importance for a given job i. The higher the value of I_i, the more important job i is. The goal is to minimize the total weighted flow time. The priority ratio is calculated by finding the ratio of processing time to priority weight, t_i/I_i. Now apply the SPT rule to t_i/I_i. That is, rank jobs in increasing order of t_i/I_i.

Suppose the priority weights of importance for Example 7.1 are (2, 1, 4, 10) for jobs 1, 2, 3, and 4, respectively, where 10 is the most important and 0 is the least important. Now, find t_i/I_i and rank alternatives.

Job, i	1	2	3	4
Priority of importance, I_i	2	1	4	10
t_i	4	9	5	2
t_i/I_i	2.0	9.0	1.25	0.2
Rank	3	4	2	1

Job, i	t_i	W_i	F_i	d_i	$|L_i|$	T_i
4	2	0	2	26	24	0
3	5	2	7	7	0	0
1	4	7	11	18	7	0
2	9	11	20	10	10	10
Total	20	20	40	—	41	10

Objectives

Average flow time $(\bar{F}) = 40/4 = 10$

Average lateness $(\bar{L}) = 41/4 = 10.25$

Average tardiness $(\bar{T}) = 10/4 = 2.50$

6. *Critical Ratio (CR) Rule for Dynamic Case*   This method can be used for dynamic cases. That is, as new jobs arrive, this method can be used to find the sequence of both old and new jobs. This method balances between minimizing the waiting time and the tardiness for the current jobs. The CR for each job is calculated as follows:

$$\mathrm{CR}_i = \frac{\text{due date for job } i - \text{current time}}{\text{processing time for job } i}$$

$$= \frac{d_i - t^0}{t_i} \tag{7.7}$$

The job with the smallest CR is sequenced first. When a ratio is negative, the job is late. The CR may change as time passes; therefore, the CR for each job in the system needs to be recalculated every time there is a new job arrival. The major advantage of this method is that it allows for dynamic sequencing. Every time the processing of a job is completed, the new current time t^0 is used to find new CR values, which are used to find the next job to be processed.

Example without Arrival of New Jobs Consider Example 7.1 where the current time t^0 is zero. The CR for each job is calculated as presented in the following table. Job 2 has the minimum CR and will be the first job to be processed. Note that $t^0 = 0$.

Job, i	t_i	d_i	$(d_i - t^0)/t_i$	CR_i	Rank
1	4	18	(18-0)/4	4.5	3
2	9	10	(10-0)/9	1.11	1
3	5	7	(7-0)/5	1.4	2
4	2	26	(26-0)/2	13	4

All CRs should be calculated as soon as job 2 is processed. Then the current time is set at $t_0 = 9$. The new CR values are presented in the following table where job 3 has the minimum CR. Therefore job 3 will be processed next. Similarly at $t_0 = 14$, job 1 is selected to be processed next. The final solution is $(2 \rightarrow 3 \rightarrow 1 \rightarrow 4)$.

Job, i	t_i	d_i	CR_i $t_0 = 0$	$t_0 = 9$	$t_0 = 14$	Order	Job, i	t_i	W_i	F_i	d_i	$\lvert L_i \rvert$	T_i
1	4	18	4.5	2.25	1	3	2	9	0	9	10	1	0
2	9	10	1.11	0.11	−0.44	1	3	5	9	14	7	7	7
3	5	7	1.4	−0.4	−1.4	2	1	4	14	18	18	0	0
4	2	26	13	8.5	6	4	4	2	18	20	26	6	0
							Total	20	41	61	—	14	7

Objectives

$$\bar{F} = 61/4 = 15.25$$
$$\bar{L} = 14/4 = 3.5$$
$$\bar{T} = 7/4 = 1.75$$

Example with Arrival of New Jobs Consider Example 7.1. Suppose that at time $t = 4$, while job 2 is being processed, a new job, job 5, arrives. Job 5 has a processing time of 6 and a due date of 15. At $t^0 = 9$, job 2 processing is finished; therefore the sequencing must be reevaluated at this point by considering jobs 1, 3, 4, and 5. Note that $t^0 = 9$.

Job, i	t_i	d_i	$(d_i - t^0)/t_i$	CR_i	Rank
1	4	18	(18-9)/4	2.25	3
3	5	7	(7-9)/5	−0.4	1
4	2	26	(26-9)/2	8.5	4
5	6	15	(15-9)/6	1	2

TABLE 7.1   Summary of Sequencing Methods for Example 7.1

Objectives	FCFS	LCFS	SPT	EDD	Weighted SPT	CR
Average flow time ($\bar{F}$) (min.)	13.75	11.25	**9.75**	14.25	10	15.25
Average lateness ($\bar{L}$) (min.)	8.50	8.00	12.50	**3.00**	10.25	3.5
Average tardiness ($\bar{T}$) (min.)	3.50	2.00	3.50	**1.00**	2.50	1.75
Sequence	1,2,3,4	4,3,2,1	4,1,3,2	3,2,1,4	4,3,1,2	2,3,1,4

Job 3 is selected to be processed next, and the algorithm is repeated once job 3 is finished. The table below summarizes the CR for all steps. The final solution is $(2\rightarrow3\rightarrow5\rightarrow1\rightarrow4)$.

Job, i	t_i	d_i	CR_i				Order
			$t_0 = 0$	$t_0 = 9$	$t_0 = 14$	$t_0 = 20$	
1	4	18	4.5	2.25	1	−0.5	4
2	9	10	1.11	0.11	−0.44	−1.1	1
3	5	7	1.4	−0.4	−1.4	−2.6	2
4	2	26	13	8.5	6	3	5
5	6	15	2.5	1	0.17	−0.83	3

Summary of All Sequencing Methods   The performances of the above sequencing methods are compared in Table 7.1. Note that the SPT provides the best average flow time and the EDD the best average tardiness (shown in boldface). In this example, the EDD coincidently also provided the best average lateness (also in bold). Depending on the problem and objectives, one of the above methods should be used.

Supplement S7.1.doc shows a new JIT method for $n \times 1$ sequencing problems.

7.2.3   Bicriteria Composite Approach for $n \times 1$

Consider the following two objectives:

Minimize average flow time, $f_1 = \bar{F}$.
Minimize average tardiness, $f_2 = \bar{T}$.

The first objective also minimizes the average waiting time for jobs in the system. The second objective may also minimize the average lateness.

For a given additive utility function, the following method can be used to generate the best alternative:

$$\text{Minimize } U = w_{\text{SPT}}\bar{F} + w_{\text{EDD}}\bar{T} \qquad (7.8)$$

where w_{SPT} and w_{EDD} are the weights of importance for f_1 and f_2, respectively.

1. Find a sequence S_{SPT} which minimizes average flow time. This sequence is obtained by using the SPT rule. Now, calculate $\bar{F}$ and $\bar{T}$ for the S_{SPT} sequence.
2. Find a sequence S_{EDD} which minimizes average tardiness. Use the EDD rule to find the solution. Calculate $\bar{F}$ and $\bar{T}$ for the S_{EDD} sequence.
3. If the two sequences are the same, that is, $S_{SPT} = S_{EDD}$, there is only one optimal solution that minimizes both objectives. Stop.
4. Consider the given weights of importance (w_{SPT}, w_{EDD}) for the two objectives, respectively. Apply the following composite MCDM approach to find the sequence of jobs.

Composite MCDM Approach for Sequencing Jobs   This approach assigns a priority index to each job for each objective. The objective weights (w_{SPT}, w_{EDD}) are then used to create a composite priority index for each job.

- First, define the priority increment as $1/(n-1)$ where n is the number of jobs.
- Assign a job priority index based on the SPT sequence, $P_i(SPT)$, from 0 to 1 to each job. Zero is assigned to the first job in the SPT sequence, and the priority increment $1/(n-1)$ is assigned to the second job. Each subsequent job in the sequence is given a priority index that is $1/(n-1)$ higher than the previous job until the last job in the sequence is assigned a value of 1.
- Assign job priorities for the EDD sequence $P_i(EDD)$ in a similar manner.
- Use weights of importance for the objectives (w_{SPT}, w_{EDD}) to find the composite priorities index Z_i for each job:

$$Z_i = w_{SPT} P_i(SPT) + w_{EDD} P_i(EDD) \tag{7.9}$$

- Sequence the jobs in ascending order of their composite priority index Z_i.

Example 7.2   Bicriteria Analysis for Sequencing n Jobs by One Processor   Consider Example 7.1. Suppose the weights of importance for the two objectives are $(0.6, 0.4)$ for f_1 and f_2, respectively. As in Example 7.1, the processing times and due dates (in days) of the four jobs are as follows:

Job, i	1	2	3	4
Process time, t_i	4	9	5	2
Due date, d_i	18	10	7	26

The solution using the SPT method (see Example 7.1) to minimize f_1 is

$$S_{SPT} = (4 \rightarrow 1 \rightarrow 3 \rightarrow 2) \quad \text{where } f_1 = \bar{F} = 9.75 \quad f_2 = \bar{T} = 3.5$$

The solution using the EDD method (see Example 7.1) to minimize f_2 is

$$S_{EDD} = (3 \rightarrow 2 \rightarrow 1 \rightarrow 4) \quad \text{where } f_1 = \bar{F} = 14.25 \quad f_2 = \bar{T} = 1$$

The above two solutions are not the same; therefore find the composite priority index Z_i for each job and sequence the jobs using this index.

Composite Priority Index   The number of jobs in this problem is four ($n = 4$). Therefore, the priority increment is $1/(n - 1) = 1/(4 - 1) = \frac{1}{3}$. The priority indices for SPT and EDD for each job are presented in the following tables. The composite priority index is then found by combining the two objectives where $Z_i = 0.6P_i(\text{SPT}) + 0.4P_i(\text{EDD})$:

SPT Sequence					EDD Sequence					Composite Priority Index				
Job, i	4	1	3	2	Job, i	3	2	1	4	Job, i	1	2	3	4
$P_i(\text{SPT})$	0	$\frac{1}{3}$	$\frac{2}{3}$	1	$P_i(\text{EDD})$	0	$\frac{1}{3}$	$\frac{2}{3}$	1	Z_i	0.47	0.73	0.4	0.4
										Rank	3	4	1 or 2	1 or 2

In this example, there are two alternative solutions because the composite priority index for jobs 3 and 4 are the same. The two solutions are ($3\rightarrow4\rightarrow1\rightarrow2$) and ($4\rightarrow3\rightarrow1\rightarrow2$). When there are multiple-sequence solutions, calculate the average flow time f_1 and average tardiness f_2 for each of sequence. Choose the sequence that best minimizes the additive utility function $U = w_1f_1 + w_2f_2$.

For example, for sequence $3\rightarrow4\rightarrow1\rightarrow2$, the solution is as follows:

Job, i	t_i	W_i	F_i	d_i	$|L_i|$	T_i
3	5	0	5	7	2	0
4	2	5	7	26	19	0
1	4	7	11	18	7	0
2	9	11	20	10	10	10
Total	20	23	43	—	38	10

Objectives

$$f_1 = \bar{F} = 10.75$$
$$f_2 = \bar{T} = 2.5$$

For sequence ($4\rightarrow3\rightarrow1\rightarrow2$), the solution is the same as the weighted SPT example solution (see Example 7.1), where $f_1 = \bar{F} = 10.00$ and $f_2 = \bar{T} = 2.5$.

Since the sequence ($4\rightarrow3\rightarrow1\rightarrow2$) dominates the sequence ($3\rightarrow4\rightarrow1\rightarrow2$), that is, it has a shorter average flow time $\bar{F}$, choose ($4\rightarrow3\rightarrow1\rightarrow2$) as the best sequence for this bicriteria problem.

7.2.4   Generating the Efficient Frontier for $n \times 1$

For small sequencing problems, such as Example 7.2, it is possible to generate all possible sequences and then identify the efficient set of alternatives. The set of efficient alternatives for Example 7.2 is shown in Table 7.2 and Figure 7.1.

TABLE 7.2 Set of Efficient Alternatives for Example 7.1

Alternatives	a_1	a_2	a_3	a_4	a_5
Min. f_1	9.75	10	11.25	13.75	14.25
Min. f_2	3.5	2.5	2	1.5	1.0
Sequence	4→1→3→2	4→3→1→2	4→3→2→1	3→2→4→1	3→2→1→4
Efficient?	Yes	Yes	Yes	Yes	Yes

Generating Efficient Alternatives Using Composite MCDM   For larger sequencing problems, generating all possible sequences is not feasible. The composite MCDM method can be used to generate a set of efficient alternatives. This can be accomplished by varying the weights w_{SPT} and w_{EDD} in Equation (7.9).

For example, consider five sets of weights (w_{SPT}, w_{EDD}): (0, 1), (0.25, 0.75), (0.5, 0.5), (0.75, 0.25), and (1, 0). Solving five different composite MCDM sequencing problems may result in five efficient alternatives. If the two sets of adjacent weights generate the same sequence, then consider using some other set of weights that is not between the two adjacent set of weights.

Example 7.3   Efficient Frontiers for Sequencing n Jobs by One Processor   Consider Example 7.2. Generate five efficient points using five different weights:

Using weights $w = (0.5, 0.5)$ for (w_{SPT}, w_{EDD}), the solution is:

Job, i	1	2	3	4	Sequence	3→4→1→2
Z_i	0.50	0.67	0.33	0.50	f_1	10.75
Rank	2	4	1	2	f_2	2.50

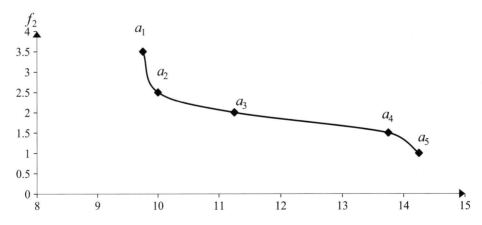

Figure 7.1   Efficient frontier for Example 7.2 where both objectives are minimized.

For weights $w = (0.75, 0.25)$, the solution is:

Job, i	1	2	3	4	Sequence	$4{\to}1{\to}3{\to}2$
Z_i	0.42	0.83	0.50	0.25	f_1	9.75
Rank	2	4	3	1	f_2	3.50

For weights $w = (0.25, 0.75)$, the MCDM solution is:

Job, i	1	2	3	4	Sequence	$3{\to}2{\to}1{\to}4$
Z_i	0.58	0.50	0.17	0.75	f_1	14.25
Rank	3	2	1	4	f_2	1.00

Table 7.3 shows the five sequences generated by using five sets of weights. The weights (1,0) and (0.75, 0.25) generated the same solution. Therefore, consider using a different set of weights, for example, use (0.65, 0.35), which is between (0.75, 0.25) and (0.5, 0.5). This generates a new solution as presented in Table 7.3. Note that only alternatives a_1, a_2, and a_5 are efficient. To generate more efficient alternatives, one can improve upon the generated inefficient points by using the pairwise exchange method (which is discussed in Chapter 13). For example, consider alternative a_7, $3{\to}1{\to}4{\to}2$. By exchanging 1 and 2, a new efficient solution a_4, $3{\to}2{\to}4{\to}1$, is found.

Supplement S7.2.doc shows how to use the pairwise exchange method for generating more efficient alternatives.

7.2.5 Tricriteria for *n* × 1[†]

Tricriteria sequencing problems can be formulated as follows:

f_1 = minimize average flow time $\bar{F}$. Use the SPT rule (minimizes avgerage waiting).
f_2 = minimize average tardiness $\bar{T}$. Use the EDD rule (customer's priority).
f_3 = maximize satisfying job priorities. Use the job priority rule (producer's priority).

TABLE 7.3 Composite MCDM Approach for Example 7.3

Weights	(1, 0) or (0.75, 0.25)	(0.65, 0.35)	(0.5, 0.5)	(0.4, 0.6)	(0.25, 0.75) or (0, 1)
Alternatives	a_1	a_2	a_6	a_7	a_5
Min. f_1	9.75	10	10.75	11.25	14.25
Min. f_2	3.5	2.5	2.5	2.5	1.0
Sequence	$4{\to}1{\to}3{\to}2$	$4{\to}3{\to}1{\to}2$	$3{\to}4{\to}1{\to}2$	$3{\to}1{\to}4{\to}2$	$3{\to}2{\to}1{\to}4$
Efficient?	Yes	Yes	No	No	Yes

Average flow times f_1 and f_2 are measured by $\bar{F}$ and $\bar{T}$, respectively. In the following, we present how to measure f_3.

Job Priority Rule (PRI)   Suppose the producer's priority or goal sequence is S_g. Let a given current sequence solution be labeled S_c. The PRI objective f_3 seeks to minimize the difference between the current sequence S_c and the goal sequence S_g; f_3 is measured by the rectilinear distance between the goal sequence and the current sequence:

$$\text{Minimize } f_3 = \sum_{i=1}^{n} |c_i - g_i| \tag{7.10}$$

where c_i is the position (rank) of job i in the current sequence and g_i is the position (rank) of job i in the goal sequence. For example, suppose the current job sequence is $(4 \rightarrow 1 \rightarrow 3 \rightarrow 2)$ and the goal (priority) sequence is $(1 \rightarrow 2 \rightarrow 3 \rightarrow 4)$. Then f_3 is calculated as presented below.

Job, i	1	2	3	4	f_3
Current position, c_i	2	4	3	1	
Goal position, g_i	1	2	3	4	
$|c_i - g_i|$	1	2	0	3	6

Note that if the current sequence is the same as the goal sequence, then, $f_3 = 0$; this is the best possible (ideal) solution for job priorities.

The approach for generating efficient alternatives for tricriteria sequencing problems is similar to the approach of bicriteria problems. For given weights of importance for the three objectives f_1, f_2, and f_3, use the composite MCDM approach for all three objectives.

Example 7.4   Tricriteria $n \times 1$ Consider Example 7.2. For the third objective, suppose the goal job priority sequence is $1 \rightarrow 2 \rightarrow 3 \rightarrow 4$. For the additive utility function, if the weights of importance for the three objectives f_1, f_2, and f_3 are $(0.3, 0.3, 0.4)$, respectively, find the solution to the MCDM problem. First, find $Z_i = 0.3P_i(\text{SPT}) + 0.3P_i(\text{EDD}) + 0.4P_i(\text{PRI})$:

For SPT, the sequence is $(4 \rightarrow 1 \rightarrow 3 \rightarrow 2)$ with SPT job priority indices of $(0, 1/3, 2/3, 1)$ for jobs 4, 1, 3, and 2, respectively.

For EDD, the sequence is $(3 \rightarrow 2 \rightarrow 1 \rightarrow 4)$ with EDD job priority indices of $(0, 1/3, 2/3, 1)$ for jobs 3, 2, 1, and 4, respectively.

For PRI, the sequence is $(1 \rightarrow 2 \rightarrow 3 \rightarrow 4)$ with PRI job priority indices of $(0, 1/3, 2/3, 1)$ for jobs 1, 2, 3, and 4, respectively.

Therefore, the composite priority index Z_i can be found as follows:

Job, i	1	2	3	4
$0.3P_i(\text{SPT})$	$0.3 \times 1/3$	0.3×1	$0.3 \times 2/3$	0.3×0
$0.3P_i(\text{EDD})$	$+ 0.3 \times 2/3$	$+ 0.3 \times 1/3$	$+ 0.3 \times 0$	$+ 0.3 \times 1$
$0.4P_i(\text{PRI})$	$+ 0.4 \times 0$	$+ 0.4 \times 1/3$	$+ 0.4 \times 2/3$	$+ 0.4 \times 1$
Z_i (sum)	$= 0.30$	$= 0.53$	$= 0.47$	$= 0.70$
Rank	1	3	2	4

Therefore, the sequence solution is $1 \rightarrow 3 \rightarrow 2 \rightarrow 4$. This solution is presented below where f_3 is measured as $|1 - 1| + |2 - 3| + |3 - 2| + |4 - 4| = 2$.

Job, i	t_i	W_i	F_i	d_i	$|L_i|$	T_i
1	4	0	4	18	14	0
3	5	4	9	7	2	2
2	9	9	18	10	8	8
4	2	18	20	26	6	0
Total	20	31	51	—	30	10

Objectives

$$f_1 = \bar{F} = 51/4 = 12.75$$
$$f_2 = \bar{T} = 10/4 = 2.5$$
$$f_3 = 2$$

7.3 SEQUENCING *n* JOBS BY TWO PROCESSORS

7.3.1 Flow Shop of *n* × 2 (Johnson's Rule)

For sequencing n jobs by two processors, Johnson's algorithm provides an optimal solution for minimizing the total time to complete all jobs (which is defined as makespan M). The algorithm is based on the SPT rule when there are two processors A and B. All jobs processed in the same order by the two processors. Label the first processor as A and the second processor as B, and denote $A \rightarrow B$ as the order of the processors. Denote

$A_i =$ processing time on processor A (first processor) for job i

$B_i =$ processing time on processor B (second processor) for job i

Johnson's algorithm is as follows:

1. List the processing times of all jobs, $i = 1, 2, \ldots, n$, on processors A and B by $\{A_1, A_2, \ldots, A_n; B_1, B_2, \ldots, B_n\}$. Consider n unassigned locations that must be assigned

to the n jobs such that only one job is assigned to each location and vice versa: 1, 2, ..., n.

2. Choose the job that has the SPT in the processing times list, and label it job j.

   (a) If the SPT is associated with processor A, place job j in the first available position on the job locations.

   (b) If the SPT is associated with processor B, place job j in the last available position on the job locations.

3. Remove both A_j and B_j from the list.

4. Repeat steps 2 and 3 until all jobs have been placed in the sequence.

Example 7.5 Sequencing n Jobs by Two Processors   Consider the following table, which presents the processing times (e.g., in days) of four jobs on processors A and B:

Job, i	1	2	3	4
Processor A processing time	4	9	5	2
Processor B processing time	3	1	7	6

Therefore, the list of processing times is $\{A_1, A_2, A_3, A_4; B_1, B_2, B_3, B_4\}$ with values of $\{4, 9, 5, 2; 3, 1, 7, 6\}$. There are four locations to be assigned by four jobs:

1. The shortest time is $B_2 = 1$. Since the shortest time is on processor B, place job 2 at the end of the sequence. Remove both A_2 and B_2 from the list. The list becomes $\{A_1, A_3, A_4; B_1, B_3, B_4)$ with values of $\{4, 5, 2; 3, 7, 6\}$.

2. The shortest time in the list is $A_4 = 2$. Place job 4 at the beginning of the sequence since the shortest time corresponds to processor A. Removing both A_4 and B_4 from the list, the list becomes $\{A_1, A_3; B_1, B_3)$ with values of $\{4, 5; 3, 7\}$.

3. The shortest time in the list is $B_1 = 3$. Place job 1 in the last available space, the third space. Removing job 1 from the list, the list becomes $\{A_3; B_3\}$.

4. Job 3 is the only remaining job. Place job 3 in the remaining unfilled space. The summary of the four iterations is listed below.

Iteration	1	2	3	4
Job Assignment	▯▯▯ 2	4 ▯▯ 2	4 ▯ 1 2	4 3 1 2

So the optimal sequence is ($4 \rightarrow 3 \rightarrow 1 \rightarrow 2$). Now, it is necessary to find the start and finish of each job as processed by each processor. One can use either a table or a Gantt chart to represent the solution. The Gantt chart for this sequence, ($4 \rightarrow 3 \rightarrow 1 \rightarrow 2$), is shown below. The shaded areas are idle time. Processor A has a total idle time of 1 day. Processor B has a total idle time of $2 + 2 = 4$ days. From the Gantt chart, it can be seen that the makespan of this sequence is 21 days.

Processor A	A_4	A_3	A_1	A_2		
Processor B		B_4	B_3	B_1	B_2	
Time Milestones	0 2	7 8	11	15	18	21

The above solution can also be presented by a table as follows:

Job, i	4	3	1	2	Makespan
A_i: Duration	2	5	4	9	—
B_i: Duration	6	7	3	1	—
A_i: Start time	0	2	7	11	—
A_i: Finish time	2	7	11	20	20
B_i: Start time	2	8	15	20	—
B_i: Finish time	8	15	18	21	21

Sometimes, it is possible to find alternative optimal solutions for the same problem. For example, for the same $(4\to 3\to 1\to 2)$ sequence, an alternative solution to the above problem is to start job 4 on processor B at time $t = 4$. The makespan for this solution is also 21 days.

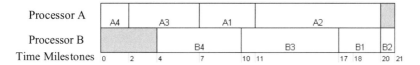

7.3.2 Job Shop of $n \times 2$

In flow shop problems, all jobs go through the processors in the same sequence. In the job shop problem, each job may have a different sequence of processors. In the case of two processors, some jobs must be processed first on processor A and then on processor B, $A \to B$, but other jobs must be processed first on processor B and then on A, $B \to A$. Although this problem is more complex than the $n \times 2$ flow shop problem covered in the last section, it can be solved by an extension of Johnson's algorithm.

Consider Johnson's algorithm covered in the last section. Use the following procedure instead of step 2 of Johnson's algorithm. Choose job j associated with the shortest processing time from the list:

(a) Assume the shortest processing time is associated with processor A.
  (i) If job j has sequence $A \to B$, then place job j in the first available space.
  (ii) If job j has sequence $B \to A$, then place job j in the last available space.

(b) Assume the shortest processing time is associated with processor B.

   (i) If job j has sequence $A{\to}B$, then place job j in the last available space.

   (ii) If job j has sequence $B{\to}A$, then place job j in the first available space.

Example 7.6   Job Shop Sequencing of Four Jobs by Two Processors   Consider the following problem:

Job, i	1	2	3	4
Processor A processing time	4	1	5	2
Processor B processing time	3	2	4	4
Processor sequence	$A{\to}B$	$B{\to}A$	$B{\to}A$	$A{\to}B$

The ordered list of processing times is $\{A_2, A_4, B_2, B_1, A_1, B_3, B_4, A_3\}$:

1. The shortest time in the list is A_2 and job 2 has sequence $B{\to}A$. Place job 2 at the end of the sequence. Remove A_2 and B_2 from the list. The list becomes $\{A_4, B_1, A_1, B_3, B_4, A_3\}$.

2. The shortest time in the list is now A_4 and job 4 has sequence $A{\to}B$. Place job 4 in the first available space. Remove A_4 and B_4 from the list. The list becomes $\{B_1, A_1, B_3, A_3\}$.

3. The next shortest time in the list is B_1 and job 1 has sequence $A{\to}B$. Place job 1 in the last available space. Removing both A_1 and B_1 from the list results in $\{B_3, A_3\}$.

4. Only job 3 remains, so it is placed in the remaining space for the sequence. The solution is $4{\to}\,3{\to}\,1{\to}\,2$, which has the following Gantt chart with a makespan of 25 days:

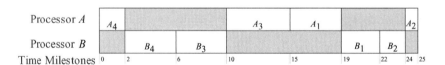

Note that the above approach is a heuristic, that is, the solution may not necessarily be optimal. This solution can be improved by assigning the first priority job on the first processor that becomes available. For example, as soon as processor A becomes available, the highest priority job that has the processing ordering of $A{\to}B$ (i.e., job 1) can be assigned to it. The order of the jobs will be $(4{\to}1{\to}3{\to}2)$. This solution has a makespan of 19 days.

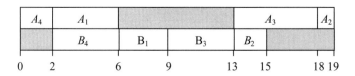

7.3.3 Multicriteria of *n* × 2

Bicriteria $n \times 2$ sequencing problem is:

Minimize f_1 = makespan
Minimize f_2 = total tardiness

The best sequence for f_1 can be obtained by using Johnson's rule. The best sequence for f_2 can be found using the EDD rule. The MCDM sequencing of n jobs by two processors is similar to the MCDM sequencing of n jobs by one processor covered in Section 7.2.3. For $n \times 2$ problems, the makespan is used instead of the average flow time, which was used in $n \times 1$ problems, because in $n \times 1$ problems the makespan is constant but in $n \times 2$ problems this value can be optimized. The following examples show the use of the composite MCDM approach for sequencing n jobs by two processors and how to generate a sample of efficient alternatives.

Example 7.7 Bicriteria of Sequencing Four Jobs by Two Processors   Consider Example 7.5 with two processors ($A{\to}B$) having the following due dates d_i (in days). Find the best sequence for the bicriteria problem, where the weights of importance for objectives f_1 and f_2 are (0.6, 0.4), respectively.

Job, i	1	2	3	4
Processor A processing time	4	9	5	2
Processor B processing time	3	1	7	6
Due date for job i, d_i	8	14	20	12

1. *Minimum Makespan*   As shown in Example 7.5, the optimal sequence using Johnson's rule is $(4{\to}3{\to}1{\to}2)$ with a makespan of 21.
2. *Minimum Tardiness*   Using the EDD rule, the optimal sequence is $(1{\to}4{\to}2{\to}3)$. The above two solutions are not the same; therefore proceed with finding the composite priority index.

Minimum Makespan: $(4{\to}3{\to}1{\to}2)$						Minimum Tardiness: $(1{\to}4{\to}2{\to}3)$					
Job, i	A_i	B_i	F_i	d_i	$	L_i	$	T_i	Job, i A_i B_i F_i d_i $	L_i	$ T_i
4	2	6	8	12	4	0	1	4	3	7	8	1	0
3	5	7	15	20	5	0	4	2	6	13	12	1	1
1	4	3	18	8	10	10	2	9	1	16	14	2	2
2	9	1	21	14	7	7	3	5	7	27	20	7	7
Objectives			$f_1=21$		$f_2=17$		Objectives $f_1=27$ $f_2=10$

3. *Composite Priority Index*   The number of jobs in this example is four, that is, $n = 4$. Therefore, the priority increment is $1/(n-1) = 1/(4-1) = \frac{1}{3}$.

SPT sequence (Johnson's rule)					EDD sequence				
Job, i	4	3	1	2	Job, i	1	4	2	3
Priority P_i(SPT)	0	$\frac{1}{3}$	$\frac{2}{3}$	1	Priority P_i(EDD)	0	$\frac{1}{3}$	$\frac{2}{3}$	1

Now find $Z_i = 0.6P_i(\text{SPT}) + 0.4P_i(\text{EDD})$. For example, for job 1, $Z_1 = 0.6(2/3) + 0.4(0) = 0.4$.

Job, i	1	2	3	4
Z_i	0.4	0.87	0.6	0.13
Rank	2	4	3	1

Therefore, for the bicriteria problem, $(4{\rightarrow}1{\rightarrow}3{\rightarrow}2)$ is the best sequence. For this sequence, the objective values are:

Job, i	A_i	B_i	F_i	d_i	$|L_i|$	T_i
4	2	6	8	12	4	0
1	4	3	11	8	3	3
3	5	7	18	20	2	0
2	9	1	21	14	7	7
Objectives			$f_1 = 21$			$f_2 = 10$

Since both objectives are minimized, alternative $(4{\rightarrow}1{\rightarrow}3{\rightarrow}2)$ with $(f_1 = 21, f_2 = 10)$ dominates the solutions of both SPT (21, 17) and EDD (27, 10).

By using the pairwise exchange method (as discussed in Chapter 13), the current alternative $(4{\rightarrow}1{\rightarrow}3{\rightarrow}2)$ may be improved. By exchanging the positions of jobs 4 and 1, sequence $(1{\rightarrow}4{\rightarrow}3{\rightarrow}2)$ can be generated with objective values $f_1 = 21$ and $f_2 = 8$. This solution, $(1{\rightarrow}4{\rightarrow}3{\rightarrow}2)$, dominates the previous solution, $(4{\rightarrow}1{\rightarrow}3{\rightarrow}2)$.

Example 7.8   Generating a Sample of Efficient Alternatives for $n \times 2$ This approach is similar to building an efficient frontier for the bicriteria of $n \times 1$. Consider Example 7.7. Generate a set of efficient alternatives by varying weights. For $W = (0.5, 0.5)$, the problem is solved. The solution is: $(4{\rightarrow}1{\rightarrow}3{\rightarrow}2)$. For this sequence, the objective values are: $f_1 = 21$ and $f_2 = 10$.

For weights $(0.75, 0.25)$, the solution is: $(4{\rightarrow}3{\rightarrow}1{\rightarrow}2)$, with $f_1 = 21$ and $f_2 = 17$.
For weights $(0.25, 0.75)$, the solution is: $(1{\rightarrow}4{\rightarrow}2{\rightarrow}3)$, with $f_1 = 27$ and $f_2 = 10$.

	Weights (0.5, 0.5)				Weights (0.75, 0.25)				Weights (0.25, 0.75)			
Job, i	1	2	3	4	1	2	3	4	1	2	3	4
Z_i	0.33	0.83	0.67	0.17	0.5	0.92	0.5	0.08	0.17	0.75	0.83	0.25
Rank	2	4	3	1	3	4	2	1	1	3	4	2

The above five sequences are summarized in the table below.

Weights	(1, 0)	(0.75, 0.25)	(0.5, 0.5)	(0.25, 0.75)	(0, 1)
Alternatives	a_1	a_1	a_2	a_3	a_3
Sequence	4→1→3→2	4→3→1→2	3→4→2→1	3→2→4→1	3→2→1→4
Min. f_1	21	21	21	27	27
Min. f_2	17	17	10	10	10
Efficient?	No	No	Yes	No	No

It can be observed that alternative a_2 dominates all other alternatives. Thus, only one efficient solution to this problem was generated so far.

Using the composite method, three different alternatives were generated. All adjacent points to the three alternatives a_1, a_2, and a_3 are generated using the pairwise exchange method. Only one new efficient solution $a_4 = \{1\rightarrow4\rightarrow3\rightarrow2\}$ with $f_1 = 21$ and $f_2 = 8$ is generated. This alternative dominates the previous efficient point a_2 (i.e., $a_4 \succ a_2$).

Example 7.9   Tricriteria for $n \times 2$ As discussed in the tricriteria for $n \times 1$ problems, the third objective can be a measure of complying with the producer's job priorities. Consider Example 7.7. Suppose the producer's job priority order is (1→2→3→4). Suppose that the weights of importance for the three objectives f_1, f_2, and f_3, are (0.3, 0.3, 0.4), respectively. Find the best multicriteria solution for this problem. First generate the composite priority index $Z_i = 0.3P_i(\text{SPT}) + 0.3P_i(\text{EDD}) + 0.4P_i(\text{PRI})$.

Job, i	1	2	3	4
Z_i	0.20	0.63	0.67	0.50
Rank	1	3	4	2

The solution of the above table is (1→4→2→3). The details of this solution are shown below where $f_3 = |1 - 1| + |2 - 3| + |3 - 4| + |4 - 2| = 0 + 1 + 1 + 2 = 4$.

Job, i	A_i	B_i	F_i	d_i	T_i	f_3
1	4	3	7	8	0	$|1 - 1|$
4	2	6	13	12	1	$|4 - 2|$
2	9	1	16	14	2	$|2 - 3|$
3	5	7	27	20	7	$|3 - 4|$
Objectives			$f_1 = 27$		$f_2 = 10$	$f_3 = 4$

7.4   SEQUENCING n JOBS BY m PROCESSORS

7.4.1   Algorithm for $n \times 3$[†]

The problem of sequencing n jobs by three processors is similar to the problem of sequencing n jobs by two processors. All jobs are processed in the same order on three processors A, B, and C: first on A, then on B, and finally on C. An optimal solution can be found by an extension of Johnson's algorithm if certain conditions are met. If the conditions are not met, a different method, such as the branch-and-bound algorithm, can be used to obtain the optimal solution.

Consider three processors A, B, and C, where:

A_i = processing time on processor A (first processor) for job i

B_i = processing time on processor B (second processor) for job i

C_i = processing time on processor C (third processor) for job i

Consider Min. $A = \text{Min}\{A_1, A_2, \ldots, A_n\}$, Max. $B = \text{Max}\{B_1, B_2, \ldots, B_n\}$, and Min. $C = \text{Min}\{C_1, C_2, \ldots, C_n\}$. Johnson's algorithm will render an optimal solution if either of the following two conditions holds:

(a) Min. $A \geq$ Max. B

(b) Min. $C \geq$ Max. B

That is, for all jobs, the minimum processing time on either processor A or processor C must be greater than or equal to the maximum processing time of processor B. If the above condition is not met, optimality cannot be guaranteed.

The following steps can be used to find the solution to $n \times 3$ problems:

1. Create two dummy processors and label them processor AB and processor BC.
   - The processing time of job i on processor AB is equal to the sum of the processing times for job i on processors A and B.
   - The processing time of job i on processor BC is equal to the sum of the processing times for job i on processors B and C.
2. Apply Johnson's algorithm to the n jobs on the dummy processors AB and BC.
3. Use the job sequence order obtained in step 2 to make a solution for the three processors A, B, and C. (This can be accomplished by using either a Gantt chart or a table.)

It should be noted that in most applications the above optimality conditions (a) or (b) may not hold, and the above method will render a heuristic solution.

Example 7.10 Sequencing Four Jobs by Three Processors   The following table shows the processing times (in days) for four jobs on three processors. The processing order is $A \rightarrow B \rightarrow C$. Find the best sequence and verify whether the solution is optimal.

Job, i	1	2	3	4	Min.	Max.
Processor A	10	6	7	10	6	NA
Processor B	3	1	5	6	NA	6
Processor C	8	8	9	9	8	NA

First, check the conditions for which the Johnson's algorithm can render an optimal solution: Min. $A = \text{Min}\{10, 6, 7, 10\} = 6$; Max. $B = \text{Max}\{3, 1, 5, 6\} = 6$; Min. $C = \text{Min}\{8, 8, 9, 9\} = 8$. Since Min. $A \geq$ Max. B, an optimal solution can be found. Now, calculate the processing times for the dummy processors AB and BC. The processing time AB_i is the sum of processing times A_i and B_i. The time BC_i is the sum of B_i and C_i. This is presented in the following table:

Job, i	1	2	3	4
Processing Time AB_i	13	7	12	16
Processing Time BC_i	11	9	14	15

Apply Johnson's algorithm to the above table. The list of processing times is $\{13, 7, 12, 16; 11, 9, 14, 15\}$. The shortest time in the list is $AB_2 = 7$. Place job 2 at the beginning of the sequence. Removing both AB_2 and BC_2 from the list, it becomes $\{13, 12, 16; 11, 14, 15\}$. The shortest time in the list is now $BC_1 = 11$. Place job 1 at the end of the sequence. Removing both AB_1 and BC_1 from the list, it becomes $\{12, 16; 14, 15\}$. The shortest time in the list is now $AB_3 = 12$. Place job 3 in the first available space, which is the second space. Only job 4 remains, so it is placed in the remaining space in the sequence. In summary the solution is:

Therefore, the optimal sequence is $2 \rightarrow 3 \rightarrow 4 \rightarrow 1$. Now show the Gantt chart for the sequence on the three processors A, B, and C. The makespan is 46.

7.4.2 Algorithm for *n* x *m*: Campbell Method

n x *2 Composite Approach for Solving n* x *m Flow Shop*   In the previous section, we show that in some cases it may be possible to find the optimal solution for $n \times 3$ problems. For $n \times m$ where $m > 3$, there is no easy way to find the optimal solution. The method often used to solve this problem is based on the results of the method in the previous section where the $n \times 3$ was reduced to a $n \times 2$ problem so that Johnson's rule can be applied to it. Such a method was developed by Campbell et al. (1970) and uses the $n \times 2$ Johnson's algorithm as a subroutine. The general approach is to reduce the $n \times m$ problem to a series of $n \times 2$ problems and then solve each problem by Johnson's algorithm in a manner similar to how $n \times 3$ problems were solved in the last section. The pseudoprocessors for the algorithm can be generated using the rules provided for the $n \times 3$ method. The idea is to generate a set of composite pseudoprocessors by incrementally combining different processors. A set of $n \times 2$ problems can be generated each having different composite processors. Each of these composite processor problems can then be solved, and the best alternative can be chosen from the given set of solutions.

The process for creating composite pseudoprocessors is as follows:

1. Set $k = 1$.
2. Consider the first k processors versus the last k processors. Find the summation of job times of the first k processors versus the summation of job times of the last k processors. Solve the generated $n \times 2$ problem by Johnson's rule. Record the solution and the makespan.
3. Set $k = k + 1$ and repeat step 2 until $k = m - 1$.
4. Choose the sequence that has the minimum makespan.

Example 7.11   Flow Shop of Four Jobs with Four Processors Using $n \times 2$ Consider the problem of sequencing four jobs by four processors (A, B, C, and D) as presented in the following table. All jobs are processed in the order ($A \rightarrow B \rightarrow C \rightarrow D$).

Job, i	A	B	C	D	Total
1	4	1	3	2	10
2	1	6	5	3	15
3	5	2	4	1	12
4	2	4	1	5	12
Total	12	13	13	11	49

The total processing times on processors A, B, C, and D are 12, 13, 13, and 11, respectively. Therefore, it can be observed that the worst (feasible) scheduling would have a makespan of 49 min, doing jobs sequentially one at a time. The ideal scheduling would have a makespan of 15, but usually such an ideal solution is infeasible.

For $k = 1$, consider just the first and last processors' processing times, that is, only consider processors A and D. Now solve this 4×2 problem. Using Johnson's algorithm, the sequence is 2→4→1→3. For this sequence the makespan is 23, calculated by using a table or a Gantt chart as discussed earlier.

For $k = 2$, consider the first two and the last two processors. Create two pseudoprocessors *AB* and *CD*. For *AB* add up the processing times of *A* and *B*. For *CD* add up the processing times of *C* and *D*. The resulting problem is shown below. Applying Johnson's algorithm gives a sequence of 1→4→2→3 with a makespan of 26.

For $k = 3$, create pseudoprocessors for the first three processors *ABC* and the last three processors *BCD*. Applying Johnson's algorithm gives a sequence of 4→2→3→1 with a makespan of 26.

	$k = 1$		$k = 2$		$k = 3$	
Job, *i*	A	D	AB	CD	ABC	BCD
1	4	2	5	5	8	6
2	1	3	7	8	12	14
3	5	1	7	5	11	7
4	2	5	6	6	7	10
	2→4→1→3, $f_1 = 23$		1→4→2→3, $f_1 = 26$		4→2→3→1, $f_1 = 26$	

Now, choose the best solution. The best sequence is 2→4→1→3 (associated with the $k = 1$ solution) with a makespan of 23.

Example 7.12   Flow Shop for Eight Jobs and Seven Processors Using *n* × 2   Consider a problem of sequencing eight jobs by seven processors with the following processing times (in minutes):

Job, *i*	A	B	C	D	E	F	G	Total
1	13	79	23	71	60	27	2	275
2	31	13	14	94	60	61	57	330
3	17	1	0	23	36	8	86	171
4	19	28	10	4	58	73	40	232
5	94	75	0	58	0	68	46	341
6	8	24	3	32	4	94	89	254
7	10	57	13	1	92	75	29	277
8	80	17	38	40	66	25	88	354
Total	272	294	101	323	376	431	437	2234

In this example, there are seven processors, so k varies from 1 to 6. For example for $k = 4$, the following table shows processing times for the two combined processors, which can be sequenced as an $n \times 2$ problem. The solution to this problem is $(3{\to}4{\to}6{\to}7{\to}2{\to}8{\to}5{\to}1)$ with a makespan of 632.

Job, i	$ABCD_i$	$DEFG_i$
1	186	160
2	152	272
3	41	153
4	61	175
5	227	172
6	67	219
7	81	197
8	175	219

The solutions of all problems are listed in the following table:

k	8 Jobs $\times$ 2 Processors	Job Sequence Solution	Makespan
1	A vs. G	$6{\to}7{\to}3{\to}4{\to}2{\to}8{\to}5{\to}1$	618
2	AB vs. FG	$3{\to}6{\to}2{\to}4{\to}7{\to}8{\to}5{\to}1$	628
3	ABC vs. EFG	$3{\to}6{\to}4{\to}2{\to}7{\to}8{\to}5{\to}1$	596
4	ABCD vs. DEFG	$3{\to}4{\to}6{\to}7{\to}2{\to}8{\to}5{\to}1$	632
5	ABCDE vs. CDEFG	$6{\to}3{\to}4{\to}7{\to}2{\to}8{\to}1{\to}5$	605
6	ABCDEF vs. BCDEFG	$3{\to}6{\to}4{\to}7{\to}8{\to}2{\to}1{\to}5$	595

The alternative associated with $k = 6$ has the lowest makespan, 595 min. This sequence is selected as the best. By considering all possible sequences, we can verify that the optimal solution for this example is $(3{\to}6{\to}4{\to}7{\to}2{\to}8{\to}1{\to}5)$, which has a makespan of 584 min. Such complete enumeration of all alternatives for larger problems is not feasible due to the excessive computational time. In this example, the error can be measured as $(595 - 584) / 584 = 1.9\%$.

7.4.3 Multicriteria of $n \times m$[†]

The MCDM of $n \times m$ problems is similar to the MCDM of $n \times 2$ problems.

Example 7.13 Bicriteria Sequencing Four Jobs by Three Processors Consider Example 7.10 with the following due dates in days:

Job, i	1	2	3	4
d_i	18	35	22	28

Minimum Makespan Using Johnson's algorithm, the optimal sequence for minimizing the makespan is $2\rightarrow3\rightarrow4\rightarrow1$ (see Example 7.10 solution). For this sequence, the makespan is 46 and the total tardiness is 43.

Minimum Total Tardiness According to the EDD rule, the sequence is $(1\rightarrow3\rightarrow4\rightarrow2)$. The solution for this sequence is presented below. The makespan is 50 days and the total tardiness is 41 days.

\multicolumn{4}{c}{Minimum Makespan: $(2\rightarrow3\rightarrow4\rightarrow1)$}				\multicolumn{4}{c}{Minimum Tardiness: $(1\rightarrow3\rightarrow4\rightarrow2)$}			
Job, i	F_i	d_i	T_i	Job, i	F_i	d_i	T_i
2	15	35	0	1	21	18	3
3	27	22	5	3	31	22	9
4	38	28	10	4	42	28	14
1	46	18	28	2	50	35	15
Objectives	$f_1 = 46$		$f_2 = 43$	Objectives	$f_1 = 50$		$f_2 = 41$

Suppose the weights of importance for the two objectives are $(0.5, 0.5)$, respectively. Using the composite priority index, the following solution is generated:

Job, i	1	2	3	4
Z_i	0.50	0.50	0.33	0.67
Rank	2	2	1	4

The best sequence is either $3\rightarrow1\rightarrow2\rightarrow4$ or $3\rightarrow2\rightarrow1\rightarrow4$. Since sequence $3\rightarrow1\rightarrow2\rightarrow4$ has $f_1 = 48$ and $f_2 = 33$ but sequence $3\rightarrow2\rightarrow1\rightarrow4$ has $f_1 = 48$ and $f_2 = 39$, choose sequence $3\rightarrow1\rightarrow2\rightarrow4$ as the best solution. This solution dominates (i.e., it is better in terms of both objectives) the EDD solution.

Example 7.14 Efficient Solutions for Bicriteria for Four Jobs by Three Processors
Consider Example 7.13. Find the efficient frontier for this example.

Phase 1: To generate the first midpoint, set the weights for the composite function Z_i to $(0.5, 0.5)$ and solve the sequencing problem. The solution is as follows:

Job, i	1	2	3	4
Z_i	0.50	0.50	0.33	0.67
Rank	2	3	1	4

The optimal sequence is $3\rightarrow1\rightarrow2\rightarrow4$ where $f_1 = 48$ and $f_2 = 33$.

Solving for weights (0.75, 0.25), the solution is:

Job, i	1	2	3	4
Z_i	0.75	0.25	0.33	0.67
Rank	4	1	2	3

The best sequence is $2{\rightarrow}3{\rightarrow}4{\rightarrow}1$ where $f_1 = 46$ and $f_2 = 43$. By searching adjacent points to this point, we find that $3{\rightarrow}2{\rightarrow}4{\rightarrow}1$ with (46, 38) dominates this point.

For weights (0.25, 0.75), the solution is:

Job, i	1	2	3	4
Z_i	0.25	0.75	0.33	0.67
Rank	1	4	2	3

The best sequence is $1{\rightarrow}3{\rightarrow}4{\rightarrow}2$ where $f_1 = 50$ and $f_2 = 41$. The MCDM alternatives are summarized in the following table:

Weights	(1, 0)	(0.75, 0.25)	(0.5, 0.5)	(0.25, 0.75)	(0, 1)
Alternatives	a_1	a_2	a_3	a_4	a_5
Sequence	$2{\rightarrow}3{\rightarrow}4{\rightarrow}1$	$3{\rightarrow}2{\rightarrow}4{\rightarrow}1$	$3{\rightarrow}1{\rightarrow}2{\rightarrow}4$	$1{\rightarrow}3{\rightarrow}4{\rightarrow}2$	$1{\rightarrow}3{\rightarrow}4{\rightarrow}2$
Min. f_1	46	46	48	50	50
Min. f_2	43	38	33	41	41

Since alternative 2 dominates alternative 1 and alternatives 4 and 5 are the same, only three efficient solutions are generated.

7.5 JOB SHOP OF TWO JOBS BY m PROCESSORS

7.5.1 Algorithm for 2 × m

In the previous sections, the sequencing of n jobs by one, two, three, and m processors was introduced. It was shown that optimality can be guaranteed only for $n \times 2$ problems. In this section, we discuss an algorithm for processing two jobs on m processors where each job may have a different processing sequence. For example, consider a job shop problem where two jobs are processed by five different operations: sawing, drilling, turning, tapping, and milling. Each job is processed by a different order of these processors.

The objective is to find a solution that minimizes the makespan. A graphical method is used to determine the sequence with the shortest processing time. The solution to this algorithm is guaranteed to be optimal. See Figure 7.2 for an example.

1. On the **x** axis, depict the processing times of job 1 on the M processors sequentially.
2. On the **y** axis, depict the processing times of job 2 on the M processors sequentially.
3. Shade the areas where the same processor is on both the **x** axis and the **y** axis. This represents both jobs being processed by the same processor, which is a conflict.

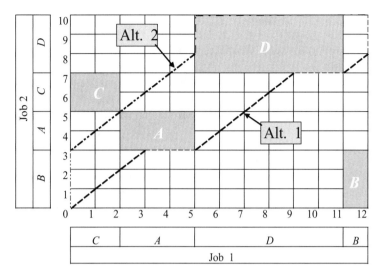

Figure 7.2 Sequencing two jobs by four processors *A, B, C,* and *D*. Alternative 1 has a makespan of 14 and alternative 2 has a makespan of 17.

4. By trial and error, try to find the best path from the origin to the upper right corner of the graph while not passing through the shaded areas. The line may travel horizontally, vertically, or diagonally (45°). The best path will move diagonally for the longest portion.

5. The solution can be interpreted as follows:
   - A diagonal line (45°) indicates that both jobs are being processed at the same time.
   - A horizontal line indicates that only job 1 is being processed while job 2 is idle.
   - A vertical line indicates that job 2 is being processed while job 1 is idle.

By maximizing the portion of the path that is diagonal, both jobs will be processed at the same time as much as possible, and therefore the makespan will be minimized. A shaded area means that both jobs cannot be processed simultaneously on the same processor, so it is an infeasible area to schedule.

Supplement S7.3.doc shows an example of sequencing two jobs on three processors.

Example 7.15 Sequencing Two Jobs on Four Processors Consider the following processing times for two jobs by four processors. Job 1 has processor sequence $C \rightarrow A \rightarrow D \rightarrow B$ and job 2 has processor sequence $B \rightarrow A \rightarrow C \rightarrow D$. Find the best job sequence for this problem and show the start and finish of each job on each processor.

	Processing Time					
	A	*B*	*C*	*D*	Total	Sequence
Job 1	3	1	2	6	12	$C \rightarrow A \rightarrow D \rightarrow B$
Job 2	2	3	2	3	10	$B \rightarrow A \rightarrow C \rightarrow D$

Job 1 has a total flow time of 12 and is depicted as the **x** axis. Job 2 has a total flow time of 10 and is depicted as the **y** axis. The rectangle size is 12×10. See Figure 7.2.

The blocked areas A, B, C, and D indicate infeasible areas. Now, find a line that connects $(0, 0)$ to $(12, 10)$. Such a line is a solution to the problem. However, note that the line that directly connects $(0, 0)$ to $(12, 10)$ passes through the A and D shaded regions, and therefore it is not a feasible solution. Now, try to find the line with the longest diagonal line portion that connects $(0, 0)$ to $(12, 10)$ without going through any of the shaded areas of A, B, C, and D. For the purpose of illustration, we present two alternatives, 1 and 2, that connect the two points $(0, 0)$ and $(12, 10)$.

Alternative 1 starts from the origin and goes diagonally toward the blocked area of processor A. After reaching the blocked area for processor A at point $(3, 3)$, the only feasible way is to move horizontally until it passes the shaded area of A. Recall that this horizontal line indicates that only job 1 is being processed on processor A while job 2 is idle, awaiting the same processor to become available. This occurs at point $(5, 3)$. At $(5, 3)$ the line runs diagonally until it reaches the shaded area of processor D at point $(9, 7)$. Once again the line must run horizontally until it passes the shaded area of processor D. At point $(11, 7)$, the line runs diagonally again. At point $(12, 8)$, the line must run vertically along the edge of the graph until it reaches the corner point. Processors B and C are not in conflict in alternative 1. The length of the diagonal portion of the line can be measured by adding the number of blocks along the path that are crossed by the diagonal line. Alternative 1's line has a diagonal length of eight.

Now, consider alternative 2. The path runs vertically from point $(0, 0)$ to point $(0, 3)$, and then it runs diagonally until it reaches the shaded area of processor D at point $(5, 8)$. It then runs vertically along the shaded area of processor D up to point $(5, 10)$. Lastly, it runs horizontally starting at point $(5, 10)$ until it reaches the corner point $(12, 10)$. This alternative has only one conflict, in using processor D. Its diagonal length is 5.

Alternative 1 has a diagonal length of 8. Alternative 2 has a diagonal length of 5. This means that alternative 1 is better than alternative 2. By trial and error, one can verify that alternative 1 has the longest feasible diagonal portion, and therefore its solution is optimal.

A Gantt chart can be used to show the solution. The Gantt charts for alternatives 1 and 2 are shown below. The shaded areas represent idle times. For alternative 1, the makespan is 14. For alternative 2, the makespan is 17.

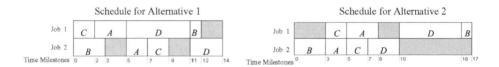

7.5.2 Multicriteria of 2 × *m*

Consider minimizing makespan as f_1 and minimizing total tardiness as f_2. The MCDM analysis of two jobs by m processors is simple. There are only two possible efficient solutions to this problem, that is, either job 1 is finished first or job 2 is finished first. To generate the two solutions, first find the solution associated with the longest diagonal line such that job 1 is finished first; then find the solution associated with the longest diagonal line such that job 2 is finished first.

Example 7.16   MCDM of Sequencing Two Jobs by m Processors   Consider Example 7.15. Suppose job 1 is due in 15 days and job 2 is due in 10 days.

To minimize f_1, makespan, solve the problem by the $2 \times m$ graphical algorithm (see the solution to Example 7.15). The minimum makespan is 14 days. Job 1 is finished first. Now, measure f_2, the total tardiness for this sequence, $f_2 = 4$ (see the following tables).

To minimize f_2, find the optimal path (by trial and error) such that job 2 is finished as close or before its due date, and then job 1 is finished as close or before its due date. This solution is shown as alternative 2 in Figure 7.2. For this solution, total tardiness $f_2 = 2$ and makespan $f_1 = 17$. Both alternatives are efficient with respect to each other. The two alternatives can then be evaluated by the DM so that the best alternative can be selected.

Alternative 1: Job 1 Finished First				Alternative 2: Job 2 Finished First				MCDM Alternative		
Job, i	F_i	d_i	T_i	Job, i	F_i	d_i	T_i		Alt. 1	Alt. 2
1	12	15	0	2	10	10	0	Min. f_1	14.00	17.00
2	14	10	4	1	17	15	2	Min. f_2	4.00	2.00
Object:	$f_1 = 14$		$f_2 = 4$	Object:	$f_1 = 17$		$f_2 = 2$			

7.6   SEQUENCING OF $n \times m$: HEAD–TAIL APPROACH

7.6.1   Overview of Head–Tail for Solving $n \times m$

There are two types of $n \times m$ sequencing problems: (a) flow shop, where the sequence of processors is the same for all jobs, and (b) job shop, where the sequence of processors may be different for different jobs. The $n \times m$ job shop problem is much more difficult to solve than the $n \times m$ flow shop problem. In this chapter, two methods for solving flow shop problems are discussed. The first method is based on the $n \times 2$ SPT covered in Section 7.4.2. The second is based on the $2 \times m$ graphical method covered in Section 7.6.1. Section 7.6.2 covers job shop problems and also uses the $2 \times m$ graphical method as its basis. The problem of sequencing n jobs by m processors is a combinatorial problem; therefore finding the optimal solution for large problems is computationally infeasible. In this section, we develop a new heuristic for solving $n \times m$ sequencing problems. The approach is based on the unidirectional head–tail method covered in Chapter 13.

To solve the $n \times m$ sequencing problem, first convert the $n \times m$ problem into a series of $2 \times m$ problems; then solve each $2 \times m$ problem by the graphical method covered in Section 7.5. Then generate an $n \times n$ flow matrix which will be used in the head–tail method.

Unidirectional Head–Tail Methods for Solving $n \times m$ Four methods are presented in Chapter 13 for solving unidirectional flow problems. They are (in order of complexity) ordinal, cardinal, simultaneous, and integer programming. When the final solution is obtained, the pairwise exchange method can be used to attempt to find an improved solution. All four methods can be used to solve sequencing problems. The ordinal method requires much less computational time than the cardinal method, but the solution may not be as good as the solution obtained from cardinal method. In this chapter, we only show the application of the cardinal unidirectional head–tail method.

Supplement S7.4a.doc shows the method and example for flow shop using ordinal head–tail.
Supplement S7.4b.doc shows the method and example for flow shop using simultaneous head–tail.

2 × m Composite Formulation of n × m Problems   Consider job i versus job j. Solve this problem by the $2 \times m$ graphical method, record the solution, and identify which job is completed first. Suppose that job i is completed first. Now solve the $2 \times m$ problem such that job j is completed first and job i is completed second. Record both these solutions and their makespans. These solutions correspond to (i, j) and (j, i) in the flow matrix, respectively. Note that it is not important which job is started first, but it is important to know which job is completed first.

Steps of Head–Tail Methods   By using a set of simple rules, this method progressively builds the solution by connecting pairs of jobs. The following notation is used for connecting pairs of sequences of jobs:

$s_p = (i \rightarrow j)$ is a candidate pair sequence where job i is completed before job j

$s_q = (\rightarrow \rightarrow \rightarrow \cdots)$ is a partially constructed sequence solution so far

1. Consider the ranking of all pairs of jobs (i, j). Rank pairs based on their makespans from the shortest to the longest makespan.
2. Consider the highest ranked unassigned pair $s_p = (i, j)$ such that:
   (a) Both i and j are not in s_q.
   (b) Either i or j is in s_q.
3. Use relevant head–tail rules to combine s_p and s_q.
4. Repeat steps 2 and 3 until all jobs are assigned to s_q.
5. Try to improve the obtained final solution by using the pairwise exchange method.

Unidirectional Cardinal Rule (Conflict Resolution)   If there is a common job i in $s_p = (i \rightarrow j)$ and $s_q = (\cdots \rightarrow i \rightarrow \cdots)$, then consider inserting j (the uncommon job) between each pair of jobs in in s_q. Choose the alternative that has the best objective function value (e.g., the minimum makespan).

For example, if $s_p = (2 \rightarrow 5)$ and $s_q = (1 \rightarrow 2 \rightarrow 4)$, the uncommon job is 5. The common job is 2. Consider all possible locations that job 5 can be inserted in sequence s_q. They are $(5 \rightarrow 1 \rightarrow 2 \rightarrow 4)$, $(1 \rightarrow 5 \rightarrow 2 \rightarrow 4)$, $(1 \rightarrow 2 \rightarrow 5 \rightarrow 4)$, and $(1 \rightarrow 2 \rightarrow 4 \rightarrow 5)$. Choose the sequence that has the lowest makespan as the new s_q.

7.6.2 Flow Shop of $n \times m$: Cardinal Head–Tail

For flow shop problems, the sequence of processors is the same for all jobs; therefore, the Gantt chart solution will generally provide the optimal solution for $2 \times m$ problems. Therefore, we suggest using the Gantt chart directly to find the makespan for a given $2 \times m$ problems, that is, it is not necessary to use the $2 \times m$ graphical method. The steps of the cardinal head–tail method are shown by solving the following example.

Example 7.17   Solving a 4 × 4 Flow Shop Problem by Cardinal Head–Tail   Consider Example 7.11, the flow shop problem of four jobs by four processors. Solve this problem by the cardinal head–tail method.

2 × m Composite Formulation of $n \times m$ for Flow Shop   First generate the $n \times n$ makespan matrix by considering each pair of jobs. For example, consider job 2 versus job 3. Solve this problem by $2 \times m$ graphical method (or the Gantt chart). The makespan is 17 when job 2 is scheduled before job 3. Now find the best solution when job 3 is scheduled to be finished before job 2. The makespan is 21 for this solution.

The makespans for all combinations of each pair of jobs are shown in the matrix below.

	1	2	3	4
1	0	19	16	16
2	17	0	17	20
3	16	21	0	17
4	14	20	14	0

Cardinal Head–Tail Methods for Solving $n \times m$ Flow Shop Problems   Now, apply the cardinal head–tail method to the matrix. First, order the pairs in ascending order:

Rank	1	2	3	4	5	6	7	8	9	10	11	12
w^0 Pair	4→1	4→3	1→3	1→4	3→1	2→1	2→3	3→4	1→2	2→4	4→2	3→2
w_{ij}	14	14	16	16	16	17	17	17	19	20	20	21

1. Consider the first ranked pair of jobs, $s_p = (4{\rightarrow}1)$; add s_p to s_q; $s_q = (4{\rightarrow}1)$.
2. Consider $s_p = (4{\rightarrow}3)$; apply cardinal rule 2 and find the best location for job 3. There are three alternatives to be considered as presented below.

Sequence	3→4→1	4→3→1	4→1→3
Makespan	19	**18**	**18**

The sequences 4→3→1 and 4→1→3 have the same makespan (indicated by boldface). Alternative $s_q = (4{\rightarrow}3{\rightarrow}1)$ is selected arbitrarily.

3. Consider $s_p = (1{\rightarrow}3)$, but ignore it due to its conflict with $(3{\rightarrow}1)$ in $s_q = (4{\rightarrow}3{\rightarrow}1)$. Ignore $(1{\rightarrow}4)$ and $(3{\rightarrow}1)$ as 1, 4, and 3 are already assigned in $s_q = (4{\rightarrow}3{\rightarrow}1)$.
4. Consider $s_p = (2{\rightarrow}1)$; as compared to $s_1 = (4{\rightarrow}3{\rightarrow}1)$, there are four possible location for job 2 as presented below.

Sequence	2→4→3→1	4→2→3→1	4→3→2→1	4→3→1→2
Makespan	**23**	26	25	26

$2\rightarrow4\rightarrow3\rightarrow1$ has the lowest makespan, 23 (bolded).

A summary of the method is provided below. The makespan is 23.

Iteration	Used Rank	s_p	Action Taken	Resulting Sequence, s_q	Candidates
1	1	$4\rightarrow1$	Add s_p as a new sequence	$(4\rightarrow1)$	2,3
2	2	$4\rightarrow3$	Insert station 3 before station 1	$(4\rightarrow3\rightarrow1)$	2
3	6	$2\rightarrow1$	Insert station 2 before station 4	$(2\rightarrow4\rightarrow3\rightarrow1)$	—

Example 7.18 Solving an 8 × 7 Flow Shop Problem by Cardinal Head–Tail Consider Example 7.12, the flow shop problem of eight jobs by seven processors. Solve this problem by the cardinal head–tail method.

2 × m Composite Formulation of n × m Flow Shop For each pair of jobs, the makespan for each pair of jobs is presented in the following table:

	1	2	3	4	5	6	7	8
1	—	458	376	417	387	456	442	425
2	332	—	384	382	383	452	390	386
3	292	315	—	249	358	271	294	371
4	309	317	318	—	360	375	315	373
5	431	436	427	408	—	478	436	373
6	294	306	340	294	478	—	299	362
7	329	369	363	361	362	431	—	277
8	359	437	440	412	421	449	357	—

Cardinal Head–Tail Method for Solving n × m Flow Shop Problems Now, apply the head–tail method. First, rank all pairs in ascending order:

	1	2	3	4	5	6	7	8
1	—	54	31	40	36	53	50	42
2	16	—	34	32	33	52	37	35
3	4	11	—	1	19	2	5	27
4	10	13	14	—	21	30	11	28
5	44	46	43	38	—	55	46	28
6	5	9	17	5	55	—	8	23
7	15	26	25	22	23	44	—	3
8	20	48	49	39	41	51	18	—

For easier reference, some of the ranking are shown below.

Rank	1	2	3	4	5	6	7	8	9	...	19
w^0 Pair	3→4	3→6	7→8	3→1	3→7	6→1	6→4	6→7	6→2	...	3→5
w_{ij}	249	271	277	292	294	294	294	299	306	...	358

1. Consider the first pair of jobs, $s_p = (3→4)$; set $s_q = (3→4)$.
2. Consider $s_p = (3→6)$. Since there is a conflict between $(3→4)$ and $(3→6)$, determine the best location for 6 in the sequence $(3→4)$.

Sequence	6→3→4	3→6→4	3→4→6
Makespan	380	311	392

The sequence $(3→6→4)$ has the lowest makespan. Therefore, $s_q = (3→6→4)$.

3. Temporarily discard the pair $(7→8)$. Consider $s_p = (3→1)$. Since there is a conflict with s_q, determine the best location for 1 in the sequence $(3→6→4)$.

Sequence	1→3→6→4	3→1→6→4	3→6→1→4	3→6→4→1
Makespan	513	513	453	339

Therefore, $s_q = (3→6→4→1)$.

4. Consider $s_p = (3→7)$. Since there is a conflict with s_q, determine the best location for 7 in the sequence $(3→6→4→1)$.

Sequence	7→3→6→4→1	3→7→6→4→1	3→6→7→4→1	3→6→4→7→1	3→6→4→1→7
Makespan	494	490	402	396*	506

Therefore, $s_q = (3→6→4→7→1)$.

5. Consider $s_p = (7→8)$. Since there is a conflict with s_q, determine the best location for 8 in the sequence $(3→6→4→7→1)$.

	8→3→6→4 →7→1	3→8→6→4 →7→1	3→6→8→4 →7→1	3→6→4→8 →7→1	3→6→4→7 →8→1	3→6→4→7 →1→8
Sequence						
Makespan	600	556	522	483	449*	546

Therefore, $s_q = (3→6→4→7→8→1)$.

6. Discard the pairs $(6\rightarrow1)$, $(6\rightarrow4)$, and $(6\rightarrow7)$. Consider $s_p = (6\rightarrow2)$. Since there is a conflict with s_q, determine the best location for 2 in the sequence $(3\rightarrow6\rightarrow4\rightarrow7\rightarrow8\rightarrow1)$.

	$2\rightarrow3\rightarrow6\rightarrow4\rightarrow7$ $\rightarrow8\rightarrow1$	$3\rightarrow2\rightarrow6\rightarrow4\rightarrow7$ $\rightarrow8\rightarrow1$	$3\rightarrow6\rightarrow2\rightarrow4\rightarrow7$ $\rightarrow8\rightarrow1$	$3\rightarrow6\rightarrow4\rightarrow2\rightarrow7$ $\rightarrow8\rightarrow1$
Sequence				
Makespan	664	651	582	544

	$3\rightarrow6\rightarrow4\rightarrow7\rightarrow2\rightarrow8\rightarrow1$	$3\rightarrow6\rightarrow4\rightarrow7\rightarrow8\rightarrow2\rightarrow1$	$3\rightarrow6\rightarrow4\rightarrow7\rightarrow8\rightarrow1\rightarrow2$
Sequence			
Makespan	538	506*	596

Therefore, $s_q = (3\rightarrow6\rightarrow4\rightarrow7\rightarrow8\rightarrow2\rightarrow1)$.

7. Discard all pairs until $(3\rightarrow5)$. Consider $s_p = (3\rightarrow5)$. Since there is a conflict with s_q, determine the best location for 5 in the sequence $(3\rightarrow6\rightarrow4\rightarrow7\rightarrow8\rightarrow2\rightarrow1)$.

	$5\rightarrow3\rightarrow6\rightarrow4\rightarrow7\rightarrow8$ $\rightarrow2\rightarrow1$	$3\rightarrow5\rightarrow6\rightarrow4\rightarrow7\rightarrow8$ $\rightarrow2\rightarrow1$	$3\rightarrow6\rightarrow5\rightarrow4\rightarrow7\rightarrow8$ $\rightarrow2\rightarrow1$	$3\rightarrow6\rightarrow4\rightarrow5\rightarrow7\rightarrow8$ $\rightarrow2\rightarrow1$
Sequence				
Makespan	732	730	657	639

	$3\rightarrow6\rightarrow4\rightarrow7\rightarrow5\rightarrow8$ $\rightarrow2\rightarrow1$	$3\rightarrow6\rightarrow4\rightarrow7\rightarrow8\rightarrow5$ $\rightarrow2\rightarrow1$	$3\rightarrow6\rightarrow4\rightarrow7\rightarrow8\rightarrow2$ $\rightarrow5\rightarrow1$	$3\rightarrow6\rightarrow4\rightarrow7\rightarrow8\rightarrow2$ $\rightarrow1\rightarrow5$
Sequence				
Makespan	597	635	596	595*

The sequence $(3\rightarrow6\rightarrow4\rightarrow7\rightarrow8\rightarrow2\rightarrow1\rightarrow5)$ is selected, with a makespan of 595. A summary of the method is provided below.

Iteration	Used Rank	s_p	Action Taken	Resulting Sequence, s_q	Candidates
1	1	$3\rightarrow4$	Add s_p as a new sequence	$(3\rightarrow4)$	1,2,5,6,7,8
2	2	$3\rightarrow6$	Insert station 6 before station 4	$(3\rightarrow6\rightarrow4)$	1,2,5,7,8
3	—	$7\rightarrow8$	Discard temporarily	$(3\rightarrow6\rightarrow4)$	1,2,5,7,8
4	4	$3\rightarrow1$	Insert station 1 before station 6	$(3\rightarrow6\rightarrow4\rightarrow1)$	2,5,7,8
5	5	$3\rightarrow7$	Insert station 7 before station 1	$(3\rightarrow6\rightarrow4\rightarrow7\rightarrow1)$	2,5,8
6	—	$7\rightarrow8$	Reconsider $7\rightarrow8$, insert 5 after 4	$(3\rightarrow6\rightarrow4\rightarrow7\rightarrow8\rightarrow1)$	2,5
7	9	$6\rightarrow2$	Insert station 1 before station 6	$(3\rightarrow6\rightarrow4\rightarrow7\rightarrow8\rightarrow2\rightarrow1)$	5
8	19	$3\rightarrow5$	Insert station 5 before station 7	$(3\rightarrow6\rightarrow4\rightarrow7\rightarrow8\rightarrow2\rightarrow1\rightarrow5)$	

At this point, apply the pairwise exchange method to the obtained final solution. By exchanging the positions of jobs 2 and 8, the optimal solution $(3{\rightarrow}6{\rightarrow}4{\rightarrow}7{\rightarrow}2{\rightarrow}8 {\rightarrow}1{\rightarrow}5)$ with a makespan of 584 min will be found.

7.6.3   Job Shop of $n \times m$: Cardinal Head–Tail[†]

As stated before, in job shop problems, different products may have different machine sequences. As a result, job shop problems are far more difficult to solve than flow shop problems. The same cardinal head–tail method presented in the previous section can be used to solve this problem. Note that it is necessary to generate an $n \times n$ flow matrix by solving $2 \times m$ problems by the graphical method (of Section 7.5) for each pair of jobs (i,j).

Supplement S7.5.doc shows the method and example for job shop using ordinal head–tail.

Example 7.19   Solving a 4 $\times$ 4 Job Shop Problem by the Cardinal Head–Tail Method Consider a job shop problem of processing four jobs by four processors as shown in the following table.

Solve this problem by the cardinal head–tail method.

Job, i	A	B	C	D	Processor Sequence
1	4	1	3	2	$D{\rightarrow}C{\rightarrow}B{\rightarrow}A$
2	1	6	5	3	$A{\rightarrow}B{\rightarrow}C{\rightarrow}D$
3	5	2	4	1	$B{\rightarrow}A{\rightarrow}C{\rightarrow}D$
4	2	4	1	5	$C{\rightarrow}A{\rightarrow}B{\rightarrow}D$

$2 \times m$ Composite Formulation of $n \times m$ Job Shop Problem   For example, consider job 2 versus job 3. Use the $2 \times m$ graphical method (of Section 7.5) to solve this problem. The solution (details not shown) indicates that the makespan for completing job 2 before job 3 is 16, and the makespan for completing job 3 before job 2 is 18. The makespans and their rankings for all pairs are presented below.

Makespan Matrix of Pairs of Jobs					Rankings of Pairs of Jobs				
	1	2	3	4		1	2	3	4
1	0	12	10	12	1	—	2	1	2
2	15	0	16	19	2	8	—	9	12
3	12	18	0	13	3	2	11	—	7
4	12	16	12	0	4	2	9	2	—

The steps of the cardinal head–tail method are presented below.

1. Consider the first pair of jobs, $s_p = (1 \rightarrow 3)$; set $s_q = (1 \rightarrow 3)$.
2. Consider $s_p = (4 \rightarrow 1)$; find the best location for job 4. The best solution is $s_q = (4 \rightarrow 1 \rightarrow 3)$.

Sequence	$4 \rightarrow 1 \rightarrow 3$	$1 \rightarrow 4 \rightarrow 3$	$1 \rightarrow 3 \rightarrow 4$
Makespan	13	20	20

3. Consider $s_p = (1 \rightarrow 2)$; this is in conflict with s_q. Find the best location for job 2 in the sequence $(4 \rightarrow 1 \rightarrow 3)$.

Sequence	$2 \rightarrow 4 \rightarrow 1 \rightarrow 3$	$4 \rightarrow 2 \rightarrow 1 \rightarrow 3$	$4 \rightarrow 1 \rightarrow 2 \rightarrow 3$	$4 \rightarrow 1 \rightarrow 3 \rightarrow 2$
Makespan	21	21	21	21

Each sequence has a makespan of 21. Therefore, each of these solutions is optimal. For the purpose of illustration, the sequence $(4 \rightarrow 1 \rightarrow 2 \rightarrow 3)$ is selected. Since all jobs are assigned, the final sequence is $(4 \rightarrow 1 \rightarrow 2 \rightarrow 3)$. The makespan is 21. The Gantt chart for this solution is presented below.

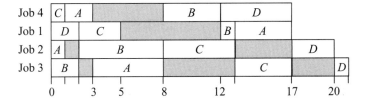

Iteration	Used Rank	s_p	Action Taken	Resulting Sequence, s_q	Candidates
1	1	$1 \rightarrow 3$	Add s_p as a new sequence	$(1 \rightarrow 3)$	2, 4
2	2	$4 \rightarrow 1$	Insert job 4 before job 1	$(4 \rightarrow 1 \rightarrow 3)$	2
3	2	$1 \rightarrow 2$	Insert job 2 before job 3	$(4 \rightarrow 1 \rightarrow 2 \rightarrow 3)$	

Example 7.20   Solving a 7 × 8 Job Shop Problem by Cardinal Head–Tail   Consider the job shop problem of eight jobs by seven processors as presented in Table 7.4.

TABLE 7.4   Job Shop Problem with Eight Jobs and Seven Processors

Job, i	A	B	C	D	E	F	G	Processor Order
1	13	79	23	71	60	27	2	$B \rightarrow A \rightarrow D \rightarrow C \rightarrow E \rightarrow G \rightarrow F$
2	31	13	14	94	60	61	57	$A \rightarrow C \rightarrow D \rightarrow F \rightarrow B \rightarrow G \rightarrow E$
3	17	1	0	23	36	8	86	$A \rightarrow B \rightarrow C \rightarrow D \rightarrow E \rightarrow F \rightarrow G$
4	19	28	10	4	58	73	40	$C \rightarrow A \rightarrow B \rightarrow G \rightarrow F \rightarrow E \rightarrow D$
5	94	75	0	58	0	68	46	$D \rightarrow C \rightarrow F \rightarrow B \rightarrow E \rightarrow A \rightarrow G$
6	8	24	3	32	4	94	89	$G \rightarrow F \rightarrow E \rightarrow D \rightarrow C \rightarrow B \rightarrow A$
7	10	57	13	1	92	75	29	$E \rightarrow D \rightarrow G \rightarrow C \rightarrow F \rightarrow B \rightarrow A$
8	80	17	38	40	66	25	88	$G \rightarrow E \rightarrow D \rightarrow B \rightarrow F \rightarrow A \rightarrow C$

$2 \times m$ Composite Formulation of $n \times m$ Job Shop   Solve a $2 \times m$ (i.e., 2×8) problem for each pair of jobs. The following table shows the summary of all solutions:

	1	2	3	4	5	6	7	8
1	—	322	275	275	290	275	276	275
2	353	—	330	382	330	374	330	341
3	233	356	—	190	220	171	222	174
4	308	253	239	—	263	318	232	263
5	341	341	341	341	—	341	341	341
6	258	289	254	254	303	—	254	384
7	277	342	278	312	305	325	—	330
8	363	354	354	354	413	414	358	—

The rankings of the above pairs in ascending order are shown below.

	1	2	3	4	5	6	7	8
1	—	30	16	16	24	16	20	16
2	45	—	32	53	32	52	32	36
3	7	49	—	3	4	1	5	2
4	27	9	8	—	14	29	6	14
5	36	36	36	36	—	36	36	36
6	13	23	10	10	25	—	10	54
7	21	44	22	28	26	31	—	32
8	51	46	46	46	55	56	50	—

Cardinal Head–Tail Method for Solving $n \times m$ Job Shop Problems

1. Consider the first ranked pair of jobs, $s_p = (3{\rightarrow}6)$; set $s_q = (3{\rightarrow}6)$.
2. Consider $s_p = (3{\rightarrow}8)$. Since there is a conflict between $(3{\rightarrow}6)$ and $(3{\rightarrow}8)$, determine the best location for 8 in the sequence $(3{\rightarrow}6)$. Possible solutions are presented below.

Sequence	$8{\rightarrow}3{\rightarrow}6$	$3{\rightarrow}8{\rightarrow}6$	$3{\rightarrow}6{\rightarrow}8$
Makespan	**428**	**428**	529

Both $(8{\rightarrow}3{\rightarrow}6)$ and $(3{\rightarrow}8{\rightarrow}6)$ have the lowest makespan, 428. Either sequence can be used. Suppose that arbitrarily $(8{\rightarrow}3{\rightarrow}6)$ is selected.

3. All remaining pairs result in conflicts, which are resolved using the cardinal rules. A summary of all iterations is provided below.

Iteration	Used Rank	s_p	Action Taken	Resulting Sequence, s_q	Candidates
1	1	3→6	Add s_p as a new sequence	3→6	1,2,4,5,7,8
2	2	3→8	Insert station 8 before station 3	8→3→6	1,2,4,5,7
3	3	3→4	Insert station 4 before station 3	8→4→3→6	1,2,5,7
4	4	3→5	Insert station 3 before station 6	8→5→4→3→6	1,2,7
5	5	3→7	Insert station 7 after station 5	8→5→7→4→3→6	1,2
6	7	3→1	Insert station 1 before station 6	8→5→7→4→3→1→6	2
7	9	4→2	Insert station 4 before station 2	8→5→7→4→2→3→1→6	

The final solution is 8→5→7→4→2→3→1→6 with a makespan of 559.

7.6.4 Multicriteria of $n \times m^{\dagger}$

This method is the same as the multicriteria $n \times m$ method of Section 7.4.3. The MCDM approach for $n \times m$ is based on the MCDM approach discussed for $n \times 2$ MCDM problems. Note that because heuristics are used in solving $n \times m$ problems, there is no guarantee that the generated solutions are efficient.

Example 7.21   MCDM of Flow Shop Sequencing Four Jobs by Four Processors
Consider Example 7.11 with the following due dates. Find the sequencing associated with the weights of importance of (0.5, 0.5) for the two objectives.

Job, i	1	2	3	4
d_i	15	25	13	22

Minimum Makespan   The best sequence for minimizing the makespan is 2→4→1→3 (see Example 7.11). For this solution, the makespan is 23 days and total tardiness is 17 days.

Minimum Tardiness   According to the EDD rule, the sequence was found to be 3→1→4→2.

For this solution, the makespan is 29 days and total tardiness is 5 days.

First Alternative: (2→4→1→3)				Second Alternative: (3→1→4→2)			
Job, i	F_i	d_i	T_i	Job, i	F_i	d_i	T_i
2	15	25	0	3	12	13	0
4	20	22	0	1	16	15	1
1	22	15	7	4	21	22	0
3	23	13	10	2	29	25	4
Objectives	$f_1 = 23$		$f_2 = 17$	Objectives	$f_1 = 29$		$f_2 = 5$

For bicriteria problems, there may exist a set of efficient alternatives that range between the two extreme alternatives. Now suppose the weights of importance for the two objectives are (0.5, 0.5). The composite priority index is $Z_i = 0.5P_i(\text{SPT}) + 0.5P_i(\text{EDD})$.

Job, i	1	2	3	4
Z_i	0.33	0.17	0.83	0.67
Rank	2	1	4	3

For the sequence (2→1→4→3), $f_1 = 24$ and $f_2 = 12$.

7.7 STOCHASTIC SEQUENCING

Sequencing problems can be classified into the following four groups:

I. *Static Problems with Deterministic Processing Times*   In these problems processing times are deterministic and the arrival of jobs is static, that is, all jobs are known and are ready to be processed at the beginning of the scheduling period. These problems are addressed in previous sections of this chapter.

II. *Dynamic Problems with Deterministic Processing Times*   Processing times are deterministic and the arrival of jobs is dynamic, that is, jobs may arrive intermittently throughout the systems operations. These problems are discussed in Section 7.2 by using the critical ratio rule.

III. *Static Problems with Stochastic Processing Times*   In the previous sections, it is assumed that the processing times of all jobs are deterministic, that is, known with certainty. In some applications, the processing time for a job may not be deterministic. This section deals with stochastic sequencing where processing times can be probabilistic. It is assumed that the probability distributions of the processing times are known and are independent from each other. The probability distribution of the processing time of a job can be presented by either a continuous function or a discrete function.

IV. *Dynamic Problems with Stochastic Processing Times*   In addition to the assumption of probabilistic processing times, the arrival of incoming jobs can be random. In this case, the interarrival and processing times are probabilistic, usually represented by exponential distributions. Queuing and/or simulation models can be used to analyze dynamic stochastic scheduling problems.

- Use single-channel, single-phase models to analyze $n \times 1$ sequencing problems.
- Use multi-channel, multi-phase models to analyze $n \times m$ (flow and job shop) sequencing problems.

7.7.1 Stochastic Sequencing of $n \times 1$[†]

Use of Beta Probability Distribution   The probability distribution for the processing time of each job can be represented by the beta distribution. The normal distribution is a special case of the beta distribution where the mode is equal to the mean. Three parameters can be used to approximate a beta distribution: a, the optimistic time; m, the most likely time; and b, the pessimistic time. Though each individual processing time is represented by a beta distribution, the total processing time (makespan) of a sequence can be represented by the normal distribution. This is due to the central limit theorem, which states that the sum of independent random variables (processing times) is approximately normal as the number of independent variables increases. We note that stochastic $n \times 1$ sequencing problems are somewhat similar to PERT and the following definitions have already been introduced for PERT problems (see Chapter 8).

The following notation is used to analyze stochastic sequencing. For each job, consider:

$m =$ most likely processing time

$a =$ optimistic processing time, where $a \leq m$

$b =$ pessimistic processing time, where $b \geq m$

$t_e =$ expected processing time, where $t_e = (a + 4m + b)/6$

$\mathrm{SD} =$ standard deviation of expected processing time, where $\mathrm{SD} = (b - a)/6$

$\mathrm{Var} =$ variance of expected processing time, where $\mathrm{Var} = \mathrm{SD}^2$

Shortest Expected Processing Time (SEPT) Rule   This method is analogous to the SPT rule in deterministic $n \times 1$ sequencing problems, which minimizes the expected (average) flow time. The SEPT rule minimizes the expected flow time of all jobs. As in the SPT rule, sequence the jobs in an increasing order of expected processing times.

Sequencing Approach   First find the expected processing times of all jobs. Then apply the SEPT rule to sequence them. Measure the following criteria for the final sequencing solution:

$$M_e = \sum_{i=1}^{N} t_{e,i} + z\,\mathrm{SD}_i \quad \text{Expected makespan with } P(Z = z) \text{ chance of completion}$$

$$V_T = \sum_{i=1}^{N} \mathrm{Var}_i \quad \text{Variance of expected duration of all jobs}$$

$$\mathrm{SD}_T = \sqrt{V_T} \quad \text{Standard deviation of expected duration of all jobs}$$

$$\bar{F}_e = \frac{1}{n} \sum_{i=1}^{n} F_{e,i} \quad \text{Expected average flow time of sequence}$$

Risk-Averse SPT Rule for Sequencing   To minimize the impact of uncertainty on the expected average flow time, jobs can be sequenced by the risk-averse SPT rule. We define this rule as follows. Sequence the jobs in ascending order of $t_{e,i} + z\,\text{SD}_i$ where z is a given value that ranges from 0 to 3; a higher z signifies higher risk aversion. Four discrete values of z are 0, 1, 2, and 3 and are associated with probabilities of 50, 84.13, 97.72, and 99.87% (respectively) that the job will be completed within $t_{e,i} + z\,\text{SD}_i$. This method sequences jobs that are less risky (in terms of the total processing time) first, which decreases the uncertainty in the average waiting time of all jobs. The DM can choose a z value (from 0 to 3) and solve the problem using the SPT rule. Alternatively, the value of z can be chosen using the bicriteria method presented in the next section.

Supplement S7.6.doc discusses the use of exponential distribution for solving stochastic $n \times 1$ sequencing problems.

Example 7.22 SEPT Rule   Consider sequencing four jobs on one processor. The most optimistic processing time (a), most likely processing time (m), and most pessimistic processing time (b) for each job are shown in the following table (in minutes):

(a) Find the best sequence using the SEPT rule and analyze the performance of the chosen sequence.

(b) Find the probability of finishing all jobs in 38 min or less.

(c) For a 84.13% probability of completion of each job, find the sequencing solution. Find the probability of completion associated with this makespan.

(a) Using the given information, find the expected processing time, standard deviation, and variance for each job. For example, for job 1, $t_{e,1} = (4 + 4 \times 6 + 14)/6 = 7$, and $\text{SD}_1 = (14 - 4)/6 = 1.667$ and $\text{Var}_1 = (1.667)^2 = 2.778$.

job, i	a	m	b	$t_{e,i}$	SD_i	Var_i	SEPT Rank
1	4	6	14	7	1.667	2.778	1
2	9	10	11	10	0.333	0.111	4
3	6	9	15	9.5	1.5	2.25	3
4	4	8	9	7.5	0.833	0.694	2
Total	—	—	—	34	—	5.833	—

According to the SEPT rule, the best sequence is $1 \rightarrow 4 \rightarrow 3 \rightarrow 2$. For this solution

$$M_e = 7 + 10 + 9.5 + 7.5 = 34 \text{ min}$$

$$V_T = 2.778 + 0.111 + 2.25 + 0.694 = 5.833$$

$$SD_T = \sqrt{5.833} = 2.415$$

$$\bar{F}_e = \frac{1}{n}\sum_{i=1}^{n} F_{e,i} = \frac{7 + 14.5 + 24 + 34}{4} = 19.9 \text{ min}$$

Therefore, the total expected processing time for this problem is 34 min, where the probability of finishing all jobs within 34 min is 50%. Note that regardless of the sequence in $n \times 1$ problems, the makespan (total processing time), total variance, and total standard deviation remain the same.

(b) To find the probability of finishing all jobs within 38 min, first find the z associated with 38 min:

$$z = \frac{38 - M_e}{\text{SD}_T} = \frac{38 - 34}{2.415} = 1.66$$

Using the normal probability distribution table at the back of the book yields

$$P(Z \leq 1.66) = \Phi(1.66) \approx 0.95$$

Therefore, the probability of finishing all jobs in 38 min or less is 95%.

(c) The sequencing solution for $P(Z \leq z) = 84.13\%$ or $z = 1$ is presented below.

job, i	$t_{e.i}$	SD_i	Var_i	$t = t_{e,i} + 1\ \text{SD}_i$	SEPT Rank
1	7	1.667	2.778	8.667	2
2	10	0.333	0.111	10.333	3
3	9.5	1.500	2.250	11	4
4	7.5	0.833	0.694	8.333	1
Total	34	—	5.833	36.415	—

For $z = 1$, the sequence of jobs can be determined by applying the SEPT rule to $t_{e,i} + 1\ \text{SD}_i$, and the solution is $4 \rightarrow 1 \rightarrow 2 \rightarrow 3$. The expected average flow time of this solution ($z = 1$) is $\bar{F}_e = (7.5 + 14.5 + 24.5 + 34)/4 = 20.125$, which is higher than $\bar{F}_e = 19.9$ when $z = 0$. However, the associated risk of completion is less for $z = 1$ than for $z = 0$.

See Supplement S7.7.xls.

7.7.2   Bicriteria Stochastic Sequencing of $n \times 1$[†*]

Consider the following two objectives:

Minimize expected average flow time, $f_1 = \bar{F}_e$.
Maximize confidence level for completing expected average flow time, $f_2 = z$.

The value of z can vary from 0 to 3 where 3 is the highest confidence. For simplicity, only consider $z = 1, 2, 3$. The bicriteria method is as follows:

1. Measure $t_{e,i}$, SD_i for each job.
2. Sequence the jobs using the SEPT rule and the risk-averse SEPT rule using $z = 1, 2, 3$.

3. If all sequences (solutions) are the same, stop. If two z values generate the same sequence, use the higher z for multicriteria ranking (it will produce a better utility value).

4. Measure f_1 and f_2 for all sequences. This is the set of bicriteria alternatives. Use a multicriteria decision-making method to choose the best alternative.

Example 7.23   Bicriteria Stochastic Scheduling   Consider Example 7.22. Generate four bicriteria solutions using $z = 0, 1, 2, 3$. Find the best sequence for the four jobs by using an additive utility function to rank alternatives, where the weights of importance of the normalized values of the two objectives are (0.6, 0.4).

Using the SEPT rule and the risk-averse rule for $z = 1, 2, 3$, the following sequences are found:

$f_2 = z$	Job, i	1	2	3	4	Sequence	$f_1 = \bar{F}_e$
$z = 0$	$t_{e,i}$	7	10	9.5	7.5	$1 \to 4 \to 3 \to 2$	19.88
	SD_i	1.67	0.33	1.5	0.83	—	—
$z = 1$	$t_{e,i} + 1\,SD_i$	8.67	10.33	11	8.33	$4 \to 1 \to 2 \to 3$	20.13
$z = 2$	$t_{e,i} + 2\,SD_i$	10.33	10.66	12.5	9.16	$4 \to 1 \to 2 \to 3$	20.13
$z = 3$	$t_{e,i} + 3\,SD_i$	12.00	10.99	14	9.99	$4 \to 2 \to 1 \to 3$	20.88

Table 7.5 shows all alternatives and their utility function values. Note that expected processing times are used in finding the expected average flow time. Note that $z = 2$ and $z = 3$ produce the same sequence, so only the higher z value ($z = 3$) is considered for multicriteria ranking. Alternative 2 ($z = 2$) is included in the table below for illustration, but it would typically be discarded. Alternative 3 is the best alternative because it has the best utility value.

See Supplement S7.8.xls.

7.7.3   Stochastic Sequencing of $n \times 2$[†]

Consider sequencing $n \times 2$ problems where the processing times of jobs are probabilistic. The same procedure used for solving $n \times 1$ stochastic sequencing (described in Section 7.7.1) can be used to solve this problem. First find the expected processing times and

TABLE 7.5   Alternatives and Their Utility Values for Example 7.23

Alternatives	1	**2**	3	4
Sequence	$1 \to 4 \to 3 \to 2$	$4 \to 1 \to 2 \to 3$	$4 \to 1 \to 2 \to 3$	$4 \to 2 \to 1 \to 3$
Min. $f_1 = \bar{F}_e$	19.88	20.13	20.13	20.88
Max. $f_2 = z$	0	1	2	3
Max. f_1'	1.00	0.75	0.75	0.00
Max. f_2'	0.00	0.33	0.67	1.00
Max. $U(0.6, 0.4)$	0.60	0.58	0.72	0.40

standard deviations of all jobs. Then apply the $n \times 2$ sequencing procedure (Johnson's rule of Section 7.3.1) to the expected processing times of the jobs. That is, apply the SEPT rule to Johnson's rule for $n \times 2$ problems. Then find the expected makespan of the obtained sequence. Note that in $n \times 2$ sequencing the makespan depends on the sequence of jobs, but in $n \times 1$ it is the same for every sequence. The risk-averse rule for stochastic $n \times 1$ sequencing from the previous section can also be applied to stochastic $n \times 2$ sequencing.

Example 7.24   Stochastic Sequencing for $n \times 2$ Consider the processing times and standard deviations for four jobs on two processors as presented in the following table:

Job, i	1	2	3	4
Processor A: $t_{e,i}$, SD$_i$	4, 2.67	8, 0.33	5, 1.5	2, 3.5
Processor B: $t_{e,i}$, SD$_i$	3, 1.33	2, 4	7, 0.67	6, 1.33

The sequence of processors is A→ B for all jobs.

(a) Find the sequence using the SEPT rule and Johnson's rule.

(b) For a 84.13% probability of completion of each job, find the sequencing solution.

(a) By applying Johnson's rule to the above expected values, the solution is (4→ 3→ 1→ 2). The makespan for this solution is 21 (e.g., using a Gantt chart). This solution is associated with a 50% probability of occurrence.

(b) For a 84.13% probability of completion of each job, use $z = 1$ to find the processing time for each job. For example, the processing time for the first job on machine A is $4 + 1 \times 2.67 = 6.67$.

Job, i	1	2	3	4
Processor A: $t_{e,i} + 1$ SD$_i$	6.67	8.33	6.5	5.5
Processor B: $t_{e,i} + 1$ SD$_i$	4.33	6	7.67	7.33

See Supplement S7.9.xls.

By applying Johnson's rule, the sequence is 4→3→2→1. The expected makespan for this sequence is 22.

Bicriteria Stochastic Sequencing of $n \times 2$[†]   The bicriteria of stochastic $n \times 2$ problems is the same as the bicriteria for stochastic $n \times 1$ problems. The following example shows the details of this method.

Example 7.25   Bicriteria Stochastic $n \times 2$ Sequencing   Consider Example 7.24. Generate four bicriteria alternatives using $z = 0, 1, 2, 3$. Find the best alternative where weights of importance for the two objectives are (0.6, 0.4), respectively. To solve the

TABLE 7.6   Alternative Sequences and Their Utility Function Values for Example 7.25

Alternatives	1	2	3	4
Sequence	4→3→1→2	4→3→2→1	3→2→4→1	2→4→3→1
Min. $f_1 = M_e$	21	22	24	26
Max. $f_2 = z$	0	1	2	3
Efficient?	Yes	Yes	Yes	Yes
Max. f_1'	1	0.8	0.4	0
Max. f_2'	0.00	0.33	0.67	1.00
Max. $U(0.6, 0.4)$	0.60	0.61	0.51	0.40

bicriteria problem, generate four sequences by applying Johnson's rule for $z = 0, 1, 2, 3$. The details for the $z = 0$ and $z = 1$ are shown in Example 7.24. Table 7.6 shows the summary of all four solutions. Sequence 4→3→2→1 has the best utility value, and therefore it is the best alternative.

See Supplement S7.10.xls.

Stochastic Sequencing of n × m[†]   The solution to this problem is similar to stochastic $n \times 2$ problems except that after finding the expected values of all jobs the methods developed in Section 7.7 can be used to find the sequence and its performance measurements.

REFERENCES

General

Akers, S. B. "A Graphical Approach to Production Scheduling Problems." *Operations Research*, Vol. 4, 1956, pp. 244–245.

Allahverdi, A., C. T. Ng, T. C. E. Cheng, and M. Y. Kovalyov. "A Survey of Scheduling Problems with Setup Times or Costs." *European Journal of Operational Research*, Vol. 187, No. 3, 2008, pp. 985–1032.

Baker, K. R. *Elements of Sequencing and Scheduling*, New York: Wiley, 1995.

Campbell, H. G., R. A. Dudek, and M. L. Smith. "A Heuristic Algorithm for the *n* Job, *m* Machine Sequencing Problem." *Management Science*, Vol. 16, No. 10, 1970, pp. 630–637.

Conway, R. W., W. L. Maxwell, and L. W. Miller. *Theory of Scheduling*. Reading, MA: Addison Wesley, 1967.

Elsayed, E. A., and T. O. Boucher. *Analysis and Control of Production Systems*, 2nd ed. Englewood Cliffs, NJ: Prentice Hall, 1994.

French, S. *Sequencing and Scheduling: An Introduction to the Mathematics of the Job Shop*, Chichester, England: Ellis Horwood, 1982.

Giglio, R. J., and H. M. Wagner. "Approximate Solutions to the Three Machine Scheduling Problem." *Operations Research*, Vol. 12, No. 2, 1964, pp. 305–324.

Ignall, E. J., and L. E. Schrage. "Application of the Branch and Bound Technique to Some Flow-Shop Scheduling Problems." *Operations Research*, Vol. 13, No. 3, 1965, pp. 400–412.

Jackson, J. R. "An Extension of Johnson's Results on Job-lot Scheduling." *Naval Research Logistics Quarterly*, Vol. 3, No. 3, 1956, pp. 201–203.

Johnson, S. M. "Optimal Two and Three Stage Production Schedules with Setup Times Included." *Naval Research Logistics Quarterly*, Vol. 1, 1954, pp. 61–68.

King, J. R., and A. S. Spachir. "Heuristics for Flow-Shop Scheduling." *International Journal of Production Research*, Vol. 18, No. 3, 1980, pp. 345–357.

Moore, J. M. "Sequencing *n* Jobs on One Machine to Minimize the Number of Tardy Jobs." *Management Science*, Vol. 17, No. 1, 1968, pp. 102–109.

Nahmias, S. *Production and Operations Analysis*, 6th ed. New York: McGraw-Hill/Irwin, 2008.

Pinedo, M. *Scheduling, Theory, Algorithms and Systems*, Englewood Cliffs, NJ: Prentice Hall, 1995.

Ramudhin, A., J. J. Bartholdi, J. M. Calvin, J. H. Vande Vate, and G. Weiss. "A Probabilistic Analysis of Two-Machine Flowshops." *Operations Research*, Vol. 44, No. 6, 1996, pp. 899–908.

Stevenson, W. J. *Operations Management*, 9th ed. New York: McGraw-Hill Irwin, 2007.

Multiobjective

Allouche, M. A., B. Aouni, J.-M. Martel, T. Loukil, and A. Rebai. "Solving Multi-Criteria Scheduling Flow Shop Problem through Compromise Programming and Satisfaction Functions." *European Journal of Operational Research*, Vol. 192, No. 2, 2009, pp. 460–467.

Arroyo, J. E. C., and V. A. Armentano. "Genetic Local Search for Multi-Objective Flowshop Scheduling Problems." *European Journal of Operational Research*, Vol. 167, No. 3, 2005, pp. 717–738

Bagchi, T. P. *Multiobjective Scheduling by Genetic Algorithms*, City: AH Dordrecht, The Netherland, Springer, 1999.

Daniels, R. L., and R. J. Chambers. "Multiobjective Flow-Shop Scheduling." *Naval Research Logistics*, Vol. 40, 1990, pp. 85–101.

Dileepan, P., and T. Sen. "Bicriterion Static Scheduling Research for a Single Machine." *Omega*, Vol. 16, No. 1, 1988, pp. 53–59.

Fry, T. D., R. D. Armstrong, and H. Lewis. "A Framework for Single Machine Multiple Objective Sequencing Research." *Omega*, Vol. 17, No. 6, 1989, pp. 595–607.

Gawiejnowicz, S., W. Kurc, and L. Pankowska. "Pareto and Scalar Bicriterion Optimization in Scheduling Deteriorating Jobs." *Computers & Operations Research*, Vol. 33, No. 3, 2006, pp. 746–767.

Haral, U., R. W. Chen, W. G. Ferrell, Jr., and M. B. Kurz. "Multiobjective Single Machine Scheduling with Nontraditional Requirements." *International Journal of Production Economics*, Vol. 106, No. 2, 2007, pp. 574–584.

Hoogeveen, H. "Multicriteria Scheduling." *European Journal of Operational Research*, Vol. 167, No. 3, 2005, pp. 592–623.

Kindt, T. V., and J.-C. Billaut. *Multicriteria Scheduling: Theory, Models and Algorithms*. Berlin: Springer, 2002.

Malakooti, B. "Multi-Objective Sequencing and Analysis and Head-Tail Approach." Case Western Reserve University, Cleveland, OH, 2013.

Malakooti, B. "Double Helix Value Functions, Ordinal/Cardinal Approach, Additive Utility Functions, Multiple Criteria, Decision Paradigm, Process, and Types." *International Journal of Information Technology & Decision Making (IJITDM)*, Forthcoming, 2014a.

Malakooti, B. "Z Utility Theory: Decisions Under Risk and Multiple Criteria" to appear, 2014b.

Malakooti, B., H. Kim, and S. Sheikh. "Bat Intelligence Search with Application to Multi-Objective Multi-Processor Scheduling Optimization." *International Journal of Advanced Manufacturing Technology*, Vol. 60, No. 9, 2012, pp. 1071–1086.

Malakooti, B., S. Sheikh, C. Al-Najjar, and H. Kim. "Multi-Objective Energy Aware Multiprocessor Scheduling Using Bat Intelligence." *Journal of Intelligent Manufacturing*, 2012, pp. 1–15. DOI 10.1007/s10845-012-0629-6.

Mosheiov, G., and A. Sarig. "A Multi-Criteria Scheduling with Due-Window Assignment Problem." *Mathematical and Computer Modelling*, Vol. 48, Nos. 5–6, 2008, pp. 898–907.

Nagara, A., J. Haddocka, and S. Heragu. "Multiple and Bicriteria Scheduling: A Literature Survey." *European Journal of Operational Research*, Vol. 81, No. 1, 1995, pp. 88–104.

Rajendran, C. "Two-Stage Flow Shop Scheduling Problem with Bicriteria." *Journal of the Operational Research Society*, Vol. 43, 1992, pp. 871–884.

Sheikh, S. "Multi-Objective Flexible Flow Lines with Due Window, Time Lag and Job Rejection." *International Journal of Advanced Manufacturing Technology*, 2012, pp. 1–11.

Wassenbove, L. N. V., and K. R. Baker. "A Bicriterion Approach to Time/Cost Trade-Offs in Sequencing." *European Journal of Operational Research*, Vol. 11, 1982, pp. 48–54.

EXERCISES

Introduction (Section 7.1)

1.1 For a specific industry (e.g., a manufacturing plant) discuss relevant sequencing problems, sequencing approaches, and sequencing objectives. How are the objectives measured? Which objectives are conflicting and competing?

1.2 Choose a specific service industry/business (e.g., bank) or government organization. Discuss relevant sequencing problems, sequencing approaches, and sequencing objectives. How are the objectives measured? Which objectives are conflicting and competing?

Sequencing Performance Measurements (Section 7.2.1) and Algorithms for $n \times 1$ (Section 7.2.2)

2.1 Compare and contrast all the sequencing rules discussed in Section 7.2. Also state a possible application for each rule.

2.2 Compare and contrast the five sequencing criteria (flow time, lateness, etc.). State a possible application where each criterion should be used.

2.3 Consider the following processing times and due dates for five jobs:

Job, i	1	2	3	4	5
t_i	3	8	6	7	10
d_i	11	9	30	25	19
I_i	4	7	2	9	10

(a) Solve the problem using the six sequencing methods.

(b) Measure the five sequencing criteria for each method.

(c) Compare the six solutions. Which ones do you select? Which ones are dominated?

2.4 Consider the following processing times and due dates for seven jobs. Respond to parts (a)–(c) of exercise 2.3.

Job, i	1	2	3	4	5	6	7
t_i	3	8	6	7	10	3	6
d_i	11	9	30	25	19	20	11
I_i	2	4	3	7	8	3	7

2.5 Consider the following process times and due dates for 10 jobs. Respond to parts (a)–(c) of exercise 2.3.

Job, i	1	2	3	4	5	6	7	8	9	10
t_i	3	5	7	9	3	4	2	10	7	12
d_i	10	22	26	30	16	28	46	12	40	31
I_i	1	3	2	4	5	3	5	2	3	1

2.6 Consider exercise 2.3. Consider a dynamic case, where a new job, job 6, arrives at time 10. This new job has a processing time of 4 units where its due date is 20. Solve the problem by the critical ratio method. State the average flow time and the average tardiness.

2.7 Consider exercise 2.4. Consider a dynamic case, where a new job, job 8, arrives at time 12. This new job has a processing time of 8 units where its due date is 21. Solve the problem by the critical ratio method. State the average flow time and the average tardiness.

Bicriteria Composite Approach for $n \times 1$ (Section 7.2.3) and Generating the Efficient Frontier for $n \times 1$ (Section 7.2.4)

2.8 Consider exercise 2.3 where the two objective function weights are $(0.7, 0.3)$.

(a) Find the best solution. State the values of f_1 (average flow time) and f_2 (average tardiness).

(b) Show three multicriteria solutions to this problem.

2.9 Consider exercise 2.4 where the two objective function weights are $(0.6, 0.4)$. Respond to parts (a) and (b) of exercise 2.8.

2.10 Consider exercise 2.5 where the two objective function weights are $(0.4, 0.6)$. Respond to parts (a) and (b) of exercise 2.8.

Multicriteria for *n* × 1 (Section 7.2.5)[†]

2.11 Consider exercise 2.3 where objective function weights are $(0.2, 0.3, 0.5)$.

    (a) For the third objective, suppose the ideal job priority sequence is $1{\rightarrow}5{\rightarrow}3{\rightarrow}4{\rightarrow}2$. Find the best solution. State the values of f_1 (average flow time), f_2 (average tardiness), and f_3 (maximize job priorities).

    (b) Generate a set efficient frontier. Determine which alternatives are inefficient.

2.12 Consider exercise 2.4 where objective function weights are $(0.3, 0.4, 0.3)$.

    (a) For the third objective, suppose the ideal job priority sequence is $7{\rightarrow}1{\rightarrow}6{\rightarrow}5{\rightarrow}3{\rightarrow}4{\rightarrow}2$. Find the best solution. State the values of f_1 (average flow time), f_2 (average tardiness), and f_3 (maximize job priorities).

    (b) Generate a set of efficient frontier. Determine which alternatives are inefficient.

Flow Shop of *n* × 2 (Section 7.3.1) and Job Shop of *n* × 2 (Section 7.3.2)

3.1 Consider the following data. Use Johnson's rule to find a sequence for this problem. Find the values of the objectives f_1 = makespan and f_2 = tardiness for parts (a)–(d) below.

Job, i	1	2	3	4
Processor A	5	9	8	7
Processor B	4	6	5	2
Due date, d_i	6	11	16	20

    (a) Each job is first processed on processor A and then on B, $A{\rightarrow}B$.

    (b) Each job is processed in $B{\rightarrow}A$ order.

    (c) Jobs 1 and 2 are processed in $A{\rightarrow}B$ order, and jobs 3 and 4 are processed in $B{\rightarrow}A$ order.

    (d) Apply the EDD rule where the processing order is $A{\rightarrow}B$.

3.2 Consider the following data. Use Johnson's rule to find a sequence for this problem. Find the values of the objectives f_1 = makespan and f_2 = tardiness for parts (a)–(d) below.

Job, i	1	2	3	4	5	6
Processor A	4	1	5	9	10	6
Processor B	7	3	8	6	4	5
Due date, d_i	7	18	16	25	14	22

    Respond to parts (a), (b), and (d) of exercise 3.1.

    (c) Jobs 1, 3, and 5 are processed in $A{\rightarrow}B$ order and jobs 2, 4, and 6 in $B{\rightarrow}A$ order.

3.3 Consider the following data. Use Johnson's rule to find a sequence for this problem. Find the values of the objectives $f_1 = $ makespan and $f_2 = $ tardiness for parts (a)–(d) below.

Job, i	1	2	3	4	5	6	7	8	9	10
Processor A	5	2	9	3	8	4	1	10	6	7
Processor B	3	7	4	2	6	4	8	5	1	9
Due date, d_i	20	13	23	12	15	41	45	52	33	61

Respond to parts (a), (b), and (d) of exercise 3.1.

(c) Jobs 1, 3, 4, 7, and 10 are processed in $A \rightarrow B$ order and jobs 2, 5, 6, 8 and 9 in $B \rightarrow A$ order.

Multicriteria of $n \times 2$ (Section 7.3.3)

3.4 Consider exercise 3.1 where the weights of two objective functions are $(0.7, 0.3)$.

(a) Find the best solution. State the values of f_1 and f_2.

(b) Show three multicriteria solutions to this problem.

3.5 Consider exercise 3.2 where the weights of two objective functions are $(0.2, 0.8)$. Respond to parts (a) and (b) of exercise 3.4.

3.6 Consider exercise 3.3 where the weights of two objective functions are $(0.6, 0.4)$. Respond to parts (a) and (b) of exercise 3.4.

Algorithm for $n \times 3^\dagger$ (Section 7.4.1)

4.1 Consider processing three jobs on three processors in $A \rightarrow B \rightarrow C$ order.

Job, i	1	2	3
Processor A	3	5	1
Processor B	11	8	9
Processor C	6	10	6
Due date, d_i	25	40	38

(a) Use Johnson's rule to find a sequence for the above data. Is this solution optimal?

(b) Show a Gantt chart for this solution. Also, measure the total tardiness for the obtained sequence.

(c) Use Johnson's rule to find a sequence for the above data where the order is $A \rightarrow C \rightarrow B$. Is this solution optimal?

4.2 Consider the following data with given processor order $A \rightarrow B \rightarrow C$:

Job, i	1	2	3	4
Processor A	11	8	9	10
Processor B	2	4	1	6
Processor C	7	12	8	9
Due date, d_i	25	40	38	35

Respond to parts (a)–(c) of exercise 4.1.

4.3 Consider the following data with given processor order $A \rightarrow B \rightarrow C$:

Job, i	Processor A	Processor B	Processor C	Due Date, d_i
1	11	2	9	29
2	9	3	6	45
3	7	6	8	22
4	12	5	3	18
5	8	4	7	38

Respond to parts (a)–(c) of exercise 4.1.

4.4 For $n \times 3$, prove that any of the two conditions stated for solving $n \times 3$ problems will yield the optimal solution?

Algorithm for $n \times m$: Campbell Method (Section 7.4.2)

4.5 Consider exercise 4.1. Find the minimum makespan using Campbell's method.

4.6 Consider exercise 4.2. Find the minimum makespan using Campbell's method.

4.7 Consider the following problem using $A \rightarrow B \rightarrow C \rightarrow D$ order. Find the minimum makespan using Campbell's method.

Job, i	A	B	C	D	d_i
1	3	8	6	10	29
2	11	9	3	7	43
3	7	5	11	6	21
4	2	9	5	8	59

4.8 Consider the following problem using $A \rightarrow B \rightarrow C \rightarrow D \rightarrow E$ order. Find the shortest processing time using Campbell's method.

Job, i	A	B	C	D	E	d_i
1	3	8	6	1	5	29
2	11	9	3	2	6	43
3	7	5	11	8	1	21
4	2	9	5	8	2	59
5	4	1	10	3	6	32

4.9 Consider the following problem using $A \to B \to C \to D \to E \to F \to G$ order. Find the minimum makespan using Campbell's method.

Job, i	A	B	C	D	E	F	G	d_i
1	3	8	6	1	5	2	9	49
2	11	9	3	2	6	2	4	83
3	7	5	11	8	1	8	4	51
4	2	9	5	8	2	13	9	99
5	7	2	4	8	10	11	7	64
6	3	7	10	6	3	6	11	102
7	8	4	7	4	8	3	7	75
8	3	6	5	3	11	7	2	37
9	9	3	6	5	5	8	4	46

4.10 What is the number of possible solutions to an $n \times m$ sequencing problem in general? In particular, what is the number of possible solutions for 20 jobs and 15 processors?

Multicriteria of $n \times m^\dagger$ (Section 7.4.3)

4.11 Consider exercise 4.1 where the weights of importance of the two objectives are $(0.3, 0.7)$. Find the best sequence for solving this problem.

4.12 Consider exercise 4.2 where the weights of importance of the two objectives are $(0.4, 0.6)$. Find the best sequence for solving this problem.

4.13 Consider exercise 4.3 where the weights of importance of the two objectives are $(0.7, 0.3)$. Find the best sequence for solving this problem.

Algorithm for $2 \times m$ (Section 7.5.1)

5.1 Show that the $2 \times m$ graphical method always generates the optimal solution. Alternatively, you can show a counterexample that the $2 \times m$ graphical method may fail to generate the optimal solution.

5.2 Consider the following data:

Job, i	A	B	C	Due Date, d_i	Processing Order
1	8	10	3	13	$B \to A \to C$
2	4	8	11	15	$C \to A \to B$

Find the minimum makespan and show a Gantt chart for this solution. Also, measure the total tardiness for the obtained sequence.

5.3 Consider the following data:

Job, i	A	B	C	D	Due Date, d_i	Processing Order
1	6	1	4	7	18	$D \to B \to A \to C$
2	5	4	9	2	17	$C \to D \to B \to A$

Find the minimum makespan and show a Gantt chart for this solution. Also, measure the total tardiness for the obtained sequence.

5.4 Consider the following data:

Job, i	A	B	C	D	E	Due Date, d_i	Processing Order
1	8	3	6	9	2	29	$B \to A \to E \to C \to D$
2	7	4	8	4	2	25	$A \to B \to C \to D \to E$

Find the minimum makespan and show a Gantt chart for this solution. Also, measure the total tardiness for the obtained sequence.

5.5 Consider the following data:

Job, i	A	B	C	D	E	F	G	Due Date, d_i	Processing Order
1	4	3	1	6	9	10	2	36	$A \to B \to C \to D \to E \to F \to G$
2	3	5	8	1	4	3	8	34	$A \to B \to E \to C \to G \to F \to D$

Find the minimum makespan and show a Gantt chart for this solution. Also, measure the total tardiness for the obtained sequence.

Multicriteria of 2 × *m* (Section 7.5.2)

5.6 Consider exercise 5.2. Generate efficient alternatives and determine inefficient ones.

5.7 Consider exercise 5.3. Generate efficient alternatives and determine inefficient ones.

5.8 Consider exercise 5.3. Generate efficient alternatives and determine inefficient ones.

Flow Shop of *n* × *m*: Cardinal Head–Tail (Section 7.6.2)

6.1 Consider exercise 4.1. Solve this problem by the cardinal head–tail method.

6.2 Consider exercise 4.2. Solve this problem by the cardinal head–tail method.

6.3 Consider exercise 4.7. Solve this problem by the cardinal head–tail method.

6.4 Consider exercise 4.8. Solve this problem by the cardinal head–tail method.

6.5 Consider exercise 4.9. Solve this problem by the cardinal head–tail method.

Job Shop of $n \times m$: Cardinal Head–Tail[†] (Section 7.6.3)

6.6 Consider the following data. Solve this problem by the cardinal head–tail method.

Job, i	A	B	C	d_i	Job Shop
1	4	9	4	29	$A \to C \to B$
2	7	11	3	43	$C \to A \to B$
3	5	6	10	21	$B \to A \to C$

6.7 Consider the following data. Solve this problem by the cardinal head–tail method.

Job, i	A	B	C	d_i	Job Shop
1	8	5	2	29	$B \to C \to A$
2	4	7	8	43	$B \to A \to C$
3	7	4	3	21	$A \to B \to C$

6.8 Consider the following data. Solve this problem by the cardinal head–tail method.

Job, i	A	B	C	D	d_i	Job Shop
1	3	8	6	10	29	$A \to C \to B \to D$
2	11	9	3	7	43	$C \to A \to B \to D$
3	7	5	11	6	21	$B \to A \to D \to C$
4	2	9	5	8	59	$D \to B \to C \to A$

6.9 Consider the following data. Solve the job shop problem by the cardinal head–tail method.

Job, i	A	B	C	D	E	F	G	d_i	Job Shop
1	3	8	6	1	5	2	9	49	$B \to A \to E \to D \to C \to G \to F$
2	11	9	3	2	6	2	4	83	$D \to F \to A \to G \to C \to E \to B$
3	7	5	11	8	1	8	4	51	$F \to C \to E \to B \to D \to G \to A$
4	2	9	5	8	2	13	9	99	$A \to B \to C \to D \to E \to F \to G$
5	7	2	4	8	10	11	7	64	$B \to D \to A \to F \to G \to C \to E$
6	3	7	10	6	3	6	11	102	$A \to D \to C \to B \to G \to F \to E$

Multicriteria of $n \times m$[†] (Section 7.6.4)

6.10 Consider exercise 4.1, suppose that the weights of importance of the two objectives are $(0.7, 0.3)$.

   (a) Find the best multicriteria solution. State the values of f_1 and f_2.

   (b) Show only three multicriteria solutions.

6.11 Consider exercise 4.2. Suppose that the weights of importance of the two objectives are (0.2, 0.8). Respond to parts (a) and (b) of exercise 6.10.

6.12 Consider the first four jobs of exercise 4.7. Suppose that the weights of importance of the two objectives are (0.6, 0.4). Respond to parts (a) and (b) of exercise 6.10.

Stochastic Sequencing of $n \times 1^\dagger$ (Section 7.7.1)

7.1 Consider sequencing four jobs on one processor for the following data in hours:

Job, i	a	m	b
1	3	9	15
2	10	13	14
3	6	8	11
4	2	6	9

(a) Find the expected processing time, variance, and standard deviation of each job.

(b) Determine the optimal sequence that minimizes the expected average flow time. Analyze the performance of the obtained sequence, that is, find expected average flow time, makespan, and variance and standard deviation of the obtained sequence.

(c) Find the probability of finishing all jobs in 39 h or less.

(d) For a 84.13% probability of completion of each job. Find the sequencing solution.

7.2 Consider sequencing five jobs on one processor for the following data in hours. Respond to parts (a)–(d) of exercise 7.1.

Job, i	a	m	b
1	1	4	8
2	8	10	10
3	3	6	11
4	4	7	12
5	2	5	10

Bicriteria Stochastic Sequencing of $n \times 1^\dagger$ (Section 7.7.2)

7.3 Consider exercise 7.1. Generate four bicriteria alternatives for the risk-averse cases $z = 0, 1, 2, 3$. Find the best sequence for the four jobs where the weights of importance of the normalized values of the two objectives are (0.5, 0.5).

7.4 Consider exercise 7.2. Generate four bicriteria alternatives for the risk-averse cases $z = 0, 1, 2, 3$. Find the best sequence for the four jobs where the weights of importance of the normalized values of the two objectives are (0.3, 0.7).

Stochastic Sequencing of $n \times 2^\dagger$ (Section 7.7.3)

7.5 Consider sequencing four jobs on two processors where all jobs are processed in the order of A first and then B. Expected processing time $t_{e,i}$ and standard deviation SD_i for each job are shown below.

    (a) Determine the optimal sequence that minimizes the expected makespan. What is the makespan for this solution.

    (b) For a 97.72% probability of completion of each job, find the optimal sequencing solution and its makespan. Compare the solutions to (a) and (b).

    (c) Generate four bicriteria alternatives for the risk-averse cases $z = 0, 1, 2, 3$.

Job, i	1	2	3	4
Processor A: $(t_{e,i}, SD_i)$	(5,3.5)	(8,0.5)	(9,0)	(7,2)
Processor B: $(t_{e,i}, SD_i)$	(4,4)	(5,3)	(6,2.5)	(2, 5)

7.6 Consider sequencing five jobs on two processors where all jobs are processed in the order of A first and then B. Expected processing time $t_{e,i}$ and standard deviation SD_i for each job are shown below. Respond to parts (a)–(c) of exercise 7.5.

Job, i	1	2	3	4	5
Processor A: $(t_{e,i}, SD_i)$	(4,3)	(1,5)	(5,2.5)	(9,0.5)	(10,0)
Processor B: $(t_{e,i}, SD_i)$	(7,1	(3,4.5)	(8,0.5)	(6, 1.5)	(4,5)

CHAPTER 8

PROJECT MANAGEMENT

8.1 INTRODUCTION

The construction of all historical man-made wonders of the world, such as the Great
Pyramids of Giza, the Roman Coliseum, the Great Wall of China, and the Persepolis of
the Persian Empire, required excellent project management in terms of organizing work,
planning, and optimizing the use of resources. In the past century, project management
tools have substantially improved in response to the need for even more complex projects.
Modern structures such as the Hoover Dam, the Chesapeake Bay Bridge, and the Three
Gorges Dam of China were also built using advanced project management tools. NASA
space exploration projects and the operation of the U.S. military (such as the management of the two U.S. wars in Iraq) were also heavily dependent on the use of project
management methods and tools. The use of project management is not only for complex
projects; the daily activities of a person can be planned and optimized by the use of project
management tools.

A project consists of several activities that must be performed in a particular sequence
and order. The primary objective of project management is to plan a project so that it
can be completed as quickly as possible and to monitor the progress of that planning.
Project management is also concerned with the allocation of available resources while
minimizing the total cost, total duration, and risk of delay in the completion of the
project. Modern project management computer software packages can be used to plan
and monitor (in real time) hundreds of thousands of activities along with their needed
resources.

Note: Advanced material that can be omitted without loss of generality will be indicated by a dagger.

Operations and Production Systems with Multiple Objectives, First Edition. Behnam Malakooti.
© 2014 John Wiley & Sons, Inc. Published 2014 by John Wiley & Sons, Inc.

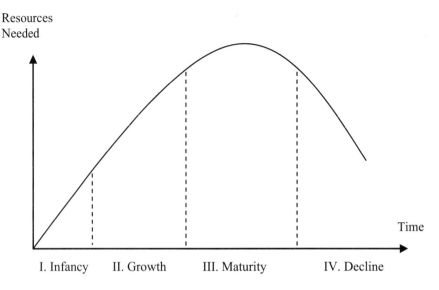

Figure 8.1   Four phases of project management life cycle.

Project Management Life-Cycle Phases   Project management has a life cycle similar to the life cycle of products and processes. It consists of four phases (see Figure 8.1):

    I. In the infancy phase, the project management problem is defined.

    II. In the growth phase, all details and information related to the project are gathered.

    III. In the maturity phase, the project management problem is planned and optimized.

    IV. In the decline phase, the project construction starts, where project management is executed and monitored.

Overview of Critical-Path Method and Its Extensions   The most important aspect of project management is the scheduling of activities. The two most popular and effective project management methods are the critical-path method (CPM) and the project evaluation and review technique (PERT). CPM is used when activity times are deterministic, while PERT is used when activity times are probabilistic. Although, historically, CPM and PERT were developed for different purposes, their analysis and objectives are very similar.

    CPM is used to find the minimum feasible duration to complete a project and to identify the start and finish time for each activity. CPM also identifies critical activities whose delay will postpone the project completion time. When the project is being implemented, the progress of the project and each activity are monitored to keep the project on schedule. Line of balance (LOB) and Gantt charts are visual methods that can be used to track the project progress and all of its activities.

Generally, it is possible to decrease the minimum project duration by incurring additional costs. An extension of CPM called the time–cost trade-off method can be used to find different reduced project durations and their associated costs. In CPM, the durations of activities are assumed to be known with certainty, that is, CPM solves deterministic project management problems. In PERT the duration of each activity is probabilistic. PERT uses three time estimates for each activity: optimistic (the shortest), most likely (the average), and pessimistic (the longest). In PERT, the CPM model is expanded to find the minimum project duration while considering probabilistic activity durations. PERT can also be used to provide probabilities associated with different project durations. The shorter the targeted project duration, the less likely the project will be completed on time.

Another aspect of project management to be considered is resource constraints, such as machines and workers needed to perform activities. The CPM model can be expanded to find the minimum feasible project duration while considering resource constraints.

Project management often involves considering multiple and conflicting objectives such as minimizing total cost, minimizing total project duration, and maximizing the probability of completing the project on time (minimizing the risk). In this chapter, we have developed multicriteria (objective) optimization procedures for solving several different multicriteria project management problems. Compared to CPM, these multicriteria procedures provide the DM with broader tools and information for choosing more satisfactory alternatives.

Project Management Software   Currently, almost all business, government, and non-profit organizations require the use of project management software for managing and monitoring their projects. There are several project management software packages that come with a variety of options. For example, Microsoft Project allows the use of Excel spreadsheets. A majority of these software packages are also connected to accounting activities, expenditure monitoring, and budget allocation.

8.2 CRITICAL-PATH METHOD

The primary objective of project management is to plan the sequence of all activities (which are also called tasks or jobs) of a project such that:

(a) Project duration is minimized.
(b) The sequence of all activities is feasible.

Project management problems can be presented by a network of activities that are sequentially related. CPM is the most popular and effective scheduling method for solving a project management problem. CPM was originally developed by Kelly and Walker (1959). CPM is used to find the minimum overall project duration and identify the critical activities. An activity is critical if a delay in its duration will result in the delay of the total project duration. Therefore, it is very important for a project manager to be able to identify and monitor critical activities. In addition to identifying critical activities, CPM also provides a schedule of the earliest and latest starting and finishing times of all activities. This gives the project manager some flexibility since the manager will know which noncritical activities can be delayed.

8.2.1 Constructing CPM Network

A project management network is a graphical representation of the sequence of all activities for a given project, in which:

- Each activity is associated with only one unique arc (arrow).
- Each activity must be uniquely identified by two nodes for its start and finish time.
- The CPM network has only one starting node and only one ending node.

The CPM network diagram is composed of three basic symbols: nodes, arrows, and pseudoarrows. These are explained below.

Node i   An event i is depicted by a node. A node signifies the start or end of an activity. More than one activity may start or finish at a given node. In a project network diagram, nodes are connected by arrows and are sequentially numbered from left to right and then from top to bottom.

Arrows   Each arrow represents one activity and connects two nodes to each other in one direction. Arrows going into the same node represent activities that finish at the same time. Arrows coming out of the same node represent activities that start at the same time. Each arrow is labeled with the activity name or a symbol. Each activity can be labeled by (i,j), which means it connects node i to node j. The duration of the activity is represented by t_{ij}. In a network diagram, only one arrow can connect any two nodes.

There are two types of arrows:

Real arrow $\xrightarrow{\;\;(i,j),t_{ij}\;\;}$ Solid-line arrow represents real (actual) activity

Pseudoarrow $\;-\;-\;\overset{(i,j),0}{-}\!-\;\rightarrow$ Dashed-line arrow represents pseudoactivity

The duration of every pseudoactivity is zero. Pseudoactivities are used to correctly represent the sequence of activities in a CPM network when a proper presentation cannot be achieved by only using real activities. For example, if two activities have the same starting and ending nodes (i.e., they cannot be distinguished from each other using CPM notation), one should use a pseudoactivity to distinguish the two activities from each other. This will be illustrated by some examples below.

To construct the network of a project, use the following steps:

1. Create the start node, node 1. All activities without predecessors start from node 1.
2. Construct the network of nodes based on the sequence of activities.

TABLE 8.1 **Predecessors and Durations of Activities for Example 2.1**

Activity	Predecessors	Duration	Activity	Predecessors	Duration
A	—	5	F	C	6
B	—	10	G	E	8
C	—	7	H	F,G	1
D	A	3	I	H	9
E	A,B	2	J	D,F,G	5

3. All activities that do not precede any other activity are connected to the final node *n*.

4. After the network is constructed, check that:
   - It completely represents the problem.
   - It has the minimum number of arrows (both for real and pseudoactivities).
   - There are no cycles.
   - All arrows are pointing forward to the end node.
   - The network has one start node and one end node.

Example 8.1 **Constructing a Project Management Network with Ten Activities**
Consider a project that has 10 activities labeled *A–J*. The activities, their durations (in days), and their predecessors are listed in Table 8.1. The CPM network for this problem is constructed as follows:

Step 1   Initialize. Based on Table 8.1, activities *A*, *B*, and *C* do not have any predecessors; therefore, they should all start from node 1. See the network below.

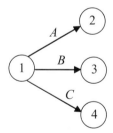

Step 2   Construct the network of nodes based on the sequence of activities that follow activities *A*, *B*, and *C*; see the following network. Note that without the use of pseudoactivities the network cannot be properly constructed. For example, one must make a pseudoactivity from node 2 to node 3 because activity *A* is a predecessor of activities *D* and *E* and activity *B* is a predecessor of activity *E* but not activity *D*.

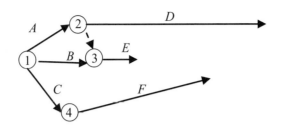

Now consider the next set of activities. Because activities D, F, and G are predecessors of activity J but activity D is not a predecessor of activity H, one must add a pseudoactivity from node 6 to node 8 to properly represent the problem.

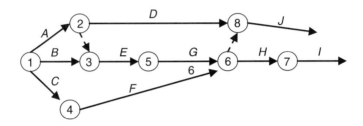

Step 3   Complete the network. Open-ended arrows must all end at the last node, that is, node 9.

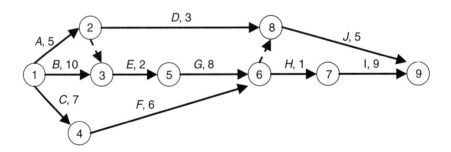

Step 4   Verify that the network is correct. Now compare this network presentation to the precedence requirements of Table 8.1. Verify that it completely represents the problem, it has the minimum number of arrows, it has no recycling, all arrows are forward, and it starts at one node and finishes with one node.

Illustration of Correct and Incorrect CPM Networks   In the following table, six different examples of correct and incorrect network formulations are provided. Read "$\prec$" as before for the precedence requirement.

	Example I	Example II	Example III	Example IV	Example V	Example VI
Constraint	$A, B \prec C$	$A \prec D, E;$ $B \prec E, F;$ $C \prec F$	$A \prec C, D;$ $B \prec C;$ $C \prec D$	$A \prec C; B \prec D;$ $A \prec D; B \prec C$	$A \prec C;$ $C \prec D;$ $B \prec C$	$A \prec C;$ $B \prec C;$ $C \prec D;$ $D \prec A$
Incorrect solution Reason	Two activities cannot have same links (1,2)	A is shown to precede F incorrectly	Link 2–4 is redundant, not necessary	Must start and end in one node	B must occur before C	Cannot have circular formulation
Correct solution						Wrong formulation, fix formulation

8.2.2 Critical-Path Method Algorithm

The purpose of CPM is to find the shortest feasible time to complete a project.

Critical Path   There are many paths that connect the starting node to the finishing node. A critical path is the path with the longest duration. The activities on the critical path are critical activities. There could be more than one critical path for the same problem, all having the same project duration. The critical path is designated on the network diagram with a bold or thick arrow. The sum of the durations of the critical activities on a critical path is the minimum project duration, which is the minimum feasible solution to the CPM problem.

Consider Example 8.1 and its final network. The path *B–E–G–H–I* has a duration of 10 $+ 2 + 8 + 1 + 9 = 30$ days, which is the minimum feasible duration for this problem. Note that path *A–D–J* has a duration of $5 + 3 + 5 = 13$ days and path *C–F–H–I* has a duration of $7 + 6 + 1 + 9 = 23$ days. However, the project cannot be completed in 13 or 23 days since there is a path that requires a longer duration.

In this section, the CPM algorithm is used to find the critical activities and paths. The following notation is used in CPM (also see Figure 2.1):

t_{ij} = activity duration for activity (i,j), associated with starting node i and finishing node j

$T_{E,j}$ = earliest event occurrence at node j

$T_{L,j}$ = latest event occurrence at node j

TF_j = total float time for occurrence at node j, $\mathrm{TF}_j = T_{L,j} - T_{E,j}$

ES_{ij} = earliest start time for activity (i, j)

EF_{ij} = earliest finish time for activity (i, j), $\mathrm{EF}_{ij} = \mathrm{ES}_{ij} + t_{ij}$

LS_{ij} = latest start time for activity (i, j)

LF_{ij} = latest finish time for activity (i, j), $\mathrm{LF}_{ij} = \mathrm{LS}_{ij} + t_{ij}$

TF_{ij} = total float time (slack) for activity (i, j), $\mathrm{TF}_{ij} = \mathrm{LF}_{ij} - \mathrm{EF}_{ij} = \mathrm{LS}_{ij} - \mathrm{ES}_{ij}$

Figure 8.2 shows the conventional placement of each symbol and its corresponding numerical value.

Note that the ES of an activity is the earliest possible time that an activity can begin, the EF is the earliest time an activity can finish taking into account the duration of preceding activities, the LS is the latest time an activity can begin without delaying the completion of the entire project, and the LF is the latest time an activity can finish taking into account a late start.

There are two assumptions made in CPM. First, the duration of each activity, t_{ij}, is assumed to be known. Second, it is assumed that ample resources are available to perform

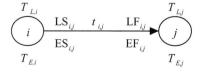

Figure 8.2   Basic network diagram for activity (i, j).

all activities. These two assumptions are relaxed in PERT (Section 8.5) and CPM with resource constraints (Section 8.7).

The CPM algorithm consists of the following three steps:

1. *Forward Pass*   In the forward-pass phase, the earliest feasible completion times for all activities and the shortest feasible project duration are found. Find the earliest feasible time to complete each activity as follows:

(a) Set $T_{E,1} = 0$ for the start node.
(b) $ES_{ij} = TE_i$ for all activities starting at node i.
(c) $EF_{ij} = ES_{ij} + t_{ij}$.
(d) $T_{E,j}$ = maximum earliest finish of all the activities finishing at node j
    $= \max\{EF_{k,j}$, for all activities ending at node $j\}$
(e) Repeat steps (a)–(d) until all nodes have been traversed. Find $T_{E,n}$.

2. *Backward Pass*   In the backward pass, the network is viewed from the end node toward the start node. The backward pass determines the latest feasible completion times of all activities, starting at the end node and moving backward toward the start node. The shortest feasible project duration (obtained from the forward pass) is used as the given project duration. Find the latest feasible time for each activity, starting from the final node as follows:

(a) Set $T_{L,n} = T_{E,n}$ for the final node.
(b) $LF_{ij} = T_{L,j}$ for all activities ending at node j.
(c) $LS_{ij} = LF_{ij} - t_{ij}$.
(d) $T_{L,j}$ = minimum latest start of all activities starting at node j
    $= \text{Min}\{LS_{j,k}$, for all activities starting at node $j\}$
(e) Repeat steps (a)–(d) until all nodes have been traversed.

3. *Critical Path*   If an activity's earliest finish time is the same as its latest finish time, it is a critical activity. The critical path consists of critical activities. To identify the critical activities, first find the TF for each activity:

$$TF_{ij} = LF_{ij} - EF_{ij} = LS_{ij} - ES_{ij} \quad \text{for each activity } (i,j)$$

Activities with a float time of zero (i.e., $TF_{ij} = 0$) are critical. The critical path consists of critical activities that are connected from the start node to the end node in one continuous forward pass. Bold arrows are used to depict the critical path. The solution of the CPM algorithm is guaranteed to be optimal. An example of the CPM algorithm is shown below.

Example 8.2   CPM Solution for Five Activities   Given the network diagram in Figure 8.3*a*, find the earliest and latest feasible times for all activities and identify the critical path. Show both the graphical and tabular solutions.

Graphical Solution   Now apply the forward pass. The start times for activities A and B are 0. The time for node 3 is Max$\{8, 9\} = 9$. The earliest completion time of the project at node 4 is Max$\{8, 10\} = 10$. See Figure 8.3*b*.

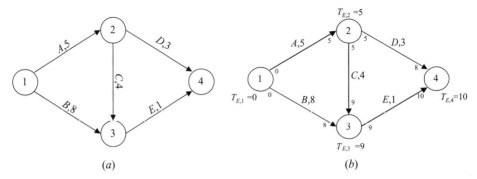

Figure 8.3   (*a*) Network diagram; (*b*) forward-pass solution.

Now apply the backward pass, starting at node 4 with a duration of 10. See Figure 8.4*a*. The latest start for activity *D* is $10 - 3 = 7$; the latest start for activity *E* is $10 - 1 = 9$. The latest time for node 2 is $\text{Min}\{5, 7\} = 5$. Also, the latest time for node 1 is $\text{Min}\{0, 1\} = 0$.

Now, critical activities can be identified. Observe that the earliest start times are the same as the latest start times for activities *A*, *C*, and *E*. That is, their float (slack) times are zero. Hence, they are critical activities. The critical path for this example is *A–C–E*, shown by bold arrows in Figure 8.4*b*.

Tabular CPM Solution   The above algorithm can also be performed and presented in a tabular format. Note that activities can be represented by their associated node numbers. For example, activity *A* can be represented by (1,2). Hence, activities *A*, *B*, *C*, *D*, and *E* can be represented by (1,2), (1,3), (2,3), (2,4), and (3,4), respectively. Table 8.2 is easy to construct using a spreadsheet or writing a computer program and illustrates the tabular CPM solution for this example. The columns labeled ES (earliest start) and EF (earliest finish) are first calculated for the forward pass. For each activity, find

$$\text{EF}_{ij} = \text{ES}_{ij} + t_{ij}$$

The ES for an activity without a precedent is set to zero. The ES for an activity with precedents is set to the maximum EF of all its precedents. For example, in Table 8.2,

$$\text{ES}_{(3,4)} = \text{Max}\{\text{EF}_{(1,3)}, \text{EF}_{(2,3)}\} = \text{Max}\{8, 9\} = 9$$

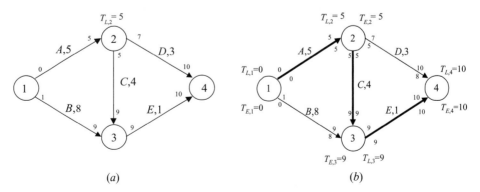

Figure 8.4   (*a*) Backward-pass solution; (*b*) critical-path diagram.

TABLE 8.2 Solution for Example 8.2

	Given Network			Forward Pass		Backward Pass		Final Step	
Activity	(i,j)	Precedents	t_{ij}	ES_{ij}	EF_{ij}	LS_{ij}	LF_{ij}	TF_{ij}	Critical?
A	(1, 2)	—	5	0	5	0	5	0	Yes
B	(1, 3)	—	8	0	8	1	9	1	No
C	(2, 3)	A	4	5	9	5	9	0	Yes
D	(2, 4)	A	3	5	8	7	10	2	No
E	(3, 4)	B,C	1	9	10	9	10	0	Yes

The maximum EF value for activities D and E, which finish at the last node (4), is the project duration, that is, Max $\{8,10\}=10$.

In the next step, the columns labeled LS (latest start) and LF (latest finish) are filled according to the backward pass. For example, activity A precedes activities C and D; therefore,

$$LF_{(1,2)} = \text{Min}\{LS_{(2,3)}, LS_{(2,4)}\} = \text{Min}\{5, 7\} = 5$$

Finally, TF_{ij} is calculated by subtracting EF_{ij} from LF_{ij}. Activities with zero in the TF column are marked as critical, and the critical path is identified. In this example, the critical path is A–C–E and the project duration is 10 days.

Example 8.3 CPM Solution for Nine Activities   Consider a project that consists of nine activities (A–I). Table 8.3 shows the activities' precedence and durations.

(a) Construct a CPM network diagram
(b) Find the optimal solution by the tabular CPM approach.

The solution is presented below.

(a) The network diagram solution is presented in Figure 8.5.
(b) The tabular-based CPM algorithm solution is given in Table 8.4. Based on Table 8.4, the critical path for this example is B–E–G–H–I. These activities are all critical and the project duration is 16 days.

TABLE 8.3 Predecessors and Durations of Activities

Activity	Predecessor	Duration
A	—	4
B	—	5
C	—	2
D	A	7
E	A, B	3
F	C	4
G	E	1
H	F, G	2
I	H	5

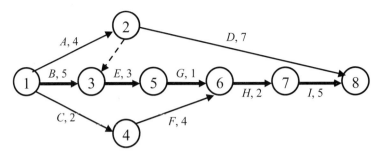

Figure 8.5   CPM network diagram for Example 8.3.

8.2.3 Project Monitoring and Gantt Chart

Line of Balance   Project planning should be conducted before the project is actually started. However, it is very important to monitor the project's progress during the project's completion. The LOB is a method for monitoring the project's progress and determining necessary changes to keep the project on track. The LOB can be used to keep track of all of the activities of the project and to identify activities that are not on schedule. The LOB technique can use charts to depict the actual status of the project versus the planned activities. In this method, all nodes are depicted on the x axis in the order of their completion (or only the critical nodes) and the cumulative project duration is depicted on the y axis.

For example, consider Example 8.3 and Figure 8.5. In this example, the critical path consists of nodes 1, 3, 5, 6, 7, and 8. These critical nodes are the ones associated with the critical activities B, E, G, H, and I. In Figure 8.6, the scheduled plan is based on the CPM solution. That is, each given node has a completion time based on the solution from the CPM. For example, at node 5, the planned finish time is day 8, which is the finish time of critical activity E. Now, observe the curve depicting the actual progress of the project in Figure 8.6. The actual completion of node 5 (activity E) is day 12. The LOB shows that by day 12 the project completion is delayed by at least four days. Realizing the project is

TABLE 8.4   CPM Algorithm Solution for Example 8.3

Activity	(i,j)	Given Network Precedents	t_{ij}	Forward Pass ES_{ij}	EF_{ij}	Backward Pass LS_{ij}	LF_{ij}	Final Step TF_{ij}	Critical?
A	(1,2)	—	4	0	4	1	5	1	No
B	(1,3)	—	5	0	5	0	5	0	Yes
C	(1,4)	—	2	0	2	3	5	3	No
D	(2,8)	A, (1,2)	7	4	11	9	16	5	No
E	(3,5)	A, (1,2); B, (1,3)	3	5	8	5	8	0	Yes
F	(4,6)	C, (1,4)	4	2	6	5	9	3	No
G	(5,6)	E, (3,5)	1	8	9	8	9	0	Yes
H	(6,7)	F, (4,6); G, (5,6)	2	9	11	9	11	0	Yes
I	(7,8)	H, (6,7)	5	11	16	11	16	0	Yes

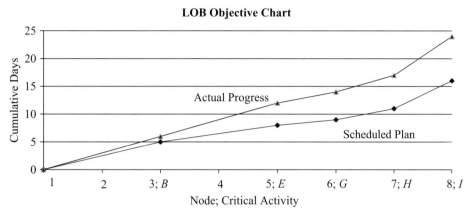

Figure 8.6 Project monitoring by the LOB for Example 8.3, nodes on critical paths.

behind at day 8, the project manager may consider some changes to expedite the project completion time. The LOB chart is useful for a quick and visual monitoring of the progress of the project as a whole and also on a daily basis. Furthermore, as the project is delayed, the critical path and activities of the project may change. Therefore, CPM may be used to solve the planning problem again to identify new critical activities.

Gantt Chart In 1910, Henry Gantt developed a visual project management tool called a Gantt chart. Today, Gantt charts are still commonly used in project management. A Gantt chart is a simple scheduling tool that is useful for monitoring a project. It shows the progress of work, concurrence of different activities, duration of each activity, as well as each activity's precedence requirements.

Figure 8.7 depicts a Gantt chart for Example 8.2. As can be seen, critical activities are placed at the top, whereas noncritical activities are placed at the bottom. The length of the bars for each activity represent the activity duration (e.g., activity A has duration of five time units). The bar for each activity begins at the earliest start time and ends at the earliest finish time for that activity (see Table 8.2 for the start and finish times for Example 8.2).

The connectors between the activities depict the precedence requirements; for example, activity A precedes both activities D and C. As a result, the slack times for each activity can also be deduced from the chart below. For example, consider activity B, which has an earliest finish time of 8 and a precedence over activity E with an earliest (and latest) start time of 9. It can be found that activity B can be delayed up to one unit (such that it finishes at time 9) without affecting the project duration; therefore, the slack time of activity B is one unit (i.e., the length of the horizontal line of the outgoing connector following activity B, as illustrated in Figure 8.7). Note that the critical activities A, C, and E have no slack time on their corresponding outgoing connector; which is why they are considered critical.

In addition, the progress of each activity can be depicted in the Gantt chart; the blank "bars" for each activity can be filled in as the activity is being completed. Figure 8.7 depicts the progress of activities at day 8 for Example 8.2. Note that although activities A, C, E, and D are on-time (i.e., their progress is as planned on the eighth day), activity B is running late by one day.

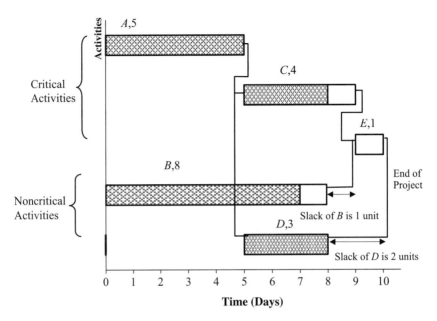

Figure 8.7   Gantt chart for Example 8.2 at day 8. The progress of each activity is shown by filled bars.

8.3 CPM TIME–COST TRADE-OFF METHOD

The total cost of the project consists of direct and indirect costs. The direct cost includes the cost of hiring or subcontracting contractors to perform different activities. Indirect costs are constant and independent of the cost of performing the activities. Examples of indirect costs are the cost of permanent employees, equipment, and overhead. Indirect costs are defined per unit of duration (e.g., per day) and they remain the same regardless of the project duration. In this section, we only consider the direct cost, that is, the direct cost of performing the activities. In the next section, we will consider both direct and indirect costs.

The project duration determined by the CPM may be too long because of deadlines and other time-related requirements. Therefore, the DM may be willing to pay an extra cost to reduce the project duration. The CPM time–cost trade-off method can be used to generate a number of alternatives for reducing the project duration which are associated with the additional cost of expediting the project. For each activity, there is a cost associated with reducing its duration per unit time. To reduce the project duration, the duration of critical activities must be reduced. An activity whose duration is reduced is called a crashed activity. The relationship of each activity cost versus its duration is presented in Figure 8.8.

In Figure 8.8a, the time–cost trade-off is linear; that is, the cost per unit time is constant. In Figure 8.8b, the time–cost trade-off function is nonlinear, which means the cost per unit time increases as time gets shorter. In this chapter, a linear time–cost trade-off is used; however, the method of this section can also be used with a nonlinear time–cost trade-off.

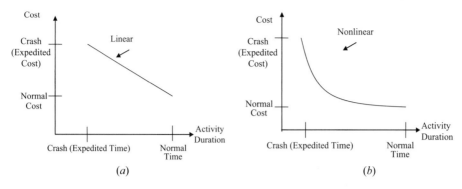

Figure 8.8   Time–cost trade-off for each activity: (*a*) linear: constant slope, cost per day; (*b*) nonlinear: varying slope, cost per day.

8.3.1   Time–Cost Trade-Off Algorithm

Consider the following notation for each activity (i, j):

$t_{n,ij}$ = normal time duration for activity (i, j) starting at node i and finishing at node j

$t_{c,ij}$ = crash time duration (minimum time possible) for activity (i, j)

$c_{n,ij}$ = normal cost for activity (i, j)

$c_{c,ij}$ = crash cost for activity (i, j)

k_{ij} = linear marginal crash cost per unit time

The linear marginal crash cost per unit time is calculated as follows:

$$k_{ij} = \frac{c_{c,ij} - c_{n,ij}}{t_{n,ij} - t_{c,ij}} \tag{8.1}$$

The steps for the time–cost trade-off method are presented below:

1. Consider the normal time of all activities. Solve the problem with CPM to find the normal project duration and the critical activities and path(s).
2. Choose a critical activity or combination of critical activities that will result in a reduction of the completion time on all critical paths by one unit of time. The selected activity (or activities) should have the following characteristics:
   (i) The activity duration can be reduced.
   (ii) The marginal cost k_{ij} is the minimum of all feasible activities whose reduction can reduce the project duration.
3. Reduce the duration of the chosen activity or activities by one time unit.
4. Calculate the additional cost and the total cost of the project.
5. Repeat step 2–4 until no more reduction (crashing) is feasible for at least one critical path.

Example 8.4   Time–Cost Trade-Off for Five Activities   Consider Example 8.2 as it is presented in Figure 8.4. Consider the normal time, crash time, normal cost, and crash cost for each activity as presented in Table 8.5.

TABLE 8.5   Activity Costs for Example 8.4

Activity (i, j)	A (1, 2)	B (1, 3)	C (2, 3)	D (2, 4)	E (3, 4)
Normal time (days)	5	8	4	3	1
Crash time (days)	3	5	3	2	1
Normal cost ($)	100	80	50	20	60
Crash cost ($)	160	110	65	40	60
Cost/day (k_{ij})	30	10	15	20	—

Determine all possible project durations and their associated costs.

See Figure 8.5 and Table 8.2 for the CPM solution with normal times. The critical path is A–C–E with a duration of 10 days. The total cost associated with the CPM solution of using activities with normal times is $100 + 80 + 50 + 20 + 60 = \$310$.

Now, the marginal cost of each activity should be calculated. For example, the cost of reducing activity A by one day is

$$\frac{160 - 100}{5 - 3} = \$30 \text{ per day}$$

The marginal costs for all activities are presented in the last row of Table 8.5. Note that activity E cannot be crashed.

Time–Cost Trade-Off Algorithm Solution   Table 8.6 is the summary of the solution to this problem. Each row presents one iteration for reducing the total project duration by one day.

Iteration 0 represents the solution for the normal time as shown in Table 8.6.

In iteration 1, activity C has the lowest marginal cost of two possible feasible candidates A and C; that is, Min$\{30, 15\} = 15$. Activity C is selected and reduced from four days to three days. In iteration 1, the total project duration is reduced by one day and the additional cost is $15. Therefore, the total cost associated with nine days is $310 + 15 = \$325$. Note that E cannot be selected because it cannot be crashed (reduced).

In iteration 2, there are two critical paths: $(A$–C–$E)$ and $(B$–$E)$. Activity C cannot be selected because it is at its minimum time. Hence, activities A and B are chosen.

TABLE 8.6   Iterations for Time–Cost Trade-Off Algorithm for Example 8.4

Iteration	Critical Path	Job Crashed, New Duration	Marginal Additional Direct Cost ($)	Total Additional Direct Cost ($)	Total Direct Project Cost ($)	Project Duration (days)	Activities at Minimum
0	A–C–E	—	—	—	310	10	E
1	A–C–E	C, 3	15	15	325	9	C, E
2	A–C–E	A, 4	30	55	365	8	C, E
	B–E	B, 7	10				
3	A–C–EB–	A, 3	30	95	405	7	A, C, E
	E	B, 6	10				

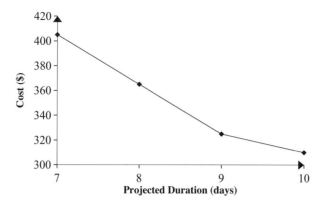

Figure 8.9   Time–cost trade-off alternatives for Example 8.4.

In iteration 3, activities A and B are crashed again to lower the duration of the critical paths A–C–E and B–E. Note that while activity D has a lower marginal crash cost than A, it is not crashed because it is not on a critical path. (Note that in this section only direct cost is considered. In the next section, we will consider both direct and indirect costs.)

In iteration 3, the project time cannot be reduced any further because all activities on the critical path A–C–E are at their minimum crash times. Note that critical path B–E could be further reduced, but this would not result in shorter project duration since critical path A–C–E cannot be reduced further. Figure 8.9 shows a plot of total cost versus total duration. Therefore, the minimum project duration is seven days, with an additional cost of $405 - 310 = \$95$.

See Supplement S8.1.xls.

Example 8.5   Time–Cost Trade-Off for Nine Activities   Consider the network diagram presented in Figure 8.10. The normal time, crash time, normal cost, and crash cost for all activities A–I are presented in the Table 8.7.

Determine all possible project durations and their associated costs.

First, find the marginal cost per day $k_{i,j}$ for all activities. These values are presented in the last row of Table 8.7.

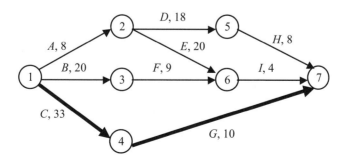

Figure 8.10   Network for Example 8.5.

TABLE 8.7   Times and Costs for Example 8.5

Activity	A	B	C	D	E	F	G	H	I
Normal time	8	20	33	18	20	9	10	8	4
Crash time	6	16	21	14	17	7	7	8	3
Normal cost	10	22	30	20	6	4	9	3	6
Crash cost	15	38	42	26	7.5	8	18	3	9.5
Cost/day, k_{ij}	2.5	4	1	1.5	0.5	2	3	—	3.5

Now solve the problem using the CPM algorithm. The critical path using normal activity times is C–G, with duration of 43 days.

Now, apply the time–cost trade-off algorithm. The solution is given in Table 8.8. At iteration 0, the total cost for normal time is $10 + 22 + 30 + 20 + 6 + 4 + 9 + 3 + 6 = 110$.

In iteration 15, the project time cannot be reduced any further as all the activities on the critical paths C–G and A–D–H are at the minimum crash times. In Example 8.5, there are 16 different alternatives, ranging from 43 to 28 days for project durations with an additional cost of $0–$48. The project cost increased from $110 to $110 + 48 = \$158$. One of the 16 alternatives can be selected and used for the actual project implementation.

8.3.2   Considering Indirect Cost

The total cost of the project is given as

$$\text{Total cost} = \text{direct cost} + \text{indirect cost}$$

By reducing the project duration, the total indirect cost decreases. In contrast to the indirect cost, the direct cost increases as the project duration decreases. Therefore, if the project duration is increased, the direct cost of the project will be decreased but the indirect cost of the project will be increased. In some cases, the decrease in the indirect cost is more than the increase in the direct cost. In these cases, it is economical to reduce the project duration up to the point that the total cost is minimized. Reducing the project duration after this point will increase the total cost. See Figure 8.11 for the relationship between project duration, indirect cost, direct cost, and total cost. The best project duration is associated with the minimum total cost.

Direct cost was considered in the last section. When considering both direct and indirect costs, the time–cost trade-off method of the last section should be modified to find the total project duration associated with the minimum total cost; calculate the total cost as the sum of the direct cost and the indirect cost in each iteration. The optimal project duration T^* is associated with the minimum total cost. In the time–cost trade-off method, the project duration starts at $T_{\max}$ (associated with normal time) and the total cost decreases as the duration is decreased in each iteration. Stop at the iteration where the total cost starts to increase.

In Figure 8.11, if only direct cost is considered, the minimum direct cost is at point $T_{\max}$ when the normal times are used. If only indirect cost is considered, $T_{\min}$ is the optimum solution. If both direct and indirect cost are considered, point T^* is the optimal duration that minimizes the total cost.

TABLE 8.8 Example 8.5 Time–Cost Trade-Off Solution

Iteration	Critical Path	Job Crashed, New Duration	Added Direct Cost ($1000)	Total Additional Direct Cost ($1000)	f_2 = Total Project Cost ($1000)	f_1 = Project Duration (days)	Activities at Minimum
0	C–G	—	—	—	110	43	H
1	C–G	C, 32	1	1	111	42	H
⋮	⋮	⋮	⋮	⋮	⋮	⋮	⋮
9	C–G	C, 24	1	9	119	34	H
10	C–G	C, 23	1	10	121.5	33	H
	A–D–H	D, 17	1.5	11.5			
11	C–G	C, 22	1	12.5	126	32	H
	A–D–H	D, 16	1.5	14			
	B–F–I	F, 8	2	16			
12	C–G	C, 21	1	17	131	31	C, F, H
	A–D–H	D, 15	1.5	18.5			
	B–F–I	F, 7	2	20.5			
	A–E–I	E, 19	0.5	21			
13	C–G	G, 9	3	24	139	30	C, D, F, H, I
	A–D–H	D, 14	1.5	25.5			
	B–F–I	I, 3	3.5	29			
	A–E–I	(I reduced above)					
14	C–G	G, 8	3	32	148.5	29	C, D, F, H, I
	A–D–H	A, 7	2.5	34.5			
	B–F–I	B, 19	4	38.5			
	A–E–I	(A reduced above)					
15	C–G	G, 7	3	41.5	158	28	A, C, D, F,
	A–D–H	A, 6	2.5	44			G, H, I
	B–F–I	B, 18	4	48			
	A–E–I	(A reduced above)					

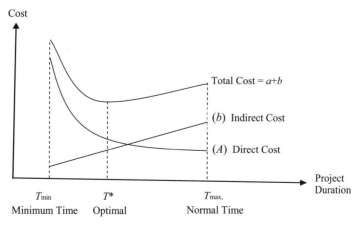

Figure 8.11 Total cost as function of indirect and direct costs.

TABLE 8.9   Costs for Project Durations in Example 8.6

f_1 = Project duration	43	42	41	40	39	38	37	36	35	34	33	**32**	**31**	30	29	28
Direct cost	110	111	112	113	114	115	116	117	118	119	121.5	**126**	**131**	139	148.5	158
Indirect cost	215	210	205	200	195	190	185	180	175	170	165	**160**	**155**	150	145	140
f_2 = Total cost	325	321	317	313	309	305	301	297	293	289	286.5	**286**	**286**	289	293.5	298
Efficient?	No	No	No	No	No	No	No	No	No	No	No	**No**	**Yes**	Yes	Yes	Yes

Example 8.6   Considering Indirect Cost for Nine Activities   Consider Example 8.5. Suppose the indirect cost is $5 per day (in $1000). Find all possible project durations using the time–cost trade-off method. Also, find the project duration that minimizes the total cost.

The summary of the solution is shown in Table 8.9. First consider Table 8.8, which is the time–cost trade-off solution of Example 8.5 where only direct cost is considered. For each iteration of Table 8.8, calculate and add the indirect cost to the direct cost, which is also presented in Table 8.8. The solutions for each iteration (i.e., for each given project duration) are shown in Table 8.9. For example, the indirect cost for a duration of 43 days is 43 days × $5 per day = $215, and the total cost is 110 + 215 = $325. It can be observed in Table 8.9 that as project duration decreases, total cost first decreases and then increases. Also see Figure 8.12.

The minimum total cost achieved is $286 (in thousands), which is associated with a project duration of 31 or 32 days. Because project duration should be minimized, choose a duration of 31 days as the best solution.

8.3.3   Bicriteria Time–Cost Trade-Off

The bicriteria time–cost trade-off project management can be formulated as follows:

Minimize project duration, f_1.

Minimize total cost, f_2.

Subject to the set of constraints on the project management problem.

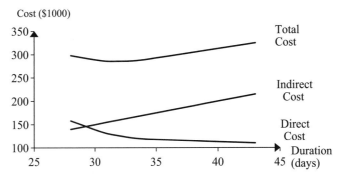

Figure 8.12   Direct, indirect, and total cost for Example 8.6. Optimal project duration is 31 days, which minimizes the total cost.

TABLE 8.10   Set of Bicriteria Alternatives for Example 8.7

Alternative (iteration)	0	1	2	3	4	5	6	7	8	9	10	11	12	13	14	15
f_1 = Project duration	43	42	41	40	39	38	37	36	35	34	33	32	31	30	29	28
f_2 = Direct cost	110	111	112	113	114	115	116	117	118	119	121.5	126	131	139	148.5	158
Efficient?	Yes	Yes	Yes	Yes	Yes	Yes	Yes	Yes	Yes	Yes	Yes	Yes	Yes	Yes	Yes	Yes
Rank	—	—	—	—	5	4	2	1	3							

The total cost can be measured by considering only the direct cost (Section 8.3.1) or by considering both direct and indirect costs (Section 8.3.2). In the multicriteria approach, first generate the set of efficient alternatives. Then use a multicriteria method (e.g., paired comparison of alternatives or an additive utility function) to choose the best alternative.

Bicriteria Time Versus Direct Cost Only   When only the direct cost is considered, the alternatives generated by the time–cost trade-off method (of Section 8.3.1) will be efficient.

Example 8.7   Bicriteria Considering Only Direct Cost   Consider Example 8.5. Find the efficient alternatives for the bicriteria problem. Ask the DM to rank the best five alternatives.

Consider the solution of Example 8.5, which was presented in Table 8.8. In this example, there are 16 efficient alternatives. The set of bicriteria alternatives for this example are summarized in Table 8.10. The ranking of the hypothetical DM is also shown.

For Example 8.5, the plot of total direct cost versus the total duration is given in Figure 8.13.

See Supplement S8.2.xls.

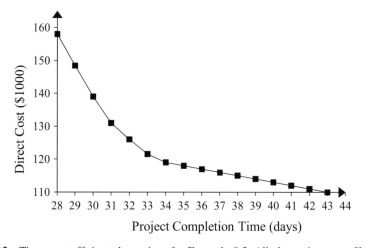

Figure 8.13   Time–cost efficient alternatives for Example 8.5. All alternatives are efficient when considering only direct cost.

Total Cost ($1000)

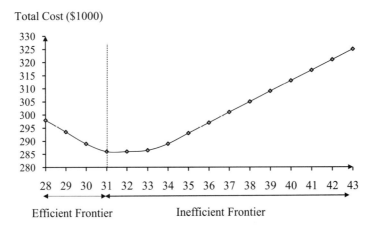

Figure 8.14   Total duration versus total direct and indirect costs; alternatives with durations less than 31 days (the minimum total cost) are efficient.

Bicriteria Time Versus Total (Direct and Indirect) Cost   When both direct and indirect costs are considered, the alternatives generated by using the method of Section 8.3.2 may not all be efficient. Only alternatives with durations less than or equal to the duration associated with the minimum total cost are efficient. In Figure 3.7 all alternatives on the left side of T^* (and including T^*) are efficient; all alternatives on the right side of it are inefficient.

Example 8.8   Bicriteria Considering Total (Direct and Indirect) Cost   Consider Example 8.5. Suppose the indirect cost is $5 per day (in $1000). Identify the efficient solutions to this example. Ask the DM to rank the efficient alternatives.

The solution to this problem is presented in Example 8.6. All the alternative solutions are presented in Table 8.9 where efficient and inefficient alternatives are identified. The same information is shown in Figure 8.14.

Because the duration associated with the minimum cost is 31, the alternatives with durations of 28, 29, 30, and 31 are efficient. Point (31, 286) dominates all alternatives whose duration is greater than 31; for example, point (32, 286) is dominated by point (31, 286). All alternatives whose durations are greater than 31 days are inefficient. Suppose that the DM's rankings are project durations 30, 31, 29, and 28.

See Supplement S8.3.xls.

Example 8.9   Selection of Best Alternative Using Additive Utility Function   Consider Example 8.4, where only the direct cost was considered. The solution for this problem was presented in Table 8.6. Suppose that an additive utility function is used to rank the alternatives. For the additive utility function, the weights of importance for the normalized values of the two objectives are 0.4 and 0.6, respectively. Find the best alternative.

TABLE 8.11 Solution to Multicriteria Example

	a_1	a_2	a_3	a_4	Minimum	Maximum
Duration, f_1	10	9	8	7	7	10
Total direct cost, f_2	310	325	365	405	310	405
Normalized duration, f_1'	1.00	0.67	0.33	0.00		
Normalized total cost, f_2'	0.00	0.16	0.58	1.00		
Minimize $U = 0.4f_1' + 0.6f_2'$	0.40	0.36	0.48	0.60		
Rank	2	1	3	4		

In this example, there are four efficient alternatives which are shown in Table 8.11. First, find normalized values of objectives f_1 and f_2 and then rank the bicriteria alternatives by minimizing the additive utility function:

$$U = 0.4f_1' + 0.6f_2'$$

For example, the normalized values of the two objectives for alternative a_2 are calculated as follows:

$$f_1' = \frac{9-7}{10-7} = 0.67$$

$$f_2' = \frac{325-310}{405-310} = 0.16$$

The ranking of the alternatives is presented in Table 8.11. Alternative a_2, with duration of nine days and cost of $325, is ranked as the best alternative.

See Supplement S8.4.xls.

8.4 LINEAR PROGRAMMING FOR PROJECT MANAGEMENT

8.4.1 Linear Programming for Solving CPM

Linear programming can be used to formulate and solve project management problems. Solving the project management problem by LP is computationally less efficient than solving it by the CPM. However, formulating the project management problem by LP has several advantages that cannot be achieved by the CPM, including sensitivity analysis and ease of incorporating constraints. With advances in computers and the development of new LP algorithms that can solve large-scale LP problems very efficiently, virtually all project management problems can be solved by LP.

The notation used for the linear programming formulation is as follows:

x_i = earliest time for node i for $i = 1, \ldots, n$ (decision variable)

t_{ij} = duration of activity (i,j) from node i to node j (known information)

Two different LP problems can be solved to determine the optimal solution:

- Find the earliest times for each node (to find the optimal project duration).
- Find the latest times for each node (to find the critical activities).

By subtracting the earliest times from the latest times for each node, the critical activities and critical path will be identified. That is, activities that have zero float (slack) time are critical. The details of these LP problems are presented below.

Step 1   The LP problem for the forward pass (i.e., finding the earliest times) is formulated as follows:

PROBLEM 8.1   FINDING THE EARLIEST TIMES
Minimize:

$$f = \sum_{i=1}^{n} x_i \tag{8.2}$$

Subject to:

$$x_j - x_i \geq t_{ij} \quad \text{for all } i, j \text{ nodes for } (i, j) \text{ activities}$$
$$x_i \geq 0 \qquad \text{for } i = 1, 2, \ldots, n$$

Solve this problem using LP software to find the optimal solution for all decision variables. The optimal value of x_n, the earliest time of the final node, represents the optimal (minimum feasible) duration of the project. From the LP solution, x_i for $i = 1, \ldots, n$ values (the earliest times for all the nodes) as well as the earliest times for all activities (i,j) can be calculated.

Step 2   The LP problem for the backward pass (i.e., to find the latest times) is formulated as follows:

PROBLEM 8.2   FINDING THE LATEST TIMES
Minimize:

$$f = nx_n - \sum_{i=1}^{n-1} x_i \tag{8.3}$$

Subject to:

$$x_j - x_i \geq t_{ij} \quad \text{for all } i, j \text{ nodes for } (i, j) \text{ activities}$$
$$x_i \geq 0 \qquad \text{for } i = 1, 2, \ldots, n$$

Note that both Problems 8.1 and 8.2 have the same set of constraints. Variable x_n is the project completion time. This value will be the same value found in Problem 8.1 because its coefficient n is larger than the coefficients of the remaining variables, which are all 1. That is, in Problem 8.2's objective function, the value of nx_n will always be greater than the

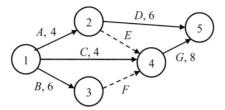

Figure 8.15 Network for Example 8.10.

value of $x_1 + x_2 + \cdots + x_{n-1}$ because $x_n \geq x_i$ for $i = 1, \ldots, n$ in the LP solution. Therefore, in Problem 8.2's objective function, the priority is to minimize x_n first, which finds and enforces the minimum project duration for the obtained solution. Then, the summation of $-x_i$ is minimized, which generates the largest feasible value of x_i for $i = 1, \ldots, n-1$; that is, it finds the latest start times for each node. Instead of nx_n, one can use Mx_n in Problem 8.2 where M is any given number greater than n.

After the values for x_i are found for both Problems 8.1 and 8.2, subtract the first solution from the second solution for each x_i to find the float time. If the difference is equal to zero for a given node, that node is on the critical path. Therefore, all critical activities can be identified.

Example 8.10 Linear Program Formulation Consider the network diagram in Figure 8.15.

Step 1 The following LP formulation (Problem 8.1) is used to find the earliest times for all nodes:

Minimize:

$$f = x_1 + x_2 + x_3 + x_4 + x_5$$

Subject to:

$$
\begin{array}{ll}
x_2 - x_1 \geq 4 & x_4 - x_3 \geq 0 \\
x_4 - x_1 \geq 4 & x_5 - x_2 \geq 6 \\
x_3 - x_1 \geq 6 & x_5 - x_4 \geq 8 \\
x_4 - x_2 \geq 0 & x_1, x_2, x_3, x_4, x_5 \geq 0
\end{array}
$$

Solving this problem by a LP computer package, the solution is as follows:

Variable	x_1	x_2	x_3	x_4	x_5	f
Node earliest time	0	4	6	6	14	30
Earliest start for activity	A, B, C	D, E	F	G	—	—

The above solution values correspond to the earliest start times for each node and their corresponding activities. In this solution, $x_5 = 14$; therefore, the minimum project duration is 14 days.

TABLE 8.12   Solution to Example 8.10

Given Network				Forward Pass		Backward Pass		Final Step	
Activity	(i, j)	Precedents	t_{ij}	ES_{ij}	EF_{ij}	LS_{ij}	LF_{ij}	TF_{ij}	Critical?
A	(1, 2)	—	4	0	$0 + 4 = 4$	2	6	2	No
B	(1, 3)	—	6	0	$0 + 6 = 6$	0	6	0	Yes
C	(1, 4)	—	4	0	$0 + 4 = 4$	2	6	2	No
D	(2, 5)	A	6	4	$4 + 6 = 10$	8	14	4	No
E	(2, 4)	A	0	4	$4 + 0 = 4$	6	6	2	No
F	(3, 4)	B	0	6	$6 + 0 = 6$	6	6	0	Yes
G	(4, 5)	A, B, C, E, F	8	6	$6 + 8 = 14$	6	14	0	Yes

Step 2   Now formulate and solve the second LP formulation (Problem 8.2):

Minimize $f = 5x_5 - x_1 - x_2 - x_3 - x_4$.

Subject to the same constraints as Problem 8.1 above.

The solution to this problem is as follows:

Variable	x_1	x_2	x_3	x_4	x_5	f
Node latest time	0	6	6	6	14	52
Latest finish for activity	—	A	B	C, E, F	D, G	—

In this solution, $x_5 = 14$ is the lowest project duration, while $x_1,$ $x_2,$ $x_3,$ x_4 indicate the latest times for nodes 1–4, respectively. For example, the earliest time for node 2 is 4 (based on the solution to Problem 8.1) and its latest time is 6 (based on the solution to Problem 8.2). We can now use the values from Problem 8.1 (the earliest times) and Problem 8.2 (the latest times) to calculate the slack time for each activity. This is summarized in Table 8.12. According to Table 8.12, critical activities are B, F, and G. The earliest and latest start times for each node are shown above each node in Figure 8.16.

See Supplement S8.5-1.lg4.
See Supplement S8.5-2.lg4.

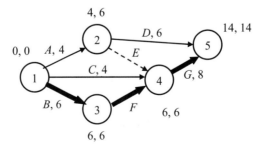

Figure 8.16   Solution of Example 8.10 by LP.

8.4.2   Linear Programming Time–Cost Trade-Off[†]

Linear programming can be used to solve the time–cost trade-off problem. The LP model will find the earliest and latest times for each node of the project network while minimizing the total marginal cost. Consider the following notation:

K = indirect cost per unit time

x_i = earliest time for node i

t_{ij} = time required to complete activity (i,j)

$t_{n,ij}$ = time required to complete activity (i,j) for normal time

$t_{c,ij}$ = minimum time to complete activity (i,j)

$c_{n,ij}$ = normal cost for activity (i,j)

$c_{c,ij}$ = crash cost for activity (i,j)

k_{ij} = crash cost per unit time for activity (i,j)

The time–cost trade-off LP formulation is presented below in Problem 8.3. This LP formulation (Problem 8.3) is based on the constraints of Problem 8.1 or Problem 8.2. In addition to their constraints, it has additional constraints and a different objective function. The objective is to minimize the total cost, which is a function of the direct cost and indirect cost. Indirect cost is represented by $K * x_n$, where x_n is the project duration.

Note that although we refer to the objective function of Problem 8.3 as the total cost, it is actually the marginal direct cost of the project. For the direct cost, only crashed costs are considered. The normal costs can be removed for optimization purposes as they are constant. Therefore, in some problems, it is possible to have a negative total (marginal) cost, which means that by considering the indirect cost the actual total cost of the project reduces when some activities are crashed.

PROBLEM 8.3   TIME–MARGINAL COST TRADE-OFF LP

Minimize:

$$\text{Total cost } f = \sum_{\text{all}(i,j)} -k_{ij}t_{ij} + Kx_n \tag{8.4}$$

Subject to:

$$
\begin{aligned}
x_j - x_i &\geq t_{ij} & \text{for all activities } (i,j) & \qquad t_{ij} \geq 0 & \text{for all activities } (i,j) \\
t_{ij} &\leq t_{n,ij} & \text{for all activities } (i,j) & \\
t_{ij} &\geq t_{c,ij} & \text{for all activities } (i,j) & \qquad x_i \geq 0 & \text{for } 1 \leq i \leq n
\end{aligned}
$$

Example 8.11   Solving Time–Cost Trade-Off by LP   Consider the problem presented in Figure 8.17. Also consider the normal time, crash time, normal cost, and crash cost for each activity as presented in Table 8.13. The indirect cost is $60 per day. All costs are in thousands of dollars.

Find the project duration that minimizes the total project cost. Also, show the optimal schedule for all activities.

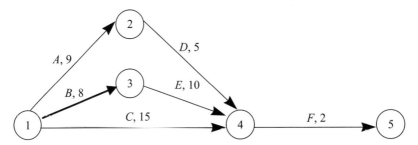

Figure 8.17 Network diagram of Example 8.11.

First, calculate the marginal (cost per day) values k_{ij} for all activities. The values for all k_{ij} are presented in Table 8.13. For example, the marginal cost for activity A is calculated as

$$k_{12} = \frac{170 - 110}{9 - 6} = \frac{60}{3} = \$20 \text{ per day}$$

The Problem 8.3 formulation for this problem is shown below.

Minimize:

$$f = -20t_{12} - 25t_{13} - 30t_{14} - 10t_{24} - 15t_{34} - 40t_{45} + 60x_5$$

Subject to:

$$x_2 - x_1 - t_{12} \geq 0 \qquad 6 \leq t_{12} \leq 9$$
$$x_3 - x_1 - t_{13} \geq 0 \qquad 6 \leq t_{13} \leq 8$$
$$x_4 - x_1 - t_{14} \geq 0 \qquad 10 \leq t_{14} \leq 15$$
$$x_4 - x_2 - t_{24} \geq 0 \qquad 3 \leq t_{24} \leq 5$$
$$x_4 - x_3 - t_{34} \geq 0 \qquad 6 \leq t_{34} \leq 10$$
$$x_5 - x_4 - t_{45} \geq 0 \qquad 1 \leq t_{45} \leq 2$$
$$x_i \geq 0 \qquad \text{for all } i$$
$$t_{ij} \geq 0 \qquad \text{for all activities } (i,j)$$

TABLE 8.13 Information for Activities A–F

Activity label (i,j)	Path i–j	$t_{n,ij}$	$t_{c,ij}$	$c_{n,ij}$	$c_{c,ij}$	k_{ij}	$t_{n,ij} - t_{c,ij}$
A	1–2	9	6	110	170	20	3
B	1–3	8	6	150	200	25	2
C	1–4	15	10	250	400	30	5
D	2–4	5	3	80	100	10	2
E	3–4	10	6	90	150	15	4
F	4–5	2	1	110	150	40	1

The LP problem above can be solved by using an LP software package. The optimal solution is shown in the table below where the optimum duration for the project is $X_5^* = 15$ days.

t_{12}	t_{13}	t_{14}	t_{24}	t_{34}	t_{45}	x_5	f
9	8	14	5	6	1	15	−80

Note that the objective function value of the optimal solution is

$$f = (-20 \times 9 - 25 \times 8 - 30 \times 14 - 10 \times 5 - 15 \times 6 - 40 \times 1) + 60 \times 15$$
$$= -980 + 900 = -80$$

Therefore, the minimum marginal cost for this project is −$80 (in thousands), which is obtained using LP. If the normal times were used, the project duration would be 20 days and the objective function would be $-20 \times 9 - 25 \times 8 - 30 \times 15 - 10 \times 5 - 15 \times 10 - 40 \times 2 + 60 \times 20 = -1110 + 1200 = 90$. This is higher than the optimal solution of −80 obtained by the LP solution.

See Supplement S8.6.lg4.

8.4.3 Multiobjective LP Time–Cost Trade-Off[†]

The multiple-objective linear programming (MOLP) problem for the time–total marginal cost trade-off problem can be stated as:

Minimize project duration: $f_1 = x_n$.

Minimize project marginal cost: $f_2 = \sum_{\text{all}(i,j)} -k_{ij}t_{ij} + Kx_n$.

Subject to: Constraints of Problem 8.3.

The project cost includes both direct and indirect costs. If only the direct cost is considered, then the indirect cost Kx_n should not be considered. We can generate all efficient points by using the following linear programming problem.

PROBLEM 8.4 TIME–MARGINAL COST TRADE-OFF MOLP PROBLEM

Minimize:

$$U = w_1 f_1' + w_2 f_2'$$

Subject to:

$$f_1 = x_n$$
$$f_2 = \sum_{\text{all}(i,j)} -k_{ij}t_{ij} + Kx_n$$
$$f_1' = \frac{f_1 - f_{1,\min}}{f_{1,\max} - f_{1,\min}}$$
$$f_2' = \frac{f_2 - f_{2,\min}}{f_{2,\max} - f_{2,\min}}$$
$$x_{n,\min} \le x_n \le x_{n,\max}$$

Constraints of Problem 8.3

Here, $x_{n,\min}$ and $x_{n,\max}$ are given lower and upper bounds on the project duration. In practice, the DM can usually provide upper and lower bounds on the project duration. Assume w_1 and w_2 are weights of importance for normalized objectives f_1 and f_2, respectively, such that $w_1 + w_2 = 1$ and $w_1, w_2 > 0$. By using different weights, different efficient alternatives can be generated.

Furthermore, for an additive utility function, if the DM provides the weights of importance for the two objectives, the best alternative can be found by solving the multiobjective LP-4 problem.

Example 8.12 MOLP for Time–Cost Trade-Off   Consider the time–cost trade-off Examples 8.5 and 8.6 where there is an indirect cost of $5 per day.

(a) Suppose the weights of importance for the normalized values of the two objective functions are $(0.25, 0.75)$, respectively. Find the optimal solution to this problem.

(b) Also, find optimal solutions for the following weights: $(0.9, 0.1)$, $(0.75, 0.25)$, $(0.55, 0.45)$, $(0.5, 0.5)$, $(0.35, 0.65)$, $(0.1, 0.9)$, and $(0, 1)$.

(a) To find the normalized values of the objective functions, first find minimum and maximum values of the two objectives by solving two different LP problems. Consider the following constraints:

$$f_1 = x_7$$
$$f_2 = -2.5t_{12} - 4t_{13} - 1t_{14} - 1.5t_{25} - 0.5t_{26} - 2t_{36} - 3t_{47} - 3.5t_{67} + 5x_7$$

$6 \le t_{12} \le 8$	$x_2 - x_1 - t_{12} \ge 0$
$16 \le t_{133} \le 20$	$x_3 - x_1 - t_{13} \ge 0$
$21 \le t_{14} \le 33$	$x_4 - x_1 - t_{14} \ge 0$
$14 \le t_{25} \le 18$	$x_5 - x_2 - t_{25} \ge 0$
$17 \le t_{26} \le 20$	$x_6 - x_2 - t_{26} \ge 0$
$7 \le t_{36} \le 9$	$x_6 - x_3 - t_{36} \ge 0$
$7 \le t_{47} \le 10$	$x_7 - x_4 - t_{47} \ge 0$
$t_{57} = 8$	$x_7 - x_5 - t_{57} \ge 0$
$3 \le t_{67} \le 4$	$x_7 - x_6 - t_{67} \ge 0$
$t_{ij} \ge 0$ for all activities (i,j)	$x_i \ge 0$ for all i

The two LP problems are presented in the table below along with their solutions.

			Solution	
Problem	Objective	Subject To	f_1	f_2
1	Min. f_1	Example 4.3 MOLP constraints	28	-40
2	Min. f_2	Example 4.3 MOLP constraints	32	-56

See Supplement S8.7.lg4.

Using these solutions and the maximum and minimum for each objective, the normalizing equations for the two objectives are

$$f_1' = \frac{f_1 - 28}{32 - 28} \quad \text{or} \quad 4f_1' + 28 = f_1$$
$$f_2' = \frac{f_2 - (-56)}{-40 - (-56)} \quad \text{or} \quad 16f_2' - 56 = f_2$$

Now the weighted MOLP problem, can be formulated:

Minimize:

$$U = 0.25 f_1' + 0.75 f_2'$$

Subject to:

$$4f_1' + 28 = f_1; \quad 16f_2' - 56 = f_2; \quad \text{and MOLP constraints}$$

Now solve the above problem using a LP software package. The solution is presented as alternative a_1 in Table 8.14, where $f_1 = 31$ and $f_2 = -56$.

TABLE 8.14   Bicriteria Alternatives for Example 8.12

Given weights (w_1, w_2)	(0.1, 0.9) or (0.25, 0.75) or (0.35, 0.65)	(0.5, 0.5)	(0.525, 0.475)	(0.55, 0.45) or (0.75,0.25) or (0.9,0.1)
Alternatives	a_1	a_2	a_3	a_4
Minimize duration, f_1	31	30	29	28
Minimize total marginal cost, f_2	-56	-53	-48.5	-44
Marginal direct cost : $\sum_{\text{all}(i,j)} -k_{ij}t_{ij}$	-211	-203	-193.5	-184
Indirect cost: Kx_n	155	150	145	140
Total cost = direct cost + indirect cost	$131 + 31 \times 5 = 286$	$139 + 30 \times 5 = 289$	$148.5 + 29 \times 5 = 293.5$	$158 + 28 \times 5 = 298$
Efficient?	Yes	Yes	Yes	Yes

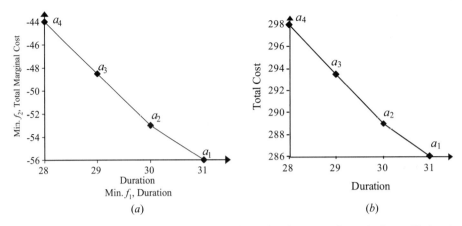

Figure 8.18   Efficient alternatives for Example 8.12: (*a*) duration vs. total marginal cost; (*b*) duration vs. total cost.

(b) Now, resolve the MOLP Example 4.3 problem with the following weights: $(0.9, 0.1)$, $(0.75, 0.25)$, $(0.55, 0.45)$, $(0.5, 0.5)$, $(0.35, 0.65)$, $(0.1, 0.9)$, and $(0.01, 0.99)$. That is, solve seven different LP problems. The solutions are presented in Table 8.14.

This problem has four efficient alternatives, $a_1, \ldots, a_4$, as presented in Table 8.14. The cost, duration of these efficient points, and corresponding total project cost for each alternative are shown in Table 8.14. Note that this solution matches the solution obtained by the time–cost trade-off method for the same example in Table 8.9.

See Supplement S8.8.lg4.

Figure 8.18*a* shows the duration versus the total marginal cost of all possible solutions of Example 8.12. Figure 8.18*b* shows the duration versus total cost (marginal direct cost and indirect cost) for the same example. Note that alternative a_3 is located on the line that connects alternatives a_2 and a_4 in Figures 8.18*a*, *b*.

See Supplement S8.9.xls.

8.5   PERT: PROBABILISTIC CPM

8.5.1   PERT Method

In CPM, the durations of activities are assumed to be known and constant. In practice, however, the durations of activities may not be known with certainty. PERT was developed by Booz, Allen, and Hamilton (1959) in a joint effort with the U.S. Navy to deal with this issue of uncertainty of durations in project management. The duration time for each activity is treated as a random variable, each having a random probability distribution.

In PERT, the probability distribution of a given activity duration is estimated by using:

a = optimistic duration of a given activity, where $a \le m$
m = most likely duration of a given activity
b = pessimistic duration of a given activity, where $b \ge m$

Using these three values, the mean and variance for the duration of each activity can be calculated. Then, the mean values of the durations are used in the CPM to find a project duration with a 0.5 probability of the project being completed within that project duration. Furthermore, the probabilities of different project durations can also be calculated.

The probability distribution of the duration of each activity is presented by a beta distribution. Beta distributions are more general than the normal distribution because the mode (the highest frequency) can be anywhere on the given interval. The mode for beta distributions is not necessarily equal to the mean, where in the normal distribution the mode is always equal to the mean. The most likely activity time in PERT, presented by m, is the same as the mode for the beta distribution. The three estimates a, b, and m can be used to approximate different forms of the beta distribution functions. Three examples of beta distributions are provided in Figure 8.19 where the optimistic time is $a = 2$ and the pessimistic time is $b = 20$. Figure 8.19 presents cases where $m = 6, 11, 18$. Note that the normal distribution (e.g., Figure 8.19b) is a special case of the beta distribution.

Although the distribution of each activity is beta, the distribution of the project duration is normal. This is true because, based on the central limit theorem, the summation of independent random variables (in this case activity durations) is approximately normal as the number of independent random variables (in this case the number of activities) increases.

In summary, in addition to CPM assumptions, PERT is based on the following assumptions:

- The duration of each activity has a beta probability distribution, and the mean and standard deviation is obtained from three estimates: optimistic, most likely, and pessimistic.
- The overall project time has a normal distribution based on the mean values of activities.
- Activities are independent of each other.

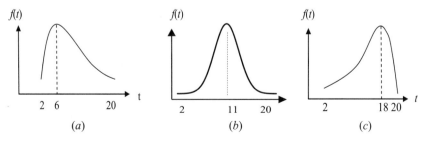

Figure 8.19   Three beta distribution probability densities: (a) m = 6; (b) m = 11; (c) m = 18.

In addition to the CPM notation, in PERT, the following notation is also used with respect to the duration of each given activity.

	Expected Duration	Variance	Standard Deviation
Each activity, (i,j)	$t_e = (a + 4m + b)/6$	$\text{Var} = \text{SD}^2$ $= (b - a)^2/36$	$\text{SD} = (b - a)/6$
Project (for only critical activities (i,j) on one critical path)	$T_e = \Sigma\, t_{e(i,j)}$	$V_T = \Sigma\, \text{Var}_{(i,j)}$	$\text{SD}_T = \sqrt{V_T}$

PERT Algorithm   PERT consists of the following steps:

Step 1   Determine t_e, SD, and Var for each activity using the three estimates a, m, and b.

Step 2   Denote the expected time as $t_{e(i,j)}$ and the variance as $\text{Var}(i,j)$ for each activity (i, j).

Apply the CPM algorithm to obtain the critical path and the critical activities using $t_{e(i,j)}$ as the duration for each activity (i,j).

Step 3   For the critical path, calculate the mean T_e, the variance V_T, and the standard deviation SD_T.

Probabilistic Analysis with PERT   In PERT, there is a 50% chance (i.e., 0.5 probability) that the project will be completed within a project duration of T_e. To determine the probability of completing a project within a given specified time t, do the following:

- Find the number of standard deviations from the mean:

$$z = \frac{t - T_e}{\text{SD}_T}$$

- Use the normal probability distribution table [also known as $\Phi(Z)$, at the end of this book] to find the corresponding cumulative probability where T is the normal random variable for project duration:

$$P\left(\frac{T - T_e}{\text{SD}_T} \le z\right) = P(Z \le z)$$

Note that if a given duration t is less than the mean value T_e, then the z value will be negative and one should use $\Phi(-Z) = 1 - \Phi(Z)$.

Example 8.13   PERT Example for Five Activities   Consider the network diagram consisting of five activities as presented in Figure 8.20. The optimistic (a), most likely (m), and pessimistic (b) durations of each activity are presented in Table 8.15.

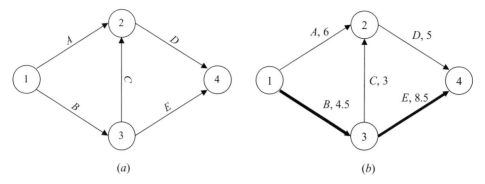

Figure 8.20   (*a*) Example 8.13 network diagram; (*b*) Critical path *B–E* using CPM.

(a) Find the critical path and project duration using mean values of activities.
(b) Find the probability of finishing the project within 14 days.
(c) Find the probability of finishing the project within 10, 13, 15, 16, and 17 days

(a) For each activity, use *a*, *m*, and *b* to generate t_e, SD, and Var. All these values are presented in Table 8.15. For example, for activity *A*,

$$t_{e(A)} = \frac{1}{6}(3 + 4 \times 6 + 9) = 6 \qquad SD_{(A)} = \frac{1}{6}(9 - 3) = 1 \qquad Var_{(A)} = (1)^2 = 1$$

Now, apply the CPM using the mean activity times t_e. The CPM solution is as follows:

Activity	t_e	ES	EF	LS	LF	TF	Critical?
A	6	0	6	2	8	2	No
B	4.5	0	4.5	0	4.5	0	Yes
C	3	4.5	7.5	5	8	0.5	No
D	5	7.5	12.5	8	13	0.5	No
E	8.5	4.5	13	4.5	13	0	Yes

TABLE 8.15   Information for Activities *A–E*

Activity	Given			Find		
	a	*m*	*b*	t_e	SD	Var.
A	3	6	9	6	1	1
B	1	5	6	4.5	0.833	0.694
C	1	2	9	3	1.333	1.777
D	2	5	8	5	1	1
E	6	8	13	8.5	1.167	1.361

The critical path is *B–E*. The mean project duration is

$$T_e = t_{e(B)} + t_{e(E)} = 4.5 + 8.5 = 13 \text{ days}$$

The probability of finishing the project within 13 days is 50%. The variance and the standard deviation of the project duration based on the mean values are

$$V_T - \text{Var}_{(B)} + \text{Var}_{(E)} = 0.694 + 1.361 = 2.055$$

$$\text{SD}_T = (V_T)^{1/2} = (2.055)^{1/2} = 1.434$$

(b) To find the probability of finishing the project within 14 days, first find the *z* associated with 14 days:

$$z = \frac{t - T_e}{\text{SD}_T} = \frac{14 - 13}{1.434} = 0.70$$

By using the normal probability distribution table at the end of the book, we find that

$$P(Z \leq 0.7) = \Phi(0.70) = 0.7580$$

Therefore, the probability of finishing the project in 14 days or less is 75.80%.

(c) The probabilities of completing the project within the given durations are presented below.

Completing Project Within	z Value	$P(Z \leq z)$	Probability
10 days	$(10 - 13)/1.434 = -2.09$	$P(Z \leq -2.09) = 0.0183$	1.83%
13 days	$(13 - 13)/1.434 = 0$	$P(Z \leq 0) = 0.50$	50%
15 days	$(15 - 13)/1.434 = 1.39$	$P(Z \leq 1.39) = 0.9171$	91.77%
16 days	$(16 - 13)/1.434 = 2.09$	$P(Z \leq 2.09) = 0.9817$	98.17%
17 days	$(17 - 13)/1.434 = 2.79$	$P(Z \leq 2.79) = 0.9974$	99.74%

The normal probability distribution and the probabilities of durations of 15, 13, and 10 are presented in Figure 8.21.

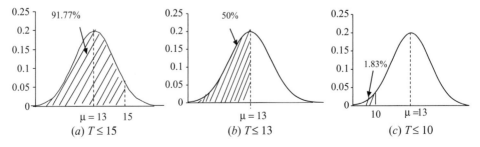

Figure 8.21   Probability of completing the project within 15, 13, and 10 days.

8.5.2 Bicriteria PERT

In PERT, there is a 0.5 probability that the project can be completed within (or beyond) the calculated mean project duration. However, suppose that the DM wishes to plan based on a different project duration, called targeted project duration.

For a targeted projection duration, the DM would like to know the probability of completion and identify critical activities based on the targeted project duration. Three cases can occur:

(a) The targeted project duration is less than the mean project duration (risk prone).

(b) The targeted project duration is more than the mean project duration (risk averse).

(c) The targeted project duration is the same as the mean project duration (risk neutral).

For case (a), more activities will be critical and the probability of completion will be less than 0.5. For case (b), fewer activities will be critical and the probability of completion will be more than 0.5. Case (c) is the same as PERT covered in last section. The bicriteria PERT problem is:

Minimize f_1: Project duration

Maximize f_2: Probability of completion

Subject to: CPM and PERT problem constraints

To generate a set of efficient bicriteria alternatives for this problem, use the method of Section 8.5.1 (PERT) to identify the probability of completion for different project durations. Then for each project duration, identify the critical path (or paths) and all critical activities. The DM can then choose one of the generated bicriteria alternatives. Depending on the selected alternative, some activities may become critical and some activities may become noncritical. For projects with durations less than the mean project duration, more activities become critical and the slack times of noncritical activities decrease. The reverse is true for alternatives with durations greater than the mean project duration. Therefore, the DM can plan according to the selected alternative. In the multicriteria PERT problem, depending on the level of risk that the DM wishes to take, different critical paths and activities may be considered.

Example 8.14 Bicriteria PERT   Consider Example 8.13. Generate seven bicriteria alternatives for this problem.

Recall that in Example 8.13 the mean for the project duration T_e was 13 and the standard deviation SD_T was 1.43. Seven different alternatives are generated as presented in Table 8.16. The graph of the project duration versus the probability of completion is shown in Figure 8.22.

For each alternative, critical activities can be identified. For example, for $T_e + 1$ $SD_T = 14.43$, none of the activities are critical. Both B and E can have slack times. So, if the DM chooses this alternative (84.13% probability of completion where the expected project duration is 13), there will be no critical activities. But for $T_e - 1$ $SD_T = 11.57$, in addition to path B–E (whose duration is 13), path B–C–D (whose duration is 12.5) also becomes critical. Therefore, in this case the critical activities will be B, C, D, and E.

TABLE 8.16 **Seven Bicriteria Alternatives for Example 8.14**

Risk Attitude	Risk Prone			Risk Neutral	Risk Averse		
Alternatives	$T_e - 3\,\mathrm{SD}_T$	$T_e - 2\,\mathrm{SD}_T$	$T_e - 1\,\mathrm{SD}_T$	T_e	$T_e + 1\,\mathrm{SD}_T$	$T_e + 2\,\mathrm{SD}_T$	$T_e + 3\,\mathrm{SD}_T$
f_1 = Project duration	8.71	10.14	11.57	13	14.43	15.86	17.29
f_2 = Probability of completion	0.13%	2.28%	15.87%	50%	84.13%	97.72%	99.87%
Critical activities	All	All	B, C, D, E	B, E	None	None	None
Efficient?	Yes	Yes	Yes	Yes	Yes	Yes	Yes

Suppose that the targeted project duration is 10 days; then all activities are critical. In this case, the durations of all activities should be reduced. But, if the target project duration is 18, then none of the critical activities are critical.

8.5.3 Tricriteria Time–Cost Trade-Off in PERT[†]

The tricriteria time–cost trade-off PERT problem is:

Minimize f_1: Project duration
Maximize f_2: Probability of completion
Minimize f_3: Project cost
Subject to: CPM and PERT problem constraints

In this problem, the time–cost trade-off method of Section 8.3 can be considered while activity durations are probabilities. By using crashed activities, the mean project duration will decrease but the cost of the project will increase.

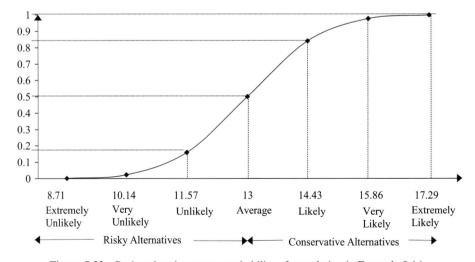

Figure 8.22 Project duration versus probability of completion in Example 8.14.

For simplicity of presentation, suppose that when an activity is crashed all the duration parameters [i.e., optimistic (a), most likely (m), and pessimistic (b)] are equally crashed with the same given cost. For example, consider activity A with $t_{e(A)}$ expected duration. If it is crashed by t units of time, then the optimistic time $a_{(A)}$ becomes $a_{(A)} - t$; the most likely time $m_{(A)}$ becomes $m_{(A)} - t$; and the pessimistic time $b_{(A)}$ becomes $b_{(A)} - t$. The mean and the variance of activity A, when it is crashed, are calculated as follows:

$$t_{e,\text{new}(A)} = \frac{a_{(A)} - t + 4(m_{(A)} - t) + b_{(A)} - t}{6} = t_{e(A)} - \frac{6t}{6} = t_{e(A)} - t$$

$$SD_{(A_{\text{new}})} = \frac{(b - t) - (a - t)}{6} = \frac{b - a}{6} = SD_{(A)}$$

Therefore, crashing activity A by t units of time will result in a decrease of t units in the expected duration but will not change the standard deviation. If A is on the critical path after being crashed, then the total project duration T_e is also reduced by the same amount t:

$$T_{e,\text{new}} = T_e - t$$

As a result, the new project duration has a normal distribution with the same standard deviation as before but with a new mean $T_{e,\text{new}}$. Therefore, the probability of the project duration T to be less or equal to the new mean $T_{e,\text{new}}$ is 50%; that is, $P_{\text{new}}(T \le T_{e,\text{new}}) = 0.5$.

Use the following procedure to generate a set of efficient alternatives for the tricriteria time–cost trade-off PERT problem.

Step 1   Solve the PERT problem for the given activity durations (use Section 8.5.1 method).

Step 2   Use the bicriteria PERT method (use Section 8.5.2 method) to generate a set of bicriteria (duration vs. probability of completion) alternatives for the given activity durations.

Step 3   Crash the eligible critical activity associated with the lowest cost (use Section 8.3.1). If no activity can be crashed, stop: All tricriteria alternatives have been generated. Otherwise, go to step 1.

Example 8.15   Tricriteria PERT   Consider Example 8.13. Suppose that the expected durations for activities B, C, and E can be crashed at $8, $12, and $16 per day, respectively. The maximum allowable crash time is one day for each of these activities. Show the efficient frontier for this problem. Also, show the probability of completion for each alternative if the targeted duration is 13.

The summary of alternatives is presented in Figure 8.23 and Table 8.17. To generate the first alternative, use mean durations and normal cost. Apply the PERT method. The solution is shown in Figure 5.6a as alternative a_1, whose duration is 13. By crashing activity B, alternative a_2 is generated, which has an additional cost of $8. By crashing activities B and E, alternative a_3 is generated, which has an additional cost of $8 + $16 = $24. By crashing activities B, C, and E, alternative a_4 is generated, which has an additional cost of $8 + $12 + $16 = $36.

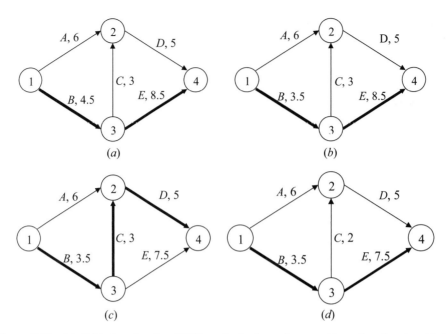

Figure 8.23 (*a*) Critical path using PERT method; alternative a_1 with project duration 13. (*b*) Activity *B* is crashed to 3.5; alternative a_4 with project duration 12. (*c*) Activities *B* and *E* are crashed to 3.5 and 7.5, respectively; alternative a_7 with project duration 11.5. (*d*) Activities *B*, *C*, and *E* are crashed to 3.5, 2, and 7.5, respectively; alternative a_{10} with project duration 11.

Figure 8.24 shows the comparison of the four different alternatives. For example, the probabilites of completion for the targeted duration of 13 are 50, 75.49, 85.31, and 91.92% for the four alternatives, respectively. Note that Figure 8.24 can be used to find the probability of completion for any given target duration. For example, if the target duration is 14 days, then the probabilities of completion for 14 days are 75.80, 91.92, 95.99, and 98.21% for alternatives a_1, a_2, a_3, and a_4, respectively.

See Supplement S8.10.xls.

TABLE 8.17 Seven Alternatives for Tricriteria PERT

Alternatives	a_1	a_2	a_2'	a_3	a_3'	a_4	a_4'
Critical activities	B, E	B,E	B,E	B,C,D	—	A,D,B,E	—
Min. f_1 = mean project duration	13	12	13	11.50	13	11	13
Max. f_2 = probability of completion	50%	50%	75.49%	50%	85.31%	50%	91.92%
Crashed activities	None	B	B	B, E	B, E	B, C, E	B, C, E
Min. f_3 = additional cost	0	8	8	24	24	36	36
Efficient?	Yes	Yes	Yes	Yes	Yes	Yes	Yes

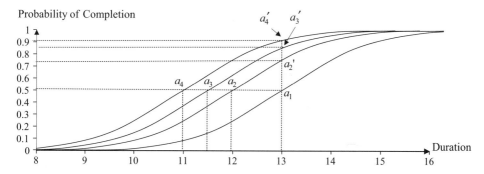

Figure 8.24 Project duration versus probability of completion in Example 8.14.

8.6 PROJECT MANAGEMENT WITH RESOURCE CONSTRAINTS

Often the resources required to perform different activities for projects are limited. Types of resources include materials, manpower, labor, space, and equipment associated with different activities. When resources are limited, it may not be possible to simply use CPM for planning. Activities that must be performed simultaneously may require the same resources; therefore, they must be scheduled sequentially. When resource constraints are taken into account, the total duration of the project usually increases. CPM can be modified to consider resource limitations. In this section, we cover CPM with single resource, CPM with multiple resources, and multicriteria CPM with resource constraints.

8.6.1 CPM with One Resource Constraint

In this section, we discuss a heuristic algorithm for solving the CPM problem with one resource constraint. It is assumed that the same amount of the resource is available throughout the project. In this method a heuristic is used to rank all activities. Then, activities are assigned according to the order of their ranked position. Possible heuristics to rank activities are:

(a) Lowest float or slack time
(b) Smallest "latest finish" time
(c) Smallest "earliest start" time
(d) Longest duration
(e) Largest resource requirement
(f) ACTIM: longest feasible path to complete a project starting at a specified activity

In the following, we use the ACTIM (ACTivity TIMe) heuristic to rank all activities. For each activity, ACTIM is the duration of the longest feasible (critical) path beginning from that activity to the last activity (end) node of the project. The duration of that path is the ACTIM value for that activity. In other words, the ACTIM value is the maximum time that an activity can control the project network considering all paths. Activities are ranked according to decreasing ACTIM values with the highest value ranked as the number 1 activity to be scheduled. The algorithm for CPM with one resource constraint is as follows:

TABLE 8.18 Information for Activities *A–F*

Activity	Predecessor	Duration	Resource	ACTIM	ACTIM Ranking
A	—	3	1	9	2
B	—	4	2	11	1
C	A	2	1	6	4
D	B	3	4	7	3
E	B	6	2	6	4
F	C, D	4	2	4	6

Step 1   Use a heuristic to rank all activities (e.g., use ACTIM).

Step 2   Find the highest ranked (feasible) activity that satisfies:

- Precedence requirements
- Resource limitations
  If such an activity exists, go to step 3.

  2a Skip to the earliest time when one of the activities currently in progress is completed.

  2b Renew (increase) resource availability for activities that are completed. Restart step 2.

Step 3   Start the activity found in step 2. Decrease the resource availability for the activity being initiated.

Go to step 2 until all activities are assigned and completed. Stop.

In each iteration, the highest ranked allowable activity that satisfies the resource constraints will be started. When all resources are used, no other activity will be allowed to start. As soon as an activity is completed, its released resource can be used to start the next ranked activity which uses the released resource subject to the precedence feasibility.

Example 8.16   CPM with Single Resource   A project consists of six activities *A–F*. The duration, activity precedence, and resource requirements for each activity are presented in Table 8.18. Figure 8.25 presents this project network. The total number of resources available at any given time is four units.

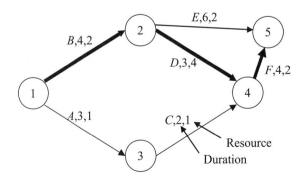

Figure 8.25 Network for Example 8.16.

TABLE 8.19   Solution for Example 8.16 (CPM with Single Resource)

Day	Activity	Duration	Start	Finish	Resource Required	Resource Available (Total = 4)	Ranked and Feasible
0	B	4	0	4	2	4 − 2 = 2	A
0	A	3	0	3	1	2 − 1 = 1	—
3	Finish A				−1	1 − (−1) = 2	C
3	C	2	3	5	1	2 − 1 = 1	—
4	Finish B				−2	1 − (−2) = 3	E
4	E	6	4	10	2	3 − 2 = 1	—
5	Finish C				−1	1 − (−1) = 2	—
10	Finish E				−2	2 − (−2) = 4	D
10	D	3	10	13	4	4 − 4= 0	—
13	Finish D				−4	0 − (−4) = 4	F
13	F	4	13	17	2	4 − 2 = 2	—

Find the shortest feasible time to complete this project using ACTIM to rank activities.

Suppose that there are no resource constraints for this problem. Then, one can use the CPM solution to this problem, which is a critical path of B–D–F with a duration of $4 + 3 + 4 = 11$ days. However, due to the limited resource constraints, the project duration will be more than 11 days. To find the solution with resource constraints, first find the ACTIM for all activities. For example, consider activity B; there are two paths starting with activity B (node 1) and ending at node 5. The first path, B–E, has a total duration of $4 + 6 = 10$. The second path, B–D–F, has a total duration of $4 + 3 + 4 = 11$. Therefore, the ACTIM value for activity B is Max$\{10,11\} = 11$. The ACTIM for all activities are given in Table 8.18. The ACTIM ranking is $\{B, A, D, C, E, F\}$.

The solution to this example is presented in Table 8.19.

Based on Table 8.19, this project will be completed in 17 days. The graphical solution of this example is also shown in Figure 8.26a, where the project duration is 17 days. It is important to realize that the above method is just a heuristic. For example, by trial and error, one can find an improved solution to this example. Figure 8.26b shows a solution where a project duration of 13 days is found. This solution of 13 days is a substantial improvement to the 17-day solution found by the method.

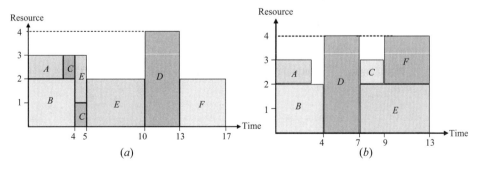

Figure 8.26   (a) Solution based on the heuristic algorithm for Example 8.16. (b) Improved solution by trial and error for Example 8.16.

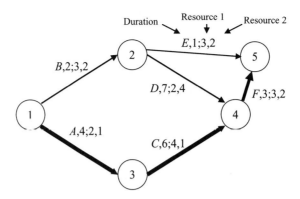

Figure 8.27 Network for Example 8.17.

8.6.2 CPM with Multiple-Resource Constraints

The algorithm for allocating multiple resources is the same as the algorithm for the single resource covered in the last section. However, in the allocation process, several resource constraints must be monitored simultaneously. For multiple resources, one can also use ACTIM to rank activities. The computation and ranking of ACTIM is the same as in the method discussed in the last section for single-resource problems.

Example 8.17   CPM with Resource for Multiple Resources   Consider a problem with six activities and two resources as shown in Figure 8.27 and Table 8.20. The first resource has a maximum of five units and the second resource has a maximum of four units. Find the shortest feasible time to accomplish the project considering both resources.

If there were no resource limitations, the critical path would be A–C–F with a duration of 13 days. The project duration when resource constraints are considered will be more than 13 days. The ACTIM ranking of activities is {A, B, D, C, F, E}. The solution is provided in Table 8.21. Figure 8.28 shows the graphical resource allocation solution of this problem.

At time 0, first assign activity A, reduce resource 1 by two units, and reduce resource 2 by one unit. The available units of resources are $5 - 2 = 3$ for resource 1 and $4 - 1 = 3$ for resource 2. This is shown in the second row of Table 8.21. At time 0, also assign activity

TABLE 8.20   Information for Activities A–F

Activity	Precedence	Duration	Resources 1	Resources 2	ACTIM	ACTIM Ranking
A	—	4	2	1	13	1
B	—	2	3	2	12	2
C	A	6	4	1	9	4
D	B	7	2	4	10	3
E	B	1	3	2	1	6
F	C, D	3	3	2	3	5
Max.	—	—	5	4	—	—

TABLE 8.21   Solution for Example 8.17

Time	Activity	Duration	Start	Finish	Resources Required		Resources Available		Ranked Feasible
					R_1	R_2	R_1	R_2	
							5	4	
0	—	—	—	—	—	—			A,B
0	A	4	0	4	2	1	$5-2=3$	$4-1=3$	B
0	B	2	0	2	3	2	$3-3=0$	$3-2=1$	—
2	Finish B	—	—	—	-3	-2	$0-(-3)=3$	$1-(-2)=3$	E
2	E	1	2	3	3	2	$3-3=0$	$3-2=1$	—
3	Finish E	—	—	—	-3	-2	$0-(-3)=3$	$1-(-2)=3$	—
4	Finish A	—	—	—	-2	-1	$3-(-2)=5$	$3-(-1)=4$	D,C
4	D	7	4	11	2	4	$5-2=3$	$4-4=0$	—
11	Finish D	—	—	—	-2	-4	$3-(-2)=5$	$0-(-4)=4$	C
11	C	6	11	17	4	1	$5-4=1$	$4-1=3$	—
17	Finish C	—	—	—	-4	-1	$1-(-4)=5$	$3-(-1)=4$	F
17	F	3	17	20	3	2	$5-3=2$	$4-2=2$	—
20	Finish F	—	—	—	-3	-2	$2-(-3)=5$	$2-(-2)=4$	—

B, because there are sufficient resources to start this activity. This is shown in the third row of Table 8.21. The availability of resource 1 is $3 - 3 = 0$ and availability of resource 2 is $3 - 2 = 1$.

At time 1, no resources are released; therefore, consider the next period, 2. At time 2, activity *B* is completed, and therefore, resources become available in period 3. Resource 1's availability becomes $0 + 3 = 3$ and resource 2's availability becomes $1 + 2 = 3$. Therefore, activity *E* can be considered; however, activity *D* cannot yet be considered because of resource requirements.

8.6.3 Bicriteria CPM with Resource Constraints[†]

In multicriteria project management with resources, two objectives are considered: project duration and units of resources available. By increasing the amount of available resources, the project duration may decrease, but this will be associated with more cost. In this section, we present the multicriteria approach for one resource. For multiple resources, each resource can be considered as a separate objective.

Suppose the lower (minimum) and the upper (maximum) bounds on the given resource are denoted by R_{min} and R_{max}, respectively. The multicriteria problem can be stated as:

Minimize: Project duration, f_1

Minimize: Units of resources R, f_2

Subject to: Problem constraints and $R_{min} \leq R \leq R_{max}$

To solve this multicriteria problem, first generate a number of efficient multicriteria alternatives. Then choose one of the alternatives as the best alternative. Consider a number of resources *R*. For each given *R*, solve the problem (by the method of Section 8.6.1), and find the associated minimum project duration. For a limited number of resources, it is possible to consider all possible values of *R*. For example, if the minimum and maximum

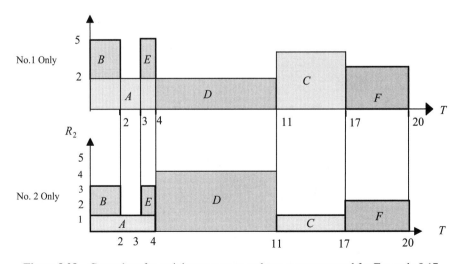

Figure 8.28 Gantt chart for activity sequences and two resources used for Example 8.17.

amounts of a resource are 8 and 12, respectively, then consider all five different amounts of the resource, that is, $R = 8, 9, 10, 11, 12$. For problems with a wide range of resources, instead of considering all values of R, a one-dimensional search (optimization) can be used to find the optimal number of resources. For bisectional one-dimensional search, five different equally distributed points (in terms of R value) can be generated. For each given R, solve the problem and find the minimum project duration. Once the best alternative is selected by the DM, proceed with the range of R values adjacent to the best selected R. Repeat the above process until no more alternatives can be considered:

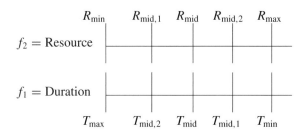

Example 8.18   Multicriteria Approach for CPM with One Resource   Consider
a project as presented in Figure 8.29. Suppose the minimum and maximum resource limitations are 8 and 16, respectively.

Generate five efficient multicriteria alternatives for this problem.

The solution is as follows: $R_{min} = 8$ and $R_{max} = 16$. Using bisectional approach, the five equally distributed R values are 8, 10, 12, 14, and 16. These R values are found as follows: $(16+8)/2 = 12$, $(8 + 12)/2 = 10$, and $(12 | 16)/2 = 14$.

Alternative a_1   For $R = 8$, solve the problem to find the minimum duration for the project using eight resources. The solution is $f_1 = 17$ and $f_2 = 8$.

Alternative a_2   For $R = 10$, find the minimum project duration. The solution is $f_1 = 15$ and $f_2 = 10$.

Alternative a_3   For $R = 12$, find the minimum project duration. The solution is $f_1 = 14$ and $f_2 = 12$.

Alternative a_4   For $R = 14$, find the minimum project duration. The solution is $f_1 = 14$ and $f_2 = 14$.

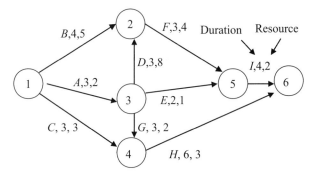

Figure 8.29   Network for Example 8.18.

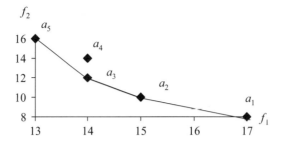

Figure 8.30   Efficient Frontier for Example 8.18.

Alternative a_5   For $R = 16$, find the minimum project duration. The critical path is A–D–F–I with a duration of 13 days. Therefore the solution is $f_1 = 13$ and $f_2 = 16$ units.

The multicriteria alternatives are summarized in the following table. The alternatives are also shown graphically in Figure 8.30.

	a_1	a_2	a_3	a_4	a_5
Min. $f_2 = R$	8	10	12	14	16
Min. $f_1 = T$	17	15	14	14	13
Efficient?	Yes	Yes	Yes	No	Yes

Notice that all generated alternatives may not necessarily be efficient. For example, a_4 is inefficient as it is dominated by a_3. Therefore, alternative a_4 can be eliminated.

Now, suppose the DM selects a_2 with $f_2 = R = 10$ and $f_1 = T = 15$ as the best alternative. In the next step, consider a new range of alternatives centered around a_2. Therefore, the new alternatives for this step range from $R = 8$ to $R = 12$; that is, $R = 8, 9, 10, 11, 12$, are considered. Their bicriteria alternatives are presented in the table below. Only two new alternatives are generated: a_6 and a_7.

	a_1	a_6	a_2	a_7	a_3
Minimum $R = f_2$	8	9	10	11	12
Minimum $T = f_1$	17	17	15	14	14
Efficient ?	Yes	No	Yes	Yes	No

Suppose the best solution is alternative a_2. The search process is then completed because there are no more feasible alternatives adjacent to the best alternative.

REFERENCES

General

Adamiecki, K. "Harmonygraph." *Polish Journal of Organizational Review*, 1931 (in polish).

Anderson, D. R., D. J. Sweeny, and T. A. Williams. *An Introduction to Management Science*, 4th ed., St. Paul, MN: West Publishing, 1985.

Archibald, R. D., and R. L. Villoria. *Network-Based Management Systems, PERT/CPM*. New York: Wiley, 1967.

Berman, E. B. "Resource Allocation in a PERT Network under Continuous Activity Time-Cost Functions." *Management Science*, Vol. 10, No. 4, 1964, 734–745.

Charnes, A., and W. W. Cooper. "A Network Interpretation and a Directed Subdual Algorithm for Critical Path Scheduling." *Journal of Industrial Engineering*, Vol. 13, No. 4, 1962, pp. 213–219.

Cleveland, D. I., and W. R. King. *Project Management Handbook*. New York: Van Nostrand Reinhold, 1984.

Deckro, R. F., J. E. Hebert, W. A. Verdini, P. H. Grimsrud, and S. Venkateshwar. "Nonlinear Time/Cost Tradeoff Models in Project Management." *Computers & Industrial Engineering*, Vol. 28, No. 2, 1995, pp. 219–229.

Deng, Z. M., H. Li, C. M. Tam, Q. P. Shen, and P. E. D. Love. "An Application of the Internet-Based Project Management System." *Automation in Construction*, Vol. 10, No. 2, 2001, pp. 239–246.

Elsayed, E. A. "Algorithm for Project Scheduling with Resource Constraints." *International Journal of Production Research*, Vol. 20, No. 1, 1982, pp. 95–103.

Evarts, H. F. *Introduction to PERT*. Boston: Allyn & Bacon, 1964.

Falk, J. E., and J. L. Horowitz. "Critical Path Problems with Concave Cost-Time Curves." *Management Science*, Vol. 19, No. 4, 1974, pp. 446–455.

Fazar, W. "Program Evaluation Review Technique," The American Statistician, Vol. 13, No. 2, (April 1959), p.10.

Fulkerson, D. R. "A Network Flow Computation for Project Cost Curves." *Management Science*, Vol. 7, No. 2, 1961, pp. 167–178.

Gelbard, R., N. Pliskin, and I. Spiegler. "Integrating System Analysis and Project Management Tools." *International Journal of Project Management*, Vol. 20, No. 6, 2002, pp. 461–468.

Goodman, L. J. *Project Planning and Management*. New York: Van Nostrand Reinhold, 1988.

Goyal, S. K. "A Note on a Simple CPM Time–Cost Trade-Off Algorithm." *Management Science*, Vol. 21, No. 6, 1975, pp. 718–722.

Heizer, J., and B. Render. *Production & Operations Management*, 4th ed. Upper Saddle River, NJ: Prentice Hall, 1996.

Kelly, J. "Critical-Path Planning and Scheduling: Mathematical Basis." *Operations Research*, Vol. 9, No. 3, 1961, pp. 296–320.

Kelly, J. E. and M. R. Walker. "Critical-Path Planning and Scheduling." In Papers presented at the December 1–3, 1959, eastern joint IRE-AIEE-ACM computer conference, New York, NY, USA, pp. 160–173. ACM, 1959.

Maccrimmon, K. R., and C. A. Ryavec. "Analytical Studies of the PERT Assumptions." *Operations Research*, Vol. 12, No. 1, 1964, pp. 16–37.

Nahmias, S. *Production and Operations Analysis*, 6th ed. New York: McGraw-Hill/Irwin, 2008.

Patterson, J. H. "A Comparison of Exact Approaches for Solving the Multiple Constrained Resource Project Scheduling Problem." *Management Science*, Vol. 30, No. 7, 1984, pp. 854–867.

Prager, W. "A Structural Method of Computing Project Cost Polygons." *Management Science*, Vol. 9, No. 3, 1963, pp. 394–404.

Siemens, N. "A Simple CPM Time–Cost Trade-Off Algorithm." *Management Science*, Vol. 17, No. 6, Application Series, 1971, pp. B354–B363.

Wagner, H. *Principles of Operations Research*, 2nd ed. Englewood Cliffs, NJ: Prentice Hall, 1975.

Multiobjective

Abbasi, B., S. Shadrokh, and J. Arkat. "Bi-Objective Resource-Constrained Project Scheduling with Robustness and Makespan Criteria." *Applied Mathematics and Computation*, Vol. 180, No. 1, 2006, pp. 146–152.

Al-Fanzin, M. A., and M. Haouari. "A Bi-Objective Model for Robust Resource-Constrained Project Scheduling." *International Journal of Production Economics*, Vol. 96, No. 2, 2005, pp. 175–187.

Al-Harbi, K. M. A.-S. "Application of the AHP in Project Management." *International Journal of Project Management*, Vol. 19, No. 1, 2001, pp. 19–27.

Al-Tabtabai, H. M., and V. P. Thomas. "Negotiation and Resolution of Conflict Using AHP: An Application to Project Management." *Engineering Construction*, Vol. 11, No. 2, 2004, pp. 90–100.

Arikan, F., and Z. Gungor. "An Application of Fuzzy Goal Programming to a Multiobjective Project Network Problem." *Fuzzy Sets and Systems*, Vol. 119, 2001, pp. 49–58.

Azaron, A., H. Katagiri, M. Sakawa, K. Kato, and A. Memariani. "A Multi-Objective Resource Allocation Problem in PERT Networks." *European Journal of Operational Research*, Vol. 172, No. 3, 2006, pp. 838–854.

Azaron, A., and R. Tavakkoli-Moghaddam. "Multi-Objective Time–Cost Trade-Off in Dynamic PERT Networks Using an Interactive Approach." *European Journal of Operational Research*, Vol. 180, No. 3, 2007, pp. 1186–1200.

Cohon, J. L. *Multiobjective Programming and Planning*. New York: Dover, 1978.

Davis, K. R., A. Stam, and R. A. Grzybowski. "Resource Constrained Project Scheduling with Multiple Objectives: A Decision Support Approach." *Computers & Operations Research*, Vol. 19, No. 7, 1992, pp. 657–669.

Deckro, R. F., and J. Hebert. "A Multiple Objective Programming Framework for Tradeoffs in Project Scheduling." *Engineering Costs and Production Economics*, Vol. 18, No. 3, 1990, pp. 255–264.

Dey, P. K. "Project Risk Management: A Combined Analytic Hierarchy Process and Decision Tree Approach." *Cost Engineering*, Vol. 44, Part 3, 2002, pp. 13–27.

Hapke, M., A. Jaszkiewicz, and R. Słowinski. "Interactive Analysis of Multiple-Criteria Project Scheduling Problems." *European Journal of Operational Research*, Vol. 107, No. 2, 1998, pp. 315–324.

Liang, T.-F. "Fuzzy Multi-Objective Project Management Decisions Using Two-Phase Fuzzy Goal Programming Approach." *Computers & Industrial Engineering*, Vol. 57, No. 4, 2009, pp. 1407–1416.

Lova, A., C. Maroto, and P. Tormos. "A Multicriteria Heuristic Method to Improve Resource Allocation in Multiproject Scheduling." *European Journal of Operational Research*, Vol. 127, No. 2, 2000, pp. 408–424.

Malakooti, B. "Multi-Objective Project Management." Case Western Reserve University, Cleveland, OH, 2013.

Mohanty, R. P., and M. K. Siddiq. "Multiple Projects—Multiple Resources Constrained Scheduling: A Multiobjective Analysis." *Engineering Costs and Production Economics*, Vol. 18, No. 1, 1989, pp. 83–92.

Moore, L. J., T. W. Bernard, E. R. Clayton, and S. M. Lee. "Analysis of a Multi-Criteria Project Crashing Model." *IIE Transactions*, Vol. 10, No. 2, 1978, pp. 163–169.

Mota, C. M. D. M., A. T. D. Almeida, and L. H. Alencar. "A Multiple Criteria Decision Model for Assigning Priorities to Activities in Project Management." *International Journal of Project Management*, Vol. 27, No. 2, 2009, pp. 175–181.

Neely, W. P., R. M. North, and J. C. Fortson. "Planning and Selecting Multiobjective Projects by Goal Programming." *Journal of the American Water Resources Association*, Vol. 12, No. 1, 1976, pp. 19–25.

Słowinski, R., B. Soniewicki, and J. Wёglarz. "DSS for Multiobjective Project Scheduling." *European Journal of Operational Research*, Vol. 79, No. 2, 1994, pp. 220–229.

Zammori, F. A., B. Marcello, and F. Marco. "A Fuzzy Multi-Criteria Approach for Critical Path Definition." *International Journal of Project Management*, Vol. 27, No. 3, 2009, pp. 278–291.

EXERCISES

Constructing CPM Network (Section 8.2.1) and Critical-Path Method Algorithm (Section 8.2.2)

2.1 Given the following table:

Activity	Precedents	Duration (weeks)
A	—	5
B	A	7
C	A	2
D	B	3
E	C,D	10

(a) Construct the network diagram.

(b) Use the CPM to find the shortest feasible time to complete this project. Show both graphical and tabular methods.

(c) What are the critical activities and critical path?

2.2 Given the following table:

Activity	Precedents	Duration (days)	Activity	Precedents	Duration (days)
A	—	5	E	B,D	4
B	A	4	F	A,C	8
C	—	5	G	E,F	16
D	A,C	7			

Respond to parts (a)–(c) from exercise 2.1.

2.3 Given the following table:

Activity	Precedents	Duration (days)	Activity	Precedents	Duration (days)
A	—	2	E	A,B	2
B	—	4	F	C,D	4
C	A	5	G	E	4
D	A,B	6	H	F	7

Respond to parts (a)–(c) from exercise 2.1.

2.4 Given the following table:

Activity	Precedents	Duration (days)	Activity	Precedents	Duration (days)
A	—	3	F	C	4
B	—	2	G	C,E	4
C	A	2	H	D,F	5
D	B	3	I	G	3
E	B	2.5			

Respond to parts (a)–(c) from exercise 2.1.

2.5 Consider the network shown below. Respond to parts (b) and (c) from exercise 2.1.

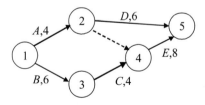

2.6 For the network shown below, respond to parts (b) and (c) from exercise 2.1.

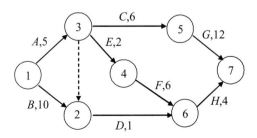

2.7 Consider the network shown below. Answer parts (b) and (c) from exercise 2.1.

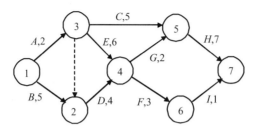

Project Monitoring and Gantt Chart (Section 8.2.3)

2.8 Consider exercise 2.1; show the Gantt chart and the line of balance where the actual project completion activity A is delayed by four days.

2.9 Consider exercise 2.2; show the Gantt chart and the line of balance where the actual project completion activity A is delayed by five days.

2.10 Consider exercise 2.3; show the Gantt chart and the line of balance where the actual project completion activities A and B are delayed by five days.

Time–Cost Trade-Off Algorithm (Section 8.3.1)

3.1 Consider the time–cost table below and its network. Suppose indirect cost is $5 per day. Consider only the direct cost presented in the table. Perform the time–cost trade-off method and identify all possible alternatives.

Activity	Normal Time	Crash Time	Normal Cost	Crash Cost
A	9	7	8	10
B	4	3	12	15
C	5	3	10	12
D	6	4	6	8
E	7	5	5	10

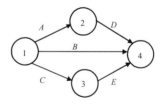

3.2 Consider the time–cost table below and its network. Suppose indirect cost is $1.50 per day. Consider only the direct cost presented in the table below. Perform the time–cost trade-off method and identify all possible alternatives.

Activity	Normal Time	Crash Time	Normal Cost	Crash Cost
A	4	3	10	15
B	6	4	9	13
C	7	5	8	12
D	9	6	15	19
E	8	2	13	22
F	4	2	8	10

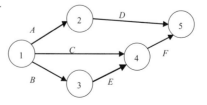

3.3 Consider the time–cost table below and its network. Suppose the indirect cost is $6 per day. Consider only the direct cost presented in the table below. Perform the time–cost trade-off method and identify all possible alternatives.

Activity	Normal Time	Crash Time	Normal Cost	Crash Cost
A	5	4	10	20
B	7	5	14	30
C	8	4	15	25
D	9	6	18	40
E	8	6	14	25
F	2	1	5	20
G	5	3	12	18
H	5	5	22	22

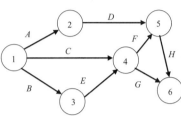

Considering Indirect Cost (Section 8.3.2)

3.4 Consider exercise 3.1. Consider both indirect cost and direct cost. Find the optimal project duration that minimizes the total of direct and indirect costs.

3.5 Consider exercise 3.2. Consider both indirect cost and direct cost. Find the optimal project duration that minimizes the total of direct and indirect costs.

3.6 Consider exercise 3.3. Consider both indirect cost and direct cost. Find the optimal project duration that minimizes the total of direct and indirect costs.

Bicriteria Time–Cost Trade-Off (Section 8.3.3)

3.7 Consider exercises 3.1 and 3.4. Suppose the weights of the normalized objective functions of an additive utility function are (0.75, 0.25), respectively.

   (a) List the set of bicriteria alternatives based on only considering direct cost. Identify efficient and inefficient alternatives. Rank the bicriteria alternatives.

   (b) List the set of bicriteria alternatives by considering both direct and indirect costs. Identify efficient and inefficient alternatives. Rank the bicriteria alternatives.

3.8 Consider exercises 3.2 and 3.5. Suppose the weights of the normalized objective functions of an additive utility function are (0.5, 0.5), respectively. Respond to parts (a) and (b) from exercise 3.7.

3.9 Consider exercises 3.3 and 3.6. Suppose the weights of the normalized objective functions of an additive utility function are (0.4, 0.6), respectively. Respond to parts (a) and (b) from exercise 3.8.

Linear Programming for Solving CPM (Section 8.4.1)

4.1 Consider exercise 3.1. Only consider the network and the normal times given in exercise 3.1. Use the LP approach to find the project duration. Also, identify earliest times and latest times for all activities and the critical path by using LP.

4.2 Consider exercise 3.2. Only consider the network and the normal times given in exercise 3.2. Use the LP approach to find the project duration. Also, identify earliest times and latest times for all activities and the critical path by using LP.

4.3 Consider exercise 3.3. Only consider the network and the normal times given in exercise 3.3. Use the LP approach to find the project duration. Also, identify earliest times and latest times for all activities and the critical path by using LP

Linear Programming for Time–Cost Trade-Off (Section 8.4.2)[†]

4.4 Consider exercise 3.1. Use LP to find the solution that minimizes the total of direct and indirect costs.

4.5 Consider exercise 3.2. Use LP to find the solution that minimizes the total of direct and indirect costs.

4.6 Consider exercise 3.3. Use LP to find the solution that minimizes the total of direct and indirect costs.

Multiobjective LP Time–Cost Trade-Off (Section 8.4.3)[†]

4.7 Consider exercise 3.1. Consider both direct and indirect costs. Solve this problem using an indirect cost of $0.50 per period.
  (a) What is the best alternative if the weights of importance are 0.6 and 0.4, respectively?
  (b) What is the best alternative if the weights of importance are 0.3 and 0.7, respectively?

4.8 Consider exercise 3.2. Consider both direct and indirect costs.
  (a) What is the best alternative if the weights of importance are 0.1 and 0.9, respectively?
  (b) What is the best alternative if the weights of importance are 0.7 and 0.3, respectively?

4.9 Consider exercise 3.3. Consider both direct and indirect costs.
  (a) What is the best alternative if the weights of importance are 0.3 and 0.7, respectively?
  (b) What is the best alternative if the weights of importance are 0.8 and 0.2, respectively?

PERT Method (Section 8.5.1)

5.1 Consider the network diagram below and three time estimates in the accompanying table.
  (a) Find the critical path and project duration using PERT. Explain what the solution means.
  (b) Find the probability of completing the project in less than 13 weeks.
  (c) Find the probability of completing the project in more than 18 weeks.

(d) How many weeks will it take to complete the project with a 84.13% probability?

Activity	Optimistic Time (weeks)	Most Likely Time (weeks)	Pessimistic Time (weeks)
A	2	4	8
B	3	4	5
C	3	6	7
D	1	5	7
E	1	6	7

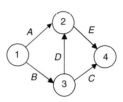

5.2 Consider the network diagram below and the three time estimates in the accompanying table.

   (a) Find the critical path and project duration using PERT. Explain what the solution means.

   (b) Find the probability of completing the project in less than 12 weeks

   (c) Find the probability of completing the project in more than 15 weeks

   (d) How many weeks will it take to complete the project with a 99.87% probability?

Activity	Optimistic Time (weeks)	Most Likely Time (weeks)	Pessimistic Time (weeks)
A	2	4	7
B	1	2	3
C	2	4	8
D	4	8	10
E	2	3	4
F	4	6	7

5.3 Consider the network diagram below and the three time estimates in the accompanying table.

   (a) Find the critical path and project duration using PERT. Explain what the solution means.

   (b) Find the probability of completing the project in less than 20 weeks

   (c) Find the probability of completing the project in less than 25 weeks

   (d) How many weeks will it take to complete the project with a 97.72% probability?

Activity	Optimistic Time (weeks)	Most Likely Time (weeks)	Pessimistic Time (weeks)
A	3	5	7
B	5	7	10
C	2	3	4
D	1	6	10
E	3	5	8
F	2	6	10
G	4	8	12

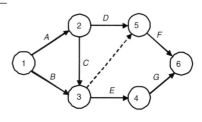

5.4 Consider the network diagram below and the three time estimates in the accompanying table.

(a) Find the critical path and project duration using PERT. Explain what the solution means.

(b) Find the probability of completing the project in more than 24 weeks

(c) Find the probability of completing the project in more than 30 weeks

(d) How many weeks will it take to complete the project with a 69.15% probability?

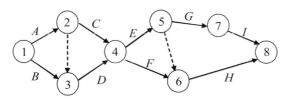

Activity	a (weeks)	m (weeks)	b (weeks)	Activity	a (weeks)	m (weeks)	b (weeks)
A	4	5	8	F	3	7	10
B	3	6	10	G	2	5	9
C	2	6	8	H	2	4	6
D	1	4	6	I	3	5	7
E	2	6	8				

Bicriteria PERT (Section 8.5.2)

5.5 Consider exercise 5.1. Generate five bicriteria alternatives associated with $z = -2$, $-1, 0, 1, 2$. List the two objective values. Identify the probability of completion, the project duration, and the critical activities associated with each alternative.

5.6 Consider exercise 5.2. Generate five bicriteria alternatives associated with $z = -2$, $-1, 0, 1, 2$. List the two objective values. Identify the probability of completion, the project duration, and the critical activities associated with each alternative.

5.7 Consider exercise 5.3. Generate five bicriteria alternatives associated with $z = -2$, $-1, 0, 1, 2$. List the two objective values. Identify the probability of completion, the project duration, and the critical activities associated with each alternative.

Tricriteria Time–Cost Trade-Off in PERT (Section 8.5.3)[†]

5.8 Consider exercise 5.1. Suppose that the expected durations for activities B and D can be crashed, each only by one day at a cost of $22 and $30 per week, respectively. Generate a set of efficient tricriteria alternatives. Show the probability of completion of all alternatives if the target duration is the project duration of the first alternative without considering any crashed activity.

5.9 Consider exercise 5.2. Suppose that the expected durations for activities A and E can be crashed, each only by one day at the cost of $8 and $12 per day, respectively. Generate a set of efficient tricriteria alternatives. Show the probability of completion of all alternatives if the target duration is the project duration of the first alternative without considering any crashed activity.

CPM with One Resource Constraint (Section 8.6.1)

6.1 Consider the following network problem where the duration and resource requirements of each activity are presented at each arrow. Use the ACTIM heuristic to rank activities. Find the minimum project duration:

(a) If there are 6 resources
(b) If there are 10 resources

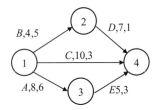

6.2 Consider the following network problem where the duration and resource requirements of each activity are presented at each arrow. Use the ACTIM heuristic to rank activities. Find the minimum project duration:

(a) If there are 4 resources
(b) If there are 9 resources

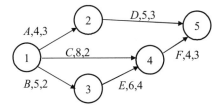

6.3 Consider the following network problem where the duration and resource requirements of each activity are presented at each arrow. Use ACTIM heuristic to rank activities. Find the minimum project duration:

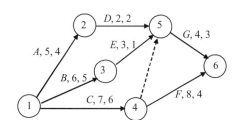

(a) If there are 7 resources

(b) If there are 11 resources

CPM with Multiple-Resource Constraints (Section 8.6.2)

6.4 Consider the following network problem where the duration, resource 1, and resource 2 requirements of each activity are presented at each arrow. Use the ACTIM heuristic to rank activities. Find the minimum project duration:

(a) If there are 8 units for the first resource and 10 for the second resource

(b) If there are 11 units for the first resource and 14 for the second resource

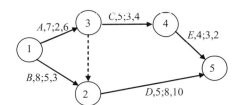

6.5 Consider the following network problem where the duration, resource 1, and resource 2 requirements of each activity are presented at each arrow. Use the ACTIM heuristic to rank activities. Find the minimum project duration:

(a) If there are 8 units for the first resource and 6 for the second resource

(b) If there are 12 units for the first resource and 15 for the second resource

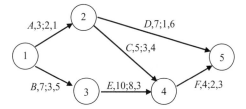

6.6 Consider the following network problem where the duration, resource 1, and resource 2 requirements of each activity are presented at each arrow. Use the ACTIM heuristic to rank activities. Find the minimum project duration:

(a) If there are 10 units for the first resource and 10 for the second resource

(b) If there are 14 units for the first resource and 16 for the second resource

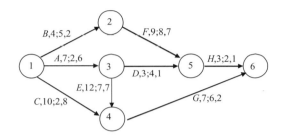

Bicriteria CPM with Resource Constraints (Section 8.6.3)[†]

6.7 Consider exercise 6.1. Resource availability can vary from 6 to 10.
   (a) Generate five bicriteria alternatives.
   (b) Identify alternatives that are efficient.

6.8 Consider exercise 6.2. Resource availability can vary from 4 to 9.
   (a) Generate five bicriteria alternatives.
   (b) Identify alternatives that are efficient.

6.9 Consider exercise 6.3. Resource availability can vary from 7 to 11.
   (a) Generate five bicriteria alternatives.
   (b) Identify alternatives that are efficient.

6.10 Consider exercise 6.4. Generate four tricriteria alternatives where resource availability for resource 1 can vary from 8 to 11.
   (a) Generate four bicriteria alternatives. (*Note*: There are only four possible multicriteria alternatives to this problem.)
   (b) Identify alternatives that are efficient.

CHAPTER 9

SUPPLY CHAIN AND TRANSPORTATION

9.1 SUPPLY CHAIN MANAGEMENT

Supply chain management is an effective integrated system that supports the management of material flow from the suppliers to the manufacturers and from the manufacturers to the customers. Supply chain management builds up and utilizes a network among the customers, retailers, distributors, suppliers, and manufacturers. Supply chain management may have the following objectives: (a) reduce inventory at all levels, (b) expedite and facilitate transportation, (c) make assignments to suppliers and vendors, and (d) maintain an online real-time computerized database for all involved parties. See Figure 1.2 of Chapter 1 for a simple example of a supply chain.

For the proper fulfillment of customer orders and the avoidance of unnecessary interruptions, it is important that demand of various parts and products be communicated properly throughout the supply chain network. To achieve this goal, one should also be aware of the lead times associated with the order of different parts and raw materials. For example, there is a certain time lapse between when the retailer orders the items from the distributor and when the orders are received. The lead time can cause accumulation of unwanted inventory due to overestimations at various levels. An efficient supply chain management network can resolve such accumulation of unwanted inventory. Successful supply chain management requires a high degree of functional and organizational integrations. The implementation of such integration can take a long time and involve many parties at each stage. To develop an integrated supply chain system, organizations generally divide the responsibilities for managing the flow of materials into three departments: (1) purchasing, (2) production, and (3) distribution.

Note: Advanced material that can be omitted without loss of generality will be indicated by a dagger.

Operations and Production Systems with Multiple Objectives, First Edition. Behnam Malakooti.
© 2014 John Wiley & Sons, Inc. Published 2014 by John Wiley & Sons, Inc.

1. *Purchasing* Purchasing is the decision-making process regarding the selection of suppliers (markets to buy from) and the levels of inventories. Inventories can be of various forms: raw materials, work-in-progress, or finished goods. Purchasing departments make the most important decisions of the three previously mentioned divisions. Two strategies for the procurement of raw materials and required goods are as follows. The first strategy is based on the competitiveness by incorporating suppliers' negotiations. Several suppliers are played against each other in an attempt to secure lowest costs. This places the responsibility of having the best quality, cost, and technical ability into the hands of each supplier. While this strategy is typically economical, long-term relationships are rarely made as a result. The second purchasing strategy is based on forming a long-term, mutually beneficial relationship with one or more suppliers. A long-term supplier is more likely to understand the business goals of a company, customize orders for the company, and be willing to work together to lower costs. Many automotive companies incorporate long-term suppliers into their supply chain.

2. *Production* Production is the management of the transformation process, which deals with all aspects of producing goods, including production quantities, machine scheduling, and employee and resource management.

3. *Distribution* Distribution deals with the management of the flow of materials from manufacturers to customers, the flow of products from warehouses to retailers, and the storage and transportation of materials and products.

Vertical Integration A different approach to material procurement is vertical integration. In vertical integration, a company develops the capability of producing and supplying their own material needs and the means to distribute the final goods. Wal Mart is an example of a firm that has successfully incorporated vertical integration. From an economic standpoint, vertical integration is beneficial, although the financial and technical resources for continual development may be lacking. Vertical integration can be either forward or backward.

Forward and Backward Integration In forward integration, the finished product is distributed to the end customer through firm-owned or firm-managed distribution centers. In backward integration, a firm internally handles all of its own raw materials or intermediate supplies. Backward integration can range from the purchase of a natural resource facility to the production of parts required for the finished product. Most automotive companies incorporate both forward and backward integration into their supply chain management.

Levels of Integration As shown in Figure 9.1, there are three levels of integration in supply chain systems. The internal supply chain refers to everything that can be directly controlled by the firm. This is depicted by the dotted line in Figure 9.1, type 2.

Relationship of Supply Chain to Other Chapters of This Book Supply chain is related and uses all techniques covered in all chapters of this book. In particular the following chapters can be used to solve supply chain problems:

Chapter 1: In supply chain systems, one must consider different criteria while making strategic planning.

Chapter 11: The models provided in this chapter are directly applicable to supply chain distributed networks.

Type 1: Independent Supply Chain Entities (Not Integrated)

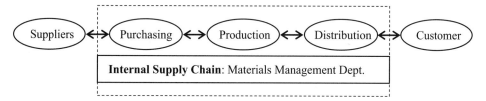

Type 2: Internal Integration (Semiintegrated)

Type 3: Supply Chain Integration (Fully Integrated)

Figure 9.1   Three types of supply chains.

Chapter 16: Different suppliers can be evaluated in terms of their cost, quality, and reliability in meeting due dates. Supplier selection optimization problems (with and without capacity limitations) are formulated and solved in the location allocation section of Chapter 16.

Chapter 6: One way to measure the performance of a supply chain system is to measure the inventory levels of each of the components of the supply chain. Then, find the total or average of the inventory levels. For example, one can find the average inventory value of each component of the supply chain and then measure the total or average inventory of the supply chain system. Other inventory performance measurements are average weeks of available supply and inventory turnover. Lower values for these performance characteristics mean better performance of the given supply chain system.

Chapter 10: It is possible to measure the effectiveness of each component (suppliers or production units) of the given supply chain and compare it to the performance of other similar components or other supply management systems.

9.1.1   Customer and Supplier Interface

Customer Interface   The customer is the entity whose demands must be fulfilled by the firm. The two main aspects of the customer interface are the order placement and order fulfillment processes. In recent years, the Internet has dramatically changed the way this interface acts. The Internet has helped firms reduce inventories and reduce transaction costs in the supply chain by improving information flow. In addition, the Internet has alleviated the need for the middleman.

The order placement process registers the need for the product or service of a manufacturer or firm and also makes sure that proper acknowledgments of these needs have been sent. This process actually generates the demands in the supply chain system. It is very important for the firm to make sure that this process is simple, smooth, fast, and accurate. The Internet and modern technology have reengineered this process to benefit both the customer and the firm. For example, previously, if a traveler needed a room in a hotel, he had to contact (e.g., by phone) various hotels to check the rates and availability of rooms. After he had obtained a suitable room, the hotel needed to enter all the information about the traveler in its register before giving the room keys to the traveler. Now, with the help of the Internet, a traveler can search for the room online and make his reservations beforehand. To check in, the traveler can present his confirmation number and the hotel employee already has all the required information. A substantial amount of customer time and employee time is saved in this process. The hotel can serve more customers more efficiently in the same amount of time. This improves the cash flow for the hotel and also guarantees more satisfaction to the customers. A similar approach can be applied in supply chains for procurement and purchasing.

The order fulfillment process deals with all the issues regarding the delivery of the product or service to the customers that were registered during the order placement process. The time between the order placement process and the order fulfillment process can vary for different supply management systems. Modern technology has facilitated the information flow, which to a large extent accelerates the order fulfillment process. A manufacturer can now have an accurate estimate of its customer demand and inventory quantities and locations at various facilities, distributors, and retailers. This helps the manufacturer to create a better production schedule, forecast demand better, and order raw materials from the suppliers more efficiently. The supply chain management system facilitates the interface of supply and demand organizations, which results in reduced inventory levels, inventory costs, and time to fulfill demand.

Supplier Interface With the advancement of the engineering world, manufacturing companies have come to realize that in order to efficiently meet the demand of their customers they need to efficiently fulfill their own demands from their own suppliers. The Internet has provided tools for virtual marketplaces, which helps firms improve their purchasing process via electronic purchasing (e-purchasing).

Different approaches are now being used for e-purchasing. The most popular one is electronic data interchange (EDI). This technology helps by transmitting routine business documents such as invoices, purchase orders, payments, and other transactions from computer to computer. This supports the data integration of different firms and suppliers with different businesses located at distant places. In an EDI system, the manufacturer places an order to a supplier electronically by using the supplier's electronic catalog. The supplier's system acknowledges the order, checks the manufacturer's credit history, checks the availability of the product, and notifies the warehouse and the shipping department. As soon as the item is ready and shipped, the supplier's accounting department sends a bill to the manufacturer. All these processes take place electronically. The implementation of an EDI system has eliminated the use of paper, reduced the stationary costs, expedited the whole process, decreased the delays caused by postal delivery of orders, improved the response times, decreased inventories, and saved the time that information previously required to travel from one point to another in the supply chain.

Two other important aspects of the supplier interface are supplier selection and supplier outsourcing. Supplier selection is a decision-making process, which is guided by four important objectives. These objectives are the price of the product or service, the quality of the product or service, the delivery time of the product or service, and the environmental impact of the product or service. Outsourcing deals with the manufacturer's decision to buy certain material from a supplier rather than producing it within the internal supply chain. This decision requires a thorough understanding by the manufacturer of its own core competencies and also depends on the degree of sourcing control and flexibility to change the supply chain. If the manufacturer can fulfill its own demand more competitively than the market, then outsourcing might not be necessary.

9.1.2 Supply Chain Performance Criteria

The supply chain management is a complex and involved process with many different intricacies. For a company to succeed, the uniqueness of each supply chain must be understood. If a company were to ignore the importance of supply chains, it would lose its competitive edge. Although supply chain management is complicated, it can be optimized. To measure the effectiveness of supply chain performance, there are several objectives (criteria) that can be measured.

Integration   The success of a supply chain system can be measured by the level of integration (vertical and horizontal) of its various components. Supply chain management utilizes all aspects of production planning methods, including forecasting, aggregate planning, material requirement planning, inventory, quality control, reliability, and maintenance. The integration of all these functions can also be measured. Also, the integration of the customer interface with the supplier interface forms another important aspect of supply chain management. The successful implementation of these aspects requires the efficient use of resources, assignment of jobs (or orders) to contractors, and transportation of goods and information throughout the system.

Agility and Flexibility   It is possible to measure the level of automation in terms of data exchange and the level of agility and flexibility in fulfilling orders. Statistics collected on cost, time, and quality can also be monitored to measure the effectiveness of supply chain performance.

Financial Measures   Several measurements can be taken to evaluate the financial effectiveness of supply chain management. Return on assets (ROA) measures the ratio of net income over total assets. By considering inventory as part of a company's investment for future use, reducing aggregate inventory will increase ROA. Working capital is another way to reflect financial investment in production. Shorter time lags between production and sales will result in better cash flow. Decreasing weeks of supply or increasing inventory turnovers will decrease inventory holding and will increase the availability of working capitals. The cash flow within a company can be used to measure the timeliness of a company in generating products.

Assignment   An important aspect of supply chain deals with the assignment of a set of given jobs (or projects) that can be performed by a number of subcontractors (or divisions). Proper assignment of jobs increases the effectiveness of the supply chain performance. This topic is covered in Sections 9.2 and 9.3.

Transportation The supply chain will rely on transportation or transshipment of materials that should be moved from a number of locations to other locations. Proper transportation planning increases the effectiveness of the supply chain performance. This topic is covered in Sections 9.4 and 9.5.

Distributed System and Network Flow Performance Supply chain management can benefit substantially by implementing the principles of distributed systems and networks. The effectiveness of a supply chain management system can also be measured by its level of distribution.

Multiple-Criteria Measurement All of the above different criteria can be used to measure or evaluate the performance of different supply chain systems. The above listed performance criteria can be presented by $f_1, f_2, \ldots, f_K$. Consider $f_1', f_2', \ldots, f_K'$ as the normalized values of $f_1, f_2, \ldots, f_K$ and $w_1, w_2, \ldots, w_K$ as the weights of importance of the normalized objectives, respectively.

When objective p is maximized, normalize it by $f_p' = (f_p - f_{\min}) / (f_{\max} - f_{\min})$.

When objective p is minimized, normalize it by $f_p' = [-f_p - (-f_{\max})] / [-f_{\min} - (-f_{\max})]$.

Therefore, all normalized objective values are maximized. The following additive utility function can be used to measure the performance of each supply chain alternative (see Chapter 16, which provides details for solving MCDM problems).

PROBLEM 9.1   MEASURING THE PERFORMANCE OF SUPPLY CHAIN SYSTEMS

Maximize $U = w_1 f_1' + w_2 f_2' + \cdots + w_K f_K'$.

9.2   ASSIGNMENT PROBLEM

In the assignment problem, individual candidates are assigned to individual positions. Examples of assignment problems include assigning contractors to jobs, facilities to cities, and salespeople to territories. This problem does not allow duplication of assignments to any one location or position; that is, only one job is assigned to one contractor and vice versa (e.g., two jobs cannot be assigned to the same contractor and vice versa). This problem can be viewed as a transportation problem requiring an equal number of supplier locations and destinations while minimizing the cost of transportation. In addition to minimizing cost, the method can be employed to optimize other objectives. Solutions to this problem can be obtained via the Hungarian algorithm (also referred to as Flood's method) or by a linear programming formulation.

Typical assignment problems will involve n candidates and m positions to which those candidates may be assigned. The cost of assigning each candidate i to each position j varies due to the circumstances of the problem. If there are an unequal number of candidates and positions, the problem can be solved by using a "dummy" candidate or position. This will be discussed later in this chapter.

9.2.1 Hungarian Method

The Hungarian algorithm is an iterative method that guarantees to find an optimal final assignment solution. This method is based on the steepest descent optimization approach (for minimization problem). This method can be used efficiently for very large problems. It utilizes row and column matrix operations to minimize the total assignment cost for the given problem. The objective is to assign n candidates i to m positions j in order to minimize the total cost of assignment. For example, assign n contractors to m contracts while minimizing the total cost of all jobs to be completed. Define variable $x_{ij} = 1$ if candidate i is assigned to position j; $x_{ij} = 0$ otherwise. The assignment matrix is shown below.

i \ j	Position 1	Position 2	$\cdots$	Position m	Total
Candidate 1	x_{11}	x_{12}	$\cdots$	x_{1m}	1
Candidate 2	x_{21}	x_{22}	$\cdots$	x_{2m}	1
$\vdots$	$\vdots$	$\vdots$	$\vdots$	$\vdots$	$\vdots$
Candidate n	x_{n1}	x_{n2}	$\cdots$	x_{nm}	1
Total	1	1	1	1	$n = m$

Suppose c_{ij} is the cost of assigning candidate i to position j. Matrix C is an $n \times m$ matrix for all candidates and positions. The procedure for assigning candidates to positions is as follows:

1. If there are an unequal number of candidates i or positions j, add dummy variables to the smaller dimension as needed (either candidates or positions) until the matrix is square (see Section 9.2.2 for details on how to add dummy variables).
2. For each row i, find the minimum c_{ij} in that row and set each $c_{ij} = c_{ij} - \text{Min}\{c_{ij}\}$.
3. Using the matrix from step 2, for each column j, find the minimum c_{ij} in that column and set each $c_{ij} = c_{ij} - \text{Min}\{c_{ij}\}$.
4. Determine if an optimal solution has been reached. Do this by drawing the minimum number of vertical and horizontal shaded lines to cover all the zeros. If the number of shaded lines is equal to n, an optimal solution has been reached. Candidate i may potentially be assigned to position j if $c_{ij} = 0$ in the current matrix. The exact assignment is determined by process of elimination and inspection (e.g., see Example 9.1).
5. Select the minimum c_{ij} of all i,j that do not have shaded lines through them. Modify the matrix in the following way:
   (a) Numbers with no shaded line through them: Subtract $\text{Min}\{c_{ij}\}$ from each c_{ij}.
   (b) Numbers with one shaded line through them: Leave numbers as is.
   (c) Numbers with two shaded lines through them: Add $\text{Min}\{c_{ij}\}$ to each c_{ij}.
   Go to step 4.

In the above algorithm, each shaded line indicates one assignment whether vertical or horizontal.

Example 9.1 Hungarian Method Consider a facility location problem in which four different facilities can be built in one of four possible sites (or locations). The cost of building facility i in location j, c_{ij}, is given in the matrix below for each i,j pair. Assign facilities 1, 2, 3, 4 to locations A, B, C, D in order to minimize total cost of building the four facilities.

Facility, i	Locations, j			
	A	B	C	D
1	50	82	60	90
2	30	100	55	100
3	80	75	65	62
4	70	62	90	80

1. There are an equal number of candidates and positions (i.e., four facilities and four locations). There is no need for dummy variables.
2. All c_{ij} in the above matrix are positive numbers. The minimum c_{ij} is found for each row and then subtracted from each element in that row.

Cost Matrix with Minimum Row

Facility, i	Locations, j				Min $c_{ij,\text{row}}$
	A	B	C	D	
1	50	82	60	90	50
2	30	100	55	100	30
3	80	75	65	62	62
4	70	62	90	80	62

Cost Matrix, $c_{ij} = c_{ij} - (\text{Min } c_{ij,\text{row}})$

Facility, i	Locations, j			
	A	B	C	D
1	0	32	10	40
2	0	70	25	70
3	18	13	3	0
4	8	0	28	18

3. The minimum c_{ij} is found for each column and is then subtracted from each element in that column.

Cost Matrix with Minimum Column

Facility, i	Locations, j			
	A	B	C	D
1	0	32	10	40
2	0	70	25	70
3	18	13	3	0
4	8	0	28	18
Min $c_{ij,\text{col}}$	0	0	3	0

Cost Matrix, $c_{ij} = c_{ij} - (\text{Min } c_{ij,\text{col}})$

Facility, i	Locations, j			
	A	B	C	D
1	0	32	7	40
2	0	70	22	70
3	18	13	0	0
4	8	0	25	18

Note: Three shaded lines cover all 0 values; solution is not optimal.

4. By inspecting the modified matrix, we see that the minimum number of shaded lines that can be drawn to cover all zeros is three (the three shaded lines are two horizontal that cover rows 3 and 4 and one vertical that covers column A). Since the number of shaded lines is less than n (i.e., $3 < 4$), an optimal solution has not been reached.

5. The Min c_{ij} for the above matrix is 7 for all elements that do not have shaded lines through them. So, we will subtract 7 from the elements that have no shaded lines through them (step 5a), add 7 to the elements that have two shaded lines through them (step 5c), and do nothing to the elements that have one shaded line through them (step 5b). The result is the matrix below.

6. Now, try to cover all 0 elements with the minimum number of shaded lines. The solution is shown below. There are four shaded lines; hence, the optimum solution is obtained.

The assignment of each facility to each location is performed by matching up facilities and locations so that each pair has a zero element in the matrix as shown above. Assign those items with a unique solution first; that is, first assign rows or columns that have only one zero (e.g., 4 to B). In some cases there may be more than one solution. Note that one facility must be assigned to only one location and vice versa. The assignment is done by inspection, as shown in the following table.

Locations, j					Assignment Steps				
						Potential			Step 3:
Facility, i	A	B	C	D	Facility	Location	Step 1	Step 2	Final	Cost
1	0	25	0	33	1	A or C	?	C	C	60
2	0	63	15	63	2	A	A	A	A	30
3	25	13	0	0	3	C or D	?	?	D	62
4	15	0	25	18	4	B	B	B	B	62
						Total				214

Note: Four shaded lines cover all 0 values; solution is optimal.

Locations are made by process of elimination. First assign the unique assignments: 2 to A and 4 to B (as they have only one zero in their row). Since 2 is already assigned to A, then 1 must be assigned to C. Lastly, since 1 has been assigned to C, then 3 must be assigned to D. Next, determine the total cost of the above assignment. This is done by adding together the cost of assigning each facility to its assigned location. The costs are obtained from the initial table of costs. The cost of assigning facility 1 to location C is 60, facility 2 to location A is 30, facility 3 to location D is 62, and facility 4 to location B is 62. Therefore, the total cost for all four assignments is $60 + 30 + 62 + 62 = \$214$.

9.2.2 Extensions of Assignment Problem

There are some assignment problems that do not fit the general formulation that was discussed in Section 9.2.1. Three different formulations and possible solutions are discussed next.

Maximization Problems   One may consider an objective that needs to be maximized instead of cost, which is usually minimized. For a maximization assignment problem, use $-c_{ij}$ for each element. Then use the assignment algorithm.

Unbalanced Facility Assignment   If the number of candidates is n and the number of positions is m, where $n \neq m$, then dummy vectors (as either rows or columns) must be added to the cost matrix:

> When candidates outnumber positions $(n > m)$, create $n - m$ dummy columns (positions).
>
> When positions outnumber candidates $(n < m)$, create $m - n$ dummy rows (candidates).

The c_{ij} associated with each dummy variable should be zero; this means that either step 2 or step 3 (not both) of the Hungarian algorithm can be skipped, since the matrix would not change. The dimensions of the matrix should now be the maximum of m and n.

Facility Assignment with Constraints   Suppose the DM (e.g., manager) does not want to assign some candidates to some positions. In this case, for each pair of such candidates and positions, set their associated $c_{ij} = M$, where M represents a very high cost relative to the other costs in the matrix. Making M large enough ensures that the assignment of this candidate to this position will not occur.

Example 9.2   Unbalanced Assignment Problem   Four contractors 1, 2, 3, and 4 have bids on three available jobs A, B, and C. The bids from each contractor are listed in the matrix below. Also, job B may not be assigned to contractor 2. Determine what assignments should be made.

| | Job | | |
Contractor	A	B	C
1	50	75	12
2	46	$M+$	11
3	68	73	15
4	55	100	13

1. The number of jobs ($m = 3$) is not equal to the number of contractors ($n = 4$). Add a dummy job D to make the two numbers equal. Adding job D produces an $n \times m$ matrix, where $n = m$. The cost of the job for each contractor for job D is assigned to 0. The M bid (a very large number) by contractor 2 on job B can be interpreted as an extremely high bid, indicating that job B will not be assigned to contractor 2. Now apply the assignment method from Section 9.2.1.

2.

Contractor	Job				Min $c_{ij,\text{row}}$
	A	B	C	D	
1	50	75	12	0	0
2	46	$M+$	11	0	0
3	68	73	15	0	0
4	55	100	13	0	0

3.

Contractor	Job, j			
	A	B	C	D
1	50	75	12	0
2	46	$M+$	11	0
3	68	73	15	0
4	55	100	13	0
Min $c_{ij,\text{col}}$	46	73	11	0

4. (I)

Contractor	Job, j			
	A	B	C	D
1	4	2	**1**	0
2	0	$M+$	0	0
3	22	0	4	0
4	9	27	2	0

Note: No. of shaded lines $= 3 < n$; not optimal.

4. (II)

Contractor	Job, j			
	A	B	C	D
1	3	1	0	0
2	0	$M+$	0	0
3	22	0	4	0
4	8	26	1	0

Note: No. of shaded lines $= 4 = n$; optimal.

5.

Contractor	Potential Jobs	Step 1	Step 2	Chosen Job	Cost
1	C or D	?	C	C	12
2	A or C	?	?	A	46
3	B	B	B	B	73
4	D	D	D	D	0

From the solution, it can be seen that since contractor 4 was assigned to the dummy job, contractor 4 does not get assigned to a real job. The total cost is $12 + 46 + 73 = \$131$.

See Supplement S9.1.xls.

9.2.3 Multicriteria Assignment Problem

Bicriteria Suppose q_{ij} represents the qualitative preference for assigning candidate i to position j. That is, the pair (of candidate and position) that has the highest q_{ij} should be assigned first. The objective is then to assign the set of candidates to the set of positions such that the total qualitative assignment is maximized. This preference information can be presented by a given preference matrix Q. Each possible assignment solution is scored by these closeness preferences. While costs should be minimized, qualitative preferences should be maximized. Consider two objectives:

Minimize total cost of assignment, f_1
Maximize total qualitative preferences for assignment f_2
Subject to the set of constraints of the assignment problem

When the additive utility function is minimized, the problem can be stated as:

Minimize $U = w_1 f_1' + w_2 f_2'$.
Subject to:

$$f_1' = \frac{f_1 - f_{\min}}{f_{\max} - f_{\min}} \qquad f_2' = \frac{-f_2 - (-f_{\max})}{-f_{\min} - (-f_{\max})}$$

Set of constraints of assignment problem

where f_1' and f_2' are normalized values of f_1 and f_2 for the minimization problem.

In order to solve this bicriteria problem by the Hungarian method, use the following steps. First convert the quantitative preference rankings to numerical values and then change the maximization objective to a minimization problem. Then generate the MCDM composite matrix (see Section 13.7.6 for the details of generating the MCDM composite matrix) to be used in the Hungarian method. The objective is to minimize the utility function.

1. Convert both the cost and preference matrices into the standard form and add any needed dummy vectors.
2. Normalize both objectives for minimization by

$$c'_{ij} = \frac{c_{ij} - c_{min}}{c_{max} - c_{min}} \qquad q'_{ij} = \frac{-q_{ij} - (-q_{max})}{-q_{min} - (-q_{max})}$$

3. Create a composite matrix Z by combining the cost and qualitative matrices using the given weights (w_1, w_2):

$$\text{Minimize } z_{ij} = w_1 c'_{ij} + w_2 q'_{ij}.$$

4. Solve the problem using the assignment method.

Note that normalized values range from 0 to 1. In order to minimize the normalized values of both objectives, the second objective is normalized by using $-q_{ij}$ because the q_{ij} are maximized. As a result, in the composite matrix, all z_{ij} components are minimized. The weights of importance for minimizing the normalized cost and minimizing the normalized qualitative preferences (w_1 and w_2, respectively) are given.

Example 9.3   Bicriteria of Facility Assignment   Consider the cost matrix and quality preference matrix listed below for assigning three facilities to three locations. Determine the optimal assignment solutions by considering each objective separately. Then use the MCDM method to determine the optimal assignment by considering both objectives, where weights of importance for total cost and total preference for closeness are $w_1 = 0.5$ and $w_2 = 0.5$. To convert qualitative preferences to quantitative values, use 5, 4, 3, 2, 1, 0 instead of A, E, I, O, U, X, respectively. Convert the qualitative preferences matrix to quantitative values as follows:

	Cost Matrix				Qualitative Preference Matrix				Qualitative Rating Matrix		
	A	B	C		A	B	C		A	B	C
1	50	82	60	1	U	E	O	1	1	4	2
2	30	100	55	2	X	A	O	2	0	5	2
3	80	75	65	3	E	I	I	3	4	3	3

This example asks for three solutions: first, for minimizing total cost, f_1; second, for maximizing qualitative preferences, f_2; and finally, for a MCDM problem using an additive utility function. First, consider only the cost matrix C and solve the problem. The solution is as follows. The qualitative preferences associated with this solution are also listed. Second, consider only the preference matrix Q and solve the problem. The solution is as follows. The cost associated with this solution is also listed.

		Optimal Solution for Cost		Optimal Solution for Qualitative Rating		
Facility	Location	Minimum Cost	Qualitative Preference	Final	Cost	Maximum Qualitative Preference
1	C	60	2	C	60	2
2	A	30	0	B	100	5
3	B	75	3	A	80	4
Total		165	5		240	11

Third, for the bicriteria problem, the matrices must first be normalized. Normalized cost and the qualitative preferences matrices (for minimization problems) are shown in the table below. For example,

$$c'_{1A} = \frac{50 - 30}{100 - 30} = 0.29 \qquad q'_{1A} = \frac{-1 - (-5)}{-0 - (-5)} = 0.8$$

Construct the composite matrix using weights (0.5, 0.5); therefore $z_{ij} = 0.5c'_{ij} + 0.5q'_{ij}$. This is given in the following:

	Normalized Values of Cost Matrix				Normalized Values of Quantitative Preference Matrix				Composite Matrix Z (Combined Cost and Qualitative)		
	A	B	C		A	B	C		A	B	C
1	0.29	0.74	0.43	1	0.8	0.2	0.6	1	0.54	0.47	0.51
2	0.00	1.00	0.36	2	1	0	0.6	2	0.50	0.50	0.48
3	0.71	0.64	0.50	3	0.2	0.4	0.4	3	0.46	0.52	0.45

Find the solution using the Hungarian algorithm as follows:

Step 1   Minimize rows:

	A	B	C
1	0.07	0.00	0.04
2	0.02	0.02	0.00
3	0.01	0.07	0.00

Step 2   Minimize columns:

	A	B	C
1	0.06	0.00	0.04
2	0.01	0.02	0.00
3	0.00	0.07	0.00

Step 3   Three shaded lines, $N = 3$, optimal:

	A	B	C
1	0.06	0.00	0.04
2	0.01	0.02	0.00
3	0.00	0.07	0.00

The final solution is

Facility	Possible	Final	Cost	Qualitative Preference
1	B	B	82	4
2	C	C	55	2
3	A or C	A	80	4
Total			217	10

The total cost and total qualitative preference are calculated using the original objective functions f_1 and f_2. In this example, three efficient alternatives are listed below:

	a_1	a_2	a_3
Min. f_1	240	217	165
Max. f_2	11	10	5

See Supplement S9.2.xls.

Tricriteria Assignment Problem[†]   The above bicriteria approach can be extended to several criteria. For example, consider a third objective, the associated time of the assignment:

Minimize total time, f_3.

We can define the objective matrix T where t_{ij} is the job time if candidate i is to be assigned to position j (e.g., different candidates can perform different jobs uniquely in terms of job duration). In the case of assigning facilities to locations, t_{ij} could be the setup time needed to build the facility in that location. In all cases, this objective is minimized. First we should normalize each element of the three matrices c_{ij}, q_{ij}, and t_{ij} as c'_{ij}, q'_{ij}, and t'_{ij}. With these values, make the composite matrix Z with given weights of importance (w_1, w_2, w_3) as follows: $z_{ij} = w_1 c'_{ij} + w_2 q'_{ij} + w_3 t'_{ij}$. The following example illustrates a tricriteria problem.

Example 9.4   Tricriteria of Facility Assignment   Consider Example 9.3. In addition, consider a third objective of minimizing the total setup time of constructing each facility to each location using the values given in the following objective matrix T. Suppose the weights of importance are 0.3, 0.5, and 0.2 for cost, preference, and time, respectively. Find the optimal assignment solution.

	Time Matrix, T				Normalized Time Matrix, T'				Composite Matrix, Z		
	A	B	C		A	B	C		A	B	C
1	11	4	5	1	1	0.3	0.4	1	0.69	0.38	0.51
2	3	8	1	2	0.2	0.7	0	2	0.54	0.44	0.41
3	9	7	2	3	0.8	0.6	0.1	3	0.47	0.51	0.37

For minimization problems, normalize c_{ij}, $-q_{ij}$, and t_{ij}. Use weighted utility $z_{ij} = 0.3c'_{ij} + 0.5q'_{ij} + 0.2t'_{ij}$ to produce the following Z matrix.

Using the Hungarian method, the solution is as follows:

Facility	Potential	Final	Cost, f_1	Qualitative Preference, f_2	Setup Time, f_3
1	B	B	82	4	4
2	C	C	55	2	1
3	A or C	A	80	4	9
Total value of each objective			217	10	14

See Supplement S9.3.xls.

9.3 OPTIMIZATION FOR ASSIGNMENT

9.3.1 Integer Linear Programming

The Hungarian algorithm works well; however, the algorithm cannot be used for assignment problems when different types of constraints are considered. This type of problem can be formulated and solved by integer linear programming (ILP).

Consider the same notation used for the assignment problem. The ILP formulation is as follows:

PROBLEM 9.2 ILP ASSIGNMENT PROBLEM

Minimize:

$$\text{Total Cost} = \sum_{i=1}^{n} \sum_{j=1}^{m} c_{ij} x_{ij}$$

Subject to:

$$\sum_{i=1}^{n} x_{ij} = 1 \quad \text{for } j = 1, 2, \ldots, m \tag{9.1}$$

$$\sum_{j=1}^{m} x_{ij} = 1 \quad \text{for } i = 1, 2, \ldots, n \tag{9.2}$$

$$x_{ij} = \begin{cases} 1 & \text{if candidate } j \text{ is assigned to position } i \\ 0 & \text{otherwise} \end{cases} \tag{9.3}$$

$$\text{Other constraints on candidates and positions} \tag{9.4}$$

where c_{ij} is the cost if candidate i is assigned to position j. The first constraint (1) ensures that each position will be assigned to exactly one candidate. The second constraint (2) also ensures that each candidate will be assigned to exactly one position.

Unbalanced ILP Problems   Note that the above problem formulation assumes a balanced assignment problem ($n = m$). In the case of an unbalanced problem, dummy variables will be added as specified in Section 9.2.2. See Example 9.2 for an illustration of this technique. In the LP formulation, the coefficients for the dummy variables in the objective function will be zero. Note that constraints must also be added for the dummy variables.

Example 9.5   Integer Linear Programming for Assignment Problem   Given the following cost matrix, formulate the ILP problem and solve it with an appropriate software package.

	Position, j				
Candidate, i	1	2	3	4	5
1	123	456	789	135	246
2	100	500	800	200	300
3	250	450	750	150	350
4	111	555	777	222	333
5	132	465	798	153	264

The constraint equations for this problem are listed below. Each x_{ij} is binary (either 0 or 1).

Minimize:

$$z = 123x_{11} + 456x_{12} + 789x_{13} + 135x_{14} + 246x_{15} + 100x_{21} + 500x_{22} + 800x_{23} + 200x_{24}$$
$$+ 300x_{25} + 250x_{31} + 450x_{32} + 750x_{33} + 150x_{34} + 350x_{35} + 111x_{41} + 555x_{42}$$
$$+ 777x_{43} + 222x_{44} + 333x_{45} + 132x_{51} + 465x_{52} + 798x_{53} + 153x_{54} + 264x_{55}$$

Subject to:

$$x_{11} + x_{21} + x_{31} + x_{41} + x_{51} = 1$$
$$x_{12} + x_{22} + x_{32} + x_{42} + x_{52} = 1$$
$$x_{13} + x_{23} + x_{33} + x_{43} + x_{53} = 1$$
$$x_{14} + x_{24} + x_{34} + x_{44} + x_{54} = 1$$
$$x_{15} + x_{25} + x_{35} + x_{45} + x_{55} = 1$$

$$x_{11} + x_{12} + x_{13} + x_{14} + x_{15} = 1$$
$$x_{21} + x_{22} + x_{23} + x_{24} + x_{25} = 1$$
$$x_{31} + x_{32} + x_{33} + x_{34} + x_{35} = 1$$
$$x_{41} + x_{42} + x_{43} + x_{44} + x_{45} = 1$$
$$x_{51} + x_{52} + x_{53} + x_{54} + x_{55} = 1$$

x_{ij} is binary $(0, 1)$ for $i, j = 1, 2, 3, 4, 5$

Solving the problem by an ILP software package yields the assignments as follows, where $x_{ij} = 1$ means candidate i is assigned to position j. Assume $x_{15} = 1$ (which means candidate 1 is assigned to position 5) and also $x_{21} = x_{32} = x_{43} = x_{54} = 1$, where all other $x_{ij} = 0$. The total cost of this solution is $246 + 100 + 450 + 777 + 153 = \$1726$.

Alternative Optimal Solution In some problems, there are several alternative optimal solutions. In the above example, the alternate optimal solution is $x_{14} = 1$ (i.e., assign candidate 1 to position 4 and also $x_{21} = x_{32} = x_{43} = x_{55} = 1$, where all other $x_{ij} = 0$. This solution has the same total cost of $z = 135 \times 1 + 100 \times 1 + 450 \times 1 + 777 \times 1 + 264 \times 1 = \$1726$.

See Supplement S9.4.lg4.

Additional Constraints Suppose (i) candidate 1 can be only assigned to positions 1, 2, or 5 (i.e., it cannot be assigned to positions 3 or 4), (ii) position 4 can be only assigned to candidates 2 or 3 (i.e., it cannot be taken by 1, 4, and 5), and (iii) if candidate 2 is assigned to position 4, then candidate 1 should be assigned to position 5, and vice versa. Hence, the following three sets of constrains are added to the set of constraints of this example:

(i) $x_{13} = 0$ and $x_{14} = 0$
(ii) $x_{14} = 0$, $x_{44} = 0$, and $x_{54} = 0$
(iii) $x_{24} = x_{15}$

Now, solve the above problem with the added constraints. The solution is $x_{12} = x_{21} = x_{34} = x_{43} = x_{55} = 1$; all other $x_{ij} = 0$. The total cost is $456 + 100 + 150 + 777 + 264 = \$1747$.

See Supplement S9.5.lg4.

9.3.2 Multiobjective Optimization of Assignment[†]

MOLP for Tricriteria Assignment Problem The multiple-objective linear programming (MOLP) formulation for tricriteria problems involves optimizing assignments with respect to cost, qualitative preference, and time objectives.

Minimize total cost, f_1
Maximize total qualitative preferences, f_2

Minimize total time, f_3

Subject to MOLP constrains

In the bicriteria MOLP problems, any two of the above three objectives can be selected.

The approach for solving the MOLP problem involves first optimizing each of the objective functions separately. There are several methods for solving MOLP problems. In the following, we present using a normalized additive utility function with given weights of importance (w_1, w_2, w_3), where f_1', f_2', and f_3' are normalized values of $f_1, f_2,$ and f_3. (For more examples of using this method see the multiple-objective optimization formulation covered in Chapter 6).

PROBLEM 9.3   MOLP FOR TRICRITERIA ASSIGNMENT PROBLEM

Minimize:

$$U = w_1 f_1' + w_2 f_2' + w_3 f_3'$$

Subject to:

$$f_1 = \sum_{i=1}^{n} \sum_{j=1}^{m} c_{ij} x_{ij} \tag{9.5}$$

$$f_2 = \sum_{i=1}^{n} \sum_{j=1}^{m} q_{ij} x_{ij} \tag{9.6}$$

$$f_3 = \sum_{i=1}^{n} \sum_{j=1}^{m} t_{ij} x_{ij} \tag{9.7}$$

$$f_1' = \frac{f_1 - f_{1\min}}{f_{1\max} - f_{1\min}} \tag{9.8}$$

$$f_2' = \frac{-f_2 - (-f_{2\max})}{-f_{2\min} - (-f_{2\max})} \tag{9.9}$$

$$f_3' = \frac{f_3 - f_{3\min}}{f_{3\max} - f_{3\min}} \tag{9.10}$$

All constraints for facility assignment (ILP)

The following example shows how this methodology can be used to solve tricriteria problems. Note that all three normalized objective values are being minimized.

Example 9.6   MOLP for Facility Assignment   Consider Example 9.4. Formulate the individual assignment MOLP problems for minimizing total cost using the cost of locations, maximizing qualitative preferences, and minimizing processing time using the processing time matrix. Find the extreme values for each objective. Then, solve the MCDM problem using weights $W = (0.35, 0.45, 0.2)$ for the three objectives, respectively.

We must first find the optimum (maximum or minimum) values of each of the objective functions and then normalize them. To do this, we solve three separate linear programming problems, each associated with one of the three objectives. For each solution we then calculate the other two objectives as a function of the objective being optimized. That is, solve the following three problems separately:

LP1 Minimize:

$$f_1 = 50x_{11} + 82x_{12} + 60x_{13} + 30x_{21} + 100x_{22} + 55x_{23} + 80x_{31} + 75x_{32} + 65x_{33}.$$

Subject to:

$$
\begin{array}{ll}
x_{11} + x_{21} + x_{31} = 1 & \quad x_{11} + x_{12} + x_{13} = 1 \\
x_{12} + x_{22} + x_{32} = 1 & \quad x_{21} + x_{22} + x_{23} = 1 \\
x_{13} + x_{23} + x_{33} = 1 & \quad x_{31} + x_{32} + x_{33} = 1 \\
\multicolumn{2}{l}{x_{ij} \text{ is binary } (0, 1) \text{ for } i, j = 1, 2, 3}
\end{array}
$$

Solving LP1 using an LP software, the solution is $x_{21} = 1$, $x_{32} = 1$, and $x_{13} = 1$ (i.e., to assign contractor 2 to job 1, contractor 3 to job 2, and contractor 1 to job 3). The objective values for f_1, f_2, and f_3 are 165, 5, and 15, respectively.

LP2 Maximize:

$$f_2 = 1x_{11} + 4x_{12} + 2x_{13} + 0x_{21} + 5x_{22} + 2x_{23} + 4x_{31} + 3x_{32} + 3x_{33}.$$

Subject to the same constraints of LP1.

LP3 Minimize:

$$f_3 = 11x_{11} + 4x_{12} + 5x_{13} + 3x_{21} + 8x_{22} + 1x_{23} + 9x_{31} + 7x_{32} + 2x_{33}.$$

Subject to the same constraints of LP1.

The solutions to LP1, LP2, and LP3 are summarized in the following table:

		f_1	f_2	f_3	$j = 1$	$j = 2$	$j = 3$
					\multicolumn{3}{c}{Assignment}
LP1:	Min. f_1	165	5	15	2	3	1
LP2:	Max. f_2	240	11	22	3	2	1
LP3:	Min. f_3	177	7	9	2	1	3
	Min.	165	5	9			
	Max.	240	11	22			

See Supplement S9.6.lg4.

Normalize $f_1, f_2,$ and f_3 using the following constraints.

$$f_1' = \frac{f_1 - 165}{240 - 165} \quad \text{or} \quad 75f_1' - f_1 = -165$$

$$f_2' = \frac{-f_2 + 11}{-5 + 11} \quad \text{or} \quad 6f_2' + f_2 = 11$$

$$f_3' = \frac{f_3 - 9}{22 - 9} \quad \text{or} \quad 13f_3' - f_3 = -9$$

Now formulate the weighted utility linear programming problem as follows:

Minimize:

$$U = 0.35 f_1' + 0.45 f_2' + 0.2 f_3'$$

Subject to:

$$f_1 = 50x_{11} + 82x_{12} + 60x_{13} + 30x_{21} + 100x_{22} + 55x_{23} + 80x_{31} + 75x_{32} + 65x_{33}$$

$$f_2 = 1x_{11} + 4x_{12} + 2x_{13} + 0x_{21} + 5x_{22} + 2x_{23} + 4x_{31} + 3x_{32} + 3x_{33}$$

$$f_3 = 11x_{11} + 4x_{12} + 5x_{13} + 3x_{21} + 8x_{22} + 1x_{23} + 9x_{31} + 7x_{32} + 2x_{33}$$

$$75f_1' - f_1 = -165; \qquad 6f_2' + f_2 = 11; \qquad 13f_3' - f_3 = -9$$

f_i for $i = 1, 2, 3$ is unrestricted

$f_i' \geq 0$ for $i = 1, 2, 3$

All constraints of LP1 of this example (presented above)

The solution to this MOLP is:

x_{ij}	1	2	3
1	0	1	0
2	1	0	0
3	0	0	1

The solution is $x_{12} = 1$, $x_{21} = 1$, and $x_{33} = 1$, where $f_1 = 177, f_2 = 7,$ and $f_3 = 9$. This is found by multiplying the cost, preference, and time matrices by the solution matrix. Note that because this is a small example, the solution happens to be the same as LP3 for minimizing f_3.

See Supplement S9.7.lg4.

9.4 TRANSPORTATION PROBLEM

The objective of the transportation problems is to minimize the total cost for delivering a certain quantity of units from a set of origin points to a set of destination points, where transportation costs are different for each origin–destination pair. Several algorithms have been developed to solve transportation problems. Vogel's method and its MCDM are discussed in this section. Linear programming approaches will be discussed in the next section.

9.4.1 Transportation Problem

In the transportation problem, there are n suppliers (origin points) with capacities of $(s_1, \ldots, s_m)$ that must ship a quantity of goods to m demands (destination points) with demands of $(d_1, \ldots, d_m)$. Each origin point is denoted by i, and each destination point is denoted by j. The cost coefficient (c_{ij}) is a transportation cost associated with shipping one unit from origin point i to demand point j. The problem is to identify x_{ij}, the quantity of goods that should flow from point i to point j, in order to minimize the total cost (TC). The transportation problem is:

Minimize:

$$\text{TC} = \sum_{i=1}^{n} \sum_{j=1}^{m} c_{ij} x_{ij}$$

Subject to: Constraints of the transportation problem

Example 9.7 Transportation Problem   Figure 9.2 is a pictorial view of a typical transportation problem. In this example, there are $n = 3$ facilities. These supply points (s_1, s_2, and s_3) can produce up to 200, 300, and 250 units, respectively, in a given period of time (e.g., one month). There are $m = 4$ customers. The demands (d_1, d_2, d_3, and d_4) for each customer are 100, 200, 300, and 150 units, respectively. Finally, the values on each route (arrow) between points are the cost coefficients c_{ij} associated with shipping one unit from

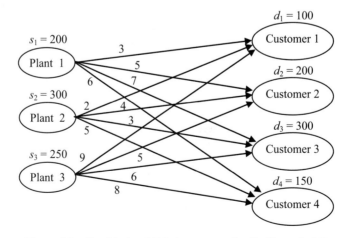

Figure 9.2   Graphical and Tabular presentation for Example 9.7.

supplier i to customer j. For example, to ship one unit from plant 1 to customer 2 incurs a cost of $C_{12} = \$5$.

The pictorial representation below can be translated into a transportation matrix as shown in the table. The suppliers are listed on the left side of the matrix with their corresponding supply listed on the right side of the matrix. The destination points are listed across the top of the matrix with their corresponding demands listed across the bottom. The cost c_{ij} is placed in the upper right-hand corner in each cell of the main body of the matrix.

The transportation problem formulation can be generalized as shown in the following table. The problem is to find the value of the unknown quantity x_{ij}, to be shipped from origin i to destination j.

		Destination $(1, ..., m)$				
		1	2	$\cdots$	m	Supply
Origin $(1, ..., n)$	1			$\cdots$		s_1
	2			$\cdots$		s_2
	$\vdots$	$\vdots$	$\vdots$	x_{ij} c_{ij}	$\vdots$	$\vdots$
	n			$\cdots$		s_n
	Demand	d_1	d_2	$\cdots$	d_m	

The General Transportation Problem

Northwest Corner Method for Finding a Basic Feasible Solution   A simple approach to finding a feasible solution to this problem (while ignoring minimizing cost) is to start assigning maximum values to meet the demand from the northwest corner and proceeding in rows, then in columns toward the southeast corner.

	Customer				
Plant	1	2	3	4	Supply
1	3 100	5 100	7 0	6 0	200
2	2 0	4 100	3 200	5 0	300
3	9 0	5 0	6 100	8 150	250
Demand	100	200	300	150	750

See Supplement S9.8.docx shows details of the northwest method.

9.4.2   Minimizing Cost by Vogel's Method[†]

In this section, Vogel's method will be covered for solving transportation problems. Vogel's method can only be applied to balanced transportation problems, meaning that total supply must equal total demand. That is,

$$\sum_{j=1}^{m} d_j = \sum_{i=1}^{n} s_i$$

Unbalanced transportation problems can be modified using dummy variables and turned into balanced transportation problems. This will be discussed in Section 9.4.3.

The final solution found by Vogel's method may not be optimal. That is, this algorithm is a heuristic method for minimizing the TC. The final solution found by the LP approach of Section 9.5 is always the optimal solution.

Finding a Basic Feasible Solution   Similar to the northwest corner method, Vogel's solution is also a basic feasible solution; however, it is usually a much better solution in terms of minimizing the total transportation cost. This method requires considerably less computational effort than linear programming (see Section 9.5). Therefore, for large-scale problems, Vogel's method is desirable. The other advantage of this method is that its solution can be used as the initial basic feasible solution in the simplex method, reducing the required computation time of the simplex method.

Penalty Cost   A penalty cost for each row P_i and column P_j is calculated and utilized in the transportation matrix. The penalty cost is calculated by taking the difference between the smallest two costs (c_{ij}) for each row and for each column:

1. Set up a transportation matrix and add a row under demand and a column next to supply for penalty cost figures.
2. Calculate penalty costs for each row P_i and for each column P_j.
3. Identify the current maximum penalty cost over the entire matrix, that is, find $\text{Max}\{\text{row}P_i, \text{col}P_j\}$. If there is a tie, arbitrarily select one.
4. Identify the minimum transportation cost, that is, $\text{Min}\{c_{ij}\}$, within the row or column associated with the current maximum penalty cost. Again, arbitrarily select one in the case of a tie. In the selection of $\text{Min}\{c_{ij}\}$, if there are ties, break the tie arbitrarily.
5. Set x_{ij} equal to the maximum number of units that can be shipped from i to j (i.e., $\text{Min}\{s_i, d_j\}$).

    If $x_{ij} = d_j$, then cross out column j and set $s_i = s_i - d_j$.
    If $x_{ij} = s_i$, then cross out row i and set $d_j = d_j - s_i$.

6. Recalculate penalty costs, $\text{row}P_i$ and $\text{col}P_j$ (excluding all rows or columns that have been eliminated).
7. Apply steps 3–6 until all demands have been met.
8. Find the total cost:

$$\text{TC} = \sum_{i=1}^{n} \sum_{j=1}^{m} c_{ij} x_{ij}$$

Example 9.8   Vogel's Method for the Transportation Problem   Consider the cost matrix given in the following table for the transportation problem, for example, $c_{12} = \$18$ for cost per unit from origin 1 to destination 2. The demand at destinations 1, 2, 3, and 4 is 300, 100, 250, and 50, respectively. The supply at origins 1, 2, and 3 is 200, 300, and 200, respectively. Find the best assignment solution to minimize the total cost for the transportation problem by using Vogel's algorithm. Calculate the total cost for this assignment.

First, determine if the problem is balanced. This problem is balanced because the total quantity supplied $(200 + 300 + 200 = 700)$ is equal to the total quantity in demand $(300 + 100 + 250 + 50 = 700)$ by the destinations.

1. Set up the transportation matrix and insert costs into the upper right-hand corner of the main body of the matrix as shown below.

Origin i	Destination j				Supply
	1	2	3	4	
1	30 x_{ij}	18	20	8	200
2	25	3	18	35	300
3	25	3	20	21	200
Demand	300	100	250	50	700

2. Calculate the penalty cost for each row and column. For example, the penalty cost for row 1 is $\text{row}P_1 = 18 - 8 = 10$ (note that 18 and 8 are the two smallest numbers in row 1). The penalty cost for column 1 is $\text{col}P_1 = 25 - 25 = 0$.

Iteration 1	1	2	3	4	Supply	Penalty
1	30	18	20	8	200	$18 - 8 = 10$
2	25	3	18	35	300	$18 - 3 = 15$
3	25	3	20	21	200	$20 - 3 = 17$
Demand	300	100	250	50		
Penalty	$25 - 25 = 0$	$3 - 3 = 0$	$20 - 18 = 2$	$21 - 8 = 13$		

3. The maximum penalty cost for all rows and columns is 17, which is associated with row 3.
4. For row 3, $\text{Min}\{c_{3j}\} = 3$. Hence choose x_{32}.
5. For row 3 and column 2, find $\text{Min}\{100, 200\} = 100$. That is, a maximum of 100 units can be shipped from origin 3 to destination 2. Now adjust the demand and supply for row 3 and column 2 accordingly. Demand for 2 is $100 - 100 = 0$, and supply at 3 is $200 - 100 = 100$. This is shown in the next table.
6. Recalculate the penalty costs while eliminating column 2 as it is completely assigned.

7. Next, iterate steps 3–6 until all supply and demand constraints have been met.

Iteration 2	1	2	3	4	Supply	Iteration 2
1	30	18	20	8	200	10
2	25	3	18	35	300	15
3	25	3 100	20	21	200 - 100	17
Demand	300	100-100	250	50		
Penalty	0	0	2	13		

Penalty Cost Max(row, column) = 17 (row 3)
$\text{Min}\{c_{3j}\} = 3$ (column 2)
Assign $\text{Min}(s_i, d_j)$ to $x_{ij} = \text{Min}(200, \underline{100})$

Iteration 3	1	3	4	Supply	Penalty
1	30	20 50	8	200 - 50	12
2	25	18	35	300	7
3	25	20	21	100	1
Demand	300	250	50-50		
Penalty	0	2	13		

Penalty Cost Max(row, column) = 13 (column 4)
$\text{Min}\{c_{i4}\} = 8$ (row 1)
Assign $\text{Min}(s_i, d_j)$ to $x_{ij} = \text{Min}(50, 200)$

Iteration 4	1	3	Supply	Penalty
1	30	20 150	150 - 150	10
2	25	18	300	7
3	25	20	100	5
Demand	300	250-150		
Penalty	0	2		

Penalty Cost Max(row, column) = 10 (row 1)
$\text{Min}\{c_{1j}\} = 20$ (column 3)
Assign $\text{Min}(s_i, d_j)$ to $x_{ij} = \text{Min}(150, 250)$

Iteration 5	1	3	Supply	Penalty
2	25	18 100	300 - 100	7
3	25	20	100	5
Demand	300	100-100		
Penalty	0	2		

Penalty Cost Max(row, column) = 7 (row 2)
$\text{Min}\{c_{2j}\} = 18$ (column 3)
Assign $\text{Min}(s_i, d_j)$ to $x_{ij} = \text{Min}(100, 300)$

Iteration 6	1	Supply	Penalty
2	25 200	200 - 200	25
3	25	100	25
Demand	300-200		
Penalty	0		

Penalty Cost Max(row, column) = 25 (row 2)

$\text{Min}\{c_{2j}\} = 25$ (column 1)
Assign $\text{Min}(s_i, d_j)$ to $x_{ij} = \text{Min}(200, 300)$

Iteration 7	1	Supply	Penalty
3	25 100	100	25
Demand	100		
Penalty	25		

Penalty Cost Max (row, column) = 25 (row 3)
$\text{Min}\{c_{3j}\} = 25$ (column 1)
Assign $\text{Min}(s_i, d_j)$ to $x_{ij} = \text{Min}(100, 100)$

Iter. 8	1	2	3	4	Supply
1	30 0	18 0	20 150	8 50	200
2	25 200	3 0	18 100	35 0	300
3	25 100	3 100	20 0	21 0	200
Demand	300	100	250	50	

The final iteration represents the best solution using Vogel's method, with the values of x_{ij} and the total cost calculated above. The solution is as follows. All supply and demands are met: $x_{13} = 150$, $x_{14} = 50$, $x_{21} = 200$, $x_{23} = 100$, $x_{31} = 100$, and $x_{32} = 100$.

Total cost is given as

$$\sum_{i,j}^{3} c_{ij}x_i = c_{13}x_{13} + c_{14}x_{14} + c_{21}x_{21} + c_{23}x_{23} + c_{31}x_{31} + c_{32}x_{32}$$

$$= (20 \times 150) + (8 \times 50) + (25 \times 200) + (18 \times 100) + (25 \times 100)$$

$$+ (3 \times 100) = \$13,000$$

See Supplement S9.9.xls.

9.4.3 Unbalanced Transportation Problem[†]

As mentioned earlier, the total supply may not necessarily equal the total demand. When the total quantity of goods available from the origins is different from the total quantity demanded by the destinations, we have an unbalanced transportation problem. For an unbalanced transportation problem, we define a "dummy" variable to account for the difference.

> *Case 1: Demand Exceeds Supply,* $\sum_{i=1}^{n} s_i < \sum_{j=1}^{m} d_j$ If the total supply is less than the total demand, insert a dummy supply row s_{dummy} which has a capacity $s_{\text{dummy}} = \sum_{j=1}^{m} d_j - \sum_{i=1}^{n} s_i$ and assign its variable costs to be an arbitrary large positive number. If the costs associated with dummy variables represent a meaningful value to the problem, such as the cost incurred for each unit of demand that cannot be met, include the dummy supply row in the total cost for the problem; otherwise, the associated dummy terms $c_{\text{dummy},j} x_{\text{dummy},j}$ should be ignored in the total-cost calculation once the problem has been solved.
>
> *Case 2: Supply Exceeds Demand,* $\sum_{i=1}^{n} s_i > \sum_{j=1}^{m} d_j$ If the supply exceeds the demand, then create a dummy demand column with $d_{\text{dummy}} = \sum_{i=1}^{n} s_i - \sum_{j=1}^{m} d_j$ and assign its costs to zero. The quantities assigned to the dummy demand point d_{dummy} will not affect the total cost after finding a solution to the problem. Alternatively, these costs could also reflect the cost of keeping those units in inventory at each supply point, in which case they would be included as part of the total cost of the solution.

Example 9.9   Unbalanced Transportation Problem   Consider the following example with two origins and four destinations. In this case, demand cannot be met ($\sum_{i=1}^{n} s_i < \sum_{j=1}^{m} d_j$), 500 < 700, that is, the total demand is greater than the total supply. A "dummy" point with an arbitrarily high cost is introduced as the third supply point. The dummy cost is set as 1000 per unit. Now add $700 - 500 = 200$ capacity for the dummy supply point. This formulation is presented in the following table. Now the problem can be solved as a balanced transportation problem.

	1	2	3	4	Supply
1	30	18	20	8	200
2	25	3	18	35	300
Demand	300	100	250	50	500 ≠ 700

	1	2	3	4	Supply
1	30	18	20	8	200
2	25	3	18	35	300
Dummy	1000	1000	1000	1000	200
Demand	300	100	250	50	700 = 700

The solution to above problem is:

	1	2	3	4
1	30 0	18 0	20 150	8 50
2	25 100	3 100	18 100	35 0
Dummy	1000 200	1000 0	1000 0	1000 0

Because the dummy row does not represent a real supply point, it is not included in the total cost calculation: $TC = 20 \times 150 + 8 \times 50 + 25 \times 100 + 3 \times 100 + 18 \times 100 = \$8000$.

See Supplement S9.10.xls.

9.4.4 Multicriteria Transportation Problem[†]

In transportation problems, several objectives can be considered. These objectives may include cost, quality of service, and transportation time. The definition of the three objectives is similar to Section 9.2.3 for multicriteria assignment problems. For qualitative preferences, q_{ij} represents the qualitative preference for assigning origin i to destination j. That is, the pairs with higher q_{ij} should be assigned first.

Minimize total cost, f_1

Maximize total qualitative preferences, f_2

Minimize total time, f_3

Similar to the MCDM approach for a tricriteria assignment problem, an additive utility function (to be minimized) can be used to generate a composite matrix of normalized values of the three objectives; then apply Vogel's method to the composite matrix. A bicriteria example is provided below.

Example 9.10   Bicriteria Vogel's Transportation Method   Consider the following information on the cost and qualitative preferences for two objectives. The supplies at origins 1, 2, and 3 are 200, 300, and 200, respectively. The demands at destinations 1, 2, 3, and 4 are 300, 100, 250, and 50, respectively. If the weights of importance of the two objectives (w_1, w_2) are $(0.6, 0.4)$, generate the best solution.

c_{ij}		To Destination j				q_{ij}		To Destination j			
		1	2	3	4			1	2	3	4
From	1	30	18	20	8	From	1	E	O	I	A
origin i	2	25	3	18	35	origin i	2	X	A	O	X
	3	25	3	20	21		3	O	A	I	O

First convert the qualitative values for preferences to their quantitative ratings. Normalize the cost and the qualitative ratings. The normalized values for c_{ij} and q_{ij} are as follows:

Normalized Matrix C'					Normalized Matrix, Q'				
	1	2	3	4		1	2	3	4
1	0.84	0.47	0.53	0.16	1	0.2	0.6	0.4	0
2	0.69	0	0.47	1	2	1	0	0.6	1
3	0.69	0	0.53	0.56	3	0.6	0	0.4	0.6

Now construct the composite matrix using $z_{ij} = 0.6c'_{ij} + 0.4q'_{ij}$ as shown in the following table. Then use Vogel's method to solve the standard transportation problem assuming the supply and demand are the same as the previous problem.

	1	2	3	4	Supply
1	0.58	0.52	0.48	0.10	200
2	0.81	0	0.52	1	300
3	0.65	0	0.48	0.58	200
Demand	300	100	250	50	700

Vogel's method is now applied to the composite matrix. The solution is as follows:

	1	2	3	4	Supply	Penalty
1	0.58	0.52	0.48	0.10	200	0.38
2	0.81	0 [100]	0.52	1	200	0.52
3	0.65	0	0.48	0.58	200	0.48
Demand	300	0	250	50		
Penalty	0.07	0	0	0.48		

	1	3	4	Supply	Penalty
1	0.58	0.48	0.10 [50]	150	0.38
2	0.81	0.52	1	200	0.29
3	0.65	0.48	0.58	200	0.10
Demand	300	250	0		
Penalty	0.07	0	0.48		

	1	3	Supply	Penalty
1	0.58	0.48	150	0.10
2	0.81	0.52 [200]	0	0.29
3	0.65	0.48	200	0.17
Demand	300	50		
Penalty	0.07	0		

	1	3	Supply	Penalty
1	0.58	0.48	150	0.10
3	0.65	0.48 [50]	150	0.17
Demand	300	0		
Penalty	0.07	0		

	1	Supply	Penalty
1	0.58 [150]	150	0.58
3	0.65 [150]	0	0.65
Demand	150		
Penalty	0.07		

Final Solution

	1	2	3	4
1	0.58 [150]	0.52 [0]	0.48	0.10 [50]
2	0.81 [0]	0 [100]	0.52 [200]	1 [0]
3	0.65 [150]	0 [0]	0.48 [50]	0.58 [0]

Now use the original cost and preference matrices to calculate the total cost and the total qualitative preferences from the assigned x_{ij} values in the flow matrix:

Total cost (to be minimized):

$$f_1 = 150 \times 30 + 50 \times 8 + 100 \times 3 + 200 \times 18 + 150 \times 25 + 50 \times 20 = \$13{,}550$$

Total qualitative preferences (to be maximized):

$$f_2 = 150 \times 4 + 50 \times 5 + 100 \times 5 + 200 \times 2 + 150 \times 2 + 50 \times 3 = 2200$$

See Supplement S9.11.xls.

9.5 OPTIMIZATION FOR TRANSPORTATION

Transportation problems can be formulated and solved by linear programming, which minimizes the total shipping cost from origins to customers. The LP can be used to find the optimal solution.

i ╲ j	Destination 1	Destination 2	$\cdots$	Destination m	Total Supply
Origin 1	x_{11}	x_{12}	$\cdots$	x_{1m}	s_1
Origin 2	x_{21}	x_{22}	$\cdots$	x_{2m}	s_2
$\vdots$	$\vdots$	$\vdots$	$\vdots$	$\vdots$	$\vdots$
Origin n	x_{n1}	x_{n2}	$\cdots$	x_{nm}	s_n
Total demand	d_1	d_2	$\cdots$	d_m	$\sum_{i=1}^{n} s_i = \sum_{j=1}^{m} d_j$

9.5.1 Linear Programming for Transportation

The LP formulation for minimizing total cost is as follows:

PROBLEM 9.4   LINEAR PROGRAMMING TRANSPORTATION PROBLEM

Minimize:

$$\text{Total cost} = \sum_{i=1}^{n} \sum_{j=1}^{m} c_{ij} x_{ij}$$

Subject to:

$$\sum_{i=1}^{n} x_{ij} = s_j \quad \text{for all } j = 1, \ldots, m$$

$$\sum_{j=1}^{m} x_{ij} = d_i \quad \text{for all } i = 1, \ldots, n$$

$$x_{ij} \geq 0 \quad \text{for all } i \text{ and } j$$

Unbalanced Transportation Problems   The above LP formulation may only be applied to balanced problems where the total supply is equal to the total demand. For unbalanced problems, the equality constraints can be satisfied by introducing dummy variables. The approach for formulating unbalanced LP transportation problems is similar to the application of dummy variables as presented in Section 9.4.3 and Example 9.9.

Example 9.11 Linear Programming for Transportation Problem Consider Example 9.8. Note that total supply equals total demand for this problem. Formulate the problem by LP to minimize total cost. The LP formulation is:

Minimize:

$$\text{Total cost} = 30x_{11} + 18x_{12} + 20x_{13} + 8x_{14}$$
$$+ 25x_{21} + 3x_{22} + 18x_{23} + 35x_{24}$$
$$+ 25x_{31} + 3x_{32} + 20x_{33} + 21x_{34}$$

Subject to:

$$x_{11} + x_{12} + x_{13} + x_{14} = 200 \qquad x_{11} + x_{21} + x_{31} = 300$$
$$x_{21} + x_{22} + x_{23} + x_{24} = 300 \qquad x_{12} + x_{22} + x_{32} = 100$$
$$x_{31} + x_{32} + x_{33} + x_{34} = 200 \qquad x_{13} + x_{23} + x_{33} = 250$$
$$x_{14} + x_{24} + x_{34} = 50$$
$$x_{ij} \geq 0 \quad \text{for all } i = 1, 2, 3, \ j = 1, 2, 3, 4$$

Solving this problem by a linear programming software package, the following solution will be obtained:

	1	2	3	4
1	$x_{11} = 0$	$x_{12} = 0$	$x_{13} = 150$	$x_{14} = 50$
2	$x_{21} = 137$	$x_{22} = 63$	$x_{23} = 100$	$x_{24} = 0$
3	$x_{31} = 163$	$x_{32} = 37$	$x_{33} = 0$	$x_{34} = 0$

See Supplement S9.12.lg4.
The total cost is $20(150) + 8(50) + 25(137) + 3(63) + 18(100) + 25(163) + 3(37) = \$13,000$. Comparing the total cost resulting from the linear programming solution to that derived using Vogel's method in Example 9.8, we can see that both produced the same total cost, although the solutions are somewhat different. Because Vogel's method is a heuristic, it may not necessarily produce an optimal solution, whereas linear programming always generates the optimal solution. However, Vogel's method is computationally much faster than LP to solve this problem.

9.5.2 Multiobjective Optimization of Transportation[†]

The MOLP formulation of the transportation problem is similar to the assignment problem. We can consider three objectives: cost, qualitative preferences, and transportation time. Transportation time functions are similar to cost functions. The total transportation time is the sum of the times required to move each unit over the associated route. The three objectives of the problem are:

Minimize total cost, f_1
Maximize total qualitative preferences, f_2
Minimize total transportation time, f_3
Subject to constraints of LP transportation problem

We can solve the formulated MOLP problem by the MCDM approach. Here, we present the solution using a normalized additive utility function. The normalized additive MOLP formulation of the transportation problem is as follows. Note that the normalized values f_1', f_2', f_3' are being minimized. Also note that because f_2 is maximized, $-f_2$ is used when it is normalized.

PROBLEM 9.5   MULTIOBJECTIVE LINEAR PROGRAMMING OF TRANSPORTATION

Minimize:

$$U = w_1 f_1' + w_2 f_2' + w_3 f_3'$$

Subject to:

$$f_1 = \sum_{i=1}^{n}\sum_{j=1}^{m} c_{ij}x_{ij} \qquad f_1' = \frac{f_1 - f_{1\,min}}{f_{1\,max} - f_{1\,min}}$$

$$f_2 = \sum_{i=1}^{n}\sum_{j=1}^{m} q_{ij}x_{ij} \qquad f_2' = \frac{-f_2 - (-f_{2\,max})}{-f_{2\,min} - (-f_{2\,max})}$$

$$f_3 = \sum_{i=1}^{n}\sum_{j=1}^{m} t_{ij}x_{ij} \qquad f_3' = \frac{f_3 - f_{3\,min}}{f_{3\,max} - f_{3\,min}}$$

All constraints for transportation LP problem

Before solving the above problem, three separate LP problems should be solved to find all f_{min} and f_{max} values to be used for normalization equations. The weights of importance w_1, w_2, w_3 are also given prior to solving the above MOLP problem.

Example 9.12   Additive Utility MOLP for the Transportation Problem   Consider the following triobjective transportation problem. Solve the MOLP problem if weights of importance for the three objectives are $(0.4, 0.3, 0.3)$.

	f_1: Cost Matrix						f_2: Qualitative Preferences Matrix					f_3: Time Matrix			
	1	2	3	4	supply		1	2	3	4		1	2	3	4
1	30	18	20	8	200	1	E	O	I	A	1	4	5	3	6
2	25	3	18	35	300	2	X	A	O	X	2	7	9	2	4
3	25	3	20	21	200	3	O	A	I	O	3	11	7	8	1
Demand	300	100	250	50											

To find maximum and minimum values of each objective, first solve three independent LP problems (P1, P2, and P3) for each objective function independently.

	f_1	f_2	f_3
P1: Minimize f_1	13,000	1600	4150
P2: Maximize f_2	13,550	2200	4250
P3: Minimize f_3	14,350	2000	2950
Maximum	14,350	2200	4250
Minimum	13,000	1600	2950

See Supplement S9.13.lg4.

Then each objective is normalized as follows:

$$f_1' = \frac{f_1 - 13,000}{14,350 - 13,000} \quad \text{or} \quad 1350 f_1' - f_1 = -13,000$$

$$f_2' = \frac{-f_2 + 2200}{-1600 + 2200} \quad \text{or} \quad 600 f_2' + f_2 = 2200$$

$$f_3' = \frac{f_3 - 2950}{4250 - 2950} \quad \text{or} \quad 1300 f_3' - f_3 = -2950$$

Continuing the MOLP method by normalizing the objective values and solving for the maximized additive utility, $U = w_1 f_1' + w_2 f_2' + w_3 f_3'$ for weights $(0.4, 0.3, 0.3)$, results in the following solution:

Minimize:

$$U = 0.4 f_1' + 0.3 f_2' + 0.3 f_3'$$

Subject to:

$$f_1 = 30x_{11} + 18x_{12} + 20x_{13} + 8x_{14} + 25x_{21} + 3x_{22} + 18x_{23} + 35x_{24} + 25x_{31}$$
$$+ 3x_{32} + 20x_{33} + 21x_{34}$$

$$f_2 = 4x_{11} + 2x_{12} + 3x_{13} + 5x_{14} + 0x_{21} + 5x_{22} + 2x_{23} + 0x_{24} + 2x_{31} + 5x_{32}$$
$$+ 3x_{33} + 2x_{34}$$

$$f_3 = 4x_{11} + 5x_{12} + 3x_{13} + 6x_{14} + 7x_{21} + 9x_{22} + 2x_{23} + 4x_{24} + 11x_{31} + 7x_{32}$$
$$+ 8x_{33} + 1x_{34}$$

$$1350 f_1' - f_1 = -13,000$$

$$600 f_2' + f_2 = 2200$$

$$1300 f_3' - f_3 = -2950$$

All constraints of LP problem in Example 9.11

The MOLP Solution is as follows:

	1	2	3	4
1	150	0	0	50
2	50	0	250	0
3	100	100	0	0

where $f_1 = \$13,450$, total cost; $f_2 = 2,050$, total closeness; and $f_3 = 3550$, total transportation time. These objective values are, of course, within the minimum and maximum value boundaries obtained earlier.

See Supplement S9.14.lg4.

9.5.3 Transshipment Problem

The transportation problem can be extended to a more general problem termed the transshipment problem. In this type of problem, goods do not necessarily flow directly from supply points to demand points but instead may move through a number of intermediate points placed between the initial supply points and the final demand points (the problem becomes similar to a network problem). Supply flows first from origins to the intermediate points, then to the destinations. The transshipment problem can be viewed as the flow of goods along two or more transportation networks chained together. As in all transportation problems, each route between points has a per-unit transportation cost, here defined as either c_{ik} or c_{kj}, where i is an origin point, k is an intermediate point, and j is a destination point. Transshipment points may have transshipment flow constraints. Here, T_k is the capacity constraint for the intermediate point k. In this case, a transshipment flow for an intermediate point means that the point "demands" a quantity T_k for the supply points and also "supplies" this quantity to the final demand points.

Figure 9.3 illustrates a graphical example of a transshipment problem showing coefficient shipping costs c_{ik} and c_{kj} along the paths and the capacity constraints (supply, demand, or transshipment flow) of the points. The following table presents this information in standard transportation matrix form. Warehouse 1 must have a flow of exactly 300 units ($T_1 = 300$), but warehouse 2 can have unlimited flow (hence $T_2 =$ any).

Transshipment Destinations, k					Final Destinations, j					
		1	2	Supply			1	2	3	4	Supply
Starting	1	5	3	200	Transshipment	1	4	2	6	1	300
Origins, i	2	2	1	300	Origins, k	2	3	5	4	2	Any
	3	7	9	250	Demand		300	200	100	150	
Demand		300	Any								

Transshipment problems are best solved using a linear programming approach. To avoid confusion, flow from origins to transshipment points will be labeled as x_{ik} and flow from

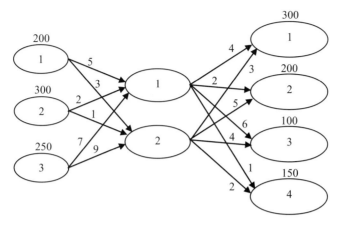

Figure 9.3 Transshipment problem.

transshipment points to destinations will be labeled as y_{kj}. As in the previous LP formulation, a set of constraints should be constructed for each point, whether it is a supply point, a demand point, or an intermediate point. For each intermediate point T_k, the flow must be balanced, meaning that the total quantity coming in must equal the total quantity going out. Since variables are placed on the left side of an equation in LP, each intermediate constraint will be of the form

$$\sum_{i=1}^{n} x_{ik} - \sum_{j=1}^{m} y_{kj} = 0$$

The LP formulation of the transshipment problem is as follows:

PROBLEM 9.6   TRANSSHIPMENT PROBLEM

Minimize:

$$z = \sum_{i=1}^{n} \sum_{k=1}^{p} c_{ik} x_{ik} + \sum_{k=1}^{p} \sum_{j=1}^{m} c_{kj} y_{kj}$$

Subject to:

$$\sum_{k=1}^{p} x_{ik} = s_i \qquad \text{for all } i = 1, \ldots, n \quad \text{(supply constraints at origins)}$$

$$\sum_{i=1}^{n} x_{ik} = T_k \qquad \text{for all } k = 1, \ldots, p \quad \text{(transshipment constraints)}$$

$$\sum_{k=1}^{p} y_{kj} = d_j \qquad \text{for all } j = 1, \ldots, m \quad \text{(demand constraints at destinations)}$$

$$\sum_{i=1}^{n} x_{ik} - \sum_{j=1}^{m} y_{kj} = 0 \quad \text{for all } k = 1, \ldots, p \quad \text{(transshipment balancing)}$$

$$x_{ik} \geq 0 \qquad y_{kj} \geq 0 \quad \text{for all } i, j, k$$

The requirements of demand and all equality constraints to be satisfied can be relaxed by allowing dummy variables as discussed in the LP formulation of transportation. Therefore, inventories or deficits can occur.

Example 9.13 Transshipment Problem Consider the transshipment problem example given earlier in this section. Formulate the linear program for this problem, and then solve it to find the minimized total cost and the associated flow matrix solution.

In this example, for simplicity of presentation, we will use $i = (1, 2, 3)$, $k = (1, 2)$, and $j = (1, 2, 3, 4)$. The LP formulation is:

Minimize:

$$z = 5x_{11} + 3x_{12} + 2x_{21} + 1x_{22} + 7x_{31} + 9x_{32} + 4y_{11}$$
$$+ 2y_{12} + 6y_{13} + 1y_{14} + 3y_{21} + 5y_{22} + 4y_{23} + 2y_{24}$$

Subject to:

Supply constraints at origins

$$x_{11} + x_{12} = 200$$
$$x_{21} + x_{22} = 300$$
$$x_{31} + x_{32} = 250$$

Transshipment constraints

$$x_{11} + x_{21} + x_{31} = 300$$
$$x_{12} + x_{22} + x_{32} = 450$$

Demand constraints at destinations

$$y_{11} + y_{21} = 300$$
$$y_{12} + y_{22} = 200$$
$$y_{13} + y_{23} = 100$$
$$y_{14} + y_{24} = 150$$

Transshipment balancing constraints

$$x_{11} + x_{21} + x_{31} - y_{11} - y_{12} - y_{13} - y_{14} = 0$$
$$x_{12} + x_{22} + x_{32} - y_{21} - y_{22} - y_{23} - y_{24} = 0$$

Positive flow values

$$x_{ik}, y_{kj} \geq 0 \quad \text{for all } i, j, k$$

Now, we solve this problem using an LP computer package to find the optimal solution for all decision variables. The flow matrix solution to the linear program is as follows:

Transshipment Destinations, k					Final Destinations, j						
	x_{ik}	1	2	Total		y_{kj}	1	2	3	4	Total, T_k
Starting	1	0	200	200	Transshipment	1	0	200	0	100	300
Origins, i	2	50	250	300	Origins, k	2	300	0	100	50	450
	3	250	0	250	Total		300	200	100	150	$750 = 750$
Total, T_k		300	450	$750 = 750$							

The total cost for the problem is given as: $\text{TC} = 3(200) + 2(50) + 1(250) + 7(250) + 2(200) + 1(100) + 3(300) + 4(100) + 2(50) = \$4600$.

See Supplements S9.15.xls, S9.15.lg4.

REFERENCES

General

Christopher, M. "Logistics and Supply Chain Management: Strategies for Reducing Cost and Improving Service." *International Journal of Logistics Research and Applications*, Vol. 2, No. 1, 1999, pp. 103–104.

Christopher, M. *Logistics and Supply Chain Management: Creating Value-Added Networks*. Harlow, England, New York: Financial Times Prentice Hall, 2005.

Cooper, M. C., D. M. Lambert, and J. D. Pagh. "Supply Chain Management: More Than a New Name for Logistics." *International Journal of Logistics Management*, Vol. 8, No. 1, 1997, pp. 1–14.

Copacino, W. C. *Supply Chain Management: The Basics and Beyond*. Boca Raton, FL: CRC Press, 1997.

Dornier, P.-P., R. Ernst, M. Fender, and P. Kouvelis. *Global Operations and Logistics: Text and Cases*, New York: Wiley, 1998.

Eilon, S., C. D. T. Watson-Gandy, and N. Christofides. *Distribution Management: Mathematical Modeling and Practical Analysis*. St. Martin's Griffin, 1976.

Handfield, R. B., and E. L. Nichols. *Introduction to Supply Chain Management*. Upper Saddle River, NJ: Prentice Hall, 1998.

Heizer, J., and B. Render. *Production & Operations Management*. 10th ed. Upper Saddle River, NJ: Prentice Hall, 2010.

Hong-Minh, S. M., S. M. Disney, and M. M. Naim. "The Dynamics of Emergency Transhipment Supply Chains." *International Journal of Physical Distribution & Logistics Management*, Vol. 30, No. 9, 2000, pp. 788–816.

Hugos, M. *Essentials of Supply Chain Management*, 2nd ed. Hoboken, NJ: Wiley, 2006.

Korpela, J., A. Lehmusvaara, and M. Tuominen. "An Analytic Approach to Supply Chain Development." *International Journal of Production Economics*, Vol. 71, Nos. 1–3, 2001, pp. 145–155.

Kumar, K. "Technology for Supporting Supply Chain Management: Introduction." *Communications of the ACM*, Vol. 44, No. 6, 2001, pp. 58–61.

Lummus, R. R., and R. J. Vokurka. "Defining Supply Chain Management: A Historical Perspective and Practical Guidelines." *Industrial Management & Data Systems*, Vol. 99, No. 1, 1999, pp. 11–17.

Martin, A. *Distribution Resource Planning*, rev. ed. Thousand Oaks, California: Oliver Wight Publications, 1990.

Mentzer, J. T. *Supply Chain Management*. Thousand Oaks, California: Sage Publications, 2000.

Nahmias, S. *Production and Operations Analysis*, 6th ed. New York: McGraw-Hill Higher Education, 2008.

Pirkul, H., and V. Jayaraman. "Production, Transportation, and Distribution Planning in a Multi-Commodity Tri-echelon System." *Transportation Science*, Vol. 30, No. 4, 1996, pp. 291–302.

Ross, D. F. *Competing through Supply Chain Management: Creating Market-Winning Strategies through Supply Chain Partnerships*. Norwell, Massachusetts: Kluwer Academic, 1999.

Schwarz, Leroy B. *Multi-Level Production/Inventory Control Systems: Theory and Practice*. New York: North-Holland Publishing Company, 1981.

Simchi-Levi, D., P. Kaminski, and E. Simchi-Levi. *Designing and Managing the Supply Chain: Concepts, Strategies, and Case Studies*. New York: McGraw-Hill/Irwin, 1999.

Stank, T. P., and T. J. Goldsby. "A Framework for Transportation Decision Making in an Integrated Supply Chain." *Supply Chain Management: An International Journal*, Vol. 5, No. 2, March 27, 2000, pp. 71–78.

Multicriteria Supply Chain

Amin, H. S., and G. Zhang. "An Integrated Model for Closed-Loop Supply Chain Configuration and Supplier Selection: Multi-Objective Approach." *Expert Systems with Applications*, Vol. 39, No. 8, 2012, pp. 6782–6791.

Aneja, Y. P., and K. P. K. Nair. "Bicriteria Transportation Problem." *Management Science*, Vol. 25, No. 1, 1979, pp. 73–78.

Clímaco, J. N., C. H. Antunes, and M. J. Alves. "Interactive Decision Support for Multiobjective Transportation Problems." *European Journal of Operational Research*, Vol. 65, No. 1, 1993, pp. 58–67.

Current, J., and H. Min. "Multiobjective Design of Transportation Networks: Taxonomy and Annotation." *European Journal of Operational Research*, Vol. 26, No. 2, 1986, pp. 187–201.

Das, S. K., A. Goswami, and S. S. Alam. "Multiobjective Transportation Problem with Interval Cost, Source and Destination Parameters." *European Journal of Operational Research*, Vol. 117, No. 1, 1999, pp. 100–112.

El-Wahed, W. F. A. "A Multi-Objective Transportation Problem under Fuzziness." *Fuzzy Sets and Systems*, Vol. 117, No. 1, 2001, pp. 27–33.

Gen, M. L., and K. Y. Ida. "Solving Multi-Objective Transportation Problem by Spanning Tree-Based Genetic Algorithm." *IEICE Transactions on Fundamentals of Electronics Communications and Computer Sciences E Series A*, Vol. 82, Part 12, 1999, pp. 2802–2810.

Gen, M., K. Ida, and Y. Li. "Bicriteria Transportation Problem by Hybrid Genetic Algorithm." *Computers & Industrial Engineering*, Vol. 35, Nos. 1–2, 1998, pp. 363–366.

Giuliano, G. "A Multicriteria Method for Transportation Investment Planning." *Transportation Research Part A: General*, Vol. 19, No. 1, 1985, pp. 29–41.

Kahraman, C., U. Cebeci, and Z. Ulukan. "Multi-Criteria Supplier Selection Using Fuzzy AHP." *Logistics Information Management*, Vol. 16, No. 6, 2003, pp. 382–394.

Lee, S. M., M. J. Schniederjans, and J. P. Cole. "A Multicriteria Assignment Problem: A Goal Programming Approach." *Interfaces*, Vol. 13, No. 4, 1983, pp. 75–81.

Liu, F.-H. F., and H. L. Hai. "The Voting Analytic Hierarchy Process Method for Selecting Supplier." *International Journal of Production Economics*, Vol. 97, No. 3, 2005, pp. 308–317.

Malakooti, B. "Multiple Objective Optimization of Assignment and Transportation Problems." Case Western Reserve University, Cleveland, OH, 2013.

Ng, W. L. "An Efficient and Simple Model for Multiple Criteria Supplier Selection Problem, *European Journal of Operational Research*, Vol. 186, No. 3, 2008, pp. 1059–1067.

Ozdemir, M. S., and R. N. Gasimov. "The Analytic Hierarchy Process and Multiobjective 0–1 Faculty Course Assignment." *European Journal of Operational Research*, Vol. 157, No. 2, 2004, pp. 398–408.

Perçin, S. "An Application of the Integrated AHP-PGP Model in Supplier Selection." *Measuring Business Excellence*, Vol. 10, No. 4, 2006, pp. 34–49.

Ringuest, J. L., and D. B. Rinks. "Interactive Solutions for the Linear Multiobjective Transportation Problem." *European Journal of Operational Research*, Vol. 32, No. 1, 1987, pp. 96–106.

Saaty, T. L. "Transport Planning with Multiple Criteria: The Analytic Hierarchy Process Applications and Progress Review." *Journal of Advanced Transportation*, Vol. 29, No. 1, 1995, pp. 81–126.

Sharma, S., S. V. Ukkusuri, and T. V. Mathew. "Pareto Optimal Multiobjective Optimization for Robust Transportation Network Design Problem." *Transportation Research Record: Journal of the Transportation Research Board*, Vol. 2090, 2009, pp. 95–104.

Weber, C. A., and J. R. Current. "A Multiobjective Approach to Vendor Selection." *European Journal of Operational Research*, Vol. 68, No. 2, 1993, pp. 173–184.

White, D. J. "A Special Multi-Objective Assignment Problem." *Journal of the Operational Research Society*, Vol. 35, No. 8, 1984, pp. 759–767.

Xia, W., and Z. Wu. "Supplier Selection with Multiple Criteria in Volume Discount Environments." *Omega* Vol. 35, No. 5, 2007, pp. 494–504.

Yang, L., and Y. Feng. "Λ Bicriteria Solid Transportation Problem with Fixed Charge under Stochastic Environment." *Applied Mathematical Modelling*, Vol. 31, No. 12, 2007, pp. 2668–2683.

Yang, X., and M. Gen. "Evolution Program for Bicriteria Transportation Problem." *Computers & Industrial Engineering*, Vol. 27, Nos. 1–4, 1994, pp. 481–484.

EXERCISES

Supply Chain Management (Section 9.1)

1.1 Briefly describe the supply chain characteristics of two major auto industries (such as Toyota) and compare their approaches. *Hint*: You may use the Internet to find the needed information.

1.2 Briefly describe the supply chain characteristics of two major retail industries (such as Wal-Mart) and compare their approaches. *Hint*: You may use the Internet to find the needed information.

1.3 Briefly describe the supply chain characteristics of two major computer industries (such as IBM) and compare their approaches. *Hint*: You may use the Internet to find the needed information.

1.4 A corporation can have five different ways to operate its supply chain system. These alternatives can be measured in terms of the following four criteria. The following table gives the cost (in thousands of dollars to be minimized), integration level (to be maximized), agility and flexibility (to be maximized), and inventory levels (in millions to be minimized). Symbols A, E, I, O, U, and X represent the level of desirability associated with 5, 4, 3, 2, 1, and 0 numerical values, respectively.

Location	Minimum Cost (f_1)	Maximum Integration Level (f_2)	Maximum Agility and Flexibility (f_3)	Mininum Inventory Level (f_4)
Plan 1	246,000	U	I	60
Plan 2	136,000	I	U	15
Plan 3	378,000	X	E	95
Plan 4	245,000	A	U	42
Plan 5	272,000	I	A	57

Suppose the weights of importance for the normalized values (all normalized values are maximized) of the four objectives are (0.2, 0.2, 0.3, 0.3), respectively. Use the additive utility function to rank alternatives and determine the best location.

Hungarian Method (Section 9.2.1) and Extensions of Assignment Problem (Section 9.2.2)

2.1 Consider the following cost information for the assignments of projects to contractors. Use the Hungarian algorithm to find the best assignment. What is its total cost?

Projects	Contractors 1	2	3
1	34	25	64
2	53	46	50
3	29	59	47

2.2 Consider the following cost information for the assignments of projects to contractors. Use the Hungarian algorithm to find the best assignment. What is its total cost?

Projects	Contractors 1	2	3	4
1	90	50	60	20
2	80	10	55	100
3	75	10	65	62
4	30	55	90	100

2.3 Consider the following cost information for the assignments of facilities to locations. Use the Hungarian algorithm to find the best assignment. What is its total cost?

		Location			
Facility	1	2	3	4	5
1	80	70	60	30	50
2	40	50	45	20	90
3	55	40	75	82	30
4	30	65	80	50	100
5	75	49	59	64	80

2.4 Consider exercise 2.1. Assume that assignments of project 1 to contractor 2 and project 3 to contractor 3 are not allowed. Solve this problem using the Hungarian algorithm.

2.5 Consider exercise 2.2. Assume there are only three contractors: 1, 2, and 4 (i.e., eliminate 3). Write the changes in the formulation for solving the Hungarian algorithm. Also assume that assignments of project 2 to contractor 2 and project 4 to contractor 4 are not allowed.

2.6 Consider exercise 2.3. There are only four locations (i.e., eliminate location 2). Write the changes in the formulation for solving the Hungarian algorithm. Also assume that assignments of facility 1 to location 4, facility 3 to location 5, and facility 4 to location 1 are not allowed.

Multicriteria Assignment Problem (Section 9.2.3)

2.7 Consider exercise 2.1. Now consider a second objective, the qualitative preferences for assignment. Generate the optimal solution using the composite objective function for additive utility (using the MCDM method based on the Hungarian method). What are the values for the given variables for the two objective MCDM problems? Assume the DM's weights for normalized values of w_1 and w_2 are 0.3 and 0.7.

		Contractors	
Projects	1	2	3
1	U	X	A
2	I	O	E
3	U	E	I

2.8 Consider exercise 2.2. Now consider a second objective, the qualitative preferences for assignment. Generate the optimal solution using the composite objective function for additive utility (using the MCDM method based on the Hungarian method). What are the values for the given variables for the two objective MCDM problems? Assume the DM's weights for normalized values of w_1 and w_2 are 0.6 and 0.4.

	Contractors			
Projects	1	2	3	4
1	U	O	I	A
2	X	A	O	U
3	O	A	I	O
4	E	O	U	X

Integer Linear Programming (Section 9.3.1)

3.1 Consider exercise 2.1.
  (a) Formulate an integer linear program. Present your formulation in standard format.
  (b) Solve the problem using a computer package. Show the solution and the value of the objective function.

3.2 Consider exercise 2.2. Respond to parts (a) and (b) of exercise 3.1.

3.3 Consider exercise 2.3. Respond to parts (a) and (b) of exercise 3.1.

3.4 Consider exercise 2.4. Respond to parts (a) and (b) of exercise 3.1.

3.5 Consider exercise 2.5. Respond to parts (a) and (b) of exercise 3.1.

3.6 Consider exercise 2.6. Respond to parts (a) and (b) of exercise 3.1.

Multiobjective Optimization of Assignment (Section 9.3.2)
MOLP Bicriteria[†]

3.7 Consider exercise 2.7.
  (a) Formulate an ILP to maximize total qualitative preferences. Solve the problem using a computer package.
  (b) Show the MOLP formulation of this problem. Solve the problem using a computer package.

3.8 Consider exercise 2.8. Respond to parts (a) and (b) of exercise 3.7.

MOLP Tricriteria

3.9 Consider exercise 2.7. Now consider a third objective, project time, as defined below. Assume the weights of importance are 0.3, 0.4, and 0.3.

	Contractors		
Projects	A	B	C
1	16	23	14
2	23	16	25
3	21	15	17

(a) Formulate an ILP to minimize time. Solve the problem using a computer package.

(b) Show the MOLP formulation of this problem. Solve the problem using a computer package.

3.10 Consider exercise 2.8. Now consider a third objective, project time, as defined below. Assume the weights of importance are 0.4, 0.2, and 0.4. Respond to parts (a) and (b) of exercise 3.9.

Projects	Contractors			
	A	B	C	D
1	39	45	46	52
2	48	50	35	43
3	55	47	38	42
4	41	49	46	50

Transportation Problem (Section 9.4.2)[†]

4.1 Consider the following transportation cost matrix. Find the best solution using Vogel's method.

		Destination				Supply
		1	2	3	4	
Origin	1	7	14	9	16	180
	2	20	7	10	17	200
	3	17	17	13	12	200
	Demand	200	150	80	150	

4.2 Consider the following transportation cost matrix. Find the best solution using Vogel's method.

		Destination				Supply
		1	2	3	4	
Origin	1	17	14	22	21	250
	2	12	10	12	19	200
	3	7	20	13	9	50
	4	10	9	11	21	250
	Demand	100	200	300	150	

4.3 Consider the following transportation cost matrix. Find the best solution using Vogel's method.

		Destination					
		1	2	3	4	5	Supply
Origin	1	7	5	10	7	9	330
	2	3	10	6	5	3	270
	3	6	8	3	11	6	190
	4	5	5	13	8	5	210
	Demand	200	140	150	300	210	

Minimizing Cost by Vogel's Method[†] (Section 9.4.4)

4.4 Consider exercise 4.1. Now consider a second objective, the qualitative preferences for transportation. Generate the optimal solution using the composite objective function for additive utility. What are the values for the given variables for the two objective MCDM problems? Assume the DM's weights for normalized values of w_1 and w_2 are 0.3 and 0.7.

		Destination		
Origin	1	2	3	4
1	A	A	O	U
2	E	X	X	E
3	X	A	I	O

4.5 Consider exercise 4.2. Now consider a second objective, the qualitative preferences for transportation. Generate the optimal solution using the composite objective function for additive utility. What are the values for the given variables for the two objective MCDM problems? Assume the DM's weights for normalized values of w_1 and w_2 are 0.6 and 0.4.

		Destination		
Origin	1	2	3	4
1	A	I	E	A
2	X	E	O	A
3	A	O	A	U
4	O	U	E	X

4.6 Consider exercise 4.3. Now consider a second objective, the qualitative preferences for transportation. Generate the optimal solution using the composite objective function for additive utility. What are the values for the given variables for the two objective

MCDM problems? Assume the DM's weights for normalized values of w_1 and w_2 are 0.5 and 0.5.

	Destination				
Origin	1	2	3	4	5
1	*O*	*I*	*E*	*I*	*E*
2	*I*	*E*	*U*	*A*	*O*
3	*A*	*U*	*X*	*I*	*U*
4	*O*	*X*	*E*	*O*	*X*

Optimization for Transportation (Section 9.5.1)

5.1 Consider exercise 4.1.

(a) Formulate the linear program. Present your formulation in standard format.

(b) Solve the problem using a computer package. Show the solution and the value of the objective function.

5.2 Consider exercise 4.2. Respond to parts (a) and (b) of exercise 5.1.

5.3 Consider exercise 4.3. Respond to parts (a) and (b) of exercise 5.1.

Multiobjective Optimization of Transportation (Section 9.5.2)[†]

5.4 Consider exercise 4.4.

(a) Formulate an LP to maximize total qualitative preferences. Solve the problem using a computer package.

(b) Show the MOLP formulation of this problem. Solve the problem using a computer package.

5.5 Consider exercise 4.5. Respond to parts (a) and (b) of exercise 5.6.

5.6 Consider exercise 4.6. Respond to parts (a) and (b) of exercise 5.6.

5.7 Consider exercise 4.4. Now consider a third objective, time, as defined below. Assume the weights of importance are 0.3, 0.4, and 0.3.

	Destination			
Origin	1	2	3	4
1	59	77	58	73
2	67	68	59	77
3	65	76	72	72

(a) Formulate an LP to minimize time. Solve the problem using a computer package.

(b) Show the MOLP formulation of this problem. Solve the problem using a computer package.

5.8 Consider exercise 4.5. Now consider a third objective, time, as defined below. Assume the weights of importance are 0.4, 0.2, and 0.4. Respond to parts (a) and (b) of exercise 5.9.

	Destination			
Origin	1	2	3	4
1	26	27	34	38
2	37	22	34	27
3	24	25	21	35
4	41	19	32	25

5.9 Consider exercise 4.6. Now consider a third objective, time, as defined below. Assume the weights of importance are 0.4, 0.2, and 0.4. Respond to parts (a) and (b) of exercise 5.9.

	Destination				
Origin	1	2	3	4	5
1	82	73	64	78	72
2	62	68	59	77	58
3	64	66	76	67	66
4	81	72	72	65	74

Transshipment Problem (Section 9.5.3)

5.10 Consider the following transportation cost matrices for transshipment. Find a transportation solution that minimizes total cost using linear programming.

		Transshipment Destination					Final Destination		
		1	2	Supply			1	2	3
Starting	1	13	15	240	Transshipment	1	11	13	15
origin	2	16	17	160	origin	2	13	10	14
	3	18	14	320		Demand	250	270	200
	Demand	Any	350						

5.11 Consider the following transportation cost matrices for transshipment. Find a transportation solution that minimizes total cost using linear programming.

		Transshipment Destination							Final Destination				
		1	2	3	Supply				1	2	3	4	5
Starting	1	7	10	8	300	Transshipment	1		9	11	14	10	15
origins	2	9	8	10	250	origins	2		10	15	12	13	11
	3	10	7	11	250		3		12	9	10	15	9
	4	8	9	8	320		Demand		280	230	320	280	200
	5	6	11	9	190								
	Demand	180	340	Any									

CHAPTER 10

PRODUCTIVITY AND EFFICIENCY

10.1 INTRODUCTION

Often productivity is correlated to profitability, which signifies a company's ability to generate revenues in comparison to expenses of inputs. However, for a given company, a high rate of profitability does not mean a high rate of productivity. A production process with low productivity can still generate high profits if a favorable market exists. Conversely, a production process with high productivity may still result in a decrease in profit if market conditions do not comply. Many factors contribute to the improvement of productivity. Such factors include the quality of materials, versatility and capability of equipment, skill of employees, design of production systems, and management approach. When all these factors work synchronously, improvement in productivity can be achieved. Such improvements, in turn, will improve the quality of produced goods and will result in a lower cost of production, which will contribute to an increase in profitability.

There are several approaches for improving productivity. The most evident way to improve productivity is to use automated equipment and processes to increase the speed of the system. The emergence and development of computers (hardware and software) has also been a main contributor to increased productivity. Furthermore, the advent of the World Wide Web, telecommunication systems, and global economy has substantially improved and will continue to improve productivity of different operational systems. From the human factor point of view, laborers working under comfortable, safe, fair, and ergonomically designed conditions tend to be more productive and responsible for quality products. Most organizations are in constant pursuit of increasing productivity. Before trying to improve

Note: Advanced material that can be omitted without loss of generality will be indicated by a dagger.
Note: Materials based on Malakooti (2013) will be indicated by asterisk.

Operations and Production Systems with Multiple Objectives, First Edition. Behnam Malakooti.
© 2014 John Wiley & Sons, Inc. Published 2014 by John Wiley & Sons, Inc.

productivity, one should identify which point of view is used to measure productivity. Alternatively, one may consider productivity as a multidimensional (criteria) by considering different criteria. We introduce the concept of multiobjective productivity and efficiency in this chapter by considering:

- Productivity measurement from cost point of view
- Productivity measurement from quality point of view
- Productivity measurement from quantity point of view

We will first show the application of multiobjective productivity to single-factor problems (problems with one input and one output). Then, we will show its application to multifactor problems (problems with several inputs and several outputs). In this chapter, we introduce a new definition called efficiency for comparing the productivity of comparable processes. We introduce the concept of multicriteria productivity and efficiency to compare different operational systems. Efficiency can be used to compare the productivity of similar operational systems by using a common denominator while productivity of a single system is measured independent of other systems. A linear programming formulation approach based on data envelopment analysis (DEA) is used to compare the productivity and efficiency of similar operational systems with multi-inputs and multioutputs. The multiple-objective optimization of productivity and efficiency is also developed. Productivity is discussed in Sections 10.2 and 10.3. Efficiency is discussed in Sections 10.4–10.6. Productivity of a network of processors is discussed in Section 10.7. In this chapter, first we define a simple productivity index consisting of one input and one output. Then, we measure productivity growth when there are multiple inputs and multiple outputs.

10.2   BASIC PRODUCTIVITY INDEXES

Productivity is defined as the amount of output per one unit of input per period for production and service processes. Productivity is commonly measured in terms of output per unit worker per unit hour, which is known as the labor productivity. Productivity growth is a rate comparing the productivity of a given period to a base period. Generally, higher productivity is associated with higher standard of living in a country. Productivity in the United States has been increasing steadily since 1890, which has contributed to the growth of the economy and the increase in the standard of living in the United States. In the United States in the 1910s, productivity grew significantly after the assembly line was developed in the auto industry. The use of assembly lines enabled the auto industry to produce a high number of cars in a short period of time by utilizing resources effectively, which resulted in a lower cost of production and increased profit.

Productivity is defined as the ratio of output to input. A higher ratio is associated with higher productivity:

$$\text{Productivity} = \frac{\text{output}}{\text{input}} \qquad (10.1)$$

Therefore, productivity is not a measure of profitability or cost. By increasing productivity, however, profitability may increase. In measuring productivity, the key question is the definition of inputs and outputs. Some examples of productivity inputs and outputs are given below.

Organization	Education	Factory	Bank	Hospital	Auto Fuel
Output	No. of credit hours taken by students per semester	No. of products produced per shift	No. of customers served per day	No. of patients per day	No. of miles
Input	No. of teachers	Labor hours	No. of staffs	No. of staffs	Gallons of gas

Productivity Growth   Productivity growth is defined as the increase in productivity from a base period to the current period with respect to the productivity of the base period:

$$\text{Productivity growth} = \frac{\text{current productivity} - \text{base productivity}}{\text{base productivity}} \times 100\% \quad (10.2)$$

For example, the productivity was 75 (items produced per hour of labor) 10 year ago, and it is now 85. The productivity growth with respect to the base year (10 years ago) is given as

$$\text{Productivity growth} = \frac{85 - 75}{75} \times 100\% = 13.3\%$$

It is expected that productivity growth will be positive. Therefore, similar to interest rates and inflation rates, productivity growth can also be compounding, as shown in the equation

$$\text{Current productivity} = (\text{base productivity}) \times [1 + (\text{productivity growth/year})]^{(\text{no. of years})} \quad (10.3)$$

In the United States, the output per work hour (of an average worker) increased at an annual average rate of about 2.5% in the past century. For example, suppose that an organization produced, on average, 26 units of products per hour 100 years ago. Using a compounding productivity growth of 2.5% per year, the current productivity of this organization is found using Equation (10.3) as follows:

$26 \times (1 + 0.025)^{(100)} = 307$ products per hour compared to base year of 100 years ago

Types of Productivity Indexes   There are three approaches to measure productivity with respect to its components (factors)—partial measures, multifactor measures, and total measures:

The partial productivity index is measured based on a single input and a single output.
The multifactor productivity index is measured based on a subset of multiple inputs and outputs.

The total productivity index is measured by considering all inputs and outputs. The general components of the outputs and inputs are shown below:

$$\text{Total productivity index} = \frac{\text{product} + \text{service}}{\text{labor} + \text{materials} + \text{capital} + \text{energy}} \qquad (10.4)$$

Example 10.1   Partial, Multifactor, and Total Productivity Measurements   Consider the table below for an example with three inputs and one output.

	Input			
Labor	Machine	Energy	Output, Profit
$50,000	$30,000	$60,000	$350,000

The partial productivity measurements for labor, machine, and energy are:

	Labor	Machine	Energy
Partial productivity	$\dfrac{\text{Output}}{\text{Labor}} = \dfrac{350}{50} = 7$	$\dfrac{\text{Output}}{\text{Machine}} = \dfrac{350}{30} = 11.7$	$\dfrac{\text{Output}}{\text{Energy}} = \dfrac{350}{60} = 5.8$

The multifactor productivity measurements for labor and machine, machine and energy, and labor and energy are:

	Labor and Machine	Machine and Energy	Labor and Energy
Multifactor productivity	$\dfrac{\text{Output}}{\text{Labor} + \text{machine}}$	$\dfrac{\text{Output}}{\text{Machine} + \text{energy}}$	$\dfrac{\text{Output}}{\text{Labor} + \text{energy}}$
	$= \dfrac{350}{50 + 30} = 4.4$	$= \dfrac{350}{30 + 60} = 3.9$	$= \dfrac{350}{50 + 60} = 3.2$

The total productivity measurement is given as

$$\text{Total productivity} = \frac{350}{50 + 30 + 60} = 2.5$$

Example 10.2   Multi-Input and Multioutput Total Productivity   Total productivity using Equation (10.4) for a corporation in the current year is given below where all factors are in millions of dollars (M):

$$\text{Total productivity index} = \frac{120M + 180M(\text{income})}{90M + 80M + 50M + 20M(\text{cost})} = \frac{300M}{240M} = 1.25$$

Suppose in the base year (e.g., five years ago), the total productivity index was 1.05. Find the productivity growth for this corporation.

The productivity growth is given as

$$\text{Productivity growth} = \frac{1.25 - 1.05}{1.05} \times 100\% = 19.0\%$$

This means that the productivity has increased compared to the base year by 19.0%.

10.3 MULTIFACTOR PRODUCTIVITY GROWTH[†]

A related definition of productivity growth was introduced by the American Productivity Center (APC). This method considers inflation rates so that all prices are compared based on their actual values. It can be calculated using the following steps:

- Find the partial productivity index of each single unit.
- Combine the partial productivity indices to generate the factor productivity of the system.
- Use the base-year price index to neutralize the inflation factor.

The *factor productivity index* for a system that consists of I outputs and J inputs is

$$\text{Factor productivity index} = \frac{\text{current period output sell/base period output sell}}{\text{current period input cost/base period input cost}}$$

$$= \frac{\sum_{i=1}^{I}(O_{2,i}P_{1,i})/\sum_{i=1}^{I}(O_{1,i}P_{1,i})}{\sum_{j=1}^{J}(I_{2,j}C_{1,j})/\sum_{j=1}^{J}(I_{1,j}C_{1,j})} \tag{10.5}$$

where

$$O_{1,i}, O_{2,i} = \text{output in base (1) and current (2) periods, respectively, for output } i$$
$$P_{1,i} = \text{price per unit of output in terms of prices in base period (1) for output } i$$
$$I_{1,j}, I_{2,j} = \text{input in base (1) and current (2) periods, respectively, for input } j$$
$$C_{1,j} = \text{cost per unit of input in terms of prices in base period (1) for input } j$$
$$i = 1,\dots,I \text{ output factors}$$
$$j = 1,\dots,J \text{ input factors}$$

In the factor productivity index, the prices of the base year are used for inputs and outputs. Two extensions of the above factor productivity index are presented below.

The *price recovery index* is an indicator of how the given organization is able to combat inflation related to the costs of inputs. In this case, the prices of the current year (2) versus the base year (1) are used for current versus base-year factors, respectively, while considering the outputs and inputs of the current period (2):

$$\text{Price recovery index} = \frac{\sum_{i=1}^{I}(O_{2,i}P_{1,i})/\sum_{i=1}^{I}(O_{1,i}P_{1,i})}{\sum_{j=1}^{J}(I_{2,j}C_{1,j})/\sum_{j=1}^{J}(I_{1,j}C_{1,j})} \quad (10.6)$$

The *cost effectiveness index* uses the prices of the current period versus the prices of the base period while considering the outputs and inputs of the current (2) versus the base year (1). This can be measured by multiplying Equation (10.5) by Equation (10.6):

$$\text{Cost effectiveness index} = \frac{\sum_{i=1}^{I}(O_{2,i}P_{1,i})/\sum_{i=1}^{I}(O_{1,i}P_{1,i})}{\sum_{j=1}^{J}(I_{2,j}C_{1,j})/\sum_{j=1}^{J}(I_{1,j}C_{1,j})} \quad (10.7)$$

Example 10.3 Multifactor Productivity Growth Indexes Consider an example with two inputs (in terms of labor hours) and one output (in terms of the number of product units). The quantities and prices for the inputs and outputs for the base year and current year are presented below. Find all three productivity growth indices.

	Period 1: Base Year		Period 2: Current Year	
Output	Quantity $O_{1,i}$	Price, $P_{1,i}$	Quantity $O_{2,i}$	Price, $P_{2,i}$
product, $i=1$	800 units	$80/unit	1300 units	$110/unit
Input	Quantity, $I_{1,j}$	Cost, $C_{1,j}$	Quantity, $I_{2,j}$	Cost, $C_{2,j}$
Labor, $j=1$	8000 h	$30/h	8500 h	$40/h
Labor, $j=2$	3000 h	$35/h	2500 h	$45/h

$$\text{Factor productivity index} = \frac{(1300 \times 80)/(800 \times 80)}{(8500 \times 30 + 2500 \times 35)/(8000 \times 30 + 3000 \times 35)} = \frac{1.63}{0.99}$$
$$= 1.64$$

$$\text{Price recovery index} = \frac{(1300 \times 110)/(1300 \times 80)}{(8500 \times 40 + 2500 \times 45)/(8500 \times 30 + 2500 \times 35)} = \frac{1.38}{1.32}$$
$$= 1.04$$

$$\text{Cost effectiveness index} = \frac{(1300 \times 110)/(800 \times 80)}{(8500 \times 40 + 2500 \times 45)/(8000 \times 30 + 3000 \times 35)} = \frac{2.23}{1.31}$$
$$= 1.70$$

We can verify that Equation (10.7) is (3.1) × (3.2), that is, $1.64 \times 1.04 = 1.70$.

See Supplement S10.1.xls.

10.4   SINGLE-FACTOR EFFICIENCY

10.4.1   Single-Factor Efficiency

Suppose that we are interested in comparing the productivity of similar processes (e.g., similar industries) labeled as Decision-making units (DMUs) on a scale of 0 to 1. Decision-making units can be simply referred to as units.

Let O_r and I_r be the output and input for unit r where there are R different units. The productivities of the units are $P_1, P_2, \ldots, P_R$. The productivity of unit r is defined as

$$P_r = \frac{O_r}{I_r} \qquad (10.8)$$

In order to compare the productivity of different units, normalize the productivity of each unit to a range of 0–1 where the unit with the highest productivity is rated as 1. The (productivity) efficiency of unit r is given as

$$E_r = \frac{\text{productivity of unit } r}{\text{productivity of most productive unit}}$$

$$= \frac{P_r}{P_{\max}} \qquad (10.9)$$

where $P_{\max} = \text{Max}\{P_1, P_2, \ldots, P_R\}$.

Example 10.4   Single-Factor Efficiency of Doctors   Consider the amount of time a doctor spends in a hospital as the input. Also, consider the actual billing (in $100 per week) issued by the doctor as the output. Suppose the doctor is paid for the amount he or she spends in the hospital and that all doctors are paid the same amount per hour. The data of inputs and outputs for five doctors are provided in Table 10.1. Find the efficiency of each doctor.

The solution to this problem is provided in Table 10.1 where doctor 5 has the highest productivity of 8.0 and therefore the highest efficiency of 1. Doctor 4 has the lowest productivity and efficiency.

TABLE 10.1   Single-Objective Productivity and Efficiency for Example 10.4

Doctor no., r	1	2	3	4	5
Outputs, O_r (billing in $100 per week)	250	350	220	320	160
Inputs, I_r (hours per week)	40	45	30	60	20
Productivity, $P_r = O_r/I_r$	6.25	7.77	7.33	5.33	8.00
Efficiency, $E_r = P_r/P_{\max}$	0.781	0.971	0.916	0.666	1
Rank	4	2	3	5	1

TABLE 10.2 Bicriteria Efficiency for Example 10.5

Doctor no., r	1	2	3	4	5
Outputs, $O_{r,1}$ (quantity, in $100 per week)	250	350	220	320	160
Outputs, $O_{r,2}$ (quality per week)	3	2	2	5	1
Inputs, I_r (hours spent in hospital per week)	40	45	30	60	20
Quantity productivity, $P_{r,1} = O_{r,1}/I_r$	6.25	7.77	7.33	5.33	8.00
Quality productivity, $P_{r,2} = O_{r,2}/I_r$	0.075	0.044	0.067	0.083	0.050
f_1 = quantity efficiency, $E_{r,1} = P_{r,1}/P_{\max,1}$	0.781	0.971	0.916	0.666	1
f_2 = quality efficiency, $E_{r,2} = P_{r,2}/P_{\max,2}$	0.9	0.533	0.800	1	0.600

10.4.2 Bicriteria Single-Factor Efficiency*

In the bicriteria problem, two different efficiencies are considered; for example:

Maximize $f_1 = E_{r,1}$ = quantity efficiency
Maximize $f_2 = E_{r,2}$ = quality efficiency

In this problem, the two criteria (objectives) for each unit are measured. Then, the definition of multiobjective efficiency can be used to identify inefficient alternatives. Furthermore, a utility function (e.g., an additive utility function) can be used to rank the alternatives.

Example 10.5 Bicriteria of Single-Factor Efficiency   Consider Example 10.4. Also, consider a second objective of maximizing the quality of patient care as shown in Table 10.2. These quality ratings could be based on the patients' ranking of the doctors, how long patients are willing to wait to get an appointment with a particular doctor, or the number of patients' complaints against a doctor. The quality ratings are from 0 to 5, where 0 is the worst and 5 is the best. Table 10.2 shows a comparison of five doctors for two outputs: f_1, billing in $100 per week, and f_2, quality ratings by patients.

In terms of the first objective, doctor 5 is the best and doctor 4 is the worst. But in terms of the second objective, doctor 4 is the best and doctor 2 is the worst. The five doctors' performances are presented in Figure 10.1. Note that doctor 2 is inferior to doctor 5 in terms of both objectives, that is, doctor 2 is inefficient.

One can rank the doctors using an additive utility function $U = w_1 f_1 + w_2 f_2$, where w_1 and w_2 are the weights of importance of the first and second objectives, respectively, $w_1 \geq 0$, $w_2 \geq 0$, $w_1 + w_2 = 1$. Suppose that the DM (e.g., the board of trustees of the hospital) places weights of importance of 0.6 and 0.4 on the quantity and the quality, respectively. The five doctors can be ranked by maximizing $U = 0.6f_1 + 0.4f_2$ as presented in the following table where doctor 3 is ranked the best and doctor 2 is ranked the worst:

Doctor no., r	1	2	3	4	5
f_1 = quantity efficiency, $E_{r,1}$	0.781	0.971	0.916	0.666	1
f_2 = quality efficiency, $E_{r,2}$	0.900	0.533	0.800	1	0.600
$U = 0.6f_1 + 0.4f_2$	0.829	0.796	0.870	0.800	0.840
Rank	3	5	1	4	2

See Supplement S10.2.xls.

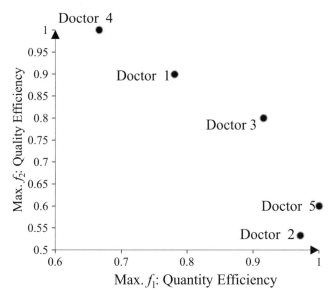

Figure 10.1 Example 10.5 presentation of efficiencies of five doctors in terms of f_1 (quantity efficiency) and f_2 (quality efficiency). Doctor 5 dominates doctor 2 in terms of both objectives.

10.5 MULTIFACTOR EFFICIENCY AND DEA

In multifactor efficiency, multiple inputs and multiple outputs are used to evaluate units. We will label units as alternatives to be consistent with our MCDM and MOO notation. The purpose of multifactor efficiency is to compare these alternatives (units) based on their inputs and outputs. To make this comparison we use the DEA method, which uses linear programming. The efficiencies of different units are compared on a scale of 0–1, where 1 is the efficiency of the most productive unit. The DEA approach is closely related to the multi-objective optimization approach in the sense that each output is maximized as an objective; therefore, the MCDM definition of efficiency can be applied to the productivity problem.

10.5.1 Data Envelopment Analysis[†]

Also known as the frontier analysis, DEA, is an optimization modeling approach that can be used to evaluate the efficiency of different units. DEA was first introduced by Charnes, Cooper, and Rhodes (1978). DEA has gained popularity among various business firms and government agencies for analyzing the efficiency of similar units within an organization or across several organizations. DEA specifically measures the efficiency of a group of organizations that have similar multiple inputs and similar multiple outputs. Examples of such organizations are banks, hospitals, governments, schools, military units, and manufacturing companies.

There are several approaches for examining the efficiency of units. A statistical method known as the central tendency method assesses units in relation to an average unit. DEA, on the other hand, is an extreme-point method, which evaluates each unit in comparison to the best unit (i.e., compares them only to one unit). Although DEA has several advantages, it also has shortcomings. Since DEA is an extreme-point method, errors in measurements can have a considerable effect on the results.

DEA employs a linear programming formulation of the problem to determine the efficiency of each unit with multiple inputs and outputs. The efficiency in DEA is defined as the ratio of the weighted sum of outputs to the weighted sum of inputs. For each inefficient unit, DEA determines efficient units that can be designated as benchmarks for improvements. The benchmarks are found using the dual of the LP.

10.5.2 Linear Programming Formulation for DEA[†]

As stated before, the problem is to compare R units (alternatives) to each other where each unit r has J inputs and I outputs. The following guidelines are used to formulate the LP objective function and its constraints:

1. Efficiency is defined as

$$\text{Efficiency} = \frac{\text{total weighted output}}{\text{total weighted input}}$$

Efficiencies are normalized from 0 to 1; the maximum efficiency for each unit is 1.

2. The input (y_j) and output (x_i) decision variables are greater than zero.

3. To compare alternatives, the total weighted input of each unit is always equal to 1. The total weighted output is between 0 and 1.

4. The optimal output values $(x_1, x_2, \ldots, x_I)$ and input values $(y_1, y_2, \ldots, y_J)$ are found by solving an LP problem as presented below. These values are used to evaluate the efficiency of each unit.

5. Comparing two alternatives, an alternative with a higher efficiency E_r is better than an alternative with a lower efficiency. That is, alternatives are ranked in descending order of their efficiency values. The alternative with the highest efficiency is the best one; its value is always 1.

The following notation is used in DEA LP formulation:

Decision variables:

x_i: Output variable i for $i = 1, \ldots, I$

y_j: Input variable j for $j = 1, \ldots, J$

Known coefficients:

a_{ri} : Coefficient of output i for unit r for $r = 1, \ldots, R, i = 1, \ldots, I$

b_{rj}: Coefficient of input j for unit r, for $r = 1, \ldots, R, j = 1, \ldots, J$

Therefore, there are R units, each having I outputs and j inputs. For each unit r, one LP problem should be solved to find its productivity (also called efficiency). Each unit has its own distinct LP objective function and constraints, but all LP problems also have a common set of constraints. The LP formulation for a given alternative r is presented below:

PROBLEM 10.1   DEA LP PROBLEM FOR UNIT r
Maximize:

$$E_r = \sum_{i=1}^{I} a_{ri} x_i \quad \text{(total output for given } r\text{)} \tag{10.10}$$

Subject to:

$$-\sum_{i=1}^{I} a_{ri} x_i + \sum_{j=1}^{J} b_{rj} y_j \geq 0 \quad \text{for } r = 1, 2, \ldots, R \quad \text{(common constraints)} \tag{10.11}$$

$$\sum_{j=1}^{J} b_{rj} y_j = 1 \quad \text{(total input for given, } r\text{)} \tag{10.12}$$

$$x_i \geq \varepsilon \quad \text{for } i = 1, \ldots, I \tag{10.13}$$

$$y_j \geq \varepsilon \quad \text{for } j = 1, \ldots, J \tag{10.14}$$

where ε is a given small positive number, for example, set $\varepsilon = 0.0001$ before solving the LP problem. Note that constraints (10.11), (10.13), and (10.14) are the same for all different LP problems associated with each alternative. However, (10.10) and (10.12) are different for each alternative r.

Example 10.6   Efficiency of Multifactor Multidecision Units by DEA LP   Consider a health clinic with three doctors with the following outputs:

a_1: Amount of billing (in $1000) charged by a doctor per week and paid by patients

a_2: Amount of extra income that a doctor creates (in $1000) per week (e.g., laboratory tests and x-rays) paid by patients

Consider the following inputs:

b_1: Amount of hours per week a doctor works

b_2: Amount of space (in square feet) of the hospital designated to a doctor

b_3: Number of hours per week the hospital staff (e.g., secretaries and nurses) spend to support a doctor's activities

Table 10.3 provides input and output information for the three doctors labeled as $r = 1$, 2, 3. Find the efficiencies of the three doctors compared to each other.

TABLE 10.3   Example 10.1 Input and Output Coefficients for Three Doctors

Doctor Number, r	Input (b_{rj})			Output (a_{ri})	
	$j = 1$	$j = 2$	$j = 3$	$i = 1$	$i = 2$
1	15	20	12	6	5
2	20	23	15	22	8
3	10	18	10	3	3

In this example, the decision variables are:

x_i: Output i for $i = 1, 2$
y_j: Input j for $j = 1, 2, 3$

The known coefficients are:

a_{ri}: Coefficient of output i for unit r for $r = 1, 2, 3, i - 1, 2$
b_{rj}: Coefficient of input j for unit r for $r = 1, 2, 3, j = 1, 2, 3$

The values of these coefficients are shown in Table 10.3.

To show the LP formulation of this problem, we show the output and input formulation for Doctor 1.

$$\text{Output} = 6x_1 + 5x_2 \quad \text{(total output of doctor 1)}$$

$$\text{Input} = 15y_1 + 20y_2 + 12y_3 \quad \text{(total input of doctor 1)}$$

We set outputs to be less than inputs for doctor 1 using $-\text{output} + \text{input} \geq 0$:

$$-6x_1 - 5x_2 + 15y_1 + 20y_2 + 12y_3 \geq 0$$

In order for the efficiency to be on a scale of 0–1, set the total input for this unit (i.e., doctor 1) to 1:

$$15y_1 + 20y_2 + 12y_3 = 1$$

$$x_1, x_2 \geq 0.001 \qquad y_1, y_2, y_3 \geq 0.001$$

Similar constraints for doctors 2 and 3 can be constructed. This set of constraints will be used in the LP formulation. The LP formulation for determining the efficiency of doctor 1 is presented below.

DEA LP Problem for Doctor 1

Maximize:

$$E_1 = 6x_1 + 5x_2 \quad \text{(efficiency for doctor 1)} \tag{10.15}$$

Subject to:

$$-6x_1 - 5x_2 + 15y_1 + 20y_2 + 12y_3 \geq 0 \quad \text{(for doctor 1)} \tag{10.16}$$

$$-22x_1 - 8x_2 + 20y_1 + 23y_2 + 15y_3 \geq 0 \quad \text{(for doctor 2)} \tag{10.17}$$

$$-3x_1 - 3x_2 + 10y_1 + 18y_2 + 10y_3 \geq 0 \quad \text{(for doctor 3)} \tag{10.18}$$

$$15y_1 + 20y_2 + 12y_3 = 1 \quad \text{(normalized input for doctor 1)} \tag{10.19}$$

$$x_i \geq 0.001 \quad \text{for } i = 1, 2 \tag{10.20}$$

$$y_j \geq 0.001 \quad \text{for } j = 1, 2, 3 \tag{10.21}$$

Constraints (10.16)–(10.18) ensure that the efficiency cannot be greater than 1 for all three units. Constraint (10.19) normalizes the total input for doctor 1 to be equal to 1. Constraints (10.20) and (10.21) ensure all decision variables are positive.

The result of solving the above LP (by using LP computer software) is shown below:

| \multicolumn{6}{c}{**Inputs, Outputs, and Efficiency of Doctor 1**} |
x_1	x_2	y_1	y_2	y_3	E_1
0.0001	0.1663	0.066	0.0001	0.0001	0.8323

Since $E_1 = 0.8323$, that is, $E_1 < 1$, this doctor is inefficient compared to some other doctors. Also, it can be stated that

$$\text{Doctor 1 efficiency} = 0.8323 \quad (\text{or } 83.23\%)$$

Now, we show the LP formulation for doctor 2.

DEA LP Problem for Doctor 2

Maximize:

$$E_2 = 22x_1 + 8x_2 \quad (\text{efficiency for doctor 2})$$

Subject to:

$$-6x_1 - 5x_2 + 15y_1 + 20y_2 + 12y_3 \geq 0 \quad (\text{for doctor 1})$$
$$-22x_1 - 8x_2 + 20y_1 + 23y_2 + 15y_3 \geq 0 \quad (\text{for doctor 2})$$
$$-3x_1 - 3x_2 + 10y_1 + 18y_2 + 10y_3 \geq 0 \quad (\text{for doctor 3})$$
$$20y_1 + 23y_2 + 15y_3 = 1 \quad (\text{normalized Input for doctor 2})$$
$$x_i \geq 0.001 \quad \text{for } i = 1, 2$$
$$y_i \geq 0.001 \quad \text{for } j = 1, 2, 3$$

The solution to the above LP problem for doctor 2 is presented in Table 10.4. Table 10.4 also presents the LP solutions for doctors 1 and 3.

Based on the results presented in Table 10.4, it can be concluded that the efficiencies of doctors 1, 2, and 3 are 83.23, 100, and 74.9%, respectively. Doctors 1 and 3 are inefficient with respect to doctor 2, who is the most efficient.

The efficiency of each doctor can be verified by using the values of x and y from the solution to each LP problem. To do this, find the output, input, and efficiency of each doctor.

TABLE 10.4 Solution to Three LP Problems for Three Doctors in Terms of Normalized Inputs, Normalized Outputs, and Efficiency

Doctor Number, r	LP	x_1	x_2	y_1	y_2	y_3	Normalized Input	Normalized Output	Efficiency, E_r
1	1	0.0001	0.1663	0.066	0.0001	0.0001	1	0.832	0.832
2	2	0.0001	0.1247	0.0498	0.0001	0.0001	1	1	1
3	3	0.0001	0.2495	0.0997	0.0001	0.0001	1	0.749	0.749

For example, for doctor 2:

Output for doctor 2: $a_{21}x_1 + a_{22}x_2 = 22(0.0001) + 8(0.1247) = 1$
Input for doctor 2: $b_{21}y_1 + b_{22}y_2 + b_{23}y_3 = 20(0.0498) + 23(0.0001) + 1(0.0001)$
$$= 1$$
Efficiency for doctor 2: Output/Input $= 1/1 = 1$

Measurement of Input and Output Factors in DEA In DEA, the measurement of different inputs and outputs do not have to be the same or specified. For example, the cost per square foot of space is different than the cost per hour of work of staff. But cost units are identical across all units for a given input or output factor. For example, the cost per hour for all doctors is the same, although not specified. Therefore, in DEA, without specifying the cost of different inputs and outputs, it is possible to compare different units. Note that if the unit costs of different input and output factors were known, then all inputs could be aggregated into one factor. Also, all outputs could be aggregated into one factor. In this case, the single-factor efficiency method of Section 10.4 can be used instead of the DEA LP method.

10.6 MULTICRITERIA EFFICIENCY AND DEA

10.6.1 Multiobjective Linear Programming of Efficiency and DEA[†]

In the last section, efficiency was defined based on productivity, that is, quantity efficiency. However, it is possible to have other types of efficiency. For example, one can consider efficiency in terms of the quality of products or services. The multiobjective linear programming (MOLP) problem for K types of efficiency is presented in the following MOLP problem where each type of efficiency (objective) is measured between 0 and 1.

PROBLEM 10.2 DEA MOLP PROBLEM FOR UNIT r

Maximize quantity efficiency, $f_1 = E_{r,1}$
Maximize quality efficiency, $f_2 = E_{r,2}$
Minimize cost efficiency, $f_3 = E_{r,3}$

$\vdots$

Maximize Kth objective efficiency, $f_K = E_{r,K}$
Subject to constraints of Problem 10.1 DEA LP for objectives $k = 1, 2, \ldots, K$

In the following, we show how to measure each objective function for each unit. We then show how to generate the set of multiobjective alternatives (units). Suppose objectives are denoted by $k = 1, \ldots, K$. Consider the following notation:

$x_{i,k}$: output i for objective k (decision variables)

$a_{r,i,k}$: output i for objective k for alternative r (given coefficients)

For each objective k and each unit r, the following LP problem can be formulated and solved.

Note that as in the single-objective DEA LP formulation, the value of all inputs for each unit is always equal to 1. The output, however, may vary from 0 to 1.

PROBLEM 10.3   DEA LP PROBLEM FOR UNIT r AND OBJECTIVE k
Maximize:

$$E_{r,k} = \sum_{i=1}^{I} a_{r,i,k} x_{i,k} \quad \text{for given } r \text{ and } k \tag{10.22}$$

Subject to:

$$-\sum_{i=1}^{I} a_{r,i,k} x_{i,k} + \sum_{j=1}^{J} b_{rj} y_j \geq 0 \quad \text{for } r = 1, \ldots, R \text{ (for given } k) \tag{10.23}$$

$$\sum_{j=1}^{J} b_{rj} y_j = 1 \quad \text{(total input for given } r) \tag{10.24}$$

$$x_{i,k} \geq 0.0001 \quad \text{for } i = 1, \ldots, I \text{ and given } k \tag{10.25}$$

$$y_j \geq 0.0001 \quad \text{for } j = 1, \ldots, J \tag{10.26}$$

Therefore, for each given unit r, we must solve K single-objective DEA LP problems. For a given unit r, the values of the K objectives are

$$\mathbf{E}_r = (E_{r,1}, E_{r,2}, \ldots, E_{r,K})$$

Therefore, the set of multiobjective alternatives (vectors) for the R units are $\mathbf{E}_1, \mathbf{E}_2, \ldots, \mathbf{E}_R$.

A MCDM method (e.g., using an additive utility function) can be used to rank the alternatives (units).

Example 10.7   Bicriteria DEA Linear Programming   Consider Example 10.6 with two objectives and the data as presented in Table 10.5. The first objective is the same as the one defined in Example 10.6, that is, quantity efficiency. The second objective is quality efficiency. This objective measures the quality of services rendered by each of the three doctors. This objective can be measured based on patient satisfaction with a doctor or based

TABLE 10.5   Coefficients for Inputs and Outputs for Two Objectives

Doctor Number, r	Input, b_{rj}			Output, $f_1(a_{r,i,k})$, $k = 1$, 1st Objective: Quantity		Output, $f_2(a_{r,i,k})$; $k = 2$, 2nd Objective: Quality	
	$j = 1$	$j = 2$	$j = 3$	$i = 1$	$i = 2$	$i = 1$	$i = 2$
1	15	20	12	6	5	0.84	0.68
2	20	23	15	22	8	0.80	0.49
3	10	18	10	3	3	0.70	0.60

on the number of claims against the given doctor. Suppose that quality can be rated from 0 to 1 where 0 is the worst and 1 is the best quality.

(a) Rank each alternative (doctor) in terms of the first objective.
(b) Rank each alternative (doctor) in terms of the second objective.
(c) List multicriteria alternatives and identify the inefficient alternatives.

(a) The solution for part (a) is given in Example 10.7.
(b) The solution for part (b) is as follows. The LP formulation for the first doctor ($r = 1$) and the second objective ($k = 2$), quality efficiency, is shown below:
Maximize:

$$E_{1,2} = 0.84x_{1,2} + 0.68x_{2,2} \quad \text{for } r = 1, k = 2 \tag{10.27}$$

Subject to:

$$-0.84x_{1,2} - 0.68x_{2,2} + 15y_1 + 20y_2 + 12y_3 \geq 0 \tag{10.28}$$

$$-0.80x_{1,2} - 0.49x_{2,2} + 20y_1 + 23y_2 + 15y_3 \geq 0 \tag{10.29}$$

$$-0.70x_{1,2} - 0.60x_{2,2} + 10y_1 + 18y_2 + 10y_3 \geq 0 \tag{10.30}$$

$$15y_1 + 20y_2 + 12y_3 = 1 \tag{10.31}$$

$$x_{1,2}, x_{2,2} \geq 0.0001 \tag{10.32}$$

$$y_1, y_2, y_3 \geq 0.0001 \tag{10.33}$$

Solving this problem by a LP software package, the optimal solution for doctor 1 in terms of the second objective is:

Input and Output Results for Doctor 1 for Objective 2

$x_{1,2}$	$x_{2,2}$	y_1	y_2	y_3	$E_{1,2}$
0.001	1.47	0.005	0.05	0.001	1

See Supplement S10.3.lg4.

TABLE 10.6   Input and Output Results for the Three Doctors for the Second Objective, Quality

	$x_{1,2}$	$x_{2,2}$	y_1	y_2	y_3	$E_{r,2}$
Doctor 1	1.190	0.0001	0.019	0.0357	0.0001	1
Doctor 2	1.0303	0.001	0.001	0.042	0.001	0.825
Doctor 3	0.001	1.666	0.006	0.05	0.001	1

TABLE 10.7   Summary of Two Objectives for Three Doctors

	Doctor 1	Doctor 2	Doctor 3
Max $f_1 = E_{r,1}$	0.8323	1	0.749
Max $f_2 = E_{r,2}$	1	0.825	1
Efficient?	Yes	Yes	No

Since $E_{1,2} = 1$, doctor 1 is efficient with regard to the second objective, quality. Similarly two other LPs can be solved for doctors 2 and 3. Table 10.6 shows the optimal solutions for all three doctors for the second objective (by solving three different LP problems).

See Supplements S10.4-1.lg4, S10.4-2.lg4.

The quality efficiencies for doctors 1, 2, and 3 are 100, 82.8, and 100%, respectively.

(c) To respond to part (c), consider the solution to the first objective, that is, Table 10.4, and the solution for the second objective, that is, Table 10.6. These two tables are presented as Table 10.7. This is the set of multiobjective vectors.

According to Table 10.7, doctor 3 is inefficient because doctor 1 is better or equal in terms of both objectives.

10.6.2   Multicriteria Ranking of Units[†]

Once all alternatives (units) are measured in terms of the different objectives, a MCDM ranking approach can be used to rank them. For the purpose of illustration, suppose that an additive utility function is used to rank the alternatives:

Maximize: $U = w_1 f_1 + w_2 f_2 + \cdots + w_K f_K$

Subject to: Set of multicriteria alternatives

where $w_1, w_2, \ldots, w_K$ are the weights of importance of objectives $1, 2, \ldots, K$, respectively.

Consider Example 10.7. Suppose that w_1 and w_2 are (0.7, 0.3) for objectives 1 and 2, respectively. The ranking of units is presented below. The ranking is doctor 2, doctor 1, and doctor 3.

Doctor r	1	2	3
Max $U = 0.7f_1 + 0.3f_2$	0.8826	0.9484	0.8243
Rank	2	1	3

10.7 PRODUCTIVITY OF A NETWORK OF PROCESSORS

10.7.1 Productivity of a Network of Processors

In a production system semifinished products are moved forward until their production is completed. A production system can be defined as a network of processors (e.g., facilities or machineries) having n sequential phases in series, where each phase can consist of up to m processors in parallel.

A system that consists of only one processor in each phase is called a serial system. A very simple example is presented in Figure 10.2 where the output of processor 1 is inputted to processor 2. Often, the outputs of a processor are not 100% good in quality, and a portion of them is rejected as scrap. In many industries, improvement in quality has caused scrap rates to be decreased. Nevertheless, in some industries, the scrap rate is still a substantial number. The system productivity can be calculated while considering the scrap rates. For example, suppose for a processor the capacity is 2000 units per hour and its scrap rate is 2%, or 0.02. Then the total amount of good units per hour produced would be $2000 \times (1 - 0.02) = 2000(0.98) = 1960$. Alternatively, if the productivity rate of a system is 98%, then its productivity is $2000(0.98) = 1960$ per hour.

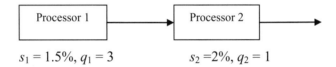

$$s_1 = 1.5\%, \ q_1 = 3 \qquad\qquad s_2 = 2\%, \ q_2 = 1$$

Figure 10.2   Serial system consisting of two processors.

The productivity for a set of processors in series can be calculated using the following notation:

s_r: Scrap rate of processor r

C_r: Capacity of processor r

q_r: Number of units needed from processor r to produce one unit of product at the next processor

P_r: Productivity of processor r that can be used for the next phase

The productivity of processor r, P_r, with scrap rate s_r is

$$P_r = \frac{C_r(1 - s_r)}{q_r} \qquad\qquad (10.34)$$

Example 10.8   Finding the Productivity of a Serial System   Consider a simple case of two processors, where processor 1 is inputted to processor 2.

Suppose the specifications of processor 1 are $C_1 = 2000$ units/h, $s_1 = 0.015$, and $q_1 = 3$ and the specifications of processor 2 are $C_2 = 750$ units/h, $s_2 = 0.02$, and $q_2 = 1$. The productivity of processor 2 needs to be found. First, the productivity of processor 1 can be calculated using Equation (10.34):

$$P_1 = \frac{2000(1 - 0.015)}{3} \quad \text{or} \quad P_1 = 656.7 \text{ units/h}$$

Since the output from processor 1 is inputted to processor 2, the usable capacity of processor 2 is limited by the productivity of processor 1:

$$C_2^* = \text{Min}\{C_2, P_1\} \quad \text{or} \quad C_2^* = \text{Min}\{750,656.7\} = 656.7 \text{ units/h}$$

This value can then be used in Equation (10.34) to find the productivity of processor 2:

$$P_2 = \frac{656.7(1 - 0.02)}{1} \quad \text{or} \quad P_2 = 643.6 \text{ units/h}$$

Therefore, the productivity of this system is 643.6 units/h.

Example 10.9   Finding the Needed Capacity of a Serial System   Consider the system in Figure 10.2 with $s_1 = 0.015$, $q_1 = 3$, $s_2 = 0.02$, and $q_2 = 1$. Consider the goal is to produce 1000 units as the output of processor 2. What is the needed capacity at processors 1 and 2?

First, the needed capacity of processor 2 must be determined using Equation (10.34):

$$1000 = \frac{C_2(1 - 0.02)}{1} \quad \text{or} \quad C_2 = 1020.4 \text{ units/h}$$

Now, since the input of processor 2 is the output of processor 1, the productivity of processor 1 must be equal to the capacity of processor 2 to meet the needed productivity of processor 2. So, the needed productivity of processor 1 is

$$P_1 = C_2 = 1020.4 \text{ units/h}$$

Equation (10.34) can be used to find the needed capacity for processor 1:

$$1020.4 = \frac{P_1(1 - 0.015)}{3} \quad \text{or} \quad P_1 = 3107.8 \text{ units/h}$$

Bottleneck Identification   When multiple processors each producing different parts are in parallel in a phase, the actual productivity of that phase is limited by the processor that has the lowest productivity. This limiting processor is called the bottleneck. The following equation is used to calculate the productivity for phase i, which consists of m processors:

$$P_i = \text{Min}\{P_{i,1}, P_{i,2}, \ldots, P_{i,m}\} \tag{10.35}$$

The processor with the minimum productivity is the bottleneck. When attempting to increase the productivity of the whole production system, the capacity of the bottleneck processor of each phase must be increased.

Example 10.10   Finding the Productivity of Parallel Systems   Figure 10.3 is a simple parallel system with three processors. Processors 2 and 3 are parallel in phase I and their outputs are inputted to processor 1. To produce one unit of output at processor 1, there must be three units from processor 2 (i.e., $q_2 = 3$) and two units of processor 3 (i.e., $q_3 = 2$). The scrap rates for processor 1, processor 2, and processor 3 are 2, 1.5, and 3%, respectively. Also, suppose that the capacity at processor 2 is 2500 units/h, the capacity at processor 3 is 1700 units/h, and the capacity at processor 1 is 850 units/h.

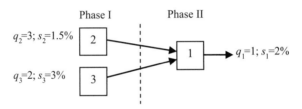

Figure 10.3   Production process for Example 10.10.

Find the productivity of processor 1 and find the bottleneck processor.
The productivity of processors 2 and 3 can be found using Equation (10.34):

$$P_2 = \frac{2500(1 - 0.015)}{3} \qquad \text{or} \qquad P_2 = 820.8 \text{ units/h}$$

$$P_3 = \frac{1700(1 - 0.03)}{2} \qquad \text{or} \qquad P_3 = 824.5 \text{ units/h}$$

The productivity of phase I can be found using Equation (10.35):

$$P_1 = \text{Min}\{P_2, P_3\} = \text{Min}\{820.8, 824.5\} = 820.8$$

Since processors 2 and 3 are parallel and their outputs are inputted to processor 1, the productivity of phase I may limit the capacity of processor 1. That is, the usable capacity of processor 1 is the minimum of the capacity of processor 1 and the productivity of phase I:

$$C_1^* = \text{Min}\{C_1, P_I\} \qquad \text{or} \qquad C_1 = \text{Min}\{850, 820.8\} = 820.8 \text{ units/h}$$

Using Equation (10.34), the productivity of processor 1 can be found:

$$P_1 = \frac{820.8(1 - 0.02)}{1} \quad \text{or} \quad P_1 = 804.4 \text{ units/h}$$

The bottleneck processor is the parallel processor with the lowest productivity; in this case the bottleneck of phase I is processor 2.

Example 10.11   Finding the Needed Capacity for a Network of Parallel Systems
Consider the system in Figure 10.3. Suppose the productivity goal at processor 1 is 980 good products per hour. Find out the needed capacity for processors 1, 2, and 3, respectively.
    Using Equation (10.34),

$$C_1 = \frac{q_1 P_1}{1 - s_1} = \frac{1 \times 980}{0.98} = 1000.00$$

Now, set $P_2 = C_1 = 1000$ and $P_3 = C_1 = 1000$. For processors 2 and 3,

$$C_2 = \frac{3 \times 1000}{1 - 0.015} = 3045.69 \quad C_3 = \frac{2 \times 1000}{1 - 0.03} = 2061.86$$

Therefore, in order to satisfy a productivity of 980 units/h at processor 1, the needed capacities for processors 1, 2, and 3 are 1000.00, 3045.69, and 2061.86, respectively.

Example 10.12   Finding Bottlenecks for Each Phase for a Network   Consider a multiphase system consisting of eight processors and four phases of production as shown in Figure 10.4.
    The specifications for the system are shown in Table 10.8. Find the productivity of the system. Also find the bottleneck for each phase and for the whole system.
    The solution is presented in Table 10.9. Each phase must be analyzed in order to determine its productivity and its bottleneck. For phase I, Equation (10.34) is used to

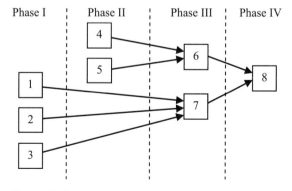

Figure 10.4   Example 10.12 multiphase network system.

TABLE 10.8   Example 10.12 Specifications for System Presented in Figure 10.4

	Phase I			Phase II		Phase III		Phase IV
Processor r	1	2	3	4	5	6	7	8
Capacity (C_r) (units/h)	3300.0	2200.0	4500.0	2300.0	3350.0	1300.0	1100.0	700.0
Scrap rate (s_r), %	2.00	2.50	1.80	2.20	1.70	2.80	2.40	1.50
No. of needed parts (q_r)	3	2	4	2	3	2	2	1

find the productivity for each processor. Equation (10.35) is then used to determine the productivity of each phase. For phase I, processor 2 is identified as the bottleneck since it has the lowest productivity. Notice that the productivity of this phase limits the productivity of the next processor, 7. These steps can be followed for each of the phases until the total productivity of the system is found in phase IV, which is 515.5 units/h.

10.7.2   Multicriteria Productivity of a Network[†]

In capacity/productivity planning, we can consider two objectives:

Maximize f_1 = system productivity

Minimize f_2 = total cost of improving system productivity

Suppose that each processor's productivity can be improved for a given cost. The problem is to find the set of efficient alternatives in terms of system productivity and the total system cost. Then the best alternative can be selected using a MCDM method. A simple example for two levels of improving productivity and the associated costs for each processor is presented below. A comprehensive method is introduced in Section 19.5.2.

Example 10.13   Generating Multicriteria Alternatives for a Productivity Problem
Consider the system illustrated in Figure 10.3. The capacities for processors 1, 2, and 3 are 850, 2500, and 1700 units/h, respectively. Suppose that each processor's scrap rate can be improved for the following given costs (all costs are in tens of thousands):

- Processor 1 scrap rate can be reduced from 2 to 1.5%, with costs of $52.

TABLE 10.9   Solution for Example 10.12

	Phase I			Phase II		Phase III		Phase IV
Processor r	1	2	3	4	5	6	7	8
Productivity of process (units/h)	1078.0	1072.5	1104.8	1124.7	1097.7	533.5	523.4	515.5
Productivity of phase (units/h)		1072.5			1097.7		523.4	515.5
Bottleneck of phase		Processor 2			Processor 5		Processor 7	—

TABLE 10.10   Multicriteria Analysis of Capacity Planning with Scrap Rates

	a_1	a_2	a_3	a_4	a_5	a_6	a_7	a_8
Processor 1, s_1	2%	1.5%	2%	2%	1.5%	1.5%	2%	1.5%
Processor 2, s_2	1.5%	1.5%	1%	1.5%	1%	1.5%	1%	1%
Processor 3, s_3	3%	3%	3%	2.5%	3%	2.5%	2.5%	2.5%
Processor 1 cost	0	52	0	0	52	52	0	52
Processor 2 cost	0	0	67	0	67	0	67	67
Processor 3 cost	0	0	0	40	0	40	40	40
Max. system productivity, f_1	804.4	808.5	808.4	804.4	812.1	808.5	808.5	812.6
Min. total cost, f_2	0	52	67	40	119	92	107	159
Efficient alternative?	Yes	Yes	No	No	Yes	No	No	Yes

- Processor 2 scrap rate can be reduced from 1.5 to 1%, with costs of $67.
- Processor 3 scrap rate can be reduced from 3 to 2.5%, with costs of $40.

Now, consider all possible alternatives in terms of different combinations of scrap rates s_1, s_2, and s_3. Table 10.10 presents all such alternatives where alternative a_1 is the original system which was solved as Example 10.10 with a productivity of 804.4. Alternative a_8 has improvements in all three processor scrap rates. Each alternative has different level of productivity, scrap rate, and total cost. For each alternative, find the system productivity f_1, and the total cost of improvements, f_2. The solution is provided in Table 10.10.

Note that alternative a_2 has a productivity of 808.52 and a total cost of $52. Alternative a_3 has a productivity of 808.4 and a total cost of $67. Therefore, alternative a_2 dominates alternative a_3. Figure 10.5 illustrates the efficient frontier for all of the alternatives.

See Supplement S10.5.xls.

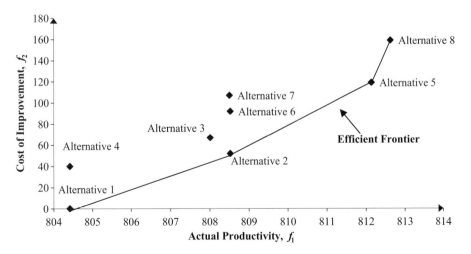

Figure 10.5   Bicriteria capacity planning with scrap rates.

REFERENCES

Productivity

Coelli, T. *An Introduction to Efficiency and Productivity Analysis*, 2nd ed. New York, NY: Springer, 2005.

Felix, G. H., and J. L. Riggs. "Productivity Measurement by Objectives." *National Productivity Review*, Vol. 2, No. 4, 1983, pp. 386–393.

Fried, H. O., C. A. K. Lovell, and S. S. Schmidt. *The Measurement of Productive Efficiency and Productivity Growth*. Oxford, NY: Oxford University Press, 2008.

Mapes, J., C. New, and M. Szwejczewski. "Performance Trade-offs in Manufacturing Plants." *International Journal of Operations & Production Management*, Vol. 17, No. 10, 1997, pp. 1020–1033.

Murphy, P. "Service Performance Measurement Using Simple Techniques Actually Works." *Journal of Marketing Practice: Applied Marketing Science*, Vol. 5, No. 2, 1999, pp. 56–73.

Prokopenko, J. *Productivity Management: A Practical Handbook*. Geneva: International Labour Organization, 1987.

Vough, C. F., and B. Asbell. *Productivity: A Practical Program for Improving Efficiency*. City; Productivity Research International, 1986.

Young, S. T. "Multiple Productivity Measurement Approaches for Management." *Health Care Management Review*, Vol. 17, No. 2, 1992, pp. 51–58.

Data Envelope Analysis

Anderson, P., and N. C. Petersen. "A Procedure for Ranking Efficient Units in Data Envelopment Analysis." *Management Science*, Vol. 39, No. 10, 1993, pp. 1261–1264.

Charnes, A., C. T. Clark, W. W. Cooper, and B. Golany. "A Development Study of Data Envelopment Analysis in Measuring the Efficiency of Maintenance Units in U.S. Air Forces." *Annals of Operational Research*, Vol. 2, No. 1, 1984, pp. 95–112.

Charnes, A., W. W. Cooper, and E. Rhodes. "Measuring the Efficiency of Decision Making Units." *European Journal of Operational Research*, Vol. 2, 1978, pp. 429–444.

Charnes, A., W. W. Cooper, and R. M. Thrall. "A Structure for Classifying and Characterizing Efficiency and Inefficiency in Data Envelopment Analysis." *Journal of Productivity Analysis*, Vol. 2, No. 3, 1991, pp. 197–237.

Cook, W. D., and L. M. Seiford. "Data Envelopment Analysis (DEA)—Thirty Years On." *European Journal of Operational Research*, Vol. 192, No. 1, 2009, pp. 1–17.

Cooper, W. W., L. M. Seiford, and K. Tone. *Introduction to Data Envelopment Analysis and its Uses*. City; Springer Science, 2006.

Winston, W. L., and M. Venkataramanan. *Introduction to Mathematical Programming*, 4th ed. City: Duxbury Press, 2002.

Zhou, P., B. W. Ang, and K. L. Poh. "A Survey of Data Envelopment Analysis in Energy and Environmental Studies." *European Journal of Operational Research*, Vol. 189, No. 1, 2008, pp. 1–18.

Multiobjective Productivity and Data Envelope Analysis

Chen, Y.-W., M. Larbani, and Y.-P. Chang. "Multiobjective Data Envelopment Analysis." *Journal of the Operational Research Society*, Vol. 60, No. 11, 2009, pp. 1556–1566.

Doyle, J., and R. Green. "Data Envelopment Analysis and Multiple Criteria Decision Making." *Omega*, Vol. 21, No. 6, 1993, pp. 713–715.

Joro, T., P. Korhonen, and J. Wallenius. "Structural Comparison of Data Envelopment Analysis and Multiple Objective Linear Programming." *Management Science*, Vol. 44, No. 7, 1998, pp. 962–970.

Li, X.-B., and G. R. Reeves. "A Multiple Criteria Approach to Data Envelopment Analysis." *European Journal of Operational Research*, Vol. 115, No. 3, 1999, pp. 507–517.

Malakooti, B. "Multi-Criteria of Productivity, Efficiency, and Data Envelope Analysis" Case Western Reserve University, Cleveland OH, December, 2013.

Malekmohammadi, N., F. H. Lotfi, and A. B. Jaafar. "Target Setting in Data Envelopment Analysis Using MOLP." *Applied Mathematical Modelling*, Vol. 35, No. 1, 2011, pp. 328–338.

Soleimani-Damaneh, M. "Optimality for Non-Smooth Fractional Multiple Objective Programming." *Nonlinear Analysis*, Vol. 68, No. 10, 2008, pp. 2873–2878.

EXERCISES

Basic Productivity Indexes (Section 10.2)

2.1 Consider an electronics company. Suppose 20 units of earphones were produced per hour in the base year and the productivity growth rate is 3.2% per year. What would be the productivity in 20 years after the base year for the same input?

2.2 Consider an automobile factory. The table below shows the values in thousands of dollars of three inputs (labor, machine, and energy) and one output (profit) factors.

Input: Costs			Output: Profit in $1000
Labor in $1000	Machine in $1000	Energy in $1000	
75	62	34	760

(a) Calculate the partial measures of productivity for labor, machine, and energy.
(b) Calculate the multifactor measures of productivity for labor and machine, machine and energy, and labor and energy.
(c) Calculate the total measure of productivity.

2.3 Consider a furniture company. Suppose all units of inputs and outputs are measured for millions of dollars. The table below shows the values of inputs and outputs in million dollars.

Input				Output	
Labor	Material	Capital	Energy	Product	Service
25	40	80	120	150	130

(a) Calculate the partial measures of productivity for labor, material, capital, and energy.

(b) Calculate the multifactor measures of productivity for labor and material, material and capital, capital and energy, and labor and energy.

(c) Calculate the total measure of productivity.

Multifactor Productivity Growth[†] (Section 10.3)

3.1 Consider a furniture company. The table below shows two inputs (in terms of labor hours) and one output (in terms of number of desks produced). The quantities and prices for inputs and outputs for the base year and current year are presented below.

	Period 1: Base Year		Period 2: Current Year	
Output	Quantity $O_{1,i}$	Price $P_{1,i}$	Quantity $O_{2,i}$	Price $P_{2,i}$
Product	450 units	$75/unit	1200 units	$100/unit
Input	Quantity $I_{1,j}$	Cost $C_{1,j}$	Quantity $I_{2,j}$	Cost $C_{2,j}$
Labor 1	7500 h	$28/h	9200 h	$42/h
Labor 2	3800 h	$40/h	3200 h	$47/h

(a) What is the factor productivity index? What does it mean?

(b) What is the price recovery index? What does it mean?

(c) What is the cost effectiveness index? What does it mean?

3.2 Consider an automobile company. Consider the table below with two inputs (in terms of labor hours) and one output (in terms of number of automobiles manufactured). The quantities and prices for inputs and outputs for the base year and current year are presented below.

	Period 1: Base Year		Period 2: Current Year	
Output	Quantity $O_{1,i}$	Price $P_{1,i}$	Quantity $O_{2,i}$	Price $P_{2,i}$
Product 1	300 units	$50/unit	800 units	$40/unit
Product 2	650 units	$55/unit	900 units	$60/unit
Input	Quantity $I_{1,j}$	Cost $C_{1,j}$	Quantity $I_{2,j}$	Cost, $C_{2,j}$
Labor 1	9500 h	$45/h	8800 h	$30/h
Labor 2	5300 h	$30/h	2900 h	$38/h

(a) What is the factor productivity index? What does it mean?

(b) What is the price recovery index? What does it mean?

(c) What is the cost effectiveness index? What does it mean?

Single-Factor Efficiency (Section 10.4.1)

4.1 Consider five furniture manufacturers. The input is the amount of furniture material in pounds and the output is the amount of furniture produced in pounds. The inputs and outputs for the five furniture manufacturers are shown in the table below.

Manufacturer no.	1	2	3	4	5
Outputs	34	42	15	25	17
Inputs	51	50	27	31	25

(a) What is the productivity for each of the five manufacturers?

(b) What is the efficiency for each of the five manufacturers?

4.2 Consider seven comparable electricity generators. The input is the amount of energy used by each generator, and the output is the amount of energy generated by each generator per given period of time. The inputs and outputs for the seven generators are shown in the table below.

Generator no.	1	2	3	4	5	6	7
Outputs	20	45	18	27	25	15	28
Inputs	45	55	30	45	31	20	32

(a) What is the productivity for each of the seven generators?

(b) What is the efficiency for each of the seven generators?

Bicriteria Single-Factor Efficiency (Section 10.4.2)

4.3 Consider five comparable companies. The efficiencies of these companies can be measured in terms of two objectives: f_1, maximize production (in terms of number of computers produced), and f_2, maximize quality. The quality is rated from 0 to 5 where 5 signifies the highest quality. Inputs are measured in terms of number of labor hours.

Computer company no.	1	2	3	4	5
Output 1, production	11	15	20	18	31
Output 2, quality	5	3	1	2	4
Inputs, labor hours	21	30	25	22	41

(a) Measure production efficiency for all five companies.

(b) Measure quality efficiency for all five companies.

(c) Identify inefficient companies.

(d) Rank alternatives if weights of importance are (0.6, 0.4) for the two objectives, respectively.

4.4 Consider exercise 4.2. Suppose another output of maximizing quality is also considered. This quality presents the environmental impact of each company where 0 is the worst and 5 is the best impact.

Computer company no.	1	2	3	4	5	6	7
Output 1, production	20	45	18	27	25	15	28
Output 2, quality	5	1	2	4	3	2	0
Inputs	45	55	30	45	31	20	32

(a) Measure quality efficiency for all seven generators.

(b) Identify inefficient generators.

(c) Rank generators if weights of importance are (0.7, 0.3) for the two objectives, respectively.

Multifactor Efficiency and DEA[†] (Section 10.5)

5.1 The table below shows three factories with one input (labor in hours) and one output (profit in $1000).

Factory	Input, $j = 1$	Productivity—output, $i = 1$
1	5	4
2	8	7
3	4	2

(a) Formulate the LP problem for factory 1. Solve the LP using an LP software package. Find the efficiency of factory 1.

(b) Find the efficiencies of factories 2 and 3. Compare the efficiencies of the three factories.

(c) Solve this problem by single-factor efficiency of Section 10.4. Compare this solution with the results obtained in parts (a) and (b).

5.2 The table below shows three automobile factories with three inputs (labor in hours, material in pounds, and energy in joules) and one output as the number of cars produced.

Factory	Input			Output
	$j = 1$	$j = 2$	$j = 3$	$i = 1$
1	5	8	7	4
2	7	9	4	5
3	4	5	9	2

(a) Formulate the LP problem for factory 1. Solve the LP problem using an LP software package. Find the efficiency of factory 1.

(b) Find the efficiencies of factories 2 and 3. Compare the efficiencies of the three factories.

5.3 The table below shows three different fast food restaurants in the same city with three inputs (labor in hours, food ingredients in pounds, and energy in joules) and two outputs as the number of sandwiches type 1 and type 2 sold in dollars.

Fast Food	Input			Output	
	$j = 1$	$j = 2$	$j = 3$	$i = 1$	$i = 2$
1	5	6	7	4	5
2	8	9	4	7	3
3	4	6	7	2	6

(a) Formulate the LP problem for fast food restaurant 1. Solve the LP problem using an LP software package. Find the efficiency of fast food restaurant 1.

(b) Find the efficiencies of fast food restaurants 2 and 3. Compare the efficiencies of the three fast food restaurant.

Multicriteria Efficiency and DEA[†] (Section 10.6)

6.1 Consider exercise 5.2. Consider an additional objective of the quality of the products being produced, f_2. Quality ratings are from 1 to 10 where 1 is the worst and 10 is the best. The table below shows the inputs and outputs for the two objectives.

Factory	Input			Quantity, Output	Quality, Output
	$j = 1$	$j = 2$	$j = 3$	$i = 1$	$k = 2$
1	5	8	7	4	7
2	7	9	4	5	5
3	4	5	9	2	3

(a) For the quality output, formulate LPs for all three factories and solve them. Identify quality efficiencies for each factory.

(b) Use the values for the first objective based on the results of exercise 5.2. List the bicriteria alternatives. Which ones are efficient or inefficient?

(c) Suppose weights of importance are (0.4, 0.6) for the first and second objectives, respectively. Rank the alternatives according to these weights.

6.2 Consider exercise 5.3. Consider an additional objective of maximizing quality of food, f_2. Quality ratings are from 1 to 10 where 1 is the worst and 10 is the best. The table below shows the inputs and outputs associated with the two objectives.

Fast Food	Labor, Input			Quantity, Output, $k = 1$		Quality, Output, $k = 2$	
	$j = 1$	$j = 2$	$j = 3$	$i = 1$	$i = 2$	$i = 1$	$i = 2$
1	5	6	7	4	5	1	5
2	8	9	4	7	3	2	8
3	4	6	7	2	6	3	4

    (a) For the quality output, formulate LPs for all three fast food restaurants and solve them. Identify the quality efficiency of each fast food restaurant.

    (b) Use the values for the first objective based on the results of exercise 5.3. List the bicriteria alternatives. Which ones are efficient or inefficient?

    (c) Suppose weights of importance are (0.8, 0.2) for the first and second objectives, respectively. Rank the alternatives according to these weights.

Productivity of a Network of Processors (Section 10.7.1)

7.1 To produce one unit of output, processor 3 requires two units of output from processor 1 and three units of output from processor 2. The scrap rate at processor 1 is 1.8%, at processor 2 is 2.0%, and at processor 3 is 2.5%. No in-process inventory can exist between processors 1 and 3 or between 2 and 3.

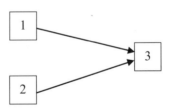

    (a) Assume the capacity at processor 1 is 180, the capacity at processor 2 is 170, and the capacity at processor 3 is 100 units/h. Determine the actual productivity of processor 3 in finished good units per hour. What is the bottleneck processor?

    (b) If the needed productivity for the final good finished product at processor 3 is 100 units/h, find the needed capacity for processors 1 and 2.

7.2 A production process consists of three processors: 1, 2, and 3. Processor 1 must be completed before processor 2, and processor 2 must be completed before processor 3. The current scrap rates are 4.5% for 3, 4% for 1, and 4% for 2. There is no inventory between any of the processes. Three units from processor 2 are needed to produce one unit of processor 3, and three units of processor 1 are needed to produce each unit of processor 2.

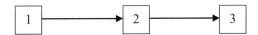

(a) The current capacity for processors 1, 2, and 3 are 3300, 3200, and 3000 units/h, respectively. Determine the actual productivity of the system in finished good units per hour.

(b) If the required good finished product should be 1000, find the needed capacity of processors 1, 2, and 3.

7.3 Consider the production process shown below with the specifications shown in the table.

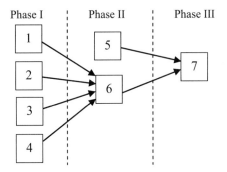

(a) Consider the information in rows 1, 2, 3, and 4 of the table. Find the productivity and bottleneck of each phase and the overall system productivity.

(b) Consider the information in rows 1, 3, and 4 of the table. Suppose that the needed productivity at processor 7 is 600 units/h. Find the needed capacity for processors 1–6.

	Phase I				Phase II		Phase III
1. Processor r	1	2	3	4	5	6	7
2. Capacity (C_r), units/h	5000	4500	3500	5800	1800	900	800
3. Scrap rate (s_r), %	1.60	1.70	2.20	2.00	1.50	1.20	1.40
4. No. needed (q_r)	6	5	4	7	4	2	1
5. New improvement investment, $1000	80	70	90	65	85	0	40
6. New improvement in scrap rates	1.40%	1.50%	1.90%	1.50%	1.40%	1.20%	1.10%

Multicriteria Productivity of a Network (Section 10.7.2)

7.4 Consider the given capacities in part (a). Suppose that with $5 (in 10,000s) processor 2 can improve its scrap rate from 2 to 1%. Also, suppose that with $40 (in 10,000s) processor 1 can improve its scrap rate from 1.8 to 1%. List all multicriteria alternatives for maximizing productivity and minimizing total cost of improvement, and identify all inefficient alternatives.

7.5 Suppose that with new investment of $250,000 for processor 1, $300,000 for processor 2, and $750,000 for processor 3 the new scrap rate at each machine can be reduced to 3, 2, and 1% for processors 3, 2, and 1, respectively. List all multicriteria alternatives for combinations of improvements, and identify inefficient alternatives.

7.6 Consider the information in rows 1–6 of the table. With new investment in each processor (as shown in row 5 of the table), the scrap rate of each processor is shown in row 6 of the table. List all multicriteria alternatives for different combinations of new improvements of processors. Identify efficient and inefficient alternatives.

CHAPTER 11

ENERGY SYSTEM DESIGN
AND OPERATION

11.1 INTRODUCTION

The same way that carpenter ants and termites can cause a magnificent mansion to collapse, humans can destroy Earth's delicate ecological system by abusing energy. Our civilization has been developed by harnessing energy, and it will cease to exist without usable energy. Energy is a scalar quantity that describes the amount of work that is performed by a force. The first law of thermodynamics states that energy cannot be created or destroyed. Rather, it is transferred from one form to another. For example, when an automobile is propelled, it converts gasoline energy into kinetic (usable) energy and waste (unusable energy). Most energy systems waste a lot of energy when producing usable energy, for example, generating heat by burning wood in a fireplace. Substantial energy is lost when energy is stored, converted into other forms of energy, or transported. Therefore, in the design of energy systems, optimally, energy sources should be directly connected and as close as possible to the users and also become available when needed. For example, most electricity is delivered to homes from power plants through cables from a long distance away.

Energy Systems Simple energy systems consist of a set of components that are used to supply, save, and use energy. For example, solar panels, wind turbines, and home electrical systems are simple energy systems. An example of a simple energy system that consists of energy generators, energy users, and energy savers is a hybrid car; see Figure 11.1.

Some hypothetical examples of energy systems consisting of generators, savers, and users are presented in Table 11.1.

Note: Advanced material that can be omitted without loss of generality will be indicated by a dagger.

Operations and Production Systems with Multiple Objectives, First Edition. Behnam Malakooti.
© 2014 John Wiley & Sons, Inc. Published 2014 by John Wiley & Sons, Inc.

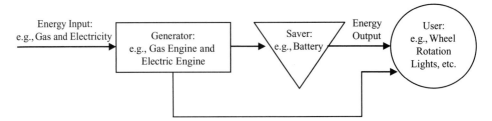

Figure 11.1   A simple hybrid energy system.

TABLE 11.1   Examples of Future Energy Systems

	Car	House	Factory	Ship	Train
Generator 1	Gas engine	Solar panels	Coal-based plant	Wind sails	Nuclear power
Generator 2	Electric engine	Electric grid	Nuclear power	Nuclear power	Electric grid
Saver 1	Chemical battery	Chemical battery	Chemical battery	Chemical battery	Chemical battery
Saver 2	Liquid battery	Water/gas wells	Water/gas wells	Liquid battery	Liquid battery
User 1	To rotate wheels	House appliances	Machinery	Propellers	Engine
User 2	Other devices	Heating/cooling	Transportation	Jet-air turbines	Jet-air turbines

Complex and Distributed Energy Systems   A complex energy system consists of many different energy components; it is usually a large-scale system. An example is a network of solar panels, wind turbines, power plants that generate energy and houses that use the energy. Another example of a complex energy system is the U.S. power grid (see Figure 11.2), which is used to deliver electricity from power plants to houses, public

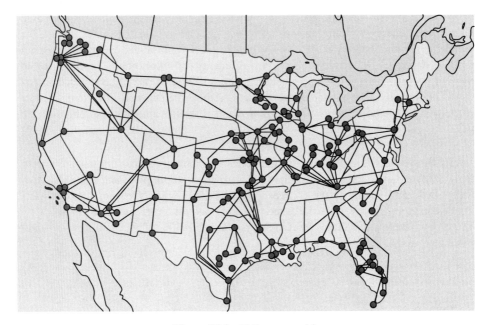

Figure 11.2   U.S. power grid.

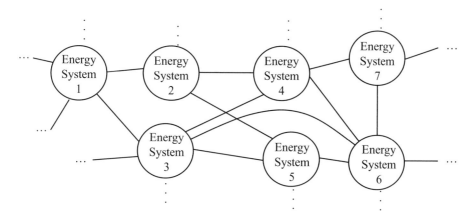

Figure 11.3 Distributed energy system of systems.

organizations, and industries. This complex system must be controlled to provide a steady and stable flow of electrical energy to different users while unpredictable turbulences and shortages may occur. In such grids it is important to provide the energy in an efficient manner to many different users while considering the safety, political, and economic concerns of different involved parties.

Energy systems can be connected to each other as a distributed network, as shown in Figure 11.3. Generally, an energy system's operation and design can be optimized with respect to the neighboring energy systems that are directly connected to it.

Multiple-Objective Energy Systems There are always different options for generating or procurement of energy for a given application, but there are trade-offs among the different options. For example, the fuel of a power plant can be nuclear, natural gas, coal, oil, or biofuel. Cables that connect long-distance transmission towers can be made out of different materials. Copper can be used; it is not an expensive material but has a large energy loss. Alternatively, superconducting wire has a small energy loss but is much more costly than copper wire. In a house, one can use energy-efficient light bulbs whose purchase price is more than incandescent bulbs or purchase energy-efficient appliances, heating, and cooling systems that cost more than their ordinary counterparts. Another option is related to sacrificing convenience for energy efficiency, for example, keeping a house's temperature lower in winter and higher in summer. The energy systems approach is based on identifying different sources and costs of energy, the availability of resources, and human preferences.

Bicriteria Trade-Off Characteristics of Energy Systems Energy systems may have conflicting criteria, such as the trade-off of speed versus energy efficiency in a car. Figure 11.4 shows the fuel efficiency of a typical car [in terms of miles per gallon (MPG)] versus different speeds [miles per hour (MPH)]. The best fuel efficiency is achieved at around 60 MPH. By maximizing speed, the total travel time is minimized. However, at speeds higher than 60 MPH, this will result in a lower MPG, which results in more gas

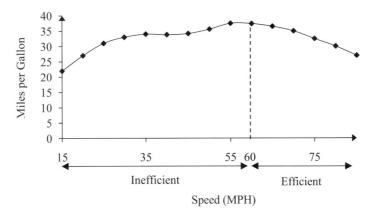

Figure 11.4 Trade-off of speed versus MPG for typical passenger car.

use. Considering only these two criteria, alternatives with speeds greater than 60 MPH are efficient. Now, consider the improved design of cars in the future. The weights of cars may be reduced considerably, better aerodynamic designs may be applied to reduce air and road friction, the use of hybrid and more efficient conversions of sources of energy may be achieved, and better roads and traffic control systems may be applied. Cars may someday have an average fuel efficiency of 100 MPG for speeds up to 75 MPH.

In this chapter, we discuss the basics of understanding and modeling energy systems by considering multiple options and using multiple-objective optimization to find the most preferred option. For example, a major use of energy is in transportation systems, where the transportation may have many different forms. In Section 11.5, we will cover routing algorithms that are used to find the shortest distance (paths or routes) from a given source to a given destination while there may exist multiple paths on a network similar to the network presented in Figure 11.3. With the use of global positioning systems, it is possible to find routes that maximize fuel efficiency by applying a routing algorithm. Energy system problems (similar to supply chain system problems) can be solved by using the topics covered in this book:

- The forecasting methods in Chapter 3 to predict energy use (demand), energy generation (supply), and energy price
- The aggregate planning method in Chapter 4 for energy planning
- The location methods in Chapter 16 for the selection of the location of energy generators, savers, and users
- The inventory methods in Chapter 6 to find the optimal amount of energy to be saved in savers for different periods (i.e., treat energy as inventory)
- The assignment methods in Chapter 9 for the optimal assignment of generators and (selling) savers to users and (buying) savers
- The transportation and transshipment methods in Chapter 9 for energy transportation problems
- The productivity and efficiency measurements in Chapter 10 to compare the productivity and efficiency of different generators, savers, and users

11.2   ENERGY PERSPECTIVE

11.2.1   Climate Change and Global Warming

Global Warming   One factor believed to be greatly influencing global climate change is the increased level of carbon dioxide in the atmosphere. The largest factor in the increase of carbon dioxide levels is the combustion of fossil fuels, the world's largest energy source. Scientists have found a direct correlation between the increase of the level of carbon dioxide and the increase in Earth's temperature. See Figure 11.5.

Population Growth   Earth's population has a direct impact on the current energy issues. Population growth is at the highest level in history. With more people on Earth and more countries becoming industrialized, the rate that people use energy is increasing. Since there are limited amounts of effective energy sources such as fossil fuels, the increasing population is a major issue in regards to energy.

Earth's Carrying Capacity   Carrying capacity is the maximum number of people that Earth can support indefinitely. This is due to limited amounts of land, water, and energy. Since most of the world's energy currently comes from fossil fuels, which are not renewable, it can be shown that Earth's carrying capacity in regards to energy is below the current population level. See Figure 11.6. If a sustainable form of energy can be used as effectively as fossil fuels, the carrying capacity will increase, and population will become less of an issue in regards to energy.

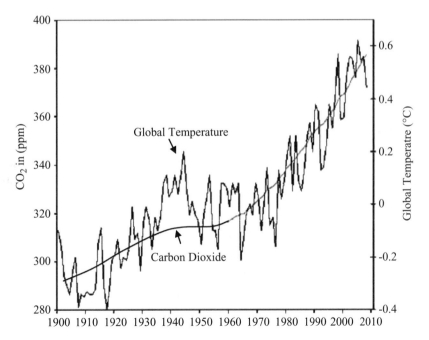

Figure 11.5   Global temperature and CO_2 levels.

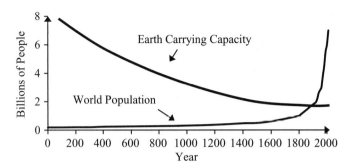

Figure 11.6   Earth's carrying capacity declines over time as world population increases (in billions).

11.2.2 Comparison of Different Types of Energy

A major source of energy in the world is fossil fuels (oil, coal, natural gas). According to the U.S. Department of Energy, fossil fuels provide over 85% of the total energy consumed in the world. This would not be an issue if fossil fuels were renewable resources and did not have negative environmental effects. However, most fossil fuels take millions of years to form, and the reserves are being depleted much faster than this. Hence, fossil fuels are not a sustainable form of energy. Current sources of worldwide energy consumption consist of:

1. Crude oil: 36%
2. Coal: 27%
3. Natural gas: 23%
4. Other: 14%

In addition to fossil fuels (total of 86%), other sources of energy (total of 14%) include nuclear energy and renewable energies such as solar, wind, biofuel, geothermal, and hydropower. All of these forms of energy are currently being used in some capacity; however, currently none of them can be effectively used to significantly reduce the world's dependence on fossil fuels. An optimal energy resource should be both *reliable* and *sustainable*. This has been the fundamental challenge in developing different energy options.

Table 11.2 specifies how each type of energy resource is transformed into energy and what type of technology typically uses it. Note that depending on the rate of energy need, a different power generator can be selected. Generally, the faster the generator, the costlier it is to use it.

TABLE 11.2   Conversion of Different Fuels into Electricity Outputs

Resource Input	Energy Converter	Output	Rate of Power Generation
Coal	Power plant	Electricity	Fast
Natural gas	Steam turbine	Electricity/heat	Fast
Uranium	Nuclear reactor	Electricity	Fast
Sunlight–solar	Photovoltaic cell	Electricity	Slow
Wind	Turbines	Electricity	Medium

TABLE 11.3   Comparison of Energy Cost Examples for Various Sources

Resource	Energy Density per Unit	Estimated Cost per Unit	Estimated Cost per Unit of Energy ($/GJ)
Coal	30 GJ/tonne	$50/tonne	$1.67
Crude oil	5.2 GJ/barrel	$70/barrel	$13.46
Natural gas	36.4 MJ/m^3	$0.44/m^3	$12.09
Nuclear	86,400 GJ/kg uranium	$210/kg uranium	$0.002
Solar	1 J/s or 1 W peak	$4/W peak	$4.00
Wind	—	$55.80/MWh	$15.50

Note: Nuclear cost does not include processing cost.

In order to compare the values of the different types of fuels, it is important to look at their energy densities. A resource's energy density is the amount of energy that can be produced per unit mass. Different types of fuel energy densities are measured using different units. In order to compare their relative value, they must be converted into units of energy. This is shown in Table 11.3.

As demonstrated in Table 11.3, with the exception of nuclear, coal is the cheapest fossil fuel in terms of cost per unit of energy. However, other costs must also be accounted for, such as transporting and storing the raw materials and the capital investment in facilities that are used for converting raw material into usable energy. For example, coal is the cheapest fossil fuel in cost per unit of energy, but it is not the cheapest fuel as it is the most difficult to extract and transport.

It is expected that advances in technology will enable the development of novel and sustainable energy generation, storage, distribution, and utilization. Breakthrough advances in these areas can be achieved through the discovery of new materials and processes and improved engineering designs. Technological advancements can be used to harvest renewable energy such as wind energy and solar (photovoltaic and/or thermal) energy. High-efficiency and low-cost electrochemical batteries using fuel cells and electrolyzers are possible advancements. New electrical grid interface solutions, controls, and sensors can be used to effectively utilize the generated energy. Also, the development of new optimization-based approaches for managing the power grid (distributed and centralized management of power transmission and delivery) can increase the productivity and efficiency of energy systems. Areas such as power electronics, aero-/hydrodynamics, structures, electromechanical systems, electrochemical energy storage, and distribution can also help to achieve this goal.

Supplement S11.1.docx describes different types of energy sources such as nuclear, solar, and geothermal in more detail.

11.2.3 Environmental Impact

The environmental impacts of different types of energy sources are shown in Table 11.4. There are two systems to measure the weight of importance of different types of environmental impacts. These are listed in Table 11.5.

TABLE 11.4 Environmental Impacts of Energy Types

Energy Type	Environmental Impacts
Biomass power	Air emissions and solid waste
Coal power	Air emissions and solid waste
Hydro power	Wildlife impacts
Natural gas power	Air emissions
Nuclear power	Radioactive waste and accidental spillage
Oil power	Air emissions and solid waste
Solar power	No significant impacts
Wind power	Wildlife impacts

Acidification is the decrease in the pH in the world's oceans. The result is more acid in Earth's bodies of water. This affects the ecology of the ocean and its inhabitants. Many fish and oceanic plants are affected by the minutest changes in acidity in their environment. Nitrification involves adding more agricultural nutrients to land and bodies of water. Examples of nitrification include adding fluoride to tap water and iodine to table salt. Solid wastes are the reactants that are left over after a chemical reaction occurs in an energy source. An example of solid waste is the nuclear waste left over from nuclear reactors.

11.2.4 Energy Measurement

In the United States, the standard unit of measurement for energy is the British thermal unit (BTU), but most other countries use the joule. A BTU is defined as the energy required to increase the temperature of one pound mass (lbm) of water from 39 to 40°F. For example, a small air conditioner or heater may use 1000 BTU/h. It is estimated that one person in the United States uses 10^6 BTU a year. Note that 10^{15} BTU is known as a quad. One BTU is equivalent to 1055 J where a joule is the energy needed to move an object one meter using one newton of force. The conversion of BTU into electric power is

$$1 \text{ kWh} = 3412 \text{ BTU} = 3.6 \times 10^6 \text{ J}$$

When BTU is used as a measure of power, the actual unit is BTU per hour, but this is assumed and simply shortened to BTU. Consider the following simple example to find the cost of energy use. Suppose that an appliance uses 2000 BTU of electricity per hour. The

TABLE 11.5 Weights of Importance of Various Environmental Impacts Caused by Energy Production

Environmental Impact Factors	EPA Weight	Harvard Weight
Global warming	0.27	0.28
Acidification	0.13	0.17
Nitrification	0.13	0.18
Natural resource depletion	0.13	0.15
Indoor air quality	0.27	0.12
Solid waste	0.07	0.10
Total	1.0	1.0

cost of electricity is $0.30 per kilowatt-hour. How much does it cost to use the appliance for 5 h?

$$\left(\frac{2000 \text{ BTU}}{h}\right)\left(\frac{1 \text{ kWh}}{3412 \text{ BTU}}\right)\left(\frac{\$0.30}{1 \text{ kWh}}\right) \times 5 \text{ h} = \$0.88$$

11.3 MULTICRITERIA ENERGY DECISIONS

11.3.1 Seven Principles for Energy Sustainability

Several factors are involved when considering the selection of an energy type. The first and most important factor is the cost per unit of energy. The second factor is the rate at which the energy source can be converted to become available for energy use. The third factor is the renewability, or the rate at which a unit of energy can be replaced. This factor may not be as important to individual users, but it has a much larger impact on the global level. The renewability of fossil fuels is very slow, as it takes millions of years for them to be renewed. Conversely, the renewability factor for wind and solar is very high. The final factor is the environmental impact of using a type of energy.

It is obvious that the solution to the energy problem is multifaceted and one should make some sacrifices to achieve energy sustainability. We propose the following seven principles for energy sustainability. But before we present these principles, we need to define the energy renewability cycle. The renewability cycle refers to the period of time that the consumed energy is replenished by nature. Figure 11.7 shows the renewability cycles for several alternative energy sources. According to the following energy principles, shorter cycle energy sources such as wind, solar, and biofuel should be used.

Seven Principles of Sustainability We define the following seven principles to characterize sustainable and efficient energy systems:

1. *Energy Renewability Cycle* All energy sources have a cycle of renewal; see Figure 11.7. To conserve energy, energy sources with shorter renewability cycles should be used.

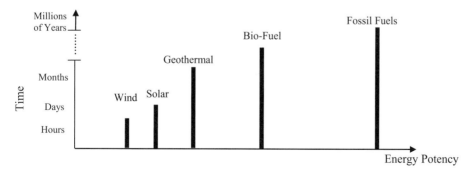

Figure 11.7 Comparison of potency of different energy options vs. their renewability cycle.

2. *Hybrid Systems*   Hybrid energy systems consist of several different sources of energy, energy generators, energy savers, and users of energy. Hybrid energy systems should be used to optimize energy conservation.

3. *Energy-Efficient Behavior*   Energy usage can be substantially reduced by changing human behaviors and preferences (this could be as simple as turning off lights or cars when not in use, or based on automation by the use of sensors and intelligent systems to minimize the use of energy).

4. *Energy Sustainability Equilibrium*   Energy sustainability equilibrium can be achieved when the energy renewal cycle is less than or equal to the energy usage cycle. Reduce the energy use until such equilibrium is achieved.

5. *Optimal Layout of Energy Systems*   Design energy systems and their components (generators, savers, and users) in such a way that generated energy is fed into energy users with the least energy loss (see the techniques developed in Chapters 9, 13, 15, and 16).

6. *Just-in-Time Energy Systems*   Apply the just-in-time (pull) system instead of the prevalent wasteful push system in the design and operation of energy systems; for example, generate energy as needed (see Chapter 5).

7. *Distributed Energy Systems*   Apply distributed system principles for connecting different energy systems for achieving efficient distribution and optimal usage of energy.

11.3.2 Bicriteria Energy Selection

Though in general fossil fuels are cheaper than other types of energy, they are also the least renewable and have the highest adverse environmental impact. The fact that each type of energy has positive and negative attributes makes it necessary to develop a multicriteria approach for energy selection.

Consider the following bicriteria problem:

Minimize $f_1 =$ cost per unit of energy

Minimize $f_2 =$ carbon dioxide (CO_2) emissions per unit of energy

Inefficient alternatives can be identified. Note that an inefficient alternative will never be the best alternative but it can be the second best alternative; therefore, when ranking a set of alternatives, inefficient alternatives should not be eliminated. An additive utility function can be used to rank different resources of energy, where different users may have different weights resulting in different rankings of alternatives.

For more complex cases, the environmental impact index of the energy source can be used instead of the carbon dioxide emissions, as it includes these. In order to calculate this, the DM must first decide whether to use the U.S. Environmental Protection Agency (EPA) or the Harvard weighting scale as presented in Section 11.2.3. If the environmental impact index is not available, CO_2 emissions may be used in its place. The emissions values for several different types of energy are listed in Table 11.6.

Note that biofuel carbon emissions can vary greatly depending on the chemical composition of the fuel.

TABLE 11.6 CO_2 Emissions from Energy Sources

Resource	Oil	Natural gas	Coal	Wind	Solar	Biofuel	Nuclear
CO_2 emissions	164,000 lb per billion BTU	117,000 lb per billion BTU	208,000 lb per billion BTU	0 lb per billion BTU	0 lb per billion BTU	80,000 lb per billion BTU	0 lb per billion BTU

Source: U.S. Department of Energy.

TABLE 11.7   Set of Alternatives for Example 11.1

Resource	Natural gas	Coal	Wind	Solar	Biofuel	Nuclear
Min. f_1 (cost) per million BTU	$19.37	$27.78	$28.42	$61.75	$32.97	$33.38
Min. f_2 (environmental impact)	0.5	0.7	0.2	0.1	0.4	0.9
Efficient?	Yes	No	Yes	Yes	No	No

Example 11.1   Bicriteria Approach for Energy Selection   Consider six common types of energy with two criteria values given in Table 11.7.

In Table 11.7, the given costs (dollars per million BTU) are rough estimates of the total cost to generate electricity for each energy resource; these costs include all costs that must be paid to generate energy, including capital, operations and maintenance, and fuel. These costs are used in this example to demonstrate the bicriteria approach. The environmental impact index is on a 0–1 scale where 0 is no environmental impact (best) and 1 is the maximum impact (worst). The set of alternatives given in Table 11.7 are presented in Figure 11.8. Note that coal, biofuel, and nuclear are inefficient, as they are dominated by either wind or natural gas where both objectives are minimized.

11.3.3   Multicriteria Energy Selection

The multicriteria energy problem can be viewed differently by different DMs. For example, the government of a country may put more emphasis on criteria such as future availably of

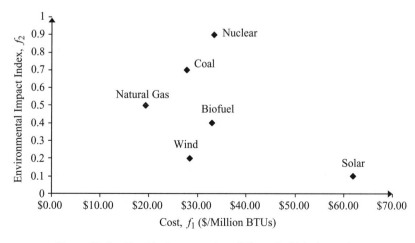

Figure 11.8   Graphical presentation of Example 11.1 alternatives.

resources and their environmental impacts. But an individual DM may be more concerned with criteria such as cost rather than indirect consequences. Multicriteria energy selection problems can be defined as follows:

Minimize f_1 = cost per unit of energy, for example, in dollars per million BTU

Minimize f_2 = environmental impact per unit of energy, for example, 0 (the best) to 1 (the worst)

Maximize f_3 = availability/accessibility rate of energy, for example, 0 (the worst) to 1 (the best)

Maximize f_4 = renewability rate of energy source, for example, 0 (the worst) to 1 (the best)

The first two criteria were discussed in the last section. The availability/accessibility rate of energy, f_3, refers to how quickly a given energy source can become available for use. For example, natural gas has the best rate because it can be immediately converted to usable energy having a very high rate of output, but solar energy rate is low. The renewability f_4 refers to how long it would take for the resource to be naturally replenished. For example, if one gallon of oil is burned, it will take millions of years for that gallon to be replenished by nature; thus its renewability rate is very low (0). Multicriteria energy alternatives can be evaluated by the following additive utility function:

$$\text{Minimize } U = w_1 f_1' + w_2 f_2' + w_3 f_3' + w_4 f_4'$$

where f_i' is the normalized (between 0 and 1) values of f_i.

Example 11.2   Multicriteria Ranking of Energy Sources   Consider six common types of energy with values provided in Table 11.8. The cost and environmental impact indexes are the same as in Example 11.1. The availability rate and renewability rate are given between 0 and 1. Minimize $U = 0.5f_1' + 0.3f_2' + 0.1f_3' + 0.1f_4'$.

TABLE 11.8   Set of Alternatives for Example 11.2

Resource, minimize	Natural gas	Coal	Wind	Solar	Biofuel	Nuclear
Min. f_1, cost($/million BTU)	$19.37	$27.78	$28.42	$61.75	$32.97	$33.38
Min. f_2, environmental impact	0.5	0.7	0.2	0.1	0.4	0.9
Max. f_3, availability/accessibility	1	0.9	0.7	0.5	0.6	0.95
Max. f_4, renewability of energy	0	0	0.6	0.5	0.75	0
Min. f_1'	0	0.20	0.21	1	0.32	0.33
Min. f_2'	0.50	0.75	0.13	0	0.38	1
Min. f_3'	0	0.20	0.60	1	0.80	0.10
Min. f_4'	1	1	0.20	0.33	0	1
Min. U	0.25	0.44	0.22	0.63	0.35	0.58
Efficient?	Yes	No	Yes	Yes	Yes	No
Rank	2	4	1	6	3	5

Nuclear Energy: The Most Controversial Yet the Most Promising Energy Source The current nuclear power is based on fission power (vs. the future fusion nuclear power). In the above example we assumed that renewability of fission nuclear power is 0; but one may argue that fusion nuclear power will have a very high renewability rate (close to 1) depending on future scientific discoveries and technologies. Also in the above example we assumed that the environmental impact of fission nuclear power is worse than all other sources of energy, at a 0.9 rating. This environmental impact of fission nuclear power can be drastically improved with the development of new technologies and storage methods for nuclear waste, which may make nuclear power the most environmentally friendly energy source. Because there appears to be no end to the insatiable human need for energy, nuclear power may be the best (and, perhaps, the only) friend that we have in the future.

See Supplement S11.2.xls.

11.4 ROUTING AND PROCUREMENT IN ENERGY SYSTEMS

In distributed networks, locations are connected to each other directly or through other locations, forming a network. For example, suppose that a package must be sent from a city (called the origin) to another city (called the destination). The package may be routed through a number of cities before reaching the destination. To minimize the energy used for transportation, the shortest path that connects the origin to the destination should be found. In energy systems, routing can be used to find the shortest path for sending generated energy from a generator through a network to a user. There are several algorithms for finding the shortest path; each method has advantages and disadvantages. In this section, Dijkstra's algorithm, the most used routing method, is covered for solving the routing problem. In addition to minimizing the total traveling distance, multicriteria routing can consider different objectives such as total travel time (by considering congestion patterns, or loads, of different paths) and traveling costs. Another method for finding the shortest path is the Bellman–Ford algorithm (which is not covered in this chapter). The advantage of the Bellman–Ford algorithm is that it has the rescaling characteristic. That is, more nodes can be added or removed from the distributed system (i.e., it produces solutions useful for designing open-ended distributed systems).

11.4.1 Shortest Route Algorithm[†]

The routing problem is concerned with finding the shortest path from a given source (origin) to a given destination. For example, in Figure 11.10, the shortest path from *B* to *E* is *B–C–E* or *B–C–D–E* with a total cost of 8. Dijkstra's algorithm finds the shortest path from one node to all other nodes in the network. In each iteration, this algorithm finds a node which has not yet been explored and has the least cost path from the source. Next, the shortest paths for each of the node's neighbors are updated. The algorithm tracks the path which will result in the least cost path (or the shortest distance path). When all nodes are explored, the algorithm stops. Dijkstra's algorithm deals with a static case, where the link costs do not change over time.

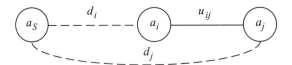

Figure 11.9   Presentation of notation.

Status of Node   The status of a node can be either temporary or permanent. At the beginning of the algorithm, all nodes have temporary status. A node's status becomes permanent when the least cost path from the source node to the given node is identified and verified. The path that connects the source to a given permanent node is the best path to the given node.

The notation used in this algorithm is presented in Figure 11.9:

a_S = source node for all distance calculations

a_j = jth node in the network

d_j = current calculated distance from the source node a_S to node a_j

u_{ij} = weight (or cost) of link between neighboring (adjacent) nodes a_i and a_j

$P(j)$ = predecessor of node j

N = set of nodes to which shortest paths from a_S have been evaluated (but not necessarily finalized)

M = set of nodes to which the shortest paths from a_S have not been evaluated

1. Initialization: N is empty and M contains all nodes in the network. Set the distance of the source node to zero and all other distances to infinity. Set

$$d_j = \begin{cases} 0 & \text{if } a_j = a_S \\ +\infty & \text{otherwise} \end{cases}$$

   Also set

$$P(j) = \begin{cases} j & \text{if } a_j = a_S \\ \text{undefined} & \text{otherwise} \end{cases}$$

2. Among all the nodes in set M, identify the node a_i to which the current distance from a_s is the shortest (i.e., $d_i = \text{Min}[d_i \mid a_i \in M]$). Remove node a_i from set M and add it to set N.

3. Update distances: Identify from set M the neighboring nodes of a_i (as found in the previous step) and update the distance from a_S to these nodes; that is,

   If $d_i + u_{ij} < d_j$, then replace d_j with $d_i + u_{ij}$ and set $P(j) = i$ for all j where a_j is a neighboring node of a_i.

4. Check the termination condition: Check if the set M is empty. If not, go to step 2. Otherwise, stop.

Note that in step 3 a_j is a neighboring node of node a_i. It is possible that a_j may belong to set N, which means that d_j (the shortest path from a_S to the node a_j) is less than or equal to $d_i + u_{ij}$. Therefore, d_j will not be changed in this step. As a result, there is no need to consider any neighbor of a_i belonging to set N.

Applications of Dijkstra's Algorithm for Solving Internet Routing Problems
Dijkstra's algorithm can be used to find the shortest path from one Internet node to other Internet nodes. It is applied for solving some of the routing problems in the Internet networks.

Supplement S11.3.docx describes the distributed link state protocol and its multiobjective.

Bellman–Ford Routing Method
Supplement S11.4.docx describes the Bellman–Ford routing approach and its multiobjective.

Example 11.3   Dijkstra's Algorithm for a Five-Node Network   Consider the network presented in Figure 11.10. Use Dijkstra's algorithm to find the shortest path from node A to all other nodes in this network.

Before solving the problem by Dijkstra's algorithm, observe that this is a small problem for which all paths can be easily enumerated to determine the shortest paths from A to all other nodes. For example, the shortest path from A to E is A–D–E with a total cost of $9 + 1 = 10$. Solving this problem by Dijkstra's algorithm is shown below.

Iteration 0

1. Initialize the system; $N = \{\ \}$, $M = \{A, B, C, D, E\}$. The distance from node A to all other nodes is ∞, while the distance of node A to itself is 0. That is, from node A, $d_A = 0$, $d_B = d_C = d_D = d_E = \infty$. The predecessor of node A is itself whereas the predecessor for all other nodes is ∞. That is, at node A, $P(A) = A$, but $P(B) = P(C) = P(D) = P(E) = \infty$.
2. Select the node with the shortest distance to the source node A, which is node A itself. In this case, node A becomes permanent.
      Now, set $N = \{A\}$ and $M = \{B, C, D, E\}$.

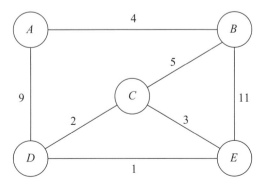

Figure 11.10   Initial network configuration for Example 11.3.

3. The nodes in set N (only node A) have neighbor nodes B and D. That is, B and D are neighbor nodes to node A. From Figure 11.10, it can be seen that link A–B is 4 and link A–D is 9. Therefore, calculate the total link cost from A to B and from A to D:

$$d_B = \text{Min}(d_B, d_A + u_{AB})$$
$$= \text{Min}(\infty, 0 + 4) = 4$$
with $P(B) = A$ (i.e., predecessor of B is A)

$$d_D = \text{Min}(d_D, d_A + u_{AD})$$
$$= \text{Min}(\infty, 0 + 9) = 9$$
with $P(D) = A$ (i.e., predecessor of D is A)

To determine the next permanent node (i.e., the node through which the minimum cost is achieved), find

$$d_j = \text{Min}(d_B, d_D) = \text{Min}(4, 9) = 4 = d_B \quad P(B) = A$$

Therefore, link A–B is the least cost path. Node B is added to set N. The table below shows the initial iteration of Dijkstra's algorithm.

Iteration	Permanent Nodes, N	Temporary Nodes, M	B Cost, Predecessor, $d_B, P(B)$	C Cost, Predecessor, $d_C, P(C)$	D Cost, Predecessor, $d_D, P(D)$	E Cost, Predecessor, $d_E, P(E)$
0	$\{A\}$	$\{B, C, D, E\}$	4, A	∞, ∞	9, A	∞, ∞

4. Since $M \neq \{\cdot\}$, the termination condition is not met. Consider the next iteration.

Iteration 1

2. In Iteration 0, node B had the smallest cost; therefore, B was removed from set M and was added to set N. That is, $N = \{A, B\}$ and $M = \{C, D, E\}$.
3. Node B has neighbor nodes C and E. The least cost paths to these nodes are calculated below. Note that d_D was determined in the previous iteration.

$$d_C = \text{Min}(\infty, 4 + 5) = 9 \quad P(C) = B$$
$$d_E = \text{Min}(\infty, 4 + 11) = 15 \quad P(E) = B$$

The next permanent node must be chosen. Since both nodes C and D have the same cost, (i.e., $d_C = d_D = 9$), then either node C or D can be chosen arbitrarily. Suppose node D is selected:

$$d_j = \text{Min}(d_C, d_D, d_E) = \text{Min}(9, 9, 15) = 9 = d_D \quad P(D) = A$$

The table below shows the current summary of Dijkstra's algorithm.

Iteration	Permanent Nodes, N	Temporary Nodes, M	B Cost, Predecessor, $d_B, P(B)$	C Cost, Predecessor, $d_C, P(C)$	D Cost, Predecessor, $d_D, P(D)$	E Cost, Predecessor, $d_E, P(E)$
0	$\{A\}$	$\{B, C, D, E\}$	4, A	∞, ∞	9, A	∞, ∞
1	$\{A, B\}$	$\{C, D, E\}$		9, B	9, A	15, B

4. Since $M \neq \{\ \}$, the termination condition is not met. Consider the next iteration.

Iteration 2

2. In Iteration 1, D is selected to become permanent and leave set M; it is added to set N. That is, $N = \{A, B, D\}$ and $M = \{C, E\}$.
3. The neighboring nodes of D in set M are C, E. The least cost path to node C is through B:

$$d_C = \text{Min}(9, 9 + 2) = 9 \quad P(C) = B$$

The cost to node E via node B is 15. The path via node D is 10:

$$d_E = \text{Min}(15, 9 + 1) = 10 \quad P(E) = D$$

Node C is selected to be added to N:

$$d_j = \text{Min}(d_C, d_E) = \text{Min}(9, 10) = d_C$$

4. Since $M \neq \{\cdot\}$, the termination condition is not met. Consider the next iteration.

Iteration 3

2. In this iteration, $N = \{A, B, D, C\}$ and $M = \{E\}$.
3. The only neighbor of node C in set M is E and it is updated as follows:

$$d_E = \text{Min}(10, 9 + 3) = 10 \quad P(E) = D$$

Since node E is the only remaining node, it will be added to N:

$$d_j = \text{Min}(d_E) = \text{Min}(10) = d_E$$

4. Since $M = \{\cdot\}$, the termination condition is met. The algorithm terminates.
   The results of Dijkstra's algorithm for each iteration are shown in Figure 11.11 and the steps are summarized in Table 11.9.

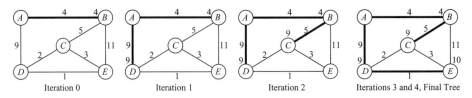

Figure 11.11   Network configuration after completion of link state algorithm.

TABLE 11.9   Summary of Dijkstra's Algorithm

Iteration	Permanent Nodes, N	Temporary Nodes, M	B Cost, Predecessor, $d_B, P(B)$	C Cost, Predecessor, $d_C, P(C)$	D Cost, Predecessor, $d_D, P(D)$	E Cost, Predecessor, $d_E, P(E)$
0	$\{A\}$	$\{B, C, D, E\}$	$4, A$	∞	$9, A$	∞
1	$\{A, B\}$	$\{C, D, E\}$		$9, B$	$9, A$	$15, B$
2	$\{A, B, D\}$	$\{C, E\}$		$9, B$		$10, D$
3	$\{A, B, D, C\}$	$\{E\}$				$10, D$
4	$\{A, B, D, C, E\}$	$\{\ \}$				

Finding Optimal Path from Table   The optimal path from node A to any given node can be identified from Table 11.9. For example, to find the optimal path to C, see the column associated with C in Table 11.9. Its predecessor is node B. Now, see the column associated with B in Table 11.9. Its predecessor is node A. Since node A is the source, stop backtracking. The best path from node A to node C is A–B–C. Similarly, to find the best path from node A to node E, see the column associated with E in Table 11.9. Its predecessor is node D. Now, see the column associated with D in Table 11.9. Its predecessor is node A. Since node A is the source, stop backtracking. The best path from node A to node E is A–D–E.

Alternate Solution   In iteration 1, both C and D had the same link cost. Suppose node C was selected instead of node D. Table 11.10 shows the alternative solution.

TABLE 11.10   Summary of the Alternative Solution of the Dijkstra's Algorithm

Iteration	Permanent Nodes, N	Temporary Nodes, M	B Cost, Predecessor, $d_B, P(B)$	C Cost, Predecessor, $d_C, P(C)$	D Cost, Predecessor, $d_D, P(D)$	E Cost, Predecessor, $d_E, P(E)$
0	$\{A\}$	$\{B, C, D, E\}$	$4, A$	∞	$9, A$	∞
1	$\{A, B\}$	$\{C, D, E\}$		$9, B$	$9, A$	$15, B$
2	$\{A, B, C\}$	$\{D, E\}$			$9, A$	$12, C$
3	$\{A, B, D, C\}$	$\{E\}$				$10, D$
4	$\{A, B, D, C, E\}$	$\{\ \}$				

11.4.2   Energy Procurement in Distributed Systems[†][*]

In energy systems, each user can purchase the needed energy from any source. The routing problem discussed in the last section can be modified to find the best energy source. The routing problem formulation will be slightly changed:

- Label the source node as the user and all other destinations as the energy generators.
- Identify the amount of energy needed by the user and the capacity of generated energy for each generator for the given period.
- Solve the routing problem by a routing algorithm (e.g., use the method of the last section).
- For each destination, add the purchasing price of the energy generator to its routing cost.
- Purchase from the generators with the lowest prices to fulfill the needed user's energy requirements. If the capacity of the generator with the lowest price is completely used, then use the next best generator.

Operations of Distributed Systems   In distributed systems, there will be many users that may purchase energy from the same energy generator. To apply the algorithm, update the network information as soon as a user purchases the given energy. The purchasing priority will be given to the first requester (bidder) to purchase the needed energy. In distributed networks, all energy capacities and routing costs can be changed by the generator at any time. For this problem, the information about new prices and capacities of generators will be periodically broadcasted to all nodes (users), and therefore periodically each user must solve the routing and the purchasing problem by using the latest information.

Example 11.4   Energy Procurement Algorithm for a Five-Node Network   Consider the network presented in Figure 11.12. Suppose that node A is the user and nodes B, C, D, and E are the energy generators. The selling price and the capacity of each generator for a given period is given above each node in Figure 11.12. The user requires 8500 units of energy for the given period. Find the purchasing plan for node A.

Consider the solution given in Table 11.9 for the routing costs from each node to A. The steps of the method are shown in Table 11.11.

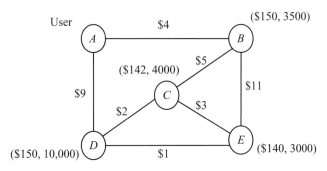

Figure 11.12   Initial network configuration for Example 11.4.

TABLE 11.11 Summary of Finding Purchase Plan where Node A is User

Energy Generator	Routing Cost	Purchase Cost	Total Cost	Ranking	Energy Capacity	Purchase Plan	Remaining Energy
B	4	150	154	3	3,500	1,500	2,000
C	9	142	151	2	4,000	4,000	0
D	9	150	159	4	10,000	0	10,000
E	10	140	150	1	3,000	3,000	0
Total	—	—	—	—	20,500	8,500	12,000

11.4.3 Multicriteria Energy Routing[†*]

Consider the following four objectives for the energy routing problem:

Minimize f_1: total energy of transportation
Minimize f_2: total time of transportation
Minimize f_3: total cost of transportation
Minimize f_4: total environmental impact of transportation

Each link connecting a pair of nodes may have these criteria values associated with it. From the source to the destination, different paths are associated with different values of the above objectives. Generally, these objectives are conflicting. Alternatives are defined as the shortest paths (in terms of the multiobjective metric) that connect the source and destination nodes. The process of choosing the best alternative (the multicriteria path) is based on the DM's preference using a utility function.

Multicriteria Composite Approach In this method, an additive utility function is used to rank alternatives. The utility function is evaluated by the normalized weight approach. Normalizing weights requires fewer computations than normalizing objectives and is especially useful for distributed and computer routing problems. In the normalized weight approach, the minimum values of all objectives for all links are assumed to be zero (i.e., set $f_{i,\min} = 0$). Note that the actual $f_{i,\min}$ may not be equal to zero, but for practical purposes it is used to simplify the procedure and calculations.

Consider each link that directly connects nodes p and q presented by (p, q). Let the value of criterion (objective) i be presented by $f_{i,pq}$ for $i = 1, 2, \ldots, k$. Assume the minimum value of all links (p, q) for all objectives is zero:

$$f_{i,\min} = 0 \quad \text{for } i = 1, 2, \ldots, k$$

The following is a summary of the steps used to solve a multicriteria routing problem.

1. Find the maximum value for all links (p, q) for all objectives; label as

$$f_{i,\max} = \text{Max}\{f_{i,pq} \text{ for all } (p, q)\} \quad \text{for } i = 1, 2, \ldots, k$$

2. Assess the desired weights for the normalized values of the objectives ($w_1, w_2, \ldots, w_k$). Find the normalized weights ($w_1', w_2', \ldots, w_k'$) using the equation below, where w_i' is the normalized weight:

$$w_i' = \frac{w_i}{f_{i,\max}} \quad \text{for } i = 1, 2, \ldots, k$$

3. Find the composite utility value z_{pq} for each link (p, q):

$$z_{pq} = \sum_{i=1}^{k} w_i' f_{pq} = w_1' f_{1,pq} + w_2' f_{2,pq} + \cdots + w_k' f_{k,pq} \qquad (11.1)$$

4. Use the composite utility values z_{pq} to solve the routing problem.

Example 11.5 MCDM Using Dijkstra's Algorithm   Consider Example 11.3 with two criteria as presented in the network in Figure 11.13a. Suppose the weights of importance for the normalized objective values are $w_1 = 0.6$ and $w_2 = 0.4$.

Use the multicriteria Dijkstra algorithm to find the shortest path from node A to all other nodes in the network. The solution follows.

Initial Phase: Finding Normalized Weights

1. By examining the network in Figure 11.13a, it can be seen that the maximum values (considering all links) are $f_{1,\max} = 11$ and $f_{2,\max} = 98$.
2. Find the normalized weights. Using $w_i' = w_i/f_{i,\max}$, the normalized weights are

$$w_1' = \frac{0.6}{11} = 0.055 \qquad w_2' = \frac{0.4}{98} = 0.004$$

3. Now use the additive utility function to generate the composite multicriteria link. For example, the composite utility value for link A–B for the given normalized weights (0.055, 0.004) is calculated below:

$$z_{AB} = w_i' f_{1,AB} + w_i' f_{2,AB} = (0.055)(4) + (0.004)(65) = 0.480$$

The composite multicriteria values for all links are shown in Figure 11.13b.

4. Now solve this network problem (as given in Figure 11.13b) by Dijkstra's algorithm.

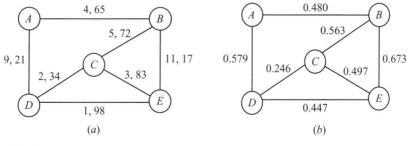

Figure 11.13   (a) Network for bicriteria problem; (b) composite network for weights $W = (0.6, 0.4)$.

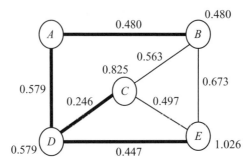

Figure 11.14 Multicriteria network and tree created by multicriteria Dijkstra algorithm for node A.

The results of applying Dijkstra's algorithm are shown in Figure 11.14 and Table 11.12. The optimal path from A to E is A–D–E with a total utility of $0.579 + 0.447 = 1.026$. Since utilities are minimized, path A–D–E is associated with the minimum utility. The two objective values for this path are

$$f_1 = 9 + 1 = 10 \qquad f_2 = 21 + 98 = 119$$

The solutions for all paths are presented in Table 11.12.

Example 11.5 Where Node E is Source Consider Example 11.5, but show the solution when node E is the source node. The same normalized values as in Figure 11.13b are used. The solution is presented in Figure 11.15 and Table 11.13.

Generating All Efficient Alternatives The set of efficient alternatives for multicriteria routing can be generated by varying the weights of importance and generating their associated composite links by using Equation (11.1) and then solving the problem for the given weights.

Example 11.6 Generating a Set of Efficient Alternatives Consider the network in Figure 11.16. Generate a set of efficient alternatives to this problem where the source node is A and the destination node is E.

By varying the weights and solving the multicriteria routing problem for each of the given weights, four efficient alternatives can be generated for paths from A to E. See Table 11.14.

TABLE 11.12 Final Results of Multicriteria Link State Algorithm

Iteration	Permanent Nodes, N	Temporary Nodes, M	B Energy, Predecessor, $B, P(B)$	C Energy, Predecessor, $C, P(C)$	D Energy, Predecessor, $D, P(D)$	E Energy, Predecessor, $E, P(E)$
0	$\{A\}$	$\{B, C, D, E\}$	0.480, A	∞	0.579, A	∞
1	$\{A, B\}$	$\{C, D, E\}$		1.043, B	0.579, A	1.153, B
2	$\{A, B, D\}$	$\{C, E\}$		0.825, D		1.026, D
3	$\{A, B, D, C\}$	$\{E\}$				1.026, D
4	$\{A, B, D, C, E\}$	$\{\ \}$				

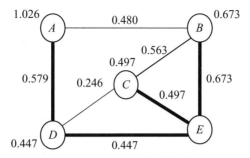

Figure 11.15   Multicriteria network and tree created by multicriteria Dijkstra algorithm at node E.

TABLE 11.13   Final Results of Multicriteria Link State Algorithm

Iteration	Permanent Nodes,	Temporary Nodes,	A Energy, Predecessor,	B Energy, Predecessor,	C Energy, Predecessor,	D Energy, Predecessor,
	N	M	$A, P(A)$	$B, P(B)$	$C, P(C)$	$D, P(D)$
0	$\{E\}$	$\{A, B, C, D\}$	∞	0.673, E	0.497, E	0.447, E
1	$\{D, E\}$	$\{A, B, C\}$	1.026, D	0.673, E	0.497, E	
2	$\{C, D, E\}$	$\{A, B\}$	1.026, D	0.673, E		
3	$\{B, D, C, E\}$	$\{A\}$	1.026, D			
4	$\{A, B, D, C, E\}$	$\{\ \}$				

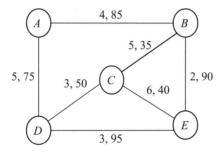

Figure 11.16   Network for Example 11.6.

TABLE 11.14   Set of Efficient Paths from Node A to Node E

Actual weights, w	(0, 1)	(0.05, 0.95)	(0.1, 0.9)	(1, 0)
$(f_{1,max}, f_{2,max})$	(6, 95)	(6, 95)	(6, 95)	(6, 95)
Normalized weights, w'	(0, 0.011)	(0.008, 0.010)	(0.017, 0.009)	(0.167, 0)
Best path	$A–B–C–E$	$A–D–C–E$	$A–D–E$	$A–B–E$
f_1	15	14	8	6
f_2	160	165	170	175

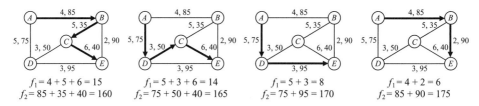

$$f_1 = 4 + 5 + 6 = 15$$
$$f_2 = 85 + 35 + 40 = 160$$

$$f_1 = 5 + 3 + 6 = 14$$
$$f_2 = 75 + 50 + 40 = 165$$

$$f_1 = 5 + 3 = 8$$
$$f_2 = 75 + 95 = 170$$

$$f_1 = 4 + 2 = 6$$
$$f_2 = 85 + 90 = 175$$

Figure 11.17   Set of all efficient paths from A to E for Example 11.6.

These four efficient alternatives are presented in Figure 11.17.

11.5   OPTIMIZATION OF ENERGY SYSTEMS

In this section, several models for solving energy system problems will be presented. The first is called energy system optimization, which is a linear programming model that can be used to represent any energy system. The second model is called energy design optimization, which is used to choose the best design for a given system. The third and fourth models discuss solving multiperiod problems. The fifth model presents the multiobjective optimization problem. All models are based on the model presented in Section 11.5.1.

11.5.1   Energy Operations Optimization

An energy system is defined based on the relationship between three energy entities: users, savers, and generators. A user is an entity that consumes energy for certain purposes. Examples of users are cars, home appliances, and factory machinery. A user can receive energy from either a generator or a saver. A saver is a device that is able to store energy for future use. If the rate of energy production is higher than the rate of energy consumption, it can be saved in the saver for later use. A generator is a device that generates energy. For fossil fuels this may be a combustion engine, whereas for wind energy this is likely a wind turbine. These three entities of energy systems can be used to describe a variety of energy applications such as automobiles, houses, and the power grid.

An energy network can have a number of generators, savers, and users. These devices are connected to each other by links. The goal is to satisfy the demand of each user and save some energy for future use. It is assumed that the generators and the savers can satisfy the total demand. There are costs associated with generating energy at a given generator as well as transportation costs through each link. There are also capacity constraints for generators and savers. A generator can only produce a certain amount of energy at a given time while a saver is used to store a certain amount of energy at a given time. A saver may also need a minimum amount of energy at all times for future use.

An energy system is a closed system where all users, generators, and savers are coordinated to minimize the total cost and to satisfy the needed demands of the users. The energy saved by the savers is paid for by the users of the system for future use. The system is defined for one period given that the energy savings of the past are available for the user. The savings at the end of the period are left for the next period's use. The amount in the saver is set depending on the prediction of the next period's energy prices. For example, if it is predicted that the generator's prices will be higher in the next period, then more energy is stored in savers for future use and vice versa. Here, Δs_p is defined as the predicted change in the future price of energy.

Hub Property of Savers   The saver can act as a hub in which all input energy from the generator is transported to the user without saving energy in the saver.

Supply-and-Demand Behavior of Savers   Depending on the price of the energy in the future, it may be economical to save or release energy from savers. If the future price of energy will be higher, then more energy should be saved in the saver now. On the other hand, if the future price of energy will be lower, then the saver should sell as much energy as it can now, and the energy level of the saver should be set to the minimum.

This section presents an optimization approach for the operation of energy systems. The method utilizes linear programming (LP) to identify:

- The optimal amount of energy that should be generated by each energy generator
- The optimal amount of energy that should be stored in each energy saver for future use
- The optimal amount of energy that should be transported on each link

In this section, single-period energy operations optimization is covered.

The decision variables and costs for formulating the energy planning problem are presented in Table 11.15. By using the notation provided in Table 11.15, the energy systems problem for a given period is formulated as follows.

TABLE 11.15   Notation Used for Energy Systems

K_i: Maximum capacity of generator i
D_j: Energy demand of user j
g_i: Cost per unit of energy for generator i

I: Total number of generators $i = 1, \ldots, I$
J: Total number of users $j = 1, \ldots, J$
P: Total number of savers $p = 1, \ldots, P$

x_{ij}: Amount of energy from generator i to user j
c_{ij}: Energy transportation cost from generator i to user j

$q_{p,0}$: Initial energy of saver p
$q_{p,\max}$: Maximum capacity of saver p
$q_{p,\min}$: Minimum required energy of saver p
q_p: Energy level of saver p (to be determined)
Δs_p: Projected gain in dollars of energy prices for the next period for saver p

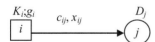

y_{ip}: Amount of energy from generator i to saver p
c_{ip}: Energy transportation cost from generator i to saver p

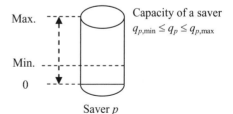

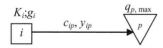

z_{pj}: Amount of energy from saver p to user j
c_{pj}: Energy transportation cost from saver p to user j

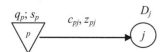

PROBLEM 11.1 ENERGY OPERATION OPTIMIZATION (FROM A BUYER'S AND SAVER'S POINT OF VIEW)

Minimize:

$$f_1 = \sum_{i=1}^{I} \sum_{j=1}^{J} c_{ij} x_{ij} \qquad \text{(transportation cost from generators to users)} \qquad (11.2a)$$

$$+ \sum_{i=1}^{I} \sum_{p=1}^{P} c_{ip} y_{ip} \qquad \text{(transportation cost from generators to savers)} \qquad (11.2b)$$

$$+ \sum_{p=1}^{P} \sum_{j=1}^{J} c_{pj} z_{pj} \qquad \text{(transportation cost from savers to users)} \qquad (11.2c)$$

$$+ \sum_{i=1}^{I} g_i \sum_{j=1}^{J} x_{ij} \qquad \text{(cost of buying from generators to users)} \qquad (11.2d)$$

$$+ \sum_{i=1}^{I} g_i \sum_{p=1}^{P} y_{ip} \qquad \text{(cost of buying from generators to savers)} \qquad (11.2e)$$

$$- G \sum_{p=1}^{P} \Delta s_p q_p \qquad \text{(total gain for future saving)} \qquad (11.2f)$$

Subject to:

$$\sum_{i=1}^{I} x_{ij} + \sum_{p=1}^{P} z_{pj} \geq D_j \qquad \begin{array}{l} \text{for all } j = 1, \ldots, J \\ \text{(required demand of each user)} \end{array} \qquad (11.3)$$

$$\sum_{j=1}^{J} x_{ij} + \sum_{p=1}^{P} y_{ip} \leq K_i \qquad \begin{array}{l} \text{for all } i = 1, \ldots, I \\ \text{(production capacity of generators)} \end{array} \qquad (11.4)$$

$$q_{p,0} + \sum_{i=1}^{I} y_{ip} - \sum_{j=1}^{J} z_{pj} = q_p \qquad \begin{array}{l} \text{for all } p = 1, \ldots, P \\ \text{(energy level of savers)} \end{array} \qquad (11.5)$$

$$q_p \leq q_{p,\max} \qquad \begin{array}{l} \text{for all } p = 1, \ldots, P \\ \text{(maximum allowable level for savers)} \end{array} \qquad (11.6)$$

$$q_p \geq q_{p,\min} \qquad \begin{array}{l} \text{for all } p = 1, \ldots, P \\ \text{(minimum required level for savers)} \end{array} \qquad (11.7)$$

$$x_{ij} \geq 0 \qquad y_{ip} \geq 0 \qquad z_{pj} \geq 0 \qquad q_p = 0 \quad \text{for all } i, j, p \qquad (11.8)$$

where G is a given large positive number (e.g., set $G = 100,000$).

The objective function (11.2) minimizes the total cost of generating, transferring, and saving energy. The marginal gains or losses for energy savers are defined as follows:

- If $\Delta s_p > 0$, then the energy price of the next period will be higher. Therefore, save (or buy) energy for future use (i.e., q_p is maximized).
- If $\Delta s_p < 0$, then the energy price of the next period will be lower. Therefore, sell the existing energy in the saver (i.e., q_p is minimized).

The demand constraints (11.3) ensure that the total amount of energy being transferred to the users from the generators and the savers is greater than or equal to the energy demand by the users. The production constraints (11.4) show that the amount of energy being transferred from each generator to users and savers is not more than the maximum capacity of the given generator. Constraints (11.5) show the amount of energy that is present in the savers at the end of the period. This amount is equal to the amount of energy that is initially in the saver plus the amount that is transferred from the generators to the saver minus the amount that is transferred from the saver to users. Constraints (11.6) show that the amount of energy in saver p at the end of the period should be less than or equal to the maximum allowable capacity of the saver. Constraints (11.7) ensure that the amount of energy in saver p at the end of the period is more than or equal to the required minimum level of the saver for emergency purposes. Based on the prediction of energy prices, either more energy is saved in the saver or it is sold to result in less energy in the saver. The nonnegativity constraints (11.8) ensure that all amounts of energy generated, transferred, or saved will be nonnegative values. See Example 11.7.

Example 11.7   Energy Systems Operations Optimization   Consider Figure 11.18. In this system, there are four users, two generators, and two savers. In this example all nodes are directly connected except the generators (to each other), the users (to each other), and the savers (to each other). The demands at users 1, 2, 3, and 4 are 400, 450, 300, and 400 units of energy, respectively. The energy cost per unit of energy for generators 1 and 2 are $59.5 and $60, respectively (generating cost). The maximum capacities of generators 1 and

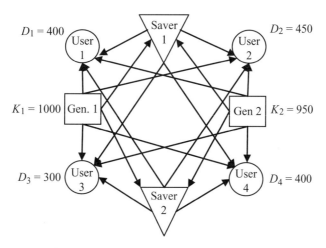

Figure 11.18   Energy System for Example 11.7.

TABLE 11.16 Data for Example 11.7

c_{ij}, Trans. Cost		User j				c_{pj}, Trans. Cost		User j			
from i to j		1	2	3	4	from p to j		1	2	3	4
Gen.	1	6	5	8	4	Saver	1	1	2	1	2
i	2	7	4	5	7	p	2	2	0.5	1.5	1
						Demand D_j		400	450	300	400

c_{ip}, Trans. Cost		Saver p			Gen.	Gen. Costs, g_i (\$)	Gen. Cap., K_i		Saver	Init. Enr $q_{p,0}$	Min. Enr $q_{p,min}$	Max. Enr $q_{p,max}$
from i to p		1	2									
Gen.	1	4.5	5		1	59.5	1000		1	100	60	200
i	2	6	4		2	60	950		2	150	24	240

Note: Enr = energy, Gen. = generator, Trans. = transportation, and Cap. = capacity.

2 are 1000 and 950 units of energy, respectively. The initial energy levels in savers 1 and 2 are 100 and 150 and the maximum capacity of savers 1 and 2 are 200 and 240 units of energy, respectively. The minimum capacity of savers 1 and 2 are 60 and 24 units of energy, respectively. The data for this problem are presented in Table 11.16. Suppose that $\Delta s_1 = +2$ and $\Delta s_2 = +3$. The next period's energy price is predicted to be higher for savers 1 and 2 (i.e., buy energy or $\Delta s_p > 0$ for all savers).

Formulate the energy systems operation problem and then solve it to find the optimal solution that minimizes the total cost for the energy operation problem.

The LP formulation of this example based on Problem 11.1 is presented below:

Minimize:

$$f_1 = [6x_{11} + 5x_{12} + 8x_{13} + 4x_{14} + 7x_{21} + 4x_{22} + 5x_{23} + 7x_{24} + 4.5y_{11} + 6y_{21}$$

$$+ 5y_{12} + 4y_{22}] + [1z_{11} + 2z_{12} + 1z_{13} + 2z_{14} + 2z_{21} + 0.5z_{22} + 1.5z_{23} + z_{24}]$$

$$+ [59.5(x_{11} + x_{12} + x_{13} + x_{14}) + 60(x_{21} + x_{22} + x_{23} + x_{24})$$

$$+ 59.5(y_{11} + y_{12}) + 60(y_{21} + y_{22})] - 100,000[(2)q_1 + (3)q_2]$$

Subject to:
Demand constraints:

$$x_{11} + x_{21} + z_{11} + z_{21} \geq 400$$
$$x_{12} + x_{22} + z_{12} + z_{22} \geq 450$$
$$x_{13} + x_{23} + z_{13} + z_{23} \geq 300$$
$$x_{14} + x_{24} + z_{14} + z_{24} \geq 400$$

Capacity limits for energy:

$$x_{11} + x_{12} + x_{13} + x_{14} + y_{11} + y_{12} \leq 1000$$
$$x_{21} + x_{22} + x_{23} + x_{24} + y_{21} + y_{22} \leq 950$$

Energy level of saver:

$$100 + (y_{11} + y_{21}) - (z_{11} + z_{12} + z_{13} + z_{14}) = q_1$$
$$150 + (y_{12} + y_{22}) - (z_{21} + z_{22} + z_{23} + z_{24}) = q_2$$

Minimum required level for savers:

$$q_1 \geq 60$$
$$q_2 \geq 24$$

Maximum required level for savers:

$$q_1 \leq 200$$
$$q_2 \leq 240$$

Nonnegativity of variables:

$$x_{ij}, y_{ip}, z_{pj}, q_p \geq 0 \quad \text{for all } i = 1, 2,$$
$$j = 1, 2, 3, 4 \quad p = 1, 2$$

Now, solve this problem by using an LP computer package to find the optimal solution for all decision variables. The solution to this problem is as follows. Note that this problem has an alternative optimal solution. However, the total cost for both solutions are the same.

Gen. i / Energy Trans. x_{ij}	User j 1	2	3	4	Total	Gen. i / Energy Trans. y_{ip}	Saver p 1	2	Total	Saver p / Energy Trans. z_{pj}	User j 1	2	3	4	Total
1	0	0	0	400	400	1	500	0	500	1	400	0	0	0	400
2	0	450	300	0	750	2	0	90	90	2	0	0	0	0	0
Total	0	450	300	400	1150	Total	500	90	590	Total	400	0	0	0	400

This solution shows that generator 1 should generate 400 and 500 units of energy for user 4 and saver 1, respectively, and generator 2 should generate 450, 300, and 90 units of energy for users 2, 3, and saver 2, respectively. Saver 1 should provide 400 units of energy for user 1. In total, generator 1 generates $400 + 500 = 900$ units of energy and generator 2 generates $450 + 300 + 90 = 840$ units. At the end of the period, the energy levels of savers 1 and 2 are $q_1 = 200$ and $q_2 = 240$, respectively. The solution is presented in Figure 11.19. The objective function value f_1 and the total benefit of this problem is

$$
\begin{aligned}
\text{Total cost} = &[6(0) + 5(0) + 8(0) + 4(400) + 7(0) + 4(450) + 5(300) + 7(0) + 4.5(500) \\
&+ 6(0) + 5(0) + 4(90)] + [1(400) + 2(0) + 1(0) + 2(0) + 2(0) + 0.5(0) \\
&+ 1.5(0) + 1(0)] + [59.5(0 + 0 + 0 + 400) + 60(0 + 450 + 300 + 0) \\
&+ 59.5(500 + 0) + 60(0 + 90)] - (2)200 - (3)240 = \$110,740
\end{aligned}
$$

See Supplement S11.5.lg4.

Example 11.8   Energy Systems Operations Optimization   In Example 11.7, the future prices were higher than the current prices. But in this example consider the future prices are predicted to be lower for savers 1 and 2 (i.e., $\Delta s_p < 0$ for both savers). In this example, we use $\Delta s_1 = -3$ and $\Delta s_2 = -2$.

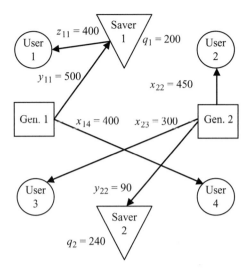

Figure 11.19 Solution for Example 11.7.

The solution is presented in Figure 11.20 where $q_1 = 60$, $q_2 = 24$, and $f_1 = \$89,367$.

Energy Trans. x_{ij}		User j					Energy Trans. y_{ip}		Saver p			Energy Trans. z_{pj}		User j				
		1	2	3	4	Total			1	2	Total			1	2	3	4	Total
Gen. i	1	0	0	0	400	400	Gen. i	1	534	0	534	Saver p	1	400	0	174	0	574
	2	0	450	0	0	450		2	0	0	0		2	0	0	126	0	126
	Total	0	450	0	400	850		Total	534	0	534		Total	400	0	300	0	700

See Supplement S11.6.lg4.

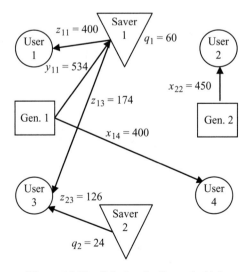

Figure 11.20 Solution for Example 11.8.

11.5.2 Energy Systems Design Optimization

In Section 11.5.1, optimal planning and operations of existing energy systems were discussed. In this section, the design of operation systems is presented. For example, in a house there are choices to buy energy generators, savers, and types of users (appliances). In this case, not only the operation costs but the purchase price costs should be considered. The optimization formulation of this problem is based on Problem 11.1 but includes the following modifications.

The additional cost of purchasing generators, savers, and users should be considered in the objective function (11.2). In the following, we show the formulation for purchasing generators. The same approach can be generalized for purchasing savers and users. Add to the objective function (11.2) the total cost of purchasing generators:

$$+ \sum_{i=1}^{I} c_i u_i \tag{11.9}$$

where u_i is a binary decision variable and c_i is the investment cost for purchasing generator i for $i = 1, \ldots, I$.

The constraints of energy systems design are the same as Problem 11.1 except that binary variables for each generator, saver, and user are added, where 1 means the item is purchased and 0 means it is not purchased. The updated generator constraints (11.10) are shown below. Constraints (11.10) replace constraints (11.4) in Problem 11.1, where u_i is a binary decision variable:

$$\sum_{j=1}^{J} x_{ij} + \sum_{p=1}^{P} y_{ip} \le K_i u_i \quad \text{for } i = 1, \ldots, I \quad \text{(generator investment constraints)} \tag{11.10}$$

where

$$u_i = \begin{cases} 1 & \text{if generator } i \text{ is chosen to be purchased} \\ 0 & \text{otherwise} \end{cases}$$

PROBLEM 11.2   ENERGY SYSTEM DESIGN OPTIMIZATION

Minimize objective function: (11.2) + (11.9)
Constraints (11.3), (11.5)–(11.8), and (11.10)

Example 11.9   Energy Systems Design for Purchasing Generators   Consider the problem presented in Figure 11.21.

Suppose that there is no generator and it is possible to buy one or more of three possible generators. The users and savers are existing entities as given in Example 11.7. The input parameters are given in Table 11.17.

Formulate the energy systems design problem, and find the optimal design and operation plan for this problem.

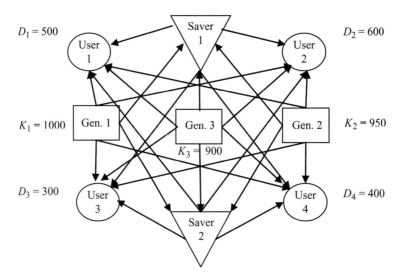

Figure 11.21 Energy systems design for Example 11.9.

TABLE 11.17 Information for Example 11.9

c_{ij}, Trans. Cost from i to j		1	2	3	4
Gen.	1	4	4	6	3
i	2	5	3	4.5	5
	3	5	4	3.5	3.5

c_{pj}, Trans. Cost from p to j		1	2	3	4
Saver p	1	3.5	3	2.5	4
	2	4	3.5	4	2.5
Demand D_j		500	600	300	400

c_{ip}, Trans. Cost from i to p		1	2
Gen.	1	1	2
i	2	2	0.75
	3	2	0.8

Gen.	Purchase Cost in $1000	Gen. Costs, g_i ($)	Gen. Cap., K_i
1	27	59.5	1000
2	26.95	59.75	950
3	27.3	59.25	900

Saver	Init. Enr $q_{p,0}$	Min. Enr $q_{p,min}$	Max. Enr $q_{p,max}$
1	60	60	300
2	24	24	240

Note: Enr = energy, Gen. = generator, Trans. = transportation, and Cap. = capacity.

The optimization problem formulation is presented below:

Minimize:

$$Z = [4x_{11} + 4x_{12} + 6x_{13} + 3x_{14} + 5x_{21} + 3x_{22} + 4.5x_{23} + 5x_{24} + 5x_{31} + 4x_{32}$$
$$+ 3.5x_{33} + 3.5x_{34} + 1y_{11} + 2y_{21} + 2y_{31} + 2y_{12} + 0.75y_{22} + 0.8y_{32}]$$
$$+ [3.5z_{11} + 3z_{12} + 2.5z_{13} + 4z_{14} + 4z_{21} + 3.5z_{22} + 4z_{23} + 2.5z_{24}]$$
$$+ [59.5(x_{11} + x_{12} + x_{13} + x_{14}) + 59.75(x_{21} + x_{22} + x_{23} + x_{24})$$
$$+ 59.25(x_{31} + x_{32} + x_{33} + x_{34}) + 59.5(y_{11} + y_{12})$$
$$+ 59.75(y_{21} + y_{22}) + 59.25(y_{31} + y_{32})] + [27{,}000\, u_1 + 26{,}950\, u_2 + 27{,}300\, u_3]$$

Subject to:

$$x_{11} + x_{21} + x_{31} + z_{11} + z_{21} \geq 500;$$
$$x_{12} + x_{22} + x_{32} + z_{12} + z_{22} \geq 600;$$
$$x_{13} + x_{23} + x_{33} + z_{13} + z_{23} \geq 300;$$
$$x_{14} + x_{24} + x_{34} + z_{14} + z_{24} \geq 400$$

$$x_{11} + x_{12} + x_{13} + x_{14} + y_{11} + y_{12} \leq 1000u_1$$
$$x_{21} + x_{22} + x_{23} + x_{24} + y_{21} + y_{22} \leq 950u_2$$
$$x_{31} + x_{32} + x_{33} + x_{34} + y_{31} + y_{32} \leq 900u_3$$

$$60 + (y_{11} + y_{21} + y_{31}) - (z_{11} + z_{12} + z_{13} + z_{14}) = q_1; \qquad q_1 \leq 300$$
$$24 + (y_{12} + y_{22} + y_{32}) - (z_{21} + z_{22} + z_{23} + z_{24}) = q_2; \qquad q_2 \leq 240$$

$$x_{ij}, y_{ip}, z_{pj}, q_p \geq 0 \quad \text{for all } i = 1, 2, 3; j = 1, 2, 3, 4; \qquad q_1 \geq 60 \quad q_2 \geq 24$$
$$\text{and } p = 1, 2; \qquad u_i: \text{binary variable for } i = 1, 2, 3$$

Now, solve this problem using an Integer LP computer package (we used LINGO) to find the optimal solution for all decision variables. The solution is shown below.

Energy Trans. x_{ij}	User j					Energy Trans. y_{ip}	Saver p			Energy Trans. z_{pj}	User j				
	1	2	3	4	Total		1	2	Total		1	2	3	4	Total
Gen. i 1	500	0	0	200	700	Gen. i 1	300	0	300	Saver p 1	0	0	300	0	300
2	0	600	0	0	600	2	0	200	200	2	0	0	0	200	200
3	0	0	0	0	0	3	0	0	0						
Total	500	600	0	200	1300	Total	300	200	500	Total	0	0	300	200	500

In the final solution, $u_1 = 1$, $u_2 = 1$, $u_3 = 0$, which means that the best decision is to purchase generators 1 and 2 but not generator 3. The total cost is \$167,350.

The solution is shown in Figure 11.22.

See Supplement S11.7.lg4.

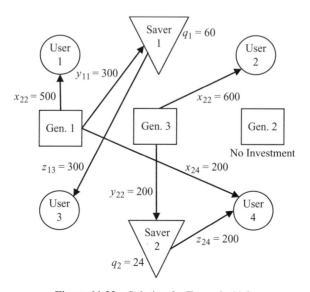

Figure 11.22   Solution for Example 11.9.

Examples of Using Different Demands   The following table shows that the optimal design solutions for different users' demands could be different even though the total demand may be the same.

Alternative		1	2	3	4
D_j, Demand of User j	D_1	500	200	200	100
	D_2	600	200	200	500
	D_3	300	700	300	200
	D_4	400	700	200	100
Total energy production by generators		1,800	1,800	900	900
Total cost ($)		167,350	167,325	83,800	83,925
Selected generators		1, 2	1, 3	1	2

11.5.3   Period-by-Period Optimization[†]

A current trend in the energy market is to sell energy at a variable rate which is mostly dependent on the demand and the supply of energy. This strategy reduces the maximum energy use at the peak demand periods. By reducing the peak demands, the overall cost for producing energy will be reduced. It also eliminates the need for power plants that are only used at peak times. As a result, the market value of energy changes periodically, for example, during a given day. In this section, we discuss formulating energy operations optimization problems for multiple periods.

Suppose there are T periods denoted by $t = 1, 2, \ldots, T$.

For a multiperiod, two different problems can be formulated:

Period-by-Period Approach   The price of the energy and demands for each period are only known at the beginning of the period. So, the energy problem is solved at the beginning of each period. This is based on the method covered in Section 11.5.1.

Aggregate Multiperiod Approach   The price of the energy and the demand of a given number of future periods are known at the beginning of the first period. This is covered in Section 11.5.4.

Example 11.10   Period-by-Period Energy Operations Optimization   Consider Example 11.7 where two different periods, day and night, are considered. For each period, there are different costs of buying and selling energy. The demands and cost information for the day period (which is the beginning period) is given in Table 11.18. Assume that the generator and saver capacities are the same as in Example 11.7. Also suppose that the initial energy levels in savers 1 and 2 are 100 and 150.

TABLE 11.18   Information for Night Period Energy Problem

Period	Demand User 1	Demand User 2	Demand User 3	Demand User 4	Predicted Change in Energy Price for Saver 1, $\Delta s_{1,t}$	Predicted Change in Energy Price for Saver 2, $\Delta s_{2,t}$	Generator 1 Costs ($g_{1,t}$)	Generator 2 Costs ($g_{2,t}$)
1, Night	200	250	100	150	+5	+10	59.5	60

TABLE 11.19 Information for Day Period Energy Problem

Period	Demand User 1	Demand User 2	Demand User 3	Demand User 4	Predicted Change in Energy Price for Saver 1, $\Delta s_{1,t}$	Predicted Change in Energy Price for Saver 2, $\Delta s_{2,t}$	Generator 1 Costs $(g_{1,t})$	Generator 2 Costs $(g_{2,t})$
2, Day	500	600	300	400	-10	-5	64	65

For this period, an optimization problem similar to Example 11.7 should be solved. The solution for this problem is provided below where for period 1 the solution is $q_{1,1} = 200$, $q_{2,1} = 240$, and $f_1 = \$53,785$.

Solution for Period 1: Night

Energy Trans. x_{ij1}	User j 1	2	3	4	Total
Gen. i 1	0	0	0	150	150
2	0	250	0	0	250
Total	0	250	0	150	400

Energy Trans. y_{ip1}	Saver p 1	2	Total
Gen. i 1	400	0	400
2	0	90	90
Total	400	90	490

Energy Trans. z_{pj1}	User j 1	2	3	4	Total
Saver p 1	200	0	100	0	300
2	0	0	0	0	0
Total	200	0	100	0	300

 See Supplement S11.8.lg4.

The solution of period 1 (night) will be used for formulating the energy optimization problem of the next period. For the next period (day), the information is provided in Table 11.19.

For period 2 the solution is $q_{1,2} = 60$, $q_{2,2} = 24$, and $f_1 = \$100,564$. The total cost for both periods is $\$53,785 + \$100,564 = \$154,349$. It can be seen from the solution that the saver will be charged during the night and used during the day.

Solution for Period 2: Day

Energy Trans. x_{ij2}	User j 1	2	3	4	Total
Gen. i 1	0	0	0	400	400
2	0	444	0	0	444
Total	0	444	0	400	844

Energy Trans. y_{ip2}	Saver p 1	2	Total
Gen. i 1	600	0	600
2	0	0	0
Total	600	0	600

Energy Trans. z_{pj2}	User j 1	2	3	4	Total
Saver p 1	500	0	240	0	740
2	0	156	60	0	216
Total	500	156	300	0	956

See Supplement S11.9.lg4.

11.5.4 Aggregate Multiperiod Optimization[†]

In the aggregate planning approach, one large problem based on Problem 11.1 is solved for $t = 1, 2, \ldots, T$ periods. These problems are connected by the saver's energy levels where the initial energy level for the next period is the energy level of the saver at the end of the current period. The formulation of this problem is similar to aggregate planning for multiperiods covered in Chapter 4.

TABLE 11.20 Notation for Multiperiod Energy Operations

T: Number of periods, $t = 1, \ldots, T$	$g_{i,t}$: Cost per unit of energy for generator i in period t
$x_{ij,t}$: Amount of energy from generator i to user j in period t	$q_{p,t}$: Energy of saver p at end of period t
$y_{ip,t}$: Amount of energy from generator i to saver p in period t	$q_{p,t,\max}$: Maximum capacity of saver p in period t
$z_{pj,t}$: Amount of energy from saver p to user j in period t	$q_{p,t,\min}$: Minimum required energy of saver p in period t
$K_{i,t}$: Maximum capacity of generator i in period t	$q_{p,0}$: Initial energy level of saver p
$D_{j,t}$: Energy demand of each user j in period t	

All variables will have an additional subscript of t. See Table 11.20.

Two consecutive periods are connected to each other by modifying constraint (11.5) in Problem 11.1, which is shown by constraint (11.14). Note that the total gain for future savings as presented in (11.2) is not used in the objective function of multiperiod energy optimization [Equation (11.11)], because in this case future prices (of the given periods) are known.

PROBLEM 11.3   MULTIPERIOD ENERGY OPERATION OPTIMIZATION

Minimize:

$$f_1 = \sum_{t=1}^{T}\sum_{i=1}^{I}\sum_{j=1}^{J} c_{ij,t}x_{ij,t} + \sum_{t=1}^{T}\sum_{i=1}^{I}\sum_{p=1}^{P} c_{ip,t}y_{ip,t} + \sum_{t=1}^{T}\sum_{p=1}^{P}\sum_{j=1}^{J} c_{pj,t}z_{pj,t}$$

$$+ \sum_{t=1}^{T}\sum_{i=1}^{I} g_{i,t}\sum_{j=1}^{J} x_{ij,t} + \sum_{t=1}^{T}\sum_{i=1}^{I} g_{i,t}\sum_{p=1}^{P} y_{ip,t} \tag{11.11}$$

Subject to:

$$\sum_{i=1}^{I} x_{ij,t} + \sum_{p=1}^{P} z_{pj,t} \geq D_{j,t} \quad \text{for all } j = 1, \ldots, J, t = 1, \ldots, T \tag{11.12}$$

$$\sum_{j=1}^{J} x_{ij,t} + \sum_{p=1}^{P} y_{ip,t} \leq K_{i,t} \quad \text{for all } i = 1, \ldots, I, t = 1, \ldots, T \tag{11.13}$$

$$q_{p,t-1} + \sum_{i=1}^{I} y_{ip,t} - \sum_{j=1}^{J} z_{pj,t} = q_{p,t} \quad \text{for all } p = 1, \ldots, P, t = 1, 2, \ldots, T \tag{11.14}$$

$$q_{p,t} \leq q_{p,t,\max} \quad \text{for all } p = 1, \ldots, P, t = 1, 2, \ldots, T \tag{11.15}$$

$$q_p \geq q_{p,t,\min} \quad \text{for all } p = 1, \ldots, P, t = 1, 2, \ldots, T \tag{11.16}$$

$$x_{ij,t} \geq 0 \quad y_{ip,t} \geq 0 \quad z_{pj,t} \geq 0 \quad q_{p,t} \geq 0 \quad \text{for all } i, j, p, t = 1, 2, \ldots, T \tag{11.17}$$

Example 11.11   Multiperiod Energy Operations Optimization: Aggregate   Consider Example 11.10. Suppose that the information for both periods is known at the beginning of the first period. The expected changes in the energy prices for the savers are shown in Tables 11.18 and 11.19. Find the optimal solution for both periods.

The formulation for this part is similar to Example 11.7 except that constraints (11.14) are used as an input–output of the different periods to connect them together and (11.2) is not used in the objective function. Constraints (11.14) and (11.17) are shown here.

Period 1	Period 2
$100 + (y_{11,1} + y_{21,1}) - (z_{11,1} + z_{12,1} + z_{13,1}$ $+ z_{14,1}) = q_{1,1}$	$q_{1,1} + (y_{11,2} + y_{21,2}) - (z_{11,2} + z_{12,2} + z_{13,2}$ $+ z_{14,2}) = q_{1,2}$
$150 + (y_{12,1} + y_{22,1}) - (z_{21,1} + z_{22,1} + z_{23,1}$ $+ z_{24,1}) = q_{2,1}$	$q_{2,1} + (y_{12,2} + y_{22,2}) - (z_{21,2} + z_{22,2} + z_{23,2}$ $+ z_{24,2}) = q_{2,2}$

$x_{ij,t}, y_{ip,t}, z_{pj,t}, q_{p,t} \geq 0$ for all $i = 1, 2, j = 1, 2, 3, 4, p = 1,2$, and $t = 1, 2$

The solution to this problem follows.

Solution for Period 1

Energy Trans. x_{ij1}	User j 1	2	3	4	Total	Energy Trans. y_{ip1}	Saver p 1	2	Total	Energy Trans. z_{pj1}	User j 1	2	3	4	Total
Gen. i 1	0	0	0	150	150	Gen. i 1	300	0	300	Saver p 1	200	0	0	0	200
2	0	250	100	0	350	2	0	90	90	2	0	0	0	0	0
Total	0	250	100	150	500	Total	300	90	390	Total	200	0	0	0	200

Solution for Period 2

Energy Trans. x_{ij2}	User j 1	2	3	4	Total	Energy Trans. y_{ip2}	Saver p 1	2	Total	Energy Trans. z_{pj2}	User j 1	2	3	4	Total
Gen. i 1	0	0	0	400	400	Gen. i 1	600	0	600	Saver p 1	500	0	240	0	740
2	0	384	60	0	444	2	0	0	0	2	0	216	0	0	216
Total	0	384	60	400	844	Total	600	0	600	Total	500	216	240	0	956

In this solution, $q_{1,1} = 200$, $q_{1,2} = 60$, $q_{2,1} = 240$, $q_{2,2} = 24$, and $f_1 = \$154{,}349$, which is the same cost found in part (a) [note that the solution in the above table is an alternate solution to the solution in part (a)]. However, the total cost for these two approaches may not necessarily be the same.

See Supplement S11.10.lg4.

11.5.5   Multiobjective Optimization of Energy Systems[†]

Energy cost is considered an important objective in energy systems. Another important objective is the environmental impact of the energy system. Different energy generators have different environmental impacts. In the following example, we introduce bicriteria energy systems operations optimization considering two objective functions: energy cost and environmental impact. Suppose that the environmental impact factor per unit of energy for each energy generator is presented by m_i for $i = 1, \dots, I$. The bicriteria problem is presented below.

PROBLEM 11.4   BIOBJECTIVES OF ENERGY SYSTEMS OPERATIONS

Minimize total cost: $f_1 =$ Use Equations (11.2a)–(11.2e).

Minimize total environmental impact:

$$f_2 = \sum_{i=1}^{I} m_i \left(\sum_{j=1}^{J} x_{ij} + \sum_{p=1}^{P} y_{ip} \right) \tag{11.18}$$

Subject to constraints (11.3)–(11.8).

Each $\sum_{j=1}^{J} x_{ij} + \sum_{p=1}^{P} y_{ip}$ in (11.18) is the amount of energy produced by generator i. This amount multiplied by the environmental factor per unit of energy (m_i) results in the total environmental impact of generator i. The constraints of Problem 11.4 are the same as the constraints of Problem 11.1. To find the best alternative for a given system, use the following MCDM process. First, a range must be found for the objectives. Solve the following problems to find the minimum and maximum values for each objective.

	Solution	
Problem 1: Minimize f_1 subject to constraints of Problem 11.1	$f_{1,\min}$	$f_{2,\max}$
Problem 2: Minimize f_2 subject to constraints of Problem 11.1	$f_{1,\max}$	$f_{2,\min}$

From these ranges, choose an evenly distributed set of values for f_2. For each value of f_2, solve the following problem.

PROBLEM 11.5   CONSTRAINED BICRITERIA ENERGY SYSTEMS OPERATIONS

Minimize: $f_1 =$ Equations (11.2a)–(11.2e)
Subject to: $f_2 =$ Equation (11.18)
          $f_2 \leq$ selected f_2 value
          Constraints of Problem 11.1: (11.3)–(11.8)

By varying f_2 over its range, many alternatives can be found with a wide range of solutions. The DM can then select the best alternative or specify a value of f_2 to be used to solve the problem. The following example illustrates this process.

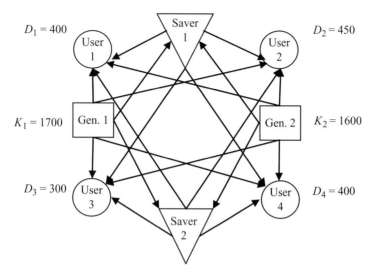

Figure 11.23   Energy system for Example 11.12.

Example 11.12   Bicriteria Energy Systems Operations   Consider an energy system with two generators. Suppose that the first generator is operated by biofuel energy and the second generator is operated by solar energy with environmental factors of $m_1 = 0.8$ and $m_2 = 0.3$, respectively. The biofuel generator is much cheaper than the solar energy generator. Generate a set of six alternatives to this problem.

The details of the problem are given in Figure 11.23. Notice that the saver energy levels in this problem are set at the minimum levels, so that the savers are only used as hubs. If the savers are used as energy sources, the environmental impact from the energy previously generated to fill them will not be measured.

c_{ij}, Trans. Cost		User j			
from i to j		1	2	3	4
Gen.	1	6	5	8	4
i	2	7	4	5	7

c_{pj}, Trans. Cost		User j			
from p to j		1	2	3	4
Saver	1	1	2	1	2
p	2	2	0.5	1.5	1
Demand D_j		400	450	300	400

c_{ip}, Trans. Cost from i to p		Saver p	
		1	2
Gen.	1	4.5	5
i	2	6	4

Gen.	Gen. Costs, g_i (\$)	Gen. Cap. K_i	Env. Imp. m_i
1	100	1700	0.8
2	200	1600	0.3

Saver	Init. Enr $q_{p,0}$	Min. Enr $q_{p,\min}$	Max. Enr $q_{p,\max}$
1	60	60	200
2	24	24	240

First, the range of the objectives must be found. The objectives must be calculated when minimizing only f_1, the total cost. This problem is as follows:

Minimize $f_1 =$ Equations (11.2a)–(11.2e)

Subject to constraints of Problem 11.1

The solution to this problem is shown as problem 1 in the table below.

Then, the objectives must be calculated when minimizing only f_2, the environmental impact. This problem is as follows:

Minimize:

$$f_2 = \sum_{i=1}^{2} m_i \left(\sum_{j=1}^{4} x_{ij} + \sum_{p=1}^{2} y_{ip} \right)$$

Subject to constraints of Problem 11.1

The solution to this problem is shown as problem 2 in the following table:

	Solution	
Problem 1: Minimize f_1 subject to constraints of Problem 11.1	162,700	1240
Problem 2: Minimize f_2 subject to constraints of Problem 11.1	318,900	465

Now that the range of f_2 is known, find an evenly distributed range of values for f_2 and then solve Problem 11.4 for each of these values. In this case, the following six evenly distributed values of f_2 are used: 465, 620, 775, 930, 1085, and 1240.

For example, if $f_2 = 775$ is selected, the following problem must be solved to generate the alternative.

Problem 11.5 for Example 11.12:

Minimize: $f_1 = $ Equations (11.2a)–(11.2e)
Subject to:

$$f_2 = \sum_{i=1}^{2} m_i \left(\sum_{j=1}^{4} x_{ij} + \sum_{p=1}^{2} y_{ip} \right)$$

$$f_2 \leq 775$$

Constraints of Problem 11.1

The solution to this bicriteria energy systems problem is as follows:

Enr. Trans. x_{ij}	User j				
Gen. i	1	2	3	4	Total
1	0	0	0	400	400
2	0	450	300	0	750
Total	0	450	300	400	1150

Enr. Trans. y_{ip}	Saver p		
Gen. i	1	2	Total
1	220	0	220
2	0	180	180
Total	220	180	400

Enr Trans. z_{pj}	User j				Total
Saver p	1	2	3	4	
1	220	0	0	0	220
2	180	0	0	0	180
Total	400	0	0	0	400

Gen. Prod., k_i	1	2
	620	930

Saver Energy Level, q_p	1	2
	60	24

The total cost (f_1) and total environmental impact (f_2) for this solution are

$$f_1 = \$255,190$$

$$f_2 = 0.8(0 + 0 + 0 + 400 + 220 + 0) + 0.3(0 + 450 + 300 + 0 + 0 + 180) = 775$$

This solution shows that generator 1 generates a total of 620 units of energy. Generator 2 generates 930 units of energy.

Generating Different Bicriteria Solutions   Consider the Environmental Impact; it varies from 465 to 1240. Five equally distributed points are 465, 620, 775, 930, 1085, and 1240. For each of these given f_2 values, solve Problem 11.5 for Example 11.12. For example, for the second alternative, use $f_2 \leq 620$. The following table shows the six generated points. All alternatives are efficient. Notice how the use of generator 1 (low cost, high environmental impact) and generator 2 (high cost, low environmental impact) vary in the alternatives. Now the best alternative can be selected by the DM.

Alternative	1	2	3	4	5	6
f_1, Total cost	317,700	286,390	255,190	224,165	193,390	162,700
f_2, environmental impact	465	620	775	930	1,085	1,240
Generator 1 energy produced	0	310	620	930	1,240	1,550
Generator 2 energy produced	1,550	1,240	930	620	310	0

See Supplement S11.11.lg4.

11.6   EFFICIENCY OF ENERGY SYSTEMS

In evaluating energy systems, it is important to compare the efficiencies of different alternative systems. Different energy generators produce different outputs of energy using different inputs. There are multiple inputs and outputs that can be used to calculate the efficiency of given comparable generators. See Table 11.21 for some examples of inputs and outputs of energy systems for generators.

TABLE 11.21   Inputs and Outputs for Energy Generators

Generator	Inputs			Outputs		
Fossil fuel plant	Fuel	Maintenance cost	Manpower	Electricity	Reliability	Environmental impact
Solar panel	Sunlight	Maintenance cost	PV cell size	Electricity	Reliability	Environmental impact
Wind turbine	Wind	Maintenance cost	Size of turbine	Electricity	Reliability	Environmental impact
Nuclear plant	Uranium	Maintenance cost	Manpower	Electricity	Reliability	Environmental impact

TABLE 11.22   Inputs and Outputs for Energy Users

User	Inputs		Outputs	
Car	Fuel	Load	Miles traveled	Environmental impact
Refrigerator	Electricity	Capacity	Temperature maintained	Environmental impact

  Similarly, different energy users produce different outputs using different inputs of energy. Multiple inputs and outputs can be used to calculate the efficiency of given comparable users. Table 11.22 shows some examples of inputs and outputs of energy systems for energy users.

11.6.1   Energy Systems with One Input and One Output

In Chapter 10, we developed methods for measuring productivity and efficiency of different operational systems. Such methods can be used to measure the productivity and efficiency of energy systems. We can categorize two classes of problems:

1. One input and one output: There is a simple method for solving this problem; see Section 10.4.
2. Multiple inputs and multiple outputs: There is a linear programming method for solving this problem; see Section 10.5.

Example 11.13   Efficiency of Single Input and Output of Five Power Plants   There are five electrical power plants in Cuyahoga County in Ohio. The manager of the plants wishes to compare the operational efficiencies of the plants and determine if any plants are inefficient. The input is fuel consumed and the output is the generated electricity. The measured input and output are shown in Table 11.23. Find the efficiency of the power plants.

  The solution for the above LP problem for plant 1 is shown in Table 11.23. It can be seen that plant 1 is the most efficient while plant 5 is least efficient. The ranking of the efficiency of the plants is also given.

11.6.2   Energy Systems with Multiple Inputs and Outputs[†]

For multiple input–multiple output energy system problems, a linear programming approach (see Section 10.6) can be used to compare the efficiencies of different energy units (generators or users). This method can be used to compare the efficiencies of energy systems and choose the most efficient units. Also, for solving the multicriteria efficiency of energy

TABLE 11.23   Inputs, Outputs, and Efficiency of Five Plants

Plant No., r	Fuel: Input (I_r)	Electricity: Output (O_r)	Productivity $P_r = O_r/I_r$	Efficiency, $E_r = P_r/P_{max}$	Efficiency, Rank
1	461	1945	4.22	1.00	1
2	444	1689	3.80	0.90	2
3	537	1956	3.64	0.86	3
4	523	1780	3.40	0.81	4
5	246	507	2.06	0.49	5

TABLE 11.24   Measured Inputs and Outputs for Five Power Plants

Unit, r	Input (b_{rj})			Output (a_{ri})		
	Fuel, $j = 1$	Costs, $j = 2$	Man-Hours, $j = 3$	Electricity, $i = 1$	Availability, $i = 2$	Positive Environmental Impact, $i = 3$
Plant 1	461	1530	214	1945	71.5	0.46
Plant 2	444	1750	249	1689	80.4	0.43
Plant 3	537	1583	226	1956	73.1	0.47
Plant 4	523	1698	276	1780	60.9	0.33
Plant 5	246	1078	151	507	43.6	0.15

systems, see Section 10.6. The following example shows how this method can be used to compare the efficiency of five power plants. Since more output is better when measuring efficiency, we use positive environmental impact (measured on a 0–1 scale where 0 is the worst and 1 is the best). Positive environmental impact = 1; note that environmental impact was defined in Section 10.3 on a 0–1 scale where 0 was the best and 1 was the worse.

Example 11.14   Efficiency of Multiple Inputs and Outputs of Five Power Plants
Consider Example 11.13, where there are multiple inputs and outputs to the plant. The inputs are fuel consumed, maintenance costs, and man-hours. The measured outputs are generated electricity, plant availability (percentage of time the plant is available to produce electricity), and positive environmental impact (measured on a 0–1 scale where 0 is the worst and 1 is the best). The measured inputs and outputs are shown in the Table 11.24. Find the efficiency of the power plants.

In the following we show the LP formulation for finding the efficiency of plant 1.

LP Problem for Plant 1

Maximize:

$$E_1 = 1945x_1 + 71.5x_2 + 0.46x_3 \quad \text{(efficiency for plant 1)}$$

Subject to:

$$-1945x_1 - 71.5x_2 - 0.46x_3 + 461y_1 + 1530y_2 + 214y_3 \geq 0 \quad \text{(for plant 1)}$$

$$-1689x_1 - 80.4x_2 - 0.43x_3 + 444y_1 + 1750y_2 + 249y_3 \geq 0 \quad \text{(for plant 2)}$$

$$-1956x_1 - 73.1x_2 - 0.47x_3 + 537y_1 + 1583y_2 + 226y_3 \geq 0 \quad \text{(for plant 3)}$$

$$-1780x_1 - 60.9x_2 - 0.33x_3 + 523y_1 + 1698y_2 + 276y_3 \geq 0 \quad \text{(for plant 4)}$$

$$-507x_1 - 43.6x_2 - 0.15x_3 + 246y_1 + 1078y_2 + 151y_3 \geq 0 \quad \text{(for plant 5)}$$

$$461y_1 + 1530y_2 + 214y_3 = 1 \quad \text{(normalized input for plant 1)}$$

$$x_i \geq 0.0001 \quad \text{for } i = 1, 2, 3$$

$$y_j \geq 0.0001 \quad \text{for } j = 1, 2, 3$$

TABLE 11.25   Inputs, Outputs, and Efficiency of Five Plants

Variables	x_1	x_2	x_3	y_1	y_2	y_3	E	Efficiency Rank
Plant 1	0.0005	0.0001	0.0001	0.0018	0.0001	0.0001	1	1
Plant 2	0.0003	0.0069	0.0001	0.0018	0.0001	0.0001	1	1
Plant 3	0.0001	0.0107	0.0001	0.0001	0.0006	0.0001	0.979	3
Plant 4	0.0005	0.0001	0.0001	0.0001	0.0005	0.0001	0.820	5
Plant 5	0.0001	0.0201	0.0001	0.0036	0.0001	0.0001	0.926	4

The solution for the above LP problem for plant 1 is shown in the first row of Table 11.25. The results of solving the LP problems for the other four plants are also shown in Table 11.25. It can be seen that plants 1 and 2 are most efficient while plants 3, 4, and 5 are less efficient. The ranking of the efficiency of plants is also given.

See Supplement S11.12.lg4.

11.7   CASE STUDY: WIND ENERGY SYSTEM

11.7.1   Wind Energy[†]

Both solar energy and wind energy are becoming increasingly popular throughout the world due to their fast renewability cycles and low environmental impacts as well as the increase in the cost of fossil fuels, which makes these types of energy economically attractive. Wind energy is collected by using a wind turbine which converts the kinetic energy of wind into mechanical energy by rotating blades. This mechanical energy can then be utilized in its current form or it can be further converted into electricity. Although wind energy has been used for centuries, dating back to a time when windmills were used to grind grains, it has never become a major source of energy production due to economic reasons and the variability associated with wind speed. The huge expense of buying and installing wind turbines is a prohibitive factor in using this type of energy. In order for a turbine to collect wind energy, the wind must be strong enough to rotate the turbine's blades, something that cannot be accomplished in all areas of the world. Even in areas that are windy, there are fluctuations in the wind; therefore, energy cannot be produced consistently. However, a combination of wind energy and other types of energy (such as solar electricity) can provide a steady and reliable supply of electricity. Also, wind is considered a great energy source for remote locations where other types of energy may not fulfill all energy needs. Currently, new wind turbines have become more efficient in generating electricity. They also have less environmental impact and generate less greenhouse gases such as carbon dioxide.

Types of Wind Turbines   There are different types of wind turbines, each with different applications. Two well-known wind turbines are horizontal-axis and vertical-axis turbines, as shown in Figures 11.24a,b. Horizontal-axis wind turbines have better efficiency than vertical-axis wind turbines in terms of generating power. In this section, we only analyze horizontal-axis wind turbines.

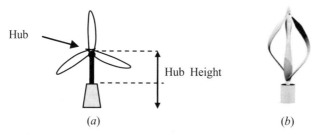

Figure 11.24 (*a*) Horizontal-axis wind turbines; (*b*) vertical-axis wind turbines.

Number of Blades Horizontal-axis wind turbines have various designs. Depending on several factors, different numbers of blades can be used in building wind turbines. Aerodynamic efficiency may increase by increasing the number of blades but the cost of making the turbine also increases. Currently, the three-blade turbine is considered the most applicable and economical wind turbine.

Noise Issue Wind turbines generate two types of noise, aerodynamic and mechanical. Aerodynamic noise is generated by blades while passing through the air. Mechanical noise is generated from the turbine's internal gears. Noise emission in turbines is also dependent on the location of the blades (either upwind or downwind of the turbine tower) and the speed of the rotor.

Wind Table Wind data such as mean speed and peak speed for some geographical locations in the United States are summarized in Table 11.26. This table also shows the percentage of time that wind speed is recorded in a given speed range. Wind speed intervals can be defined with different width so as to cover the whole range of wind speed (see first row of Table 11.26). Wind speed can be recorded in miles per hour or meters per hour. For example, in Cleveland, there are wind speeds of 8–12 MPH for 35% of the time in a year. Mean (average) and peak wind speeds in Cleveland are 11.6 and 78.1 MPH, respectively.

Wind Turbine Classes Depending on the mean speed and wind turbulence characteristics, different types of wind turbines and sizes of wind blade lengths are used. They are classified into four classes (I, II, III, and IV) associated with 112, 95.1, 83.9, and 67.1 peak speed (miles per hour), respectively.

TABLE 11.26 Annual Percentage of Wind Speed for Different Geographical Locations

Station	State	0–3 MPH	4–7 MPH	8–12 MPH	13–18 MPH	19–24 MPH	25–31 MPH	32–38 MPH	Mean Speed (MPH)	Peak Speed (MPH)
Boston	MA	3	12	33	35	12	4	1	13.3	87
Cleveland	OH	7	18	35	29	9	2	NA	11.6	78.1
Denver	CO	11	27	34	22	5	2	NA	10	64.9
Honolulu	HI	9	17	27	32	12	2	NA	12.1	67.1
Los Angeles	CA	28	33	27	11	1		NA	6.8	49

Source: Climatography of the United States, Series 82: Decennial Census of the United States Climate.

11.7.2 Wind Power Characteristics[†]

The wind power equation shows how much power P (in watts) a wind turbine can generate from wind with a certain speed. The wind power is proportional to the area swept by the blade (A), air density (ρ), and wind speed (V) to the power of 3. Air density is a function of temperature and altitude. A default value for ρ is 1.225 kg/m^3, which is equal to the density of dry air with temperature 32°F at sea level. The wind power equation is

$$P = 0.5\eta\rho c_p AV^3 \tag{11.19}$$

where

P = power (W)
η = mechanical and electrical efficiency
ρ = air density (kg/m^3)
c_p = power coefficient (aerodynamic coefficient)
$A = \pi(d/2)^2$ = area swept by blades (m^2)
d = rotor diameter of wind turbine (m)
V = average wind speed (m/s)

In practice, the kilowatt-hour (1000 Wh) is used instead of the watt as the unit of generated power. Betz's law proves that a hypothetical ideal wind turbine cannot capture more than 59.3% of the wind energy. Modern turbines can only approach to 70–80% of this theoretical limit. For example, for 78% of the Betz limit the power coefficient will be $c_p = 0.593 \times 0.78 = 0.463$. Here η is the mechanical and electrical efficiency where $\eta < 1$.

Example 11.15   Wind Power Equation   A company plans to install a wind turbine at a place close to sea level (the air density at sea level is $\rho = 1.225$ kg/m^3). The wind is coming mainly from one direction with annual average speed of 10 m/s. Suppose that there is no wind obstacle in the vicinity of the wind farm. Suppose that all turbines have the same mechanical and electrical efficiency of 0.8. Calculate generated power for five different rotor diameter sizes of 1, 30, 60, 90, and 125.

An interval value is considered for possible c_p values of each turbine, where c_p can be any value within these intervals, but for simplicity of analysis, we will only work with the average value of the interval. Power coefficient intervals with respect to each rotor diameter are shown in Table 11.27.

For example, for turbines with rotor diameter of 30 m the swept area by the rotor is

$$A = \pi \left(\tfrac{1}{2}d\right)^2 = 3.14 \left(\tfrac{30}{2}\right)^2 = 706.8 \text{ m}^2$$

TABLE 11.27   Power Coefficient for Turbines of Different Sizes

Rotor (d)	1	30	60	90	125
c_p	[0.20,0.30]	[0.30,0.40]	[0.38,0.48]	[0.25,0.40]	[0.25,0.38]
Avg c_p	0.25	0.35	0.43	0.33	0.32

TABLE 11.28   Generated Power for Five Wind Turbine Diameters

Rotor (d) m	1	30	60	90	125
Area A (m^2) $= \pi (d/2)^2$	0.8	706.8	2,827.4	6,361.5	12,271.5
Power P (W) $= 0.5 \eta \rho\, c_p A V^3$	96	121,223	595,723	1,013,075	1,894,104

The converted wind power is

$$P = 0.5 \times 0.8 \times 1.225 \text{ kg/m}^3 \times 0.35 \times 706.8 \text{ m}^2 \times 10^3 \text{ m/s}^3 = 121{,}223 \text{ W}$$

(or 121.223 kW)

See Table 11.28.

Operable Ranges of Wind Speed   The plot of power versus wind speed is shown in Figure 11.25. The turbine starts turning at cutting speed V_C and reaches its maximum power capacity at V_R. As soon as the wind speed reaches its maximum limit V_F, the turbine shuts down to prevent damage to the turbine due to very high wind. The general function for the power output of wind turbine is shown in Figure 11.25.

Different blades have different operable signatures such as the one presented in Figure 11.26. One should select a blade that maximizes average generated power for the given

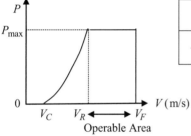

If	$V < V_C$	$V_C < V < V_R$	$V_R < V < V_F$	$V > V_F$
$P =$	0	Equation (11.19)	P_{max}	0

Figure 11.25   Wind speed versus generated power.

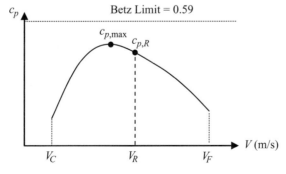

Figure 11.26   Power coefficient c_p versus wind speed V, where $V_C < V < V_F$.

TABLE 11.29   Physical Property Notation

β: Pitch angle (rad)	V: Undisturbed wind speed (m/s)
λ: Tip speed ratio (rad)	ω: Angular speed of turbine (rad/s)
β_i: Pitch angle i, where $i = 1, \ldots, k$	$r\omega$: Tip speed (m∗rad/s)
and $\beta_1 < \beta_2 < \cdots < \beta_k$	V_c: Cutting speed for wind (turbine starts turning)
r: Radius of rotating turbine (m)	V_R: Rated speed (turbine reaches maximum capacity)
n: Rotational speed (cycles/min)	V_F: Maximum operable speed for wind turbine
$C_{p,max}$: Maximum power coefficient	$c_{p,R}$: Rated power coefficient

wind power for the given geographical area. All examples in this section consider wind speed V in the operable area $V_C < V < V_R$.

Betz Limit   Equation (11.19) shows that the generated power P is a function of the power coefficient c_p. However, c_p is also dependent on wind speed V. Figure 11.26 shows the relationship between wind speed and the power coefficient. As wind speed increases from V_C, the power coefficient c_p increases. This trend continues until c_p reaches its maximum value $c_{p,max}$. After $c_{p,max}$, the generated power continues to increase while c_p starts to decrease until it reaches the rated power coefficient $c_{p,R}$ at wind speed V_R. After this point, c_p decreases with a faster rate until the wind speed reaches the maximum operable speed V_F, which makes the turbine shut down.

Physical Properties of Blades in Wind Turbine   Equation (11.19) shows the power equation for the physical properties of wind turbine blades given in Table 11.29. Figure 11.27a shows the "airfoil," which is the cross-sectional view of the turbine blade. One important parameter of the airfoil is pitch angle β. The pitch angle is automatically adjusted by the wind turbine to control the speed of the turbine in order to maximize the amount of produced energy. Pitch angle is the angle between the chord line and the rotation plane. The chord line is the straight line that connects the head and the tail of the airfoil. The rotation plane is the plane in which the blade is rotating. The angular speed ω is the

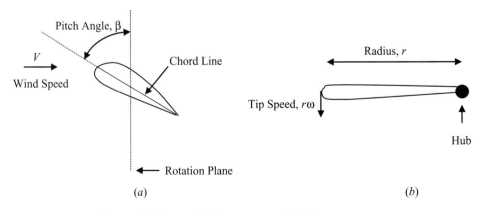

Figure 11.27   (a) Airfoil parameters; (b) blade parameters.

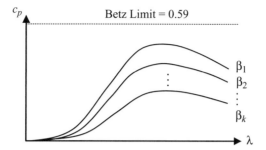

Figure 11.28 Power coefficient c_p versus tip speed ratio λ for different pitch angle β, where $\beta_1 < \beta_2 < \cdots < \beta_k$.

vector that shows the angular speed of the blade and the vector in which the blade rotates. The angular speed ω is calculated from the rotational speed n (cycle/min) using

$$\omega = \frac{2\pi n}{60} \quad (\text{rad/s})$$

The speed of the blade with angular speed ω and radius r is $r\omega$, called the "tip speed"; see Figure 11.27b. The ratio of tip speed to undisturbed wind speed, V, is the tip speed ratio λ, where

$$\lambda = \frac{r\omega}{V}$$

Figure 11.28 shows that for a given fixed speed ratio λ, as the pitch angle increases from β_1 to β_k, the power coefficient increases. The given physical properties will result in different turbine power generation such as the ones presented in Figure 11.28.

Example 11.16 Tip Speed Ratio   Suppose that we are designing a turbine blade with a radius of 20 m and a rotational speed of 50 cycles/min. Calculate the speed and tip speed ratio of this blade for wind speeds of 10 and 15 m/s.
  The angular speed $\omega = 2\pi n/60 = 2 \times 3.14 \times 50/60 = 5.23$ rad/s.
  The speed of this blade is $r\omega = 5.23 \times 20 = 104.67$.
  Therefore, for $V = 10$ m/s, the tip speed ratio $\lambda = r\omega/V = 104.67/10 = 10.47$, and for $V = 15$ m/s it is $\lambda = r\omega/V = 104.67/15 = 6.98$.

11.7.3 Break-Even Analysis[†]

Economies of scale are the most significant considerations regarding feasibility study and capacity growth. As the number of operational years increases, the average cost per year decreases. The selling price of generated power per year, B, is set based on the power purchase agreement (PPA) between the seller and customer. Suppose the cost of generating power per year is given by C_V. Note that this is a variable cost dependent on the number of years n; C_F is the initial fixed cost needed to install the generator.

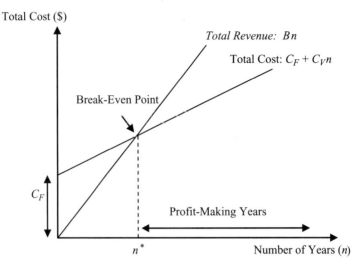

Figure 11.29 Break-even analysis for finding break-even point.

There are two types of costs involving the cost–benefit analysis of a wind turbine, fixed costs and variable costs. Fixed costs include the price of purchasing the turbine, installation, and other related initial costs. Variable costs include operation, maintenance, and insurance costs. Following is the break-even analysis notation:

n = number of years
n^* = break-even point (years)
C_F = fixed cost
C_V = variable cost per year
B = selling price of generated power per year
$B - C_V$ = net revenue per year

The break-even point (in years) is the point at which the total costs are paid off by total revenue. The fixed cost for installing a wind turbine plus the variable cost of generated power equals total revenue:

$$C_F + C_V n = Bn \qquad (11.20)$$

Figure 11.29 illustrates the break-even lines. After the break-even point n^* the wind turbine will be profitable.

Equation (11.20) can be solved to find the break-even point n^*. In the following equation $B - C_V$ is the net revenue per year:

$$n^* = \frac{C_F}{B - C_V} \qquad (11.21)$$

Example 11.17 Break-Even Analysis for Wind Turbine   Consider Example 11.15. Suppose that the average price of 1 kWh is 11 cents. Note that the turbines are only operable

TABLE 11.30   Revenue and Costs for Five Wind Turbine Diameters

Rotor (d), m	1	30	60	90	125
Selling price of generated power ($/year); B	53	66,672	327,647	557,191	1,041,757
Fixed costs ($); C_F	200	400,000	1,500,000	3,000,000	5,500,000
Variable costs per year ($); C_V	35	15,000	25,000	40,000	200,000
Break-even point (in years); n^* using Equation (11.21)	11.2	7.7	5.0	5.8	6.5

for a portion of the year due to turbine maintenance, low wind speeds (when the wind speed is not enough to turn the turbine blades), or gusty winds (when turbines are shut down to prevent damage). Suppose that available hours per year are 5000 h. Find the revenue per year and the break-even point for all five rotor diameter sizes and show them graphically. Table 11.30 shows the revenue, fixed cost, and variable costs of these turbines. Break-even points are shown in the last row of Table 11.30.

Example 11.15 showed that the converted wind power for a turbine with rotor diameter of 30 m is 121.223 kW. Therefore, the revenue (in dollars) of generated energy per year for this turbine is

$$B = 121.223 \text{ kW} \times 0.11 \text{ \$/kWh} \times 5000 \text{ h/year} = 66,672 \text{ \$/year}$$

Therefore, the break-even point for this turbine is

$$n^* = \frac{400,000}{66,672 - 15,000} = 7.7 \text{ years}$$

The cost–benefit analysis for the 60-m wind turbine is illustrated in Figure 11.30a. For this turbine, the total costs are paid off after five years of installation. Figure 11.30b shows the relationship between diameter size and the break-even year. In this figure, the wind turbine with a 60-m rotor has the lowest break-even point, five years.

See Supplement S11.13.xls.

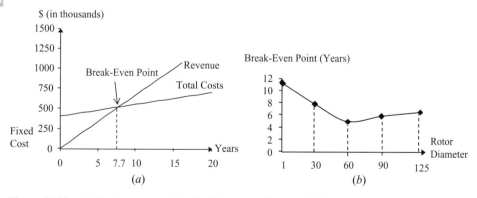

Figure 11.30   (a) Break-even analysis for 60-m rotor diameter; (b) break-even point (years) for five different rotor diameters.

REFERENCES

General

Albright, L., and F. Vanek. *Energy Systems Engineering*. New York: McGraw-Hill, 2008.

Endrenyi, J. *Reliability Modeling in Electric Power Systems*. New York: Wiley, 1978.

Groscurth, H. M., T. Bruckner, and R. Kümmel. "Modeling of Energy Services Supply Systems." *Energy*, Vol. 20, No. 9, 1995, pp. 941–958.

Jebaraj, S., and S. Iniyan. "A Review of Energy Models." *Renewable and Sustainable Energy Reviews*, Vol. 10, No. 4, 2006, pp. 281–311.

Khator, S. K., and L. C. Leung. "Power Distribution Planning: A Review of Models and Issues." *IEEE Transactions on Power Systems*, Vol. 12, No. 3, 1997, pp. 1151–1159.

Malakooti, B. "A Personal Computer Model for Energy Conservation Selection." *Microcomputer Applications*, Vol. 10, No. 1, 1987, pp. 41–54.

Nijcamp, P., and A. Volwahsen. "New Directions in Integrated Energy Planning." *Energy Policy*, Vol. 18, No. 8, 1990, pp. 764–773.

Ostergaard, P. A. "Reviewing Optimisation Criteria for Energy Systems Analyses of Renewable Energy Integration." *Energy*, Vol. 34, No. 9, 2009, pp. 1236–1245.

Multiobjective Energy Systems

Akash, B. A., R. Mamlook, and M. S. Mohsen. "Multi-Criteria Selection of Electric Power Plants Using Analytical Hierarchy Process." *Electric Power Systems Research*, Vol. 52, 1999, pp. 29–35.

Georgopoulou, E., D. Lalas, and L. Papagiannakis. "A Multicriteria Decision Aid Approach for Energy Planning Problems: The Case of Renewable Energy Option." *European Journal of Operational Research*, Vol. 103, No. 1, 1997, pp. 38–54.

Kavrakoglu, I. "Multi-Objective Strategies in Power System Planning." *European Journal of Operations Research*, Vol. 12, No. 2, 1983, pp. 159–170.

Malakooti, B. "Implementation of MCDM for the Glass Industry Energy System." *IIE Transactions*, Vol. 18, No. 4, 1986, pp. 374–379.

Malakooti, B., S. Sheikh, C. Al-Najjar, and H. Kim. "Multi-Objective Energy Aware Multiprocessor Scheduling Using Bat Intelligence." *Journal of Intelligent Manufacturing*, 2012, pp. 1–15.

Malakooti, B. "Multi-Objective Energy Systems Design and Optimization." *Case Western Reserve University*, 2013.

Polatidis, H., D. A. Haralambopoulos, G. Munda, and R. Vreeker. "Selecting an Appropriate Multi-Criteria Decision Analysis Technique for Renewable Energy Planning." *Energy Sources, Part B*, Vol. 1, 2006, pp. 181–193.

Psarras, J., P. Capros, and J. E. Samoulidis. "Multi-Criteria Analysis Using a Large Scale Energy Supply LP Model." *European Journal of Operations Research*, Vol. 44, No. 3, 1990, pp. 383–394.

Schulz, V., and H. Stehfest. "Regional Energy Supply Optimization with Multiple Objectives." *European Journal of Operations Research*, Vol. 17, Issue 3, 1984, pp. 302–312.

Sheikh, S., and B. Malakooti. "Integrated Energy Systems with Multi-Objective." IEEE Energy Tech Conference, Cleveland, OH, May 25–26, 2011, pp. 1–5.

Sheikh, S., and B. Malakooti. "Design, Operation, and Efficiency of Energy Systems." IEEE Energy Tech Conference, Cleveland, OH, May 29–31, 2012.

Sheikh, S., M. K. Komaki, and B. Malakooti. "Energy Planning Problem Using Multiple Objective Optimization." Forthcoming 2014.

Sheikh, S., M. K. Komaki, and B. Malakooti. "Integrated Risk and Multi-Objective Optimization of Energy Systems." *Engineering Optimization*, Forthcoming 2014.

Wang, J.-J., Y.-Y. Jing, C.-F. Zhang, and J.-H. Zhao. "Review on Multi-Criteria Decision Analysis Aid in Sustainable Energy Decision-Making." *Renewable and Sustainable Energy Reviews*, Vol. 13, No. 9, 2009, pp. 2263–2278.

Routing

Bazaraa, M., J. Jarvis, and H. D. Sherali. *Linear Programming and Network Flows*, 2nd ed. New York: Wiley, 1990.

Coulouris, G., J. Dollimore, and T. Kindberg. *Distributed Systems Concepts and Design*, 4th ed. Reading, HA: Addison-Wesley, 2005.

Hiller, F. S., and G. J. Lieberman. *Introduction to Operations Research*, 8th ed. New York: McGraw-Hill, 2005.

Kepaptsoglou, K. "Transit Route Network Design Problem: Review." *Journal of Transportation Engineering*, Vol. 135, No. 8, 2009, pp. 491–505.

Kurose, J. F., and K. W. Ross. *Computer Networking*. New York: Pearson Addison Wesley, 2005.

Malakooti, B. "Unidirectional Loop Network Layout by a LP Heuristic and Design of Tele-Communications Networks." *Journal of Intelligent Manufacturing*, Vol. 15, 2004, pp. 117–125.

Nahmias, S. *Production and Operations Analysis*, 6th ed., New York: McGraw-Hill, 2008.

Taha, H. *Operations Research: An Introduction*, 8th ed. Upper Saddle River, NJ: Prentice Hall, 2006.

Multiobjective/Energy Routing

Barritt, B. J., S. Sheikh, C. Al-Najjar, and B. Malakooti. "Mobile Ad-Hoc Network Broadcasting: A Multi-Criteria Approach." *International Journal of Communication Systems*, Vol. 24, No. 4, 2011, pp. 438–460.

Boffey, B., F. R. F. García, G. Laporte, J. A. Mesa, and B. P. Pelegrín. "Multiobjective Routing Problems." *Top*, Vol. 3, No. 2, 1995, pp. 167–220.

Climaco, J. C. N., J. M. F. Craveirinha, and M. M. B. Pascoal. "A Bicriterion Approach for Routing Problems in Multimedia Networks." *Networks*, Vol. 41, No. 4, 2003, pp. 206–220.

Coutinho-Rodrigues, J. M., J. C. N. Clímaco, and J. R. Current. "An Interactive Bi-Objective Shortest Path Approach: Searching for Unsupported Nondominated Solutions." *Computer and Operation Research*, Vol. 26, No. 8, 1999, pp. 789–798.

Current, J. R., C. S. Revelle, and J. L. Cohon. "An Interactive Approach to Identify the Best Compromise Solution for Two Objective Shortest Path Problems." *Computers & Operations Research*, Vol. 17, No. 2, 1990, pp. 187–198.

Guo, Z., S. Sheikh, C. Al-Najjar, H. Kim, and B. Malakooti. "Mobile Ad Hoc Network Proactive Routing with Delay Prediction Using Neural Network." *Wireless Networks*, Vol. 16, No. 6, 2010, pp. 1601–1620.

Guo, Z., S. Sheikh, C. Al-Najjar, M. Lehman, and B. Malakooti. "Energy Aware Proactive Optimized Link State Routing in Mobile Ad-hoc Networks." *Applied Mathematical Modeling*, Vol. 35, No. 10, 2011a, pp. 4715–4729.

Guo, Z., S. Sheikh, C. Al-Najjar, and B. Malakooti. "Multi-Objective OLSR for Proactive Routing in MANET with Delay, Energy, and Link Lifetime Predictions." *Applied Mathematical Modeling*, Vol. 35, No. 3, 2011b, pp. 1413–1426.

Henig, M. I. "The Shortest Path Problem with Two Objective Functions." *European Journal of Operational Research*, Vol. 25, No. 2, 1986, pp. 281–291.

Malakooti, B., and K. Bhasin. "Roadmap for Developing Reconfigurable Intelligent Internet Protocol for Space Communication Networks." 2005 IEEE International Conference on Networking, Sensing and Control, Tucson, AZ, March 2005.

Malakooti, B., and I. Thomas. "A Distributed Composite Multiple Criteria Routing Using Distance Vector." IEEE International Conference on Networking, Sensing and Control (ICNSC06), Ft. Lauderdale, FL, April 23–25, 2006.

Malakooti, B., I. Thomas, S. K. Tanguturi, S. Gajurel, H. Kim, and K. Bhasin. "Multiple Criteria Network Routing with Simulation Results." IERC, May 2006.

Warburton, A. "Approximation of Pareto Optima in Multiple-Objective, Shortest-Path Problems." *Operational Research*, Vol. 35, No. 1, 1987, pp. 70–79.

EXERCISES

Introduction (Section 11.1)

1.1 Discuss two examples of simple energy systems which were not discussed in this chapter.

1.2 Discuss two examples of complex energy systems which were not discussed in this chapter.

1.3 Discuss energy bicriteria trade-off characteristics of two examples which were not discussed in this chapter.

Energy Perspective (Section 11.2)

2.1 Based on the information provided in Section 11.2.1, predict global warming, population growth, and Earth's carrying capacity for years 2025, 2050, and 2100. What are the approaches to control such trends and what are the positive consequences of such control action?

2.2 Answer the following questions in regard to natural gas versus biomass:
   (a) What are the advantages of using these energy sources?
   (b) What are the disadvantages of using these energy sources?
   (c) For what applications would these energy sources be the most beneficial?

2.3 Consider oil and wind energy. Answer parts (a)–(c) of exercise 2.2.

2.4 Consider coal and solar energy. Answer parts (a)–(c) of exercise 2.2.

2.5 Consider nuclear energy and hydro power. Answer parts (a)–(c) of exercise 2.2.

2.6 A machine uses 1000 J/second. The cost of electricity to run the machine is $0.50/kWh. How much will it cost to run the machine for 10 h?

2.7 A country uses 51.8 quad of energy per year. If energy in that country costs $2.00/kWh, how much money does the country spend on energy each year?

Multicriteria Energy Decisions (Section 11.3)

3.1 Discuss one example (or case study) for each of the seven principles for sustainable energy systems. Show how these principles save energy for these cases.

3.2 Consider the following table for four different sources of energy:

Resource, minimize	Natural Gas	Coal	Wind	Solar
Min. f_1, cost ($/million BTU)	$25	$30	$35	$65
Min. f_2, environmental impact	0.5	0.8	0.3	0.2
Max. f_3, availability/accessibility	0.8	0.85	0.60	0.45
Max. f_4, renewability of energy	0	0	0.8	0.9

(a) Consider the bicriteria problem of cost versus environmental impact and graphically show if any of the given four sources are inefficient. Suppose that the weights of importance for the bicriteria problem of cost versus environmental impact are (0.7,0.3), respectively. Rank all alternatives.

(b) Consider the four-criteria problem and identify if any of these resources are inefficient. Suppose that the weights of importance for the four-criteria problem are (0.4,0.2, 0.2,0.2), respectively. Rank all alternatives.

3.3 Consider the following table for five different sources of energy:

Resource, minimize	Natural Gas	Coal	Wind	Solar	Nuclear
Min. f_1, cost ($/million BTU)	$40	$10	$55	$70	$35
Min. f_2, environmental impact	0.4	0.9	0.2	0.2	0.7
Max. f_3, availability/accessibility	0.6	0.8	0.4	0.3	0.6
Max. f_4, renewability of energy	0.1	0.1	0.7	0.8	0.3

Respond to parts (a) and (b) of exercise 3.2.

Shortest Route Algorithm[†] (Section 11.4.1)

4.1 Consider the network shown below. Use Dijkstra's algorithm to:
(a) Find the shortest path from node A to all other nodes (show numerically and graphically).
(b) Find the shortest path from node C to all other nodes (show numerically and graphically).

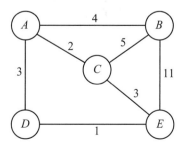

4.2 Consider the network shown below. Use Dijkstra's algorithm to:

(a) Find the shortest path from node A to all other nodes (show numerically and graphically).

(b) Find the shortest path from node C to all other nodes (show numerically and graphically).

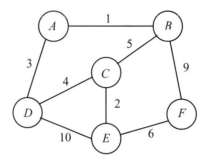

4.3 Consider the network shown below. Use Dijkstra's algorithm to:

(a) Find the shortest path from node D to all other nodes (show numerically and graphically).

(b) Find the shortest path from node E to all other nodes (show numerically and graphically).

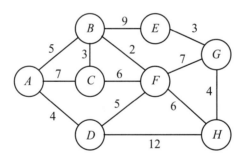

Energy Procurement in Distributed Systems[†] (Section 11.4.2)

4.4 Consider exercise 4.1. Suppose node A is the energy user and other nodes are energy generators. The user demand is 1500 for the given period. The selling price and the capacity of each generator for the given period for each node are as follows: B, $100, 500; C, $110, 600; D, $105, 800; E, $90, 400. Find the energy purchasing plan for node A.

4.5 Consider exercise 4.2. Suppose node A is the energy user and other nodes are energy generators. The user demand is 20,000 for the given period. The selling price and the capacity of each generator for the given period for each node are as follows: B, $50, 5,000; C, $42, 6,000; D, $48, 4,000; E, $36, 9,000; F, $40, 3,500. Find the energy purchasing plan for node A.

4.6 Consider exercise 4.3. Suppose node A is the energy user and other nodes are energy generators. The user demand is 500 for the given period. The selling price and the capacity of each generator for the given period for each node are as follows: B, $350, 100; C, $341, 85; D, $342, 95; E, $345, 200; F, $344, 150; G, $340, 110; H, $337, 90. Find the energy purchasing plan for node A.

Multicriteria Energy Routing[†] (Section 11.4.3)

4.7 Consider the network shown below. Suppose the weights of importance are (0.75, 0.25), respectively. Use the multicriteria Dijkstra algorithm to:

(a) Find the normalized weights

(b) Find the best path from node C to all other nodes

(c) Show two efficient paths from C to D

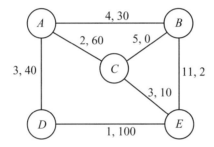

4.8 Consider the network shown below. Suppose the weights of importance are $(0.6, 0.4)$. Use the multicriteria Dijkstra algorithm to:

(a) Find the normalized weights

(b) Find the best path from node A to all other nodes

(c) Show two efficient paths from A to F

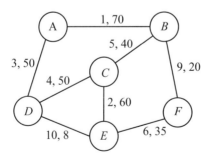

4.9 Consider the network shown below. Suppose the weights of importance are $(0.2, 0.8)$. Use the multicriteria Dijkstra algorithm to:

(a) Find the normalized weights

(b) Find the best path from node A to all other nodes

(c) Show two efficient paths from A to H

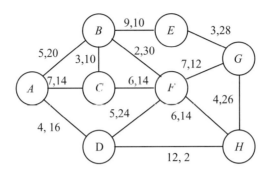

Energy Operations Optimization (Section 11.5.1)

5.1 Consider a house with four users, two generators, and one saver. The following tables and figure show the required information, including transportation costs, energy demand, cost per unit of energy for generators, maximum capacity of generators, initial energy in saver 1, and minimum and maximum saver energy levels.

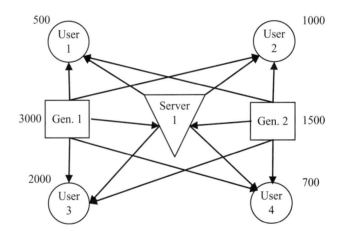

c_{ij}, Trans. Cost		User j			
from i to j		**1**	**2**	**3**	**4**
Gen.	**1**	1.5	2	0.75	1
i	**2**	2	1.8	0.6	0.4

c_{pj}, Trans. Cost		User j			
from p to j		**1**	**2**	**3**	**4**
Saver	**1**	1	0.5	1	0.5
Demand D_j (BTU)		500	1000	2000	700

c_{ip}, Trans. Cost from i to p		Saver p
		1
Gen.	**1**	1
i	**2**	1.2

Gen.	Gen. Costs, g_i (\$)	Gen. Cap., K_i (BTU)
1	30	3000
2	40	1500

Saver	Init. Enr $q_{p,0}$ (BTU)	Min. Enr $q_{p,min}$	Max. Enr $q_{p,max}$
1	200	90	720

(a) Suppose that the predicted change in energy prices is +$5. Formulate and solve the energy operation program for this problem. Explain the meaning of the result.

(b) Assume that saver 1 can only deliver energy to users 1 and 3. Formulate and solve the energy operations problem.

5.2 Consider exercise 5.1. Answer parts (a) and (b) of exercise 5.1 given that the predicted change in energy prices is −$5.

5.3 Consider an energy system problem where three generators and two savers can serve the four users. The following figure and tables show required information, including transportation costs, energy demand for users, cost per unit of energy for generators, maximum capacity of generators, initial energy in savers, and minimum and maximum savers energy levels.

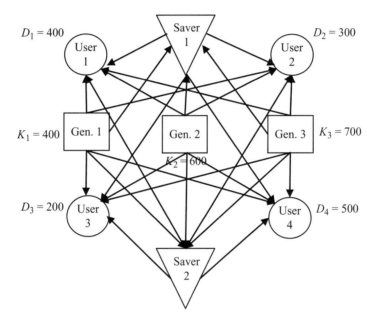

c_{ij}, Trans. Cost		User j			
from i to j		**1**	**2**	**3**	**4**
Gen.	**1**	2	4	3	5
i	**2**	3	4	5	7
	3	6	2	1	5

c_{pj}, Trans. Cost		User j			
from p to j		**1**	**2**	**3**	**4**
Saver p	**1**	2	1	6	4
	2	5	2	3	5
Demand D_j		400	300	150	800

c_{ip}, Trans. Cost		Saver p	
from i to p		**1**	**2**
Gen.	**1**	0.5	0.75
i	**2**	0.25	1
	3	0.5	0.5

Gen.	Gen. costs, g_i ($)	Gen. Cap., K_i (BTU)
1	0.5	400
2	0.6	600
3	0.4	700

Saver	Init. Enr $q_{p,0}$	Min. Enr $q_{p,min}$	Max. Enr $q_{p,max}$
1	40	10	80
2	35	5	100

(a) Suppose that the predicted change in energy price is +$1 for both savers. Formulate and solve the energy operation program for this problem. Explain the meaning of the result.

(b) Assume that saver 1 can only deliver energy to users 1 and 2. Formulate and solve the energy operations problem.

5.4 Consider exercise 5.3. Answer parts (a) and (b) of exercise 5.3 given that the predicted change in energy prices is −$1.

Energy Systems Design Optimization (Section 11.5.2)

5.5 Consider exercise 5.1. The investment costs of the generators are $20,000 and $30,000, respectively. For this problem, the capacity of generators 1 and 2 are both 4200. Formulate and solve the energy design problem. Explain the meaning of the result.

5.6 Consider exercise 5.3. The investment costs of the generators are $2000, $3000, and $1000, respectively. Suppose the capacity of generators 1, 2, and 3 have been increased to 1200, 1000, and 1300, respectively. Formulate and solve the energy design problem. Explain the meaning of the result.

Period-by-Period Optimization† (Section 11.5.3)

5.7 Consider exercise 5.1. Suppose that two periods are considered. The demands and cost information for this problem are given in the following table:

Period	Demand User 1	Demand User 2	Demand User 3	Demand User 4	Predicted Change in Energy Price for Saver 1, $\Delta s_{1,t}$
1, Night	500	1000	2000	700	+10
2, Day	800	1000	2200	800	−5

Suppose that the information for each of the two periods is known only at the beginning of each period. Find the optimal solution for both periods.

5.8 Consider exercise 5.3. Suppose that two periods are considered. The demands and cost information for this problem are given in the following table:

Period	Demand User 1	Demand User 2	Demand User 3	Demand User 4	Predicted Change in Energy Price for Saver 1, $\Delta s_{1,t}$	Predicted Change in Energy Price for Saver 2, $\Delta s_{2,t}$
1, Night	500	250	200	600	+0.4	+0.7
2, Day	400	300	150	500	−0.3	−6

Suppose that the information for each of the two periods is known only at the beginning of each period. Find the optimal solution for both periods.

Aggregate Multiperiod Optimization[†] (Section 11.5.4)

5.9 Consider exercise 5.7. Suppose that information for both periods is known at the beginning of the first period. Find the optimal solution for both periods.

5.10 Consider exercise 5.8. Suppose that information for both periods is known at the beginning of the first period. Find the optimal solution for both periods.

Multiobjective Optimization of Energy Systems[†] (Section 11.5.5)

5.11 Consider the energy system presented in exercise 5.1. Replace the generators in exercise 5.1 with two new generators as presented in the following table:

Generators	Energy Source	Generating Costs ($)	Capacity (K_i)	Environmental Factor
Biofuel	Solar Energy	30	4500	0.5
Solar energy	Oil	70	4000	0.1

(a) Generate five bicriteria alternatives evenly distributed in terms of environmental impact.

(b) Choose the best alternative using an additive utility function where weights of importance are 0.4 and 0.6 for the two normalized objectives of cost and environmental impact, respectively.

5.12 Consider the energy system presented in exercise 5.3. Replace the generators in exercise 5.2 with three new generators as presented in the following table. The initial levels of savers 1 and 2 are 10 and 5, respectively.

Generators	Energy Source	Generating Costs ($)	Capacity (K_i)	Environmental Factor
1	Solar Energy	80	1700	0.2
2	Oil	20	1600	0.9
3	Nuclear Energy	55	1800	0.5

Respond to parts (a) and (b) of exercise 5.11.

Energy Systems with One Input and One Output (Section 11.6.1)

6.1 Consider three power plants with different energy generators (fossil fuel, solar panel, and wind turbine). The inputs and outputs for the three generators are shown in the table below. The manager of the plants wishes to compare the energy operational efficiencies of the plants and determine inefficient plants.

Generator no.	1	2	3
Output: energy	895	760	710
Input: cost	120	236	197

(a) What is the energy productivity of the three power plants?

(b) What is the energy efficiency of the three power plants?

6.2 Compare the energy efficiencies of four different cars and determine if any cars are inefficient. The inputs and outputs for four new cars are shown in the table below.

Generator no.	1	2	3	4
Output: miles traveled for given input	1.5	1.7	1.3	18
Input: amount of Gas in gallons	0.1	0.15	0.17	1.8

(a) What is the energy productivity of the four cars?

(b) What is the energy efficiency of the four cars?

Energy Systems with Multiple Inputs and Outputs[†] (Section 11.6.2)

6.3 A laundromat has five different drying machines. The owner of the lanundromat wishes to compare the operational efficiencies of five drying machines and determine if any of them are inefficient. The inputs are electricity and maintenance costs. The outputs are percentage of clothes dried and smoothness of machine operations (measured on a 0–1 scale where 0 is the worst and 1 is the best). The inputs and outputs are shown in the following table. Find the efficiencies of all five drying machines.

	Input		Output	
Unit r	Electricity Consumed, $j = 1$	Maintenance Costs, $j = 2$	Percentage of Clothes Dried, $i = 1$	Smoothness of Machine Running, $i = 2$
Drier 1	150	60	89	0.15
Drier 2	102	55	85	0.30
Drier 3	180	78	80	0.42
Drier 4	90	40	90	0.25
Drier 5	200	120	83	0.18

6.4 There are three power plants with different generators (fossil fuel, solar panel, and wind turbine) in an industrial zone. The three inputs and three outputs for the these generators are shown in the table below. The inputs are fuel consumed in BTU, maintenance costs, and manpower. The outputs are generated electricity, plant availability (the percentage of time the power plant is available to produce electricity), and positive environmental impact (measured on a 0–1 scale where 0 is the worst and 1 is the best). Find the efficiencies of the three power plants.

Unit r	Input (b_{rj})			Output (a_{ri})		
	Fuel (BTU), $j = 1$	Maintenance Costs ($), $j = 2$	Operational Man-Hours, $j = 3$	Electricity, $i = 1$	Percentage Availability, $i = 2$	Positive Environmental Impact, $i = 3$
Plant 1 (fossil fuel)	120	287	214	895	61.5	0.26
Plant 2 (solar panel)	236	410	95	760	80.4	0.65
Plant 3 (wind turbine)	197	316	110	710	70.9	0.67

Wind Power Characteristics[†] (Section 11.7.2)

7.1 A company plans to install a wind turbine 1000 m above sea level (where the air density is 1.0 kg/m^3). The wind is coming mainly from one direction with an annual speed of 9 m/s. Suppose there is no wind obstacle in the vicinity of the wind farm. Also suppose that the mechanical efficiency is 0.8. Calculate the generated power for four different rotor diameter sizes of 20, 40, 60, and 80 m.

Rotor (d)	20	40	60	80
c_p	0.35	0.45	0.47	0.41

7.2 Consider installing a wind turbine at sea level (where the air density is 1.225 kg/m^3). The wind is coming mainly from one direction with an annual speed of 12 m/s. Suppose there is no wind obstacle in the vicinity of the wind farm. Also suppose that the mechanical efficiency is 0.8. Calculate the generated power for four different rotor diameter sizes of 5, 25, 50, and 75 m.

Rotor (d)	5	25	50	75
c_p	0.32	0.44	0.48	0.43

Break-Even Analysis[†] (Section 11.7.3)

7.3 Consider exercise 7.1. Suppose that the average price of 1 kWh is 9 cents and there are 5000 h of operations per year. Variable costs per year and fixed costs for four different rotor diameter sizes of 20, 40, 60, and 80 m are shown in the following table. Find the break-even point for each of the four rotor diameter sizes. Also show the solutions graphically.

Rotor (d)	20	40	60	80
Variable costs per year ($)	13,000	20,000	35,000	50,000
Fixed costs ($)	250,000	500,000	1,500,000	2,500,000

7.4 Consider exercise 7.2. Suppose that the average electricity price is 11 cents/kWh. There are 5000 hours of operation per year. Variable costs per year and fixed costs for four different rotor diameter sizes of 5, 25, 50, and 75 m are shown in the following table. Find the break-even point for each of the four rotor diameter sizes. Also show the solutions graphically.

Rotor (d)	5	25	50	75
Variable costs per year ($)	250	20,000	10,000	12,000
Fixed costs ($)	25,000	350,000	1,200,000	2,300,000

CHAPTER 12

CLUSTERING AND GROUP TECHNOLOGY

12.1 INTRODUCTION

A "cluster" is a group of objects that are "similar" to each other and are "dissimilar" to members of other clusters in terms of their attributes. Clustering is an approach for identifying clusters. Clustering is an essential part of the human cognitive, learning, and discovering processes. Clustering has many applications, including classification of parts and processors in manufacturing systems, categorization of behavior in social sciences, identfication of market segments for marketing, classification of diseases in medicine, classification of animals or plants, and pattern recognition in artificial intelligence. In this chapter, the group technology problem is used to illustrate different clustering methods. However, the clustering methods of this chapter are applicable to a variety of clustering problems. In group technology, different machines (as objects) are clustered for the purpose of processing similar parts (as attributes). Group technology simplifies and streamlines complex operational systems (see Section 12.3). Ideally, each cluster of machines would process an exclusive group of parts (i.e., parts belonging to one cluster are not processed by machines of other clusters).

An object is a single entity having a number of attributes. Based on the similarity of attributes, objects are clustered into different groups. For example, in classifying insects, the objects could be grasshoppers, bees, and ants. The attributes could be the number of legs and whether the object has wings. Generally, in clustering problems, the number of objects is much higher than the number of attributes. In most applications, the resulting clusters will be different depending on the designation of objects or attributes. In some problems, objects and attributes can be arbitrarily decided. For example, human races can be objects

Note: Advanced material that can be omitted without loss of generality will be indicated by a dagger.

Operations and Production Systems with Multiple Objectives, First Edition. Behnam Malakooti.
© 2014 John Wiley & Sons, Inc. Published 2014 by John Wiley & Sons, Inc.

and skin colors can be attributes, or vice versa; also in group technology, machines can be labeled as attributes and parts as objects or vice versa. In ideal situations, objects based on their attributes should be clustered into mutually exclusive clusters. That is, ideally each cluster should have its own objects and these objects are not common with other clusters. However, exclusive separation of attributes does not occur in many clustering problems. In fact, the clustering problem will be trivial (i.e., it can be easily solved) if such mutual exclusiveness of clusters occurs.

Clustering can be used to provide additional information besides the classification of objects. These include:

- *Predicting Attributes of Object*   If it is determined that an object belongs to a certain cluster, then the object should have the attributes of that group. For example, one can classify plants as green objects having cellulose. Then any object that is declared a plant will have the attributes of a plant.
- *Reduction of Data*   Clustering can be used to reduce the amount of data for describing an object. For example, instead of providing a detailed description of an oak tree, one can refer to it by its category, "tree".
- *Determining Outliers in Group*   If an object does not have all the attributes of the group it is assigned to, then the object can be declared as an outlier to that group. For example, a green animal may be wrongly attributed to be a plant because of its color; however, it does not have other plant attributes.

1. *Partitional Clustering*   In partitional clustering, the set of objects can be partitioned so that the most similar objects are grouped in the same cluster. There are three types of partitional clustering:

- *Exclusive Clustering*   In this method, each object can belong to only one cluster, that is, the clusters are mutually exclusive.
- *Overlapping (Fuzzy) Clustering*   The clusters formed by these methods are fuzzy and some objects may belong to two or more clusters. Fuzzy clustering performs a minimization of distance from cluster centers (similar to K-means) but also allows partial membership.
- *Probabilistic Clustering*   Probabilistic clustering uses a certain probability distribution for the formation of clusters that can result in cluster shapes other than the circular one, for example, elliptical Gaussian.

2. *Hierarchical Clustering and Similarity Coefficient Approaches*   The objects are arranged in a hierarchy based on their dissimilarity (distances) or their similarities (closeness). Different levels in the hierarchy represent the clustering of the objects at different similarity levels.

3. *Two-Modal and One-Modal Clustering*   Clustering can be done based on using the similarity of objects in terms of their attributes (which is called two modal) or it can be done based on the similarity of objects to each other (which is called one modal). Usually after similarities are measured as a two modal (object to attribute), the problem is converted into a one modal (object to object).

4. *Multiple Clustering*   In multiple clustering, the same set of objects can be clustered differently based on which attributes are considered. For example, consider the set of six

Attribute 2

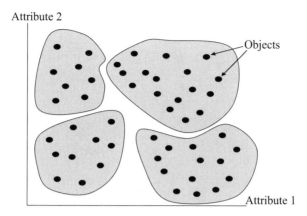

Figure 12.1   Two-dimensional clustering where four clusters are identified by visual inspection.

people: {Anna, Bob, Carey, David, Elsa, Frank}. A clustering based on age may result in two separate clusters: children {Anna, Elsa, Frank} and adults {Bob, Carey, David}. However, a clustering with respect to gender may result in different cluster sets: female {Anna, Carey, Elsa} and male {Bob, David, Frank}.

5. *Binary or Continuous Attributes*   The attributes of objects can be represented by binary or continuous real numbers. In binary, each attribute value can be either one or zero. In continuous, the attributes can assume any real value.

6. *Two-Dimensional Clustering*   When there are two attributes, the clustering problem becomes easy because it can be presented graphically. Each object is graphed in a two-dimensional space where each axis represents a given attribute. Figure 12.1 demonstrates an example of multiple objects having two attributes; for example, people are clustered based on their income and their level of education. Four clusters are identified. For the two-dimensional case, clustering can be performed by visual inspection graphically separating objects into exclusive sets.

In this chapter, we will cover several different clustering methods as listed in Table 12.1. Each method can be used for specific types of data and problems. For binary attributes, one can use rank order clustering, similarity coefficient (binary), or P-median. For continuous data, one can use the K-means or similarity coefficient (continuous) methods. The multicriteria clustering approach can be used for the selection of the best multicriteria clustering alternative. It is covered for all clustering methods.

A summary of the clustering methods covered in this chapter is provided in Table 12.1.

Multiple-Criteria Clustering and Optimal Number of Clusters   The quality and benefits of clustering depend on the number of clusters. An important question in clustering problems is to determine the best number of clusters for the given problem. To respond to this question, we have also developed a multicriteria approach for each clustering method presented in this chapter. Most clustering problems can be shown by an elbow graph where the trade-off between maximizing the number of clusters (to maximize the usage of clustering) and the distinguishability of clustering occurs. See Figure 12.2. Distinguishability of clusters is defined in Section 12.7.

TABLE 12.1 Classification of Clustering Methods

Method	Type of Attribute Measurement	One or Two Modal	Solution Approach	Number of Clusters	Type of Similarity Measurement	Covered in Section
Visual	Continuous or binary	Both	Intuition	TBD	Intuitive	12.3
Rank order cluster	Binary (0 or 1)	Two	Rearrange columns and rows	TBD	Jaccard index (similarity)	12.4
Similarity coefficient: binary	Binary (0 or 1)	One	Hierarchical, add members to clusters	Given	Jaccard index (similarity)	12.5
Hierarchical continuous	Continuous (real no.)	One	Hierarchical, add members to clusters	Given	Euclidean distance	12.5
P-median	Binary (0 or 1)	One	Optimization; use centers	Given	Rectilinear distance	12.6
P-median	Continuous	One	Optimization; use centers	Given	Euclidean distance	12.6
K-Means	Continuous (real no.)	One	Iteratively change membership; use centers	Given	Euclidean distance	12.7
Multicriteria clustering	Binary or continuous	Both	All methods	TBD	All methods	12.4–12.8

TBD, To Be Decided.

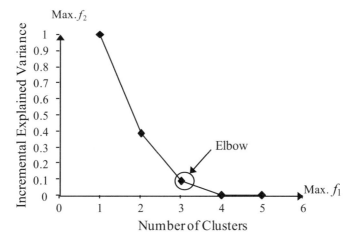

Figure 12.2   Elbow criterion where the highest trade-off for distinguishability of clustering vs. number of clusters occurs.

Supplement S12.1.docx describes how to find the optimal number of clusters using bicriteria clustering method. It also covers applications of clustering.

Types and Scaling of Attributes   Attributes can be binary, nominal, ordinal, continuous linear scaled, or continuous ratio scaled. Attributes can be scaled using 0 to 1, or −1 to 1 or using mean and standard deviation values. Depending on scaling values the clustering results may vary considerably.

Supplement S12.2.docx describes types and scaling of attributes used in clustering.

12.2   CLUSTERING DATA AND MEASUREMENTS

12.2.1   Types of Data Representation

Two-Modal Formulation   The two-modal formulation refers to a matrix with columns corresponding to the objects to be clustered and the rows corresponding to a set of attributes that describe the objects. Suppose there are m objects and n attributes; then the data are presented as an $m \times n$ matrix. This matrix is not symmetric. Rank order clustering (Section 12.4) uses the two-modal formulation, but methods of Sections 12.5 and 12.6 convert the two-modal data into one-modal data.

One-Modal Formulation   In one-modal formulation, the attribute information is presented as pairwise measures of either the similarity or dissimilarity of the objects. In this case, the data for the m objects will be shown in an $m \times m$ matrix. The value of element (q, j) in the matrix represents a comparison between the qth object and the jth object. This matrix is always symmetric. The K-means method (Section 12.7) uses one-modal formulation.

Example 12.1   One- and Two-Modal Continuous Attributes—Dissimilarity   Consider the data points in the table below. Find a dissimilarity matrix for objects a_1, a_2, a_3, and a_4.

Alternative	a_1	a_2	a_3	a_4
Attribute 1	13	23	15	20
Attribute 2	60	40	33	55

Two Modal Representation

The above table represents the two-modal formulation of the data. Now suppose that the Euclidean distance (see next section) is used to measure the distance between each pair of objects [e.g., $d(a_1, a_2)$ is the Euclidean distance between a_1 and a_2]. The resulting matrix of the above table is shown in the table below as a one-modal representation. The lower these values, the more similar are the objects. For example, alternatives a_1 and a_4 are similar and alternatives a_1 and a_3 are not similar.

	a_1	a_2	a_3	a_4
a_1	0	22.4	27.1	8.6
a_2	22.4	0	10.6	15.3
a_3	27.1	10.6	0	22.6
a_4	8.6	15.3	22.6	0

Example 12.2   One- and Two-Modal Binary Attributes—Similarity   Consider four people (as objects). Table 12.2 presents the set of four attributes for each of the four people. Find the similarity index for the four people (objects). Show the one-modal representation of this problem.

The data provided in Table 12.2 give the two-modal representation of this problem. To generate the one-modal representation, convert the data to a similarity matrix using the simple matching coefficient approach by comparing each pair of objects presented in the columns of Table 12.2. The results are shown in Table 12.3. Note that the one-modal data are symmetric, where $a_{ij} = a_{ji}$.

TABLE 12.2   Two-Modal Binary Data For Four People

Attributes \ Objects	Anne	Laura	Tom	Scott
Fair hair	1	0	1	0
Wears glasses	0	1	1	0
Left-handed	0	0	1	1
Married	1	0	0	1

TABLE 12.3   One-Modal Binary Similarity Matrix for Four People

Attributes / Objects	Anne	Laura	Tom	Scott
Anne	1.00	0.25	0.25	0.50
Laura	0.25	1.00	0.50	0.25
Tom	0.25	0.50	1.00	0.25
Scott	0.50	0.25	0.25	1.00

The simple matching coefficient approach for finding the similarity is presented by the following example. The similarity between Anne and Laura can be determined by finding how many identical row entities they have in common where $1 \neq 0$, $0 \neq 1$, $0 = 0$, and $1 \neq 0$ for rows 1–4, respectively, in Tables 12.2 and 12.3. Since Anne and Laura have only one of the four attributes in common (they are both not left-handed), the similarity matching coefficient of (Anne, Laura) $= (0 + 0 + 1 + 0)/4 = 1/4 = 0.25$. The higher this value is, the more similar the objects are.

12.2.2   Distance Measurement Metrics

Clustering seeks to group objects into different subsets according to some kind of similarity among the objects. The concept of similarity is not as simple as it appears. In order to be objective, quantitative measures are needed. Also, a similarity criterion is needed to determine the final clustering of objects. The similarity measurement and the number of clusters are two factors that should be considered in a clustering process. The DM can specify the number of clusters before solving the problem. However, some clustering methods support finding the best number of clusters while identifying the number of members of each cluster which is directly related to how similarity (or dissimilarity) of groups of objects is measured.

The distance metric (as presented below) is a popular choice for measuring similarity. However, this method is not suitable for all applications. For example, when using attributes with nominal values, the distance metric has no meaning. Also, in the last section, we used the simple matching coefficient approach (which is not a distance metric) for measuring the similarity of humans as objects, and in Section 12.5.1 we will use the Jaccard metric to measure similarity for specific types of binary data. However, the most common distance metric (to measure dissimilarity) is presented below. Given a set of objects, the distance metric is defined as a real-valued function d which satisfies the following three properties. Consider two vectors (of objects) $\mathbf{x}$, $\mathbf{y}$, where $d(\mathbf{x}, \mathbf{y})$ is the distance between $\mathbf{x}$ and $\mathbf{y}$. For three objects ($\mathbf{x}$, $\mathbf{y}$, and $\mathbf{z}$), the components of each object (x_i, y_i, and z_i) represent the value of the attribute i associated with each object, respectively:

1. Distinguishability of nonidenticals, $d(\mathbf{x}, \mathbf{y}) \neq 0$, and indistinguishability of identicals, $d(\mathbf{x}, \mathbf{x}) = 0$

2. Symmetry and nonnegativity: $d(\mathbf{y}, \mathbf{z}) = d(\mathbf{z}, \mathbf{y}) \geq 0$

3. Triangle inequality: $d(\mathbf{w}, \mathbf{z}) \leq d(\mathbf{w}, \mathbf{y}) + d(\mathbf{y}, \mathbf{z})$ (i.e., going directly from $\mathbf{w}$ to $\mathbf{z}$ is shorter than or equal to a path detouring through $\mathbf{y}$)

The most common similarity measure is the distance metric. A generalization of the distance metric for n-dimensional data (n attributes) is given by the Minkowski equation:

$$d\,(\mathbf{x}, \mathbf{y}) = \left(\sum_{i=1}^{n} |x_i - y_i|^L\right)^{1/L} \tag{12.1}$$

The Manhattan (or rectilinear) distance results when $L = 1$:

$$d\,(\mathbf{x}, \mathbf{y}) = \sum_{i=1}^{n} |x_i - y_i| \tag{12.2}$$

The Euclidean metric is a special case of the Minkowski equation where $L = 2$:

$$d\,(\mathbf{x}, \mathbf{y}) = \left(\sum_{i=1}^{n} |x_i - y_i|^2\right)^{1/2} \tag{12.3}$$

In the Minkowski equation, as L goes to infinity, the limit of $d(\mathbf{x}, \mathbf{y})$ becomes

$$d(\mathbf{x}, \mathbf{y}) = \text{Max}\{|x_i - y_i| \quad \text{for } i = 1, 2, \ldots, n\} \tag{12.4}$$

This is also known as the infinity norm.

Figure 12.3 illustrates a two-dimensional case where $\mathbf{x} = (3,5)$, $\mathbf{y} = (6,4)$, and the following Minkowski equations can be used:

Type	Distance Measurement
When $L = 1$ (Manhattan)	$d(\mathbf{x}, \mathbf{y}) = |3 - 6| + |5 - 4| = 4$
When $L = 2$ (Euclidean)	$d(\mathbf{x,y}) = \sqrt{(3 - 6)^2 + (5 - 4)^2} = 3.16$
When $L \to \infty$ (infinity norm)	$d(\mathbf{x}, \mathbf{y}) = \text{Max}\{|3 - 6|, |5 - 4|\} = 3$

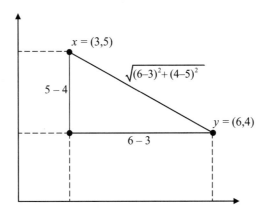

Figure 12.3 Distance metric in two-dimensional case.

12.3 GROUP TECHNOLOGY CLUSTERING

12.3.1 Group Technology Problem

Group technology is the most important application of clustering in manufacturing systems. Group technology provides methods of clustering machines (as objects) based on the parts (as attributes) that they process. Groups of parts that are primarily processed by a given cluster of machines are called a part family. A facility may consist of several cells where each cell represents a cluster of machines (objects) to produce or assemble one family of parts.

Group technology is a strategy that combines the productivity of mass production with the flexibility of a job shop environment. One of the goals of group technology is to streamline production operations to achieve a flexible environment. Setup time reduction is an obvious payoff from grouping parts into families because of the reduction in machine setup and materials handling. Ideally, the cellular environment would be conducive to just-in-time (JIT) manufacturing where items can be processed at the needed time without finishing early or late. Smoother production flow can also be achieved by cellular manufacturing. Consider a job shop layout based on groups of similar machines. Parts may be processed by machines at different locations creating a complex routing, as shown in Figure 12.4a. With a cellular system, a group of machines (which are not usually similar) are arranged to handle a group of similar parts. Each part remains in its cell throughout the manufacturing process, as shown in Figure 12.4b.

Some of the objectives of cellular manufacturing are:

1. Maximized space utilization
2. Minimized material handling volume/flow
3. Minimized work-in-process inventory
4. Minimized job throughput time
5. Maximized flexibility
6. Maximized machine utilization

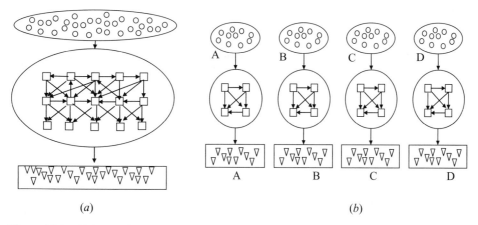

Figure 12.4 (a) Processing with one cluster, which has 15 machines; (b) processing four types of items by four clusters of processors, each cluster with four machines.

There are compromises that must be considered when adopting cellular manufacturing. Increased flexibility may come at the expense of higher equipment costs for having an increase in the number of machines. Also, there will be an increase in operations and labor costs for running several cells. Additionally, if clusters are not totally exclusive, some parts may be routed between cells. This may cause queuing problems and/or bottlenecks.

12.3.2 Classification and Coding

Part classification and coding refer to the process of identifying similarities between parts and categorizing them accordingly through a coding system. Typically, we separate part similarities into two categories. The first is design attributes, which refer to concrete attributes like size, shape, and material. The second category is process (manufacturing) attributes, which refer to characteristics that define how the part is processed and the sequence of processors required to make the part.

Classification and coding have a variety of applications and uses. For example, they can be used in design retrieval, automated process planning, and machine cell design. To develop a process plan for a new part, one can retrieve process plans of similar parts and use the existing plans as the basis for the new part. Thus, coding and classification can save man-hours of developing a process plan for similar parts.

There are several different systems of classification, and each must be customized to the particular application in which the system is used. In order to code components into appropriate categories, tables with a listing of symbols and characters that represent particular attributes can be created and used. The result of this should be a unique identification code that is assigned to each part. The code may be (a) numerical, (b) alphabetical, or (c) alphanumeric. There are three structures used in classification and coding.

Chain-Type Structure (Polycode)   In a chain-type structure, the meaning of each symbol of the code is always the same. It is completely independent of the sequence. This method is also known as polycode, meaning that there are possibly different combinations of codes for the same item. For illustration, consider presenting a person by (hair, eye, height); regardless of the position of each attribute, the same identification is obtained. For example, (green eyes, brown hair, short height) is the same as (brown hair, short height, green eyes).

Hierarchical Structure (Monocode)   In a hierarchical structure, each symbol of the code may have a different interpretation depending on the preceding symbol. This approach is also termed monocode. This system is much more structured than polycode. An example of this method is the Zip code used by the U.S. postal office, where beginning digits refer to the broader geographical areas and subsequent digits identify specific locations within it. Figure 12.5 represents a monocode system which consists of integers starting at 1. Each digit depends on the previous digit for its meaning. For example, the value of digit 3 can mean gold or silver. For example, 111 represents a large, square, gold part, but 121 represents a large, circle, gold part.

Hybrid Structure (Hybrid Code)   A hybrid or mixed-mode structure is a combination of the chain (polycode) and hierarchical (monocode) structures. One commonly used industrial classification system, known as Optiz, is an example of hybrid structures. Figure 12.6 shows how Optiz uses nine digits and four letters. This is split into sections. Digits 1–5

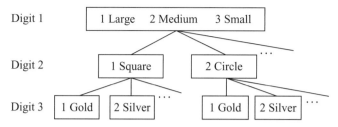

Figure 12.5   Example of monocode system.

represent the physical form of the object. Digits 6–9 are the supplementary code that represents features such as the actual manufacturing processes that the specific part undergoes. Finally there is a secondary code of four letters that depicts even more detailed aspects of the production and sequence of the part. The basic form for the Optiz classification is given below.

Field explanation	Form code	Supplementary code	Secondary code
Digits location	12345	6789	*ABCD*

Figure 12.6 shows the basic structure of the form and supplementary codes for the Optiz system; L and D are used to represent length and diameter, respectively.

12.3.3   Clustering by Visual Inspection and Intuition

Visual inspection and intuition make up a powerful technique for solving small clustering problems or problems that can be presented by two modal. To use the method,

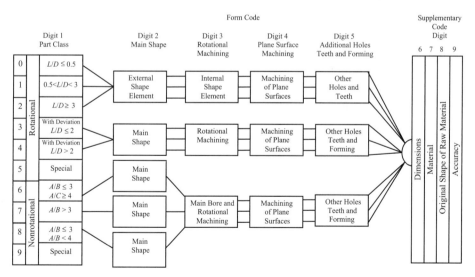

Figure 12.6   Optiz form and supplementary codes. (*Source:* From H. Optiz, *A Classification System to Describe Workpieces*, Permagon Press, 1970.)

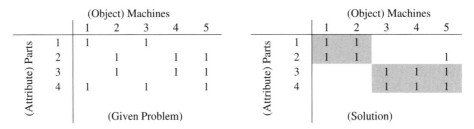

Figure 12.7 Visual inspection for five machines with four attributes.

first create an objects (machines)–attributes (parts) matrix by listing machines as columns and parts as rows. Place a 1 in each machine–part cell if the corresponding part visits (uses) the corresponding machine. Then, cluster the similar parts based on using similar machines by visual inspection of the matrix. Rearrange the rows and columns until there are clusters of 1's together for given sets of machines and parts. Then form cells of machines and families of parts by selecting logical groups of similar parts and machines. Make additional rearrangements of rows and columns for better part–machine grouping until no further improvements are apparent. To reduce material handling among different cells, minimize the number of exceptional elements. Exceptional elements are parts that must visit another cell because a machine in that cell is also required to process the given part. Visual inspection can be an easy, straightforward approach for small problems; however, as complexity increases, it can become overwhelming. This method provides no guarantee of optimality. The methods of the next sections can solve larger and more complex problems.

Example 12.3 Visual Inspection and Intuition Consider four parts that are processes by five machines as presented on the left side of Figure 12.7. Cluster these machines and parts.

The solution is given on the right side of Figure 12.7. The following solution for the machine–parts matrix depicts a logical grouping of machines into cells. There are two machine cells, $S_1 = \{1, 3\}$ and $S_2 = \{2, 4, 5\}$, and two parts families, $PF_1 = \{1, 4\}$ and $PF_2 = \{2, 3\}$.

The above solution can be summarized as follows:

Two-machine cells	$S_1 = \{M1, M3\}$	$S_2 = \{M2, M4, M5\}$
Two-part families	$PF_1 = \{P1, P4\}$	$PF_2 = \{P2, P3\}$
Exceptional element(s)?	Yes, part 4 visits cell 2	No

In this method, rearranging the rows and columns is completely up to the designer's discretion. In this example, there is only one exceptional element, part 4, which must visit machine 5 in cell S_2.

Objects (Machines) (*j*)

		1	2	...	*j*	...	*m*
Attributes Parts (*i*)	1	a_{11}	a_{12}	...	a_{1j}	...	a_{1m}
	2	a_{21}	a_{22}	...	a_{2j}	...	a_{2m}
	⋮	⋮	⋮	⋮	⋮	⋮	⋮
	i	a_{i1}	a_{i2}	...	a_{ij}	...	a_{im}
	⋮	⋮	⋮	⋮	⋮	⋮	⋮
	n	a_{n1}	a_{n2}	...	a_{nj}	...	a_{nm}

Figure 12.8 Machine–part matrix notation.

12.4 RANK ORDER CLUSTERING

12.4.1 ROC Method for Binary Data

The rank order clustering (ROC) method is based on rearranging the rows and columns of the object–attribute matrix such that clustering is visually apparent in the form of a diagonally ordered matrix. In the group technology problem, machines are treated as objects and parts are treated as attributes. However, the solution is the same if machines are treated as attributes and parts are treated as objects.

In group technology, there are a total of *m* objects (machines) represented by columns and *n* attributes (parts) represented by rows. A single element a_{ij} refers to part *i* visiting (being processed by) machine *j*. The value of a_{ij} is either 1 or 0:

$$a_{ij} = \begin{cases} 1 & \text{if part } i \text{ visits machine } j \\ 0 & \text{otherwise} \end{cases} \quad \text{for } i = 1, \ldots, n, \quad j = 1, \ldots, m$$

The part–machine matrix is shown in Figure 12.8.

The ROC method, also known as King's method, rearranges columns and rows of a given object–attribute (machine–part) matrix such that the matrix appears as a diagonal block from the top left to the bottom right. The ideal rearranged matrix will have values of zero on the upper right side and the lower left side of the diagonal block but will have values of unity in the diagonal area. An example of an ideal clustering is shown in Figure 12.9 for five parts and eight machines. In Figure 12.9, there are three completely separate clusters.

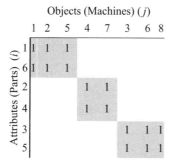

Figure 12.9 Ideal machine–part matrix.

This is an ideal clustering because each cell can operate without having interactions with other cells. The three sets of cell machine clusters are $S_1 = \{1, 2, 5\}$, $S_2 = \{4, 7\}$, and $S_3 = \{3, 6, 8\}$. Their corresponding part families (PF) are $PF_1 = \{1, 6\}$, $PF_2 = \{2, 4\}$, and $PF_3 = \{3, 5\}$, respectively.

In practice, ideal arrangements such as Figure 12.9 usually do not occur. The goal of ROC is to generate the closest shape to the ideal diagonal. The ROC method is based on giving very high scores to elements that appear in the top left corner and very low scores to elements that appear in the bottom right corner. In each iteration, rows and columns are rearranged toward the ideal diagonal shape until no more improvements can occur. Any element that has a value of 1 outside the blocked diagonal areas is called an exceptional element. The purpose of the ROC (King's) algorithm is to minimize the number of exceptional elements as the objective function of the optimization problem. Because ROC is heuristic, it may not always find the optimal solution for the stated objective.

Initial Phase Assign a binary weight (BW) to each column j in the matrix: $BW_j = 2^{m-j}$. Assign a binary weight to each row i in the matrix: $BW_i = 2^{n-i}$. These binary weight fixed values are associated with the column and row numbers; they do not change through the algorithm. For example, when $n = 9$, BW_i values are:

i	1	2	3	4	5	6	7	8	9
$BW_i = 2^{9-i}$	256	128	64	32	16	8	4	2	1

For example, $BW_2 = 2^{9-2} = 128$.

Algorithm

0. Start with an initial given (or arbitrarily arranged) machine–part matrix.
1. Calculate the decimal weight W_i of each row i using

$$W_i = \sum_{j=1}^{m} BW_j a_{ij} \qquad (12.5)$$

   where a_{ij} equals 1 if part i visits machine j and 0 otherwise.
2. Rearrange the rows in descending order of their decimal weight value. In case of a tie, order is arbitrary.
3. Calculate the decimal weight W_j of column j using

$$W_j = \sum_{i=1}^{n} BW_i a_{ij} \qquad (12.6)$$

4. Rearrange the columns in descending order of their decimal weight value. In case of a tie, order is arbitrary.
5. Repeat steps 1–4 until all row and column weights are in descending order, that is, no improvement is possible.

Final Phase Decide on the number of clusters and assign objects (machines) to each cluster and parts to each part family. If possible, rearrange rows and columns to reduce the number of exceptional elements. If possible, use the visual inspection and intuition method to further refine and improve the solution.

Example 12.4   ROC Method   Consider the part–machine matrix below. Use the ROC method to determine the optimal grouping of machines and parts so that the exceptional elements are minimized.

	Machines (j)				
Parts (i)	1	2	3	4	5
1	1	1			
2		1	1	1	1
3		1		1	1
4			1	1	1
5	1			1	1
6			1	1	1
7	1	1			

The BWs are assigned to each column and row. See Figure 12.10. Then, the row weights W_i are calculated for each part i. Note that the BW values remain the same for all subsequent iterations, but W_i and W_j change in each iteration and their values depend on the current iteration.

Then row weights are ranked in descending order as shown in Figure 12.10.

The details for calculating row weights are given below:

$$W_1 = 16 \times 1 + 8 \times 1 + 4 \times 0 + 2 \times 0 + 1 \times 0 = 24$$
$$W_2 = 16 \times 0 + 8 \times 1 + 4 \times 1 + 2 \times 1 + 1 \times 1 = 15$$
$$W_3 = 16 \times 0 + 8 \times 1 + 4 \times 0 + 2 \times 1 + 1 \times 1 = 11$$
$$W_4 = 16 \times 0 + 8 \times 0 + 4 \times 1 + 2 \times 1 + 1 \times 1 = 7$$
$$W_5 = 16 \times 1 + 8 \times 0 + 4 \times 0 + 2 \times 1 + 1 \times 1 = 19$$
$$W_6 = 16 \times 0 + 8 \times 0 + 4 \times 1 + 2 \times 1 + 1 \times 1 = 7$$
$$W_7 = 16 \times 1 + 8 \times 1 + 4 \times 0 + 2 \times 0 + 1 \times 0 = 24$$

	BW$_j$	16	8	4	2	1		
BW$_i$		1	2	3	4	5	W_i	Rank
64	1	1	1				24	1
32	2		1	1	1	1	15	4
16	3		1		1	1	11	5
8	4			1	1	1	7	6
4	5	1			1	1	19	3
2	6			1	1	1	7	7
1	7	1	1				24	2

Figure 12.10   Ranking of row of initial data (first iteration of Example 12.4).

	BW$_j$	16	8	4	2	1		
BW$_i$		1	2	3	4	5	W_i	Rank
64	1	1	1				24	1
32	7	1	1				24	2
16	5	1			1	1	19	3
8	2		1	1	1	1	15	4
4	3		1		1	1	11	5
2	4			1	1	1	7	6
1	6			1	1	1	7	7
	W_j	112	108	11	31	31		
	Rank	1	2	5	3	4		

Figure 12.11 Ranking rows.

Now, rearrange the location of rows according to the ranking of row weights. For example, row 7 is ranked as second and therefore should be moved to the second row in the new matrix. This is shown in Figure 12.11.

Now calculate the weights for each column W_j and rank each column. See Figure 12.11. Since the rankings for the columns are not in descending order, reorder the columns in descending order. Now calculate part weights W_j and rank them. This is shown in Figure 12.12.

Because both the row and the column weights are in descending order, no more improvements can be made. Figure 12.12 is the final solution. Now, by visual inspection, group machines and parts into the desired number of cells while minimizing the number of exceptional elements. Suppose two clusters are considered. The matrix in Figure 12.13 shows the two cells as shaded areas.

Cell 1 consists of machines 1 and 2, that is, $S_1 = \{1, 2\}$, and parts 1 and 7, that is, $PF_1 = \{1, 7\}$. Cell 2 consists of machines 4, 5, and 3, that is, $S_2 = \{4, 5, 3\}$, and parts 5, 2, 3, 4, and 6, that is, $PF_2 = \{5, 2, 3, 4, 6\}$. There are three exceptional elements that lie outside these two cells (outside the shaded areas in the matrix in Figure 12.13). This means that although parts 5, 2, and 3 belong to cell 2 (i.e., primarily processed by cell 2), they must also be processed by cell 1.

	BW$_j$	16	8	4	2	1		
BW$_i$		1	2	4	5	3	W_i	Rank
64	1	1	1				24	1
32	7	1	1				24	2
16	5	1		1	1		19	3
8	2		1	1	1	1	15	4
4	3		1	1	1		11	5
2	4			1	1	1	7	6
1	6			1	1	1	7	7
	W_j	112	108	31	31	11		
	Rank	1	2	3	4	5		

Figure 12.12 Ranking columns. This is the final tableau because all rows and columns are ranked in descending order.

$\,^{j}_{i}$	1	2	4	5	3
1	1	1			
7	1	1			
5	1		1	1	
2		1	1	1	1
3		1	1	1	
4			1	1	1
6			1	1	1

Figure 12.13 Clustering assignments for two clusters.

12.4.2 Bicriteria Rank Order Clustering

In the part–machine family formulation, it is usually assumed that the number of clusters (cells) is given, and the objective is to group parts and machines to minimize the number of exceptional elements (outliers). In bicriteria group technology, the following two objectives are considered:

Maximize f_1 = number of cells (to maximize flexibility)

Minimize f_2 = number of exceptional elements (to maximize productivity)

Subject to $f_{1,\min} \le f_1 \le f_{1,\max}$

where $f_{1,\min}$ and $f_{1,\max}$ are given lower and upper bounds on the number of cells. By maximizing the number of cells, the flexibility of the group technology is maximized. By minimizing the exceptional elements, the intratravel among cells is reduced, which means that the productivity is increased because there is less movement of parts among cells. Note that the maximum number of cells (f_1) is Min$\{m, n\}$. In Example 12.4, Min$\{5, 7\} = 5$. In reality, the number of cells is usually much less than the number of machines or parts. Ranking of multicriteria alternatives can be accomplished by using one of the multicriteria ranking methods. Additive utility functions can be used to rank alternatives, where f_1' and f_2' are the normalized values of f_1 and f_2, respectively. The weights of importance for normalized objective values are w_1 and w_2, respectively. Do not confuse the weights of importance (w_1, w_2 in small letters) with the weights of rows and columns (W_i, W_j in capital letters) in the ROC.

Example 12.5   Bicriteria Rank Order Clustering   Consider the final clustering shown in Figure 12.13 for Example 12.4. This final solution can be used to generate the following four clustering alternatives, having one, two, three, and four clusters as shown in Figure 12.14. Note that because machines 4 and 5 have identical vectors, they will always be in the same cell; therefore the maximum number of cells for this problem is four. Their alternatives are labeled F_1, F_2, F_3, and F_4, for one to four cells, respectively. For example, F_2 is the same as the solution to Example 12.4, two cells and three exceptional elements. In alternative F_3, there are three clusters (cells) but the number of exceptional elements is five.

All multiobjective alternatives for this problem are presented in Figure 12.15. They are all efficient.

For an additive utility function suppose the weights of importance for normalized objective values are $w_1 = 0.55$ and $w_2 = 0.45$, respectively. The rankings of alternatives are presented in Table 12.4 where f_1' and f_2' are the normalized values of f_1 and f_2, respectively.

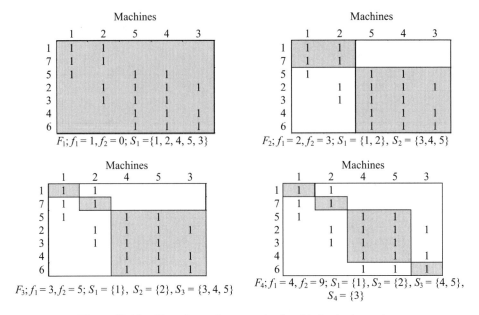

Figure 12.14 Clustering assignments for four bicriteria alternatives.

For example, f_1' for alternative F_2 is calculated by using the normalization equation as follows:

$$f_1' = \frac{2-1}{4-1} = 0.33$$

Since f_2 is minimized, consider maximizing $-f_2$ for the normalization. Then f_2' for alternative F_2 is calculated as

$$f_2' = \frac{-3-(-9)}{0-(-9)} = 0.67$$

For the given weights of importance (0.55, 0.45), the optimal solution is F_3, that is, having three cells which had five exceptional elements. Note that by using different weights of importance alternatives F_1, F_3, or F_4 can be selected; however, F_2 can only be selected if a nonlinear utility function is used.

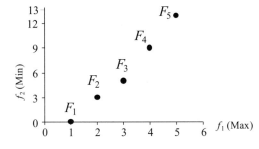

Figure 12.15 Bicriteria efficient frontier of Example 12.5.

TABLE 12.4   Ranking of Bicriteria Alternatives Using Weights (0.55,0.45)

	F_1	F_2	F_3	F_4	Minimum	Maximum
Max. f_1	1	2	3	4	1	4
Min. f_2	0	3	5	9	0	9
Max. f_1'	0	0.33	0.67	1		
Max. f_2'	1	0.67	0.44	0		
Max. $U = 0.55 f_1' + 0.45 f_2'$	0.45	0.48	0.57	0.55		
Rank	4	3	1	2		

12.4.3   Selection of Machines to Be Duplicated[†*]

For a given clustering solution, it is possible to reduce the number of exceptional elements by duplicating some of the machines (i.e., purchasing additional identical machines). The two identical (duplicated) machines can be used in two different cells, reducing the number of exceptional elements. In this case, the machine that minimizes the total number of exceptional elements is duplicated. In practice, there are always limitations on the number of duplicated machines based on the costs and the availability of physical space to place the duplicated machines. For the purpose of simplicity, suppose that the costs of all machines are the same. If the costs of machines are different, then the machine that has the lowest ratio of cost to number of exceptional elements is duplicated.

The approach for the selection of the best machine to be duplicated is as follows:

1. Solve the clustering problem by the chosen method (e.g., ROC).
2. Choose the machine which has the most exceptional elements in its column associated with one cluster.
3. Duplicate the chosen machine and add the duplicated machine to the cluster that has the most exceptional elements associated with this duplicated machine.
4. Repeat steps 2 and 3 until all duplicated machines are assigned.

Consider Example 12.4 and its final solution presented in Figure 12.13. In this example, machine 2 has two exceptional elements associated with the second cluster. Duplicate this machine, and add the duplicated machine 2 to the second cluster. Label this duplicated machine as 2′. Cell 1 consists of machines 1 and 2, that is, $S_1 = \{1, 2\}$, serving parts 1 and 7. Cell 2 consists of machines 2′, 4, 5, and 3, that is, $S_2 = \{2', 4, 5, 3\}$, serving parts 5, 2, 3, 4, and 6. This solution has only one exceptional element.

12.4.4   Tricriteria Rank Order Clustering[†*]

The number of duplicated machines can be considered as a third criterion. The three-criteria clustering problem can be formulated as:

Maximize f_1 = number of cells (to maximize flexibility)

Minimize f_2 = number of exceptional elements (to maximize productivity)

Minimize f_3 = number of duplicated machines (to minimize cost of duplication)

For a given number of cells, by increasing the number of duplicated machines, the number of exceptional elements will decrease. The downfall is the cost of purchasing and

Initial Problem

i \ j	1	2	3	4	5	6	7	8
1			1		1			
2				1			1	1
3			1	1				
4				1	1			
5	1	1						
6		1						
7			1	1	1			
8	1							
9				1		1		
10						1		1
11	1	1					1	
12							1	1
13							1	

ROC Solution

i \ j	1	2	7	8	4	6	3	5
11	1	1	1					
5	1	1						
8	1							
6		1						
2			1	1	1			
12			1	1				
13			1					
10				1		1		
9					1	1		
7					1		1	1
3					1		1	
4					1			1
1							1	1

(a)   (b)

Figure 12.16 (*a*) Initial data, Example 12.6; (*b*) ROC solution, Example 12.6.

maintaining the duplicated machines. To generate efficient alternatives, consider each solution generated by the bicriteria problem of Section 12.4.2. For each bicriteria alternative, apply the procedure of selecting the best duplicated machine as discussed in Section 12.4.3. In the above problem, we assumed that the costs of all machines are the same; if machine costs are different, then the total cost of duplicated machines should be used instead of the number of duplicated machines as the third objective.

Example 12.6   Tricriteria Rank Order Clustering   Consider the problem presented in Figure 12.16*a*. Suppose the number of cells can vary from two to three. Also, suppose that only a maximum of two machines may be duplicated. Find all efficient alternatives for this problem.

The ROC solution is presented in Figure 12.16*b*. In this problem, $f_1 = 2, 3$ and $f_3 = 0, 1, 2$.

First consider the bicriteria problem. The cell formation for two and three cells is presented in Figures 12.17*a,b*. For both solutions no machines are duplicated.

Now consider the tricriteria problem for each of the above two alternatives if one machine is duplicated.

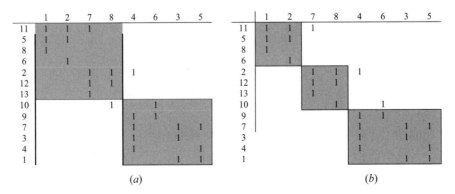

Figure 12.17 (*a*) $F_1; f_1 = 2, f_2 = 2, f_3 = 0, S_1 = \{1, 2, 7, 8\}, S_2 = \{4, 6, 3, 5\}$; (*b*) $F_2; f_1 = 3, f_2 = 3, f_3 = 0; S_1 = \{1, 2\}, S_2 = \{7, 8\}, S_3 = \{4, 6, 3, 5\}$.

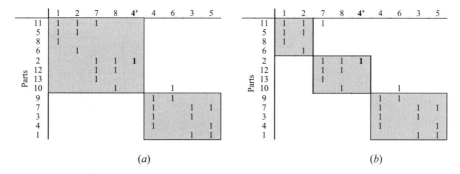

Figure 12.18 (a) F_3; $f_1 = 2, f_2 = 1, f_3 = 1$; $S_1 = \{1, 2, 7, 8, 4'\}$, $S_2 = \{4, 6, 3, 5\}$; (b) F_4; $f_1 = 3$, $f_2 = 2, f_3 = 1$ $S_1 = \{1, 2\}$, $S_2 = \{7, 8, 4'\}$, $S_3 = \{4, 6, 3, 5\}$.

Use Method Covered in Section 12.4.3 to Find Best Machine to Duplicate Consider alternative F_1, shown in Figure 12.17a. Duplicate the machine which has the most exceptional elements in its column associated with one cluster. In this case, either machine 4 or 8 can be selected, both having only one exceptional element. Arbitrarily select machine 4 to be duplicated. Denote the duplicated machine 4 by $4'$. The solution is presented in Figure 12.18a, labeled as F_3.

Now consider alternative F_2, shown in Figure 12.17b. Machines 4, 6, and 7 have one exceptional element. Suppose that machine 4 is arbitrarily selected to be duplicated. The solution is F_4, which is shown in Figure 12.18b.

Now consider allowing two machines to be duplicated. Consider alternative F_1. By duplicating machines 4 and 8, that is, having $4'$ and $8'$, alternative F_5, shown in Figure 12.19a, can be generated. This solution does not have any exceptional elements.

Now consider alternative F_2. Machines 4, 6, or 7 can be duplicated. Suppose machines 4 and 6 are selected for duplication. By duplicating machines 4 and 6, alternative F_6, shown in Figure 12.19b, can be generated. This solution has one exceptional element.

The summary of these six alternatives is presented in Table 12.5. All the multicriteria alternatives are efficient. Suppose an additive utility function is used to rank the alternatives with given weights of importance for normalized objectives $w_1 = 0.2$, $w_2 = 0.3$, $w_3 = 0.5$ for the three objectives, respectively. The ranking of alternatives is provided in Table 12.5.

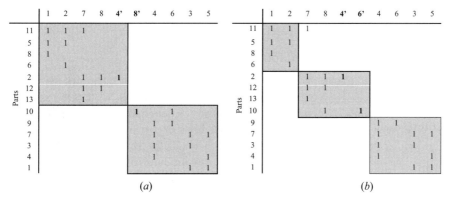

Figure 12.19 (a) F_5; $f_1 = 2, f_2 = 0, f_3 = 2$; $S_1 = \{1, 2, 7, 8, 4'\}$, $S_2 = \{8', 4, 6, 3, 5\}$; (b) F_6; $f_1 = 3, f_2 = 1, f_3 = 2$; $S_1 = \{1, 2\}$, $S_2 = \{7, 8, 4', 6'\}$, $S_3 = \{4, 6, 3, 5\}$.

TABLE 12.5 Ranking of Tricriteria Alternatives for Example 12.6

	F_1	F_3	F_5	F_2	F_4	F_6	Minimum	Maximum
Max. f_1, no. of cells	2	2	2	3	3	3	2	3
Min. f_2, no. of exceptional elements	2	1	0	3	2	1	3	0
Min. f_3, no. of duplicated machines	0	1	2	0	1	2	2	0
Max. f_1'	0.00	0.00	0.00	1.00	1.00	1.00		
Max. f_2'	0.33	0.67	1.00	0.00	0.33	0.67		
Max. f_3'	1.00	0.50	0.00	1.00	0.50	0.00		
$U = 0.2f_1' + 0.3f_2' + 0.5f_3'$	0.60	0.45	0.30	0.70	0.55	0.40		
Rank	2	4	6	1	3	5		

12.5 SIMILARITY COEFFICIENT–HIERARCHICAL CLUSTERING

Similarity coefficient (used for binary data) and hierarchical (used for continuous data) clustering algorithms result in nested partitions of objects rather than just a single clustering set. The similarity coefficient (SC) and hierarchical algorithms generally follow one of two approaches in creating the clusters, agglomerative and divisive:

- Agglomerative clustering begins with each object as a cluster and merges clusters together until the algorithm converges.
- Divisive clustering begins with a single cluster containing all objects and splits the cluster successively.

In this section, we will discuss the agglomerative approach to clustering. There are several variants of the agglomerative clustering algorithm based on how similarity (or dissimilarity) is measured between clusters (linkage methods are discussed in Section 12.5.4). In the SC method (Sections 12.5.1–12.5.3) we use the Jaccard metric for binary data to measure similarities of pairs of objects. In hierarchical clustering (Sections 12.5.4 and 12.5.5), we use Euclidean distance for continuous data to present dissimilarity of pairs of objects.

12.5.1 Similarity Coefficient for Binary Data

In the SC approach, the similarity of pairs of objects in terms of their attributes is used for clustering objects. For the group technology problem, machines are labeled as objects and parts as attributes. A popular distance metric for continuous data is the Euclidean distance between pairs of objects. If the distance between a pair of objects is small, then they are similar and can be clustered in the same cluster. In the SC approach for binary data, the Jaccard metric (Jaccard, 1901) is used to measure the similarity of pairs of objects. The higher this value is, the more similar the two objects are. For a given number of clusters, the objective of the SC method is to maximize the summation of the Jaccard coefficients for each cluster. The Jaccard coefficient is preferred to the Minkowski distance measure [(12.4), e.g., Euclidean distance (2.3)] for many applications where binary data (such as the group technology) is used because attributes having zero values do not carry any information in terms of similarity between pairs of machines.

Jaccard Similarity Coefficient for Pair of Objects Consider n attributes, $i = 1$, $2, \ldots, n$. The similarity index for a pair of objects q and j is

$$\text{SC}_{qj} = \frac{\sum_{i=1}^{n} a_{iq} a_{ij}}{\sum_{i=1}^{n} (a_{iq} + a_{ij} - a_{iq} a_{ij})} = \frac{\text{no. of common attributes in both objects } q \text{ and } j}{\text{no. of attributes in at least one of two objects}}$$

(12.7)

where

$$a_{ij} = \begin{cases} 1 & \text{if object } j \text{ has attribute } i \\ 0 & \text{otherwise} \end{cases}$$

In group technology, where objects are defined as machines and attributes are defined as parts, the SC is defined as the number of parts that are processed by both machines divided by the number of parts that are processed by at least one of the two machines. That is, for a given pair of machines (q and j), the value of the numerator indicates the number of parts that are processed by both machines q and j. The value of the denominator is the total number of parts that are processed by either machine q or machine j. For example, consider two objects with five attributes, $\mathbf{a}_1 = (0, 0, 1, 0, 1)$ and $\mathbf{a}_2 = (0, 1, 1, 0, 1)$. Their similarity coefficient is

$$\text{SC}_{12} =$$

$$\frac{(0 \times 0) + (0 \times 1) + (1 \times 1) + (0 \times 0) + (1 \times 1)}{(0 + 0 - 0 \times 0) + (0 + 1 - 0 \times 1) + (1 + 1 - 1 \times 1) + (0 + 0 - 0 \times 0) + (1 + 1 - 1 \times 1)}$$

$$= \frac{2}{3} = 0.66$$

This means that objects $\mathbf{a}_1$ and $\mathbf{a}_2$ are 0.66 or 66% similar in terms of their positive attributes. The following table shows how to simplify measuring the SC. First, ignore the attributes that have zero values in both objects. In this example, the first and fourth attributes have zero values for both objects, so they are ignored. Now, measure the similarity of the attributes that have values of unity. In this case, two out of three components are identical.

Attribute i	1	2	3	4	5		i	2	3	5
Object $\mathbf{a}_1$	0	0	1	0	1	$\Rightarrow$	$\mathbf{a}_1$	0	1	1
Object $\mathbf{a}_2$	0	1	1	0	1		$\mathbf{a}_2$	1	1	1
	Ignore	Use	Use	Ignore	Use					

Two out of three attributes have the same values

$$\text{SC}_{1,2} = \tfrac{2}{3} = 0.66$$

Complete Linkage   After SC of pairs of all objects are obtained, the two-modal (n object versus m attributes) problem is presented as one modal (n objects versus n objects). Linkage refers to measuring the similarity (or closeness) of two clusters in one-modal presentation. In this section, we use complete linkage for similarity which is the "maximum" of SC values when comparing columns of objects. For example, consider three objects $a = (0.4,0,0.3)$, $b = (0.1,0.7,0.6)$, and $c = (0.5,0.25,0.2)$. If the three objects a, b, and c are assigned to the same cluster labeled as (a,b,c), this cluster is represented by a vector which is associated with the maximum of each entry, so the complete linkage of group (a,b,c) is $(0.5,0.7, 0.6)$. Section 12.5.4 discusses three linkage approaches. Single linkage in Section 12.5.5 is based on dissimilarity (distance), which is the same as complete linkage in this section, which is based on similarity (closeness).

SC Method: Combining Only One Pair of Objects in Each Iteration   Before using the SC method, one should first identify the number of desired clusters. The SC method starts with the maximum number of clusters and then reduces the number of clusters in each iteration until the number of desired clusters is achieved. The algorithm begins by calculating the SC values for all possible pairs of objects (machines). In this approach, in each iteration only one pair of clusters is combined in each iteration, and the approach can continue until the desired number of clusters is achieved. A tree structure can be used to represent many possible clusterings based on the use of different SC thresholds. This method is shown in Example 12.7.

Example 12.7   Finding Similarity Coefficient Matrix   Consider the matrix for five objects (machines) and seven attributes (parts) in Figure 12.20. This example is the same as Example 12.4, which was solved in Section 12.4 by the ROC method. For the matrix, first find the SC index for each pair of objects. For example, the SC for objects 1 and 2 is

$$SC_{12} = \frac{(1 \times 1) + (0 \times 1) + (0 \times 1) + (0 \times 0) + (1 \times 0) + (0 \times 0) + (1 \times 1)}{\begin{aligned}&(1 + 1 - 1 \times 1) + (0 + 1 - 0 \times 1) + (0 + 1 - 0 \times 1) + (0 + 0 - 0 \times 0)\\&+ (1 + 0 - 1 \times 0) + (0 + 0 - 0 \times 0) + (1 + 1 - 1 \times 1)\end{aligned}}$$

$$= \frac{2}{5} = 0.4$$

		Objects (Machines)				
i \ j		1	2	3	4	5
	1	1	1			
	2		1	1	1	1
	3		1		1	1
Attributes (Parts)	4			1	1	1
	5	1			1	1
	6			1	1	1
	7	1	1			

Figure 12.20   Machine–part matrix for Example 12.7.

	Similarity of Five Objects				
	1	2	3	4	5
1	—	0.4	0	0.14	0.14
2	0.4	—	0.17	0.29	0.29
3	0	0.17	—	0.6	0.6
4	0.14	0.29	0.6	—	1
5	0.14	0.29	0.6	1	—

Figure 12.21   Similarity coefficients.

This can be interpreted as two parts visiting both machines for every five parts that visit either machine 1 or 2. The SC of machines 4 and 5 is 1 because both machines are visited by the same parts. The SC for all pairs of objects is shown in an $m \times m$ matrix as presented in Figure 12.21. Note that in all SC methods the SC matrix is symmetrical around its main diagonal. But for purpose of clarity, the whole matrix is shown here.

Example 12.8   Hierarchical Tree (Combining One Pair in Each Iteration)   Consider Example 12.7. In this example, only two objects are combined in each iteration.

Iteration 1

Step 1   SC values between pairs of objects are found as follows:

(Machine) Pairs, (q, j)	(1,2)	(1,3)	(1,4)	(1,5)	(2,3)	(2,4)	(2,5)	(3,4)	(3,5)	(4,5)	SC_{avg}
SC_{qj}	0.40	0.00	0.14	0.14	0.17	0.29	0.29	0.60	0.60	1.00	0.36
Combine the pair?	No	No	No	No	No	No	No	No	No	Yes	

Step 2   Identify the pair of objects associated with the maximum SC ($SC_{max} = 1$).
Step 3   Merge objects 4 and 5 into (4,5). There are four clusters: $\{1, 2, 3, (4,5)\}$.

Iteration 2

Step 1   Find the SC of the four clusters $\{1, 2, 3, (4,5)\}$ (see Figure 12.22).

	Similarity of Five Objects				
	1	2	3	4	5
1	—	0.4	0	0.14	0.14
2	0.4	—	0.17	0.29	0.29
3	0	0.17	—	0.6	0.6
4	0.14	0.29	0.6	—	1
5	0.14	0.29	0.6	1	—
Max.	0.14	0.29	0.6		

$\Rightarrow$

	Similarity of Four Clusters			
	1	2	3	(4,5)
1	—	0.4	0	0.14
2	0.4	—	0.17	0.29
3	0	0.17	—	0.6
(4,5)	0.14	0.29	0.6	—

Figure 12.22   Similarity coefficient for step 1 of iteration 2.

	Similarity of Four Clusters						Similarity of Three Objects		
	1	2	3	(4,5)			1	2	(3,4,5)
1	—	0.4	0	0.14		1	—	0.4	0.14
2	0.4	—	0.17	0.29	⇨	2	0.4	—	0.29
3	0	0.17	—	0.6		(3,4,5)	0.14	0.29	—
(4,5)	0.14	0.29	0.6	—					

Figure 12.23   Similarity coefficient for step 1 of iteration 3.

The linkage of objects (4,5) is (0.14, 0.29, 0.6) and is associated with the maximum values of rows 4 and 5 (in this case both rows are identical).

Step 2 SC_{max} is 0.6 for objects 3 and (4, 5). Therefore, 3 and (4, 5) need to be merged.

Step 3   Merge 3 and (4, 5) into (3, 4, 5). There are three clusters: {1, 2 and (3, 4, 5)}.

Iteration 3

Step 1   Find the SC of the three objects (see Figure 12.23).

Step 2 SC_{max} is 0.40 for objects 1 and 2.

Step 3   Merge 1 and 2 into (1,2). There are two clusters: {(1,2), (3,4,5)}.

Iteration 4

Step 1   Find the SC of the two clusters (see Figure 12.24).

Therefore, SC_{max} is 0.29. Combine (1, 2) and (3, 4, 5). There is only one cluster: {(1,2,3,4,5)}. The following table shows how the selection of different SC thresholds leads to different clustering. Figure 12.25 shows the hierarchical tree for use of the SC threshold for this example. The selection of different SC threshold values will cause different objects to be merged, which results in different clusters. This is shown in the table below.

Steps	1	2	3	4
Used SC_{thresh}	1	0.6	0.4	0.29
No. of clusters	4	3	2	1
Cluster members	{(1),(2),(3),(4,5)}	{(1),(2),(3,4,5)}	{(1,2),(3,4,5)}	{(1,2,3,4,5)}

	Similarity of Three Objects					Similarity of Two Clusters	
	1	2	(3,4,5)			(1,2)	(3,4,5)
1	—	0.4	0.14		(1,2)	—	0.29
2	0.4	—	0.29	⇨	(3,4,5)	0.29	—
(3,4,5)	0.14	0.29	—				

Figure 12.24   Similarity coefficient for step 1 of iteration 4.

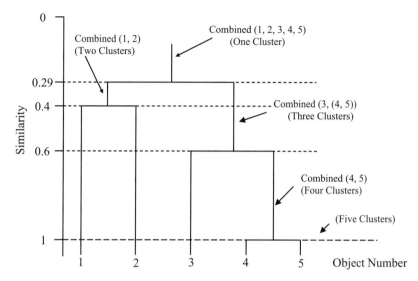

Figure 12.25   Hierarchical tree structure resulting from hierarchical clustering for Example 12.8.

The explanation of Figure 12.25 and the above table are as follows when an SC threshold SC_{thresh} is used.

1. At a SC_{thresh} of 1, objects 4 and 5 are combined. Therefore, there will be four clusters: $\{(1), (2), (3), (4,5)\}$.
2. At a SC_{thresh} of 0.6, there are three clusters: $\{(1), (2), (3,4,5)\}$.
3. At a SC_{thresh} of 0.4, there are two clusters: $\{(1,2), (3,4,5)\}$.
4. At a SC_{thresh} of 0.29, there is one cluster: $\{(1,2,3,4,5)\}$.

Note that Figure 12.25 is based on the combinations of only pairs of objects in hierarchical order. However, the above solutions may be improved by considering more objects to be combined. In this case, the problem can be solved iteratively using the combined (reduced) sets of objects.

SC Threshold Method to Assign More Than One Pair of Objects in Each Iteration[†]   The SC method can have a maximum of m iterations when there are m objects. To expedite the process of clustering (for larger problems), that is, to have less than m iterations, in each iteration, an SC threshold SC_{thresh} which is lower than the maximum SC value should be selected. To do this in each iteration a threshold value of SC should be selected. In each iteration, each pair of objects whose SC is higher than or equal to the SC threshold are combined (are assigned to the same cluster). A linkage approach is used to present the combined objects. The process is repeated until the desired number of clusters are achieved or all SC values are lower than the given SC threshold.

Choosing the SC threshold SC_{thresh} value is subjective; different values can be selected in different iterations. A high value of SC_{thresh} results in choosing fewer objects. A smaller threshold value will allow more objects to be assigned to the same cluster. For example, one may use the average of all SC scores, SC_{avg}, as the threshold in each iteration.

The SC algorithm when using SC thresholds is as follows:

Step 0   Identify the desired number of clusters.

Step 1   Find the similarity coefficient for all pairs of objects.

Step 2   Identify the SC threshold for each iteration.

Step 3   Assign objects to appropriate clusters if their SC is more than the SC threshold.

Step 4   Find the linkage of the assigned objects. This linkage represents the cluster by one column (or one row as the matrix is symmetric).

Repeat steps 1–4 until the given number of clusters is achieved.

Supplement S12.3.docx describes details and examples of the SC for binary data using SC thresholds.

12.5.2   Simultaneous SC Clustering[†]

In this section, we develop the simultaneous SC method, which is based on the simultaneous head–tail approach discussed in Section 13.7.6. In this method, a number of alternatives (e.g., Q solutions) are developed simultaneously. Since the clustering problem is combinatorial, by generating Q alternatives instead of one, there is a better chance of finding the optimal or better clustering solution. As the number of simultaneous alternatives increases, the likelihood of this method finding better solutions increases. The computational effort of solving even larger problems is reasonable.

The simultaneous SC method presented here uses the SC method covered in Section 12.5.1 where only one object is assigned in each iteration. However, the SC threshold method can assign more than one object in each iteration. In the first iteration, the Q best candidates (i.e., pairs of objects in terms of their SC values) are selected and each of these candidates are treated as one separate solution. In each iteration, each of the Q solutions is developed. But, if any of the developed solutions is the same as one of the previously developed solutions (i.e., it is a replica), the next best pair of objects is selected to avoid developing identical solutions simultaneously. In the following, first a small example is presented to show the basics of the method, then a bigger example is described to show that this method can find better alternatives than the one generated by the original SC method of Section 12.5.1.

Example 12.9   Simultaneous Hierarchical Clustering—A Small Example   Consider Example 12.7. Apply simultaneous clustering for three ($Q = 3$) simultaneous solutions.

Iteration 1   Consider the ranking of SC scores given in Example 12.7. The best three pairs are (4,5), (3,4), and (3,5). For each of these pairs, apply the SC method. The best candidates that do not generate replicas are shaded in Figure 12.26.

The best three pairs in Figure 12.26 are {3,(4,5)}, {(3,4),5}, and {(3,5),4}. Since these three pairs generate the same clustering solutions for the next iteration, select the best pair for alternative I, the second best pair for alternative II, and the second best pair for alternative III. The first best pair for alternative I is {3,(4,5)} shown in Figure 12.27a. The second best pair for alternative II is {(1,2)} shown in Figure 12.27b. The second best pair for alternative III is {(1,2)} shown in Figure 12.27c.

Iteration 2   The SC for each of the three alternatives are shown in Figure 12.27.

	1	2	3	(4,5)
1	—	0.4	0	0.14
2	0.4	—	0.17	0.29
3	0	0.17	—	0.6
(4,5)	0.14	0.29	**0.6**	—

(a)

	1	2	5	(3,4)
1	—	0.4	0.14	0.14
2	**0.4**	—	0.29	0.29
5	0.14	0.29	—	1
(3,4)	0.14	0.29	1	—

(b)

	1	2	4	(3,5)
1	—	0.4	0.14	0.14
2	**0.4**	—	0.29	0.29
4	0.14	0.29	—	1
(3,5)	0.14	0.29	1	—

(c)

Figure 12.26 Three simultaneous solutions for iteration 1: SCs for (a) {(1),(2),(3,4,5)}, (b) {(1,2),(3,4),(5)}, and (c) {(1,2),4,(3,5)}.

	1	2	(3,4,5)
1	—	0.4	0.14
2	**0.4**	—	0.29
(3,4,5)	0.14	0.29	—

(a)

	(1,2)	(3,4)	5
(1,2)	—	0.29	0.29
(3,4)	**0.29**	—	1
5	0.29	1	—

(b)

	(1,2)	4	(3,5)
(1,2)	—	0.29	0.29
4	**0.29**	—	1
(3,5)	0.29	1	—

(c)

Figure 12.27 Three simultaneous solutions for iteration 2: SCs for (a) {(1),(2),(3,4,5)}, (b) {(1,2),(3,4),(5)}, and (c) {(1,2),4,(3,5)}.

The best three pairs (that do not generate replica solutions) are {(1),(2)}, {(1,2),(3,4)}, and {(1,2),(4)} in Figure 12.27a,b,c, respectively. Since any further clustering will result in a single cluster consisting of all objects, the method is stopped.

Summary of All Solutions   Table 12.6 shows the summary of all the solutions for this example. The set of alternatives can be screened out by identifying the efficient alternatives, while considering the bicriteria problem of maximizing the number of clusters and minimizing the number of exceptional elements. The inefficient alternatives can be eliminated. See Table 12.6.

As the number of objects increases, the simultaneous method might find better solutions than the SC method as presented in the next example.

See Supplement S12.4.xls.

TABLE 12.6   Summary of All Clustering Solutions by Simultaneous Clustering Method

Iteration	f_1 = No. of Cluster	Alternatives	f_2 = No. of Exceptional Elements	Efficient?
1	$K = 4$	I. {(1),(2),(3),(4,5)}	9	Yes
		II. {(1),(2),(3,4),(5)}	10	No
		III. {(1),(2),(4),(3,5)}	10	No
2	$K = 3$	I. {(1),(2),(3,4,5)}	5	Yes
		II. {(1,2),(3,4),5}	8	No
		III. {(1,2),4,(3,5)}	8	No
3	$K = 2$	I. {(1,2),(3,4,5)}	3	Yes
		II. {(5),(1,2,3,4)}	6	No
		III. {(1,2,4),(3,5)}	6	No
4	$K = 1$	(1,2,3,4,5)	0	Yes

TABLE 12.7   Summary of Alternatives for Different Cluster Numbers Using Simultaneous Method

Iteration	$f_1 =$ No. of Clusters	Alternatives Cluster Assignments	$f_2 =$ No. of Exceptional Elements	Efficient?
1	$K = 7$	I. {(1,2),(3),(5),(4),(6),(7),(8)}	11	Yes
		II. {(2),(1),(3,5),(4),(6),(7),(8)}	11	Yes
		III. {(3),(1),(2),(4),(5),(6),(7,8)}	11	Yes
2	$K = 6$	I. {(1,2),(3,5),(4),(6),(7),(8)}	9	Yes
		II. {(2),(1),(3,5),(4),(6),(7,8)}	9	Yes
		III. {(3),(1,2),(4),(5),(6),(7,8)}	9	Yes
3	$K = 5$	I. {(1,2),(3,5),(4),(6),(7,8)}	7	No
		II. {(2),(1),(3,4,5),(6),(7,8)}	6	Yes
		III. {(3),(1,2),(4,5),(6),(7,8)}	7	No
4	$K = 4$	I. {(1,2),(3,4,5),(6),(7,8)}	6	Yes
		II. {(2),(1),(3,4,5,6),(7,8)}	6	Yes
		III. {(3),(1,2),(4,5,6),(7,8)}	6	Yes
5	$K = 3$	I. {(1,2),(3,4,5,6),(7,8)}	3	Yes
		II. {(2),(1),(3,4,5,6,7,8)}	5	No
		III. {(3),(1,2),(4,5,6,7,8)}	6	No
6	$K = 2$	I. {(1,2),(3,4,5,6,7,8)}	2	Yes
		II. {(2),(1,3,4,5,6,7,8)}	2	Yes
		III. {(3),(1,2,4,5,6,7,8)}	3	No
7	$K = 1$	(1,2,3,4,5,6,7,8)	0	Yes

Example 12.10   Simultaneous Hierarchical Clustering—A Larger Example   Consider Example 12.6.3. Apply simultaneous clustering for three ($Q = 3$) simultaneous solutions. Table 12.7 shows the summary and the final solutions for each iteration.

Solutions found by the SC method (Section 12.5.1) are shaded in Table 12.7. The solution found by the SC method for $K = 5$ is inefficient compared to the solution generated by the simultaneous method. The hierarchical tree structure for all three solutions for $K = 5$ is presented in Figure 12.28. For alternatives II and III, the dashed vertical lines show the SC values at which a delayed object (which could have been clustered sooner) is clustered.

See Supplement S12.5.xls.

12.5.3   Multicriteria Similarity Coefficient[†]

Bicriteria Similarity Coefficient   As discussed in Section 12.4.2, the bicriteria problem is:

> Maximize $f_1 =$ number of clusters
> Minimize $f_2 =$ exceptional elements

Consider the SC method discussed in Section 12.5.1. In the first iteration, the maximum number of clusters are known (which is the same as the number of objects). So, the approach starts with the maximum number of clusters, and it reduces the number of objects by one in

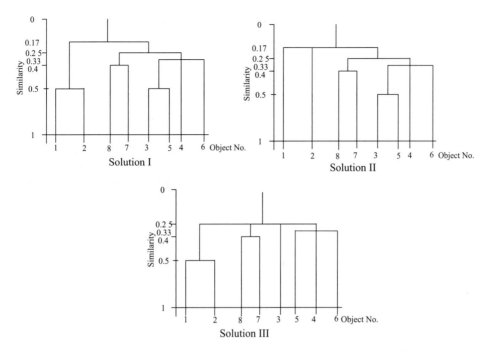

Figure 12.28   Three different hierarchical tree solutions for Example 12.10, for $K = 5$.

each iteration. For the clustering solution generated in each iteration, measure the number of exceptional elements. The method is summarized as follows:

1. In each iteration of the SC algorithm, for a given clustering solution, measure the number of exceptional elements.
2. Make a list of bicriteria alternatives and identify the inefficient ones.

Example 12.11   Bicriteria SC Algorithm   Consider Example 12.7 and the solution by adding one object at a time as presented in Example 12.9. The set of solutions based on Example 12.9 are provided in Table 12.8. (See Figure 12.14 for details of the solutions.)

TABLE 12.8   First Iteration of Bicriteria Alternatives for Example 12.9

Iteration	1	2	3	4
Alternative labels	F_4	F_3	F_2	F_1
f_1 = No. of clusters	4	3	2	1
f_2 = No. of exceptional elements	9	5	3	0
Clusters	(1),(2),(3),(4,5)	(1),(2),(3,4,5)	(1,2),(3,4,5)	(1,2,3,4,5)
Efficient?	Yes	Yes	Yes	Yes

Tricriteria Similarity Coefficient with Duplicated Machines   The tricriteria clustering of the SC is similar to tricriteria clustering of ROC (see Sections 12.4.3 and 12.4.4) but using the bicriteria SC algorithm for generating efficient alternatives for the bicriteria problem. This method is presented through Example 12.12.

Example 12.12   Tricriteria Similarity Coefficient   Consider Example 12.7. Suppose the maximum number of clusters is three and suppose there can be up to two duplicated machines.

The set of efficient alternatives for the bicriteria of this problem was generated in Example 12.9. See Table 12.7. Alternative F_4 is eliminated because it exceeds the maximum of three clusters. Therefore, the efficient feasible solutions for the bicriteria problems are

$$F_1 = \{(1,2,3,4,5)\} \quad F_2 = \{(1,2),(3,4,5)\} \quad F_3 = \{(1),(2),(3,4,5)\}$$

Consider each bicriteria alternative (as presented in Example 12.9). For each given solution, duplicate the machine with the highest number of exceptional elements in its column.

Here, we show the details of how to generate efficient solutions associated with alternative F_2. Consider the solution for alternative F_2. In this clustering, machine 2 has the most exceptional elements, that is, two. Duplicate machine 2 and label it as $2'$. Machine $2'$ is placed in the second cell, that is, each cell has one machine 2. This is shown as alternative F_5, in Figure 12.29. Since the maximum number of allowable duplicated machines is two, in the next iteration, consider duplicating machine 1 in addition to machine 2, which was already duplicated in alternative F_5. This is presented as alternative F_6 in Figure 12.29.

Now consider alternative F_3 (Example 12.9), which has three clusters. Two more alternatives, F_7 and F_8, can be generated by considering duplicated machines for alternative F_3. The details of alternatives F_3, F_7, and F_8 are not shown, but they are simple variations

No Duplicated Machines; Two Cells					
	1	2	3	4	5
1	1	1			
7	1	1			
5	1			1	1
3		1		1	1
2		1	1	1	1
6			1	1	1
4			1	1	1

$F_2; f_1 = 2, f_2 = 3, f_3 = 0; s_1 = \{1,2\}, s_2 = \{3,4,5\}$

One Duplicated Machine, $2'$; Two Cells						
	1	2	$2'$	3	4	5
1	1	1				
7	1	1				
5	1				1	1
3			1		1	1
2			1	1	1	1
6				1	1	1
4				1	1	1

$F_5; f_1 = 2, f_2 = 1, f_3 = 1; s_1 = \{1,2\}, s_2 = \{2',3,4,5\}$

Two Duplicated Machines, $1'$ and $2'$; Two Cells							
	1	2	$1'$	$2'$	3	4	5
1	1	1					
7	1	1					
5			1			1	1
3				1		1	1
2				1	1	1	1
6					1	1	1
4					1	1	1

$F_6; f_1 = 2, f_2 = 0, f_3 = 2; s_1 = \{1,2\}, s_2 = \{1',2',3,4,5\}$

Figure 12.29   Three alternatives for Example 12.12.

of F_3. Note that for one family, F_1, there cannot be a duplicated machine as there are no exceptional elements. All alternatives are shown below.

Bicriteria	F_1	F_2			F_3		
Tricriteria	F_1	F_2	F_5	F_6	F_3	F_7	F_8
f_1 (Max.), no. of cells	1	2	2	2	3	3	3
f_2 (Min.), no. of exceptional elements	0	3	1	0	5	3	2
f_3 (Min.), no. of duplicated machines	0	0	1	2	0	1	2
Efficient?	Yes	Yes	Yes	Yes	Yes	Yes	Yes

12.5.4 Linkage Approaches[†]

Linkage refers to the similarity (or closeness) of two clusters. There are several approaches for defining linkage; they include complete linkage, single linkage, and average linkage. Consider the members of two clusters:

- *Complete Linkage*   Uses the maximum distance (least similarity) of all pairs of objects belonging to two different clusters.
- *Single Linkage*   Uses the minimum distance (most similarity) of all pairs of objects belonging to two different clusters.
- *Average Linkage*   Uses the average distance of all pairs of objects belonging to two different clusters.

In *complete-link* or *complete-linkage clustering*, the similarity of two clusters is the distance of their two *most dissimilar* members. In *single-link* or *single-linkage clustering*, the similarity of two clusters is the distance of their two *most similar* members. The complete-linkage approach produces clusters which are close and compact. Single linkage produces clusters consisting of chains of objects. The average linkage uses the average of the similarity of all objects in the two clusters. Each of these methods has its applications, advantages, and disadvantages. For two-attribute problems, linkages can be shown graphically. Examples of single linkage and complete linkage are shown for two clusters in the Figure 12.30. The above linkage definitions are based on dissimilarity (distance) measurements while Sections 12.5.1–12.5.3 use similarity (closeness); therefore, complete linkage as used in Sections 12.5.1–12.5.3 is the same as single linkage in this section.

Example 12.13   Linkage Measures   Consider four objects with three attributes as given in the following matrix:

(a) Find the Euclidean distance (dissimilarity) for all pairs of objects.
(b) Find complete, single, and average linkages of clusters $(\mathbf{a}_1)$, $(\mathbf{a}_2)$ and $(\mathbf{a}_2, \mathbf{a}_3)$.
(c) Find the complete linkages for clusters $(\mathbf{a}_2, \mathbf{a}_3)$ and $(\mathbf{a}_1, \mathbf{a}_4)$.

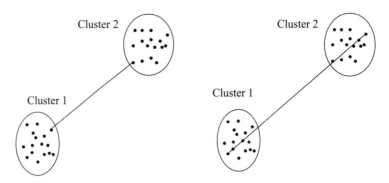

(a) Single Linkage most similar   (b) Complete Linkage least similar

Figure 12.30   Single and complete linkage for two-attribute example.

(a) The Euclidean distance matrix for this problem is as follows:

		Two Modal					One Modal using Euclidean Distance			
		Objects						Dissimilarity (Distance) of Four Objects			
		a_1	a_2	a_3	a_4			a_1	a_2	a_3	a_4
Attributes	1	2	1	2	4	Objects	a_1	0	5.09	3.16	4.12
	2	0	5	3	2		a_2	5.09	0	2.45	5.20
	3	5	5	4	2		a_3	3.16	2.45	0	3
							a_4	4.12	5.20	3	0

For example, the Euclidean distance between a_2 and a_3 is

$$d_{23} = \sqrt{(2-1)^2 + (5-3)^2 + (5-4)^2} = 2.45$$

(b) Objects a_2 and a_3 have the closest distance (are the most similar) and they can be grouped in one cluster (a_2, a_3). Table 12.9 shows how all three linkages are measured between (a_2, a_3), (a_1), and (a_4).

TABLE 12.9   Measuring Three Types of Linkages between (a_2, a_3), (a_1), and (a_4)

	a_1	a_2	a_3	a_4	Complete Linkage (Max.) $a_2 a_3$	Single Linkage (Min.) $a_2 a_3$	Average Linkage (Average) $a_2 a_3$
a_1	0	5.09	3.16	4.12	5.09	3.16	4.13
a_4	4.12	5.20	3	0	5.20	3	4.10

For example, the linkage of clusters ($\mathbf{a}_1$) and ($\mathbf{a}_2, \mathbf{a}_3$) is calculated as follows.

Complete linkage (($\mathbf{a}_2, \mathbf{a}_3$), $\mathbf{a}_1$) = Max{5.09, 3.16} = 5.09
Single linkage (($\mathbf{a}_2, \mathbf{a}_3$), $\mathbf{a}_1$) = Min{5.09, 3.16} = 3.16
Average linkage (($\mathbf{a}_2, \mathbf{a}_3$), $\mathbf{a}_1$) = Average{5.09, 3.16} = 4.13

(c) The following table shows complete linkage for clusters ($\mathbf{a}_2, \mathbf{a}_3$) and ($\mathbf{a}_1, \mathbf{a}_4$), which is Max{5.09, 5.20}= 5.20.

	$\mathbf{a}_1$	$\mathbf{a}_2\mathbf{a}_3$	$\mathbf{a}_4$	Complete Linkage (Max) $\mathbf{a}_1\mathbf{a}_4$
($\mathbf{a}_2, \mathbf{a}_3$)	5.09	0	5.20	5.20

12.5.5 Hierarchical Clustering for Continuous Data[†]

The hierarchical clustering approach is similar to the SC approach covered in Section 12.5.1 except that here the Euclidean distance is used to show the dissimilarity of pairs of objects for continuous data. In this approach only two objects (or an object and a cluster) are combined in each iteration. Complete linkage is used to find the similarity of two clusters but one can use other linkage methods. The multicriteria approach of this problem is similar to the multicriteria approach covered in Sections 12.5.2 and 12.5.3 for binary data.

Example 12.14   Hierarchical Algorithm for Continuous Data   Consider objects *A–H* shown in Table 12.10. Cluster these objects using the hierarchical method. Cluster the two most similar objects (or clusters) in each iteration. Also, use hierarchical complete linkage in each step. Use the Euclidean distance to measure distance. Use the complete-linkage approach for combining objects. The solution is as follows:

Iteration 1

Step 1   Calculate the Euclidean distance for all pairs of objects. For example, the Euclidean distance between *B* and *E* is calculated as

$$d(B, E) = \sqrt{(2 - 5)^2 + (1 - 5)^2} = 5.0$$

TABLE 12.10   Objects Used in Example 12.14

Objects	A	B	C	D	E	F	G	H
Attribute 1	1	2	5	7	5	11	14	15
Attribute 2	1	1	1	1	5	11	10	14

Iteration 1

	A	B	C	D	E	F	G	H
A	—	1	4	6	5.7	14.1	15.8	19.1
B	1	—	3	5	5	13.5	15	18.4
C	4	3	—	2	4	11.7	12.7	16.4
D	6	5	2	—	4.5	10.8	11.4	15.3
E	5.7	5	4	4.5	—	8.5	10.3	13.5
F	14.1	13.5	11.7	10.8	8.5	—	3.2	5
G	15.8	15	12.7	11.4	10.3	3.2	—	4.1
H	19.1	18.4	16.4	15.3	13.5	5	4.1	—
Max.			4	6	5.7	14.1	15.8	19.1

Iteration 2

	AB	C	D	E	F	G	H
A,B	—	4	6	5.7	14.1	15.8	19.1
C	4	—	2	4	11.7	12.7	16.4
D	6	2	—	4.5	10.8	11.4	15.3
E	5.7	4	4.5	—	8.5	10.3	13.5
F	14.1	11.7	10.8	8.5	—	3.2	5
G	15.8	12.7	11.4	10.3	3.2	—	4.1
H	19.1	16.4	15.3	13.5	5	4.1	—
Max. 6			4.5	11.7	12.7	16.4

Figure 12.31   Complete linkage for step 3.

See the matrix in Figure 12.31 labeled iteration 1 for the Euclidean distances of pairs of all objects.

Step 2 A and B have the lowest distance value of 1 (i.e., the highest similarity). Therefore, they are selected to be in the same cluster.

Step 3   The complete linkage of A and B is shown in Figure 12.31, iteration 1.

In the iteration 1 table, the distance between object A and itself is zero, which is not shown in the matrix. The highest distance between objects A and B and all other objects (i.e., C, D, E, F, G, H) is then selected as the combined distance. For example, in the column corresponding to object C, the distance from C to A is $d(A,C) = 4$, and the distance from B to C is $d(B,C) = 3$. So, the distance from combined object AB to C is $d(AB,C) =$ Max$\{4,3\} = 4$.

At this iteration, there are seven clusters. To generate the distance matrix for this iteration, combine A and B into one column vector (AB) by using the complete-linkage approach. See the iteration 2 matrix in Figure 12.32 where the AB combined vector is shown. In iteration 2, objects C and D have the lowest value (2); therefore, C and D will be clustered together. Their combined column vector, CD, is presented in the matrix for iteration 3. CD is generated by simply choosing the higher values for each row of columns AB, E, F, G, and H in iteration 2. The remaining iterations follow the same approach.

For iteration 7, since there will be only one cluster for the next iteration, the highest distance will be zero.

Table 12.11 shows the summary of the hierarchical method for Example 12.14. Depending on the application, one of these clustering alternatives can be selected. Bicriteria clustering can be used to choose the best alternative.

Figure 12.33 shows the tree associated with this clustering example. The labels on the axis are the distances at which the grouping of clusters occurred. It can be seen that the tree is labeled with the level of similarity on the right side (or dissimilarity on the left side) at which the clustering changes. Note that the similarity value is equal to the maximum distance minus the dissimilarity value. Cutting the tree at different levels will result in different clustering sets at different distance levels. For example, cutting the tree at a dissimilarity value of 6 will result in the following two clusters: $\{(A, B, C, D, E), (F, G, H)\}$.

Iteration 2

	A,B	C	D	E	F	G	H
A,B	0	4	6	5.7	14.1	15.8	19.1
C	4	0	2	4	11.7	12.7	16.4
D	6	2	0	4.5	10.8	11.4	15.3
E	5.7	4	4.5	0	8.5	10.3	13.5
F	14.1	11.7	10.8	8.5	0	3.2	5
G	15.8	12.7	11.4	10.3	3.2	0	4.1
H	19.1	16.4	15.3	13.5	5	4.1	0
Max.	6			4.5	11.7	12.7	16.4

Iteration 3

	A,B	C,D	E	F	G	H
A,B	0	6	5.7	14.1	15.8	19.1
C,D	6	0	4.5	11.7	12.7	16.4
E	5.7	4.5	0	8.5	10.3	13.5
F	14.1	11.7	8.5	0	3.2	5
G	15.8	12.7	10.3	3.2	0	4.1
H	19.1	16.4	13.5	5	4.1	0
Max.	15.8	12.7	10.3			5

⇒

Iteration 4

	A,B	C,D	E	F,G	H
A,B	0	6	5.7	15.8	19.1
C,D	6	0	4.5	12.7	16.4
E	5.7	4.5	0	10.3	13.5
F,G	15.8	12.7	10.3	0	5
H	19.1	16.4	13.5	5	0
Max.	6			12.7	16.4

⇒

Iteration 5

	A,B	C,D,E	F,G	H
A,B	0	6	15.8	19.1
C,D,E	6	0	12.7	16.4
F,G	15.8	12.7	0	5
H	19.1	16.4	5	0
Max.	19.1	16.4		

Iteration 6

	A,B	C,D,E	F,G,H
A,B	0	6	19.1
C,D,E	6	0	16.4
F,G,H	19.1	16.4	0
Max.			19.1

⇒

Iteration 7

	A,B,C,D,E	F,G,H
A,B,C,D,E	0	19.1
F,G,H	19.1	0
Max.	0	

Iteration 8

	A,B,C,D,E,F,G,H
(A,B,C,D,E,F,G,H)	0

Figure 12.32   Details of all iterations of Example 12.14.

For purpose of illustration, suppose that the dissimilarity threshold is set at a value of 4. In iteration 4, the minimum threshold is 4.5. Therefore, the algorithm stops with five clusters. Figure 12.34 shows the graphical clustering for iterations 4 and 7.

12.6   *P*-MEDIAN OPTIMIZATION CLUSTERING

12.6.1   *P*-Median Method for Binary Data[†]

The *P*-median method uses optimization to solve the clustering problem. The objective of the optimization problem is to minimize the total dissimilarity coefficient between objects in the same cluster. The desired number of clusters is given as a constraint in the optimization problem. The optimization solution provides the clustering of objects where each cluster has one object as its center. For the group technology problem, machines are treated as

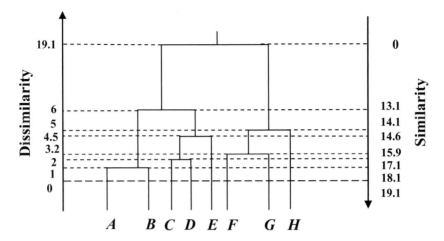

Figure 12.33   Hierarchical tree structure resulting from hierarchical clustering.

objects and parts are treated as attributes. For each cluster of machines, its associated part family can be identified based on the solution of optimization.

The dissimilarity for objects q and j is defined as follows, which is the same as the rectilinear distance:

$$d_{qj} = \sum_{i=1}^{n} |a_{iq} - a_{ij}| \tag{12.8}$$

$$a_{ij} = \begin{cases} 1 & \text{if object } j \text{ has value "1" for attribute } i \\ 0 & \text{otherwise} \end{cases} \tag{12.9}$$

Note that the dissimilarity between any object and itself is zero. That is, $d_{jj} = 0$ for $j = 1, 2, \ldots, m$. The dissimilarity matrix d_{qj} is a square matrix with m rows and m columns when there are m objects. This matrix is symmetric with respect to its diagonal axis, that is, $d_{qj} = d_{jq}$ and $d_{jj} = 0$. It is generally assumed that there are more objects than attributes and the number of desired clusters is less than the number of objects. With respect to the group technology problem, this method does not directly minimize the number of exceptional elements and therefore may not always provide an optimal solution for group technology

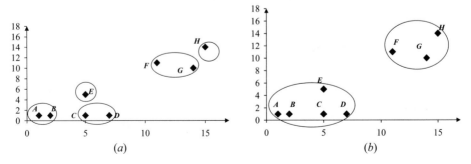

Figure 12.34   Graphical clustering for (a) iteration 4 and (b) iteration 7.

TABLE 12.11 Summary of Hierarchical Method for Example 12.14

Iteration	No. of Clusters	Distance Threshold	Cluster Members
1	8	0	{A,B,C,D,E,F,G,H}
2	7	1	{(A,B),C,D,E,F,G,H}
3	6	2	{(A,B),(C,D),E,F,G,H}
4	5	3.2	{(A,B),(C,D),E,(F,G),H}
5	4	4.5	{(A,B),(C,D,E),(F,G),H}
6	3	5	{(A,B),(C,D,E),(F,G,H)}
7	2	6	{(A,B,C,D,E),(F,G,H)}
8	1	19.1	{(A,B,C,D,E),(F,G,H)}

problems. The *P*-median model is formulated by the following 0–1 integer linear program (ILP) where $d_{q,j}$ are known coefficients and *P* is the given number of clusters.

PROBLEM 12.1   *P*-MEDIAN PROBLEM

Minimize:

$$z = \sum_{q=1}^{m}\sum_{j=1}^{m} d_{qj}x_{qj} \tag{12.10}$$

Subject to:

$$\sum_{j=1}^{m} x_{qj} = 1 \quad \text{for } q = 1, 2, \ldots, m \tag{12.11}$$

$$\sum_{j=1}^{m} x_{jj} = P \tag{12.12}$$

$$x_{qj} \leq x_{jj} \quad \text{for } q = 1, 2, \ldots, m, \ j = 1, 2, \ldots, m \tag{12.13}$$

$$x_{qj} = \begin{cases} 1 & \text{if object } q \text{ belongs to cluster } j \\ 0 & \text{otherwise} \end{cases} \tag{12.14}$$

The objective function (12.9) of the above ILP represents the total dissimilarity cost within the clusters, which is minimized for the given number of clusters, *P*. It is straightforward to change *P* and rerun the computer software to find the solution for different numbers of clusters.

Constraint (12.10) enforces that an object will be assigned to only one cluster. Constraint (12.11) requires that *P* clusters are generated. Constraint (12.12) ensures that object *q* will belong to a cluster *j* whose center is object *j* where $x_{jj} = 1$.

In the final ILP solution, when $x_{qj} = 1$, objects *q* and *j* are in the same cluster. When $x_{jj} = 1$, object *j* is the center of cluster *j* and there is only one cluster center for each cluster. The final solution may be further improved by visual inspection and intuition.

Example 12.15   *P*-Median Method   Cluster the following seven objects (machines) each having five attributes (parts). Use the *P*-median method to determine the best clustering when there are three clusters (Note that this is not the same as Example 12.4.)

		Objects (Machines)						
		1	2	3	4	5	6	7
Attributes (Parts)	1	1				1		1
	2	1	1	1				1
	3		1		1		1	
	4		1	1	1	1	1	
	5		1	1	1	1	1	

In this problem, $P = 3$. First, find the dissimilarities for all pairs of objects. For example, the dissimilarity between object 1 and object 2 for attributes $i = 1, 2, 3, 4, 5$ is calculated as follows:

Attributes	1	2	3	4	5	Sum
$\mathbf{a}_1$	1	1	0	0	0	
$\mathbf{a}_2$	0	1	1	1	1	
d_{12}	$|1-0|$	$|1-1|$	$|0-1|$	$|0-1|$	$|0-1|$	4

Finding d_{qj} is fairly straightforward. For example, consider objects $\mathbf{a}_1$ and $\mathbf{a}_2$. Only attribute 2 is the same for both objects. This means that four attributes (1, 3, 4, and 5) are not the same for both objects. Therefore, the dissimilarity of objects 1 and 2 is 4 (out of five attributes).

For objects 1 and 7, $d_{17} = 0$, which means that the two objects are identical, that is, their dissimilarity is zero. Note that dissimilarity is symmetric, that is, $d_{qj} = d_{jq}$, for example, $d_{12} = d_{21} = 4$. The dissimilarity (distance) matrix for the above example is given as follows:

$$d_{qj} = \begin{bmatrix} 0 & 4 & 3 & 5 & 3 & 5 & 0 \\ 4 & 0 & 1 & 1 & 3 & 1 & 4 \\ 3 & 1 & 0 & 2 & 2 & 2 & 3 \\ 5 & 1 & 2 & 0 & 2 & 0 & 5 \\ 3 & 3 & 2 & 2 & 0 & 2 & 3 \\ 5 & 1 & 2 & 0 & 2 & 0 & 5 \\ 0 & 4 & 3 & 5 & 3 & 5 & 0 \end{bmatrix}$$

Using the above d_{qj} values, the *P*-median ILP problem is formulated as follows for three clusters, that is, $P = 3$:

Minimize:

$$z = \sum_{q=1}^{7} \sum_{j=1}^{7} d_{qj} x_{qj} \tag{12.15}$$

$$\begin{aligned}
= \ & 4x_{12} + 3x_{13} + 5x_{14} + 3x_{15} + 5x_{16} + 0x_{17} \\
& + 4x_{21} + x_{23} + x_{24} + 3x_{25} + x_{26} + 4x_{27} \\
& + 3x_{31} + x_{32} + 2x_{34} + 2x_{35} + 2x_{36} + 3x_{37} \\
& + 5x_{41} + x_{42} + 2x_{43} + 2x_{45} + 0x_{46} + 5x_{47} \\
& + 3x_{51} + 3x_{52} + 2x_{53} + 2x_{54} + 2x_{56} + 3x_{57} \\
& + 5x_{61} + x_{62} + 2x_{63} + 0x_{64} + 2x_{65} + 5x_{67} \\
& + 0x_{71} + 4x_{72} + 3x_{73} + 5x_{74} + 3x_{75} + 5x_{76}
\end{aligned}$$

Subject to:

$$x_{11} + x_{12} + x_{13} + x_{14} + x_{15} + x_{16} + x_{17} = 1; \quad x_{51} + x_{52} + x_{53} + x_{54} + x_{55} + x_{56} + x_{57} = 1$$

$$x_{21} + x_{22} + x_{23} + x_{24} + x_{25} + x_{26} + x_{27} = 1; \quad x_{61} + x_{62} + x_{63} + x_{64} + x_{65} + x_{66} + x_{67} = 1$$

$$x_{31} + x_{32} + x_{33} + x_{34} + x_{35} + x_{36} + x_{37} = 1; \quad x_{71} + x_{72} + x_{73} + x_{74} + x_{75} + x_{76} + x_{77} = 1$$

$$x_{41} + x_{42} + x_{43} + x_{44} + x_{45} + x_{46} + x_{47} = 1 \tag{12.16}$$

$$x_{11} + x_{22} + x_{33} + x_{44} + x_{55} + x_{66} + x_{77} = 3 \tag{12.17}$$

$x_{12} - x_{22} \le 0;$	$x_{24} - x_{44} \le 0;$	$x_{36} - x_{66} \le 0;$	$x_{51} - x_{11} \le 0;$	$x_{64} - x_{44} \le 0$
$x_{13} - x_{33} \le 0;$	$x_{25} - x_{55} \le 0;$	$x_{37} - x_{77} \le 0;$	$x_{52} - x_{22} \le 0;$	$x_{65} - x_{55} \le 0$
$x_{14} - x_{44} \le 0;$	$x_{26} - x_{66} \le 0;$	$x_{41} - x_{11} \le 0;$	$x_{53} - x_{33} \le 0;$	$x_{67} - x_{77} \le 0$
$x_{15} - x_{55} \le 0;$	$x_{27} - x_{77} \le 0;$	$x_{42} - x_{22} \le 0;$	$x_{54} - x_{44} \le 0;$	$x_{71} - x_{11} \le 0$
$x_{16} - x_{66} \le 0;$	$x_{31} - x_{11} \le 0;$	$x_{43} - x_{33} \le 0;$	$x_{56} - x_{66} \le 0;$	$x_{72} - x_{22} \le 0$
$x_{17} - x_{77} \le 0;$	$x_{32} - x_{22} \le 0;$	$x_{45} - x_{55} \le 0;$	$x_{57} - x_{77} \le 0;$	$x_{73} - x_{33} \le 0$
$x_{21} - x_{11} \le 0;$	$x_{34} - x_{44} \le 0;$	$x_{46} - x_{66} \le 0;$	$x_{61} - x_{11} \le 0;$	$x_{74} - x_{44} \le 0$
$x_{23} - x_{33} \le 0;$	$x_{35} - x_{55} \le 0;$	$x_{47} - x_{77} \le 0;$	$x_{62} - x_{22} \le 0;$	$x_{75} - x_{55} \le 0$
$x_{63} - x_{33} \le 0;$	$x_{76} - x_{66} \le 0$			

$$\tag{12.18}$$

$$x_{qj} = \{0 \text{ or } 1\} \quad \text{for } q, \ j = 1, 2, 3, 4, 5, 6, 7 \tag{12.19}$$

See Supplement S12.6.lg4.

Solving the problem by an integer linear programming computer package (we used LINGO), the optimal solution is

$$x_{17} = x_{77} = x_{23} = x_{33} = x_{53} = x_{44} = x_{64} = 1$$

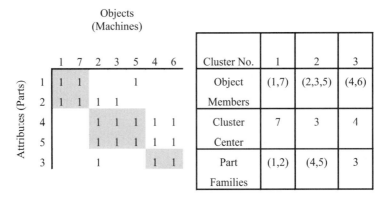

Figure 12.35 Part assignment for Example 12.15.

All other decision variables are zero.

The objective value is $z = 1x_{23} + 2x_{53} + 0 + 0 \cdots + 0 = 3$. The interpretation of variables whose values are 1 is as follows. Variable $x_{qj} = 1$ implies that object q is in cluster j, which means that object j is also a member and the center of this cluster. Variables that have at least one common q or j belong to the same cluster. For example, x_{17} means that object 1 is in cluster 7. The grouping of variables are $\{(x_{17}, x_{77}), (x_{23}, x_{33}, x_{53}), \text{ and } (x_{44}, x_{64})\}$. This means the clusters of objects are $\{(1,7), (2,3,5), \text{ and } (4,6)\}$. Variable $x_{jj} = 1$ implies that objects j is in cluster j and it is its center. Therefore, objects 7, 3, and 4 are the centers of the three clusters, respectively. The objective value is 3, which is the total dissimilarity of this solution. The matrix solution is provided in the following. This clustering solution has eight exceptional elements.

Based on Figure 12.35, the part families are (1,2), 3, and (4,5) corresponding to the three clusters.

In this example, there were five attributes and seven objects. One can measure the dissimilarity of pairs of attributes (instead of objects) and use these values to solve the group technology problem. The solution will provide a grouping of attributes for which the grouping of objects can be determined. When the number of objects is substantially higher than the number of attributes, it is computationally advantageous to group based on attributes and then assign objects to given clusters of attributes.

12.6.2 Bicriteria *P*-Median Clustering[†]

The *P*-median approach can be extended to solve the bicriteria clustering problem. To generate the set of efficient alternatives, solve the *P*-median problem for all possible values of *P*. That is, for each given value of *P*, solve the *P*-median problem and record its clustering solution. Also, find the total number of exceptional elements.

Example 12.16   Bicriteria *P*-Median Method   Consider Example 12.15. Generate all efficient bicriteria solutions for this example.

In this problem, there can be up to five clusters. Therefore, $P = 1, 2, 3, 4, 5$. For each value of *P*, solve the *P*-median ILP problem. The solution for each problem is presented in Table 12.12. For $P = 1$, that is, one cluster, the solution is obvious as presented in the

TABLE 12.12   Five Different Problems Solved by *P*-Median

Min $z =$ ($\sum\sum d_{qj} x_{qj}$)	$f_1 = P$ (P Given)	Clusters with Centers x_{jj} and $x_{qj} = 1$	Objects (Machine) Clusters	Attributes (Part) Families	Min $f_2 =$ No. of Exceptional Elements	Alternatives
13	1	$(x_{13}, x_{23}, x_{33}, x_{43}, x_{53}, x_{63}, x_{73})$	(1,2,3,4,5,6,7)	(1, 2, 3, 4, 5)	0	F_1
5	2	(x_{17}, x_{77}) $(x_{26}, x_{36}, x_{46}, x_{56}, x_{66})$	(1,7) (2,3,4,5,6)	(1,2) (3,4,5)	3	F_2
3	3	(x_{17}, x_{77}) (x_{23}, x_{33}, x_{53}) (x_{44}, x_{64})	(1,7) (2,3,5) (4,6)	(1,2) (4,5) (3)	8	F_3
1	4	(x_{17}, x_{77}) (x_{23}, x_{33}) (x_{44}, x_{64}) (x_{55})	(1,7) (2,3) (4,6) (5)	(1,2) (4) (3) (5)	11	F_4
0	5	(x_{17}, x_{77}) (x_{22}) (x_{33}) (x_{44}, x_{64}) (x_{55})	(1,7) (2) (3) (4,6) (5)	(1) (2) (4) (3) (5)	13	F_5

For $P = 2$, the cluster centers are 6 and 7. There are three exceptional elements.

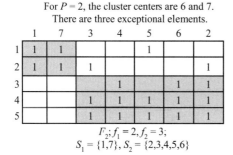

$F_2; f_1 = 2, f_2 = 3;$
$S_1 = \{1,7\}, S_2 = \{2,3,4,5,6\}$

For $P = 3$, the cluster centers are 3, 4, and 7. There are eight exceptional elements.

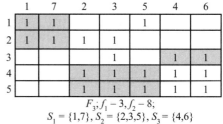

$F_3; f_1 = 3, f_2 = 8;$
$S_1 = \{1,7\}, S_2 = \{2,3,5\}, S_3 = \{4,6\}$

For $P = 4$, the cluster centers are 3, 4, 5, and 7. There are 11 exceptional elements.

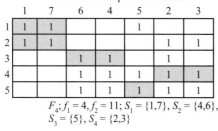

$F_4; f_1 = 4, f_2 = 11; S_1 = \{1,7\}, S_2 = \{4,6\},$
$S_3 = \{5\}, S_4 = \{2,3\}$

For $P = 5$, the cluster centers are 1, 2, 3, 4, and 5. There are 13 exceptional elements.

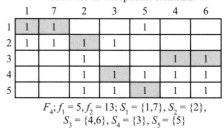

$F_4; f_1 = 5, f_2 = 13; S_1 = \{1,7\}, S_2 = \{2\},$
$S_3 = \{4,6\}, S_4 = \{3\}, S_5 = \{5\}$

Figure 12.36 Assignment of four solutions of P-median problems.

first row of Table 12.12. For $P = 2$, that is, two clusters, set $P = 2$ and solve the P-median problem. The ILP solution is:

Cluster 1: Machines 1 and 7 with parts 1 and 2.

Cluster 2: Machines 2, 3, 4, 5, and 6 with parts 3, 4, and 5.

This grouping has only three exceptional elements.

For $P = 3$, the solution was presented in Example 12.15. Solutions for $P = 4$ and $P = 5$ are presented in Table 12.12.

For $P = 1$, there is one object cluster [in this case, object (machine) 3 is the cluster center] and all parts are members. There is only one cell and no exceptional elements. The matrix solutions for $P = 2, 3, 4, 5$ are presented in Figure 12.36.

The list of bicriteria alternatives are summarized in the following table; a multicriteria method (e.g., an additive utility function) can be used to find the best alternative.

	F_1	F_2	F_3	F_4	F_5
Max. f_1 = No. of cells (P)	1	2	3	4	5
Min. f_2 = No. of exceptional element	0	3	8	11	13
Efficient?	Yes	Yes	Yes	Yes	Yes

Tricriteria P-Median   The tricriteria problem for the P-median is similar to the tricriteria problem of rank order clustering of Section 12.4.4.

Supplement S12.7.docx describes tricriteria P-median clustering and gives an example.

12.6.3 P-Median for Continuous Data[†]

The *P*-median approach discussed in Section 12.6.1 can be used to solve problems with continuous data. For continuous data, find the Euclidean distances for all pairs of objects where the distance for each pair of objects *q* and *j* is defined as follows:

$$d_{qj} = \left(\sum_{i=1}^{n} (a_{iq} - a_{ij})^2 \right)^{1/2}$$

The *P*-median ILP formulation is the same as the *P*-median ILP problem presented in Section 12.6.1.

Example 12.17   P-Median for Continuous Example   Consider the following set of biattribute objects A–G (labeled as 1–7). Use the *P*-median for continuous data to determine the best clustering when there are three clusters.

Object	A,1	B, 2	C,3	D,4	E,5	F,6	G,6
Attribute 1	1	2	2	8	9	8	10
Attribute 2	10	9	8	3	3	8	8

In this problem, $P = 3$. First, find the dissimilarities for all pairs of objects. The complete dissimilarity matrix for the above example is given in the following matrix. For example,

$$d_{A,B} = d_{1,2} = \sqrt{(1-2)^2 + (10-9)^2} = 1.41$$

$$d_{qj} = \begin{bmatrix} 0 & 1.41 & 2.24 & 9.90 & 10.63 & 7.28 & 9.22 \\ 1.41 & 0 & 1 & 8.49 & 9.22 & 6.08 & 8.06 \\ 2.24 & 1 & 0 & 7.80 & 8.60 & 6 & 8 \\ 9.90 & 8.49 & 7.80 & 0 & 1 & 5 & 5.39 \\ 10.63 & 9.22 & 8.60 & 1 & 0 & 5.10 & 5.10 \\ 7.28 & 6.08 & 6 & 5 & 5.10 & 0 & 2 \\ 9.22 & 8.06 & 8 & 5.39 & 5.10 & 2 & 0 \end{bmatrix}$$

Using the above d_{qj} values, the *P*-median ILP problem is formulated as follows where $P = 3$.

Constraint (12.10) enforces each object is assigned to only one cluster. Constraint (12.11) shows the required number of clusters. Constraint (12.12) ensures that object *q* will belong to a cluster *j*, whose center is object *j*, where $x_{jj} = 1$:

Minimize:

$$z = \sum_{q=1}^{7} \sum_{j=1}^{7} d_{qj} x_{qj}$$

$$= 1.41x_{12} + 2.24x_{13} + 9.90x_{14} + 10.63x_{15} + 7.28x_{16} + 9.22x_{17}$$

$$+ 1.41x_{21} + x_{23} + 8.49x_{24} + 9.22x_{25} + 6.08x_{26} + 8.06x_{27}$$

$$+ 2.24x_{31} + x_{32} + 7.80x_{34} + 8.60x_{35} + 6x_{36} + 8x_{37}$$

$$+ 9.90x_{41} + 8.49x_{42} + 7.80x_{43} + x_{45} + 5x_{46} + 5.39x_{47}$$

$$+ 10.63x_{51} + 9.22x_{52} + 8.60x_{53} + x_{54} + 5.10x_{56} + 5.10x_{57}$$

$$+ 7.28x_{61} + 6.08x_{62} + 6x_{63} + 5x_{64} + 5.10x_{65} + 2x_{67}$$

$$+ 9.22x_{71} + 8.06x_{72} + 8x_{73} + 5.39x_{74} + 5.10x_{75} + 2x_{76}$$

Subject to:

$$x_{11} + x_{12} + x_{13} + x_{14} + x_{15} + x_{16} + x_{17} = 1; \qquad x_{51} + x_{52} + x_{53} + x_{54} + x_{55} + x_{56} + x_{57} = 1$$

$$x_{21} + x_{22} + x_{23} + x_{24} + x_{25} + x_{26} + x_{27} = 1; \qquad x_{61} + x_{62} + x_{63} + x_{64} + x_{65} + x_{66} + x_{67} = 1$$

$$x_{31} + x_{32} + x_{33} + x_{34} + x_{35} + x_{36} + x_{37} = 1; \qquad x_{71} + x_{72} + x_{73} + x_{74} + x_{75} + x_{76} + x_{77} = 1$$

$$x_{41} + x_{42} + x_{43} + x_{44} + x_{45} + x_{46} + x_{47} = 1 \tag{12.20}$$

$$x_{11} + x_{22} + x_{33} + x_{44} + x_{55} + x_{66} + x_{77} = 3 \tag{12.21}$$

$x_{12} - x_{22} \le 0;$	$x_{24} - x_{44} \le 0;$	$x_{36} - x_{66} \le 0;$	$x_{51} - x_{11} \le 0;$	$x_{64} - x_{44} \le 0$
$x_{13} - x_{33} \le 0;$	$x_{25} - x_{55} \le 0;$	$x_{37} - x_{77} \le 0;$	$x_{52} - x_{22} \le 0;$	$x_{65} - x_{55} \le 0$
$x_{14} - x_{44} \le 0;$	$x_{26} - x_{66} \le 0;$	$x_{41} - x_{11} \le 0;$	$x_{53} - x_{33} \le 0;$	$x_{67} - x_{77} \le 0$
$x_{15} - x_{55} \le 0;$	$x_{27} - x_{77} \le 0;$	$x_{42} - x_{22} \le 0;$	$x_{54} - x_{44} \le 0;$	$x_{71} - x_{11} \le 0$
$x_{16} - x_{66} \le 0;$	$x_{31} - x_{11} \le 0;$	$x_{43} - x_{33} \le 0;$	$x_{56} - x_{66} \le 0;$	$x_{72} - x_{22} \le 0$
$x_{17} - x_{77} \le 0;$	$x_{32} - x_{22} \le 0;$	$x_{45} - x_{55} \le 0;$	$x_{57} - x_{77} \le 0;$	$x_{73} - x_{33} \le 0$
$x_{21} - x_{11} \le 0;$	$x_{34} - x_{44} \le 0;$	$x_{46} - x_{66} \le 0;$	$x_{61} - x_{11} \le 0;$	$x_{74} - x_{44} \le 0$
$x_{23} - x_{33} \le 0;$	$x_{35} - x_{55} \le 0;$	$x_{47} - x_{77} \le 0;$	$x_{62} - x_{22} \le 0;$	$x_{75} - x_{55} \le 0$
$x_{63} - x_{33} \le 0;$	$x_{76} - x_{66} \le 0$			

$$\tag{12.22}$$

$$x_{qj} = \{0 \text{ or } 1\} \text{ for } q, \ j = 1, 2, 3, 4, 5, 6, 7 \tag{12.23}$$

Solving the problem by an ILP computer package, the optimal solution is as follows:

$$x_{12} = x_{22} = x_{32} = x_{45} = x_{55} = x_{66} = x_{76} = 1 \qquad z = 5.41$$

All other decision variables are zero. The objective function is 5.41, which means that the total Euclidean distance between all objects in the same cluster is 5.41. The three clusters of variables are $(x_{12}, x_{22}, x_{32}), (x_{45}, x_{55})$, and (x_{66}, x_{76}) whose centers are x_{22}, x_{55}, and x_{66}, respectively. Therefore, the following three groups of objects are formed: $\{(1,2,3), (4,5), (6,7)\}$, which can also be labeled as $\{(A, B, C), (D, E), (F, G)\}$. As we will see in the next section, this solution is the same as the solution found by the K-means method (which also uses centers) for the same problem.

12.7 *K*-MEANS CLUSTERING

In this section, *K*-means is presented for continuous linear-scaled data sets; however, it can also be used for binary or discrete data. *K*-means is a partitional clustering method where all objects are initially partitioned into a given number of clusters rather than progressively assigning objects to clusters as in the SC approach (Section 12.5). In partitional clustering, the number of clusters and the members of each cluster (or the centers of each cluster) must be given before solving the problem, and then the *K*-means method is used in attempt to find an improved solution. In comparison, in the *P*-median method (Section 12.6) only the number of clusters should be given but not their membership (or the cluster centers). Also, in *P*-median, the final cluster centers are actual objects while in the *K*-means this may not necessarily be true. Also, for large problems, solving ILP (for *P*-median) is computationally very time consuming, but *K*-means can solve very large problems fairly quickly.

The method is called *K*-means because *K* different centers are considered in each iteration. In the *K*-means method, the closeness of objects with respect to given centers is measured by the Euclidean distance. The objects are clustered based on their closeness to the given centers. In the initial phase of the method, the number of clusters and cluster centers (or each cluster membership) is identified. Then *K*-means clustering is used to find improved cluster centers and cluster memberships. In each iteration, the centers are changed, resulting in different (improved) clustering. The method is stopped when new improved centers cannot be generated and/or the members of clusters remain the same in two consecutive iterations.

It is important to note that the initial assignment of the cluster centers has a large effect on the final solution of the clustering. That is, the results of the algorithm are sensitive to the initial selection of the cluster centers and may run into a local minimum. Similar to all other clustering methods covered in this chapter, *K*-means is a heuristic method. Therefore, the *K*-means clustering method can be performed several times, each time using different initial cluster centers. Then, the best clustering solution can be chosen. *K*-means works best with compact and isolated clusters. *K*-means performs weakly when the clusters have very large variances or when the clusters are elongated (rather than being distributed in a circular manner).

12.7.1 *K*-Means Method

The most common objective in a partitional clustering algorithm is to minimize the distance between members of a cluster to their associated cluster center. This objective is known as the squared-error criterion, which is the same as the Euclidean distance. The distance between a given object $\mathbf{a}_j$ and a cluster center $\mathbf{c}_k$ is defined in Equation (12.14) where there are n attributes denoted by i:

$$d(\mathbf{a}_j, \mathbf{c}_k) = \left(\sum_{i=1}^{n} \left(a_{ij} - c_{ik} \right)^2 \right)^{1/2} \tag{12.24}$$

where $\mathbf{c}_k = (c_{1k}, c_{2k}, \ldots, c_{nk})$ is the center of cluster k. There are K clusters $k = 1, 2, \ldots, K$.

Deviation   Deviation is defined as the total distance of all cluster members to their cluster centers. Deviation can be measured by the within sum of squares (WSS), which is defined as

$$\text{WSS} = \sum_{k=1}^{K} \sum_{a_j \in S_k} (\mathbf{a}_j - \mathbf{c}_k)^2 \tag{12.25}$$

where S_k is the set of objects that belong to cluster k.

The purpose of the K-means algorithm is to find a clustering solution such that for a given K clusters WSS is minimized. That is:

Minimize WSS

Subject to the set of data and the number of clusters K

Separation   Separation is defined as the total distance of all cluster centers to their grand cluster center, $\mathbf{c} = (c_1, c_i, \ldots, c_n)$, where c_i is the average of attribute i for all objects, calculated as

$$c_i = \frac{\sum_{j=1}^{m} a_{ij}}{m} \quad \text{for } i = 1, \ldots, n$$

Separation can be measured by the between sum of squares (BSS), which is defined as

$$\text{BSS} = \sum_{k=1}^{K} |S_k|(\mathbf{c} - \mathbf{c}_k)^2 \tag{12.26}$$

It can be proven that the summation of the BSS and WSS for any given number of clusters in any iteration of K-means is constant by using the centers generated at the end of each iteration:

Total sum of squares (TSS) = WSS + BSS   for any number of clusters

Since BSS requires less computational effort than WSS, maximizing BSS can be used instead of minimizing WSS as the objective of the K-means method; that is:

Maximize BSS

Subject to the set of data and the number of clusters K

The algorithm begins by assigning initial clusters. This is usually accomplished by selecting K random cluster centers $\mathbf{c}_k$, $k = 1, \ldots, K$. Then, the objects are assigned to the cluster whose center is the closest to the given object. The cluster centers are recalculated in each iteration as the mean of their members.

The algorithm consists of the following steps:

Step 0 (Initialization) Select the K centers $\mathbf{c}_k$ either provided by the analysis or randomly generated. (Alternatively, if a clustering solution is given, go to step 2.)

Step 1 Measure the distance of each object to each center. Assign each object to the closest cluster center (in terms of distance), that is, identify which object $\mathbf{a}_j$ belongs to which cluster S_k. In case of a tie, assign the object arbitrarily to one of the closest clusters.

Step 2 Recalculate each cluster center $\mathbf{c}_k$ to be the mean (average) in terms of the attributes of all objects that belong to the given cluster k, where

$$c_{ik} = \frac{1}{|S_k|} \sum_{a_j \in S_k} a_{ij} \quad \text{for} \quad i = 1, 2, \ldots, n \tag{12.27}$$

where $|S_k|$ is the number of objects in cluster k.
If a cluster is empty, then do not update its associated cluster center.

Repeat steps 1 and 2 until the cluster memberships do not change.

Example 12.18 Using *K*-Means to Cluster a Two-Attribute Problem Consider the following set of biattribute objects *A–G*:

Objects	A	B	C	D	E	F	G
Attribute 1	1	2	2	8	9	8	10
Attribute 2	10	9	8	3	3	8	8

Cluster these objects into three groups using the K-means algorithm. Suppose the initial cluster centers are $\mathbf{c}_1 = (2,5)$, $\mathbf{c}_2 = (6,5)$, and $\mathbf{c}_3 = (6,10)$. (Note that, alternatively, if the initial clustering is given as $S_1 = \{B, C\}$, $S_2 = \{D, E\}$, and $S_3 = \{A, F, G\}$, then go to step 2 in the initial iteration, that is, skip step 1.)

Initial Iteration

Step 1 Table 12.13 shows the Euclidean distance between each object and the initial cluster centers $\mathbf{c}_1$, $\mathbf{c}_2$, and $\mathbf{c}_3$. For example, the Euclidean distance between object A and $\mathbf{c}_1$ is

$$d(A, \mathbf{c}_1) = \sqrt{(1-2)^2 + (10-5)^2} = \sqrt{1+25} = 5.1$$

Each object becomes a member of the cluster whose center is closest. That is, for each row (object), find the minimum distance value to the three cluster centers. Then assign the object to the cluster corresponding to the chosen center. For example, object A's Euclidean distances to the three centers are 5.1, 7.1, and 5, respectively. Therefore, A is assigned to cluster 3. Objects B and C are closest to cluster center $\mathbf{c}_1$, so B and C are assigned to cluster 1. The memberships for the three sets of clusters, based on Table 12.13, are

$$S_1 = \{B, C\} \qquad S_2 = \{D, E\} \qquad S_3 = \{A, F, G\}$$

TABLE 12.13 Distances from Each Cluster Center to Each Object for Initial Iteration

Objects	$c_1 = (2,5)$	$c_2 = (6,5)$	$c_3 = (6,10)$	Minimum	Assign To
A	5.1	7.1	5.0	5.0	c_3
B	4.0	5.7	4.1	4.0	c_1
C	3.0	5.0	4.5	3.0	c_1
D	6.3	2.8	7.3	2.8	c_2
E	7.3	3.6	7.6	3.6	c_2
F	6.7	3.6	2.8	2.8	c_3
G	8.5	5.0	4.5	4.5	c_3
Members	B, C	D, E	A,F,G		

Step 2   The new cluster centers are calculated as follows: $c_1 = (2,8.5)$, $c_2 = (8.5,3)$, and $c_3 = (6.3,8.7)$. For example for the first cluster center, the value of the first attribute is $(2+2)/2 = 2$, and the value of the second attribute is $(9+8)/2 = 8.5$.

	First Cluster			Second Cluster			Third Cluster		
Object	B	C	c_1 avg.	D	E	c_2 avg.	A	F	G	c_3 avg.
Attribute 1	2	2	2	8	9	8.5	1	8	10	6.3
Attribute 2	9	8	8.5	3	3	3	10	8	8	8.7

Iteration 1   The new cluster memberships are determined using the cluster centers found at the end of the initial iteration; see Table 12.14A. Now recalculate each cluster center. The new cluster centers are calculated as follows: $c_1 = (1.7,9)$, $c_2 = (8.5,3)$, and $c_3 = (9,8)$.

Iteration 2   The new cluster memberships are determined using the cluster centers found at the end of iteration 1; see Table 12.14B. The clustering method converges in iteration 2 because the members of each cluster remain the same. The final clusters are

$$S_1 = \{A, B, C\} \qquad S_2 = \{D, E\} \qquad S_3 = \{F, G\}$$

Figure 12.37a illustrates the convergence of the cluster centers where cluster k is encircled around c_k.

As the following table shows, the TSS does not change throughout the iterations of the K-means algorithm. This property holds for any K-means problem.

	Minimize WSS	Maximize BSS	TSS
Iteration 1	48.33	89.10	137.43
Iteration 2	5.17	132.26	137.43

See Supplements S12.8.xls, S12.8.xlsx.

TABLE 12.14 Summary of Iterations 1 and 2 for Example 12.18

A. Iteration 1

Objects	$c_1 = (2,8.5)$	$c_2 = (8.5,3)$	$c_3 = (6.3,8.7)$	Minimum	Assign To
A	1.8	10.3	5.5	1.8	c_1
B	0.5	8.8	4.3	0.5	c_1
C	0.5	8.2	4.4	0.5	c_1
D	8.1	0.5	5.9	0.5	c_2
E	8.9	0.5	6.3	0.5	c_2
F	6.0	5.0	1.8	1.8	c_3
G	8.0	5.2	3.7	3.7	c_3
Members	A,B,C	D,E	F,G		

B. Iteration 2

Objects	$c_1 = (1.7,9)$	$c_2 = (8.5,3)$	$c_3 = (9,8)$	Minimum	Assign To
A	1.8	10.3	5.5	1.8	c_1
B	0.5	8.8	4.3	0.5	c_1
C	0.5	8.2	4.4	0.5	c_1
D	8.1	0.5	5.9	0.5	c_2
E	8.9	0.5	6.3	0.5	c_2
F	6.0	5.0	1.8	1.8	c_3
G	8.0	5.2	3.7	3.7	c_3
Members	A,B,C	D,E	F,G		

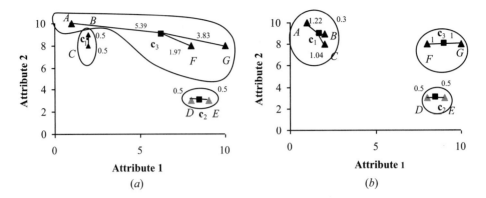

Figure 12.37   Euclidean distances of each member to its center: (*a*) Initial clustering; (*b*) iteration 2. The distances are closer in the second iteration.

Example 12.19   Using Different Initial Centers and Different Numbers of Clusters
Consider Example 12.18. Show the solution for 1, 2, 3, and 4 clusters. Show the impact of using different initial centers. Also, show the BSS, WSS, and TSS for each cluster. Verify the TSS is the same for all different cluster numbers.

Consider $K = 1$ Cluster   In this case, there is just one cluster. The cluster center for this alternative is computed in the following table. The membership for one cluster is $S_1 = \{A,B,C,D,E,F,G\}$. The center (based on the average for all objects for each attribute) is (5.71,7). Using this center, the BSS and WSS for $K = 1$ are

$$\text{BSS} = (5.71 - 5.71)^2 + (7 - 7)^2 = 0$$
$$\text{WSS} = (1 - 5.71)^2 + (10 - 7)^2 + \cdots + (10 - 5.71)^2 + (8 - 7)^2 = 137.43$$

where TSS = WSS + BSS = 137.43 + 0 = 137.43. Note that the TSS is the same for all alternatives in Table 12.15.

Object	A	B	C	D	E	F	G	c (Average)
Attribute 1	1	2	2	8	9	8	10	5.71
Attribute 2	10	9	8	3	3	8	8	7

Consider $K = 2$ Clusters   Table 12.15 shows the solution for this problem for initial cluster centers $c_1 = (2, 5)$ and $c_2 = (6, 5)$:

$$\text{BSS}\,(K = 2) = 3[(5.71 - 1.67)^2 + (7 - 9)^2] + 4[(5.71 - 8.75)^2 + (7 - 5.5)^2] = 106.93$$
$$\text{WSS}\,(K = 2) = (1 - 1.67)^2 + (10 - 9)^2 + \cdots + (10 - 8.75)^2 + (8 - 5.5)^2 = 30.50$$

TABLE 12.15 Clustering Alternatives for $K = 1, 2, 3, 4$

No. of clusters	$K = 1$	$K = 2$	$K = 3$		$K = 4$	
Solution	F_1	F_2	F_3	F_5	F_4	F_6
Initial cluster centers	$c_1 = (2, 5)$	$c_1 = (2, 5)$ $c_2 = (6, 5)$	$c_1 = (2, 5)$ $c_2 = (8, 2)$ $c_3 = (8, 8)$	$c_1 = (1, 10)$ $c_2 = (2, 9)$ $c_3 = (6, 5)$	$c_1 = (2, 9)$ $c_2 = (8, 2)$ $c_3 = (8, 8)$ $c_4 = (9, 7)$	$c_1 = (1, 10)$ $c_2 = (2, 9)$ $c_3 = (8, 2)$ $c_4 = (8, 8)$
Final cluster centers	$c_1 = (5.71, 7)$	$c_1 = (1.67, 9)$ $c_2 = (8.75, 5.5)$	$c_1 = (1.67, 9)$ $c_2 = (8.5, 3)$ $c_3 = (9, 8)$	$c_1 = (1, 10)$ $c_2 = (2, 8.5)$ $c_3 = (8.75, 5.5)$	$c_1 = (1.67, 9)$ $c_2 = (8.5, 3)$ $c_3 = (8, 8)$ $c_4 = (10, 8)$	$c_1 = (1, 10)$ $c_2 = (2, 8.5)$ $c_3 = (8.5, 3)$ $c_4 = (9, 8)$
Cluster sets	$S_1 = \{A, B, C, D, E, F, G\}$	$S_1 = \{A, B, C\}$ $S_2 = \{D, E, F, G\}$	$S_1 = \{A, B, C\}$ $S_2 = \{D, E\}$ $S_3 = \{F, G\}$	$S_1 = \{A\}$ $S_2 = \{B, C\}$ $S_3 = \{D, E, F, G\}$	$S_1 = \{A, B, C\}$ $S_2 = \{D, E\}$ $S_3 = \{F\}$ $S_4 = \{G\}$	$S_1 = \{A\}$ $S_2 = \{B, C\}$ $S_3 = \{D, E\}$ $S_4 = \{F, G\}$
BSS (max.)	0	106.93	132.26	109.18	134.26	134.43
WSS (min.)	137.43	30.50	5.17	28.25	3.17	3.00
TSS	137.43	137.43	137.43	137.43	137.43	137.43
Better solution for given K?	Yes	Yes	Yes	No	No	Yes

Consider $K = 3$ Clusters   For two different sets of initial centers, two solutions are generated, as shown in Table 12.15. Note that the BSS for solutions F_3 and F_5 are 132.26 and 109.18, respectively. Therefore, solution F_3 is better than solution F_5 because it has a higher BSS.

Consider $K = 4$ Clusters   For two different sets of initial centers, solutions F_4 and F_6 are generated as shown in Table 12.15.

Comparisons of K-Means to ROC, SC Clustering, and P-Median for Binary Data Consider Example 12.4. In this example, there are five objects and seven attributes. This example was solved by K-means with initial cluster centers for $K = 2$ of $\mathbf{c}_1$: (1, 0, 1, 0, 1, 0, 1) and $\mathbf{c}_2$: (0, 1, 0, 1, 1, 0, 0). The initial cluster centers for $K = 3$ were $\mathbf{c}_1$: (1, 0, 1, 0, 1, 0, 1), $\mathbf{c}_2$: (0, 1, 0, 1, 1, 0, 0), and $\mathbf{c}_3$: (1, 1, 1, 1, 0, 0, 1).

This example was solved by the ROC method in Section 12.4, the SC method in Section 12.5, and the P-median method in Section 12.6 (the details of the solutions are not shown). The solutions of all four clustering methods are identical. The four clustering approaches are heuristics and therefore may not generate the optimal solution. But, for a small problem such as Example 12.4, all generated solutions are optimal. For larger problems, the solution of these four clustering methods may vary greatly. Each method can be used effectively for a specific type of problem and application.

 See Supplement S12.9.xlsx.

12.7.2   Bicriteria K-Means[†]

The bicriteria for K-means (and all other methods with continuous data) is:

Maximize $f_1 = K$ (number of clusters)

Maximize $f_2 = \text{BSS}_k - \text{BSS}_{k-1}$ (distinguishability of clusters)

The first objective maximizes the flexibility of clustering (as discussed in previous multi-criteria sections). The second objective, incremental BSS, measures incremental explained variance. It measures the usefulness of increasing the number of clusters. This objective is measured by using Equation (12.16) for BSS of k and BSS of $k - 1$ number of clusters. To generate all efficient alternatives, find the best clustering solution for each given K, the number of clusters. Then tabulate the bicriteria solutions. There are always K efficient solutions for this problem.

Example 12.20 K-Means Bicriteria Clustering   Consider Example 12.18. Generate bicriteria alternatives for up to four clusters (i.e., $K = 1, 2, 3, 4$). For each alternative, measure the two objective functions.

The method for generating efficient solutions is the same as the approach used for solving Example 12.19. See Table 12.15 for all solutions. Only four solutions are efficient for $K = 1, 2, 3, 4$. They are shown in Table 12.16.

Finding the incremental explained variance is very simple. Use the BSS values obtained in Table 12.15. For example, the incremental BSS for alternative F_3 is

$$\text{BSS}_k - \text{BSS}_{k-1} = 132.26 - 106.93 = 25.33$$

TABLE 12.16   Summary of Objective Functions for Example 12.19

No. of clusters	$K = 1$	$K = 2$	$K = 3$	$K = 4$
Alternative label	F_1	F_2	F_3	F_4
BSS	0	106.93	132.26	134.43
Max. $f_1 = K$	1	2	3	4
Max. $f_2 = BSS_k - BSS_{k-1}$	137.43	106.93	25.33	2.17
Efficient?	Yes	Yes	Yes	Yes

We define $BSS_k - BSS_{k-1}$ for $K = 1$ as the TSS since there is no BSS_0.

See Supplement S12.10.xls.

12.8   MULTIPERSPECTIVE MULTICRITERIA CLUSTERING

12.8.1   Multicriteria Clustering Problems[†]

Clustering analysis is concerned with the grouping of alternatives into clusters based on their attributes. Multiple-criteria clustering is concerned with the clustering and subsequent ranking of alternatives that belong to each cluster. In some real-world problems, the range of consequences of different alternatives are considerably different, and the selection of a group of alternatives (instead of only one best alternative) is necessary. Traditional MCDM approaches treat the set of alternatives with the same method of analysis, but multicriteria clustering allows the set of alternatives to be partitioned so that different MCDM methods can be applied to different clusters. Examples of multiple-criteria clustering are:

- The selection of team members in sports, business, academia, research institutes, and so on.
- The selection of groups of machines in manufacturing. Machines that can perform similar jobs are grouped together, and then a group of machines is selected.
- The selection among groups of investment (portfolios) (e.g., choose among different types of stocks, bonds, and certificates of deposit based on income and risks).
- The selection of foods for different diet purposes.

Consider the example of choosing team members for a new football team. Suppose there are 100 candidate players (objects which in a multicriteria problem are labeled as alternatives) who can be clustered into three clusters; each cluster represents a certain role (e.g., defense, offense). Players can then be ranked compared to other players in their respective clusters and one or more players will be selected from each cluster.

What distinguishes multicriteria clustering from other clustering methods covered in this chapter is that in multicriteria clustering alternatives are selected for the purpose of ranking, but they will be ranked within their own clusters. Therefore, alternatives belonging to different clusters will be evaluated differently. It is interesting to note that a member may be ranked very low or very high depending on to which cluster it is assigned. For example, in the investment portfolio example, the DM would like to use the different utility functions to rank the alternatives in different clusters. Using the same additive utility function, all alternatives in certificates of-deposit clusters can be ranked, but a different utility function should be used to rank alternatives in stock clusters which are associated with much higher risks. These two clusters, although evaluated by the same criteria, would have different utility functions. The selection of the best alternative of each cluster completes the selection process.

Multiple-criteria clustering compared to MCDM may have the following benefits:

- It decreases the set of alternatives. The DM may be interested in some of the clusters and therefore discard alternatives belonging to other clusters.
- It decreases the number of criteria. The DM may only need to consider those criteria that are applicable to the given cluster.
- It provides a basis for a more in-depth evaluation of very different alternatives. Each cluster of alternatives can be explored and analyzed in more depth.
- It provides an approach for solving group decision-making problems. Each DM may rank alternatives in each cluster differently, so the differences for each cluster can be separately negotiated.
- It provides a practical approach for solving nonlinear utility functions where each cluster ranking can be solved by a simple linear additive utility function.

Clustering Attributes Versus Criteria Used in MCDM In some applications, the set of attributes for clustering are the same as the criteria used for evaluating alternatives, but clustering attributes may not necessarily be the same as the criteria used for evaluating alternatives. However, they can overlap; for example, three out of four clustering attributes could be the same as three out of five criteria for evaluating alternatives.

Multicriteria Clustering Approach The multicriteria clustering approach can be summarized as follows:

1. Identify the attributes for clustering objects (alternatives).
2. Cluster alternatives using a given clustering method. Assume there are K clusters.
3. Identify criteria to be used for evaluating alternatives (if possible, use clustering attributes as criteria).
4. For each cluster, assess its utility function. Rank alternatives of each cluster using its utility function.
5. Choose the final set of selected alternatives.

12.8.2 Z Theory Clustering Using *K*-Means†

In this section, we use K-means clustering for solving multicriteria clustering problems. In this approach, objects are clustered based on their attributes, which are the same as the criteria used for evaluating alternatives. Furthermore, alternatives in each cluster are ranked by using ideal goal utility functions for measuring distances to the Ideal alternative of each cluster where each cluster has its own weights of importance for criteria [using

Equation (12.18)]. For each cluster, the ideal alternative is determined by the DM for each cluster where alternatives closer to the ideal alternative are ranked higher.

Z theory (introduced by Malakooti, 2014) is based on using nonlinear utility functions for ranking multicriteria alternatives. An extension of Z theory for solving multicriteria K-means clustering is introduced here. Suppose for each cluster the k the center is $\mathbf{c}_k = (c_{1,k}, c_{2,k}, \ldots, c_{n,k})$. The ideal alternative for cluster k is $F_{k,G} = (f_{1,k,G}, f_{2,k,G}, \ldots, f_{n,k,G})$. It is assumed that the ideal alternative of each cluster dominates (or at least is efficient with respect to) all alternatives belonging to that cluster. Each cluster has weights of importance $\mathbf{w}_k = (w_{1,k}, w_{2,k}, \ldots, w_{n,k})$, where $w_{i,k}$ is the weight associated with the ith attribute for objects in cluster $k = 1,2,\ldots, K$. The weights of all attributes (criteria) are positive, that is, $w_{ik} > 0$ for $i = 1,\ldots, n$ and $k = 1,2,\ldots, K$. Each cluster has its own ideal alternative and weights of importance. The distance to the ideal alternative of each cluster k is defined as follows:

$$\text{Minimize DV}_{k,G} = \sqrt{\sum_{i=1}^{n} w_{i,k}(a_{i,j} - f_{i,k,G})^2} \qquad \text{for cluster } k \qquad (12.28)$$

Note that this MCDM method (12.18) is related to goal programming where the distance to goals are minimized. In multicriteria clustering, the K ideal goal problems must be solved. After clustering is completed, then for each cluster the (12.18) utility function is used to rank its members.

Note that (12.18) is a generalization of (12.14), which is used in the K-means method. That is, if we set all weights equal to one, that is, $w_{ik} = 1$, and also set goals to be the same as the centers of each cluster, that is, $F_{k,G} = \mathbf{c}_k$, then (12.18) becomes (12.14).

Ideal Goal Versus Center of a Cluster   In K-means, the center of each cluster is the mean of the cluster. Then, if some of the alternatives are inefficient, the center may also be inefficient with respect to some of the alternatives. However, in multicriteria K-means, we require that the ideal alternative of each cluster dominate (or at least be efficient with respect to) all members of the cluster.

Nonlinear Z Utility Function for Each Cluster   The utility function for each cluster k is defined as $U_k = \text{LV}_k + z_k\,\text{DV}_{k,G}$ where LV_k is a linear value (similar to additive utility function) for each cluster k, z_k is a coefficient (the weight of importance of closeness to the ideal goal usually between -1 and 0), and $\text{DV}_{k,G}$ is the distance to the ideal goal. For each object in each cluster $k = 1,2,\ldots, K$ the Z utility function (12.19) is

$$\text{Maximize } U_{k,j} = \text{LV}_{k,j} + z_{k,j}\text{DV}_{k,j,G} \quad \text{for object } j \text{ in cluster } k \qquad (12.29)$$

where

$$\text{LV}_{k,j} = w_{1,k}f_{1,k,j} + w_{2,k}f_{2,k,j} \cdots + w_{n,k}f_{n,k,j} \quad \text{for object } j \text{ in cluster } k$$

$$(12.30)$$

To apply this utility function to K-means, first cluster alternatives using (12.18), then rank alternatives of each cluster using (12.19).

Example 12.21   Z Theory Bicriteria Clustering of Example 12.18   Consider Example 12.18. The final solution of clustering this problem using K-means is given in Example 12.18. Suppose that the DM agrees with $F_{1,G}$:(2, 9), $F_{2,G}$:(9, 3), $F_{3,G}$:(10, 8) as the ideal goals for the three clusters, respectively. The weights of importance for three clusters are (0.2, 0.8), (0.6, 0.4), and (0.3, 0.7), respectively. Also, suppose that z_1, z_2, and z_3 are -0.3, -0.4, and -0.1, respectively. Rank alternatives for each cluster.

In this example, all members of each cluster are efficient with respect to their cluster centers. The ideal goal distance [Equation (12.18)] can be applied for each cluster. The ideal goal functions for each of the three clusters are presented below:

$$DV_{1,G} = \sqrt{0.2(a_{ij} - 2)^2 + 0.8(a_{ij} - 9)^2}$$

$$DV_{2,G} = \sqrt{0.6(a_{ij} - 9)^2 + 0.4(a_{ij} - 3)^2}$$

$$DV_{3,G} = \sqrt{0.3(a_{ij} - 10)^2 + 0.7(a_{ij} - 8)^2}$$

For example, the ideal goal function for object A in cluster 1 is

$$DV_{1,G}(A) = \sqrt{0.2(1 - 2)^2 + 0.8(10 - 9)^2} = 1$$

The Z utility function for object A in cluster 1 is

$$U_{1,1} = 0.2a_{11} + 0.8a_{21} - 0.3DV_{1,1,G} = 0.2 \times 1 + 0.8 \times 10 - 0.3 \times 1 = 8.2 - 0.3 = 7.9$$

The distances and utilities for all objects are summarized in the table below.

Goals	$F_{1,G} = (2,9)$			$F_{2,G} = (9,3)$		$F_{3,G} = (10,8)$	
	$W_1 = (0.2, 0.8)$,			$W_2 = (0.6, 0.4)$,		$W_3 = (0.3, 0.7)$,	
Weights	$z_1 = -0.3$			$z_2 = -0.4$		$z_3 = -0.1$	
Alternatives	A	B	C	D	E	F	G
$DV_{k,G}$	1.00	0.00	0.89	0.77	0.00	1.10	0.00
LV_k	8.20	7.60	6.80	6.00	6.60	8.00	8.60
$U_k = LV_k + z_k\,DV_{k,G}$	7.90	7.60	6.53	5.69	6.60	7.89	8.60
Rank U_k	2	4	6	7	5	3	1

Therefore, the best alternative for clusters 1, 2, and 3 are A, E, and G, respectively.

See Supplement S12.11.xls.

Example 12.22   Z Utility Bicriteria Clustering Using K-Means   Consider the multicriteria alternatives given below with 10 alternatives and 2 attributes (criteria).

	F_1	F_2	F_3	F_4	F_5	F_6	F_7	F_8	F_9	F_{10}
Attribute 1	35	37	40	43	45	48	50	52	55	60
Attribute 2	80	75	70	65	60	55	50	45	40	30

Suppose there are three clusters, and the DM provides the following initial cluster centers: $c_1 = (47, 80)$, $c_2 = (50, 75)$, and $c_3 = (56, 70)$. Suppose the DM assigns the following weights of importance for each cluster: $w_1 = (0.3, 0.7)$, $w_2 = (0.6, 0.4)$, and $w_3 = (0.5, 0.5)$, respectively. Suppose that z_1, z_2, and z_3 are $-0.5, -0.2$, and -0.1, respectively. Use multicriteria clustering to cluster and rank alternatives.

The clustering of alternatives using the K-means method is performed as follows.

Iteration 1   Using Equation (12.14), the distances from the cluster centers are:

	F_1	F_2	F_3	F_4	F_5	F_6	F_7	F_8	F_9	F_{10}
c_1	6.57	6.89	9.20	12.74	16.77	20.92	25.15	29.41	33.75	42.43
c_2	12.04	10.07	8.37	8.33	10.25	12.74	15.81	19.04	22.47	29.50
c_3	16.45	13.89	11.31	9.85	10.51	12.02	14.76	17.90	21.22	28.43
Best, Min.	c_1	c_1	c_2	c_2	c_2	c_3	c_3	c_3	c_3	c_3

For example, the distance for F_1 and c_1 is DV $(F_1, c_1) = \sqrt{0.3(35 - 47)^2 + 0.7(80 - 80)^2} = 6.57$.

Each alternative is then assigned to the closest cluster center. For example, F_1 is assigned to the cluster c_1.

The clusters are: cluster 1, $\{F_1, F_2\}$; cluster 2, $\{F_3, F_4, F_5\}$; cluster 3, $\{F_6, F_7, F_8, F_9, F_{10}\}$. Now, calculate the new cluster centers using Equation (12.17). The new cluster centers are $c_1 = (36, 77.5)$, $c_2 = (42.67, 65)$, and $c_3 = (53, 44)$.

The new centers are all efficient with respect to the set of alternatives.

Iteration 2   Using Equation (12.14), the distances from the new cluster centers are:

	F_1	F_2	F_3	F_4	F_5	F_6	F_7	F_8	F_9	F_{10}
c_1	2.16	2.16	6.65	11.14	15.45	19.94	24.25	28.57	33.06	41.86
c_2	11.19	7.70	3.78	0.26	3.64	7.55	11.06	14.57	18.47	25.89
c_3	28.46	24.67	20.55	16.45	12.65	8.54	4.74	1.00	3.16	11.07
Best, Min.	c_1	c_1	c_2	c_2	c_2	c_2	c_3	c_3	c_3	c_3

Each alternative is then assigned to the closest cluster center.

The clusters are: cluster 1, $\{F_1, F_2\}$; cluster 2, $\{F_3, F_4, F_5, F_6\}$; cluster 3, $\{F_7, F_8, F_9, F_{10}\}$.

TABLE 12.17 Final Clustering and Ranking for Example 12.22

Goal	$F_{1,G} = (40, 80)$			$F_{2,G} = (50, 70)$				$F_{3,G} = (60, 50)$		
Weight	$\mathbf{w}_1 = (0.3, 0.7), z_1 = -0.5$			$\mathbf{w}_2 = (0.6, 0.4), z_2 = -0.2$				$\mathbf{w}_3 = (0.5, 0.5), z_3 = -0.1$		
Alternatives	F_1	F_2	F_3	F_4	F_5	F_6	F_7	F_8	F_9	F_{10}
$DV_{k,G}$	2.74	4.49	7.75	6.28	7.42	9.61	7.07	6.67	7.91	14.14
LV_k	66.5	63.6	52	51.8	51	50.8	50	48.5	47.5	45
$U_k = LV_k + z_k DV_{k,G}$	65.13	61.35	50.45	50.54	49.52	48.88	49.29	47.83	46.71	43.59
Rank U_k	1	2	4	3	5	7	6	8	9	10

Now, the new cluster centers can be calculated. They are $c_1 = (36, 77.5)$, $c_2 = (44, 62.5)$, and $c_3 = (54.25, 41.25)$.

The clustering process is now repeated using the new cluster centers.

Iteration 3 Using Equation (12.14), the distances from the cluster centers are:

	F_1	F_2	F_3	F_4	F_5	F_6	F_7	F_8	F_9	F_{10}
c_1	2.16	2.16	6.65	11.14	15.45	19.94	24.25	28.57	33.06	41.86
c_2	13.08	9.59	5.67	1.76	1.76	5.67	9.17	12.68	16.59	24.00
c_3	30.60	26.80	22.69	18.58	14.78	10.68	6.88	3.09	1.03	8.93
Best, Min.	c_1	c_1	c_2	c_2	c_2	c_2	c_3	c_3	c_3	c_3

Each alternative is then assigned to the closest cluster center. Now, the new cluster centers are calculated to be $c_1 = (36, 77.5)$, $c_2 = (44, 62.5)$, and $c_3 = (54.25, 41.25)$. The new cluster centers are the same as the previous centers; therefore, the cluster membership will not change.

At this point, ask the DM to verify if the centers are acceptable as ideal alternatives (goals) or if they should be reassessed. Suppose that the DM's ideal alternatives (goals) for the three centers are $F_{1,G} = (40, 80)$, $F_{2,G} = (50, 70)$, and $F_{3,G} = (60, 50)$.

Now the ideal goal approach is used to rank alternatives in each cluster:

$$DV_{1,G} = \sqrt{0.3(a_{ij} - 40)^2 + 0.7(a_{ij} - 80)^2}$$

$$DV_{2,G} = \sqrt{0.6(a_{ij} - 50)^2 + 0.4(a_{ij} - 70)^2}$$

$$DV_{3,G} = \sqrt{0.5(a_{ij} - 60)^2 + 0.5(a_{ij} - 50)^2}$$

The ideal goal approach generates the rankings in Table 12.17 using Equation (12.18).

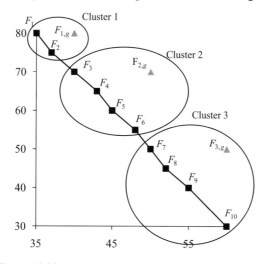

Figure 12.38 Graphical representation of Example 12.22.

As shown in the table, the optimal alternatives for clusters c_1, c_2, and c_3 are F_1, F_4, and F_7, respectively. The solution is shown graphically in Figure 12.38.

See Supplement S12.12.xls.

REFERENCES

Clustering and Classification

Aldenderfer, M. S., R. K. Blashfield, R. Blashfield, and R. K. Blashfield. *Cluster Analysis*. London, UK: SAGE Publications, 1984.

Arabie, P., L. J. Hubert, and G. D. Soete. *Clustering and Classification*. London, UK: World Scientific, 1996.

Askin, R. G., and C. R. Standridge. "Flexible Manufacturing Systems." In *Models of Analysis of Manufacturing Systems*. New York: Wiley, 1993, pp. 125–162.

Berkhin, P. "Survey of Clustering Data Mining Techniques." *Grouping Multidimensional Data*, 2006, pp. 25–71.

Brandimarte, P., and A. Villa. *Advanced Models for Manufacturing Systems Management*. Boca Raton, FL: CRC Press, 1995.

Everitt, B., S. Landau, and M. Leese. *Cluster Analysis*. West Sussex, UK: Oxford University Press, 2001.

Gan, G., C. Ma, and J. Wu. *Data Clustering: Theory, Algorithms, and Applications*, SIAM, Society for Industrial and Applied Mathematics, 2007.

Gordon, A. D. *Classification*, 2nd ed. Chapman & Hall, 1999.

Groover, M. P. *Automation and Production Systems and Computer-Integrated Manufacturing*, 3rd ed. Upper Saddle River, NJ: Prentice Hall, 2007.

Heragu, S. S. "Group Technology and Cellular Manufacturing." *IEEE Transactions on Systems, Man, and Cybernetics*, Vol. 24, No. 2, 1994, pp. 203–215.

Heragu, S. S. *Facilities Design*, 3rd ed. Boca Raton, FL: CRC Press, 2008.

Jaccard, P. "Étude comparative de la distribution florale dans une portion des Alpes et des Jura." *Bulletin del la Société Vaudoise des Sciences Naturelles*, Vol. 37, 1901, pp. 547–579.

Jain, A. K., and R. C. Dubes. *Algorithms for Clustering Data*. Englewood Cliffs, NJ: Prentice Hall, 1988.

Jain, A. K., M. N. Murty, and P. J. Flynn. "Data Clustering: A Review." *ACM Computing Surveys*, Vol. 31, No. 3, 1999, pp. 264–323.

Kalpakjian, S., and S. Schmid. *Manufacturing Engineering and Technology*, 5th ed. Upper Saddle River, NJ: Prentice Hall, 2006.

Kaufman, L., and P. J. Rousseeuw. *Finding Groups in Data: An Introduction to Cluster Analysis*, New York: Wiley, 1990.

Kogan, J. *Introduction to Clustering Large and High-Dimensional Data*. New York, NY: Cambridge University Press, 2006.

MacKay, D. J. C. *Information Theory, Inference, and Learning Algorithms*, Cambridge, UK: Cambridge University Press, 2003.

Malakooti, B., and Z. Yang. "A Variable-Parameter Unsupervised Learning Clustering Neural for Machine Part Group Formation." *International Journal of Production Research*, Vol. 33, No. 9, 1995, pp. 2395–2413.

Mirkin, B. G. *Mathematical Classification and Clustering*. Springer, 1996.

Romesburg, H. C. *Cluster Analysis for Researchers*. Lulu Press, North Carolina, 1989.

Sen, Sumit, and R. N. Dave. "Application of noise clustering in group technology." In Fuzzy Information Processing Society, 1999. NAFIPS. 18th International Conference of the North American, pp. 366–370. IEEE, 1999.

Tibshirani, R., G. Walther, and T. Hastie. "Estimating the Number of Clusters in a Dataset via the Gap Statistic." Techical Report 208, Dept. of Statistics, Standford University, 2000.

Wu, W., H. Xiong, and S. Shekhar. *Clustering and Information Retrieval*, New York: Springer, 2007.

Xu, R., and D. Wunsch. "Survey of Clustering Algorithms." *IEEE Transactions on Neural Networks*, Vol. 16, No. 3, 2005, pp. 645–678.

Multiobjective Clustering and Classification

Belacel, N., H. B. Raval, and A. P. Punnen. "Learning Multicriteria Classification Method PROAFTN from Data." *Computers and Operations Research*, Vol. 34, 2007, pp. 1885–1898.

Brusco, M. J., and D. Steinley. "Cross Validation Issues in Multiobjective Clustering." *British Journal of Mathematical and Statistical Psychology*, Vol. 62, No. 2, 2009, pp. 349–368.

Doumpos, M., Y. Marinakis, M. Marinaki, and C. Zopounidis. "An Evolutionary Approach to Construction of Outranking Models for Multicriteria Classification: The Case of the ELECTRE TRI Method." *European Journal of Operational Research*, Vol. 199, No. 2, 2009, pp. 496–505.

Doumpos, M., and C. Zopounidis. *Multicriteria Decision Aid Classification Methods*. Dordrech–Boston–London: Kluwer Academic, 2002.

Ferligoj, A., and V. Batagelj. "Direct Multicriteria Clustering Algorithms." *Journal of Classification*, Vol. 9, 1992, pp. 43–61.

Fernandez, E., J. Navarro, and S. Bernal. "Handling Multicriteria Preferences in Cluster Analysis." *European Journal of Operational Research*, Vol. 202, No. 3, 2010, pp. 819–827.

Han, C., and I. Ham. "Multiobjective Cluster Analysis for Part Family Formations." *Journal of Manufacturing Systems*. Vol. 5, No. 4, 1986, pp. 223–230.

Handl, J., and J. Knowles. "Exploiting the Trade-Off—The Benefits of Multiple Objectives in Data Clustering." *Lecture Notes in Computer Science*, Vol. 3410, 2005, pp. 547–560.

Handl, J., and J. Knowles. "An Evolutionary Approach to Multiobjective Clustering." *IEEE Transactions on Evolutionary Computation*, Vol. 11, No. 1, 2007, pp. 56–76.

Law, M. H., A. P. Topchy, and A. K. Jain. "Multiobjective Data Clustering." *IEEE Computer Society Conference on Computer Vision and Pattern Recognition*, 2004. CVPR 2004. Proceedings of the 2004 IEEE Computer Society Conference on, Vol. 2, 2004, pp. 11–424.

Malakooti, B. "Multiple Criteria Clustering of ROC, Similarity Coefficient, P-median for Binary and Continuous Data." EECS Department, Case Western Reserve University, Cleveland, OH, 2012.

Malakooti, B. "Multiple Criteria Clustering of Randk Order, Similarity Coefficient, P-median, K-Means for Binary and Continuous Data." Case Western Reserve University, Cleveland, OH, 2013.

Malakooti, B. "Z Utility Theory: Decisions Under Risk and Multiple Criteria" to appear, 2014.

Malakooti, B., and V. Raman. "Clustering and Selection of Multiple Criteria Alternatives Using Unsupervised and Supervised Neural Networks." *Journal of Intelligent Manufacturing*, Vol. 11, 2000, pp. 435–453.

Malakooti, B., and Z. Yang. "Multiple Criteria Approach and Generation of Efficient Alternatives for Machine-Part Family Formation Group Technology." *IIE Transactions*, Vol. 34, No. 9, 2002, pp. 837–846.

Malakooti, B., and Z. Yang. "Clustering and Group Selection of Multiple Criteria Alternatives with Application to Space Networks." *IEEE Transactions on Systems, Man, and Cybernetics Part B: Cybernetics*, Vol. 33, No. 9, 2004, pp. 40–51.

Malakooti, B., J. Youchul, K. Bhasin, A. Teber, J. Mathewson, D. Wang, and Z. Xu. "Multi-Criteria Clustering for Intelligent Tele-Communications Systems." *IEEE International Conference on Systems, Man and Cybernetics*, Vol. 1, 2002, pp. 421–426.

Saha, S., and S. Bandyopadhyay. "A New Multiobjective Clustering Technique Based on the Concepts of Stability and Symmetry." *International Journal of Knowledge and Information Systems*, Vol. 23, No. 1, 2010a.

Saha, S., and S. Bandyopadhyay. "A Symmetry Based Multiobjective Clustering Technique for Automatic Evolution of Clusters." *Pattern Recognition*, Vol. 43, No. 3, 2010b, pp. 738–751.

Smet, Y. D., and L. M. Guzman. "Towards Multicriteria Clustering: An Extension of the *K*-Means Algorithm." *European Journal of Operational Research*, Vol. 158, 2004, pp 390–398.

Valls, A. "Using Classification as an Aggregation Tool in MCDM." *Fuzzy Sets and Systems*, Vol. 115, No. 1, 2000, pp. 159–168.

Won, Y., and S. Kim. "Multiple Criteria Clustering Algorithm for Solving the Group Technology Problem with Multiple Process Routings." *Computers and Industrial Engineering*, Vol. 32, No. 1, 1997, pp. 207–220.

Zopounidis, C., and M. Doumpos. "Multicriteria Classification and Sorting Methods: A Literature Review." *European Journal of Operational Research*, Vol. 138, Issue 2, 2000, pp. 229–246.

EXERCISES

Types of Data Representation (Section 12.2.1)

2.1 Consider the following data points. Find a dissimilarity matrix for objects *A, B, C, D, E*, and *F*. Show two and one modal.

	A	*B*	*C*	*D*	*E*	*F*
Weight (kg)	17	68	45	15	12	85
Height (cm)	93	170	155	95	88	180

2.2 Find the similarity matrix for the following binary data. Show two and one modal.

Objects / Attributes	Ted	Abby	Barney	Bob	Lily	Carly
Fair hair	0	1	1	1	1	0
Wears glasses	0	1	0	1	0	1
Male	1	0	1	1	0	0
Left-handed	0	1	0	0	1	1
Married	0	1	0	1	1	0
Employed	1	0	1	0	0	0

Distance Measurement Metrics (Section 12.2.2)

2.3 Consider two points $\mathbf{x} = \{2,3,0\}$ and $\mathbf{y} = \{5,1,2\}$. Calculate the distance between points $\mathbf{x}$ and $\mathbf{y}$ using the Minkowski metric for $Q = 1, 2, 3, \infty$.

2.4 Consider the six points presented in exercise 2.1.

    (a) Calculate the distance between all pairs of points using the Minkowski metric for $Q = 1, 2, \infty$.

    (b) Present the set of data for each metric graphically. For two sets of clusters, comment on how using different metrics may change the clustering of alternatives.

Group Technology Clustering[†] (Section 12.3)

3.1 Provide the best part family and machine cell formations by rearranging columns and rows of the following part–machine matrix. Use the visual inspection and intuition approach.

Parts	Machines				
	1	2	3	4	5
1		1		1	
2	1		1		1
3			1	1	1
4	1	1		1	

3.2 Provide the best part family and machine cell formations by rearranging columns and rows of the following part–machine matrix. Use the visual inspection and intuition approach.

Parts	Machines				
	1	2	3	4	5
1	1	1			1
2			1	1	1
3	1	1	1	1	
4			1		1
5	1	1		1	
6	1	1		1	1

3.3 Provide the best part family and machine cell formations by rearranging columns and rows of the following part–machine matrix. Use the visual inspection and intuition approach.

	Machines						
Parts	1	2	3	4	5	6	7
1					1		1
2		1		1		1	
3	1	1					
4		1	1	1		1	
5	1		1			1	
6		1	1	1			
7				1	1		1
8	1	1					

3.4 Provide the best part family and machine cell formations by rearranging columns and rows of the following part–machine matrix. Use the visual inspection and intuition approach.

	Machines								
Parts	1	2	3	4	5	6	7	8	9
1					1		1		1
2				1		1		1	
3	1		1	1					
4				1	1	1		1	
5			1		1			1	
6	1			1	1	1			
7		1				1	1		1
8			1	1					
9	1	1				1			1
10	1		1						
11		1					1		1
12	1						1		1
13				1	1			1	

ROC Method for Binary Data (Section 12.4.1)

4.1 Consider the matrix in exercise 3.1.

    (a) Use the ROC method to rearrange the rows and columns.

    (b) Based on the visual inspection of the solution you found in part (a), suggest the number of cells that should be used for this problem. For this number of cells, identify the clustering of machines and parts.

    (c) Suppose there are three clusters. How many exceptional elements are in your solution for three clusters? Can you find a better solution that has less exceptional elements?

4.2 Consider the matrix in exercise 3.2. Respond to parts (a)–(c) of exercise 4.1.

4.3 Consider the matrix in exercise 3.3. Respond to parts (a)–(c) of exercise 4.1.

4.4 Consider the matrix in exercise 3.4. Respond to parts (a)–(c) of exercise 4.1.

4.5 Does King's method obtain an optimal solution? Explain why or why not.

Bicriteria Rank Order Clustering (Section 12.4.2)

4.6 Consider exercise 3.1. Suppose the weights of importance for the normalized values of the two objectives are 0.6 and 0.4 for the number of clusters and number of exceptional elements, respectively.

    (a) Consider two criteria. Using the ROC method and generate three efficient clustering alternatives for one, two, and three cells.

    (b) Find the best alternative for the multicriteria problem using an additive utility function.

4.7 Consider exercise 3.2. Suppose the weights of importance for the normalized values of the two objectives are $\mathbf{w} = (0.5, 0.5)$ for number of clusters and number of exceptional elements, respectively. Respond to parts (a) and (b) of exercise 4.6.

4.8 Consider exercise 3.3. Suppose the weights of importance for the normalized values of the two objectives are $\mathbf{w} = (0.7, 0.3)$ for the number of clusters and number of exceptional elements, respectively.

    (a) Consider two criteria. Using the ROC method, generate five efficient clustering alternatives for one, two, three, four, and five cells.

    (b) Find the best alternative for the MCDM problem using an additive utility function.

Selection of Machines To Be Duplicated (Section 12.4.3)[†] and Tricriteria Rank Order Clustering (Section 12.4.4)[†]

4.9 Consider exercise 3.1.

    (a) Consider three criteria. Only allow up to two machine duplications and a maximum of two clusters. Find all efficient alternatives.

    (b) Rank the alternatives using an additive utility function where weights of importance for normalized objective values are (0.3, 0.3, 0.4) for number of clusters, number of exceptional elements, and number of duplicated machines, respectively.

4.10 Consider exercise 3.2.

    (a) Consider three criteria. Only allow up to one machine duplication and a maximum of three clusters. Find all efficient alternatives.

(b) Rank the alternatives using an additive utility function where weights of importance for normalized objective values are (0.6, 0.3, 0.1) for number of clusters, number of exceptional elements, and number of duplicated machines, respectively.

4.11 Consider exercise 3.3.

(a) Consider three criteria. Only allow up to one machine duplication and a maximum of three clusters. Find all efficient alternatives.

(b) Rank the alternatives using an additive utility function where weights of importance for normalized objective values are (0.2, 0.4, 0.4) for number of clusters, number of exceptional elements, and number of duplicated machines, respectively.

Similarity Coefficient for Binary Data (Section 12.5.1)

5.1 Consider exercise 3.1. For three clusters, use the similarity coefficient algorithm.

(a) Provide the best clustering (part family and machine cell) formations by using the SC method, that is, in each iteration add only one object. Identify the list of parts and machines for each cell.

(b) Show the hierarchical tree structure for part (a).

5.2 Consider exercise 3.2. Use the similarity coefficient algorithm. Respond to parts (a) and (b) of exercise 5.1.

5.3 Consider exercise 3.3. Use the similarity coefficient algorithm. Respond to parts (a) and (b) of exercise 5.1.

5.4 Can the SC algorithm generate the optimal grouping? Explain why or why not.

5.5 Explain the effect of using different SC thresholds in the similarity coefficient algorithm. What is the result of using a high SC_{thresh} compared to a low SC_{thresh}?

Simultaneous SC Clustering[†] (Section 12.5.2)

5.6 Consider exercise 3.1.

(a) Use simultaneous head–tail SC clustering. Use two ($Q = 2$) simultaneous solutions and find the best clustering for (1) two clusters, (2) three clusters, and (3) four clusters.

(b) Present the results of the clustering for each iteration (and their similarity coefficients).

5.7 Consider exercise 3.2. Respond to parts (a) and (b) of exercise 5.6.

5.8 Consider exercise 3.3. Respond to parts (a) and (b) of exercise 5.6.

Multicriteria Similarity Coefficient[†] (Section 12.5.3)

5.9 Consider exercise 5.1.

(a) List the set of bicriteria alternatives.

(b) Use the bicriteria clustering with SC algorithm to generate efficient alternatives.

(c) Use the tricriteria clustering with SC algorithm to generate efficient alternatives. Allow up to a total of two machines to be duplicated.

5.10 Consider exercise 5.2. Respond to parts (a)–(c) of exercise 5.6.

5.11 Consider exercise 5.3. Respond to parts (a)–(c) of exercise 5.6.

Linkage Approaches[†] (Section 12.5.4)

5.12 Consider four objects with four attributes given in the following matrix. Use the Euclidean distance to find the distance for each pair of objects. Consider two sets of clusters $\{A,C\}$ and $\{B, D\}$. Show the linkage between the two clusters using the three linkage methods.

		Objects			
		A	B	C	D
Attributes	1	5	3	4	2
	2	3	2	5	1
	3	6	2	1	5
	4	1	7	3	4

5.13 Consider five objects with three attributes given in the following matrix. Use the Euclidean distance to find the distance for each pair of objects. Consider three sets of clusters $\{A,B\}$, $\{C\}$, and $\{D,E\}$. Show the linkage between the three clusters using complete- and single-linkage methods.

		Objects				
		A	B	C	D	E
Attributes	1	7	6	1	2	1
	2	2	3	9	3	3
	3	1	2	1	1	2

Hierarchical Clustering for Continuous Data[†] (Section 12.5.5)

5.14 Consider exercise 5.13. Cluster objects A–E using the SC method for continuous data. Use complete linkage and the Euclidean distance metric. Show the clusters and the hierarchical tree for the solution obtained.

5.15 Cluster objects *A–H* using the SC method for continuous data. Use complete linkage and the Euclidean distance metric. Show the clusters and the hierarchical tree for the solution obtained.

Objects	*A*	*B*	*C*	*D*	*E*	*F*	*G*	*H*
Attribute 1	0	4	3	10	14	9	12	20
Attribute 2	17	10	14	10	5	4	10	4

5.16 Cluster objects *A–K* using the SC method for continuous data. Use complete linkage and the Euclidean distance metric. Show all clusters. Also, show the hierarchical tree for the solution obtained.

Objects	*A*	*B*	*C*	*D*	*E*	*F*	*G*	*H*	*I*	*J*
Attribute 1	145	109	184	134	152	174	134	170	144	126
Attribute 2	27	25	37	26	20	39	21	32	25	23
Attribute 3	60	98	58	98	62	64	99	53	56	80

P-Median Method for Binary Data[†] (Section 12.6.1)

6.1 Consider exercise 3.1.

(a) Formulate the *P*-median ILP.

(b) Suppose the desired number of cells is two. Solve the ILP. Show the detailed solution. What are the families of machines and parts? Identify the number of exceptional elements.

(c) Suppose the desired number of cells is three. Solve the ILP. Show the detailed solution.

6.2 Consider exercise 3.2. Respond to parts (a)–(c) of exercise 6.1.

6.3 Consider exercise 3.3. Respond to parts (a)–(c) of exercise 6.1.

6.4 Does the *P*-median method generate an optimal solution? Explain why. If the *P*-median method is not optimal, how can its solution be improved?

Bicriteria *P*-Median Clustering[†] (Section 12.6.2)

6.5 Consider exercise 6.1. List the set of efficient alternatives when the number of clusters is 1, 2, and 3.

6.6 Consider exercise 6.2. List the set of efficient alternatives when the number of clusters is 1, 2, and 3.

6.7 Consider exercise 6.3. List the set of efficient alternatives when the number of clusters is 2, 3, and 4.

P-Median for Continuous Data[†] (Section 12.6.3)

6.8 Consider exercise 5.13.

  (a) Formulate the *P*-median ILP.

  (b) Suppose the desired number of cells is 2. Solve the ILP. Show the detailed solution.

  (c) Suppose the desired number of cells is 3. Solve the ILP. Show the detailed solution.

6.9 Consider exercise 5.15. Respond to parts (a)–(c) of exercise 6.8.

K-Means Method (Section 12.7.1)

7.1 Consider the 10 objects with two attributes (x, y) below.

	1	2	3	4	5	6	7	8	9	10
x	1	3	10	8	1	4	5	3	7	0
y	2	3	1	2	7	8	9	4	8	3

  (a) Cluster the objects using the *K*-means algorithm. Suppose the initial cluster centers are $c_1 = (3, 9)$, $c_2 = (2, 3)$, and $c_3 = (8, 2)$.

  (b) Cluster the objects using the *K*-means algorithm. Suppose the initial cluster centers are $c_1 = (2, 5)$, $c_2 = (7, 9)$, and $c_3 = (6, 1)$.

  (c) Explain how the different starting centers affect the final clustering. Show the best solution graphically.

  (d) Compute BSS and TSS for the final solutions of parts (a)–(c). Discuss why their TSS is the same and what solution should be selected.

7.2 Consider the 20 objects shown below. Determine the cluster centers and cluster membership for each cluster using 0–1 normalized values for each attribute.

		Objects																			
		1	2	3	4	5	6	7	8	9	10	11	12	13	14	15	16	17	18	19	20
Attributes	1	3	6	1	9	5	3	8	9	3	7	8	3	1	6	9	3	8	3	2	5
	2	55	36	73	23	67	43	26	36	94	41	17	83	31	8	91	21	49	59	71	62
	3	11	15	1	12	20	17	19	2	6	13	7	5	9	14	19	16	5	4	20	12
	4	0	1	1	0	1	1	1	0	0	1	0	0	1	0	1	0	0	1	0	1

  (a) Cluster the objects using the *K*-means algorithm where the four initial cluster centers are $c_1 = (5, 50, 10, 0)$, $c_2 = (8, 20, 10, 1)$, $c_3 = (10, 30, 0, 0)$, and $c_4 = (2, 90, 15, 1)$.

(b) Cluster the objects using the K-means algorithm where the five initial cluster centers are $c_1 = (10, 50, 20, 0)$, $c_2 = (2, 15, 15, 0)$, $c_3 = (5, 37, 10, 1)$, $c_4 = (8, 70, 1, 1)$, and $c_5 = (4, 95, 7, 0)$.

(c) Compare and contrast results of parts (a) and (b).

7.3 Consider the 16 objects shown below. For three clusters, the initial cluster centers are $c_1 = (1, 1, 128, 128, 128)$, $c_2 = (2, 4, 128, 128, 128)$, and $c_3 = (4, 2, 128, 128, 128)$. Use the K-means algorithm to cluster the pixels in the image.

	1	2	3	4	5	6	7	8	9	10	11	12	13	14	15	16
x	1	1	1	1	2	2	2	2	3	3	3	3	4	4	4	4
y	1	2	3	4	1	2	3	4	1	2	3	4	1	2	3	4
R	228	225	128	64	247	0	40	128	255	53	74	187	250	245	228	147
G	105	128	0	0	187	0	40	0	128	70	48	6	92	16	27	19
B	22	0	128	64	9	0	40	255	64	77	99	187	64	10	133	232

(a) Determine cluster centers and cluster membership using the K-means algorithm.

(b) Determine cluster centers and cluster membership for each cluster using 0–1 normalized values for each criteria.

(c) Do the solutions in parts (a) and (b) differ? Why? Which one is better?

7.4 Consider the data in exercise 7.3. Suppose the initial cluster centers are $c_1 = (2, 3, 32, 8, 26)$, $c_2 = (1, 4, 100, 7, 98)$, and $c_3 = (3, 1, 225, 75, 40)$. Respond to parts (a)–(c) of exercise 7.3.

Multicriteria K-Means[†] (Section 12.7.2)

7.5 Consider the seven objects given below.

Attribute	F_1	F_2	F_3	F_4	F_5	F_6	F_7
1	25	23	33	12	25	26	19
2	91	62	55	91	98	48	68

(a) Generate three bicriteria alternatives for one, two, and three clusters. Use the following initial cluster centers. For two clusters, use $c_1 = (10, 75)$ and $c_2 = (25, 50)$. For three clusters, use $c_1 = (12, 80)$, $c_2 = (30, 55)$, and $c_3 = (20, 72)$.

(b) Identify efficient alternatives and show them graphically.

7.6 Consider the seven objects given below.

Attribute	F_1	F_2	F_3	F_4	F_5	F_6	F_7
1	76	70	55	89	66	87	56
2	14	1	18	17	23	9	2
3	140	181	142	177	158	131	114

(a) Generate three bicriteria alternatives for one, two, and three clusters. Use the following initial cluster centers. For two clusters, use $c_1 = (74, 18, 150)$ and $c_2 = (50, 25, 112)$. For three clusters, use $c_1 = (70, 10, 120)$, $c_2 = (65, 13, 150)$, and $c_3 = (70, 15, 200)$.

(b) Identify efficient alternatives.

Multiperspective Multiple-Criteria Clustering[†] (Section 12.8)

8.1 Consider the multicriteria clustering alternatives given below. Suppose there are three clusters, whose initial cluster centers (goals) are $c_1 = (20, 90)$, $c_2 = (35, 80)$, and $c_3 = (30, 61)$. Suppose the weights of importance for each cluster are $w_1 = (0.7, 0.3)$, $w_2 = (0.5, 0.5)$, and $w_3 = (0.4, 0.6)$.

Attribute	F_1	F_2	F_3	F_4	F_5	F_6	F_7	F_8	F_9	F_{10}
f_1	27	20	38	29	34	16	27	30	16	28
f_2	77	98	93	73	56	96	57	82	72	94

(a) Use multicriteria clustering to cluster the alternatives.

(b) Rank alternatives in each clustering using the given weights and the final center of each cluster as its ideal alternative.

(c) Determine the optimal alternative for each of the clusters.

(d) For each cluster, assume the z coefficient is -0.3 and rank alternatives using Z theory clustering.

8.2 Consider the multicriteria clustering alternatives given below. Suppose there are four clusters whose initial cluster centers are $c_1 = (12, 80)$, $c_2 = (25, 55)$, $c_3 = (46, 72)$, and $c_4 = (37, 90)$. Suppose the weights of importance for each cluster are $w_1 = (0.6, 0.4)$, $w_2 = (0.3, 0.7)$, $w_3 = (0.5, 0.5)$, and $w_4 = (0.8, 0.2)$.

Attribute	F_1	F_2	F_3	F_4	F_5	F_6	F_7	F_8	F_9	F_{10}	F_{11}	F_{12}	F_{13}	F_{14}	F_{15}
f_1	25	23	33	12	25	26	32	16	10	43	16	21	40	54	50
f_2	91	62	55	91	98	48	75	100	61	90	33	69	99	63	37

Respond to parts (a)–(c) of exercise 8.1.

8.3 Consider the multicriteria clustering alternatives given below. Suppose there are three clusters whose initial cluster centers are $c_1 = (74, 18, 150)$, $c_2 = (50, 25, 112)$, and $c_3 = (90, 7, 186)$. Suppose the weights of importance for each cluster are $w_1 = (0.6, 0.3, 0.1)$, $w_2 = (0.3, 0.3, 0.4)$, and $w_3 = (0.2, 0.6, 0.2)$.

Attribute	F_1	F_2	F_3	F_4	F_5	F_6	F_7	F_8	F_9	F_{10}	F_{11}	F_{12}
f_1	76	70	55	89	66	87	56	78	66	52	60	94
f_2	14	1	18	17	23	9	2	10	18	13	20	19
f_3	140	181	142	177	158	131	114	108	166	172	186	109

Respond to parts (a)–(c) of exercise 8.1.

CHAPTER 13

CELLULAR LAYOUTS AND NETWORKS

13.1 INTRODUCTION

Many layout and network problems can be formulated as a single-row layout problem, a network that has only one layer of adjacent nodes (either physically or conceptually) in which each node can be connected to its two adjacent nodes. Single-row layouts are easy to monitor, maintain, and utilize. Also, the independence of separate cells improves the work flow and reliability of the system. However, to reduce the movement of parts among different cells, sometimes duplication of certain equipment in different cells may become necessary. Some applications of cellular systems are presented below.

Cellular Manufacturing For practical purposes, many manufacturing facilities, corporations, and service industries divide their operations into separate areas called cells. Each cell has independent operations from other cells. In cellular manufacturing, units may enter the facility from the load station. Then, the units flow along a one-dimensional path to visit the needed workstations. The units may be recycled through the same loop of workstations for further work. Finally, the units exit at the unloading station. The layout problem can be defined as finding the linear (or circular) order of workstations that will optimize the given objectives, such as minimizing the total flow, total cost, and cycle time.

Flexible Manufacturing Systems In manufacturing systems, in order to produce or assemble products, materials are moved from one workstation to another. Materials travel between adjacent workstations either unidirectionally (one way) or bidirectionally (two ways). The traditional job shop design (i.e., grouping the same processors in one area) has a

Note: Advanced material that can be omitted without loss of generality will be indicated by a dagger.

Operations and Production Systems with Multiple Objectives, First Edition. Behnam Malakooti.
© 2014 John Wiley & Sons, Inc. Published 2014 by John Wiley & Sons, Inc.

717

very poor performance for solving flexible flow problems where groups of products visit the same sequence of processors. Job shop design can be very time consuming and may create many in-process inventories for solving this type of problem. In flexible manufacturing systems, a variety of products can be produced using automation. Each line of products may go through a different sequence of processors. In some assembly systems semifinished products may be recycled (i.e., returned through a loop) to be further processed by the same station. Also, automated guided vehicles can be used for transportation of inventory and semifinished products to various stations (processing departments) within the factory facility. The layout of flexible assembly systems and the routing of automated guided vehicles can be represented by the cellular network presented in this chapter.

Service Industries   Different operational and service systems can be effectively modeled and optimized using single-row layout techniques. Some service industries, such as airports and fast food restaurants, can be designed using single-row network models, where the flow of customers or food through the facilities is optimized.

Routing, Transportation, and Scheduling Problems   A number of routing problems (such as the traveling salesman problem) can be represented by the single-row network problem where the locations are given (and fixed) but the best route must be decided. For example, certain transportation problems, such as cross-country shipping routes for trucks, can be modeled by a single-row layout. Also, many scheduling and production planning problems can be expressed as a single-row (or loop) network problem; see Chapter 7.

Types of Cellular Layouts and Networks   There are five types of single-row network layout problems:

(a) **Unidirectional as Single-Row Network**   In unidirectional single-row layouts, the flow is only one way, but different products with different sequences for using stations are processed. A product may go through several loops to be processed by stations arranged differently than the sequence needed by different products. See Figures 13.1a,b for linear and circular examples. The sequence of stations is $2 \rightarrow 3 \rightarrow 1 \rightarrow 4$.

(b) **Bidirectional as Single-Row Network**   Bidirectional single-row loop layouts are similar to unidirectional, except that the flow can be two ways. Hence, there is no need for a returning loop; the loop is implicit. See Figures 13.2a,b for linear and circular examples. The sequence of stations is $2 \leftrightarrow 3 \leftrightarrow 1 \leftrightarrow 4$ (read $i \leftrightarrow j$ as bidirectional movement between stations i and j). Note that bidirectional linear and circular layouts are different and should

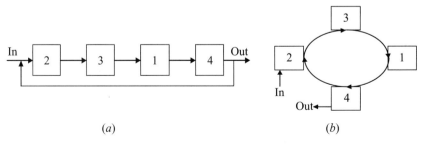

Figure 13.1   Two types of unidirectional layout: (*a*) linear; (*b*) circular.

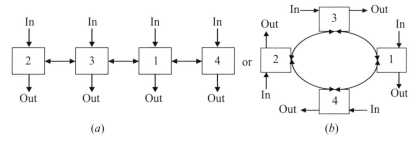

Figure 13.2 Two types of bidirectional layout: (*a*) linear; (*b*) circular.

be solved differently. For example, the distance from station 2 to station 4 is three units in Figure 13.2*a* but one unit in Figure 13.2*b*.

(c) Hybrid Single-Row Network Figure 13.3 illustrates a hybrid single-row loop layout for unidirectional flow. Parts can enter and leave the system at any given station without being forced to go through a complete loop. This is a more flexible system than the unidirectional (Figure 13.3*a*) and it can be used for many different applications. This problem can also be formulated and solved by either uni- or bidirectional algorithms; however, the objective function (minimizing total flow) should be formulated for this problem.

(d) Traveling Salesman Problem as Single-Row Unidirectional Network In the above three layout problems, the problem is to identify the best location for each node (e.g., workstation), where the flow information for each pair of nodes is given. The same type of information can be used to define an entirely different problem. Suppose the node locations are fixed and distance information for each pair of nodes is given. The routing problem can be defined as finding the shortest total distance while visiting all of the nodes. In the well-known combinatorial traveling salesman problem (TSP), the salesman must visit all of the nodes (cities) and return to the point of origin while minimizing the total distance traveled. In Figure 13.4, both layouts (linear and circular) have the same total distance. For the linear view, the first station is repeated at the end of the sequence to represent the salesman returning to the starting city. In Section 13.7, it will be shown that TSP problems can also be solved by the head–tail methods developed in this chapter.

(e) Combinatorial Problems as Single-Row Network Many combinatorial problems (such as TSP) can be formulated as a single-row network problem presented in this chapter. The methods of Section 13.7 are presented for solving generic combinatorial optimization problems.

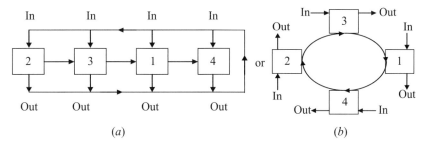

Figure 13.3 Two types of hybrid layout: (*a*) linear; (*b*) circular.

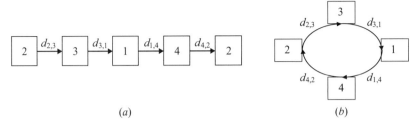

(a)   (b)

Figure 13.4   Two types of TSP: (*a*) linear; (*b*) circular.

Ordinal (Qualitative) versus Cardinal Data   In many applications, the data can only be presented as ordinal. For example, one can present his or her preferences for paired comparisons of two objects (or alternatives) by A, E, I, O, U, and X, where A is the highest preference, U is the lowest preference, and X is indifferent. The amount of preference, however, is not known or cannot be assessed. For example, a preference of A is higher than E, but the difference in preference cannot be determined. To solve this problem, ordinal information can be represented by a set of discrete numbers, for example, A, E, I, O, U, and X can be represented by use of 5, 4, 3, 2, 1, and 0, respectively. The ordinal head–tail method presented in this chapter can be used to solve this type of ordinal information. On the other hand, the cardinal head–tail method can be used to solve cardinal preference information.

Head–Tail Approaches: Overview   To find exact optimal solutions to any of the above problems, one might generate all possible solutions or use the well-known branch-and-bound procedure. But these methods are very demanding in terms of computational time and computer memory for larger problems. For example, for a layout problem with 100 stations, the calculation time to find an exact optimal solution is astronomical. Because exact algorithms require such high computational time to solve large combinatorial problems, heuristics are often used to solve them. It is important to note that because of inaccuracies in the data and problem formulation, the benefit of an optimal solution over an approximate one may be negligible. For example, market demand fluctuations, unexpected breakdowns, shortages, and product defects during production can all yield inaccurate flow data. In this chapter, we develop a new heuristic, called the head–tail method, to solve single-row layout problems (unidirectional loop, bidirectional loop) and traveling salesman combinatorial problems. We also introduce MCDM approaches for solving the above problems. Finally, we provide approaches for solving unidirectional and bidirectional problems by integer programming optimization methods. The applications of the four methods covered in this section are presented below:

- *Ordinal head–tail* can be used to solve extremely large problems because it is computationally very efficient. It can be also used to solve qualitative (ordinal) information. It renders acceptable solutions.
- *Cardinal head–tail* can be used to solve very large problems; it renders good solutions.
- *Simultaneous head–tail* can be used to solve large problems; it renders very good solutions.
- *Integer programming optimization* can be used to solve medium-sized problems; it renders excellent solutions.

13.2   UNIDIRECTIONAL NETWORK PROBLEM

A principal concern of operational layout problems (such as manufacturing systems) is the minimization of the total flow of materials that are moved through the system. Every part is processed by machines according to its processing sequence. The layout problem is to find the optimal arrangement of stations that minimizes the total flow of all parts. *A poorly designed layout will increase the materials flow cost and work-in-process.*

Consider a single-row layout in the form of a line with a loop feedback or a circle. In the linear layout, the input and output gates are in different locations (Figure 13.1*a*), but there is a return line to the input gate. In a circular layout, input and output gates of the loop are adjacent (Figure 13.1*b*). If a product passes through the loop once, from in to out as in either Figure 13.1*a* or Figure 13.1*b*, it counts as one loop. The cell is designed to process several products at the same time. Each different product may have a different sequence of machines. As a result, different products will require a different number of loops. Assembly lines (see Chapter 14) have only one loop; they process only one product with no return loops.

The sequences represented in Figures 13.1*a,b* are identical; only the orientation (circular vs. linear) is different. The distance traveled by the product from input to output in both of these layouts is considered constant. Hence, distances among pairs of machines have no effect on the solution. The distinction between loop and linear layouts will not affect the solution procedure of this section. Therefore, one is only concerned with the sequence of stations regardless of distances. Note that inputs may directly enter each station, but it counts as one loop when they arrive at the out gate.

Label stations (or machines) by $i = 1, 2, 3, \ldots, n$ and products by $j = 1, 2, 3, \ldots, m$, where there are n machines and m products. Let s_j be the sequence of stations for product j. That is, product j is processed according to sequence s_j. Two methods are presented below for measuring total flow; the two methods use different total-flow measurements in their algorithms. The first method measures the total flow based on all individual product sequences. The second method is based on known from–to flow for each pair of stations. The latter approach is used in the unidirectional algorithms presented in this chapter.

13.2.1   Total Flow Based on Product Sequence

For a given layout, the total flow for one product is the total number of loops that the product has to go through to be processed by all stations multiplied by the lot size of the product. The total flow for all products is the summation of the total flow of each product.

Example 13.1   Total Flow Based on Loops for Four-Product, Four-Station Problem
Consider the following example of four products and their production sequence requirements along with lot sizes:

Product (j)	Sequence of Stations (s_i)	Lot Size
1	$2 \rightarrow 3 \rightarrow 1 \rightarrow 4$	176
2	$3 \rightarrow 1 \rightarrow 2 \rightarrow 4$	92
3	$2 \rightarrow 3 \rightarrow 4 \rightarrow 1$	35
4	$1 \rightarrow 4 \rightarrow 3$	73

Find out which of the following two layouts has a better total flow.

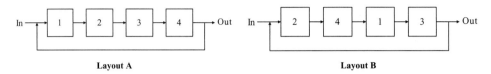

Layout A   **Layout B**

In this example, $n = 4$, $m = 4$, $s_1 = \{2{\rightarrow}3{\rightarrow}1{\rightarrow}4\}$, $s_2 = \{3{\rightarrow}1{\rightarrow}2{\rightarrow}4\}$, $s_3 = \{2{\rightarrow}3{\rightarrow}4{\rightarrow}1\}$, and $s_4 = \{1{\rightarrow}4{\rightarrow}3\}$. Layout A has a sequence of $\{1{\rightarrow}2{\rightarrow}3{\rightarrow}4\}$ and layout B has a sequence of $\{2{\rightarrow}4{\rightarrow}1{\rightarrow}3\}$. Note that product 4 only requires station 1, 3, and 4.

Given this information, we would like to calculate the total flow for each of the layouts. To calculate the total flow for a single product, we multiply the lot size by the number of loops required to finish the production sequence. We then sum the total flow for all products to find the total flow for the layout. Using the facility layouts shown in the above figures, we calculate:

Layout A: Total flow based on loops $= 2 \times 176 + 2 \times 92 + 2 \times 35 + 2 \times 73 = 752$.
Layout B: Total flow based on loops $= 3 \times 176 + 3 \times 92 + 2 \times 35 + 2 \times 73 = 1020$.

Layout A results in less total flow for the products; therefore, layout A has a better flow than layout B.

Converting Product Sequence Flow to From–To Matrix   In order to solve the product sequencing problem using the methods developed in this chapter, it is necessary to convert the sequence flows into from–to flow format. An approach for converting the flow of different part sequences with different lot sizes is as follows. Only consider the flow of adjacent pairs and aggregate them for each pair. In Example 13.1, the sequence 1–4 is used by product 1 and product 4; so the total flow of 1–4 is the sum of the lot sizes for these two products, that is, $176 + 73 = 249$. Similarly, the total flow from 4 to 1 is 35. The complete conversion of the information in Example 13.1 is shown in Example 13.2.

13.2.2 Total Flow Based on From–To Matrix

For all unidirectional methods, we will use the total-flow measurement based on the from–to flow information. In a from–to flow matrix, each element (w_{ij}) represents the flow from station i to station j. Therefore, the total flow (in loops) between station i and station j for a given layout depends on the order of stations i and j in the sequence. If station i comes before station j, then products moving from i to j will traverse a single loop through the sequence layout; the flow for this pair of stations is w_{ij}. However, when station i is located after station j, then products moving from i to j will traverse two loops through the sequence layout; the flow for this pair of stations is $2w_{ij}$. Equation (13.1) can be used to represent the total flow for all pairs of stations. This can be used as the objective of the given problem:

Minimize total flow $= 1 \times$ sum of all one-loop flows $+ 2 \times$ sum of all two-loop flows

$$(13.1)$$

Example 13.2   Total Flow Based on From–To Information for Unidirectional   Consider the following from–to flow matrix information for four stations. Based on the from/to information, calculate the total flow for facility layouts A and B. Find out which layout has a better total flow.

Workflow Matrix

From, (i) \ Station	To (j) 1	2	3	4
1	0	92	0	249
2	0	0	211	110
3	268	0	0	35
4	35	18	73	0

Facility Layout A

Facility Layout B

The table below shows the loop requirements for each pair (i, j). Based on the above information, the total flow for layouts A and B is computed and presented below.

Layout Sequence in Layout A		Layout Sequence in Layout B	
1 loop pairs	2 loop pairs	1 loop pairs	2 loop pairs
1→2	2→1	1→3	1→2
1→3	3→2	2→1	1→4
1→4	3→1	2→3	3→1
2→3	4→3	2→4	3→2
2→4	4→2	4→1	3→4
3→4	4→1	4→3	4→2

Layout A: Total flow $= 1 \text{ loop} \times (92 + 249 + 211 + 110 + 35) + 2 \text{ loops} \times (35 + 18 + 73 + 268) = 1485$.

Layout B: Total Flow $= 1 \text{ loop} \times (211 + 110 + 35 + 73) + 2 \text{ loops} \times (92 + 249 + 268 + 35 + 18) = 1753$.

Hence, layout *A* is better in terms of total flow. In the following method, we use total flow based on from–to flow information for solving unidirectional problems.

13.2.3   Concordance and Discordance Measurement

Concordance and Discordance   For a given layout, we can define the following:

Concordance: $v_1 = $ sum of all one loop flows.
Discordance: $v_2 = $ sum of all two loop flows.

For different sequencing solutions, the summation of the concordance and discordance scores $(v_1 + v_2)$ is always a constant number; it is equal to the summation of all matrix entries. As defined in the previous section, total flow $= v_1 + 2v_2$. Therefore, the problem is to:

$$\text{Minimize total flow} = v_1 + 2v_2.$$

This is equivalent to minimizing $(v_1 + v_2) + v_2$. However, since $v_1 + v_2$ is constant, this problem is equivalent to:

$$\text{Minimizing discordance} = v_2.$$

So, the problem is to find a solution that minimizes discordance; this in turn maximizes concordance.

Measuring Total Flow of Given Solution   To simplify the calculations of total flows for problems with many stations, the following approach can be used: Consider the from–to flow matrix information. Reorder the columns and rows of the from–to flow matrix in order of the sequence solution. The total score of the upper right triangle is the concordance and the total score of the lower left triangle is the discordance. Then measure total flow as $\text{TF} = v_1 + 2v_2$.

Example 13.3   Concordance, Discordance, and Total-Flow Measurement   Consider the from–to flow matrix from Example 13.2. For the layout B solution, that is, $2 \rightarrow 4 \rightarrow 1 \rightarrow 3$, find the concordance, discordance, and total flow:

Reordered Workflow Matrix

	Station	To (*j*)			
		2	4	1	3
From (*i*)	2	0	110	0	211
	4	18	0	35	73
	1	92	249	0	0
	3	0	35	268	0

Facility Layout B

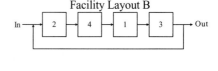

Concordance (summation of upper triangle) $= (110 + 211 + 35 + 73) = 429$.
Discordance (summation of lower triangle) $= (18 + 92 + 249 + 35 + 268) = 662$.
Total flow $= 1 \text{ loop} \times (429) + 2 \text{ loops} \times (662) = 1753$.

13.3   UNIDIRECTIONAL HEAD–TAIL METHODS

13.3.1   Ordinal (Qualitative) Head–Tail for Unidirectional Flow

In this section, we develop a heuristic to find the best layout for a single-row unidirectional layout using from–to flow information. Note that in this case best does not mean optimal as heuristics are used. This heuristic is computationally very fast and easy to implement. The heuristic starts with the pair of stations that have the highest flow as the initial partial layout. More stations are progressively added to the existing partial layout. To build the layout, four simple rules are used. In each iteration, one station is added to the partial layout being developed. For a given layout, all parts (or products) must flow through the same loop. This means that the distances between pairs of stations do not matter, but the number of loops (one or two) that the parts must go through is minimized. Suppose in a layout, station i is before station j, that is, $i \rightarrow j$. The flow will then be one loop for w_{ij} and two loops for w_{ji}. The problem is to find the order of stations that minimizes the total flow as measured in Equation (13.1).

Largest Difference Rule   Consider all pairs of stations:

- For all pairs of stations (i, j), calculate the flow difference, $[w_{ij} - w_{ji}]$.
- Rank all $[w_{ij} - w_{ji}]$ in descending order; label them by vector $\mathbf{w}_0$.
- If two or more $[w_{ij} - w_{ji}]$ values are equal, rank the (i, j) pair with the larger single w_{ij} value first.

Example 13.4   Largest Difference Rule   Consider Example 13.2.

		Workflow: $[w_{ij}]$						Difference for Workflow: $[w_{ij} - w_{ji}]$			
		To (j)						To (j)			
	Station	1	2	3	4		Station	1	2	3	4
From (i)	1	0	92	0	249	From (i)	1	0	92	−268	214
	2	0	0	211	110		2	−92	0	211	92
	3	268	0	0	35		3	268	−211	0	−38
	4	35	18	73	0		4	−214	−92	38	0

The ranking of all pairs of stations of $[w_{ij} - w_{ji}]$ is presented below.

			To (j)			
		Station	1	2	3	4
From (i)		1	—	5	12	2
		2	8	—	3	4
		3	1	10	—	7
		4	11	9	6	—

The pairs $(2, 4)$ and $(1, 2)$ have equal difference $[w_{ij} - w_{ji}]$. However, because $w_{24} = 110 > w_{12} = 92$, pair $(2, 4)$ is ranked higher than $(1, 2)$.

Consider the following notations for constructing partial sequences of workstations.

$s_p = (i \rightarrow j)$ is a candidate pair sequence where station i is before station j.

$s_q = (\rightarrow \rightarrow \rightarrow \cdots)$ is a partially constructed sequence solution so far.

Generic Steps and Rules of All Head–Tail Methods

1. Consider the ranking of all pairs (i, j). Set the highest ranked pair as s_q.
2. Consider the highest ranked unassigned pair, $s_p = (i, j)$, such that:
   (a) Both i and j are not in s_q.
   (b) Either i or j is in s_q.
3. Use relevant head–tail rules to combine s_p and s_q.
4. Repeat steps 2 and 3 until all stations are assigned to s_q.
5. Try to improve upon the obtained solution by use of the pairwise exchange method.

Temporarily Discarding Unassigned Pairs   If a pair $s_p = (i, j)$ does not have any common element with $s_q = (\cdots)$, then $s_p = (i, j)$ is discarded temporarily only in the current iteration, but it is reconsidered in the next iteration. For example, suppose that $s_p = (5,3)$ and $s_q = (6,1,2,4)$. Temporarily discard $s_p = (5,3)$ and reconsider it in the next iteration.

Permanently Discarding Assigned Pairs   If a pair $s_p = (i, j)$ has two common elements with $s_q = (\ldots i \ldots, j \ldots)$, then $s_p = (i, j)$ is discarded permanently (because both i and j are already assigned). For example, suppose that $s_p = (4,6)$ and $s_q = (6,1,2,4)$. Both stations 4 and 6 have already been placed in s_q; permanently discard $s_p = (4,6)$.

Ordinal Head–Tail Rule for Unidirectional   In the ordinal head–tail method, the best location for the conflicting (uncommon) station is found by inserting it in the first found location without causing any additional loop penalty to the objective function. To simplify the presentation of this rule, consider two cases:

Unidirectional Ordinal (Conflict Resolution) Rule (j after i) When there is one common station i in $s_p = (i{\rightarrow}j)$ and $s_q = (\ldots \rightarrow i \rightarrow \cdots m \ldots)$, insert j (the uncommon station of s_p) after i in s_q and move j to the right until it is before a station m such that $m{\rightarrow}j$ flow is negative, that is, it is better to have sequence $j{\rightarrow}m$ than $m{\rightarrow}j$.

Unidirectional Ordinal (Conflict Resolution) Rule (i after j) When there is one common station i in $s_p = (j{\rightarrow}i)$ and $s_q = (\ldots m \ldots \rightarrow i \rightarrow)$, insert j (the uncommon station of s_p) before i in s_q and move j to the left until it is after a station m such that $j{\rightarrow}m$ flow is negative, that is it is better to have sequence $m{\rightarrow}j$ than $j{\rightarrow}m$.

Example of Ordinal Rule (j after i) Consider $s_q = (7{\rightarrow}1{\rightarrow}2{\rightarrow}4)$ and $s_p = (7{\rightarrow}5)$. In this case, 5 must be inserted after 7 in s_q. Possible choices are presented below. Note that throughout the correct placement is indicate by the shading.

Sequence	a. 7→5?→1→2→4	b. 7→1→5?→2→4	c. 7→1→2→5?→4	d. 7→1→2→4→5?
Flow, decision	7→5? Positive, yes	1→5? Positive, yes	2→5? Positive, yes	4→5? Negative, no

(a) First suppose $1{\rightarrow}5$ flow is positive. Then place 5 after 1 in s_q. (b) Now, suppose $2{\rightarrow}5$ flow is positive. Then place 5 after 2. (c) Suppose $4{\rightarrow}5$ flow is negative. Then do not place 5 after 4. The best solution is $s_q = (7{\rightarrow}1{\rightarrow}2{\rightarrow}5{\rightarrow}4)$.

Example of Ordinal Rule (i after j) Consider $s_q = (7{\rightarrow}1{\rightarrow}2{\rightarrow}4)$ and $s_p = (5{\rightarrow}4)$. In this case, 5 must be inserted before 4 in s_q. Possible choices are presented below.

Sequence	a. 7→1→2→5?→4	b. 7→1→5?→2→4	c. 7→5?→1→2→4	d. 5?→7→1→2→4
Flow, decision	5?→4, Positive, yes	5?→2, Positive, yes	5?→1, Positive, yes	5?→7, Negative, no

(a) First suppose $5{\rightarrow}2$ flow is positive. Then place 5 before 2 in s_q. (b) Now, suppose $5{\rightarrow}1$ flow is positive. Then place 5 before 1. (c) Suppose $5{\rightarrow}7$ flow is negative. Then do not move 5 before 7. The best solution is: $s_q = (7{\rightarrow}5{\rightarrow}1{\rightarrow}2{\rightarrow}4)$.

Example 13.4 for Ordinal Head–Tail Consider Example 13.4. A summary of the method is provided below.

Iteration/ Rank	s_p	Action	Resulting sequence, s_q	Candidates
1	3→1	Add s_p to s_1	$\{s_1 = (3{\rightarrow}1)\}$	2,4
2	1→4	Add s_p to s_1	$\{s_1 = (3{\rightarrow}1{\rightarrow}4)\}$	2
3	2→3	Add s_p to s_1	$\{s_1 = (2{\rightarrow}3{\rightarrow}1{\rightarrow}4)\}$	0, Stop

All elements have been assigned, and the solution is $(2{\rightarrow}3{\rightarrow}1{\rightarrow}4)$.

$$\text{Total flow} = (211 + 110 + 268 + 35 + 249) + 2(35 + 73 + 18 + 92)$$

$$= 873 + 2(218) = 1309$$

Concordance is 873 and discordance is 218. Note that total concordance and discordance, that is, $873 + 218 = 1091$, is constant regardless of the chosen solution, but the total-flow value may be different for different solutions.

Example 13.5 Ordinal Head–Tail for Eight Stations

Original Unidirectional Flow (W)									Difference Flow Matrix							
	1	2	3	4	5	6	7	8		1	2	3	4	5	6	7	8
1	0	0	97	0	0	0	0	0	1	0	−84	97	−88	0	−109	0	0
2	84	0	185	182	5	101	63	0	2	84	0	23	182	5	101	63	0
3	0	162	0	103	0	0	0	160	3	−97	−23	0	−32	0	−122	−63	160
4	88	0	135	0	29	0	20	0	4	88	−182	32	0	29	0	20	0
5	0	0	0	0	0	171	123	60	5	0	−5	0	−29	0	171	−46	31
6	109	0	122	0	0	0	0	0	6	109	−101	122	0	−171	0	0	0
7	0	0	63	0	169	0	0	103	7	0	−63	63	−20	46	0	0	103
8	0	0	0	0	29	0	0	0	8	0	0	−160	0	−31	0	−103	0

Rank the pairs of stations using the largest difference method.

Ranking of Difference Flow Matrix								
	1	2	3	4	5	6	7	8
1	—	47	8	48	20	52	21	22
2	10	—	17	1	19	7	11	23
3	49	40	—	43	24	53	45	3
4	9	56	14	—	16	25	18	26
5	27	38	28	41	—	2	44	15
6	5	50	4	29	55	—	30	31
7	32	46	12	39	13	33	—	6
8	34	35	54	36	42	37	51	—

For ease of referencing, the first 19 ranked flow differences are listed below.

Rank	1	2	3	4	5	6	7	8	9	10	11	12	13	14	15	16	17	18	19
w_o	(2,4)	(5,6)	(3,8)	(6,3)	(6,1)	(7,8)	(2,6)	(1,3)	(4,1)	(2,1)	(2,7)	(7,3)	(7,5)	(4,3)	(5,8)	(4,5)	(2,3)	(4,7)	(2,5)
w_{ij}-w_{ji}	182	171	160	122	109	103	101	97	88	84	63	63	46	32	31	29	23	20	5

A summary of the above method is provided in the following table:

Iteration	Used Rank	s_p	Action Taken	Resulting Sequence, S	Candidates
1	1	2→4	Add s_p as a new sequence	(2→4)	1,3,5,6,7,8
2	—	5→6	Discard temporarily	(2→4)	1,3,5,6,7,8
3	—	3→8	Discard temporarily	(2→4)	1,3,5,6,7,8
4	—	6→3	Discard temporarily	(2→4)	1,3,5,6,7,8
5	—	6→1	Discard temporarily	(2→4)	1,3,5,6,7,8
6	—	7→8	Discard temporarily	(2→4)	1,3,5,6,7,8
7	7	2→6	Insert station 6 after station 4	(2→4→6)	1,3,5,7,8
8	2	5→6	Reconsider 5→6, insert 5 after 4	(2→4→5→6)	1,3,7,8
9	4	6→3	Reconsider 6→3, insert 3 after 6	(2→4→5→6→3)	1,7,8
10	3	3→8	Reconsider 3→8, insert 8 after 3	(2→4→5→6→3→8)	1,7
11	5	6→1	Reconsider 6→1, insert 1 after 6	(2→4→5→6→1→3→8)	7
12	6	7→8	Reconsider 7→8, insert 7 after 4	(2→4→7→5→6→1→3→8)	0, Stop

In iteration 7, station 6 could be inserted before or after station 4. It is inserted after 4 because 6–4 has zero flow, that is, it does not give any penalty to move it further to the front. In iteration 8, 5 can be inserted before 2 or 4; it is not moved before 4 because 5–4 flow is negative. In iteration 12, 7 can be inserted anywhere before 8 in $(2{\to}4{\to}5{\to}6{\to}1{\to}3{\to}8)$; it is not moved before 4 because 7–4 flow is negative.

The total flow of the obtained final sequence $(2{\to}4{\to}7{\to}5{\to}6{\to}1{\to}3{\to}8)$ is

$$\text{Total flow} = 1 \times (182 + 63 + 5 + 101 + 84 + 185 + 20 + 29 + 88 + 135 + 169 + 63$$
$$+ 103 + 171 + 60 + 109 + 122 + 97 + 160) + 2 \times (162 + 103 + 123 + 29)$$
$$= 1946 + 2 \times 417 = 2780$$

For this solution, concordance $= 1946$ and discordance $= 417$.

Note that by reordering columns and rows of the original flow of Example 13.5, in order of $(2{\to}4{\to}7{\to}5{\to}6{\to}1{\to}3{\to}8)$, the total score of the upper right triangle is the concordance and the total score of the lower left triangle is the discordance. This simplifies total calculations for large problems.

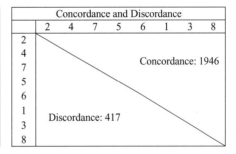

See Supplement S13.1.xls.

Supplement S13.2.docx shows details of the head-tail method for solving single-row unidirectional layout with simultaneous growth.

Variations of Head–Tail Methods   There are several different heuristics based on head–tail methods that can be used for solving unidirectional and bidirectional flow problems.

13.3.2 Cardinal Head–Tail Method for Unidirectional Flow

In the cardinal head–tail method, the best location for the conflicting (uncommon) station is found by considering its insertion in all possible locations in the partially built sequence solution. For each possible location, the objective function value is calculated and the one with the lowest total flow is selected.

Unidirectional Cardinal Rule (Conflict Resolution)   If there is a common station i, in $s_p = (i{\to}j)$ and $s_q = (\ldots \to i \to \ldots)$, then consider inserting j (the uncommon station) between each pair of stations in s_q. Choose the alternative that has the best objective function value (e.g., the minimum total flow).

For example, for $s_p = (2 \rightarrow 5)$ and $s_q = (1 \rightarrow 2 \rightarrow 3 \rightarrow 4)$, the common station is 2 and the uncommon station is 5. Consider all possible locations that station 5 can be inserted in sequence s_q. They are listed below.

Sequence	$5 \rightarrow 1 \rightarrow 2 \rightarrow 3 \rightarrow 4$	$1 \rightarrow 5 \rightarrow 2 \rightarrow 3 \rightarrow 4$	$1 \rightarrow 2 \rightarrow 5 \rightarrow 3 \rightarrow 4$	$1 \rightarrow 2 \rightarrow 3 \rightarrow 5 \rightarrow 4$	$1 \rightarrow 2 \rightarrow 3 \rightarrow 4 \rightarrow 5$
Total flow	560	622	584	668	**520**

In this example, sequence $(1 \rightarrow 2 \rightarrow 3 \rightarrow 4 \rightarrow 5)$ has the lowest flow of 520, which is selected as the best alternative. (Throughout boldface numbers indicate the minimum value.)

Example 13.6   Cardinal Head–Tail Method   Consider Example 13.5. Suppose it is a unidirectional problem. Solve it by the cardinal head–tail method.

The steps of the method are presented below.

0. Consider the list of ranked pairs of stations based on the largest difference given in Example 13.5.
1. Consider $s_p = (2 \rightarrow 4)$; label it as $s_q = (2 \rightarrow 4)$.
2–6. Temporarily discard pairs $(5 \rightarrow 6)$, $(3 \rightarrow 8)$, $(6 \rightarrow 3)$, $(6 \rightarrow 1)$, and $(7 \rightarrow 8)$.
7. Consider $s_p = (2 \rightarrow 6)$, which has a common station, 2, with $s_q = (2 \rightarrow 4)$. Find the best location of uncommon station 6 in $s_q = (2 \rightarrow 4)$. The solutions are listed in the following table.

Sequence	$6 \rightarrow 2 \rightarrow 4$	$2 \rightarrow 6 \rightarrow 4$	$2 \rightarrow 4 \rightarrow 6$
Total flow	384	**283**	**283**

Since there are two sequences with the same minimum total flow of 283, arbitrarily select one. For this example, the sequence $s_q = (2 \rightarrow 4 \rightarrow 6)$ is selected.

8. Reconsider $s_p = (5 \rightarrow 6)$; there is a conflict with $s_q = (2 \rightarrow 4 \rightarrow 6)$. The location of station 5 should be determined.

Sequence	5,2,4,6	2,5,4,6	2,4,5,6	2,4,6,5
Total flow	522	517	**488**	659

The sequence with the lowest total flow is $2 \rightarrow 4 \rightarrow 5 \rightarrow 6$. Therefore choose $s_q = (2 \rightarrow 4 \rightarrow 5 \rightarrow 6)$.

9. Reconsider $s_p = (6 \rightarrow 3)$; find the best location for station 3. The five alternatives and their total flows are listed as the columns in the table below. Here, $s_q = (2 \rightarrow 4 \rightarrow 5 \rightarrow 6 \rightarrow 3)$ has the lowest total flow and it is selected.

Sequence	3,2,4,5,6	2,3,4,5,6	2,4,3,5,6	2,4,5,3,6	2,4,5,6,3
Total flow	1637	1614	1582	1582	**1460**

10. Reconsider $s_p = (3\rightarrow8)$; find the best location for station 8. The six alternatives and their total flows are listed as the columns in the table below. Here, $s_q = (2\rightarrow4\rightarrow5\rightarrow6\rightarrow3\rightarrow8)$ is the best alternative.

Sequence	88,2,4,5,6,3	2,8,4,5,6,3	2,4,8,5,6,3	2,4,5,8,6,3	2,4,5,6,8,3	2,4,5,6,3,8
Total flow	1929	1929	1929	1898	1898	**1738**

11. Reconsider $s_p = (6\rightarrow1)$; there is a conflict. Find the best location for station 1. The seven possible sequences are presented below.

Sequence	1,2,4,5, 6,3,8	2,1,4,5, 6,3,8	2,4,1,5, 6,3,8	2,4,5,1, 6,3,8	2,4,5,6, 1,3,8	2,4,5,6, 3,1,8	2,4,5,6, 3,8,1
Total flow	2397	2473	2385	2416	**2307**	2404	2404

The sequence with the lowest total flow is $s_q = (2\rightarrow4\rightarrow5\rightarrow6\rightarrow1\rightarrow3\rightarrow8)$.

12. Consider $s_p = (7\rightarrow8)$; Find the best location for station 7.

Sequence	7,2,4,5, 6,1,3,8	2,7,4,5, 6,1,3,8	2,4,7,5, 6,1,3,8	2,4,5,7, 6,1,3,8	2,4,5,6, 7,1,3,8	2,4,5,6, 1,7,3,8	2,4,5,6, 1,3,7,8	2,4,5,6, 1,3,8,7
Total flow	2863	**2800**	2843	2909	2863	2863	2926	2966

The sequence with the lowest total flow is $s_q = (2\rightarrow7\rightarrow4\rightarrow5\rightarrow6\rightarrow 1\rightarrow3\rightarrow8)$. Since all stations are assigned, this is the final solution with a total flow of 2800. A summary of the above method is provided in the following table:

Iteration	Used Rank	s_p	Action Taken	Resulting sequence, S	Candidates
1	1	$2\rightarrow4$	Add s_p as a new sequence	$(2\rightarrow4)$	1,3,5,6,7,8
2	—	$5\rightarrow6$	Discard temporarily	$(2\rightarrow4)$	1,3,5,6,7,8
3	—	$3\rightarrow8$	Discard temporarily	$(2\rightarrow4)$	1,3,5,6,7,8
4	—	$6\rightarrow3$	Discard temporarily	$(2\rightarrow4)$	1,3,5,6,7,8
5	—	$6\rightarrow1$	Discard temporarily	$(2\rightarrow4)$	1,3,5,6,7,8
6	—	$7\rightarrow8$	Discard temporarily	$(2\rightarrow4)$	1,3,5,6,7,8
7	7	$2\rightarrow6$	Insert station 6	$(2\rightarrow4\rightarrow6)$	1,3,5,7,8
8	2	$5\rightarrow6$	Reconsider discarded $5\rightarrow6$	$(2\rightarrow4\rightarrow5\rightarrow6)$	1,3,7,8
9	4	$6\rightarrow3$	Reconsider discarded $6\rightarrow3$	$(2\rightarrow4\rightarrow5\rightarrow6\rightarrow3)$	1,7,8
10	3	$3\rightarrow8$	Reconsider discarded $3\rightarrow8$	$(2\rightarrow4\rightarrow5\rightarrow6\rightarrow3\rightarrow8)$	1,7
11	5	$6\rightarrow1$	Reconsider discarded $6\rightarrow1$	$(2\rightarrow4\rightarrow5\rightarrow6\rightarrow1\rightarrow3\rightarrow8)$	7
12	6	$7\rightarrow8$	Reconsider discarded $7\rightarrow8$	$(2\rightarrow7\rightarrow4\rightarrow5\rightarrow6\rightarrow1\rightarrow3\rightarrow8)$	0, Stop

The total for the final solution is

$$
\begin{aligned}
\text{Total flow} =\ & 1 \times (63 + 182 + 5 + 101 + 84 + 185 + 169 + 63 + 103 + 29 + 88 + 135 \\
& + 171 + 60 + 109 + 122 + 97 + 160) + 2 \times (162 + 20 + 123 + 103 + 29) \\
=\ & 1926 + 2 \times 437 = 2800
\end{aligned}
$$

Simultaneous Head–Tail Method for Unidirectional[†]   It is possible to generate a number of alternatives simultaneously in order to increase the chances of finding the optimal solution. See Sections 13.7.4 and 13.7.7 for more details.

Supplement S13.3.docx shows details of simultaneous head–tail for unidirectional layouts.

13.3.3   Pairwise Exchange Improvement[†]

It is possible to try to improve the solutions of head–tail methods. To do this, apply the pairwise exchange (PWE) method to the obtained solution in each iteration of the head–tail method. Then use the improved solution in the next iteration of the head–tail method. A simpler approach is to only apply the PWE method to the final solution (last iteration) of the head–tail method. The PWE method, also known as steepest descent, is a well-known heuristic used for solving single-objective facility layout problems. For more details see the PWE method covered in Section 15.4. The steps for Pair-wise Exchange (improvement) method for cell and combinatorial problems are as follows:

1. Swap all possible and feasible pairs of elements (e.g., stations) of the current alternative. These are called adjacent alternatives.
2. Find the objective function (e.g., the total flow) of each generated adjacent alternative.
3. Compare the objective function values of the current and all the generated adjacent alternatives.
4. If the current alternative objective function value is better (e.g., lower for the minimization problem) than all generated adjacent alternatives, stop and label it as the best.
5. Choose an adjacent alterative which has the best (e.g., lowest) objective function value given that this alternative is not the same as one of the previously generated alternatives. If a new equivalent or better adjacent alternative cannot be found, stop and label the current as the best.
6. Label the selected adjacent alternative as the current alternative. Go to step 1.

In this method we allow equivalent alternatives (in terms of the value of the objective function) to be selected; therefore, in step 5, it is verified that the selected alternative is not the same as one of the previously selected alternatives so that the algorithm will not go through a cycle.

Example 13.6 Revisited   Consider the final solution obtained by the cardinal head–tail method in Example 13.6, $2 \rightarrow 7 \rightarrow 4 \rightarrow 5 \rightarrow 6 \rightarrow 1 \rightarrow 3 \rightarrow 8$, as the current alternative. This problem has eight stations; therefore in each iteration of PWE method, $(8 \times 7)/2 = 28$ alternatives should be considered. In the following table, for simplicity of presentation, only the selected best pair of stations of each iteration is shown. In iteration 0, the best adjacent solution is obtained by swapping the locations of stations 4 and 7. In the next iteration, an improved solution cannot be found; therefore the final solution is $(2 \rightarrow 4 \rightarrow 7 \rightarrow 5 \rightarrow 6 \rightarrow 1 \rightarrow 3 \rightarrow 8)$ with a total flow of 2780.

Iteration	Pairwise Exchange	Solution	Objective Value
0	—	2→7→4→5→6→1→3→8	2800
1	(7,4)	2→4→7→5→6→1→3→8	2780

See Supplement S13.4.xls.

13.3.4 Multicriteria Unidirectional by Composite Approach[†]

Bicriteria In addition to the total flow as the only objective function (Section 13.3.1), several other objectives (criteria) can be used in evaluating alternatives. For example, from a quality and/or operations point of view, it may be preferred to have one station before another station. In this case, qualitative preferences for all pairs of stations can be assessed. Qualitative preferences can be expressed by A, E, I, O, U, and X. Such qualitative measures can be converted to quantitative measures by using $A = 5$, $E = 4$, $I = 3$, $O = 2$, $U = 1$, and $X = 0$. The preferences are given only for the unidirectional precedence. The rating for each pair of stations is presented by p_{ij} and their matrix is presented by P. For bicriteria consider:

Minimize f_1 = total flow (based on quantitative information w_{ij} of matrix W)

Minimize f_2 = total qualitative preferences (based on qualitative information p_{ij} of matrix P)

Tricriteria For tricriteria, in addition to the above two criteria, the following third criterion can also be considered:

Minimize f_3 = Total flow time.

The third objective, f_3, means that the processing (and/or setup) times of stations can vary depending on the precedence of stations. For example, a station setup time can be different depending on the previous station. This processing time can be presented as t_{ij} for pairs of stations of i first and then j. Its matrix is presented by T.

Composite MCDM Approach for Solving Tricriteria Unidirectional Problem
This method is based on the composite MCDM approach:

1. Normalize the matrices W, P, and T to obtain W', P', and T', where w'_{ij}, p'_{ij}, and t'_{ij} represent the three criteria values for each pair of stations i and j.
2. Assign weights of importance w_1, w_2, and w_3 for minimizing the normalized values of objectives 1, 2, and 3, respectively.
3. Generate a composite matrix Z where $z'_{ij} = w_1 w'_{ij} + w_2 p'_{ij} + w_3 t'_{ij}$.
4. Find the best solution for matrix Z using one of the head–tail methods.
5. Measure the values of the three objectives f_1, f_2, and f_3.

Example 13.7   Tricriteria Problem   Consider the weights of importance for the three objectives to be $\mathbf{w} = (0.3, 0.3, 0.4)$. Flow quantity, qualitative preferences, and flow time for each pair of stations are given below. Use the cardinal head–tail method to solve this problem.

Flow Quantity W				Preferences P				Flow Time T			
1	2	3	4	1	2	3	4	1	2	3	4
1 —	92	0	249	1 —	U/1	E/4	U/1	1 —	2	7	1
2 0	—	211	92	2 A/5	—	E/4	X/0	2 3	—	6	2
3 268	0	—	35	3 X/0	E/4	—	A/5	3 4	5	—	10
4 35	0	73	—	4 E/4	A/5	U/1	—	4 9	8	3	—

1. For flow, the minimum is 0 and the maximum is 268. For qualitative preferences, the minimum is 0 and the maximum is 5. Therefore, 0, 0.2, 0.4, 0.6, 0.8, 1 can be used for X, U, O, I, E, A, respectively. For flow time, use 0 as the minimum and 10 as the maximum. Normalized matrices are as follows:

Flow W'				Preferences P'				Flow Time T'			
1	2	3	4	1	2	3	4	1	2	3	4
1 —	0.3	0	0.9	1 —	0.2	0.8	0.2	1 —	0.2	0.7	0.1
2 0	—	0.8	0.3	2 1	—	0.8	0	2 0.3	—	0.6	0.2
3 1	0	—	0.1	3 0	0.8	—	1	3 0.4	0.5	—	1
4 0.1	0	0.27	—	4 0.8	1	0.2	—	4 0.9	0.8	0.3	—

2. The composite matrix is constructed by using $z_{ij} = 0.3w'_{ij} + 0.3p'_{ij} + 0.4t'_{ij}$.
3. Compute the differences matrix of $z_{ij} - z_{ji}$.

Composite Matrix z_{ij}					Composite Differences $z_{ij} - z_{ji}$			
Step 2	1	2	3	4	Step 3	1	2	3	4
1	—	0.2	0.5	0.4	1	—	-0.2	0	-0.2
2	0.4	—	0.7	0.1	2	0.2	—	0.3	-0.5
3	0.5	0.4	—	0.7	3	0	-0.3	—	0.4
4	0.6	0.6	0.3	—	4	0.2	0.5	-0.4	—

$\Rightarrow$

The Ranking of composite differences $[z_{ij} - z_{ji}]$ is presented below.

		To (j)			
	Station	1	2	3	4
From (i)	1	—	9	6	8
	2	5	—	3	12
	3	6	10	—	2
	4	4	1	11	—

4. Solve the problem using the chosen head–tail method (here the cardinal method is used):

   (a) $s_p = (4{\rightarrow}2)$; set $s_q = (4{\rightarrow}2)$.
   (b) Consider $s_p = (3{\rightarrow}4)$; find the best location for station 3 in $s_q = (4{\rightarrow}2)$; the best solution is $s_q = (3{\rightarrow}4{\rightarrow}2)$.
   (c) $s_p = (2{\rightarrow}3)$; stations 2 and 3 already exist in s_q. Discard this pair.
   (d) $s_p = (4{\rightarrow}1)$; consider s_p versus s_q. Find the best location for object (station) 1.

Sequence	$1{\rightarrow}3{\rightarrow}4{\rightarrow}2$	$3{\rightarrow}1{\rightarrow}4{\rightarrow}2$	$3{\rightarrow}4{\rightarrow}1{\rightarrow}2$	**$3{\rightarrow}4{\rightarrow}2{\rightarrow}1$**
Total flow	8	8	7.8	**7.6**

Therefore, $s_q = (3{\rightarrow}4{\rightarrow}2{\rightarrow}1)$ is the best solution. The three objective values for this solution are:

f_1, Total flow $= (35 + 268 + 35) + 2(92 + 249 + 211 + 92 + 73) = 1772$.
f_2, Total qualitative preferences $= (5 + 4 + 5 + 4 + 5) + 2(1 + 4 + 1 + 4 + 1) = 45$.
f_3, Total flow time $= (4 + 5 + 10 + 8 + 9 + 3) + 2(2 + 7 + 1 + 2 + 6 + 3) = 81$.

See Supplement S13.5.xls.

13.4 BIDIRECTIONAL HEAD–TAIL METHODS

The bidirectional flow problem is similar to the unidirectional flow problem discussed in the last section, except that the object (e.g., the product) can flow in both directions. Automated guided vehicles, flexible manufacturing systems, jobshops, and cell systems often employ bidirectional flow for processing different parts. Token-ring computer networks are a nonmanufacturing example of a system that uses bidirectional flow. In a bidirectional layout, every station has its own input and output gates that is, the incoming raw material can directly enter the starting station and leave from the finishing station when the job is completed. The conceptual layout of bidirectional flow is shown in Figure 13.2, where the sequence of machines is $2\leftrightarrow3\leftrightarrow1\leftrightarrow4$. The layout could be linear or circular. The distances between pairs of stations are calculated differently depending on whether the layout is linear or circular.

Total-Flow Measurement Based on Distances of Pairs of Locations For a problem with n stations, there will be n locations. Each station will be assigned to one location. The distances between pairs of locations can be used in calculating the total flow of a given bidirectional layout solution. For simplicity, it can be assumed that stations are equidistant from their adjacent stations in a linear or a circular form. For example, in Figure 13.2a, a linear layout, the distance between stations 2 and 3 is one unit, and the distance between stations 2 and 4 is three units. However, in Figure 13.2b, a circular layout, the distance between stations 2 and 4 is one unit. The distances can be presented by a from–to distance matrix (see the following example). The distance matrix is symmetric; that is, the distance from station i to station j, d_{ij}, is equal to the distance from station j to location i, d_{ji}. In this section, the distances are measured based on a linear layout similar to Figure 13.2a; however, the developed procedures can also be applied to circular distances as well. The total flow for a given sequence solution (alternative) is

$$\text{TF} = \sum_{i=1}^{n}\sum_{j=1}^{n} w_{ij}d_{ij} \qquad (13.2)$$

Combined-Flow Matrix Because the flow is bidirectional, for every pair of stations (i, j) where $i < j$, create a new w_{ij} equal to $w_{ij} + w_{ji}$; then set $w_{ji} = 0$. These values can be used in calculating the total flow.

Example 13.8 Bidirectional Total-Flow Calculation Based on Distances Consider the following given flow matrix. Find the total flow for two different alternatives, $(2\leftrightarrow3\leftrightarrow1\leftrightarrow4)$ and $(1\leftrightarrow2\leftrightarrow3\leftrightarrow4)$.

First, find the combined-flow matrix by calculating $[w_{ij} + w_{ji}]$. The distances for both alternatives are also given in the following matrices assuming that the distance between each pair of adjacent stations is 1:

	Flow Matrix					Combined-Flow Matrix			
	1	2	3	4		1	2	3	4
1	0	92	0	249	1	0	92	268	284
2	0	0	211	92	2		0	211	92
3	268	0	0	35	3			0	108
4	35	0	73	0	4				0

D: Distances for $(2\leftrightarrow3\leftrightarrow1\leftrightarrow4)$					D: Distances for $(1\leftrightarrow2\leftrightarrow3\leftrightarrow4)$			
	1	2	3	4		1	2	3	4
1	0	2	1	1	1	0	1	2	3
2	—	0	1	3	2	—	0	1	2
3	—	—	0	2	3	—	—	0	1
4	—	—	—	0	4	—	—	—	0

The total flow for the first alternative, $(2\leftrightarrow3\leftrightarrow1\leftrightarrow4)$, is

$$\text{TF} = \sum_{i=1}^{n}\sum_{j=1}^{n} w_{ij}d_{ij} = 92 \times 2 + 268 \times 1 + 284 \times 1 + 211 \times 1 + 92 \times 3 + 108 \times 2$$

$$= 1439$$

The total flow for the second alternative, $(1\leftrightarrow2\leftrightarrow3\leftrightarrow4)$, is

$$\text{TF} = \sum_{i=1}^{n}\sum_{j=1}^{n} w_{ij}d_{ij} = 92 \times 1 + 268 \times 2 + 284 \times 3 + 211 \times 1 + 92 \times 2 + 108 \times 1$$

$$= 1983$$

Observe that $(2\leftrightarrow3\leftrightarrow1\leftrightarrow4)$ has a better total flow than $(1\leftrightarrow2\leftrightarrow3\leftrightarrow4)$.

13.4.1 Ordinal (Qualitative) Head–Tail for Bidirectional Flow

Ordering All Pairs by Largest Combined Rule   Consider the combined-flow matrix. Order all pairs (i, j) in descending order of w_{ij}. Label the ordered vectors of w_{ij} by $\mathbf{w}_0$. Consider $s_p = (i \leftrightarrow j)$, $s_q = (\leftrightarrow \leftrightarrow \leftrightarrow \dots)$, and $\mathbf{w}_0 = $ vector of w_{ij} in descending order of w_{ij}.

Bidirectional Ordinal (Conflict Resolution) Rule   If there is a common station in sequences s_p and s_q, that is, $s_p = (i \leftrightarrow j)$ and $s_q = (\dots n \leftrightarrow i \leftrightarrow m \dots)$, then place the new station j next to the common station i in s_q on the side that has a higher ranking to its adjacent station, that is, compare $j \leftrightarrow n$ and $j \leftrightarrow m$ flows and insert j on the side that has higher flow.

For example, suppose that $s_p = (6 \leftrightarrow 1)$ and $s_q = (7 \leftrightarrow 2 \leftrightarrow 6 \leftrightarrow 5 \leftrightarrow 4)$. Station 6 is common between the two sequences. Decide if station 1 should be on the left or on the right side of station 6 in s_q. This is shown by the following graphical representation:

$$(7 \longleftrightarrow 2 \leftrightarrow 6 \leftrightarrow 5 \longleftrightarrow 4)$$
$$? \quad \updownarrow \quad ?$$
$$1$$

Suppose that $(1, 5)$ is ranked higher than $(1, 2)$ (e.g., flow between 1 and 5 is more than 1 and 2). Therefore the solution is $(7\leftrightarrow2\leftrightarrow6\leftrightarrow1\leftrightarrow5\leftrightarrow4)$.

Attaching Head–Tail Sequences Note that the above rule means that when the common station of $s_p = (i\leftrightarrow j)$ is at the head or tail of s_q, simply attach s_p to s_q at the common station. For example, suppose that $s_q = (3\leftrightarrow2\leftrightarrow5)$ and $s_p = (1\leftrightarrow3)$. Connect the two sequences at the common station 3, $s_q = (1\leftrightarrow3\leftrightarrow2\leftrightarrow5)$.

Linear Mirror Property The mirror image of sequence s_q, that is, its reverse sequence, is equivalent to s_q. Both sequences will have the same total flow. For example, $s_q = (6\leftrightarrow1\leftrightarrow2\leftrightarrow4)$ is equivalent to $s_q = (4\leftrightarrow2\leftrightarrow1\leftrightarrow6)$.

Circular Layouts Bidirectional layout problems can be linear or circular, for example, see Figures 1.2a,b. Linear layouts are presented in this section. The approaches covered for solving linear layouts can be extended for solving circular layouts. In circular problems, the conflict resolution rule should be applied in every iteration because the head and the tail of the given sequence are always connected.

Example 13.9   Ordinal Head–Tail for Bidirectional for Four Stations   Consider the information in Example 13.8. The combined-flow matrix and its ranking are shown below.

Combined-Flow Matrix					Ranking of Flow Matrix			
	1	2	3	4		1	2	3	4
1	0	92	268	284	1	0	5	2	1
2		0	211	92	2		0	3	6
3			0	108	3			0	4
4				0	4				0

A summary of the method is provided below.

Iteration	Used Rank	s_p	Action Taken	Resulting sequence, s_q	Candidates
1	1	$1\leftrightarrow4$	Add s_p as a new sequence	$(1\leftrightarrow4)$	2, 3
2	2	$1\leftrightarrow3$	Insert station 3	$(3\leftrightarrow1\leftrightarrow4)$	2
3	3	$2\leftrightarrow3$	Insert station 3	$(2\leftrightarrow3\leftrightarrow1\leftrightarrow4)$	—

The final layout is $(2\leftrightarrow3\leftrightarrow1\leftrightarrow4)$. This layout is the same as $(4\leftrightarrow1\leftrightarrow3\leftrightarrow2)$. The distance between each pair of adjacent stations is one unit. For example, the distance between stations 4 and 2 is three units, but between stations 4 and 3 it is two units. The total flow for this example is

$$\text{Total flow} = 2 \times 92 + 1 \times 268 + 1 \times 284 + 1 \times 211 + 3 \times 92 + 2 \times 108 = 1439$$

Example 13.10   Ordinal Head–Tail for Bidirectional for Seven Stations   Consider the flow matrix shown below. The combined-flow matrix for the bidirectional heuristic is as follows:

	Flow Matrix								Combined-Flow Matrix					
	1	2	3	4	5	6	7		1	2	3	4	5	6	7
1	0	24	15	0	0	0	0	1	0	69	78	5	51	76	59
2	45	0	4	18	0	66	0	2		0	42	58	31	88	97
3	63	38	0	0	15	13	0	3			0	6	30	22	29
4	5	40	6	0	54	0	0	4				0	99	98	26
5	51	31	15	45	0	0	0	5					0	99	2
6	76	22	9	98	99	0	45	6						0	85
7	59	97	29	26	2	40	0	7							0

Rank all w_{ij} pairs. In the above table, both (4,5) and (5,6) have a flow of 99. The tie is arbitrarily broken by choosing (4,5) first.

The ranking of the flow matrix is as follows:

	1	2	3	4	5	6	7
1	0	9	7	20	12	8	10
2		0	13	11	14	5	4
3			0	19	15	18	16
4				0	1	3	17
5					0	2	21
6						0	6
7							0

A summary of the method is provided below.

Iteration	Used Rank	s_p	Action Taken	Resulting Sequence, s_q	Candidates
1	1	4↔5	Add s_p as a new sequence	(4↔5)	1,2,3,6,7
2	2	5↔6	Insert station 6	(4↔5↔6)	1,2,3,7
3	3	4↔6	Discard permanently	(4↔5↔6)	1,2,3,7
4	—	2↔7	Discard temporarily	(4↔5↔6)	1,2,3,7
5	5	2↔6	Insert station 2	(4↔5↔6↔2)	1,3,7
6	4	2↔7	Reconsider discarded 2→7	(4↔5↔6↔2↔7)	1,3
7	6	6↔7	Discard permanently	(4↔5↔6↔2↔7)	1,3
8	—	1↔3	Discard temporarily	(4↔5↔6↔2↔7)	1,3
9	8	1↔6	Insert 1 between 6 and 2	(4↔5↔6↔1↔2↔7)	3
10	7	1↔3	Reconsider 1↔3, insert 3	(4↔5↔6↔1↔3↔2↔7)	0, Stop

In iteration 8, $s_p = (1\leftrightarrow6)$. Station 6 is common between the two sequences. Find the location for station 1 next to 6. Apply the ordinal conflict resolution rule. Since (1,2) is ranked higher than (1,5), the new sequence is $s_q = (4\leftrightarrow5\leftrightarrow6\leftrightarrow1\leftrightarrow2\leftrightarrow7)$.

In iteration 10, reconsider $s_p = (1\leftrightarrow3)$. Find the best location for station 3 next to station 1. Compare (2,3) and (3,6); because (2,3) has a higher ranking, put 3 next to 2. Then the new sequence is $s_q = (4\leftrightarrow5\leftrightarrow6\leftrightarrow1\leftrightarrow3\leftrightarrow2\leftrightarrow7)$. Stop since all stations 1–7 have already been placed in the sequence.

The distance matrix is presented below where the distance between each pair of adjacent stations is 1.

	1	2	3	4	5	6	7
1	—	2	1	3	2	1	3
2			1	5	4	3	1
3				4	3	2	2
4					1	2	6
5						1	5
6							4
7							

$$\text{Total flow} = 2 \times 69 + 1 \times 78 + 3 \times 5 + 2 \times 51 + 1 \times 76 + 3 \times 59 + 1 \times 42 + 5 \times 58$$
$$+ 4 \times 31 + 3 \times 88 + 1 \times 97 + 4 \times 6 + 3 \times 30 + 2 \times 22 + 2 \times 29 + 1$$
$$\times 99 + 2 \times 98 + 6 \times 26 + 1 \times 99 + 5 \times 2 + 4 \times 85 = 2519$$

Improving Bidirectional Solutions by Pairwise Exchange[†]   The pairwise Exchange method can be used to possibly improve the solutions obtained in each iteration of the head–tail method. For simplicity of presentation, we will only apply it to the final solution of the head–tail method. The summary of the pairwise exchange method is shown below.

Iteration	Pairwise Exchange	Solution	Objective Value
0	—	$4\leftrightarrow5\leftrightarrow6\leftrightarrow1\leftrightarrow3\leftrightarrow2\leftrightarrow7$	2519
1	(3,7)	$4\leftrightarrow5\leftrightarrow6\leftrightarrow1\leftrightarrow7\leftrightarrow2\leftrightarrow3$	2447
2	(1,2)	$4\leftrightarrow5\leftrightarrow6\leftrightarrow2\leftrightarrow7\leftrightarrow1\leftrightarrow3$	2285

Therefore, the solution can be improved to 2285 using the pairwise exchange method.

Supplement S13.6.docx shows details of pairwise exchange for bidirectional solutions.

13.4.2   Cardinal Head–Tail for Bidirectional Flow

This method is the same as the cardinal head–tail for unidirectional networks (Section 13.3.2). The method is based on the following rule.

Bidirectional Cardinal (Conflict Resolution) Rule   If there is a common station i in $s_p = (i \leftrightarrow j)$ and $s_q = (\ldots \leftrightarrow i \leftrightarrow \ldots)$ insert j (the uncommon station) between each pair of adjacent stations in s_q, then choose the alternative that has the best objective function value (e.g., the minimum total flow).

For example, suppose $s_p = \{2 \leftrightarrow 4\}$ and $s_q = \{1 \leftrightarrow 2 \leftrightarrow 3\}$. The common station is 2 and the uncommon station is 4; there is a conflict with station 4. Consider all possible locations that station 4 can be inserted in s_q. Possible sequences are

$(4 \leftrightarrow 1 \leftrightarrow 2 \leftrightarrow 3)$	$(1 \leftrightarrow 4 \leftrightarrow 2 \leftrightarrow 3)$	$(1 \leftrightarrow 2 \leftrightarrow 4 \leftrightarrow 3)$	$(1 \leftrightarrow 2 \leftrightarrow 3 \leftrightarrow 4)$

Then, choose the sequence which has the lowest total flow.

Example 13.11   Using Cardinal Head–Tail to Solve Example 13.10   Consider the information given in Example 13.10. The steps of the method are as follows:

1. $s_p = (4 \leftrightarrow 5)$ and $s_q = (4 \leftrightarrow 5)$.
2. $s_p = (5 \leftrightarrow 6)$. Station 5 is common between the two sequences. Find the location for station 6 in $s_q = (4 \leftrightarrow 5)$. Apply the cardinal conflict resolution rule.

Sequence	6,4,5	4,6,5	4,5,6
Total flow	395	395	**394**

   Therefore the best sequence is $s_q = (4 \leftrightarrow 5 \leftrightarrow 6)$.
3. $s_p = (4 \leftrightarrow 6)$. Discard s_p since 4 and 6 are already assigned.
4. $s_p = (2 \leftrightarrow 7)$. Discard pair $(2 \leftrightarrow 7)$ temporarily.
5. $s_p = (2 \leftrightarrow 6)$. There is a conflict.

Sequence	2,4,5,6	4,2,5,6	4,5,2,6	4,5,6,2
Total flow	778	856	826	**718**

   New $s_q = (4 \leftrightarrow 5 \leftrightarrow 6 \leftrightarrow 2)$.
6. $s_p = (2 \leftrightarrow 7)$. There is a conflict.

Sequence	7,4,5,6,2	4,7,5,6,2	4,5,7,6,2	4,5,6,7,2	4,5,6,2,7
Total flow	1391	1462	1337	1159	**1095**

   New $s_q = (4 \leftrightarrow 5 \leftrightarrow 6 \leftrightarrow 2 \leftrightarrow 7)$.
7. $s_p = (6 \leftrightarrow 7)$. Discard s_p since 6 and 7 are already assigned.
8. $s_p = (1 \leftrightarrow 3)$. Discard this temporarily.

9. $s_p = (1 \leftrightarrow 6)$. There is a conflict.

Sequence	1,4,5,6,2,7	4,1,5,6,2,7	4,5,1,6,2,7	4,5,6,1,2,7	4,5,6,2,1,7	4,5,6,2,7,1
Total flow	2001	2027	1861	**1667**	1758	1749

New $s_q = (4 \leftrightarrow 5 \leftrightarrow 6 \leftrightarrow 1 \leftrightarrow 2 \leftrightarrow 7)$.

10. $s_p = (1 \leftrightarrow 3)$. There is a conflict with s_q.

Sequence	3,4,5,6, 1,2,7	4,3,5,6, 1,2,7	4,5,3,6, 1,2,7	4,5,6,3, 1,2,7	4,5,6,1, 3,2,7	4,5,6,1, 2,3,7	4,5,6,1, 2,7,3
Total flow	2593	2678	2597	2536	2519	2477	**2386**

New $s_q = (4 \leftrightarrow 5 \leftrightarrow 6 \leftrightarrow 1 \leftrightarrow 2 \leftrightarrow 7 \leftrightarrow 3)$.

Stop since all stations 1–7 have already been placed in the sequence.
A summary of the method is provided below.

Iteration	Used Rank	s_p	Action Taken	Resulting Sequence, s_q	Candidates
1	1	$4 \leftrightarrow 5$	Add s_p as a new sequence	$(4 \leftrightarrow 5)$	1,2,3,6,7
2	2	$5 \leftrightarrow 6$	Insert station 6	$(4 \leftrightarrow 5 \leftrightarrow 6)$	1,2,3,7
3	3	$4 \leftrightarrow 6$	Discard permanently	$(4 \leftrightarrow 5 \leftrightarrow 6)$	1,2,3,7
4	—	$2 \leftrightarrow 7$	Discard temporarily	$(4 \leftrightarrow 5 \leftrightarrow 6)$	1,2,3,7
5	5	$2 \leftrightarrow 6$	Insert station 2	$(4 \leftrightarrow 5 \leftrightarrow 6 \leftrightarrow 2)$	1,3,7
6	4	$2 \leftrightarrow 7$	Reconsider discarded 2→7	$(4 \leftrightarrow 5 \leftrightarrow 6 \leftrightarrow 2 \leftrightarrow 7)$	1,3
7	6	$6 \leftrightarrow 7$	Discard permanently	$(4 \leftrightarrow 5 \leftrightarrow 6 \leftrightarrow 2 \leftrightarrow 7)$	1,3
8	—	$1 \leftrightarrow 3$	Discard temporarily	$(4 \leftrightarrow 5 \leftrightarrow 6 \leftrightarrow 2 \leftrightarrow 7)$	1,3
9	8	$1 \leftrightarrow 6$	Insert station 1	$(4 \leftrightarrow 5 \leftrightarrow 6 \leftrightarrow 1 \leftrightarrow 2 \leftrightarrow 7)$	3
10	7	$1 \leftrightarrow 3$	Reconsider discarded $1 \leftrightarrow 3$	$(4 \leftrightarrow 5 \leftrightarrow 6 \leftrightarrow 1 \leftrightarrow 2 \leftrightarrow 7 \leftrightarrow 3)$	0, Stop

The final sequence is $(4 \leftrightarrow 5 \leftrightarrow 6 \leftrightarrow 1 \leftrightarrow 2 \leftrightarrow 7 \leftrightarrow 3)$ with a total flow of 2386.

$$\text{Total flow} = 1 \times 69 + 3 \times 78 + 3 \times 5 + 2 \times 51 + 1 \times 76 + 2 \times 59 + 2 \times 42 + 4 \times 58$$
$$+ 3 \times 31 + 2 \times 88 + 1 \times 97 + 6 \times 6 + 5 \times 30 + 4 \times 22 + 1 \times 29 + 1 \times 99$$
$$+ 2 \times 98 + 5 \times 26 + 1 \times 99 + 4 \times 2 + 3 \times 85 = 2386$$

This solution is better (has a lower objective function value) than the solution generated by the ordinal head–tail bidirectional method.

See Supplement S13.7.xls.

Pairwise Exchange Improvement[†]   The summary of the pairwise exchange method applied to the final solution of the head–tail method is shown below.

Iteration	Pairwise Exchange	Solution	Objective Value
0	—	$4\leftrightarrow5\leftrightarrow6\leftrightarrow1\leftrightarrow2\leftrightarrow7\leftrightarrow3$	2386
1	(1,7)	$4\leftrightarrow5\leftrightarrow6\leftrightarrow7\leftrightarrow2\leftrightarrow1\leftrightarrow3$	2326
2	(2,7)	$4\leftrightarrow5\leftrightarrow6\leftrightarrow2\leftrightarrow7\leftrightarrow1\leftrightarrow3$	2285

See Supplement S13.8.xls.

Simultaneous Head–Tail Method for Bidirectional[†]   It is possible to generate a number of alternatives simultaneously in order to increase the chances of finding the optimal solution. See Sections 13.7.4 and 13.7.7 for more details.

Supplement S13.9.docx shows details of simultaneous head–tail for bidirectional layouts.

13.4.3   Multicriteria Bidirectional by Composite Approach[†]

This method is similar to the bicriteria and tricriteria unidirectional problems of Section 13.3.4. However, the second criterion, qualitative preferences, for bidirectional layouts is defined as follows. For bidirectional, the qualitative preferences (A,E,I,O,U,X) for pairs of stations indicate the amount of preference that the two stations should be close to each other. The desired closeness of each pair of stations could be based on the use of common resources in operations and maintenance. For example, the rating of $A = 5$ indicates that the two stations should be very close to each other. The problem is then solved in the same manner as in Section 13.3.4.

13.5   UNIDIRECTIONAL OPTIMIZATION

13.5.1   Optimization Method for Unidirectional Flow[†]

The head–tail methods for solving the unidirectional layout problem are computationally very fast and may generate optimal solutions in many cases. However, optimality cannot be guaranteed. Problem 13.1, ILP can be used to present a formulation of the unidirectional cell problem. For this problem formulation, the optimal solution can be found if such a solution exists. However, solving ILP problems requires substantial computational time, and therefore this approach can only be used for solving small cell layout problems.

PROBLEM 13.1   INTEGER LINEAR PROGRAMMING FOR UNIDIRECTIONAL PROBLEMS

Minimize:

$$f_1 = \sum_{i=1}^{n} \sum_{j=1, i\neq j}^{n} \left(2w_{ij}y_{ij} + w_{ij}y_{ji}\right) \tag{13.3}$$

Subject to:

$$x_i - x_j + My_{ji} \geq 1 \qquad \text{for } i, j = 1, \ldots, n \quad \text{where } i \neq j \qquad (13.4)$$

$$x_i, x_j \geq 0 \qquad \text{for } i, j = 1, \ldots, n \quad \text{where } i \neq j \qquad (13.5)$$

$$y_{ij} = 0 \text{ or } 1 \qquad \text{for } i, j = 1, \ldots, n \quad \text{where } i \neq j \qquad (13.6)$$

where M is a given large number, for example set $M = 999$ before solving the problem. Note that the objective function represents total flow, which is minimized.

The solution for the x_i variables is interpreted as follows: if $x_i < x_j$, station i is located before station j in the single-row layout sequence. Variables y_{ij} are binary; when $y_{ij} = 1$, station j is located before station i in the layout sequence.

Example 13.12   Solving Example 13.4 Using Optimization Method   Consider Example 13.4. The from–to matrix is shown again for simplicity. Formulate and solve the ILP problem.

w_{ij}	1	2	3	4
1	0	92	0	249
2	0	0	211	110
3	268	0	0	35
4	35	18	73	0

Minimize:

$$\begin{aligned} f_1 &= 2 \times (92y_{12} + 249y_{14} + 211y_{23} + 110y_{24} + 268y_{31} + 35y_{34} + 35y_{41} \\ &\quad + 18y_{42} + 73y_{43}) + (92y_{21} + 249y_{41} + 211y_{32} + 110y_{42} + 268y_{13} \\ &\quad + 35y_{43} + 35y_{34} + 18y_{24} + 73y_{34}) \end{aligned}$$

Subject to:

$$x_1 - x_2 + 999y_{21} \geq 1; \qquad x_2 - x_4 + 999y_{42} \geq 1; \qquad x_4 - x_2 + 999y_{24} \geq 1$$

$$x_1 - x_3 + 999y_{31} \geq 1; \qquad x_3 - x_1 + 999y_{13} \geq 1; \qquad x_4 - x_3 + 999y_{34} \geq 1$$

$$x_1 - x_4 + 999y_{41} \geq 1; \qquad x_3 - x_2 + 999y_{23} \geq 1; \qquad x_i \geq 0 \quad \text{for } i = 1, 2, 3, 4$$

$$x_2 - x_1 + 999y_{12} \geq 1; \qquad x_3 - x_4 + 999y_{43} \geq 1; \qquad y_{ij} = 0, 1 \quad \text{for } i, j = 1, 2, 3, 4 \; i \neq j$$

$$x_2 - x_3 + 999y_{32} \geq 1; \qquad x_4 - x_1 + 999y_{14} \geq 1;$$

The following tables are the results of solving the problem by an ILP software package:

y_{ij}	1	2	3	4
1	—	1	1	0
2	0	—	0	0
3	0	1	—	0
4	1	1	1	—

Decision variable	x_1	x_2	x_3	x_4	Objective Function Value
Solution of ILP	3	1	2	4	1309
Sequence	3	1	2	4	

Since in the optimal solution $x_2 < x_3 < x_1 < x_4$ (i.e., $1 < 2 < 3 < 4$), it means the order of the sequence is $2 \rightarrow 3 \rightarrow 1 \rightarrow 4$, where

$$\text{Total flow} = 1 \times (249 + 211 + 110 + 268 + 35) + 2 \times (92 + 35 + 18 + 73) = 1309$$

See Supplement S13.10.lg4.

Example 13.13   Using Unidirectional Optimization Method   Consider Example 13.5. The from–to matrix is shown again for simplicity. Formulate and solve the ILP problem.

			Original Unidirectional Flow (W)					
	1	2	3	4	5	6	7	8
1	0	0	97	0	0	0	0	0
2	84	0	185	182	5	101	63	0
3	0	162	0	103	0	0	0	160
4	88	0	135	0	29	0	20	0
5	0	0	0	0	0	171	123	60
6	109	0	122	0	0	0	0	0
7	0	0	63	0	169	0	0	103
8	0	0	0	0	29	0	0	0

Minimize:

$$\begin{aligned}
f_1 = {} & 2 \times (97y_{13} + 84y_{21} + 185y_{23} + 182y_{24} + 5y_{25} + 101y_{26} + 63y_{27} + 162y_{32} \\
& + 103y_{34} + 160y_{38} + 88y_{41} + 135y_{43} + 29y_{45} + 20y_{47} + 171y_{56} + 123y_{57} \\
& + 60y_{58} + 109y_{61} + 122y_{63} + 63y_{73} + 169y_{75} + 103y_{78} + 29y_{85}) \\
& + (97y_{31} + 84y_{12} + 185y_{32} + 182y_{42} + 5y_{52} + 101y_{62} + 63y_{72} + 162y_{23} \\
& + 103y_{43} + 160y_{83} + 88y_{14} + 135y_{34} + 29y_{54} + 20y_{74} + 171y_{65} + 123y_{75} \\
& + 60y_{85} + 109y_{16} + 122y_{36} + 63y_{37} + 169y_{57} + 103y_{87} + 29y_{58})
\end{aligned}$$

Subject to:

$$x_1 - x_2 + 999y_{21} \geq 1; \qquad x_3 - x_1 + 999y_{13} \geq 1$$
$$x_1 - x_3 + 999y_{31} \geq 1; \qquad x_3 - x_2 + 999y_{23} \geq 1$$
$$x_1 - x_4 + 999y_{41} \geq 1; \qquad x_3 - x_4 + 999y_{43} \geq 1$$
$$x_1 - x_5 + 999y_{51} \geq 1; \qquad x_3 - x_5 + 999y_{53} \geq 1$$
$$x_1 - x_6 + 999y_{61} \geq 1; \qquad x_3 - x_6 + 999y_{63} \geq 1$$
$$x_1 - x_7 + 999y_{71} \geq 1; \qquad x_3 - x_7 + 999y_{73} \geq 1$$
$$x_1 - x_8 + 999y_{81} \geq 1; \qquad x_3 - x_8 + 999y_{83} \geq 1$$
$$x_2 - x_1 + 999y_{12} \geq 1; \qquad x_4 - x_1 + 999y_{14} \geq 1$$
$$x_2 - x_3 + 999y_{32} \geq 1; \qquad x_4 - x_2 + 999y_{24} \geq 1$$
$$x_2 - x_4 + 999y_{42} \geq 1; \qquad x_4 - x_3 + 999y_{34} \geq 1$$
$$x_2 - x_5 + 999y_{52} \geq 1; \qquad x_4 - x_5 + 999y_{54} \geq 1$$
$$x_2 - x_6 + 999y_{62} \geq 1; \qquad x_4 - x_6 + 999y_{64} \geq 1$$
$$x_2 - x_7 + 999y_{72} \geq 1; \qquad x_4 - x_7 + 999y_{74} \geq 1$$
$$x_2 - x_8 + 999y_{82} \geq 1; \qquad x_4 - x_8 + 999y_{84} \geq 1$$
$$x_5 - x_1 + 999y_{15} \geq 1; \qquad x_7 - x_1 + 999y_{17} \geq 1$$
$$x_5 - x_2 + 999y_{25} \geq 1; \qquad x_7 - x_2 + 999y_{27} \geq 1$$
$$x_5 - x_3 + 999y_{35} \geq 1; \qquad x_7 - x_3 + 999y_{37} \geq 1$$
$$x_5 - x_4 + 999y_{45} \geq 1; \qquad x_7 - x_4 + 999y_{47} \geq 1$$
$$x_5 - x_6 + 999y_{65} \geq 1; \qquad x_7 - x_5 + 999y_{57} \geq 1$$
$$x_5 - x_7 + 999y_{75} \geq 1; \qquad x_7 - x_6 + 999y_{67} \geq 1$$
$$x_5 - x_8 + 999y_{85} \geq 1; \qquad x_7 - x_8 + 999y_{87} \geq 1$$
$$x_6 - x_1 + 999y_{16} \geq 1; \qquad x_8 - x_1 + 999y_{18} \geq 1$$
$$x_6 - x_2 + 999y_{26} \geq 1; \qquad x_8 - x_2 + 999y_{28} \geq 1$$
$$x_6 - x_3 + 999y_{36} \geq 1; \qquad x_8 - x_3 + 999y_{38} \geq 1$$
$$x_6 - x_4 + 999y_{46} \geq 1; \qquad x_8 - x_4 + 999y_{48} \geq 1$$
$$x_6 - x_5 + 999y_{56} \geq 1; \qquad x_8 - x_5 + 999y_{58} \geq 1$$
$$x_6 - x_7 + 999y_{76} \geq 1; \qquad x_8 - x_6 + 999y_{68} \geq 1$$
$$x_6 - x_8 + 999y_{86} \geq 1; \qquad x_8 - x_7 + 999y_{78} \geq 1$$
$$x_i \geq 0 \quad \text{for all } i$$
$$y_{ij} = 0, \text{ for all } i, j; i \neq j$$

The following tables are the results of solving the problem by an ILP software package:

Variable	x_1	x_2	x_3	x_4	x_5	x_6	x_7	x_8	f_1
Value	996	0	997	1	994	995	2	998	2780
Sequence	6	1	7	2	4	5	3	8	

y_{ij}	1	2	3	4	5	6	7	8
1	0	1	0	1	1	1	1	1
2	0	0	0	0	0	0	0	1
3	1	1	0	1	1	1	1	0
4	0	1	0	0	0	1	0	1
5	1	1	1	1	0	0	1	0
6	0	1	0	1	1	0	1	1
7	1	1	0	1	0	1	0	0
8	1	1	1	1	1	1	1	0

Since in the optimal solution $x_2 < x_4 < x_7 < x_5 < x_6 < x_1 < x_3 < x_8$, it means the order of the sequence is $2\rightarrow4\rightarrow7\rightarrow5\rightarrow6\rightarrow1\rightarrow3\rightarrow8$, where

$$\begin{aligned} \text{Total flow} &= 1 \times (97 + 84 + 185 + 182 + 5 + 101 + 63 + 160 + 88 + 135 + 29 \\ &\quad + 20 + 171 + 60 + 109 + 122 + 63 + 169 + 103) + 2 \\ &\quad \times (162 + 103 + 123 + 29) = 2780 \end{aligned}$$

This is the same solution generated by the ordinal head–tail method (with pairwise exchange) (Example 13.5), the cardinal head–tail method (Example 13.6), and the simultaneous head–tail method (Example 13.7).

See Supplement S13.11.lg4.

13.5.2 Multicriteria Optimization of Unidirectional Flow[†*]

For tricriteria of multiple-objective optimization of this problem, consider the same three objectives, f_1 = total flow, f_2 = total qualitative preferences, and f_3 = total flow time, defined in the tricriteria head–tail method in Section 13.3.5. Note that the composite coefficient approach used in Section 13.3.5 is a heuristic, whereas the following approach renders the optimal solution for the multicriteria problem. Formulate the weighted multiple-objective integer linear programming (MOILP) problem as follows.

PROBLEM 13.2   MULTIPLE-OBJECTIVE INTEGER LINEAR PROGRAMMING FOR UNIDIRECTIONAL

Minimize:

$$U = w_1 f_1' + w_2 f_2' + w_3 f_3'$$

Subject to:

$$f_1 = \sum_{i=1}^{n} \sum_{j=1, j \neq i}^{n} 2w_{ij} y_{ij} + w_{ij} y_{ji} \qquad f_1' = \frac{f_1 - f_{1,\min}}{f_{1,\max} - f_{1,\min}}$$

$$f_2 = \sum_{i=1}^{n} \sum_{j=1, j \neq i}^{n} 2p_{ij} y_{ij} + p_{ij} y_{ji} \qquad f_2' = \frac{f_2 - f_{2,\min}}{f_{2,\max} - f_{2,\min}}$$

$$f_3 = \sum_{i=1}^{n} \sum_{j=1, j \neq i}^{n} 2t_{ij} y_{ij} + t_{ij} y_{ji} \qquad f_3' = \frac{f_3 - f_{3,\min}}{f_{3,\max} - f_{3,\min}}$$

Constraints of Problem 13.1

Note that in the above problem weights of objectives (w_1, w_2, w_3); flow information (w_{ij}), qualitative information (p_{ij}), and time information (t_{ij}) for all pairs i, j are known before solving the problem. The only unknown variables are y_{ij} and x_i. The minimum and maximum of each objective can be obtained by solving three single objective-function problems; these values are used to derive the normalized objective constraints (f_1', f_2', f_3'). In this W-MOILP problem, constraints are linear and the problem can be solved by an available ILP package.

Example 13.14   Optimization Method for Tricriteria Problem for Example 3.4
Consider the tricriteria problem in Example 3.4 solved by the unidirectional head–tail method. Suppose the weights of importance are $\mathbf{w} = (0.3, 0.3, 0.4)$ for the normalized objective functions, flow, closeness, and flow time, respectively. Note that based on the maximum and the minimum values of each objective function, normalized objective functions are derived as follows:

$$f_1' = \frac{f_1 - 1255}{1772 - 1255} \quad \text{or} \quad f_1 - 517 f_1' = 1255$$

$$f_2' = \frac{f_2 - 45}{55 - 45} \quad \text{or} \quad f_2 - 10 f_2' = 45$$

$$f_3' = \frac{f_3 - 81}{94 - 81} \quad \text{or} \quad f_3 - 13 f_3' = 81$$

Now formulate the multiple-objective ILP problem:

Minimize:

$$U = 0.3 f_1' + 0.3 f_2' + 0.4 f_3'$$

Subject to:

$$
\begin{aligned}
f_1 = {} & 2(92 y_{12} + 0 y_{13} + 249 y_{14} + 0 y_{21} + 211 y_{23} + 92 y_{24} + 268 y_{31} + 0 y_{32} \\
& + 35 y_{34} + 35 y_{41} + 0 y_{42} + 73 y_{43}) + (92 y_{21} + 0 y_{31} + 249 y_{41} + 0 y_{12} \\
& + 211 y_{32} + 92 y_{42} + 268 y_{13} + 0 y_{23} + 35 y_{43} + 35 y_{14} + 0 y_{24} + 73 y_{34}) \\
f_2 = {} & 2(1 y_{12} + 4 y_{13} + 1 y_{14} + 5 y_{21} + 4 y_{23} + 0 y_{24} + 0 y_{31} + 4 y_{32} + 5 y_{34} \\
& + 4 y_{41} + 5 y_{42} + 1 y_{43}) + (1 y_{21} + 4 y_{31} + 1 y_{41} + 5 y_{12} + 4 y_{32} + 0 y_{42} \\
& + 0 y_{13} + 4 y_{23} + 5 y_{43} + 4 y_{14} + 5 y_{24} + 1 y_{34}) \\
f_3 = {} & 2(2 y_{12} + 7 y_{13} + 1 y_{14} + 3 y_{21} + 6 y_{23} + 2 y_{24} + 4 y_{31} + 5 y_{32} + 10 y_{34} \\
& + 9 y_{41} + 8 y_{42} + 3 y_{43}) + (2 y_{21} + 7 y_{31} + 1 y_{41} + 3 y_{12} + 6 y_{32} + 2 y_{42} \\
& + 4 y_{13} + 5 y_{23} + 10 y_{43} + 9 y_{14} + 8 y_{24} + 3 y_{34})
\end{aligned}
$$

$$f_1 - 517 f_1' = 1255 \qquad f_2 - 10 f_2' = 45 \qquad f_3 - 13 f_3' = 81$$

$$x_1 - x_2 + 999y_{21} \geq 1; \qquad x_3 - x_1 + 999y_{13} \geq 1$$
$$x_1 - x_3 + 999y_{31} \geq 1; \qquad x_3 - x_2 + 999y_{23} \geq 1$$
$$x_1 - x_4 + 999y_{41} \geq 1; \qquad x_3 - x_4 + 999y_{43} \geq 1^\dagger$$
$$x_2 - x_1 + 999y_{12} \geq 1; \qquad x_4 - x_1 + 999y_{14} \geq 1$$
$$x_2 - x_3 + 999y_{32} \geq 1; \qquad x_4 - x_2 + 999y_{24} \geq 1$$
$$x_2 - x_4 + 999y_{42} \geq 1; \qquad x_4 - x_3 + 999y_{34} \geq 1$$

$$x_i \geq \text{ for } i = 1, 2, 3, 4$$
$$y_{ij} = 0, 1 \quad \text{for } i, j = 1, 2, 3, 4, \ i \neq j$$

The following tables show the solution:

Variables	x_1	x_2	x_3	x_4	U
Value	3	2	0	1	0.3
Sequence	4	3	1	2	

y_{ji}	1	2	3	4
1	0	1	1	1
2	0	0	1	1
3	0	0	0	0
4	0	0	1	0

Hence, the final sequence is $3 \rightarrow 4 \rightarrow 2 \rightarrow 1$.

$$f_1 = \text{total flow} = 1(268 + 35 + 35) + 2(92 + 249 + 211 + 92 + 73) = 1772$$
$$f_2 = \text{total qualitative preferences} = 1(4 + 5 + 4 + 4 + 5 + 1) + 2(1 + 1 + 4 + 5) = 45$$
$$f_3 = \text{total flow time} = 1(3 + 4 + 9 + 5 + 8 + 10) + 2(2 + 7 + 1 + 6 + 2 + 3) = 81$$

Note that this solution is different than the one obtained in Example 3.4, which is (1469, 50, 86) by the head–tail method. The optimization ILP method provides a better solution because the composite MCDM approach is a heuristic, but ILP provides an exact optimal solution.

See Supplement S13.12.lg4.

13.6 BIDIRECTIONAL OPTIMIZATION

13.6.1 Optimization Method for Bidirectional Flow[†]

In this model, it is assumed that all departments have the same size, and the distance between any two adjacent departments is the same (one unit). Define x_i as the relative location of department i from the origin (e.g., zero). Hence, the layout is determined by the solution for finding values of $x_1, x_2, \ldots, x_n$. The binary variable y_{ij} equals 1 when station i is located before station j. Note that this y_{ij} definition is the opposite of the definition used in the unidirectional optimization method of the last section. Define the distance between stations i and j as x_{ij}^+ and x_{ij}^-. Hence, the objective function represents total flow. The integer linear program is shown below.

PROBLEM 13.3   INTEGER LINEAR PROGRAMMING FOR BIDIRECTIONAL PROBLEMS

Minimize:

$$f_1 = \sum_{i=1}^{n-1} \sum_{j=i+1}^{n} w_{ij}(x_{ij}^+ + x_{ij}^-) \tag{13.7}$$

Subject to:

$$x_i - x_j + My_{ij} \geq 1 \qquad \text{for } i = 1, 2, \ldots, n-i, \; j = i+1, \ldots, n \tag{13.8}$$

$$-(x_i - x_j) + M(1 - y_{ij}) \geq 1 \qquad \text{for } i = 1, 2, \ldots, n-i, \; j = i+1, \ldots, n \tag{13.9}$$

$$x_{ij}^+ - x_{ij}^- = x_i - x_j \qquad \text{for } i = 1, 2, \ldots, n-i, \; j = i+1, \ldots, n \tag{13.10}$$

$$x_i, x_j, x_{ij}^+, x_{ij}^- \geq 0 \qquad \text{for } i = 1, 2, \ldots, n-i, \; j = i+1, \ldots, n \tag{13.11}$$

$$y_{ij} = 0, 1 \qquad \text{for } i = 1, 2, \ldots, n-i, \; j = i+1, \ldots, n \tag{13.12}$$

In this model, w_{ij} is the total flow between station i and j and can also be defined as $w_{ij} = c_{ij}f_{ij}$, where c_{ij} is the cost per unit and f_{ij} is the total flow for a pair of stations i and j. The above formulation is based on Heragu and Kusiak (1990).

The method can be summarized as follows:

1. Consider the given from–to flow matrix (W). Generate the combined-flow matrix, where the new w_{ij} is equal to $w_{ij} + w_{ji}$. Then set $w_{ji} = 0$.
2. Formulate the ILP problem.
3. Solve the problem by an appropriate available computer program for ILP.
4. Arrange the location of the stations according to the increasing order of the optimal x_i solutions. The lowest value of x_i will be the first station.

Example 13.15   Using Optimization Method for Four Objects   Consider Example 13.8. The from–to matrix is shown again for simplicity. The combined-flow matrix is calculated as follows:

	From–To Matrix					Combined–Flow Matrix			
	1	2	3	4		1	2	3	4
1	0	92	0	249	1	0	92	268	284
2	0	0	211	92	2	—	0	211	92
3	268	0	0	35	3	—	—	0	108
4	35	0	73	0	4	—	—	—	0

Minimize:

$$Z = 92(x_{12}^+ + x_{12}^-) + 268(x_{13}^+ + x_{13}^-) + 284(x_{14}^+ + x_{14}^-) + 211(x_{23}^+ + x_{23}^-)$$
$$+ 92(x_{24}^+ + x_{24}^-) + 108(x_{34}^+ + x_{34}^-)$$

Subject to:

$x_1 - x_2 + 999y_{12} \geq 1;$	$x_1 - x_2 + 999y_{12} \leq 998;$	$x_1 - x_2 - x_{12}^+ + x_{12}^- = 0$
$x_1 - x_3 + 999y_{13} \geq 1;$	$x_1 - x_3 + 999y_{13} \leq 998;$	$x_1 - x_3 - x_{13}^+ + x_{13}^- = 0$
$x_1 - x_4 + 999y_{14} \geq 1;$	$x_1 - x_4 + 999y_{14} \leq 998;$	$x_1 - x_4 - x_{14}^+ + x_{14}^- = 0$
$x_2 - x_3 + 999y_{23} \geq 1;$	$x_2 - x_3 + 999y_{23} \leq 998;$	$x_2 - x_3 - x_{23}^+ + x_{23}^- = 0$
$x_2 - x_4 + 999y_{24} \geq 1;$	$x_2 - x_4 + 999y_{24} \leq 998;$	$x_2 - x_4 - x_{24}^+ + x_{24}^- = 0$
$x_3 - x_4 + 999y_{34} \geq 1;$	$x_3 - x_4 + 999y_{34} \leq 998;$	$x_3 - x_4 - x_{34}^+ + x_{34}^- = 0$

$$x_i \geq 0 \quad \text{for } i = 1, 2, 3, 4$$
$$x_{ij}^+, x_{ij}^- \geq 0 \quad \text{for } i = 1, 2, 3, j = 2, 3, 4$$
$$y_{ij} = 0, 1 \quad \text{for } i = 1, 2, 3, j = 2, 3, 4$$

Note that the constraints for Equation (13.9) in the second column have been simplified. The results of the solution are shown in the following tables:

x_1	x_2	x_3	x_4	f_1
2	0	1	3	1439

y_{ij}	1	2	3	4	x_{ij}^+	1	2	3	4	x_{ij}^-	1	2	3	4
1		0	0	1	1		2	1	0	1		0	0	1
2			1	1	2			0	0	2			1	3
3				1	3				0	3				2
4					4					4				

Since $x_2 < x_3 < x_1 < x_4$, the facility sequence is $2 \leftrightarrow 3 \leftrightarrow 1 \leftrightarrow 4$ or its mirror $4 \leftrightarrow 1 \leftrightarrow 3 \leftrightarrow 2$. This layout has the following distances for each pair of stations, where the distances are symmetric, for example, the distance from 1 to 4 is the same as the distance from 4 to 1.

d_{ij}	1	2	3	4
1		2	1	1
2			1	3
3				2
4				

Total flow can be calculated based on the distances as follows:

$$\text{Total flow} = \sum_{i=1}^{4}\sum_{j=1}^{4} w_{ij}d_{ij} = 2 \times 92 + 1 \times 268 + 1 \times 284 + 1 \times 211$$

$$+ 3 \times 92 + 2 \times 108 = 1439$$

This total flow is the same as the total flow found in Example 13.9.

 See Supplement S13.13.lg4.

Example 13.16   Using Optimization Method for Seven Stations   Consider Example 13.10, a seven-station problem with the following combined-flow matrix:

	1	2	3	4	5	6	7
1	0	69	78	5	51	76	59
2		0	42	58	31	88	97
3			0	6	30	22	29
4				0	99	98	26
5					0	99	2
6						0	85
7							0

The bidirectional optimization problem is formulated as follows.

Minimize:

$$f_1 = 69(x_{12}^{+} + x_{12}^{-}) + 78(x_{13}^{+} + x_{13}^{-}) + 5(x_{14}^{+} + x_{14}^{-}) + 51(x_{15}^{+} + x_{15}^{-}) + 76(x_{16}^{+} + x_{16}^{-})$$
$$+ 59(x_{17}^{+} + x_{17}^{-}) + 42(x_{23}^{+} + x_{23}^{-}) + 58(x_{24}^{+} + x_{24}^{-}) + 31(x_{25}^{+} + x_{25}^{-})$$
$$+ 88(x_{26}^{+} + x_{26}^{-}) + 97(x_{27}^{+} + x_{27}^{-}) + 6(x_{34}^{+} + x_{34}^{-}) + 30(x_{35}^{+} + x_{35}^{-}) + 22(x_{36}^{+} + x_{36}^{-})$$
$$+ 29(x_{37}^{+} + x_{37}^{-}) + 99(x_{45}^{+} + x_{45}^{-}) + 98(x_{46}^{+} + x_{46}^{-}) + 26(x_{47}^{+} + x_{47}^{-})$$
$$+ 99(x_{56}^{+} + x_{56}^{-}) + 2(x_{57}^{+} + x_{57}^{-}) + 85(x_{67}^{+} + x_{67}^{-})$$

Subject to:

$$
\begin{array}{lll}
x_1 - x_2 + 999y_{12} \geq 1; & x_1 - x_2 + 999y_{12} \leq 998; & x_1 - x_2 - x_{12}^{+} + x_{12}^{-} = 0 \\
x_1 - x_3 + 999y_{13} \geq 1; & x_1 - x_3 + 999y_{13} \leq 998; & x_1 - x_3 - x_{13}^{+} + x_{13}^{-} = 0 \\
x_1 - x_4 + 999y_{14} \geq 1; & x_1 - x_4 + 999y_{14} \leq 998; & x_1 - x_4 - x_{14}^{+} + x_{14}^{-} = 0 \\
x_1 - x_5 + 999y_{15} \geq 1; & x_1 - x_5 + 999y_{15} \leq 998; & x_1 - x_5 - x_{15}^{+} + x_{15}^{-} = 0 \\
x_1 - x_6 + 999y_{16} \geq 1; & x_1 - x_6 + 999y_{16} \leq 998; & x_1 - x_6 - x_{16}^{+} + x_{16}^{-} = 0 \\
x_1 - x_7 + 999y_{17} \geq 1; & x_1 - x_7 + 999y_{17} \leq 998; & x_1 - x_7 - x_{17}^{+} + x_{17}^{-} = 0
\end{array}
$$

$$x_2 - x_3 + 999y_{23} \geq 1; \qquad x_2 - x_3 + 999y_{23} \leq 998; \qquad x_2 - x_3 - x_{23}^+ + x_{23}^- = 0$$
$$x_2 - x_4 + 999y_{24} \geq 1; \qquad x_2 - x_4 + 999y_{24} \leq 998; \qquad x_2 - x_4 - x_{24}^+ + x_{24}^- = 0$$
$$x_2 - x_5 + 999y_{25} \geq 1; \qquad x_2 - x_5 + 999y_{25} \leq 998; \qquad x_2 - x_5 - x_{25}^+ + x_{25}^- = 0$$
$$x_2 - x_6 + 999y_{26} \geq 1; \qquad x_2 - x_6 + 999y_{26} \leq 998; \qquad x_2 - x_6 - x_{26}^+ + x_{26}^- = 0$$
$$x_2 - x_7 + 999y_{27} \geq 1; \qquad x_2 - x_7 + 999y_{27} \leq 998; \qquad x_2 - x_7 - x_{27}^+ + x_{27}^- = 0$$
$$x_3 - x_4 + 999y_{34} \geq 1; \qquad x_3 - x_4 + 999y_{34} \leq 998; \qquad x_3 - x_4 - x_{34}^+ + x_{34}^- = 0$$
$$x_3 - x_5 + 999y_{35} \geq 1; \qquad x_3 - x_5 + 999y_{35} \leq 998; \qquad x_3 - x_5 - x_{35}^+ + x_{35}^- = 0$$
$$x_3 - x_6 + 999y_{36} \geq 1; \qquad x_3 - x_6 + 999y_{36} \leq 998; \qquad x_3 - x_6 - x_{36}^+ + x_{36}^- = 0$$
$$x_3 - x_7 + 999y_{37} \geq 1; \qquad x_3 - x_7 + 999y_{37} \leq 998; \qquad x_3 - x_7 - x_{37}^+ + x_{37}^- = 0$$
$$x_4 - x_5 + 999y_{45} \geq 1; \qquad x_4 - x_5 + 999y_{45} \leq 998; \qquad x_4 - x_5 - x_{45}^+ + x_{45}^- = 0$$
$$x_4 - x_6 + 999y_{46} \geq 1; \qquad x_4 - x_6 + 999y_{46} \leq 998; \qquad x_4 - x_6 - x_{46}^+ + x_{46}^- = 0$$
$$x_4 - x_7 + 999y_{47} \geq 1; \qquad x_4 - x_7 + 999y_{47} \leq 998; \qquad x_4 - x_7 - x_{47}^+ + x_{47}^- = 0$$
$$x_5 - x_6 + 999y_{56} \geq 1; \qquad x_5 - x_6 + 999y_{56} \leq 998; \qquad x_5 - x_6 - x_{56}^+ + x_{56}^- = 0$$
$$x_5 - x_7 + 999y_{57} \geq 1; \qquad x_5 - x_7 + 999y_{57} \leq 998; \qquad x_5 - x_7 - x_{57}^+ + x_{57}^- = 0$$
$$x_6 - x_7 + 999y_{67} \geq 1; \qquad x_6 - x_7 + 999y_{67} \leq 998; \qquad x_6 - x_7 - x_{67}^+ + x_{67}^- = 0$$

$$x_i \geq 0, \quad \text{for all } i \; x_{ij}^+, x_{ij}^- \geq 0 \quad \text{for all } i, j \; y_{ij} = 0, 1 \quad \text{for all } i, j$$

The solution is shown in the following tables:

x_1	x_2	x_3	x_4	x_5	x_6	x_7	f_1
1	3	0	6	5	4	2	2285

y_{ij}	1	2	3	4	5	6	7	x_{ij}^+	1	2	3	4	5	6	7	x_{ij}^-	1	2	3	4	5	6	7
1		1	0	1	1	1	1	1		0	1	0	0	0	0	1		2	0	5	4	3	1
2			0	1	1	1	0	2			3	0	0	0	1	2			0	3	2	1	0
3				1	1	1	1	3				0	0	0	0	3				6	5	4	2
4					0	0	0	4					1	2	4	4					0	0	0
5						0	0	5						1	3	5						0	0
6							0	6							2	6							0
7								7								7							

Since $x_3 < x_1 < x_7 < x_2 < x_6 < x_5 < x_4$, the facility sequence is $3 \leftrightarrow 1 \leftrightarrow 7 \leftrightarrow 2 \leftrightarrow 6 \leftrightarrow 5 \leftrightarrow 4$ (or $4 \leftrightarrow 5 \leftrightarrow 6 \leftrightarrow 2 \leftrightarrow 7 \leftrightarrow 1 \leftrightarrow 3$). For this layout the distances for each pair of stations are presented below.

d_{ij}	1	2	3	4	5	6	7
1	0	2	1	5	4	3	1
2		0	3	3	2	1	1
3			0	6	5	4	2
4				0	1	2	4
5					0	1	3
6						0	2
7							0

The total flow can be calculated using the distances and flows as follows:

$$
\begin{aligned}
\text{Total flow} = {} & (2 \times 69) + (1 \times 78) + (5 \times 5) + (4 \times 51) + (3 \times 76) + (1 \times 59) \\
& + (3 \times 42) + (3 \times 58) + (2 \times 31) + (1 \times 88) + (1 \times 97) + (6 \times 6) \\
& + (5 \times 30) + (4 \times 22) + (2 \times 29) + (1 \times 99) + (2 \times 98) + (4 \times 26) \\
& + (1 \times 99) + (3 \times 2) + (2 \times 85) = 2285
\end{aligned}
$$

 See Supplement S13.14.lg4.

13.6.2 Multicriteria Optimization for Bidirectional Flow[†]

This method is similar to the bicriteria and tricriteria optimization method for unidirectional layout problems covered in Section 13.5.2.

Supplement S13.15.docx shows details of multicriteria optimization for bidirectional layout.

13.7 COMBINATORIAL OPTIMIZATION BY HEAD–TAIL

Generally, it is not possible to find verifiable optimal solutions for single-objective combinatorial problems; therefore, one cannot also guarantee finding the best alternative for combinatorial multiobjective optimization problems. The methods used for solving single–objective combinatorial optimization problems are usually heuristics. The most known heuristics are:

- Genetic algorithm
- Tabu search
- Simulated annealing

The head–tail methods developed in this chapter can also be used for solving generic single- and multiple-objective combinatorial problems. The key in using the head–tail method is the ability to formulate the given combinatorial problem in the form of pairwise comparisons of their decision variables. In Sections 13.3 and 13.4, we showed how to present cell layout (uni- and bidirectional) problems by pairwise comparisons of their decision variables. Several combinatorial single and multiobjective optimization (MOO) problems are also solved by the head–tail method in this book:

- Cellular (single-row) unidirectional—rank in descending order of flow difference of pairs
- Cellular (single-row) bidirectional—rank in descending order of flow summation of pairs
- Layout design (multirow) bidirectional—rank in descending order of flow summation of pairs
- Scheduling and sequencing, used unidirectional formulation—rank in ascending order of flows of pairs

Rules of Ranking Pairs   Depending on the application problems different ranking of pairs of object may apply; rules include largest difference rule, largest summation rule, largest flow rule, and smallest flow rule.

Types of Flows   Depending on the application problem, different types of flow may apply; flows may be unidirectional, bidirectional, linear, circular, or some combinations of them.

Traveling Salesman Problem Demonstration   To demonstrate how the head–tail method can be used to solve combinatorial problems, first we review its steps. Then, we illustrate how to formulate and solve a well-known combinatorial problem, the TSP, and its multiobjective formulation by the head–tail method.

13.7.1   Traveling Salesman Problem: Head–Tail Approach[†]

Traveling Salesman Problem with Single Objective   In the TSP, a traveling sales-man must visit n cities and return to the starting city (point of origin). The objective function is to minimize the total distance traveled by the salesman to visit all the cities. The distances between pairs of connected (also called adjacent) cities are provided. For example, suppose that one wishes to drive through all of the capital cities in Europe (on the mainland). That is, all capital cities must be visited, and the end of the trip must be the city where the trip was started. The TSP is a combinatorial problem and the approach for solving it can be used to solve other combinatorial problems. In the following, we demonstrate how to solve the TSP using the head–tail methods.

Suppose d_{ij} is the distance between the two connected (adjacent) cities i and j. The problem is to find a sequence s which visits all the cities only once and minimizes the total distance (TD):

$$\text{Minimize TD}_s = \sum_{(i,j)}^{n} d_{(i,j)} \quad \text{for all connected } (i, j) \text{ nodes for given sequence } s \quad (13.13)$$

Consider s_p = pairsequence $(i \leftrightarrow j)$, where city i is adjacent (connected) to city j. Suppose that s_q is a partially constructed sequence of cities, $s_q = (\ldots \leftrightarrow m \leftrightarrow n \leftrightarrow p \ldots)$. Note that for the TSP objective function (13.13) is used, but for a different combinatorial problem different objective functions should be used.

General Steps of Head–Tail Procedure

1. Rank all pairs (i, j) of elements (e.g., cities, nodes, or objects) for TSP rank in ascending order of distances d_{ij}. Present them by vector $\mathbf{d}_0$.
2. Start with the best pair of unassigned elements (e.g., the two cities with the short-est distance). In each iteration, add one new element (e.g., city) to the partially constructed sequence until all elements (cities) are assigned. Use one of the three head–tail approaches to construct the solution.
3. Try to improve the generated solutions by using the pairwise exchange method.

13.7.2   Ordinal (Qualitative) Head–Tail Combinatorial[†]

This method is based on the Bidirectional method covered in Section 13.4.1, which uses the bidirectional ordinal (conflict resolution) Rule.

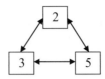

Figure 13.5   Circular presentation of three equivalent linear sequences, $(3\leftrightarrow2\leftrightarrow5)$, $(5\leftrightarrow2\leftrightarrow3)$, and $(2\leftrightarrow5\leftrightarrow3)$.

Circular and Mirroring Property   Present the circular sequence of the TSP by a linear sequence, where the two ends of the linear sequence are assumed to be connected. All linear presentations of a circular sequence are equivalent. For example, in the circular sequence $s_q = (3\leftrightarrow2\leftrightarrow5)$, elements 3 and 5 are connected. This is equivalent to its mirror, $s_q = (5\leftrightarrow2\leftrightarrow3)$. It is also equivalent to $s_q = (2\leftrightarrow5\leftrightarrow3)$. The circular view of these three linear sequences is illustrated in Figure 13.5.

Example 13.17   Traveling Salesman Problem: Six Cities Using Ordinal Head–Tail
Consider six cities A, B, C, D, E, and F, with the distances given in Figure 13.6. Note that X signifies that the two cities are not directly connected; cities A and E and also cities B and F are not connected.

See Supplement S13.16.lg4.

Solve this problem by the ordinal head–tail method. The ranking of all pairs of the Distance Matrix (d_{ij}) is shown below.

	A	B	C	D	E	F
A		1	2	3	—	10
B			5	4	7	—
C				6	8	12
D					9	11
E						13
F						

The bidirectional ordinal head–tail steps are as follows:

1. $s_p = (A\leftrightarrow B)$; then the constructed partial sequence is $s_q = (A\leftrightarrow B)$:
2. Consider the next ranked pair, $s_p = (A\leftrightarrow C)$; the new partial sequence is $s_q = (B\leftrightarrow A\leftrightarrow C)$:

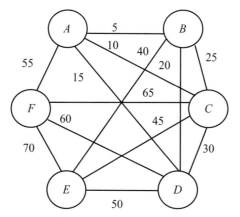

Figure 13.6   Network distances for Example 13.17

3. Consider the next ranked pair, $s_p = (A \leftrightarrow D)$; use the conflict resolution rule:

$$B \longleftrightarrow A \longleftrightarrow C$$
$$_? \nwarrow \updownarrow \nearrow _?$$
$$D$$

BD's distance is less than CD's distance; therefore the solution is $s_q = (B \leftrightarrow D \leftrightarrow A \leftrightarrow C)$.

4–6. The next three ranked pairs, $(B \leftrightarrow D)$, $(B \leftrightarrow C)$, and $(C \leftrightarrow D)$, are redundant; discard them.

7. Cyesonsider $s_p = (B \leftrightarrow E)$; use the conflict resolution rule. Show $s_q = (B \leftrightarrow D \leftrightarrow A \leftrightarrow C)$ as $(C \leftrightarrow B \leftrightarrow D \leftrightarrow A)$:

$$C \longleftrightarrow B \longleftrightarrow D \longleftrightarrow A$$
$$_? \nwarrow \updownarrow \nearrow _?$$
$$E$$

CE's distance is less than DE's distance; therefore $s_q = (C \leftrightarrow E \leftrightarrow B \leftrightarrow D \leftrightarrow A)$.

8 and 9. The next two pairs, $(C \leftrightarrow E)$, and $(D \leftrightarrow E)$, are redundant; discard them.

10. Consider $s_p = (A \leftrightarrow F)$; use conflict resolution. Show $s_q = (C \leftrightarrow E \leftrightarrow B \leftrightarrow D \leftrightarrow A)$ as $(E \leftrightarrow B \leftrightarrow D \leftrightarrow A \leftrightarrow C)$:

$$E \longleftrightarrow B \longleftrightarrow D \longleftrightarrow A \longleftrightarrow C$$
$$_? \nwarrow \updownarrow \nearrow _?$$
$$F$$

DF's distance is less than CF's distance; therefore $s_q = (E \leftrightarrow B \leftrightarrow D \leftrightarrow F \leftrightarrow A \leftrightarrow C)$.

All cities (nodes) are assigned. Stop the final solution of the ordinal head–tail method is $s_q = (E \leftrightarrow B \leftrightarrow D \leftrightarrow F \leftrightarrow A \leftrightarrow C)$. Note that E and C are connected. The total distance for this sequence is $40 + 20 + 60 + 55 + 10 + 45 = 230$.

A summary of the method is provided below.

Iteration	Used Rank	s_p	Action Taken	Resulting Sequence, s_q	Candidates
1	1	$A \leftrightarrow B$	Add s_p as a new sequence	$(A \leftrightarrow B)$	C,D,E,F
2	2	$A \leftrightarrow C$	Insert station C	$(B \leftrightarrow A \leftrightarrow C)$	D,E,F
3	3	$A \leftrightarrow D$	Insert station D	$(B \leftrightarrow D \leftrightarrow A \leftrightarrow C)$	E,F
4	—	$B \leftrightarrow D$	Discard permanently	$(B \leftrightarrow D \leftrightarrow A \leftrightarrow C)$	E,F
5	—	$B \leftrightarrow C$	Discard permanently	$(B \leftrightarrow D \leftrightarrow A \leftrightarrow C)$	E,F
6	—	$C \leftrightarrow D$	Discard permanently	$(B \leftrightarrow D \leftrightarrow A \leftrightarrow C)$	E,F
7	7	$B \leftrightarrow E$	Insert station D	$(C \leftrightarrow E \leftrightarrow B \leftrightarrow D \leftrightarrow A)$	F
8	—	$C \leftrightarrow E$	Discard permanently	$(C \leftrightarrow E \leftrightarrow B \leftrightarrow D \leftrightarrow A)$	F
9	—	$D \leftrightarrow E$	Discard permanently	$(C \leftrightarrow E \leftrightarrow B \leftrightarrow D \leftrightarrow A)$	F
10	10	$A \leftrightarrow F$	Insert station F	$(E \leftrightarrow B \leftrightarrow D \leftrightarrow F \leftrightarrow A \leftrightarrow C)$	0, Stop

13.7.3 Cardinal Head–Tail Combinatorial[†]

This method is based on the bidirectional cardinal head–tail method covered in Section 13.4.2. It uses the bidirectional cardinal (conflict resolution) rule. The following shows an example of applying this rule for TSP.

For $s_p = (2 \leftrightarrow 5)$ and $s_q = (1 \leftrightarrow 2 \leftrightarrow 3 \leftrightarrow 4)$, find the best location for 5. To do this, evaluate all four possible sequences shown in Figure 13.7 and then choose the alternative with the least total distance. The linear and circular views of the four different sequences are shown in Figure 13.7.

Example 13.18 Traveling Salesman Problem: Six Cities Using Cardinal Head–Tail
Consider Example 13.17. Solve this problem by the Cardinal head–tail method. The steps of the method are as follows:

1. $s_p = (A \leftrightarrow B)$; the constructed partial sequence is $s_q = (A \leftrightarrow B)$
2. Consider new pair $s_p = (A \leftrightarrow C)$; find the best location for C with respect to $(A \leftrightarrow B)$. The solution is $s_q = (B \leftrightarrow A \leftrightarrow C)$.
3. Consider $s_p = (A \leftrightarrow D)$; apply the cardinal conflict resolution rule and find the best location for node D. The three possible solutions are:

Sequence	$B \leftrightarrow D \leftrightarrow A \leftrightarrow C$	$B \leftrightarrow A \leftrightarrow D \leftrightarrow C$	$B \leftrightarrow A \leftrightarrow C \leftrightarrow D$
Objective value	$20 + 15 + 10 +$ $25 = 70$	$5 + 15 + 30 +$ $25 = 75$	$5 + 10 + 30 +$ $20 = \mathbf{65}$

The best solution is $B \leftrightarrow A \leftrightarrow C \leftrightarrow D$.

4–6. The next three lowest pairs, $(B \leftrightarrow D)$, $(B \leftrightarrow C)$, and $(C \leftrightarrow D)$, are redundant; discard them.

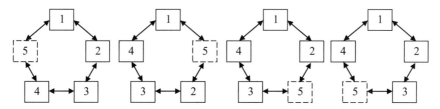

Figure 13.7 Four different linear and circular sequences for inserting node 5.

7. Consider $s_p = (B \leftrightarrow E)$. Apply the cardinal conflict resolution rule to find the best location for E; the four generated solutions are presented below. Note that $(B \leftrightarrow A \leftrightarrow C \leftrightarrow D \leftrightarrow E)$ is the same as $(E \leftrightarrow B \leftrightarrow A \leftrightarrow C \leftrightarrow D)$.

Sequence	$B \leftrightarrow E \leftrightarrow A$ $\leftrightarrow C \leftrightarrow D$	$B \leftrightarrow A \leftrightarrow E$ $\leftrightarrow C \leftrightarrow D$	$B \leftrightarrow A \leftrightarrow C$ $\leftrightarrow E \leftrightarrow D$	$B \leftrightarrow A \leftrightarrow C$ $\leftrightarrow D \leftrightarrow E$
Objective value	Infeasible	Infeasible	$5 + 10 + 45 + 50$ $+ 20 = \mathbf{130}$	$5 + 10 + 30 + 50$ $+ 40 = 135$

8 and 9. The next two pairs, $(C \leftrightarrow E)$ and $(D \leftrightarrow E)$, are redundant. Discard them.

10. Consider $s_p = (A \leftrightarrow F)$. Apply the cardinal conflict resolution rule to find the best location for F in $B \leftrightarrow A \leftrightarrow C \leftrightarrow E \leftrightarrow D$. The three feasible alternatives are shown in the following table. Note that $(B \leftrightarrow F \leftrightarrow A \leftrightarrow C \leftrightarrow E \leftrightarrow D)$ and $(F \leftrightarrow B \leftrightarrow A \leftrightarrow C \leftrightarrow E \leftrightarrow D)$ are infeasible and cannot be considered.

Sequence	$B \leftrightarrow A \leftrightarrow F \leftrightarrow C \leftrightarrow E \leftrightarrow D$	$B \leftrightarrow A \leftrightarrow C \leftrightarrow F \leftrightarrow E \leftrightarrow D$	$B \leftrightarrow A \leftrightarrow C \leftrightarrow E \leftrightarrow F \leftrightarrow D$
Objective value	$5 + 55 + 65 + 45 + 50$ $+ 20 = 240$	$5 + 10 + 65 + 70 + 50$ $+ 20 = 220$	$5 + 10 + 45 + 70 + 60$ $+ 20 = \mathbf{210}$

Because all nodes (cities) are assigned, stop. The final solution is $s_q = (B \leftrightarrow A \leftrightarrow C \leftrightarrow E \leftrightarrow F \leftrightarrow D)$. A summary of the method is provided below.

Iteration	Used Rank	s_p	Action Taken	Resulting sequence, s_q	Candidates
1	1	$A \leftrightarrow B$	Add s_p as a new sequence	$(A \leftrightarrow B)$	C,D,E,F
2	2	$A \leftrightarrow C$	Insert station C	$(B \leftrightarrow A \leftrightarrow C)$	D,E,F
3	3	$A \leftrightarrow D$	Insert station D	$(B \leftrightarrow A \leftrightarrow C \leftrightarrow D)$	E,F
4	—	$B \leftrightarrow D$	Discard permanently	$(B \leftrightarrow A \leftrightarrow C \leftrightarrow D)$	E,F
5	—	$B \leftrightarrow C$	Discard permanently	$(B \leftrightarrow A \leftrightarrow C \leftrightarrow D)$	E,F
6	—	$C \leftrightarrow D$	Discard permanently	$(B \leftrightarrow A \leftrightarrow C \leftrightarrow D)$	E,F
7	7	$B \leftrightarrow E$	Insert station E	$(B \leftrightarrow A \leftrightarrow C \leftrightarrow E \leftrightarrow D)$	F
8	—	$C \leftrightarrow E$	Discard permanently	$(B \leftrightarrow A \leftrightarrow C \leftrightarrow E \leftrightarrow D)$	F
9	—	$D \leftrightarrow E$	Discard permanently	$(B \leftrightarrow A \leftrightarrow C \leftrightarrow E \leftrightarrow D)$	F
10	10	$A \leftrightarrow F$	Insert station F	$(B \leftrightarrow A \leftrightarrow C \leftrightarrow E \leftrightarrow F \leftrightarrow D)$	0, Stop

The solution is $(B \leftrightarrow A \leftrightarrow C \leftrightarrow E \leftrightarrow F \leftrightarrow D)$ with total flow $= 5 + 10 + 45 + 70 + 60 + 20 = 210$. This is the optimal solution. Therefore, applying the pairwise exchange method will not result in a better alternative.

13.7.4   Simultaneous Head–Tail Combinatorial[†]

Since the head–tail method is a heuristic, by generating different alternative solutions, it is possible to increase the chances of finding the optimal alternative. In this method, m different alternatives are simultaneously developed. In the first iteration, start with the m highest ranked pairs of elements (e.g., cities or objects) and label them as m seeds. Augment each seed by inserting in it the best possible unassigned element. Use the best augmented seed as the seed solution for the next iteration. Repeat this process until all elements are assigned. At the end of this process there should be m different alternatives solutions to the problem. Then choose the best final solution. It is possible that some of the solutions generated by different seeds will be replicas of each other. To avoid generating replica solutions, the following rule is used. This method is based on the following two rules.

Simultaneous Optimal Insertion (Conflict Resolution) Rule   Consider element j which is not in sequence $s_q = (\dots \rightarrow i \rightarrow n \rightarrow \dots)$. Find the best location for element j by inserting j between each pair of adjacent stations in s_q. Choose the best location based on the value of the objective function (e.g., choose the alternative associated with the minimum total flow or total distance in TSP).

For example, consider inserting element 7 into $s_q = (2 \rightarrow 3 \rightarrow 4)$. Consider all possible locations that element 7 can be inserted in sequence s_q. There are four possible solutions as presented below; the best solution is $2 \rightarrow 7 \rightarrow 3 \rightarrow 4$, which will be used as the seed in the next iteration.

Sequence	$7 \rightarrow 2 \rightarrow 3 \rightarrow 4$	$2 \rightarrow 7 \rightarrow 3 \rightarrow 4$	$2 \rightarrow 3 \rightarrow 7 \rightarrow 4$	$2 \rightarrow 3 \rightarrow 4 \rightarrow 7$
Total flow	387	**261**	373	286

Simultaneous Replica Avoidance Rule   In each iteration of the simultaneous head–tail method, if a generated alternative of a given seed is a replica of one of the generated alternatives by other seeds, consider the next best new (which is not a replica) alternative with respect to the given seed. For example, suppose that with respect to the first seed solution $s_1 = (6 \leftrightarrow 1 \leftrightarrow 2)$ the best alternative is $(6 \leftrightarrow 1 \leftrightarrow 4 \leftrightarrow 2)$, that is, element 4 is inserted between 1 and 2. Also, suppose that with respect to the second seed solution $s_2 = (4 \leftrightarrow 6 \leftrightarrow 1)$ the best solution is $(6 \leftrightarrow 1 \leftrightarrow 4 \leftrightarrow 2)$, that is, element 2 is inserted after 4. This latter solution is a replica of the first solution. Therefore, instead of the second solution, the next best alternative which should not be a replica should be selected. Suppose that the next best solution with respect to the second seed, $s_2 = (4 \leftrightarrow 6 \leftrightarrow 1)$, is $(4 \leftrightarrow 6 \leftrightarrow 3 \leftrightarrow 1)$; that is, element 3 is inserted between 6 and 1. This solution is then selected for the second seed.

Example 13.19   TSP: Six Cities Using Simultaneous Head–Tail Method   Consider Example 13.17 Solve this problem by the simultaneous ordinal head–tail method by considering three ($m = 3$) simultaneous alternative solutions.

TABLE 13.1   Iteration 1 of Example 13.19

Elements		Seeds		
		(A,B)	(A,C)	(A,D)
B	Seq.	—	(A,C,B)	**(A,D,B)**
	TF	—	40	**40**
C	Seq.	**(A,B,C)**	—	(A,D,C)
	TF	**40**	—	55
D	Seq.	(A,B,D)	**(A,C,D)**	—
	TF	40	**55**	—
E	Seq.	(A,B,E)	(A,C,E)	(A,D,E)
	TF	45	55	65
F	Seq.	—	(A,C,F)	(A,D,F)
	TF	—	130	130
Best seed	Seq.	**(A,B,C)**	**(A,C,D)**	**(A,D,B)**
	TF	**40**	**55**	**40**

The first step is to determine the three initial seeds. This is accomplished by selecting the first three ranked pairs of nodes. The initial three seeds are (A, B), (A, C), and (A, D).

Now construct a table of seeds and elements as presented in Table 13.1. In iteration 1, for the first seed (A, B), the new pair is (A, C) and the new partial sequence is $s_q = (B \leftrightarrow A \leftrightarrow C)$. For the second seed (A, C), the new pair is (A, B), which is a replica. Therefore, the next available pair is (A, D) and the new partial sequence is $s_q = (C \leftrightarrow A \leftrightarrow D)$. Note that because element A is common in all three seeds, it is not shown as a row in the tables. Also, note that element E cannot be connected to A, and B cannot be connected to F, so they cannot be inserted next to each other in the solutions generated in Tables 13.1 and 13.2.

Note that in Table 13.2, (A,B,D,C) from the first column and (A,C,D,B) from the second column have the same sequence. Therefore, (A,C,E,D) is selected to avoid replica solutions. Table 13.4 below shows the final iteration, where the best solution is (A,C,E,F,D,B) with 210 total flow objective function. Note that because elements A and D are common in all

TABLE 13.2   Iteration 2 of Example 13.19

Elements		Seeds		
		(A,B,C)	(A,C,D)	(A,D,B)
B	Seq.	—	(A,C,D,B)	—
	TF	—	65	—
C	Seq.	—	—	(A,C,D,B)
	TF	—	—	65
D	Seq.	(A,B,D,C)	—	—
	TF	65	—	—
E	Seq.	(A,B,E,C)	(A,C,E,D)	(A,D,E,B)
	TF	100	120	110
F	Seq.	(A,B,C,F)	(A,C,D,F)	(A,F,D,B)
	TF	150	155	140
Best seed	Seq.	**(A,B,D,C)**	**(A,C,E,D)**	**(A,D,E,B)**
	TF	**65**	**120**	**110**

TABLE 13.3   Iteration 1 of Example 13.19

Elements		Seeds		
		(A,B,D,C)	(A,C,E,D)	(A,D,E,B)
B	Seq.	—	(A,C,E,D,B)	—
	TF	—	130	—
C	Seq.	—	—	(A,C,D,F,B)
	TF	—	—	135
E	Seq.	(A,B,D,E,C)	—	—
	TF	130	—	—
F	Seq.	(A,B,D,F,C)	(A,C,E,F,D)	(A,D,F,E,B)
	TF	160	200	190
Best seed	Seq.	(A,B,D,E,C)	(A,C,E,F,D)	(A,C,D,E,B)
	TF	130	200	135

three seeds, they are not shown as rows in Table 13.3. Similarly, A, C, D, and E a re not shown as rows in Table 13.4.

It can be verified that the obtained final solution ($A \leftrightarrow C \leftrightarrow E \leftrightarrow F \leftrightarrow D \leftrightarrow B$) is the optimal solution to this problem. Note that this solution is identical to the cardinal solution ($B \leftrightarrow A \leftrightarrow C \leftrightarrow E \leftrightarrow F \leftrightarrow D$) since the sequences are circular networks where the head and tail are connected.

13.7.5   Pairwise Exchange Improvement Method[†]

Consider the solution obtained in each iteration of the head–tail method. The solution of each iteration of the head–tail method may be improved by using the pairwise exchange method. Then the improved solution can be used in the next iteration of the head–tail method.

The pairwise exchange method was summarized in Section 13.3.3. For the TSP, consider all pairs of cities in a given solution (alternative). Exchange the locations of each pair of cities and measure the resulting objective function. Choose the solution (alternative) with the minimum objective function value. If this objective value is the same as or lower than the given (current) alternative, choose this alternative as the best alternative and repeat the process. Otherwise, stop the process.

TABLE 13.4   Iteration 2 of Example 13.19

Elements		Seeds		
		(A,B,D,E,C)	(A,C,E,F,D)	(A,C,D,E,B)
B	Seq.	—	(A,C,E,F,D,B)	—
	TF	—	210	—
F	Seq.	(A,B,D,E,C,F)	—	(A,C,D,F,E,B)
	TF	185	—	215
Best seed	Seq.	(A,B,D,E,C,F)	(A,C,E,F,D,B)	(A,C,D,F,E,B)
	TF	240	210	215

Example 13.20   Applying Pairwise Exchange to Improve a Given Solution   Consider Example 13.17, where the final solution is $E{\leftrightarrow}B{\leftrightarrow}D{\leftrightarrow}F{\leftrightarrow}A{\leftrightarrow}C$ with objective function value of 40+20+60+55+10+45=230. Apply the pairwise exchange method in an attempt to find a better solution.

There are six possible (feasible) pair exchanges in this example. These are shown in the following table. Note that when measuring the total distance, the distance from the last city to the first city must be included.

Pair of nodes	Current solution	EB	ED	BF
Generated sequence	$E{\leftrightarrow}B{\leftrightarrow}D{\leftrightarrow}$ $F{\leftrightarrow}A{\leftrightarrow}C$	$B{\leftrightarrow}E{\leftrightarrow}D{\leftrightarrow}$ $F{\leftrightarrow}A{\leftrightarrow}C$	$D{\leftrightarrow}B{\leftrightarrow}E{\leftrightarrow}$ $F{\leftrightarrow}A{\leftrightarrow}C$	$E{\leftrightarrow}F{\leftrightarrow}D{\leftrightarrow}$ $B{\leftrightarrow}A{\leftrightarrow}C$
Objective function	40+20+60+55 +10+45 = 230	40+50+60+55 +10+25 = 240	20+40+70+55 +10+30 = 225	**70+60+20+5 +10+45 = 210**

Pair of nodes	DC	BC	FA
Generated sequence	$E{\leftrightarrow}B{\leftrightarrow}C{\leftrightarrow}$ $F{\leftrightarrow}A{\leftrightarrow}D$	$E{\leftrightarrow}C{\leftrightarrow}D{\leftrightarrow}$ $F{\leftrightarrow}A$ ${\leftrightarrow}B$	$E{\leftrightarrow}B{\leftrightarrow}D{\leftrightarrow}$ $A{\leftrightarrow}F{\leftrightarrow}C$
Generated sequence	40+25+65+55 +15+50 = 250	45+30+60+55 +5+40 = 235	40+20+15+55 +65+45 = 240

Therefore, the best sequence is $E{\leftrightarrow}F{\leftrightarrow}D{\leftrightarrow}B{\leftrightarrow}A{\leftrightarrow}C$, which has the lowest objective function of 210. Now, this alternative is selected as the best solution for this iteration. The pairwise exchange method should be once again applied to this solution where $E{\leftrightarrow}F{\leftrightarrow}D{\leftrightarrow}B{\leftrightarrow}A{\leftrightarrow}C$ is used as the current alternative. The next iteration does not generate a better solution, and the pairwise exchange method is stopped. Therefore, $E{\leftrightarrow}F{\leftrightarrow}D{\leftrightarrow}B{\leftrightarrow}A{\leftrightarrow}C$ is the best alternative. Note that this solution is identical to $B{\leftrightarrow}A{\leftrightarrow}C{\leftrightarrow}E{\leftrightarrow}F{\leftrightarrow}D$ (solution of the cardinal method) and also identical to $A{\leftrightarrow}C{\leftrightarrow}E{\leftrightarrow}F{\leftrightarrow}D{\leftrightarrow}B$ (solution of the simultaneous head–tail method) because the heads and tails of sequences are connected.

13.7.6   Multiobjective Combinatorial Optimization[†]

This method can be used to solve a variety of combinatorial MOO problems. This method uses existing (available) single-objective heuristic optimization methods after it converts the MOO problem into a single-objective problem. For this method, it is assumed that objective functions are homogeneous; that is, all objective functions have the same mathematical form, but their coefficients are different. For an additive utility function, the composite approach is as follows:

1. For each objective function, find the highest and lowest coefficients and normalize all coefficients based on these values.
2. Assess the weights of importance for the normalized values of the k objective functions $w_1, w_2, \ldots$, and w_k.
3. Generate the weighted composite of objective coefficients for the given weights of importance.
4. Use an available single-objective heuristic or optimization method to solve the weighted composite problem.

The weighted composite coefficient is

$$z_{qj} = w_1 c_{1,q'j} + w_2 c_2, q'_j \cdots + w_k c_k, q'_j \quad \text{for all pairs of objects, } q, j$$

where $c_{i,qj}$ is the coefficient for objective $i = 1, 2, \ldots k$. Now use one of the head–tail methods to solve the problem. In the following example, we use the ordinal head–tail to solve this problem.

For more examples and details of this method, see the MOO approach for solving multiobjective problems for uni- and bidirectional problems in this chapter. This approach is used in several chapters of this book, including Chapters 7, 9, 15, and 16.

Example 13.21 Bicriteria Traveling Salesman Problem: Using Ordinal Head–Tail
Consider Example 13.17. Consider that there are two criteria ($k=2$) for selection of the best route: total time of travel and total cost of travel (e.g., suppose that better roads have tolls and/or their traveling time is less). The distance objective ($i=1$) and cost objective ($i=2$) matrices are shown in the following tables.

	Distance Matrix, $c_{1,qj}$						Cost Matrix, $c_{2,qj}$					
	A	*B*	*C*	*D*	*E*	*F*	*A*	*B*	*C*	*D*	*E*	*F*
A	—	5	10	15	—	55	—	13	15	2	—	14
B			25	20	40	—			9	11	8	—
C				30	45	65				3	7	4
D					50	60					20	19
E						70						20
F												

We apply the composite approach to formulate this problem given that the weights of importance for normalized coefficients are (0.6, 0.4).

The values of distance and cost coefficients are first normalized. Note that the maximum and minimum distances are max $c_{1,qj} = 70$ and min $c_{1,qj} = 5$, and the minimum and maximum costs are max $c_{2,qj} = 20$ and min $c_{2,>qj} = 2$. For example, if the distance is 15, it is normalized as $(15 - 5)/(70-5) = 0.15$.

	Distance Matrix $c_{1,qj'}$						Cost Matrix $c_{2,qj'}$					
	A	*B*	*C*	*D*	*E*	*F*	*A*	*B*	*C*	*D*	*E*	*F*
A	—	0.00	0.08	0.15	—	0.77	—	0.61	0.72	0.00	—	0.67
B			0.31	0.23	0.54	—			0.39	0.50	0.33	—
C				0.38	0.62	0.92				0.06	0.28	0.11
D					0.69	0.85					1.00	0.94
E						1.00						1.00
F												

Use $(0.6, 0.4)$ as weights of importance for the objectives and generate each component of the composite matrix by

$$z_{qj} = 0.6c'_{1,qj} + 0.4c'_{2,qj}$$

For example, $z_{23} = 0.6c_{1,23'} + 0.4c_{2,23'} = 0.6 \times 0.31 + 0.4 \times 0.39 = 0.34$. The composite matrix for this example follows:

Composite Matrix z_{qj}							Ranking of Composite Matrix (z_{qj})					
	A	B	C	D	E	F		A	B	C	D	E	F
A	—	0.24	0.34	0.09	—	0.73	A	—	2	4	1	—	10
B			0.34	0.34	0.46	—	B			5	6	7	—
C				0.25	0.48	0.60	C				3	8	9
D					0.82	0.89	D					11	12
E						1.00	E						13
F							F						

Now, this problem can be solved by any head–tail method or other heuristics. The final solution using the ordinal head–tail heuristic is $B \leftrightarrow A \leftrightarrow D \leftrightarrow F \leftrightarrow C \leftrightarrow E$. For this solution:

f_1, total distance $= 5 + 15 + 60 + 65 + 45 + 40 = 230$

f_2, total cost $= 13 + 2 + 19 + 4 + 7 + 8 = 53$

Now try to improve the solution, $B \leftrightarrow A \leftrightarrow D \leftrightarrow F \leftrightarrow C \leftrightarrow E$, by the pairwise exchange method. There are six possible pairwise exchanges as shown in the following table:

Pair of nodes	Current solution	CE	AE	CF
Generated sequence $U = 0.6f'_1 + 0.4f'_2$ (f_1, f_2)	$B \leftrightarrow A \leftrightarrow D \leftrightarrow$ $F \leftrightarrow C \leftrightarrow E$ $0.6 \times 3.08 + 0.4 \times$ $2.27 = 2.76$ $(230,53)$	$B \leftrightarrow A \leftrightarrow D \leftrightarrow$ $F \leftrightarrow E \leftrightarrow C$ $0.6 \times 2.93 + 0.4 \times$ $3.22 = 3.05$ $(220,70)$	$B \leftrightarrow E \leftrightarrow D \leftrightarrow$ $F \leftrightarrow C \leftrightarrow A$ $0.6 \times 3.08 + 0.4 \times$ $3.71 = 3.33$ $(230,79)$	$B \leftrightarrow A \leftrightarrow D \leftrightarrow$ $C \leftrightarrow F \leftrightarrow E$ **$0.6 \times 2.99 + 0.4 \times$** **$2.11 = 2.64$** **$(225,50)$**

Pair of nodes	BF	AD	DF
Generated sequence $U = 0.6f'_1 + 0.4f'_2$ (f_1, f_2)	$F \leftrightarrow A \leftrightarrow D \leftrightarrow B \leftrightarrow$ $C \leftrightarrow E$ $0.6 \times 3.08 + 0.4 \times$ $2.84 = 2.98$ $(230,43)$	$B \leftrightarrow D \leftrightarrow A \leftrightarrow F \leftrightarrow$ $C \leftrightarrow E$ $0.6 \times 3.23 + 0.4 \times$ $1.89 = 2.69$ $(240,46)$	$B \leftrightarrow A \leftrightarrow F \leftrightarrow D \leftrightarrow$ $C \leftrightarrow E$ $0.6 \times 3.16 + 0.4 \times$ $2.89 = 3.05$ $(235,64)$

The best (minimum) utility function is 2.64, associated with $B \leftrightarrow A \leftrightarrow D \leftrightarrow C \leftrightarrow F \leftrightarrow E$.

In an attempt to find a better solution to this problem, one can use the Pairwise exchange method. However, in this example, the obtained solution cannot be improved by using the Pairwise Exchange method.

13.7.7   Computational Efficiency of Head–Tail Methods[†]

Comparisons of Solved Examples   In general, the ranking of the four methods in order of their ability to find better solutions is optimization, simultaneous, cardinal, and ordinal. This ranking is the reverse in terms of the computational time required to solve given problems. In all examples that we solved, the four head–tail solutions were either optimal or lead to optimal solutions by using the pairwise exchange method. Consider unidirectional Example 13.5 (with eight stations), which was solved by all four methods. The following table is a summary of results where boldface signifies the optimal solution.

Method	Ordinal	Cardinal	Simultaneous	Optimization
Sequence	$2\to4\to7\to5\to6$ $\to1\to3\to8$	$2\to7\to4\to5\to6$ $\to1\to3\to8$	$2\to4\to7\to5\to6$ $\to1\to3\to8$	$2\to4\to7\to5\to6$ $\to1\to3\to8$
Total flow	**2780**	**2800**	**2780**	**2780**
Pairwise exchange	—	2780	—	—

The bidirectional Example 13.10 (with seven stations) was also solved by all four methods; the summary of the solutions is presented below.

Method	Ordinal	Cardinal	Simultaneous	Optimization
Sequence	$4\leftrightarrow5\leftrightarrow6\leftrightarrow1\leftrightarrow$ $3\leftrightarrow2\leftrightarrow7$	$4\leftrightarrow5\leftrightarrow6\leftrightarrow1\leftrightarrow$ $2\leftrightarrow7\leftrightarrow3$	$4\leftrightarrow5\leftrightarrow6\leftrightarrow2\leftrightarrow$ $7\leftrightarrow1\leftrightarrow3$	$4\leftrightarrow5\leftrightarrow6\leftrightarrow2\leftrightarrow$ $7\leftrightarrow1\leftrightarrow3$
Total flow	2519	2386	**2285**	**2285**
Pairwise exchange	**2285**	**2285**	—	—

Number of Problems Solved by Each Head–Tail Method   One way to compare the three head–tail approaches is by considering the number of problems that must be solved by each method, where the location of objects and the objective function (e.g., the total flow) value for a given alternative is calculated. In some applications (e.g., see Chapter 7) measuring each objective function value requires relatively high computational time.

1. **Ordinal Head–Tail**   The ordinal head–tail method uses the least computational time (excluding the sorting computational times). For a problem with n stations, the ordinal head–tail method solves one problem in each iteration in order to find the best location for the insertion of a new station. The maximum number of iterations is $n-1$; therefore, $n-1$ problems should be solved where the size of the problems varies from 2 to n. To solve each problem, however, it only checks the sign of the flow difference.

2. **Cardinal Head–Tail** The cardinal head–tail method also solves $n - 1$ problems but it requires more computation time than the ordinal method because to find the location for each insertion one must calculate the objective function (e.g., total flow), which may be a complex function.

3. **Simultaneous Head–Tail** Consider a problem with n stations and m simultaneous solutions. For example, for $n = 7$ and $m = 3$, in the first iteration $(7 - 2) \times 3 = 15$ problems and in the second iteration $(7 - 3) \times 3 = 12$ problems need to be solved, and so on. In general, in the first iteration $(n - 2)m$ and in the second iteration $(n - 3)*m$, problems need to be solved, and so on. Therefore, the total number of problems is $m \sum_{i=2}^{n-1} (n - i)$. Each problem involves calculating the objective function as in the cardinal head–tail problem, which may be complex functions.

4. **Optimization** This method finds the optimal solution for the given formulated integer programming problem. Only small-sized problems can be solved by integer programming algorithms due to their substantial computational requirements.

Complexity Analysis of Head–Tail Methods Computational efficiency shows how much computer memory space and time are required by each iteration of the algorithm. Big-O notation is used to show the time complexity of an algorithm which usually depends on the size of inputs presented by n. Suppose that the input size of the algorithm approaches infinity. Therefore, the limiting behavior of the algorithm can be described when it applies to very large problems. In this case, multiplicative constants and lower order terms are ignored in evaluating time complexity.

There are different classes of time complexities, for example, $O(n)$ represents linear time, $O(\log n)$ represents logarithmic time, and $O(n \times \log n)$ presents linear–logarithmic time. Note that $O(\log n)$ uses much less time than $O(n \log n)$. Note that $O(n^m)$ is still considered polynomial regardless of how large m is. However, $O(e^n)$ has exponential order because as n increases, the computational time increases exponentially. Three head–tail methods presented in this chapter require very small polynomial time as shown in the following. But, it was not verified that they obtain the optimal solution, that is, they are heuristics.

The complexity analysis for the head–tail methods are based on $n \times n$ number of inputs (the size of the from–to matrix) where n is the number of stations, so the problem size is $k = n^2$. Sorting the flow matrix needs $O(k \log k) = O(n^2 \log (n))$; this is the time needed by the best sorting algorithms. Calculation time for the objective function (e.g., total flow) is $O(k)$ for problems presented in this chapter.

- The ordinal method (bi- or unidirectional) is $O(k)$ without sorting, but with sorting the total is $O(k \log k)$. Note that $O(k) + O(k \log k)$ has the same order of $O(k \log k)$ since the $(k \log k)$ is a higher order term than k and it takes over the order. (Only the highest order terms are needed to present the order.)
- The cardinal method (bi- or unidirectional) is $O(k^2)$ without sorting, but with sorting the total is $O(k^2 + k \log k)$, which is the same as $O(k^2)$.
- The simultaneous method (bi- or unidirectional) is $O(mn^3)$ or $O(mk^{1.5})$ without sorting, but with sorting the total is $O(mk(k^{0.5} + \log k))$ which is $O(mk^{1.5})$ because $O(k \log k)$ is less than $O(k^{1.5})$ but it is greater than $O(n)$.

REFERENCES

General

Afentakis, P. "A Loop Layout Design Problem for Flexible Manufacturing Systems." *International Journal of Flexible Manufacturing Systems*, Vol. 1, 1989, pp. 175–196.

Buzacott, J. A. and D. D. Yao. "Flexible Manufacturing Systems: A Review of Analytical Models." *Management Science*, Vol. 32, No. 7, 1986, pp. 890–905.

Francis, R. L., F. M. Ginnis, and J. A. White. *Facility Layout and Location: An Analytical Approach*, 2nd ed., Englewood Cliffs. NJ: Prentice Hall, 1991.

Heragu, S. S. *Facilities Design*, 3rd ed. Boca Raton, FL: CRC Press; 2008.

Hyer, L., and K. H. Brown. "The Discipline of Real Cells." *Journal of Operations Management*, Vol. 17, No. 5, 1999, pp. 557–574.

Kouvelis, P., and M. W. Kim. "Unidirectional Loop Network Layout Problem in Automated Manufacturing Systems." *Operations Research*, Vol. 40, No. 3 1992, pp. 533–550.

Mahdavi, I., and B. Mahadevan. "CLASS: An Algorithm for Cellular Manufacturing System and Layout Design Using Sequence Data." *Robotics and Computer-Integrated Manufacturing*, Vol. 24, No. 3, 2008, pp. 488–497.

Malakooti, B. "Unidirectional Loop Network Layout by a LP Heuristic and Design of Telecommunications Networks." *Journal of Intelligent Manufacturing*, Vol. 15, No. 1, 2004, pp. 117–124.

Malakooti, B., and N. Malakooti. "Some Insights into Malakooti et al. *Integrated Group Technology, Cell Formation and Process Planning' Approach*." *International Journal of Production Research*, Vol. 45, No. 8, 2007, pp. 1933–1936.

Malakooti, B., and Z. Yang. "An Unsupervised Neural Network Approach for Machine-Part Cell Design." *ICNN'94 Proceedings of IEEE International Conference on Neural Networks*, Vol. II, 1994, pp. 665–670.

Malakooti, B., Z. Yang, and N. Malakooti. "Integrated Group Technology, Cell Formation, Process and Production Planning with Application to the Emergency Room." *International Journal of Production Research*, Vol. 42, 2004, pp. 1769–1786.

Malakooti, B., "Solving Combinatorial Uni & Bi-Directional, and Traveling Salesman Problems by Head-Tails Methods: Single and Multiple Objective Optimization ." *Case Western Reserve University, Cleveland, Ohio 2013*.

Morris, J. S., and R. J. Tersine. "A Comparison of Cell Loading Practices in Group Technology." *Journal of Manufacturing and Operations Managememt*, Vol. 2, No. 4, 1989, p. 299.

Morris, J. S., and R. J. Tersine. "A Simulation Analysis of Factors Influencing the Attractiveness of Group Tech-nology Cellular Layout." *Management Science*, Vol. 36, 1990, pp. 1567–1578.

Nahmias, S. *"Production and Operations Analysis,"* 6th ed. New York: McGraw-Hill/Irwin 2008.

Offodile, O. F., A. Mehrez, and J. Grznar. "Cellular Manufacturing: A Taxonomic Review Framework." *Journal of Manufacturing Systems*, Vol. 13, No. 3, 1994, pp. 196–225.

Sassani, F. "A Simulation Study on Performance Improvement of Group Technology Cells." *International Journal of Production Research*, Vol. 28, 1990, pp. 293–300.

Sunderesh, S. H., and K. Andrew. "Machine Layout: An Optimization and Knowledge-Based Approach." The International Journal of production Research, Vol. 28, No. 4, 1990, pp. 615–635.

Selim, H. M., R. G. Askin, and A. J. Vakharia. "Cell Formation in Group Technology: Review, Evaluation and Directions for Future Research." *Computers & Industrial Engineering*, Vol. 34, No. 1, 1998, pp. 3–20.

Seo, Y., D. Sheen, C. Moon, and T. Kim. "Integrated Design of Workcells and Unidirectional Flowpath Layout." *Computers & Industrial Engineering*, 51, 2006, pp. 142–153.

Singh,N. "Design of Cellular Manufacturing Systems: An Invited Review." *European Journal of Operational Research*, 69, 1993, pp. 284–291.

Stevenson, W. J., *Operations Management*, 9th ed. New York: McGraw-Hill Irwin, 2007.

Suresh, N. C., and J. M. Kay. *Group Technology and Cellular Manufacturing: A State~of-the-Art Synthesis of Research and Practice*. Boston: Kluwer Academic, 1997.

Tompkins, J. A., and J. A. White. *Facilities Planning*. New York: Wiley, 1984.

Yang, Z., and B. Malakooti. "Machine-Part Cell Formation & Operation Allocation in Cellular Manufacturing." 5th IE Research Conference, 1996, pp. 31–36.

Multiobjective Cellular Systems

Agarwal, D., S. Sahu, and P. K. Ray. "Cell Formation in Cellular Manufacturing: An AHP-Based Framework for Evaluation of Various Techniques." *International Journal of Manufacturing Technology and Management*, Vol. 5, Nos. 5–6, 2003, pp. 521–535.

Arkan, F., and Z. Gungor. "Modeling of a Manufacturing Cell Design Problem with Fuzzy Multi-objective Parametric Programming." *Mathematical and Computer Modelling*, Vol. 50, Nos. 3–4, 2009, pp. 407–420.

Bajestani, M. A., M. Rabbani, A. R. Rahimi-Vahed, and G. B. Khoshkhou. "A Multi-Objective Scatter Search for a Dynamic Cell Formation Problem." *Computers & Operations Research*, Vol. 36, No. 3, 2009, pp. 777–794.

Chan, F. T. S., and K. Abhary. "Design and Evaluation of Automated Cellular Manufacturing Systems with Simulation Modelling and AHP Approach: A Case Study." *Integrated Manufacturing Systems*, Vol. 7, No. 6, 1996, pp. 39–52.

Jayakrishnan, G. N., and T. T. Narendran. "A Bicriterion Algorithm for Cell Formation Using Ordinal and Ratio-Level Data." *International Journal of Production Research*, Vol. 37, No. 3, 1999, pp. 539– 556.

Lei, D., and Z. Wu. "Tabu Search for Multiple-Criteria Manufacturing Cell Design." *International Journal of Advanced Manufacturing Technology*, Vol. 28, 2006, pp. 950–956.

Low, C., Y. Yip, and T. H. Wu. "Modelling and Heuristics of FMS Scheduling with Multiple Objectives." *Computers & Operations Research*, Vol. 33, 2006, pp. 674–694.

Mahesh, O., and G. Srinivasan. "Multi-Objectives for Incremental Cell Formation Problem." *Annals of Operations Research*, Vol. 143, No. 1, 2006, pp. 157–170.

Malakooti, B. "Z Utility Theory : Decisions Under Risk and Multiple Criteria" to appear, 2014b.

Malakooti, B., and Z. Yang. "Machine-Part Cell Formation by Clustering Multiple Criteria Approach." *IIE Transactions*, Vol. 34, No. 9, 2002, pp. 837–846.

Mansouri, S. A., S. M. Moattar Husseini, and S. T. Newman. "A Review of the Modern Approaches to Multi-Criteria Cell Design." *International Journal of Production Research*, Vol. 38, No. 5, 2000, pp. 1201–1218.

Mansouri, S. A., S. M. Moattar-Husseini, and S. H. Zegordi. "A Genetic Algorithm for Multiple Objective Dealing with Exceptional Elements in Cellular Manufacturing." *Production Planning & Control*, Vol. 14, No. 5, 2003, pp. 437–446.

Neto, A. R. P., and E. V. G. Filho. "A Simulation-based Evolutionary Multiobjective Approach to Manufacturing Cell Formation." *Computers & Industrial Engineering*, Vol. 59, No. 1, August 2010, pp. 64–74.

Stam, A., and M. Kuula. "Selecting a Flexible Manufacturing System Using Multiple Criteria Analysis." *International Journal of Production Research*, Vol. 29, No. 4, 1991, pp. 803–820.

Yang, Z., and B. Malakooti. "A Multiple Criteria Neural Network Approach for Layout of Machine-Part Group Formation in Cellular Manufacturing." Third IE Research Conference, 1994, pp. 249–254.

Yasuda, K., L. Hu, and Y. Yin. "A Grouping Genetic Algorithm for the Multi-Objective Cell Formation Problem." *International Journal of Production Research*, Vol. 43, No. 4, 15, 2005, pp. 829–853.

EXERCISES

Unidirectional Network Problem (Section 13.2)

2.1 Consider the following two products and their production sequence requirements on three stations along with the product lot sizes given below. The given layout is $2 \rightarrow 3 \rightarrow 1$.

Product (i)	Sequence of Machines	Lot Size
I	$2 \rightarrow 1 \rightarrow 3$	15
II	$1 \rightarrow 2 \rightarrow 3$	20

   (a) Find the total flow based on loops.

   (b) Find the total flow based on from–to information.

2.2 Consider the following two products and their production sequence requirements on four stations along with the product lot sizes given below. The given layout is $4 \rightarrow 3 \rightarrow 1 \rightarrow 2$. Respond to parts (a) and (b) of exercise 2.1.

Product (i)	Sequence of Machines	Lot Size
I	$3 \rightarrow 2 \rightarrow 4 \rightarrow 1$	50
II	$2 \rightarrow 3 \rightarrow 1 \rightarrow 4$	100
III	$4 \rightarrow 1 \rightarrow 2 \rightarrow 3$	60
IV	$1 \rightarrow 3 \rightarrow 4$	45

2.3 Consider the following problem of five delivery schedules between four towns along with the number of packages to be delivered. The given layout is $4 \rightarrow 2 \rightarrow 1 \rightarrow 3$. Respond to parts (a) and (b) of exercise 2.1.

Schedule (i)	Sequence of Cities (S_i)	Delivery Size
I	$3 \rightarrow 2 \rightarrow 1$	20
II	$2 \rightarrow 4 \rightarrow 3 \rightarrow 1$	100
III	$4 \rightarrow 1 \rightarrow 2 \rightarrow 3$	35
IV	$2 \rightarrow 3 \rightarrow 4 \rightarrow 1$	50
V	$1 \rightarrow 3 \rightarrow 4$	25

Unidirectional Head–Tail Methods (Sections 3.3.1 and 3.3.2)

3.1 Consider the following flow matrix. The layout is unidirectional.

   (a) Find the best layout to minimize the total flow using the ordinal head–tail method. Then find concordance, discordance, and total flow for the obtained solution.

   (b) Find the best layout to minimize the total flow using the cardinal head–tail method. Then find concordance, discordance, and total flow for the obtained solution.

	1	2	3	4
1	—	92	0	249
2	0	—	211	110
3	268	0	—	35
4	35	18	73	—

3.2 Consider the following flow matrix. The layout is unidirectional. Respond to parts (a) and (b) of exercise 3.1.

	1	2	3	4	5
1	—	10	0	100	60
2	70	—	20	0	70
3	80	20	—	15	80
4	25	0	0	—	0
5	0	40	0	200	—

3.3 Consider the following flow matrix. The layout is unidirectional. Respond to parts (a) and (b) of exercise 3.1.

	1	2	3	4	5	6
1	—	77	83	49	0	73
2	58	—	91	52	85	69
3	44	58	—	0	55	36
4	88	94	152	—	49	0
5	143	68	88	107	—	37
6	74	83	100	148	99	—

3.4 Consider the following qualitative preferences matrix. The layout is unidirectional. Respond to parts (a) and (b) of exercise 3.1.

	1	2	3	4
1	—	A	U	E
2	E	—	I	X
3	U	I	—	U
4	X	A	O	—

3.5 Consider the following qualitative preferences matrix. The layout is unidirectional. Respond to parts (a) and (b) of exercise 3.1.

	1	2	3	4	5
1	—	U	X	E	O
2	I	—	U	X	I
3	E	U	—	U	E
4	O	X	X	—	X
5	X	O	X	A	—

Pairwise Exchange Improvement (Section 13.3.3)[†]

3.6 Consider part (b) of exercise 3.2. Consider the final solution obtained by the cardinal head–tail method. Apply the pair-wise exchange method to the final solution. Is the obtained solution better? Hint: Only perform one iteration of PWE.

3.7 Consider part (b) of exercise 3.3. Consider the final solution obtained by the cardinal head–tail method. Apply the pairwise exchange method to the final solution. Is the obtained solution better? Hint: Only perform one iteration of PWE.

Multicriteria Unidirectional by Composite Approach[†] (Section 13.3.4)

3.8 Consider exercises 3.1 and 3.4. Find the best solution for the MCDM problem when the weights are (0.4, 0.6) for flow and qualitative preferences, respectively. (Use the ordinal head–tail method.)

3.9 Consider exercises 3.2 and 3.5. Find the best solution for the MCDM problem when the weights are (0.6, 0.4) for flow and qualitative preferences, respectively. (Use the ordinal head–tail method.)

3.10 Consider exercises 3.1 and 3.4 along with the flow time matrix below. Assume the weights of importance are (0.4, 0.3, 0.3) for flow, qualitative preferences, and flow time, respectively. Find the best solutions for the tricriteria problem. (Use the ordinal head–tail method.)

	1	2	3	4
1	—	4	7	1
2	5	—	2	3
3	1	7	—	5
4	6	8	5	—

3.11 Consider exercise 3.2 and 3.5 along with the flow time matrix below. Assume the weights of importance are (0.2, 0.5, 0.3) for flow, qualitative preferences, and flow time, respectively. Find the best solutions for the tricriteria problem. (Use the ordinal head–tail method.)

	1	2	3	4	5
1	—	2	4	1	4
2	0	—	3	1	1
3	1	4	—	0	7
4	4	6	3	—	1
5	0	5	2	1	—

Ordinal and Cardinal Head–Tail for Bidirectional Flows (Sections 13.4.1 and 13.4.2)

4.1 Consider exercise 3.1. Assume the layout is bidirectional.

   (a) Find the best layout to minimize the total flow using the ordinal head–tail method.

   (b) Find the best layout to minimize the total flow using the cardinal head–tail method.

4.2 Consider exercise 3.2. Assume the layout is bidirectional. Respond to parts (a) and (b) of exercise 4.1.

4.3 Consider exercise 3.3. Assume the layout is bidirectional. Respond to parts (a) and (b) of exercise 4.1.

4.4 Consider exercise 3.4. Assume the layout is bidirectional. Respond to parts (a) and (b) of exercise 4.1.

4.5 Consider exercise 3.5. Assume the layout is bidirectional. Respond to parts (a) and (b) of exercise 4.1.

4.6 Consider exercise 4.1. Assume that the layout is circular.

   (a) Determine the best layout using the ordinal head–tail method.

   (b) Determine the best layout using the cardinal head–tail method

   (c) Compare the results from parts (a) and (b). Which method produced a better result?

4.7 Consider exercise 4.2. Assume the layout is circular. Respond to parts (a)–(c) of exercise 4.6.

4.8 Consider exercise 4.3. Assume the layout is circular. Respond to parts (a)–(c) of exercise of Problem 4.6.

Multicriteria Bidirectional by Composite Approach[†] (Section 13.4.3)

4.9 Consider exercises 3.1 and 3.4. Assume the layout is bidirectional. Find the best solution for the MCDM problem when the weights of importance of two objectives are (0.4, 0.6) for flow and qualitative preferences, respectively.

4.10 Consider exercises 3.2 and 3.5. Assume the layout is bidirectional. Find the best solution for the MCDM problem when the weights of importance of two objectives are (0.6, 0.4) for flow and qualitative preferences, respectively.

Optimization Method for Unidirectional Flow† (Section 13.5.1)

5.1 Consider exercise 3.1.

(a) Formulate the problem by integer linear programming for the unidirectional layout.

(b) Solve part (a) with an appropriate computer software.

(c) Explain the solution obtained in part (b).

5.2 Consider exercise 3.2. Respond to parts (a)–(c) of exercise 5.1.

5.3 Consider exercise 3.3. Respond to parts (a)–(c) of exercise 5.1.

Multicriteria Optimization of Unidirectional Flow† (Section 13.5.2)

5.4 Consider exercise 3.8. Find the best solution for the MCDM problem when weights of importance are (0.4, 0.6).

5.5 Consider exercise 3.9. Find the best solution for the MCDM problem when weights of importance are (0.6, 0.4).

Optimization Method for Biddirectional Flow† (Section 13.6.1)

6.1 Consider exercise 3.1. Assume the layout is bidirectional.

(a) Formulate the problem by integer linear programming.

(b) Solve part (a) with an appropriate computer software.

(c) Explain the solution obtained in part (b)

6.2 Consider exercise 3.2. Assume the layout is bidirectional. Respond to parts (a)–(c) of exercise 6.1.

6.3 Consider exercise 3.3. Assume the layout is bidirectional. Respond to parts (a)–(c) of exercise 6.1.

Ordinal (Qualitative) Head–Tail Combinatorial† (Section 13.7.2)

7.1 Consider six cities A, B, C, D, E, and F with the following Matrix (d_{ij}) of the distances between cities where X means the two cities cannot be directly connected.

d_{ij}	A	B	C	D	E	F
A	0	15	12	6	10	8
B		0	16	X	2	2
C			0	2	5	3
D				0	X	3
E					0	4
F						0

Find the minimum distance for this TSP using the ordinal head–tail method.

7.2 Consider eight cities A, B, C, D, E, F, G, and H with the following Matrix (d_{ij}) of the distances between cities where X means the two cities cannot be directly connected.

d_{ij}	A	B	C	D	E	F	G	H
A	0	18	9	19	4	19	13	X
B		0	14	10	5	2	1	19
C			0	10	18	1	8	16
D				0	17	13	17	8
E					0	18	20	14
F						0	4	X
G							0	13
H								0

Find the minimum distance for this TSP using the ordinal head–tail method.

Cardinal Head–Tail Combinatorial[†] (Section 13.7.3)

7.3 Solve exercise 7.1 using the cardinal head–tail method.

7.4 Solve exercise 7.2 using the cardinal head–tail method.

Simultaneous Head–Tail Combinatorial[†] (Section 13.7.4)

7.5 Solve exercise 7.1 using the simultaneous ordinal cardinal head–tail method. Consider two simultaneous alternatives ($m=2$).

7.6 Solve exercise 7.2 using the simultaneous ordinal cardinal head–tail method. Consider two simultaneous alternatives ($m=2$).

Pairwise Exchange Improvement Method[†] (Section 13.7.5)

7.7 Consider the final solution obtained in solving exercise 7.1. Apply two iterations of pairwise exchange to improve the final solution.

7.8 Consider the final solution obtained in solving exercise 7.2. Apply two iterations of pairwise exchange to improve the final solution.

Multiobjective Combinatorial Optimization[†] (Section 13.7.6)

7.9 Consider exercise 7.1 with the following distances and cost matrices. Suppose the weights of importance for normalized coefficients for distance and costs are $(0.3, 0.7)$, respectively.

	Distance Matrix							Cost Matrix					
d_{qj}	A	B	C	D	E	F	c_{qj}	A	B	C	D	E	F
A	0	15	12	6	10	8	A	0	5	1	1	4	3
B		0	16	X	2	2	B		0	2	X	1	1
C			0	2	5	3	C			0	4	1	2
D				0	X	3	D				0	X	1
E					0	4	E					0	3
F						0	F						0

Solve this bicriteria problem using the ordinal head–tail method. Do one iteration of pairwise exchange to improve the result. Show the final solution of the two objectives.

7.10 Consider exercise 7.2 with the following distances and cost matrices. Suppose the weights of importance for normalized coefficients for distance and costs are (0.8, 0.2), respectively.

	Distance Matrix									Cost Matrix							
d_{qj}	A	B	C	D	E	F	G	H	c_{qj}	A	B	C	D	E	F	G	H
A	0	18	9	19	4	19	13	X	A	0	5	2	4	1	4	5	X
B		0	14	10	5	2	1	19	B		0	3	2	4	3	0	5
C			0	10	18	1	8	16	C			0	4	4	4	1	3
D				0	17	13	17	8	D				0	3	3	0	3
E					0	18	20	14	E					0	5	1	5
F						0	4	X	F						0	1	X
G							0	13	G							0	5
H								0	H								0

Solve this bicriteria problem using the ordinal head–tail method. Do one iteration of pairwise exchange to improve the result. Show the final solution of the two objectives.

CHAPTER 14

ASSEMBLY SYSTEMS

14.1 INTRODUCTION

Most mass production systems have assembly layouts. Automobile factories and the clothing industry are examples of assembly systems. As the unfinished product is moved along a one-directional assembly line, the needed components are added to the semifinished product until the product assembly is completed. In this type of layout, usually only one or very few types of products are assembled through the system. Assembly system layouts are the simplest form of facility layouts as compared to single-row and multirow layouts. Assembly line balancing involves allocating various tasks of a production line to specific stations in order to minimize bottlenecks, maximize resource utilization, maximize productivity (by minimizing cycle time), minimize investment cost, and maximize qualitative preferences.

Traditionally, assembly line balancing (ALB) techniques have been used for designing assembly system layouts. Currently, the emerging technological growth and rapid changes in the market require assembly systems to be designed in such a way that frequent changes and the restructuring of assembly systems are feasible and economical to meet the changing demands of customers. For this new era of dynamic environments, new assembly system layout methods are needed.

Assembly layout systems are unidirectional single-row layouts where the flow moves in only one direction and there are no loops. Cellular system layouts are also single row layouts but products may go through several loops (see Chapter 13). An example of assembly system layout is given in Figure 14.1a where the sequence of stations is $1\rightarrow2\rightarrow3\rightarrow4$ (read $i\rightarrow j$ as unidirectional movement from station i to station j). Because assembly lines are usually long, the linear lines are connected by U-shape lines. Figure 14.1b shows a layout of 15 stations in the form of connected U shapes.

Note: Advanced material that can be omitted without loss of generality will be indicated by a dagger.

Operations and Production Systems with Multiple Objectives, First Edition. Behnam Malakooti.
© 2014 John Wiley & Sons, Inc. Published 2014 by John Wiley & Sons, Inc.

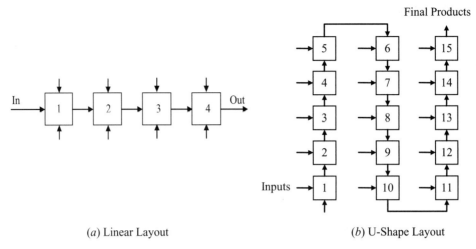

(*a*) Linear Layout   (*b*) U-Shape Layout

Figure 14.1   Unidirectional single-row assembly system layout.

14.2   ASSEMBLY LINE PROBLEM

The approach for solving the ALB problem consists of the following steps:

1. Identify all the tasks needed to assemble the product.
2. Determine the time it takes to perform each task.
3. Establish the sequence in which the tasks must be done to make a product.
4. Define the target cycle time, that is, the needed production rate.
5. Find the minimum number of stations needed to perform the tasks for the targeted cycle time.

Three different methods for solving ALB problems will be discussed in the next section. The following notation and measurements are used in ALB problems:

$$ST_j = \sum_{i \in S_j} t_i \qquad \text{station time } j$$

$$LE = \frac{\left(\sum_{j=1}^{K} ST_j\right)}{(K * CT)} \times 100\% \qquad \text{line efficiency}$$

$$SI = \sqrt{\sum_{j=1}^{K} (ST_{max} - ST_j)^2} \qquad \text{smoothness index}$$

$$K_{min} = \left\lceil \frac{\sum_{i=1}^{N} t_i}{CT} \right\rceil \qquad \begin{array}{l} \text{theoretical min. no.} \\ \text{of stations} \end{array}$$

where
N = total number of tasks
i = individual task number
    ($i = 1, \ldots, N$)
t_i = time required to complete task i
K = number of stations
j = individual station number
    ($j = 1, \ldots, K$)
S_j = set of tasks in station
    j ($j = 1, \ldots, K$)
$\lceil\ \rceil$ means round to the next highest integer

The cycle time CT is the time that elapses between two consecutively finished products at the end of the line, CT is equal to the maximum station time of all K stations, that is, the maximum ST_j.

The line efficiency is the total processing time needed to assemble the product divided by the total time the product spends on the assembly line; the ideal value of LE is 100%.

The smoothness index shows the fluctuation in idle times of the stations on the assembly line; the ideal value for SI is zero.

Multiple Objectives of ALB   There are five possible objectives for the ALB problem:

- Minimize the number of stations (K).
- Minimize cycle time (CT).
- Maximize total qualitative closeness for pairs of tasks.
- Maximize net profit.
- Minimize total cost.

In the following sections, we will discuss methods for solving ALB problems considering each of the above objectives. In this chapter, we will also discuss multiple-criteria approaches to consider multiple objectives simultaneously.

14.3   ASSEMBLY LINE BALANCING METHODS

In ALB, each station is usually supervised by one worker; therefore the total cost can be minimized by minimizing the total number of stations. Given a production requirement per period, one can find the required cycle time to meet the demand. The purpose of solving the assembly line balancing problem is to find the minimum number of stations to meet the demand (or cycle time).

ALB Problem

Minimize:	number of stations
Subject to:	given cycle time and precedence of tasks

The ALB is a combinatorial problem; therefore heuristics are used for solving this problem. In order to assign tasks to stations, different heuristics are used to rank different tasks. Three historically well-known heuristics for solving the ALB problem are the largest candidate rule, column precedence, and positional weight method. Because heuristics may not render an optimal solution, one may try all three methods. Other heuristics can also be used. The common steps of the three ALB heuristics are as follows:

1. Rank all tasks according to the given heuristic.
2. Assign the highest ranked unassigned task to the first available station if:
   (a) It is feasible with respect to the precedence constraints.
   (b) It is feasible with respect to the given cycle time, that is, the station time does not exceed the CT.

Task, i	Task Time, t_i	Precedence
1	3	None
2	4	1
3	1	1
4	4	1
5	3	2
6	1	3 and 4
7	2	5 and 6

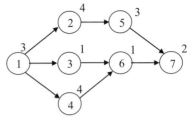

Precedence Diagram

Figure 14.2   Precedence information and precedence diagram for Example 14.1.

3. If it is not possible to add any task to the current station, start a new station.
4. Go to step 2 until all tasks are assigned.

After all assignments are completed, one can try to improve the obtained solution.

14.3.1 Largest Candidate Rule

This method was developed by Moodie and Young (1965). In this method tasks are ranked in descending order of their duration; that is, the task with the longest duration is ranked first.

Example 14.1   Assembly Line Balancing Problem (Largest Candidate Rule)   Consider a product whose assembly requires seven different tasks. The precedence information and task times (durations) as well as the precedence diagram for this problem are given in Figure 14.2. In the precedence diagram, the task number is written inside each node and the processing time is provided on the top of each node. Suppose that the cycle time is 7 min. Find the minimum number of stations using the largest candidate rule.

For this problem the theoretical minimum number of stations is three, meaning that the acutal number of stations will be three or more for the given cycle time of 7:

$$K_{\min} = \frac{\sum t_i}{\mathrm{CT}} = \frac{18}{7} = \lceil 2.57 \rceil = 3$$

The ranking of tasks based on the largest candidate rule is listed below. When there is a tie, first use the task associated with the smallest index i. For example, tasks 2 and 4 both have the same duration, but task 2 has a smaller index; therefore tasks 2 and 4 are ranked first and second. Task 6 is ranked last because it has the lowest time.

Now consider solving the ALB problem. To assign the first task to the first station, consider task 2, which is ranked the highest. However, this task cannot be assigned because its predecessor, task 1, has not yet been assigned. Therefore, in the first iteration task 1 is assigned to station 1. In the next iteration, task 2 is also assigned to station 1 because the total duration of tasks 1 and 2 does not exceed the cycle time. The assignment of the remaining tasks is shown in the table below.

Ranking			Solution				
Task i	t_i	Rank	Station j	Task i	t_i	ST_j	$CT - ST_j$
1	3	3	1	1	3	3	
2	4	1		2	4	7	$7-7=0$
3	1	6	2	4	4	4	
4	4	2		5	3	7	$7-7=0$
5	3	4	3	3	1	1	
6	1	7		6	1	2	
7	2	5		7	2	4	$7-4=3$

In this example there are three stations, so the largest candidate rule produced the minimum possible number of stations for CT = 7. Note that the maximum station time (7) is equal to the cycle time. The line efficiency and smoothness index for this solution are

$$LE = \frac{18}{3 \times 7} \times 100\% = 85.7\% \qquad SI = \sqrt{0^2 + 0^2 + 3^2} = 3$$

For this example, the layout is presented in Figure 14.3, where the flow is shown by an arrow.

14.3.2 Column Precedence

This method was developed by Kilbridgem and Wester (1961). In this method, the precedence diagram is presented by the formation of columns that have the same number of predecessors. The steps of separating a precedence diagram into columns are as follows:

1. Place all tasks with no predecessors in the first column.
2. In the next column, place all tasks whose predecessors have already been placed in the preceding columns.
3. Repeat step 2 until all tasks are assigned.

For example, Figure 14.2 is presented as Figure 14.4 where there are four columns.

When constructing the precedence diagram, if any task can also be placed in the next column without violating precedence requirements, also show this task in the next column but use a dashed circle to represent it as a possible option.

The ranking rule in this method is based on the order of columns and then in descending order of task times within each column.

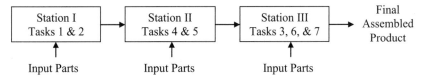

Figure 14.3 Layout of ALB for Example 14.1.

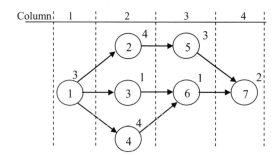

Figure 14.4 Figure 14.2 structured into four columns.

Example 14.2 Assembly Line Balancing Problem (Column Precedence Method)
Consider Example 14.1. The ranking of tasks by the column precedence method is presented below. For example, for the second column the ranking of tasks is 2, 4, and 3. Also, solving this example by the column precedence method is shown below.

Ranking			Solution				
Task i	t_i	Rank	Station j	Task i	t_i	ST_j	$CT - ST_j$
1	3	1	1	1	3	3	
2	4	2		2	4	7	$7 - 7 = 0$
3	1	4	2	4	4	4	
4	4	3		3	1	5	
5	3	5		6	1	6	$7 - 6 = 1$
6	1	6	3	5	3	3	
7	2	7		7	2	5	$7 - 5 = 2$

$$LE = \frac{18}{3 \times 7} \times 100\% = 85.7\% \qquad SI = \sqrt{0^2 + 1^2 + 2^2} = 2.24$$

This solution also has three stations; however, the assignment of this method is different than the largest candidate rule and the SI of 2.24 is better than the SI of 3 generated by the largest candidate rule.

14.3.3 Positional Weight Method

This method was developed by Helgeson and Birnie (1961). In this method each task has a positional weight. The positional weight (PW) of a task is the sum of the time required to complete the task being considered and the times of all tasks following that task that cannot be done until the current task is finished. That is,

$$PW_i = \sum_{p=i}^{n} t_p \quad \text{for all tasks } p \text{ on all paths following task } i$$

In this method tasks are ordered in descending order of their positional weights. If the rankings of two tasks are equal, then the task with the larger task time is ranked first.

Example 14.3   Assembly Line Balancing Problem (Positional Weight Method)
Consider Example 14.1. Ranking of tasks and the ALB solution by the PW method is presented below.

Task i	t_i	PW		Rank	Station j	Task i	t_i	ST_j	$CT - ST_j$
		Ranking			Solution				
1	3	$3+4+1+4+3+1+2 = 18$		1	1	1	3	3	
2	4	$4+3+2 = 9$		2		2	4	7	$7-7=0$
3	1	$1+1+2 = 4$		5	2	4	4	4	
4	4	$4+1+2 = 7$		3		5	3	7	$7-7=0$
5	3	$3+2 = 5$		4	3	3	1	1	
6	1	$1+2 = 3$		6		6	1	2	
7	2	2		7		7	2	4	$7-4=3$

In this example, the PW method solution is identical to the largest candidate rule solution of Example 14.1.

Comparison of the Three Heuristics   For Example 14.1, all three solutions produced the same number of stations ($K = 3$) and therefore the same line efficiency (LE $= 85.7\%$). The column precedence method produced a superior smoothness index (SI $= 2.24$) when compared to the other two heuristics, which produced SI $= 3$. Therefore, the solution produced by the column precedence method is the best solution for a cycle time of 7.

Supplement S14.1.docx shows an example of a larger ALB problem which is solved by the three heuristics presented in this section.

14.4   QUALITATIVE ALB: HEAD–TAIL APPROACH[†]

The objective of the classical ALB methods is to minimize the number of stations subject to the given cycle time. In many applications, however, there are different objectives that must be considered when solving ALB problems. Consider a pair of tasks. Because of several factors, it may be preferred that both tasks be assigned to the same station because they may use the same machines, materials, or workers. Such closeness preferences are qualitative in nature. The qualitative closeness for each pair of tasks can be assessed using qualitative ratings of A, E, I, O, U, and X. When solving the ALB, the qualitative ratings can be converted into numerical ratings, for example, use 5, 4, 3, 2, 1, and 0 for A, E, I, O, U, and X, respectively. Suppose for each pair of tasks (i,k), p_{ik} represents their desired closeness rating. In this section, we develop a new heuristic method called qualitative head–tail ALB to maximize the total qualitative preferences subject to a given cycle time. This method uses the head–tail heuristic in terms of ranking pairs of tasks and connecting pairs of tasks that are close to each other. The qualitative head–tail ALB method is as follows:

1. Consider the network precedence information, the given cycle time, and the qualitative closeness rating for all pairs of tasks.

2. Rank all pairs of tasks according to their qualitative closeness.
3. Assign the highest ranked unassigned task to the first available station if:
   (a) It is feasible with respect to the precedence constraints.
   (b) It is feasible with respect to the given cycle time, that is, the station time does not exceed the CT.
   (c) It has the highest closeness to the tasks in the station.
4. If it is not possible to add any task to the current station, start a new station.
5. Go to step 3 until all tasks are assigned.

Total Closeness Objective Function   The objective function of this problem, the total closeness, is calculated as follows. First, find the closeness value of each station as the total p_{ik} for all pairs of tasks i,k assigned to the station. Then find the total closeness score as the summation of the scores of all stations.

Example 14.4   Qualitative ALB Problem   Consider Example 14.1 and Figure 14.2 with a cycle time of 7 min. Also, consider the following closeness qualitative information and its converted numerical assignments for each pair of tasks i and k:

	Qualitative Closeness								Qualitative Closeness Ratings p_{ik}						
	1	2	3	4	5	6	7		1	2	3	4	5	6	7
1		X	A	U	U	U	A	1		0	5	1	1	1	5
2			E	I	A	U	U	2			4	3	5	1	1
3				X	A	I	U	3				0	5	3	1
4					X	U	U	4					0	1	1
5						O	A	5						2	5
6							A	6							5
7								7							

Station 1

1. Consider station 1. The only feasible task is task 1. Assign task 1 to station 1.
2. Assign the next feasible task with the highest rating p_{ik} to station 1. The closeness ranking of all feasible tasks is presented in the following table.
3. Assign task 3 to station 1.

Ranking: Feasible Pairs for Station 1			Current Solution				
Feasible Pairs	Closeness	Rank	Station j	Task i	t_i	ST_j	$CT - ST_j$
1,2	0	3	1	**1**	3	3	
1,3	5	1		**3**	1	4	$7 - 4 = 3$
1,4	1	2					

Station 2

4. According to the cycle time, it is not possible to assign more tasks to station 1; therefore consider the next station, 2.
5. According to precedence requirements, the only feasible tasks are 2 and 4. According to the largest candidate rule, task 2 is assigned to station 2.
6. Rank all feasible pairs that can be assigned to station 2. See the following table.
7. Assign task 5 to station 2.

Ranking: Feasible Pairs for Station 2			Current Solution				
Feasible Pairs	Closeness	Rank	Station j	Task i	t_i	ST_j	$CT - ST_j$
2,4	3	2	1	1	3	3	
2,5	5	1		3	1	4	$7 - 4 = 3$
2,6	1	3	2	2	4	4	
				5	3	7	$7 - 7 = 0$

Station 3

8. No other task is feasible to be assigned to station 2; consider station 3. Since the only feasible task is task 4, assign it to station 3.
9. Similarly, tasks 6 and 7 are the only remaining feasible tasks; assign them to station 3.

The final solution is presented below with three stations, $K = 3$, and a cycle time of $CT = 7$.

Station j	Task i	t_i	ST_j	$CT - ST_j$	$\sum p_{ik}$
1	1	3	3		
	3	1	4	$7 - 4 = 3$	**5**
2	2	4	4		
	5	3	7	$7 - 7 = 0$	**5**
3	**4**	4	4		
	6	1	5		
	7	2	7	$7 - 7 = 0$	$1 + 1 + 5 = 7$
Total closeness score					**17**

Total Closeness Calculation In this example, the total closeness is calculated as follows. In station 1, tasks 1 and 3 have a closeness score of 5. In station 2, tasks 2 and 5 have a score of 5. In station 3, for tasks 4, 6, and 7, consider all possible combinations.

That is, $p_{46} + p_{47} + p_{67} = 1 + 1 + 5 = 7$. Therefore, this layout has a total (qualitative) closeness score of:

$$5 + 5 + 7 = 17$$

Note that for this solution

$$\text{LE} = \frac{18}{3 \times 7} = 0.8571 = 85.71\% \qquad \text{SI} = \sqrt{3^2 + 0^2 + 0^2} = 3$$

The LE and SI of this solution are the same as the solution obtained in the positional weight method of Example 14.3, but their assignments are different and have different closeness scores. The total closeness for Example 14.3 is $(0 + 0 + (3 + 1 + 5)) = 9$, which is lower (worse) than the Example 14.1 solution.

As a point of reference, one can find the maximum total score by considering only one station and adding the ratings of all pairs of tasks. For the above Example 4.1, if the cycle time is 18, all tasks will be assigned to only one station. In this case, the total closeness score is 50. Therefore, the score of 17 of the above example can be compared to the ideal score of 50.

14.5 MULTICRITERIA ALB

14.5.1 Minimizing Cycle Time for Given Number of Stations

For a given number of stations, K, the minimum cycle time can be found. This can be done by a simple enumeration of all cycle times that give the same K solution and then choosing the solution that has the lowest CT. This search process can be improved by using a one-dimensional search on CT.

Consider Example 14.1, and suppose that the given cycle time is 6. For $\text{CT} = 6$, the minimum number of stations, K, is 4. Now, consider $\text{CT} = 5$; the solution to this ALB problem is $K = 4$. Therefore, $\text{CT} = 5$ and $K = 4$ is better than $\text{CT} = 6$ and $K = 4$. Now, consider $\text{CT} = 4$; the solution is $K = 5$. This means that the best (minimum) CT for $K = 4$ is $\text{CT} = 5$. The set of all efficient solutions for Example 14.6 is shown in Table 14.1.

Example 14.5   Improving Given ALB Solution   Consider the set of tasks and corresponding precedence diagram in Figure 14.5.

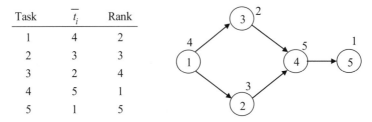

Task	$\bar{t}_i$	Rank
1	4	2
2	3	3
3	2	4
4	5	1
5	1	5

Figure 14.5   Set of tasks and precedence diagram for Example 14.5.

The given cycle time is CT = 9. Solving this problem by the largest candidate rule gives the following solution:

Station j	Task i	t_i	ST_j
1	1	4	4
	2	3	7
	3	2	9
2	4	5	5
	5	1	6

However, using the method from Section 14.5.1, it may be possible to find a two-station solution with a lower cycle time. Solving the problem for a cycle time of CT = 8, an improved solution with two stations is found as follows:

Station j	Task i	t_i	ST_j
1	1	4	4
	2	3	7
2	3	2	2
	4	5	7
	5	1	8

Supplement S14.2.docx shows details of minimizing cycle time for a given number of stations.

14.5.2 Optimizing ALB Profit

In designing the layout of an ALB problem, both the CT and the number of stations K may be unknown. That is, the long-term future demand for the product may be unknown. But information related to production expenses and profits may be available. In this method, the objective function is to maximize the total profit (TP):

$$TP = (R - O)Q - (G + L)K \qquad (14.1)$$

where R is the sale price per unit, O is the input material cost per unit, Q is the production volume per period, G is the average equipment cost per period per station, and L is the labor cost per period per station.

Suppose the production duration of each period is T^0. Then, the production volume per period, Q, and the TP are defined as follows:

$$Q = \frac{T^0}{CT} \qquad (14.2)$$

$$TP = \frac{(R - O)T^0}{CT} - (G + L)K \qquad (14.3)$$

Therefore, total profit is a function of the number of stations K and the CT. For each given K, it is possible to find the minimum CT, which corresponds to the maximum total profit. This problem can be solved by enumerating the solutions for all K values and choosing the solution associated with the highest total profit. To expedite this search, a one-dimensional search approach (e.g., bisectional method) can be used on K as the single variable. For small problems, it is easier to generate all solutions associated with different values of K and then choose the best solution that maximizes the total profit.

The method for enumeration of all alternatives for maximizing profit is as follows:

1. Set the minimum and the maximum cycle time for the problem; note that the the-oretical minimum and maximum cycle times are $CT_{min} = Max\{t_1, \ldots, t_n\}$ and $CT_{max} = $ summation for all tasks, t_i.
2. Find K_{max}, which is associated with CT_{min}, and K_{min}, which is associated with CT_{max}.
3. Set $CT_1 = CT_{min}$ and set $i = 1$.
4. For a given CT_i, solve the ALB problem to find its K_i. If a lower CT is found for the same K, only use the solution with the lowest CT.
5. Find TP_i (total profit for alternative i) for the given CT_i and K_i using Equation (14.3).
6. Set $i = i + 1$ and $CT_i = CT_{i-1} + 1$. Go to step 4.
7. Stop when $CT_i = CT_{max}$; the alternative with the highest TP_i is the optimal solution.

Example 14.6 Iterative Method for Maximizing Profit Consider the assembly system in Example 14.1 with precedence diagram given in Figure 14.2. Suppose that for each assembled product the sale price is $15 and operation cost is $5. The station (equipment) cost per period is $3000, and labor cost per period is $1000 per station. That is, $R = \$15$, $O = \$5$, $G = \$3000$, and $L = \$1000$. Suppose that each period is four weeks, and there is a total of 35 h/week. Then the production period in minutes is $T^0 = $ (4 weeks)(35 h/week) (60 min/h) = 8400 min. Therefore, the production per period is $Q = 8400/CT$.

1. The minimum cycle time is $CT_{min} = 4$ and the maximum cycle time is $CT_{max} = 18$.
2. Find the optimal solution for $CT_{min} = 4$ and $CT_{max} = 18$. For $CT_{min} = 4$, $K_{max} = 5$, and for $CT_{max} = 18$, $K_{min} = 1$. The solutions are presented below.

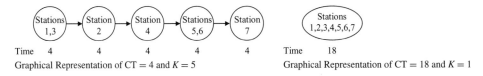

Time 4   4   4   4   4   Time 18
Graphical Representation of CT = 4 and K = 5   Graphical Representation of CT = 18 and K = 1

3. Set $CT_1 = 4$ and set $i = 1$.
4. Solving the ALB problem for $CT_1 = 4$ produces $K_1 = 5$.
5. Find total net profit for $CT_1 = 4$ and $K_1 = 5$
   $TP = (R - O)Q^0/CT_1 - (G + L)K_1 = 10 \times 8400/4 - (3000 + 1000 \times 5 = 1000$
   The total cost is $5 \times 8400/4 + 4000 \times 5 = 30{,}500$.
   The production quantity is $8400/4 = 2100$ units.
6. Set $i = 2$ and $CT_2 = 4 + 1 = 5$ and continue the algorithm.

Repeat this process until $CT_i = CT_{max} = 18$. Five efficient alternatives are generated as presented in Table 14.1.

TABLE 14.1 Set of All Efficient Solutions (for CT vs. K) Maximizing TP and Minimizing TC

	a_1	a_2	a_3	a_4	a_5
Cycle time, CT	4	5	7	9	18
No. of stations, K	5	4	3	2	1
Total net profit, TP	1,000	800	0	1,333	667
Total cost, TC	30,500	24,400	18,000	12,667	6,333

In the above example, a_4 has the largest total profit, 1333. This alternative is the optimal solution for maximizing total profit. A plot of CT versus TP is presented in Figure 14.6. According to Figure 14.5, the total net profit function is not unimodal, that is, it may have several local optima. Therefore, to find the optimal solution, all possible alternatives in terms of K must be enumerated.

14.5.3 Bicriteria ALB Problem

The Bicriteria ALB problem can be formulated as follows:

Minimize number of stations: $f_1 = K$
Minimize cycle time: $f_2 = \text{CT}$
Subject to the constraints of the ALB problem

Objectives f_1 and f_2 are in conflict with each other. That is, as the number of stations increases, the cycle time decreases.

Efficient Alternatives for K versus CT   Consider Example 14.6 and the solutions presented in Table 14.1. Figure 14.7 shows the efficient frontier for this example. For example, the alternative associated with CT $= 6$ and $K = 4$ is dominated by alternative a_2 (CT $= 6$ and $K = 4$), and therefore it is inefficient and can be eliminated. Also, the alternatives associated with CT $= 8, 10, 11, \ldots, 16, 17$ (all associated with $K = 2$) are inefficient; they are dominated by a_4 (CT $= 9$ and $K = 2$) and can be eliminated.

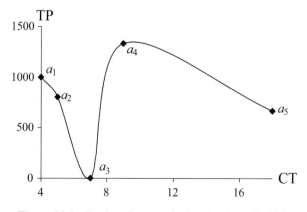

Figure 14.6   Total profit vs. cycle time for Example 14.5.

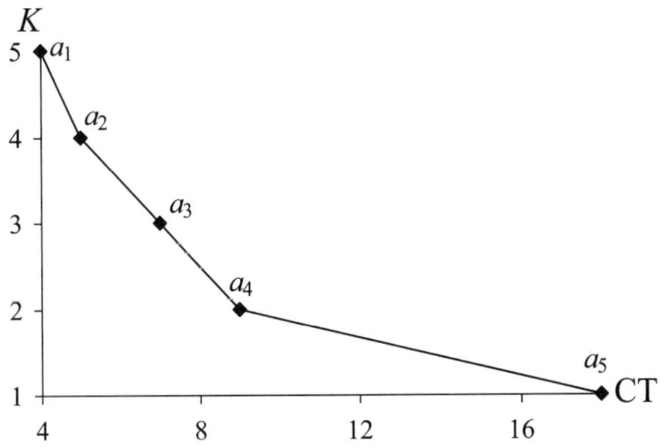

Figure 14.7 Number of stations vs. cycle time for Example 14.6.

Supplement S14.3.docx shows how to find efficient alternatives for the bicriteria ALB problem of cycle time versus total cost.

Generating All Efficient Alternatives To generate all efficient alternatives, first generate the efficient frontier in terms of CT and K and then calculate the total cost associated with the given CT and K.

Choosing Best Alternative Use a MCDM method for the selection of the best efficient alternative. For example, an additive utility function can be used to rank all alternatives as shown in the following example. If the set of efficient alternatives is very large, instead of generating all efficient alternatives, use a one-dimensional search on the number of stations, K. In this case, use the DM's utility function as the objective function for the one-dimensional search problem.

Example 14.7 Bicriteria ALB Problem Consider Example 14.6. The following table shows all five efficient alternatives. Rank alternatives using an additive utility function with weights of importance (0.45, 0.55) for the normalized objective function values f_1' and f_2', respectively.

$$\text{Minimize } U = 0.45 f_1' + 0.55 f_2'$$

The efficient MC alternatives and their ranking are presented below.

	a_1	a_2	a_3	a_4	a_5	Minimum	Maximum
Min. f_1, K	5	4	3	2	1	1	5
Min. f_2, CT	4	5	7	9	18	4	18
Min. f_1'	1	0.75	0.5	0.25	0	$w_1 = 0.45$	
Min. f_2'	0	0.07	0.21	0.36	1	$w_2 = 0.55$	
Min. U	0.45	0.38	0.34	0.31	0.55		
Rank	4	3	2	1	5		

14.5.4   Tricriteria ALB with Qualitative Closeness

In addition to the bicriteria problem covered in Section 14.3, an objective that represents qualitative closeness for pairs of tasks can be considered. The measurement of total qualitative closeness was discussed in Section 14.4. The tricriteria ALB problem is

Minimize f_1, number of stations
Minimize f_2, cycle time
Maximize f_3, qualitative closeness
Subject to the constraints of the ALB problem

This approach for solving tricriteria ALB is similar to the bicriteria method of Section 5.2.

Supplement S14.4.docx shows an example of tricriteria ALB with qualitative preferences and contains homework problems.

14.6   MIXED-PRODUCT ALB

In previous sections, the ALB problem was formulated for only one product. In this section we discuss the ALB problem where several products can be assembled on the same line. This problem is referred to as the mixed-product ALB where each product may have a different set of tasks and task precedence. Typically, products that share the same production line will have similar tasks but different precedence diagrams. To solve this problem, a common precedence diagram for all products can be created. Then the ALB methods discussed in the previous sections can be applied to solve the problem associated with the combined precedence. If in the combined precedence diagram there are two different durations for the same task, the larger value should be used, ignoring the smaller one. The details of this approach are shown in the following example.

Example 14.8   Mixed-Product ALB   Consider the precedence diagram for two products as shown in Figure 14.8. The two-product common tasks are 1, 2, 4, and 7. For a cycle time of 3 min, solve the ALB problem.

First, we create the combined precedence diagram of products 1 and 2 as shown in Figure 14.9. Note that because the task 2 duration is 1 min for product 1 and 2 min for product 2, use 2 min (the larger amount) in the combined precedence diagram.

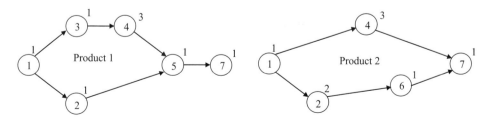

Figure 14.8   Product 1 and 2 precedence diagram.

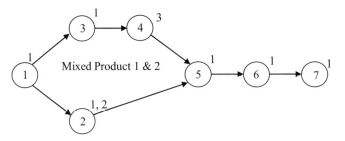

Figure 14.9   Mixed (combined) product 1 and 2 precedence diagram.

Now, use any of the ALB heuristics that were covered in previous sections to solve this problem. For the purpose of illustration, we use the positional weight method to solve this problem.

The following table shows the positional weight for each task based on Figure 14.9:

Task	1	3	4	2	5	6	7
Positional weight	10	7	6	5	3	2	1

The assignment of tasks to stations is shown in the following table:

Station j	Task i	Product 1		Product 2	
		t_i	ST_j	t_i	ST_j
1	1	1	1	1	1
	3	1	2	0	1
	2	1	3	2	3
2	4	3	3	3	3
3	5	1	1	0	0
	6	0	1	1	1
	7	1	2	1	2

If the two products were each assembled on a separate line, then for a cycle time of 3, product 1 would have three stations and product 2 would also have three stations. By sharing the same production line, either product 1 or product 2 is produced every cycle time, that is, every 3 min. Note that the production rate or cycle time remains the same for different proportions of products 1 and 2. In the above solution, when product 1 is in station 1, tasks 1, 2, and 3 are performed, but when product 2 is in station 1, tasks 1 and 3 are performed. In station 2, either product 1 or 2 is processed. Also, when product 1 is in station 3, tasks 5 and 7 are completed, but when product 2 is in station 3, tasks 6 and 7 are completed.

14.7 STOCHASTIC ALB

14.7.1 Single-Objective Stochastic ALB

In the ALB problems discussed so far, the processing time for each task was assumed to be known and fixed. That is, we considered a deterministic ALB problem. In some situations, task times may vary due to the nature of the work, type of materials, machinery, and the worker. In these cases, using stochastic times is more realistic than using deterministic times. The stochastic time accounts for variation in the duration of each task. A stochastic ALB method can be used to solve this type of ALB problem. In this method the mean (expected or average) time of each task is used to solve the problem. Then the variance of each task (assuming a normal distribution) and each station is calculated. The method establishes confidence intervals for the completion of the assigned tasks at each station. If the station time variance is large, then the approach allows more time for such variations.

Suppose $\bar{t}_i$ is the mean of task i time, V_i is the variance of task i time, and SD_i is the standard deviation of task i time. The standard deviation for each station j is

$$SD_j = \left[\sum_{i \in S_j} V_i \right]^{1/2} \quad \text{for tasks } i \text{ belonging to set of tasks in station } j, S_j \qquad (14.4)$$

For each station j, the station time is calculated as

$$ST_j = \sum_{i \in S_j} t_i + z\, SD_j \quad \text{for } j = 1, \ldots, K \qquad (14.5)$$

where S_j is the set of tasks in station j, z is the coefficient value from the normal standardized distribution associated with a certain confidence level that the actual time of task i will be within the range $\bar{t}_i - z\, SD_i$ to $\bar{t}_i + z\, SD_i$. For example, $z = 3$ is associated with a 99.74% confidence level, $z = 2$ for 95.44%, $z = 1$ for 68.26%, and $z = 0$ for 50%. For other values of z, see the normal distribution table at the end of the book. For example, if $\bar{t}_i = 5, z = 3$, and $SD_j = 2$, then there is 99.74% confidence that task i time will be within $5 - 3 \times 2$ to $5 + 3 \times 2$, or -1 to 11 min. Because the minimum time cannot be less than 0, the range is 0–11.

The method for solving stochastic ALB problems is as follows:

1. Determine the mean time $\bar{t}_i$ and variance V_i for all tasks 1, 2, ... , N. Also, identify the cycle time and the desired confidence level z for the cycle time.
2. Use one of the ALB methods (covered in Section 4.3) to rank each task using their mean values $\bar{t}_i$.
3. Assign tasks to stations according to the ALB method for the given cycle time. The station time is determined by Equation (14.5). Repeat step 3 until all tasks are assigned to all stations.

Example 14.9 Stochastic ALB   Consider Figure 14.10 where for each task the first number is the mean time $\bar{t}_i$ and the second number is its variance V_i. Given that the cycle time CT $= 10$ and the required confidence level $z = 2.5$ (i.e., 98.76%), solve the ALB problem. Use the largest candidate rule to rank the tasks based on their mean time $\bar{t}_i$.

Task	$\bar{t}_i$	V_i	Rank
1	4	2	2
2	3	1	3
3	2	2	4
4	5	1	1
5	1	0	5

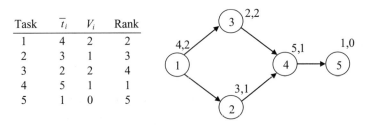

Figure 14.10   Set of tasks and precedence diagram for Example 14.9.

The ranking of each task by using $\bar{t}_i$ is given in Figure 14.10. The ALB solution is presented in the following table:

Station j	Task i	$\bar{t}_i$	$\sum_{i \in S_j} \bar{t}_i$	V_i	$\sum_{i \in S_j} V_i$	$SD_j = [\sum_{i \in S_j} V_i]^{1/2}$	ST_j
1	1	4	4	2	2	1.41	$(4) + 2.5 \times 1.41 = 7.53$
2	2	3	3	1	1	1.00	$(3) + 2.5 \times 1 = 5.50$
	3	2	5	2	3	1.73	$(3 + 2) + 2.5 \times 1.73 = 9.33$
3	4	5	5	1	1	1.00	$(5) + 2.5 \times 1 = 7.50$
	5	1	6	0	1	1.00	$(5 + 1) + 2.5 \times 1 = 8.50$

In this example, for the three stations, the cycle time is $Max\{7.53, 9.33, 8.5\} = 9.33$ min, with 98.76% confidence that it will occur within this duration. As a matter of illustration, if a confidence level of 50% is used, that is, $z = 0$, then only two stations are needed as presented in the following table:

Station j	Task i	t_i	ST_j
1	1	4	4
	2	3	7
2	3	2	2
	4	5	7
	5	1	8

In this case, only two stations are needed to achieve the cycle time of 8 with a 50% confidence level versus three stations with a cycle time of 9.33 with a 98.76% confidence level. Such a trade-off between the two solutions is further explained in the next section.

14.7.2   Multicriteria Stochastic ALB[†*]

In multicriteria stochastic ALB, the objectives are:

Minimize $f_1 =$ number of stations
Minimize $f_2 =$ cycle time
Maximize $f_3 =$ confidence level

The first objective minimizes the total cost. The second objective maximizes the productivity. The third objective maximizes the confidence level that the chosen cycle time will occur. To generate a set of efficient alternatives to this problem, consider a set of different z values. For each given z, solve the stochastic ALB problem as presented in the previous section and report the three objectives. Commonly used z values are 0, 1, 2, and 3 and are associated with 50, 68.26, 95.44, and 99.74%. The DM can then choose the best efficient alternative.

Example 14.10   Multicriteria Stochastic ALB   Consider Example 14.9. Suppose the target cycle time is 10. For $z = 0, 1, 2, 3$, four alternatives can be generated. They are listed below. For a given z and CT, the number of stations is found, so alternatives are presented in the order of f_3, f_2, and f_1.

z	0	1	2	3
Max. f_3 = confidence level	50%	68.26%	95.44%	99.74%
Min. f_2 = cycle time	8	9.73	8.46	9.00
Min. f_1 = no. of stations	2	2	3	4
Efficient ?	Yes	Yes	Yes	Yes

See Supplement S14.5.xls on the book companion site.

In this problem the given cycle time is 10, but after solving the problem for each alternative, a better CT was generated for the obtained number of stations, which is the CT reported in the above table. Comparing the alternatives associated with $z = 0$ and $z = 1$, both alternatives have two stations. For $z = 0$, the cycle time is 8 with a confidence level of 50%. For $z = 1$, that is, increasing the confidence level to 68.28%, the cycle time increases to 9.73.

REFERENCES

General

Baybars, I. "A Survey of Exact Algorithms for the Simple Assembly Line Balancing Problem." *Management Science*, Vol. 32, No. 8, 1986, pp. 909–932.

Becker, C. "A Survey on Problems and Methods in Generalized Assembly Line Balancing." *European Journal of Operational Research*, Vol. 168, No. 3, 2006, pp. 694–715.

Chow, W.-M. *Assembly Line Design, Manufacturing Engineering and Materials Processing, New York:* Marcel Dekker, 1990.

Erel, E., and S. C. Sarin. "A Survey of the Assembly Line Balancing Procedures." *Production Planning and Control*, Vol. 9, No. 5, 1998, pp. 414–434.

Held, M., R. M. Karp, and R. Shareshian. "Assembly-Line Balancing–Dynamic Programming with Precedence Constraints." *Operations Research*, Vol. 11, No. 3, 1963, pp. 442–459.

Helgeson, W. P., and D. P. Birnie. "Assembly Line Balancing Using the Ranked Positional Weight Technique." *Journal of Industrial Engineering*, Vol. 12, No. 6, 1961, pp. 394–398.

Kilbridgem, M. D., and L. Wester. "A Heuristic Method of Assembly Line Balancing." *Journal of Industrial Engineering*, Vol. 12, No. 4, 1961, pp. 292–298.

Linhart, R. *The Assembly Line*. City: University of Massachusetts Press, 1981.

Nahmias, S. *Production and Operations Analysis*, 8th ed. New York: McGraw-Hill/Irwin, 2008.

Pinto, P. A., D. G. Dannenbring, and B. M. Khumawala. "Assembly Line Balancing with Processing Alternatives: An Application." *Management Science*, Vol. 29, No. 7, 1983, pp. 817–830.

Rekiek, B., and A. Delchambre. *Assembly Line Design: The Balancing of Mixed-Model Hybrid Assembly Lines with Genetic Algorithms*. New York: Springer, 2005.

Sawik, T. "An LP-Based Approach for Loading and Routing in a Flexible Assembly Line." *International Journal of Production Economics*, Vol. 64, 2000, pp. 49–58.

Scholl, A. *Balancing and Sequencing of Assembly Lines*. New York: Physica, 1999.

Scholl, A., and C. Becker. "State-of-the-Art Exact and Heuristic Solution Procedures for Simple Assembly Line Balancing." *European Journal of Operational Research*, Vol. 168, No. 3, 2006, pp. 666–697.

Tonge, F. M. *A Heuristic Program for Assembly Line Balancing*, Englewood Cliffs, NJ: Prentice-Hall, 1961.

Traven, B. *Assembly Line*. City: New England Free Press, 1977.

Multiobjective

Al-Anzi, F. S., and A. Allahverdi. "Heuristics for a Two-Stage Assembly Flowshop with Bicriteria of Maximum Lateness and Makespan." *Computers and Operations Research*, Vol. 36, No. 9, 2009, pp. 2682–2689.

Ayağ, Z., and R. G. Ozdemir. "A Combined Fuzzy AHP-Goal Programming Approach to Assembly-Line Selection." *Journal of Intelligent and Fuzzy Systems*, Vol. 18, No. 4, 2007, pp. 345–362.

Bukchin, J., and M. Masin. "Multi-Objective Design of Team Oriented Assembly Systems." *European Journal of Operational Research*, Vol. 156, No. 2, 2004, pp. 326–352.

Celano, G., S. Fichera, V. Grasso, U. L. Commare, and G. Perrone. "An Evolutionary Approach to Multi-Objective Scheduling of Mixed Model Assembly Lines." *Computers & Industrial Engineering*, Vol. 37, Nos. 1–2, 1999, pp. 69–73.

Chen, R.-S., K.-Y. Lu, and S.-C.Yu. "A Hybrid Genetic Algorithm Approach on Multi-Objective of Assembly Planning Problem." *Engineering Applications of Artificial Intelligence*, Vol. 15, No. 5, 2002, pp. 447–457.

Choi, Y.-K., D. M. Lee, and Y. B. Cho. "An Approach to Multi-Criteria Assembly Sequence Planning Using Genetic Algorithms." *The International Journal of Advanced Manufacturing Technology*, Vol. 42, Nos. 1–2, 2009, pp. 180–188.

Deckro, R. F., and S. Rangachari. "A Goal Approach to Assembly Line Balancing." *Computers & Operations Research*, Vol. 17, No. 5, 1990, pp. 509–521.

Gokcen, H., and K. Agpak. "A Goal Programming Approach to Simple U-line Balancing Problem." *European Journal of Operational Research*, Vol. 171, No. 2, 2006, pp. 577–585.

Jolai, F., M. J. Rezaee, and A. Vazifeh. "Multi-Criteria Decision Making for Assembly Line Balancing." *Journal of Intelligent Manufacturing*, Vol. 20, No. 1, 2009, pp. 113–121.

Kara, Y., H. Gokcen, and Y. Atasagun. "Balancing Parallel Assembly Lines with Precise and Fuzzy Goals." *International Journal of Production Research*, Vol. 48, No. 6, 2010, pp. 1685–1703.

Kilbridge, M., and L. Wester. "The Balance Delay Problem." *Management Science*, Vol. 8, No. 1, 1961, pp. 69–84.

Li, T. "Applying TRIZ and AHP to Develop Innovative Design for Automated Assembly Systems." Vol. 46, Nos. 1–4, 2010, pp. 301–313.

Moodie, C. L., and H. H. Young. "A Heuristic Method of Assembly Line Balancing for Assumptions of Constant or Variable Work Element Times." PhD diss., Purdue University, 1964.

Malakooti, B. "A Multiple Criteria Decision Making Approach for the Assembly Line Balancing Problem." *International Journal of Production Research*, Vol. 29, 1991, pp. 1979–2001.

Malakooti, B. "Assembly Line Balancing with Buffers by Multiple Criteria Optimization." *International Journal of Production Research*, Vol. 32, No. 9, 1994, pp. 2159–2178.

Malakooti, B., and A. Kumar. "A Knowledge-Based System for Solving Multi-Objective Assembly Line Balancing Problems." *International Journal of Production Research*, Vol. 34, 1996, pp. 2533–2552.

Malakooti, B., "Head Tail Qualitative Assembly Line Balancing with Multiple Objectives." *Case Western Reserve University, Cleveland, Ohio* 2013.

McMullen, P. R., and G. V. Frazier. "Using Simulated Annealing to Solve a Multiobjective Assembly Line Balancing Problem with Parallel Workstations." *International Journal of Production Research*, Vol. 36, No. 10, 1998, pp. 2717–2741.

McMullen, P. R., and P. Tarasewich. "Multi-Objective Assembly Line Balancing via a Modified Ant Colony Optimization Technique." *International Journal of Production Research*, Vol. 44, No. 1, 2006, pp. 27–42.

Nourmohammadia, A., and M. Zandieh. "Assembly Line Balancing by a New Multi-Objective Differential Evolution Algorithm Based on TOPSIS." *International Journal of Production Research*, Vol. 49, No. 10, 2011, pp. 2833–2855.

Pastor, R., C. Andrés, A. Duran, and M. Pérez. "Tabu Search Algorithms for an Industrial Multi-Product and Multi-Objective Assembly Line Balancing Problem, with Reduction of the Task Dispersion." *Journal of the Operational Research Society*, Vol. 53, No. 12, 2002, pp. 1317–1323.

Suwannarongsri, S., and D. Puangdownreong. "Optimal Balancing of Multi-Objective Assembly Lines via Metaheuristic Approach." Proceedings of the 10th WSEAS International Conference on Evolutionary Computing, 2009, pp. 52–56.

EXERCISES

Assembly Line Balancing Methods (Section 14.3)

3.1 Consider an ALB problem with six tasks and the following time requirements and precedence relationships:

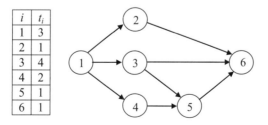

i	t_i
1	3
2	1
3	4
4	2
5	1
6	1

For a cycle time of 6:

(a) Solve ALB by the largest candidate rule. Find the line efficiency and smoothness index.

(b) Solve ALB by the column precedence Method. Find the line efficiency and smoothness index.

(c) Solve ALB by the positional weight method. Find the line efficiency and smoothness index.

(d) Determine which of the above methods provides the best solution.

3.2 Consider an ALB problem with seven tasks and the following time requirements and precedence relationships. Suppose the cycle time is 7. Respond to parts (a)–(d) of exercise 3.1.

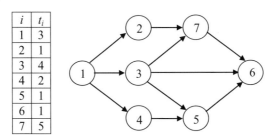

i	t_i
1	3
2	1
3	4
4	2
5	1
6	1
7	5

3.3 Consider an ALB problem with 11 tasks and the following time requirements and precedence relationships. Suppose the cycle time is 9. Respond to parts (a)–(d) of exercise 3.1.

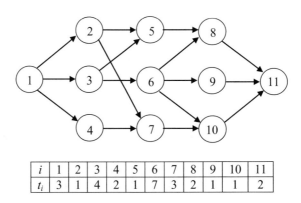

i	1	2	3	4	5	6	7	8	9	10	11
t_i	3	1	4	2	1	7	3	2	1	1	2

3.4 Consider an ALB problem with 15 tasks and the following time requirements and precedence relationships. Suppose the cycle time is 10. Respond to parts (a)–(d) of exercise 3.1.

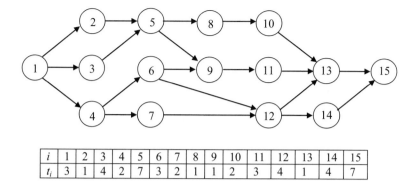

i	1	2	3	4	5	6	7	8	9	10	11	12	13	14	15
t_i	3	1	4	2	7	3	2	1	1	2	3	4	1	4	7

Qualitative ALB: Head–Tail Approach (Section 14.4)[†]

4.1 Consider exercise 3.1. Consider the following qualitative closeness information for each pair of tasks. For a cycle time of 6, use the qualitative ALB approach to assign the tasks to the minimum number of stations. Measure the total closeness of the final solution and compare it to the ideal total closeness.

	1	2	3	4	5	6
1		X	E	U	O	U
2			U	A	U	O
3				O	E	O
4					E	X
5						U

4.2 Consider exercise 3.2. Consider the following qualitative closeness information for each pair of tasks. For a cycle time of 9, use the qualitative ALB approach to assign the tasks to the minimum number of stations. Measure the total closeness of the final solution and compare it to the ideal total closeness.

	1	2	3	4	5	6	7
1		X	E	U	O	U	A
2			U	A	U	O	E
3				O	E	O	I
4					E	X	U
5						U	X
6							A

4.3 Consider exercise 3.3. Consider the following qualitative closeness information for each pair of tasks. For a cycle time of 13, use the qualitative ALB approach to assign the tasks to the minimum number of stations. Measure the total closeness of the final solution and compare it to the ideal total closeness.

	1	2	3	4	5	6	7	8	9	10	11
1	X	E	U	O	U	U	U	U	U	U	U
2		U	A	U	O	U	U	U	U	U	O
3			O	E	O	U	U	U	U	U	O
4				E	X	I	U	U	U	U	X
5					U	I	U	U	U	U	A
6						E	A	O	U	U	E
7							I	U	U	U	I
8								X	I	U	
9									O	X	
10										E	

Minimizing Cycle Time for Given Number of Stations (Section 14.5.1) and Optimizing ALB Profit (Section 14.5.2)

5.1 Consider exercise 3.1. Suppose that the sale price is $30 per unit, operation cost is $5 per unit, station cost per period is $5000, and labor cost per period is $1000. Suppose that there is a total of four weeks in each period and each week has 40 h. Find the best number of stations and cycle time for maximizing profit.

5.2 Consider exercise 3.2. Suppose that the sale price is $30 per unit, operation cost is $5 per unit, station cost per period is $500, and labor cost per period is $100. Suppose that there is a total of four weeks in each period and each week has 40 h. Find the best number of stations and cycle time for maximizing profit.

5.3 Consider exercise 3.3. Suppose that the sale price is $36 per unit, operation cost is $1 per unit, station cost is $5000 per period, and labor cost per period is $1000. Suppose that there is a total of four weeks in each period and each week has 40 h. Find the best number of stations and cycle time for maximizing profit.

Bicriteria ALB Problem (Section 14.5.3)

5.4 Consider exercise 3.1. Also consider the cost information given in exercise 5.1.

(a) Generate the efficient points associated with number of stations and cycle time. Rank alternatives using 0.6 and 0.4 as weights of importance for number of stations and cycle time, respectively.

(b) Generate the efficient points associated with number of stations, cycle time, and total cost. Rank alternatives using 0.3, 0.3, and 0.4 as weights of importance for number of stations, cycle time, and total cost, respectively.

5.5 Consider exercise 3.2. Also consider the cost information given in exercise 5.2.

(a) Generate the efficient points associated with number of stations and cycle time. Rank alternatives using 0.2 and 0.8 as weights of importance for number of stations and cycle time, respectively.

(b) Generate the efficient points associated with number of stations, cycle time, and total cost. Rank alternatives using 0.4, 0.2, and 0.4 as weights of importance for number of stations, cycle time, and total cost, respectively.

5.6 Consider exercise 3.3. Also consider the cost information given in exercise 5.3.

(a) Generate the efficient points associated with number of stations and cycle time. Rank alternatives using 0.5 and 0.5 as weights of importance for number of stations and cycle time, respectively.

(b) Generate the efficient points associated with number of stations, cycle time, and total cost. Rank alternatives using 0.2, 0.5, and 0.3 as weights of importance for number of stations, cycle time, and total cost, respectively.

Mixed-Product ALB (Section 14.6)

6.1 Consider exercises 3.1 and 3.2. Suppose these two products are assembled on the same production line. Given the cycle time of 8 min, solve the mixed ALB problem and identify the best number of stations.

6.2 Consider exercises 3.1 and 3.3. Suppose these two products are assembled on the same production line. Given the cycle time of 13 min, solve the mixed ALB problem and identify the best number of stations.

6.3 Consider exercises 3.1, 3.2, and 3.3. Suppose these three products are assembled on the same production line. Given the cycle time of 11 min, solve the mixed ALB problem and identify the best number of stations.

Single-Objective Stochastic ALB (Section 14.7.1)

7.1 Consider exercise 3.1. Consider the following mean task times and variances. Use the stochastic ALB to solve the problem. Suppose the cycle time is 9.

Task	1	2	3	4	5	6
Mean time, variance	(3,1)	(1,1)	(4,2)	(2,1)	(1,1)	(1,0)

(a) Use confidence level of 68.26% (i.e., $z = 1$). Find the number of stations.
(b) Use confidence level of 95.44% (i.e., $z = 2$). Find the number of stations.
(c) Use confidence level of 99.74% (i.e., $z = 3$). Find the number of stations.

7.2 Consider exercise 3.2. Consider the following mean task times and variances. Use the stochastic ALB to solve the problem. Suppose the cycle time is 12.

Task	1	2	3	4	5	6	7
Mean time, variation	(3,1)	(1,0)	(4,2)	(2,1)	(1,1)	(1,0)	(5,3)

Respond to parts (a)–(c) of exercise 7.1.

7.3 Consider exercise 3.3. Consider the following mean task times and variances. Use the stochastic ALB to solve the problem. Suppose the cycle time is 14.

Task	1	2	3	4	5	6	7	8	9	10	11
Mean time, variation	(3,1)	(1,0)	(4,2)	(2,1)	(1,1)	(7,4)	(3,1)	(2,1)	(1,0)	(1,1)	(2, 1)

Respond to parts (a)–(c) of exercise 7.1.

Multicriteria Stochastic ALB (Section 14.7.2)[†]

7.4 Consider exercise 7.1. List a set alternatives for $z = 0, 1, 2, 3$. Identify efficient alternatives.

7.5 Consider exercise 7.2. List a set of alternatives for $z = 0, 1, 2, 3$. Identify efficient alternatives.

7.6 Consider exercise 7.3. List a set of alternatives for $z = 0, 1, 2, 3$. Identify efficient alternatives.

CHAPTER 15

FACILITY LAYOUT

15.1 INTRODUCTION

Effective layout designs are necessary in many different businesses such as manufacturing facilities, assembly systems, warehouses, production and operations, retail stores, schools, office buildings, airports, health care (e.g., hospitals), utility providers (e.g., water and sewer facilities), energy-generating facilities (e.g., nuclear power plants), service industries, and government operational facilities. Historically, however, layout design techniques were developed to solve manufacturing facility layout problems to identify the best locations for different manufacturing departments. Since the 1950s, facility layout methods have been expanded to solve a variety of applications, including the design of computer motherboards and integrated circuits (ICs). Diminishing physical space availability, high cost of relayouts, safety issues, and substantial long-term investments are good reasons for getting the design right the first time.

Recently, with manufacturing facilities taking on a whole new level of complexity, flexibility, and capital investment, it has become imperative to develop a systems approach for solving complex layout problems. Modern layout planning relies on gathering critical information from multiple sources with the goal of designing a layout which can provide most flexibility for future modifications and expansions while optimizing utilization, efficiency, and total cost.

Facility layout design is a multiobjective problem where one must consider different criteria such as operational costs, the construction and installation costs for future relayouts, the organizational leadership goals, employee job satisfaction, utilization of space, workforce management, equipment utilization, ease and simplicity of materials handling,

Note: Advanced material that can be omitted without loss of generality will be indicated by a dagger.

Operations and Production Systems with Multiple Objectives, First Edition. Behnam Malakooti.
© 2014 John Wiley & Sons, Inc. Published 2014 by John Wiley & Sons, Inc.

803

materials flow costs, inventory work in process, inventory of incoming materials and inventory of outgoing final products, ease of future expansion, and safety factors. Most of these objectives are difficult to quantify. In this chapter, new concepts and methods are introduced for solving multiple-objective layout problems.

Minimizing materials handling cost is one of the key objectives in layout design of manufacturing systems because of the following reasons:

1. Materials handling costs tend to be high (30–70% of total production cost).
2. Materials handling costs can be substantially reduced by proper layout design.
3. Materials handling costs are quantifiable, for example, they are often proportional to distance.
4. Minimizing material handling costs will also minimize work in process and cycle times.

Often the optimal layout design cannot be found due to the complexity of the layout problem. Layout design is a combinatorial problem, where the number of possible alternatives increases exponentially as the number of departments increases. For example, a facility layout with 25 departments has $25! \approx 1.55 \times 10^{25}$ possible layout solutions. Even with the fastest computers, many months may be needed to consider all possible options for even a small layout problem. Therefore, heuristic approaches are used to solve layout problems in which the resulting layout is satisfactory but not necessarily optimal. For example, one can use a heuristic approach to minimize the materials handling costs. On the other hand, certain layout design aspects are somewhat of an art and rely on human experience, intuition, and judgment. Layout design is an evolving process that requires both optimization and human insight. It is therefore important to critique, revise, and improve proposed layout designs before determining the final layout.

The approaches for solving layout problems can be categorized as follows:

- Construction approaches that do not require starting with an initial layout solution. This is covered under systematic layout planning (SLP, Section 15.2), rule-based layout (RBL, Section 15.3), and quadratic assignment problem optimization (QAP, Section 15.7).
- Improvement approaches that require starting with an initial feasible layout solution. This is covered under pairwise exchange (Sections 15.4 and 15.5) and facility relayout (Section 15.6).

In this chapter, we develop three new approaches for solving layout design problems: (a) solving single-objective layout problems by a new construction approach called rule-based layout which uses the head–tail methods, (b) solving multiobjective layout problems, and (c) solving relayout problems, which in practice occur far more frequently than new layout designs.

Layout problems can be represented by a network connecting different departments as nodes. Such layout design presentations can be classified based on the number of rows and columns in the layout:

(I) *Single-row layouts* have a single row allowing loops of return; they can be linear or circular. In this layout only adjacent departments (e.g., stations) can be connected. This type of layout is covered in Chapters 13 and 14.

(II) *Multirow layouts* have multiple rows and columns where all departments can be connected. This type of layout is more complex than the single-row layout. Methods for this layout are addressed in this chapter.

15.1.1 Layout Classification

Manufacturing layout problems can be classified based on their production volume and flexibility (or product diversity). Figure 15.1 demonstrates the classification of four types of manufacturing layouts based on volume and flexibility. For example, a job shop is associated with a high-flexibility and low-volume environment, but product layout is associated with a low-flexibility and high-volume environment.

In the following, several types of layout problems are discussed. The first four layout problems are related to manufacturing industries. Other layouts are related to service and distribution industries and computer industries.

(a) *Product Layout (Assembly Systems)* Product layouts are used in mass production systems where volumes of products are high and products are fairly standardized. In mass production systems, production processes are organized in a sequential form to expedite the flow of materials. Resources are organized to support the flow of materials. Product layouts require a large initial capital investment in procurement of the necessary equipment. Mass fabrication and assembly lines are two examples of product layouts. Fabrication involves manufacturing processes such as metal cutting, welding, casting, extrusion, forming, and machining. In assembly lines, off-the-shelf parts and components are assembled to create the final product. This type of layout is common in many industries, including automobile, computer assembly, and fast food industries. See Figure 15.2 for an illustration of an assembly line. This topic is covered in Chapter 14.

(b) *Process Layout (Job Shop)* In process or job shop layout, processes and machines are grouped based on their common functions. This type of a layout is typically used for manufacturing a high variety of low- and medium-volume products. This type of layout is ideal for products with varying manufacturing requirements and routings. For example, a manufacturer of metal parts may divide the fabrication facility into processor type 1: casting

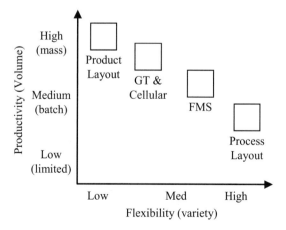

Figure 15.1 Classification of manufacturingg layouts based on productivity and flexibility.

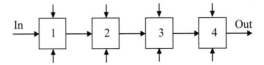

Figure 15.2   Example of a product layout assembly line with four stations.

(three groups of casting machines with each group having five machines), processor type 2: heat treating (two groups each with nine machines), and processor type 3: machining (milling, drilling, turning, with each group having two machines). See Figure 15.3 for an illustration.

(c) *Cellular Layout (Group Technology)* In this type of layout, parts are grouped into part families based on similarities. These families are manufactured in cells that consist of equipment and resources dedicated to producing that family of parts. Group technology is particularly valuable for facilities with medium- to high-volume manufacturing and relatively varying products. By grouping parts into families, cells are designed to handle the entire production process for the part family. Group technology has been implemented extensively in a variety of industries. See Figure 15.4 for an illustration of group technology. In Figure 15.4, each cell has several different machines in it. Clustering and group procedures are covered in Chapter 12. The set up (layout) of each cell is covered in Chapter 13.

(d) *Flexible Manufacturing Systems* In flexible manufacturing systems (FMSs), computers are used to automatically set up machines and change the tools of manufacturing equipment. A machine operator inputs the specifications for a particular part and the computer numerical control (CNC) automatically sets up the machines and tooling needed for production. An FMS is highly automated and requires far less human intervention than other layout methods. An FMS works best when there are a variety of similar parts that have medium volumes (e.g., batches of 50 identical parts).

(e) *Semiflexible Assembly Systems* We define semiflexible assembly systems as systems whose layout can be modified with reasonable time and cost. Examples of semiflexible

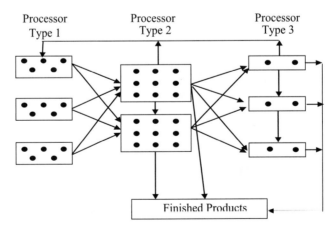

Figure 15.3   Example of process layout where different parts may be processed by different processors: Various parts use various routes. Each processor type has similar machines, where each machine is shown as a filled circle.

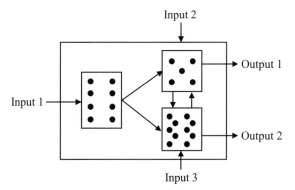

Figure 15.4 Group technology layout with three cells: Each cell consists of different types of machines.

systems are service industries, repair industries, labor-intensive mass production systems (e.g., garment factories where sewing machines can be moved easily), and cellular manu-facturing systems (see Chapter 13). In these systems, departments can be relocated without causing substantial cost or interruption of work.

(f) *Retail and Service* Retail layouts are designed to provide maximum service and access for the customer to the needed service or products. In retail layouts, products are placed in such a way that sales per square foot of floor space are maximized. Also, products are placed in a way that increases the probability of selling items based on the dynamic behavior of customers, that is, the customer is exposed to more products while looking for specific items. In service layouts, critical resources are placed in such a way that employees can most effectively render the needed services while giving the customer easy access and minimal waiting time and minimizing noise and other unpleasant interactions. This type of layout fosters better communication, teamwork, and responsibility.

(g) *Warehouse* For a warehouse layout, utilization of space must be balanced against the materials handling costs. Space utilization must be maximized while ensuring easy access to the goods. Issues dealing with loading and unloading, placement, stock picking, shelving heights, damages, safety, and spoilage are all important. Warehousing methods may utilize advanced conveyors, robotics, and computerized automated guided vehicles. More automation will result in significant productivity improvements but may be very costly.

(h) *Fixed Position* This type of layout is used when the project or product is too large to be moved or the product is in a fixed position. The product stays in a fixed location and the workstations are positioned around the product. Ships, buildings, dams, and oil drilling rigs are examples of fixed-position layouts. See Figure 15.5. Chapter 8 provides a systematic way to organize the work sequences and scheduling of needed resources for this type of product.

(i) *Very-Large-Scale Integration (VLSI)* VLSI is an example of a complex (system-of-systems) layout problem. In VLSI thousands of transistor-based circuits are combined onto a single chip, creating ICs such as a microprocessor. Due to the limited size of a single chip, a multilayer approach is used to find the best location for each component. Also, because of the vast amount of transistors placed on a given chip, a hierarchical approach for the layout can also be used.

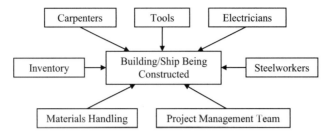

Figure 15.5   Example of fixed-position layout.

15.1.2 Hierarchical Layout Planning

Most real-world layout problems are a combination of several types of layouts and require several different approaches and strategies for solving different subproblem layout problems as presented in Chapters 13 and 14. A related approach for formulating and solving hierarchical multiple objective problems is presented in Chapter 1.

The hierarchical approach for solving layout problems consists of the aggregation and decomposition of a complex and/or large-scale layout problem into smaller problems that are solvable by different methods. The approach is top down and consists of three levels:

Level I Superdepartment Layout Planning This is the highest level of aggregation. Simplify and aggregate many related departments into single superdepartments. Ignore all detailed relationships.

Level II Intermediate Aggregate Department Layout Planning This is the intermediate level of aggregation. Each aggregated department consists of several departments. Ignore the lowest level of relationships and departments. Solve the layout problem for each department.

Level III Department Layout Planning This is the lowest level of aggregation, that is, it has the highest level of detail in the relationships between departments. For each department, design the needed layout.

The hierarchical method is iterative, meaning that levels I, II, and III will be modified and improved until a satisfactory solution to all levels is found.

GE High-Intensity Discharge (HIO) Bulbs Case Study GE's division of lighting manufactures a variety of bulbs. One group of bulbs consists of various high-intensity discharge bulbs. These bulbs have a high level of illumination and are usually used for commercial, industrial, and large-scale buildings. The GE lamp plant in Ravenna, Ohio, has a mixed/hierarchical layout. A substantially simplified layout is presented in Figure 15.6. (The author of this book was in charge of facility layout planning for expansion and renovation of HID lines.)

In this layout, there are:

- Three different groups of products: I, II, and III
- Three quality control inspection groups: Q-I, Q-II, and Q-III
- One common finishing station (FS)
- Three packaging stations: P-I, P-II, and P-III

- One storage area for shipment and outgoing inventory
- One storage area for incoming parts and raw materials
- One cafeteria
- One area for technical support
- One department for management offices

Each assembly and production line produces a group of high-intensity discharge bulbs that are grouped based on size and/or the type of common machinery being used. In each line, there is some bidirectional movement, that is, there are some loops in the line, but in general the movement is unidirectional. Each line has several quality control stations; the most important one is located at the end of each line. The assembled parts from all three lines go through a common finishing station (a glass heating treatment and sealing process). Finally, different types of bulbs are assembled, packaged, and labeled for shipment to vendors or warehouses.

The application of the hierarchical approach to make a layout design for this problem is as follows:

Level I Superdepartment Layout Planning Start at the highest level: that is, consider administrative offices, Lines I, II, and III, the finishing station, packaging stations I, II, and III, incoming storage, and outgoing storage. This is a total of 10 superdepartments, which is the maximum level of aggregation recommended in this approach. Set up the best layout for these 10 superdepartments. This layout is presented in Figure 15.6.

Level II Intermediate Aggregate Department Layout Planning Make the layout for each of the intermediate aggregate level departments. Provide layouts for each of the above 10 superdepartments. For example, the administrative offices consist of management offices, the cafeteria, and technical support. The technical support area consists of design, engineering, quality, operations management, and maintenance offices.

Level III Department Layout Planning This is the most detailed layout planning for each department. For example, the details of each quality control station (Q-I, Q-II, and Q-III) are provided in this phase. Similarly, each assembly and production line consists of several stations. The layout of each office and station also occurs in this level.

Depending on the characteristics of each of the above layout problems, different layout methods should be used. For example, for the highest level of layout design, use the methods

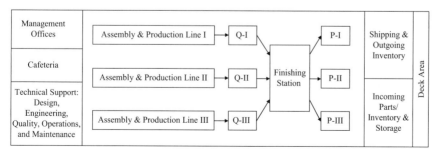

Figure 15.6   Layout of high-intensity discharge bulb plant.

covered in this chapter. Layout design approaches for each station, each assembly, each production line, each cell, and each station are covered in Chapters 12–14.

15.2   SYSTEMATIC LAYOUT PLANNING

Systematic layout planning is a technique that considers quantitative and qualitative factors and restrictions to determine a facility layout. SLP uses a problem-solving approach that consists of collecting information, analyzing the problem, generating alternative solutions, evaluating the solutions, and selecting the best layout. Developed by Muther, Richard, "Systematic layout planning." (1973). SLP utilizes a qualitative graphical network approach to construct facility layouts. SLP is a construction method that does not require an initial layout (vs. improvement methods that require an initial layout solution). Figure 15.7 shows a block diagram of the SLP procedure. The three major steps of the SLP method are explained below.

1. *Analysis of Problem*   Data are collected for activities (functions and departments) and relationships are determined for each pair of activities. There are three basic types of data

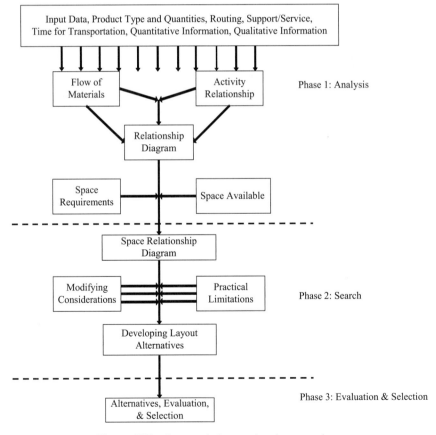

Figure 15.7   Systematic layout planning procedure.

that need to be collected: (a) numeric, such as space requirements and flow; (b) graphic, including sketches or drawings of the facility, equipment, and so on; and (c) qualitative information such as preferences of the designer. The number of departments and their functions, the costs of material flow between departments, and the desired closeness of departments are determined. The costs for material flow between each pair of departments can be based on the volume and difficulty of product movement. The designer then creates a materials flow matrix (from–to chart) and an activity RELationship diagram (REL chart) that shows the desired closeness between departments. The materials flow matrix consists of entries that show the typical number of trips and the volumes that occur between pairs of departments. The REL chart uses letters *A, E, I, O, U, X* to represent the importance of having two departments in close proximity to one another. For example, a rating of *A* indicates a preference for the two departments to be very close. Table 15.1 shows the notation used in REL charts. One can use 4, 3, 2, 1, 0, −10 or 5, 4, 3, 2, 1, 0 for *A, E, I, O, U, X* notation, respectively.

2. *Search Stage* The search stage of the SLP process begins by constructing a departmental relationship diagram. First a network of the layout is presented. Figure 15.7 is an example of the graphical depiction of a facility layout produced by using SLP. Each pair of departments is connected by a set of lines based on Table 15.1. The line representations can be thought of as similar to rubber bands and springs. For example, two departments with an *E* rating have three rubber bands connecting them, which create a stronger pull than an *O* rating, which has only one rubber band. An *X* rating is a spring, which repels the departments instead of drawing them together. The network should then be readjusted by moving nodes to different positions such that pairs of departments which have stronger connections are closer to each other. Many different alternatives can be constructed by rearranging department locations. In this stage, constraints and limitations are used to modify layout alternatives into feasible and practical solutions. Space requirements and available space must then be considered. A base unit for measuring areas is determined. Each department's size is specified by the number of these base units. Feasible alternatives are then generated by drawing a layout of the departments and letting the connections "push" and "pull" each other into a practical solution.

TABLE 15.1   Qualitative Notation, Graphical Symbols, and Two Different Quantitative Values Used for Showing Desirability of Closeness of Each Pair of Departments

Qualitative Notation	Meaning for Two Departments	Number of Lines	Line Representation	Ordinal Rating Method I	Ordinal Rating Method II
A	MUST be together (absolutely)	4	≡≡≡	4	5
E	SHOULD be together (especially)	3	≡≡	3	4
I	COULD be together (important)	2	=	2	3
O	MIGHT be together (ordinary)	1	—	1	2
U	Proximity is unimportant (unimportant)	0	None	0	1
X	SHOULD NOT be together (undesirable)	Dashed	- - - - - - - -	−10	0

3. *Evaluation and Selection* Alternatives generated during the search stage are critiqued in the evaluation stage. In this stage the layout designer presents the alternatives to the DMs to find out which layout best satisfies their objectives and needs. In this stage, multiple-criteria decision making (MCDM) approaches can be used to evaluate alternatives and select the best alternatives. The MCDM evaluation of alternatives is our modification to SLP; it will be covered in more depth later in this chapter.

Example 15.1   SLP Procedure   Consider a facility consisting of seven departments, each with individual space requirements as shown in the table below. Each department can be broken up into units or blocks of 100 ft². For example, department 1, which is 600 ft², will require six blocks of space when divided into 100-ft² blocks. The total allotted area for the new facility is 2400 ft² and must be laid within a rectangular space with a dimension of 4 × 6 blocks. Each department can fit into a square or a rectangle (e.g., a department can be presented by a 1 × 6 or 2 × 3 block shape).

Department	1	2	3	4	5	6	7	Sum
Square feet	600	200	400	200	500	100	400	2400
No. of blocks	6	2	4	2	5	1	4	24

The matrix below shows the given desired proximity or closeness for each pair of departments. For example, departments 1 and 2 must be very close together (*A* rating), departments 1 and 3 might be together (*O* rating), and departments 3 and 4 should not be together (*X* rating). Note that the desired closeness rating between two departments is symmetric. It is the same in either direction (e.g., closeness of department 1 to 2 is the same as 2 to 1). Use SLP to find a layout solution to this problem for (a) a 400 × 600 facility and (b) a 200 × 1200 facility.

	1	2	3	4	5	6	7
1		*A*	*O*	*A*	*I*	*A*	*A*
2			*U*	*E*	*O*	*E*	*X*
3				*X*	*U*	*U*	*E*
4					*U*	*O*	*O*
5						*I*	*I*
6							*X*
7							

Figure 15.8 shows the initial solution where each department is presented by a node and the desired closeness for each pair of departments is presented by lines that connect the two departments. Therefore, the layout problem is presented by a network of nodes which are connected by a set of lines. For example, the closeness rating of *X* for department 2 and department 7 is presented by a dotted line. The arrangement of the departments in this first figure is arbitrary. Now imagine that each line is a rubber band and that the more lines there are between two departments, the more location "pull" there will be for the two departments. The dotted lines are springs, indicating that the two departments should be

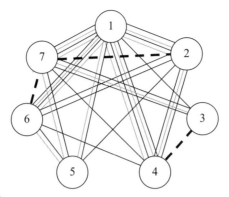

Figure 15.8   Initial graph to present the problem.

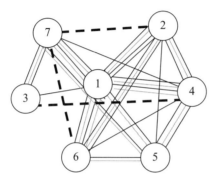

Figure 15.9   First iteration to improve total closeness.

located as far apart as possible. Figure 15.9 shows the first iteration of the rearrangement of the relationship diagram to minimize the strain on these bands. In the example, department 1 has the strongest relationships with other departments so it is centered and the other departments are arranged around it.

Figure 15.10 shows the next iteration. Rearrangement of the network should be repeated until no more improvements can be achieved. Once the final graphical relationship is

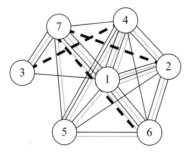

Figure 15.10   Further rearrangement to improve total closeness.

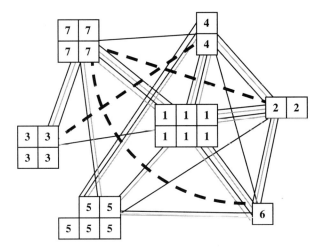

Figure 15.11   Relationship diagram showing departmental space requirements.

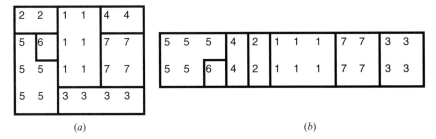

(a) (b)

Figure 15.12 (a) Relationship diagram for 4 × 6 area; (b) layout solution for a 2 × 12 area.

achieved, the space requirements for each department are considered. Figure 15.11 shows the area of each department on the network.

(a) Consider the layout for a 4 × 6 block facility. Collapse and rotate the spatial relationship diagram shown in Figure 15.5 to fit into the 4 × 6 space. This is shown in Figure 15.12*a*.

(b) For the 2 × 12 block facility, the solution is shown in Figure 15.12*b*.

The above solutions (a) and (b) generated by the SLP method, based on space require-ments, availability, closeness preferences, and other relative factors, are acceptable solutions but may not be optimal. The following two methods may provide better solutions than SLP. However, it may be useful to first use SLP, as it provides a good formulation of the problem, and then use the Rule Based Layout and Pairwise Exchange Method.

15.3   RULE-BASED LAYOUT

SLP is useful since it provides a systematic approach for defining and formulating the layout problem. However, the process of improving the initial graph of the layout becomes very difficult as the number of departments increases. Also, the evaluation process in SLP is subjective and depends on the layout designer. A different approach for solving layout problem is by use of expert systems, which are powerful tools for solving ill-structured problems such as layouts. In this section, we develop a new rule-based constructive approach called rule-based layout (Malakooti, 2013). In RBL, the starting point is the two departments that are the most central to all departments. Then, in each iteration, one department is added to the partially constructed layout until all departments are assigned. The selection of the location of each department is based on qualitative or quantitative rules and other pertinent information. RBL can use head–tail rules (developed in Chapter 13) which can solve very large layout problems with minimal computational efforts. Because this method is rule based, it can incorporate complicated constraints while developing the layout solution. Furthermore, it can be used in conjunction with expert systems approaches (which also use rules) to solve layout problems.

15.3.1 Calculating Total Cost and Flow

Preferences for departmental proximity can be subjective (qualitative) or objective (quantitative) presented for all pairs of departments. Consider a distance matrix D consisting of elements d_{ij} where d_{ij} is the distance between pairs of departments i and j. Also consider a volume matrix V consisting of elements v_{ij} where v_{ij} is the volume of movement between pairs of departments i and j during a specified period of time. Similarly, a cost matrix C consists of elements c_{ij} where c_{ij} is the cost per unit of volume per unit of distance for each pair of departments. The flow cost between each pair of departments i and j is calculated by multiplying the distance, the load volume, and the cost per unit of volume per unit of distance. The total flow cost (TC) for a given layout is calculated as

$$\text{Total Cost, TC} = \sum_{i=1}^{n} \sum_{j=1}^{n} d_{ij} v_{ij} c_{ij} \tag{15.1}$$

where n is the number of departments. In Equation (15.1), if $c_{ij} = 1$ for all i and j (e.g., using equal costs for all types of movements), the total flow can be written as in Equation (15.2)

$$\text{Total Flow, TF} = \sum_{i=1}^{n} \sum_{j=1}^{n} d_{ij} v_{ij} \tag{15.2}$$

To simplify the presentation of facility layout problems, each department can be represented by one square and distance between each pair of departments can be measured from the center of each department.

Example 15.2 Calculation of Total Cost for Four-Department Problem   The flow matrix for the four departments (A, B, C, and D) is shown in Figure 15.13. The flows for each pair of departments can be aggregated because the material movement is bidirectional. For example, for departments B and C, the flow from B to C is 25 and the flow from C to B is 15. Therefore, 40 is the aggregate flow between departments B and C. The cost matrix C shows the cost in dollars of moving one unit of material per unit of distance for the pair of departments. For example, the material handling cost of B to D is \$2 per one unit of distance per one unit of volume. See Figure 15.13 for these matrices. Find the total

Original Flow Matrix

	A	B	C	D
A	—	20	15	6
B	5	—	25	20
C	35	15	—	5
D	4	40	10	—

Aggregated Flow, V

	A	B	C	D
A	—	25	50	10
B		—	40	60
C			—	15
D				—

Cost, C

	A	B	C	D
A	—	2	3	1
B		—	5	2
C			—	6
D				—

Given Locations

1	2
3	4

Distance, D

	1	2	3	4
1	—	1	1	2
2		—	2	1
3			—	1
4				—

Solution Layout Department

A	B
1	2
C	D
3	4

Location

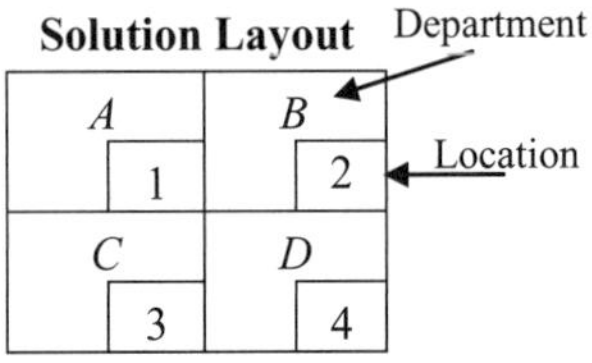

Figure 15.13   Distance, cost, and flow matrices.

cost for the solution layout assuming the materials are moved only in one axis (x or y) at a time.

Consider a facility with four departments of equal size in a 2×2 square area shown as "given locations" in Figure 15.13. Suppose the solution to this layout problem is assigning departments A, B, C, and D to locations 1, 2, 3, and 4, respectively, shown as "solution layout." Also, since the materials can only be moved in one axis at a time, it is appropriate to use rectilinear distance measurements. This is shown as distance matrix D. The calculation of the distance matrix is fairly simple in this example. Each pair of departments that share an edge has a distance of 1. Each pair of departments that share only one corner has a distance of 2. See Figure 15.13 for all the information.

To determine the total cost of the layout given in Figure 15.13, for each pair of departments multiply the distance by the aggregate flow by the cost to get the flow cost for that pair of departments; then add the costs of all pairs of departments. For example, the flow cost of moving between departments B and C is $2 \times 40 \times 5 = \$400$. The total cost of the initial layout is calculated using Equation (15.1) as follows:

$$TC = 1 \times 25 \times 2 + 1 \times 50 \times 3 + 2 \times 10 \times 1 + 2 \times 40 \times 5 + 1$$
$$\times 60 \times 2 + 1 \times 15 \times 6 = \$830$$

15.3.2   Identical Layouts

For computational efficiency, generation of identical layouts should be avoided. Identical layouts have the same distance matrix. For example, consider a single-row layout with four departments: A, B, C, and D. Read$\leftrightarrow$as bidirectional movement. Consider layout I: ($A\leftrightarrow B\leftrightarrow C\leftrightarrow D$). Suppose the size of the departments is the same and the distance between each pair of adjacent locations is 1. For example, the distance from department A to department B is 1, and the distance from department B to department D is 2. Therefore, by flipping layout I, an identical layout II ($D\leftrightarrow C\leftrightarrow B\leftrightarrow A$) can be generated.

Consider Example 16.1, which has four departments in the form of a 2×2 square layout. For this problem several layouts can be generated, but many of them are identical.

Rectilinear Distances

	A	B	C	D
A	—	2	1	1
B		—	1	1
C			—	2
D				—

A	C
D	B

,

B	C
D	A

,

A	D
C	B

,

B	D
C	A

Figure 15.14   Four equivalent layouts with identical distances generated by rotating and mirroring the first layout.

This occurs when the swapping of a pair of departments results in a layout with the same distance matrix. By rotating or mirroring layouts, identical layouts (in terms of distances) can be generated. Four identical layouts are presented in Figure 15.14. The distances are presented on the right side.

15.3.3   Rule-Based Layout for Single Row

Layouts with a single row (linear or circular in $1 \times n$ form) can be solved more easily than multirow layouts ($m \times n$). Design of single-row layouts is covered in-depth in Sections 13.4 and 13.6 by using the head–tail bidirectional methods (ordinal, cardinal, and optimization). To use these methods, for any pair of departments (i, j), create a new aggregated flow v_{ij} and set $v_{ji} = 0$. Order all pairs (i, j) in descending order of v_{ij}; ties are broken arbitrarily. Label the ordered vector of v_{ij} as $\mathbf{v}_o$.

> *Ordinal Head–Tail for Single-Row Layouts*   See Section 13.4.1 for details and examples. The following example illustrates this method. The ordinal head–tail method is expanded in the next section for multirow layout problems.
>
> *Cardinal Head–Tail for Single-Row Layouts*   See Section 13.4.2 for details and examples.
>
> *Optimization for Single-Row Layouts*   Integer linear programming can also be used to solve this problem. See Section 13.6.1.

In the following, we solve a single-row layout using the ordinal head–tail method for bidirectional networks based on the following rule:

> *Bidirectional Ordinal (Conflict Resolution) Rule for Single Row*   If there is a common department (or station) in sequences $s_p = (i \leftrightarrow j)$ and $s_q = (\ldots n \leftrightarrow i \leftrightarrow m \ldots)$, then place the new department j next to the common station i in s_q on the side that has a higher ranking to its adjacent station, that is, compare $j \leftrightarrow n$ and $j \leftrightarrow m$ flows and insert j on the side that has higher flow.

Example 15.3   Single-Row Layout of Six Departments by RBL—Ordinal Head–Tail

Consider the information given in Figure 15.15. Suppose the layout should be in the form of a 1×6 rectangle.

1. $s_p = (A \leftrightarrow F)$; then set $s_q = (A \leftrightarrow F)$.
2. $s_p = (C \leftrightarrow F)$. The mirror image is $s_p = (F \leftrightarrow C)$. Therefore, $s_q = (A \leftrightarrow F \leftrightarrow C)$.

Distance D

	1	2	3	4	5	6
1	—	1	2	3	4	5
2		—	1	2	3	4
3			—	1	2	3
4				—	1	2
5					—	1
6						—

Locations

1	2	3	4	5	6

Flow V

	A	B	C	D	E	F
A	—	10	20	30	10	100
B		—	50	60	10	20
C			—	10	20	80
D				—	50	60
E					—	10
F						—

Ranking of Flow V

	A	B	C	D	E	F
A	—	11	8	7	11	1
B		—	5	3	11	8
C			—	11	8	2
D				—	5	3
E					—	11
F						—

Figure 15.15   Initial data for Example 15.3.

3. Discard $(B \leftrightarrow D)$ temporarily as it does not have a common element with s_q.
4. $s_p = (D \leftrightarrow F)$; there is conflict for location of D.
   Use the conflict resolution rule to find the location for D. This is shown as

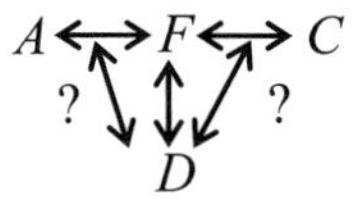

$$A \longleftrightarrow F \longleftrightarrow C$$
$$? \quad D \quad ?$$

Since the flow of $(D, A) = 30$ is more than the flow of $(D,C) = 10$, locate D between A and F; therefore, $s_q = (A \leftrightarrow D \leftrightarrow F \leftrightarrow C)$.

   4-1. Reconsider $(B \leftrightarrow D)$. The location of B should be determined. This is shown as

$$(A \longleftrightarrow D \longleftrightarrow F \longleftrightarrow C)$$
$$? \quad B \quad ?$$

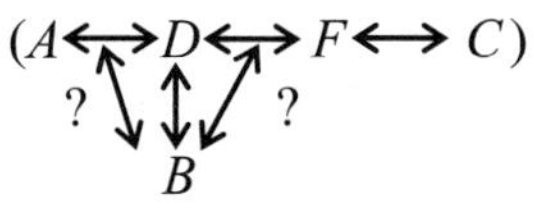

The flow of $(B,F) = 20$ is more than the flow of $(B,A) = 10$; locate B on the right side of D. Therefore, $s_q = (A \leftrightarrow D \leftrightarrow B \leftrightarrow F \leftrightarrow C)$.

5. Consider $(B \leftrightarrow C)$; discard it because it is a duplicate of elements in the sequence.
6. Consider $(D \leftrightarrow E)$. The location of E should be determined. This is shown as

$$(A \longleftrightarrow D \longleftrightarrow B \longleftrightarrow F \longleftrightarrow C)$$
$$? \quad E \quad ?$$

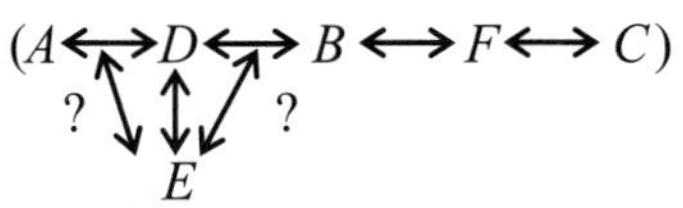

The flow of $(E,A) = 10$ is equal to the flow of $(E,B) = 10$. Therefore E can be put on either side of D; arbitrarily, locate it on the right side of D. Therefore, $s_q = (A{\leftrightarrow}D{\leftrightarrow}E{\leftrightarrow}B{\leftrightarrow}F{\leftrightarrow}C)$.

The final layout is $(A{\leftrightarrow}D{\leftrightarrow}E{\leftrightarrow}B{\leftrightarrow}F{\leftrightarrow}C)$. Using the above given distances and Equation (15.2), the total flow of this layout is given as

$$
\begin{aligned}
\text{Total flow} = {} & 10 \times 3 + 20 \times 5 + 30 \times 1 + 10 \times 2 + 100 \times 4 + 50 \times 2 + 60 \times 2 \\
& + 10 \times 1 + 20 \times 1 + 10 \times 4 + 20 \times 3 + 80 \times 1 + 50 \times 1 \\
& + 60 \times 3 + 10 \times 2 = 1260
\end{aligned}
$$

The summary of the method is shown in the following table.

Iteration	Used Rank	s_p	Action Taken	Resulting Sequence, s_q	Candidates
1	1	$A{\leftrightarrow}F$	Add s_p as a new sequence	$(A{\leftrightarrow}F)$	B,C,D,E
2	2	$C{\leftrightarrow}F$	Insert station C	$(A{\leftrightarrow}F{\leftrightarrow}C)$	B,D,E
3	—	$B{\leftrightarrow}D$	Discard temporarily	$(A{\leftrightarrow}F{\leftrightarrow}C)$	B,D,E
4	3	$D{\leftrightarrow}F$	Insert station D	$(A{\leftrightarrow}D{\leftrightarrow}F{\leftrightarrow}C)$	B,E
4–1	3	$B{\leftrightarrow}D$	Reconsider $B{\leftrightarrow}D$	$(A{\leftrightarrow}D{\leftrightarrow}B{\leftrightarrow}F{\leftrightarrow}C)$	E
5	—	$B{\leftrightarrow}C$	Discard permanently	$(A{\leftrightarrow}D{\leftrightarrow}B{\leftrightarrow}F{\leftrightarrow}C)$	E
6	5	$D{\leftrightarrow}E$	Insert station E	$(A{\leftrightarrow}D{\leftrightarrow}E{\leftrightarrow}B{\leftrightarrow}F{\leftrightarrow}C)$	0, Stop

Cardinal Head–Tail for Single-Row Layouts We can solve the single-row layout problem of this section by using the cardinal head–tail method (see details in Section 13.4.2). Solving Example 15.3 by this method will result in the final layout solution of $A{\leftrightarrow}D{\leftrightarrow}E{\leftrightarrow}F{\leftrightarrow}C{\leftrightarrow}B$ with 1170 total flow. This solution is better (has a lower total flow) than the solution obtained by using the ordinal head–tail method for Example 15.3. Generally, the cardinal head–tail method generates better solutions than the ordinal head–tail method, but such improvements are at the expense of more computational burden.

15.3.4 Rule-Based Layout for Multiple Rows

We can extend the single-row RBL head–tail method (for bidirectional flow) to solve multiple-row layout problems. The single-row approach expands horizontally. The multirow layout approach uses all rules of the single-row approach but allows both horizontal and vertical expansions.

The following rules show how to progressively augment s_q by combining existing $s_q = (\ldots \leftrightarrow i \leftrightarrow \ldots)$, the partially built layout, with a new $s_p = (i{\leftrightarrow}j)$; that is, find a location for department j. By using the following rules, all stations will be assigned and a complete solution will be obtained.

Expanding Phase

Ordinal Rule 1 (Attaching Sequences, Horizontal and Vertical)

(a) *Horizontal* If there is one common department in the pair of sequences s_p and s_q such that the common department is either at the head or the tail of sequence s_p and s_q, then form a new sequence by attaching s_p to s_q by their common department. For example, given $s_q = (C \leftrightarrow B \leftrightarrow E)$ and $s_p = (A \leftrightarrow C)$, the new sequence is $s_q = (A \leftrightarrow C \leftrightarrow B \leftrightarrow E)$.

(b) *Vertical* A sequence can be attached in the middle of an existing sequence at a $90°$ angle if the new sequence has one common department with the existing sequence. For example, adding the sequence $(B \leftrightarrow F)$ to the sequence $(A \leftrightarrow B \leftrightarrow C \leftrightarrow D)$ in Figure 15.16a results in the sequence in Figure 15.16b.

Ordinal Rule 2 (Attaching Sequences, Diagonal) If there is a common department in sequences s_p and s_q, for example, $s_p = (i \leftrightarrow j)$ and $s_q = (\ldots \leftrightarrow i \leftrightarrow \ldots)$, and ordinal rule 1 fails, that is, department j cannot be placed next to i in s_q, then insert it at a $45°$ angle on the side which is available and has the highest ranking priority with respect to its adjacent departments.

For example, consider step 5 of Figure 15.9. Suppose that in the next step the highest ranked pair is $(1 \leftrightarrow 5)$. Department 5 cannot be located next to department 1, horizontally or vertically. But there are four possible locations for 5 at a $45°$ angle. Suppose that pairs (5,2) and (5,6) have the highest ranking; then 5 will be located next to 2 and 6, as shown in step 6 of Figure 15.9.

Ordinal Rule 3 (Extending) Keep the existing sequence as open (extended) as possible by stretching all connected lines as straight as possible. That is, add the new sequence to the old sequence in the form of a straight line if possible. For example, adding $(E \leftrightarrow F)$ to the sequence in Figure 15.16b results in the sequence in Figure 15.16c.

Reshaping Phase

Ordinal Rule 4 (Rotation) To fit the partial constructed layout tree into the given layout locations, if possible, use vertical, horizontal, or diagonal rotation of branches while satisfying closeness preferences.

In Example 15.4, in Figure 15.16c, E is rotated around D, resulting in Figure 15.16d to allow B to fold down to C in Figure 15.16e.

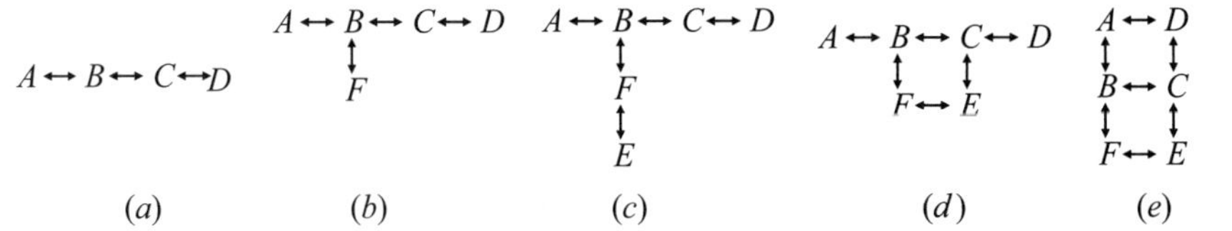

Figure 15.16   (a–c) Sequences expanding; (d, e) sequences folding.

Ordinal Rule 5 (Folding)   To fit the partially constructed layout tree into the given layout locations, if possible, bend the branches of the layout tree while satisfying closeness preferences. For example, suppose the given layout is 2×3 where the given constructed sequence is shown in Figure 15.16c. Consider folding branch *B–F–E*. It can be folded to the right or to the left. Because the ranking of (*E,C*) is higher than (*E,A*), fold the branch to the right, connecting *E* to *C*, as shown in Figure 15.16d. Figure 15.16d uses the folding rule to fit into the given layout space. To do this, bend *A* and *D* upward and connect them as shown in Figure 15.16e. This is the final layout which now fits in the 2×3 space.

Fitting Phase

Ordinal Rule 6 (Conflict Resolution for Multirow Layout)   If the generated solution using the reshaping rules does not fit into the given layout space, cut low-priority (ranking) departments and paste them into the closest empty spaces based on closeness preferences.

Suppose that Figure 15.17a shows the sequence obtained from phase I where the layout space is 2×3. In phase II folding results in Figure 15.17b. The sequence given in Figure 15.17b cannot be folded to fit the layout shape of 2×3. Suppose that the departments (*A,B*) have the lowest flow of all attached departments in the layout; then *A* is the best choice to be cut. Apply ordinal rule 6 by cutting department *A* and pasting it to the open space. The result is shown in Figure 15.17c.

Cardinal RBL Head–Tail Method for Multirow Layout   The method covered in this section is based on the RBL ordinal head–tail method. However, it is also possible to use the cardinal head–tail method to solve this problem. To do this, follow all the rules presented in this section, but the location decisions are made by measuring the total flow for each possible insertion of the candidate department in the current alternative (partially built layout) and then choosing the best alternative in each iteration.

Improving Obtained Solution by Pairwise Exchange Method of Section 15.4
It may be possible to improve the RBL final solution by applying the Pairwise Exchange method which will be covered in the next section.

Example 15.4   RBL Head–Tail Multirow Layout: Six Departments   Consider the information given in Figure 15.18. Suppose the layout should be in the form of a 2×3 rectangle.

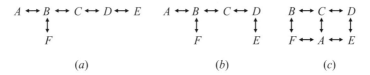

(a)   (b)   (c)

Figure 15.17   (a) Given solution; (b) folding; (c) cut and paste *A*.

Locations

1	2	3
4	5	6

Distance D

	1	2	3	4	5	6
1	—	1	2	1	2	3
2		—	1	2	1	2
3			—	3	2	1
4				—	1	2
5					—	1
6						—

Flow V

	A	B	C	D	E	F
A	—	10	20	30	10	100
B		—	50	60	10	20
C			—	10	20	80
D				—	50	60
E					—	10
F						—

Ranking of Flows

	A	B	C	D	E	F
A	—	11	8	7	11	1
B		—	5	3	11	8
C			—	11	8	2
D				—	5	3
E					—	11
F						—

Figure 15.18   Initial data for Example 15.4.

The rules for the RBL head–tail heuristic are applied as follows:

Expanding Steps

1. $s_p = (A \leftrightarrow F)$; $s_q = (A \leftrightarrow F)$.
2. $s_p = (C \leftrightarrow F)$ or $s_p = (F \leftrightarrow C)$; $s_q = (A \leftrightarrow F \leftrightarrow C)$.
3. $s_p = (B \leftrightarrow D)$. Ignore s_p temporarily.
4. $s_p = (D \leftrightarrow F)$; s_p can be attached to s_q; see Figure 15.7a.
5. $s_p = (B \leftrightarrow D)$; B can be attached to D; see Figure 15.7b.
6. $s_p = (D \leftrightarrow E)$; E can be attached to D. Since (E,C) has a higher flow than (E,A), E is attached to the right side of D. See Figure 15.7c.

All departments are assigned; reshape to fit the layout to the 2×3 block requirement.

Reshaping Steps

7. Consider Figure 15.19c. (E,C) has a flow of 20 compared to a flow of 10 for (E,A). (B,C) has a flow of 50 compared to a flow of 10 for (B,A). Therefore, the advantage of having B next to C (an improvement of 40) outweighs the advantage of having E next to C (an improvement of 10). Therefore, in Figure 15.19c, E can be rotated from left to the right side. This rotation will allow B to fold toward (closer to) C instead of A. The result of this rotation rule is shown in Figure 15.19d.
8. In Figure 15.19e, B is folded down toward C.

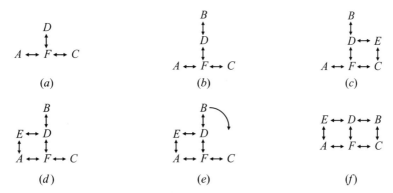

Figure 15.19 (*a–c*) Sequence expanding; (*d*) rotation of *E*; (*e*) sequence folding; (*f*) solution.

The final solution layout is shown in Figure 15.19*f*. For this solution

$$\text{Total flow} = 3(10) + 2(20) + 2(30) + 1(10) + 1(100) + 1(50) + 1(60) + 2(10) + 2(20)$$
$$+ 2(10) + 3(20) + 1(80) + 1(50) + 1(60) + 2(10) = 700$$

The summary of the method is shown in the following table.

Iteration	Used Rank	s_p	Action Taken		Resulting Sequence	Candidates
1	1	$A \leftrightarrow F$	Add s_p as a new sequence		$\{(A \leftrightarrow F)\}$	B,C,D,E
2	2	$C \leftrightarrow F$	Insert station C		$\{(A \leftrightarrow F \leftrightarrow C)\}$	B,D,E
3	—	$B \leftrightarrow D$	Discard temporarily		$\{(A \leftrightarrow F \leftrightarrow C)\}$	B,D,E
4	3	$D \leftrightarrow F$	Insert station D		Figure 15.19*a*	B,E
4	3	$B \leftrightarrow D$	Consider $B \leftrightarrow D$		Figure 15.19*b*	E
5	—	$B \leftrightarrow C$	Discard permanently		—	E
6	5	$D \leftrightarrow E$	Insert station E		Figure 15.19*c*	—
7	—	—	Rotate E		Figure 15.19*d*	—
7	—	—	Fold B		Figures 15.19*e*, 15.19*f*	

Example 15.5   RBL Head–Tail Multirow Layout: Nine Departments   Consider the information given below. Suppose the layout is in the form of the 3×3 square below. The distances are rectilinear. For example, the distance between locations 1 and 2 is 1, and the distance between locations 1 and 9 is 4.

1	2	3
4	5	6
7	8	9

					Flow V										Rankings of Pairs (i,j) of Flow V			
	A	B	C	D	E	F	G	H	I		A	B	C	D	E	F	G	H	I
A		30	85	44	79	89	52	28	48	A		26	6	25	9	5	18	27	20
B			80	48	2	28	90	61	55	B			8	20	35	27	4	14	16
C				66	16	47	68	91	97	C				12	31	22	10	3	2
D					61	55	50	46	83	D					14	16	19	23	7
E						64	15	67	46	E						13	32	11	23
F							8	13	25	F							34	33	29
G								17	1	G								30	36
H									99	H									1
I										I									

The rules of the RBL head–tail approach are applied as follows:

Expanding Steps

1. Consider $s_p = (H\leftrightarrow I)$; $s_q = (H\leftrightarrow I)$.
2. Consider $s_p = (C\leftrightarrow I)$; $s_q = (H\leftrightarrow I\leftrightarrow C)$.
3. Consider $s_p = (C\leftrightarrow H)$. Discard this pair as C and H are already assigned and extended.
4. Temporarily discard $(B\leftrightarrow G)$.
5. Temporarily discard $(A\leftrightarrow F)$.
6. Consider $s_p = (A\leftrightarrow C)$; $s_q = (H\leftrightarrow I\leftrightarrow C\leftrightarrow A)$.
7. Reconsider $s_p = (A\leftrightarrow F)$; $s_q = (H\leftrightarrow I\leftrightarrow C\leftrightarrow A\leftrightarrow F)$.
8. Consider $s_p = (D\leftrightarrow I)$. Add s_p to s_q as shown in Figure 15.20a.
9. Consider $s_p = (B\leftrightarrow C)$. Add s_p to s_q as shown in Figure 15.20b.

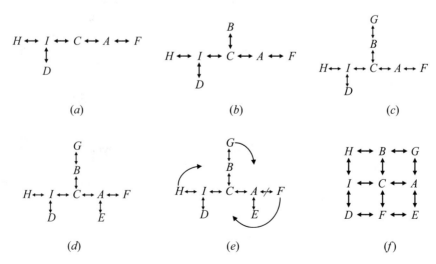

Figure 15.20 (a) Expanding, add D; (b) expanding, add B; (c) expanding, add G; (d) expanding, add E; (e) reshaping and fitting; (f) final solution.

10. Reconsider $s_p = (B \leftrightarrow G)$. Add s_p to s_q as shown in Figure 15.20c.

11. Consider $s_p = (A \leftrightarrow E)$. Add s_p to s_q as shown in Figure 15.20d.

All elements have been added, but the solution does not fit in the 3×3 layout. Go to the next phase.

Reshaping Step

12. Branches will be folded according to the highest preferences. Pairs (B,H), (E,F), and (A,G) are highly ranked. Therefore, H should be folded toward B and G should be folded toward A, as shown in Figure 15.20e.

Fitting Step

13. Department F cannot be folded into the empty space; use the cut-and-paste rule to move it into the empty space, as shown in Figure 15.20e.

Figure 15.20 is the final layout. At this point, the pairwise exchange method may be applied in an attempt to improve the solution.

The summary of the method is shown in the following table.

Iteration	Used Rank	s_p	Action Taken	Resulting Sequence, S	Candidates
1	1	$H \leftrightarrow I$	Add s_p as a new sequence	$(H \leftrightarrow I)$	A,B,C,D,E,F,G
2	2	$C \leftrightarrow I$	Insert station C	$(H \leftrightarrow I \leftrightarrow C)$	A,B,D,E,F,G
3	3	$C \leftrightarrow H$	Discard permanently	—	A,B,D,E,F,G
4	—	$B \leftrightarrow G$	Discard temporarily	—	A,B,D,E,F,G
5	—	$A \leftrightarrow F$	Discard temporarily	—	A,B,D,E,F,G
6	6	$A \leftrightarrow C$	Insert station A	$(H \leftrightarrow I \leftrightarrow C \leftrightarrow A)$	B,D,E,F,G
7	5	$A \leftrightarrow F$	Consider $A \leftrightarrow F$	$(H \leftrightarrow I \leftrightarrow C \leftrightarrow A \leftrightarrow F)$	B,D,E,G
8	7	$D \leftrightarrow I$	Insert station D	Figure 15.20a	B,E,G
9	8	$B \leftrightarrow C$	Insert station B	Figure 15.20b	E,G
10	4	$B \leftrightarrow G$	Consider $B \leftrightarrow G$	Figure 15.20c	E
11	9	$A \leftrightarrow E$	Insert station E	Figure 15.20d	—
12	—	—	Fold	Figure 15.20e	—
13	—	—	Cut and paste	Figure 15.20f	—

The final solution is shown in Figure 15.20f, where the total flow is found as the sum of the flows for each pair (i, j). The flow for pair (i, j) is the product of Distance

(i, j) and Flow (i, j). See the following table for the distance and flow × distance for each pair.

	Distances per Pair										Flow × Distance per Pair								
	A	B	C	D	E	F	G	H	I		A	B	C	D	E	F	G	H	I
A		2	1	3	1	2	1	3	2	A		60	85	132	79	178	52	84	96
B			1	3	3	2	1	1	2	B			80	144	6	56	90	61	110
C				2	2	1	2	2	1	C				132	32	47	136	182	97
D					2	1	4	2	1	D					122	55	200	92	83
E						1	2	4	3	E						64	30	268	138
F							3	3	2	F							24	39	50
G								2	3	G								34	3
H									1	H									99
I										I									

$$\text{Total flow} = \sum_{i=1}^{n}\sum_{j=1}^{n}\text{flow per pair}(i,j) = \sum_{i=1}^{n}\sum_{j=1}^{n}(\text{Distance}(i,j) \times \text{Flow}(i,j)) = 3240$$

15.3.5 RBL for Qualitative and Odd Shapes

The RBL approach can be applied to departments of unequal sizes and also to qualitative ratings. For unequal-size departments, first assume the departments are equal in size and then solve the expansion and folding phases of the layout problem. Once the folding phase is complete, insert the non-equal-sized departments. Fit the layout tree into the desired layout. For qualitative information, convert the qualitative preferences into quantitative information as presented in the SLP method.

Example 15.6 RBL Head–Tail for Unequal Department Sizes   Consider Example 15.1, which has qualitative preferential closeness and non-equal-sized departments. The A, E, I, O, U, and X ratings can be converted to the numerical values 5, 4, 3, 2, 1, and 0, respectively. Then the ranking of the pairs of departments i, j will be performed and are shown below. Note that pairs that tie (i.e., with the same ratings) have the same ranking.

	1	2	3	4	5	6	7
1		1	11	1	8	1	1
2			15	5	11	5	X
3				X	15	15	5
4					15	11	11
5						8	8
6							X
7							

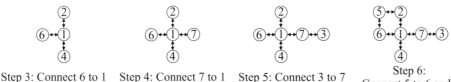

Step 3: Connect 6 to 1   Step 4: Connect 7 to 1   Step 5: Connect 3 to 7   Step 6: Connect 5 to 6 and 2

Figure 15.21 Graphical representation of layout tree steps 3, 4, 5, and 6.

The construction of the layout by the RBL head–tail method is as follows:

Expanding Steps

Step 1 $s_p = (1 \leftrightarrow 2)$; $s_q = (1 \leftrightarrow 2)$.

Step 2 $s_p = (1 \leftrightarrow 4)$; $s_q = (4 \leftrightarrow 1 \leftrightarrow 2)$.

Steps 3, 4, and 5   Departments 6, 7, and 3 are added; see Figure 15.21.

Step 6   Next consider $s_p = (1 \leftrightarrow 5)$. Department 5 can be located in four possible locations diagonally from department 1, as shown in Figure 15.21, step 6. Because pairs (5,2) and (5,6) have the highest ranking, it will be located next to 2 and 6.

Since all departments are assigned, the expansion phase is complete.

Reshaping Steps

Step 7   Consider the next pair, $(4 \leftrightarrow 6)$. Departments 4 and 6 cannot be folded. Consider the next pair, $(4 \leftrightarrow 7)$. Departments 4 and 7 cannot be folded. Consider the next pair, $(2 \leftrightarrow 3)$. Fold 3 closer to 2 (see step 7 in Figure 15.22).

Fitting Steps

Step 8   Present the actual size and shape of the departments.

Step 9   Move the departments to fit in the layout size of 4×6 units.

The final layout is shown in step 9 of Figure 15.22. The center of each department is shown by a filled circle. The distances for the final RBL head–tail solution between each

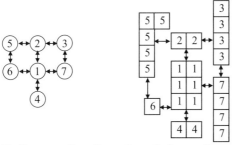

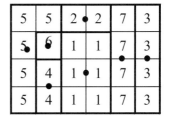

Step 7: Connect 3 to 2   Step 8: Insert Spaces   Step 9: Final Layout

Figure 15.22 Graphical representation of steps 7, 8, and 9.

pair of departments are shown in the following table:

	1	2	3	4	5	6	7
1		200	300	200	310	250	200
2			400	400	350	250	300
3				500	510	450	100
4					210	150	400
5						80	410
6							350
7							

Now, calculate the total preferential closeness using these distances:

$$\text{Total closeness} = \sum_{i=1}^{n} \sum_{j=1}^{n} (\text{Closeness}(i, j) \times \text{Distance}(i, j)) = 13{,}620$$

The summary of the method is shown in the following table:

Iteration	Used Rank	s_p	Action Taken	Resulting Sequence, S	Candidates
1	1	1↔2	Add s_p as a new sequence	1↔2	3,4,5,6,7
2	1	1↔4	Insert station 4	4↔1↔2	3,5,6,7
3	1	1↔6	Insert station 6	Figure 15.21, step 3	3,5,7
4	1	1↔7	Insert station 7	Figure 15.21, step 4	3,5
—	—	2↔4	Discard permanently	—	3,5
—	—	2↔6	Discard permanently	—	3,5
5	5	3↔7	Insert station 3	Figure 15.21, step 5	5
6	8	1↔5	Insert station 5	Figure 15.21, step 6	—
7	8	2↔3	Reshape: Fold 3 to 2	Figure 15.22, step 7	—
9	—	—	Reshape: Show spaces	Figure 15.22, step 8	—
10	—	—	Fit	Figure 15.22, step 9	—

Comparison of RBL and SLP Solutions   The above example was solved by the SLP method in Section 15.2. The layout and the distances for the SLP solution are shown in Figure 15.23 and the table below. The total closeness for the SLP solution is 15,500. The

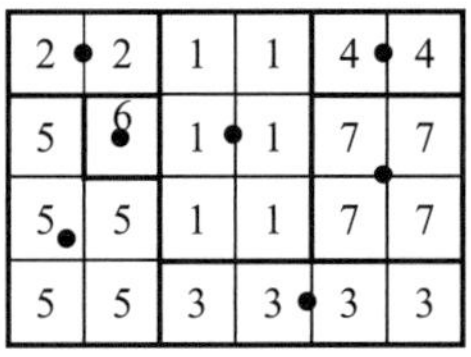

Figure 15.23   SLP layout solution (Example 15.1).

rule-based planning solution (Example 15.6) was 13,620, which is considerably better than the solution found by the SLP method in Example 15.1.

	1	2	3	4	5	6	7
1		300	300	300	330	150	250
2			600	400	230	150	550
3				400	390	450	250
4					630	450	150
5						180	480
6							400
7							

See Supplement S15.1.xls.

15.3.6 Rule-Based Layout with Constraints

In layout problems, often certain conditions and constraints must be considered. Such constraints can be enforced when the given department constraints are considered in the expansion, folding, and fitting phases (see Figure 15.24). We show this through the following example.

Consider Example 15.5. Suppose that the following constraints should be satisfied:

(a) Department F should have two exits to the outside; therefore, it should be at a corner.
(b) For safety reasons department G should be connected to both departments B and I.

Consider the solution of Example 15.5 up to iteration 12.

1. Rotate E and move it above A to allow for F to take position in the corner under A.
2. Fold G to the left side above I to satisfy condition (b) above.
3. Cut H and paste it between D and F.

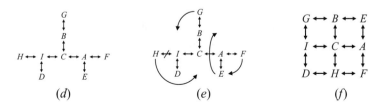

(d)   (e)   (f)

Figure 15.24 (*a*) Sequence expanding; (*b*) folding and fitting; (*c*) final solution.

15.4 PAIRWISE EXCHANGE METHOD

15.4.1 Pairwise Exchange Procedure

Once an initial layout is designed, it can be improved by using the pairwise exchange method, which involves swapping all possible pairs of departments in order to generate a better solution. The objective for this layout method is to minimize the total interdepartmental flow of all products among all pairs of departments. The swap of the two departments that results in the lowest total flow cost is selected as the best layout for the given iteration. Then the pairwise exchange method is repeated using the selected alternative as the current layout solution. The process is repeated until no more improvements are possible. This method is a heuristic, that is, it is not guaranteed that the optimal solution can be found. For large layout problems, the final solution will greatly depend on the initial solution.

Pairwise exchange is based on a class of optimization methods called steepest descent where in each iteration the best direction for improving the objective function is identified and a step in that direction is taken. Pairwise exchange has been applied in several facility layout software packages such as CRAFT (Computer Relative Allocation of Facilities Technique); see Section 15.8.

The steps of this method are as follows:

1. Generate all possible adjacent layout alternatives by swapping each possible pair of departments of the current layout.
2. For each generated adjacent layout, find the total cost TC.
3. Compare the TC values of the current and all generated adjacent layouts. If the current layout has the lowest TC, stop and accept it as the best layout. Otherwise, choose the generated layout (which has not been considered before) associated with the lowest TC. Label it as the current layout. Go to step 1.

In step 3, it is possible to choose an adjacent layout which may have the same TC value of the current layout. To avoid cycling, the new selected alternative should always be a new one, that is, not explored before. By choosing alternatives that have the same TC as the current alternative, the possibility of finding better new alternative layouts may increase in future iterations.

Example 15.7 Pairwise Exchange for Four Departments Consider a facility with four departments of equal sizes. The initial layout (assignment of departments to locations) is shown in Figure 15.25. The Figure 15.25 below gives the rectilinear distances and flow between departments.

In this example, assume that the cost per unit movement for all pairs of departments are the same, that is, use $c_{ij} = 1$ for all i, j. Use pairwise exchange to find better layout solutions.

Consider the following table, which shows the calculation of total flow for exchanging each pair of departments. To find the total flow for each exchange, first find the distance column for the given exchange. Then find the total flow by multiplying the flow and distance columns. For example, the calculation of total flow (TF) for exchanging departments A and C is

$$TF(A, C) = 1 \times 2 + 14 \times 1 + 14 \times 1 + 10 \times 1 + 13 \times 1 + 13 \times 2 = 79$$

Locations	
1	2
3	4

Distances of Locations D				
	1	2	3	4
1	—	1	1	2
2		—	2	1
3			—	1
4				—

Flow V				
	A	B	C	D
A	—	1	14	14
B		—	10	13
C			—	13
D				—

Initial Assignments

A	B
C	D

Figure 15.25   Initial data for pairwise exchange.

The alternatives generated in the first iteration of the pairwise exchange method are shown below. The goal is to minimize the total flow.

Flow	Facility Pairs	Current Layout Distances	Pairwise Exchange—Distances					
			AB	AC	AD	BC	BD	CD
1	AB	1	1	2	1	1	2	1
14	AC	1	2	1	1	1	1	2
14	AD	2	1	1	2	2	1	1
10	BC	2	1	1	2	2	1	1
13	BD	1	2	1	1	1	1	2
13	CD	1	1	2	1	1	2	1
	Total flow	89	92	79	89	89	79	92

The total flow of the current layout is 89; by switching A–C or B–D, the total flow can be reduced to 79. Suppose A and C are selected to be exchanged. Then the improved layout is as follows:

C	B
A	D

Label this layout as the current layout and repeat the pairwise exchange procedure. No better alternatives can be found, declare the current layout as the best alternative, and stop.

Example 15.8   Pairwise Exchange for Six Departments   Consider six departments A–F. The location positions are 1–6 as shown in Figure 15.26. Flow and distance matrices and the current layout (initial assignments) are shown below. The total flow for this initial layout is 1030. Find better alternative solutions by the pairwise exchange (PWE) method.

Locations

1	2	3
4	5	6

Initial Assignments

A 1	*D* 2	*C* 3
B 4	*E* 5	*F* 6

Flow, *V*

	A	*B*	*C*	*D*	*E*	*F*
A	—	10	20	30	10	100
B		—	50	60	10	20
C			—	10	20	80
D				—	50	60
E					—	10
F						—

Distance of Locations, *D*

	1	2	3	4	5	6
1	—	1	2	1	2	3
2		—	1	2	1	2
3			—	1	2	1
4				—	1	2
5					—	1
6						—

Figure 15.26 Initial data for pairwise exchange.

At the current alternative, adjacent alternatives generated by the PWE method are shown below.

			Pairwise Exchange—Distances														
Flows	Facility Pairs	Current Layout	*AB*	*AC*	*AD*	*AE*	*AF*	*BC*	*BD*	*BE*	*BF*	*CD*	*CE*	*CF*	*DE*	*DF*	*EF*
10	*AB*	1	1	3	2	1	2	2	1	2	3	1	1	1	1	1	1
20	*AC*	2	3	2	1	2	1	1	2	2	2	1	2	3	2	2	2
30	*AD*	1	2	1	1	1	2	1	1	1	1	2	1	1	2	3	1
10	*AE*	2	1	2	1	2	1	2	2	1	2	2	2	2	1	2	3
100	*AF*	3	2	1	2	1	3	3	3	3	1	3	3	2	3	1	2
50	*BC*	3	2	1	3	3	3	3	1	2	1	2	1	2	3	3	3
60	*BD*	2	1	2	1	2	2	1	2	1	2	3	2	2	1	2	2
10	*BE*	1	2	1	1	1	1	2	1	1	1	1	3	1	2	1	2
20	*BF*	2	3	2	2	2	1	1	2	1	2	2	2	3	2	2	1
10	*CD*	1	1	1	2	1	1	2	3	1	1	1	1	2	2	1	1
20	*CE*	2	2	2	2	2	2	1	2	3	2	1	2	1	1	2	1
80	*CF*	1	1	3	1	1	2	2	1	1	3	2	1	1	1	1	2
50	*DE*	1	1	1	2	1	1	1	1	2	1	2	1	1	1	1	2
60	*DF*	2	2	2	3	2	1	2	2	2	2	1	2	1	1	2	1
10	*EF*	1	1	1	1	3	2	1	1	2	1	1	1	2	2	1	1
Total flow		1030	890	910	970	**850**	1050	1020	950	980	910	1100	950	860	940	890	980

The best adjacent alternative is generated by swapping departments *A* and *E*, which has a total flow of 850 (shown in bold). This is selected as the best alternative; therefore, for the next iteration, the current layout is

E	*D*	*C*
B	*A*	*F*

The following table shows the next iteration of the PWE method. The details of pairwise exchanges are not shown.

Current Layout	Pairwise Exchange														
	AB	AC	AD	AE	AF	BC	BD	BE	BF	CD	CE	CF	DE	DF	EF
TF	850	870	**770**	870	1030	870	840	770	820	910	920	910	860	850	930	980

The best adjacent alternative is generated by swapping departments A and C, with a total flow of 770. Thus, the current layout will become

E	D	A
B	C	F

The following table shows the next iteration of the PWE method. No better alternative is generated. The current layout remains the best with a total flow of 770. The details of pairwise exchanges are not shown.

Current Layout	Pairwise Exchange														
	AB	AC	AD	AE	AF	BC	BD	BE	BF	CD	CE	CF	DE	DF	EF
TF	**770**	850	850	890	950	810	780	810	800	950	820	910	840	780	850	900

The total flow of the initial layout was 1030, and the total flow of the final layout is 770, that is, almost 25% improvement on the total flow. This demonstrates the PWE method can be a powerful tool to improve the given initial layout solution. But its effectiveness depends highly on the initial layout. The final solution can often be a local optima.

See Supplement S15.2.xls.

15.4.2   Pairwise Exchange for Qualitative Data

It is possible to use qualitative information for desired closeness of pairs of departments, for example, A, E, I, O, U, and X instead of quantitative data in the PWE method. To do this, first convert the qualitative information into quantitative information by assigning ordinal numerical scores to the qualitative information (such as A, E, I, O, U, and X). For each pair of departments, let p_{ij} be the desired closeness and P be the qualitative matrix. There are two scoring options for converting the qualitative ratings (A, E, I, O, U, X) to quantitative values, as shown in Table 15.1. The final solution will significantly depend on the quantitative ratings method I (4, 3, 2, 1, 0, −10) or II (5, 4, 3, 2, 1, 0). By using

Location

1	2	3	4

Locations Distance Matrix

	1	2	3	4
1	—	1	2	3
2		—	1	2
3			—	1
4				—

Initial Assignment

A	B	C	D

Qualitative Maxtrix

	A	B	C	D
A	—	X	E	A
B		—	U	E
C			—	X
D				—

Converted Qualitative Matrix

	A	B	C	D
A	—	−10	3	4
B		—	0	3
C			—	−10
D				—

Figure 15.27   Initial data for PWE example.

qualitative closeness matrix P, the total closeness for a given layout can be calculated instead of its total flow:

$$\text{Total closeness} = \sum_{i=1}^{n}\sum_{j=1}^{n} d_{ij} p_{ij} \tag{15.3}$$

where d_{ij} is the distance from department i to department j and p_{ij} is the closeness rating for departments i and j. The closeness ratings can be viewed as the importance of locating pairs of departments close to each other. Therefore, the total closeness can be treated in the same way that total flow is used in the PWE method, that is, it should be minimized.

Example15.9   Pairwise Exchange Method for Qualitative Information   Consider four departments A, B, C, and D, four locations, the initial assignment, and the qualitative matrix as shown in Figure 15.27. Find improved solutions by the PWE method.

The converted qualitative matrix using rating I of Table 15.1 for this example is shown in Figure 15.28. The total closeness score for the initial layout is given as

$$\text{Total closeness} = 1 \times (-10) + 2 \times (3) + 3 \times (4) + 1 \times (0) + 2 \times (3) + 1 \times (-10) = 4$$

After applying the PWE method, in the fourth iteration, no more improvement is possible and the final layout is C, A, D, B with total closeness $= -30$.

Location	1	2	3	4	Total Closeness
Initial layout	A	B	C	D	4
Iteration 1	C	B	A	D	−24
Iteration 2	C	A	B	D	−26
Iteration 3	C	A	D	B	−30

Supplement S15.3.doc shows the details of Example 15.9.

Location

1	2	3	4

Assignment

A	B	C	D

Flow Matrix

	A	B	C	D
A	—	100	40	10
B		—	60	30
C			—	90
D				—

Qualitative Matrix

	A	B	C	D
A	—	X	E	A
B		—	U	E
C			—	X
D				—

Converted Qualitative Matrix

	A	B	C	D
A	—	0	4	5
B		—	1	4
C			—	0
D				—

Figure 15.28   Initial data for bicriteria layout problem.

15.4.3   Hybrid Rule-Based and Pairwise Exchange[†]

It is possible to combine PWE and RBL methods in order to create a more powerful layout algorithm that has the advantages of both methods. Here are some possible options:

1. Apply PWE at the end of each iteration of RBL in an attempt to improve the partially constructed solution.
2. Apply PWE to the final solution of the RBL method.

The advantages of combining the RBL method with the PWE method are: RBL does not require an initial feasible solution; it requires substantially less calculations; it may not get trapped in a local optima; it can easily use non-equal-sized departments; and it can incorporate nonquantitative restrictions (constraints) and preferences during the construction process. The advantage of PWE to RBL is that it is designed to improve a given solution; therefore, it may be used to improve upon the final solution of RBL.

The following examples discuss some of these points.

(a) Example 15.7 (four departments) can be solved by the RBL method as follows.
  (i) $(A \leftrightarrow C)$.
  (ii) Add $(A \leftrightarrow D)$ to generate $(D \leftrightarrow A \leftrightarrow C)$.
  (iii) Add $(B \leftrightarrow D)$ to generate $(B \leftrightarrow D \leftrightarrow A \leftrightarrow C)$.
  (iv) Fold the sequence to fit into the given location space. The final layout is:

C	A
B	D

This solution is the same as the final solution obtained in Example 15.7, but it was generated with much less computational effort.

(b) Example 15.8 (six departments) was solved in three iterations by the PWE method with a total flow of 770. Example 15.8 was also solved by the RBL method as

Example 15.4 in Section 15.3, which had a total flow of 700. The final solution using RBL was better than the final solution using PWE, while many more computations were needed for PWE. For this example, when PWE was applied to the final solution of RBL, no better solution could be found.

See Supplement S15.4.xls.

(c) Example 15.9 (four departments) was solved in four iterations by the PWE method. This example can be solved by the RBL method as shown below:

    (i) Consider $(A \leftrightarrow D)$.

    (ii) Add $(A \leftrightarrow C)$ to $(A \leftrightarrow D)$. This results in $(C \leftrightarrow A \leftrightarrow D)$.

    (iii) Add $(B \leftrightarrow D)$ to $(C \leftrightarrow A \leftrightarrow D)$. This results in $(C \leftrightarrow A \leftrightarrow D \leftrightarrow B)$.

The final solution of RBL is the same as the final solution of Example 15.9 but RBL uses much less computation.

(d) Example 15.5 (nine departments) was solved in Section 15.3 by the RBL method. The final solution is presented in Figure 15.8*f* with a total flow of 3240. Now, one can try to improve the final solution of RBL by using PWE. But, even for small problems of this size, it takes a substantial amount of time to apply even one iteration of PWE. In the first iteration of PWE, no improved solution can be found.

Supplement S15.5.docx shows the details of the PWE method for Example 15.5.

15.5 MULTICRITERIA LAYOUT PLANNING

15.5.1 Bicriteria Layout

In bicriteria facility layout problem, the two objectives are:

$$\text{Minimize total flow:} \quad f_1 = \sum_{i=1}^{n} \sum_{j=1}^{n} d_{ij} v_{ij}$$

$$\text{Minimize total closeness:} \quad f_2 = \sum_{i=1}^{n} \sum_{j=1}^{n} d_{ij} p_{ij}$$

For example, suppose that the flow between departments A and B is very high and from a total-flow point of view should be located very close to each other. However, from a safety (or maintenance) point of view, they should be very far from each other. The above bicriteria problem can be used to handle such conflicting objectives.

    Suppose that an additive utility function can be used to rank multicriteria layout alternatives. In the following we show how to normalize each objective function and generate an efficient solution that optimizes the given additive utility function. Let w_1 and w_2 be the weights of importance of the two objective functions, respectively.

    First find the normalized values of the two objective functions. To normalize the flow information, find the largest flow ($v_{ij,\text{max}}$) and smallest flow ($v_{ij,\text{min}}$), and then use the formula shown below to normalize all flow values presented by v_{ij}: they will be between 0 and 1. Similarly, closeness information can be normalized between 0 and 1 using the following formula where $p_{ij,max}$ and $p_{ij,min}$ are the largest and smallest p_{ij} values, respectively:

$$v'_{ij} = \frac{v_{ij} - v_{ij,min}}{v_{ij,max} - v_{ij,min}} \qquad p'_{ij} = \frac{p_{ij} - p_{ij,min}}{p_{ij,max} - p_{ij,min}}$$

Then for each pair of departments (i, j), calculate

$$z_{ij} = w_1 v'_{ij} + w_2 p'_{ij}$$

The matrix of z_{ij} values is the composite matrix Z. This composite multicriteria matrix can be used to solve one of the facility layout methods covered in this chapter.

Note that the additive utility function (U) can be defined as the weighted sum of the flow and the closeness values using weights of importance of w_1 and w_2 for the flow and closeness, respectively. In order to minimize the additive utility function $U = w_1 f_1' + w_2 f_2$:

$$\text{Minimize } U = \sum_{i-1}^{n} \sum_{j=1}^{n} d_{ij}(w_1 v'_{ij} + w_2 p'_{ij}) = \sum_{i=1}^{n} \sum_{j=1}^{n} d_{ij} z_{ij}$$

The value of $z_{ij} = w_1 v_{ij}' + w_2 p_{ij}'$ is constant for i, j and given weights. Therefore, z_{ij} only need to be calculated once for each pair of departments. Therefore, for the generated Z matrix, the bicriteria layout problem can be solved by RBL or PWE methods. The alternative with the smallest U value is the best layout.

Example 15.10 Bicriteria Layout   For the four-department facility layout problem presented in Figure 15.28, find the best bicriteria layout solution. Suppose that the DM's weights of importance for flow and closeness objectives are $w_1 = 0.3$ and $w_2 = 0.7$, respectively. Consider using 0–5 numerical ratings for X–A qualitative ratings.

First convert qualitative closeness data to quantitative data as presented in Figure 15.28. Then normalize the flow matrix and closeness matrix values as shown in Figure 15.29.

Normalized Flow Matrix
$(v'_{ij}) = (v_{ij} - 10)/(100 - 10)$

	A	B	C	D
A	—	1	0.333	0
B		—	0.556	0.222
C			—	0.889
D				—

Normalized Qualitative Matrix
$(p'_{ij}) = (p_{ij} - 0)/(5 - 0)$

	A	B	C	D
A	—	0	0.8	1
B		—	0.2	0.8
C			—	0
D				—

Figure 15.29 Normalized flow and qualitative closeness values.

A composite matrix Z can be generated by using $z_{ij} = w_1 v_{ij}' + w_2 p_{ij}' = 0.3 v_{ij}' + 0.7 p_{ij}'$:

	A	B	C	D
A		0.3	0.66	0.7
B			0.307	0.627
C				0.267
D				–

Solving Bicriteria Problem by RBL Method   First rank all pairs in descending order of z_{ij} values: (A,D), (A,C), (B,D), (B,C), (A,B), and (C,D).

The steps of RBL for assigning departments are

1. $(A \leftrightarrow D)$.
2. $(C \leftrightarrow A \leftrightarrow D)$.
3. $(C \leftrightarrow A \leftrightarrow D \leftrightarrow B)$.

This is the final solution, and its mirror is $(B \leftrightarrow D \leftrightarrow A \leftrightarrow C)$. The objective values for this layout are

$$f_1 = 2 \times 100 + 1 \times 40 + 1 \times 10 + 3 \times 60 + 1 \times 30 + 2 \times 90 = 640$$
$$f_2 = 2 \times 0 + 1 \times 4 + 1 \times 5 + 3 \times 1 + 1 \times 4 + 2 \times 0 = 16$$

No improved solution can be found by applying the PWE method to the final solution of this example.

Supplement S15.6.docx shows the details of Example 15.10.

15.5.2 Efficient Frontier

By systematically varying weights (w_1, w_2) in the method presented in Section 15.5.1, a set of efficient bicriteria alternatives can be generated. To do this, first generate the Z matrix for each given set of weights and then use either the RBL or PWE method to solve the problem.

Note that for a given set of weights, when the PWE method is used, alternatives generated in different iterations of the PWE method may also be efficient and should be recorded. That is, each PWE solution in each iteration of the PWE method can be efficient.

Consider Example 15.10. Consider the set of four weights given in the first row of the following table. For each given set of weights, solve the problem by the RBL method. The generated solution is recorded under each set of weights. The set of generated four alternatives are all efficient.

(w_1, w_2)	(0.99, 0.01)	(0.6, 0.4)	(0.5, 0.5)	(0.01, 0.99)
Layout	*ABCD*	*BACD*	*DBAC*	*CADB*
Min. f_1	420	460	580	640
Min. f_2	32	28	20	16
Efficient?	Yes	Yes	Yes	Yes

15.6   FACILITY RELAYOUT[†]

There are two types of layout problems: design of a new facility and redesign of an existing facility. The first problem was addressed in previous sections of this chapter. However, in practice, the majority of layout problems are concerned with modifying existing facilities to accommodate changes in the business strategies and coping with new technologies or renovations. Also, relayout could be a substantial concern because many existing layouts are inefficient since they were designed based on the flow information of discontinued products. That is, the layout was optimal for the old flow of materials but is inefficient for the new flow of materials. The objectives of relayout problems can be similar to the objectives of new layouts. The designer must consider the desired qualitative closeness, materials flow, and other objectives for relayout problems. The relayout problem is usually more complex than the layout problem because it is far more difficult to relocate existing facilities than of replace an old machine with a new machine. The relayout approach developed in this section can be particularly useful for solving relayout of semiflexible assembly systems.

Benefit of Relayout   The benefit of relayout is the reduction in total flow. The calculation of total flow is shown in Section 15.3.1.

Cost of Relayout   The cost of relayout is the total cost of relocating the departments. The relayout costs include removing, moving (transportation), and installing facilities in new locations. The total cost of relayout should be within a given budget as most companies have capital budget restrictions. In the following section, we develop an approach based on the PWE method for solving the relayout problem.

15.6.1   Calculating Relayout Cost

In order to move a piece of equipment (or a department), several costs must be considered. First, there will be the cost of removing the machine from its current location. Then, there is the cost of transporting the machine to its new location. Finally, there is an installation cost for the machine. Suppose that m out of n departments should be relocated where $m < n$. For relocating m departments, there are m disconnecting costs, m moving costs, and m installation costs. Consider the following three costs associated with relocating department i:

Disconnection cost for department i, D_i

Moving cost per unit distance for department i, M_i

Installation cost for department i, I_i

The total moving cost is a function of the distance (distance times M_i), while the other two costs (D_i and I_i) are fixed. See Figure 15.30.

Dept.	1	2	...	m
D_i	D_1	D_2	...	D_m
M_i	M_1	M_2	...	M_m
I_i	I_1	I_2	...	I_m

Figure 15.30   Table showing disconnecting, moving, and installing costs for department i.

Consider the relocation cost for switching two departments (i,j):

$$R(ij \rightarrow ji) = D_i + I_i + D_j + I_j + M_i d_{ij} + M_j d_{ij} \qquad (15.4)$$

where d_{ij} is the rectilinear distance that the two departments are moved. Similarly, for three departments (i,j,k),

$$R(ijk \rightarrow kij) = D_i + I_i + D_j + I_j + D_k + I_k + M_i d_{ik} + M_j d_{ij} + M_k d_{jk} \qquad (15.5)$$

where $R(ijk \rightarrow kij)$ is the cost of moving departments i, j, and k to the locations of departments k, i, and j respectively.

A general formulation for relocating m departments is presented below, where X_{ij} is a binary variable: It is 1 if department i is moved to the location of department j and 0 otherwise. The general m department formula is shown below:

$$R(ij \cdots m \rightarrow \cdots) = \sum_{i=1}^{m}(D_i + I_i) + \sum_{i=1}^{m}\sum_{j=1}^{m}(M_i d_{ij} X_{ij}) \qquad (15.6)$$

Example 15.11 Calculating Relocation Cost for Four Departments   Consider a layout with four departments. The relocation cost of each department (e.g., in thousands of dollars) and the initial layout are shown in Figure 15.31.

Relocation of Two Departments   Suppose that exchanging departments A and D provides the most improvement in the layout. By exchanging departments A and D, the new layout will be

D	B	C	A

The relocation costs are found by adding the disconnecting, moving, and installation costs:

$$R(AD \rightarrow DA) = D_A + D_D + I_A + I_D + d_{AD}M_A + d_{AD}M_D$$

Dept.	A	B	C	D
D_i	5	12	10	15
M_i	5	10	8	5
I_i	15	10	30	20

A	B	C	D

Figure 15.31   Relocation costs and initial layout.

The distance between A and D is 3 units. The cost of exchanging departments A and D is

$$R(AD \rightarrow DA) = 5 + 15 + 15 + 20 + 3 \times 5 + 3 \times 5 = \$85$$

Relocation of Three Departments   Suppose departments A, C, and D are moved but B is not moved. Suppose the best new layout is

C	B	D	A

The cost of relocation for departments A, C, and D is

$$R(ACD \rightarrow DAC) = D_A + D_C + D_D + I_A + I_C + I_D + d_{AD}M_A + d_{CA}M_C + d_{DC}M_D$$
$$R(ACD \rightarrow DAC) = 5 + 10 + 15 + 15 + 30 + 20 + 3 \times 5 + 2 \times 8 + 1 \times 5 = \$131$$

The final relocation cost is $131.

15.6.2   Relayout with Given Budget

The Relayout problem is concerned with selecting the best layout alternative within the given budget. This problem can be solved by minimizing either the total qualitative closeness (TQC) or TF of all pairs of all departments subject to the given budget limitation for relayout:

$$\text{TQC} = \sum_{i=1}^{m} \sum_{j=1}^{m} d_{ij} p_{ij} \tag{15.7a}$$

$$\text{TF} = \sum_{i=1}^{m} \sum_{j=1}^{m} d_{ij} v_{ij} \tag{15.7b}$$

One can consider either qualitative closeness preferences or materials flow for exchange of pairs of departments. In the following, we show how to optimize the TQC while maintaining costs within a certain budget constraint. The following information is needed before starting the method: the desired qualitative closeness matrix for pairs of departments in terms of A, E, I, O, U, X, the total budget, the size of the departments, and the existing layout, labeled as S^0. Then, convert the qualitative closeness information into quantitative closeness information as discussed in Section 15.4.2. To solve this problem, the PWE method can be used iteratively until the best TQC (or TF) improvement is achieved within the given budget. The algorithm for solving this problem is as follows:

0. Label the initial layout S^0 as the current alternative and measure its TQC or TF.
1. At the current alternative, follow the PWE algorithm to find the pair of departments (i, j) that minimizes the TQC or TF. Label the pair $(i, j)^*$. If there is a tie, choose the one with the lowest cost of relocation.
2. Check if the cost of relocation for all of the selected departments is less than the allocated budget.

(a) If yes, exchange $(i, j)^*$. Label the alternative as the current one and go to step 1.

(b) If no, consider the next (i, j) that decreases the TC. Label it as $(i, j)^*$ and repeat step 2.

3. When no more changes are possible, stop.

Example 15.12   Relayout to Minimize Total Closeness within Budget   Given four departments and the distance and closeness relationship information given below, find the best layout for a budget of $150,000. The moving costs are also provided and are in thousands of dollars. The initial layout is shown in Figure 15.32.

Step 1   Convert the qualitative closeness matrix to a quantitative closeness matrix shown below based on the DM's preference. In this example, qualitative values of 4, 3, 2, 1, 0, -10 are used for A, E, I, O, U, X, respectively. These values use a penalty of -10 for two departments with a qualitative value of X.

	A	B	C	D
A		-10	3	4
B			0	3
C				-10
D				

Step 2   The existing layout is $S^0 = (A{\rightarrow}1, B{\rightarrow}2, C{\rightarrow}3, D{\rightarrow}4)$. Read $\rightarrow$ as assigned; that is, department A is assigned to location 1, B to 2, C to 3, and D to 4. For this assignment

$$\text{TQC} = \sum_i \sum_j d_{ij} p_{ij} = 1 \times (-10) + 2 \times 3 + 3 \times 4 + 1 \times 0$$
$$+ 2 \times 3 + 1 \times (-10) = 4$$

The cost of relocation is $R = 0$ since this is the initial matrix.

Qualitative Closeness Matrix

	A	B	C	D
A	—	X	E	A
B		—	U	E
C			—	X
D				—

Distance Matrix

	A	B	C	D
A	—	1	2	3
B		—	1	2
C			—	1
D				—

Location

1	2	3	4

Initial Layout

A	B	C	D

Cost of Relocation Matrix

	A	B	C	D
D_i	5	12	10	15
M_i	5	10	8	5
I_i	15	10	30	20

Figure 15.32   Initial data for relocation problem.

Step 3 Following the PWE procedure, the adjacent alternatives to the current layout are shown in following table:

Closeness	Facility Pairs	Current Distances	Pairwise Exchange					
			AB	AC	AD	BC	BD	CD
−10	AB	1	1	1	2	2	3	1
3	AC	2	1	2	1	1	2	3
4	AD	3	2	1	3	3	1	2
0	BC	1	2	1	1	1	1	2
3	BD	2	3	2	1	1	2	1
−10	CD	1	1	3	2	2	1	1
Total closeness		4	0	−24	−22	−22	−24	0
Layout		ABCD	BACD	CBAD	DBCA	ACBD	ADCB	ABDC

Exchanging A and C (or exchanging B and D) would provide the largest improvement in the total closeness. The two possible improved layouts are

| C | B | A | D |

| A | D | C | B |

The associated relocation costs can be calculated as follows:

$$R(AC \rightarrow CA) = 5 + 15 + 10 + 30 + (2 \times 5) + (2 \times 8) = \$86$$
$$R(BD \rightarrow DB) = 12 + 15 + 10 + 20 + (2 \times 10) + (2 \times 5) = \$87$$

Since $R(AC \rightarrow CA)$ relocation cost ($\$86$) is less than $R(BD \rightarrow DB)$ relocation cost ($\$87$), choose A and C to be the candidates for relocation. Since the relocation cost of $\$86$ is less than $\$150$ (the budget), exchanging A and C is acceptable. The improved layout is $S^1 = (C \rightarrow 1, B \rightarrow 2, A \rightarrow 3, D \rightarrow 4)$.

Total qualitative closeness improvement is $4 - (-24) = 28$.

Repeat step 3 for the new layout.

Closeness	Facility Pairs	Current Distances	Pairwise Exchange					
			AB	AC	AD	BC	BD	CD
−10	AB	1	1	1	2	2	1	1
3	AC	2	1	2	3	1	2	1
4	AD	1	2	3	1	1	1	2
0	BC	1	2	1	1	1	3	2
3	BD	2	1	2	1	3	2	1
−10	CD	3	3	1	2	2	1	3
Total closeness		−24	−26	4	−24	−24	−4	−26
Layout		CBAD	CABD	ABCD	CBDA	BCAD	CDAB	DBAC

Exchanging A and B (or C and D) would provide the largest improvement in total closeness. The improved layouts are as follows:

C	A	B	D

or

D	B	A	C

The total relocation costs compared to the initial layout, S^0, are calculated as follows:

$$R(ABC \to BCA) = 5 + 15 + 12 + 10 + 10 + 30 + (1 \times 5) + (1 \times 10) + (2 \times 8) = \$113$$
$$R(ACD \to CDA) = 5 + 10 + 15 + 15 + 30 + 20 + (2 \times 5) + (1 \times 8) + (3 \times 5) = \$128$$

Since $R(ABC \to BCA) < R(ACD \to CDA)$, that is, $\$113 < \$128$, choose $ABC \to BCA$. Now, since the cost of relocation, $\$113$, is less than the budget, $\$150$, accept the selected improved layout, $S^2 = (C \to 1, A \to 2, B \to 3, D \to 4)$.

Total qualitative closeness improvement is $4 - (-26) = 30$.

Repeat step 3 for the new layout.

			Pairwise Exchange					
Closeness	Facility Pairs	Current Distances	AB	AC	AD	BC	BD	CD
−10	AB	1	1	2	1	1	2	1
3	AC	1	2	1	3	1	1	2
4	AD	2	1	3	2	2	1	1
0	BC	2	1	1	2	2	3	1
3	BD	1	2	1	1	3	1	2
−10	CD	3	3	2	1	1	2	3
Total closeness		−26	−24	−22	0	0	**−30**	−24
Layout		CABD	CBAD	ACBD	CDBA	BACD	CADB	DABC

BD has the lowest total closeness of -30 (shown in bold), so B and D are candidates for exchange. The relocation cost is calculated for the change of this layout compared to the existing layout:

$$R(ABCD \to BDAC) = 5 + 12 + 10 + 15 + 15 + 10 + 30 + 20 + 5 + 2$$
$$\times 10 + 2 \times 8 + 5 = \$163$$

Since $\$163$ is more than $\$150$ (the budget), this exchange cannot be made. No more improvement is possible. Therefore, S^2 is the optimal solution to this problem and its total cost is within the given budget. The final layout is

C	A	B	D

The total qualitative closeness is -26 for this layout and the total relocation cost is $\$113$.

Total qualitative closeness improvement is $4 - (-26) = 30$ for the final layout compared to the existing layout.

Bicriteria Relayout Problem   The bicriteria relayout problem is:

Minimize f_1 = total qualitative closeness (or total flow)
Minimize f_2 = total relocation cost

The method covered in this section can be used to generate efficient alternatives in terms of the above two objective functions. To do this, list the values of both objective functions for each generated layout in all iterations and then eliminate the inefficient alternatives. For Example 15.12 some efficient alternatives are listed below.

Layout	ABCD	CBAD	CABD	CADB
Min. f_1	4	−24	−26	−30
Min. f_2	0	86	113	163
Efficient?	Yes	Yes	Yes	Yes

Tricriteria Relayout Problem   The tricriteria relayout problem is presented below and is based on considering both total qualitative closeness and total flow in addition to the total cost of relocation.

Minimize f_1 = total qualitative closeness
Minimize f_2 = total flow
Minimize f_3 = total relocation cost

15.7   QUADRATIC ASSIGNMENT OPTIMIZATION†

The quadratic assignment problem (QAP) is a mathematical programming formulation of the facility layout problem. Mathematically the QAP is difficult to solve, but its formulation provides a good insight of the complexity of the layout problem. The following coefficients and decision variables are used to define the objective function and constraints of the QAP:

a_{ij} = net revenue generated by moving department i to location j
d_{jl} = distance between location j and location l
f_{ik} = cost of material flow between department i and department k per unit of distance
n = number of departments and locations
$$x_{ij} = \begin{cases} 1 & \text{if department } i \text{ is assigned to location } j \\ 0 & \text{otherwise} \end{cases}$$

PROBLEM 15.1 QAP

Maximize:

$$\sum_{i=1}^{n}\sum_{j=1}^{n}a_{ij}x_{ij} - \sum_{i=1}^{n}\sum_{j=1}^{n}\sum_{k=1, i\neq k}^{n}\sum_{l=1, j\neq l}^{n} f_{ik}d_{jl}x_{ij}x_{kl} \tag{15.8}$$

Subject to:

$$\sum_{j=1}^{n}x_{ij} = 1 \quad \text{for departments } i = 1, 2, \ldots, n \tag{15.9}$$

$$\sum_{i=1}^{n}x_{ij} = 1 \quad \text{for locations } j = 1, 2, \ldots, n \tag{15.10}$$

$$x_{ij} = 0, 1 \quad i = 1, 2, \ldots\ldots, n, \quad j = 1, 2, \ldots\ldots, n \tag{15.11}$$

The objective function (15.8) maximizes the total revenue minus the materials handling (transportation) cost for all pairs of departments. The term $f_{ik}d_{jl}$ represents the flow between departments i and k if they are located in locations j and l, respectively. This term is considered in the total flow only if $x_{ij}x_{kl} = 1$, meaning that department i is in location j ($x_{ij} = 1$) and department k is in location 1 ($x_{kl} = 1$). The terms in (15.8) consider all pairs of departments (i, k) in all possible locations (j, l). Constraint (15.9) enforces that only one location is assigned to each department. Constraint (15.10) enforces that only one department is assigned to each location. Equation (15.11) is a binary (0 or 1) constraint. For each assignment variable, x_{ij}, 1 means it is assigned, and 0 means it is not assigned.

In the case where $a_{ij} = 0$, the objective function (15.8) can be simplified to (15.12) in the QAP:

$$\text{Minimize} \sum_{i=1}^{n}\sum_{j=1}^{n}\sum_{k=1, i\neq k}^{n}\sum_{l=1, j\neq l}^{n} f_{ik}d_{jl}x_{ij}x_{kl} \tag{15.12}$$

In objective function (15.12), from department i (the origin), materials are moved to department k (the destination). Thus, the materials handling cost per unit of distance from department i to department k is f_{ik}. Similarly, location j is the origin location and location l is the destination location. The distance between location j and location l is d_{jl}.

Example 15.13 Quadratic Assignment Problem Consider a layout problem with three departments 1, 2, and 3. Use the QAP to formulate the facility layout problem. The information for this problem is given below. The material flows for each pair of departments are symmetric; therefore, flow information is only provided in the upper right corner of the interflow of departments matrix.

Locations			Distance of Locations				Interflow of Departments			
?		?		l=1	l=2	l=3		k=1	k=2	k=3
1		2	j=1	—	1	1	i=1	—	50	10
?			j=2	1	—	2	i=2		—	20
3			j=3	1	2	—	i=3			—

The QAP is formulated as follows:

Minimize:

$$f = \sum_{i=1}^{3}\sum_{j=1}^{3}\sum_{k=1, i\neq k}^{3}\sum_{l=1, j\neq l}^{3} f_{ik}d_{jl}x_{ij}x_{kl}$$

$$= (50 \times 1 \times x_{11}x_{22} + 50 \times 1 \times x_{11}x_{23} + 10 \times 1 \times x_{11}x_{32} + 10 \times 1 \times x_{11}x_{33}$$

$$+ 50 \times 1 \times x_{12}x_{21} + 50 \times 2 \times x_{12}x_{23} + 10 \times 1 \times x_{12}x_{31} + 10 \times 2 \times x_{12}x_{33}$$

$$+ 50 \times 1 \times x_{13}x_{21} + 50 \times 2 \times x_{13}x_{22} + 10 \times 1 \times x_{13}x_{31} + 10 \times 2 \times x_{13}x_{32}$$

$$+ 20 \times 1 \times x_{21}x_{32} + 20 \times 1 \times x_{21}x_{33} + 20 \times 1 \times x_{22}x_{31} + 20 \times 2 \times x_{22}x_{33}$$

$$+ 20 \times 1 \times x_{23}x_{31} + 20 \times 2 \times x_{23}x_{32})$$

Subject to:

$$x_{11} + x_{12} + x_{13} = 1; \qquad x_{11} + x_{21} + x_{31} = 1$$
$$x_{21} + x_{22} + x_{23} = 1; \qquad x_{12} + x_{22} + x_{32} = 1$$
$$x_{31} + x_{32} + x_{33} = 1; \qquad x_{13} + x_{23} + x_{33} = 1$$

$$x_{ij} = 0,\ 1 \quad \text{for } i = 1, 2, 3, \text{ and } j = 1, 2, 3$$

This problem can be solved by an optimization software package that can handle 0–1 integer programming and also quadratic objective functions. The solution for this example is given in Figure 15.33. The meaning of the x_{ij} variables are as follows:

$x_{12} = 1$: Department 1 is assigned to location 2.
$x_{21} = 1$: Department 2 is assigned to location 1.
$x_{33} = 1$: Department 3 is assigned to location 3.

Note that the objective function for this solution is $50 \times 1 \times x_{12}x_{21} + 10 \times 2 \times x_{12}x_{33} + 20 \times 1 \times x_{21}x_{33} = 90$. Also, note that this problem has alternate optimal solution $x_{13} = 1$, $x_{21} = 1$, $x_{32} = 1$, which also has a total objective function value of 90, or $50 \times 1 \times x_{13}x_{21} + 10 \times 2 \times x_{13}x_{32} + 20 \times 1 \times x_{21}x_{32} = 90$.

See Supplements S15.7.xls, S15.7.lg4.

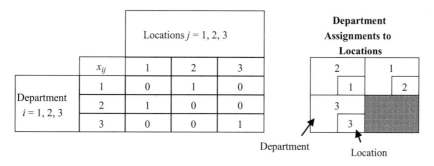

Figure 15.33   Solution for Example 15.13.

15.8 LAYOUT SOFTWARE PACKAGES

As the number of departments increases, creating layouts without the aid of a computer software program becomes extremely difficult. Generating different alternatives by hand is impractical. Using computers, thousands of alternatives can be generated and analyzed very quickly. The most important characteristic of a good software package for solving layout problems is to allow for qualitative restrictions and preferences to be implemented. We recommend first using a construction method such as RBL or SLP, then using this solution as an initial layout for an improvement method, such as CRAFT. In addition, it is essential to use a commercial CAD package (e.g., AutoCAD) to finalize the layout solutions. However, none of layout software programs (as presented below) can compete with the creativity of a human layout planner. Therefore, all solutions generated by these software programs should be reviewed and improved on by the designer.

Some well-known and traditional software packages are listed below.

CRAFT Computerized Relative Allocation of Facilities Technique (Buffa, Armour, and Vollman, 1964) is a software based on the PWE method covered in this chapter. It can also deal with exchanging three departments at the same time. CRAFT is an improvement method that takes an initial layout and tries to improve it via PWE of the centroids of each department. Because PWE is heuristic, CRAFT's final solution is not guaranteed to be the optimal solution; this comment is also true for other layout software packages listed in this section. However, the effectiveness of PWE and CRAFT very much depends on the quality of the given initial layout. Further, CRAFT treats transportation costs as a linear function of the distance, which may not be the case in some applications. Usually, the cost of initial movement is high but the cost of moving per unit of distance decreases as the total distance traveled increases.

ALDEP Automated Layout DEsign Program (Seehof and Evans, 1967) is a constructive method that uses a "REL chart" as an input; see the SLP method covered in this chapter. It places a random department in the upper left corner of the layout and proceeds to fill in the remaining areas. A department with a high closeness to the placed department is assigned next, and the process continues until all departments are placed. Several alternatives are generated, and the one with the best overall closeness score is chosen. This method always generates rectangular strip layouts.

CORELAP COmputerized RElationship LAyout Planning (Lee and Moore, 1967) is also a constructive method based on SLP. It places departments according to closeness similar to ALDEP. However, CORELAP does not use a random start. Instead, it begins by placing the department with the highest overall closeness in the center. Then, similar to ALDEP, it places departments with the highest closeness next to the placed departments until the facility is completely designed.

COFAD COmputerized FAcility Design (Tompkins and Reed, 1973) is an improvement method, like CRAFT. However, unlike CRAFT, COFAD first tries to find a better initial layout and then tries to improve upon the initial layout. Similar to CRAFT, it stops when no further improvements can be made.

FADES FAcilities Design Expert System (Fisher and Nof, 1984) is a software package that attempts to incorporate layout expert opinions in developing layouts.

CAFLAS Computer Aided Facility LAyout Selection (Malakooti, 1987; Malakooti and D'Souza, 1987) is the first method that addresses the multiple-criteria (objectives) facility layout problems. It can consider both linear and nonlinear utility functions. It is based on the PWE improvement method.

MOFLES Multi-Objective Facility Layout Expert System (Malakooti and Tsurushima, 1989) is a construction artificial intelligence software package that integrates layout logical expert rules with quantitative and qualitative information. The rules, preferences, and restrictions are based on a multiple-objective knowledge base engine. The expert system consists of four modules: (1) database (raw information), (2) knowledge base (domain knowledge by experts), (3) inference engine (solution control strategies), and (4) priority base (information on priority and objectives).

VIP-PLANOPT Visually Interfaced Package-PLANOPT (1995) is a general-purpose facility layout optimization software which is used to solve large real-world industrial facility layout design problems. VIP-PLANOPT uses both "hard" and "soft" modules where a hard module has fixed dimensions (width and length) for a department and a soft module has a fixed area but unknown dimensions.

FLEXPERT Facility Layout EXPERT (Badiru, 1996) uses fuzzy logic and a multicriteria concept of the layout problem at the same time as an expert system. It generates the best layout that satisfies both qualitative and quantitative constraints.

RBL Rule Based Layout (Malakooti, 2013) is based on the rule-based approach summarized in Section 15.3. This packages integrates both rule-based and PWE optimization for solving large-scale, ill-structured, and complex layout problems. RBL also uses MOFLES (expert system) and CAFLAS (multiobjective layout planning).

REFERENCES

General

Apple, J. M. *Plant Layout and Material Handling.* 3rd Ed. New York: Wiley, 1977.

Badiru, A. B., and A. Arif. "FLEXPERT: Facility Layout Expert System Using Fuzzy Linguistic Relationship Codes." *IIE Transactions*, Vol. 28, No. 4, 1996, pp. 295–308.

Buffa, E. S., G. C. Armour, and T. E. Vollman. "Allocating Facilities with CRAFT." *Harvard Business Review*, Vol. 42, 1964, pp. 136–158.

Coleman, D. R. "Plant Layout: Computers versus Humans." *Management Science*, Vol. 24, 1977, pp. 107–112.

Drira, A., H. Pierreval, and S. H. Gabouj. "Facility Layout Problems: A Survey." *Annual Reviews in Control*, Vol. 31, 2007, pp. 255–267.

Fisher, E. L., and S. Y. Nof. "FADES: Knowledge-Based Facility Design." *Annual International Industrial Engineering Conference Proceedings*, 1984, pp. 74–82.

Francis, R. L., L. F. McGinnis, and J. A. While. *Facility Layout and Location*, 2nd ed. Englewood Cliffs. NJ: Prentice Hall, 1992.

Heragu, S. S. *Facilities Design*, 3rd ed. Boca Raton, FL: CRC Press, 2008.

Heragu, S. S., and A. Kusiak. "Efficient Models for the Facility Layout Problem." *European Journal of Operational Research*, Vol. 53, No. 1, 1991, pp. 1–13.

Hillier, F. S. "Quantitative Tools for Plant Layout Analysis." *Journal of Industrial Engineering*, Vol. 14, 1963, pp. 33–40.

Lee, R. C., and J. M. Moore. "CORELAP-Computerized Relationship Layout Planning." *Journal of Industrial Engineering*, Vol. 18, No. 3, 1967, pp. 195–200.

Loiola, E. M., N. M. M. Abreu, P. O. B. Netto, P. Hahn, and T. Querido. "A Survey for the Quadratic Assignment Problem." *European Journal of Operational Research*, Vol. 176, 2007, pp. 657–690.

Muther, R. *Practical Plant Layout*. New York: McGraw-Hill, 1955.

Nahmias, S. *Production and Opertatioins Analysis*, 6th ed. New York: McGraw-Hill/Irwin, 2008.

Nicol, L. M., and R. H. Hollier. "Plant Layout in Practice." *Material Flow*, Vol. 1, No. 3, 1983, pp. 177–188.

Raman, D., S. V. Nagalingam, and G. C. I. Lin. "Towards Measuring the Effectiveness of a Facilities Layout." *Robotics and Computer-Integrated Manufacturing*, Vol. 25, No. 1, 2009, pp. 191–203.

Rosenblatt, M. J. "The Dynamics of Plant Layout." *Management Science*, Vol. 32, No. 1, 1986, pp. 76–86.

Rubadeux, K. L. PLANOPT—A Fortran Optimization Program for Planetary Transmission Design, MS Thesis, The University of Akron, 1995.

Seehof, Jerrold. M., and W. O. Evans. "Automated layout disign program." *Journal of Industrial Engineering* 18, no. 12 (1961): 690–695.

Singh, S. P., and R. R. K. Sharma, "A Review of Different Approaches to the Facility Layout Problems." *International Journal of Advanced Manufacturing Technology*, Vol. 30, 2006, pp. 425–433.

Snyder, L. V. "Facility Location under Uncertainty: A Review." *IIE Transactions*, 2006, Vol. 38, No. 7, pp. 547–554.

Tompkins, J. A., and J. A. White. *Facilities Planning*. New York: Wiley, 1984.

Tompkins, J. A., and R. Reed. "COFAD-A New Approach to Computerized Layout." Modern Materials Handling 30, No. 5 (1975): 40–43.

Xiea, W., and N. V. Sahinidis. "A Branch-and-Bound Algorithm for the Continuous Facility Layout Problem." *Computers & Chemical Engineering*, Vol. 32, Nos. 4–5, 2008, pp. 1016–1028.

Multiobjective Optimization

Aiello, G., M. Enea, and G. Galante. "A Multi-Objective Approach to Facility Layout Problem by Genetic Search Algorithm and Electre Method." *Robotics and Computer-Integrated Manufacturing*, Vol. 22, Nos. 5–6, 2006, pp. 447–455.

Chen, C.-W., and D. Y. Sha. "Heuristic Approach for Solving the Multi-Objective Facility Layout Problem." *International Journal of Production Research*, Vol. 43, No. 21, 2005, pp. 4493–4507.

Chiang, W.-C., P. Kouvelis, and T. L. Urban. "Single and Multi-Objective Facility Layout with Workflow Interference Considerations." *European Journal of Operational Research*, Vol. 174, No. 3, 2006, pp. 1414–1426.

Fortenberry, J. C., and J. F. Cox. "Multiple Criteria Approach to the Facilities Layout Problem." *International Journal of Production Research*, Vol. 23, No. 4, 1985, pp. 773–782.

Jacobs, F. R. "A Layout Planning System with Multiple Criteria and a Variable Domain Representation." *Management Science*, Vol. 33, No. 8, 1987, pp. 1020–1034.

Malakooti, B. "Computer Aided Facility Layout Selection with Applications to Multiple Criteria Manufacturing Planning Problems." *Large Scale Systems: Theory and Applications, Special Issue on Complex Systems Issues in Manufacturing*, Vol. 12, 1987, pp. 109–123.

Malakooti, B. "Multi-Objective Facility Layout: A Heuristic Method to Generate All Efficient Alternatives." *International Journal of Production Research*, Vol. 27, No. 7, 1989, pp. 1225–1238.

Malakooti, B. "Rule Based Layout Planning, Multiple Objective Optimization, and Re-layout" Case Western Reserve University, Cleveland, OH, 2013.

Malakooti, B., and G. D'Souza. "Multiple Objective Programming for the Quadratic Assignment Problem." *International Journal of Production Research*, Vol. 25, No. 2, 1987, pp. 285–300.

Malakooti, B., and A. Tsurushima. "An Expert System Using Priorities for Solving Multiple Criteria Facility Layout Problems." *International Journal of Production Research*, Vol. 27, No. 5, 1989, pp. 793–808.

Pelinescu, D. M., and M. Y. Wang. "Multi-Objective Optimal Fixture Layout Design." *Robotics and Computer-Integrated Manufacturing*, Vol. 18, 2002, pp. 365–372.

Rosenblatt, M. J. "The Facility Layout Problem: A Multi-Goal Approach." *International Journal of Production Research*, Vol. 17, 1979, pp. 323–332.

Şahin, R. "A Simulated Annealing Algorithm for Solving the Bi-objective Facility Layout Problem." *Expert Systems with Applications*, Vol. 38, No. 4, 2011, pp. 4460–4465.

Sha, D., and C. Chen. "A New Approach to the Multiple Objective Facility Layout Problem." *Integrated Manufacturing Systems*, Vol. 12, No. 1, 2001, pp. 59–66.

Singh, S. P., and V. K. Singh. "An Improved Heuristic Approach for Multi-Objective Facility Layout Problem." *International Journal of Production Research*, Vol. 48, No. 4, 2010, pp. 1171–1194.

Singh, S. P., and V. K. Singh. "Three-Level AHP-Based Heuristic Approach for a Multi-Objective Facility Layout Problem." *International Journal of Production Research*, Vol. 49, No. 4, 2011, pp. 1105–1125.

Urban, T. L. "A Multiple Criteria Model for the Facilities Layout Problem." *International Journal of Production Research*, Vol. 25, No. 12, 1987, pp. 1805–1812.

Yang, T., and Ch. Kuo. "A Hierarchical AHP/DEA Methodology for the Facilities Layout Design Problem." *European Journal of Operational Research*, Vol. 147, 2003, pp. 128–136.

Yang, Z., and B. Malakooti. "A Multiple Criteria Neural Network Approach for Layout of Machine—Part Group Formation in Cellular Manufacturing." Third IE Research Conference, 1994, pp. 249–254.

Ye, M., and G. Zhou. "A Local Genetic Approach to Multi-Objective Facility Layout Problems with Fixed Aisles." *International Journal of Production Research*, Vol. 45, No. 22, 2007, pp. 5243–5264.

EXERCISES

Systematic Layout Planning (Section 15.2)

2.1 A machining job shop company is planning to construct a new shop to handle increased demand. The company has developed the following space requirements and closeness ratings for the four departments. The company would like the layout of the facility to be 20 ft by 50 ft.

Space Requirements

Department	1	2	3	4
Square feet	200	200	200	400

Desired Closeness

	1	2	3	4
1		I	O	A
2			E	X
3				U
4				

(a) Use the SLP approach to generate the best layout

(b) Provide an alternative to the layout generated in part (a). Compare the two layouts.

2.2 A company is planning to construct a factory to handle increased demand. The company has developed the following space requirements and closeness ratings for the five departments. The company would like the layout of the facility to be 40 ft by 50 ft. Respond to parts (a) and (b) of exercise 2.1.

Space Requirements

Department	1	2	3	4	5
Square feet	300	400	300	400	600

Desired Closeness

	1	2	3	4	5
1		*I*	*O*	*A*	*U*
2			*E*	*X*	*U*
3				*U*	*I*
4					*U*
5					

2.3 LX Computer Corp. is planning to construct a facility to build computers. This facility will have six departments. LX would like the layout of the facility to be 40 ft by 35 ft. The requirements are shown below. Respond to parts (a) and (b) of exercise 2.1.

Space Requirements

Department	1	2	3	4	5	6
Square feet	150	300	150	350	250	200

Desired Closeness

	1	2	3	4	5	6
1		*A*	*E*	*U*	*X*	*I*
2			*I*	*I*	*O*	*U*
3				*E*	*O*	*A*
4					*E*	*U*
5						*I*
6						

2.4 QRD Systems Inc. is constructing a new widget factory. QRD would like the layout of the facility to be 50 ft by 50 ft. The requirements are given below. Respond to parts (a) and (b) of exercise 2.1.

Space Requirements

Department	1	2	3	4	6	7	8	9	10	Total
Square feet	200	400	300	200	500	100	200	400	200	2500

Desired Closeness

	1	2	3	4	5	6	7	8	9	10
1		*I*	*O*	*A*	*E*	*I*	*O*	*U*	*X*	*X*
2			*E*	*X*	*A*	*O*	*I*	*U*	*O*	*I*
3				*U*	*E*	*I*	*U*	*O*	*E*	*I*
4					*I*	*E*	*O*	*U*	*X*	*O*
5						*E*	*O*	*A*	*A*	*I*
6							*I*	*O*	*E*	*U*
7								*I*	*O*	*U*
8									*E*	*I*
9										*A*
10										

Rule-Based Layout for Single Row (Section 15.3.3)

3.1 Consider the following flow matrix. Find the best layout to minimize the total flow using the RBL head–tail method, and then find the total flow for the layout.

Location Distances

1	2	3	4	5

Dept.	A	B	C	D	E
A	—	10	0	100	60
B	70	—	20	0	70
C	80	20	—	15	80
D	25	0	0	—	0
E	0	40	0	200	—

3.2 Consider the following flow matrix. Find the best layout to minimize the total flow using the RBL head–tail method, and then find the total flow for the layout.

Location Distances

1	2	3	4	5	6

Dept.	A	B	C	D	E	F
A	—	77	83	49	0	73
B	58	—	91	52	85	69
C	44	58	—	0	55	36
D	88	94	152	—	49	0
E	143	68	88	107	—	37
F	74	83	100	148	99	—

Rule-Based Layout for Multiple Rows (Section 15.3.4)

3.3 Consider the following flow matrix. Find the best layout to minimize the total flow using the RBL head–tail method, and then find the total flow for the layout.

Locations

1	2	3
4	5	6

Interdepartmental Flow

Dept.	A	B	C	D	E	F
A		25	14	8	5	2
B			29	17	14	6
C				4	31	18
D					10	30
E						19
F						

3.4 Consider the following flow matrix. Find the best layout to minimize the total flow using the RBL head–tail method, and then find the total flow for the layout.

Interdepartmental Flow

	A	B	C	D	E	F	G	H	I	J	K	L
A		35	24	15	5	13	15	20	12	18	21	15
B			19	12	14	19	5	25	14	8	5	2
C				4	31	11	21	19	14	17	6	29
D					10	29	30	4	31	18	14	24
E						29	20	15	5	27	10	30
F							18	24	18	12	9	19
G								14	11	30	26	8
H									21	13	18	26
I										19	24	11
J											15	21
K												12
L												

Locations

1	2	3	4
5	6	7	8
9	10	11	12

RBL for Qualitative and Odd Shapes (Section 15.3.5)

3.5 Solve the layout problem given in exercise 2.1 by the RBL head–tail method.

3.6 Solve the layout problem given in exercise 2.2 by the RBL head–tail method.

3.7 Solve the layout problem given in exercise 2.3 by the RBL head–tail method.

3.8 Solve the layout problem given in exercise 2.2 by the RBL head–tail method.

Pairwise Exchange Method (Section 15.4)

4.1 Consider the following layout. You cannot assign a department to the shaded location.

Locations

1	2
3	

Layout

A	B
C	

Interdepartmental Flow

	1	2	3
1		13	14
2			11
3			

(a) Calculate the total flow for initial layout.

(b) Compute two iterations of the PWE method. Show the improved layout and the new total flow.

(c) Is the solution in part (b) the best solution? Can you suggest a better solution?

(d) Solve this problem using the RBL head–tail method and compare the solution to the solution obtained in part (b).

(e) Continue to compute iterations from the layout found in part (b) until the final solution is found (i.e., the layout cannot be further improved).

4.2 The ABC Company wishes to construct a new facility. Suppose the initial layout is $A \rightarrow 1$, $B \rightarrow 2$, $C \rightarrow 3$, and $D \rightarrow 4$. They have also anticipated interdepartmental flow in this facility. Respond to parts, (a)–(e) of exercise 4.1.

Location

1	2
3	4

Layout

A	B
C	D

Interdepartmental Flow

	A	B	C	D
A		15	5	27
B			18	12
C				30
D				

4.3 LX Computer Corp. is planning to construct a facility to build computers. This facility will have six departments. The initial layout and interdepartmental flows are shown below. Respond to parts (a)–(e) of exercise 4.1.

Location

1	2	3
4	5	6

Layout

A	B	C
D	E	F

Interdepartmental Flow

	A	B	C	D	E	F
A		25	14	8	5	2
B			29	17	14	6
C				4	31	18
D					10	30
E						19
F						

4.4 Vertical Computer Co. is planning to construct a facility to build computers. This facility will have seven departments. The initial layout and interdepartmental flows are given below. Respond to parts (a)–(e) of exercise 4.1.

Location

1	2	3	4
5		6	7

Layout

A	B	C	D
E		F	G

Interdepartmental Flow

	A	B	C	D	E	F	G
A		35	24	15	5	13	15
B			19	12	14	19	5
C				4	31	11	21
D					10	29	30
E						29	20
F							18
G							

4.5 Consider exercise 3.1. Solve it by the PWE method. The initial layout is

1	2	3	4	5

   (a) Use the algorithm for only two iterations.
   (b) Continue the algorithm for the solution found in part (a) until no more improvement is possible.

4.6 Consider exercise 3.2. Solve it by the PWE method. The initial layout is

1	2	3	4	5	6

   Respond to parts (a) and (b) of exercise 4.5.

4.7 Consider exercise 3.3. Solve it by the PWE method. The initial layout is

A	B	C
D	E	F

   Respond to parts (a) and (b) of exercise 4.5.

Bicriteria Layout (Section 15.5.1) and Efficient Frontier (Section 15.5.2)

5.1 Consider exercise 4.1. The closeness and travel time for the pairs of departments are given below. For part (c), the weights of importance are 0.6 and 0.4 for flow and closeness, respectively. For part (d), the weights of importance are 0.2, 0.4, and 0.4 for flow, closeness, and travel time, respectively.

Desired Closeness

	A	B	C
A		I	A
B			U
C			

Travel Time

	A	B	C
A		4	2
B			5
C			

   (a) Calculate the total closeness for the initial layout.
   (b) Calculate the total travel time for the initial layout.
   (c) Use the bicriteria approach, where total flow and total closeness are the two objectives. Find the best layout using the RBL head–tail method.
   (d) Use the tricriteria approach and find the best layout using the RBL head–tail method.

5.2 Consider exercise 4.2. The closeness and travel time for the pairs of departments are given below. For part (c), the weights of importance are 0.7 and 0.3 for flow and closeness, respectively. For part (d), the weights of importance are 0.3, 0.2, and 0.5 for flow, closeness, and travel time, respectively. Respond to parts (a)–(d) of exercise 5.1.

Desired Closeness

	A	B	C	D
A		E	A	X
B			I	U
C				E
D				

TravelTime

	A	B	C	D
A		6	1	5
B			3	2
C				4
D				

5.3 Consider exercise 4.3. The closeness and travel time for the pairs of departments are given below. For part (c), the weights of importance are 0.4 and 0.6 for flow and closeness, respectively. For part (d), the weights of importance are 0.4, 0.2, and 0.4 for flow, closeness, and travel time, respectively. Respond to parts (a)–(d) of exercise 5.1.

Desired Closeness

A	B	C	D	E	F	F
A		A	X	U	O	E
B			I	E	O	E
C				X	A	E
D					U	X
E						I
F						

Travel Time

	A	B	C	D	E	F
A		4	2	5	2	9
B			1	4	8	1
C				6	3	3
D					2	5
E						7
F						

5.4 Consider exercise 4.4. The closeness and travel time for the pairs of departments are given below. For part (c), the weights of importance are 0.5 and 0.5 for flow and closeness, respectively. For part (d), the weights of importance are 0.3, 0.4, and 0.3 for flow, closeness and travel time, respectively. Respond to parts (a)–(d) of exercise 5.1.

Desired Closeness

	A	B	C	D	E	F	G
A		E	A	I	X	O	A
B			U	E	I	E	O
C				O	X	A	U
D					E	X	I
E						U	O
F							A
G							

TravelTime

	A	B	C	D	E	F	G
A		9	6	4	8	2	3
B			4	1	6	3	5
C				3	4	5	9
D					7	3	4
E						2	8
F							7
G							

Calculating Relayout Cost (Section 15.6.1) and Relayout with Given Budget (Section 15.6.2)[†]

6.1 Consider the information given in exercises 4.1 and 5.1. Suppose that the initial layout is to be improved with a budget of $70 (in thousands) and relocation costs are given below.

	A	B	C
Di	6	15	15
Mi	6	13	12
Ii	20	32	28

(a) Find the best layout for minimizing total flow within the budget. Perform a maximum of three iterations.

(b) List the set of bicriteria alternatives based on the solution to part (a).

6.2 Consider exercises 4.2 and 5.2. The ABC Company plans to change the layout. Their budget is $120 (in thousands) and relocation costs are given below. Respond to parts (a) and (b) of exercise 6.1.

	A	B	C	D
Di	5	12	10	15
Mi	5	10	8	5
Ii	18	22	30	20

6.3 Consider the exercises 4.3 and 5.3. The LX Computer Corp. plans to change the layout. Their budget is $150 and relocation costs are given below. Respond to parts (a) and (b) of exercise 6.1.

	A	B	C	D	E	F
Di	7	5	12	15	8	20
Mi	5	4	10	6	8	5
Ii	20	25	27	24	25	30

6.4 Consider exercises 4.4 and 5.4. The Vertical Computer Co. plans to change the layout. Their budget is $270 and relocation costs are given below. Respond to parts (a) and (b) of exercise 6.1.

	A	B	C	D	E	F	G
Di	8	12	7	6	15	9	10
Mi	9	5	10	7	4	8	13
Ii	23	28	32	30	30	25	29

Quadratic Assignment Optimization (Section 15.7)[†]

7.1 Consider exercise 4.1.

    (a) Formulate it by the quadratic assignment problem.

    (b) Solve the problem by an available nonlinear integer (0–1) program software package.

    (c) Compare the solution generated by the quadratic assignment problem with the solution obtained by PWE and the RBL head–tail method.

7.2 Consider exercise 4.2. Respond to parts (a)–(c) of exercise 7.1.

7.3 Consider exercise 4.3. Respond to parts (a)–(c) of exercise 7.1.

CHAPTER 16

LOCATION DECISIONS

16.1 INTRODUCTION

The location problem is concerned with identifying the best locations for given objects in relation to other objects based on some metric such as the distances between the different objects. The location of any operational organization, such as a facility, is very important because it greatly affects operations, costs, and revenues. Generally, both the distances to and the amount of interaction among facilities are considered when choosing the location for a new facility. For instance, a new cafeteria should be located within the range of all the buildings on a campus but should be closest to the buildings with the highest numbers of people. In addition to distances and costs, many other criteria, including qualitative factors, must also be considered when choosing a location.

Historically, location decision techniques were developed to determine the location of facilities. However, location decisions are applicable to many design, engineering, management, and business problems. Purchasing a residential house is an example of a location problem where many factors are considered. In purchasing a house, many purchasers consider the American proverb to choose a house based on three criteria: "Location, location, location."

Facility location is a long-term decision and is influenced by many quantitative and qualitative factors. For a given location, several types of cost, such as fixed costs, energy, raw materials, transportation, and labor, may not drastically change for a long time. Location selection is also a long-term decision due to the prohibitive expense of moving an established facility.

The location problem can have different scales. Location problems can be small scale, such as deciding the location of a microprocessor on a computer's motherboard, or large

Note: Advanced material that can be omitted without loss of generality will be indicated by a dagger.

Operations and Production Systems with Multiple Objectives, First Edition. Behnam Malakooti.
© 2014 John Wiley & Sons, Inc. Published 2014 by John Wiley & Sons, Inc.

TABLE 16.1 Global, Regional, and Civic Site Selection Factors

Global Factors	Regional Factors	Civic Factors
Government laws, rules, and regulations	Community and support services	Construction and land costs
Exchange rates	Desirability of location	Access to airports/highways
Cost, skill, and work ethics of labor	Environmental factors	Utilities availability
Tax structure and incentives	Regional law and regulations	Educational facilities
Proximity to raw materials	Proximity to raw materials	Transportation
Proximity to markets	Proximity to markets	Community support
Available infrastructure	Global factors	Regional factors

scale, such as deciding the location of a new facility within a city, region, state, and country. The location criteria in Table 16.1 are divided into three scales: global, regional, and civic. All these factors should be considered when choosing a facility location.

Regardless of the scale, there are two general types of location problems: continuous and discrete. In the continuous-location problem, one or more locations from a continuous set of locations can be chosen for the intended use. For example, consider the location for a new distribution center that can be built almost anywhere with respect to an existing network of factories across the country. In continuous-location problems, it is assumed that there exists ample space and virtually any point in that ample space can be selected. Alternatively, one can think of a location problem where the distances between the objects are so large that the exact location of objects does not matter.

In the discrete-facility-location problem, one or more locations from a finite set of locations are chosen for the intended use. For example, consider the selection of one of five given cities for the headquarters of a new corporation. This type of problem is most often encountered where a set of possible locations are identified and the best candidate needs to be selected. For example, when constructing a new factory, warehouse, or distribution center, one of several potential locations should be selected. Similar to continuous–location problems, many criteria must be considered, such as distances, materials flow, and qualitative factors in conjunction with the existing facilities. As stated before, the location decision problem is about the identification of the best location for objects with respect to other objects. Each object is assumed to have a required area (or space) of zero; therefore, each object can be presented as a point having zero area. The main difference between location decision problems and facility layout problems is that in the layout problem each object has a specific area and all objects should fit into a given total area. Figure 16.1a shows a location problem where the location of a new object, E, should be determined with respect to the existing objects, A, B, C, and D. Note that the total area in a location problem does not need to be known; only distances between pairs of points are needed. For example, the distance between points B and D is 125 miles. Figure 16.1b shows a layout problem where departments A, B, C, and D are assigned to certain locations where each department has a given area and the total area for all departments must be considered.

Section 16.2 discusses the break-even graphical approach for the selection of the best location considering fixed and variable costs of potential new sites. Sections 16.3 and 16.4 cover the continuous-location problem of finding the best location compared to n existing locations. For this problem, two methods are used for measuring distances: rectilinear

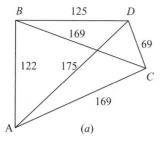

Figure 16.1 (*a*) Location problem for four objects as points; (*b*) layout problem of four departments in a 5 × 3 area.

(Section 16.3) and Euclidean (Section 16.4). Section 16.5 covers the discrete-location problem where given locations are evaluated by use of a multiple-criteria decision making (MCDM) method. In particular, the ordinal/cardinal approach (OCA) is presented for assessing criteria values and ranking alternatives. Section 16.6 presents single- and multiple-objective optimization approaches for finding suppliers to satisfy the demands of a set of locations. Section 16.7 expands the approaches of Section 16.6 for problems with capacity limitations on the supplier.

16.2 BREAK-EVEN ANALYSIS

The selection of a site for a new facility can be based on minimizing the total cost, which consists of fixed and variable costs. Fixed costs include the overhead of operating a facility, including building expenses, fixed salaries, capital investments, and fixed taxes. Fixed costs are independent of the production quantity, but variable costs depend on the production and distribution quantities. Examples of production costs are hired labor, raw materials, and inventory. Distribution costs are the expenses involved in transporting the finished product to given destinations such as to consumers and/or warehouses.

In considering a set of locations, if the total revenue of each location can be predicted, the location with the maximum profit is selected as the best location alternative. Otherwise, the location with the minimum total cost is selected. The total cost consists of the sum of the fixed, distribution, and production costs. Distribution costs, production costs, and total revenue are all functions of production volume, while fixed costs are constant. In Figure 16.2, one given location is analyzed with respect to its different costs and benefits.

Suppose that the total revenue can be predicted. In Figure 16.2, if the demand quantity Q is between Q_1 and Q_2, then the net profit (i.e., total revenue minus total cost) is positive. Hence, this range is the feasible solution range. The maximum profit is at point Q^*, where the difference between total revenue and total cost is the largest. Points Q_1 and Q_2 are break-even points. At a break-even point, the total revenue is equal to the total cost. By making a graph similar to Figure 16.2 for each given location, one can identify the range of quantities for which the net profit is positive. Each location may have different fixed costs, production costs, and distribution costs and also different total revenue. Using the above approach, one can identify the location that has the maximum profit for the given expected demand quantity.

Now, consider several locations that are compared in terms of their total costs. In this case, the location that has the minimum total cost for the given expected demand quantity

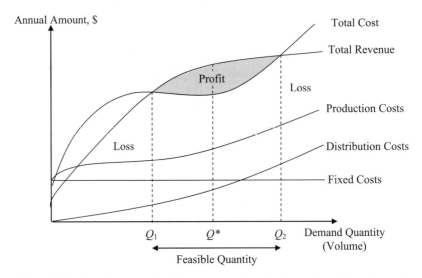

Figure 16.2 Break-even analysis for one given location for net benefit vs. demand.

should be selected. For example, a plant can be built in one of three possible locations: Boston, Chicago, and Phoenix. Figure 16.3 shows the costs for each city as a function of demand quantity. Using break-even analysis for a given demand quantity, the alternative with the lowest cost can be found:

For $0 \leq Q \leq Q_1$: Boston will have the lowest cost.

For $Q_1 \leq Q \leq Q_2$: Chicago will have the lowest cost.

For $Q_2 \leq Q$: Phoenix will have the lowest cost.

Assume Q_1 and Q_2 are break-even points. At point Q_1, the cost of Boston and Chicago are equal. Therefore, either of the two facilities can be selected. At Q_2, the cost of Chicago and Phoenix are equal.

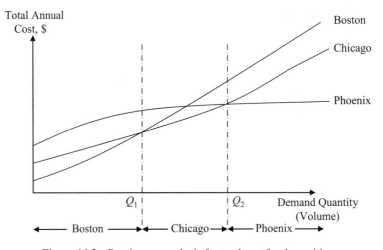

Figure 16.3 Break-even analysis for total cost for three cities.

16.3 RECTILINEAR METHOD

16.3.1 Rectilinear Method

The distance between two locations can be determined by rectilinear or Euclidean measurements. Rectilinear distance is the total length of the straight lines connecting two points while traveling in only one axial (horizontal or vertical) direction at a time. Euclidean distance is the length of the straight line that connects the two points where traveling simultaneously in both the horizontal and vertical directions is allowed. Distances in a town or a building complex are best approximated by rectilinear measurement since it is rarely possible to travel directly between two locations because of obstacles. Streets or walkways are often laid out in grid form allowing rectilinear traveling. Euclidean measurement, on the other hand, can be used for open spaces or for large-scale distances, such as the distance between two cities, where it is generally possible to travel the shortest path between the two locations. These two types of distance measurements are presented in Figure 16.4 for two points, (x_1, y_1) and (x_2, y_2). In Figure 16.4, the rectilinear distance is the summation of the two straight lines, a (vertical) and b (horizontal). The Euclidean distance is the straight line, c (diagonal), between the two points:

- Rectilinear distance = $|x_1 - x_2| + |y_1 - y_2| = |5 - 3| + |1 - 4| = 2 + 3 = 5$.
- Euclidean distance = $\sqrt{(x_1 - x_2)^2 + (y_1 - y_2)^2} = \sqrt{(5 - 3)^2 + (1 - 4)^2} = 3.61$.

Rectilinear Optimization Problem   Consider finding an optimal location (x, y) in order to minimize the total distances from this point to n given locations (x_i, y_i) for $i = 1, 2, \ldots, n$. The new location is selected from an infinite number of possible points. In this section, distances are measured using the rectilinear method. That is,

$$\text{Minimize total rectilinear distances} = \sum_{i=1}^{n} (|x - x_i| + |y - y_i|) \qquad (16.1)$$

Now consider the volume of materials that flow between the new facility and each of the existing facilities. Let the volume of materials of each of the n existing facilities located at $(x_1, y_1), (x_2, y_2), \ldots, (x_n, y_n)$ be presented by $v_1, v_2, \ldots, v_n)$, respectively. Now consider

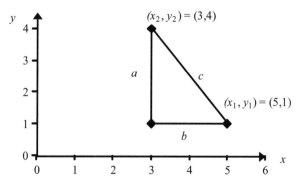

Figure 16.4   Illustration of rectilinear and euclidean distances.

finding the location of the new facility (x, y) by minimizing the total flow of materials based on rectilinear distances. Then (16.1) can be modified as

$$\text{Minimize total flow:} \quad f(x, y) = \sum_{i=1}^{n} v_i(|x - x_i| + |y - y_i|) \tag{16.2}$$

The optimal x, y location can be determined by minimizing $f(x, y)$. This function can be expressed as two separate functions:

$$f(x, y) = g(x) + h(y) \tag{16.3}$$

$$g(x) = \sum_{i=1}^{n} v_i|x - x_i| \tag{16.4}$$

$$h(y) = \sum_{i=1}^{n} v_i|y - y_i| \tag{16.5}$$

Therefore, $f(x, y)$ can be minimized by minimizing the two separate functions $g(x)$ and $h(y)$.

Example 16.1   Rectilinear Method for Two Points with Equal Flows   Find the optimal location with respect to two existing locations $(x_1, y_1) = (2, 6)$ and $(x_2, y_2) = (10, 1)$, where the volume of materials v_1 and v_2 are equal to unity.

In this case, use Equation (16.1). The optimal location is any solution that satisfies $2 \leq x \leq 10$ and $1 \leq y \leq 6$; see Figure 16.5. That is, there are an infinite number of optimal solutions to this problem; that is, any point in the shaded area in Figure 16.5 is an optimal solution. For example, in Figure 16.5, paths 1, 2, and 3 all have the same distance. Note that in Figure 16.5 all location points (x, y) in the rectangular shaded area have the same distance to the two given points; that is, $(|x_1 - x| + |x_2 - x|, |y_1 - y| + |y_2 - y|) = (|x_1 - x_2|, |y_1 - y_2|)$.

Finding Optimal Location by Use of Median Definition   Consider the rectilinear optimization problem presented in Equation (16.1). When there are only two given locations, Example 16.1 demonstrated that any point in the given rectangular in Figure 16.5 is the

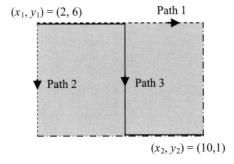

Figure 16.5   All points in the rectangular area have the same minimum distance to the two points (x_1, y_1) and (x_2, y_2).

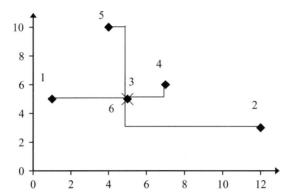

Figure 16.6   Rectilinear distances from existing facilities 1–5 to the new location 6 at (5, 5) for Example 16.2.

optimal solution. However, when there are more than two points, it is shown that the median approach can be used to find the optimal location. This method is based on the median definition for each coordinate. Consider n points $(x_1, y_1), (x_2, y_2), \ldots, (x_n, y_n)$. The optimal value for x is the median value of $(x_1, x_2, \ldots, x_n)$. Similarly, the optimal value for y is the median value of $(y_1, y_2, \ldots, y_n)$.

Example 16.2   Rectilinear Method for Five Points   Consider building a new cafeteria where there is ample land available such that a new facility can be built practically anywhere. There are five existing facilities, each presented as a point, whose coordinates are presented in the following table. A graphical representation of these five points is shown in Figure 16.6.

i	Building	x_i	y_i
1	Accounting	1	5
2	Warehouse	12	3
3	Shipping	5	5
4	Engineering	7	6
5	Marketing	4	10

To solve the problem, rank each coordinate (x and y) in ascending order, and then find the median.

	x Coordinates in Ascending Order				y Coordinates in Ascending Order	
i	Building	x_i		i	Building	y_i
1	Accounting	1		2	Warehouse	3
5	Marketing	4		1	Accounting	5
3	**Shipping**	**5**		**3**	**Shipping**	**5**
4	Engineering	7		4	Engineering	6
2	Warehouse	12		5	Marketing	10

Since there are five points, the median coordinate is found by dividing $n/2$, that is, $5/2 = 2.5$, which is rounded up to 3. Therefore, the third point is the median. By listing the x coordinates in ascending order as shown in the above tables, the median is found to be $x = 5$. The median of the y coordinates is $y = 5$. Therefore, the optimal location is (5, 5). The total distance can be calculated for this new location in terms of two separate functions, Equations (16.4) and (16.5):

$$g(x) = |1 - 5| + |4 - 5| + |5 - 5| + |7 - 5| + |12 - 5| = 14$$
$$h(y) = |3 - 5| + |5 - 5| + |5 - 5| + |6 - 5| + |10 - 5| = 8$$
$$f(x, y) = g(x) + h(y) = 14 + 8 = 22$$

This means that the total distance from the given locations to this new location is 22. Coincidentally, the new location (5, 5) is at the same location as the shipping building. The distances of all of the existing facilities to this new facility are shown in Figure 16.6.

Finding Optimal Location by Use of Median Approach When Considering Volumes   Consider the rectilinear optimization problem presented in Equation (16.2). The procedure to find the optimal location follows.

For each coordinate (x, y):

1. Order the locations in ascending order of that coordinate.
2. Consider the locations' volume (v_i). Calculate the cumulative sum of the volumes, one at a time, in the order obtained in step 1.
3. Find half of the highest cumulative volume.
4. Find the coordinate value of the location with a cumulative volume that is equal to or the next highest than half of the total cumulative volume. This coordinate is the optimum value.

Example 16.3   Rectilinear Method for Five Points with Unequal Volumes   Consider Example 16.2. Find the optimal coordinates for the new facility based on the number of employees working in each of the existing buildings. The objective is to minimize the total distance that all employees travel to the cafeteria. The following table shows the coordinates and the number of employees. Find the optimal location (x, y) for the cafeteria.

i	Building	x_i	y_i	Employees, v_i
1	Accounting	1	5	35
2	Warehouse	12	3	25
3	Shipping	5	5	10
4	Engineering	7	6	45
5	Marketing	4	10	40

Following the procedure, order the buildings in ascending order of each coordinate (x, y) separately. Note that the number of "employees" is considered to be the volume for each location. Now find the cumulative values of the volumes for each row.

i	Building	x_i	Volume, v_i	Cumulative Volume, Σv_i	i	Building	y_i	Volume, v_i	Cumulative Volume, Σv_i
1	Accounting	1	35	35	2	Warehouse	3	25	25
5	Marketing	4	40	75	1	Accounting	5	35	60
3	Shipping	**5**	10	**85**	3	Shipping	5	10	70
4	Engineering	7	45	130	4	Engineering	**6**	45	**115**
2	Warehouse	12	25	155	5	Marketing	10	40	155
	Half of cumulative			$155/2 = 77.5$		Half of cumulative			$155/2 = 77.5$

If two location facilities have the same coordinate values (e.g., both accounting and shipping have $y = 5$ values), their order can be chosen arbitrarily (e.g., for the y coordinate, either accounting or shipping can be put first).

To determine the optimal coordinate, divide the highest cumulative flow by 2. For both x and y, this value is 77.5. Now for each coordinate separately, find the first row in the table with a cumulative flow that exceeds or is equal to this value. The coordinate corresponding to that row is the optimal coordinate. For x, the optimal value is 5 (indicated in bold), because the cumulative value of the shipping building is the first row that has a cumulative flow higher than 77.5. For y, the optimal coordinate is 6 because the engineering building row is the first row that has a cumulative flow higher than 77.5. Therefore, the new location should be built at point (5,6). Using Equation (16.3), the total flow between the new point (5,6) and the given locations is calculated as follows:

$$f(x, y) = g(x) + h(y)$$
$$= [35|5 - 1| + 40|5 - 4| + 10|5 - 5| + 45|5 - 7| + 25|5 - 12|] + [25|6 - 3|$$
$$+ 35|6 - 5| + 10|6 - 5| + 45|6 - 6| + 40|6 - 10|] = 445 + 280 = 725$$

16.3.2 Multicriteria Rectilinear Method

In the previous section, one criterion (e.g., total flow) was used to determine the location of a new facility. In this section, multiple criteria are used to determine the location of a new facility.

Bicriteria Location Problem   Consider the following two criteria (objectives):

Minimize total flow:

$$f_{1(x,y)} = \sum_{i=1}^{n} v_i(|x - x_i| + |y - y_i|)$$

Minimize total qualitative closeness:

$$f_2(x, y) = \sum_{i=1}^{n} q_i(|x - x_i| + |y - y_i|)$$

The value of f_1 can be based on quantitative objectives, such as transportation cost, and f_2 can be based on qualitative objectives, such as closeness preferences. The qualitative closeness is presented by A, E, I, O, U, and X, which can be converted into numerical ratings of 5, 4, 3, 2, 1, and 0, respectively. Such a numerical rating is presented by q_i for location i. The higher the q_i rating, the closer the new facility should be to location i.

The bicriteria location problem can be solved by the composite MCDM approach (see Section 13.7.6 for details). For an additive utility function, the multicriteria composite index for each location i is

$$z_i = w_1 v_i' + w_2 q_i'$$

where v_i' and q_i' are normalized values of v_i and q_i; v_i is the volume of department i. When minimizing f_1 and f_2, the utility function is also minimized; that is, minimize $U = w_1 f_1' + w_2 f_2'$.

Tricriteria Location Problem   Similarly, a third objective can be considered. For example, the third objective could be the transportation time.

$$\text{Minimize transportation time:} \quad f_3(x, y) = \sum_{i=1}^{n} t_i (|x - x_i| + |y - y_i|)$$

where t_i is the transportation time per unit of distance for location i. The tricriteria problem can be solved using the same composite MCDM approach.

Example 16.4   Bicriteria Location Problem Using Rectilinear Method   Consider Example 16.3 with the following two criteria: number of employees in each location and qualitative closeness given by the employees.

i	Building	x_i	y_i	No. of Employees, v_i	Qualitative Closeness, q_i
1	Accounting	1	5	35	I
2	Warehouse	12	3	25	E
3	Shipping	5	5	10	A
4	Engineering	7	6	45	U
5	Marketing	4	10	40	O

Suppose the weights of importance for the normalized values of the two objectives, flow and qualitative closeness, are 0.65 and 0.35, respectively. Find the optimal location for the new cafeteria taking both criteria into account.

First, convert the qualitative values of A, E, I, O, U, and X to the ratings of 5 to 0 as shown in the table below. Then, normalize each of the criteria. Recall that each value is normalized by using $f' = (f - f_{min})/(f_{max} - f_{min})$. For example, the normalized value for the volume of the accounting building is $(35 - 10)/(45 - 10) = 0.71$. Once the data are normalized, the next step is to calculate the composite objective value for each department by using $z_i = w_1 v_i' + w_2 q_i' = 0.65 v_i' + 0.35 q_i'$. For example, $z_1 = 0.65(0.71) + 0.35(0.5) = 0.64$.

i	Building	x_i	y_i	v_i	q_i	v_i'	q_i'	z_i
1	Accounting	1	5	35	3	0.71	0.50	0.64
2	Warehouse	12	3	25	4	0.43	0.75	0.54
3	Shipping	5	5	10	5	0.00	1.00	0.35
4	Engineering	7	6	45	1	1.00	0.00	0.65
5	Marketing	4	10	40	2	0.86	0.25	0.64
	Minimum			10	1	0	0	
	Maximum			45	5	1	1	

Now apply the rectilinear median method using the composite value z_i as the weight for each location. This is shown in the following tables.

i	Building	x_i	Composite, z_i	Cumulative Composite	i	Building	y_i	Composite, z_i	Cumulative Composite
1	Accounting	1	0.64	0.64	2	Warehouse	3	0.54	0.54
5	Marketing	4	0.64	1.28	1	Accounting	5	0.64	1.18
3	Shipping	5	0.35	**1.63**	3	Shipping	5	0.35	**1.53**
4	Engineering	7	0.65	2.28	4	Engineering	6	0.65	2.18
2	Warehouse	12	0.54	2.83	5	Marketing	10	0.64	2.83
	Half of cumulative			$2.83/2 = 1.41$		Half of cumulative			$2.83/2 = 1.41$

Half of the cumulative composite is 1.41. Now find the first row in the table where the cumulative composite exceeds this value. For x, the optimal coordinate is 5, because shipping is the first department that has a cumulative composite value greater than 1.41. Similarly, for y, the optimal coordinate is 5, because shipping is the first department that has a cumulative composite value that exceeds 1.41. The optimal location for the new facility is (5,5). Use Equation (16.3) to find the value of each of the two objective functions f_1 and f_2:

$$f_1 = 35|5 - 1| + 40|5 - 4| + 10|5 - 5| + 45|5 - 7| + 25|5 - 12| + 25|5 - 3|$$
$$+ 35|5 - 5| + 10|5 - 5| + 45|5 - 6| + 40|5 - 10|$$
$$= 445 + 295 = 740$$

$$f_2 = 3|5 - 1| + 2|5 - 4| + 5|5 - 5| + 1|5 - 7| + 4|5 - 12| + 4|5 - 3| + 3|5 - 5|$$
$$+ 5|5 - 5| + 1|5 - 6| + 2|5 - 10|$$
$$= 44 + 19 = 63$$

Generating Set of Efficient Alternatives By varying the weights of importance, it may be possible to generate different efficient alternatives. Three alternatives are presented below.

	a_1	a_2	a_3	a_3	a_3
(w_1, w_2)	(1, 0)	(0.75, 0.25)	(0.5, 0.5)	(0.25, 0.75)	(0, 1)
(x, y)	(4, 6)	(5, 6)	(5, 5)	(5, 5)	(5, 5)
f_1	730	725	740	740	740
f_2	77	72	63	63	63
Efficient?	No	Yes	Yes	Yes	Yes

Alternative a_1 is dominated by a_2 because its corresponding two objectives have higher values; that is, 725 is better than 730 and 72 is better than 77. Note that the weights (0.5, 0.5), (0.25, 0.75), and (0, 1) generate the same solution, a_3. For small problems, there are usually only a few efficient alternatives because central locations usually satisfy both objectives at the same time.

See Supplement S16.1.xls.

16.4   EUCLIDEAN METHOD

16.4.1   Center-of-Gravity Method

The center-of-gravity method is similar to the rectilinear method of the last section, but it uses Euclidean distance instead of rectilinear distance to find the optimal location; see Figure 16.4. For example, consider Figure 16.7 with four points that form a rectangle. The four points are A: (1, 1), B: (1, 5), C: (7, 1), and D: (7, 5). Suppose the volumes of materials

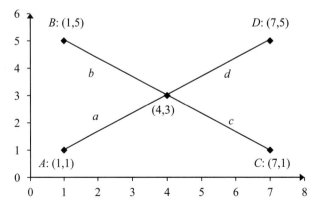

Figure 16.7   For Euclidean distance, point (4, 3) is the center of gravity when the volumes of all four locations, A, B, C, and D, are identical.

at these four points are the same. The center of gravity is at the center of the rectangle, which is the point (4, 3). This point is closest to all four points in terms of total distance if the distances are measured by straight lines. That is, the total distance for this example is $a + b + c + d = 3.6 + 3.6 + 3.6 + 3.6 = 14.4$.

Now consider that each existing point (i) has a different volume, v_i. To find the optimal solution (x, y), there are two methods: the gravity method [based on Equation (16.6)] and the Euclidean distance method [based on Equation (16.7)]:

Minimize total flow:

$$f(x, y) = \sum_{i=1}^{n} v_i[(x - x_i)^2 + (y - y_i)^2] \quad \text{(gravity)} \quad (16.6)$$

Minimize total flow:

$$f(x, y) = \sum_{i=1}^{n} v_i\sqrt{(x - x_i)^2 + (y - y_i)^2} \quad \text{(Euclidean)} \quad (16.7)$$

The solution to Equation (16.6) is called the center of gravity. The gravity equation (16.6) is much simpler to solve than Equation (16.7). The optimal solution for the Euclidean distance method (16.7) cannot be obtained using the derivative method. An iterative optimization method developed by Weiszfeld (1936) finds the solution to this problem. For the purpose of simplicity, only the center-of-gravity method will be presented here. It is a widely used method which guarantees the optimal solution for (16.6).

To determine the optimal solution (x, y) for Equation (16.6), take the partial derivatives of objective function $f(x, y)$ [Equation (16.6)], with respect to x and y:

$$\frac{\partial f(x, y)}{\partial x} = 2\sum_{i=1}^{n} v_i(x - x_i) = 0 \qquad \frac{\partial f(x, y)}{\partial y} = 2\sum_{i=1}^{n} v_i(y - y_i) = 0$$

The optimal solution is found by solving each equation separately. The solution is

$$x = \frac{\sum_{i=1}^{n} (v_i x_i)}{\sum_{i=1}^{n} v_i} \qquad y = \frac{\sum_{i=1}^{n} (v_i y_i)}{\sum_{i=1}^{n} v_i} \qquad (16.8)$$

Example 16.5 Finding Location for New Facility A regional telephone company needs to decide the location of a training facility for its operators. The operators are located at operating centers in four different towns. The training facility has to serve these four centers. The coordinates of the four towns and the number of operators (in hundreds) working in each town are given below.

(a) Determine the optimal coordinates of the new training facility using the center-of-gravity method.

(b) Determine the optimal coordinates of the new training facility using Euclidean distance:

Town	x_i	y_i	No. of Operators, v_i
1	42	65	7.2
2	51	50	4.6
3	18	75	6.6
4	49	35	4.1

(a) The number of operators of each town can be treated as the volume v_i. Now apply the center-of-gravity formulas (16.8) to find the coordinates of the new location (see Figure 16.8):

$$x = \frac{42 \times 7.2 + 51 \times 4.6 + 18 \times 6.6 + 49 \times 4.1}{7.2 + 4.6 + 6.6 + 4.1} \approx 38$$

$$y = \frac{65 \times 7.2 + 50 \times 4.6 + 75 \times 6.6 + 35 \times 4.1)}{7.2 + 4.6 + 6.6 + 4.1} \approx 59$$

Therefore, for the center-of-gravity method, the best location for the new facility is (38, 59), and the total flow is

$$f(x, y) = 7.2[(38 - 42)^2 + (59 - 65)^2] + 4.6[(38 - 51)^2 + (59 - 50)^2]$$
$$+ 6.6[(38 - 18)^2 + (59 - 75)^2] + 4.1[(38 - 49)^2 + (59 - 35)^2] = 8712$$

(b) Using optimization software (e.g., LINGO) to solve Equation (16.7) (the Euclidean distance), the optimal location is (42, 65), where $f(x, y) = 378.37$. Note that the two solutions are different because their objective functions [Equations (16.6) and (16.7)] are defined differently.

See Supplements 16.2.xls, 16.2.lg4.

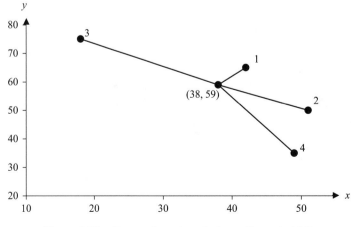

Figure 16.8   Center-of-gravity solution to Example 16.5.

16.4.2 Multicriteria Center-of-Gravity-Method

This method is similar to the MCDM of the rectilinear method except that center-of-gravity distances are used (see Section 16.3.2 for details):

Minimize f_1 = total flow or cost
Minimize f_2 = total qualitative closeness
Minimize f_3 = total transportation time

Each objective function is presented by a center-of-gravity function, that is, Equation (16.6). To rank alternatives, find the composite values $z = w_1 f_1' + w_2 f_2' + w_3 f_3'$, and then use this value as the volume in Equation (16.6).

Example 16.6   Tricriteria Example for Center of Gravity   Suppose a telephone company has three criteria (objectives) in considering the location of a new training facility: number of operators (v_i), qualitative closeness (q_i), and number of skilled teachers in each town (t_i). The data for this problem are shown in the table below. Use an additive utility function to find the best location for the new facility where the weights of importance for the normalized values of the three objectives are 0.3, 0.35, and 0.35, respectively. Use the center-of-gravity method.

To find the optimal solution, first convert the qualitative closeness into quantitative values. Normalize all criteria values and use the composite MCDM approach as shown below. After the normalization, the composite value for each existing location is calculated using $z_i = 0.3v_i' + 0.35q_i' + 0.35t_i'$.

Town	x_i	y_i	v_i	q_i	t_i	v_i'	q_i'	t_i'	z_i
1	42	65	7.2	X=0	6	1.00	0.00	0.60	0.51
2	51	50	4.6	A=5	0	0.16	1.00	0.00	0.40
3	18	75	6.6	I=3	8	0.81	0.60	0.80	0.73
4	49	35	4.1	E=4	10	0.00	0.80	1.00	0.63
Minimum	18	35	4.1	0	0				
Maximum	51	75	7.2	5	10				

Now use the composite values z_i, as the volume v_i in the center-of-gravity equation (16.8):

$$x = \frac{42 \times 0.51 + 51 \times 0.40 + 18 \times 0.73 + 49 \times 0.63}{0.51 + 0.40 + 0.73 + 0.63} \approx 38$$

$$y = \frac{65 \times 0.51 + 50 \times 0.40 + 75 \times 0.73 + 35 \times 0.63}{0.51 + 0.40 + 0.73 + 0.63} \approx 57$$

The best location for the new training facility is (38, 57) when considering the three criteria. Using the gravity equation (16.6) to calculate the value of each of the three objective functions, the solution associated with (38, 57) is $f_1 = 8,809, f_2 = 5,745$, and $f_3 = 12,332$.

A Set of Efficient Alternatives   A set of efficient alternatives can be generated by varying the weights of importance and solving the MCDM location problem. Four efficient alternatives are shown below and in Figure 16.9.

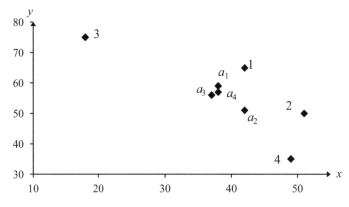

Figure 16.9   A set of Efficient alternatives for Example 16.6.

w	a_1	a_2	a_3	a_4
(w_1, w_2, w_3)	$(1,0,0)$	$(0,1,0)$	$(0,0,1)$	$(0.3, 0.35, 0.35)$
(x, y)	$(38, 59)$	$(42, 51)$	$(37, 56)$	$(38, 57)$
f_1	8,712	10,642	8,994	8,809
f_2	6,006	5,086	5,666	5,745
f_3	12,530	13,442	12,262	12,332
Efficient?	Yes	Yes	Yes	Yes

See Supplement S16.3.xls.

16.5   MULTICRITERIA LOCATION SELECTION

In facility location selection problems, the best location should be selected based on a set of criteria. The multicriteria location problem is identical to generic MCDM problems, and therefore, the methods developed for solving MCDM problems can be directly applied to solve the multicriteria location selection problem. In Section 16.5.1, we show how to apply the OCA method for solving this problem; OCA is based on additive (linear) utility functions. In Section 16.5.2, we show how to apply the Z theory approach for solving this problem; Z theory can solve nonlinear utility functions which are broader than additive (linear) utility functions.

16.5.1   Ordinal/Cardinal Approach for Location Selection

In almost all real-world problems, the selection of a location is based on several criteria. A list of location criteria was provided in Table 16.1. In this section, OCA is used to rank all alternative locations where the DM's preferences have an additive utility function (otherwise the Z theory approach of the next section should be used). In the following example, OCA is used to rank five alternatives when there are four criteria.

TABLE 16.2   Criteria Assessment for Example 16.7

	Criteria	Boston	Cleveland	Pittsburgh	Phoenix	LA	Minimum	Maximum
Min. f_1	Initial cost (millions)	$780	$650	$680	$530	$800	$530	$800
Max. f_2	Educational institutes (0–10)	10	6	7	4	8	4	10
Max. f_3	Resources availability (0–100)	25	85	75	56	44	25	85
Max. f_4	Government support (0–10)	5	8	7	10	6	5	10

Example 16.7   Multicriteria Location Problem Using OCA   A manufacturing company is planning to build a new plant by considering four criteria in one of the five possible locations: Boston, Cleveland, Pittsburgh, Phoenix, and Los Angeles; see Table 16.2. Use OCA to find the utility of these five locations and rank them.

Step 1: Assessing Criteria   First, order the criteria in accordance with their importance. Suppose that the order of importance of the criteria is f_1, f_2, f_3, and f_4. Then, the values of each criterion for each alternative should be assessed; the assessed values are shown in Table 16.2. The second and fourth objectives are assessed on a rating scale from 0 to 10, where 0 is the worst and 10 is the best. The third objective is assessed on a rating scale from 0 to 100, where 0 is the worst and 100 is the best.

   Normalizing Criteria Values: The criteria values in Table 16.2 are normalized on a scale of 0–1 for maximizing all objectives (criteria). The normalized values are shown in Table 16.3. Note that because f_1 (cost) is minimized, then one should normalize-f_1 (which is maximized); for example, for Boston,

$$f_1' = \frac{-780 - (-800)}{-530 - (-800)} = 0.07$$

However, f_2, f_3, and f_4 are maximized; for example, for Boston,

$$f_2' = \frac{10 - 4}{10 - 4} = 1.$$

Step 2: Assessing Weights of Importance   With respect to normalized values of criteria (0–1 scale), weights of importance of each criterion should be assessed. The DM is asked to provide the ratings of the difference of importance of two given weights on a given scale. Here, we use a scale of 0–10 where 0 means that the two weights are equally important and 10 means that the more important weight is extremely preferred to the less important weight (see OCA in Chapter 2 for more explanations). Note that w_0 is a point of reference

TABLE 16.3   Normalized Criteria for Example 16.7

	Criteria	Boston	Cleveland	Pittsburgh	Phoenix	LA
Max. f_1'	Initial cost (millions)	$0.07	$0.56	$0.44	$1	$0
Max. f_2'	Educational institutes (0–10)	1	0.33	0.50	0	0.67
Max. f_3'	Resources availability (0–100)	0	1	0.83	0.52	0.32
Max. f_4'	Government support (0–10)	0	0.60	0.40	1	0.20

which is equal to zero (i.e., it is the worst possible value for any weights of importance). Table 16.4A shows the DM's assessments of the paired comparison of the four weights. For example, weights of cost on a scale of 0–10, is 10 (this is shown by comparing w_1 to w_0, but w_1 to w_2 is 4. Note that w_1 to w_3 and w_1 to w_4 are both 8.

Ordinal Consistency Consider Table 16.4A assessed ratings, and note that all assessed ratings are in ascending order (from left to right and from bottom to top), which is consistent with the order of their importance, that is, w_1, w_2, w_3 w_4, and w_0. Therefore, these responses are ordinal consistent.

Finding Relative Ratings Now, we can move on to generate their relative rating values on the given scale of 0–10. This is accomplished in Table 16.4B. Consider each column of Table 16.4A, and find the difference of each entry with respect to the minimum value of that column. For example, for the first column, the relative rating of w_2 with respect to w_1 is calculated as $-4 - (-10) - 6$.

Finding Weights of Importance Weights of importance are shown in Table 16.4C. Consider each column of Table 16.4B, and then divide each entry by its sum. For example for the first column, the weight of w_2 with respect to w_1 is calculated as $6/20 = 0.3$. The assessed weights are the average weights of all five values for each weigh. For example $\overline{w}_1 = (0.5 + 0.53 + 0.50 + 0.48 + 0.5) = 0.5$.

Cardinal Consistency The responses are cardinal inconsistent because the columns of Table 16.4 are not identical. The consistency index ratio (CIR) can be used to measure the inconsistency for each weight (each row) in Table 16.4C. Each CIR is calculated by its mean absolute deviation (MAD) with respect to its average.

Acceptable Weight Assessment To find out whether the assessed weights are acceptable, consider the total of CIR, which is 0.04. Because this value (0.04) is less than the

TABLE 16.4   Weight Assessment by OCA for Example 16.7

A. Paired Comparison of Four Weights						B. Relative Ratings				
R_{ij}	w_1	w_2	w_3	w_4	w_0	R_{ij}^*	w_1	w_2	w_3	w_4	w_0
w_1	0	4	8	8	10	w_1	10	10	10	10	10
w_2	−4	0	4	5	6	w_2	6	6	6	7	6
w_3	−8	−4	0	0	2	w_3	2	2	2	2	2
w_4	−8	−5	0	0	2	w_4	2	1	2	2	2
w_0	−10	−6	−2	−2	0	w_0	0	0	0	0	0
$R_{\min,j}$	−10	−6	−2	−2	0	Sum	20	19	20	21	20

C. Assessed Weights

	w_1	w_2	w_3	w_4	w_0	$\overline{w}_i$	CIR_i	Acceptable Consistency?
w_1	0.5	0.53	0.5	0.48	0.5	0.5	0.01	Yes
w_2	0.3	0.32	0.3	0.33	0.3	0.31	0.01	Yes
w_3	0.1	0.11	0.1	0.1	0.1	0.1	0	Yes
w_4	0.1	0.05	0.1	0.1	0.1	0.09	0.01	Yes
Sum	1	1	1	1	1	1	0.04	<0.05, Yes

TABLE 16.5   Normalized Criteria Values and Ranking of Alternatives by OCA for Example 16.7

	Boston	Cleveland	Pittsburgh	Phoenix	LA	Weights
Max. f_1'	0.07	0.56	0.44	1	0	0.50
Max. f_2'	1	0.33	0.5	0	0.67	0.31
Max. f_3'	0	1	0.83	0.52	0.32	0.10
Max. f_4'	0	0.6	0.4	1	0.2	0.09
Max. U	0.35	0.54	0.49	0.64	0.26	—
Rank	4	2	3	1	5	—

minimum acceptable threshold (0.05), then assessed values are acceptable. Therefore, for the additive utility function, the weights of importance for normalized objectives f_1', f_2', f_3', f_4' are 0.50, 0.31, 0.10, 0.09, respectively. The utility function (to be maximized) is $U = 0.5f_1' + 0.31f_2' + 0.1f_3' + 0.09f_4'$. Note that all normalized values of criteria in Table 16.3 are maximized.

Step 3. Ranking Alternatives   Alternatives are ranked by maximizing the utility function $U = 0.5f_1' + 0.31f_2' + 0.1f_3' + 0.09f_4'$. Table 16.5 shows the utility values and ranking of all alternatives, where Phoenix has the highest utility function value; that is, it is the best location.

See Supplement S16.4.xlsx.

16.5.2   Z Theory for Location Selection

Suppose that the assessed additive utility of Section 16.5.1 does not present the DM's preferences and therefore a nonlinear utility function should be used. This can be identified by asking the DM to verify the assessed additive utility function values of a subset of alternatives. If the DM does not agree with the assessed utility values, then a nonlinear utility function should be used. Consider the following two objectives for the selection of the best location:

Minimize total cost (or objective factors), f_1

Maximize qualitative preferences (or subjective factors), f_2^A

This bicriteria problem can be used to capture the essential aspects of a multiple-criteria location selection problem. All cost-related factors are measured and aggregated by the first objective, f_1, and all noncost factors are aggregated and presented by f_2^A. The process of quantification of both these objectives is very critical to correctly presenting this problem. The aggregation of the first objective (total cost) is rather easy as all its components are measurable (e.g., in dollars). However, the aggregation of the second objective, f_2^A, is not trivial as it consists of several noncommensurate criteria.

Suppose that f_2^A is an additive function of all subjective factors. A well-known approach that provides a structure for aggregation of such a criterion is the analytic hierarchy process (AHP) (Saaty, 1980). Unfortunately, despite its appeal, the AHP has been shown to have limitations (it may cause rank reversal of alternatives). Therefore, we recommend using the OCA method for aggregating all subjective factors. It should be noted that when the method of the last section for using OCA fails [(i.e., when an additive (linear) utility function cannot present the DM's preferences], one should attempt to apply the method of this section, which uses a nonlinear utility function with respect to f_1 and f_2^A.

Consider the normalized value of the two objectives (on a scale of 0–1) presented by f_2' and f_2^A, respectively. The Z-nonlinear utility function can be used to rank all alternatives:

$$U = \text{LV} + z * \text{DV}_M$$
$$\text{LV} = w_1 f_1' + w_2 f_2^A$$
$$\text{DV}_M = \left| \text{LV} - f_1'^{w_1} \times f_2^{Aw_2} \right|$$

where, $w_1 > 0$, $w_2 > 0$, and $w_1 + w_2 = 1$. The value of f_2^A can be assessed by using OCA with respect to all subjective factors. This will be explained is the following example.

Example 16.8   Bicriteria Location Problem Using Z Theory and OCA   Consider Example 16.7. Suppose that the assessed additive utility of Example 16.7 does not present the DM's preferences and therefore a nonlinear utility function should be used. Solve this problem using the Z theory utility function where all subjective factors (objectives f_2, f_3, and f_4 in Table 16.1) can be aggregated and presented by an additive utility function and presented by f_2^A. Suppose that the weights of importance for normalized criteria (f_1 and f_2^A) are 0.7 and 0.3, respectively; and the Z utility function coefficient is -0.4, $z = -0.4$.

Aggregation of Subjective Criteria, f_2^A   Criteria f_2, f_3, and f_4 can be aggregated and presented by f_2^A using the OCA method. The weights of importance of these three criteria are assessed as presented in Table 16.6.

TABLE 16.6   Weight Assessment by OCA for Example 16.8

A. Paired Comparision of Three Weights					B. Relative Ratings				
R_{ij}	w_2	w_3	w_4	w_0	R_{ij}^*	w_2	w_3	w_4	w_0
w_2	0	4	5	8	w_2	8	8	8	8
w_3	−4	0	1	4	w_3	4	4	4	4
w_4	−5	−1	0	3	w_4	3	3	3	3
w_0	−8	−4	−3	0	w_0	0	0	0	0
$R_{\min,j}$	−8	−4	−3	0	Sum	15	15	15	15

C. Assessed Weights

	w_2	w_3	w_4	w_0	$\bar{w}_i$	CIR_i	Acceptable Consistency?
w_2	0.53	0.53	0.53	0.53	0.53	0	Yes
w_3	0.27	0.27	0.27	0.27	0.27	0	Yes
w_4	0.2	0.2	0.2	0.2	0.2	0	Yes
Sum	1	1	1	1	1	0	Yes

TABLE 16.7 **Normalized Bicriteria Values and Ranking of Alternatives by Z Theory Nonlinear Utility Function for Example 16.8**

	Boston	Cleveland	Pittsburgh	Phoenix	LA	Weights
Max. f_1'	0.07	0.56	0.44	1	0	0.7
Max. f_2^A	0.53	0.56	0.57	0.34	0.48	0.3
LV	0.21	0.56	0.48	0.80	0.14	—
DV_M	0.080	0.000	0.003	0.079	0.145	—
Max. U	0.18	0.56	0.48	0.77	0.09	—
Rank	4	2	3	1	5	—

Now, we find the aggregate values of three subjective factors for each alternative using

$$f_2^A = w_2 f_2' + w_3 f_3' + w_4 f_4' = 0.53 f_2' + 0.27 f_3' + 0.20 f_4'$$

The Z utility function is

$$LV = w_1 f_1' + w_2^A f_2^A = 0.7 f_1' + 0.3 f_2^A$$

$$MV_M = f_1'^{0.7} * f_2^{A0.3}$$

$$U = LV + z * |LV - MV_M| 0.7 f_1' + 0.3 f_2^A - 0.4 * |0.7 f_1' + 0.3 f_2^A - f_1'^{0.7} * f_2^{A0.3}|$$

The set of bicriteria alternatives are presented in Table 16.7. The following shows the details of finding the utility value for Boston:

$$f_2^A = w_2 f_2' + w_3 f_3' + w_4 \times f_4' = 0.53 \times 1 + 0.27 \times 0 + 0.20 \times 0 = 0.53$$

$$\text{(using Table 16.5)}$$

$$LV = 0.7 f_1' + 0.3 f_2^A = 0.7 \times 0.07 + 0.3 \times 0.53 = 0.21$$

$$MV_M = f_1'^{0.7} * \left(f_2^A\right)^{0.3} = 0.07^{0.7} * 0.53^{0.3} = 0.13$$

$$U = LV - 0.4 * |LV - MV_M| = 0.21 - 0.4 * |0.21 - 0.13| = 0.18$$

See Supplement **S16.5.xlsx.**

16.6   LOCATION ALLOCATION–SUPPLIER SELECTION

16.6.1   Single-Objective Location Allocation

Suppose there are m given locations whose demands in terms of given volumes must be fulfilled by one or more of n given suppliers. In this section, we assume that suppliers have infinite capacity to satisfy the demands of given locations. However, in practical situations, suppliers have limited capacities. (The location allocation problem when considering limited resources is discussed in Section 16.6.3.) The notation for formulating this problem is as follows:

Known coefficients:

i = demand locations; $i = 1, 2, \ldots, m$ (e.g., existing market or warehouse demand)

j = supply locations; $j = 1, 2, \ldots, n$ (e.g., existing supplier or plants to be built)

m = number of demand locations

n = number of supply locations

c_{ij} = cost of satisfying demand i by supply location j for one unit of product

K_j = fixed cost of each supply location (e.g., construction cost) for $j = 1, 2, \ldots, n$

Decision variables (to be determined):

x_{ij} = fraction of demand location i being supplied by supply location j ranging from 0 to 1

Y_j = supply location selection decision variable, where $Y_j = 1$ means supply location j is selected; otherwise, $Y_j = 0$ means supply location j is not selected, that is, its values will be either 0 or 1

This optimization problem can be formulated by the following integer linear programming (ILP) problem. This problem assumes that each supplier has unlimited capacity.

PROBLEM 16.1   LOCATION ALLOCATION OPTIMIZATION WITHOUT CAPACITY

Minimize:

$$\text{Total cost} = \sum_{i=1}^{m} \sum_{j=1}^{n} c_{ij} x_{ij} + \sum_{j=1}^{n} K_j Y_j \tag{16.9}$$

Subject to:

$$\sum_{i=1}^{m} x_{ij} \leq mY_j \quad \text{for } j = 1, 2, \ldots, n \tag{16.10}$$

$$\sum_{j=1}^{n} x_{ij} = 1 \quad \text{for } i = 1, 2, \ldots, m \tag{16.11}$$

$$Y_j = 0, \ 1 \quad \text{for } j = 1, 2, \ldots, n \tag{16.12}$$

$$x_{ij} \geq 0 \quad \text{for all } i, j \tag{16.13}$$

Equation (16.9) is the total-cost objective function, which is minimized. The total cost is the sum of transportation costs (the first term) plus the initial costs of the selected suppliers (the second term). Equation (16.10) ensures that each supplier j can satisfy the total demand of all demand locations. Equation (16.4) ensures that the demand of each location i is satisfied. Equation (16.12) ensures that either a supplier j is selected or it is not. For example, at location j, either a facility is constructed ($Y_j = 1$) or it is not ($Y_j = 0$). Equation (16.13) is the nonnegativity constraint on all decision variables where x_{ij} represents the fraction of the demand of location i supplied by location j.

The fixed cost of the selection of a supplier refers to all related fixed costs for the given planning period. That is, by choosing a supplier location (whether a new building is constructed or leased), there is a specific fixed initial cost to use the supplier location. It can be shown that in the optimal solution to Problem 16.1 the values of x_{ij} will always be either 0 or 1, that is, the demand location will buy all its needed items only from one supplier. This is shown in the following example.

Example 16.9   Single-Objective Optimization for Supplier Selection   Suppose there are four existing warehouses (demand locations) and three potential sites that can be selected to satisfy the demands of the four warehouses. The table below shows the cost (including the transportation cost) of buying items for each warehouse from each supply site. The initial cost of using each of the suppliers is also given. Costs are in thousands of dollars. For example, Cleveland's demand can be satisfied by any of the three suppliers (Reno, Flagstaff, and Rochester), with costs of 81, 71, and 110, respectively. For Cleveland the best choice is Flagstaff because it has the lowest cost in terms of shipment and also in terms of the initial cost of 66.

Identify how much should be shipped from each of the three suppliers to each of the four warehouses such that the total cost is minimized.

		From: Plant (Supplier) Site (j)		
		1. Reno	2. Flagstaff	3. Rochester
To: Warehouse (Demand) Site (i)	1. Cleveland	81	71	110
	2. Los Angeles	101	95	46
	3. Pittsburgh	72	79	192
	4. Chicago	12	18	59
Initial cost for each supplier		85	66	92

To solve this problem, first define decision variables as presented below.

	From (j) / To (i)	1. Reno	2. Flagstaff	3. Rochester	Total Demand
Fraction of demand (i) supplied by supplier (j)	1. Cleveland	x_{11}	x_{12}	x_{13}	1
	2. Los Angeles	x_{21}	x_{22}	x_{23}	1
	3. Pittsburgh	x_{31}	x_{32}	x_{33}	1
	4. Chicago	x_{41}	x_{42}	x_{43}	1
	If	$Y_1 = 1$, Use Reno	$Y_2 = 1$, Use Flagstaff	$Y_3 = 1$, Use Rochester	Total: 4

The total demand at each of the four demand locations is 1, which must be satisfied by one or more of the three supply locations. Therefore, the total amount needed to satisfy the demand of all four cities is 4. That is, the capacity of each supplier is sufficient to satisfy the demand of all four cities.

This problem is formulated by the following ILP problem:

Minimize:

$$\text{Total cost} = 81x_{11} + 71x_{12} + 110x_{13} + 101x_{21} + 95x_{22} + 46x_{23} + 72x_{31}$$
$$+ 79x_{32} + 192x_{33} + 12x_{41} + 18x_{42} + 59x_{43} + 85Y_1 + 66Y_2 + 92Y_3$$

Subject to:

$$x_{11} + x_{21} + x_{31} + x_{41} \leq 4Y_1$$
$$x_{12} + x_{22} + x_{32} + x_{42} \leq 4Y_2$$
$$x_{13} + x_{23} + x_{33} + x_{43} \leq 4Y_3$$

$$x_{11} + x_{12} + x_{13} = 1$$
$$x_{21} + x_{22} + x_{23} = 1$$
$$x_{31} + x_{32} + x_{33} = 1$$
$$x_{41} + x_{42} + x_{43} = 1$$

$$Y_j = 0, 1 \text{ for } j = 1, 2, 3$$
$$x_{ij} \geq 0 \quad \text{for all } i, j$$

Solving this problem using ILP software, the optimal solution is as follows:

x_{ij}	1	2	3
1	0	1	0
2	0	1	0
3	0	1	0
4	0	1	0
Y_j	0	1	0

In the above solution, $Y_2 = 1$ means that only the Flagstaff supplier should be used to supply all four warehouses. The total cost is $71 + $95 + $79 + $18 + $66 = $329. Note that $66 is the initial cost of using the Flagstaff supplier, and the remaining cost is the total shipment (purchase and transportation) cost from Flagstaff to the four existing warehouses.

See Supplements 16.6.xls, 16.6.lg4.

16.6.2   Multiple-Objective Location Allocation[†]

There are several criteria that can be used for the selection of suppliers (or locations). For example, consider the following two objectives:

Minimize total cost, f_1
Maximize total qualitative preferences, f_2

For the second objective define q_{ij} as the coefficient of the qualitative preference matrix for demand location i to be satisfied by supplier j. A high qualitative preference, q_{ij}, means that it is preferred that demand location i be supplied by supplier j. Using an additive utility function, the multiple-objective integer linear programming (MOILP) problem can be formulated as follows.

To solve this problem, first identify the maximum and minimum values for each objective ($f_{1\max}, f_{1\min}, f_{2\max}, f_{2\min}$). Then, use these values to normalize each objective function. Because f_1 is minimized, use maximizing $-f_1$; therefore, its normalized value, f_1', is maximized. The MOILP problem is presented below.

PROBLEM 16.2 TWO-OBJECTIVE LOCATION ALLOCATION OPTIMIZATION

Maximize:

$$U = w_1 f_1' + w_2 f_2' \tag{16.14}$$

Subject to:

$$f_1 = \sum_{i=1}^{m} \sum_{j=1}^{n} c_{ij} x_{ij} + \sum_{j=1}^{n} K_j Y_j \tag{16.15}$$

$$f_2 = \sum_{i=1}^{m} \sum_{j=1}^{n} q_{ij} x_{ij} \tag{16.16}$$

$$f_1' = \frac{-(f_1 - f_{1\max})}{f_{1\max} - f_{1\min}} \tag{16.17}$$

$$f_2' = \frac{f_2 - f_{2\min}}{f_{2\max} - f_{2\min}} \tag{16.18}$$

$$\sum_{i=1}^{m} x_{ij} \leq m Y_j \qquad \text{for } j = 1, 2, \ldots, n \tag{16.10}$$

$$\sum_{j=1}^{n} x_{ij} = 1 \qquad \text{for } i = 1, 2, \ldots, m \tag{16.11}$$

$$Y_j = 0, 1 \qquad \text{for } j = 1, 2, \ldots, n \tag{16.12}$$

$$x_{ij} \geq 0 \qquad \text{for all } i, j \tag{16.13}$$

The above bicriteria problem can be generalized to three or more criteria. For example, for a tricriteria problem, the third objective could be minimizing transportation time (f_3) as presented in the following problem, where t_{ij} represents the transportation time between demand location i and supplier j.

PROBLEM 16.3 THREE-OBJECTIVE LOCATION ALLOCATION OPTIMIZATION

Maximize:

$$U = w_1 f_1' + w_2 f_2' + w_3 f_3' \tag{16.19}$$

Subject to:
Equations (16.15)–(16.18)

$$f_3 = \sum_{i=1}^{m} \sum_{j=1}^{n} t_{ij} x_{ij} \tag{16.20}$$

$$f_3' = \frac{-(f_3 - f_{3\max})}{f_{3\max} - f_{3\min}} \tag{16.21}$$

All constrains of bicriteria MOILP, i.e., Equations (16.10)–(16.13).

Example 16.10 Two-Objective Location Allocation Optimization Consider Example 16.9. In addition to the cost information, also consider the following qualitative preferences for each pair of demand sites and supplier.

Additive Utility Approach Use the additive utility function to determine the optimal solution if the weights of importance are (0.3, 0.7) for normalized values of the two objectives, respectively.

		From: Plant (Supplier) Site (j)		
		Reno	Flagstaff	Rochester
To: Warehouse	Cleveland	$I = 3$	$E = 4$	$O = 2$
(Demand) Site (i)	Los Angeles	$X = 0$	$X = 0$	$A = 5$
	Pittsburgh	$E = 4$	$O = 2$	$U = 1$
	Chicago	$A = 5$	$A = 5$	$E = 4$

For example, it is most desirable, with an A rating, to supply Los Angeles from Rochester. However, it is less desirable, with an X rating, to supply Los Angeles from Reno and/or Flagstaff.

Finding Normalized Objective Functions In Example 16.9, the minimum value for f_1 (cost) was found. Now, solve the problem by maximizing f_2 subject to the constraints of Example 16.9.

The solutions for both problems are listed below.

	Min. f_1	Max. f_2	Min.	Max.
f_1	329	450	329	450
f_2	11	18	11	18

The normalized objective function equations for f_1 and f_2 are presented below where both normalized values are maximized:

$$f_1' = \frac{-f_1 + 450}{-329 + 450} \quad \text{or} \quad 121f_1' + f_1 = 450$$

$$f_2' = \frac{f_2 - 11}{18 - 11} \quad \text{or} \quad 7f_2' - f_2 = -11$$

The two-objective optimization problem is formulated as follows:

Maximize:

$$Z = 0.3f_1' + 0.7f_2'$$

Subject to:

$$f_1 = 81x_{11} + 71x_{12} + 110x_{13} + 101x_{21} + 95x_{22} + 46x_{23} + 72x_{31} + 79x_{32}$$
$$+ 192x_{33} + 12x_{41} + 18x_{42} + 59x_{43} + 85Y_1 + 66Y_2 + 92Y_3$$
$$f_2 = 3x_{11} + 4x_{12} + 2x_{13} + 0x_{21} + 0x_{22} + 5x_{23} + 4x_{31} + 2x_{32} + 1x_{33}$$
$$+ 5x_{41} + 5x_{42} + 4x_{43}$$
$$121f_1' + f_1 = 450$$
$$7f_2' - f_2 = -11$$

All constraints of Example 16.9.

The solution of this problem by an ILP computer software package is given below. The solution is to use two suppliers: one in Reno and one in Rochester. Supply Cleveland, Pittsburgh, and Chicago from the plant at Reno and supply Los Angeles from the plant at Rochester.

x_{ij}	1	2	3
1	1	0	0
2	0	0	1
3	1	0	0
4	1	0	0
Y_j	1	0	1

Total cost (f_1) = \$81 + \$72 + \$12 + \$85 + \$46 + \$92 = \$388
Total qualitative preference (f_2) = 3 + 5 + 4 + 5 = 17

See Supplement S16.7.lg4.

Example 16.11   Three-Objective Location Allocation Optimization   The tricriteria problem can be presented as:

Minimize total cost, f_1
Maximize total qualitative preferences, f_2
Minimize total transportation time, f_3

Consider Example 16.10 with a third criterion, transportation time. Consider the following distance matrix in minutes, which represents the transportation time:

		Potential Plant Sites (j)		
		1. Reno	2. Flagstaff	3. Rochester
Existing	1. Cleveland	2240	1980	260
Warehouse	2. Los Angeles	520	460	2660
Sites (i)	3. Pittsburgh	2360	1960	280
	4. Chicago	1910	1660	600

Consider three separate MOILP problems using each of the three objective functions. The solutions for each of the three MOILP problems are presented below.

	Min. f_1	Max. f_2	Min. f_3	Min.	Max.
f_1	329	450	699	329	699
f_2	11	18	7	7	18
f_3	6340	8660	1320	1320	8660

Use the above minimum and maximum values of each objective to generate the normalized objective function. The solution procedure is similar to Example 16.10. The normalized objective function equations for f_1, f_2, and f_3 are as follows. The normalized objective function equation for f_3 is maximized:

$$f_1' = \frac{-f_1 + 699}{-329 + 699} \qquad \text{or} \qquad 370 f_1' + f_1 = 699$$

$$f_2' = \frac{f_2 - 7}{18 - 7} \qquad \text{or} \qquad 11 f_2' - f_2 = -7$$

$$f_3' = \frac{-f_3 + 8660}{-1320 + 8660} \qquad \text{or} \qquad 7340 f_3' + f_3 = 8660$$

Suppose the weights of importance for the normalized values of the three objectives are $(0.3, 0.2, 0.5)$:

Maximize: $U = 0.3 f_1' + 0.2 f_2' + 0.5 f_3'$
Subject to: Constraints of Example 16.9.

The final solution is:

x_{ij}	1	2	3
1	0	0	1
2	0	0	1
3	0	0	1
4	0	0	1
Y_j	0	0	1

Total cost (f_1) = \$499
Total qualitative preference (f_2) = 12
Total time (f_3) = 3520 min

In this solution, only the Flagstaff supplier is used to supply all four warehouses.

See Supplements 16.8.xls, 16.8.lg4.

16.6.3   Location Allocation with Capacity Constraints

Suppose there are m given locations whose demands in terms of given volumes must be fulfilled by one or more of n given suppliers where each supplier has a capacity limit. This problem is a generalization of the problem in the previous section using the following additional notations:

c_{ij} = cost of satisfying demand i by supplier j for one unit of demand
s_j = capacity of supplier j for $j = 1, 2, \ldots, n$
D_i = demand of location i for $i = 1, 2, \ldots, m$

The decision variables (to be determined) are:

x_{ij} = amount of demand of location i to be supplied by supplier j
Y_j = supply location selection, where $Y_j = 1$ means supply location j is selected; otherwise, $Y_j = 0$ means supply location j is not selected

This optimization problem can be formulated by the following ILP problem.

PROBLEM 16.4   LOCATION ALLOCATION OPTIMIZATION WITH LIMITED CAPACITY

Minimize:

$$\text{Total cost} = \sum_{i=1}^{m} \sum_{j=1}^{n} c_{ij} x_{ij} + \sum_{j=1}^{n} K_j Y_j \qquad (16.22)$$

Subject to:

$$\sum_{i=1}^{m} x_{ij} \le s_j Y_j \quad \text{for } j = 1, 2, \dots, n \tag{16.23}$$

$$\sum_{j=1}^{n} x_{ij} = D_i \quad \text{for } i = 1, 2, \dots, m \tag{16.24}$$

$$Y_j = 0, 1 \quad \text{for } j = 1, 2, \dots, n \tag{16.25}$$

$$x_{ij} \ge 0 \quad \text{for all } i, j \tag{16.26}$$

Equation (16.23) ensures that the total amount assigned to each supplier j does not exceed its capacity. Equation (16.24) ensures that the demand of each location i is satisfied. It can be shown that Problem 16.1 is a special case of Problem 16.4.

Example 16.12   Single-Objective Optimization for Supplier Selection   Consider Example 16.9 where the demand and supply for each location is given in the following table. Suppose that the initial costs to use the suppliers at Reno, Flagstaff, and Rochester are 85,000, 66,000, and 92,000, respectively.

	From (j) / To (i)	1. Reno	2. Flagstaff	3. Rochester	Demand
Amount of demand (i) supplied by supplier (j)	1. Cleveland	x_{11}	x_{12}	x_{13}	800
	2. Los Angeles	x_{21}	x_{22}	x_{23}	3500
	3. Pittsburgh	x_{31}	x_{32}	x_{33}	1000
	4. Chicago	x_{41}	x_{42}		2200
	Capacity	5500	3000	5000	—

Identify how much should be shipped from each of the three suppliers to each of the four warehouses such that the total cost is minimized.

This is formulated by the following ILP problem:

Minimize:

$$\text{Total cost} = 81x_{11} + 71x_{12} + 110x_{13} + 101x_{21} + 95x_{22} + 46x_{23} + 72x_{31} + 79x_{32}$$
$$+ 192x_{33} + 12x_{41} + 18x_{42} + 59x_{43} + 85{,}000Y_1 + 66{,}000Y_2$$
$$+ 92{,}000Y_3$$

Subject to:

$$x_{11} + x_{12} + x_{13} = 800$$
$$x_{11} + x_{21} + x_{31} + x_{41} \le 5500Y_1 \qquad x_{21} + x_{22} + x_{23} = 3500 \qquad Y_j = 0, 1 \quad \text{for } j = 1, 2, 3$$
$$x_{12} + x_{22} + x_{32} + x_{42} \le 3000Y_2 \qquad x_{31} + x_{32} + x_{33} = 1000 \qquad x_{ij} \ge 0 \quad \text{for all } i, j$$
$$x_{13} + x_{23} + x_{33} + x_{43} \le 5000Y_3 \qquad x_{41} + x_{42} + x_{43} = 2200$$

Solving this problem by ILP software, the optimal solution is as follows:

x_{ij}	1	2	3	Demand
1	800	0	0	800
2	0	0	3500	3500
3	1000	0	0	1000
4	2200	0	0	2200
Y_j	1	0	1	–
Used capacity	4000	0	3500	Total = 7500

In the above solution, $Y_1 = 1$, $Y_2 = 0$, $Y_3 = 1$. This solution means that two suppliers, Reno and Rochester, should be used to supply the four warehouses. The total cost is $501,200. Note that $85,000 and $92,000 are the initial costs of using the two suppliers, and the remaining cost is the total shipment (purchase and transportation) cost from suppliers to the four existing warehouses.

See Supplement S16.9.lg4.

REFERENCES

General

Atkins, R. J., and R. H. Shriver. "New Approaches to Facilities Location." *Harvard Business Review*, 1968, pp. 70–79.

Ballou, R. H. "Dynamic Warehouse Location Analysis." *Journal of Marketing Research*, Vol. 5, No. 3, 1968, pp. 271–276.

Brandeau, M. L. "An Overview of Representative Problems in Location Research." *Management Science*, Vol. 35 No. 6, 1989, pp. 645–674.

Drezner, Z. *Facility Location: A Survey of Applications and Methods*. New York: Springer, 1995.

Drezner, Z., and H. W. Hamacher. *Facility Location: Applications and Theory*, New York: Springer, 2004.

Efroymson, M. A., and T. L. Ray. "A Branch and Bound Algorithm for Plant Location." *Operations Research*, Vol. 14, No. 3, 1966, pp. 361–368.

Farahani, R. Z., and M. Hekmatfar. *Facility Location: Concepts, Models, Algorithms and Case Studies*. City: Physica, 2009.

Francis, R. L., and J. A. White. *Facility Layout and Location: An Analytical Approach*, Englewood Cliffs, NJ: Prentice Hall, 1974.

Heragu, S. S. *Facilities Design*, 3rd ed., Boca Raton, FL: CRC press, 2008.

Konforty, Y., and A. Tamir. "The Single Facility Location Problem with Minimum Distance Constraints." *Location Science*, Vol. 5, No. 3, 1997, pp. 147–163.

Nahmias, S. *Production and Operations Analysis*, 6th ed. New York: McGraw-Hill/Irwin, 2008.

Nugent, C. E, T. E Vollmann, and J. Ruml. "An Experimental Comparison of Techniques for the Assignment of Facilities to Locations." *Operations Research*, Vol. 16, No.1, 1968, pp. 150–173.

Ohsawa, Y., and A. Imai. "Degree of Locational Freedom in a Single Facility Euclidean Minimax Location Model." *Location Science*, Vol. 5, No. 1, 1997, pp. 29–45.

Owen, S. H., and M. S. Daskin. "Strategic Facility Location: A Review." *European Journal of Operational Research*, Vol. 111, No. 3, 1998, pp. 423–447.

Partovi, F. Y. "An Analythic Model for Locating Facilities Strategically." *Omega*, Vol. 34, No. 1, 2006, pp. 41–55.

ReVelle, C. S., H. A. Eiselt, and M. S. Daskin. "A Bibliography for Some Fundamental Problem Categories in Discrete Location Science." *European Journal of Operational Research*, Vol. 184, No. 3, 2008, pp. 817–848.

Sale, D. R. *Manufacturing Facilities: Location. Planning and Design*, Boston: PWS Publishing, 1994.

Stevenson, W. J. *Operations Management*, 10th ed. New York: Mc Graw-Hill Irwin, 2008.

Sule, D. R. *Manufacturing Facilities: Location, Planning and Design*, 3rd ed. Boca Raton, FL: CRC Press, 2008.

Weber, A. *Alfreds Weber's Theory of the Location of Industries*. Chicago: University of Chicago Press, 1929.

Weiszfeld, E. "Sur le point pour lequel la somme des distances de n points donnes et minimum." *Tohoku Mathematical journal*, Vol. 34, pp. 355–386, 1937.

White, J. A. "Locational Analysis." *European Journal of Operational Research*, Vol. 12, 1983, pp. 220–252.

Wilamowsky, Y., S. Epstein, and B. Dickman. "How the Oldest Recorded Multiple Facility Location Problem Was Solved." *Location Science*, Vol. 3, No. 1, 1995. pp. 55–60.

Yang, Z., and B. Malakooti. "Machine-Part Cell Formation & Operation Allocation in Cellular Manufacturing." 5th IE Research Conference, 1996, pp. 31–36.

Multiobjective Optimization

Aktar, D. E., and U. Ozden. "An Integrated Multiobjective Decision Making Process for Supplier Selection and Order Allocation." *Omega*, Vol. 36, No. 1, 2008, pp. 76–90.

Badri, M. A. "Combining the Analytic Hierarchy Process and Goal Programming for Global Facility Location-Allocation Problem." *International Journal of Production Economics*, Vol. 62, No. 3, 1999, pp. 237–248.

Chuang, P.-T. "Combining the Analytic Hierarchy Process and Quality Function Deployment for a Location Decision from a Requirement Perspective." *The International Journal of Advanced Manufacturing Technology*, Vol. 18, No. 11, 2001, pp. 842–849.

Current, J., H. Min, and D. Schilling. "Multiobjective Analysis of Facility Location Decisions." *European Journal of Operational Research*, Vol. 49, No. 3, 1990, pp. 295–307.

Farahani, R. Z., and N. Asgari. "Combination of MCDM and Covering Techniques in a Hierarchical Model for Facility Location: A Case Study." *European Journal of Operational Research*, Vol. 176, No. 3, 2007, pp. 1839–1858.

Farahani, R. Z., M. SteadieSeifi, and N. Asgari. "Multiple Criteria Facility Location Problems: A Survey." *Applied Mathematical Modelling*, Vol. 34, No. 7, 2010, pp. 1689–1709.

Fernández, E., and J. Puerto. "Multiobjective Solution of the Uncapacitated Plant Location Problem." *European Journal of Operational Research*, Vol. 145, No. 3, 16, 2003, pp. 509–529.

Hamacher, H. W., and S. Nickel. "Multicriteria Planar Location Problems." *European Journal of Operational Research* Vol. 94, No. 1, 1996, pp. 66–86.

Karkazis, J. "Facilities Location in a Competitive Environment: A Promethee Based Multiple Criteria Analysis." *European Journal of Operational Research*, Vol. 42, No. 3, 1989, pp. 294–304.

Lee, S. M., G. I. Green, and C. S. Kim. "A Multiple Criteria Model for the Location-Allocation Problem." *Computers & Operations Research*, Vol. 8, No. 1, 1981, pp. 1–8.

Malakooti, B. "Computer Aided Facility Layout Selection with Applications to Multiple Criteria Manufacturing Planning Problems," *Large Scale Systems: Theory and Applications, Special Issue on Complex Systems Issues in Manufacturing*, Vol. 12, 1987, pp. 109–123.

Malakooti, B. "Multi-Objective Location Decision Problem." Case Western Reserve University, Cleveland, Ohio, 2013.

Malczewksi, J., and W. Ogryczak. "The Multiple Criteria Location Problem: 2. Preference-Based Techniques and Interactive Decision Support." *Environment and Planning*, 1996, Vol. 28, No. 1, pp. 69–98.

McGinnis, L. F., and J. A. White. "A Single Facility Rectilinear Location Problem with Multiple Criteria." *Transportation Science*, Vol. 12, No. 3, 1978, pp. 217–231.

Saaty, T. *The Analytic Hierarchy Process: Planning, Priority Setting, Resource Allocation.* London: McGraw-Hill, 1980.

Şahin, G., and H. Süral. "A Review of Hierarchical Facility Location Models." *Computers & Operations Research*, Vol. 34, No. 8, 2007, pp. 2310–2331.

Yang, J., and H. Lee. "An AHP Decision Model for Facility Location Selection." *Facilities*, Vol. 15, No. 9/10, 1997, pp. 241–254.

EXERCISES

Break-Even Analysis (Section 16.2)

2.1 Consider the following annual cost information for three cities. Show the volume ranges for which each city is optimal. All costs are in $10,000.

City	A	B	C
Fixed cost	10	22	35
Cost/volume	2.0	1.7	1.5

2.2 For the following three cities, show the volume ranges for which each alternative is optimal using break-even analysis. The volume ranges from 0 to 1000. All costs are in $10,000.

City	A	B	C
Annual cost	$e^{(volume/100)} + 10$	$50 \ln (volume) + 30$	$4 \times volume + 20$

2.3 Consider the cost information provided in the following table for four cities. Show the volume ranges for which each alternative is optimal. All costs are in $10,000.

City	A	B	C	D	E
Fixed cost	100	120	280	350	400
Cost/volume	17	10	9	8	5

Rectilinear Method (Section 16.3.1)

3.1 An amusement park wants to build a restroom facility. The park is laid out in a grid fashion with paths running north to south and west to east. They would like the restroom facility to be close to the most five popular rides as presented in the following table:

Ride	X	Y	Avg. No. of Riders (f_1)	Qualitative Closeness (f_2)
1	4	17	50	A
2	7	5	65	I
3	23	10	45	U
4	14	13	20	E
5	8	8	80	E

(a) Find the optimal location using the rectilinear distance method and only considering the first objective (f_1), quantitative information.

(b) Find the optimal location using the rectilinear distance method and only considering the second objective (f_2), qualitative closeness information.

3.2 The ABC Co. wants to build a new service/repair facility within its corporate offices. There are six departments that will utilize this new facility. Find the optimal location for the new cafeteria. Respond to parts (a) and (b) of exercise 3.1.

Facility	X	Y	No. of Workers (f_1)	Qualitative Closeness (f_2)
1	1	8	10	A
2	20	16	50	O
3	4	9	30	U
4	6	8	20	E
5	10	13	23	I
6	15	5	17	A

3.3 A department store would like to know the optimal coordinates to build a new store in a metropolitan area. There are six potential markets that have the following populations of customers. The locations of these six markets, along with their populations, are listed in the table below. Respond to parts (a) and (b) of exercise 3.1.

Locality	Latitude	Longitude	Population (f_1)	Qualitative Closeness (f_2)
1	45.50	75.50	120,000	U
2	45.12	75.78	345,000	O
3	45.89	75.02	165,000	A
4	45.34	75.95	85,000	A
5	45.20	75.45	450,000	E
6	45.03	75.67	110,000	X

3.4 Consider a large warehouse facility such that by the use of storage shelves locations can be utilized in three-dimensional space (x, y, z). Respond to parts (a) and (b) of exercise 3.1.

Facility	X	Y	Z	No. of Users (f_1)	Qualitative Closeness (f_2)
1	20	34	25	200	E
2	40	21	30	150	U
3	80	50	10	30	O
4	55	43	90	300	E
5	61	60	74	175	I
6	33	49	80	90	A

Multicriteria Rectilinear Method (Section 16.3.2)

3.5 Consider both criteria presented in exercise 3.1. Suppose that the weights of importance of the normalized objectives for the additive utility function are $(0.3, 0.7)$. Find the best location for the bicriteria problem.

3.6 Consider both criteria presented in exercise 3.2. Suppose that the weights of importance of the normalized objectives for the additive utility function are $(0.6, 0.4)$. Find the best location for the bicriteria problem.

3.7 Consider both criteria presented in exercise 3.3. Suppose that the weights of importance of the normalized objectives for the additive utility function are $(0.5, 0.5)$. Find the best location for the bicriteria problem.

3.8 Consider both criteria presented in exercise 3.4. Suppose that the weights of importance of the normalized objectives for the additive utility function are $(0.4, 0.6)$ Find the best location for the bicriteria problem.

Center-of-Gravity Method (Section 16.4.1)

4.1 Consider exercise 3.1.

   (a) Determine the optimal coordinates of the new restroom facility using the center-of-gravity method. Solve the problem using the first and second objectives, separately.

   (b) Determine the optimal coordinates of the new restroom facility using Euclidean distance. Solve the problem using the first and second objectives, separately.

4.2 Consider exercise 3.2.

   (a) Determine the optimal coordinates of the new service/repair facility using the center-of-gravity method. Solve the problem using the first and second objectives, separately.

   (b) Determine the optimal coordinates of the new service/repair facility using Euclidean distance. Solve the problem using the first and second objectives, separately.

4.3 Consider exercise 3.3.

   (a) Determine the optimal coordinates of the new store using the center-of-gravity method. Solve the problem using the first and second objectives, separately.

   (b) Determine the optimal coordinates of the new store using Euclidean distance. Solve the problem using the first and second objectives, separately.

4.4 Consider exercise 3.4.

   (a) Determine the optimal coordinates of the new facility using the center-of-gravity method. Solve the problem using the first and second objectives, separately.

   (b) Determine the optimal coordinates of the new facility using Euclidean distance. Solve the problem using the first and second objectives, separately.

Multicriteria Center-of-Gravity Method (Section 16.4.2)

4.5 Solve exercise 3.5. Use the center-of-gravity method to find the optimal coordinates of the new training facility.

4.6 Solve exercise 3.6. Use the center-of-gravity method to find the optimal coordinates of the new training facility.

4.7 Solve exercise 3.7. Use the center-of-gravity method to find the optimal coordinates of the new training facility.

4.8 Solve exercise 3.8. Use the center-of-gravity method to find the optimal coordinates of the new training facility.

Ordinal/Cardinal Approach for Location Section (Section 16.5.1)

5.1 A company wants to open a new office in one of five possible cities. The following table gives the cost (to be minimized) (in hundreds of dollars) and qualitative preference (to be maximized) for each of the five cities.

Location	Cost (f_1)	Qualitative Preference (f_2)
Olympia	125000	E
Nashville	200000	E
Charleston	155000	I
Atlanta	210000	A
Green Bay	145000	U

   (a) Suppose the weights of importance are $(0.6, 0.4)$ for the normalized criteria values for an additive utility function. Rank all the cities and determine the best city in which to open the new office.

   (b) Suppose that the weights of importance are not given but the following responses are given for the comparison of pairs of weights where $w_0 = 0$ is a reference point. Assess the weights and rank all alternatives.

(c) Determine if the given information on weights are ordinal and cardinal consistent and estimate the CIR. Do you recommend using the additive utility function?

R_{ij}	w_1	w_2	w_0
w_1	0	6	9
w_2	−6	0	4
w_0	−9	−4	0

5.2 A university is planning to open a new cafeteria in one of its seven existing buildings. The following table gives the cost (in thousands of dollars), qualitative preference (to be maximized), and environmental preference (to be maximized, on a scale of 0–100) for each of the locations.

Location	Cost (f_1)	Qualitative Preference (f_2)	Environment Preference (f_3)
A	5200	E	60
B	6300	O	70
C	4600	A	20
D	5700	I	30
E	4700	E	10
F	6000	U	100
G	5900	I	80

(a) Suppose the weights of importance for the normalized values of the three objectives are $(0.3, 0.4, 0.3)$, respectively. Use the additive utility function to rank alternatives and determine the best location.

(b) Suppose that the weights of importance are not given but the following responses are given for the comparison of pairs of weights where $w_0 = 0$ is a reference point. Assess the weights and rank all alternatives.

(c) Determine if the given information on weights are ordinal and cardinal consistent and estimate the CIR. Do you recommend using the additive utility function?

R_{ij}	w_1	w_2	w_3	w_0
w_1	0	3	7	9
w_2	−3	0	3	6
w_3	−7	−3	0	3
w_0	−9	−6	−3	0

5.3 A corporation is planning to open a new manufacturing facility in one of five possible cities. The following table gives the cost (in thousands of dollars), educational institutes, government support, and skilled labor (on a scale of 0–100) for each city. All four criteria are maximized except cost.

Location	Cost (f_1)	Educational Institutes (f_2)	Government Support (f_3)	Skilled Labor (f_4)
Olympia	225,000	U	I	70
Nashville	100,000	I	O	10
Charleston	355,000	X	E	100
Atlanta	210,000	E	U	30
Green Bay	245,000	I	A	50

(a) Suppose the weights of importance for the normalized values of the four objectives are (0.2, 0.2, 0.3, 0.3), respectively. Use the additive utility function to rank alternatives and determine the best location.

(b) Suppose that the weights of importance are not given but the following responses are given for the comparison of pairs of weights where $w_0 = 0$ is a reference point. Assess the weights and rank all alternatives.

(c) Determine if the given information on weights are ordinal and cardinal consistent and estimate the CIR. Do you recommend using the additive utility function?

R_{ij}	w_1	w_3	w_2	w_4	w_0
w_1	0	2	4	5	9
w_3	−2	0	2	3	6
w_2	−4	−2	0	1	4
w_4	−5	−3	−1	0	1
w_0	−9	−6	−4	−1	0

5.4 A pharmaceutical company wants to open a new laboratory in one of seven cities. The following table gives the cost (in thousands of dollars), educational institutes, government support, desirability of location, and skilled labor (on a scale of 0–100) for each city.

Location	Cost (f_1)	Educational Institutes (f_2)	Government Support (f_3)	Desirability of Location (f_4)	Skilled Labor (f_5)
A	6200	A	U	E	80
B	5300	I	A	O	70
C	4600	X	I	U	50
D	4700	E	X	E	60
E	5700	I	I	A	70
F	5000	U	E	A	65
G	4900	X	A	I	30

(a) Suppose the weights of importance for the normalized values of the five objectives are (0.2, 0.2, 0.3, 0.1, 0.2), respectively. Use the additive utility function to rank all alternatives and determine the best location.

(b) Suppose that the weights of importance are not given but the following responses are given for the comparison of pairs of weights where $w_0 = 0$ is a reference point. Assess the weights and rank all alternatives.

(c) Determine if the given information on weights are ordinal and cardinal consistent and estimate the CIR. Do you recommend using the additive utility function?

R_{ij}	w_1	w_3	w_2	w_4	w_5	w_0
w_1	0	2	4	5	9	10
w_3	−2	0	2	4	7	8
w_2	−4	−2	0	0	1	4
w_4	−5	−4	0	0	1	3
w_5	−9	−7	−1	−1	0	1
w_0	−10	−8	−4	−3	−1	0

Single-Objective Location Allocation (Section 16.6.1)

6.1 The following table shows the cost of purchasing items for each demand location i from each supplier j. The table also shows the initial cost of using each supplier.

		From: Plant (Supplier) Site (j)			
		A	B	C	D
To: Warehouse (Demand) Site (i)	Beijing	34	33	37	32
	Tokyo	39	40	33	37
	San Diego	61	43	40	51
Initial cost for each supplier		49	52	65	58

(a) Determine the amount of shipment from each supplier to each demand site. Which suppliers should be selected and how much is the total cost?

(b) If site B cannot serve Beijing and site C cannot serve Tokyo, what is the solution to part (a)?

6.2 Consider the transportation costs from each potential plant site j, to each existing warehouse i, in the table below. The initial cost at each supplier location is also given.

		From: Plant (Supplier) Site (j)			
		A	B	C	D
To: Warehouse (Demand) Site (i)	Seoul	60	81	80	100
	New York	90	110	95	35
	New Delhi	85	101	79	180
	London	25	34	29	65
Initial Cost for each supplier		85	80	65	92

(a) Determine the amount of shipment from each supplier to each demand site. Which suppliers should be selected and how much is the total cost?

(b) If site A cannot serve New York and site C cannot serve London, what is the solution to part (a)?

6.3 Consider the transportation costs from each potential plant site j to each existing market i in the table below. The initial cost of each supplier location is also given.

		From: Plant (Supplier) Site (j)					
		A	B	C	D	E	F
	Bangkok	100	75	160	170	150	50
	Austin	256	88	30	180	90	150
To: Warehouse	Cairo	90	73	170	150	160	130
(Demand) site (i)	Hong Kong	35	110	185	200	95	100
	Quebec	110	130	210	100	190	155
	Sydney	220	110	105	25	220	190
Initial Cost for each Supplier		110	90	120	69	95	115

(a) Determine the amount of shipment from each supplier to each demand site. Which suppliers should be selected and how much is the total cost?

(b) If site D cannot serve Quebec and site B cannot serve Bangkok, what is the solution to part (a)?

Multiple-Objective Location Allocation[†] (Section 16.6.2)

6.4 Consider exercise 6.1. For this problem consider a second objective of qualitative preference as shown in the table below. Assume the weights of importance are (0.6, 0.4) for the two objectives.

		From: Plant (Supplier) Site (j))			
		A	B	C	D
To: Warehouse	Beijing	X	O	E	U
(Demand) Site (i)	Tokyo	I	A	X	O
	San Diego	O	E	I	A

(a) What is the optimal solution considering only the qualitative preference?

(b) Use the MOILP approach to determine the best solution using the normalized values for the additive utility function.

6.5 Consider exercise 6.2. For this problem consider a second objective of qualitative preference as shown in the table below. Assume the weights of importance are (0.5, 0.5) for the two objectives. Respond to parts (a) and (b) of exercise 6.4.

		From: Plant (Supplier) Site (j)			
		A	B	C	D
To: Warehouse (Demand) Site (i)	Seoul	A	I	E	O
	New York	U	X	X	A
	New Delhi	O	E	O	U
	London	I	A	A	E

6.6 Consider exercise 6.3. For this problem consider a second objective of qualitative preference as shown in the table below. Assume the weights of importance are (0.3, 0.7) for the two objectives. Respond to parts (a) and (b) of exercise 6.4.

		From: Plant (Supplier) Site (j)					
		A	B	C	D	E	F
To: Warehouse (Demand) Site (I)	Bangkok	I	X	A	I	E	U
	Austin	A	I	X	E	U	O
	Cairo	O	U	U	O	O	E
	Hong Kong	X	I	I	A	X	X
	Quebec	I	A	E	O	E	I
	Sydney	E	E	O	X	A	A

Location Allocation with Capacity Constraints (Section 16.6.3)

6.7 Consider exercise 6.1. Suppose that the capacity limitations for plants A, B, C, and D are 650, 350, 500, and 600, respectively. The demands required by Beijing, Tokyo, and San Diego warehouses are 500, 600, and 750, respectively. Respond to all parts of exercise 6.1.

6.8 Consider exercise 6.2. Suppose that the capacity limitations for potential plant sites A, B, C, and D are 2000, 1400, 1800, and 2500, respectively. The demands required by Seoul, New York, New Delhi, and London warehouses are 1500, 1100, 960, and 800, respectively. Respond to all parts of exercise 6.2.

6.9 Consider exercise 6.3. Suppose that the capacity limitations for potential plant sites A, B, C, D, E and F are 2000, 1500, 2200, 1000, 900 and 1200, respectively. The demands required by Bangkok, Austin, Cairo, Hong Kong, Quebec and, Sydney warehouses are 800, 500, 400, 600, 350, and 650, respectively. Respond to all parts of exercise 6.3.

CHAPTER 17

QUALITY CONTROL AND ASSURANCE

17.1 INTRODUCTION

This chapter provides approaches for measuring the quality of processes and products. Defining quality, however, is difficult due to its multidimensionality. A simple definition of quality, from a producer's point of view, is conformance to specifications while such specifications are measurable; for example, the length of an object can be clearly measured. However, this definition of quality may not be of much interest to the consumer. From a consumer's point of view, quality may be defined based on its usefulness relative to the price paid to acquire such a product.

Table 17.1 illustrates our classification of a multiperspective multicriteria view of quality control (QC) of processes and products. From a system's point of view, an optimal trade-off between the quality of processes and the quality of products is considered. In addition, broader impacts of quality control, such as environmental consequences, are also considered.

Quality of System In Section 17.2, we introduce a system's approach for solving the total quality management problem while considering several criteria. In this problem, there are two conflicting and competing DMs, consumers and producers. These DMs have several objectives which are usually conflicting. We develop a multicriteria quality function deployment (QFD) approach for solving this problem. The optimal solution will be feasible in terms of the product design and the manufacturing process capabilities, and it will also improve the system's performance.

Note: Advanced material that can be omitted without loss of generality will be indicated by a dagger.

Operations and Production Systems with Multiple Objectives, First Edition. Behnam Malakooti.
© 2014 John Wiley & Sons, Inc. Published 2014 by John Wiley & Sons, Inc.

TABLE 17.1 View of Multicriteria QC Problem from Three Different Perspectives

View	QC Problem	QC Approach	QC Multicriteria Analysis
System's	Quality of the system	Quality function deployment (Section 17.2)	• Conformance to specifications • Suitable for intended use • Cost of QC
Producer's	Quality of the process	Process control (Sections 17.3–17.6)	• Duration of QC
Consumer's	Quality of the product	Acceptance sampling Plans (Sections 17.7 and 17.8)	• Environmental impact of QC • Cost of product

Quality of Process In Sections 17.2–17.6, statistical process control is presented, which includes methods for measuring and monitoring the quality of a process during production. Quality control charts are utilized to depict the performance of the process over time. These charts can be used to determine whether a process is under control and behaving as expected. Every process is expected to have some inherent random variability, which is also called random noise. Processes also have nonrandom variations. The control chart can be used to separate random noise variations from nonrandom variations of a process. Nonrandom variations are typically referred to as assignable causes. These causes must be identified and corrected to keep the process under control. Reliable QC systems reduce overall costs by accurately identifying patterns of variations in the process and making corrections before nonconforming products are produced. Statistical process control provides the needed mathematics for making quality observations, measurements, and judgments. While the products are being produced, the capability of the process is being measured. Assuming a normal distribution, the mean μ plus or minus three standard deviations σ includes 99.74% of the total population of items being produced within the required design specifications. This is presented by $\mu \pm 3\sigma$, where the term three-sigma was coined for the high quality of a process where 99.74% of its products fall within the required specifications that is, allowing 2.6 defects out of 1000 products. Currently some industries, such as the automobile and aerospace industries, are striving to achieve six-sigma quality performance; $\mu \pm 6\sigma$ for 0.9999966, that is, allowing 3.4 defects for every million products.

Quality of Product In Sections 17.7 and 17.8, we discuss acceptance sampling plans, which are used after a "lot" of products are produced. In sampling plans, only a small number of products randomly selected from a lot are inspected for quality. The reason for inspecting only a sample instead of the whole lot is that a 100% inspection is expensive, time consuming, and sometimes impossible (e.g., the inspection tests could be destructive). In sampling plans for a lot size of N, one should determine the sample size n, where n is much smaller than N. Based on the observations about the quality of the n samples, inferences would be made about the quality of the whole lot, N. The decision could be to reject or accept the lot. Rejection means the products are not submitted to the next stage of production or to the consumer, but certain actions to remedy the defects should be taken. Acceptance sampling plans are necessary to assure the consumer of the outgoing quality.

Economic Analysis of Quality Control: Bicriteria Problem   The economic analysis of QC is based on the relationship between the percentage of defective products produced and the cost of the QC system. A higher cost for a QC system is associated with a lower percentage of defective items, and this cost increases exponentially to attain a very low percentage of defective items. Figure 17.1 demonstrates an example of the basic relationship between the total cost, which includes the cost of producing poor-quality products, and the cost of quality process control.

The total QC cost consists of (a) the cost of poor-quality products submitted to the consumers and (b) the cost of measuring and controlling the quality of processes. A QC approach that allows many poor-quality products to be released to the consumer is very costly because products will be rejected and returned by the consumers. In Figure 17.1, as the percentage of defective products (p) increases, the cost associated with the rejection of defective items increases. On the other hand, a QC system designed to produce a low percentage of defective products will cost more to implement. The summation of these two costs (Figures 17.1a,b) is the total QC cost whose minimum is associated with a percentage of defective products, p^*. In this example, the minimum total cost is associated with $p^* = 3\%$. To minimize the total cost, one must design a QC system that allows p^* percent of defective products. However, in bicriteria QC problems, total cost is not the only objective.

The two objectives of bicriteria QC problems are:

Minimize the total cost of the QC system.
Minimize the percent of defective products, p.

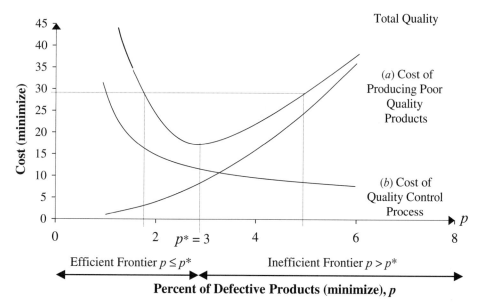

Figure 17.1   Example of optimal percent of defective, p^*, for minimizing total QC and bicriteria problem of cost versus quality.

In Figure 17.1, in bicriteria QC, alternatives associated with $p \leq p^*$ are efficient, and alternatives associated with $p > p^*$ are inefficient. Alternatives associated with $p > p^*$ are inefficient because, for a higher total cost, their percent of defective products (p) is worse than the alternative associated with $p = p^*$; that is, they are all dominated by the alternative at $p = p^*$. At $p = p^* = 3\%$, the total cost is 20 [i.e., alternative (3, 20)]. For example, at $p = 5\%$ the total cost is 30; that is, alternative (5, 30) is inefficient with respect to alternative (3, 20). Inefficient alternatives can be discarded. Therefore, the best alternatives are associated with $p \leq p^*$.

17.2   MULTIPLE-CRITERIA QUALITY FUNCTION DEPLOYMENT

17.2.1   Three Perspectives of QFD

When designing new products, companies must ensure the marketability of their products. A high-quality product with exceptional features and functionalities will still have no value if no demand exists for it. For instance, a high-performance and well-built sports car would not be marketable in an area where it snows most of the year. Therefore, before designing a new product, a company needs to discover the customers' preferences in order to ensure that the designed products satisfy the needs of the customers.

The product development process consists of four phases:

(a) Acquirement of data from customers
(b) Breakdown of customer's requirements
(c) Prioritizing customer's requirements
(d) Translation of customer's requirements into the product design

The data acquisition can be accomplished in several ways. Prominent approaches for obtaining data from customers include interviews and surveys. Once sufficient data have been gathered, the needs and requirements of customers should be categorized into different groups. Then, the importance of different groups of requirements should be assessed. In this process, a customer may consider some requirements to be more important than others. For instance, suppose a customer desires an automobile with high horsepower and also with high fuel efficiency. However, these two objectives are conflicting; high-powered automobiles generally consume more fuel. Due to these conflicting criteria, the customer's needs must be prioritized when considering conflicting and competing criteria. If the customer values fuel efficiency more than horsepower, then greater emphasis or weight should be assigned to fuel efficiency than to horsepower. In this case, when presented with two different automobiles, one with better horsepower and one with better fuel efficiency, the customer will choose the automobile with better fuel efficiency. When the categorized requirements are prioritized, the company needs to design a product that corresponds to the needs of customers while adhering to the needs and requirements of the company design process.

House of Quality   The House of Quality is a conceptual graphical tool that can be employed to bridge between customer needs and design process. The construction of the House of Quality consists of six steps:

1. Define customer needs.
2. Define product features or requirements.

3. Rate the strength of the relationship between customer needs and product requirements.

4. Perform evaluation of competing products.

5. Identify target performance specifications.

6. Determine which product requirements to deploy in the design process.

The generic configuration of the House of Quality is illustrated in Figure 17.2.

Example 17.1 illustrates how to construct the House of Quality.

Example 17.1 Constructing a House of Quality for QFD Consider designing a car where for the purpose of illustration the car features and the manufacturing processes are simplified. Figure 17.3 is the House of Quality employed to relate the customer needs to the product development process.

In Figure 17.3, the customer's needs are specified in terms of the qualities or attributes that are desired in a product. In this example, customers desire an automobile that is fuel efficient, safe, reliable, and powerful. The weights of importance are rated from 0 to 9 where 9 means the most important and 0 means the least important. For example, fuel efficiency is rated with a 9 and is therefore the most important quality in an automobile.

The technical requirements of a product are set by the producer. Some of these technical requirements are correlated with each other and also with customers' needs. For instance, the weight of the car and the horsepower both affect the power and fuel efficiency of an automobile. These requirements are also weighted from 0 to 9 corresponding to their importance. The targets specify the goal values of each technical requirement. The competitive evaluation of customer needs signifies how other companies' products satisfy the needs of the customers in a rating from 0 to 5. The technical evaluation also describes how other companies satisfy the technical requirements. After developing the House of Quality, a company will develop a product that incorporates the needs of both the customers and the company's process capability based on the House-of-Quality matrix.

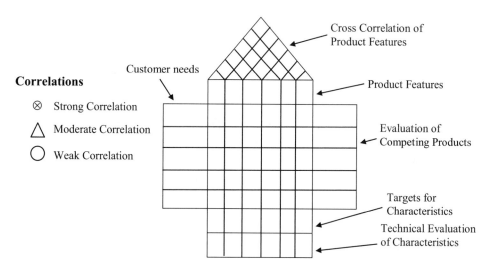

Figure 17.2 Generic configuration of House of Quality.

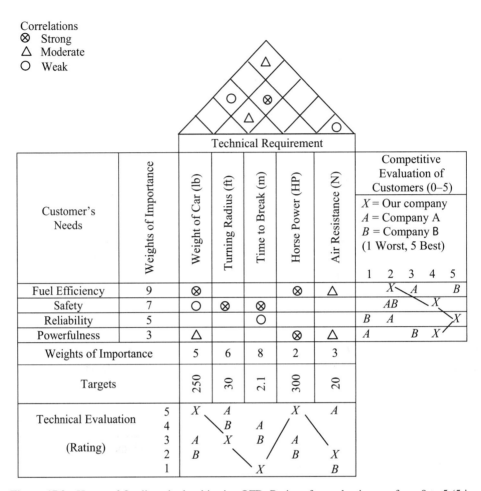

Figure 17.3 House of Quality: single-objective QFD. Ratings for evaluation are from 0 to 5 (5 is the best). Weights of importance are from 0 to 9 (9 is the best).

17.2.2 Multicriteria Views of Customer and Producer

Three Perspectives (Views) of QFD There are three perspectives that are usually involved in accepting or managing changes.

(a) *Customer's View* Usually represented by sales, marketing, and design departments.
(b) *Producer's View* Usually represented by manufacturing, production, assembly, and QC departments.
(c) *System's View* Usually represented by the engineering and management departments.

These perspectives are shown in Figure 17.4. Engineering and management departments should provide a flexible and feasible bridge to connect the left side (the customer's view) to the right side (the producer's view).

Product design, product engineering, manufacturing process, and quality management are interrelated. For example, designing a product in a certain way may be associated with a

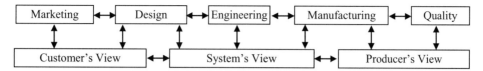

Figure 17.4   How different departments interact to improve quality.

better manufacturing process and quality management system. In practice, both the product design and the manufacturing process may change over time. There are two approaches for improving quality, design, and engineering:

Adaptive Approach (Sequential and Evolutionary)   In this approach changes are based on improving existing conditions on a small (micro) scale. Marketing, sales, design, engineering, manufacturing, production planning, and QC teams must interact to allow feasible microchanges (presented by $\Delta\mu$) to take place that will positively impact the product design, the manufacturing process, and the quality management system. These changes are usually sequential and evolutionary. This is presented in Figure 17.5a.

Generative Approach (Simultaneous and Revolutionary)   In this approach changes are based on developing new ideas through direct interactions of marketing, sales, design, engineering, manufacturing, production planning, and QC teams. Such major changes (presented by ΔM) are usually simultaneous and revolutionary. This is presented in Figure 17.5b.

17.2.3   Multicriteria Views of Customer and Producer

In multicriteria QFD, objectives of both the producer and the customer are considered in the design of the product and the design of the manufacturing process:

Multicriteria Customer's View   The customer's view of quality is based on several criteria for the given design parameters and cost for the product. See Table 17.2.

Multicriteria Producer's View   The producer's view is also based on several criteria. See Table 17.2.

It can be observed that the producer's criteria are measurable, but some of the customer's criteria are not easily measurable. For example, aesthetics and impressions are subjective and require an understanding of societal values.

The House of Quality can be modified and expanded to allow a multicriteria (MC) analysis to evaluate several product and process design alternatives. In MC-QFD, the

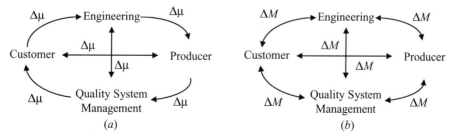

Figure 17.5   System's view of the interaction of four major players in terms of (a) feasible microchanges $\Delta\mu$ and (b) macrochanges ΔM.

TABLE 17.2 Customer's and Producer's Criteria

Customer's Criteria	Producer's Criteria
1. Cost of product	1. Cost of manufacturing
2. Reliability and durability	2. Conformance to design ranges
3. Performance	3. Ease of manufacturing and assembly
4. Comfort of use	4. Ease of QC and assurance
5. Aesthetics and impressions	5. Productivity (production rate)
6. Ease of serviceability	

correlation symbols are replaced by numerical values so that alternatives can be evaluated. Ratings of 1–9 are used instead of symbolic ratings [e.g., weak (∘), moderate (△), and strong (⊗)].

Multicriteria QFD Approach The steps of the MC-QFD procedure are as follows:

1. Identify the measurable objectives for both the customer and producer: $(f_1^C, f_2^C, \ldots f_k^C)$ and $(f_1^P, f_2^P, \ldots f_q^P)$, respectively.
2. Identify all possible feasible alternatives. Measure the objective values of these alternatives for both customer and producer.
3. Assess the weights of importance for the customer's objectives $(w_1^C, w_2^C, \ldots w_k^C)$, and assess the weights of importance for the producer's objectives $(w_1^P, w_2^P, \ldots w_q^P)$.
4. Calculate the utility function for the customer. The utility function for the customer, U^C, is calculated as follows where there are k criteria:

$$U^C = w_1^C f_1^C + w_2^C f_2^C + w_3^C f_3^C + \cdots + w_k^C f_k^C \qquad (17.1)$$

5. Calculate the utility function for the producer. The utility function for the producer, U^P, is calculated as follows where there are q criteria:

$$U^P = w_1^P f_1^P + w_2^P f_2^P + w_3^P f_3^P + \cdots + w_q^P f_q^P \qquad (17.2)$$

6. Assess the importance given to the customer, δ, versus the importance given to the producer, $1 - \delta$, where $0 < \delta < 1$.
7. Rank alternatives by maximizing the utility of the system utility function. The system utility U^S of each alternative is a composite function of both customer and producer utility functions, U^C and U^P:

$$U^S = \delta U^C + (1 - \delta)U^P \qquad (17.3)$$

where δ is a parameter that represents a compromise between the customer's and producer's interests. Higher values of δ give a higher priority to the customer while lower values of δ give a higher priority to the producer, where $0 < \delta < 1$. For example, $\delta = 0.3$ gives 30% importance to customer and 70% to producer. The default value of δ is 0.5. If needed, one can use nonlinear (e.g., Z) utility functions instead of (17.1), (17.2), and (17.3).

Example 17.2   Multiple Criteria of QFD: House of Quality   Consider Example 17.1 where there are four alternatives. The design alternatives for the customers and the company for Example 17.1 are illustrated in the table below where N denotes newtons of air resistance.

Alternative Designs	1. Weight (lb)	2. Turning Radius, (ft)	3. Time to Break from 60 to 0 MPH (s)	4. Horsepower	5. Air Resistance at 60 MPH (N)
I	2500	31	3.2	175	35
II	3500	26	3.5	200	25
III	3000	24	2.6	300	20
IV	3750	28	2.1	250	32

The customer and the company objectives are presented below:

Customer Objectives	Producer's Objectives	Decision Variables
f_1^C = fuel efficiency	f_1^P = production cost	Weight
f_2^C = safety	f_2^P = ease of manufacturing	Turning radius
f_3^C = reliability	f_3^P = ease of assembly	Time to break
f_4^C = powerfulness	f_4^P = materials cost	Horsepower
		Air resistance

The multiple-criteria House of Quality has a different configuration from the conventional House of Quality matrix presented in the last section. Figure 17.6 shows a multiple-criteria House of Quality. The ratings are between 0 and 9, where 0 is the worst and 9 is the best. The weights of importance are between 0 and 1, where their total is 1.

In Example 17.1, the ratings of alternatives are different for the customer and the producer. This is due to the fact that the customer and the producer have different objectives. The multiple-criteria House of Quality, shown in Figure 17.6, can be used to decide which alternative offers the best compromise for both the customer and the producer.

Based on the given weights for the customer,

$$U^C = 0.5 f_1^C + 0.2 f_2^C + 0.1 f_3^C + 0.2 f_4^C$$

Based on the given weights for the producer,

$$U^P = 0.3 f_1^P + 0.4 f_2^P + 0.2 f_3^P + 0.1 f_4^P$$

The importance given to the customer is 0.6 and to the producer is 0.4. Therefore,

$$U^S = \delta U^C + (1 - \delta) U^P = 0.6 U^C + 0.4 U^P$$

Design Alt.	Customer's Weights (Product) Customer's Needs (Objectives)				Customer's Utility, U^c	Technical Requirement					Producer's Utility, U^P	Producer's Weights (Process) Producer's Needs (Objectives)				System's Utility, $U^s = 0.6U^c + 0.4U^P$	System's Ranking
	0.5	0.2	0.1	0.2		Weight of Car (lb)	Turning Radius (ft)	Time to Break (m)	Horse Power (HP)	Air Resistance (N)		0.3	0.4	0.2	0.1		
	f_1^c	f_2^c	f_3^c	f_4^c								f_1^P	f_2^P	f_3^P	f_4^P		
I	8	6	5	3	6.3	2500	31	3.2	175	35	4.7	3	7	2	6	5.7*	1
II	7	3	7	7	6.2	3500	26	3.5	200	25	3.9	5	2	4	8	5.3	2
III	3	7	6	9	5.3	3000	24	2.6	300	20	3.9	8	1	3	5	4.7	5
IV	4	6	4	8	5.2	3750	28	2.1	250	32	4.8	2	9	1	4	5.0	3
A	6	4	4	2	4.3	2800	24	3.1	250	30	5.0	6	3	8	4	4.8	4
B	5	4	2	4	4.3	3000	30	2.7	190	23	5.7	3	7	2	6	4.5	6
Ideal	9	9	9	9	9	2500	24	2.1	300	35	9	9	9	9	9	9	N/A

Figure 17.6 Multiple-Criteria QFD House of Quality with four design alternatives (I, II, III, IV) compared to competitor A and B products and processes (NA = not applicable).

Now the system utility value for each alternative can be calculated. For example, for design alternative I, the utility function is

$$U^S = 0.6U^C + 0.4U^P = (0.60)(0.5 \times 8 + 0.2 \times 6 + 0.1 \times 5 + 0.2 \times 3)$$

$$+(0.4)(0.3 \times 3 + 0.4 \times 7 + 0.2 \times 2 + 0.1 \times 6) = 5.7$$

In this example, alternative I has the highest utility function compared to the utility functions of all other alternatives. Therefore, it is the best solution for this problem. It must be noted that in the above method the ranking of alternatives was obtained by the use of additive utility functions. Other multiple-criteria methods can also be used to rank alternatives.

17.3   PROCESS CONTROL BACKGROUND

17.3.1   Classification of Types of Measurements

From the consumer's perspective, quality can be defined based on a product's performance related to its cost. This view of quality is also adopted by marketing departments. From the producer's point of view, however, quality is a measure of conformance to standards. Thus, if a product is manufactured within all of its design specifications, then it is a good-quality product, regardless of whether it is appealing to the consumer. Statistical process control seeks to improve the latter type of quality to ensure that units are produced with respect to their design specifications.

Statistical process control is an approach for measuring and monitoring certain qualities of processes to meet prespecified standards for products. There are two causes for products not meeting the required specifications: natural (or random) causes and assignable (deterministic) causes:

- Natural causes are the result of natural process variation and cannot be rectified, so there is no solution for such causes.
- Assignable causes are due to an abnormality or identifiable variations in the process and can be detected and rectified by changing either the input materials or the setup of the operation.

Statistical process control provides effective and powerful statistical and graphical tools for solving problems associated with assignable causes. To solve QC problems, one should first understand the principles behind the different statistical methods used in statistical process control. Then, the problem should be categorized and solved by the appropriate class of statistical process control tools.

Types of Measurement   There are two types of measurements:

1. *Variable*   A variable measurement parameter can assume an infinite set of individual values. For example, length, diameter, and temperature can be measured within a needed accuracy. These variable values usually fall within a finite range. Variables show specifically how good or bad a particular part is. Using a micrometer to measure the diameter of a rod, D_{rod}, is an example of a variable. This information allows one to numerically compare D_{rod} to the target diameter of the rod (D_{target}). The acceptable range of the rod diameter has an upper tolerance (D_U) and lower tolerance (D_L). That is, all rods with a diameter $D_L \leq D_{rod} \leq D_U$ are acceptable, and rods with diameters outside of this range should be rejected. For example, suppose $D_U = 1.005$ and $D_L = 0.995$. If a rod is measured to have $D_{rod} = 1.003$, it is acceptable. With variable measurement, the exact variation from the target value, that is, $|D_{rod} - D_{target}|$, is known.

2. *Attribute*   An attribute measurement parameter can only assume one of two values. This is a binary classification (0 or 1, on or off, go or no-go, etc.). Attributes provide information on whether or not the item meets the requirement. For example, a go/no-go gauge, as shown in Figure 17.7, can be used to determine if a rod is good or bad. If the rod

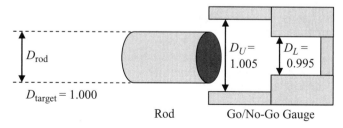

Figure 17.7   Go/no-go gauge for measuring the diameter of a rod where $D_{target} = 1.000$.

fits into the first section and does not fit into the second section, it passes inspection. The gauge is used to identify whether the rod is

(a) within control ($D_L \leq D_{\text{rod}} \leq D_U$),

(b) too large ($D_{\text{rod}} > D_U$), or

(c) too small ($D_{\text{rod}} < D_L$).

But the diameter D_{rod} is not measured and hence $|D_{\text{rod}} - D_{\text{target}}|$ is not known. That is, the degree to which the part meets the specifications is not obtained. Although a direct comparison to the target is not possible, attributes are useful for studying large numbers of samples quickly. Attributes are also faster and less expensive to obtain than variables. It is possible to convert variable data into attribute data, but not vice versa.

Types of Control Charts   There are two types of control charts: variable and attribute. Each type of chart is based on underlying statistical principles. The method of calculating the acceptable limits of these charts is based on empirical evidence and supported by relevant statistical theory. There are lower and upper limits for each chart. Values within the lower and upper control limits are acceptable and values outside of the limits are not. The data for a control chart are collected in regular sampling intervals over a period of time. The time sequence order in which data are collected must be preserved when constructing all control charts. All charts also require the data to be collected in rational groups that allow for identification of changes in the process. Grouping (or the sampling approach for the selection of groups) should be conducted randomly and all groups should be treated equally, allowing for the identification of trends for assignable causes. For example, if there is a variation in performance from one work shift to another work shift, then the sampling should be conducted for similar shifts.

Table 17.3 presents a summary of four well-known control charts. They will be discussed in depth later in this section.

Interpreting Control Charts   A process is under control if all of its samples are within the given lower and upper control limits and have no discernable patterns. When reading control charts, one should first check that all data points are within the control limits. Samples that lie outside these limits should be investigated carefully and individually.

TABLE 17.3   Summary of Four Well-Known Control Charts

	Key Characteristics of Different Charts			
Control Charts	Measurement Sampling Name	Measurement Units	Measurement Output	Probability Distribution
$\bar{X}$	Variable groups	Exact no.	Process mean	Normal
R	Variable groups	Exact no.	Process variance	Normal
p	Attribute—number of defective units in the group	Binary (0 or 1)	Mean number of defects per group	Binomial (or Poisson)
c	Attribute—number of defects in one product unit in the group	Binary (0 or 1)	Number of defects per product unit	Binomial (or Poisson)

Also, there should be no patterns in the data; thus if the data show an upward trend, downward trend, or other recognizable pattern, the entire process needs to be studied to determine what is causing the pattern.

17.3.2 Statistical Background

Histograms   A histogram is a graphical representation of a frequency distribution. It is used in statistical analysis to show the number or proportion of objects that fall into different categories. A category is a nonoverlapping interval of certain observations. The categories are also known as bins. For a given set of data, different numbers and widths of bins can be used to categorize the data. Selecting an appropriate bin width is critical in correctly identifying the general shape and important features of the frequency distribution.

By using a histogram, one can observe several attributes of a given data set. Some of these attributes are (a) overall silhouette of the frequency distribution, (b) presence of outliers, (c) symmetry of the distribution, and (d) modality. The overall silhouette provides information on the type of data distribution, for example, normal, triangular, uniform, and exponential. Outliers are points that do not correspond to the overall shape of the distribution. The symmetry conveys whether the distribution is symmetric or skewed to the right or left. The modality indicates the number of modes or peaks in the graph. A unimodal histogram contains one peak, a bimodal histogram contains two peaks, and a multimodal histogram contains multiple modes in the plot. Figure 17.8 shows unimodal, bimodal, and multimodal histograms where the modes are shaded.

The next example illustrates how to construct a histogram using given data.

Example 17.3   Constructing a Histogram Using Given Data   Consider the data provided in Table 17.4, which shows the weights of 150 male high school students in pounds. Construct a histogram that displays the overall weight distribution of these students.

First, select the appropriate bin width. The minimum and the maximum values of the data in Table 17.4 are 111.0 and 178.9 lb, respectively. We can set the minimum and maximum values of the x axis as 110 and 180, respectively. After several trials, we choose a bin width of 10. Now, count the number of data values that fall into each interval. Table 17.5 shows the frequency for each interval. The mean (average) and standard deviation (SD) of the data are 148.22 and 14.75, respectively. Calculation of the mean and standard deviation is explained later in this section.

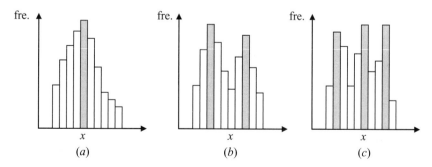

Figure 17.8   Histogram of (*a*). unimodal, (*b*). bimodal, and (*c*). multimodal plots where fre. is the frequency and x is the value of the data falling within bins.

TABLE 17.4 Weights of Male High School Students

154.3	166.8	160.5	156.3	163.7	146.1	142.8	158	174.4	157.4
142.8	160.1	138.5	137.5	155.6	135.5	145.5	153.2	156.8	175.1
134.9	147.2	160	151.9	158.3	137.2	135.7	144.6	111	136.6
133.4	157.2	136.2	149.7	136.3	164.3	145.7	173.9	136.4	143.4
152.7	145.1	137.1	167.9	171	167.3	138.7	176.3	141.7	117.3
152.6	157.5	166.2	127.9	137.3	166.6	143.6	147.7	156.8	156.2
160.1	134.4	141.4	161.7	149.1	124.6	149.4	163.6	161	175
178.8	151.2	161.6	178.2	118.7	140	123.2	119.9	159.4	156.3
156.6	156	178.9	157.7	177.5	139.1	140.7	142.8	139.5	129.4
172.2	132.4	120.6	142.6	137.4	123.9	138.2	148.2	149.7	133.6
163.7	125.3	131.9	154.4	159.4	145.4	153.1	146.6	118	155
134.5	135.6	135.7	150.6	141.7	160.1	143.3	161.6	148.8	131.7
138.4	138.3	136.2	150.7	159.6	157.4	141.7	171.4	146.2	160.3
143.7	164.9	140.3	151.6	124.7	137.5	159.8	132	147.8	133.4
154.3	150.3	161.6	159.7	150.3	119.2	134.6	142.5	174.4	136.7

Figure 17.9 has a resemblance to a normal distribution curve. The range between 150 and 159.9 lb contains the highest number of students.

Supplement S17.1.doc discusses the importance of selecting appropriate bin width.

Population Versus Sample

1. *Population Mean (Average)* If the data about the whole population are known, then the population average μ is used to represent the average of the population. Suppose there are N items in the population where the measurement of each is presented by i for $i = 1, \ldots, N$:

$$\mu = \frac{1}{N} \sum_{i=1}^{N} X_i \tag{17.4}$$

2. *Sample Average* In most applications, it is not possible to measure the values of all the members of the population. In this case, a sample is used to represent the whole population. When the population average is not known, the sample average is used instead. The sample average $\bar{X}$, or "X bar," is the most well-known measure of central tendency. It is obtained by the summation of observations divided by the number of observations. Suppose there are n observations in the sample. Then

$$\bar{X} = \frac{1}{n} \sum_{i=1}^{n} X_i$$

TABLE 17.5 Frequency Distribution of Weights, Mean, and Standard Deviation

Bin	110–119.9	120–129.9	130–139.9	140–149.9	150–159.9	160–169.9	170–179.9	Mean	SD
Frequency	6	8	34	33	35	21	13	148.29	14.66

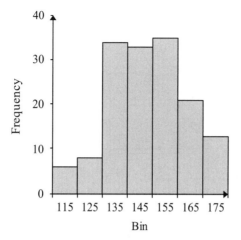

Figure 17.9   Histogram of weights for Table 17.4 where the bin width is 10.

3. *Range of Each Sample*   Range is a measure of dispersion, which is defined as the difference between the largest and the smallest value of observations:

$$R = \text{range} = X_U - X_L$$

where X_U is the largest value and X_L is the smallest value of the measured values.

4. *Population Standard Deviation*   Standard deviation σ is another measure of dispersion that shows how spread-out the observations are. The standard deviation for the population is calculated as

$$\sigma = \sqrt{\dfrac{\sum\limits_{i=1}^{N}(X_i - \mu)^2}{N}} \qquad (17.5)$$

The variance, σ^2, is another measure of dispersion.

5. *Sample Standard Deviation*   As was the case with averages, there exists a population standard deviation and a sample standard deviation. The sample standard deviation for a sample of n observations is calculated by

$$SD = \sqrt{\dfrac{\sum\limits_{i=1}^{n}(X_i - \bar{X})^2}{n-1}} \qquad (17.6)$$

Note that the sample variance is SD^2.

Since we usually do not have access to the entire population, the sample mean $\bar{X}$ is used instead of the population mean μ. Therefore, in the above formulation we substitute $\bar{X}$ for μ. Also, the sample standard deviation SD is used instead of the population standard deviation σ.

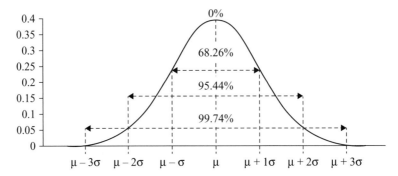

Figure 17.10   Example of normal distribution with mean μ and standard deviation σ, where $z = 1$, 2, 3.

Normal Distribution   The most common probability distribution function is the normal probability distribution, also known as the bell-shaped curve or bell curve. Its popularity is based on the fact that most natural phenomena can be validly approximated using this distribution. The normal distribution has two parameters, μ (the mean) and σ (the standard deviation). By making a simple transformation, one can generate standard normal distribution values by using the standard normal table presented at the back of this book. Note that capital Z is a random variable and small z is a given value when using standard normal distribution. For a continuous random variable, Z, the probability of the random variable Z being any constant value a, is always zero, that is, $P(Z = a) = 0$ for all a. In other words, the probability of Z being in the range $[a, a]$ is zero, or $P(a \leq Z \leq a) = 0$. Therefore, for a continuous random variable, $P(Z \leq a)$ and $P(Z < a)$ are the same, that is, $P(Z \leq a) = P(Z < a)$.

An example of a normal distribution is presented in Figure 17.10 for $\mu \pm z\sigma$. The probability that a random variable X falls between $\mu - z\sigma$ and $\mu + z\sigma$ can be easily calculated. Figure 17.10 presents values for $z = 1, 2, 3$.

Table 17.6 and Figure 17.10 show an example of values of z for several desired confidences. For example, a 95.44% confidence interval (two tails) has a z value of about 2. Some well-known and commonly used values of z are presented in Table 17.6. In this chapter we use both the one-tail, that is, $P\{Z \leq z\}$, and the two-tail confidence intervals, that is, $P\{-z \leq Z \leq z\}$.

In most QC applications, $z = 3$, known as three-sigma, is used, for which 99.74% of the data falls within the control limits (± 3). That is, $P\{-3 \leq Z \leq 3\} = 0.9974$. This can be verified by checking the normal distribution table at the back of the book.

Supplement S17.2.doc shows how to use the standard normal distribution table.

TABLE 17.6   Examples of Probabilities Associated with Different z Values for Two-tail, $P\{-z \leq Z \leq z\}$ and One Tail, $P\{Z \leq z\}$

	z	0	0.5	1.0	1.5	2.0	2.5	3.0
Two tail	$P\{-z \leq Z \leq z\}$	0%	38.30%	68.26%	86.64%	95.44%	98.76%	99.74%
One tail	$P\{Z \leq z\}$	50%	69.15%	84.13%	93.32%	97.73%	99.38%	99.87%

Central Limit Theorem The central limit theorem is an important probability result which is used in many QC and statistical applications. It guarantees that the distribution of the sum of a large number of independent random variables will be approximately a normal distribution. For example, consider random variables $X_1, X_2, \ldots, X_n$, where each random variable has a uniform distribution on the interval $(0, 1)$. The central limit theorem guarantees that the density function of $\sum_{i=1}^{n} X_i$ is a normal distribution for a large n. This means that by considering more independent random variables, regardless of each of their distributions, the sum will resemble a normal distribution. In practice, usually a small value of n (e.g., five members in a group) is used.

The central limit theorem can be used to find the sampling distribution of means. Therefore, the sample distribution of means can be easily approximated by a normal distribution. The mean and standard deviation for a normal distribution of means are defined as follows for n samples:

$$\bar{\bar{X}} = \mu = \frac{1}{n}\sum_{i=1}^{n} \bar{X}_i \qquad \text{(mean of mean of samples)} \qquad (17.7)$$

$$\sigma_{\bar{X}} = \frac{\sigma}{\sqrt{n}} \qquad \text{(standard deviation of samples)} \qquad (17.8)$$

where σ is the standard deviation of the population.

Application of Central Limit Theorem in Quality Control The central limit theorem is used as the theoretical base of the $\bar{X}$-chart, which will be explained in detail in the following section. The central limit theorem states the following:

1. The probability distribution of the average of identically distributed sample distributions follows a normal distribution
2. The average of the mean of the sample distributions, denoted by $\bar{\bar{X}}$, is the same as the mean of the population from which samples were taken, denoted by μ [Equation (3.4)].
3. The population standard deviation, denoted by σ, is related to the standard deviation of sample distributions, denoted by $\sigma_{\bar{X}}$ [Equation (17.8)].

Upper and Lower Control Limits (LCL and UCL) In quality control, the LCL and UCL are set up for the purpose of accepting good products and rejecting bad products. The range of the population is generally wider than the range of the sampling; see Figure 17.11. For this reason, control limits based on the sampling range are tighter (more discriminatory) than control limits based on the population. Therefore, if the population is out of control, very likely its sampling distribution will be identified as out of control.

Types of Errors There are two possible types of error when analyzing a process:

Type I Error Declaring a process is out of control when it is actually under control. In this case, the producer rejects good products. The producer unnecessarily must bear the cost of readjusting a good process. This is shown as α, the probability of error, in Figure 17.12. Note that there is a probability of $\alpha/2$ for lower than the LCL and a probability of $\alpha/2$ for higher than the UCL; that is, $\alpha/2 + \alpha/2 = \alpha$. Type I error is also known as an error

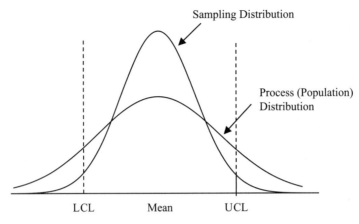

Figure 17.11 Comparison of sampling distribution and process (population) distribution.

of the first kind, an α error, or a false positive, the error of rejecting a null hypothesis when it is actually true.

Type II Error Declaring a process is under control when it is actually out of control. In this case, the consumer must unnecessarily bear the cost of accepting defective (bad) products. This is represented by β, the probability of accepting defective products.

In using the LCL and UCL, both type I and type II errors should be considered. Type II error is also known as an error of the second kind, a β error, or a false negative, the error of failing to reject a null hypothesis when it is in fact not true.

Supplement S17.3.docx explains the importance of selecting the appropriate bin width.

17.4 PROCESS CONTROL VARIABLE CHARTS

17.4.1 Control Charts for Variables

As discussed before, variables are defined for exact measurements of continuous values such as age, diameter, length, and weight. There are two types of statistical QC charts

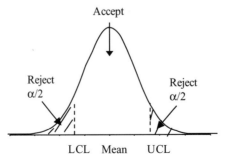

Figure 17.12 Type I error: $\alpha/2 + \alpha/2 = \alpha$; probability of declaring a good process (which is actually under control) as out of control.

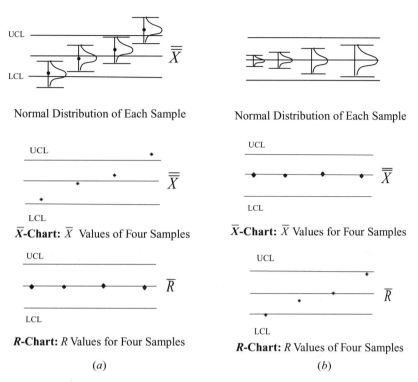

Figure 17.13 Two examples of $\bar{X}$-charts and R-charts. a: an increasing mean value with stable range. b: a stable mean value with increasing range.

for variables, $\bar{X}$-charts and R-charts. $\bar{X}$-charts show changes in averages and R-charts demonstrate changes in ranges. The causes for changes are typically different in averages and ranges. Both $\bar{X}$-charts and R-charts should be used while monitoring a process.

Figure 17.13 provides two examples of $\bar{X}$- and R-charts. In Figure 17.13a, the $\bar{X}$-chart and R-chart are shown. In this example, the mean values vary with an upward trend but the range of values is constant. In Figure 17.13b, the mean values are constant, but the values vary with an upward trend. For each example, after the $\bar{X}$-chart and R-chart, the normal distribution for each sample is shown where for each sample $\bar{X}$ is its mean and R is its range.

As stated before, the $\bar{X}$-chart and R-chart are most commonly used for variable-type data. The $\bar{X}$-chart can be used to determine a shift in the process mean. The R-chart can be used to determine whether there are variations in the range of values. Both $\bar{X}$- and R-charts are based on normal distribution assumptions, but they work relatively well for nonnormal distributions. In practice, a sample size of 4–6 per group is used where there could be about 20–30 groups.

In this section, we consider two cases: when the standard deviation of the population is known and when it is unknown. In most applications, the latter case is used.

17.4.2 Mean Chart with Known Standard Deviation

In some applications, it is possible to have an estimation of the standard deviation of the population (i.e., the process under study). In this case, such estimation is used to find the standard deviation of the sampling. As discussed before, the relationship of the population standard deviation σ and the standard deviations of the sampling is

$$\sigma_{\bar{X}} = \frac{\sigma}{\sqrt{n}} \qquad \text{where } n = \text{sample size of each group} \qquad (17.9)$$

The limits for the $\bar{X}$-charts are set as follows:

$$\text{UCL} = \bar{\bar{X}} + z\sigma_{\bar{X}}$$

$$\text{LCL} = \bar{\bar{X}} - z\sigma_{\bar{X}}$$

where z is a given coefficient. In most applications LCL cannot assume values less than zero; for example, the length of an object cannot be a negative value. In this case, if LCL $<$ 0, set LCL $=$ 0. However in some applications, LCL can be less than zero, for example, temperature.

The following information is used for $\bar{X}$-charts:

X_{ij} = value of variable for item j in group i, where $j = 1, 2, \ldots, n$, $i = 1, 2, \ldots, m$
m = number of groups under consideration, $i = 1, 2, \ldots, m$
n = number of items in each group, $j = 1, 2, \ldots, n$
σ = standard deviation of population

The following values are calculated for building $\bar{X}$-charts:

$\bar{X}_i$ = mean value of each given group (sample), $i = 1, 2, \ldots, m$
$\bar{\bar{X}}$ = grand average of all sample means [use Equation (3.4)]
$\sigma_{\bar{X}}$ = standard deviation of samples [use Equation (3.5)]

Steps for Creating X̄-Charts with Known Population Standard Deviation

When the population standard deviation σ of the process is known, control limits are calculated using the following method.

1. Compute $\bar{X}_i$ for each group.
2. Calculate

$$\bar{\bar{X}} = \frac{1}{m} \sum_{i=1}^{m} \bar{X}_i$$

Plot $\bar{\bar{X}}$ and $\bar{X}_i$ for $i = 1, \ldots, m$ on the chart.

3. Calculate the sampling distribution standard deviation: $\sigma_{\bar{X}} = \sigma/\sqrt{n}$.

4. Compute UCL and LCL and plot them on the chart for the given z value:

$$\text{UCL} = \bar{\bar{X}} + z\sigma_{\bar{X}} \qquad \text{LCL} = \bar{\bar{X}} - z\sigma_{\bar{X}}$$

5. Determine if any special cases exist and investigate. Remove exceptional data points and redo steps 1–4. Do not repeat this step.

Example 17.4   Control Limits with Given Population Standard Deviation (σ)   Suppose the estimation of the population standard deviation is 1.67. Calculate the UCL and LCL for a 99.74% confidence level where there are seven items in each sample and the average of all sample means is 5.

In this example, $\sigma = 1.67$, $z = 3$ (for 99.74%), $n = 7$, and $\bar{\bar{X}} = 5$. The solution is as follows. First, find the standard deviation of the samples:

$$\sigma_{\bar{X}} = \frac{\sigma}{\sqrt{n}} = \frac{1.67}{\sqrt{7}} = 0.631$$

The upper and lower control limits are calculated as follows:

$$\text{UCL} = 5.00 + 3 \times 0.631 = 6.89 \qquad \text{LCL} = 5.00 - 3 \times 0.631 = 3.11$$

17.4.3   Mean Chart with Unknown Standard Deviation

If the population standard deviation is not known, the following procedure can be used to find the upper and lower control limits by using the range instead of the population standard deviation. Consider:

R_i = range of values of samples for each given group, $i = 1, 2, \ldots, m$
$\bar{R}$ = mean range of groups
A_2 = given factor for computing $\bar{X}$-chart control limits for $z = 3$ (see Table 17.7)

For each given group i, the mean and range is found as follows:

$$\bar{X}_i = \frac{1}{n}\sum_{j=1}^{n} X_{ij} \qquad \text{for } i = 1, 2, \ldots m$$
$$R_i = \text{Max}\{X_{ij}; \; j = 1, 2, \ldots, n\} - \text{Min}\{X_{ij}; \; j = 1, 2, \ldots, n\}$$

Steps for Creating $\bar{X}$-Charts with Unknown Population Standard Deviation
In this procedure, sample ranges are used:

1. Compute $\bar{X}_i$ and R_i for each group i.
2. Calculate

$$\bar{\bar{X}} = \frac{1}{m}\sum_{i=1}^{m} \bar{X}_i$$

Plot $\bar{X}_i$ for $i = 1, \ldots, m$ and $\bar{\bar{X}}$ on the chart.

TABLE 17.7   Standard Table of Factors for calculating $\bar{X}$ and R Control Chart Limits for Variables Based on 99.74% Confidence ($z = 3$)

n	A_2 (Mean Factor)	D_3 (Lower Range)	D_4 (Upper Range)
2	1.880	0	3.268
3	1.023	0	2.574
4	0.729	0	2.282
5	0.577	0	2.114
6	0.483	0	2.004
7	0.419	0.076	1.924
8	0.373	0.136	1.864
9	0.337	0.184	1.816
10	0.308	0.223	1.777
11	0.285	0.256	1.744
12	0.266	0.283	1.717
13	0.249	0.307	1.693
14	0.235	0.328	1.672
15	0.223	0.347	1.653
16	0.212	0.364	1.635
17	0.203	0.379	1.621
18	0.194	0.392	1.608
19	0.187	0.404	1.596
20	0.180	0.414	1.586
21	0.173	0.425	1.576
22	0.167	0.434	1.566
23	0.162	0.443	1.557
24	0.157	0.452	1.548
25	0.153	0.459	1.541
Over 25	$3/\sqrt{n}$	—	—

3. Calculate

$$\bar{R} = \frac{1}{m} \sum_{i=1}^{m} R_i$$

4. Find values of A_2 for the given n (sample size of each group) from Table 17.7 (this is given for $z = 3$).

5. Compute UCL and LCL and plot them on the chart:

$$\text{UCL} = \bar{\bar{X}} + A_2 \bar{R}$$

$$\text{LCL} = \bar{\bar{X}} - A_2 \bar{R}$$

6. Determine if any special cases exist and investigate. Remove exceptional data points and redo steps 1–5. Do not repeat this step.

Standard Table for X̄- and R-Charts for 99.74% Confidence   Each factor is based on three standard deviations, that is, $z = 3$ for 99.74% confidence level. To find the lower and upper control limits for the $\bar{X}$- and R-charts, use the coefficient factors A_2, D_3, and D_4, readily available in Table 17.7, to calculate the lower and upper control limits. The coefficient values for different sample sizes are given in Table 17.7. Note that, generally, the sample size group n is less than 26.

Example 17.5 $\bar{X}$-Chart for Rod Weight Measurement   Five tire treads are randomly sampled from each hour of production and their weights are measured in pounds. Data for a 10-h period is summarized below. Show the upper and lower control, limits for 99.74% of the data. Is the process under control?

Group, i			Values of Sample, X_{ij}				
	$j=1$	$j=2$	$j=3$	$j=4$	$j=5$	$\bar{X}_i$	R_i
1	2.05	2.08	2.08	2.11	2.1	2.084	0.06
2	2.13	2.08	2.05	2.09	2.1	2.09	0.08
3	2.03	2.12	2.06	2.09	2.11	2.082	0.09
4	2.09	2.07	2.1	2.09	2.07	2.084	0.03
5	2.1	2.06	2.06	2.07	2.08	2.074	0.04
6	2.03	2.13	2.08	2.06	2.1	2.08	0.1
7	2.1	2.06	2.06	2.08	2.08	2.076	0.04
8	2.07	2.07	2.11	2.07	2.08	2.08	0.04
9	2.07	2.09	2.08	2.09	2.09	2.084	0.02
10	2.07	2.06	2.11	2.09	2.07	2.08	0.05
Avg.						$\bar{\bar{X}} = 2.081$	$\bar{R} = 0.055$

The number of groups is $m = 10$. The average for group $i = 1$ is calculated as

$$\bar{X}_1 = \frac{1}{n}\sum_{j=1}^{n} X_{ij} = \frac{2.05 + 2.08 + 2.08 + 2.11 + 2.10}{5} = 2.084$$

The range for group 1 is

$$R_1 = \text{Max}\{X_{ij}\} - \text{Min}\{X_{ij}\} = 2.11 - 2.05 = 0.06$$

Now, find the average range for all groups:

$$\bar{\bar{X}} = \frac{1}{10}(2.084 + 2.09 + 2.082 + 2.084 + 2.074 + 2.08 + 2.076 + 2.08 + 2.084 + 2.08)$$
$$= 2.081$$

$$\bar{R} = \frac{1}{10}(0.06 + 0.08 + 0.09 + 0.03 + 0.04 + 0.01 + 0.04 + 0.04 + 0.02 + 0.05) = 0.055$$

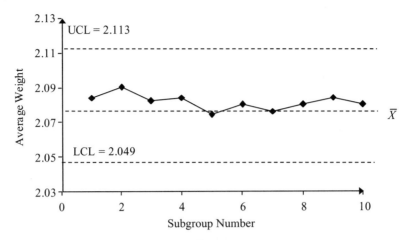

Figure 17.14 Mean $\bar{X}$-chart for Example 17.5.

From Table 17.7, for $n = 5$, $A_2 = 0.577$.

The control limits for the $\bar{X}$-chart are presented below and plotted in Figure 17.14.

$$UCL = 2.081 + 0.577 \times 0.055 = 2.113 \qquad LCL = 2.081 - 0.577 \times 0.055 = 2.049$$

This process is under control, as all data points lie within the control limits. There are no abnormalities in the data. The process capability is $2.049 < \text{weight} < 2.113$.

17.4.4 Range Chart

As stated before, in many applications, one should also monitor the changes in the variations. It is possible that two samples may have the same mean values but considerably different range values. Such changes in variations should be monitored closely as they may indicate major quality problems.

In practice, when $z = 3$, the following notation is used for constructing range charts using the values given in Table 17.7.

$D_3 =$ factor for computing LCL of R-chart

$D_4 =$ factor for computing UCL of R-chart

Steps for Creating R-charts

1. Compute R_i for each group i and plot the groups on the chart.
2. Find

$$\bar{R} = \frac{1}{m} \sum_{i=1}^{m} R_i$$

Plot $\bar{R}$.

3. Compute the UCL and LCL and plot them on the chart:

$$UCL = D_4 \bar{R} \qquad LCL = D_3 \bar{R}$$

If the calculated LCL value is less than zero, set LCL $= 0$.

4. Determine if any special cases exist and investigate. Remove exceptional data points and redo steps 1–3. Do not repeat this step.

Example 17.6   *R*-Chart for Rod Weight Measurement   Consider Example 17.5 Compute and plot the *R*-chart. Interpret the meaning of the chart.

The average range of the process is calculated as

$$\bar{R} = \frac{1}{10}(0.06 + 0.08 + 0.09 + 0.03 + 0.04 + 0.1 + 0.04 + 0.04 + 0.02 + 0.05) = 0.055$$

For $n = 5$, using Table 17.7, the parameters for the lower and upper control limits (assuming $z = 3$) are $D_3 = 0$ and $D_4 = 2.114$. Therefore, the control limits are calculated as

$$UCL = 2.114 \times 0.055 = 0.116 \qquad LCL = 0 \times 0.055 = 0$$

See Figure 17.15. The *R*-chart has random behavior, but the process is within the control limits. That is, the process is under control but variations are too much.

17.4.5   Sampling Accuracy for Mean Chart[†]

A key question in the design of statistical control charts is the selection of the sample size *n*.

The size of the sampling plan plays a vital role in achieving the desired accuracy.

The desired accuracy *h* represents the closeness of the sample average $\bar{X}$ to the true population average μ.

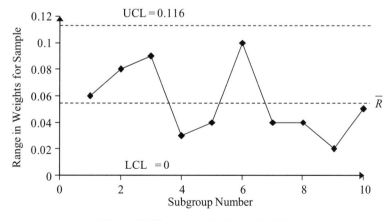

Figure 17.15   *R*-chart for Example 17.6.

A smaller h signifies a more accurate estimate of the true mean μ. As the sample size increases, the desired accuracy improves. Hence, a larger n leads to a smaller h. Also, as the sample size increases, the cost of sampling increases. Therefore, it is necessary to find the relationship of the desired accuracy h with the sample size n. The basic relationship between h and n is

$$n = \left(\frac{z\sigma_{\bar{X}}}{h\bar{X}}\right)^2 \tag{17.10}$$

where $\bar{X}$ is the average (mean) of the initial estimation (e.g., based on the initial sampling) and $\sigma_{\bar{X}}$ is its standard deviation. As was explained before, a smaller value of h is associated with a higher accuracy. Thus, to achieve high accuracy, a larger sample size n should be chosen. As for selecting the z value to be used, the commonly used industry standard is $z = 3$, which is associated with a 99.74% confidence level in the analysis.

For a given n, one can find its accuracy by

$$h = \frac{z\sigma_{\bar{X}}}{\bar{X}\sqrt{n}} \tag{17.11}$$

There is an alternative formulation for Equation (17.10) where $h\bar{X}$ is the maximum acceptable error, which is denoted by e:

$$n = \left(\frac{z\sigma_{\bar{X}}}{e}\right)^2 \tag{17.12}$$

Therefore, when the average, $\bar{X}$, is not known, one can assess the maximum acceptable error e and then find the needed sample. Note that, by definition, $e = h\bar{X}$.

Example 17.7 Finding Sampling Size for $\bar{X}$-Chart Using Desired Accuracy h Suppose the initial estimate for the average value of the length of a product is 50 in. and the estimate of its standard deviation is 0.631 in. that is, $\bar{X} = 50$, $\sigma_{\bar{X}} = 0.631$. When the required confidence level is 99.74%, that is, $z = 3$, consider the following:

(a) Find the sample size n for an accuracy of 1%, that is, $h = 0.01$.
(b) Find the accuracy when the sample size is 10, that is, $n = 10$.
(c) Suppose that the length of the product, $\bar{X}$, is unknown but the maximum acceptable error for the length of a product is 0.2 in. Find the sample size.

The solution is as follows:

(a) Using Equation (17.10),

$$n = \left(\frac{z\sigma_{\bar{X}}}{h\bar{X}}\right)^2 = \left(\frac{3 \times 0.631}{0.01 \times 50}\right)^2 = (3.786)^2 = 14.33$$

This means that 14.33 (rounded up to 15) samples are needed to achieve 1% accuracy for this problem.

(b) Applying Equation (17.11),

$$h = \frac{z\sigma_{\bar{X}}}{\bar{X}\sqrt{n}} = \frac{3(0.631)}{50\sqrt{10}} = 0.01197 \quad \text{(or } 1.197\%\text{)}$$

This means that the level of accuracy for sample size $n = 10$ is 1.197%.

(c) In this example, $e = 0.2$, $\sigma_{\bar{X}} = 0.631$, and $z = 3$. Using Equation (17.12), then

$$n = \left(\frac{z\sigma_{\bar{X}}}{e}\right)^2 = \left(\frac{3 \times 0.631}{0.2}\right)^2 = 89.59 \quad \text{(rounded up to 90)}$$

17.4.6 Bicriteria Sampling for Mean Chart[†]

The accuracy of sampling increases as n (sample size) increases. However, the cost of sampling also increases as n increases. Therefore, it is important to consider the cost of sampling when determining the sample size. Note that the minimum cost is associated with zero sample size, that is, $n = 0$, which has the worst accuracy. Consider the total cost of sampling versus the accuracy of sampling for a given confidence level z. The bicriteria problem can be defined as:

Minimize total cost of sampling, $f_1 = C_o n$
Minimize inaccuracy of sampling, $f_2 = h$ [use Equation (17.11)]
Subject to the given confidence value z and $n_{min} \leq n \leq n_{max}$

where C_o is the cost of sampling one item.

As h decreases towards zero, the cost of sampling increases exponentially. That is, it becomes very expensive to decrease h when h becomes very small. In the bicriteria problem, bounds on the sample size n are given. That is, $n_{min} \leq n \leq n_{max}$. For different values of n, the objective values f_1 and f_2 can be calculated. A multicriteria approach (e.g., using an additive utility function) can be used to rank alternatives.

Optimal Sample Size for Additive Utility Function For additive utility functions, consider minimizing

$$U = w_1 f_1' + w_2 f_2' = w_1 \left(\frac{f_1 - f_{1,min}}{f_{1,max} - f_{1,min}}\right) + w_2 \left(\frac{f_2 - f_{2,min}}{f_{2,max} - f_{2,min}}\right) \quad (17.13)$$

where f_1' and f_2' are the normalized values of objectives f_1 and f_2, respectively and w_1 and w_2 are the weights of importance of the two normalized objective functions, respectively.

Equation (17.13) can be rewritten as

$$U = w_1 \left(\frac{C_o n - f_{1,min}}{f_{1,max} - f_{1,min}}\right) + w_2 \left(\frac{z\sigma_{\bar{X}}/(\sqrt{n}\bar{X}) - f_{2,min}}{f_{2,max} - f_{2,min}}\right) \quad (17.14)$$

Consider the additive utility function (17.13). Function f_1 is linear (therefore convex) and f_2 is a convex function; f_1' and f_2' are linear functions of f_1 and f_2, respectively. Therefore, f_1'

and f_2' are both convex functions and their convex combination $U = (w_1\, f_1' + w_2\, f_2')$ is also a convex function. For convex functions, the optimal solution can be found by setting the first derivative of $U(n)$ equal to zero. (Note that we can verify that this function is convex because its second derivative is positive.) Therefore, the solution to $\partial U(n)/\partial n = 0$ is a global minimum. The optimal minimum solution can be found as follows:

$$\frac{\partial U(n)}{\partial n} = w_1\left(\frac{C_o}{f_{1,max} - f_{1,min}}\right) - w_2\left(\frac{0.5Z\sigma_{\bar{X}}n^{-3/2}\bar{X}^{(-1)}}{f_{2,max} - f_{2,min}}\right) = 0$$

$$n^* = \left(\frac{(f_{1,max} - f_{1,min})w_2 Z\sigma_{\bar{X}}}{(f_{2,max} - f_{2,min})2w_1 C_o \bar{X}}\right)^{2/3} \tag{17.15}$$

Instead of using an additive utility function $w_1 f_1' + w_2 f_2'$, it is possible to use a nonlinear utility function. For a nonlinear utility function, minimize $U(n)$ to find the optimal sampling plan n^*. This can be done by performing a one-dimensional search on n (e.g., use the bisectional search method while the objective values f_1 and f_2 are considered for each given value of n). The nonlinear utility function will not be demonstrated in this chapter.

Example 17.8   Bicriteria for $\bar{X}$-Chart to Find Best Sampling Size   Consider an example where the inspection cost per unit is $1.12, the required confidence level is 99.74%, the sample average $\bar{X}$ is 5, and the sample standard deviation $\sigma_{\bar{X}}$ is 1.83. Suppose the sample size can vary from 2 to 50. Assume the weights of importance for the additive utility function are $w_1 = 0.4$ and $w_2 = 0.6$ for the two normalized objectives.

(a) Find five bicriteria alternatives equally distributed in terms of n, that is, {2, 14, 26, 38, 50}. Which alternatives are efficient?

(b) Rank the five alternatives obtained in part (a) using the additive utility function.

(c) Identify the optimal sampling plan for the given additive utility function.

The solution is presented below.

(a) Five equally distributed alternatives in terms of n are {2, 14, 26, 38, 50}. For each n, find f_1 and f_2. For example, if $n = 38$,

$$f_1 = 1.12 \times 38 = 42.56$$
$$f_2 = h = \frac{Z\sigma_{\bar{X}}}{\sqrt{n}\bar{X}} = \frac{3 \times 1.83}{\sqrt{38} \times 5} = 0.178 \quad (\text{or } 17.8\%)$$

The five multicriteria alternatives are presented below:

Alternatives	a_1	a_2	a_3	a_4	a_5	Minimum	Maximum
n	2	14	26	38	50	2	50
$f_1 = C_o$	2.24	15.68	29.12	42.56	56.00	2.24	56.00
$n = 1.12n$							
$f_2 = h$	0.78	0.29	0.22	0.18	0.16	0.16	0.78
Efficient?	Yes	Yes	Yes	Yes	Yes	—	—

(b) The two objectives of the multicriteria alternatives should first be normalized. The normalized values are presented below. Alternatives are ranked by minimizing $U = 0.4 f_1' + 0.6 f_1'$.

n	2	14	26	38	50
Min. f_1'	0.00	0.25	0.50	0.75	1.00
Min. f_2'	1.00	0.22	0.10	0.04	0.00
Min. $U = 0.4 f_1' + 0.6 f_2'$	0.70	0.23	0.22	0.25	0.3
Ranking	5	2	1	3	4

(c) The optimal number of samples can be directly obtained by using Equation (17.15):

$$ n^* = \left(\frac{53.76 \times 0.7 \times 3 \times 1.83}{0.62 \times 2 \times 0.3 \times 1.12 \times 5} \right)^{2/3} = 21.42 \sim 22 \quad \text{(rounded up to 22)} $$

Note that, if all different values of n are considered, that is, $n = 2, 3, 4, \ldots, 50$, the same optimal solution of $n^* = 21.42$ will be obtained. The objectives values for $n = 22$ are

$$ f_1 = 1.12 \times 22 = \$24.64 \qquad f_2 = h = \frac{z \sigma_{\bar{X}}}{\sqrt{n} \bar{X}} = \frac{3 \times 1.83}{\sqrt{22} \times 5} = 0.23 $$

Figure 17.16a shows the set of all efficient bicriteria alternatives for different values of n. Figure 17.16b illustrates the additive utility function $U = 0.4 f_1' + 0.6 f_2'$ versus the sample size n.

WWW **Supplement S17.4.docx shows the alternative approach to the tricriteria sampling plan.**

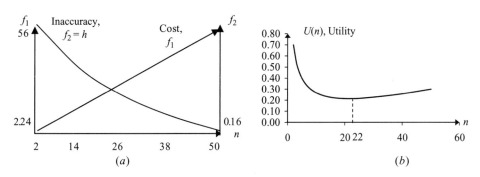

Figure 17.16 Graphical representation of Example 17.8: (a) efficient frontier, f_1 vs. f_2; (b) utility function for different sample size n. The optimal sample size $n^* = 21.42$, at which the utility function is at its minimum.

17.5 PROCESS CONTROL CAPABILITY

17.5.1 Process Capability Versus Design Tolerances

Process capability refers to the extent that a process is capable of meeting the design specifications. There are two cases that can occur:

(a) Process variability is less than or equal to the given design specifications. In this case, the process is capable of meeting the requirements. See Figure 17.17a.

(b) Process variability is more than the given design specifications. In this case, the process is not capable of meeting the requirements. See Figure 17.17b.

Process variability can be measured using the standard deviations. It is common to use $\pm 3SD$ for 99.74% confidence for the range of values that fall within the limits of mean $-$ 3SD and mean $+$ 3SD.

The three-sigma upper and lower limits (i.e., $\bar{X} \pm 3\sigma$) of the control charts are calculated based on the observed data variability. Recall that three-sigma represents the natural process capability. The natural process capability must conform to the constraint of design tolerances. However, the natural process variability is not statistically related to the design tolerances. The given design tolerances must be used to ensure that the product meets specifications. Some manufacturing companies, such as Motorola, are moving toward a six-sigma (6σ) quality standard (i.e., $\bar{X} \pm 6\sigma$) requirement for natural process capability. The 6σ process requires that nondefective products fall within control limits associated with $\pm 6\sigma$, that is, within 0.9999966%. This is the equivalent of having 3.4 defects per million opportunities (DPMO). Such perfection in quality radically changes the nature of QC inspections and methods. In practice, however, for many companies, this high-quality standard may not be economical or practical. The multicriteria sampling of the next section allows one to find the best sampling plan based on the optimal trade-off between the cost and the quality of inspection.

17.5.2 Process Capability for Symmetric Tolerances

Suppose that a processor, for example, a machine, is set up in such a way that its mean is equal to the design target. Also, suppose that the design specifications are symmetric

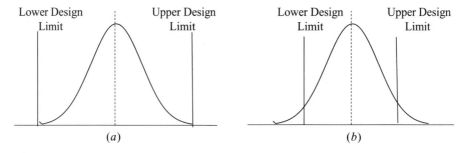

Figure 17.17 Design target is the same as the mean of the process: (a) process variability (natural capability) is within design specifications; (b) process variability (natural capability) is not within design specifications.

with respect to the mean. The process variations can be presented by a normal distribution. Process capability C_p is defined as

$$C_p = \frac{\text{upper design limit} - \text{lower design limit}}{\text{upper control limit} - \text{lower control limit of natural process variations}}$$

$$= \frac{\text{UDL} - \text{LDL}}{\text{UCL} - \text{LCL}}$$

For $z = 3$, that is, 99.74% capability,

$$C_p = \frac{\text{UDL} - \text{LDL}}{6\sigma}$$

The higher the value of C_p, the higher the process capability. Generally, if the process capability is higher than unity, it is under control. Otherwise, it is out of control.

Example 17.9   Process Capability for Symmetric Design Tolerances   The design process of a product has a target value of 15 but values between 11 and 19 are acceptable. Therefore, the acceptable design range for this product is $19 - 11 = 8$. Therefore, the design tolerance (deviation) from the target is $8/2 = 4$, that is, accept the values that range from $15 - 4$ to $15 + 4$.

Consider three different processors (I, II, III) that produce the same product. The standard deviations of the three processors are 0.5, 1, and 1.5, respectively. All three processors have a mean value of 15 (in 1/1000 of inch).

Determine the process capability of each processor.
The solution is presented below.

Processor	Standard Deviation of Process, σ	Process Capability Deviation for $z = 3$	6σ	Design Tolerance from Target X	Design Tolerance Range	C_p	Capability
I	0.5	$3 \times 0.5 = 1.5$	3	4	8	$8/3 = 2.67$	Capable
II	1	$3 \times 1 = 3$	6	4	8	$8/6 = 1.33$	Capable
III	1.5	$3 \times 1.5 = 4.5$	9	4	8	$8/9 = 0.89$	Not capable

17.5.3   Process Capability for Nonsymmetric Tolerances

For nonsymmetric design specifications, the process capability for $z = 3$ (i.e., 99.74% capability) is defined as

$$C_{pk} = \text{Min} \left\{ \left[\frac{\text{(upper design tolerance)}}{3 \, \sigma \text{ of process}} \right], \left[\frac{\text{(lower design tolerance)}}{3 \, \sigma \text{ of process}} \right] \right\} \quad \text{or}$$

$$\text{Min} \left\{ \frac{\text{(UDL} - \text{design target)}}{\text{(UCL} - \text{mean)}}, \frac{\text{(design target} - \text{LDL)}}{\text{(mean} - \text{LCL)}} \right\} \quad \text{or}$$

$$\text{Min} \left\{ \frac{X^+}{3\sigma}, \frac{X^-}{3\sigma} \right\}$$

For symmetric design specifications, the process is under control if the process capability is greater than unity. The process is out of control if the process capability is less than unity. For nonsymmetric design specifications, the minimum process capability for upper and lower design tolerances is used. This is shown in the next example.

Example 17.10   Process Capabilities for Nonsymmetric Design Tolerances   Consider three different processors: I, II, and III. The design target is 15 (in 1/1000 of inch). The upper design tolerance is 19 and the lower design tolerance is 13. All processors have a mean of 15 with standard deviations of 0.5, 1, and 1.5 for processors I, II, and III, respectively. Find each processor's capability.

The solution is presented below.

Processor	Standard Deviation of Process, σ	Process Capability Deviation for $z = 3$	Upper Design Limit Tolerance, X^+ $19 - 15 = 4$	Lower Design Limit Tolerance, X^- $15 - 13 = 2$	Upper Limit Process Capability	Lower Limit Process Capability	$C_{p,k}$ Minimum	Capability
I	0.5	$3 \times 0.5 = 1.5$	4	2	$4/1.5 = 2.67$	$2/1.5 = 1.33$	1.33	Capable
II	1	$3 \times 1 = 3$	4	2	$4/3 = 1.33$	0.67	0.67	Not capable
III	1.5	$3 \times 1.5 = 4.5$	4	2	$4/4.5 = 0.89$	0.44	0.44	Not capable

In this example, for processor II and processor III, the process capability is less than unity. Therefore, processors II and III are out of control and have poor capabilities to satisfy the design tolerances.

17.5.4   Practical Limits on Upper and Lower Control Limits

Guidelines for Setting Process UCL and LCL   In practice, as a rule of thumb, a process is designed in such a way that its natural UCL and LCL from the target design are half of those of upper design and lower design tolerances. Note that the process mean is set equal to the target design value:

$$C_p = \frac{\text{upper design limit} - \text{lower design limit}}{\text{upper control limit} - \text{lower control limit}} = \frac{\text{UDL} - \text{LDL}}{\text{UCL} - \text{LCL}} = 2$$

where $C_p = 2$ is known as a six-sigma process which results in less than 0.0018 defects in one million items. This is a relatively high expectation for quality of the process. For most industries, $C_p = 1.33$ is acceptable.

Consider Figure 17.18 for a three-sigma example.

Suppose the lower and upper design limits are 3 and 15 respectively and the design target is 9. The QC charts for the process are usually set to be half of the design limits, that is, $15 - 3 = 12$; $12/2 = 6$. Now, use $6/2 = 3$ as the design tolerance to generate the UCL and LCL for the process:

$$\text{UCL} = \text{mean} + \text{upper design tolerance} = 9 + 3 = 12$$
$$\text{LCL} = \text{mean} - \text{lower design tolerance} = 9 - 3 = 6$$

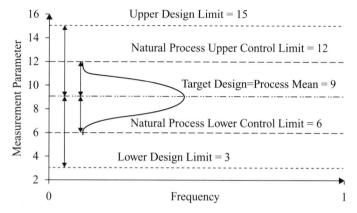

Figure 17.18   Example comparing process capability control limits and design tolerance; the process is capable of meeting tolerances for $C_p = 2$.

Note that

$$C_p = \frac{\text{UDL} - \text{LDL}}{\text{UCL} - \text{LCL}} = \frac{15 - 3}{12 - 6} = 2$$

In this example, the process mean is the same as the design target and the design tolerances are twice as large as the natural control limits. This represents an in-control, three-sigma process.

Guidelines for Identifying Out-of-Control Process   There are four ways to identify if a process is out of control for three-sigma control limits:

   (a) One point is outside the three-sigma control limits.

   (b) Two out of three successive points are on the same side and outside of two-sigma control limits.

   (c) Four out of five successive points are on the same side and outside of one-sigma control limits.

   (d) Eight successive points are above or below the center line.

If any of the above guidelines are observed on a control chart, then the process is out of control and should be investigated.

    The highest level of process capability is ±6SD, which yields 99.99966%. This is known as six-sigma (for deviation), a term used by many companies, including Motorola.

17.6   PROCESS CONTROL ATTRIBUTE CHARTS

Control charts for attributes are utilized for monitoring and tracking a process performance where the measurement is presented by binary values, 1 or 0, which represent acceptable or unacceptable. Compared to variables, attributes are faster and less costly to measure, but they provide less information about the sample. To provide reliable statistical measurements, many more samples are required for attributes than variables.

Two common types of attribute control charts are p-charts and c-charts. The p-charts are used when the process being monitored produces individual units that are either good or bad, such as a light bulb working or not. So, p-charts monitor the number of defective units per subgroup of samples, for example, 2 defective bulbs out of 1000 sample bulbs. On the other hand, c-charts are used when the product can have multiple defects and still be acceptable. In this case, it is possible to count the number of defects for the given product. An example would be the number of minor flaws in a car's paint. The c-charts measure the number of defects per unit. For a c-chart, one product is counted as one subgroup. Both p- and c-charts use a number of groups, but in a c-chart each subgroup is one unit of a product.

The following table shows some examples of p- and c-chart applications:

p-Chart Applications	c-Chart Applications
No. of defective units per group	No. of defects per unit
Two classifications:	No. of occurrences can be counted:
• Good or bad	• No. of paint chips per square foot
• Accept or reject	• No. of calls per hour
• On or off	• No. of defects in one mile of asphalt
Examples:	Examples:
• No. of plastic trash bags that have holes in them	• No. of defects in a trash bag
• No. of student loan whose applications have errors	• No. of errors per student loan application
• No. of shipments that arrive damaged	• No. of dents in each shipment package

17.6.1 Control Chart for Attributes: p-Chart

The p-chart is the most used control chart in industry as it is applicable to many situations, it is easy to use, and it is rather inexpensive to conduct. In p-charts, the value of each sample is either 1 (acceptable) or 0 (unacceptable). Suppose the random variable X represents the total number of defects in a sample of a subgroup where each subgroup consists of n items. Subgroups should be selected randomly but consistently over a period of time. For example, if different operators are used during various daytime and nighttime shifts, the daytime and nighttime shift's performance measurements should be analyzed independently.

Since each measurement can assume values of either 1 or 0 and n is a given finite discrete number, the underlying probability distribution for X can be represented by a binomial. When the sample size $n \geq 20$ and the percent error is less than 5%, that is, $p \leq 0.05$, the Poisson distribution can be used instead of the binomial distribution. The Poisson distribution is easier to use than the binomial distribution because of the extensive calculation associated with the binomial distributions. Both of these distributions will be discussed later in the sampling plans section of this chapter.

In p-charts, the percent of defective items, X/n, instead of X, is used for statistical analysis. The expected value (mean) and the variance (Var) for p-charts are

$$E\left(\frac{X}{n}\right) = p \qquad \text{[expected (mean) value]}$$
$$\text{Var}\left(\frac{X}{n}\right) = \frac{p(1-p)}{n} \qquad \text{(variance)}$$

The estimate for the true proportion of defective units in the population is denoted by $\bar{p}$, which is based on the average of the proportion of defectives in m subgroups.

Consider the following notation:

n = number of units in each subgroup (sample size)
m = number of samples (subgroups)
p_i = percent of defective units in each subgroup i for $i = 1, 2, \ldots, m$
$\bar{p}$ = average percent of defective units in subgroups
$\sigma_{\bar{p}}$ = standard deviation of sampling distribution
z = number of standard deviations for control limit calculations

As discussed before, because the measured values are binary (i.e., 0 or 1), a binomial distribution is used for the p-chart. For a binomial distribution, the standard deviation of the sampling distribution is

$$\sigma_{\bar{p}} = \sqrt{\frac{\bar{p}(1-\bar{p})}{n}} \tag{17.16}$$

For p-charts, the normal distribution approximation is used to find the control limit:

$$\text{UCL} = \bar{p} + z\sigma_{\bar{p}} \qquad \text{LCL} = \bar{p} - z\sigma_{\bar{p}}$$

Steps for Creating p-Charts

1. Compute

$$\bar{p} = \frac{1}{m}\sum_{i=1}^{m} p_i$$

2. Plot $\bar{p}$ and each subgroup p_i for $i = 1, \ldots, m$ on the control chart.
3. Compute the UCL and LCL and plot them on the control chart for the given z value:

$$\text{UCL} = \bar{p} + z\sqrt{\frac{\bar{p}(1-\bar{p})}{n}} \qquad \text{LCL} = \bar{p} - z\sqrt{\frac{\bar{p}(1-\bar{p})}{n}}$$

If the LCL is less than zero, it should be set equal to zero.
4. After plotting the data, determine if any special cases exist and investigate their causes. Remove exceptional cases and repeat steps 1–3. Do not repeat this step.

Example 17.11   *p*-Chart for Defective Light Bulbs   Each day, a sample of 200 light bulbs is tested for a standard level of illumination. If a bulb does not illuminate at the standard level, it is labeled as defective. The results of tests for 20 consecutive days are given in the following table:

Subgroup, i (Day)	Number of Defectives in Subgroup i	Subgroup, i (Day)	Number of Defectives in Subgroup i
1	12	11	24
2	0	12	20
3	8	13	28
4	20	14	16
5	12	15	12
6	8	16	32
7	24	17	24
8	20	18	28
9	16	19	40
10	20	20	36

Plot the appropriate control chart for 99.74% control limits for this problem. Find the percent defective of each subgroup. Then, determine if the process is under control.

The percent defective, p_i, for each group is calculated as follows:

Subgroup, i	Number of Defectives in Subgroup i	Percent Defective, p_i	Subgroup, i	Number of Defectives in Subgroup i	Percent Defective, p_i
1	12	$12/200 = 0.06$	11	24	$24/200 = 0.12$
2	0	$0/200 = 0.00$	12	20	$20/200 = 0.1$
3	8	$8/200 = 0.04$	13	28	$28/200 = 0.14$
4	20	$20/200 = 0.1$	14	16	$16/200 = 0.08$
5	12	$12/200 = 0.06$	15	12	$12/200 = 0.06$
6	8	$8/200 = 0.04$	16	32	$32/200 = 0.16$
7	24	$24/200 = 0.12$	17	24	$24/200 = 0.12$
8	20	$20/200 = 0.1$	18	28	$28/200 = 0.14$
9	16	$16/200 = 0.08$	19	40	$40/200 = 0.20$
10	20	$20/200 = 0.1$	20	36	$36/200 = 0.18$
			Total	400	2

In this example, $m = 20$ and $n = 200$. The average percentage for all the 20 subgroups is

$$\bar{p} = \frac{1}{20}(0.06 + 0.00 + 0.04 + \cdots + 0.18) = \frac{2}{20} = 0.1$$

That is, there are 0.1, or 10%, defective bulbs.

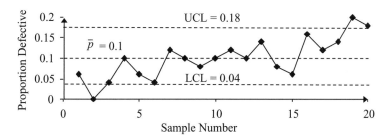

Figure 17.19   *p*-Chart for Example 17.11.

The sample standard deviation for the 20 subgroups is

$$\sigma_{\bar{p}} = \sqrt{\frac{\bar{p}(1-\bar{p})}{n}} = \sqrt{\frac{0.1(1-0.1)}{200}} = 0.02$$

For 99.74% confidence level, use $z = 3$; the control limits are calculated as follows:

$$UCL = 0.1 + 3(0.02) = 0.160 \quad LCL = 0.1 - 3(0.02) = 0.04$$

The *p*-chart for this example is presented in Figure 17.19.

The process shows an out-of-control point on day 2 with an unusually low number of defects. This may indicate problems with the inspection method or defect reporting. Days 19 and 20 also show an out-of-control condition on the high side. This also indicates a problem in the process. The upward trend of the data suggests that there is a nonrandom cause creating the process to produce more defective units. The process should be analyzed to determine the assignable causes of producing these defects.

17.6.2   Control Chart for Attributes: *c*-Chart

The *c*-chart is mostly used when the quality is measured over a continuous extent of a product (such as an area or volume) which can tolerate several inconsistencies up to a certain limit. For example, when spray painting cabinets, if the number of imperfect spots, for example, very small bubbles, in a square foot of sprayed material is less than a certain number, the product can be accepted as good. In *c*-charts, the number of defects per unit of product is measured. Each of these defects individually is considered acceptable from a QC point of view. However, when the number of defects in one unit of product increases over the limit, the product should be rejected. As another example, consider a square yard of a plastic bag. Suppose the thickness of it varies in a few spots. In this case, the plastic bag will be acceptable. However, if there are many inconsistencies in the thickness of the plastic bag, it will be rejected. Suppose X is a random variable that represents the number of defects per unit of product. Then, the Poisson distribution can be applied to represent the underlying probability distribution of X.

The mean and the variance for the Poisson distribution is $\bar{c}$, and the standard deviation is

$$\sigma_c = \sqrt{\bar{c}} \tag{17.17}$$

The Poisson distribution will be discussed in more depth in the next section of this chapter. When the number of defects $c \geq 20$, the normal distribution can be used because it provides a good approximation of the Poisson distribution.

The control limits for the c-chart are derived as

$$\text{UCL} = \bar{c} + z\sqrt{\bar{c}} \qquad \text{LCL} = \bar{c} - z\sqrt{\bar{c}}$$

Steps for Creating c-Charts

1. Compute

$$\bar{c} = \frac{1}{m}\sum_{i=1}^{m} c_i$$

  where c_i is the number of defects in unit i.
2. Plot $\bar{c}$ and groups c_i for $i = 1, \ldots, m$ values on the chart.
3. Compute the UCL and LCL and plot them on the chart:

$$\text{UCL} = \bar{c} + z\sqrt{\bar{c}} \qquad \text{LCL} = \bar{c} - z\sqrt{\bar{c}}$$

  If the LCL is less than zero, set it to zero.
4. Determine if any special cases exist and investigate. Remove all special/exceptional cases and redo steps 1–3. Do not repeat this step.

Example 17.12 *c*-Chart for Number of Flaws in Rolls of Carpet   Rolls of carpet are inspected as they come off of a production line. The following table summarizes the data for 10 rolls of carpet inspected during a 10-day period.

Apply the appropriate control chart and report whether the process is under control or not.

Roll of carpet, i	1	2	3	4	5	6	7	8	9	10
Number of defects in roll, c_i	30	28	25	29	31	27	30	30	32	26

First, find the average number of defects $\bar{c}$:

$$\bar{c} = \frac{1}{10}(30 + 28 + 25 + 29 + 31 + 27 + 30 + 30 + 32 + 26) = 28.8$$

Now, find the sample standard deviation σ_c:

$$\sigma_c = \sqrt{\bar{c}} = \sqrt{28.8} = 5.37$$

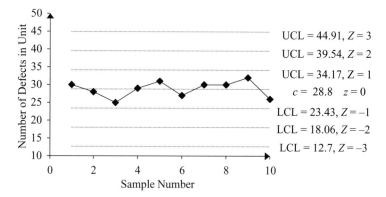

Figure 17.20 c-Chart of Example 17.12 for $z = 1, 2, 3$.

For $z = 1, 2, 3$, associated with 68.3, 95.4, and 99.74, the control limits are calculated as follows:

For $z = 1$	$\text{UCL} = 28.8 + 1 \times 5.37 = 34.17$	$\text{LCL} = 28.8 - 1 \times 5.37 = 23.43$
For $z = 2$	$\text{UCL} = 28.8 + 2 \times 5.37 = 39.54$	$\text{LCL} = 28.8 - 2 \times 5.37 = 18.06$
For $z = 3$	$\text{UCL} = 28.8 + 3 \times 5.37 = 44.91$	$\text{LCL} = 28.8 - 3 \times 5.37 = 12.69$

The control chart for $z = 1, 2, 3$ is presented in Figure 17.20. According to the control c-chart, the process is under control for $z = 1, 2, 3$.

17.6.3   Bicriteria Sampling for p-Chart[†]

The multiobjective problem for p-charts is similar to the multiobjective problem for $\bar{X}$-charts. The two objectives are to minimize cost, $f_1 = C_o n$, and minimize inaccuracy, $f_2 = h$. For p-charts the relationship between sample size n and average percent of defects $\bar{p}$ is presented below. The value of $f_2 = h$ can be obtained by the equation

$$n = \frac{z^2 \bar{p}(1 - \bar{p})}{h^2} \quad \text{or} \quad h = \frac{z\sqrt{\bar{p}(1 - \bar{p})}}{\sqrt{n}} \tag{17.18}$$

The multicriteria approach is similar to the method used for $\bar{X}$-charts. For additive utility functions, the optimal sample size can be found by setting $\partial U(n)/\partial n = 0$ and solving for the value of n. For minimizing $U = w_1 f_1' + w_2 f_2'$, the optimal solution n^* is given in Equation (17.19). The additive utility function can be written as

$$U = w_1 \left(\frac{C_o n - f_{1,\min}}{f_{1,\max} - f_{1,\min}} \right) + w_2 \left(\frac{z\sqrt{\bar{p}(1 - \bar{p})}/\sqrt{n} - f_{2,\min}}{f_{2,\max} - f_{2,\min}} \right)$$

$$\frac{\partial U(n)}{\partial n} = w_1 \left(\frac{C_o}{f_{1,\max} - f_{1,\min}} \right) - w_2 \left(\frac{0.5 n^{-3/2} z\sqrt{\bar{p}(1 - \bar{p})}}{f_{2,\max} - f_{2,\min}} \right) = 0$$

$$n^* = \left(\frac{(f_{1,\max} - f_{1,\min}) w_2 z \sqrt{\bar{p}(1 - \bar{p})}}{(f_{2,\max} - f_{2,\min}) 2 w_1 C_o} \right)^{2/3} \tag{17.19}$$

Example 17.13   Bicriteria for p-Charts   Consider a p-chart sampling problem where the initial sampling average is 10% and the sample size can vary from 2 to 50. Suppose a 99.74% confidence level is desired and the inspection cost per unit is $1.12. Suppose the weights of importance for the two normalized objectives are $w_1 = 0.3$ and $w_2 = 0.7$.

  (a) Generate five equally distributed multicriteria alternatives in terms of n.
  (b) Rank the five alternatives.
  (c) Find the optimal sample size.

In this example, $C_o = \$1.12$, $z = 3$, $\bar{p} = 0.1$, $n_{\min} = 2$, and $n_{\max} = 50$. Using Equation (17.19), one can find h for different values of n.

  (a) Five equally distributed alternatives in terms of n are $\{2, 14, 26, 38, 50\}$. Multicriteria alternatives are illustrated in the following table.
  (b) Ranking of alternatives for minimizing $U = 0.3 f_1' + 0.7 f_2'$ is also presented in the table.

Alternatives	a_1	a_2	a_3	a_4	a_5	Minimum	Maximum
n	2	14	26	38	50	2	50
Min. $f_1 = C_o n$	2.24	15.68	29.12	42.56	56	2.24	56
Min. $f_2 = h$	0.64	0.24	0.18	0.15	0.13	0.13	0.64
Min. f_1'	0.00	0.25	0.50	0.75	1.00		
Min. f_2'	1.00	0.22	0.10	0.04	0.00		
Min. $U = 0.3 f_1' + 0.7 f_2'$	0.70	0.23	0.22	0.25	0.30		
Ranking	5	2	1	3	4		

  (c) The optimal number of samples can be directly obtained by using Equation (17.19) (see Figure 17.21b):

$$n^* = \left(\frac{53.76 \times 0.7 \times 3 \times \sqrt{0.1 \times 0.9}}{0.51 \times 2 \times 0.3 \times 1.12} \right)^{2/3} = 21.38 \quad \text{(rounded to 22)}$$

Note that for $n = 22$

$$f_1 = 1.12 \times 22 = 24.64 \quad f_2 = 3 \times \left(\frac{0.1 \times 0.9}{22} \right)^{0.5} = 0.19 \quad U(n = 22) = 0.21$$

The set of all efficient bicriteria alternatives are presented in Figure 17.21a. The set of utility values for different sample sizes n is presented in Figure 17.21b.

See Supplement S17.5.xls.

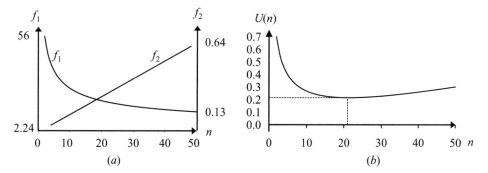

Figure 17.21 Graphical representation of Example 17.13: (*a*) efficient frontier, f_1 vs. f_2; (*b*) utility function for different sample sizes n, where at $n = 21.38$ the utility function is at its minimum, 0.21.

17.6.4 Bicriteria Sampling for *c*-Chart[†]

The multicriteria problem for *c*-charts is similar to multicriteria $\bar{X}$- and *p*-chart problems. The basic relationship between the sample size and the desired accuracy for *c*-charts is

$$n = \left(\frac{z\bar{c}}{h}\right)^2 \quad \text{or} \quad h = \frac{z\bar{c}}{\sqrt{n}} \tag{17.20}$$

For additive utility functions $w_1 f_1' + w_2 f_2'$, the optimum sample size n^* can be found as

$$U = w_1 \left(\frac{C_o n - f_{1,\min}}{f_{1,\max} - f_{1,\min}}\right) + w_2 \left(\frac{z\bar{c}/\sqrt{n} - f_{2,\min}}{f_{2,\max} - f_{2,\min}}\right)$$

$$\frac{\partial U(n)}{\partial n} = w_1 \left(\frac{C_o}{f_{1,\max} - f_{1,\min}}\right) + \left(\frac{zn^{-3/2}\bar{c}}{2(f_{2,\max} - f_{2,\min})}\right) w_2 = 0$$

$$n^* = \left(\frac{(f_{1,\max} - f_{1,\min})w_2 z\bar{c}}{(f_{2,\max} - f_{2,\min})2 w_1 C_o}\right)^{2/3} \tag{17.21}$$

Example 17.14 Bicriteria for *c*-Charts   Consider a *c*-chart sampling problem where the initial sampling average is 8.8. The sample size can vary from 2 to 50. Suppose 99.74% confidence is desired and the cost per sample is \$1.12. Find the best multicriteria alternative where the weights of importance are $w_1 = 0.4$ and $w_2 = 0.6$ for the normalized objective functions.

In this example, $C_o = \$1.12$, $z = 3$, $\bar{c} = 8.8$, $n_{\min} = 2$, and $n_{\max} = 50$.

Using Equation (17.21), the optimal sample size is as follows (see Figure 17.22):

$$n = \left(\frac{(f_{1,\max} - f_{1,\min})w_2 z\bar{c}}{(f_{2,\max} - f_{2,\min})2 w_1 C_o}\right)^{2/3} = \left(\frac{53.76 \times 0.6 \times 3 \times 8.8}{14.94 \times 2 \times 0.4 \times 1.12}\right)^{2/3} = 15.94$$

We can round 15.94 to 16. For $n = 16$, we have

$$f_1 = C_o n = 1.12 \times 16 = 17.92 \qquad f_2 = \frac{3 \times 8.8}{(16)^{(0.5)}} = 6.6 \qquad U = 0.23$$

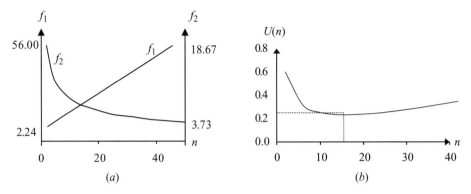

Figure 17.22   Graphical representation of Example 17.14 for utility function U vs. sample size n: (a) efficient frontier; (b) utility function values.

17.7   ACCEPTANCE SAMPLING CHARACTERISTICS

The statistical process control methods covered in the previous section are concerned with on-going quality during the production process. They are used to monitor and adjust the process. Acceptance sampling plans are concerned with the outgoing products. Instead of performing inspections on all products, sampling plans are used to inspect only a small portion of the products. Sampling plans are widely used in different industries for the following reasons:

- Cost and time of inspecting all items are usually substantial.
- Inspecting all items is impractical.
- Queuing, delays, and bottlenecks may occur when in-process inspections are conducted.
- Certain types of products could be damaged or destroyed by the type of inspection (e.g., inspecting delicate products, testing explosives, lighting matches, and tasting a chocolate bar).

The purpose of using an acceptance sampling plan is to render judgment about the quality of a lot while minimizing the time and expense of inspection. Usually, inspection is performed on a randomly selected small sample of products from a lot to make a judgment about the quality of the lot. The problem is to choose an optimal sampling plan that has minimal inspection costs while satisfying the required confidence level in the quality of the outgoing products. Acceptance sampling plans can also be used to inspect the incoming (or outgoing) materials or components for each phase of production.

Sampling plans can be used for variables (continuous measurement) or attributes (binary measurement). But in most applications attributes are used because attributes are easier and less expensive to measure. We only consider attributes in this section.

Sampling plans may be conducted in one phase (called single-phase sampling) or many phases (called sequential sampling). We will first cover the single-sampling plans and then we will discuss sequential sampling. Also, we develop methods for the selection of sampling plans using single- and multiple-criteria decision making.

17.7.1 Statistical Distributions for Sampling

Consider an attribute measurement of quality, where a binary value (0 or 1) indicates an unacceptable or acceptable quality. The following notation is used in sampling plans:

N = lot size

n = sample size

X = random variable representing the number of defectives in the lot

x = number of defectives in the sample

c = rejection level for sample n (that is, if $x > c$, reject the lot; otherwise, accept it)

p = true proportion of defective in a lot, that is, actual proportion of defects present in a lot and varies from lot to lot

p' = estimate of true proportion of defectives, p, in the lot

In sampling plans, two commonly used probability distributions are the binomial distribution and the Poisson distribution.

Binomial Probability Distribution In acceptance sampling planning, items being inspected will have only one of two possible states: good (acceptable) and bad (unacceptable). In practice, the lot size is much larger than the sample size, so a binomial approximation is sufficient to provide an accurate approximation of the probability distribution.

In the binomial probability distribution, the probability of x defects occurring is calculated as

$$P\{X = x\} = \frac{n!}{x!(n-x)!} p^x (1-p)^{n-x} \qquad \text{for} \qquad 0 \le x \le n \qquad (17.22)$$

Thus, the probability that X is less than or equal to a given rejection level c can be found as follows:

$$P\{X \le c\} = \sum_{x=0}^{c} \frac{n!}{x!(n-x)!} p^x (1-p)^{n-x} \qquad (17.23)$$

Binomial distributions have the following properties:

$$\text{Mean} = np$$

$$\text{Variance} = np(1-p)$$

$$\text{Standard deviation} = \sqrt{np(1-p)}$$

The values for binomial probabilities can be directly calculated [Equation (17.22) or (17.23)]. They can also be looked up in most statistical tables, such as the one presented at the back of this book.

Poisson Probability Distribution When the proportion of defective items in the lot, p, is relatively small and n is relatively large, the Poisson distribution is a good approximation for the binomial distribution. Typically, if $n \ge 25$ and $np' \le 5$, the Poisson distribution can be used instead of the binomial distribution. The Poisson distribution measurements are much easier to calculate and use than those of the binomial distributions. Also, the Poisson table at the back of the book covers more ranges of parameters than the corresponding binomial table. Poisson probability distributions are calculated as follows:

$$P\{X = x\} = \frac{e^{-\mu}\mu^x}{x!} \tag{17.24}$$

$$P\{X \le c\} = \sum_{x=0}^{c} \frac{e^{-\mu}\mu^x}{x!} \tag{17.25}$$

The Poisson distribution has the following properties:

$$\text{Mean } \mu = np$$
$$\text{Variance } \mu = np$$
$$\text{Standard deviation} = \sqrt{np}$$

One can use the cumulative Poisson table presented at the back of the book to find

$$P\{X \ge c \,|\, \mu = np\}$$

Note that $P\{X \le c\} = 1 - P\{X \ge (c + 1)\}$. In order to use the tables in the back of the book, one may need to round μ up to the nearest given number in the tables.

Example 17.15   Binomial and Poisson Distributions for a Light Bulb   Suppose there are samples of 50-bulbs from a lot size of 500 bulbs. If more than 2 bulbs (out of the 50 bulb sample) are defective, the lot should be rejected. It is estimated that the percentage of defective bulbs is 4%.

Find the probability distribution and the probability that a given lot will be accepted based on a sample size of 50 bulbs using (a) a binomial distribution and (b) a Poisson distribution. Compare the results of parts (a) and (b).

In this example, we are given the following information:

$$N = 500 \qquad n = 50 \qquad c = 2 \qquad p' = 0.04$$

The lot will be rejected if the number of defects in a sample of 50 is greater than 2. Therefore, the probability of acceptance, p_{acc}, is the probability that the number of defects in the sample is less than or equal to 2.

(a) *Binomial Distribution*   Using Equation (17.23),

$$p_{acc} = P\{X \le 2\} = \sum_{x=0}^{2} \frac{n!}{x!(n-x)!}(p')^x(1-p')^{n-x}$$

$$p_{acc} = \frac{50!}{0!(50-0)!}(0.04)^0(1-0.04)^{50-0} + \frac{50!}{1!(50-1)!}(0.04)^1(1-0.04)^{50-1}$$

$$+ \frac{50!}{2!(50-2)!}(0.04)^2(1-0.04)^{50-2}$$

$$= 0.1299 + 0.2706 + 0.2762 = 0.6767$$

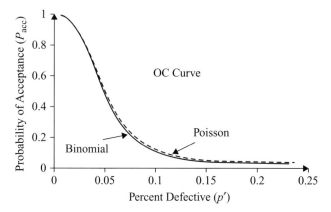

Figure 17.23 Operating characteristic curve for Example 17.16, $n = 50$, $c = 2$, using binomial and Poisson Distributions.

Therefore, there is a 67.67% chance that a given lot will be accepted for the given sampling strategy, that is, for $n = 50$ and $c = 2$. See Figure 17.23 and Table 17.8.

(b) *Poisson Distribution*   In this example, $n = 50 \geq 25$ and $np' = 50 \times 0.04 = 2 \leq 5$. Therefore, a Poisson distribution can be used instead of a binomial distribution.

Using Equation (17.7) for Poisson distributions, the probability of acceptance p_{acc} is given as

$$p_{acc} = P\{X \leq 2\} = \sum_{x=0}^{2} P\{X = x\} = \frac{e^{-2} \cdot 2^0}{0!} + \frac{e^{-2} \cdot 2^1}{1!} + \frac{e^{-2} \cdot 2^2}{2!}$$
$$= 0.1353 + 0.2707 + 0.2707 = 0.6766$$

TABLE 17.8   Finding Probability Acceptance by Binomial and Poisson Distributions

p'	n	$\mu = np'$	P_{acc} (Binomial)	P_{acc} (Poisson)
0.01	50	0.5	0.9862	0.9856
0.02	50	1.0	0.9216	0.9197
0.03	50	1.5	0.8108	0.8088
0.04	50	2.0	0.6767	0.6767
0.05	50	2.5	0.5405	0.5438
0.06	50	3.0	0.4162	0.4232
0.07	50	3.5	0.3108	0.3208
0.08	50	4.0	0.2260	0.2381
0.09	50	4.5	0.1605	0.1736
0.10	50	5.0	0.1117	0.1247
0.15	50	7.5	0.0142	0.0203
0.20	50	10.0	0.0013	0.0028
0.25	50	12.5	0.0001	0.0003

Alternatively, one can use the Poisson table at the back of the book:

$$p_{acc} = P\{X \leq 2\} = 1 - P\{X \geq 3|\mu\} = 1 - P\{X \geq 3|(50 \times 0.04)\}$$

or

$$p_{acc} = 1 - P\{X \geq 3|2\} = 1 - 0.3233 = 0.6767$$

In this example, the binomial solution of 0.6767 is almost equivalent to the Poisson solution of 0.6766. This verifies that if $n \geq 25$ and $np' \leq 5$, the Poisson distribution agrees very well with the binomial distribution. As stated before, it is computationally much easier to use a Poisson distribution, especially when statistical tables are not available.

17.7.2   Operating Characteristic Curve

The operating characteristic (OC) curve provides a graphical representation of the probability of acceptance for a given sample size n and rejection level c. The OC curve displays the probability of accepting a lot versus the probability of the number of defects in the sample. This curve is a function of n and c. Different sampling plans (n, c) have different OC curves.

The steps for creating the OC curve can be summarized as follows:

1. Select the sample size n and the rejection level c.
2. For the given n and c, find the probability of acceptance p_{acc} for a range of values of p' (estimated probability of defects). Usually p' is in the range of 0.01–0.25.
3. Graph the results of the probability of acceptance p_{acc} (y axis) versus p' (x axis). The resulting graph is the OC curve.

Example 17.16   Operating Characteristic Curve of a Light Bulb   Consider the information provided in Example 17.15. Determine the OC curve for a light bulb sampling plan. For purpose of illustration, use both binomial and Poisson distributions.

The probabilities of acceptance p_{acc} using both binomial and Poisson distributions are shown in Table 17.8. The results of p_{acc} versus p' are depicted in Figure 17.23 where the Poisson curve is shown with a dashed line and the binomial curve is shown with a solid line.

In this example, the estimated percentage of defective bulbs is 4%. Since $n = 50$, which is greater than 25, and $np' = 50 \times 0.04 = 2 \leq 5$, it is justified to use a Poisson distribution instead of a binomial distribution; also see Table 17.8.

Using Poisson Probability Table   To use the table at the back of the book, first find the mean value, that is, $\mu = np'$. Check the table of Poisson distributions at the back of the book for the given mean μ. Note that the table provides values for $P\{X \geq c\}$. If the problem requires finding a value for $P\{X < c\}$, then use

$$P\{X < c\} = 1 - P\{X \geq c\}$$

Similarly, if the problem requires finding a value for $P\{X \leq c\}$, then use

$$P\{X \leq c\} = 1 - P\{X \geq (c + 1)\}$$

For example, consider the data in the first row of Table 17.8, where $p' = 0.01$, $n = 50$, and $\mu = np' = 50 \times 0.01 = 0.5$. Find

$$P\{X \le 2\} = 1 - P\{X \ge 3\}$$

From the cumulative poisson probability table at the back of the book, find the values of $P\{X \ge 3\}$ by referring to the intersection of column $\mu = 0.5$ and row $c = 3$. This value is shown as 0.0144. Then

$$P\{X \le 2\} = 1 - 0.0144 = 0.9856$$

The values found using the Poisson distribution are listed in the last column of Table 17.8, and it can be seen that they agree very well with the binomial values also presented in Table 17.8, especially when $np' \le 5$.

17.7.3 Producer's and Consumer's Risks

The sampling plan decision problem is concerned with finding the optimal values of the sample size n and rejection size c that result in accepting a good (satisfactory) lot and in rejecting a bad lot. Since only a small sample of each lot is used, there is a possibility of error, that is, a wrong decision would be made about the lot. There are two types of errors that can occur:

Type I Error (Producer's Risk, α)   The error of rejecting a good (satisfactory) lot. Type I error is the risk that the producer may reject an acceptable lot.

Type II Error (Consumer's Risk, β)   The error of accepting a bad (unsatisfactory) lot. Type II error is the risk that a bad lot may be accepted and passed on to the consumer.

The table below outlines the basic sampling plan decision problem for the two types of errors.

	Actual Condition	
Sampling Plan Decision	Good Lot	Bad Lot
Reject lot	Type I error (producer's risk, α), $P(X > c)$	Correct decision
Accept lot	Correct decision	Type II error (consumer's risk, β), $P(X \le c)$

The following definitions are used in conjunction with the operating characteristic curve for a given n and c which can be used to measure the producer's risk (α) and the consumer's risk (β).

Acceptable Quality Level (AQL)   The AQL is presented by probability p_0. A lot is acceptable if its proportion of defects p is less than or equal to p_0. That is, accept a lot if $p \le p_0$.

Lot Tolerance Percent Defective (LTPD)   The LTPD is also called rejectable quality level (RQL). LTPD is presented by probability p_1. If the proportion of defects p is greater than or equal to p_1, then the lot is rejected. That is, reject a lot if $p > p_1$.

The following figure shows the relationships of AQL and LTPD and the acceptable ranges.

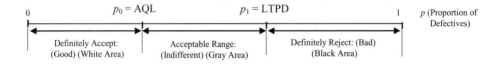

The sampling problem is to find n and c such that

$$p_0 \le p \le p_1 \quad \text{or} \quad \text{AQL} \le p \le \text{LTPD}$$

AQL must be set to be less than or equal to LTPD; for example, see Figure 7.2.

The following notation is used for risk analysis for the two different types of errors:

α = producer's risk of rejecting a satisfactory lot. This is calculated as 1 minus the probability of accepting a lot at $p' = \text{AQL}$; that is, find $1 - p_{\text{acc}}$ for given n, c, AQL, and LTPD.

β = consumer's risk of accepting an unsatisfactory lot. This is the probability of accepting a lot; that is, p_{acc} at $p' = \text{LTPD}$ for the given values of n, c, AQL, and LTPD.

Example 17.17   Calculating Producer's (α) and Consumer's Risks (β)   Consider a sampling plan where $n = 25$, $c = 4$, AQL $= 0.1$, and LTPD $= 0.3$.

Find the producer's and consumer's risk for percent of defects ranging from 0.01 to 0.30. Also show the OC curve for this problem.

The probability of acceptance for ($n = 25$, $c = 4$) is presented and depicted in Figure 17.24.

The probability of acceptance is 0.9020 when $p' = \text{AQL} = 0.1$. Therefore, the producer's risk is

$$\alpha = 1 - p_{\text{acc}} = 1 - 0.9020 = 0.0980$$

That is, there is a 9.80% chance that a good lot be rejected.

The probability of acceptance when $p' = \text{LTPD} = 0.3$ is 0.0905. Therefore, the consumer's risk is

$$\beta = p_{\text{acc}} = 0.0905$$

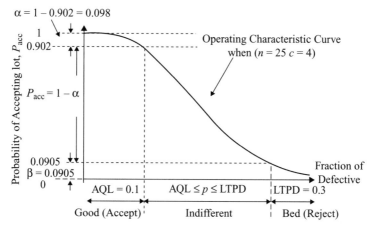

$\alpha = 1 - 0.902 = 0.098$

Figure 17.24 Relationship between AQL, LTPD, α, β, and probability of acceptance for $n = 25$, $c = 4$.

That is, there is 9.05% chance that a bad lot will be accepted.

p'	P_{acc} for $n = 25$, $c = 4$
0.01	1.0000
0.02	0.9999
0.03	0.9992
0.04	0.9972
0.05	0.9928
0.1 = AQL	0.9020
0.15	0.6821
0.2	0.4207
0.25	0.2137
0.30 = LTPD	0.0905
Producer's, α	1-0.902 = 0.0980
Consumer's, β	0.0905

Binomial Probability Distribution for Finding α and β   To find the producer's risk α, first find the probability of acceptance p_{acc} for the given values of n, c, and $p = p_0$, where p_0 is defined as the AQL:

$$\alpha = P\{X > c | p = p_0\} = \sum_{x=c+1}^{n} \frac{n!}{x!(n-x)!} p_0^x (1 - p_0)^{n-x} \qquad (17.26)$$

To find the consumer's risk β, find the probability of acceptance for the given values of n, c, and $p = p_1$ where p_1 is defined as the LTPD:

$$\beta = P\{X \le c | p = p_1\} = \sum_{x=0}^{c} \frac{n!}{x!(n-x)!} p_1^x (1 - p_1)^{n-x} \qquad (17.27)$$

Poisson's Probability Distribution for Finding α and β   As discussed before, for ease of calculation, when $n \geq 25$ and $np \leq 5$, the Poisson distribution can be used instead of the binomial distribution (it gives approximately the same solution). In the Poisson distribution,

$$P\{X = c | \mu = np_0\} = \frac{e^{-\mu}\mu^x}{x!}$$

$$\alpha = P\{X > c | \mu = np_0\} = \sum_{x=c+1}^{n} \frac{e^{-\mu}\mu^x}{x!} \tag{17.28}$$

$$\beta = P\{X \leq c | \mu = np_1\} = \sum_{x=0}^{c} \frac{e^{-\mu}\mu^x}{x!} \tag{17.29}$$

Alternatively, α can be derived as

$$\alpha = P\{X > c | \mu = np_0\} = 1 - P\{X \leq c | \mu = np_0\} = 1 - \sum_{x=0}^{c} \frac{e^{-\mu}\mu^x}{x!}$$

17.7.4 Finding Optimal Sampling Size, n, $c^{\dagger}$

Effect of Increasing n and Decreasing c on OC Curve   For a given rejection level c, as the sample size n increases and approaches the lot size N, the shape of the OC curve becomes steeper; see Figure 17.25. That is, the OC curve will become more precise in determining if a lot is good or bad. When $n = N$, the curve is vertical and provides 100%

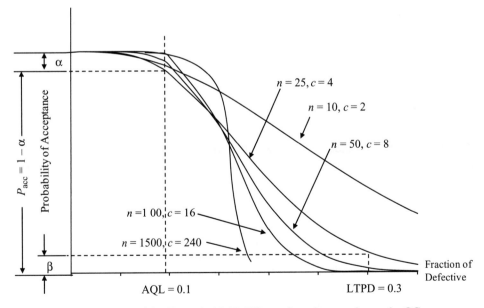

Figure 17.25 p_{acc} versus p' for Example 17.18. Effects of varying n and c on the OC curve, p_{acc}. As sample size increases, the OC curve becomes more accurate (vertical); i.e., α and β get smaller; α and β for ($n = 25$, $c = 4$) are shown.

TABLE 17.9   Example 17.18: Probability of Acceptance for Different Values of p', n, and c OC Curves are Presented in Figure 17.25

p'	$n = 10$, $c = 2$	$n = 25$, $c = 4$	$n = 50$, $c = 8$	$n = 100$, $c = 16$	$n = 1500$, $c = 240$
0.01	0.9999	1.0000	1.0000	1.0000	1.0000
0.02	0.9991	0.9999	1.0000	1.0000	1.0000
0.03	0.9972	0.9992	1.0000	1.0000	1.0000
0.04	0.9938	0.9972	0.9999	1.0000	1.0000
0.05	0.9885	0.9928	0.9992	1.0000	1.0000
AQL $= 0.1$	0.9298	0.9020	0.9421	0.9794	1.0000
0.15	0.8202	0.6821	0.6681	0.6725	0.8684
0.2	0.6778	0.4207	0.3073	0.1923	0.0000
0.25	0.5256	0.2137	0.0916	0.0211	0.0000
LTPD $= 0.30$	0.3828	0.0905	0.0183	0.0010	0.0000
Producer's, α	0.0702	0.0980	0.0579	0.0206	0
Consumer's, β	0.3828	0.0905	0.0183	0.0010	0

accuracy that the consumer is not receiving more defective products than the agreed-upon level, and there is no gray area for error.

Example 17.18   Effect of Varying n and c   Suppose that the AQL is 0.1 and the LTPD is 0.3. Consider five different sampling plans (n,c): (10, 2), (25, 4), (50, 8), (100, 16), and (1500, 240). For each plan, find the OC curve and its producer's and consumer's risks.

The solutions for the five different sampling plans are presented in Table 17.9 and illustrated in Figure 17.25. In Figure 17.25 different OC curves for different n and c values are presented, where the c/n ratio is roughly about 20%. As the number of samples n increases, the probability of a type I or type II error is reduced, although the ratio of c/n remains almost the same. Notice that for different n and c values the values of α and β are different.

Finding Optimal n and c for given AQL, LTPD, α, and β   Consider the following two given points, where AQL $\leq$ LTPD:

Point 1   Producer agrees with AQL $= p_0$ and type I error of α. Note that $1 - \alpha = p_{acc}$.
Point 2   Consumer agrees with LTPD $= p_1$ and type II error of β. Note that $\beta = p_{acc}$.

Ideally, one wishes to find n and c such that the OC curve perfectly matches the given AQL, LTPD, α, and β. Such a perfect match may not exist. Therefore, the closest OC curve can be used.

The problem is to find an (n,c) whose OC curve is the closest curve that passes through points 1 and 2:

Point 1 $(p_0, (1 - \alpha))$
Point 2 (p_1, β)

In practice, a trial-and-error approach can be used by considering different values of n and c until the best (the closest) OC curve that passes through both points is found. It is also possible to perform an exhaustive search by considering different values of n and c. However, the exhaustive method may be computationally intensive. Currently, there are standard tables that provide the best approximate values of n and c for the given two points.

Example 17.19 Finding Optimal n and c   Suppose that AQL $= 0.1$, LTPD $= 0.3$, $\alpha = 0.1$, and $\beta = 0.1$. Find the best sampling plan (n,c) from the following five sampling plans: $(10, 2)$, $(25, 4)$, $(50, 8)$, $(100, 16)$, and $(1500, 240)$. The two points are point 1: $(0.1, (1 - 0.1 = 0.9))$ and point 2: $(0.3, 0.1)$.

We would like to identify which of these five plans is the best fit. The probability of acceptance for each of these plans is presented in Table 17.9. The last two rows of Table 17.9 show the producer's risk (α) and consumer's risk (β) for each of the five plans for the given AQL $= 0.1$ and LTPD $= 0.3$. The closest α and β to the given target of $\alpha = 0.1$ and $\beta = 0.1$ is plan $n = 25$, $c = 4$ where $\alpha = 0.0980$ and $\beta = 0.0905$.

The OC curve of these five plans are presented in Figure 17.25. One can verify that the OC curve of $n = 25$, $c = 4$ is the best fit for the two given points. Therefore, choose $n = 25$ and $c = 4$ for this problem.

Supplement S17.6.docx explains a heuristic for finding an OC curve.

17.8   ACCEPTANCE SAMPLING OUTGOING QUALITY

17.8.1   Average Outgoing Quality

Average outgoing quality (AOQ) is a measure of the expected number of defective items in a lot under a given sampling plan (n,c). It is assumed that if a defective item is found in the sample, it will be replaced with a good item. If the lot is rejected, then every single item in the lot is inspected, and all defective items are replaced with good items. The AOQ measurement is as follows:

$$\text{AOQ} = \frac{E\{\text{outgoing number of defective items}\}}{E\{\text{outgoing number of items}\}} = \frac{p_{\text{acc}}\, p'(N - n)}{N} \qquad (17.30)$$

where N is lot size, n is sample size, x is number of defectives in the sample size n, c is rejection level for sample n (that is, if $x > c$, reject the lot; otherwise, accept it), and p' is the fraction of defectives. Note that for a given sampling plan the probability of acceptance, p_{acc}, should first be calculated based on the given values of p', n and c. Then, the AOQ is calculated using p_{acc}, p', n, and N.

Example 17.20   Find AOQ   Consider lots of 500 items where the sample size is 50. A lot is rejected if there are more than two defective items in each sample. The average number of defective items per lot is about 5%. That is, $N = 500$, $n = 50$, $c = 2$, and $p' = 0.05$. Find the AOQ for this sampling plan. Also, identify the number of defectives that will be passed on to the consumer with and without the sampling.

First, find p_{acc} where $n = 50$, $c = 2$, and $p' = 0.05$. The result using the Poisson table is

$$p_{acc} = P(X \leq 2|\mu = 2.5) = 1 - P(X \geq 3|\mu = 2.5) = 0.5438$$

where X is the random variable representing the number of defectives in the lot. Now use Equation (17.30) to find the AOQ:

$$\text{AOQ} = \frac{0.5438 \times 0.05(500 - 50)}{500} = 0.0245$$

In this plan, 2.45% of items are defective and will be passed on to the consumer. That is, $0.0245 \times 500 = 12.24$ items per lot of 500 are defective after inspection is conducted. Without the sampling plan, the number of defectives would be $500 \times 0.05 = 25$.

Average Outgoing Quality Limit (AOQL) In some applications, the average number of defectives in a lot (p') is not known. When p' is not known, the AOQL can be used. The AOQL represents the poorest level of the AOQ for a given (n, c) sampling plan for all possible proportions of defective items in the lot, that is, considering all different values of p'. That is, the AOQL is the maximum of all possible values of the AOQ as the value of p' varies:

$$\text{AOQL} = \text{Max}\{\text{AOQ}(p') \quad \text{for} \quad p' = 0, \ldots, 1\} \qquad (17.31)$$

The procedure for finding the AOQL is as follows. Consider different values of p'. For each given p', find its AOQ. Then choose the highest AOQ value. The results of the calculations can be shown by a table. It can also be graphically presented by plotting AOQ values on the y axis and plotting p' values on the x axis, then choosing the peak of the curve.

Example 17.21   Finding the AOQL   Suppose a lot size is 500. Find the AOQL for two sampling plan alternatives: (a) $n = 25$, $c = 1$, and (b) $n = 50$, $c = 2$.
    The solutions for the two alternatives are presented in Table 17.10 and shown in Figure 17.26. The maximum AOQ is boldface in the tables below.
    The AOQL is found by finding the maximum AOQ subject to the constraint $0 \leq p' \leq 1$:

(a) For the first alternative ($n = 25$, $c = 1$), AOQL $= 0.0315$, which is associated with $p' = 0.06$.
    That is, the maximum AOQ is AOQL $= 0.0315$.
(b) For the second alternative ($n = 50$, $c = 2$), AOQL $= 0.0244$, which is associated with $p' = 0.04$.

Both plans have the same ratio of quality, $c/n = 1/25 = 2/50$. But the second alternative ($n = 50$, $c = 2$), has a lower AOQL because as the sample size increases, for the same ratio of quality, the AOQL decreases; that is, more sampling results in a better AOQL.

TABLE 17.10 AOQL Calculation for Example 17.21 for Two alternatives

(a) $n = 25, c = 1$					(b) $n = 50, c = 2$				
p'	n	c	p_{acc}	AOQ	p'	n	c	p_{acc}	AOQ
0.01	25	1	0.9742	0.0093	0.01	50	2	0.9862	0.0089
0.02	25	1	0.9114	0.0173	0.02	50	2	0.9216	0.0166
0.03	25	1	0.8280	0.0236	0.03	50	2	0.8108	0.0219
0.04	25	1	0.7358	0.0280	**0.04**	**50**	**2**	**0.6767**	**0.0244**
0.05	25	1	0.6424	0.0305	0.05	50	2	0.5405	0.0243
0.06	**25**	**1**	**0.5527**	**0.0315**	0.06	50	2	0.4162	0.0225
0.07	25	1	0.4696	0.0312	0.07	50	2	0.3108	0.0196
0.08	25	1	0.3947	0.0300	0.08	50	2	0.2260	0.0163
0.09	25	1	0.3286	0.0281	0.09	50	2	0.1605	0.0130
0.10	25	1	0.2712	0.0258	0.10	50	2	0.1117	0.0101
0.15	25	1	0.0931	0.0133	0.15	50	2	0.0142	0.0019
0.20	25	1	0.0274	0.0052	0.20	50	2	0.0013	0.0002
0.25	25	1	0.0070	0.0017	0.25	50	2	0.0001	0.0000

17.8.2 Bicriteria Acceptance Sampling

When determining a sampling plan, one must consider the trade-offs between the cost of inspection and the average outgoing quality. The bicriteria acceptance sampling is similar to bicriteria statistical process control (Section 17.2). The bicriteria problem for the acceptance sampling can be formulated as:

Minimize cost of inspection: $f_1 = C_o n$

Minimize average outgoing quality: $f_2 = $ AOQ [use Equation (17.30)]

Subject to $n_{min} \leq n \leq n_{max}$

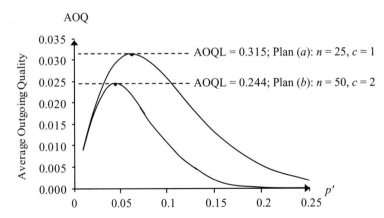

Figure 17.26 Graphical representation of AOQ and AOQL for Example 17.21 for two sampling plans (n,c): (a) (25, 1) and (b) (50, 2).

In this method, generate a set of efficient alternatives as a function of n and c. Then choose the best alternative. The notation used for this problem is summarized in the following table:

Given	Find
C_o = inspection cost per item	n = sample size, $n_{min} \leq n \leq n_{max}$
p_0 = acceptance quality level (AQL)	c = rejection level for sample size n
n_{min} = lower bound on n	f_1 = cost of inspection
n_{max} = upper bound on n	f_2 = AOQ

In this method, we assume that $c/n \geq p_0$. Therefore, for a given p_0, set $c = \lceil p_0 n \rceil$, where $\lceil . \rceil$ means that the given value is rounded up to the nearest integer.

Bicriteria Procedure for AOQ   For the given range of n_{min} to n_{max}, generate a set of equally distributed points in terms of n (e.g., generate five values of n).
For each given n:

1. Find $c = \lceil p_0 n \rceil$.
2. Find the probability of acceptance for given (n, c) and p_0.
3. Calculate the cost of inspection, f_1, for each alternative (n, c).
4. Find the average outgoing quality, f_2, for given (n, c) and p_0.
5. List the set of alternatives and identify efficient alternatives. Then use a multicriteria approach to rank alternatives.

Example 17.22   Bicriteria Procedure for AOQ   Consider an example where the cost of inspection per item is $20 and the lot size is 1000. The given AQL is 4%. The sample size can range from 5 to 100. Generate the set of alternatives for sample sizes 5, 10, 20, 40, 60, 80, and 100. Identify efficient alternatives and rank multicriteria alternatives.
In this example, $C_o = \$20$, $N = 1000$, $p_0 = AQL = 0.04$, $n_{min} = 5$, and $n_{max} = 100$.
In this example, we use the Poisson distribution to generate the AOQ values.
Table 17.11 shows the summary for calculating c, P_{acc}, f_1, and f_2 for each given value.

TABLE 17.11   Summary of Bicriteria Alternative for Example 17.22

Alternative no.	1	2	3	4	5	6	7	Min.	Max.
n	5	10	20	40	60	80	100	5	100
$c = \lceil p_0 n \rceil = \lceil 0.04n \rceil$	1	1	1	2	3	4	4	1	4
P_{acc}	0.9825	0.9384	0.8088	0.7834	0.7787	0.7806	0.6288	0.6288	0.9825
Min. $f_1 = C_o\, n = 20n$	$100	$200	$400	$800	$1,200	$1,600	$2,000	$100	$2,000
Min. $f_2 = $ AOQ	0.0391	0.0372	0.0317	0.0301	0.0293	0.0287	0.0226	0.0226	0.0391

TABLE 17.12   Ranking of Bicriteria Alternatives

Alternative no.	1	2	3	4	5	6	7
(n,c)	(5,1)	(10,1)	(20,1)	(40,2)	(60,3)	(80,4)	(100,4)
Min. f_1'	0.0000	0.0526	0.1579	0.3684	0.5789	0.7895	1.0000
Min. f_2'	1.0000	0.8820	0.5507	0.4522	0.4034	0.3698	0.0000
Min. $U = 0.7f_1' + 0.3f_2'$	0.3000	0.3014	0.2757	0.3935	0.5263	0.6636	0.7000
Rank	2	3	1	4	5	6	7

For purpose of illustration, we show the details of how to generate the two objective values for $n = 20$. First find the c value:

$$c = \lceil p_0 n \rceil = \lceil 0.04 \times 20 \rceil = \lceil 0.8 \rceil = 1$$

The closest higher integer value to 0.8 is 1. Therefore, $c = 1$ is used. Now, find the probability of acceptance for $n = 20$, $c = 1$ and use $p' = p_0 = 0.04$. The probability of acceptance is as follows using the Poisson tables at the back of the book:

$$p_{\text{acc}} = P(X \leq 1 | \mu = 0.8) = 1 - P(X \geq 2 | \mu = 0.8) = 0.8088$$

The cost of this plan is

$$f_1 = \$20n = 20 \times 20 = \$400.$$

The second objective, f_2, average outgoing quality (AOQ), using Equation (17.30), is

$$f_2 = \text{AOQ} = \frac{0.8088 \times 0.04 \times (1000 - 20)}{1000} = 0.0317$$

Table 17.12 and Figure 17.27a show the set of bicriteria alternatives for this problem. As n increases, AOQ decreases, and the cost of sampling increases. All alternatives are efficient.

For an additive utility function, suppose the weights of importance for the normalized values of the two objectives are $w_1 = 0.7$ and $w_2 = 0.3$. Therefore, the additive utility function $U = 0.7f_1' + 0.3f_2'$ can be used to rank alternatives. The rankings are shown in Table 17.12, where the best alternative is $n = 20$, $c = 1$, $f_1 = \$400$, and $f_2 = \text{AOQ} = 0.0317$.

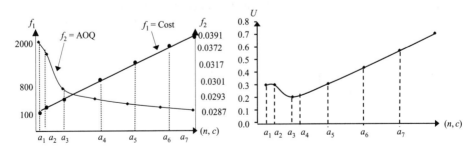

Figure 17.27   Graphical representation of Example 17.22, alternatives 1–7 present given $(n.c)$ in Table 17.12: (a) set of bicriteria alternatives; (b) additive utility function values.

Figure 17.27*b* shows the utility function values for different alternative sample sizes. For more precision, if needed, more bicriteria alternatives can be considered around the selected best alternative $n = 20$; that is, search $10 \leq n \leq 40$.

17.8.3 Sequential Sampling Plans

Sequential sampling plans are used to optimize the sampling plan procedure. In this approach, a small sample size is used initially. If the results from the initial sample are conclusive, then the lot is either accepted or rejected. However, if the results from the initial sample are inconclusive, additional samples are drawn sequentially until a conclusive judgment can be made on the acceptability of the lot. Such plans reduce the average number of samples required to determine the quality of a given lot. Therefore, sequential sampling plans are more cost effective than the single sampling plans covered in the last section. The sequential sampling plans also reduce the total time of inspections, making them an overall better option.

Double-Sampling Plans   A special case of sequential sampling is double sampling where only two consecutive sampling plans are used.

Consider the following notation for double sampling:

n_1, n_2 = no. of samples in sequential samplings 1 and 2, respectively

x_1, x_2 = actual no. of defectives in sequential samplings 1 and 2, respectively

c_1 = maximum no. of defective items in first sampling to accept lot

c_2 = minimum no. of defective items in first sampling to reject lot

c_3 = maximum cumulative no. of defective items in first and second lots for accepting lot

Figure 17.28 summarizes the double-sampling method.

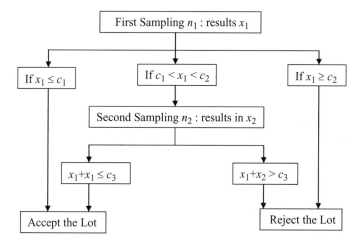

Figure 17.28   Decision pyramid for double-sampling plan.

Example 17.23   Double-Sampling Plan   Consider $c_1 = 4$, $c_2 = 7$, $c_3 = 8$, and sample sizes $n_1 = 40$ and $n_2 = 30$. For the following four examples, make a decision on whether to accept or reject by applying the double-sampling plan.

Condition	First Sample: x_1 If $x_1 \leq 4$, accept If $x_1 \geq 7$, reject	Second Sample: x_2 $4 < x_1 < 7$	Second Sample: If $x_1 + x_2 \leq 8$, accept If $x_1 + x_2 > 8$, reject	Decision
Example I	$x_1 = 3$	N/A	N/A	Accept the lot
Example II	$x_1 = 8$	N/A	N/A	Reject the lot
Example III	$x_1 = 6$	$x_2 = 2$	8	Accept the lot
Example IV	$x_1 = 5$	$x_2 = 4$	9	Reject the lot

Note that examples I and II are decided after the first sampling while examples III and IV are decided after the second sampling.

Multiple Sequential Sampling   The higher order sequential sampling plan algorithms are simple extensions of the double sampling presented in Figure 17.28. Consider the given values of $c_{i,min}$ and $c_{i,max}$ for each level of inspection i. Use the following procedure:

1. Add the number of defects to the previous number of defects.
2. If the number of defects is less than or equal to $c_{i,min}$ for level i, accept the lot. Stop.
3. If the number of defects is greater than or equal to $c_{i,max}$ for level i, reject the lot. Stop.
4. If the number of defects is greater than $c_{i,min}$ but less than the cumulative $c_{i,max}$ for level i, continue to the next level of sampling.
   Repeat steps 1–4.

Three examples of multiple sequential sampling are presented in Figure 17.29.

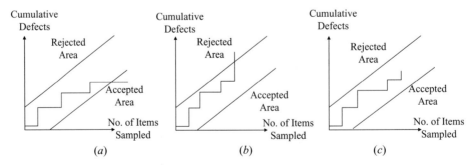

Figure 17.29   Examples of sequential sampling: (*a*) accept the lot in level 4; (*b*) reject the lot in level 4; (*c*) continue sampling in level 4.

Multiple-criteria decision making of sequential sampling is similar to the multiple-criteria approach of the single-sampling plan covered in Section 17.3.5.

REFERENCES

General

Bai, D. S., and K. T. Lee. "An Economic Design of Variable Sampling Interval X Control Charts." *International Journal of Production Economics*. Vol. 54, No. 1, 1998. pp. 57–64.

Besterfield, D. H., C. Besterfield, G. Besterfield, and M. Besterfield, 3rd ed. *Total Quality Management*. New York, NY. Pearson Education International, 2002.

Chiu, W. K. "Comments on the Economic Design of X-Charts." *Journal of American Statistical Association*. Vol. 68, 1973, pp. 919–921.

Cohen, L. *Quality Function Deployment: How to Make QFD Work for You*. Reading, MA: Addison-Wesley, 1995.

Dehnad, K. *Quality Control, Robust Design, and the Taguchi Method*. New York, NY. Kluwer Academic, 1988.

Dodge, H. F., and H. G. Roming. "A Method for Obtaining and Analyzing Sensitivity Data." *Bell System Technical Journal*, Vol. 8, 1929, pp. 613–631.

Dodge, H. P. "A Sampling Plan for Continuous Production." *Annals of Mathematical Statistics*, No. 3, Vol. 14, 1943, pp. 264–279.

Duncan, A. J. "The Economic Design of X Charts Used to Maintain Current Control of a Process." *Journal of the American Statistical Association*, Vol. 51, 1956, pp. 228–242.

Duncan, A. J. *Quality Control and Industrial Statistics*, 5th ed. New York: McGraw-Hill/Irwin, 1986.

Feigenbaum, A. V. *Total Quality Control*, 3rd ed. New York: McGraw-Hill, 1991.

Garvin, D. A. *Managing Quality*, New York, NY. Free Press, 1988.

Gibra, I. N. "Economically Optimal Determination of the Parameters of X-control Charts." *Management Science*, Vol. 17, 1971, pp. 635–646.

Grant, E. L., and R. S. Leavenworth, *Statistical Quality Control*, 7th ed. New York; McGraw-Hill, 1996.

Lin, Y., and C. Chou. "Non-Normality and the Variable Parameters $\bar{X}$ Control Charts." *European Journal of Operational Research*, Vol. 176, No. 1, 2007, pp. 361–373.

Montgomery, D. C. *Introduction to Statistical Quality Control*, 5th ed. Hoboken, NJ: Wiley, 2005.

Nahmias, S. *Production and Operation Analysis*, 6th ed. New York: McGraw-Hill, 2008.

Schmidt, J. W., and G. K. Bennett. "Economic Multiattribute Acceptance Sampling." *IIE Transactions*, Vol. 4, No. 3, 1972, pp. 194–199.

Shewhart, W.A. *Economic Control of Quality of Manufactured Product*. New York: Van Nostrand, 1931.

Taguchi, G. *On-Line Quality Control During Production*. Tokyo: Japanese Standard Association, 1981.

Taguchi, G., E. A. Elsayed, and T. Hsiang. *Quality Engineering in Production Systems*. New York: McGraw-Hill, 1989.

Ulrich, K. T., and S. D. Eppinger. *Product Design and Development*, 3rd ed. New York: McGraw-Hill, 2005.

Wald, A. "Foundations of a General Theory of Sequential Decision Functions." *Econometrica*, Vol. 15, 1947, pp. 279–313.

Multiobjective

Asadzadeh, S., and F. Khoshalhan. "Multiple-Objective Design of an $\bar{X}$ Control Control Chart with Multiple Assignable Causes." *International Journal of Advance Manufacturing Technology*, Vol. 43, Nos. 3–4, 2009, pp. 312–322.

Badri, M. A. "A Combined AHP-GP Model for Quality Control Systems." *International Journal of Production Economics*, Vol. 72, No. 1, 2001, pp. 27–40.

Bakir, M. A., and B. Altunkaynak. "The Optimization with the Genetic Algorithm Approach of the Multi-Objective, Joint Economical Design of the $\bar{X}$ and R Control Charts." *Journal of Applied Statistics*, Vol. 31, No. 7, 2004, pp. 753–772.

Bhattacharya, A., J. Geraghty, and P. Young. "Supplier Selection Paradigm: An Integrated Hierarchical QFD Methodology under Multiple-Criteria Environment." *Applied Soft Computing*, Vol. 10, No. 4, 2010, pp. 1013–1027.

Castillo, E. D., P. Mackin, and D. C. Montgomery. "Multiple-Criteria Optimal Design of X Control Charts." *IIE Transactions*, Vol. 28, No. 6, 1996, pp. 467– 474.

Celano, G., and S. Fichera. "Multiobjective Economic Design of an $\bar{X}$ Control Chart." *Computers and Industrial Engineering*, Vol. 37, No. 1, 1999, pp. 129–132.

Chen, Y., and H. C. Liao. "Multi-Criteria Design of an $\bar{X}$ Control Chart." *Computers & Industrial Engineering*, Vol. 46, No. 4, 2004, pp. 877–891.

Drezner, Z., and G. O. Wesolowsky. "Design of Multiple Criteria Sampling Plans and Charts." *International Journal of Production Research*, Vol. 29, No. 1, 1991, pp. 155–163.

Evans, G. W., and S. M. Alexander. "Multiobjective Decision Analysis for Acceptance Sampling Plans." *IIE Transactions*, Vol. 19, No. 3, 1987, pp. 308–316.

Karsak, E. E., S. Sozer, and S. E. Alptekin. "Product Planning in Quality Function Deployment Using a Combined Analytic Network Process and Goal Programming Approach." *Computers & Industrial Engineering*, Vol. 44, No. 1, 2003, pp. 171–190.

Malakooti, B. "Multi-Criteria Quality Control and Assurance." Case Western Reserve University, Cleveland, Ohio, 2013.

Malakooti, B., and W. H. Balhorn, "Selection of Acceptance Sampling Plans with Multi-Attribute Defects in Computer-Aided Quality Control." *International Journal of Production Research*, Vol. 25, No. 6, 1987, pp. 869–887.

Moskowitz, H. "Risk Preference, Multiple Attributes, and Multiple Criteria in Bayesian Acceptance Sampling Plans." *Naval Research Logistics Quarterly*, Vol. 32, No. 1, 1985, pp. 81–94.

Safaei, A. S., R. B. Kazemzadeh, and S. T. Akhavan Niaki. Multi-Objective Economic Statistical Design of X-Bar Control Chart Considering Taguchi Loss Function," *International Journal of Advanced Manufacturing Technology*, Vol. 59, issue 9-12, 2012, pp. 1091–1101.

Yanlai, L., T. Jiafu, and Y. Jianming, "Multi-Object Decision-Making Methodology for Selecting Engineering Characteristics in Quality Function Deployment." *Computer Integrated Manufacturing Systems*, Vol. 14, No. 6, 2008, pp. 1363–1369.

EXERCISES

Multiple-Criteria Quality Function Deployment (Section 17.2)

2.1 Consider the multiple-criteria House of Quality shown below.

  (a) Consider the single-objective QFD approach. Rank all alternatives.

  (b) Consider the multiple-criteria QFD approach. Rank all alternatives where the weights of importance are 0.6 for the customer and 0.4 for the producer.

Design Alt.	Customer's Weights		Customerr's Utility	Technical Requirement				Producer's Utility	Producer's Weights		System's Utility	System's Ranking
	0.4	0.6		Length	Width	Height	Weight		0.7	0.3		
	Customer's Needs								Designer Objectives			
	f_1^C	f_2^C							f_1^P	f_2^P		
1	4	7			⊗	⊗	⊗		7	3		
2	6	3		⊗	⊗		⊗		2	8		
3	5	5		⊗					6	5		
4	2	8			⊗	⊗			9	1		
Ideal	9	9		6	4	2	5		9	9		

2.2 Consider the House of Quality below for a cellular phone. Suppose the weights of importance are 0.7 for the customer and 0.3 for the producer. Consider technical requirements and objectives values given below for each design alternative.

Customer Objectives	Company Objectives	Decision Variables
f_1 = easy to carry	f_1 = low production cost	Weight
f_2 = high battery life	f_2 = ease of manufacture	Battery life (talk time)
f_3 = high reliability	f_3 = ease of assembly	Battery life (standby time)
f_4 = esthetic	f_4 = materials cost	

  (a) Consider the single-objective QFD approach. Rank all alternatives.

  (b) Consider the multiple-criteria QFD approach. Rank all alternatives where the importance given to the customer is 0.3 and to the producer is 0.7.

Design Alt.	Customer's Weights (Product) Customer's Needs (Objectives)				Customer's Utility, U^c	Technical Requirement			Producer's Utility, U^p	Producer's Weights (Process) Producer's Needs (Objectives)				System's Utility, $U^s = 0.6U^c + 0.4U^p$	System's Ranking
	0.4	0.1	0.4	0.1		Weight of Car (lb)	Turning Radius (ft)	Time to Break (m)		0.4	0.1	0.2	0.3		
	f_1^c	f_2^c	f_3^c	f_4^c						f_1^p	f_2^p	f_3^p	f_4^p		
I	8	3	5	4		2.9	4.5	300		3	5	7	2		
II	6	5	3	2		4	5.5	250		8	1	4	6		
III	3	4	7	8		5.8	5.1	450		9	5	6	2		
IV	2	8	4	1		7.5	8.2	700		7	4	3	8		
A	1	3	5	4		2.2	3.1	420		5	3	2	6		
B	2	4	7	1		2.5	2.7	270		6	1	2	1		
Ideal	9	9	9	9		2.5	9	800		9	9	9	9		

Statistical Background (Section 17.3.2)

3.1 Consider Table 17.4 of Example 17.3. Generate the histogram and find the mean and standard deviation:

(a) Bin width of 5

(b) Bin width of 20

3.2 Consider a pomegranate orchard. In the harvest season, suppose one randomly picks up nine pomegranates from nine random locations. The diameter of each pomegranate is presented in the table below. The measurements are based on a continuous scale.

Leaf no.	1	2	3	4	5	6	7	8	9
Length, in.	6	2.8	5.9	6.2	6.6	5.6	6.1	5.9	6.2

(a) Find the average (mean), standard deviation, and range of the sample of pomegranates.

(b) Is it valid to use normal distribution for this problem? Why?

(c) Assume a normal distribution, where the mean of the population and the standard deviation are 5.8 and 1.2, respectively. What is the probability that the diameter of a pomegranate will be (i) exactly 6 in. (ii) less than 5.5 in., or (iii) more than 6.5 in.

Process Control Variable Charts (Sections 17.4.1–17.4.4)

4.1 Consider the following data where there are 10 groups and each group consists of four samples.

	Sample No.			
Group	1	2	3	4
1	5.30	5.20	5.43	5.80
2	5.40	5.20	5.79	5.99
3	5.11	5.21	5.25	5.66
4	5.42	5.30	5.55	5.70
5	5.90	5.43	5.49	5.60
6	5.10	5.33	5.61	5.90
7	5.59	5.14	5.96	5.83
8	5.41	5.66	5.10	5.81
9	5.57	5.91	5.66	5.04
10	5.21	5.92	5.24	5.68

(a) If the population standard deviation is known and it is 0.2, calculate the control limits for a three-sigma $\bar{X}$-chart.

(b) If the population standard deviation is unknown, calculate the control limits for a three-sigma $\bar{X}$-chart.

(c) Draw the $\bar{X}$ and R control chart based on your answer for part (b). Analyze this chart.

(d) If the design upper control limit is 5.81 and the design lower control limit is 5.21, determine the process capability of this system.

4.2 Sampling four pieces of cut wire every hour for the past 12 h has produced the following results for the length of each cut wire

Hour	Average	Range	Hour	Average	Range
1	2.97	0.93	7	3.04	1.00
2	3.78	1.10	8	3.12	0.75
3	3.44	1.18	9	3.10	0.67
4	2.89	0.32	10	3.22	0.35
5	2.99	0.43	11	3.97	0.69
6	3.01	0.90	12	2.83	0.99

(a) If the population standard deviation is known and it is 0.26, calculate the control limits for a three-sigma $\bar{X}$-chart.

(b) If the population standard deviation is unknown, calculate the control limits for a three-sigma $\bar{X}$-chart.

(c) Draw the combined $\bar{X}$ and R control chart based on part (b). Analyze this chart.

4.3 The following table shows the resistance measured in ohms of a resistor from 10 groups where each group consists of five samples:

	Sample No.				
Subgroup	1	2	3	4	5
1	207	191	203	202	201
2	200	189	205	198	203
3	198	201	198	201	202
4	190	201	188	200	199
5	201	205	196	198	199
6	200	203	201	203	198
7	200	198	200	204	210
8	199	199	198	201	201
9	189	200	195	203	199
10	201	211	187	197	196

(a) If the population standard deviation is known and it is 2.6, calculate the control limits for a three-sigma $\bar{X}$-chart.

(b) If the population standard deviation is unknown, calculate the control limits for a three-sigma $\bar{X}$-chart.

(c) Draw the $\bar{X}$ and R control chart based on part (b). Analyze this chart.

Sampling Accuracy for Mean Chart (Section 17.4.5)[†]

4.4 Consider the width of a product produced by a processor. The initial sampling plan measurement indicates the average width is 3.5 in. and its standard deviation is 0.02 in. Suppose 99.74% confidence level is required.

(a) Suppose the initial sampling was conducted on 50 items. What is the percent accuracy of the sampling mean to the true mean of the population?

(b) Suppose the required percent accuracy is 1.5%. How many samples should be used to achieve this accuracy?

4.5 The initial sampling of the diameter of a product indicates that the standard deviation of its diameter is 0.02 in. The maximum acccpetable error is 0.04 in. from the mean of the product.

(a) If the required confidence is 95.44%, how many samples should be taken?

(b) If the required confidence is 99.74%, how many samples should be taken?

Bicriteria Sampling for Mean Chart (Section 17.4.6)[†]

4.6 The mean of the length of a product is 5 in. and its standard deviation is 0.02 in. Suppose that the required confidence is 99.74% and a feasible range for the number of sampling size can range from 30 to 120. The cost of each sample is $12.60. Suppose the weights of importance for the normalized values of the two objective (i.e., cost and inaccuracy) are 0.7 and 0.3, respectively.

(a) Generate three equally distributed multicriteria alternatives.

(b) What is the optimal sampling size?

4.7 For a sampling problem, the sampling cost per item is $1.30, the required confidence is 99.74%, the sample average length of the item is 12 in. and its sample standard deviation is 2.68 in.

(a) Suppose the number of samples can vary from 15 to 75. Generate three efficient bicriteria alternatives equally distributed in terms of the sample size.

(b) Suppose the weights of importance for the cost and inaccuracy are 0.4 and 0.6, respectively. Find the best (optimal) sampling size for the given weights for the additive utility function.

4.8 Consider a sampling plan for a new line of parts. The initial sampling indicates that the average part diameter is 2.5 in. with a standard deviation of 0.3 in. The cost of inspecting each rod is $1.5 and there is a required confidence level of 95.44%. The sample size can range from 20 to 50.

(a) Determine five efficient bicriteria alternatives which are equally distributed in terms of the number of samples.

(b) Suppose the weights of importance for the cost and inaccuracy are 0.5 and 0.5, respectively. Find the best (optimal) alternative for the given weights for the additive utility.

Process Control Capability (Section 17.5)

5.1 In the manufacturing of screws, the natural process has a mean diameter of 0.125 in. and a standard deviation of 0.001 in. The design mean is 0.125. For the following three cases, find the process capability for 3σ. Also determine if this process capability is acceptable.

(a) The upper design tolerance is 0.127 in. and the lower design tolerance is 0.123 in.

(b) The upper design tolerance is 0.129 in. and the lower design tolerance is 0.121 in.

(c) The upper design tolerance is 0.126 in. and the lower design tolerance is 0.121 in.

5.2 In the manufacturing of metal rods, the natural process has a mean length of 36 in. and a standard deviation of 0.20 in. The design mean is 36. For the following three cases, find the process capability for 3σ. Also determine if the obtained process capability is acceptable.

(a) The upper design tolerance is 36.45 in. and the lower design tolerance is 35.75 in.

(b) The upper design tolerance is 36.25 in. and the lower design tolerance is 35.75 in.

(c) The upper design tolerance is 39.75 in. and the lower design tolerance is 32.85 in.

Process Control Attribute Charts (Sections 17.6.1 and 17.6.2)

6.1 Consider the following data for a sample of 10 tires. Each tire was inspected for possible minor defects.

Number, i	1	2	3	4	5	6	7	8	9	10
Number of defects in each tire	5	4	5	6	3	1	4	3	6	7

(a) Calculate the control limits for 99.74%. Use the appropriate control chart.

(b) Draw the control chart. Analyze this chart.

6.2 Consider the following data for the number of minor defects in 10 samples of plastic tarps. The size of each tarp is 20 ft by 20 ft.

Tarp No.	No. of Defects in Each Tarp	Tarp No.	No. of Defects in Each Tarp
1	0	6	5
2	6	7	5
3	7	8	8
4	2	9	9
5	8	10	7

Respond to parts (a) and (b) of exercise 6.1.

6.3 Consider the following data for a sample of 15 buckets where each bucket contains white paint. Each bucket was inspected for different properties. The number of defects per bucket are presented below.

Five-gallon bucket no., i	1	2	3	4	5	6	7	8	9	10	11	12	13	14	15
No. of defects	2	6	2	5	2	3	3	2	0	1	3	5	3	8	4

Respond to parts (a) and (b) of exercise.

6.4 Consider 12 groups of a sampling of tires where each group consists of 250 tires. Each tire was either rejected (defective) or accepted (nondefective). The result of the sampling is provided below.

Group no.	1	2	3	4	5	6	7	8	9	10	11	12
No. of defective tires	1	2	3	4	5	6	7	8	9	6	6	1

Respond to parts (a) and (b) of exercise 6.1.

6.5 Consider a golf shooting range where golfers buy baskets of golf balls. The target is to have 100 golf balls per basket. Twelve baskets were randomly selected. The number of balls in each basket are presented below.

Basket number, i	1	2	3	4	5	6	7	8	9	10	11	12
No. of balls in basket	100	103	97	59	102	105	99	101	107	95	105	95

(a) Based on these observations, compute the three-sigma limits. Use the appropriate control chart. Is the process under control?

(b) Does any basket need to be eliminated? Why? If so, recompute the three-sigma limits after eliminating the out-of-control basket and determine if the process is under control.

6.6 A farmer picks up oranges from his orange orchard where he inspects 150 oranges every day. He either accepts the orange as good or rejects it as bad. Bad oranges are rejected and disposed. See the table below.

Day	Number of bad oranges	Day	Number of bad oranges
1	6	9	0
2	3	10	5
3	6	11	4
4	4	12	5
5	3	13	5
6	4	14	4
7	6	15	6
8	5		

(a) Based on these observations, compute the three-sigma limits. Use the appropriate chart. Is the process under control?

(b) Should any point not be considered? If so, recalculate the three-sigma limits after eliminating the out-of-control points.

Bicriteria Sampling for p-Charts (Section 17.6.3)[†]

6.7 Consider a p-chart sampling problem. The average percent of defective units in a group is 2%. The inspection cost per sampling is $1.24. Suppose the feasible sample size can vary from 10 to 50 and that 99.74% confidence is desired. Suppose the weights of importance for the cost and inaccuracy are 0.2 and 0.8, respectively.

(a) Determine three efficient bicriteria alternatives which are equally distributed in terms of the number of sampling plan.

(b) Find the best alternative for the given weights for additive utility. What is the optimal feasible solution?

6.8 The Laiho Company is looking into developing a sampling plan for a new line of parts using p-charts. The average percent of defective units in a group is 5%. Suppose 68.26% confidence is desired and the inspection cost per part is $2.7 per sampling. The sample size can range from 100 to 500. Suppose the weights of importance for the normalized values of the two objectives are 0.4 and 0.6, respectively. Respond to parts (a) and (b) of exercise 6.7.

Bicriteria Sampling for c-Charts (Section 17.6.4)[†]

6.7 For an initial sampling of a c-chart problem, the sampling average is 1.5. Suppose 99.74% confidence is desired and the sample size can vary from 15 to 45. The inspection cost per unit is $5.9. Assume the weights of importance for the normalized values of the two objectives are 0.6 and 0.4, respectively. Respond to Parts (a) and (b) of Exercise 6.7.

6.8 For a process, the average number of errors per unit is 1.72. Suppose 95.44% confidence is desired and the cost for inspecting each unit is $3.8. The sample size can vary from 40 to 120. Assume the weights of importance for the normalized values of the two objectives are 0.7 and 0.3, respectively. Respond to Part (a) and (b) of Exercise 6.7.

Statistical Distributions for Sampling (Section 17.7.1)

7.1 Consider a sample of 10 items from a lot of 100 items. If more than 2 items out of 10 samples fail, the lot is rejected. It is estimated that the lot tolerance percent defective is 5%.
  (a) Calculate the probability of acceptance using a binomial distribution and the binomial table.
  (b) Calculate the probability of acceptance using Poisson's approximation formula and the Poisson table.

7.2 Consider a sample of 25 screws from a lot size of 250. If more than 4 items out of 25 samples fail, the lot is rejected. It is estimated the lot tolerance percent defective is 10%. Respond to parts (a) and (b) of exercise 7.1.

Operating Characteristic Curve (Section 17.7.2)

7.3 Consider exercise 7.1. Find the operations characteristic curve.

7.4 Consider exercise 7.2. Find the operations characteristic curve.

7.5 A company samples 50 bolts out of a lot size of 750 bolts. If more than 3 bolts are defective, the lot is rejected.
  (a) If an estimated 4% of the samples fail, what is the probability for a lot to be accepted?
  (b) If an estimated 2% of the samples fail, what is the probability for a lot to be accepted?
  (c) Construct the operating characteristic curve of this problem

Producer's and Consumer's Risks (Section 7.7.3)

7.6 Consider a sampling plan where the required average quality level is 0.05 and the lot tolerance percent defective is 0.2. The sampling size is 40. If there are five or more defective items in a sample, the lot is rejected.

(a) Find the producer's and consumer's risks. What do they mean?

(b) If rejecting a good lot costs $10,000 to the producer and accepting a bad lot costs $12,000 to the consumer, what are the expected losses to the producer and the consumer?

7.7 A semiconductor company produces a special type of microprocessor. The company has established the value of 0.02 for acceptance quality level and the value of 0.15 for lot tolerance percent defective. Each lot consists of 300 items, and 50 items from each lot are randomly selected for inspection. If there are 6 or more defective items in a sample, the lot is rejected. Compute the producer's risk and consumer's risk. What do they mean?

Finding Optimal Sampling Size (Section 7.7.4)[†]

7.8 A producer and a consumer have agreed upon the following conditions. Acceptance quality level is 0.04, producer's risk is 0.12, lot tolerance percent defective is 0.25, and consumer's risk is 0.07.

(a) Which of three sampling plans, (a) (15,1), (b) (20,2), and (c) (30,3), is the best fit for this problem? (*Hint:* Find the OC curve for each alternative and graphically choose the closest best fit.)

(b) What sampling plan do you recommend?

Average Outgoing Quality (Section 17.8.1)

8.1 Consider a sampling plan where the lot size is 2000, the sampling plan sample size is 50, and the rejection level is 2.

(a) Find the average outgoing quality where the percent defective is 4%.

(b) Suppose the percent defective ranges from 1 to 15%. What is the average outgoing quality level for this problem? At what percent defective does the average outgoing quality level occur? What does it mean?

8.2 Consider a sampling plan where the sample size is 150, the lot size is 1000, and the rejection level is 8.

(a) What is average outgoing quality for the sampling plan if the percent of defective is about 5%?

(b) Find the average outgoing quality level for the sampling plan where the percent defective can range from 1 to 8%.

Bicriteria Acceptance Sampling (Section 17.8.2)

8.3 Suppose the cost of inspection per item is $2.8, lot size is 1200, and acceptance quality level is 4%. The sample size cannot be less than 2 and cannot be more than 40. Consider two criteria of cost and average outgoing quality.

(a) Consider three alternatives where sample size can be 20, 30, and 40. Generate the bicriteria alternatives for this problem. Are they efficient?

(b) Suppose the weights of importance for the cost of inspection and average outgoing quality are 0.2 and 0.8. Find the best alternative for additive utility?

8.4 Suppose the cost of inspection per item is $3.7, lot size is 300, and average quality level is 5%. Sample size can range from 10 to 30. Consider two criteria of cost and average outgoing quality.

(a) Consider three alternatives where sample size can be 10, 20, and 30. Generate the bicriteria alternatives for this problem. Are they efficient?

(b) Suppose the weights of importance for the cost of inspection and average outgoing quality are 0.5 and 0.5. Find the best alternative for additive utility?

Sequential Sampling Plans (Section 17.8.3)

8.5 Consider sequential sampling for four levels. The number of defects and rejection levels are listed below. For each level, determine if the lot should be accepted, rejected, or considered for the next level.

Sample Level, i	Number of Defects, x_i	$c_{i,min}$	$c_{i,max}$
1	4	2	6
2	5	7	10
3	2	10	12
4	3	12	14

8.6 Consider sequential sampling for six levels. The number of defects and rejection levels are listed below. For each level, determine whether the lot should be accepted, rejected, or considered for the next level.

Sample Level, i	Number of defects, x_i	$c_{i,min}$	$c_{i,max}$
1	5	3	7
2	2	6	9
3	4	9	12
4	1	11	14
5	3	14	17
6	8	17	22

CHAPTER 18

WORK MEASUREMENT

18.1 INTRODUCTION

The majority of mass production assembly industries use work measurement approaches to design and improve the assembly of components and also to achieve synchronized and smooth production. Even in the modern mass production assembly systems, manual assembly is still a significant part of the work and cost. Work measurement methods have applications in manufacturing (e.g., manual assembly line systems, machining, and sewing), operations, service industries, education, and construction managements. In service industries work measurement is applied to customer service representatives, technical support specialists, banking tellers, banking specialists, cleaning services, restaurants, butcher shops, and delivery services. The following areas are closely related to work measurement:

- Assembly systems design and line balancing
- Quality control and total quality improvement
- System productivity and efficiency

Each of the above topics is covered in different chapters of this book. Work measurement methods can also be used to measure and improve the productivity of individual workers. These methods can typically predict how long it should take for an average worker to perform a particular task. By establishing labor standards and analyzing these standards,

Note: Advanced material that can be omitted without loss of generality will be indicated by a dagger.

Operations and Production Systems with Multiple Objectives, First Edition. Behnam Malakooti.
© 2014 John Wiley & Sons, Inc. Published 2014 by John Wiley & Sons, Inc.

management can continuously improve productivity, efficiency, quality, and operational costs. Labor standards are commonly grouped into two types:

- Engineered standards that use established work measurement principles. These standards define the time required by a qualified worker to complete a specified task under normal conditions of fatigue, supervision, delay, and quality demand.
- Nonengineered standards for jobs/tasks for which engineered standards do not exist or apply.

Using engineered standards is much less expensive than developing nonengineered standards. In deciding which type of work measurement systems to use, a balance among effective work measurement, quality of measurement, and cost of the system must be considered. Management must then consider the benefits of each type of work measurement system and decide which system will produce the most accurate results at the lowest cost and the lowest duration.

Work measurement methods are based on statistical approaches to estimate the true (mean) value of the job time. As the number of observations (e.g., the number of samples) increases, the accuracy of estimation and the confidence level in the analysis increase, but at the same time, the cost and duration of the time study increases. Therefore, in selecting the number of samples, one should consider the following objectives:

- Minimize the total cost of sampling.
- Maximize the desired level of confidence in the study.
- Maximize the desired degree of accuracy to be close to the true (mean) values.

Historical Background of Work Measurement   The history of work measurement is covered in Chapter 1.

Productivity and Performance Measurement   Work measurement can be used to measure productivity of individual workers. It can also be used to establish the work standard time of a given job which allows for needed extra time allowances and performance ratings compared to an average worker. Therefore, the productivity of a worker (e.g., the number of parts produced per shift) can be compared to the productivity based on the work standard time of the job. Many corporations and firms focus on improving productivity and reducing cost where manual labor constitutes a significant percentage of the total cost of producing a product or providing a service. Work measurement is a scientific approach for measuring productivity and identifying a fair and acceptable amount of time needed to perform a specific task. Decisions such as the development of a new product, purchasing new equipment, and setting up due dates can all be based on the estimation of productivity and work measurement. In addition to this, productivity and standard work measurements can be used to set proper wages and incentives. A core problem in work measurement is to define universal standards that apply to the whole labor force and not just a few productive or unproductive workers. Workers are then expected to meet the given standards.

Overview of Work Measurement Methods to Find Standard Time   Standard time refers to a constant time during which a certain task should be performed while maintaining a certain level of quality and work flexibility. Some organizations use their past experiences to find standard times for their operations. This technique has the advantages of minimal costs and minimal time to implement. However, this method does not promote improvement and innovation and is typically inaccurate and not scientifically based. The pace of work defined by these standards may be slower than what could possibly be achieved. This technique is worth mentioning for background as one of the origins of work measurement. The three scientific approaches for work measurements are overviewed below and will be discussed in more depth in each section of this chapter:

1. *Time studies* (covered in Section 18.3) are used when the cycle time to produce an item is relatively short. An experienced observer is chosen to define starting and ending points, break down the work into basic work elements, measure the time to perform each element, and rate experienced operators over several trials. This technique is considered a continuous-time technique because the process is observed continuously during an operation.

2. *Work samplings* (covered in Section 18.6) are used when the cycle time to produce an item is relatively long and the process has complex elements. It is also used when labor unions simply will not permit continuous observations of the workers. In this situation, stop-watch methods are not used. Random observations of a person, machine, or process are made by an observer for a set of activities that constitutes a job assigned to a worker. At each random visit, the observer marks which activity is being performed. After a sufficient number of observations, the percentage of time that a worker is performing an activity can be measured. This method is considered discrete time because data are not gathered during continuous observations.

3. *Predetermined time standards* (covered in Section 18.7) are usually used in mass production systems where each simple task can be broken down into its simplest moves such as reach, lift, and place for predetermined distances and weights. Currently, there are several established databases. By combining (adding) the time of simple moves, the time for a simple task can be predetermined.

Relationship to Quality Control Statistical Methods   In Chapter 17 we discussed two types of measurement variables (for continuous observations) and attributes (for discrete observations). The statistical methods of work measurement are based on Chapter 17. Specifically:

- Variables methods (Section 17.4) are used in Sections 18.3, 18.4, and 18.5.
- The attribute method (*p*-chart) (Section 17.6.1) is used in Section 18.6.

18.2 WORK ANALYSIS

18.2.1 Analysis of Methods

Analysis of methods deals with evaluating how a task is completed. One can think of analysis of methods as *ergonomics*, which means "the study of work." Analysis of methods can be conducted at a general level and then can be evolved into a more detailed analysis. A technical way of examining analysis of methods is the process of converting inputs to outputs. The first step of analysis is to identify the overall input and overall output of the system. The next step is to break the system down into subcomponents and isolate the

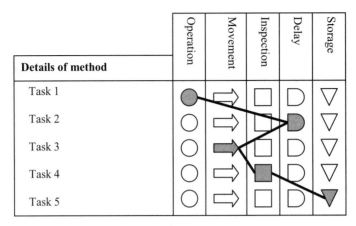

Figure 18.1   Process chart for five tasks.

different inputs and outputs of each subcomponent. Then, we try to identify the efficiency of each subcomponent and how each can be improved. Analysis of methods can arise when there is a need to save costs to a business, a factory, a government, and so on. Analysis of methods can also be used to improve safety, quality, and productivity. Analysis of methods should be an ongoing process to keep the system lean and efficient.

Process Charts   Analysis of methods often includes a process chart depicting symbols that illustrate the flow of a process. A list of conventional symbols is presented in Figure 18.1. For example, a circle depicts any operation such as driving a nail, mixing materials, and word processing. Similarly, an arrow depicts any form of transportation. Using these symbols, one can create a process flow chart with each row showing one task. The process chart starts out with an operation, then there is a delay, then a transportation, then inspection, and eventually the process ends with storage. After the process flow chart is analyzed, it can be modified to make it more efficient. After the operation system is altered, adjusting to the new operation system may take some time since workers may need to be trained to use the new operation system and training them may take some time. As a result, the system may not be functioning as efficiently during the initial stages of use. There is a learning curve that occurs. Understanding and quantifying the learning curve are important aspects of analysis of methods. Learning curves are covered in Chapter 1. A process chart allows the engineer to view the entire process and discover inefficiencies. For example, an unnecessary inspection or delay may be discovered or similar activities may be combined. Figure 18.1 shows a simple example of a job that consists of five tasks. These tasks are performed sequentially. For example, after task 1 is completed, the job is delayed (presented as task 2). Then it is moved (task 3) and inspected (task 4). Finally it is stored (task 5). In real-world applications, a job (e.g., manufacturing a part) may consist of hundreds of tasks for which identifying and understanding the flow process can be very complicated.

18.2.2   Study of Motion

The study of motion involves studying one specific process on the process flow chart. Simply put, the study of motion involves studying a specific manual motion, that is, a "circle," on the process flow chart. The study of motion can involve recording the process on a video, audio medium, digital camcorder, and/or digital sound recorder. Moreover, for smaller motions, a microscopic-level recorder can be used. Additionally, one can use micromotion study to analyze motions that are too rapid to see with the naked eye. An example of a micromotion study would be a slow-motion replay of a swimmer's stroke to determine whether there are any inefficient movements or improvable movements. Any minor improvement in the mechanics of the swimmer's stroke may result in a drastic overall improvement in time over the course of an entire event.

Often, charts are used to study motions. A popular chart used to study motion is a simultaneous motion (SIMO) chart. SIMO charts are valuable for professionals that use hands as a vital part of their work, such as surgeons, tailors, typists, video game players, pilots, dentists, and pianists. An example of a SIMO chart is shown at http://mcu.edu. tw/~ychen/op_mgm/notes/part3.html.

The objective of studying motion is to find more efficient ways for a person or machine to move certain parts. More specifically, studying motion involves determining if there is any wasted motion, fatigue from certain types of motion, altering motions to be more efficient, joining motions, improving the equipment used in the motion, and/or improving the layout of the facility. Proper improvement of motion can result in higher productivity, less fatigue, lower costs, increased safety, and higher quality.

18.3 STANDARD TIME

18.3.1 Stopwatch Studies

Historically, stopwatch time studies were the first and most accepted work measurement system. As discussed earlier, time study procedures were first proposed by Frederick W. Taylor (1883) to perform work measurement by direct observation of workers performing jobs. Although time studies have been popular with management and industrial engineers, labor unions have presented resistance to its implementation. In industries with strong labor unions, approval of the worker and their supervisor is required before time studies can be performed. A problem associated with this requirement is that random observations, which are at the core of statistically valid work measurement, cannot be conducted.

Before using the stopwatch method, one must identify and define the scope of the job, the worker's tools, workplace layout, materials, and work elements of the job being observed. Then, an experienced operator should be selected for the observation. In this method, the job should be divided into the simplest tasks that are controlled by the worker.

There are four different types of work elements:

- *Operator versus Machine-Controlled Work Elements* The time it takes for a person to set up a machine versus the time it takes for the machine to complete a task. This is the most common type of work element.
- *Regular versus Irregular Work Elements* Regular work elements occur in each cycle, but irregular work elements occur without a pattern. For example, changing a broken drill bit is an irregular element.
- *Constant versus Variable-Duration Work Elements* Repetitive identical tasks have constant duration, but tasks that deal with varying size, shape, and weight have variable durations.

- *Accidental versus Nonaccidental Work Elements* Accidental work elements have unpredictable delays caused by factors such as power failures, machine failures, and defective materials.

18.3.2 Standard Time Formulas

A job or work consists of many tasks which are also referred to as work elements. To find the standard time of a job, first find the standard time for each of its work elements. Then, measure the standard time of a job by adding the standard time of all its work elements (tasks). For each work element (task), measure its (a) time study, (b) selected time, (c) normal time, and (d) standard time by incorporating the needed allowances. Each of the above work measurement times are described below.

(a) *Time Study* Time study is the time for performing a specified task. Time studies can be obtained by three different methods: (a) conducting stopwatch studies, (b) using the given databases (tables) for the predetermined time standards (covered in Section 18.7), or (c) conducting sampling plans (covered in Sections 18.4–18.6). In all methods, all superfluous, extraneous, out-of-range, and erroneous time (data) should be removed from the data set.

(b) *Selected Time or Average Actual Cycle Time (AACT)* $\bar{t}_j$ For a given task j, the selected time is the sum of all its measured times divided by the number of observed cycles, n:

$$t_{ij} = \text{time observed for task } j \text{ in cycle } i \quad \text{for } i = 1, 2, \ldots, n \text{ (observations)}$$

$$\bar{t}_j = \frac{1}{n}\sum_{i=1}^{n} t_{ij} \qquad\qquad \text{for } j = 1, 2, \ldots, m \text{ (tasks)}$$

$$\text{(18.1)}$$

(c) *Normal Time (NT)* Normal time is the average actual cycle time multiplied by a percentage representing the operator's performance rating (R_j) for a task j. This number is assigned by the observer for the given task. Performance ratings of greater than 100% mean that the operator completed the task faster than the normal time whereas ratings less than 100% mean that the operator completed the tasks slower than the normal time:

$$\text{NT}_j = \bar{t}_j R_j \quad \text{for } j = 1, 2, \ldots, m \qquad (18.2)$$

The same task conducted by different operators may have different performance ratings but will have the same normal time.

(d) *Standard Time (ST)* Standard time is calculated by considering the needed allowances for personal time, delays, and fatigue for a given task:

$$\text{ST}_j = \frac{NT_j}{1 - \text{allowance factor for task } j} \quad \text{for } j = 1, \ldots, m \qquad (18.3)$$

These allowances are measured as a sum of percentages, usually defined by company policy. The allowance factor is a number less than unity, usually between 0.05 and 0.2.

The standard time for a job that consists of m tasks is the sum of all its standard times:

$$ST = \sum_{j=1}^{m} ST_j \qquad (18.4)$$

If the allowance factors are the same for all different tasks, then use

$$ST = \frac{\sum_{j=1}^{m} NT_j}{1 - \text{allowance factor}} \qquad (18.5)$$

Personal, Fatigue, and Delay (PFD) Allowance   Different businesses or industries have different PFD allowances for different tasks. There are also standard tables, such as Table 18.1 presented, that can be used.

Consider Table 18.1. For example, in "1. constant allowances," 5% for personal allowance and 4% for basic fatigue are allowed, that is, a total of $5\% + 4\% = 9\%$ is a standard allowance. However, in "2. Variable allowances," the total allowance should be calculated based on applicable conditions in the table. For example, the allowance for a task that consists of "(b) (ii) very awkward," "(c) 60 pounds," and "(d) (i) bad light (well below recommended)" is $7\% + 17\% + 2\% = 26\%$.

18.3.3 Standard Time Procedure

Time studies are usually conducted with a handheld computer-based (or palm) timing device and/or camera. The observer records the time when each task starts and completes

TABLE 18.1   Rest Allowances (%)

Personal, Fatigue, and Delay Allowances			
1. Constant allowances:			
(a) Personal allowance......................	7	(b) Fatigue allowance......................	5
2. Variableallowances:			
(a) Standing allowance......................	1	(f) Close attention:	
		(i) Average......................	3
		(ii) Extreme......................	6
(b) Abnormal position allowance		(g) Mental strain:	
(i) Averageawkward......................	2	(i) Complex attention	4
(ii) Extreme awkward (lying,..)	7	(ii) Verycomplex......................	10
(c) Lifting pulling, pushing Weightlifted (pounds):		(h) Noise level:	
		(i) Intermittent—loud......................	2
25...		(ii) Intermittent—very loud or high.....	
45...	3		6
60...	9		
	17		
(d) Bad light:		(i) Tediousness:	
(i) Below average......................	2	(i) Tedious......................	2
(ii) Inadequate...................................	5	(ii) Verytedious......................	5
(e) Atmospheric conditions Heat............................... Humidity......................	0–10		

for each given cycle in a continuous and uninterrupted manner. In each cycle, all different tasks are observed. To find the standard time, the following steps should be taken:

Step 1 Determine the number of cycles (sample size) n to be observed. Make an educated guess at this number or use the next section to find it.

Step 2 For each task, identify the cycle at which the time is out of the normal range. This can be identified by measuring the difference of a given task time value from its average. Remove the task time which is out of the normal range.

Step 3 For each task j, measure its $\bar{t}_j$, the average time for completing the task based on the n observations.

Step 4 For each task j, find its normal time NT_j using its performance rating R_j.

Step 5 For each task, j, find its standard time ST_j, by considering allowance factors.

Step 6 Find the standard time for the job by finding the summation of the standard times of all tasks.

Example 18.1   Finding Standard Time   Consider a job that consists of three tasks. Each task is performed by a different operator. The results of six observations and their performance ratings for each operator are shown in Table 18.2. For all tasks, the allowance factors are 5% for personal allowance, 4% for basic fatigue, and 5% for very tedious task. Find the standard time for this job.

Consider each task. For task 1, the value of 20 for cycle 4 is significantly different from the other cycle times. This value should be omitted and not used for calculations. Therefore, only five cycles will be used for the first task, that is, $n = 5$.

Now calculate the average (selected) time for each task. For example, for first task

$$\bar{t}_1 = \frac{14 + 13 + 14 + 14 + 13}{5} = \frac{68}{5} = 13.6$$

Since each task is performed by a different worker, convert the selected time to normal time. To find the standard time for each task, first find the total allowance factor. The total allowance is $5\% + 4\% + 5\% = 14\%$. Therefore, use 0.14 as the allowance factor. The standard time for job 1 is

$$ST_1 = \frac{NT_1}{1 - \text{allowance factor}} = \frac{14.96}{1 - 0.14} = \frac{14.96}{0.86} = 17.4 \text{ min}$$

TABLE 18.2   Observed Times (min) for each of Three Tasks Each of Six Cycles

		Cycle Observed (i), Minutes, t_i						
j	Task j for Cycle i	1	2	3	4	5	6	Percent Performance Rating, R_j
1	Assemble a product	14	13	14	20[a]	14	13	110%, or 1.1
2	Package the product	2	2	3	2	2	3	100%, or 1
3	Store the product	0.5	1	1	0.75	0.5	1	95%, or 0.95

[a]This value is significantly different from the other cycle times.

TABLE 18.3   Calculations for Example 18.1

Task, j	$\bar{t}_j$	R_j	$NT_j = \bar{t}_j R_j$	$ST_j = NT_j/(1 - 0.14)$
1	13.6	1.1	$13.6 \times 1.1 = 14.96$	$14.96/0.86 = 17.4$
2	2.33	1.0	$2.33 \times 1 = 2.33$	$2.33/0.86 = 2.7$
3	0.79	0.95	$0.79 \times 0.95 = 0.75$	$0.75/0.86 = 0.87$
Total	16.72	—	18.04	20.98

The calculations for standard time of all three tasks are presented in Table 18.3. The standard time for this job is

$$\text{ST} = 17.4 + 2.7 + 0.87 = 20.98 \text{ min/job}$$

18.4   SAMPLE SIZE FOR STANDARD TIME

18.4.1   Finding Sample Size

The initial n observations (cycles) can be used to find N, the sample size for a needed accuracy. As n increases, a better estimation for N can be obtained. The number of samples needed, N, is determined based on the desired accuracy, acceptable confidence level, and initial n observations. Because a job consists of m different tasks, the number of samples for each task, N_j, should be determined. Then, the sample size for all tasks will be set equal to the maximum value of all N_j, that is, $N = \text{Max}\{N_j, j = 1, 2, \ldots, m\}$.

Assume the following information:

n = initial observation sample size

z = desired confidence level

h = desired degree (level) of accuracy, percent deviation ($\pm\%$) of sample mean

t_{ij} = time measured to complete task j in cycle i where $i = 1, \ldots, n, j = 1, \ldots, m$

We can find $\bar{t}_j$ = mean value of task j and then

$$s_j = \sqrt{\frac{\sum_{i=1}^{n}(t_{ij} - \bar{t}_j)^2}{n - 1}} \quad \text{(standard deviation for task } j) \tag{18.6}$$

$$N_j = \left(\frac{zs_j}{h\bar{t}_j}\right)^2 \quad \text{(number of samples to be decided for task } j) \tag{18.7}$$

The sample size for the entire job is the maximum N_j, that is,

$$N = \text{Max}\{N_1, N_2, \ldots, N_m\}$$

It can be observed that the maximum N_j (labeled as N) is associated with the highest $s_j/\bar{t}_j$ ratio because the values of z and h are constant for all tasks, $j = 1, \ldots, m$. For the chosen element, use $\bar{t}$ and s associated with this work element.

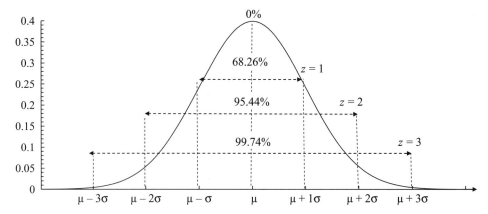

Figure 18.2   General normal distribution with mean μ, standard deviation σ, and confidence levels for $z = 1, 2, 3$.

Alternative Approach for Finding Sampling Size   There is an alternate formulation for Equation (18.7) where $h\bar{t}_j$ can be replaced with the maximum acceptable error, which is denoted by e_j:

$$N_j = \left(\frac{zs_j}{e_j} \right)^2 \tag{18.8}$$

Therefore, when the average time $\bar{t}_j$ is not known, one can assess the maximum acceptable error e and find the needed sample size. Note that, by definition, $e_j = h\bar{t}_j$.

Desired Level of Accuracy (h)   Let h be the maximum allowable percentage error for which the estimated mean value can deviate from its true value μ. As N increases, a better estimation of the true mean μ can be found.

Desired Level of Confidence (z)   The desired level of confidence (z) represents the percentage of data that will fall within the expected range of $\mu \pm zs$ where μ is the true mean and s is the standard deviation. Higher confidence levels require a higher sample size N. For example, if $z = 3$, then 99.74% of the data will fall within ± 3 standard deviations of the mean for the normal distribution. Figure 18.2 provides an example for three values of z.

Table 18.4 shows some examples of z values at several desired confidences. For example, a 95% confidence interval has a z value of about 1.96. The normal distribution table at the back of the book can be used to find confidence levels for all values of z.

TABLE 18.4   Selected z Values for Desired Confidence Levels

z	0	0.5	1	1.5	2	2.5	3
Confidence level	0%	38.30%	68.26%	86.64%	95.44%	98.76%	99.74%

Example 18.2   Finding Sample Size N   Consider Example 18.1. Suppose that the DM requires the following values:

Desired confidence level of 99.74% and therefore $z = 3$
Desired level of accuracy of $\pm 5\%$ of the sample mean, $h = 0.05$
Initial number of cycles observed in Example 18.1, that is, $n = 6$

Find the needed sample size for this problem.

In Example 18.1, the values of $\bar{t}_j$ were calculated as $\bar{t}_1 = 13.6$, $\bar{t}_2 = 2.33$, and $\bar{t}_3 = 0.79$.

First, find the standard deviation for each task j, s_j. Using the standard deviation formula in Equation (18.6), the following values can be calculated:

Task 1	Task 2	Task 3
$s_1 = \sqrt{\dfrac{\sum_{i=1}^{5}(\bar{t}_{i,1} - 13.6)^2}{5-1}}$	$s_2 = \sqrt{\dfrac{\sum_{i=1}^{6}(\bar{t}_{i,2} - 2.33)^2}{6-1}}$	$s_3 = \sqrt{\dfrac{\sum_{i=1}^{6}(\bar{t}_{i,3} - 0.79)^2}{6-1}}$
$= 0.548$	$= 0.516$	$= 0.246$

Then, the needed sample size for each task j, N_j, is found:

Task 1	Task 2	Task 3
$N_1 = \left[\dfrac{3 \times 0.548}{0.05 \times 13.6} \right]^2$	$N_2 = \left[\dfrac{3 \times 0.516}{0.05 \times 2.33} \right]^2$	$N_3 = \left[\dfrac{3 \times 0.246}{0.05 \times 0.79} \right]^2$
$= 5.85$	$= 176.56$	$= 349.07$

Therefore, the sample size N is

$$\text{Max}\{N_1, N_2, N_3\} = \{5.85, 176.56, 349.07\} = 349.07 \quad \text{(or 350)}$$

In this example, 350 samples are needed to satisfy the accuracy of 5% and the confidence level of 99.74%. A summary of the above calculations is presented in Table 18.5.

See Supplement 18.1.xls.

TABLE 18.5   Summary of Results for Example 18.2

Task, j	\multicolumn{6}{c}{Cycle Observed, Minutes, t_j}	$\bar{t}_j$	s_j	N_j
	1	2	3	4	5	6			
Task 1	14	13	14	20[a]	14	13	13.6	0.548	5.85
Task 2	2	2	3	2	2	3	2.33	0.516	176.56
Task 3	0.5	1	1	0.75	0.5	1	0.79	0.246	349.07

[a]Maximum observed time in minutes.

Example 18.3   Finding Sample Size *N* Using Alternate Formulation   Consider three tasks with standard deviations of 0.548, 0.516, and 0.246. Also, the maximum acceptable errors, for the three tasks are 0.5, 0.1, and 0.05, and the required confidence level for all tasks is 99.74%.

The needed sample size for this problem can be found using Equation (18.8):

Task 1	Task 2	Task 3
$N_1 = \left[\dfrac{3 \times 0.548}{0.5}\right]^2 = 10.81$	$N_2 = \left[\dfrac{3 \times 0.516}{0.1}\right]^2 = 239.60$	$N_3 = \left[\dfrac{3 \times 0.246}{0.05}\right]^2 = 217.86$

Therefore, the sample size *N* is

$$\text{Max}\{N_1, N_2, N_3\} = \{10.81, 239.6, 217.86\} = 239.6 \quad (\text{or } 240)$$

18.4.2   Finding Accuracy of Given Sample Size

To find the accuracy of a given sample size *N*, calculate *h* using Equation (18.7):

$$h = \frac{zs_j}{\sqrt{N_j}\bar{t}_j} \quad \text{for task } j \tag{18.9}$$

Example 18.4   Finding Accuracy for Given Sample Sizes   Consider Example 18.2. Find the accuracy if the sample size is:

$N = 6$ (i.e., the initial sample size)
$N = 239.60$ (as obtained in Example 18.3)
$N = 1000$

In Example 18.2, the sample size $N = 349.07$ was associated with task 3. Therefore, task 3 with mean $\bar{t} = 0.79$, confidence level $z = 3$, and standard deviation $s = 0.246$ is used for the analysis.

(a) For $N = 6$, the accuracy is

$$h = \left[\frac{3 \times 0.246}{\sqrt{6} \times 0.79}\right] = 0.380 \quad (\pm 38.0\%)$$

Therefore, if we use $N = 6$, we will have an accuracy of $\pm 38.0\%$, which is a very poor accuracy.

(b) For $N = 239.60$,

$$h = \left[\frac{3 \times 0.246}{\sqrt{239.60} \times 0.79}\right] = 0.06 \quad (\pm 6\%)$$

(c) Find *h* for the given $N = 1000$:

$$h = \left[\frac{3 \times 0.246}{\sqrt{1000} \times 0.79}\right] = 0.03 \quad (\pm 3\%)$$

For $N = 1000$, we will have an accuracy of $\pm 3\%$, which is much better than $N = 6$ with an accuracy of $\pm 38\%$ and also better than $N = 348.52$, which has an accuracy of $\pm 5\%$. It should be observed that $348.52 - 6 = 342.52$ additional samples improved the accuracy by $0.38 - 0.05 = 0.33\%$. But having $1000 - 348.52 = 651.48$ additional samples only improves the accuracy by $5 - 2.9 = 2.1\%$. In the multiple-criteria method of the next section, the trade-offs among the cost of sampling, its accuracy, and its confidence level are discussed.

Example 18.5 Comprehensive Time Study Example   Consider a time study for assembling a package. An observation of nine cycles is provided in Table 18.6.

Note that the bold numbers for tasks 1 and 4 are ignored as they greatly differ from other values. Tasks 4 and 7 are machine controlled.

To find the optimal sample size, the following information is also given:

- Desired level of accuracy is $\pm 1\%$, $h = 0.01$.
- Desired level of confidence is 95.44%, $z = 2$.

Using the above information, find:

(a) Standard time for each task and the entire job
(b) Find the needed number of samples
(c) Find the accuracy of the given nine samples

Breakdown of Job into Its Tasks

1. Place an empty crate under the conveyor belt on the scale.
2. Remove the lid and place it in the holder position.
3. Push the Tare button to zero the scale. (Tare refers to the weight of an empty container.)
4. Push the start button to begin transporting finished material down the conveyor belt into the crate.
5. Press the stop button when the material in the crate is at the desired weight.
6. Place lid on crate.
7. A robot removes the crate.

The time study observation sheet is provided below.

Time Study Observation Sheet

Operation:	Package finished product
Operator name:	John Doe I
Operator number:	49
Job experience:	10 Yearson this line
Part no.:	91464AJ
Part name:	Trico
Equipment name:	Banbury
Equipment number:	5
Analyst:	John Doe II
Material:	Polymer Wig Wag

Continued:

Begin Time	10:00
Finish Time	11:15
Elapsed Time	1:15
No of Machines Operated	1
Units Finished	9

TABLE 18.6   Time Study Observation Results of Nine Cycles and Seven Tasks

Task	Observed Time per Product, min									n	Performance Rating ($R\%$)
	1	2	3	4	5	6	7	8	9		
1	0.62	0.6	0.58	**4**	0.57	0.6	0.62	0.56	0.58	8	120
2	0.15	0.13	0.14	0.16	0.16	0.18	0.16	0.17	0.18	9	110
3	1	0.8	0.9	0.9	0.8	0.9	0.9	0.8	0.9	9	115
4^a	4	4	4	4	**15**	4	4	4	4	8	100
5	0.15	0.18	0.16	0.12	0.18	0.2	0.16	0.14	0.18	9	110
6	0.4	0.41	0.49	0.48	0.41	0.42	0.48	0.49	0.43	9	90
7^b	0.25	0.25	0.25	0.25	0.25	0.25	0.25	0.25	0.25	9	100

^{a}Machine controlled.

The following observations are made for this example:

- All of the assembly components will be in designated places and all machines are running normally.
- Elements 1, 2, 3, 5, and 6 are operator-controlled elements, and therefore performance ratings will be applied to them. Elements 4 and 7 are machine-controlled elements that have a standard time as provided by the machine manufacturer.
- The number of cycles that was considered for the initial sample planning was nine, that is, $n = 9$.
- Table 18.6 shows the values for each element during each cycle. In Table 18.6, two measurements are highlighted in bold for tasks 1 and 4. The first extraneous time, in cycle 4 of task 1, was due to the operator's waiting for the stock of empty crates to be replenished; this element qualifies as an accidental element and should be ignored. The second extraneous time, in cycle 5 of task 4, was due to a machine power failure that required replacement of circuit breakers and restarting of machine; this also qualifies as an accidental element and should be ignored.

For all elements, the same following allowances are used:

- Personal allowance $= 5\%$
- Basic fatigue allowance $= 4\%$

The operator performance ratings for each task are provided in the last column of Table 18.6. Note that for tasks 4 and 7 the ratings are 100% as they are performed by machines.

(a) First we find the average actual cycle time for each task, $\bar{t}_j$. Note that due to the removal of accidental time elements, the number of cycles for tasks 1 and 4 are 8 instead of 9. Then find the normal time and the standard time for each task. Since the allowance is $5\% + 4\% = 9\%$, use $1 - 0.09$ to calculate the standard times. These values are presented in Table 18.7.

The standard time for each part is

$$ST = \frac{\sum_{j=1}^{7} NT_j}{1 - \text{allowance factor}} = \frac{6.72}{1 - 0.09} = 7.39 \text{ min}$$

TABLE 18.7   Work Measurement Analysis of Example 18.5

		Part (a) Solution				Part (b) Solution	
Task, j	n	Average Actual Cycle Time, $\bar{t}_j$	Performance Rating, R_j (%)	Normal Time, $NT_i = \bar{t}_j R_i$	Standard Time, $ST_j = NT_j/(1-0.09)$	Standard Deviation s_j	Sample Size, N_j
1^a	8	0.59	120	0.71	0.78	0.022	56
2	9	0.16	110	0.17	0.19	0.017	453
3	9	0.88	115	1.01	1.11	0.067	231
4	8	4.00	100	4.00	4.40	0.000	0
5	9	0.16	110	0.18	0.20	0.024	900^a
6	9	0.45	90	0.40	0.44	0.038	285
7	9	0.25	100	0.25	0.27	0.000	0
Totals	—	6.49	—	6.72	7.39	—	Max. = 900

^{a}needed sample size.

(b) First, find the standard deviation for each task using Equation (18.6). These values are presented in Table 18.7. For example, the standard deviation for task 1 is

$$s_1 = \sqrt{\frac{(0.62-0.59)^2 + (0.6-0.59)^2 + \cdots + (0.58-0.59)^2}{8-1}}$$

$$= \sqrt{\frac{0.0035}{7}} = 0.022$$

Now, the values of z, h, s_j, and $\bar{t}_j$ and Equation (18.7) are used to calculate the sample size for each task, N_j. These values are also shown in Table 18.7.

For example, the calculation of N_1 for the first work element is presented as

$$N_1 = \left(\frac{2s_1}{0.01\bar{t}_1}\right)^2 = \left(\frac{2 \times 0.022}{0.01 \times 0.59}\right)^2 = 56$$

Now, find the needed sample size by using the maximum of the sample sizes for all the tasks:

$$\text{Max}\{56, 453, 231, 0, 900, 298, 0\} = 900$$

Therefore $N = 900$ is needed to achieve the $\pm 1\%$ level of accuracy.

(c) To find the accuracy (h) of the existing initial sample size $N = 9$, use Equation (18.8) for the given $z = 2$. Also, use the mean and standard deviation associated with the highest sample size, N, that is, use task 5 with values of $\bar{t} = 0.16$ and $s = 0.024$:

$$h = \frac{2 \times 0.024}{\sqrt{9} \times 0.16} = 0.1 \quad (\text{or } \pm 10\%)$$

See Supplement 18.2.xls.

18.5   MULTICRITERIA SAMPLING FOR STANDARD TIME

18.5.1   Generation of Set of Efficient Alternatives[†]

When choosing a sample size N, the two objectives are:

Minimize total cost of sampling, $f_1 = CN$
Minimize percent error, $f_2 = h$ [see Equation 18.8)]

Consider the following notation for the multicriteria problem. Note that C, total cost for observation of one cycle, is defined as the sum of labor and equipment expenses incurred per observation.

Given	Find
n = initial sample size	f_1 = total cost of sampling = CN
C = cost per observation for one cycle	f_2 = percent error of accuracy =
s = standard deviation of initial sample size	$h = zs/(\sqrt{N}\bar{t})$
N_{min} = minimum sample size	N = sample size
N_{max} = maximum sample size	
z = confidence level	

The following are the steps for generating a set of efficient multicriteria sampling plan alternatives:

1. Consider a number of given values of N. (For example, generate five discrete alternatives of N that are equally distributed, from N_{min} to N_{max}.)
2. Measure the values of the two objectives f_1, and f_2 for each given N. For nonlinear utility functions, the optimal solution can be found by enumerating all values of N, that is, generating f_1, f_2, and utility values U for each N from N_{min} to N_{max}. For additive utility functions, the methods of Sections 18.2 or 18.3 can be used.

Example 18.6   Generating Efficient Alternatives for Sampling Plan   Consider the following information:

$50 \leq N \leq 150$
Average time per observation, $\bar{t} = 25$ min
Standard deviation, $s = 3$
Required confidence level of 99.74%, that is, $z = 3$
Cost per observation, $C = \$50$

Generate five efficient alternatives, equally distributed in terms of sample size.

First find the five equally distributed values for N ranging from 50 to 150. That is, use $N = \{50, 75, 100, 125, 150\}$. Then find the values of f_1 and f_2 for each given N. For example, for $N = 100$,

$$f_1 = CN = 50 \times 100 = 5000$$

$$f_2 = h = \frac{zs_j}{\sqrt{N\bar{t}_j}} = \frac{3 \times 3}{\sqrt{100 \times 25}} = 0.036$$

All five alternatives $\{a_1, a_2, a_3, a_4, a_5\}$ are generated as follows:

	a_1	a_2	a_3	a_4	a_5	Minimum	Maximum
N	50	75	100	125	150	50	150
$f_1 = CN$	2500	3750	5000	6250	7500	2,500	7,500
$f_2 = h$	0.051	0.042	0.036	0.032	0.029	0.029	0.051

18.5.2 Ranking with Additive Utility Functions[†]

The given multicriteria alternatives can be ranked using an additive utility function. Consider the additive utility function with normalized objective values. For a two-objective problem, the additive utility function is

$$\text{Minimize} \quad U(N) = w_1 f_1' + w_2 f_2'$$

where f_2' and f_2' are normalized values (0–1) for the two objectives, respectively, and w_1 and w_2 are weights of importance of normalized values of the two objectives, respectively, such that $w_1 + w_2 = 1$.

Example 18.7   Ranking Alternatives   Consider Example 18.6. Suppose the weights of importance for the normalized objective functions are 0.6 and 0.4, respectively. Rank alternatives generated in Example 18.6:

$$\text{Minimize } U = 0.6 f_1' + 0.4 f_2'$$

Alternative	a_1	a_2	a_3	a_4	a_5
N	50	75	100	125	150
Min. f_1'	0.000	0.250	0.500	0.750	1.000
Min. f_2	1.000	0.566	0.307	0.130	0.000
Min. $U = 0.6 f_1' + 0.4 f_2'$	0.400	0.376	0.423	0.502	0.600
Rank	2	1	3	4	5

Therefore, for the given weights of importance, alternative a_2 with a sample size of $N = 75$ should be selected.

18.5.3 Optimal Sampling Size for Additive Utility Functions[†]

For the additive utility function, the optimal sampling size solution can be found directly by setting

$$\frac{\partial U(N)}{\partial N} = 0$$

and solving for the value of N. Since f_1 and f_2 are both convex functions, their convex combination $w_1 f_1 + w_2 f_2$ and $w_1 f_1' + w_2 f_2'$) will also be convex. Then, any local optimum will be global. The second derivative of the above convex combination is positive. Therefore, the solution for $\partial U(N)/\partial N = 0$ is also a global minimum.

The optimal value of N can be calculated as

$$U = w_1 \left(\frac{f_1 - f_{1,\min}}{f_{1,\max} - f_{1,\min}} \right) + w_2 \left(\frac{f_2 - f_{2,\min}}{f_{2,\max} - f_{2,\min}} \right)$$

$$= w_1 \left(\frac{CN - f_{1,\min}}{f_{1,\max} - f_{1,\min}} \right) + w_2 \left(\frac{zs/\sqrt{N\bar{t}} - f_{2,\min}}{f_{2,\max} - f_{2,\min}} \right)$$

$$U'(N) = \frac{\partial U}{\partial N} = w_1 \left(\frac{C}{f_{1,\max} - f_{1,\min}} \right) - w_2 \left(\frac{0.5 zs N^{-3/2}\bar{t}^{-1}}{f_{2,\max} - f_{2,\min}} \right) = 0$$

Solving for N produces the equation

$$N^* = \left(\frac{(f_{1,\max} - f_{1,\min})w_2 zs}{(f_{2,\max} - f_{2,\min})2w_1 C\bar{t}} \right)^{2/3} \tag{18.10}$$

Now, find the values of f_1, f_2, and U using the obtained optimal N^*.

Example 18.8   Optimal Solution for Additive Utility Function   Consider Example 18.6. Find the optimal sample size where the weights of importance of the two normalized objectives are 0.6 and 0.4, respectively. That is, minimize

$$U = 0.6 f_1' + 0.4 f_2'$$

The minimum and maximum values of the objectives for the given range of N are

$$f_{1,\min} = CN_{\min} = 50 \times 50 = 2500 \qquad f_{1,\max} = CN_{\max} = 50 \times 150 = 7500$$

$$f_{2,\min} = h = \frac{3 \times 3}{\sqrt{150 \times 25}} = 0.029 \qquad f_{2,\max} = h = \frac{3 \times 3}{\sqrt{50 \times 25}} = 0.051$$

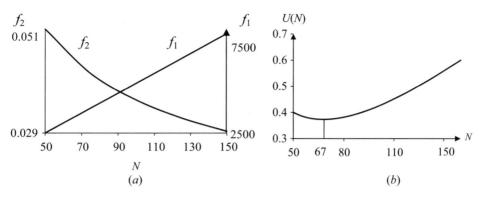

Figure 18.3 (*a*) Efficient frontier, f_1 vs. f_2. (*b*) Utility function for different sample sizes, N. The optimal sample size $N^* = 67$ occurs where the utility function is at its minimum.

Using Equation (18.10), we have

$$N^* = \left(\frac{(7500 - 2500) \times 0.4 \times 3 \times 3}{(0.051 - 0.029) \times 2 \times 0.6 \times 50 \times 25} \right)^{2/3} = 66.75 \quad \text{(rounded to 67)}$$

For this example, the utility function for N is presented in Figure 18.3, where the optimal N is 66.75 and $U = 0.373$. The values of f_1 and f_2 for the optimal N^* are

$$f_1 = cN = 50 \times 67 = 3350$$

$$f_2 = h = \frac{zs_j}{\sqrt{N\bar{t}_j}} = \frac{3 \times 3}{\sqrt{67} \times 25} = 0.044$$

Supplement 18.3.xls shows an alternative approach for the bicriteria sampling plan.

18.6 WORK SAMPLING

Originally developed in England by L. H. C. Tippet in the 1930s, work sampling is a discrete-time method of randomly observing a person, machine, or process to determine the percentage of the time the person, machine, or process is performing an activity. Work sampling is used for defining labor standards, measuring worker productivity, and comparing accidental delays. This method can be used in many applications, including construction, manufacturing, and hospitals, to measure time studies for crews or equipment. The limitation of this method is the accuracy of the random sampling. It is possible that the sampled times could produce distorted or biased results if an inadequate sampling size is chosen. Sampling size selection based on one criterion and multicriteria is covered in this section.

p-*Chart* In order to use the work sampling technique, there are three assumptions that must be made. The first assumption is that the short duration of the random sample is enough time to make a generalization of the performance. For example, consider a light bulb was randomly observed to be "on" for 75 out of 100 observations during each 24-h period. This means that the light bulb was observed being on 75% of the time. This ratio can be used to describe its usage over long periods of time. The second assumption is that the accuracy increases as the sampling size increases. For example, taking 200 instead of 100 observations of the light bulb will increase the accuracy of measurement. Because the status of the element being observed is binary, for example, it is either on or off, the third assumption is that the probability distribution is assumed to be binomial or its approximation, the Poisson distribution. Therefore, a p-chart can be used to solve this sampling problem. For the theory of p-charts, refer to Chapter 17, which covers p-charts in more depth.

18.6.1   Finding Sample Size for p-Chart

The p-chart sampling procedure consists of the following steps:

1. Take a small sample n to determine the probability of the initial estimate for the average probability $\bar{p}$.
2. Determine a sample size using Equation (18.11) where the confidence level z and the desired accuracy level h are given. That is, find N for the following given values:

$$\bar{p} = \text{Probability of initial observations}$$

$$z = \text{Desired confidence level}$$

$$h = \text{Accuracy level (\%) of sampling mean}$$

$$N = \left(\frac{z}{h}\right)^2 [\bar{p}(1 - \bar{p})] \tag{18.11}$$

Example 18.9   Work Sampling Examples: Finding Sample Size   A manager at a local hospital is concerned with the utilization of an expensive MRI (magnetic resonance imaging) device that has recently been purchased. The manager's initial estimate of how often the MRI is idle is 25%, or $\bar{p} = 0.25$. Determine the number of random observations needed to estimate the probability of idle time if:

(a) The desired accuracy is 2% and the confidence level is 95.44%.
(b) The desired accuracy is 2% and the confidence level is 99.74%.

The solution is as follows:

(a) Since the confidence level is 95.44%, $z = 2$. Also, the desired accuracy of 2% means that $h = 0.02$. Therefore, $\bar{p} = 0.25$, $h = 0.02$, and $z = 2$. Now use Equation (18.11) to find N:

$$N = \left(\frac{2}{0.02}\right)^2 (1 - 0.25)(0.25) = 1875 \text{ observations}$$

(b) Since the confidence level is 99.74%, $z = 3$. Therefore, $\bar{p} = 0.25$, $h = 0.02$, and $z = 3$. Using Equation (18.11),

$$N = \left(\frac{3}{0.02}\right)^2 (1 - 0.25)(0.25) = 4218.75 \text{ (or 4219) observations}$$

So, increasing the confidence level from 95.44 to 99.74% requires an additional $4219 - 1875 = 2344$ observations.

18.6.2   Finding Standard Time per Part

Consider the problem of finding the standard time to produce one part. The method is as follows:

1. Prepare a random schedule for observations, for example, by using a random generator method or a given random table.
2. Observe the activity. Record if the activity is on or off (or is busy or idle).
3. Provide the performance rating R of the activity for the period of study.
4. Determine the number of units produced during the entire period of study.
5. Compute the normal time per part:

$$NT = \frac{\text{total study time} \times \% \text{ of time worker performs activity} \times R}{\text{number of parts produced during period}}$$

6. Assess the allowance factor for the given activity.
7. Compute the standard time per part:

$$ST = \frac{NT}{1\text{-allowance factor}}$$

In step 5, the observation is usually very quick (i.e., either busy or idle), so usually a neutral rating of $R = 1$ is used.

Example 18.10   Work Sampling: Finding Standard Time per Part   Consider Example 18.9, part (a). As before, the percent idle time is 25%. The performance rating R is 100%, and the study is conducted for an 8-h period. Given that:

Accuracy level is 2%, or $h = 0.02$.
Confidence level is 95.44%, or $z = 2$.
Allowances: personal allowance of 6% plus basic fatigue allowance of 3% gives total allowance of $6\% + 3\% = 9\%$.
Percent idle time is 25%, or $\bar{p} = 0.25$.
Length of study is 8 h, or 480 min.
Number of units produced is 15.

Find the normal and standard times for each part produced.

The solution is as follows:

$N = (2/0.02)^2(1 - 0.25)(0.25) = 1875$ observations [using Equation (18.11)]

$\bar{p}(\text{idle}) = 25\%$, $\bar{p}(\text{working}) = 75\%$, $R = 100\%$

Fifteen units were produced during the 480 min of the study

Normal time $NT = (480)(0.75)(1.00)/15 = 24$ min/part

Standard time $ST = 24 /(1 - 0.09) = 26.37$ min/part

Therefore, the standard time for this job is 26.37 min/part.

Scheduling For Random Visits In order to develop a random schedule for visits, use the following steps:

1. Determine the number of observations n.
2. Generate n random numbers.
3. Sort the random numbers from smallest to largest.
4. Scale the random numbers according to the time frame.

Example 18.11 Scheduling for Random Visits For a work period of 9 am–5 pm (i.e., an 8-h period), generate a schedule of visits for a $\bar{p}$-chart study. Use an approach to pick up numbers from this table.

Use a random generator system to generate the random numbers. Consider Table 18.8, which is generated by a random-number system. For example, start at the first number, 409, and pick the numbers as they appear, going across the first row from left to right. Then, go to the next row and continue the process. Using this method, the stream of random numbers is 409, 947, 776, 714, ..., 544.

Now convert the selected random numbers into the time scale. An example of the calculations of three random times is shown below. Since the work time is 8 h and the range of random numbers in Table 18.8 is from 0 to 1000, then the random numbers are scaled by multiplying each number by $8/1000 = 0.008$. For example, the number 409 is converted to hours, $409 \times 0.008 = 3.27$ h. Therefore its start time is $9 + 3.27 = 12.16$ p.m.

TABLE 18.8 Random-Number Table

4 0 9	9 4 7	7 7 6	7 1 4	9 0 5	6 8 6	0 7 2	2 1 0	9 4 0	5 5 8
6 0 9	7 0 9	3 4 3	3 5 0	5 0 0	7 3 9	9 8 1	1 8 0	5 0 5	4 3 1
3 9 8	0 8 2	7 7 3	2 5 0	7 2 5	6 8 2	4 8 2	9 4 0	5 2 4	2 0 1
5 2 7	7 5 6	7 8 5	1 8 3	4 5 2	9 9 6	3 4 0	6 2 8	8 9 8	0 8 3
1 3 7	4 6 7	0 0 7	8 1 8	4 7 5	4 0 6	1 0 6	8 7 1	1 7 7	8 1 7
8 8 6	8 5 4	0 2 0	0 8 6	5 0 7	5 8 4	0 1 3	6 7 6	6 6 7	9 5 1
9 0 3	4 7 6	4 9 3	2 9 6	0 9 1	1 0 6	2 9 9	5 9 4	6 7 3	4 8 8
7 5 1	7 6 4	9 6 9	9 1 8	2 6 0	8 9 2	8 9 3	7 8 5	6 1 3	6 8 2
3 4 7	8 3 4	1 1 3	8 6 2	4 8 1	1 7 6	7 4 1	7 4 6	8 5 0	9 5 0
5 8 0	4 7 7	6 9 7	4 7 3	0 3 9	5 7 1	8 6 4	0 2 1	8 1 6	5 4 4

The orders of visits are shown as ranked.

Example	Random Number	Scaled Random Number	Start Time Plus Scaled Random No. = Observation Time	Visit Order
1	409	3.27	9 am + 3.27 = 12:16 pm	1
2	947	7.58	9 am + 7.58 = 4:35 pm	3
3	776	6.21	9 am + 6.21 = 3:13 pm	2

18.6.3   Illustration of *p*-Control Charts

In most situations, work samples take place over several periods of study (e.g., several days) where the probability for each period can be calculated. (The approach is similar to the *p*-chart covered in Chapter 17. See that chapter for details). Using control charts, upper and lower control limits can be established for the observations. This allows nonrepresentative data points to be identified and eliminated. To construct the *p*-chart and identify nonrepresentative data points, the following steps are used:

1. Find the mean probability for each period $i = 1, 2, \ldots, n$, where p_i is the probability measurement for period i. Then find the average of all periods:

$$\bar{p} = \frac{1}{n} \sum_{i=1}^{n} p_i \quad \text{(mean probability of all periods)} \tag{18.12}$$

2. Calculate the mean standard deviation (σ) of all periods:

$$\sigma = \sqrt{\frac{\bar{p}(1 - \bar{p})}{n}} \tag{18.13}$$

3. Set the upper and lower control chart limits for a given confidence level using $\bar{p}$, z, and σ:

$$\text{UCL} = \bar{p} + z\sigma \tag{18.14}$$

$$\text{LCL} = \bar{p} - z\sigma \tag{18.15}$$

4. Determine which observations fall outside the control limits.

Identifying Whether System Is Under Control   *p*-Control charts are used to graphically demonstrate whether the system or process being observed is under control or not. The required upper and lower control limits (UCL and LCL) are set and the average data observed for all periods are plotted. If all points are within the UCL and LCL, the process is under control. The control chart can also show patterns of change in the behavior of a system.

Example 18.12   Sampling Using *p*-Chart Example   Consider a piece of equipment being observed which can have one of two statuses: busy or idle. Suppose the selected sampling plan is for 12 periods. Everyday 100 observations are made. The following table shows the probability of having an idle status for each of the 12 periods.

Construct the p-chart and identify if any trends exist. Determine which points are out of the control limits and find the average probability of having an idle status. Use a 95.44% confidence level.

Period	1	2	3	4	5	6	7	8	9	10	11	12
No. of observations	100	100	100	100	100	100	100	100	100	100	100	100
No. of idle status	10	20	10	15	25	30	10	15	25	20	20	10
p_i	0.1	0.2	0.1	0.15	0.25	0.3	0.1	0.15	0.25	0.2	0.2	0.1

The solution is as follows. The mean and standard deviation for these given 12 periods are

$$\bar{p} = \frac{1}{12} \sum_{i=1}^{12} p_i = 0.175 \qquad \sigma = \sqrt{\frac{(0.175 \times (1 - 0.175))}{12}} = 0.1097$$

A confidence level of 95.44% is associated with $z = 2$. Therefore, the upper and lower control limits are

$$\text{UCL} = 0.175 + 2 \times 0.11 = 0.395 \qquad \text{LCL} = 0.175 - 2 \times 0.11 = -0.044$$

Since $\text{LCL} = -0.044$, which is less than zero, then set $\text{LCL} = 0$. (Negative values are not meaningful in this example).

The p-control chart for the above data is presented in Figure 18.4.

Based on Figure 18.4, all p_i values are within the control limits. They can be used for further analysis. There is no significant pattern in these data.

The average probability of having an idle status is $\bar{p} = 0.175$, or 17.5% of the time.

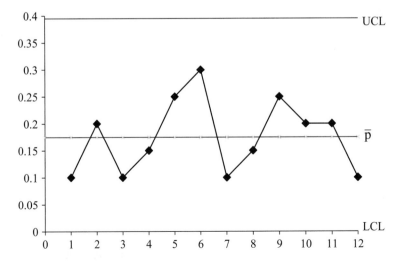

Figure 18.4   Graph showing upper and lower control limit lines.

18.6.4   Multicriteria *p*-Chart Sampling Plan†

The multicriteria *p*-chart sampling approach is similar to the multicriteria sampling covered for continuous time in Section 18.4. Equation (18.11) can be written as

$$h = z\sqrt{\frac{\bar{p}(1 - \bar{p})}{N}} \qquad (18.16)$$

where $\bar{p}$ is the average probability of observing an idle status. Consider the following two objectives:

Minimize total cost of sampling, $f_1 = CN$

Minimize the error, $f_2 = h$ [Equation (18.16)]

where C is cost (in dollars) per observation.

Generating Efficient Alternatives   Consider a range of N values, for example, five equally distributed N values between N_{min} and N_{max}. Find f_1 and f_2 for each alternative. This will be a set of efficient alternatives.

Finding Optimal N for Additive Utility Functions   For additive utility functions, the optimal solution can be obtained analytically by setting $\partial U(N)/\partial N = 0$ and solving for the value of N:

$$U(N) = w_1 f_1' + w_2 f_2' = w_1 \left(\frac{CN - f_{1,min}}{f_{1,max} - f_{1,min}} \right) + w_2 \left(\frac{z\sqrt{\bar{p}(1 - \bar{p})/N} - f_{2,min}}{f_{2,max} - f_{2,min}} \right)$$

$$\frac{\partial U(N)}{\partial(N)} = \frac{Cw_1}{f_{1,max} - f_{1,min}} - \frac{0.5 w_2 z [\bar{p}(1 - \bar{p})]^{1/2} N^{3/2}}{f_{2,max} - f_{2,min}} = 0$$

$$N^* = \left(\frac{z w_2 (f_{1,max} - f_{1,min})}{2 w_1 C (f_{2,max} - f_{2,min})} \right)^{2/3} [\bar{p}(1 - \bar{p})]^{1/3} \qquad (18.17)$$

Example 18.13   Multicriteria *p*-Chart Sampling   Consider an example where the cost of observation per period is \$1.50, the average idle time is 0.2, and the required level of confidence is 99.74%. The sample size should be between 50 and 150.

(a) Generate five efficient alternatives equally distributed by sample size.
(b) Suppose weights of importance for the normalized objectives are 0.4 and 0.6. Rank the five alternatives generated for part (a).
(c) Find the best alternative using the additive utility function. What is its utility value?

In this problem, $C = \$1.50$, $\bar{p} = 0.2$, $z = 3$, and the range for the number of samples is $50 \le N \le 150$.

(a) **Generate Five Efficient Alternatives** Five equally distributed alternatives in terms of N are $\{50, 75, 100, 125, 150\}$. For each N, find f_1 and f_2. For example, if $N = 100$,

$$f_1 = 1.5 \times 100 = 150$$

$$f_2 = h = 3\sqrt{\frac{0.2(1 - 0.2)}{100}} = 0.12 \quad \text{(or 12\%)}$$

Then find the multicriteria alternatives as presented below:

Alternatives	a_1	a_2	a_3	a_4	a_5	Minimum	Maximum
N	50	75	100	125	150	50	150
$f_1 = CN = 1.5\,N$	75	112.5	150	187.5	225	75	225
$f_2 = h$	0.170	0.139	0.120	0.107	0.098	0.098	0.170

(b) **Ranking Alternatives with the Additive Utility Function** The multicriteria alternatives should first be normalized. The alternatives obtained in part (a) are ranked below using weights of importance $W = (0.4, 0.6)$.

Alternatives	a_1	a_2	a_3	a_4	a_5
Min. f_1'	0	0.25	0.5	0.75	1
Min. f_2'	1	0.566	0.307	0.130	0
Min. $U = 0.4 f_1' + 0.6 f_2'$	0.600	0.439	0.384	0.378	0.400
Ranking	5	4	2	1	3

(c) The optimal solution is found using Equation (18.17):

$$N^* = \left(\frac{zw_2(f_{1,\max} - f_{1,\min})}{2w_1 C(f_{2,\max} - f_{2,\min})}\right)^{2/3} [\bar{p}(1 - \bar{p})]^{1/3}$$

$$= \left(\frac{3 \times 0.6 \times (225 - 75}{2 \times 0.4 \times 1.5 \times (0.17 - 0.098)}\right)^{2/3} (0.2 \times 0.8)^{1/3} = 116.03$$

In this example, for $N = 1.5, f_1 = 1.5 \times 116.03 = 174.045, f_2 = z\sqrt{\bar{p}(1 - \bar{p})/N} = (3 \times \sqrt{(0.2 \times 0.8)/116.03} = 0.111, f_1' = 0.66, f_2' = 0.187,$ and $U = 0.4 \times 0.66 + 0.6 \times 0.187 = 0.376.$ Figures 18.5a and 18.5b show efficient frontier and utility value for different sample sizes in Example 18.13.

See Supplement 18.4.xls.

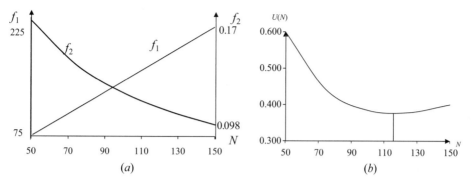

Figure 18.5 (*a*) Efficient frontier for Example 18.13. (*b*) Utility values for different sample sizes N, where $N^* = 116$.

18.7 PREDETERMINED TIME STANDARDS

Predetermined time standards have been used for many years. The method involves dividing a manual task into several elementary tasks such as reach, grab, pick, or place. These elementary tasks have predetermined times for different variables, such as the weight of an object, the distance to an object, the type of motion, conditions under which the task is performed, and the required accuracy for the task being performed. The numbers are presented in tables and are scaled in TMUs (time measurement units). Each TMU is 0.0006 min, which is equal to 0.00001 h.

Some large corporations such as UPS (United Parcels Service) have developed their own databases of predetermined time standards which are considered among the most accurate labor standards in the world. Predetermined times have been used for tasks in many different industries and businesses, such as riveting sheet metal, manually installing computer components, packaging boxes, and making hamburgers at fast food restaurants. Several commercial predetermined time standard systems are available. The most common ones are:

- MTM (Methods Time Measurement, MTM Association).
- CSD (Computerized Standard Data)
- BMT (Basic Motion Time study)
- MODAPTS (MODular Arrangement of Predetermined Time Standards)

As discussed before, predetermined time standards are used by dividing a task into its simplest elementary task components and then finding the time for each elementary task by using applicable time standard tables. The times of elementary tasks are then added together to determine the standard time for the given task. Table 18.9 shows an example of predetermined time standards by MTM.

Example 18.14 Predetermined Time Standard Method   Consider a simple GET and PLACE task of an assembly line worker who is seated in a chair. The worker must retrieve an electronic component from a basket whose center is about 12 in. away and install it in a circuit board. The circuit board arrives to the worker by a conveyor that stops exactly at the same location every time. Once the task is completed, the assembly line worker

TABLE 18.9 MTM Table for Predetermined Time Standards in TMUs or 0.00001 h

GET and PLACE			Distance Range, in.	Distance <8	8<Distance <20	20<Distance<32
Weight	**Conditions of GET**	**PLACE Accuracy**	Code	1	2	3
Weight <2 lb	Easy	Approximate	AA	20	35	50
	Easy	Loose	AB	30	45	60
	Easy	Tight	AC	40	55	70
	Difficult	Approximate	AD	20	45	60
	Difficult	Loose	AE	30	55	70
	Difficult	Tight	AF	40	65	80
	Handful	Approximate	AG	40	65	80
2 lb<Weight<18 lb		Approximate	AH	25	45	55
		Loose	AJ	40	65	75
		Tight	AK	50	75	85
18 lb<Weight<45 lb		Approximate	AL	90	106	115
		Loose	AM	95	120	130
		Tight	AN	120	145	160

Source: Reprinted with permission of the MTM Association for Standards and Research.
Note: The data should not be used without proper training in the use of the MTM-UAS System.

pushes the Continue button and the conveyor removes the finished circuit board. Then, the process is repeated. The breakdown of the assembly task into its work elements (subtasks) is presented below.

1. GET a component from the basket (this is a GET task where reaching is required).
2. PLACE the component into the designated slot on the circuit board.
3. PUSH the button to index the assembly line conveyor, which pushes away the assembled board and brings in the new circuit board (this is a GET task).

Conditions for Steps 1 and 2	Conditions for Step 3
• Weight (<2 lb)	• Weight (<2 lb)
• Place accuracy (tight)	• Place accuracy (approximate)
• Conditions of GET (easy)	• Conditions of GET (easy)
• Distance range (8–20 in.)	• Distance range (8–20 in.)

The predetermined time standards of Table 18.9 for the steps 1, 2, and 3 are explained below:

• For steps 1 and 2, the element from Table 18.9 is AC2, which has 55 TMU (row AC and column 2). The frequency of doing this task is one per one cycle.

- For step 3, the element from Table 18.9 is AA2, which has 35 TMU, and its frequency is also units.

    Therefore, the total TMU is $55 + 55 + 35 = 145$. The total standard time is

$$ST = \text{total TMU} \times 0.0006 = 145 \times 0.0006 = 0.087 \text{ min}$$

Therefore, the standard time for this operation is 0.087 min (or 5.22 s) per assembly.

REFERENCES

Amanda, Y., G. Pransky, and W. V. Mechelen. "Introduction to the Special Issue on Measurement of Work Outcomes." *Journal of Occupational Rehabilitation*, Vol. 12, No. 3, 2002, pp. 115–117.

Backman, J. *Wage Determination: An Analysis of Wage Criteria*. Princeton, NJ: Van Nostrand, 1960.

Baines, A. "Work Measurement, The Basic Principles Revisited." *Work Study*, Vol. 44, No. 7, 1995, pp. 10–14.

Barnes, R. M. *Motion and Time Study: Design and Measurement of Work*, 7th ed. New York: Wiley, 1980.

Coelli, T. J., D. S. P. Rao, C. J. O'Donnell, and G. E. Battese. *An Introduction to Efficiency and Productivity Analysis*, 2nd ed. New York: Springer, 2005.

Gantt, H. L. *Work, Wages and Profit*. New York: Engineering Management, 1913.

Gilbreth, F. B., and L. M. Gilbreth, *Fatigue Study*, 2nd ed. New York: Macmillan, 1919.

Hamill, B. J. *Work Measurement in the Office: An MTM Systems Workbook*, Epping: Gower Press, 1973.

Konz, S. *Work Design*, Columbus, OH: Grid Publishing, 1979.

Malakooti, B. "Multiple Criteria Sampling Plans for Work Measurement." Case Western Reserve University, Cleveland, Ohio, 2013.

Maxwell, S. E., K. Kelley, and J. R. Rausch. "Sample Size Planning for Statistical Power and Accuracy in Parameter Estimation." *Annual Review of Psychology*, Vol. 59, 2008, pp. 537–563.

Miller, M. J., D. J. Woehr, and N. Hudspeth. "The Meaning and Measurement of Work Ethic: Construction and Initial Validation of a Multidimensional Inventory." *Journal of Vocational Behavior*, Vol. 60, No. 3, 2002, pp. 451–489.

Mundel, M. E. *Motion and Time Study*, 5th ed. Englewood Cliffs, NJ: Prentice-Hall, 1978.

Murphy, P. "Service Performance Measurement Using Simple Techniques Actually Works." *Journal of Marketing Practice: Applied Marketing Science,* Vol. 5, No. 2, 1999. pp. 56–73.

Nahmias, S. *Production and Operations Analysis*, 6th ed. New York: McGraw-Hill, 2008.

Nakayama, S. I., K. I. Nakayama, and H. Nakayama. "A Study on Setting Standard Time Using Work Achievement Quotient." *International Journal of Production Research*, 2002, Vol. 40, No. 15, pp. 3945–3953.

Neely, A., M. Gregory, and K. Platts. "Performance Measurement System Design, A Literature Review and Research Agenda." *International Journal of Operations & Production Management*, Vol. 15, No. 4, 1995, pp. 80–116.

Nembhard, D. A., and S. M. Shafer. "The Effects of Workforce Heterogeneity on Productivity in an Experiential Learning Environment." *International Journal of Production Research*, Vol. 46, No. 14, 2008, pp. 3909–3929.

Niebel, B. W. *Motion and Time Study*, 7th ed. Homewood, IL: Irwin, 1982.

Rice, R. S. "Survey of Work Measurement and Wage Incentives." *Industrial Engineering*, Vol. 9, No. 7, 1997, pp. 18–31.

Sarkis, J. "Quantitative Models for Performance Measurement Systems—Alternate Considerations." *International Journal of Production Economics*, Vol. 86, No. 1, 2003, pp. 81–90.

Stevenson, W. J. *Operations Management*, 9th ed. New York: McGraw-Hill/Irwin, 2007.

Sudit, E. F. "Productivity Measurement in Industrial Operations." *European Journal of Operational Research*, Vol. 85, No. 3, 1995, pp. 435–453.

Suwignjo, P., U. S. Bititci, and A. S. Carrie. "Quantitative Models for Performance Measurement System." *International Journal of Production Economics*, Vol. 64, Nos. 1–3, 2000, pp. 231–241.

Wen-Hsien, T. "A Technical Note on Using Work Sampling to Estimate the Effort on Activities under Activity-Based Costing." *International Journal of Production Economics*, Vol. 43, No. 1, 1996, pp. 11–16.

Young, A., G. Pransky, and W. van Mechelen. "Introduction to the Special Issue on Measurement of Work Outcomes." *Journal of Occupational Rehabilitation*, Vol. 12, No. 3, 2002, pp. 115–117.

Young, S. T. "Multiple Productivity Measurement Approaches for Management." *Health Care Manage Rev.*, Vol. 17, No. 2, 1992, pp. 51–58.

EXERCISES

Analysis of Methods (Section 8.2.1) and Study of Motion (Section 8.2.2)

2.1 List two industries (or organizations) where analysis of methods would benefit a person working in that specific industry. Why would analysis of methods benefit that person?

2.2 Draw a flow process chart for making a cake with eggs, sugar, milk, cake mix, frosting which involves using an oven and stirring (use the symbols from Figure 2.1). For each activity also show its duration and comment on how the total process can be improved.

2.3 Draw a flow process chart for the assembly of a simple product (use the symbols from Figure 2.1). For each activity also show its duration and comment on how the total process can be improved.

Standard Time (Section 18.3)

3.1 A job consist of three tasks. The total allowance factor has been established as 12% for each of three tasks. A study of seven cycles is presented below. Performance ratings for each task are also given in the table. The demand is for 500 items to be produced in a 4-h shift. Use the information in the table below to determine:

(a) Average actual cycle time of each task

(b) Normal time of each task

(c) Standard time of the job

(d) Number of workers needed for each task

	Cycle Observed, t_j, Min							
Task, j	1	2	3	4	5	6	7	R_j
1	12	14	13	14	22	12	13	105%, or 1.05
2	0.48	0.47	3.00	0.75	0.71	0.53	0.63	95%, or 0.95
3	1.825	1.961	1.563	1.983	1.590	1.585	1.706	99%, or 0.99

3.2 Consider a job that consists of four tasks. The performance rating for each task is given in the table. The total allowance factor has been established as 17% for each of the four tasks. The demand is for 5000 items to be produced in an 8-h shift. Task 2 is machine operated. Respond to parts (a)–(d) of exercise 3.1.

Task, j	Cycle Observed, t_j, min						R_j
	1	2	3	4	5	6	
1	22	25	22	27	21	26	90%, or 0.90
2	2	2	2	2	2	2	100%, or 1
3	0.294	0.387	0.290	0.422	0.273	0.403	112%, or 1.12
4	5.5	6.6	5.5	10.6	5.3	6.8	95%, or 0.95

3.3 Consider a job that consists of five tasks. The performance rating for each task is given in the table. The total allowance factor has been established as 10% for each of the four tasks. The demand is for 150 items to be produced in a 360-min period. Task 3 is machine operated. Respond to parts (a)–(d) of exercise 3.1.

Task, j	Cycle Observed, t_j, min								R_j
	1	2	3	4	5	6	7	8	
1	0.276	0.465	0.316	0.430	0.484	0.383	0.298	0.344	95%, or 0.95
2	103	122	107	118	123	113	105	109	110%, or 1.10
3	20	20	20	20	20	20	20	20	100%, or 1
4	2.33	2.56	2.38	2.52	2.58	2.46	2.36	2.41	115%, or 1.15
5	5.3	7.6	5.8	7.2	7.8	6.6	5.6	6.1	105%, or 1.05

3.4 Consider a job that consists of eight tasks. The performance rating for each task is given in the table. The total allowance factor has been established as 15% for each of the four tasks. The demand is for 25,000 items to be produced in a 24-h day. Task 6 is machine operated. Respond to parts (a)–(d) of exercise 3.1.

Task, j	Cycle Observed, t_j, min											R_j
	1	2	3	4	5	6	7	8	9	10	11	
1	12.20	12.09	12.16	12.22	12.15	12.15	12.10	12.25	12.23	12.14	12.20	93%
2	16.6	15.7	16.3	16.8	16.2	16.2	15.8	17.0	16.8	16.1	16.6	104%
3	54	52	53	54	53	53	52	55	55	101	54	97%
4	0.653	0.421	0.570	0.689	0.552	0.554	0.451	0.749	0.708	0.531	0.647	112%
5	9.0	6.7	8.2	9.4	8.0	8.0	7.0	10.0	9.6	7.8	9.0	96%
6	1	1	1	1	1	1	1	1	1	1	1	100%
7	54	51.7	53.2	54.4	53.0	53.0	52.0	55.0	54.6	52.8	54	101%
8	1.81	1.34	1.64	1.88	1.60	1.61	1.40	2.00	1.92	1.56	1.79	110%

Sample Size for Standard Time (Section 18.4)

4.1 Consider the following information for a sampling problem. The desired confidence level is 95.44%. The level of accuracy is 10%. The mean value is 12 min. The standard deviation is 2 min.

(a) Find the sampling size required to achieve the desired values.

(b) Suppose the desired confidence level is 99.74%. Find the sampling plan.

(c) Suppose the desired level of accuracy is 5%. Find the sampling size.

(d) Suppose the level of accuracy is unknown but the sample size is known, and it is 100. What would be the desired level of accuracy?

(e) Suppose the level of accuracy is unknown but the maximum level of error is 0.5. Find the sampling size.

4.2 Consider the following information for a sampling problem. The desired confidence level is 95.44%. The level of accuracy is 6%. The mean value is 25 min. The standard deviation is 5 min.

(a) Find the sampling size required to achieve the desired values.

(b) Suppose the desired confidence level is 68.26%. Find the sampling plan.

(c) Suppose the desired level of accuracy is 3%. Find the sampling size.

(d) Suppose the level of accuracy is unknown but the sample size is known, and it is 300. What would be the desired level of accuracy?

(e) Suppose the level of accuracy is unkown but the maximum level of error is 1.5. Find the sampling size.

4.3 Consider the observed cycle times in exercise 3.1. Suppose the decision maker agrees on the following:

(i) The desired confidence level is 99.74%.

(ii) The desired level of accuracy is $h = 8\%$.

(iii) The initial number of cycles observed is 7 (see exercise 3.1).

Determine the needed sample size to meet these requirements.

4.4 Consider the observed cycle times in exercise 3.2. Suppose the decision maker agrees on the following:

(i) The desired confidence level is 95.44%.

(ii) The desired level of accuracy is 5%.

(iii) The initial number of cycles observed is 6 (see exercise 3.2).

Determine the needed sample size to meet these requirements.

4.5 Consider the observed cycle times in exercise 3.3. Suppose the decision maker agrees on the following:

(i) The desired confidence level is $Z = 2.5$.

(ii) The desired level of accuracy is 6%.

(iii) The initial number of cycles observed is 8 (see exercise 3.3).

Determine the needed sample size to meet these requirements.

Multicriteria Sampling for Standard Time (Section 18.5)[†]

5.1 An assembly plant wishes to conduct a time study on one of its assembly tasks. Due to budget limitations, the number of observations should be between 100 and 200 samples.

Each observation costs $25. From previous records, it is known that the average time to complete the task is normally distributed with a mean of 5 min and a standard deviation of 1. Their required confidence level is 99.74%.

(a) Generate five equally distributed efficient alternatives.

(b) Rank the alternatives using weights of importance (0.4, 0.6) for the normalized values of objectives 1 and 2, respectively.

(c) Compute the optimal N value for the given weights of importance.

5.2 A company wishes to do a time study on one of its operation's tasks. They wish to take between 500 and 1000 observations. Each observation costs $5. The average time to complete the task is normally distributed with a mean of 35 min and a standard deviation of 5. Their required confidence level is 95.44%.

(a) Generate five equally distributed efficient alternatives.

(b) Rank the alternatives using weights of importance (0.55, 0.45) for the normalized values of objectives 1 and 2, respectively.

(c) Compute the optimal N value.

5.3 Consider a job whose completion time is normally distributed with a mean of 20 s and a standard deviation of 2. The required confidence level is 99.74%. The cost of observation is $1.50. Suppose that the number of observations can range from 100 to 500.

(a) Generate five equally distributed efficient alternatives.

(b) Rank the alternatives using weights of importance (0.4, 0.6) for the normalized values of objectives 1 and 2, respectively.

(c) Compute the optimal N value.

5.4 Consider a job whose completion time is normally distributed with a mean of 103 s and a standard deviation of 7. The required confidence level is 99.74%. The cost of observation is $0.75. Suppose that the number of observations can range from 10,000 to 50,000.

(a) Generate five equally distributed efficient alternatives.

(b) Rank the alternatives using weights of importance (0.4, 0.6) for the normalized values of objectives 1 and 2, respectively.

(c) Compute the optimal N value.

Finding Sample Size for p-Chart (Section 18.6.1) and Finding Standard Time per Part (Section 18.6.2)

6.1 A machine is idle 20% of time. The desired accuracy is 5% and the desired confidence level is 99.74%. The performance rating is 100% and the study was conducted over a 4-h period. In this period, the machine produces 1000 units. The total allowance factor is 10%. Determine:

(a) Number of discrete random observations that must be performed

(b) Average actual cycle time per unit

(c) Normal time per unit

(d) Standard time per unit

6.2 A machine is idle 12% of time. The desired accuracy is 2.5% and the desired confidence level is 95.44%. The performance rating is 110% and the study was conducted over a

8-h period. In this period, the machine produces 2500 units. The total allowance factor is 8%. Respond to parts (a)–(d) of exercise 6.1.

6.3 A machine is idle 8% of time. The desired accuracy is 3% and the desired confidence level is 99.74%. The performance rating is 95% and the study was conducted over a 12-h period. In this period, the machine produces 5000 units. The total allowance factor is 12%. Respond to parts (a)–(d) of exercise 6.1.

Illustration of *p*-Control Charts (Section 18.6.3)

6.4 The following table shows the probability of a job having an idle status for each day of a 10-day working period. The required confidence level is 99.74%.

(a) Determine the average probability of the job having an idle status.
(b) Determine upper and lower control limits.
(c) Construct the *p*-chart.
(d) Determine if the process is out of control.

Day, i	1	2	3	4	5	6	7	8	9	10
p_i	0.23	0.16	0.15	0.22	0.20	0.15	0.10	0.16	0.17	0.22

6.5 The following table shows the probability of a job having an idle status for each day of a 12-day working period. The required confidence level is 95.44%. Respond to parts (a)–(d) of exercise 6.4.

Day	1	2	3	4	5	6	7	8	9	10	11	12
p_i	0.224	0.185	0.222	0.199	0.191	0.217	0.170	0.178	0.212	0.245	0.239	0.224

6.6 The following table shows the probability of a job having an idle status for each day of a 15-day working period. The required confidence level is 68.26%. Respond to parts (a)–(d) of exercise 6.4.

Day	1	2	3	4	5	6	7	8	9	10	11	12	13	14	15
p_i	0.19	0.18	0.23	0.17	0.23	0.15	0.18	0.21	0.23	0.23	0.21	0.19	0.16	0.18	0.24

Multicriteria *p*-Chart Sampling Plan (Section 18.6.4)[†]

6.7 The probability for a task to be idle is 15%. The cost of each observation is $20. The required confidence level is 95.44%. Suppose that the number of observations can range from 100 to 500. Suppose the weights of importance for the normalized values of the two objectives are (0.7,0.3) for the total cost and the accuracy, respectively.

(a) Generate five equally distributed alternatives in terms of the sample size.
(b) Rank the alternatives generated in part (a) using the given weights of importance.
(c) Find the optimal sample size for the given weights of importance.

6.8 The probability for a task having an idle status is 8%. The cost of each observation is $25. The required confidence level is 99.74%. Suppose that the number of observations can range from 1000 to 2500. Suppose the weights of importance for the normalized values of the two objectives are $(0.4, 0.6)$ for the total cost and the accuracy, respectively. Using this information, respond to parts (a)–(c) of exercise 6.7.

6.9 The probability for a task having an idle status is 12%. The cost of each observation is $2.50. The required confidence level is 95.44%. Suppose that the number of observation can range from 500 to 1000. Suppose the weights of importance for the normalized values of the two objectives are $(0.8, 0.2)$ for the total cost and the accuracy, respectively. Using this information, respond to parts (a)–(c) of exercise 6.7.

Predetermined Time Standards (Section 18.7)

7.1 Consider an assembly line worker is seated in a chair. The worker must retrieve a 1-lb handful of screws from a container whose center is about 16 in. away and place them in a box. The box arrives to the worker by a conveyor and stops exactly at the same location every time. Once the task is completed, the assembly line worker places the box on a different conveyor belt. Then, the above process is repeated. The breakdown of the assembly job into its work elements (tasks) is presented below.
  (i) GET the screws from the container (handful, approximate).
 (ii) PLACE the screws into the box (handful, approximate).
(iii) PLACE the box onto the belt (easy, loose).
Use Table 18.9 to determine the standard time for this operation.

7.2 An assembly line worker is seated in a chair (see Table 18.9). First, the worker must retrieve a 4-lb gear from a container whose center is about 36 in. away and install it. The product arrives to the worker by a conveyor and stops at exactly the same location every time. Once the gear is installed, the assembly line worker must install another 4-lb gear from the same container. Once both gears are installed, the worker pushes the Next button and the conveyor removes the finished part. Then, the above process is repeated. The breakdown of the assembly job into its work elements (tasks) is presented below.
  (i) GET a gear from the basket (easy, loose).
 (ii) PLACE the gear into the part (difficult, tight).
(iii) GET a gear from the basket (easy, loose).
(iv) PLACE the component into the part (difficult, tight).
 (v) PUSH (use GET to approximate PUSH) button to index the assembly line conveyor that pushes away the assembled board and brings in the new part (easy, approximate).
Use Table 18.9 to determine the standard time for this operation.

CHAPTER 19

RELIABILITY AND MAINTENANCE

19.1 INTRODUCTION

Reliability is the characteristic that decides how effectively a system operates over a certain period of time. More specifically, reliability refers to the likelihood that a system will perform up to its specifications for an expected period of time. In many applications, it is critical to measure and predict the reliability of a product, process, or system. Reliability tactics are used to improve the system's performance.

Maintenance is the process that keeps a system operational and reduces the variability and failure in its operational availability, performance, and quality. Maintenance can also be used to increase the reliability of a system. There are two types of maintenance: preventive and breakdown. Preventive maintenance is a process that ensures a system stays operational, while breakdown maintenance is performed after the system has failed and needs to be repaired or replaced. Maintenance tactics include implementing preventive maintenance and decreasing repair requirements.

Reliability and maintenance models are useful in predicting a system's performance and developing an optimal, economical plan that allows for maintaining smooth operations, preventing disruptions, and gaining financial advantages. They are also used to design systems to perform the required capabilities while costs are kept at a minimum. Proper reliability measurement and maintenance planning for components of a system will have an enormous impact on the overall operation of the system. However, the effects of reliability and maintenance planning are not limited to the machinery. Employee morale will also be positively affected by the ability to perform jobs based upon a high confidence level that the system is reliable. Furthermore, the use of reliability and maintenance models has a

Note: Advanced material that can be omitted without loss of generality will be indicated by a dagger.

Operations and Production Systems with Multiple Objectives, First Edition. Behnam Malakooti.
© 2014 John Wiley & Sons, Inc. Published 2014 by John Wiley & Sons, Inc.

significant impact in terms of customer satisfaction and meeting expected performance and life of products. These models are used by industries, businesses, service industries, and other types of organizations. The economic well-being of an institution or a business could very well hinge upon its approach for identifying and applying proper reliability and maintenance programs. Therefore, the amount of money spent on reliability and maintenance must be carefully examined.

Reliability measurements are concerned with the performance of a component, a product, or a system over a period of time. The clear distinction between quality control systems and reliability systems is that in quality control the quality of the process and the quality of the produced items are measured during the manufacturing process or right after the products are produced. In contrast, for reliability after a product (a simple component or a system) is put to use, its performance over its useful life is measured. However, a system's failure occurs due to interactions among several components. The system failure measurement and prediction are often complex and require the use of advanced analytical or simulation tools. For example, the major disasters of the two NASA space shuttles, *Challenger* and *Columbia*, were blamed on poor reliability of certain components and systems. These components and systems successfully passed the testing procedures but failed in the actual application.

This chapter covers the basics of reliability, failure rates, maintenance, and replacement measurements and methods. Furthermore, we introduce our multicriteria approaches for reliability, maintenance, and replacement policies.

19.2   RELIABILITY OF SINGLE UNITS

The reliability of a single component can be represented by a probability distribution because the exact time of failure for a component cannot be predicted. That is, the lifetime of a component, T, is a random variable. Consider a given point in time t. The cumulative distribution function for random variable T is defined as $F(t) = P\{T \leq t\}$, which is the probability of failure at or before a given time t. This failure function is shown in Equation (19.1).

The survival cumulative function is the probability that the component does not fail up to the given time t. The survival function is defined as $R(t) = 1 - F(t)$, as shown in Equation (19.2). The cumulative distribution function $F(t)$ is a differentiable function whose derivative is called a probability density function, or $f(t) = \delta F(t)/\delta t$.

The failure rate function $r(t)$, represents the likelihood that the unit, which has been operational up to a point in time t, will fail in the next instant, $t+s$, where s is a very small amount of time; see Equation (19.3). Therefore, the failure rate equation (19.3) is defined as the conditional probability $P\{t < T \leq t+s \mid T > t\}$, where the failure density function $f(t)$ is known. Important measurements related to reliability are:

(a) Failure (cumulative) function:

$$F(t) = P\{T \leq t\} \tag{19.1}$$

(b) Survival (cumulative) function:

$$R(t) = 1 - F(t) = P\{T > t\} \tag{19.2}$$

(c) Failure rate function:

$$r(t) = \frac{f(t)}{R(t)} = P\{t < T = t + s | T > t\} \tag{19.3}$$

Depending on the probability functions, the failure rate may decrease, increase, first decrease and later increase, or be independent of time. For example, as time increases, the failure rate may increase for many consumer products (e.g., automobiles). This might be due to simple wear and tear of a product (e.g., a car battery fails after a certain amount of time).

19.2.1 Exponential Reliability Functions

For many practical cases, the exponential distribution is suitable for describing the failure characteristics of an operational component or a single processor.

In the exponential distribution,

$$f(t) = \lambda e^{-\lambda t} \quad \text{(failure probability density function)} \tag{19.4}$$

$$F(t) = 1 - e^{-\lambda t} \quad \text{(failure cumulative function)} \tag{19.5}$$

where $e = 2.7183$. This function depends on the value of the parameter λ, where $1/\lambda$ is defined as the expected failure time. The standard deviation of the failure time is also $1/\lambda$ for exponential distribution. For exponential distribution, the survival function is

$$R(t) = 1 - F(t) = 1 - (1 - e^{-\lambda t}) = e^{-\lambda t} \quad \text{(survival cumulative function)} \tag{19.6}$$

The failure rate is

$$r(t) = \frac{f(t)}{R(t)} = (\lambda e^{-\lambda t})/(e^{-\lambda t}) = \lambda$$

on

$$r(t) = \lambda \quad \text{for} \quad \lambda > 0 \quad \text{(failure rate function)} \tag{19.7}$$

This means that for the exponential distribution the failure rate is constant and independent of time.

Example 19.1   Calculating *f(t)*, *F(t)*, and *R(t)*   Consider a company that produces a specific type of car battery. The company's records show that the expected failure time (expected life) of this car battery is five years.

(a) Find the failure and survival functions for a five-year-old battery.
(b) Determine the probability that a random car battery lasts longer than $t = 1, 2, 3, 4, 5, 10, 15, 20$ years.

The solution is as follows:

(a) Since the expected failure time is 5 years, $1/\lambda = 5$. This means that the failure rate is $r(t) = \lambda = 0.2$. Therefore, for an exponential function

$$
\begin{aligned}
f(t) &= 0.2e^{-0.2t} & \text{(failure probability density function)} \\
F(t) &= 1 - e^{-0.2t} & \text{(failure cumulative function)} \\
R(t) &= 1 - F(t) = e^{-0.2t} & \text{(survival cumulative function)} \\
r(t) &= 0.2 & \text{(failure rate)}
\end{aligned}
$$

The probability that a battery fails in five years or less, $P(T \leq 5)$, is represented by the failure function

$$
F(5) = 1 - e^{-0.2(5)} = 0.632
$$

That is, there is a 63.2% probability that the battery will fail up to the end of year 5. Now, consider the survival function, the probability that it will last more than five years, $P(T > 5)$:

$$
R(5) = e^{-0.2(5)} = 0.368
$$

That is, there is 36.8% probability that the battery will not fail during the first five years.

   Figure 19.1 illustrates the functions of this example. The area under the density function for $0 \leq T \leq 5$ is $F(5) = 0.632$ and the area for $T > 5$ is 0.368. Note that the total area is $0.632 + 0.368 = 1$.

(b) Therefore, the probability that a random car battery lasts longer than $t = 1, 2, 3, 4, 5, 10, 15, 20$ years is presented in the following table. Note that the probability of failure before year 10 is 0.865, or 86.5%, which is higher than year 5 with 63.2% probability.

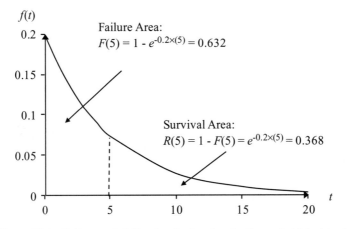

Figure 19.1   Failure probability density function for Example 19.1 at $t = 5$.

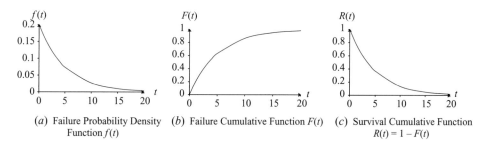

(a) Failure Probability Density Function $f(t)$ (b) Failure Cumulative Function $F(t)$ (c) Survival Cumulative Function $R(t) = 1 - F(t)$

Figure 19.2   Probability functions for Example 19.1, where $\lambda = 0.2$ for exponential distribution.

The three functions are presented in Figure 19.2.

Time	t	1	2	3	4	5	10	15	20
Failure probability density function	$f(t)$	0.164	0.134	0.110	0.090	0.074	0.027	0.010	0.004
Failure cumulative function	$F(t)$	0.181	0.330	0.451	0.551	0.632	0.865	0.950	0.982
Survival cumulative function	$R(t) =$ $1 - F(t)$	0.819	0.670	0.549	0.449	0.368	0.135	0.050	0.018

Example 19.2   Calculating $f(t), F(t), R(t),$ and $r(t)$ Consider Example 19.1. Determine:

(a) The probability that a random battery survives longer than eight years
(b) The proportion of all batteries that fail during the third year (i.e., $2 \leq T \leq 3$)
(c) The proportion of batteries which fail during the sixth year given that they have survived at least five years
(d) The proportion of batteries that fail during the fourth and fifth years

The solution is as follows:

(a) Use $R(t)$, that is, $P\{T > 8\} = e^{-0.2(8)} = 0.202$.
(b) Use $F(t)$, that is, $P\{2 \leq T < 3\} = P\{T < 3\} - P\{T \leq 2\} = 0.451 - 0.330 = 0.121$.
(c) Use $F(t)$ and $R(t)$, that is,

$$P\{5 \leq T < 6 | T > 5\} = \frac{P\{5 \leq T < 6\}}{P\{T > 5\}} = \frac{F(6) - F(5)}{R(5)}$$

$$= \frac{0.699 - 0.632}{0.367} = \frac{0.067}{0.367} = 0.182$$

(d) Use $F(t)$, that is, $P\{T \leq 5\} - P\{T \leq 3\} = 0.632 - 0.451 = 0.181$. This is shown in Figure 19.3.

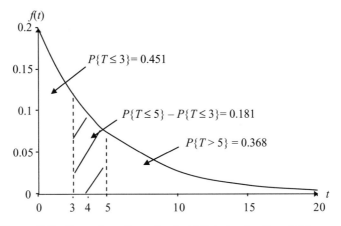

Figure 19.3   Proportion of batteries that fail during fourth and fifth years.

19.2.2   Multicriteria Reliability of Single Unit

The total cost of the reliability of a product consists of:

(a) Manufacturing costs (including design, materials, and quality control)
(b) Failure and warranty costs (including cost of returned items due to failures and liability cost)

Suppose that there are several alternatives for designing a product. As the expected life of a product increases, its manufacturing cost also increases, but its failure and warranty costs decrease. The total cost first decreases and then increases. See Figure 19.4, where the expected life that minimizes the total cost is t^*.

Suppose that a given industry has a specific life goal for the product, for example, it provides warranty up to a certain life, called the target life t^o. Therefore, the probability for

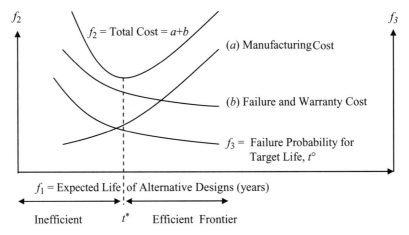

Figure 19.4   Total cost of different design alternatives with different expected lives. The design alternative associated with t^* minimizes the total cost.

failure at the target life decreases as the expected life increases. In multicriteria evaluation of design alternatives, the following three objectives can be considered. See Figure 19.4.

Maximize expected life, f_1
Minimize total cost per unit, f_2
Minimize the probability of failure for a given target life t^o, $f_3 = F(t = t^o)$
Subject to the set of design alternatives

Suppose that there are K alternative designs denoted by $k = 1, 2, \ldots, K$. In exponential distribution, for the target life t^o, the third objective can be calculated as

$$f_3 = F(t = t^o) = 1 - e^{-\lambda_k t^o} \quad \text{for design alternative } k$$

where $1/\lambda_k$ is the expected life of design alternative k.

In Figure 19.4, all design alternatives to the left of t^* are inefficient, and alternatives to its right are efficient. The best alternative will be one of the efficient alternatives.

In order to rank the multicriteria alternatives, a MCDM method can be used. A simple MCDM method is to ask the DM to directly rank alternatives based on presented objectives for each alternative.

Example 19.3 Multicriteria Reliability of Single Unit   Consider a company that produces car batteries. For a specific battery, the company is considering nine possible design alternatives. Each design has a different expected life (from one to nine years) and different total costs (from $24 to $60) per unit. See Table 19.1. In this example, suppose the target life is five years, $t^o = 5$. Thus, the warranty is given for only five years. Generate multicriteria alternatives for this problem and identify the efficient alternatives. Ask the DM to rank alternatives and choose the best one.

The expected life versus the total cost of Table 19.1 is presented in Figure 19.5 for nine different design alternatives. Suppose the DM ranks alternatives as presented in the last column of Table 19.1. The best design alternative is number 6. Figure 19.5 shows that the expected life for alternative 4 is greater than alternatives 1, 2, and 3, and the cost and probability of failure for target life $t^o = 5$ is less. Therefore, design alternative 4 dominates design alternatives 1, 2, and 3.

TABLE 19.1   Data and Solution for Example 19.2

Design Alternative, k	Max. f_1 = Expected Life f_1 (Given)	Min. f_2 = Cost ($) (Given)	Failure Rate, $\lambda_k = 1/$Exp. Life	Min. $f_3 = F$ ($t^o = 5$)	Efficient?	Ranking
1	1	50	$1/1 = 1$	0.993	No	9
2	2	39	$1/2 = 0.5$	0.918	No	7
3	3	30	$1/3 = 0.333$	0.811	No	5
4	4	24	$1/4 = 0.25$	0.713	Yes	3
5	5	26	$1/5 = 0.2$	0.632	Yes	2
6	6	30	$1/6 = 0.167$	0.565	Yes	1
7	7	34	$1/7 = 0.143$	0.510	Yes	4
8	8	40	$1/8 = 0.125$	0.465	Yes	6
9	9	60	$1/9 = 0.111$	0.426	Yes	8

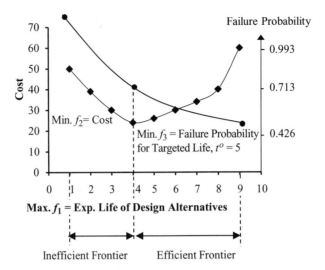

Figure 19.5   Graphical presentation of three objectives for Example 19.2 and Table 19.1.

To generate alternatives presented in Table 19.1, find $\lambda = 1$/expected life for alternative k for each given expected life. Then, find the failure probability for the given target life $t^o = 5$. For example, for an expected life of three years, $\lambda_3 = 1/3 = 0.333$:

$$F(t) = 1 - e^{-0.333t} \qquad \text{failure cumulative probability function}$$
$$F(t = 5) = 1 - e^{-0.333(5)} = 0.811 \quad \text{failure probability before target life } t^o = 5$$

Therefore, the failure probability before year 5 for a battery that has an expected life of three years is 0.811. Similarly, for an expected life of five years, $\lambda_5 = 0.2$, and the failure probability for the target life is $F(t = 5) = 0.632$.

19.3   MEAN TIME BETWEEN FAILURE

19.3.1   Mean Time between Failure for Single Units

Consider a system that consists of only one processor (one piece of equipment). Based on past observations, suppose the durations between failures are $T_1, T_2, \ldots, T_n$ and the durations to repair each failure are $D_1, D_2, \ldots, D_n$. The total time of this system is

$$\text{Total time} = (T_1 + D_1) + (T_2 + D_2) + \cdots + (T_n + D_n)$$

Consider the following two definitions:

$$\text{MTBF} = \frac{1}{n} \sum_{i=1}^{n} T_i \qquad\qquad \text{MTTR} = \frac{1}{n} \sum_{i=1}^{n} D_i$$

where MTBF is the mean time between failure for the unit and MTTR is the mean repair time. The availability of a system is calculated as

$$\text{System availability} = \frac{\text{MTBF}}{\text{MTBF} + \text{MTTR}} \tag{19.8}$$

Discrete Probability Function For a discrete probability function, find the expected values and then find the system availability using Equation (19.8). The following example demonstrates this concept.

Example 19.4   Finding System Availability   Consider a system with the following recorded running and repair times in days. In the first period, the system is operational for 12 days; then it is under repair for 2 days. Then at time $12 + 2 = 14$, it becomes operational. Find its system availability. In this example, the system was studied for 122 days.

Event number, i	1	2	3	4	5	Total	Expected value
Operational time, T_i	12	18	2	50	10	92	MTBF $= 92/5 = 18.4$
Repair time, D_i	2	6	5	8	9	30	MTTR $= 30/5 = 6$

$$\text{System availabilty} = \frac{18.4}{18.4 + 6} = 0.754$$

Therefore, this system is available 75.40% of the time.

Exponential Distribution Application to MTBF Exponential distribution can be used for the analysis of MTBF for single units. It is assumed that after each repair the unit has the same expected probability of failure and the same probability distribution function. In this case,

$$\text{Mean (MTBF)} = \bar{T} = \frac{1}{\lambda}$$
$$F(t) = 1 - e^{-\lambda t} \qquad \text{(failure probability)} \qquad (19.9a)$$
$$R(t) = 1 - F(t) = e^{-\lambda t} \quad \text{(survival probability)}$$

Example 19.5   System Availability Using Exponential Distribution   Consider Example 19.4. For an exponential distribution function, find the probability that the system does not fail in the first 10 days.

In this example, $\bar{T} = \text{MTBF} = 18.4$; therefore, $\lambda = 1/\bar{T} = 1/18.4 = 0.054$. The survival probability after 10 days is

$$P\{T > 10\} = R(t = 10) = e^{-\lambda t} = e^{-0.054 \times 10} = 0.583 \quad \text{(or 58.3\%)}$$

Normal Distribution Application to MTBF A normal distribution can also be used for MTBF applications where a unit after repair will have the same expected probability of failure and the same normal probability distribution function. For a normal distribution,

$$\text{Mean (MTBF)} = \bar{T} = \mu$$
$$\text{Standard deviation (SD)} = \sigma = \sqrt{\frac{\sum_{i=1}^{n}(T_i - \bar{T})^2}{n - 1}} \qquad (19.9b)$$

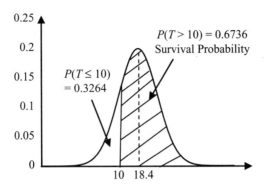

Figure 19.6   Example 19.6: Probability that the system does not fail in the first 10 days is 0.6736.

Then, find probabilities for a given period using Z values from the standard normal probability tables given at the back of the book. This is shown in the following example.

Example 19.6   System Availability Using Normal Distribution   Consider Example 19.4. For a normal distribution, find the probability that the system does not fail in the first 10 days.

In this example, we need to find $P\{T > 10\}$. Note that MTBF $= \bar{T} = 18.4$ days and the standard deviation is

$$
\text{SD} = \sqrt{\frac{(12 - 18.4)^2 + (18 - 18.4)^2 + (2 - 18.4)^2 + (50 - 18.4)^2 + (10 - 18.4)^2}{5 - 1}} = 18.56
$$

First obtain $F(T = 10) = P\{T \leq 10\}$ by finding

$$
P\left\{\frac{T - \mu}{\sigma} \leq \frac{10 - 18.4}{18.56}\right\} \quad \text{or} \quad P\{Z \leq -0.45\}
$$

Now, find $P(Z \leq -0.45)$ from the normal distribution table at the back of the book. Note that because the normal distribution is symmetric, $P\{Z \leq -0.45\} = P\{Z \geq 0.45\}$. Using the normal distribution table, we find that $P\{Z \leq 0.45\} = 0.6736$. Therefore,

$$
P\{Z \geq 0.45\} = 1 - P\{Z \leq 0.45\} = 1 - 0.6736 = 0.3264
$$

Therefore, the unit has a 32.64% chance of failure before 10 days. The probability that the unit does not fail in the first 10 days is given as (also see Figure 19.6.)

$$
P\{T > 10\} = 1 - P\{T \leq 10\} = 1 - 0.3264 = 0.6736 \quad \text{(or 67.36\%)}
$$

Using reliability notation,

$$
R(T = 10) = 1 - F(T = 10) = 1 - 0.3264 = 0.6736
$$

19.3.2 Mean Time between Failure for Multiple Units

A typical measure of reliability of a system of multiple units is the failure rate. The failure rate (FR) can be measured in two ways:

(a) Number of failures divided by total number of units tested: FR(%)
(b) Number of failures divided by operating time: FR(N)

That is,

$$FR(\%) = \frac{\text{number of failures}}{\text{number of units tested}} \times 100\% \quad \text{(percent of failures per units tested)}$$

(19.10)

$$FR(N) = \frac{\text{number of failures}}{\text{actual total operating time}} \quad \text{(failures per unit time)} \tag{19.11}$$

where FR(%) provides a measure for the percentage of failures based on the total number of units (processors) and FR(N) provides a measure of the number of failures per unit of time.

A very important measurement in the analysis of reliability of multiunit systems is the MTBF of the system. The MTBF can be interpreted as the average time between failures for a group of units running simultaneously and constantly. This term is defined as the inverse of the failure rate, FR(N), as shown in Equation (19.11):

$$\text{MTBF of system} = \frac{1}{FR(N)} \tag{19.12}$$

Example 19.7   Calculating FR and MTBF of a System (MTBF of System)   Consider 15 boilers that are used to heat up an apartment building complex which has a central heating system. All units were observed for 2000 h. Two units failed; the first after 250 h and the second after 1900 h. Suppose that once a failure occurs, the failed boiler cannot be used for the remainder of the given observed period.

Find the FR(%), FR(N), and MTBF for this system.

The failure rate percentage, FR(%), can be found as

$$FR(\%) = \frac{2}{15} \times 100 = 13.3\%$$

Therefore, 13.3% of the units failed out of the 15 units.

Now find the failure rate number, FR(N), based on the total hours in operation.

To find the total operation time, simply add together the operation times for each unit. Thirteen of the units did not fail, so their operation time is $2000 \times 13 = 26{,}000$. The two units that did fail were operational for 250 and 1900 h. Thus

$$\text{Total operation time} = 2000 \times 13 + 250 + 1900 = 28{,}150$$

Therefore,

$$FR(N) = \frac{2}{28,150} = 0.000071 \text{ failure/h}$$

Therefore, 0.000071 units will fail in every operating hour. The MTBF for the system can now be calculated:

$$MTBF = \frac{1}{0.000071} = 14,075 \text{ h}$$

That is, the average time between two consecutive system failures will be 14,075 h. (or about 1.6 years) for units running constantly.

19.3.3   Failure Rate Signature

Each system has different characteristics in terms of the relationship of its age (life) versus its failure rate. Figure 19.7 presents the failure rate curve, which shows the failure rate of a system at different time intervals. The system has different failure rates at different periods. This type of graph is known as a bath tub curve because of its shape. Most companies use this curve to determine the warranty period of their new products.

   In the early stages of a bath tub curve, the failure rate is high. This is known as the infant mortality stage. As time elapses, the failure rate decreases. Eventually, the failure rate becomes constant. At this stage the system is stable and the failure rate remains constant until the system reaches a certain age. At this point, the failure rate increases as the unit wears out.

   Figure 19.8 shows the cumulative failure rate (or probability of failure) of the same system. The slope at any point determines the failure rate at that particular instant. The slope of the curve (failure rate) is initially high. It becomes constant during the useful life and finally increases during wear-out.

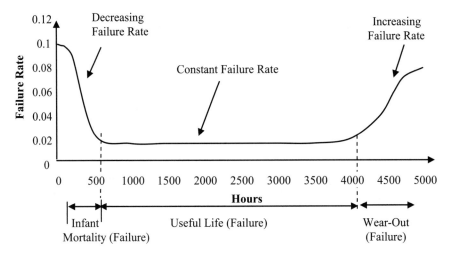

Figure 19.7   Failure rate over time.

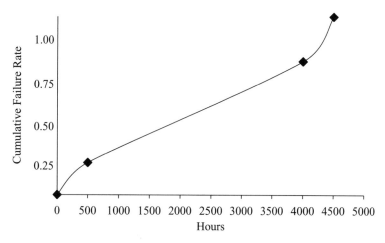

Figure 19.8   Cumulative failure rate over time.

19.4   RELIABILITY OF MULTIPLE UNITS

Consider a system that is comprised of n processors where each processor performs different tasks. Each processor (also called a component) is independent of other processors and has its own reliability. The organization of processors and the form of sequences in which tasks are processed may vary for different systems. The reliability of a system is defined by the probability of all processors running (successfully) simultaneously. There are three ways to organize processors: (a) series (linearly sequential), (b) parallel, or (c) hybrid (a mix of series and parallel). These three organizations are discussed below.

19.4.1   Series Systems

In series systems, all processors are connected to each other sequentially in linear form similar to a chain. If one processor fails, the whole system fails. Figure 19.9 shows an example of processors in series. Suppose the reliability of each processor is represented by R_i for $i = 1, 2, \ldots, n$. The reliability of a series system, R_S, is the product of the reliabilities of the various processors:

$$R_s = R_1 \times R_2 \times \cdots \times R_n \qquad (19.13)$$

The value of R_S is independent of the order of the sequence. For example, the same R_S value is calculated for the reverse sequence of n processors ($n, n - 1, \ldots, 2, 1$); that is $R_S = R_n \times R_{n-1} \times \ldots \times R_2 \times R_1$.

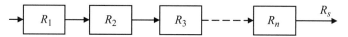

Figure 19.9   Series system of n processors.

Example 19.8 Series Systems Consider a series system consisting of three processors where the reliabilities of the processors are $R_1 = 0.95$, $R_2 = 0.90$, and $R_3 = 0.85$. Find the reliability of the system:

$$R_s = 0.95 \times 0.90 \times 0.85 = 0.727 \quad (\text{or } 72.7\%)$$

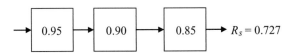

The reliability of this system is 0.727, or 72.7%. This value is smaller than the reliability of each individual component in the series. For processors in series, the reliability of each individual component is always higher than the reliability of the system.

19.4.2 Parallel Systems

The reliability of a series system can be improved by duplicating its processors. This duplication is also called redundancy. If one processor fails, the duplicated processor will be used to avoid a system failure. In other words, the system will be operational as long as at least one of the duplicated processors is working. Figure 19.10 shows an example of a parallel system where n duplicated processors are used.

The reliability of a system with n parallel processors, R_P, is calculated as follows:

$$R_P = 1 - [(1 - R_1) \times (1 - R_2) \times \cdots \times (1 - R_n)] \tag{19.14}$$

Note that R_p is the same regardless of the order of processors.

Example 19.9 Parallel Systems Consider a system consisting of three processors as shown below. The reliabilities of the processors are $R_1 = 0.7$, $R_2 = 0.9$, and $R_3 = 0.9$. The three processors can be substituted for each other to achieve the same performance. Find the reliability of this parallel system:

$$R_P = 1 - [(1 - 0.70) \times (1 - 0.90) \times (1 - 0.90)] = 0.997$$

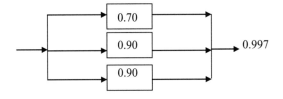

This system can successfully run 99.7% of the time. Due to redundancy, the total system reliability is always higher than the reliability of components.

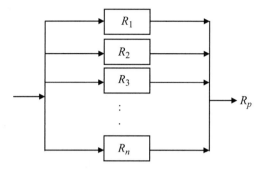

Figure 19.10   Parallel system with n processors.

19.4.3   Hybrid Systems

A hybrid system is a combination of series and parallel systems. See the following example.

Example 19.10   Combination of Parallel and Series systems   Consider a hybrid system consisting of five processors as shown below. The reliabilities of the processors are $R_1 = 0.9, R_2 = 0.9, R_3 = 0.7, R_4 = 0.9$, and $R_5 = 0.8$. Find the reliability of the hybrid system.

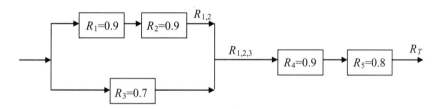

First, calculate the reliability for each of the subsystem series. The reliability of processors 1 and 2 is

$$R_{1,2} = R_1 R_2 = 0.9 \times 0.9 = 0.81$$

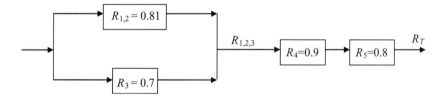

Now, use the above result to find the reliability of processors 1, 2, and 3 together:

$$R_{1,2,3} = 1 - [(1 - R_{1,2}) \times (1 - R_3)] = 1 - [(1 - 0.81) \times (1 - 0.7)] = 0.943$$

Now, consider a series system consisting of $R_{1,2,3}$, R_4, and R_5. Let R_T be the total reliability of the system:

$$R_T = R_{1,2,3}R_4R_5 = 0.943 \times 0.9 \times 0.8 = 0.679 \quad \text{(or 67.9\%)}$$

19.5   MULTICRITERIA RELIABILITY OF MULTIPLE UNITS[†]

In this section, we provide multicriteria analysis for the following three problems for improving the reliability of a multiunit system. Define R_i as the reliability of a subsystem and R'_i as the reliability of the subsystem after it is improved by one of the following three methods:

(a) **Use of Improved Units** For example, in Figure 19.11 there are three units. Each unit can be replaced with a unit that has a better reliability at the expense of a higher cost. In this example, the reliability of the original three units are R_1, R_2, and R_3, respectively. The reliabilities of improved units are R_1', R_2', and R_3'. Therefore, the system reliability increases from $R_T = R_1 R_2 R_3$ to $R_T = R_1' R_2' R_3'$.

(b) **Use of Emergency Spares or Buffers** For example, in Figure 19.12 a spare unit is added for each of the three units. In this case, the reliability of spare units (R_1^o, R_2^o, R_3^o) are less than the reliabilities of the original units. But the total system reliability increases to

$$R_T = (1 - (1 - R_1)(1 - R_1^o))[1 - (1 - R_2)(1 - R_2^o)][1 - (1 - R_3)(1 - R_3^o)]$$

(c) **Use of Duplicated Units** For example, in Figure 19.13 the original system has three units. Suppose each unit can be duplicated up to two times. In this case, the reliability of each duplicated unit is the same as the reliability of the original unit. The system reliability will improve to

$$R_T = [1 - (1 - R_1)^3][1 - (1 - R_2)^3][1 - (1 - R_3)^3]$$

It is possible to use a combination of the above three methods to improve the system reliability. In general (for the above three problems), we can consider the following two objectives:

Maximize system' reliability, $f_1 = R_T$

Minimize cost of improving system' reliability, f_2

Subject to reliability constraints and equations: budget limitations and space constraints

In the following section, we develop an algorithm called the reliability cost trade-off method that can be used to solve multicriteria problems. We only show details for the first problem.

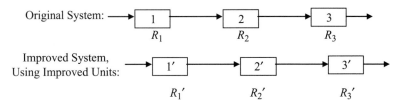

Figure 19.11   Use of improved units to increase system reliability.

Original Units R_1 1 R_2 2 R_3 3

Spare or Emergency Units 1° 2° 3°

Subsystem Reliability R_1' R_2' R_3'

Figure 19.12   Use of spare units to increase system reliability.

Supplement S19.1.docx provides a detailed approach for the following:

    (a) MCDM—use of emergency spares or buffers

    (b) MCDM—use of duplicated units

    (c) Multiple-objective optimization of reliability

Enumeration of All Alternatives   For small problems, all alternatives can be generated and their objectives can be measured. Then, the best alternative can be selected. For larger problems, use the reliability–cost trade-off method provided in the next section.

Example 19.11   Multicriteria Reliability: Enumerating All Alternatives   Consider a system consisting of two units in series with reliabilities of $R_1 = 0.9$ and $R_2 = 0.95$. The reliability of the two units can be improved to $R_1' = 0.99$, and $R_2' = 0.998$. The cost of improving unit 1 is \$120 and the cost of improving unit 2 is \$140. Find all possible multicriteria alternatives and identify efficient alternatives.

In this example, the reliability of the basic (or existing) system is $R_T = 0.9 \times 0.95 = 0.855$.

The four bicriteria alternatives are summarized in the table below. For example, for alternative a_4, the system reliability is $R_T = 0.99 \times 0.998 = 0.988$, and its additional cost compared to the original system is $120+140 = \$260$.

	a_1	a_2	a_3	a_4
R_1	0.90	0.99	0.90	0.99
R_2	0.95	0.95	0.998	0.998
Max. f_1 = system reliability; R_T	0.855	0.941	0.898	0.988
Min. f_2 = additional cost	0	120	140	260
Efficient?	Yes	Yes	No	Yes

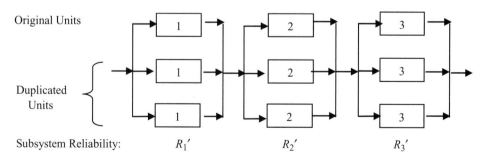

Original Units 1 2 3

Duplicated Units 1 2 3 / 1 2 3

Subsystem Reliability: R_1' R_2' R_3'

Figure 19.13   Use of duplicated units to increase system reliability.

In this example, a_2 dominates a_3 because a_2 has a lower cost but a higher reliability. Therefore, a_3 is inefficient and can be eliminated from consideration. Now, one can use a multicriteria method to choose the best alternative.

19.5.1 Reliability–Cost Trade-off Algorithm[†]

As the number of units in a system and/or the number of possible improvements for each unit increases, the number of alternatives increases exponentially. However, many of these alternatives will be inefficient. In the following, we develop a method that generates a set of efficient alternatives. In each iteration, a unit is selected that has the lowest cost–reliability ratio. This method generates an efficient frontier within a limited number of iterations.

The steps of the method are as follows. Consider a system with n units.

1. Calculate the reliability of the current system, R_T.
2. Calculate the system reliability if each unit i is improved, defined as $R_{T,i}$.
3. For each unit i, calculate the difference in the system reliability if the unit is improved, that is, find $R_{\text{dif},i} = (R_{T,i} - R_T)$ for $i = 1, 2, \ldots, n$.
4. Rank the units in ascending order of $c_i/R_{\text{dif},i}$ where c_i is the additional cost for improving unit i.
5. Choose the highest ranked unit which is feasible in terms of budget and space. Label this the current solution.
6. Measure the system reliability R_T, label it as f_1, and find the total additional cost f_2 for the given alternative.
7. Repeat steps 2–6 until the maximum improvements for all units are achieved (or the budget or space limitations do not allow any more improvement).

19.5.2 Use of Improved Units to Increase System Reliability[†]

The system's reliability can be improved by increasing the reliability of each unit. For this problem, consider the following notations:

$k = 1, 2, \ldots, K$: Number of successive improvement alternatives for each unit

$R_{i,k}$: Reliability of unit i for alternative k in ascending order that is, $R_{i,k} > R_{i,k-1}$

$c_{i,k}$: Additional cost of unit i for alternative k; if the component is upgraded from alternative $k - 1$ to alternative k.

The initial system units are labeled with $k = 1$, where it has the lowest reliability and the lowest cost for each unit i and for the system. Now, apply the reliability–cost trade-off procedure where the system reliability R_T is calculated based on the given reliability problem. The system can be in a series, parallel, or hybrid configuration. However, it may not guarantee generating an efficient solution for configurations other than series. In series, the increase in the reliability of the system depends only on the reliability of the component itself, which is increased, but in parallel (also hybrid) it depends on the values of the reliability of other components too, so it depends on the choices made in previous steps.

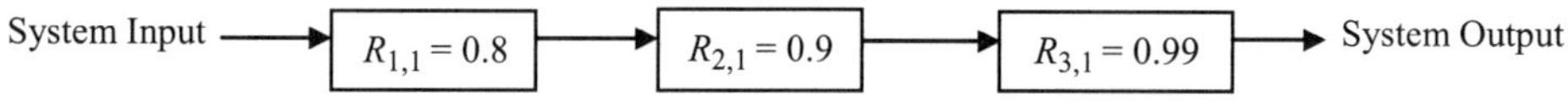

System Input → $\boxed{R_{1,1} = 0.8}$ → $\boxed{R_{2,1} = 0.9}$ → $\boxed{R_{3,1} = 0.99}$ → System Output

Figure 19.14 Example 19.12, initial system with three original units with system reliability $R_T = 0.8 \times 0.9 \times 0.99 = 0.713$.

Example 19.12 Use of Improved Units Consider a system that consists of three components in series as shown in Figure 19.14. All the needed information is presented in Table 19.2. The budget for this project is \$65,000. There are five possible cases for each unit. For example, for unit 1, the cost of improving the existing unit to the next level is \$10 (in thousands), which improves its reliability from 0.8 to 0.85. However, to improve its reliability from 0.85 to 0.9 (alternative $k = 3$), the cost is $10 + 12 = \$22$ (in thousands of dollars).

Generate a set of efficient bicriteria alternatives where the total cost does not exceed the given budget.

Now, apply the reliability–cost trade-off method. At iteration 0, reliability $f_1 = R_T = 0.9 \times 0.8 \times 0.99 = 0.713$. The solution is provided in Table 19.3 and $c_1/R_{\text{diff},1} < c_2/R_{\text{diff},2}$. Therefore, unit $i^* = 1$ is selected for reliability improvement in the text iteration. Note that unit 3 is not considered because it cannot be improved.

For example, the details of calculations for iteration 0 are presented in Table 19.4.

See Supplement S19.2.xlsx.

In iteration 0, unit 1 with the lowest ratio of 227.27 is selected (bolsface in the table). The new system reliability for the next iteration is calculated as

$$R_T = 0.85 \times 0.9 \times 0.99 = 0.757$$

Table 19.3 shows all iterations. In iteration 5, the cost of improving units 1 and 2 are 50 and 20, respectively. Therefore, the total cost would be either $62 + 50 = 112$ or $62 + 20 = 82$. Both these costs are more than the budget of 65. Therefore, stop at iteration 5. The best solution for the \$65,000 budget is to improve the reliability of unit 1 to 95% and unit 2 to 94%. The system reliability will be $R_T = 0.884$. The total additional cost would be \$62,000. The set of six efficient alternatives for iterations 0–5 (for the budget of 65) are presented in Table 19.3. Figure 19.15 shows all multicriteria alternatives regardless of the budget limitation.

The efficient frontier shows that with very small initial investment substantial gain in reliability can be achieved. But for higher reliability substantial investment is needed. The highest reliability is

$$R_T = 0.99 \times 0.99 \times 0.99 = 0.97$$

TABLE 19.2 Example 19.12 Reliability and Cost Information

Alternative, k	1	2	3	4	5	Total Additional Cost
Unit $i = 1$, $R_{1,k}$ reliability	0.8	0.85	0.9	0.95	0.99	—
$c_{1,k}$, Additional cost, \$ (thousands)	0	10	12	20	50	92
Unit $i = 2$, $R_{2,k}$ reliability	0.9	0.92	0.94	0.96	0.99	—
$c_{2,k}$, Additional cost, \$ (thousands)	0	5	15	20	40	80
Unit $i = 3$, $R_{3,k}$ reliability	0.99	0.99	0.99	0.99	0.99	—
$c_{3,k}$, Additional cost, \$ (thousands)	0	0	0	0	0	0

TABLE 19.3 Alternatives for Example 19.12

Iteration alternative	R_1, R_1'	R_2, R_2'	R_3, R_3'	$R_{T,1}, R_{\text{dif},1}$	$R_{T,2}, R_{\text{dif},2}$	$c_1, c_1/R_{\text{dif},1}$	$c_2, c_2/R_{\text{dif},2}$	i^*	Maximum $f_1 = R_T$	Mininum f_2
0	0.8, 0.85	0.9, 0.92	0.99, N/A	0.757, 0.044	0.729, 0.016	10, 227.27	5, 312.50	1	0.713	0
1	0.85, 0.9	0.9, 0.92	0.99, N/A	0.802, 0.045	0.774, 0.017	12, 266.67	5, 294.12	1	0.757	10
2	0.9, 0.95	0.9, 0.92	0.99, N/A	0.846, 0.044	0.820, 0.018	20, 454.55	5, 277.78	2	0.802	22
3	0.9, 0.95	0.92, 0.94	0.99, N/A	0.865, 0.045	0.838, 0.018	20, 444.44	15, 833.33	1	0.820	27
4	0.95, 0.99	0.92, 0.94	0.99, N/A	0.902, 0.037	0.884, 0.019	50, 1351.35	15, 789.47	2	0.865	47
5	0.95, 0.99	0.94, 0.96	0.99, N/A	0.921, 0.037	0.903, 0.019	50, 1351. 35	20, 1052.63	–	0.884	62
6				Violates budget limitation in next iteration					0.903	82

Note: All alternatives are efficient. N/A = not applicable.

TABLE 19.4   Details for Iteration 0 of Table 19.3

i	Improved	System Reliability, $R_T = R_1 R_2 R_3$	System Improvement, $R_{\mathrm{dif},i} = R_{T,i} - R_T$	$c_i/R_{\mathrm{dif},i}$
	Original	$R_T = 0.8 \times 0.9 \times 0.99$ $= 0.713$	—	—
1	R_1'	$R_{T,1} = 0.85 \times 0.9 \times 0.99$ $= 0.757$	$0.757 - 0.713 = 0.044$	$10/0.044 =$ **227.27**
2	R_2'	$R_{T,2} = 0.8 \times 0.92 \times 0.99$ $= 0.729$	$0.729 - 0.713 = 0.016$	$5/0.016 = 312.5$

The cost of this system is $92 + 80 = \$172,000$. One of the multicriteria alternatives can be selected as the best alternative.

See Supplement S19.3.xlsx.

19.6   MAINTENANCE POLICIES

19.6.1   Single-Objective Maintenance Policy

Maintenance is the upkeep of a system by performing repairs and services. Its purpose is to reduce the likelihood of system breakdowns and performance variability. Maintenance falls under two categories:

Preventive Maintenance The routine checkup and servicing of the system.

Breakdown Repair Maintenance The maintenance that occurs after the system fails.

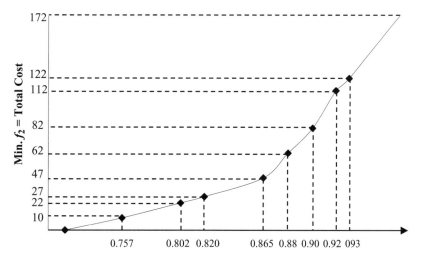

Figure 19.15   Efficient frontier for Example 19.12: total reliability vs. total additional cost for all alternatives.

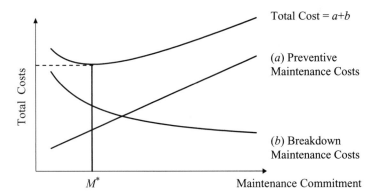

Figure 19.16   Maintenance curve with respect to preventive and breakdown costs.

Figure 19.16 illustrates the relationship of total cost versus maintenance commitments. Maintenance commitments consist of routine maintenance performed. For example, one maintenance commitment plan could offer performance maintenance six times per year versus another commitment plan of four maintenances per year. The total cost consists of preventive maintenance costs and breakdown maintenance costs. High-maintenance commitments are associated with high preventive maintenance costs and low breakdown maintenance costs. That is, as the system is maintained more frequently, there is less chance of breakdown. The minimum total cost is associated with maintenance commitment point M^*.

To decrease the number of breakdowns, it is necessary to increase periodic maintenance up to a certain point. As the breakdown cost decreases, the preventive maintenance cost increases. Therefore the total-cost curve has a minimum M^* as illustrated in Figure 19.16. Assume i = number of breakdowns, $i = 1, \ldots, n$; q_i = number of periods in which i breakdowns occurred; and

$$p_i = \frac{q_i}{\sum\limits_{i=1}^{n} q_i} \quad \text{(probability of having } i \text{ breakdowns)}$$

The expected number of breakdowns $E(i)$ is calculated as

$$E(i) = \sum_{i=1}^{n} i p_i \tag{19.15}$$

Then

Expected breakdown cost = cost per breakdown × expected number of breakdowns

$$= C * E(i) \tag{19.16}$$

where C is the average cost per breakdown. Total cost is calculated as the sum of expected breakdown costs and preventive maintenance costs:

$$\text{Total cost} = C * E(i) + \text{preventive maintenance cost} \tag{19.17}$$

Example 19.13 Finding Optimal Maintenance Policy   Consider a pump motor. The number of breakdowns for this pump motor for the past 24 months is shown in the table below. There were four months in which no breakdowns occurred, six months in which one breakdown occurred, and so on The average repair cost per breakdown is $2700.

Number of breakdowns, i	0	1	2	3	4	Total
Number of months breakdowns occurred, q_i	4	6	8	4	2	24

The company has the option to sign a contract for preventive maintenance. This contract guarantees to reduce the number of expected breakdowns to only one breakdown per month at a monthly cost of $1500. Is it beneficial for the company to sign the preventive maintenance contract?

Cost without Maintenance Contract First, calculate the expected number of breakdowns per month using (19.15).

Number of breakdowns, i	0	1	2	3	4	Total
Probability, p_i	$4/24 = 0.17$	$6/24 = 0.25$	$8/24 = 0.33$	$4/24 = 0.17$	$2/24 = 0.08$	1.00
Expected value, $E(i)$	$0 \times 0.17 = 0$	$1 \times 0.25 = 0.25$	$2 \times 0.33 = 0.67$	$3 \times 0.17 = 0.5$	$4 \times 0.08 = 0.33$	1.75

The expected number of breakdowns is 1.75 per month. Now find the cost of the current system. The total cost is equal to the expected breakdown costs as there are no preventive costs using (19.17):

$$\text{Total cost} = \$2700 \times 1.75 + 0 = \$4725 \text{ per month}$$

Cost with Maintenance Contract Consider the cost of the maintenance contract. Calculate the preventive maintenance costs as the sum of the cost of expected breakdowns with preventive maintenance and the costs of the maintenance agreement using (19.17):

$$\text{Total cost} = 1 \times \$2700 + \$1500 = \$4200$$

The total maintenance costs with the contract is lower than the total costs without the contract on a monthly basis ($4725 − $4200 = $525). Therefore, the company should sign the contract for the maintenance service.

19.6.2 Multiobjective Maintenance Policy

Consider several possible different maintenance commitment plans. Choose one of these maintenance alternatives by considering the following two objectives:

Minimize expected number of breakdowns, $f_1 = E(i)$ [use Equation (19.15)]
Minimize total cost, f_2 [use Equation (19.17)]

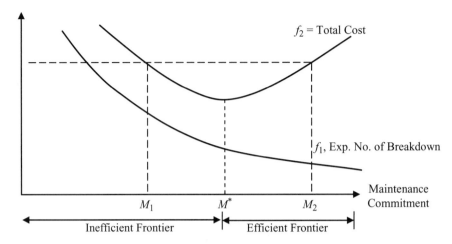

Figure 19.17   Total cost versus total number of breakdowns.

By minimizing the expected number of breakdowns, the system's reliability increases. There will be less interruptions and better customer (or user) satisfaction. This objective also minimizes the risk of failure associated with the systems performance.

Different maintenance policies are associated with different number of expected breakdowns and total maintenance costs. Higher maintenance commitments are associated with lower number of breakdowns. This relationship is presented in Figure 19.17.

Consider two maintenance commitments: M_1 and M_2 (see Figure 19.17). They both have the same total costs, but M_2 has a lower number of breakdowns. This means that point M_1 is inefficient compared to point M_2 in terms of the two objectives. Therefore, points to the right side of M^* are efficient. Points to the left of M^* are inefficient and do not need to be considered.

Example 19.14   Bicriteria Maintenance Example   The breakdown probabilities corresponding to the number of breakdowns for five different maintenance policies are given in the table below. Each breakdown costs \$100. The costs for preventive maintenance for policies $a_1, \ldots, a_5$ are \$730, \$700, \$625, \$525, and \$520, respectively.

(a) Generate the set of efficient alternatives for this problem.
(b) Choose the best alternative by minimizing an additive utility function with weights of importance $W = (0.7, 0.3)$ for the two normalized objectives, respectively.

Alternative maintenance policy	a_1	a_2	a_3	a_4	a_5
Probability (no. of breakdowns = 0)	0.1	0.2	0.2	0	0
Probability (no. of breakdowns = 1)	0.7	0.4	0.2	0.1	0
Probability (no. of breakdowns = 2)	0.2	0.3	0.4	0.4	0.5
Probability (no. of breakdowns = 3)	0	0.1	0.2	0.5	0.5
Preventive maintenance cost	730	700	625	525	520

First calculate the expected number of breakdowns for each alternative. Then, find the expected breakdown cost for each alternative. For example, for a_2:

Expected number of breakdowns:

$$f_1 = E(i) = \sum_{i=1}^{n} i p_i = 0 \times 0.2 + 1 \times 0.4 + 2 \times 0.3 + 0.1 \times 3 = 1.3$$

Expected breakdown costs: $C * E(i) = 1.3 \times 100 = \$130$.
Total cost: $f_2 = \$130 + \$700 = \$830$.

The set of multicriteria alternatives are listed below. If only cost is considered, then a_4 is the best alternative.

	a_1	a_2	a_3	a_4	a_5	Minimum	Maximum
Min. f_1 = expected no. of breakdowns	**1.1**	1.3	1.6	2.4	**2.5**	1.1	2.5
Expected breakdown costs, $C * E(i)$	110	130	160	240	250	110	250
Min. f_2 = total costs	**840**	830	785	**765**	770	765	840
Min. f_1'	0	0.143	0.357	0.929	1	0	1
Min. f_2'	1	0.867	0.267	0	0.067	0	1
Min. $U = 0.7f_1' + 0.3f_2'$	0.3	0.36	0.33	0.65	0.72		
Rank	1	3	2	4	5		
Efficient?	Yes	Yes	Yes	Yes	No		

Alternative a_5 is dominated by a_4. Similar to Figure 19.17, the first four alternatives are to the right of M^* and are efficient. Alternative a_5 is to the left of M^*, and therefore it is inefficient. For any given weights, a_4 is preferred to a_5.
    The best multicriteria alternative for the given weights is a_1.

See Supplement S19.4.xls.

19.7 REPLACEMENT POLICIES

When the cost of repairing exponentially increases or its breakdown could be disastrous, it is necessary to retire an existing functioning unit, product, or system and replace it with a new one before it could possibly fail. For example, it is necessary to replace susceptible airplanes or nuclear reactor facilities that may become defective when their failure would be catastrophic, even if the probability of failure is very low. Also, replacement of a group of products can be conducted simultaneously or during a shutdown period. For example, replacing one single street light bulb is expensive, but replacement of all light bulbs on the same street while the lift equipment and manpower are available can be much less expensive in the long term.

19.7.1 Single-Objective Replacement

Consider a unit that is being used continuously. When the unit fails and cannot be maintained, it is replaced with an identical new unit. Suppose the salvage value of the unit is negligible. The cost of maintenance per visit (C_M) is always the same, but as the unit gets older, the frequency of maintenance increases. The cost of replacing the unit per visit (C_R) is always the same. The replacement problem is to find optimal replacement age t^* when the unit should be replaced so that the total cost over a long period of time is minimized.

Let the age of a unit be denoted by t. As the age of a unit increases, its total maintenance cost increases at a higher rate. Suppose the total maintenance cost at age t is $C_M t^2/2$. Therefore, the total cost for maintenance and replacement of one unit over its lifetime t is

$$\text{Total cost (TC)} = \frac{1}{2} C_M t^2 + C_R$$

Now, consider the total cost per unit of time, that is, by dividing the above equation by t:

$$\text{TC}(t) = \left(\frac{1}{t}\right)\left(\frac{C_M t^2}{2} + C_R\right)$$

This equation can be simplified as to

$$\text{TC}(t) = \left(\frac{C_M}{2}\right) t + \frac{C_R}{t} \tag{19.18}$$

where C_M and C_R are given positive values and $\text{TC}(t)$ is a function of the replacement decision variable t. See Figure 19.18a. To find the optimal replacement time t^* that minimizes the total cost and the total cost over time, find the first derivative of Equation (19.18) with respect to t and then set it to zero:

$$\text{TC}'(t) = \frac{C_M}{2} - \frac{C_R}{t^2} = 0$$

or

$$t^* = \sqrt{\frac{2C_R}{C_M}} \tag{19.19}$$

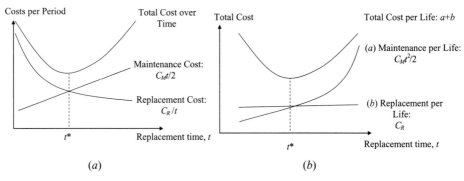

Figure 19.18 Optimal age t^* to replace item to minimize total costs: (*a*) total cost over time t; (*b*) total cost per life t.

The second derivative of TC(t) is given as

$$\text{TC}''(t) = \frac{C_R}{t^3}$$

Since both C_R and t are positive, $\text{TC}''(t) > 0$. Therefore $\text{TC}(t)$ is a convex function for which the global minimum solution is t^* in Equation (19.19).

Now, we show an alternative approach for finding the value of t^*. Consider Figure 19.18b for the total cost of replacing the item at age t. It can also be observed that at age t^* the unit should be replaced because after point t^* the maintenance costs become higher than the replacement costs. That is, t^* is the break-even point. Therefore

$$\frac{C_M t^2}{2} = C_R \qquad \text{or} \qquad t^* = \sqrt{\frac{2C_R}{C_M}}$$

This is the same solution as Equation (19.19).

Example 19.15   Single-Objective Replacement   A new copy machine costs $7500. The maintenance cost is $300 per year. The company has had the copy machine for six years.

Determine whether or not the company should replace its copy machine.

In this example, $C_M = \$300$ and $C_R = \$7500$. Note that for this example the total cost [Equation (19.18)] is

$$\text{TC}(t) = \left(\frac{300}{2}\right) t + \frac{7500}{t}$$

From Equation (19.19), the optimal time for replacement is

$$t^* = \sqrt{\frac{2C_R}{C_M}} = \sqrt{\frac{2(7500)}{300}} = \sqrt{50} = 7.07 \text{ years}$$

Since $7.07 > 6$, the company should not replace its current machine.

Figure 19.19 shows the total cost versus the number of years for this example.

19.7.2   Multiobjective Replacement Policy

In multicriteria replacement, consider two objectives. For failure probability, assume an exponential distribution function:

Minimize total cost per period:

$$f_1 = \text{TC}(t) = \left(\frac{C_M}{2}\right) t + \frac{C_R}{t} \qquad [\text{see Equation (19.18)}]$$

Minimize failure probability:

$$f_2 = F(t) = 1 - e^{-\lambda t} \tag{19.20}$$

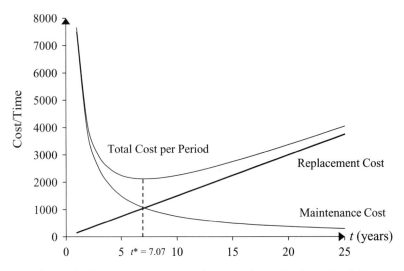

Figure 19.19   Total cost over time for Example 19.15, where $t^* = 7.07$.

The optimal replacement life t^* is associated with minimizing cost with some level of failure. As the item ages, the chance of failure increases. By reducing the replacement time to less than t^*, the failure probability decreases. Therefore, choose a lifetime between the minimum lifetime t_{min} and t^*. Alternatives associated with $t^* > t$ are inefficient and are dominated by alternatives associated with $t \leq t^*$. See Figure 19.20 for an example.

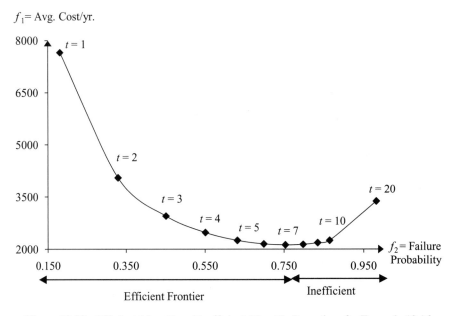

Figure 19.20   Efficient ($t^* \leq 7$) and inefficient ($t^* > 7$) alternatives for Example 19.16.

TABLE 19.5   Set of Multicriteria Alternatives for Example 19.16

Alternatives	a_1	a_2	a_3	a_4	a_5	a_6	a_7	a_8	a_9	a_{10}	a_{20}
Replacement age, t	1	2	3	4	5	6	7	8	9	10	20
Min. $f_1 =$ cost /yr (TC/yr)	7650	4050	2950	2475	2250	2150	2121	2138	2183	2250	3375
Min. $f_2 = F(t)$, failure prob.	0.181	0.330	0.451	0.551	0.632	0.699	0.753	0.798	0.835	0.865	0.982
Efficient?	Yes	Yes	Yes	Yes	Yes	Yes	Yes	No	No	No	No

Example 19.16   Multiobjective Replacement Policy   Consider Example 19.15. Suppose that the expected failure time for the copy machine is 5 years. Consider 10 alternatives, that is, lifetime of 1–10 years. Determine cost over time for each given alternative and find its probability of failure. Identify efficient and inefficient alternatives.

Since the expected failure time is 5 years, the failure rate is $\lambda = \frac{1}{5} = 0.2$. The set of alternatives is presented in Table 19.5. For the purpose of illustration, we show calculations for alternative a_2, where replacement is at age $t = 2$. The cost per period and probability of failure are

$$\text{Min. } f_1 = \text{TC}(t = 2) = \left(\frac{300}{2}\right)2 + \frac{7500}{2} = \$4050 \quad \text{(average cost over time)}$$

$$\text{Min. } f_2 = F(t) = P\{T \le 2\} = 1 - e^{-0.2(2)} = 0.330 \quad \text{(probability of failure before age 2)}$$

The calculations for all 10 alternatives and also for age 20 are shown in Table 19.5. Note that alternatives a_8, a_9, a_{10}, and a_{20} are inefficient. They are dominated by alternative a_7, which has a lower probability and failure. Figure 19.20 shows the graphical representation of these alternatives.

Now, a multicriteria method (e.g., paired comparison of alternatives or an additive utility function) can be used to choose the best alternative. Suppose the alternative chosen for this decision problem is to replace the machine at age 4 (alternative a_4).

See Supplement S19.5.xls.

19.7.3   General Replacement Model[†]

The assumptions for a general replacement model are as follows:

1. Maintenance cost increases as the equipment ages.
2. Uncertainty is explicitly considered in the failure model.
3. Equipment is operating continuously.
4. The downtime for maintenance or replacement is negligible and therefore is ignored.
5. Planning period is infinite.
6. All new equipment has the same features and cost.
7. There are only two costs, maintenance and replacement.
8. Salvage value is considered a negative value in the total cost function.
9. The objective is to minimize the total cost over time.

Let $C_M(t)$ be the instantaneous cost of a unit at age t. Suppose the maintenance cost (as an increasing function of t) and salvage value (as a decreasing function of t) both obey the exponential distribution as presented below:

$$C_M(t) = ue^{\lambda_1 t} \quad \text{(maintenance cost at time } t) \tag{19.21}$$

$$S(t) = ve^{-\lambda_2 t} \quad \text{(salvage value at time } t) \tag{19.22}$$

where μ, λ_1, v, and λ_2 are given parameters and u, λ_1, v, $\lambda_2 \geq 0$.

The optimal replacement life can be obtained by solving the following equations. Consider t_0, the replacement life, that is, the duration of one cycle. The average cost per unit time is $TC(t_0)$. The maintenance cost is the integral (cumulative) of the instantaneous maintenance cost. That is,

$$\text{Maintenance cost (cumulative)} = \int_0^{t_0} C_M(t) \, dt$$

The total cost per period for the general replacement model is

$$TC(t_0) = \left(\frac{1}{t_0}\right) C_R - \left(\frac{1}{t_0}\right) S(t_0) + \left(\frac{1}{t_0}\right) \int_0^{t_0} C_M(t) \, dt \tag{19.23}$$

To find the optimal age t^*, set the first derivative of Equation (19.23) to zero:

$$TC'(t_0) = \left(\frac{-1}{t_0^2}\right) C_R - \left(\frac{1}{t_0^2}\right) [S'(t_0)t_0 - S(t_0)] + \frac{t_0 \left(\int_0^{t_0} C_M(t) \, dt\right)' - \int_0^{t_0} C_M(t) \, dt}{t_0^2}$$

$$= 0 - C_R - S'(t_0)t_0 + S(t_0) + t_0 C_M(t_0) - \int_0^{t_0} C_M(t) \, dt = 0$$

Now, substitute $C_M(t_0)$ and $S(t_0)$ using Equations (19.21) and (19.22) respectively:

$$-C_R - v\lambda_2 t_0 e^{-\lambda_2 t_0} + v e^{-\lambda_2 t_0} + t_0 u e^{\lambda_1 t_0} - \left(\frac{u}{\lambda_1} e^{\lambda_1 t_0} - \frac{u}{\lambda_1}\right) = 0$$

$$TC'(t_0) = u e^{\lambda_1 t_0} \left(t_0 - \frac{1}{\lambda_1}\right) + v e^{-\lambda_2 t_0}(1 - t_0\lambda_2) + \frac{u}{\lambda_1} - C_R = 0 \tag{19.24}$$

In Equation (19.24), the only unknown is t_0, that is, all parameters are known.

Note that Equation (19.23), total cost per cycle t_0, is a unimodal function of t; that is, it has only one minimum global optimal point which is the best time for replacement. Therefore, Equation (19.24) has only one solution. Finding the solution to Equation (19.24) directly is not trivial. Therefore, an iterative single-variable optimization method can be used to find optimal age t^*.

Note that the total cost per period [Equation (19.23)] can be written in expanded form by using Equations (19.21) and (19.22):

$$\mathrm{TC}(t_0) = \left(\frac{1}{t_0}\right) C_R - \left(\frac{1}{t_0}\right) v e^{-\lambda_2 t} + \left(\frac{1}{t_0}\right) \int_0^{t_0} u e^{\lambda_1 t} \, dt \qquad (19.25)$$

Example 19.17   General Replacement Model   Suppose a machine replacement cost is $200,000. The salvage value is calculated based on the assumption that a machine's value decreases 10% per year. First-year machine maintenance cost is $300, but maintenance cost increases at a rate of 30% per year as the machine ages. Find the optimal replacement life.

The salvage value is calculated as follows. At year 0, the salvage value of the machine is the same as its purchase price:

$$S(0) = v e^{-\lambda_2 \times 0} \qquad \text{or} \qquad 200,000 = v e^{-\lambda_2 \times 0} \qquad \text{or} \qquad v = 200,000$$

Since, the machine loses 10% of its value each year, the salvage values for years t and $t - 1$ are related as follows:

$$\frac{S(t)}{S(t-1)} = \frac{e^{\lambda_2 t}}{e^{\lambda_2(t-1)}} = e^{\lambda_2} = 0.9 \qquad \text{or} \qquad \lambda_2 = 0.105$$

For maintenance cost, we have

$$C_M(t) = u e^{\lambda_1 t} \qquad \text{where} \qquad C_M(1) = 300$$

Since there is 30% cost increase per year, we have

$$\frac{C_M(t)}{C_M(t-1)} = \frac{e^{\lambda_1 t}}{e^{\lambda_1(t-1)}} = e^{\lambda_1} = 1.3 \qquad \text{or} \qquad \lambda_1 = 0.26$$

Now, we find the value of u for the maintenance cost function:

$$C_M(1) = 300 \qquad \text{or} \qquad u e^{(0.26) \times 1} = 300 \qquad \text{or} \qquad u = 231.32$$

Now, substitute the above obtained values into Equation (19.24):

$$\mathrm{TC}'(t_0) = 231.32 e^{0.26 t_0} \left(t_0 - \frac{1}{0.26} \right) + 200,000 e^{-0.105 t_0} (1 - t_0 \times 0.105)$$
$$+ \frac{231.32}{0.26} - 200,000 = 0$$

Now find a value that solves the above equation. This equation can be solved using a one-dimensional search method. Alternatively, one can use a trial-and-error method with a spreadsheet.

Some of the values of Equation (19.24) for different values of t_0 are shown in Table 19.6 and Figure 19.21.

TABLE 19.6   Equation (19.24) Values for some t_0

t_0	5	10	15	16	16.5	16.6	16.656	17
Total cost, Equation (19.23)	16812.8	14110.1	13444.1	13677.5	13858.2	13899.9	13924.2	14087.1
Derivative of total cost, Equation (19.24)	−141933	−183444	−95453	−44325	−11438	−4177	0	27387

That is, we find total cost by varying the value of t. The optimal time is about 16.656 years. The total cost using Equation (19.25) is

$$
\mathrm{TC}(t_0) = \left(\frac{1}{16.656} \right) \times 200{,}000 - \left(\frac{1}{16.656} \right) \times 200{,}000 e^{-0.105 \times 16.656}
$$

$$
+ \left(\frac{1}{16.656} \right) \int_0^{16.656} 231.32 e^{0.26t} \, dt = 12007.68 - 2088.96 + 4005.48 = 13924.2
$$

See Supplement S19.6.xls.

19.7.4   Multicriteria of General Replacement Model[†]

In using multicriteria of general replacement, consider two objectives. For failure probability, assume an exponential distribution function:

Minimize cost per period:

$$
f_1 = \mathrm{TC}(t_0) = \left(\frac{1}{t_0} \right) C_R - \left(\frac{1}{t_0} \right) S(t_0) + \left(\frac{1}{t_0} \right) \int_0^{t_0} C_M(t) \, dt \qquad (19.23)
$$

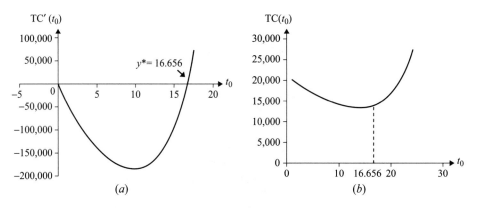

(a)   (b)

Figure 19.21   Optimal replacement age for Example 19.17; (a) first derivative of total cost [Equation 19.24)]; (b) total cost [Equation 19.23)], where $t^* = 16.656$.

TABLE 19.7 Set of Multicriteria Alternatives for Example 19.18

Alternatives	a_1	a_2	a_3	a_4	a_5	a_6	a_7	a_8
Replacement age, t	1	2	5	10	13.84	15	20	25
Min. $f_1 = $ cost /yr, $	20199	19245	16813	14110	13357	13444	16795	31056
Min. $f_2 = $ F(t), failure probability	0.181	0.330	0.632	0.865	0.937	0.950	0.982	0.993
Efficient?	Yes	Yes	Yes	Yes	Yes	No	No	No
Ranking by DM	7	3	1	2	4	5	6	8

Minimize failure probability:

$$f_2 = F(t_0) = 1 - e^{-\lambda t_0} \qquad (19.20)$$

The optimal replacement life t^* minimizes cost over time, which is associated with a probability of failure. As the equipment ages, the probability of failure increases. By reducing the replacement life to less than t^*, the failure probability decreases. Therefore, choose a replacement life that is between the minimum life time $t_{\min}$ and the optimal replacement life t^*. Alternatives associated with $t^* > t_0$ are inefficient and are dominated by the alternative associated with $t_0 = t^*$. See Table 19.7 for an example.

Example 19.18 Multicriteria of General Replacement Model   Consider Example 19.17. Suppose that the expected failure age for the machine is 5 years. Consider eight alternatives whose lives are 1, 2, 5, 10, 13.84, 15, 20, and 25 years. Determine cost per period for each given alternative and find its probability of failure. Identify efficient and inefficient alternatives. Use paired comparison to choose the best alternative and rank them. The solution is presented below.

Since the expected failure time is 5 years, the failure rate is $\lambda = \frac{1}{5} = 0.2$. That is, the probability of failure is

$$f_2 = F(t_0) = 1 - e^{-0.2 t_0}$$

The set of alternatives is presented in Table 19.7. For the purpose of illustration, we show calculations for a_2, where replacement is at age $t_0 = 2$. The total cost over time and probability of failure are

$$f_1 = \left(\frac{1}{2}\right) \times 200,000 + \left(\frac{1}{2}\right) \times \left(\int_0^2 231.32 e^{0.26t}\, dt\right) - \left(\frac{1}{2}\right) \times 200,000 e^{-0.1052} = 19245$$

$$f_2 = F(t) = P\{T \leq 2\} = 1 - e^{-2(0.20)} = 0.330$$

The summary of the resulting eight alternatives are shown in Table 19.7. They are also presented in Figure 19.22. Note that, in Example 19.17, we found that the minimum total cost occurred at year 13.84. Therefore, alternatives whose age is greater than 13.84 are

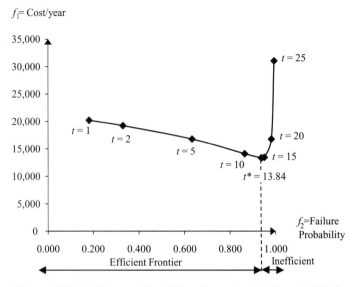

Figure 19.22   Efficient and inefficient alternatives for Example 19.18.

inefficient. The DM can now be asked to make paired comparisons of alternatives and rank them. A ranking is provided in the last row of Table 19.7. Therefore, the best alternative for this ranking is alternative a_3.

See Supplement S19.7.xls.

REFERENCES

General

Amstadter, B. L. *Reliability Mathematics*. New York: McGraw-Hill, 1971.

Barlow, R. E., and F. Proschan. *Mathematical Theory of Reliability*. New York: Wiley, 1965.

Barlow, R. E., and F. Proschan. *Statistical Theory of Reliability and Life Testing*, New York: Holt, Rinehart & Winston, 1975.

Blischke, W. R., and E. M. Scheuer. "Calculation of the Cost of Warranty Policies as a Function of Estimated Life Distributions." *Naval Research Logistics Quarterly*, Vol. 22, 1975, pp. 681–696.

Dai, Y. S., M. Xie, K. L. Poh, and G. Q. Liu, "A Study of Service Reliability and Availability for Distributed Systems." *Reliability Engineering and System Safety*, Vol. 79, No. 1, 2003, pp. 103–112.

Denson, W. "The History of Reliability Prediction." *IEEE Transactions on Reliability*, Vol. 47, No. 3, 1998, pp. 321–329.

Devarun, G., and R. Sandip. "A Decision-Making Framework for Process Plant Maintenance." *European Journal of Industrial Engineering*, Vol. 4, No. 1, 2009, pp. 78–98.

Gertsbakh, I. B. *Models of Preventive Maintenance*, Amsterdam: North Holland, 1977.

Jelinski, Z., and P. Moranda. Software Reliability Research," *In Statistical Computer Performance Evaluation*, W. Freiberger, Ed. New York: Academic, 1972.

Lotka, A. J. "A Contribution to the Theory of Self-Renewing Aggregates with Special References to Industrial Replacement." *Annals of Mathematical Statistics*, Vol. 10, 1939, pp. 1–25.

Mamer, J. "Cost Analysis of Pro Rata and Free Replacement Warranties." *Naval Reaserch Logistics Quarterly*, Vol. 29, 1982, pp. 345– 356.

Mann, L. L. *Maintenance Management*. Lexington, MA: Lexington Books, 1976.

McCall, J. J. "Maintenance Policies for Stochastically Failing Equipment: A Survey." *Management Science*, Vol. 11, 1965, pp. 493–524.

Nahmias, S. *Production and Operations Analysis*, 6th ed., New York: McGraw-Hill, 2008.

Nakagawa, T., and Q. C. Hua. "Note on Reliabilities of Series-Parallel and Parallel-Series System." *Journal of Quality in Maintenance Engineering*, Vol. 8, No. 3, 2002, pp. 274–280.

O'Connor, P. D. T. *Practical Reliability Engineering*, 4th ed. New York: Wiley, 2002.

Sarhan, A. M. "Reliability Estimations of Components from Masked System Life Data." *Reliability Engineering and System Safety*, Vol. 74, No. 1, 2001, pp. 107–113.

Shooman, M. L. "Probabilistic Models for Software Reliability Prediction." In *Statistical Computer Performance Evaluation*, W. Freiberger, Ed. New York: Academic, 1972, pp. 485–502.

Tarelko, W. "Control Model of Maintainability Level." *Reliability Engineering and System Safety*, Vol. 47, No. 2, 1995. pp. 85–91.

Turban, E. "The Use of Mathematical Models in Plant Maintenance Decision Making." *Management Science*, Vol. 13, 1967, pp. 20–27.

Utkin, Lev V. "A Second-Order Uncertainty Model for Calculation of the Interval System Reliability." *Reliability Engineering and System Safety*, Vol. 79, No. 3, 2003, pp. 341–351.

Wani, M. F., and O. P. Gandhi. "Development of Maintainability Index for Mechanical Systems." *Reliability Engineering and System Safety*, Vol. 65, No. 3, 1999, pp. 259–270.

Wilkerson, J. J. "How to Manage Maintenance." *Harvard Business Review*, Vol. 46, 1968, pp. 100–111.

Zhao, Y. X. "On Preventive Maintenance Policy of a Critical Reliability Level for System Subject to Degradation." *Reliability Engineering and System Safety*, Vol. 79, No. 3, 2003, pp. 301–308.

Multiobjective Reliability

Almeida, A. T. D. "Multicriteria Decision Making on Maintenance: Spares and Contracts Planning." *European Journal of Operational Research*, Vol. 129, No. 2, 2001, pp. 235–241.

Arunraj, N. S., and J. Maiti. "Risk-Based Maintenance Policy Selection Using AHP and Goal Programming." *Safety Science*, Vol. 48, No. 2, 2010, pp. 238–247.

Berrichi, A., L. Amodeo, F. Yalaoui, E. Châtelet, and M. Mezghiche, "Bi-Objective Optimization Algorithms for Joint Production and Maintenance Scheduling: Application to the Parallel Machine Problem." *Journal of Intelligent Manufacturing,* Vol. 20, No. 4, pp. 389–400.

Bertolini, M., and M. Bevilacqua. "A Combined Goal Programming—AHP Approach to Maintenance Selection Problem." *Reliability Engineering & System Safety*, Vol. 91, No. 7, 2006, pp. 839–848.

Bevilacqua, M., and M. Braglia. "The Analytic Hierarchy Process Applied to Maintenance Strategy Selection." *Reliability Engineering & System Safety*, Vol. 70, No. 1, 2000, pp. 71–83.

Cascales, M. S. G., and M. T. Lamata. "Multi-Criteria Analysis for a Maintenance Management Problem in an Engine Factory: Rational Choice." *Journal of Intelligent Manufacturing*, Vol. 22, No. 5, 2011, pp.779–788.

Chaichan, C., N. Nagen, and M. T. Tabucanon, "A Multicriteria Approach to the Selection of Preventive Maintenance Intervals." *International Journal of Production Economics*, Vol. 49, No. 1, 1997, pp. 55–64.

Chareonsuk, C., N. Nagarur, and M. T. Tabucanon. "A Multicriteria Approach to the Selection of Preventive Maintenance Intervals." *International Journal of Production Economics*, Vol. 49, No. 1, 1997, pp. 55–64.

Coit, D. W., J. Tongdan, and N. Wattanapongsakorn. "System Optimization with Component Reliability Estimation Uncertainty: A Multi-Criteria Approach." *IEEE Transactions on Reliability*, Vol. 53, No. 3, 2004, pp. 369–380.

Dhingra, A, K. "Optimal Apportionment of Reliability and Redundancy in Series Systems under Multiple Objectives." *IEEE Transactions on Reliability*, Vol. 41, No. 4, 1992, pp. 576–582.

Fwa, T. F., W. T. Chan, and K. Z. Hoque. "Multiobjective Optimization for Pavement Maintenance Programming." *Journal of Transportation Engineering*, Vol. 126, No. 5, 2000, pp. 367–374.

Kralj, B., and R. Petrovic. "A Multiobjective Optimization Approach to Thermal Generating Units Maintenance Scheduling." *European Journal of Operational Research*, Vol. 84, No. 2, 1995, pp. 481–493.

Labib, A. W., R. F. O'Connor, and G. B. Williams, "An Effective Maintenance System Using the Analytic Hierarchy Process." *Integrated Manufacturing System*, Vol. 9, No. 2, 1998, pp. 87–98.

Malakooti, B. "Multi-Criteria Reliability and Maintainability Analysis and Optimization." Case Western Reserve University, Cleveland, Ohio, 2013.

Quan, G., G. W. Greenwood, D. Liu, and S. Hu, "Searching for Multiobjective Preventive Maintenance Schedules: Combining Preferences with Evolutionary Algorithms." *European Journal of Operational Research*, Vol. 177, No. 3, 2007, pp. 1969–1984.

Safaei, N., D. Banjevic, and A. K. S. Jardine. "Bi-Objective Workforce-Constrained Maintenance Scheduling: A Case Study." *Journal of the Operational Research Society*, 2010, Vol. 62, 2011, pp. 1005–1018.

Salazar, D., C. M. Rocco, and B. J. Galván. "Optimization of Constrained Multiple-Objective Reliability Problems Using Evolutionary Algorithms." *Reliability Engineering & System Safety*, Vol. 91, No. 9, 2006, pp. 1057–1070.

Toshiyuki, I., I. Koichi, and A. Hajime. "Interactive Optimization of System Reliability under Multiple Objectives." *IEEE Transactions on Reliability*, Vol. R-27, No. 4, 1978, pp. 264–267.

Vaidogas, E. R., and E. K. Zavadskas. "Introducing Reliability Measures into Multi-Criteria Decision-Making." *International Journal of Management and Decision Making*, Vol 8, No. 5/6, 2007, pp. 475–496.

Wang, L., J. Chu, and J. Wu. "Selection of Optimum Maintenance Strategies Based on a Fuzzy Analytic Hierarchy Process." *International Journal of Production Economics*, Vol. 107, No. 1, 2007, pp. 151–163.

Yadav, O. P., S. S. Bhamare, and A. Rathore. "Reliability-Based Robust Design Optimization: A Multi-Objective Framework Using Hybrid Quality Loss Function." *Quality and Reliability Engineering International*, Vol. 26, No. 1, 2010, pp. 27–41.

EXERCISES

Reliability of Single Units (Section 19.2)

2.1 Consider a company that produces guitar strings. Its records show that the expected failure time of a guitar string is 15 h. To meet the current demand, the company is going to produce 75,000 guitar strings. Assume an exponential probability distribution.

(a) Determine the probability that a random guitar string lasts longer than $t = 10, 12, 14, 16, 18, 20, 25, 30$ h.

(b) Graph the survival function versus time.

(c) Determine the proportion of the 75,000 guitar strings that fail during the 17th hour.

(d) Determine the proportion of guitar strings that fail during the 13th hour given that they have survived at least 12 h.

2.2 Consider a company that produces light bulbs. Its records show that the expected failure time of a light bulb is 1500 h. To meet the current demand, the company is going to produce 25,000 light bulbs. Assume an exponential probability distribution.

(a) Determine the probability that a random light bulb lasts longer than $t = 1000$, 1100, 1200, 1300, 1400, 1500, 1600, 1800, 2000, 2500, and 3000 h.

(b) Graph the survival function versus time.

(c) Determine the proportion of the 25,000 light bulbs that fail during the 13th hundered hours.

(d) Determine the proportion of light bulbs that fail during the 16th hundered hour given that they have survived 15 hundered hours.

2.3 Consider a company that produces a machine part. Its records show that the expected failure time of the part is 8 years. To meet the current demand, the company is going to produce 700 parts. Assume an exponential probability distribution.

(a) Determine the probability that a random machine part lasts longer than $t = 5, 6, 7$, 8, 9, 10, 11, 12, 13, 14, 15, 20 years.

(b) Graph the survival function versus time.

(c) Determine the proportion of the 700 parts that fail during the 7th year.

(d) Determine the proportion of parts that fail during the 9th year given that they have survived 8 years.

2.4 Consider five design alternatives for guitar strings. The cost and expected life of each design are presented below. The target life is 16h.

Design	1	2	3	4	5
Expected life (h)	12	15	18	19	20
Cost per item, $	0.16	0.1	0.12	0.14	0.15

Generate five tricriteria alternatives (expected life, total cost, and probability of failure before target life), and identify efficient and inefficient alternatives. (Assume an exponential distribution function.)

2.5 Consider seven alternatives for high-efficiency light bulbs. The cost and expected life of each design are presented below. The target life is 19000 h.

Design	1	2	3	4	5	6	7
Expected life (in 1000 h)	8	12	15	18	20	21	22
Cost per units, $	12	9	7	8	9	12	15

Generate seven tricriteria alternatives (expected life, total cost, and probability of failure before target life). Identify efficient and inefficient alternatives. (Assume an exponential distribution function.)

Mean Time between Failure (Section 19.3)

3.1 Consider a machine with the following recorded times between failure and repair times in days.

i	1	2	3	4	5
Time between failure, T_i	12	18	2	50	10
Repair times, D_i	2	6	5	8	9

(a) Calculate MTBF and MTTR.

(b) Calculate the system availability.

(c) For an exponential distribution, find the probability that the product does not fail in the first 30 days.

(d) For a normal distribution, find the probability that the product does not fail in the first 30 days.

3.2 Consider a machine with the following recorded times between failure and repair times in days.

i	1	2	3	4	5	6	7
Time between failure, T_i	30	21	23	28	27	24	29
Repair times, D_i	3	6	6	5	3	7	4

(a) Calculate MTBF and MTTR.

(b) Calculate the system availability.

(c) For an exponential distribution, find the probability that the product does not fail in the first 30 days.

(d) For a normal distribution, find the probability that the product does not fail in the first 28 days.

3.3 Consider a machine with the following recorded times between failure and repair times in days.

i	1	2	3	4	5	6	7	8	9
Time between failure, T_i	71	93	59	69	52	70	73	75	91
Repair times, D_i	19	19	16	25	23	29	23	21	23

(a) Calculate MTBF and MTTR.

(b) Calculate the system availability.

(c) For an exponential distribution, find the probability that the product does not fail in the first 70 days.

(d) For a normal distribution, find the probability that the product does not fail in the first 70 days.

3.4 A company tested 20 simultaneously running motors for 1000 h each. During the test, three of the motor broke down. The first broke down after 100 h, the second after 690 h, and the third after 750 h. Find the percent failure rate, the number of failures per hour, and the mean time between failures for the system.

3.5 A taxi cab company has a total of 300 cabs in a particular city. The 300 cabs were observed for a period of 100 operating hours. In this period, 5 cabs failed at the following times: 2 after 10 h, 2 after 50 h, and 1 after 75 h. Find the percent failure rate, the number of failures per hour, and the mean time between failures for the system.

Reliability of Multiple Units (Section 19.4)

4.1 What is the reliability of the system below?

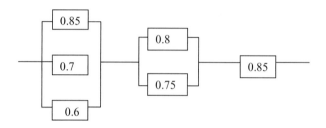

4.2 What is the reliability of the system below?

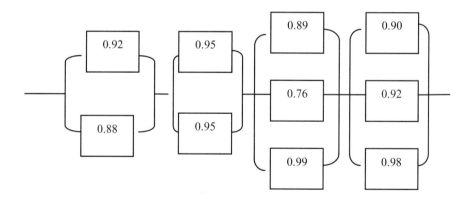

4.3 The reliability of the given system is to be improved to 0.75. An element is added as shown to accomplish this. What is the needed reliability of the added component?

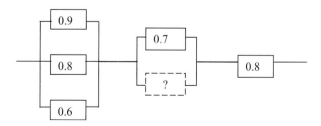

4.4 In the following system, determine how many of the four elements each with a reliability of 0.75 must be operational for the system to have a reliability of at least 0.85.

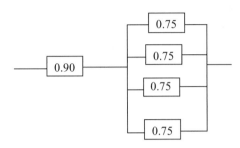

Multicriteria Reliability of Multiple Units (Section 19.5)[†]

5.1 Consider a system that consists of two units in series where the reliabilities of units 1 and 2 are 0.8 and 0.7, respectively. All the needed information is presented in the table below. The budget for this problem is $25,000. Generate a set of efficient alternatives within the given budget using the reliability–cost trade-off algorithm.

Alternative, k	1	2	3	4
$R_{1,k}$ Reliability	0.8	0.85	0.9	0.95
$c_{1,k}$, Additional cost, $ (thousands)	0	5	8	10
$R_{2,k}$ Reliability	0.7	0.8	0.85	0.90
$c_{2,k}$, Additional cost, $ (thousands)	0	5	10	15

5.2 Consider a table system that consists of three units in series. All the needed information is presented in the table below. The budget for this problem is $70,000. Generate a set of efficient alternatives within the given budget using the reliability–cost trade-off algorithm.

Alternative, k	1	2	3	4	5
$R_{1,k}$ Reliability	0.7	0.75	0.8	0.85	0.90
$c_{1,k}$, Additional cost, $ (thousands)	0	10	15	20	25
$R_{2,k}$ Reliability	0.7	0.8	0.85	0.90	0.95
$c_{2,k}$, Additional cost, $ (thousands)	0	5	10	15	20
$R_{3,k}$ Reliability	0.6	0.65	0.7	0.8	0.9
$c_{3,k}$, Additional cost, $ (thousands)	0	10	15	20	40

Single-Objective Maintenance Policy (Section 19.6.1)

6.1 Consider the data given in the table below. Each breakdown costs $600. An available service contract for preventive maintenance costs $150 per week with the guarantee of only two breakdowns per week.

Number of breakdowns	0	1	2	3
Number of weeks breakdowns occurred	14	16	13	9

(a) Find the expected breakdown cost?

(b) Should the maintenance policy be bought?

(c) If the maintenance is changed to guarantee only one breakdown, should the maintenance policy be bought?

6.2 Consider the breakdown table given below. The number of breakdowns from the past two years is shown. Each breakdown costs $3000. A service contract for preventive maintenance costs $300 per month and guarantees only two breakdowns per month.

Number of breakdowns	0	1	2	3	4	5
Number of months breakdowns occurred	3	4	7	6	3	1

(a) Find the expected breakdown cost?

(b) Should the maintenance policy be bought?

(c) If the price of fixing the machine changes to $650, should the maintenance policy be bought?

6.3 Consider the breakdown table given below. Preventive maintenance costs are $700 monthly for only two breakdowns per month. If the machine breaks down, it will cost $6500 to repair each time.

Number of breakdowns	0	1	2	3	4	5	6
Number of months breakdowns occurred	7	8	9	6	5	4	2

(a) Find the expected breakdown cost?

(b) Should the maintenance policy be bought?

(c) If the price of fixing the machine changes to $4000 and the price of the maintenance plan changes to $1500 per month, should the maintenance policy be bought?

Multiobjective Maintenance Policy (Section 19.6.2)

6.4 Consider the following five maintenance alternatives, where each alternative would result in a different daily breakdown frequency. The repair cost of each breakdown is $2000. Assume the weights of importance for the two normalized objectives are 0.6 and 0.4, respectively.

	Probability				
Number of breakdowns	a_1	a_2	a_3	a_4	a_5
0	0.2	0.2	0.2	0	0
1	0.4	0.3	0.2	0.1	0
2	0.3	0.3	0.3	0.4	0.4
3	0.1	0.2	0.3	0.5	0.6
Maintenance policy cost	4500	3000	2300	2000	1900

(a) Find the number of expected breakdowns for each alternative.

(b) Find the total maintenance cost for each alternative.

(c) Find the efficient alternatives.

(d) For the additive utility function, find the best policy.

6.5 Consider the following six different maintenance alternatives where each alternative would result in a different breakdown daily frequency. The repair cost of each breakdown is $700. Assume the weights of importance for the two normalized objectives are 0.5 and 0.5, respectively. Respond to all parts of exercise 6.4.

Number of breakdowns	Probability					
	a_1	a_2	a_3	a_4	a_5	a_6
0	0.5	0.4	0	0.1	0	0
1	0	0.3	0.1	0.2	0.1	0
2	0.5	0.2	0.4	0.2	0.2	0
3	0	0	0.4	0.3	0.3	0.4
4	0	0.1	0.1	0.2	0.4	0.6
Maintenance policy cost	800	700	550	120	40	30

6.6 Consider the following five different maintenance alternatives where each would result in a different breakdown frequency. The cost of each breakdown is $5000. Assume the weights of importance for the two normalized objectives are 0.6 and 0.4, respectively. Respond all parts of exercise 6.4.

Number of breakdowns	Probability				
	a_1	a_2	a_3	a_4	a_5
0	0.7	0.5	0.4	0.1	0.1
1	0.1	0.3	0.1	0.2	0
2	0.1	0.1	0.1	0.2	0.3
3	0.1	0.1	0.4	0.5	0.6
Maintenance policy cost	4500	3000	1800	700	200

Single-Objective Replacement (Section 19.7.1)

7.1 Consider a machine that costs $5000 to replace. The maintenance cost is $50 per year. The machine is three years old. Determine whether the machine should be replaced.

7.2 The expected life of a machine is seven years. The maintenance cost is $300 per year. How much the owner would be willing to pay to replace the machine?

7.3 A company would like to determine how much to spend on the yearly maintenance of a machine. The expected life of the machine is four years. The cost of replacement for this machine is $3000?

Multiobjective Replacement Policy (Section 19.7.2)

7.4 Consider a machine with maintenance cost of $300 per year. The machine costs $2500 to replace. The machine's expected life is eight years.

(a) Find the optimal time to replace the machine.

(b) Generate alternatives for times $t = 2, 3, 4, 5, 6$ years. Identify efficient alternatives.

(c) Rank the multicriteria alternatives generated in part (b) using an additive utility function, where the weights of importance for the two normalized objectives are 0.4 and 0.6, respectively.

7.5 Consider a machine that costs $5000 to replace and its maintenance cost is $500 per year. The expected failure time of the machine is five years.

(a) Find the optimal time to replace the machine.

(b) Generate multicriteria alternatives for times $t = 2, 3, 4, 5, 6$ years.

(c) Rank the alternatives generated in part (a) using an additive utility function, where the weights of importance for the two normalized objectives are 0.7 and 0.3, respectively.

General Replacement Model[†] (Section 19.7.3)

7.6 Consider an oil pump. The cost of the new pump is $50,000. The pump loses its sale value at a rate of 15% per year. The cost of maintenance is $2000 for the first year. Maintenance costs increases at a rate of 10% per year as it ages. Find the optimal life replacement (in years) for this pump. (Maintenance costs and salvage value have exponential functions.)

7.7 Consider a stamping die that is used to form the fender of an automobile body. The cost of new stamping die is $98,000. The die loses its sale value at a rate of 5% per month. The cost of maintenance is $2000 for the first month. The maintenance cost for the stamping die increases 2% per month. Find the optimal life replacement (in months) for this die. (Maintenance costs and salvage value have exponential functions.)

Multicriteria of General Replacement Model[†] (Section 19.7.4)

7.8 Consider exercise 7.6. Suppose the expected life of the pump is 3.5 years.

(a) Show the bicriteria alternatives for five consecutive alternatives (whose age differences are within two years), three to be efficient and two to be inefficient. *Hint:* Based on exercise 7.6, you know the optimal replacement life; choose two alternatives less than that value and two alternatives more than that value.)

(b) Rank the alternatives generated in part (a) using an additive utility function, where the weights of importance for the two normalized objectives are 0.7 and 0.3, respectively.

7.9 Consider exercise 7.7. Suppose the expected life of the die is 42 months.
Respond to parts (a) and (b) of exercise 7.8.

THE STANDARD NORMAL DISTRIBUTION

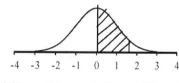

$$P(0 \leq Z \leq z) = \int_{0}^{z} \frac{1}{\sqrt{2\pi}} \, e^{\frac{-x^2}{2}} \, dx$$

Area from mean to z = 1.65

P(0≤ Z≤ 1.65) = 0.4505

z	0.00	0.01	0.02	0.03	0.04	0.05	0.06	0.07	0.08	0.09
0.00	0.0000	0.0040	0.0080	0.0120	0.0160	0.0199	0.0239	0.0279	0.0319	0.0359
0.10	0.0398	0.0438	0.0478	0.0517	0.0557	0.0596	0.0636	0.0675	0.0714	0.0753
0.20	0.0793	0.0832	0.0871	0.0910	0.0948	0.0987	0.1026	0.1064	0.1103	0.1141
0.30	0.1179	0.1217	0.1255	0.1293	0.1331	0.1368	0.1406	0.1443	0.1480	0.1517
0.40	0.1554	0.1591	0.1628	0.1664	0.1700	0.1736	0.1772	0.1808	0.1844	0.1879
0.50	0.1915	0.1950	0.1985	0.2019	0.2054	0.2088	0.2123	0.2157	0.2190	0.2224
0.60	0.2257	0.2291	0.2324	0.2357	0.2389	0.2422	0.2454	0.2486	0.2517	0.2549
0.70	0.2580	0.2611	0.2642	0.2673	0.2704	0.2734	0.2764	0.2794	0.2823	0.2852
0.80	0.2881	0.2910	0.2939	0.2967	0.2995	0.3023	0.3051	0.3078	0.3106	0.3133
0.90	0.3159	0.3186	0.3212	0.3238	0.3264	0.3289	0.3315	0.3340	0.3365	0.3389
1.00	0.3413	0.3438	0.3461	0.3485	0.3508	0.3531	0.3554	0.3577	0.3599	0.3621
1.10	0.3643	0.3665	0.3686	0.3708	0.3729	0.3749	0.3770	0.3790	0.3810	0.3830
1.20	0.3849	0.3869	0.3888	0.3907	0.3925	0.3944	0.3962	0.3980	0.3997	0.4015

(continued)

Operations and Production Systems with Multiple Objectives, First Edition. Behnam Malakooti.
© 2014 John Wiley & Sons, Inc. Published 2014 by John Wiley & Sons, Inc.

z	0.00	0.01	0.02	0.03	0.04	0.05	0.06	0.07	0.08	0.09
1.30	0.4032	0.4049	0.4066	0.4082	0.4099	0.4115	0.4131	0.4147	0.4162	0.4177
1.40	0.4192	0.4207	0.4222	0.4236	0.4251	0.4265	0.4279	0.4292	0.4306	0.4319
1.50	0.4332	0.4345	0.4357	0.4370	0.4382	0.4394	0.4406	0.4418	0.4429	0.4441
1.60	0.4452	0.4463	0.4474	0.4484	0.4495	0.4505	0.4515	0.4525	0.4535	0.4545
1.70	0.4554	0.4564	0.4573	0.4582	0.4591	0.4599	0.4608	0.4616	0.4625	0.4633
1.80	0.4641	0.4649	0.4656	0.4664	0.4671	0.4678	0.4686	0.4693	0.4699	0.4706
1.90	0.4713	0.4719	0.4726	0.4732	0.4738	0.4744	0.4750	0.4756	0.4761	0.4767
2.00	0.4772	0.4778	0.4783	0.4788	0.4793	0.4798	0.4803	0.4808	0.4812	0.4817
2.10	0.4821	0.4826	0.4830	0.4834	0.4838	0.4842	0.4846	0.4850	0.4854	0.4857
2.20	0.4861	0.4864	0.4868	0.4871	0.4875	0.4878	0.4881	0.4884	0.4887	0.4890
2.30	0.4893	0.4896	0.4898	0.4901	0.4904	0.4906	0.4909	0.4911	0.4913	0.4916
2.40	0.4918	0.4920	0.4922	0.4925	0.4927	0.4929	0.4931	0.4932	0.4934	0.4936
2.50	0.4938	0.4940	0.4941	0.4943	0.4945	0.4946	0.4948	0.4949	0.4951	0.4952
2.60	0.4953	0.4955	0.4956	0.4957	0.4959	0.4960	0.4961	0.4962	0.4963	0.4964
2.70	0.4965	0.4966	0.4967	0.4968	0.4969	0.4970	0.4971	0.4972	0.4973	0.4974
2.80	0.4974	0.4975	0.4976	0.4977	0.4977	0.4978	0.4979	0.4979	0.4980	0.4981
2.90	0.4981	0.4982	0.4982	0.4983	0.4984	0.4984	0.4985	0.4985	0.4986	0.4986
3.00	0.4987	0.4987	0.4987	0.4988	0.4988	0.4989	0.4989	0.4989	0.4990	0.4990
3.10	0.4990	0.4991	0.4991	0.4991	0.4992	0.4992	0.4992	0.4992	0.4993	0.4993
3.20	0.4993	0.4993	0.4994	0.4994	0.4994	0.4994	0.4994	0.4995	0.4995	0.4995
3.30	0.4995	0.4995	0.4995	0.4996	0.4996	0.4996	0.4996	0.4996	0.4996	0.4997
3.40	0.4997	0.4997	0.4997	0.4997	0.4997	0.4997	0.4997	0.4997	0.4997	0.4998
3.50	0.4998	0.4998	0.4998	0.4998	0.4998	0.4998	0.4998	0.4998	0.4998	0.4998
3.60	0.4998	0.4998	0.4999	0.4999	0.4999	0.4999	0.4999	0.4999	0.4999	0.4999
3.70	0.4999	0.4999	0.4999	0.4999	0.4999	0.4999	0.4999	0.4999	0.4999	0.4999
3.80	0.4999	0.4999	0.4999	0.4999	0.4999	0.4999	0.4999	0.4999	0.4999	0.4999
3.90	0.5000	0.5000	0.5000	0.5000	0.5000	0.5000	0.5000	0.5000	0.5000	0.5000
4.00 $\geq$	0.5000	0.5000	0.5000	0.5000	0.5000	0.5000	0.5000	0.5000	0.5000	0.5000

APPENDIX B

CUMULATIVE BINOMIAL PROBABILITIES

$$P(X \leq x) = \sum_{k=0}^{x} \frac{n!}{k!(n-k)!} p^n (1-p)^{n-k} \text{ for } 0 \leq x \leq n, \text{ x is the number of successes in n trials.}$$

n	x	p 0.01	0.05	0.10	0.20	0.30	0.40	0.50	0.60	0.70	0.80	0.90	0.95	0.99
2	0	0.980	0.903	0.810	0.640	0.490	0.360	0.250	0.160	0.090	0.040	0.010	0.003	0.000
	1	1.000	0.998	0.990	0.960	0.910	0.840	0.750	0.640	0.510	0.360	0.190	0.098	0.020
3	0	0.970	0.857	0.729	0.512	0.343	0.216	0.125	0.064	0.027	0.008	0.001	0.000	0.000
	1	1.000	0.993	0.972	0.896	0.784	0.648	0.500	0.352	0.216	0.104	0.028	0.007	0.000
	2	1.000	1.000	0.999	0.992	0.973	0.936	0.875	0.784	0.657	0.488	0.271	0.143	0.030
4	0	0.961	0.815	0.656	0.410	0.240	0.130	0.063	0.026	0.008	0.002	0.000	0.000	0.000
	1	0.999	0.986	0.948	0.819	0.652	0.475	0.313	0.179	0.084	0.027	0.004	0.000	0.000
	2	1.000	1.000	0.996	0.973	0.916	0.821	0.688	0.525	0.348	0.181	0.052	0.014	0.001
	3	1.000	1.000	1.000	0.998	0.992	0.974	0.938	0.870	0.760	0.590	0.344	0.185	0.039
5	0	0.951	0.774	0.590	0.328	0.168	0.078	0.031	0.010	0.002	0.000	0.000	0.000	0.000
	1	0.999	0.977	0.919	0.737	0.528	0.337	0.188	0.087	0.031	0.007	0.000	0.000	0.000
	2	1.000	0.999	0.991	0.942	0.837	0.683	0.500	0.317	0.163	0.058	0.009	0.001	0.000
	3	1.000	1.000	1.000	0.993	0.969	0.913	0.813	0.663	0.472	0.263	0.081	0.023	0.001
	4	1.000	1.000	1.000	1.000	0.998	0.990	0.969	0.922	0.832	0.672	0.410	0.226	0.049
10	0	0.904	0.599	0.349	0.107	0.028	0.006	0.001	0.000	0.000	0.000	0.000	0.000	0.000
	1	0.996	0.914	0.736	0.376	0.149	0.046	0.011	0.002	0.000	0.000	0.000	0.000	0.000
	2	1.000	0.988	0.930	0.678	0.383	0.167	0.055	0.012	0.002	0.000	0.000	0.000	0.000
	3	1.000	0.999	0.987	0.879	0.650	0.382	0.172	0.055	0.011	0.001	0.000	0.000	0.000
	4	1.000	1.000	0.998	0.967	0.850	0.633	0.377	0.166	0.047	0.006	0.000	0.000	0.000
	5	1.000	1.000	1.000	0.994	0.953	0.834	0.623	0.367	0.150	0.033	0.002	0.000	0.000
	6	1.000	1.000	1.000	0.999	0.989	0.945	0.828	0.618	0.350	0.121	0.013	0.001	0.000
	7	1.000	1.000	1.000	1.000	0.998	0.988	0.945	0.833	0.617	0.322	0.070	0.012	0.000
	8	1.000	1.000	1.000	1.000	1.000	0.998	0.989	0.954	0.851	0.624	0.264	0.086	0.004
	9	1.000	1.000	1.000	1.000	1.000	1.000	0.999	0.994	0.972	0.893	0.651	0.401	0.096

(continued)

Operations and Production Systems with Multiple Objectives, First Edition. Behnam Malakooti.
© 2014 John Wiley & Sons, Inc. Published 2014 by John Wiley & Sons, Inc.

								p							
n	x	0.01	0.05	0.10	0.20	0.30	0.40	0.50	0.60	0.70	0.80	0.90	0.95	0.99
15	0	0.860	0.463	0.206	0.035	0.005	0.000	0.000	0.000	0.000	0.000	0.000	0.000	0.000
	1	0.990	0.829	0.549	0.167	0.035	0.005	0.000	0.000	0.000	0.000	0.000	0.000	0.000
	2	1.000	0.964	0.816	0.398	0.127	0.027	0.004	0.000	0.000	0.000	0.000	0.000	0.000
	3	1.000	0.995	0.944	0.648	0.297	0.091	0.018	0.002	0.000	0.000	0.000	0.000	0.000
	4	1.000	0.999	0.987	0.836	0.515	0.217	0.059	0.009	0.001	0.000	0.000	0.000	0.000
	5	1.000	1.000	0.998	0.939	0.722	0.403	0.151	0.034	0.004	0.000	0.000	0.000	0.000
	6	1.000	1.000	1.000	0.982	0.869	0.610	0.304	0.095	0.015	0.001	0.000	0.000	0.000
	7	1.000	1.000	1.000	0.996	0.950	0.787	0.500	0.213	0.050	0.004	0.000	0.000	0.000
	8	1.000	1.000	1.000	0.999	0.985	0.905	0.696	0.390	0.131	0.018	0.000	0.000	0.000
	9	1.000	1.000	1.000	1.000	0.996	0.966	0.849	0.597	0.278	0.061	0.002	0.000	0.000
	10	1.000	1.000	1.000	1.000	0.999	0.991	0.941	0.783	0.485	0.164	0.013	0.001	0.000
	11	1.000	1.000	1.000	1.000	1.000	0.998	0.982	0.909	0.703	0.352	0.056	0.005	0.000
	12	1.000	1.000	1.000	1.000	1.000	1.000	0.996	0.973	0.873	0.602	0.184	0.036	0.000
	13	1.000	1.000	1.000	1.000	1.000	1.000	1.000	0.995	0.965	0.833	0.451	0.171	0.010
	14	1.000	1.000	1.000	1.000	1.000	1.000	1.000	1.000	0.995	0.965	0.794	0.537	0.140

CUMULATIVE POISSON PROBABILITIES

$$P\{X \geq c\} = \sum_{x=c}^{\infty} \frac{e^{-\mu}\mu^x}{x!} \text{ where } \mu = np$$

					μ					
c	0.1	0.2	0.3	0.4	0.5	0.6	0.7	0.8	0.9	1
0	1.0000	1.0000	1.0000	1.0000	1.0000	1.0000	1.0000	1.0000	1.0000	1.0000
1	0.0952	0.1813	0.2592	0.3297	0.3935	0.4512	0.5034	0.5507	0.5934	0.6321
2	0.0047	0.0175	0.0369	0.0616	0.0902	0.1219	0.1558	0.1912	0.2275	0.2642
3	0.0002	0.0011	0.0036	0.0079	0.0144	0.0231	0.0341	0.0474	0.0629	0.0803
4	0.0000	0.0001	0.0003	0.0008	0.0018	0.0034	0.0058	0.0091	0.0135	0.0190
5	0.0000	0.0000	0.0000	0.0001	0.0002	0.0004	0.0008	0.0014	0.0023	0.0037

					μ					
c	1.1	1.2	1.3	1.4	1.5	1.6	1.7	1.8	1.9	2
0	1.0000	1.0000	1.0000	1.0000	1.0000	1.0000	1.0000	1.0000	1.0000	1.0000
1	0.6671	0.6988	0.7275	0.7534	0.7769	0.7981	0.8173	0.8347	0.8504	0.8647
2	0.3010	0.3374	0.3732	0.4082	0.4422	0.4751	0.5068	0.5372	0.5663	0.5940
3	0.0996	0.1205	0.1429	0.1665	0.1912	0.2166	0.2428	0.2694	0.2963	0.3233
4	0.0257	0.0338	0.0431	0.0537	0.0656	0.0788	0.0932	0.1087	0.1253	0.1429
5	0.0054	0.0077	0.0107	0.0143	0.0186	0.0237	0.0296	0.0364	0.0441	0.0527
6	0.0010	0.0015	0.0022	0.0032	0.0045	0.0060	0.0080	0.0104	0.0132	0.0166
7	0.0001	0.0003	0.0004	0.0006	0.0009	0.0013	0.0019	0.0026	0.0034	0.0045
8	0.0000	0.0000	0.0001	0.0001	0.0002	0.0003	0.0004	0.0006	0.0008	0.0011

(continued)

Operations and Production Systems with Multiple Objectives, First Edition. Behnam Malakooti.
© 2014 John Wiley & Sons, Inc. Published 2014 by John Wiley & Sons, Inc.

					μ					
c	2.1	2.2	2.3	2.4	2.5	2.6	2.7	2.8	2.9	3
0	1.0000	1.0000	1.0000	1.0000	1.0000	1.0000	1.0000	1.0000	1.0000	1.0000
1	0.8775	0.8892	0.8997	0.9093	0.9179	0.9257	0.9328	0.9392	0.9450	0.9502
2	0.6204	0.6454	0.6691	0.6916	0.7127	0.7326	0.7513	0.7689	0.7854	0.8009
3	0.3504	0.3773	0.4040	0.4303	0.4562	0.4816	0.5064	0.5305	0.5540	0.5768
4	0.1614	0.1806	0.2007	0.2213	0.2424	0.2640	0.2859	0.3081	0.3304	0.3528
5	0.0621	0.0725	0.0838	0.0959	0.1088	0.1226	0.1371	0.1523	0.1682	0.1847
6	0.0204	0.0249	0.0300	0.0357	0.0420	0.0490	0.0567	0.0651	0.0742	0.0839
7	0.0059	0.0075	0.0094	0.0116	0.0142	0.0172	0.0206	0.0244	0.0287	0.0335
8	0.0015	0.0020	0.0026	0.0033	0.0042	0.0053	0.0066	0.0081	0.0099	0.0119
9	0.0003	0.0005	0.0006	0.0009	0.0011	0.0015	0.0019	0.0024	0.0031	0.0038
10	0.0001	0.0001	0.0001	0.0002	0.0003	0.0004	0.0005	0.0007	0.0009	0.0011

					μ					
c	3.1	3.2	3.3	3.4	3.5	3.6	3.7	3.8	3.9	4
0	1.0000	1.0000	1.0000	1.0000	1.0000	1.0000	1.0000	1.0000	1.0000	1.0000
1	0.9550	0.9592	0.9631	0.9666	0.9698	0.9727	0.9753	0.9776	0.9798	0.9817
2	0.8153	0.8288	0.8414	0.8532	0.8641	0.8743	0.8838	0.8926	0.9008	0.9084
3	0.5988	0.6201	0.6406	0.6603	0.6792	0.6973	0.7146	0.7311	0.7469	0.7619
4	0.3752	0.3975	0.4197	0.4416	0.4634	0.4848	0.5058	0.5265	0.5468	0.5665
5	0.2018	0.2194	0.2374	0.2558	0.2746	0.2936	0.3128	0.3322	0.3516	0.3712
6	0.0943	0.1054	0.1171	0.1295	0.1424	0.1559	0.1699	0.1844	0.1994	0.2149
7	0.0388	0.0446	0.0510	0.0579	0.0653	0.0733	0.0818	0.0909	0.1005	0.1107
8	0.0142	0.0168	0.0198	0.0231	0.0267	0.0308	0.0352	0.0401	0.0454	0.0511
9	0.0047	0.0057	0.0069	0.0083	0.0099	0.0117	0.0137	0.0160	0.0185	0.0214
10	0.0014	0.0018	0.0022	0.0027	0.0033	0.0040	0.0048	0.0058	0.0069	0.0081
11	0.0004	0.0005	0.0006	0.0008	0.0010	0.0013	0.0016	0.0019	0.0023	0.0028

					μ					
c	4.1	4.2	4.3	4.4	4.5	4.6	4.7	4.8	4.9	5
0	1.0000	1.0000	1.0000	1.0000	1.0000	1.0000	1.0000	1.0000	1.0000	1.0000
1	0.9834	0.9850	0.9864	0.9877	0.9889	0.9899	0.9909	0.9918	0.9926	0.9933
2	0.9155	0.9220	0.9281	0.9337	0.9389	0.9437	0.9482	0.9523	0.9561	0.9596
3	0.7762	0.7898	0.8026	0.8149	0.8264	0.8374	0.8477	0.8575	0.8667	0.8753
4	0.5858	0.6046	0.6228	0.6406	0.6577	0.6743	0.6903	0.7058	0.7207	0.7350
5	0.3907	0.4102	0.4296	0.4488	0.4679	0.4868	0.5054	0.5237	0.5418	0.5595
6	0.2307	0.2469	0.2633	0.2801	0.2971	0.3142	0.3316	0.3490	0.3665	0.3840
7	0.1214	0.1325	0.1442	0.1564	0.1689	0.1820	0.1954	0.2092	0.2233	0.2378
8	0.0573	0.0639	0.0710	0.0786	0.0866	0.0951	0.1040	0.1133	0.1231	0.1334
9	0.0245	0.0279	0.0317	0.0358	0.0403	0.0451	0.0503	0.0558	0.0618	0.0681
10	0.0095	0.0111	0.0129	0.0149	0.0171	0.0195	0.0222	0.0251	0.0283	0.0318
11	0.0034	0.0041	0.0048	0.0057	0.0067	0.0078	0.0090	0.0104	0.0120	0.0137
12	0.0011	0.0014	0.0017	0.0020	0.0024	0.0029	0.0034	0.0040	0.0047	0.0055
13	0.0003	0.0004	0.0005	0.0007	0.0008	0.0010	0.0012	0.0014	0.0017	0.0020

					μ					
c	5.1	5.2	5.3	5.4	5.5	5.6	5.7	5.8	5.9	6
0	1.0000	1.0000	1.0000	1.0000	1.0000	1.0000	1.0000	1.0000	1.0000	1.0000
1	0.9939	0.9945	0.9950	0.9955	0.9959	0.9963	0.9967	0.9970	0.9973	0.9975
2	0.9628	0.9658	0.9686	0.9711	0.9734	0.9756	0.9776	0.9794	0.9811	0.9826
3	0.8835	0.8912	0.8984	0.9052	0.9116	0.9176	0.9232	0.9285	0.9334	0.9380
4	0.7487	0.7619	0.7746	0.7867	0.7983	0.8094	0.8200	0.8300	0.8396	0.8488
5	0.5769	0.5939	0.6105	0.6267	0.6425	0.6578	0.6728	0.6873	0.7013	0.7149
6	0.4016	0.4191	0.4365	0.4539	0.4711	0.4881	0.5050	0.5217	0.5381	0.5543
7	0.2526	0.2676	0.2829	0.2983	0.3140	0.3297	0.3456	0.3616	0.3776	0.3937
8	0.1440	0.1551	0.1665	0.1783	0.1905	0.2030	0.2159	0.2290	0.2424	0.2560
9	0.0748	0.0819	0.0894	0.0973	0.1056	0.1143	0.1234	0.1328	0.1426	0.1528
10	0.0356	0.0397	0.0441	0.0488	0.0538	0.0591	0.0648	0.0708	0.0772	0.0839
11	0.0156	0.0177	0.0200	0.0225	0.0253	0.0282	0.0314	0.0349	0.0386	0.0426
12	0.0063	0.0073	0.0084	0.0096	0.0110	0.0125	0.0141	0.0159	0.0179	0.0201
13	0.0024	0.0028	0.0033	0.0038	0.0045	0.0051	0.0059	0.0068	0.0078	0.0088
14	0.0008	0.0010	0.0012	0.0014	0.0017	0.0020	0.0023	0.0027	0.0031	0.0036
15	0.0003	0.0003	0.0004	0.0005	0.0006	0.0007	0.0009	0.0010	0.0012	0.0014

					μ					
c	6.1	6.2	6.3	6.4	6.5	6.6	6.7	6.8	6.9	7
0	1.0000	1.0000	1.0000	1.0000	1.0000	1.0000	1.0000	1.0000	1.0000	1.0000
1	0.9978	0.9980	0.9982	0.9983	0.9985	0.9986	0.9988	0.9989	0.9990	0.9991
2	0.9841	0.9854	0.9866	0.9877	0.9887	0.9897	0.9905	0.9913	0.9920	0.9927
3	0.9423	0.9464	0.9502	0.9537	0.9570	0.9600	0.9629	0.9656	0.9680	0.9704
4	0.8575	0.8658	0.8736	0.8811	0.8882	0.8948	0.9012	0.9072	0.9129	0.9182
5	0.7281	0.7408	0.7531	0.7649	0.7763	0.7873	0.7978	0.8080	0.8177	0.8270
6	0.5702	0.5859	0.6012	0.6163	0.6310	0.6453	0.6594	0.6730	0.6863	0.6993
7	0.4098	0.4258	0.4418	0.4577	0.4735	0.4892	0.5047	0.5201	0.5353	0.5503
8	0.2699	0.2840	0.2983	0.3127	0.3272	0.3419	0.3567	0.3715	0.3864	0.4013
9	0.1633	0.1741	0.1852	0.1967	0.2084	0.2204	0.2327	0.2452	0.2580	0.2709

					μ					
c	6.1	6.2	6.3	6.4	6.5	6.6	6.7	6.8	6.9	7
10	0.0910	0.0984	0.1061	0.1142	0.1226	0.1314	0.1404	0.1498	0.1595	0.1695
11	0.0469	0.0514	0.0563	0.0614	0.0668	0.0726	0.0786	0.0849	0.0916	0.0985
12	0.0224	0.0250	0.0277	0.0307	0.0339	0.0373	0.0409	0.0448	0.0490	0.0533
13	0.0100	0.0113	0.0127	0.0143	0.0160	0.0179	0.0199	0.0221	0.0245	0.0270
14	0.0042	0.0048	0.0055	0.0063	0.0071	0.0080	0.0091	0.0102	0.0115	0.0128
15	0.0016	0.0019	0.0022	0.0026	0.0030	0.0034	0.0039	0.0044	0.0050	0.0057
16	0.0006	0.0007	0.0008	0.0010	0.0012	0.0014	0.0016	0.0018	0.0021	0.0024

(continued)

					μ					
c	7.1	7.2	7.3	7.4	7.5	7.6	7.7	7.8	7.9	8
0	1.0000	1.0000	1.0000	1.0000	1.0000	1.0000	1.0000	1.0000	1.0000	1.0000
1	0.9992	0.9993	0.9993	0.9994	0.9994	0.9995	0.9995	0.9996	0.9996	0.9997
2	0.9933	0.9939	0.9944	0.9949	0.9953	0.9957	0.9961	0.9964	0.9967	0.9970
3	0.9725	0.9745	0.9764	0.9781	0.9797	0.9812	0.9826	0.9839	0.9851	0.9862
4	0.9233	0.9281	0.9326	0.9368	0.9409	0.9446	0.9482	0.9515	0.9547	0.9576
5	0.8359	0.8445	0.8527	0.8605	0.8679	0.8751	0.8819	0.8883	0.8945	0.9004
6	0.7119	0.7241	0.7360	0.7474	0.7586	0.7693	0.7797	0.7897	0.7994	0.8088
7	0.5651	0.5796	0.5940	0.6080	0.6218	0.6354	0.6486	0.6616	0.6743	0.6866
8	0.4162	0.4311	0.4459	0.4607	0.4754	0.4900	0.5044	0.5188	0.5330	0.5470
9	0.2840	0.2973	0.3108	0.3243	0.3380	0.3518	0.3657	0.3796	0.3935	0.4075
10	0.1798	0.1904	0.2012	0.2123	0.2236	0.2351	0.2469	0.2589	0.2710	0.2834
11	0.1058	0.1133	0.1212	0.1293	0.1378	0.1465	0.1555	0.1648	0.1743	0.1841
12	0.0580	0.0629	0.0681	0.0735	0.0792	0.0852	0.0915	0.0980	0.1048	0.1119
13	0.0297	0.0327	0.0358	0.0391	0.0427	0.0464	0.0504	0.0546	0.0591	0.0638
14	0.0143	0.0159	0.0176	0.0195	0.0216	0.0238	0.0261	0.0286	0.0313	0.0342
15	0.0065	0.0073	0.0082	0.0092	0.0103	0.0114	0.0127	0.0141	0.0156	0.0173
16	0.0028	0.0031	0.0036	0.0041	0.0046	0.0052	0.0059	0.0066	0.0074	0.0082
17	0.0011	0.0013	0.0015	0.0017	0.0020	0.0022	0.0026	0.0029	0.0033	0.0037
18	0.0004	0.0005	0.0006	0.0007	0.0008	0.0009	0.0011	0.0012	0.0014	0.0016

					μ					
c	8.1	8.2	8.3	8.4	8.5	8.6	8.7	8.8	8.9	9
0	1.0000	1.0000	1.0000	1.0000	1.0000	1.0000	1.0000	1.0000	1.0000	1.0000
1	0.9997	0.9997	0.9998	0.9998	0.9998	0.9998	0.9998	0.9998	0.9999	0.9999
2	0.9972	0.9975	0.9977	0.9979	0.9981	0.9982	0.9984	0.9985	0.9986	0.9988
3	0.9873	0.9882	0.9891	0.9900	0.9907	0.9914	0.9921	0.9927	0.9932	0.9938
4	0.9604	0.9630	0.9654	0.9677	0.9699	0.9719	0.9738	0.9756	0.9772	0.9788
5	0.9060	0.9113	0.9163	0.9211	0.9256	0.9299	0.9340	0.9379	0.9416	0.9450
6	0.8178	0.8264	0.8347	0.8427	0.8504	0.8578	0.8648	0.8716	0.8781	0.8843
7	0.6987	0.7104	0.7219	0.7330	0.7438	0.7543	0.7645	0.7744	0.7840	0.7932
8	0.5609	0.5746	0.5881	0.6013	0.6144	0.6272	0.6398	0.6522	0.6643	0.6761
9	0.4214	0.4353	0.4493	0.4631	0.4769	0.4906	0.5042	0.5177	0.5311	0.5443
10	0.2959	0.3085	0.3212	0.3341	0.3470	0.3600	0.3731	0.3863	0.3994	0.4126
11	0.1942	0.2045	0.2150	0.2257	0.2366	0.2478	0.2591	0.2706	0.2822	0.2940
12	0.1193	0.1269	0.1348	0.1429	0.1513	0.1600	0.1689	0.1780	0.1874	0.1970
13	0.0687	0.0739	0.0793	0.0850	0.0909	0.0971	0.1035	0.1102	0.1171	0.1242
14	0.0372	0.0405	0.0439	0.0476	0.0514	0.0555	0.0597	0.0642	0.0689	0.0739
15	0.0190	0.0209	0.0229	0.0251	0.0274	0.0299	0.0325	0.0353	0.0383	0.0415
16	0.0092	0.0102	0.0113	0.0125	0.0138	0.0152	0.0168	0.0184	0.0202	0.0220
17	0.0042	0.0047	0.0053	0.0059	0.0066	0.0074	0.0082	0.0091	0.0101	0.0111
18	0.0018	0.0021	0.0023	0.0027	0.0030	0.0034	0.0038	0.0043	0.0048	0.0053
19	0.0008	0.0009	0.0010	0.0011	0.0013	0.0015	0.0017	0.0019	0.0022	0.0024
20	0.0003	0.0003	0.0004	0.0005	0.0005	0.0006	0.0007	0.0008	0.0009	0.0011

	μ									
c	9.1	9.2	9.3	9.4	9.5	9.6	9.7	9.8	9.9	10
0	1.0000	1.0000	1.0000	1.0000	1.0000	1.0000	1.0000	1.0000	1.0000	1.0000
1	0.9999	0.9999	0.9999	0.9999	0.9999	0.9999	0.9999	0.9999	0.9999	1.0000
2	0.9989	0.9990	0.9991	0.9991	0.9992	0.9993	0.9993	0.9994	0.9995	0.9995
3	0.9942	0.9947	0.9951	0.9955	0.9958	0.9962	0.9965	0.9967	0.9970	0.9972
4	0.9802	0.9816	0.9828	0.9840	0.9851	0.9862	0.9871	0.9880	0.9889	0.9897
5	0.9483	0.9514	0.9544	0.9571	0.9597	0.9622	0.9645	0.9667	0.9688	0.9707
6	0.8902	0.8959	0.9014	0.9065	0.9115	0.9162	0.9207	0.9250	0.9290	0.9329
7	0.8022	0.8108	0.8192	0.8273	0.8351	0.8426	0.8498	0.8567	0.8634	0.8699
8	0.6877	0.6990	0.7100	0.7208	0.7313	0.7416	0.7515	0.7612	0.7706	0.7798
9	0.5574	0.5704	0.5832	0.5958	0.6082	0.6204	0.6324	0.6442	0.6558	0.6672
10	0.4258	0.4389	0.4521	0.4651	0.4782	0.4911	0.5040	0.5168	0.5295	0.5421
11	0.3059	0.3180	0.3301	0.3424	0.3547	0.3671	0.3795	0.3920	0.4045	0.4170
12	0.2068	0.2168	0.2270	0.2374	0.2480	0.2588	0.2697	0.2807	0.2919	0.3032
13	0.1316	0.1393	0.1471	0.1552	0.1636	0.1721	0.1809	0.1899	0.1991	0.2084
14	0.0790	0.0844	0.0900	0.0958	0.1019	0.1081	0.1147	0.1214	0.1284	0.1355
15	0.0448	0.0483	0.0520	0.0559	0.0600	0.0643	0.0688	0.0735	0.0784	0.0835
16	0.0240	0.0262	0.0285	0.0309	0.0335	0.0362	0.0391	0.0421	0.0454	0.0487
17	0.0122	0.0135	0.0148	0.0162	0.0177	0.0194	0.0211	0.0230	0.0249	0.0270
18	0.0059	0.0066	0.0073	0.0081	0.0089	0.0098	0.0108	0.0119	0.0130	0.0143
19	0.0027	0.0031	0.0034	0.0038	0.0043	0.0048	0.0053	0.0059	0.0065	0.0072
20	0.0012	0.0014	0.0015	0.0017	0.0020	0.0022	0.0025	0.0028	0.0031	0.0035

	μ									
c	11	12	13	14	15	16	17	18	19	20
0	1.0000	1.0000	1.0000	1.0000	1.0000	1.0000	1.0000	1.0000	1.0000	1.0000
1	1.0000	1.0000	1.0000	1.0000	1.0000	1.0000	1.0000	1.0000	1.0000	1.0000
2	0.9998	0.9999	1.0000	1.0000	1.0000	1.0000	1.0000	1.0000	1.0000	1.0000
3	0.9988	0.9995	0.9998	0.9999	1.0000	1.0000	1.0000	1.0000	1.0000	1.0000
4	0.9951	0.9977	0.9989	0.9995	0.9998	0.9999	1.0000	1.0000	1.0000	1.0000
5	0.9849	0.9924	0.9963	0.9982	0.9991	0.9996	0.9998	0.9999	1.0000	1.0000
6	0.9625	0.9797	0.9893	0.9945	0.9972	0.9986	0.9993	0.9997	0.9998	0.9999
7	0.9214	0.9542	0.9741	0.9858	0.9924	0.9960	0.9979	0.9990	0.9995	0.9997
8	0.8568	0.9105	0.9460	0.9684	0.9820	0.9900	0.9946	0.9971	0.9985	0.9992
9	0.7680	0.8450	0.9002	0.9379	0.9626	0.9780	0.9874	0.9929	0.9961	0.9979
10	0.6595	0.7576	0.8342	0.8906	0.9301	0.9567	0.9739	0.9846	0.9911	0.9950
11	0.5401	0.6528	0.7483	0.8243	0.8815	0.9226	0.9509	0.9696	0.9817	0.9892
12	0.4207	0.5384	0.6468	0.7400	0.8152	0.8730	0.9153	0.9451	0.9653	0.9786
13	0.3113	0.4240	0.5369	0.6415	0.7324	0.8069	0.8650	0.9083	0.9394	0.9610
14	0.2187	0.3185	0.4270	0.5356	0.6368	0.7255	0.7991	0.8574	0.9016	0.9339
15	0.1460	0.2280	0.3249	0.4296	0.5343	0.6325	0.7192	0.7919	0.8503	0.8951
16	0.0926	0.1556	0.2364	0.3306	0.4319	0.5333	0.6285	0.7133	0.7852	0.8435
17	0.0559	0.1013	0.1645	0.2441	0.3359	0.4340	0.5323	0.6249	0.7080	0.7789
18	0.0322	0.0630	0.1095	0.1728	0.2511	0.3407	0.4360	0.5314	0.6216	0.7030
19	0.0177	0.0374	0.0698	0.1174	0.1805	0.2577	0.3450	0.4378	0.5305	0.6186

(continued)

	μ									
c	11	12	13	14	15	16	17	18	19	20
20	0.0093	0.0213	0.0427	0.0765	0.1248	0.1878	0.2637	0.3491	0.4394	0.5297
21	0.0047	0.0116	0.0250	0.0479	0.0830	0.1318	0.1945	0.2693	0.3528	0.4409
22	0.0023	0.0061	0.0141	0.0288	0.0531	0.0892	0.1385	0.2009	0.2745	0.3563
23	0.0010	0.0030	0.0076	0.0167	0.0327	0.0582	0.0953	0.1449	0.2069	0.2794
24	0.0005	0.0015	0.0040	0.0093	0.0195	0.0367	0.0633	0.1011	0.1510	0.2125
25	0.0002	0.0007	0.0020	0.0050	0.0112	0.0223	0.0406	0.0683	0.1067	0.1568

INDEX

Operations and Production Systems with Multiple Objectives, First Edition. Behnam Malakooti.
© 2014 John Wiley & Sons, Inc. Published 2014 by John Wiley & Sons, Inc.

**WILEY SERIES IN SYSTEMS ENGINEERING
AND MANAGEMENT**

Andrew P. Sage, Editor